K. Kurihara · N. Suzuki · H. Ogawa (Eds.)

Olfaction and Taste XI

Proceedings of the 11th International Symposium on Olfaction and Taste
and of the 27th Japanese Symposium on Taste and Smell

Joint Meeting held at Kosei-nenkin Kaikan, Sapporo, Japan
July 12–16, 1993

With 355 Illustrations

Springer Japan KK

KENZO KURIHARA, D.S.
Professor, Faculty of Pharmaceutical Sciences, Hokkaido University, N-13 W-6,
Kita-ku, Sapporo, 060 Japan

NORIYO SUZUKI, D.S.
Associate Professor, Animal Behavior and Intelligence, Division of Biological Sciences,
Graduate School of Science, Hokkaido University, N-10 W-8, Kita-ku, Sapporo,
060 Japan

HISASHI OGAWA, M.D., D.M.S.
Professor, Department of Physiology, Kumamoto University School of Medicine,
2-2-1 Honjo, Kumamoto, 860 Japan

ISBN 978-4-431-68357-5   ISBN 978-4-431-68355-1 (eBook)
DOI 10.1007/978-4-431-68355-1

Printed on acid-free paper

© Springer Japan 1994
Originally published by Springer-Verlag Tokyo Berlin Heidelberg New York in 1994
Softcover reprint of the hardcover 1st edition 1994

Typesetting: Best-set Typesetter Ltd., Hong Kong

Preface

The first meeting of the International Symposium on Olfaction and Taste (ISOT) was held at the Wenner Gren Center in Stockholm in 1962 under Professor Yngve Zotterman's chairmanship. Ever since the International Union of Physiological Sciences (IUPS) held its first congress, ISOT has been one of the satellite symposia of the Congress. To date, ISOT meetings have taken place as follows:

I	1962	Stockholm, Sweden
II	1965	Tokyo, Japan
III	1968	New York, USA
IV	1971	Starnberg, Germany
V	1974	Melbourne, Australia
VI	1977	Paris, France
VII	1980	Noordwijkerhout, Netherlands
VIII	1983	Melbourne, Australia
IX	1986	Snowmass, USA
X	1989	Oslo, Norway
XI	1993	Sapporo, Japan

After the first ISOT, three research organizations for chemoreception science—the Japanese Association for the Study of Taste and Smell (JASTS), the European Chemoreception Organization (ECRO), and the Association for Chemoreception Sciences (AChemS)—were founded in Japan, Europe, and the USA. Each organization has elected two representative members to the International Commission of Olfaction and Taste (ICOT) of IUPS. At the 1989 meeting held in Oslo, ICOT decided that future ISOT meetings would be convened in principle by each of the three organizations in turn. ISOT is a satellite symposium of IUPS, and hence ISOT is to be held every 4 years, since the interval between IUPS Congresses was changed from 3 to 4 years. The conference site for ISOT is, however, independent of the site for the IUPS Congress. Therefore, the next ISOT will be organized by AChemS in 1997.

ISOT XI was held at Kosei-nenkin Kaikan in Sapporo from July 12 to 16, 1993. The meeting was held as a joint congress with JASTS XXVII (President, T. Shibuya). About 630 persons participated in the congress. Four plenary lectures, 136 papers in 21 session symposia, 59 papers in an oral session, and 170 papers in a poster session were presented. The total number of papers was 369, which represents a marked increase over that of previous ISOTs. In particular, there were significantly more papers on the molecular biology of chemoreception at this conference. The Organizing Committee scheduled 21 session symposia to address every current topic in various fields of chemoreception. Many front-line scientists in each field participated in the conference and actively exchanged their scientific information. The meeting was held from early morning to late evening and active discussion was carried out at each session. Especially the audience at the poster session was in a fever of discussion. Participants in the congress also had a chance to meet informally at three evening parties and one afternoon excursion. This was an additional fruitful product of the meeting.

In compiling the proceedings of this conference into the book entitled "Olfaction and Taste XI", the papers presented at the conference were completely rearranged. That is, all short and long papers submitted to the editors were systematically arranged according to the fields of chemoreception. We hope that this book may serve as a handbook for all researchers and students who wish to study current advances in many different fields of chemoreception.

The congress was successfully coordinated by the members of the Organizing Committee. The Committee would like to extend its thanks to all participants for having contributed to the success of the meeting. The Committee also wishes to express its gratitude to the many organizations and companies that provided financial assistance to this congress. Their support was essential to the success of the conference.

KENZO KURIHARA
Chairperson of ISOT XI
Editor-in-Chief for
Olfaction and Taste XI

Contents

2. Taste Transduction

2.1 BIOCHEMICAL EVENTS

Contents

2.2 IONIC MECHANISMS

3. Olfactory Transduction

3.1 BIOCHEMICAL EVENTS

3.2 IONIC MECHANISMS

4. Genetics and Molecular Neurobiology of Chemoreception

5. Structure and Function of Flavor Compounds

Contents xiii

6. Psychophysics of Taste and Olfaction

6.1 TASTE

Contents

7. Umami: Physiology, Nutrition and Food Science

8. Central Coding Mechanisms of Taste and Olfaction

8.1 TASTE

8.2 OLFACTION

Contents

9. Mechanisms of Learning and Memory Involving the Chemical Senses

10. Nutrition and Brain Mechanisms of Food Recognition

11. Clinical Assessment of Taste

11.1 Aging in Taste

12. Clinical Assessment of Olfaction

12.1 AGING IN OLFACTION

12.2 EVALUATION AND TREATMENT OF OLFACTORY DISORDERS

13. Noninvasive Measurements of Human Chemosensory Response

14. Artificial Sensing Devices

Contents

15. Chemoreception in Aquatic Organisms

15.1 FISH

15.2 OTHERS

16. Chemoreception in Insects

16.1 Transduction

16.2 Neuroethology

Contents

1. Structure and Function of the Taste Bud and Olfactory Epithelium

Gustatory Cells as Paraneurons

TSUNEO FUJITA[1]

Key words. Gustatory cell—Paraneuron—Gut endocrine cell—Synaptic vesicles—Peptidergic vesicles—*Hydra*

Introduction

A large group of cells previously called endocrine cells and sensory cells are now categorized as "paraneurons" because they share biological features with nerve cells, or neurons [1,2]. Paraneurons are receptosecretory cells possessing a receptive site and a secretory site on their plasma membrane. They receive adequate stimuli at the former site; and at the latter they release, by an exocytotic mechanism, membrane-bound granules containing peptidic, aminic, and other signal substances that are common to neuronal secretions. Immunocytochemical studies have demonstrated that various neuron-specific proteins are shared by certain paraneurons [2,3]. Since we proposed the concept of the paraneuron in 1975, we have counted the gustatory cell as one of the most typical sensory paraneurons [1].

Gustatory Cells as Sensory Paraneurons

Despite the endeavors of previous electron microscopists [5–8], morphological identification of the gustatory cell (one of several cell types comprising mammalian taste buds) had been controversial until Yoshie et al. [9] determined that the type III cell is the sole and constant candidate to be the gustatory cell. In the guinea pig [9] and other mammalian species (unpublished data), only this type of cell possesses synaptic vesicles associated with nerve terminals [9].

As a sensory paraneuron, the type III, or gustatory, cell is bipolar with an apical receptive site provided with microvilli and a basal secretory area synaptically associated with nerves. The synaptic vesicles morphologically correspond to the structures of the same name possessed by neurons [1,2].

Synaptic Vesicles and Transmitter Candidates

Although the synaptic vesicles in murine gustatory cells have been reported to be predominantly small, clear vesicles, our recent observations in the guinea pig [9], dog [10], cat, and some other mammals indicate that they are mostly large-cored vesicles with dark, round granules, 70 to 150 nm in diameter, that well deserve to be called peptidergic granules [9,10].

Secretion of a peptidic signal substance or substances is one of the essential features of a paraneuron. From this viewpoint, we immunocytochemically investigated such peptide(s) in gustatory cells from various mammals. Our results suggest that enkephalins might be one of the products of gustatory cells in the taste bud on vallate papillae of the guinea pig, rat, and cat [11]. As the type III cells immunoreacted for Met-enkephalin-8, it is likely that preproenkephalin A is produced in the cell, and its derivative peptides are stored in the peptidergic-type vesicles. It seems possible that monoamines, including dopamine and serotonin [12,13], may be co-localized in these vesicles, but we suggest that they are present in insignificant amounts, if at all.

The content of the small, clear vesicles is unknown. If the hypothesis holds true that norepinephrine represents a transmitter in gustatory cells [14,15], this type of vesicle may include it.

Vesicle Release and Its Implications

Based on our studies of intestinal paraneurons, we conceived that stimulation of the taste bud paraneurons might cause exocytotic release of their synaptic vesicles. We obtained electron microscopic images of the vesicles opening to the synaptic cleft in an Ω shape [11,16; unpublished data]. We placed various taste substances, i.e., sweet (sucrose), umami (mixture of monosodium glutamate and guanosine monophosphate), sour (citric acid), and salty (NaCl), on the vallate papillae of guinea pigs [16] and dogs and fixed the tissue within a few minutes. Exocytotic images, seen by electron micro-

---

[1] Department of Anatomy, Niigata University School of Medicine, Asahi-machi 1, Niigata, 951 Japan

2

scopy, occurred either exclusively or mainly in the large-cored vesicles [10,11,16].

Another finding in gustatory cells was a hitherto unknown large number of the large-cored vesicles filling the basal portion of the cell; they were especially conspicuous in the guinea pig, cat, and dog [9–11]. These vesicles by far exceeded the number presumed to be used for synaptic transmission [10,11]. They tended to gather at the basal and lateral borders of the cell. When the gustatory cell was apically stimulated, as mentioned above, some of those marginal vesicles underwent exocytotic opening [10,11].

This finding strongly supports a novel view that the gustatory cell is endocrine/paracrine in nature as well as synaptocrine. In the dog we observed fenestrated blood capillaries close to the basal portion of the gustatory cell, a finding that supports its endocrine function [10]. Of more importance is the paracrine release of the vesicle contents. We hypothesize that they might stimulate Ebner's and other taste-bud-associated salivary glands.

Comparison with "Gustatory Cells in the Gut"

We have investigated the endocrine cells in the gut for about 20 years and during this period established that they are chemoreceptive sensor cells and deserve the name of paraneurons because of their neuronal nature [1,2,17]. We differentiated more than 12 cell types containing as many (or more) different signal substances, mainly peptidic and partly aminic in nature. By stimulating the paraneurons from the apical (luminal) side, we succeeded in provoking exocytosis in specific types of cell.

Comparing the gut sensor cells with the gustatory cells, it became clear that they share the essential paraneuronal features, although there are some discrepancies [18]. That the lingual gustatory cell is synaptically connected with nerves and that the intestinal paraneuron is not is an unimportant difference in terms of the function of the cell. A more serious difference is that the gustatory cells seem to consist of a single cell type, as determined by electron microscopic and immunocytochemical studies, whereas gut paraneurons comprise many types that respond to different chemical stimuli. Results of our experiment that stimulated the gustatory cells, as mentioned above, suggested that different tastes can induce vesicle release in a single gustatory cell. The electrophysiologic data by Tonosaki [19] also support the concept that a single cell may be receptive to different taste stimuli. The mechanism as to how the gustatory cell–nerve complex distinguishes the different tastes remains a mystery.

Lessons from *Hydra*

Chemosensitive cells that qualify as gustatory cells are known to be distributed at ectodermal sites away from the tongue in various animals. That taste buds may be found on the barbels, perioral area, and even the fin of certain fish is well known. The lamprey possesses microvillous paraneurons in the skin whose fine structure closely resembles that of the gustatory cells in higher vertebrates, though a possible photoreceptive function of those cells has been proposed [20].

Xenopus, living in water for life and lacking a tongue, possesses hair cells (corresponding to the lateral line system of the fish) on the sides of the body. Japanese physiologists demonstrated that these hair cells, which represent typical paraneurons and function as mechanoreceptors, are engaged also in chemoreception, being sensitive to amino acids including monosodium glutamate and certain salt solutions [21].

In *Hydra*, one of the most primitive metazoa, chemoreceptive paraneurons are distributed in the ectoderm covering the perioral area and tentacles as well as in the entoderm lining the gut. The cells in the ectoderm are selectively sensitive to glutathione [22] and much less to monosodium glutamate (preliminary data). The entodermal cells' stimuli are not well understood; the cells are suggested to be sensitive to elevated osmotic pressure.

It is possible that the ectodermal paraneurons of *Hydra* represent the prototype of the gustatory cells on the tongue and body surface of vertebrates, whereas the entodermal paraneurons of *Hydra* represent the prototype of the gut endocrine cells of vertebrates.

Conclusion

The view regarding the gustatory cell as a paraneuron (i.e., a cell possessing structural and functional features common to neurons and endocrine cells) has stimulated advances in our understanding of the biology of this cell. Comparing the cell with paraneurons (sensor/endocrine cells) in the gut proves especially worthwhile for understanding its hitherto unknown nature.

The gustatory cell possesses, in addition to small, clear vesicles, large-cored (peptidergic) vesicles. In some mammals these vesicles are so numerous and extensive in the cell base that we postulate an endocrine/paracrine function of this cell in addition to its role in synaptic transmission. Exocytotic granule release can be evoked by apically stimulating the cell with taste substances. Enkephalins are proposed as the signal peptides probably secreted by the gustatory cell, according to our immunocytochemical study. Phylogenetic studies indicate that the gustatory cells (for food uptake) and the sensor cells in the gut (for digestion) are the most primitive and essential regulatory cells inherited from the ancestor of metazoa, *Hydra*.

References

1. Fujita T (1976) The gastro-enteric endocrine cell and its paraneuronic nature. In: Fujita T, Coupland RE (eds) Chromaffin, enterochromaffin and related cells. Elsevier, Amsterdam, pp 191–208
2. Fujita T, Kanno T, Kobayashi S (1988) The paraneuron. Springer, Berlin Heidelberg New York Tokyo
3. Fujita T, Iwanaga T, Nakajima T (1983) Immunohistochemical detection of nervous system-specific proteins in normal and neoplastic paraneurons in the gut and pancreas. In: Miyoshi A (ed) Gut peptides and ulcer. Biomedical Research Foundation, Tokyo, pp 81–88
4. Iwanaga T, Takahashi-Iwanaga H, Fujita T, Yamakuni T, Takahashi Y (1985) Immunohistochemical demonstration of a cerebellar protein (spot 35 protein) in some sensory cells of guinea pig. Biomed Res 6:329–334
5. Farbman AI (1965) Fine structure of the taste bud. J Ultrastr Res 12:328–350
6. Murray RG, Murray A, Fujimoto S (1969) Fine structure of gustatory cells in rabbit taste buds. J Ultrastr Res 27:444–461
7. Takeda M, Hoshino T (1975) Fine structure of taste buds in the rat. Arch Histol Jpn 37:395–413
8. Delay RJ, Kinnamon JC, Roper SD (1986) Ultrastructure of mouse vallate taste buds. II: Cell types and cell lineage. J Comp Neurol 253:242–252
9. Yoshie S, Wakasugi C, Teraki Y, Fujita T (1990) Fine structure of taste bud in guinea pig. I: Cell characterization and innervation patterns. Arch Histol Cytol 53:103–119
10. Kanazawa H, Yoshie S, Fujita T (1993) Ultrastructure of canine circumvallate taste buds. This volume
11. Yoshie S, Wakasugi C, Kanazawa H, Fujita T (1993) Receptosecretory nature of the gustatory cell. This volume
12. Nada O, Hirata K (1977) The monoamine-containing cell in the gustatory epithelium of some vertebrates. Arch Histol Jpn Suppl 40:197–206
13. Uchida T (1985) Serotonin-like immunoreactivity in the taste bud of the mouse circumvallate papilla. Jpn Oral Biol 27:132–139
14. Morimoto K, Sato M (1982) Noradrenaline as a chemical transmitter from taste cells to sensory nerve terminals in frog. Poc Jpn Acad 51:347–352
15. Nagahama S, Kurihara K (1985) Norepinephrine as a possible transmitter involved in synaptic transmission in frog taste organs and Ca dependence of its release. J Gen Physiol 85:431–442
16. Yoshie S, Wakasugi C, Teraki Y, Kanazawa H, Iwanaga T, Fujita T (1991) Response of the taste receptor cell to the umami-substance stimulus: an electron-microscopic study. Physiol Behav 49:887–889
17. Fujita T, Kobayashi S (1977) Structure and function of gut endocrine cells. Int Rev Cytol Suppl 6:187–233
18. Fujita T (1991) Taste cells in the gut and on the tongue: their common, paraneuronal features. Physiol Behav 49:883–885
19. Tonosaki K (1988) Generation mechanisms of mouse taste cell responses. In: Miller IJ Jr (ed) The Breidler-symposium on taste and smell. Book Service Associates, Winston-Salem, pp 93–102
20. Whitear M, Lane EB (1983) Multivillous cells: epidermal sensory cells of unknown function in lamprey skin. J Zool Lond 201:259–272
21. Yanagisawa K, Shiozawa K (1980) Stimulus reception in hair cells of the lateral-line organ. Biomed Res Suppl 1:128–129
22. Lenhoff HM (1961) Activation of the feeding reflex in *Hydra littoralis*. I: Role played by reduced glutathione and quantitative assay of the feeding reflex. J Gen Physiol 45:331–344

Receptosecretory Nature of the Gustatory Cell

Sumio Yoshie[1], Chikashi Wakasugi[1], Hiroaki Kanazawa[2], and Tsuneo Fujita[2]

Key words. Gustatory stimulation—Taste bud—Gustatory cell—Synapse—Dense-cored vesicle—Guinea pig

Introduction

The taste bud is a chemoreceptive sensory organ functioning primarily for gustatory sensation. This organ, existing in the oral region of all vertebrate classes, is composed of specialized epithelial cells and nerve fibers. Extensive morphologic studies have revealed that taste buds comprise gustatory cells that form synapses with afferent sensory fibers along with other nonsynapsing sustentacular and precursor cells [e.g., 1–4]. The gustatory cells contain membrane-bound vesicles in the cytoplasm. Although these vesicles are variable in appearance and size among species, they are fundamentally categorized into small, clear and large dense-cored types. As both types of vesicles accumulate around the synaptic zone of the cytoplasm, the vesicles are considered to contain a transmitter or transmitters for the nerve.

Our previous taste stimulation study on the guinea pig taste bud revealed that the umami substances monosodium L-glutamate (MSG) and guanosine 5′-monophosphate (GMP) make the gustatory cells discharge the contents of the dense-cored vesicles not only into the synaptic cleft but also into the intercellular space [5]. The present study extends to other basic tastes in order to clarify whether different qualities of stimuli cause variable responses of the gustatory cells in guinea pigs.

Materials and Methods

Five male guinea pigs were used. After being anesthetized by an intraperitoneal injection of sodium pentobarbital (50 mg/kg), the dorsal surface of each tongue was flashed in situ with one of the following stimulants for 20 seconds: 0.5 M sucrose (sweet), 0.1 M citric acid (sour), 10 mM quinine hydro-chloride (bitter), 0.3 M sodium chloride (salty), and a mixture of 20 mM MSG and 1 mM GMP (umami). Immediately after stimulation, the circumvallate papillae were removed, cut into small pieces, and immersed overnight in 2.5% glutaraldehyde in 0.1 M cacodylate buffer at pH 7.3. They were then post-fixed in 2% osmium tetroxide in the buffer for 2 h. After dehydration through an ascending ethanol series and propylene oxide, the tissue blocks were embedded in Epon 812. Ultrathin sections were double-stained with uranyl acetate and lead citrate and then examined with a JEOL 1200EX transmission electron microscope under an accelerating voltage of 80 kV.

Results

In response to each stimulus the gustatory cells exhibited definite ultrastructural modifications. All other types of cells and the nerves distributed in the taste buds remained unaffected. Before describing the changes of the gustatory cells in detail, the ultrastructural features of the nonstimulated gustatory cells are reviewed briefly (for details, see [6]).

The gustatory cells in guinea pigs contain numerous round, dense-cored vesicles with a mean diameter of 90 nm and with a moderately electron-dense material. The vesicles are in close association with the Golgi apparatus, scatter in the perinuclear cytoplasm, and accumulate in the synaptic zone (Fig. 1). Two types of vesicles occur in the synaptic area: small, clear vesicles (40 nm mean diameter) and dense-cored vesicles (Fig. 1b). As for their localizations at the synapse, the dense-cored vesicles tend to be closely juxtaposed to the synaptic membrane, whereas the small, clear vesicles are located at a distance from the membrane.

Although every stimulant was effective in causing a response by the gustatory cells, not all the cells responded to each stimulus. Moreover, the responsive cells displayed common changes for all the stimuli as follows.

The synaptic membrane of certain gustatory cells invaginated toward the cytoplasm (Fig. 2a–d). Most invaginations were empty, but some contained a dense material comparable of that appearing in the dense-cored vesicles (Fig. 2c). Hence these structures indicate that the dense-cored vesicles are being

---

[1] Department of Oral Anatomy, Nippon Dental University, 1-8, Hamaura-cho, Niigata, 951 Japan
[2] Department of Anatomy, Niigata University School of Medicine, Asahi-machi 1, Niigata, 951 Japan

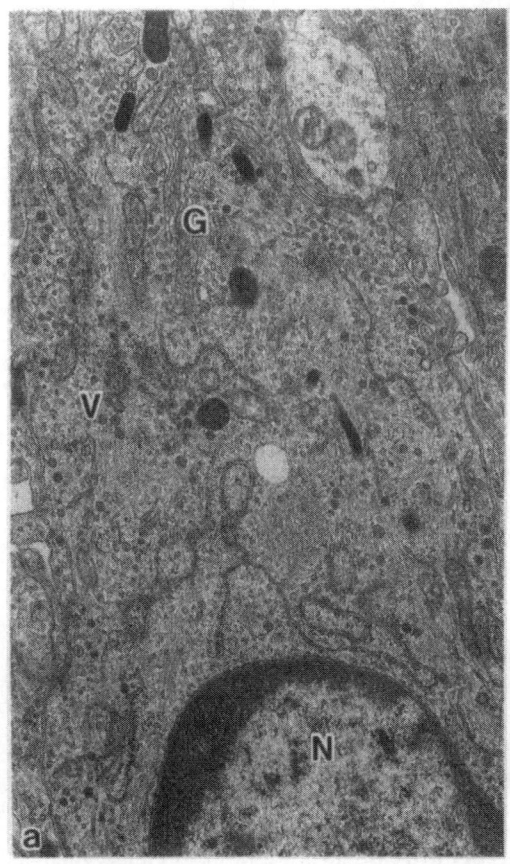
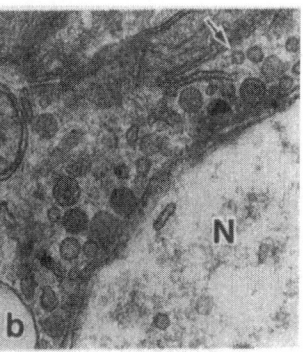

FIG. 1. Nonstimulated gustatory cell and synaptic contact with a nerve. **a** Cytoplasm contains dense-cored vesicles (*V*). Note that the vesicles are in close association with the Golgi apparatus (*G*). *N*, nucleus. **b** At the synapse, dense-cored vesicles and smaller, clear vesicles (*arrow*) are situated along the synaptic membrane of the gustatory cell. N, nerve terminal. **a** ×14 000; **b** ×48 000

exocytosed toward the nerves. Probably as a result of their rapid release, the dense-cored vesicles have decreased in number at the synapses (Fig. 2a). The small, clear vesicles, another component at the synapses, increased in number but exhibited no exocytotic signs (Fig. 2b).

In addition to their release at the synapses, the dense-cored vesicles were discharged at non-synaptic sites into the intercellular space (Fig. 2e). None of the small, clear vesicles appeared at those sites. No significant structural changes were recognized in any other organelles, e.g., endoplasmic reticulum and Golgi apparatus of the responding gustatory cells as well as the nonresponding cells.

Discussion

This study shows that the guinea pig gustatory cells, but not all, release the contents of the large dense-cored vesicles into the synaptic cleft and inter-cellular space; the small, clear vesicles at the synapses in these cells accumulate in response to every stimulus. With regard to the umami taste, the present data confirm our previous findings [5].

Although the sense of taste has a wide spectrum, sweet, sour, bitter, salty, and umami sensations are generally accepted as basic taste modalities. Indeed, electrical responses have been recorded from mouse gustatory cells stimulated with the corresponding taste substances [7]. Individual gustatory cells reportedly were variable in the manner of response: Certain cells responded to single basic taste stimuli and others to two or more stimuli. The gustatory cells receiving more than two stimuli showed no regularity in the response modalities. These electro-physiologic data signify that all the gustatory cells do not always respond to each basic taste stimulus and, accordingly, coincide with the present results.

The present findings that gustatory cells respond to different stimuli in the same way (i.e., exocytotic discharges of the dense-cored vesicles) indicate that identical transmitter substance(s) might be involved in the transduction of different taste information.

Any gustatory cells in submammalian and mammalian species comprise intrinsically dense-cored vesicles, which resemble those contained in autonomic neurons and other paraneuronal cells (e.g., Merkel cells, small granule-containing cells in the adrenal medulla, chief cells in the carotid body, and

FIG. 2. Changes of gustatory cell after stimulation. Gustatory cells are stimulated with sucrose (**a,e**), citric acid (**b**), sodium chloride (**c**), and a mixture of monosodium L-glutamate and guanosine 5′-monophosphate (**d**). **a–d** Note synapses between gustatory cells and nerve terminals (*N*). The membrane invaginations of the gustatory cells (*arrows*) indicate the dense-cored vesicles being released by exocytosis toward the nerves. **e** Dense-cored vesicle (*arrow*) is exocytosed also at a nonsynaptic site into the intercellular space. **a** ×40 000; **b** ×60 000; **c** ×63 000; **d** ×49 000; **e** ×48 000

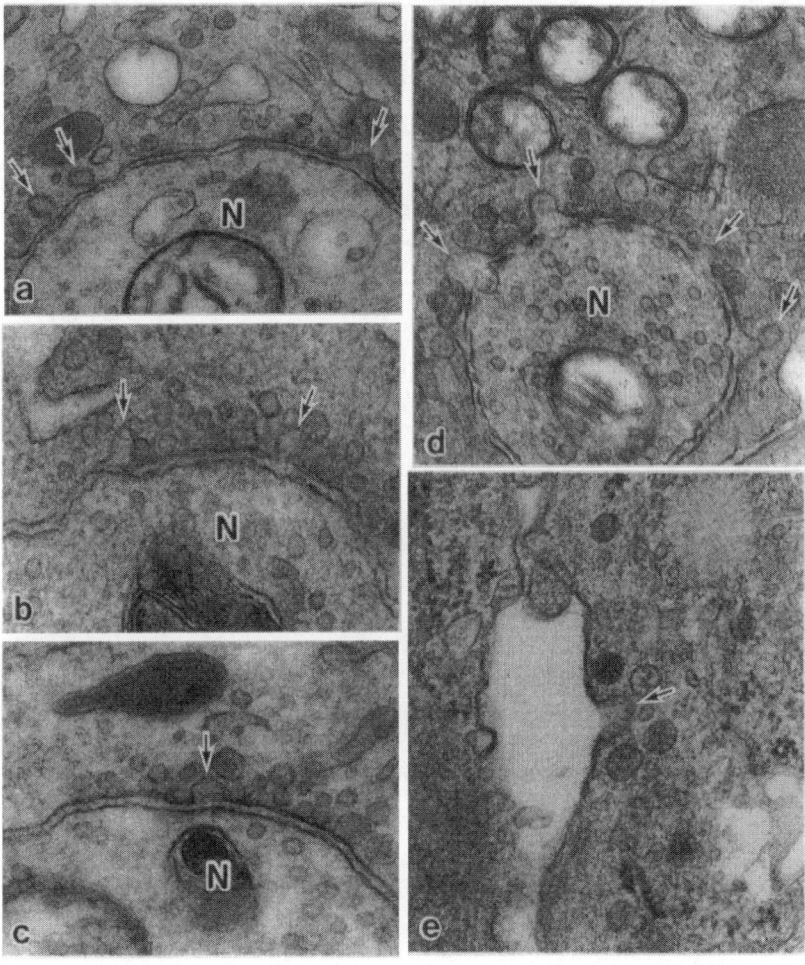

pinealocytes) [8]. In certain neurons and paraneurons, furthermore, the dense-cored vesicles have been demonstrated to include bioactive peptides [8–10]. Hence it is probable that in the gustatory cells such peptides are contained in the vesicles to transduce the excitement of the cells to the nerves.

Our preliminary immunohistochemical examination indicates the occurrence of cells immunoreactive for Met-enkephalin-Arg^6-Gly^7-Leu^8, a component peptide of preproenkephalin A [11], in the taste buds of the guinea pig and some other mammalian species (Yoshie et al., manuscript in preparation). This peptide has been demonstrated to be located in the granules or vesicles of adrenomedullary cells [12], chief cells in the carotid body [13], and Merkel cells [14]. Thus enkephalins are likely candidates for the transmitters of the gustatory cells.

Although small, clear vesicles have been reported to predominate in the nonstimulated gustatory cells in the rabbit, rat, and some other species [4,15–18], our observations in the guinea pig indicate that they accumulate densely at the synapses of the gustatory cells only after stimulation. Moreover, the present examination failed to demonstrate any evidence of their exocytotic opening. Despite these enigmatic findings, it seems reasonable to consider that the clear vesicles are also synaptic vesicles containing transmitter(s). Further studies are required to disclose how the small, clear vesicles participate in taste transduction.

Another conspicuous feature of the responding gustatory cells is the release of the dense-cored vesicles other than at the synaptic site. This action of the gustatory cells possibly causes paracrine effects of the messenger substances as local hormones on the surroundings. The Ebner's salivary gland may be considered one of the targets [19].

Acknowledgments. This work was supported by Umami Manufacturers Association of Japan.

References

1. Hirata Y (1966) Fine structure of the terminal buds on the barbels of some fishes. Arch Histol Jpn 26:507–523

2. Graziadei PPC, Dehan RS (1971) The ultrastructure of frogs' taste organs. Acta Anat (Basel) 80:563–603
3. Uchida T (1980) Ultrastructural and histochemical studies on the taste buds in some reptiles. Arch Histol Jpn 43:459–478
4. Murray RG, Murray A, Fujimoto S (1969) Fine structure of gustatory cells in rabbit taste buds. J Ultrastruct Res 27:444–461
5. Yoshie S, Wakasugi C, Teraki Y, Kanazawa H, Iwanaga T, Fujita T (1991) Response of the taste receptor cell to the umami-substance stimulus. An electron-microscopic study. Physiol Behav 45:887–889
6. Yoshie S, Wakasugi C, Teraki Y, Fujita T (1990) Fine structure of the taste bud in guinea pigs. I. Cell characterization and innervation patterns. Arch Histol Cytol 53:103–119
7. Tonosaki K, Funakoshi M (1984) Intracellular taste cell responses of mouse. Comp Biochem Physiol [A] 78:651–656
8. Fujita T, Kanno T, Kobayashi S (1988) The paraneuron. Springer, Berlin Heidelberg New York Tokyo
9. Fujita T, Kobayashi S, Uchida T (1984) Secretory aspect of neurons and paraneurons. Biomed Res Suppl 5:1–8
10. Hökfelt T, Johansson O, Ljungdahl Å, Lundberg JM (1980) Peptidergic neurons. Nature 284:515–521
11. Noda M, Furutani Y, Takahashi H, Toyosato M, Hirose T, Inayama S, Nakanishi S, Numa S (1982) Cloning and sequence analysis of cDNA for bovine adrenal preproenkephalin. Nature 295:202–206
12. Kobayashi S, Ohashi T, Fujita T, Nakao K, Yoshimasa T, Imura H, Mochizuki T, Yanaihara C, Yanaihara N, Verhofstad AAJ (1983) An immunohistochemical study on the co-storage of Met-enkephalin-Arg[6]-Gly[7]-Leu[8] and Met-enkephalin-Arg[6]-Phe[7] with adrenaline and/or noradrenline in the adrenal chromaffin cells of the rat, dog and cat. Biomed Res 4:433–442
13. Kobayashi S, Uchida T, Ohashi T, Fujita T, Nakao K, Yoshimasa T, Imura H, Mochizuki T, Yanaihara C, Yanaihara N, Verhofstad AAJ (1983) Immuno-cytochemical demonstration of the co-storage of noradrenalin with Met-enkephalin-Arg[6]-Phe[7] and Met-enkephalin-Arg[6]-Gly[7]-Leu[8] in the carotid body chief cells of the dog. Arch Histol Jpn 46:713–722
14. Hartschuh W, Weihe E, Buchler M, Helmstaedter V, Feurle GE, Forssmann WG (1979) Met-enkephalin-like immunoreactivity in Merkel cells. Cell Tissue Res 201:343–348
15. Takeda M, Hoshino T (1975) Fine structure of taste buds in the rat. Arch Histol Jpn 37:395–413
16. Takeda M (1976) An electron microscopic study on the innervation in the taste buds of the mouse circumvallate papillae. Arch Histol Jpn 39:257–269
17. Idé C, Munger BL (1980) The cytologic composition of primate laryngeal chemosensory corpuscles. Am J Anat 158:193–209
18. Paran N, Mattern CFT, Hankin RI (1975) Ultrastructure of the taste bud of the human fungiform papilla. Cell Tissue Res 161:1–10
19. Fujita T, Kobayashi S (1979) Aspects of paraneuron. Cell 11:45–55

Structure/Function Correlates in Taste Buds

JOHN C. KINNAMON[1], MARTHA M. MCPHEETERS[2], and SUE C. KINNAMON[2]

Key words. Gustation—Mudpuppy—*Necturus*—Electron microscopy—Patch-clamp

Introduction

The taste bud is an end-organ specialized for the detection of aqueous chemical stimuli. A typical taste bud comprises approximately 50 to 150 cells. Taste bud receptor cells (= taste cells) are spindle-shaped cells that terminate apically in one to several microvillar processes. The process of taste transduction involves the interaction of taste stimuli with those apical microvilli, followed by membrane conductance changes, depolarization of the taste cell membrane, and exocytosis of transmitter onto gustatory afferent neurons at synaptic contacts. Studies suggest that taste cells use a variety of mechanisms for transduction, including apically located ion channels, ligand-gated channels, and G-protein-coupled receptors [1].

Although considerable information is emerging about the biophysical properties of taste cells, little is known about the specificity of taste cells for particular taste stimuli. Early intracellular recordings revealed that most taste cells respond to all taste modalities. However, afferent nerve recordings indicate that single nerve fibers usually respond best to one or a few taste modalities [2,3]. Caprio and coworkers have shown separate amino-acid-sensitive nerve fibers in the catfish [2,4–6]. Studies using giga-seal whole cell recording suggest that particular subsets of taste cells may respond to particular taste modalities [7]. Do these subsets of taste cells correspond to a particular morphologic taste cell type?

A variety of taste cell types have been identified using ultrastructural criteria, including basal cells, dark (type I) cells, and light (type II) cells [for review see 8]. Basal cells are small, round cells found at the basolateral margin of the taste bud. These cells lack processes reaching the taste pore and are unlikely to participate directly in the transduction process. In contrast, both dark cells and light cells have processes that reach the taste pore; in addition, both cell types form chemical synapses with primary afferent neurons. Thus both dark cells and light cells possess the morphologic features requisite of taste receptor cells.

Taste stimuli elicit action potentials in taste receptor cells [9–16]. Although the role of the action potential is unclear, the capability of cells to generate action potentials is a defining feature of taste receptor cells. Another characteristic of receptor cells is the presence of voltage-activated Ca^{2+} channels [14,17], which are necessary for initiating transmitter release. Thus the presence of voltage-activated Ca^{2+} channels and the ability to generate action potentials in putative receptor cells is good evidence of their direct involvement with the taste transduction process.

As a first step in defining the functional roles of morphologically identifiable taste receptor cells, we have combined structural and functional studies of individual *Necturus* taste cells. Using giga-seal whole cell recording of isolated taste cells, we have characterized the electrophysiologic properties of taste cells and later identified the cell type with electron microscopy. All taste cells classified as dark cells had voltage-activated Na^+, K^+, and Ca^{2+} currents, suggesting that these cells can serve as taste receptor cells. Taste cells classified as light cells fell into two distinct classes based on electrophysiologic criteria. One class had only voltage-activated K^+ currents. It is unlikely that these cells are receptor cells. The second class had voltage-activated Na^+, K^+, and Ca^{2+} currents, with properties similar to those of dark cells. Thus both dark cells and a subset of light cells are capable of generating action potentials and releasing transmitter in response to taste stimulation.

Material and Methods

Taste Cell Isolation and Recording

Tongues were excised from mudpuppies (*Necturus maculosus*), and taste cells were isolated from the surrounding nongustatory epithelium according to the method of Kinnamon et al. [18]. Briefly, the epithelium was stripped from the tongue, and the mucosal surface was treated with fluorescein isothiocyanate-conjugated wheat germ agglutinin (FITC-WGA) to label the apical membranes of

[1] Department of Biological Sciences, University of Denver, Denver, CO 80208, USA
[2] Department of Anatomy & Neurobiology, Colorado State University, Fort Collins, CO, USA

mature taste cells. This procedure allowed us to select for recording only those mature cells with microvillar processes that extended into the taste pore. The epithelium was then incubated in amphibian physiological saline (APS) containing collagenase, albumin, and glucose until the mucosal layers of the nongustatory epithelium could be gently peeled free from the underlying lamina propria, leaving the taste buds attached to their papillae. The remaining tissue was treated with Ca^{2+}-free APS-containing BAPTA (Molecular Probes, Eugene, OR) to dissociate the taste buds. Isolated cells were plated into plastic culture dishes coated with Cell-Tak (Collaborative Research, Bedford, MA). Cells were viewed at 400× with a Nikon Diaphot (Nikon, Garden City, NY) inverted microscope equipped with epifluorescence. Membrane current or voltage was recorded and analyzed using the whole-cell configuration of the patch-clamp technique [19].

Electron Microscopy

After electrophysiologic recording, each cell was photographed using the light microscope to facilitate subsequent identification of the location of the cell. Cells were then fixed in 2% glutaraldehyde in 0.05 M sodium cacodylate (buffered to pH 7.2 with NaOH). The cells were postfixed with 1% OsO_4 and then dehydrated through an ethanol series and embedded in Luft's medium [20]. The cells of interest were then cut out of the embedding medium, and the embedding wafer containing the cell was glued onto a blank block and trimmed for ultramicrotomy. Semithin (0.25 μm) sections were cut using a Diatome diamond knife (Diatome, Zurich, Switzerland) on a Reichert Ultracut E ultramicrotome (Leica, Inc., Deerfield, IL) and collected onto Formvar-coated slot grids. The sections were poststained with alcoholic uranyl acetate and Reynolds lead stain [21]. The sections were examined and photographed using a Hitachi H-7000 transmission electron microscope (Hitachi, Tokyo, Japan) at an accelerating voltage of 100 kV.

Results

Previous ultrastructural studies have demonstrated that *Necturus* taste buds contain both dark cells and light cells, with light cells constituting approximately 30% of the mature receptor cell population [22,23]. Light cells were initially distinguished from dark cells based on their staining properties after treatment with quinacrine, a naturally fluorescent molecule that is taken up and retained in the apical granules of dark cells [24]. With practice, however, it became possible to select light cells based on their size and appearance in the light microscope. Isolated

light cells were ovoid or spindle-shaped with stubby processes and were considerably smaller than the dark cells. They also tended to have blebs present on the membrane surface. After isolation, light cells were found to comprise only 5% to 10% of the receptor cell population. Thus light cells either did not survive the dissociation procedure as well as dark cells, or light cells were lost at an earlier stage in the dissociation procedure.

Dark Cells

Dark cells were identified after recording by the presence of dense granular packets in the supranuclear and apical regions of the cytoplasm. Even after recording, these cells retained their ultrastructural features to a remarkable degree. All dark cells were characterized by the presence of voltage-activated inward Na^+ and Ca^{2+} currents and outward K^+ currents in response to depolarizing voltage steps from a holding potential of −80 mV. The Na^+ current activated at −20 mV, peaked at 0 mV, and decreased with further depolarization. This current activated and inactivated rapidly and was tetrodotoxin (TTX)-sensitive. The K^+ current activated at +10 mV and increased almost linearly with further depolarization. This current, which was tetraethylammonium (TEA)-sensitive, activated more slowly than the Na^+ current and exhibited little or no inactivation. The voltage-dependent Ca^{2+} current was revealed by bathing the cells in a high Ba^{2+} solution containing both TEA and TTX. This Ca^{2+} (Ba^{2+}) current activated at 0 mV, peaked at +20 mV, and decreased with further depolarization. This current inactivated slowly.

Light Cells

Light cells were identified following recording by the absence of the characteristic membrane-bound "granular packets" observed in the apical cytoplasm of dark cells. Light cells were also characterized by electron-lucent, empty-appearing vesicular inclusions distributed throughout the cytoplasm. Two general morphotypes of light cells were observed: ovoid light cells and the more spindle-shaped light cells. All light cells were significantly smaller than dark cells, as estimated from membrane capacitance measurements.

Light cells were divided into two functional populations based on electrophysiologic criteria: those cells with inward and outward currents, and those cells with outward currents only. The current profiles of the two cells with inward and outward currents resembled those of dark cells; that is, depolarizing voltage commands from a holding potential of −80 mV elicited Na^+, Ca^{2+}, and K^+ currents. Both of these cells were of the spindle-shaped morphotype. In contrast, the eight light cells with outward currents only were ovoid in shape.

Depolarizing voltage commands from a holding potential of $-80\,mV$ elicited only sustained outward K^+ currents in these cells. Bathing the cells in the high Ba^{2+} solution reduced the K^+ current, but there was no evidence of a voltage-activated Ca^{2+} current. In general, the K^+ currents of both types of light cells activated more slowly than the K^+ currents of dark cells.

Discussion

This study provides direct evidence that the different taste bud cell types have different populations of ion channels. From this observation we speculate that different taste bud cell types may respond differently to applied chemical stimuli. Dark cells possess voltage-activated Na^+, Ca^{2+}, and K^+ currents, whereas light cells belong to one of two classes: those with inward and outward currents, and those with outward K^+ currents only.

This study provides the first information about the functional properties of light cells. Some, but not all, light cells have the ability to generate action potentials in response to gustatory stimuli. It is not yet clear what role the action potential plays in taste transduction. It does appear, however, that the action potential is a common feature of taste bud receptor cells [9,10,15,25,26]. Both dark cells and some light cells contain voltage-dependent Na^+ and K^+ currents, providing a basis for regenerative potentials. These cells also contain a voltage-dependent Ca^{2+} current, which may provide a mechanism for exocytosis of transmitter at synapses. Thus given the known capabilities of these cells, we believe that both light cells and dark cells serve as receptor cells for taste stimulation.

Our data corroborate the findings of Bigiani and Roper [27], who studied membrane currents of *Necturus* taste cells in a lingual slice preparation. Two populations of receptor cells were identified, based on their current profiles. One population possessed voltage-activated Na^+, K^+, and Ca^{2+} currents, and the other population had only voltage-activated K^+ currents. Similar electrophysiologically distinct subpopulations of taste cells have been observed in rat taste buds [7,15].

We believe that it will eventually be possible to make predictions concerning the functional properties of a taste cell based on its morphotype. It is tempting to speculate that the major morphologic differences between these cell types will be mirrored in their functional properties. Based on the outward and inward currents that we have observed in both cells types, it does not appear to be the case. Satisfactory answers concerning the relation between the structure and function of taste cell types remain to be found. The present study provides evidence for functional differences among taste cells. McPheeters

and Roper [28] have shown that mudpuppies respond to a variety of sapid stimuli, ranging from salts and amino acids to complex mixtures of components from their normal diet. Our future goals are to apply appropriate sapid stimuli to isolated receptor cells, determine their responses, and finally correlate these results with the morphologic features of the receptor cells.

Acknowledgments. We thank Ms. Janet Meehl for her help in initiating this study, as well as Mr. Andrew Barber, Mr. Dan Harris, Mr. Andrew Bowerman, and Ms. Hildegarde Crowley for excellent technical asistance. We also thank Dr. Stephen D. Roper for his valuable comments on the manuscript. This work was supported in part by NIH grants DC00285 (JCK), DC00766 (SCK), and PO00244 (J.C.K., S.C.K.).

References

1. Kimura K, Beidler LM (1961) Microelectrode study of taste receptors of rat and hamster. J Comp Cell Physiol 58:131–139
2. Davenport CJ, Caprio J (1982) Taste and tactile recordings from the ramus recurrens facialis innervating flank taste buds in catfish. J Comp Physiol 147:217–229
3. Frank ME, Bieber SL, Smith DV (1988) The organization of taste sensibilities in hamster chorda tympani nerve fibers. J Gen Physiol 91:861–896
4. Caprio J (1975) High sensitivity of catfish taste receptors to amino acids. Comp Biochem Physiol [A] 52:247–251
5. Caprio J (1978) Olfaction and taste in the channel catfish: an electrophysiological study of the responses to amino acids and derivatives. J Comp Physiol 123:357–371
6. Kanwal JS, Caprio J (1983) An electrophysiological investigation of the oropharyngeal (IX–X) taste system in the channel catfish, *Ictalurus punctatus*. J Comp Physiol 150:345–357
7. Akabas M, Dodd J, Al-Awqati Q (1990) Identification of electrophysiologically distinct subpopulations of rat tast cells. J Membr Biol 114:71–78
8. Kinnamon J (1987) Organization and innervation of taste buds. In: Finger TE, Silver WL (eds) Neurobiology of taste and smell. Wiley, New York, pp 277–297
9. Roper S (1983) Regenerative impulses in taste cells. Science 220:1311–1312
10. Kashiwayanagi M, Miyake M, Kurihara K (1983) Voltage-dependent Ca^{2+} channel and Na^+ channel in frog taste cells. Am J Physiol 244:C82–C88
11. Avenet P, Lindemann B (1988) Amiloride-blockable sodium currents in isolated taste receptor cells. J Membr Biol 105:245–255
12. Avenet P, Lindemann B (1991) Non-invasive recording of receptor cell action potentials and sustained currents from single taste buds maintained in the tongue: the response to mucosal NaCl and amiloride. J Membr Biol 124:33–41
13. Avenet P, Kinnamon SC (1991) Cellular basis of taste reception. Curr Opin Neurobiol 1:198–203

14. Kinnamon S, Roper SD (1988) Evidence for a role of voltage-sensitive apical K^+ channels in sour and salt taste transduction. Chem Senses 13:115–121
15. Béhé P, DeSimone JA, Avenet P, Lindemann B (1990) Membrane currents in taste cells of the rat fungiform papilla: evidence for two types of Ca currents and inhibition of K currents by saccharin. J Gen Physiol 96:1061–1084
16. Gilbertson TA, Avenet P, Kinnamon SC, Roper SD (1992) Proton currents through amiloride-sensitive Na channels in hamster taste cells. J Gen Physiol 100: 803–824
17. Kinnamon SC, Cummings TA, Roper SD, Beam KG (1989) Calcium currents in isolated taste receptor cells of the mudpuppy. Ann NY Acad Sci 560:112–115
18. Kinnamon SC, Cummings TA, Roper SD (1988) Isolation of single taste cells from lingual epithelium. Chem Senses 13:355–366
19. Hamill OP, Marty A, Neher E, Sakmann B, Sigworth FJ (1981) Improved patch-clamp techniques for high-resolution current recording from cell and cell-free membrane patches. Pfluegers Arch 391:85–100
20. Luft JH (1961) Improvements in epoxy resin embedding methods. J Biophys Biochem Cytol 9:409–414
21. Reynolds ES (1963) The use of lead citrate at high pH as an electron-opaque stain in electron microscopy. J Cell Biol 17:208–212
22. Farbman AI, Yonkers JD (1971) Fine structure of the taste bud in the mudpuppy, *Necturus maculosus*. Am J Anat 131:353–370
23. Delay RJ, Roper SD (1988) Ultrastructure of taste cells and synapses in the mudpuppy *Necturus maculosus*. J Comp Neurol 277:268–280
24. Delay RJ, Ruiz C, Kinnamon SC, Roper SD (1990) Determination of cell type in acutely isolated taste cells using quinacrine fluorescence. Proc Soc Neurosci 16:878
25. Kinnamon SC, Roper SD (1987) Passive and active membrane properties of mudpuppy taste receptor cells. J Physiol (Lond) 383:601–614
26. Avenet P, Lindemann B (1987) Action potentials in epithelial taste receptor cells induced by mucosal calcium. J Membr Biol 95:265–269
27. Bigiani A, Roper SD (1993, in press) Electrically coupled and noncoupled taste cells in *Necturus*: identification of electrophysiologically distinct cellular subpopulations. J Gen Physiol
28. McPheeters M, Roper SD (1985) Amiloride does not block taste transduction in the mudpuppy, *Necturus maculosus*. Chem Senses 10:341–352

Role of Merkel Cells in the Taste Organ Morphogenesis of the Frog

KUNIAKI TOYOSHIMA[1]

Key words. Taste organ—Merkel cell—Ultrastructure — Immunohistochemistry — Morphogenesis — Frog

Introduction

It has long been believed that there exists an intimate association between taste buds and gustatory nerve fibers during embryogenesis. The formation of taste buds is induced only after gustatory nerves innervate the embryonic epithelium. Furthermore, the integrity of taste buds also depends on a trophic influence of the gustatory nerves. It is well known that the taste bud disappears rapidly after denervation and reappears following regeneration of the nerve fibers. Still unclear, however, is how the gustatory nerves entering the tongue find the way to their appropriate target sites where the taste buds will develop. Farbman and Mbiene [1] suggested, in their study on the development of rat fungiform papillae, that the growth of sensory nerve fibers toward the gustatory epithelium needs the existence of a chemotrophic factor originating from the target epithelium.

Evidence suggests that the Merkel cell has a trophic influence on the nerve fibers [2,3]. According to this theory, the Merkel cell may act as a target site for the developing nerve fibers. The Merkel cell may be trophic for growing nerve terminals, possibly via the secretion of diffusible substances [4]. Since a cell type containing monoamine was discovered by Reutter [5] in the taste buds of the catfish barbel, special attention has been paid to the morphology and physiology of this distinct cell type in the taste buds of various animal species [6]. Immunohistochemistry has shown that monoamine-containing (MC) cells in vertebrate taste buds are immunoreactive for both serotonin [7,8] and neuron-specific enolase (NSE) antisera [9,10].

The MC cell in mammalian taste buds exposes its apical process directly to the oral environment and has been considered to be a gustatory transducer cell. In contrast, the MC cell in amphibian and teleostean taste buds is located exclusively at the basal part and fails to reach the free surface of the taste buds. In this context, the MC cell in the taste buds of such lower vertebrates has been called the basal cell, and its role in the taste bud remains enigmatic.

This chapter describes briefly the development of the frog taste organ, particularly focusing on the origin and possible fuction of the basal cells. Our results indicate that the basal cells are modified Merkel cells, which may be responsible for initiation of taste organ morphogenesis.

Materials and Methods

Bullfrogs (*Rana catesbeiana*), ranging from metamorphic stage to climax stages A to D [11], were used as materials. The tongues were fixed in a trialdehyde-DMSO mixture [12] buffered with phosphate. The specimens were postfixed in 2% osmium tetroxide in the same buffer and then processed for electron microscopy.

For immunohistochemistry, the tongues were fixed in 4% paraformaldehyde buffered with phosphate. The tissues were then forzen with Dry Ice-isopentane and cut in a cryostat. The sections were immunostained with anti-serotonin (Chemicon, USA) and anti-NSE (Dakopatts, Denmark).

Results and Discussion

The taste organ of the adult bullfrog contains characteristic basal cells, which show immunoreactivities of both NSE and serotonin (Figs. 1 and 2). Ultrastructurally, the basal cell is characterized by the presence of numerous dense-cored granules in the cytoplasm and microvillous projections at the cell surface. These morphologic features of the basal cell in the frog taste organ conform in many aspects to the Merkel cell. Where the basal cell faces the epithelial basal lamina, intimate apposition of the nerve fibers is observed occasionally on the opposite side of the basal lamina. The detail morphology of the basal cells in the frog taste organs has been described [13–16].

[1] Department of Oral Anatomy, Kyushu Dental College, 2-6-1 Manazuru, Kokurakita-ku, Kitakyushu, 803 Japan

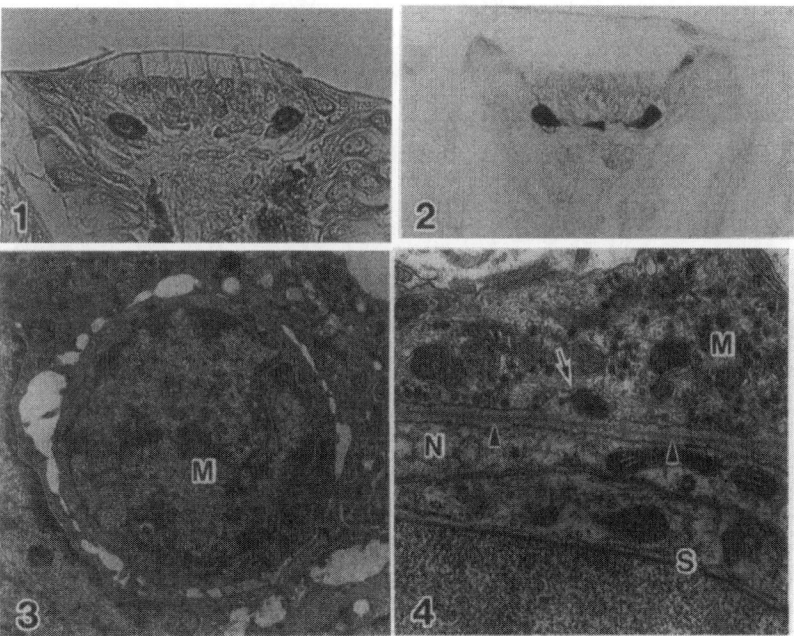

FIG. 1. Light micrograph showing NSE-immunoreactive Merkel cells in the basal periphery of the bullfrog taste organ. ×430

FIG. 3. Electron micrograph showing a Merkel cell (*M*) at the upper layer of the epithelium. ×6200

FIG. 2. Serotonin immunoreactivity is also present in Merkel cells. ×400

FIG. 4. Exocytotic release of the Merkel cell (*M*) granules (*arrow*) toward the nerve fiber (*N*) through the basal lamina (*arrowheads*). *S*, Schwann cell. ×24 500

During the metamorphic stage of the developing bullfrog, the primodium of the tongue is already recognizable in the anterior portion of the lower jaw in the oral cavity. A few Merkel cells appear at the uppermost layer of the epithelium just above the connective tissue papilla (Fig. 3). The Merkel cells are spherical and small, about 10 μm in diameter. The Merkel cell is covered by attenuated epithelial cells and is never exposed to the oral environment. The Merkel cell connects with adjacent epithelial cells by small desmosomes and lacks axon terminals. No neuronal elements are recognizable in the epithelium.

As development progresses, Merkel cells move down toward the basal lamina and finally become basal cells, which make a taste organ primodium. Where the Merkel cell borders on the basal lamina, the nerve fibers are closely in contact with the Merkel cell over a basal lamina. Occasionally, massive exocytotic releases of dense-cored granules of the Merkel cell toward the nerve fibers through the epithelial basal lamina are observed in these regions (Fig. 4). The present results clearly show that the basal cells in the frog taste organ differentiate from the Merkel cells, which appear in advance of the taste organ morphogenesis (Fig. 5).

What do Merkel cells do during morphogenesis of the taste buds? As described above, it has been supposed that the Merkel cell does not involve in mechanoreception but has a trophic influence on nerve fibers. It is of interest to suppose that the Merkel cells in the lingual epithelium of the developing frog also can act as target sites for growing gustatory nerves by secreting or leaking neurotrophic substances. These nerve fibers then invade the epithelium and may induce it to become taste organ cells by neural induction. Furthermore, even after formation of taste organs, the Merkel cells

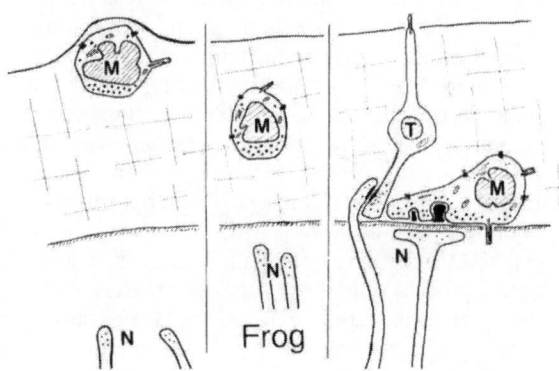

FIG. 5. Development of Merkel (*M*) and taste (*T*) cells in the gustatory epithelium of the bullfrog fungiform papillae. *N*, nerve terminal

(basal cells) in the taste organs can still secrete or leak neurotrophic substances, which may influence the survival and maintenance of nearby gustatory nerve fibers.

References

1. Farbman AI, Mbiene J-P (1991) Early development and innervation of taste bud-bearing papillae on the rat tongue. J Comp Neurol 304:172–186
2. Scott SA, Cooper E, Diamond J (1981) Merkel cells as targets of the mechanosensory nerves in salamander skin. Proc R Soc Lond [Biol] 211:453–470
3. Diamond J, Mills LR, Mearow KM (1988) Evidence that the Merkel cell is not the transducer in the mechanosensory Merkel cell-neurite complex. Progr Brain Res 74:51–56
4. Pasche F, Mérot Y, Carraux P, Saurat J-H (1990) Relationship between Merkel cells and nerve endings during embryogenesis in the mouse epidermis. J Invest Dermatol 95:247–251
5. Reutter K (1971) Die Geschmacksknospen des Zwergwelses *Amiurus nebulosus* (Lesueur): morphologische und histochemische Untersuchungen. Z Zellforsch 120:280–308
6. Nada O, Hirata K (1977) The monoamine-containing cell in the gustatory epithelium of some vertebrates. Arch Histol Jpn (Suppl) 40:197–206
7. Uchida T (1985) Serotonin-like immunoreactivity in the taste bud of the mouse circumvallate papilla. Jpn J Oral Biol 27:132–139
8. Kuramoto H (1988) An immunohistochemical study of cellular and nervous elements in the taste organ of the bullfrog, *Rana catesbeiana*. Arch Histol Cytol 51:205–221
9. Toyoshima K, Shimamura A (1988) An immunohistochemical demonstration of neuron-specific enolase in the Merkel cells of the frog taste organ. Arch Histol Cytol 51:295–297
10. Yoshie S, Wakasugi C, Teraki Y, Iwanaga T, Fujita T (1988) Immunohistochemical localization of neuron-specific proteins in the taste bud of the guinea pig. Arch Histol Cytol 51:379–384
11. Hirakawa T, Honda E, Toyoshima K, Tomo S, Nakahara S (1993) Glossopharyngeal-hypoglossal nerve reflex of the frog in metamorphosis. Arch Oral Biol 38:123–129
12. Kalt MR, Tandler B (1971) A study of fixation of early amphibian embryos for electron microscopy. J Ultrastruct Res 36:633–645
13. Düring MV, Andres KH (1976) The ultrastructure of taste and touch receptors of the frog taste organ. Cell Tissue Res 165:185–190
14. Toyoshima K, Honda E, Nakahara S, Shimamura A (1984) Ultrastructural and histochemical changes in the frog taste organ following denervation. Arch Histol Jpn 47:31–42
15. Sbarbati A, Franceschini F, Zancanaro C, Cecchini T, Ciaroni S, Osculati F (1988) The fine morphology of the basal cell in the frog's taste organ. J Submicrosc Cytol Pathol 20:73–79
16. Witt M (1993) Ultrastructure of the taste disc in the red-bellied toad *Bombina orientalis* (Discoglossidae, Salientia). Cell Tissue Res 272:59–70

Keratin Polypeptides and Taste Buds

BRUCE OAKLEY[1], ANNE LAWTON[1], LIANNA WONG[2], and CHUNXIAO ZHANG[1]

Key words. Antibody—Basal cells—Cytoskeleton —Intragemmal cells—Perigemmal cells—Salivary ducts

Introduction

Subsets of 20 soft keratin polypeptides have been identified in animal tissues by two-dimensional gel electrophoresis [1–6]. Differences among sets of keratins detected by immunocytochemistry distinguish types of epithelia or even specific regions of an epithelium [1,7–10]. From morphologic considerations we hypothesized that taste buds were islets of simple epithelium embedded in stratified squamous epithelium. If this hypothesis is correct, the keratins in taste buds ought to be distinctive from those of surrounding cells. We examined several gustatory epithelia with antikeratin antibodies to compare the immunoreactivity of fusiform, perigemmal, and basal gustatory cells and to identify useful cell markers. The present report concludes that antibodies against keratins 8 and 19 may be used as general differentiation markers for taste receptor cells.

Materials and Methods

Sprague-Dawley albino rats, Mongolian gerbils, *Meriones unguiculatus*, and New Zealand white rabbits were used.

Denervation of the Vallate Papilla

The vallate papilla was denervated by bilaterally avulsing the IX nerve in normal rats anesthetized with a mixture of 170 mg ketamine and 7 mg xylazine per kilogram body weight. All vallate taste buds degenerated and disappeared within the 15- to 21-day survival times.

Tissue Preparation

Animals were deeply anesthetized with an intraperitoneal injection of sodium pentobarbital. Rats and gerbils were perfused intracardially will Trisbuffered mammalian Ringer's solution [11] containing 0.02% sodium heparin and 0.5% procaine hydrochloride, followed by acid-alcohol fixative (70% ethanol, 10% acetic acid). Tissues of interest were excised and immersed in fixative for at least 1 h. In addition to various internal tissues, the following gustatory epithelia were examined: fungiform, foliate, and vallate papillae of the tongue and the epiglottis, nasopalatine papilla, and soft palate. Tissue sections 10 µm thick were cut in a cryostat at −20°C, mounted on gelatin-coated slides, and stored at −20°C.

Immunocytochemistry

The monoclonal antibodies (mAbs) used included LE41 (specific for cytokeratin no. 8), LE65 (no. 18), and Q3CK7 (no. 7) (Amersham, Arlington Heights, Ill.); 170.2.14 (no. 19) (Boehringer Mannheim, Indianapolis, IN); 20.5 (no. 20) (IBL Research, Cambridge, MA); AE-2 (nos. 1, 2, 10, 11) (Janssen, Piscataway, NJ; Chicago, IL); PKK3 (no. 18), (Labsystems, Raleigh, NC); LP2K (no. 19), a gift of Dr. E.B. Lane; and 4.62 (no. 19), 6B10 (no. 4), DK80.20 (no. 8), and LDS68 (no. 7) (Sigma Chemical, St. Louis, Mo.). Immunocytochemistry was carried out using either indirect immunofluorescence (FITC-, rhodamine-, or Texas Red-tagged antimouse secondary antibodies) or an avidin-biotin peroxidase method (ABC kit, Vector Laboratories, Burlingame, CA). No staining was observed when either the primary or the secondary antibody was omitted.

Results

Taste receptor cell renewal is believed to follow a lineage path in which basal (stem) cells generate suprabasal (perigemmal) daughters that in turn differentiate into receptor (intragemmal) cells [reviewed in 12]. We used a panel of mAbs against keratins in a search for suitable markers for these three classes of taste cells. This chapter emphasizes the evaluation of keratins in the cells within taste buds (intragemmal cells).

[1] Department of Biology, 3127 Natural Science Building, University of Michigan, Ann Arbor, MI 48109, USA
[2] Department of Molecular and Cell Biology, University of California at Berkeley, Berkeley, CA 94720, USA

Intragemmal (Receptor) Cells

From a survey of antikeratin mAbs we have found that the elongated taste receptor cells have keratin 7, 8, 18, 19, and 20-like immunoreactivity. A mAb against human keratin 20 reacted specifically with human fungiform and vallate taste buds (tissue courtesy of Inglis Miller) but was ineffective in other species. For the rat, gerbil, and rabbit we used at least two mAbs each for keratins 7, 8, 18, and 19. Examples are shown in Figure 1A–D. Keratin 19 was present in all mammalian taste buds examined, which included fungiform, foliate, and vallate taste buds in rat, gerbil, and rabbit, and nasopalatine, epiglottal (Fig. 1D), and palatine taste buds in rat.

All intragemmal fusiform cells appeared to be immunopositive for keratins 8 and 19 in longitudinal sections of taste buds (Fig. 1B,D), and it was confirmed by cryostat sections that grazed the surface of the vallate trench wall and sliced through the taste pits of several protruding taste buds evident as rings of keratin 8- and 19-positive cells (Fig. 1E). The center was unstained because the central cells are recessed and extend microvilli into the taste pit [13]. Sections immediately underneath the taste pit encroached on the keratin-positive cytoplasm of centrally located cells (Fig. 1F). Keratin 7-like immunoreactivity was present in many receptor cells. Keratin 18-like immunoreactivity was probably present in less than one-half of the vallate receptor cells, as judged from longitudinal sections (Fig. 1C) and some unstained cells around the taste pit.

Perigemmal Cells

Surrounding the keratin 19-positive intragemmal cells was a shell of perigemmal cells that were negative for keratin 19 (Fig. 1G, H).

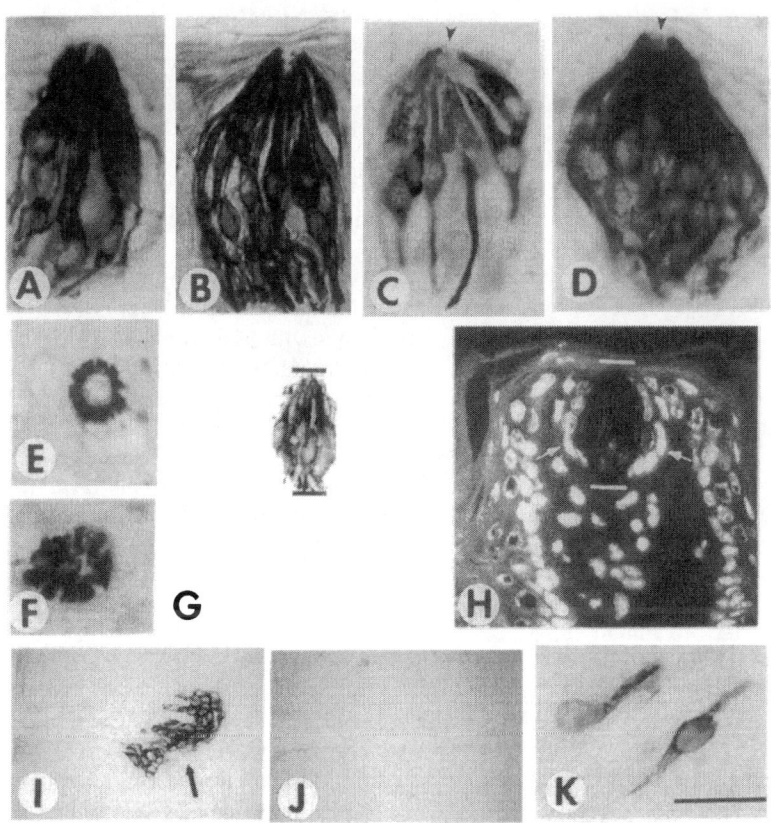

FIG. 1. **A–D** Keratin 7, 8, 18, and 19-like immunoreactivity, respectively, in rat vallate, vallate, foliate, and epiglottis, respectively. *Arrowheads* indicate the taste pit. **E,F** Keratin 19-like immunoreactivity in cells of the taste pit of two tangentially sectioned rat vallate taste buds. **G** ABC peroxidase staining of rat fungiform taste bud. **H** Hoechst nuclear stain of the same tissue section as in **G**. *Arrows* indicate keratin 19-negative perigemmal cells. The peroxidase reaction product obscures most intragemmal nuclei. *Horizontal lines* indicate the position of the taste bud visible in **G**. **I** At 21 days after bilateral denervation of the rat vallate papilla there were no taste buds or associated immunoreactivity. *Arrow* indicates residual keratin 19-positive salivary duct cells. The trench opening is at the left in **I** and **J**. **J** Absence of immunoreactivity in a control vallate trench after omission of the primary antibody. **K** Two rat vallate taste cells at postnatal day 3, antikeratin 18-like immunoreactivity. **A–K** ABC method (with added Hoechst dye in **H**). Scale bar in **K**: 19 μm for **A** and **C**; 23 μm for **B**; 15 μm for **D–F**; 32 μm for **G** and **H**; 120 μm for **I** and **J**; 17 μm for **K**

Basal Cells

Double staining and comparisons between adjacent sections indicated that basal cells, including those adjacent to or within the base of rat fungiform taste buds, were immunonegative for keratins 8, 18, and 19. Denervation of the rat vallate papilla completely eliminated taste buds and keratins 8, 18, and 19-like immunoreactivity (Fig. 1I). This finding is added evidence that gustatory basal cells lack keratins 8, 18, and 19. The residual staining in Figure 1I is of cells from the ducts of von Ebner's salivary glands where the ducts fuse with the base of the vallate trench. Omission of the primary antibody eliminated all staining (Fig. 1J).

Keratin antibodies can be used as markers to track the loss of taste receptor cells after denervation (Fig. 1I) and to examine the progression of development. Preliminary observations of developing taste buds indicate that keratin 18-like immunoreactivity is the last of these four taste bud keratins to appear (Fig. 1K).

Staining of Nongustatory Cells

The only nongustatory cells in rat lingual gustatory epithelium consistently immunoreactive for cyto-keratins 8, 18, and 19 were the cells associated with salivary ducts at the base of foliate and vallate trenches (Fig. 1I). Salivary ducts were also keratin 7-positive.

Discussion

Keratin 18-like immunoreactivity was first reported in human fungiform buds by Lane [14] and subsequently in mouse vallate taste buds by Takeda et al. [15]. Keratin 19-like immunoreactivity has been reported for human fungiform taste buds [4,16] and rat vallate taste buds [17]. Keratin 8-like immuno-reactivity [16] and keratin 20-like immunoreactivity [6] have been reported in human fungiform taste buds. Keratin 7-like immunoreactivity has not been reported previously in taste buds.

The present results indicate that fusiform taste cells have at least five keratins (7, 8, 18, 19, and 20). In contrast, nearby perigemmal cells and basal cells were keratin 7-, 8-, 18-, and 19-negative, as determined by direct observations of normal tissue and by the absence of any immunoreactivity for these keratins in the denervated vallate papilla. That taste buds and their surrounding cells have nonoverlapping sets of keratins agrees with our morphologic assessment of taste buds as islets of simple epithelium embedded in stratified squamous epithelium.

The functions of the 20 cytokeratins and the reasons for their diverse expression in subsets remain mysterious. Most tissues, including most keratinizing stratified epithelia, are keratin 19-negative [3,4,18]. Examples of rat tissues [the present work] and human tissues [3,4,18] with moderate or intense keratin 19-like immunoreactivity include taste cells, salivary ducts, tracheal mucosa, liver bile ducts, intestinal mucosae, bladder, urinary collecting ducts, and uterus. We noted that keratin 19-positive cells in rat tissues had two generalized traits. These cells were luminal cells apically bathed by specialized secretions in a duct or cavity, and they had an apical-basal polarization. Cells such as these, including taste cells, share structural features, such as apical microvilli and tight junctions, and are functionally polarized by different apical and basal ion conductance and transport mechanisms [13,20–22]. Hence the keratin 19-like immunoreactivity of taste receptor cells and their known structural and functional features suggest that taste receptor cells should be considered members of the large class of polarized epithelial cells separating ionic compartments in many nongustatory tissues [20,21]. (In contrast to studies on human tissues, we did not observe regions of keratin 19-positive basal cells in rats. Stasiak et al. [19] have made a different proposal for the significance of keratin 19 in human epithelial cells.)

Summary and Conclusion

In our research program we examined gustatory epithelia in human, rat, gerbil, and rabbit oral tissues with 21 mAbs specific for one or more of the 20 soft keratin polypeptides. Elongated cells within all types of taste buds examined in rat, gerbil, and rabbit were keratin 7-, 8-, 18-, and 19-positive. An ideal immunologic marker of taste receptor cells would be reactive to all taste receptor cells and unreactive to cells adjacent to taste buds. Specific antibodies against keratin 8 and 19 met these requirements for several gustatory papillae. In vitro studies should be able to obtain taste buds from the soft palate, fungiform papillae, or shallow excision of vallate tissue without contamination by immuno-reactive salivary cells. More generally, mAbs against keratins 7, 8, 18, and 19 can help evaluate taste bud development, renewal, degeneration, and regeneration. The application of keratin markers for taste buds is allowing us to probe the cell biology of taste cell renewal and to evaluate the role of the nerve in the formation and maintenance of taste buds. For example, in ongoing experiments we are determining the age of the youngest mature taste receptor cells, the functional life-span of a given taste receptor cell, and whether denervation hastens taste cell death or simply eliminates cell replacement.

Acknowledgments. Supported in part by NIH grant DC-00083.

References

1. Moll R, Franke WW, Schiller D, Geiger B, Krepler R (1982) The catalog of human cytokeratins: patterns of expression in normal epithelia, tumors and cultured cells. Cell 31:11–24
2. Achtstätter T, Moll R, Moore B, Franke WW (1985) Cytokeratin polypeptide patterns of different epithelia of the human male urogenital tract: immunofluorescence and gel electrophoretic studies. J Histochem 33:415–426
3. Quinlan RA, Schiller DL, Hatzfeld M, Achtstätter T, Moll R, Jorcano JL, Magin TM, Franke WW (1985) Patterns of expression and organization of cytokeratin intermediate filaments. Ann NY Acad Sci 455:282–306
4. Bartek J, Bartkova J, Taylor-Papadimitriou J, Rejthar A, Kovarik J, Lukas Z, Vojtesek B (1986) Differential expression of keratin 19 in normal human epithelial tissues revealed by monospecific monoclonal antibodies. Histochem J 18:565–575
5. O'Guin WM, Schermer A, Lynch M, Sun TT (1990) Differentiation-specific expression of keratin pairs. In: Goldman RD, Steinert PM (eds) Cellular and molecular biology of intermediate filaments. Plenum, New York, pp 301–334
6. Moll R, Lowe A, Laufer J, Franke WW (1992) Cytokeratin 20 in human carcinomas. Am J Pathol 140:427–447
7. Franke WW, Schmid E, Schiller DL, Winter S, Jarash ED, Moll R, Denk H, Jackson B, Illmensee K (1982) Differentiation-related patterns of expression of proteins and intermediate-sized filaments in tissues and cultured cells. Cold Spring Harbor Symp Quant Biol 46:431–453
8. Tseng SCG, Jarvinen MJ, Nelson WG, Huang JW, Woodcock-Mitchell J, Sun TT (1982) Correlation of specific keratins with different types of epithelial differentiation: monoclonal antibody studies. Cell 30:361–372
9. Cooper D, Schermer A, Sun TT (1985) Biology of disease classification of human epithelia and their neoplasms using monoclonal antibodies to keratins: strategies, applications and limitations. Lab Invest 52:243–256
10. Sun TT, Tseng SCG, Huang AJ-W, Cooper D, Schermer A, Lynch MH, Weiss R, Eichner R (1985) Monoclonal antibody studies of mammalian epithelial keratins: a review. Ann NY Acad Sci 445:307–329
11. Oakley B, Schafer R (1978) Experimental Neurobiology. University of Michigan Press, Ann Arbor, p 355
12. Roper SD (1987) The cell biology of vertebrate taste receptors. Annu Rev Neurosci 12:329–353
13. Murray RG (1974) The ultrastructure of taste buds. In: Friedmann I (ed) The ultrastructure of sensory organs. Elsevier, New York, pp 1–81
14. Lane EB (1982) Monoclonal antibodies provide specific intramolecular markers for the study of epithelial tonofilament organization. J Cell Biol 92:665–673
15. Takeda M, Obara N, Suzuki Y (1990) Keratin filaments of epithelial and taste-bud cells in the circumvallate papillae of adult and developing mice. Cell Tissue Res 260:41–48
16. Sawaf MH, Ouhayoun JP, Shabana AHM, Forest N (1990) Cytokeratin expression in human tongue epithelium. Am J Anat 189:155–166
17. Oakley B (1991) Neuronal-epithelial interaction in mammalian gustatory epithelium. In: Whelan J (ed) Regeneration of vertebrate sensory receptor cells (vol 160). Wiley, Chichester, pp 277–287
18. Gigi-Leitner O, Geiger B (1986) Antigenic interrelationship between the 40-kilodalton cytokeratin polypeptide and desmoplakins. Cell Motil Cytoskel 6:628–639
19. Stasiak PC, Purkis PE, Leigh IM, Lane EB (1989) Keratin 19: predicted amino acid sequence and broad tissue distribution suggest it evolved from keratinocyte keratins. J Invest Dermatol 92:707–716
20. Simmons K, Fuller SD (1985) Cell surface polarity in epithelia. Annu Rev Cell Biol 1:243–288
21. Matlin KS, Caplan MJ (1992) Epithelial cell structure and polarity. In: Seldin DW, Giebish G (eds) The kidney: physiology and pathophysiology (2nd ed.). Raven, New York, pp 447–473
22. Cummings TA, Kinnamon SC (1992) Apical K^+ channels in *Necturus* taste cells: modulation by intracellular factors and taste stimuli. J Gen Physiol 99:591–613

Frog Taste Cells After Denervation

Eiko Honda[1], Kuniaki Toyoshima[2], Teruyuki Hirakawa[1], and Satoshi Nakahara[1]

Key words. Frog taste cell—Degeneration

The structure of the taste organ and the physiological properties of the taste cell of the bullfrog (*Rana catesbeiana*) were investigated after glossopharyngeal nerve transection.

The taste organ was composed of at least three distinct cell types, i.e., taste, supporting, and basal cells. After denervation, the intragemmal nerve terminals within the taste organ disappeared completely in about 7 days in summer (April to September) and in 15 days in winter (October to March). However, the structure of the taste, supporting, and basal cells persisted for 140 days (April to September) after denervation, as observed by electron microscopy. Each cell within the taste organ was dissociated by enzymatic dissociation of fungiform papilla. Putative taste cells were identified by the in characteristic shape, i.e., they had slim apical processes. The areas (mean ± SD) of normal and denervated taste cells 120 days after surgery were 12.9 ± 4.1 and 12.1 ± 3.7 ($\times 10^2 \mu m^2$), respectively; the differences between them were not significant.

As time elapsed after denervation, impulse discharges from the glossopharyngeal nerve in response to mechanical and chemical stimuli declined, ceasing within about 7 days in summer (April to September) and within about 15 days in winter (October to March). These periods were about the same as those required for the disappearance of nerve terminals within the taste organ after denervation. Intracellular recordings were carried out in summer (April to September) to analyze the time-course of degeneration in taste cells. Statistical analysis was performed using one-way ANOVA and Fisher's protected least significant difference test for post-hoc determination. The mean magnitude of the resting potential of normal taste cells was −28.9 ± 7.0 mV, and those of denervated taste cells recorded 5, 7, 14, 21, 28, 35, 70, 105, and 140 days after surgery were −24.7 ± 6.9 mV, −26.8 ± 6.8 mV, −29.6 ± 8.2 mV, −28.5 ± 7.0 mV, −31.0 ± 6.9 mV, −23.0 ± 4.2 mV, −24.3 ± 5.6 mV, −26.6 ± 4.7 mV,

and −23.6 ± 2.9 mV, respectively. Glossopharyngeal nerve transection significantly affected the magnitude of the resting potential of the taste cells (F 9, 169 = 2.15, $P < 0.05$). In normal and denervated taste cells, receptor potentials were generated by 0.5 M NaCl, 0.01 M HCl, and 0.2 M quinine HCl. At 5, 21, 35, 70, and even 105 days after denervation, receptor potentials were elicited in each taste cell in response to NaCl, HCl, and quinine HCl. The mean magnitude of receptor potentials elicited by 0.5 M NaCl in normal taste cells was 12.1 ± 4.3 mV, and those elicited in denervated taste cells at 5, 7, 14, 21, 28, 35, 70, 105, and 140 days after surgery were 9.4 ± 4.9 mV, 10.5 ± 2.5 mV, 10.5 ± 4.0 mV, 8.5 ± 3.7 mV, 9.8 ± 4.1 mV, 7.0 ± 3.3 mV, 6.8 ± 4.3 mV, 7.1 ± 4.2 mV, and 1.5 ± 0.9 mV respectively. Glossopharyngeal nerve transection significantly affected the magnitude of receptor potential in the taste cells elicited by 0.5 M NaCl (F 9, 105 = 4.94, $P < 0.001$).

In a variety of mammals, i.e., rats, dogs, rabbits, and hamsters, taste buds degenerate and finally disappear after the transection of gustatory sensory nerves, reappearing after regeneration of the nerve fibers. These results have led to the proposal of the neurohumoral theory, i.e., that transport of trophic substances by afferent axons maintains the structure of taste organs and taste responses in mammals. The morphological and functional integrity of mammalian taste cells depends on an intact nerve supply.

However, in the present study, it was demonstrated that the frog taste organ survived for 140 days following glossopharyngeal nerve transection, and that the structure of the taste, supporting, and basal cells was maintained without obvious change during this period. Denervated taste cells also had resting potentials and generated receptor potentials in response to the four basic taste stimuli. However, the glossopharyngeal nerve transection had a significant effect on the magnitude of resting potentials and receptor potentials elicited by 0.5 M NaCl in the frog taste cells. These results indicate that glossopharyngeal nerve transection could affect the function of the taste cells, but that the structure of the frog taste organ persisted long-term following neurectomy. Consequently, we concluded that the neural dependency of the frog taste organ was not so great as that of mammals.

Departments of Physiology[1] and Oral Anatomy[2], Kyushu Dental College, 2-6-1 Manazuru, Kokurakita-ku, Kitakyushu, 803 Japan

Parasympathetic Postganglionic Nerve Fibers in the Fungiform Papillae of the Frog

Katsuhiro Inoue[1] and Yasuyuki Kitada[2]

Key words. Frog taste disk—Parasympathetic innervation

The nerve bundle in each fungiform papilla of the frog consists of afferent, and sympathetic and parasympathetic efferent nerve fibers. Efferent control of the gustatory system has been postulated from electronphysiological evidence, but there is inadequate morphological evidence for such a conclusion. Some unmyelinated nerve fibers in the nerve bundle of each fungiform papilla originate from parasympathetic postganglionic neurons within the lingual branch of the glossopharyngeal nerve [1,2]. Particular nerve endings, designated parasympathetic efferent according to morphological criteria, namely, the presence of accumulations of clear synaptic vesicles and mitochondria, besides a specific membrane system, at the regions of contact between the nerve endings and taste cells, have been observed in the taste organs of mammals, fish, and frogs. However, the presence of sensory nerve endings that contained many typical, clear synaptic vesicles of the type usually associated with cholinergic fibers has also been reported. Thus, an experimental study, designed to single out efferent fibers, is needed if we are to determine whether parasympathetic efferent nerve endings do exist in the taste disk. In this study, 20 bullfrogs (*Rana catesbeiana*), weighing 250–470 g, were anesthetized by immersing them in a solution of MS-222 (tricaine methanesulfonate, 2 g/liter). The glossopharyngeal and hypoglossal nerves were cut at the level of the angle of the mouth. The frogs were killed on day 28 or 56 after surgery (group A), or, alternatively, on day 28 after the surgery, the lingual branches of the left glossopharyngeal nerve were cut in the body of the tongue and the frogs were killed 3, 7, 14, and 21 days after the additional nerve transection (group B). For electron microscopy, the frogs were anesthetized with MS-222 and perfused through the heart with 100 ml of a mixture of 1% glutaraldehyde and 4% paraformaldehyde in 0.1 M phosphate buffer, pH 7.4, for 10 min. After perfusion, the left side of the tongue was removed and cut into small pieces. Tissues were then immersed for 2 h in the same fixative at 4°C and postfixed for 1 h at room temperature in 1% OsO_4 dissolved in 0.1 M phosphate buffer, pH 7.4. The tissues were then treated for conventional electron microscopy. In group A, in which preganglionic parasympathetic nerve fibers had been cut, although myelinated nerve fibers had disappeared or showed degeneration, no signs of degeneration were found on the unmyelinated fibers in the nerve bundles of the fungiform papillae. Just under the taste disk, the unmyelinated nerve fibers that persisted were enlarged into irregular shapes of varying size and distinct features. Except at the very terminal regions, the axonal profiles were largely embedded in infoldings of cytoplasm supporting Schwann cells. The enlargement of the axoplasm, nerve endings, was filled with large dense-core vesicles (100–120 nm in diameter) and small, clear synaptic vesicles (50–60 nm in diameter). Some naked axons penetrated the basal lamina of the taste disk and made close contact with the supporting or Merkel-like basal cells in the taste disk. In group B, in which both preganglionic and postganglionic parasympathetic nerve fibers had been cut, clear signs of degeneration were observed in most of the unmyelinated nerve fibers in the nerve bundle of the fungiform papillae. Definitive signs of degeneration were also recognized in many nerve endings. Dense lamellae, dense masses, a large dense body, the accumulation of large dense-core vesicles, and the disappearance of synaptic vesicles were also observed. These results strongly suggest that some nerve endings originate from the parasympathetic postganglionic cells; the results also seem to provide morphological evidence for the existence of an efferent control system in the taste disk.

References

1. Inoue K, Kitada Y (1991) Parasympathetic postganglionic cells in the glossopharyngeal nerve trunk and their relationship to unmyelinated nerve fibers in the fungiform papillae of the frog. Anat Rec 230:131–135
2. Inoue K, Yamaai T, Kitada Y (1992) Parasympathetic postganglionic nerve fibers in the fungiform papillae of the bullfrog, *Rana catesbeiana*. Brain Res 596:299–304

[1] Institute for Dental Science, Matsumoto Dental College, Shiojiri, 399-07 Japan
[2] Department of Oral Physiology, Okayama University Dental School, 2-5-1 Shikata-cho, Okayama, 700 Japan

Lamellar Bodies of Mouse Taste Buds

Masako Takeda, Yuko Suzuki, Nobuko Obara, and Yasuko Nagai[1]

Key words. Taste bud—Phospholipid

It has been reported that fixation with tannic acid preceding osmication preserves saturated phospholipids. Tannic acid reacts with the choline base of phosphatidylcholine and sphingomyelin to form a complex, which is then stabilized by treatment with OsO_4 and survives during embedding procedures. Thus, the ordered lamellar structures of phospholipids can be demonstrated under an electron microscope [1].

Adult dd-mice (7–15 weeks) were perfused through the left ventricle with a mixture of 0.5% tannic acid, 2% glutaraldehyde, and 4% paraformaldehyde in 0.1 M phosphate buffer. Small blocks of the tongues, containing circumvallate papillae, were excised and immersed in the above-mentioned fixative overnight at 4°C. The tissues were postfixed in 1% OsO_4 for 2 h at 4°C. After the tissues were embedded in Epon 812, thin sections were cut, stained with uranyl acetate followed by lead citrate, and then observed under an electron microscope.

Aggregations of many globular electron-dense bodies (70–500 nm in diameter) were found in the apical cytoplasm of the three differentiated types (type I, -II, and -III) of slender taste bud cells. The bodies were localized in a belt-like form below the level of the tight junction, surrounding the taste pores. High-resolution micrographs of the bodies revealed an ordered lamellar structure with a repeating period of 5 nm, forming concentric circles, characteristic of phospholipids. Most of the bodies were free in the cytoplasm rather than inside vacuoles or smooth endoplasmic reticulum. The lamellar bodies were occasionally observed in the perinuclear and basal area or in the intracellular spaces. In the taste pores, dense bodies with lamellar structure were observed among the microvilli, and occasionally on the cell membranes of the taste bud cells, as if the bodies were being released into the taste pores through the cell membrane. Epithelial cells near the taste pores also had many lamellar bodies, while, on the other hand, epithelial cells that were situated far from the taste pores rarely contained these bodies. The lamellar bodies were rarely found in taste buds treated with the fixative lacking tannic acid.

It seems that, in vivo, most phospholipids involved in the creation of lamellar bodies are diffusely distributed in the taste bud cells, and are released into the taste pores through the cell membrane by diacrine secretion.

Type II pneumocytes possess secretory bodies with a lamellar structure containing a high proportion of saturated phospholipids in a lipid-protein complex. These complexes are released into the lumens of alveoli, acting as a surfactant that reduces the surface tension at the air-liquid interface. In lung tissue adequately treated with tannic acid before osmication, highly ordered lamellar arrays of the bodies have been observed not only in type II pneumocytes and alveolar lumens, but also in Clara, ciliated, and goblet cells; the secretory products of all these cells are presumed to act as surfactants [2]. Bile, the exocrine secretion of hepatocytes, contains phospholipids and acts as a surfactant which emulsifies fat so that the action of the enzyme, lipase, is more effective in the small intestine.

Our findings suggest that phospholipids in the taste pores act as a surfactant to promote the solubility of sapid materials, thus facilitating their adsorption onto the membranes of taste receptor cell microvilli.

References

1. Kalina M, Pease DC (1977) The preservation of ultrastructure in saturated phosphatidyl cholines by tannic acid in model systems and type II pneumocytes. J Cell Biol 74:726–741
2. Ueda S (1990) Ultrastructural studies of the lung surfactant and its production cells (in Japanese). In: Yoshida S (ed) Lung surfactants. Shinkoukoeki, Tokyo, pp 27–47

[1] Department of Oral Anatomy, Higashi Nippon Gakuen University School of Dentistry, Tobetsu, Ishikari, Hokkaido, 061-02 Japan

Ultracytochemical Localization of Enzymes and Substances Associated with Transmission and Transduction in Mouse Taste Buds

Michio Kudoh[1]

Key words. Cytochemistry—Transmission

The ultracytochemical localization of monoamines, acetylcholinesterase (AChE), and Ca^{2+}-pumping ATPase, which localization is presumably related to the regulation of cytosolic Ca^{2+} concentration, was investigated in the taste buds of mouse vallate papilla.

The distribution of monoamines was examined by the glyoxylic acid-Mg^{2+}-$KMnO_4$ method [1]. In taste buds of the mouse administered an intraperitoneal injection of 5-hydroxytryptophan (5-HTP), numerous small dense vesicles appeared throughout the cytoplasm of type III cells, whereas only a few vesicles were found in type I and II cells. These vesicles aggregated especially around the Golgi apparatus and in the presynaptic regions of nerve terminals. The electron density of the vesicles was variable, perhaps reflecting differences in the amounts of monoamines stored within the vesicles. By immunohistochemical investigation with a confocal laser-scanning microscope, we identified the monoamine contained within the vesicles as serotonin. This indicates that only type III cells were capable of taking up the monoamine precursor, that they decarboxylated it into serotonin and stored it in small vesicles. However, whether these vesicles are involved in nerve transmission is questionable, since serotonin was not detected in taste buds without 5-HTP treatment.

The ultracytochemical localization of AChE was investigated by the highly sensitive method of Tago et al. [2]. AChE activity was even around the intragemmal nerves, especially at the typical synaptic regions where type III cells were in contact with the afferent nerve fibers. The activity was also extensively found in the cisternae of the rough endoplasmic reticulum (rER) and in the perinuclear space of type III cells, whereas little activity was observed in the cytoplasm of type I and II cells. These observations suggest that the type III cell is the primary gustatory receptor and that the cholinergic process might be involved in taste perception. However, further studies will be required to eluci-

date whether the type of nerve transmission is cholinergic, adrenergic, or peptidergic. Occasionally, the periphery of the nerve fiber was studded with a number of AChE-positive granules, the appearance and size of which were much like exocytotic vesicles. Thus, there might be an exporting system that carries AChE from the rER cisternae to the interspace between the cell and the nerve fiber by exocytosis [3].

There is increasing evidence that Ca^{2+}, as well as cyclic AMP, is an important intracellular regulator functioning as a second messenger for certain extracellular signaling molecules [4]. Ultracytochemical investigation revealed that Ca^{2+}-ATPase activity was intensively located on the plasma membrane of various cell types and around nerves, but not in subcellular organelles such as rER, mitochondria, and the Golgi apparatus. Plasma membrane Ca^{2+}-ATPase, therefore, may play a primary role in determining the cytosolic Ca^{2+} concentration of taste bud cells and nerves at a very low range, about 10^{-7} M. In the developing taste bud during the early postnatal period, on the other hand, Ca^{2+}-ATPase appeared to be located in the lumen and cisternae of the rER, in the Golgi apparatus, and in small vesicles in the gustatory epithelial cells. These results suggest that Ca^{2+}-ATPase synthesized in the rER passes through the Golgi apparatus to the plasma membrane, mediated by transport vesicles [5].

References

1. Koda LY, Bloom FE (1977) A light and electron microscopic study of noradrenergic terminals in the rat dentate gyrus. Brain Research 120:327–335
2. Tago H, Kimura H, Maeda S (1986) Visualization of detailed acetylcholinesterase fiber and neuron staining in rat brain by a sensitive histochemical procedure. J Histochem Cytochem 34:1431–1438
3. Kudoh M (1988) Ultrastructural and histochemical localization of acetylcholinesterase in the taste bud of mouse vallate papilla. Fukushima J Med Sci 34:27–44
4. Carafoli E, Penninston JT (1985) The calcium signal. Sci Am 253:50–58
5. Kudoh M (1990) Ultracytochemical localization of Ca^{2+}-ATPase in the mouse taste bud during the early postnatal period. Fukushima J Med Sci 36:41–57

[1] Department of Biology, Fukushima Medical College, Fukushima, 960-12 Japan

Electron Microscopic Demonstration of Guanylate Cyclase Activity in Rabbit Taste Bud Cells

Naokazu Asanuma and Hiromichi Nomura[1]

Key words. Taste—Guanylate cyclase

The possible involvement of cyclic nucleotide in sweet taste transduction has been reported [1]. The present study was undertaken to explore the possibility of cyclic GMP involvement in taste transduction by cytochemically localizing guanylate cyclase activity in taste bud cells.

On incubating rabbit foliate papillae with 5'-guanylylimidodiphosphate (substrate) and Sr^{2+} (capture ion), we observed enzymatic activity exclusively in the apical portion (microvilli and neck) of all taste bud cells with one end protruding into a taste pore. The microvillous membrane of dark cells showed especially strong activity. The long blunt process of cells that lacked microvilli often showed strong activity. The enzymatic activity was not inhibited, but rather was enhanced by 5 mM dithiothreitol; this suggested that the activity was that of guanylate cyclase [2] and not nucleotide pyrophosphatase, another guanylylimidodiphosphate-hydrolyzing enzyme [3]. Enzymatic activity was inhibited by 1 mM $CdCl_2$, $ZnCl_2$, or $HgCl_2$, and was enhanced by 1% Triton X-100, all in congruence with the reported nature of guanylate cyclase [2,4]. We have not yet succeeded in demonstrating the stimulation of the present enzymatic activity by sweet substances, probably because the experimental procedure caused some damage to a presumed mediator between the receptor and the catalytic site of the enzyme. However, the localization of guanylate cyclase in taste bud cells supports the hypothesis that cyclic GMP is involved in taste reception.

References

1. Tonosaki K, Funakoshi M (1988) Cyclic nucleotides may mediate taste transduction. Nature 331:354–356
2. Hardman JG, Sutherland EW (1969) Guanyl cyclase, an enzyme catalyzing the formation of guanosine 3',5'-monophosphate from guanosine triphosphate. J Biol Chem 244:6363–6370
3. Johnson RA, Welden J (1977) Characteristics of the enzymatic hydrolysis of 5'-adenylylimidodiphosphate: Implications for the study of adenylate cyclase. Arch Biochem Biophys 183:216–227
4. Kimura H, Murad F (1975) Localization of particulate guanylate cyclase in plasma membranes and microsomes of rat liver. J Biol Chem 250:4810–4817

[1] Department of Oral Physiology, Matsumoto Dental College, Shiojiri, 399-07 Japan

Ultrastructure of Canine Circumvallate Taste Buds

Hiroaki Kanazawa[1], Sumio Yoshie[2], and Tsuneo Fujita[1]

Key words. Dog—Sucrose stimulation

This study demonstrated the fine structure of the taste bud of the dog, with special reference to the identification of the gustatory cell and to the morphological changes in this cell due to stimulation with a sweet substance.

A 0.1 M solution of sucrose, at 36°C, was placed for 1 min on the circumvallate papillae of three male dogs (6–15 kg bw) that had been lightly anesthetized with ketamine. Immediately after this stimulation, the tongue was removed; pieces containing circumvallate papillae were excised and fixed in a solution containing 2.5% glutaraldehyde and 1% paraformaldehyde (0.05 M sodium cacodylate buffer, pH 7.3). The specimens were treated by routine techniques and examined under a transmission electron microscope. Control specimens were obtained from dogs that had received no taste stimulation.

Under the electron microscope, it could be seen that the taste bud of the dog consisted of five types of epithelial cells. Four of them essentially corresponded in fine structure to previously reported cell types, in the rabbit [1]; in the rat and mouse [2,3]; and in the guinea pig [4], whereas type V corresponded to that shown [5] in the rat. Type III cells made synaptic contacts with nerves. Not only a single profile, but also two or more synapse profiles could be counted in either longitudinal or transverse sections of a type III cell. The presynaptic membrane was densely associated with synaptic vesicles. These were predominantly a large-core type measuring approximately 100 nm, only a few being of the small clear type, approximately 30 nm in diameter. The nerve terminal of the synapses consistently revealed a considerable number of small clear vesicles, 30 nm in diameter, some of which were gathered close to the postsynaptic membrane.

The canine type III cells were characterized by the numerous large-core vesicles in their basal cytoplasm. Blood capillaries located beneath the taste bud were fenestrated.

After stimulation with 0.1 M sucrose, we recognized morphological changes in the type III cells. In some cells of this type, we recognized a few large-core vesicles localized in the presynaptic region discharging their contents into the synaptic cleft by exocytosis. Further, large-core vesicles localized at the cell base also revealed exocytotic images, opening to the intercellular spaces.

These findings suggest that type III cells are gustatory cells and that they release transmitter(s) in the large-core vesicles in response to a chemical stimulus at the cell apex. Yoshie et al. [6] showed the same findings in guinea pig gustatory cells after stimulation with a mixture of monosodium L-glutamate and guanosine 5′-monophosphate. Besides being released into the synaptic cleft, it appears that the substances are liberated in a paracrine fashion at the base of the cell. Presumably, Ebner's salivary glands may be one of the targets of the paracrine effects of the gustatory cell transmitter(s).

The substances may enter blood capillaries as hemocrine (endocrine) agents, a hypothesis supported by our observation of fenestrated capillaries near the base of the taste bud cell.

References

1. Murray RG, Murray A, Fujimoto S (1969) Fine structure of gustatory cells in rabbit taste buds. J Ultrastruct Res 27:444–461
2. Takeda M, Hoshino T (1975) Fine structure of taste buds in the rat. Arch Histol Jpn 37:395–413
3. Takeda M (1976) An electron microscopic study of the innervation in the taste buds of the mouse circumvallate papillae. Arch Histol Jpn 39:257–269
4. Yoshie S, Wakasugi C, Teraki Y, Fujita T (1990) Fine structure of the taste bud in guinea pigs. I: Cell characterization and innervation patterns. Arch Histol Cytol 53:103–119
5. Farbman AI (1965) Fine structure of the taste bud. J Ultrastruct Res 12:328–350
6. Yoshie S, Wakasugi C, Teraki Y, Kanazawa H, Iwanaga T, Fujita T (1991) Response of the taste receptor cell to the umami-substance stimulus. An electron-microscopic study. Physiol Behav 49:887–889

[1] Department of Anatomy, Niigata University School of Medicine, Asahi-machi 1, Niigata, 951 Japan
[2] Department of Oral Anatomy, Nippon Dental University, 1-8 Hamaura-cho, Niigata, 951 Japan

Structural and Functional Features of Human Fungiform Papillae

Inglis J. Miller, Jr.[1]

Key words. Human—*Fungiform papillae*—Taste buds

Human fungiform papillae have been studied in order to understand how individual variations in taste perception relate to anatomical features of the tongue. Several studies [1–3] have shown that taste thresholds and taste intensity ratings are correlated with the number of taste buds that a person has. The purpose of this paper is to characterize attributes of human fungiform papillae which are associated with the prevalence of taste buds. These attributes may be useful for characterizing papillae in studies of taste perception and in the diagnosis of taste dysfunctions. Papillae have been studied from human cadaver tongues, using paraffin sections, and from living human subjects, using videomicroscopy. The size, shape, location on the tongue, and number of taste buds on fungiform papillae are the features which were measured.

The topology of papillae from living male and female students ($n = 16$; 17–35 years) yielded 81 papillae, using videomicroscopy, as previously described [1]. Pores of taste buds were identified by the application of methylene blue. Serial reconstructions were made from samples of human cadaver tongues from donors ($n = 18$) aged 22–90 years at the time of death. Materials and methods for these papillae have been reported [4,5]. A total of 81 papillae from cadaver tongues are included here. Measurements were made with videoimages, using NIH Image 1.32–1.49 program. Simulations of papillae were made with Cricket Graph and Swivel 3-D software (Paracomp, Inc.).

Macrophotography of cadaver tongues showed papillae in a variety of shapes and sizes. Digitized video images of fungiform papillae on living tongues were measured by four dimensions: height (H); diameter (A) at $\frac{3}{4}$ of height; diameter (B) at $\frac{1}{2}$ of height; and the base (C) at the bottom of the papillae. Two-dimensional profiles of the papillae were plotted from the four measurements, and three-dimensional simulations of the papillae yielded four predominant shapes. These papillae were characterized as "flat", "derby-like", "conical", and "mushroom-like", respectively. The respective shapes of papillae were defined in terms of ratios of their heights and diameters.

The *numbers of taste buds (tb) per fungiform papilla* were significantly related to papillary *diameters* (df = 1, 160, F = 14.2; $P < 0.005$), but not to the *heights* (df = 1, 160, F = 2.2; $P > 0.10$). Taste buds (or pores) ranged from 0 to 15 per papilla, with an average of 3.6 ± 3.7 tb/pap. Fifty-seven percent (57%) of papillae had 0–3 tb, and 8.0% had 10 tb of more. Papillary diameters averaged 0.72 ± 0.18 mm (SD, $n = 162$, range 0.35–1.43), and heights averaged 0.56 ± 0.35 mm (SD, range 0–1.6). Papillary heights and diameters were not significantly related (df = 1, 160, F = 2.1; $P > 0.1$). The numbers of taste buds per papilla did not differ significantly by papillary shape or by the relative location on the tongue tip.

An important conclusion from this study is that "gustatory papillae" (i.e., those with taste buds) vary substantially from the classical "mushroom-shape". Combined studies of anatomy and taste perception are needed to sort out whether there are functional distinctions among the variant sizes and shapes of gustatory papillae.

Acknowledgments. Support for this project was provided by NIH Grant DCD 00230 from the National Institute of Deafness and Other Communicative Disorders. David Black and David Sink made important contributions to the analysis of human fungiform papillae. Ruoyu Xiao provided photography of living papillae. Frank Reedy, Jr. performed the videomicroscopy procedures.

References

1. Miller IJ Jr, Reedy FE Jr (1990) Variations in human taste bud density and taste intensity perception. Physiol Behav 47:1213–1219
2. Reedy FE Jr, Bartoshuk LM, Miller IJ Jr, Duffy VB, Yanagisawa K (1993) Relationships among papillae, taste pores, and 6-n-propylthiouracil (PROP) suprathreshold taste sensitivity. Chem Senses 18:218–219
3. Zuniga J, Davis S, Englehardt R, Miller IJ Jr, Schiffman S, Phillips C (1993) Taste performance on the anterior human tongue varies with fungiform taste bud density. Chem Senses 18:449–460
4. Miller IJ Jr (1986) Variation in human fungiform taste bud densities among regions and subjects. Anat Rec 216:474–482
5. Miller IJ Jr (1988) Human taste bud density across adult age groups. J Gerontol: Biolog Sci 43:B26–B30

[1] Wake Forest University, BGSM, Department of Neurobiology and Anatomy, Winston-Salem, NC 27157-1010, USA

Investigations on the Dynamics of Chemosensory Epithelia

Winrich Breipohl[1], Tomonori Naguro[1,2], Daniel Grandt[1,3], Alan Mackay-Sim[4], Oliver Leip[1], and Klaus Reutter[5]

Key words. Olfactory epithelium—Masera organ—Vomeronasal organ—Taste buds—Mitosis—Cell death

Introduction

Many histologic and cytologic investigations have documented a remarkable systemic uniformity of the receptor epithelia within the vertebrate olfactory (OS) and gustatory (GS) systems. However, with more detailed consideration, one is confronted with the diversity of the receptive components in the OS and GS. Some phyla have, at least temporarily, up to five peripheral olfactory sensory modalities—three sensory epithelia and two cranial nerves—qualifying for sensory perception. Similarly, in the GS, peripheral sensory diversity is established by the concomitant involvement of three cranial nerves and their respective receptor cell populations. So far, nobody has sufficiently explained this complexity and to what extent the entity of those peripheral sensory subsystems cooperates or functions independently.

One can speculate that the involvement of more than one cranial nerve is to ensure stable chemoreception even under conditions when insults have switched off major parts of the sensory system. However, securing chemoreception by the involvement of multiple cranial nerves seems to clash with data documenting extensive neuronal plasticity (i.e., morphologic instability in both systems). Neuronal plasticity follows from a continual receptor cell turnover and, related to this point, permanent reestablishment of the synapses between the sensory cells and consecutive neuronal structures.

The functional implications of uniformity, diversity, and plasticity of the two sensory systems are still poorly understood. The aims of this report are therefore:

1. To evaluate our own, mainly unpublished data on the regulation of neuronal proliferation, maturation, and cytochemical identity in order to better understand the biologic features underlying the uniformity, diversity, and plasticity of the OS (and GS) at the level of the peripheral sensory receptor epithelia
2. To encourage further investigations in these areas

Results and Discussion

Speculation on the cooperation of the various olfactory receptive components have been based on comparisons of developmental features. The vomeronasal organ (VNO) and the Masera organ (MO), but not the main olfactory epithelium (OEP), are believed to lose importance with adulthood. In rodents the VNO and MO seem to maintain functional and structural importance after puberty. In rodents the VNO is developed during intrauterine life and, as with the other two sensory epithelia, develops olfactory marker protein (OMP)-positive neuronal cells. However, during the early postnatal stages, OMP reaction is still less intensive in the VNO neuroepithelium than in the OEP and the MO neuroepithelium. The MO has been associated with the perception of suckling odors. Our investigations document, however, that in rodents an MO is also present in adult animals. The existence of OMP-positive neurons even in aged mice suggests that this olfactory receptor region maintains functional importance beyond the suckling period at least in this species. Comparable investigations in humans and other vertebrates are lacking, as are investigations on the overall regional density and first apperance of OMP-positive olfactory neurons during intrauterine life.

Developmental investigations on the peripheral olfactory receptor epithelia are of basic neurobiologic interest for our understanding of neuro-

[1] Institut für experimentelle Ophthalmologie, R.F.W.-Universität, Sigmund-Freud-Str. 25, D53105 Bonn, Germany
[2] Department of Anatomy, Tottori University, Yonago, Japan
[3] Zentrum für Innere Medizin, Universität Essen, Essen, Germany
[4] Department of Biological Sciences, Griffith University, Brisbane, Australia
[5] Anatomisches Institut, E.K.-Universität Tübingen, Tübingen, Germany

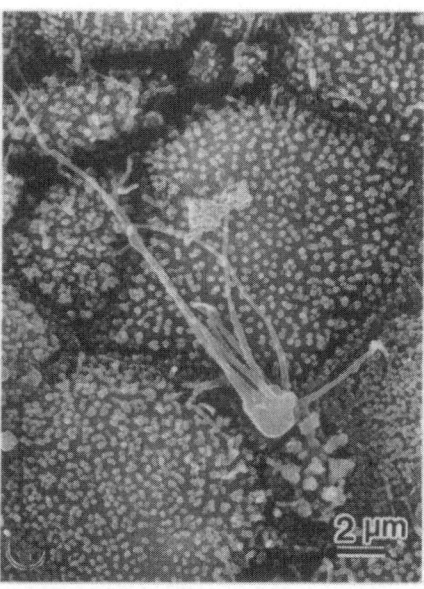

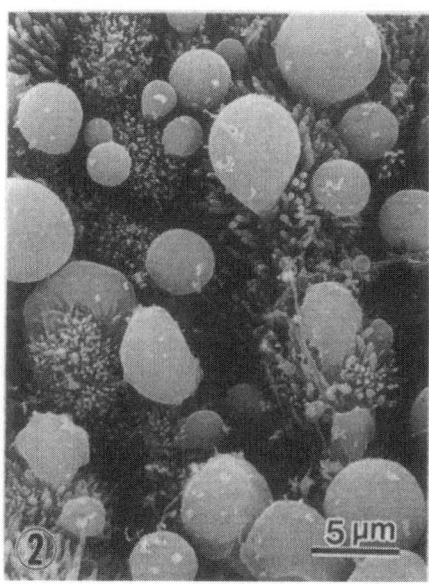

FIG. 1. Scanning electron microscopic documentation of a mature olfactory knob at the surface of the receptor-free epithelium of the vomeronasal organ. Wistar rat, 0.5 day postnatally

FIG. 2. Scanning electron microscopic documentation of the anterior ventral border of the Masera organ with many prominent apical cell protrusions not observed at later developmental stages. Wistar rat, 0.5 day postnatally

genesis and neuronal maturation. In the OEP of mice and chicken we found, during the last quarter of intrauterine and in ovo development, respectively, single mature olfactory receptor cells situated between immature olfactory receptor cells. We believe that these receptor cells may represent pioneering neuronal cells.

Similarly, receptor cells were detected during the first postnatal week in Wistar rats and NMRI mice at two atypical locations: anterior to the nasal septal olfactory area within the dorsal respiratory epithelium and within the receptor free epithelium of the VNO (Fig. 1). We believe that these temporary appearances of additional olfactory sensory epithelial spots represent typical examples of neuronal overdevelopment, a phenomenon that has so far mainly been shown for various developmental periods of higher central nervous system (CNS) structures, including the olfactory bulb [1,2]. Temporary overexpression of neuronal receptive features during normal development and at atypical locations qualifies for more attention in the future. It may also serve as a model for studies on receptor cell death due to the unsuccessful establishment of synaptic connections with other neuronal structures. Another feature, eventually indicating some kind of overdevelopment, can be found at the rostroventral edge of the MO neuroepithelium. During the first few postnatal days we regularly observed voluminous epithelial cell projections, which were never found in concomitantly investigated older animals (Fig. 2).

Additional features in favor of neuronal overdevelopment in the sensory periphery originated from investigations on olfactory receptor cell replacement.

Classically, the production of new receptor cells had hitherto been associated with the idea that an internal death clock would end the life of the receptor cells. The key investigation for this hypothesis was performed in the laboratory of Moulton in 1974 [3]. On the basis of ^{3}H-thymidine injection studies Moulton concluded (later confirmed by several other authors) that olfactory receptor cells of warm-blooded vertebrates were replaced about every 30 days according to their inborn death schedule. What had not been considered, however, was that some 30% of the receptor cells were still labeled 90 days after a single thymidine injection. Regarding the low doses applied and the injection scheme, we assumed that ^{3}H-thymidine reutilization could not be responsible for the phenomenon of labeled receptor cells 90 days after a single isotope injection.

This assumption was the starting point for systematic morphometric and metaphase arrest technique investigations in our laboratory concerning the ratio of progenitor and receptor cells within a defined area of the OEP in the nasal dome of postnatal mice at different developmental stages. Investigations comparable to those here reported for the OEP, unfortunately, have not yet been performed for the VNO and MO. Comparing the data for 2- and 8-week-old mice obtained in our morpho-

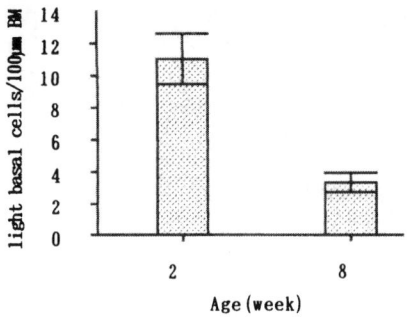

FIG. 3. Age-related frequency of light basal cells per 100 µm basal membrane in the dorsoposterior region of the septal main olfactory epithelium of 2- and 8-week-old mice

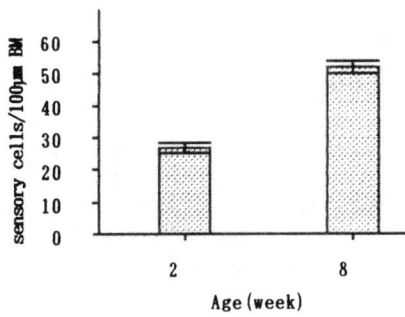

FIG. 4. Age-related frequency of sensory cell perikarya per 100 µm basal membrane in the dorsoposterior region of the septal main olfactory epithelium of 2- and 8-week-old mice

metric studies will help broaden the hypothetical implications on olfactory neuronal cell turnover and their maturational behavior published already [4–8].

The OEP height in 2- and 8-week-old animals did not differ significantly, provided samples were obtained from strictly comparable epithelial locations. Considerable changes occurred, however, in the numerical ratios between the various cell types of the OEP. The light (globose) basal cells (also referred to as immediate olfactory progenitor or precursor cells) declined by about 60% between 2 and 8 weeks of age (Fig. 3), whereas the number of olfactory receptor cell perikarya per unit basal membrane increased by nearly 90% (Fig. 4). Thus it follows that the ratio of olfactory progenitor and receptor cells declines considerably (from about 0.40 at 2 weeks to about 0.07 at 8 weeks). In other words, a reduced number of mitotically active cells would have to supply an increased number of receptor cells if it was true that programmed cell death is killing the sensory cells every 30 days. On the other hand and not shown here, the number of cell deaths in the receptor cell perikaryal layer per unit basal lamina decreased between 2 and 8 weeks of age in the OEP covering the ethmoidals (Breipohl and Darrelmann,

unpublished). Additional data (not shown here) indicate that, for the dorsoposterior septal OEP during the same period, the number of dark basal cells quadrupled, and the number of supporting cells per unit basal membrane remained almost constant.

The latter phenomenon suggests considerable changes in the relation between sensory dendrites and the supporting cells. However, preliminary studies on the mean distance between olfactory knobs and the ratio between dendrites and apical poles of the supporting cells surprisingly indicated only an insignificant decrease and increase, respectively (Breipohl and Grandt, manuscript in preparation). All these features cannot be explained by the still dominant hypothesis of an inherent life-span of 30 days for the receptor cells.

To explain the incompatibility of our data with the hypothesis of a programmed receptor cell turnover, we carefully checked the homogeneity of the sensory cell population and the respective ratio of mitoses. Mature receptor cells, by definition, do possess a dendrite ending in the form of an olfactory knob at the epithelial surface. Since Nagahara [9] published the first evidence on mitotic replacement of olfactory sensory cells, one has had to assume that within the population of sensory cells at least two fractions exist: one that has already differentiated an olfactory knob and another that has not yet done so. Strictly speaking, we should consider yet a third fraction of cells: those that undergo degeneration (for whatever reason) and that have already lost their olfactory knobs although their perikaryal regions are still prominent. As the process of cell death is probably more rapid than cell differentiation and is difficult to quantify on the basis of mere histologic evaluation, this third possibility is not considered here but is recommended for future investigations.

Between 2 and 8 weeks the number of olfactory knobs per unit basal membrane increased by about 19%, accompanied by a mean decrease of the distance between adjacent knobs by about 30%. Yet the ratio of olfactory knobs to sensory cell perikarya decreased by about 37% (from 0.90 at 2 weeks to 0.57 at 8 weeks). In other words, in 2-week-old animals about 90% and in 8-week-old animals only 57% of all receptor cells were equipped with an olfactory knob. However, the amount of mature sensory cells (i.e., with knobs) per unit basal membrane remained almost constant (24.55 ± 0.69 and 26.54 ± 0.86, respectively). The percentual increase of sensory cells without an olfactory knob, i.e., immature sensory cells (but see comments above), the concomitant increase of receptor cell perikarya and decline in those cells replacing the sensory cells would leave us with many insurmountable problems for our understanding if we adhered to the old hypothesis of an inborn receptor cell death programmed for every 30 days or so in warm-blooded animals.

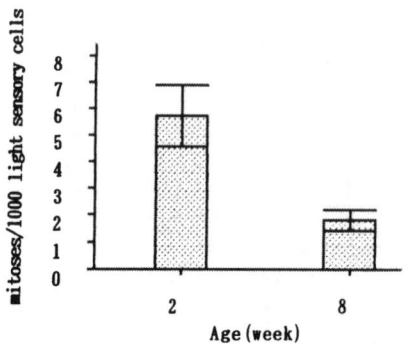

FIG. 5. Age-related frequency of mitoses per 1000 light basal cells in the dorsoposterior region of the septal main olfactory epithelium of 2- and 8-week-old mice

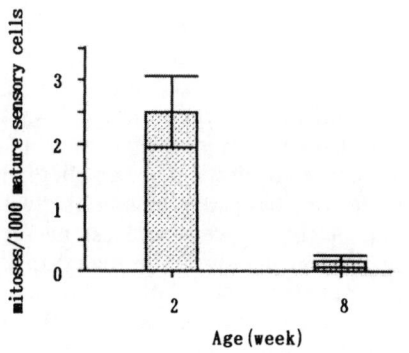

FIG. 6. Age-related frequency of mitoses per 1000 mature sensory cells in the dorsoposterior region of the septal main olfactory epithelium of 2- and 8-week-old mice

The relative and absolute increases in knobless receptor cells could not be explained. The decline in the ratio of progenitor cells to daughter (olfactory receptor) cells could only be explained by the rather dubious assumptions that the diminished number of progenitor cells must have dramatically enhanced and speeded up their mitotic activity with age. Parallel to these investigations, we showed in the OEP of the tiger salamander that mitotic activity is high in epithelial areas with few receptor cell perikarya but low in areas with densely packed receptor cells [10,11]. Analyzing the frequency of mitoses in the progenitor cell compartment of the OEP with the metaphase arrest technique and relating the obtained numbers (from 10:00 a.m. to 2:00 p.m.) to 1000 light globose basal cells and olfactory receptor cells, we found remarkable reductions between the data for 2- and 8-week-old animals. The reduced ratios of mitoses versus light basal cells and mature olfactory sensory cells contradict the already dubious assumption of enhanced mitotic activity with age (Figs. 5 and 6). As we showed for the mouse and tiger salamander [4–6,8,10–13] with two inde-

pendent methods, the ratio of mitoses per 1000 progenitor cells and receptor cells declined, and we postulate that a programmed cell turnover of the olfactory receptor cells could no longer be assumed.

Summary and Perspectives

The morphometric investigations led us to assume that the leading principle underlying the dynamics of the replacement of OEP receptor cells is a tendency of the latter to live as long as possible and to reach an optimal numerical density of mature sensory cells per epithelial unit [4,6,11–13]. Cell death phenomena probably reflect mainly dying receptor cells due to external insults. Analogous to the physiologic cell death during normal development of the CNS, the peripheral olfactory neuroepithelium could show cell death for these reasons. A careful evaluation of cell death numbers (pyknosis versus apoptosis) and location (progenitor cell compartment versus receptor cell compartment) could most likely specify these features. Investigation on this topic would, however, have to differentiate between developing growth areas and stable, mature areas. In other words, one would have to consider the age-related changes in the epithelial area of the nasal chamber covered by OEP and the total number of sensory cells therein.

We further conclude that the progenitor of receptor cells, the basal cell, provides an ample supply of immature, prospective olfactory receptor cells. These immature cells would have no dendritic knob and probably no functional synaptic contact with the olfactory bulb (although the establishment of silent synapses cannot be excluded). From the immature cells sufficient numbers of functional receptor cells could be generated within a short time after the death of one to many thousands of mature receptor cells with the outgrowth of olfactory knobs and cilia to the epithelial surface.

A variety of questions regarding the underlying processes to the above documented principles of diversity, stability, and plasticity in both the OS and GS could be tackled by multiple research strategies, some of which are discussed here. Metaphase arrest techniques together with careful morphometric, immunocytologic techniques and evaluation of DNA reduplication phenomena are further strategies used to resolve the questions in relation to neuronal cell replacement and plasticity in the two chemoreceptive systems. Cell cycle phase evaluations versus the respective locations of cells in the sensory cell epithelia together with lectin-histochemical studies and analysis of gender and hormone effects on the sensory epithelium would be strongly recommended as well. In unpublished investigations on the development of the olfactory epithelia of *Xenopus borealis*, for example, we have observed that group III lectins are especially helpful for comparing species differences and for following the maturation of the OS (Leip and Breipohl, unpublished data) and GS [14–16]. Detailed lectin-histochemical and immunocytochemical studies (e.g., to differentiate between various types of basal cells and different receptor sites on the sensory cilia and microvilli) should therefore be linked with such investigations. In addition, more detailed studies, including experimental sensory deprivation or stimulation, chemical destruction of olfactory receptor cells (e.g., by zinc sulfate and N-methylformimino-methylester), and careful analysis of cell ratios, cell death phenomena, and mitoses are required for the variety of receptor cell epithelia in both the OS and GS to characterize the dynamics of the sensory epithelia and their age-related regenerative capacities.

Including investigations on growth factors, associated glands, and the underlying extracellular matrix proteins would further help bridge the gap between descriptive morphologic studies and molecular biologic research strategies. Finally, concomitant morphometric evaluations of the receptor cell epithelia and their projection sites could definitely help to deepen our insights into the regulation of the dynamics of chemoreceptive epithelia. One comment must be added, however: in addition to the above results and features, circadian shifts of mitotic activity in the OEP in mice have been observed (Grandt and Breipohl, unpublished) and must be kept in mind as an inherent systemic feature when the above recommended research strategies and others are applied to investigate the diversity, stability, and plasticity of the OS and GS.

References

1. Rehn B, Breipohl W, Mendoza AS, Apfelbach R (1986) Changes in granule cells of the ferret olfactory bulb associated with imprinting on prey odours. Exp Brain Res 373:114–125
2. Lü ZB, Breipohl W, Rehn B (1986) Morphometric studies on the effect of postnatal occlusion on the structure of higher olfactory centers. In: Breipohl W (ed) Ontogeny of Olfaction. Springer, Berlin Heidelberg New York, pp 119–209
3. Moulton DG (1974) Dynamics of cell populations in the olfactory epithelium. Ann NY Acad Sci 237:52–61
4. Breipohl W, Grandt D, Rehn B, Mackay-Sim A, Hierche H (1985) Investigations of cell replacement in the olfactory epithelium. Neurosci Lett [Suppl] 19:7
5. Breipohl W, Rehn B, Molyneux GS, Grandt D (1985) Plasticity of neuronal cell replacement in the main olfactory epithelium of mouse. In: Proceedings of the XII International Anatomy Congress, London, p 85
6. Breipohl W, Mackay-Sim A, Rehn B, Walker C (1986) Neurogenesis in the olfactory epithelium. Chem Senses 12:180
7. Breipohl W, Rehn B, Laing D, Panhuber H (1986) Postnatal maturation of the olfactory system. Chem Senses 11:1
8. Breipohl W, Mackay-Sim A, Grandt D, Rehn B, Darrelmann C (1986) Neurogenesis in the vertebrate main olfactory epithelium. In: Breipohl W (ed) Ontogeny of olfaction. Springer, Berlin Heidelberg New York, pp 21–34
9. Nagahara Y (1940) Experimentelle Studien über die histologischen Veränderungen des Geruchsorgans nach der Olfactoriusdurchschneidung: Beiträge zur Kenntnis des feineren Baus des Geruchsorgans. Jpn J Med Sci Pathol 5:165–199
10. Mackay-Sim A, Patel U (1984) Regional differences in cell density and cell genesis in the olfactory epithelium of the salamander, Ambystoma tigrinum. Exp Brain Res 57:99–106
11. Mackay-Sim A, Breipohl W, Kremer M (1988) Cell dynamics in the olfactory epithelium: a morphometric study. Exp Brain Res 71:189–198
12. Mackay-Sim A, Kittel P (1991a) On the life span of olfactory receptor neurons. Eur J Neurosci 3:209–215
13. Mackay-Sim A, Kittel P (1991b) Cell dynamics in the adult mouse olfactory epithelium: a quantitative autoradiographic study. J Neurosci 11:979–984
14. Witt M, Reutter K (1988) Lectin histochemistry on mucous substances of the taste buds and adjacent epithelia of different vertebrates. Histochemistry 88:453–461
15. Witt M, Reutter K (1990) Electron microscopic demonstration of lectin binding sites in the taste buds of the European catfish Silurus glanis (Teleostei). Histochemistry 94:617–628
16. Reutter K, Witt M (1993) Morphology of vertebrate taste organs and their nerve supply. In: Simon SA, Roper SD (eds) Mechanisms in taste transduction. CRC Press, Boca Raton, FL

Intercellular Communications Via Gap Junctions in the Olfactory System

FERNANDO MIRAGALL, MARIAN KREMER, and ROLF DERMIETZEL[1]

Key words. Gap junctions—Connexins—Olfactory system—Olfactory ensheathing cells—Olfactory neurons—Neuronal plasticity

Introduction

Gap junctions are aggregations of membrane channels connecting adjacent cells. They form intercellular pathways for the diffusion of ions and small molecules and have been shown to be involved in important biologic events such as maintenance of intercellular buffering, synchronization of cellular behavior and cell-to-cell coordination [1]. There is also evidence that gap junctions play a regulative role in development and regeneration [2]. Isolation and biochemical analysis of gap junctions have led to the identification of a group of related proteins generically termed connexins [1]. Connexins show tissue and species specificity and are developmentally regulated. Thus among the three known major connexins of mammalian nervous tissues (Cx43, Cx32, Cx26), Cx32 is absent at early stages of development, whereas Cx26 and Cx43 are strongly expressed in embryonic nervous tissues [3]. Whereas expression of Cx32 increases during maturation of the nervous system, Cx26 mostly disappears with the advent of maturity [3]. Expression of Cx43 remains high throughout development and into adulthood [3].

The olfactory system is a chemosensory system that, together with the vomeronasal system, demonstrates plasticity and regenerative capacities unique in the adult mammalian nervous system in that its sensory neurons continuously turn over [4] and are able to reconstitute their nerves after ablation [5]. To gain further insight into the subcellular and molecular organization of the olfactory system, we have investigated the presence of gap junctions and the expression of Cx43, Cx32, and Cx26 in the olfactory epithelium and bulb of the adult mouse. For this purpose we used immunocytochemical methods at both light and electron microscopic levels as well as freeze-fracture electron microscopy.

Material and Methods

Adult NMRI mice were used in this study. Animals were deeply anesthetized with ether.

For freeze-fracture electron microscopy mice were fixed by intravascular perfusion using 3.2% glutaraldehyde and 2.6% paraformaldehyde in cacodylate buffer (0.09M, pH 7.35). Olfactory tissues were processed for freeze-fracture as previously reported [6].

For the immunolocalization of Cx26, Cx32, and Cx43, specific polyclonal antibodies against these connexins were used. The production and specificity of these antibodies have been described elsewhere [7–9]. Primary antibodies were made visible for immunofluorescent detection by goat antibodies to rabbit or mouse immunoglobulins coupled to fluorescein isothiocyanate (FITC), tetramethylrhodamine (TRITC), or Texas red (TR). For immunogold electron microscopy, primary antibodies were visualized by anti-immunoglobulin antibodies adsorbed to colloidal gold (5 or 10nm in diameter; Amersham, Braunschweig, Germany).

Indirect immunofluorescence, postembedding immunogold electron microscopy, and Northern bolt analyses were carried out as previously described [9].

Results

Freeze-Fracture Electron Microscopy

Freeze-fracture replicas of olfactory tissues showed gap junctions as characteristic dense aggregates of intermembrane particles on the P-fracture face (Figs. 1 and 2) with a complementary array of pits on the E-fracture face (not shown). In the olfactory epithelium no gap junctions were observed on olfactory sensory neurons. Gap junctions were observed associated with apical tight junctional strands (Figs. 1 and 2) or in deeper regions of the epithelium, but always on supporting cells. In the olfactory nerves, gap junctions were observed to connect adjacent olfactory ensheathing cells, the characteristic glial cells of the olfactory nerves (now shown). In the olfactory bulb, gap junctions appeared in all layers, including the leptomeninges. They were localized

[1] Institut für Anatomie, Universität Regensburg, 93040 Regensburg, Germany

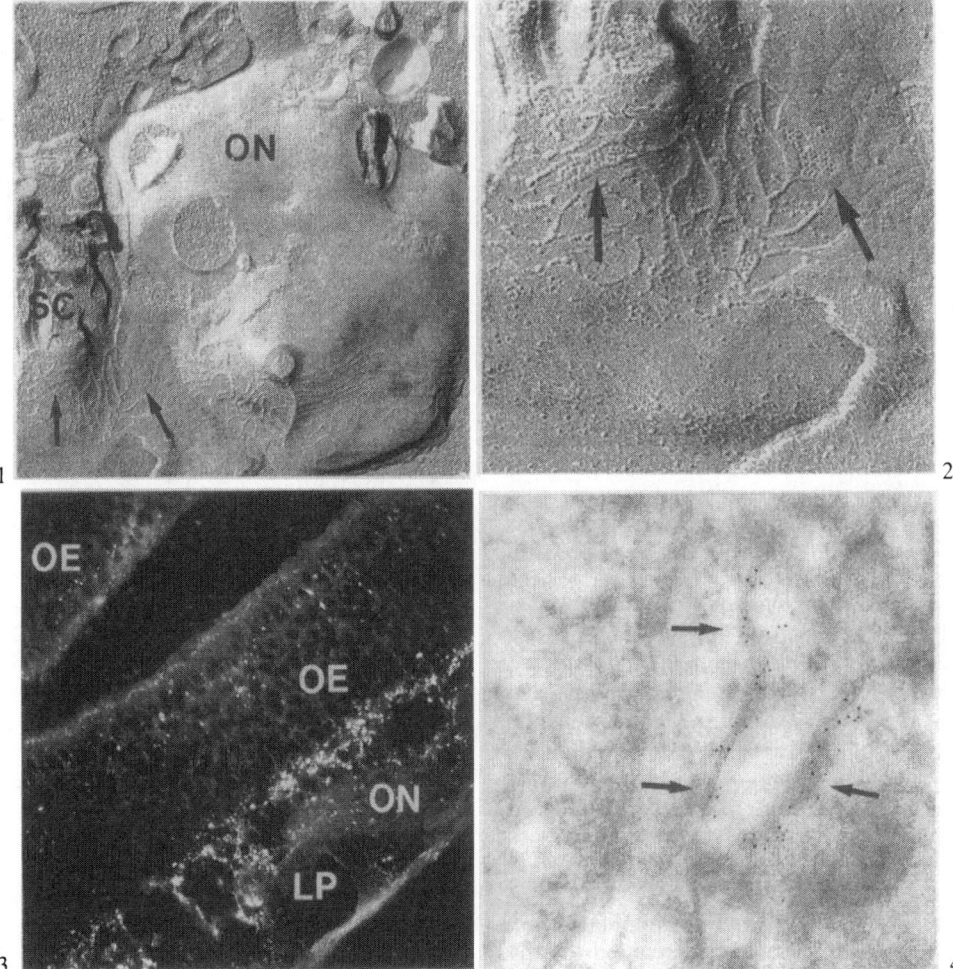

FIG. 1. Freeze-fracture electron microscopy of the apical region of the olfactory epithelium. Gap junctions (*arrows*) are localized on the supporting cell (*SC*) in association with the tight junctional complex. *ON*, olfactory neuron

FIG. 2. Higher magnification micrograph of gap junctions (*arrows*) from Fig. 1

FIG. 3. Connexin 43 immunofluorescence of the olfactory epithelium. Strong immunoreactivity is observed on the epithelium (*OE*) and the lamina propria (*LP*) including the fila olfactoria (*ON*)

FIG. 4. Immunogold electron microscopy using antibodies against Cx43 shows immunoreactive gap junctions (*arrows*) between olfactory ensheathing cells

here on olfactory ensheathing cells, astrocytes, and mitral cells (not shown).

Immunocytochemistry

To analyze the localization of Cx43, Cx32, and Cx26, immunocytochemical methods using specific polyclonal antibodies against these three proteins were used.

Cx43

Immunofluorescence analysis of the localization of Cx43 yielded clear punctate or plaque-shaped immunolabeling (Fig. 3). Cx43 immunoreactivity was mostly observed in the apical third of the olfac-tory neuroepithelium and in association with the apical junctional complex (Fig. 3). Strong immuno-fluorescence was detected in the underlying lamina propria. This fluorescence appeared as small puncta or short fibrils (Fig. 3). Cx43 immunolabeling was also observed within the olfactory nerves (Fig. 4) and in the olfactory bulb. This strong reaction was localized in all layers of the bulb (not shown). As in the olfactory epithelium and nerves, bulb immuno-labeling appeared punctate or in the form of short fibrils, the latter corresponding to the characteristic gap junctional plaques. Cx43-positive short fibrils were mostly localized in the olfactory nerve layer and at blood vessels. Double immunofluorescence analysis using polyclonal antibodies against Cx43 and monoclonal antibodies against GFAP showed strong co-localization of Cx43 immunolabeling and GFAP-positive structures, indicating association of Cx43 expression with astrocytes and in the olfactory nerve layer with ensheathing cells.

Immunogold electron microscopy confirmed the light microscopic observations. Cx43-positive gap junctions within the olfactory nerves were clearly localized between processes of the glial ensheathing cells (Fig. 4). Cx43-positive gap junctions were also observed in the bulb between astrocytic processes (not shown). In addition, the undifferentiated cells of the subependymal layer of the bulb were frequently coupled with Cx43-positive gap junctions (not shown). From the electron microscopic analysis, we can conclude that neither gap junctions nor Cx43 immunoreactivity are localized between primary olfactory sensory neurons.

Cx32

No Cx32 expression was detected immunocytochemically in the olfactory epithelium or nerves. Some Cx32-positive cellular elements were observed in the olfactory bulb. They were localized mostly at the mitral cell layer and internal plexiform layer. Scattered Cx32-positive cells were observed in association with myelinated fibers, suggesting that these cells are oligodendrocytes.

Cx26

Connexin 26 was not detected in the olfactory epithelium or nerves despite observation of clear immunoreactivity in glands of the nasal respiratory region. In the olfactory bulb Cx26 immunoreactivity was detected in the subependymal layer and periglomerular region. The leptomeninges also showed clear Cx26 expression.

Northern Blot Analysis

Northern blot analyses of total RNA from olfactory epithelia and bulbs were carried out using cDNA probes for Cx26, Cx32, and Cx43. Northern blot analyses of total mRNA from adult olfactory epithelium and bulb clearly hybridizes with Cx43 probes, giving bands at 3 kb that correspond to the size of Cx43 mRNA. Cx26 mRNA was detected at 2.6 kb in RNA preparations from olfactory epithelium and bulbs. Northern blot analysis of total mRNA from the olfactory bulb using Cx32 probes gave a band at 1.6 kb, corresponding to the size of Cx32 mRNA.

Discussion

Three major facts were determined in our study.

1. *Gap junctions are present in the peripheral olfactory system connecting nonneuronal cells.*

Numerous gap junctions are present in the olfactory epithelium and nerves. Although the presence of these intercellular junctions in the olfactory system has been previously reported [10,11], the combination of freeze-fracture electron microscopy and immunocytochemistry used in this study provides evidence of the magnitude and relevance of gap junctions in this system. Gap junctions largely connect supporting cells in the olfactory epithelium. In the olfactory nerves, ensheathing cells are strongly coupled with another via gap junctions. It therefore seems that although olfactory sensory neurons do not possess gap junctions these neurons, including their axonal processes, are enveloped by a functional syncytium that provides a homogeneous environment not only for the processes of olfactory reception and transcription but also for transmission of the olfactory impulse.

2. *Connexin 43 is the dominant one in olfactory tissues.* Among the three connexins investigated, Cx43 has been found to be dominant in olfactory tissues. Cx43 has been described to be characteristic of excitable tissues, such as heart or brain [12]. Our findings regarding Cx43 expression in the olfactory system supports this notion, despite the fact that Cx43 is localized between nonneuronal cells. Expression of Cx43 is high in the olfactory epithelium, whereas Cx32 and Cx26 are not detected. However, because in most epithelia at least two connexins are expressed concomitantly, we cannot exclude the existence of other members of the connexin family in the olfactory epithelium. Moreover, Cx26, Cx32, and Cx43 are present in the olfactory bulb, although no immunoreaction for any of these connexins was detected on mitral cells despite the fact that gap junctions are observed on these cells in freeze-fracture replicas. These observations clearly suggest the existence of other connexins (perhaps still unidentified members of the family) in the olfactory bulb.

3. *Connexin 26 is detected in the adult olfactory bulb.* In the nervous system Cx26 is strongly expressed during development, and it largely disappears with the advent of maturity [3]. Cx26 therefore represents the embryonic member of the molecular connexin family. Interestingly, we have detected Cx26 expression at the subependymal layer and at the periglomerular region of the adult olfactory bulb. These portions of the bulb have been described to retain significant embryonic features such as the expression of embryonic N-CAM, laminin, and 04 antigen [13]. The retention of Cx26 expression in the adult olfactory bulb can be considered another of these embryonic attributes.

Acknowledgments. The authors thank Drs. O. Traub, E.L. Hertzberg, E.C. Beyer, and B.J. Nicholson for antibodies and cDNA probes; I. Hertting for technical assistance; and H. Grist-Paleologo for typing the manuscript. The photographic work of K. Daßler and A. Pieringer is also acknowledged. This work has been supported by a grant of the Deutsche Forschungsgemeinschaft (Schwerpunkt Glia) to R.D.

References

1. Dermietzel R, Hwang TK, Spray DC (1990) The gap junction family: structure, function and chemistry. Anat Embryol 182:517–528

2. Warner AE, Guthrie SC, Gilula NB (1984) Antibodies to gap junction protein selectively disrupt junctional communication in early amphibian embryo. Nature 311:127–131

3. Dermietzel R, Spray DC (1993) Gap junctions in the brain, what type, how many and why? Trends Neurosci 16:186–192

4. Graziadei PPC, Monti Graziadei GA (1978) Continuous nerve cell renewal in the olfactory system. In: Jacobson M (ed) Handbook of Sensory Physiology (vol 9). Springer, Berlin Heidelberg New York, pp 55–82

5. Monti Graziadei GA, Graziadei PPC (1979) Neurogenesis and neuron regeneration in the olfactory system of mammals. II: Degeneration and reconstitution of the olfactory sensory neurons after axotomy. J Neurocytol 8:197–213

6. Miragall F (1983) Evidence for orthogonal arrays of particles in the plasma membranes of olfactory and vomeronasal sensory neurons of vertebrates. J Neurocytol 12:567–576

7. Traub O, Janssen-Timmen U, Drüge PM, Dermietzel R, Willecke K (1982) Immunological properties of gap junctions from mouse liver. J Cell Biochem 19:27–44

8. Traub O, Look J, Dermietzel R, Brümmer, Hülser D, Willecke K (1989) Comparative characterization of the 21 kDa and 26 kDa gap junction proteins in murine liver and cultured hepatocytes. J Cell Biol 108:1039–1051

9. Miragall F, Hwang TK, Traub O, Hertzberg EL, Dermietzel R (1992) Expression of connexins in the developing olfactory system of the mouse. J Comp Neurol 325:359–378

10. Kerjaschki D, Hörandner H (1976) The development of mouse olfactory vesicles and their contacts: a freeze-etching study. J Ultrastruct Res 54:420–444

11. Mendoza AS, Breipohl W, Miragall F (1980) Intercellular junctions during the development of the olfactory epithelium in the chick: a freeze-etching study. J Submicrosc Cytol 12:29–41

12. Kadle R, Zhang JT, Nicholson BJ (1991) Tissue-specific distribution of differentially phosphorylated forms of Cx43. Mol Cell Biol 11:363–369

13. Miragall F, Kadmon G, Faissner A, Antonicek H, Schachner M (1990) Retention of J1/tenascin and the polysialylated form of the neural cell adhesion molecule (N-CAM) in the adult olfactory bulb. J Neurocytol 19:899–914

Generation, Differentiation, and Maturation of Olfactory Receptor Neurons In Vitro

ANNE L. CALOF[1], MURALI D. ADUSUMALLI[1], MELINDA K. DEHAMER[1], JOSE L. GUEVARA[1], JEFF S. MUMM[1], SARA J. WHITEHEAD[1], and ARTHUR D. LANDER[2]

Key words. Olfactory epithelium—Neurogenesis—Neuronal precursor—Polypeptide growth factor—Cell migration—Extracellular matrix

Introduction

In the mammalian olfactory epithelium (OE), proliferation of neuronal precursor cells and differentiation of their progeny into olfactory receptor neurons (ORNs) are ongoing processes [1]. The OE is thus an ideal system for exploring how neurogenesis can control neuron number. There is also evidence that cell interactions regulate neurogenesis in the OE: Degeneration of ORNs following removal of the olfactory bulb (their central target) or lesioning of ORN axons leads to increased proliferation of neuronal precursor cells in the epithelium and subsequent production of new receptor neurons [2].

Although new ORNs are clearly generated in bulbectomized animals, the full complement of biochemically mature neurons is never reconstituted [3]. That immature ORNs are generated at all in bulbectomized animals suggests that neurogenesis and neuronal differentiation occur independently of target tissue [4]. However, these in vivo experiments also suggest that subsequent biochemical maturation and survival of ORNs are to some extent target tissue-dependent [3].

We have developed a culture system for OE purified from mouse embryos [4]. During our initial characterization of OE explants, we identified three cell types: (1) basal cells, which grow as epithelial sheets and express keratins; (2) postmitotic olfactory receptor neurons (ORNs), which express the neural cell adhesion molecule N-CAM; and (3) immediate neuronal precursors (INPs), which do not express keratins or N-CAM. INPs are migratory round cells that rapidly sort out from basal cells, synthesizing DNA and dividing as they migrate. ³H-Thymidine

(^{3}H-TdR) incorporation analysis indicates that INPs are the direct precursors of ORNs [4].

Interestingly, the events that occur during bulbectomy-induced neuronal degeneration and regeneration in vivo in the OE appear to be mimicked in vitro in OE cultures: Proliferation of INPs and differentiation of ORNs occur in the absence of target tissue [4]. However, genesis of ORNs from INPs continues for only 1 to 2 days and then ceases. Cessation of neurogenesis in vitro may reflect inappropriate or inadequate mitogenic stimulation of INPs, as studies suggest that INPs may be able to undergo as many as three rounds of division in vivo [5]. With time in culture, not only do the INPs cease dividing and disappear (owing to their generation of ORNs), but ORNs begin to die and by 7 days are no longer detectable [4]. In vivo, contact with the olfactory bulb appears to enable ORNs to survive for prolonged periods [6]. Because no central nervous system (CNS) cells are present in our OE cultures, any trophic factors or cell interactions normally supplied by the CNS are necessarily absent.

Our previous studies in this system suggest that olfactory neurogenesis and neuron survival in vitro are subject to regulatory influences such as those that operate in vivo, in that extrinsic agents are clearly required to permit long-term neuronal production and survival. Here we describe results from experiments designed to identify and characterize molecules that regulate generation, differentiation, and maturation of ORNs. Our experiments focus on two classes of molecules: polypeptide growth factors and glycoproteins of the extracellular matrix (ECM).

Materials and Methods

Materials

Recombinant human nerve growth factor (NGF), brain-derived neurotrophic factor (BDNF), neurotrophin 3 (NT3), and recombinant rat ciliary neurotrophic factor (CNTF) were obtained from Genentech (San Francisco, CA). Recombinant human acidic fibroblast growth factor (FGF1), basic fibroblast growth factor (FGF2), transforming

[1] Department of Biological Sciences, Room 138, Biology Building, University of Iowa, Iowa City, IA, 52242, USA
[2] Departments of Biology and Brain & Cognitive Sciences, Massachusetts Institute of Technology, Cambridge, MA, USA

growth factor-α (TGF-α), epidermal growth factor (EGF), and platelet-derived growth factors AA and BB (PDGF-AA, PDGF-BB) were obtained from US Biochemicals (Cleveland, OH). Recombinant human transforming growth factor-β2 (TGF-β2) was from Genzyme (Cambridge, MA). Laminin, fibronectin, merosin, and poly-D-lysine (PDL) were prepared or purchased as described [7]. For explant cultures, acid-cleaned glass coverslips were prepared and coated with ECM molecules as described [4,7]. For multiwell cultures, 96-well tissue culture trays (Costar no. 3596) were coated overnight with PDL (1 mg/ml in water), washed, and sterilized by ultraviolet (UV) light. AG1D5 monoclonal anti-N-CAM antibody was prepared as described [4]. Anti-MASH1 hybridoma 24B72D11 was the gift of D. Anderson (California Institute of Technology). Monoclonal anti-Thy 1.1 (Ab22) was the gift of W. Matthew (Duke University Medical School).

Tissue Culture and Analysis

For assessing growth factor effects on neuronal precursor proliferation, explants of purified OE from embryonic day 14–15 (E14–15) CD-1 mice were cultured on glass coverslips coated with PDL followed by merosin (10 μg/ml) for 3 h at 37°C [4,7]. Cultures were grown in serum-free, low calcium (0.1 mM) medium with modified N2 additives (LCM) [4], except that bovine serum albumin (BSA; Clinical Reagent Grade, ICN) was reduced to 1 mg/ml. Cultures were grown for 24 h in the tested growth factor, then growth factor was replenished, and ^{3}H-TdR (0.1 μCi/ml; 80 Ci/mmol) was added. After another 24 h, cultures were fixed and processed for autoradiography [4]. Quantitative analysis of ^{3}H-TdR incorporation was performed using explants as statistical units. For each explant analyzed, the number of migratory cells incorporating ^{3}H-TdR was counted and this number normalized to the explant area. This labeling index was measured for a minimum of 20 to 40 explants per experimental condition. The proliferation factor (PF) was then calculated as a ratio: mean labeling index for that condition/mean labeling index for explants grown in no growth factor. Percent error of the ratio was calculated as the square root of the sum of the squares of percent errors (from SEMs) of the two labeling indices being compared; it was generally around 20%.

For determining adhesion and migratory responses of olfactory neuronal cells, either E14–15 explants, or the "neuronal cell" fraction (ORNs plus INPs) of OE suspension cultures (prepared as described [7]) were cultured in LCM containing crystalline BSA 5 mg/ml [7]. For assessing effects of growth factors and pharmacologic agents on survival and maturation of ORNs, the neuronal cell fraction of OE suspension cultures was also used, except that

OE was purified from E16–17 homozygous OT-2 transgenic mice [8], and neuronal cell fractions were prepared after 6 h of suspension culture without the use of proteolytic enzymes. Neuronal cells were plated onto PDL-coated 96-well tissue culture trays at 2×10^4 to 5×10^4 cells per well in LCM with crystalline BSA 5 mg/ml.

Results and Discussion

Mitogenic Effects of Polypeptide Growth Factors on Olfactory Neuron Precursors

Compelling evidence suggests that two families of polypeptide growth factors—neurotrophins (NTs) and fibroblast growth factors (FGFs)—are important regulators of neuron number: Both families have widespread expression in the nervous system [9,10], both exert trophic effects in vivo and in vitro on a wide variety of neurons [11], and there is evidence for their mitogenic effects on neuronal precursors as well [12].

Receptors for NTs and FGFs appear to be expressed by OE cells in vivo: Affinity cross-linking experiments have demonstrated cross-linking of both BDNF and NT3 to membranes from E15 olfactory turbinates (E. Escandon, K. Nikolics, A. Calof, unpublished data). Reverse transcriptase-polymerase chain reaction analysis of RNA from E15 OE indicates that FGF receptors FR1 and 2 are also expressed [13]. To determine whether NTs, FGFs, or other families of polypeptide growth factors affect proliferation of olfactory neuron precursors, we measured the relative proliferation of migratory olfactory neuronal cells grown for 48 hours in the presence of various factors. Proliferation was assessed by measuring ^{3}H-TdR incorporation from 24 to 48 h in culture, a time when neurogenesis has greatly diminished in OE cultures grown in defined, serum-free medium [4]. Growth factors tested were FGF1 (100 ng/ml), FGF2 (10 ng/ml), PDGFaa (10 ng/ml), PDGFbb (10 ng/ml), EGF (20 ng/ml), TGF-α (20 ng/ml), BDNF (50 ng/ml), NT3 (50 ng/ml), NGF (50 ng/ml), CNTF (10 ng/ml), and TGF-β2 (10 ng/ml). Only FGF1 and FGF2 caused a statistically significant increase in the proliferation of olfactory neuronal cells. FGF2 showed the most pronounced effect, giving a three- to fourfold increase in proliferation ($\overline{PF}$ = 3.23); the effect of FGF1 was slightly less ($\overline{PF}$ = 2.82) [13].

To verify that labeled migratory cells in these assays were ORN precursors, it was demonstrated that 85% to 90% of ^{3}H-TdR-labeled cells expressed immunologically detectable N-CAM (a marker for ORNs) within 24 to 36 h following a brief (2-h) pulse of ^{3}H-TdR (5 μCi/ml; 80 Ci/mmol). This proportion was similar for cultures pulsed at t = 10 and t = 22 h in culture and for cultures grown in the presence or absence of FGF2. Thus most proliferating migratory

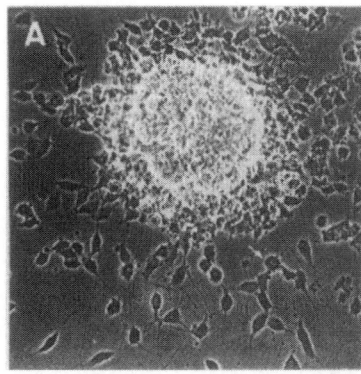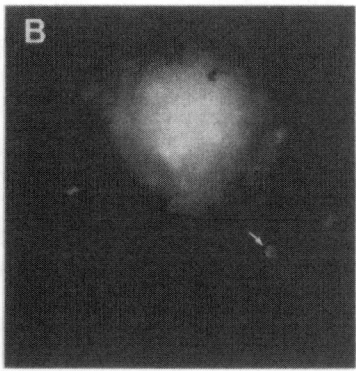

FIG. 1. Anti-MASH1 immunostaining of olfactory epithelium cultures. E14 OE explants were grown for 10 h, then fixed and stained with anti-MASH1 hybridoma supernatant followed by rhodamine-conjugated goat anti-mouse IgG (Zymed, 1:30). **A** Phase contrast. **B** Rhodamine optics. *Arrow* points to a MASH1$^+$ nucleus

cells in these cultures are INPs, and FGF2 apparently acts on this population [13].

Immortalization and Immunologic Analysis of Cultured Olfactory Epithelium Cells: New Classes of Precursor Cells

Although INPs may be identified by their morphology, lack of keratin and N-CAM immunoreactivity, and ability to migrate and incorporate ^{3}H-TdR, they may consist of more than one kind of precursor cell. We have found that a monoclonal antibody to the transcription factor MASH1 [14], a homolog of *Drosophila achaete-scute* proteins, recognizes a subset of migratory cells in early (t = 10 h) explant cultures of E14–15 OE (Fig. 1). MASH1-immunoreactive cells appear to be dividing, as they incorporate ^{3}H-TdR (not shown), but only a small proportion (3%–4%) of total migratory, ^{3}H-TdR-incorporating cells are MASH1$^+$. This finding suggests that MASH1$^+$ cells are a subset of INPs, perhaps at a specific stage of differentiation. The possibility that MASH1$^+$ cells in OE cultures are in fact progenitors of INPs is suggested by evidence that targeted disruption of the mouse MASH1 gene causes loss of most of the ORNs (F. Guillemot, personal communication).

Generation of immortalized cell lines from proliferating cells in OE cultures has revealed that progenitors for olfactory ensheathing cells are present in cultures of OE purified from E14–15 mice [15]. These cell lines express markers characteristic of olfactory ensheathing cells, including glial fibrillary acidic protein (GFAP) and S-100 [16], but no growth condition we have yet tried has caused them to express markers of neuronal differentiation, suggesting that ensheathing cells are in a lineage separate from that of ORNs.

Effects of Extracellular Matrix Molecules on Cell Migration and Axon Outgrowth by Olfactory Neuronal Cells

In vivo, neurons derived from OE exhibit two types of motile behavior. Differentiating receptor neurons

extend axons that grow through the olfactory nerve to targets in the olfactory bulb; and in the embryo some cells also leave the epithelium entirely and migrate into the brain, where they become luteinizing hormone releasing hormone-secreting neurons of the hypothalamus [17]. In vitro, neurons and INPs of the OE exhibit both types of behavior [4,7]. Interestingly, cell migration and neurite outgrowth occur on substrata treated with the ECM protein laminin or its homolog merosin but not on substrata treated with other ECM molecules, such as fibronectin or collagens [7]. The effects of laminin can be mimicked by an elastase fragment of laminin known as E8 and can be blocked by antibodies to integrin receptor α6β1, which is known to be a receptor for E8 [18]. Purified OE neuronal cells are also stimulated to migrate and extend neurites by laminin, but laminin is not adhesive for these cells [7]. In contrast, fibronectin supports strong neuronal adhesion but does not promote cell migration [7]. These results suggest that laminin, working through an integrin receptor, stimulates olfactory neuron motility via a mechanism other than promotion of cell-substratum adhesion.

Regulation of Olfactory Receptor Neuron Maturation and Survival

Using OT-2 transgenic mice, we are able to identify —and potentially purify—OE cells that can be unambiguously identified as mature ORNs (MORNs). In the OT-2 strain, regulatory elements of the olfactory marker protein gene drive expression of the Thy 1.1 cDNA in mice that are otherwise homozygous for Thy 1.2. Therefore MORNs are the only cells in OT-2 mice that express Thy 1.1 [8].

Initial experiments have examined effects of NTs and pharmacologic agents on cultures of ORNs prepared from E16–17 OT-2 embryos. We have found that aurintricarboxylic acid (AT; 100 μM), which prevents programmed cell death in other neuronal cells in vitro [19], promotes survival of cultured ORNs as well. NTs can also promote survival of ORNs in vitro, with AT > BDNF > NT3 > NGF in effectiveness (not shown). Interestingly, the number

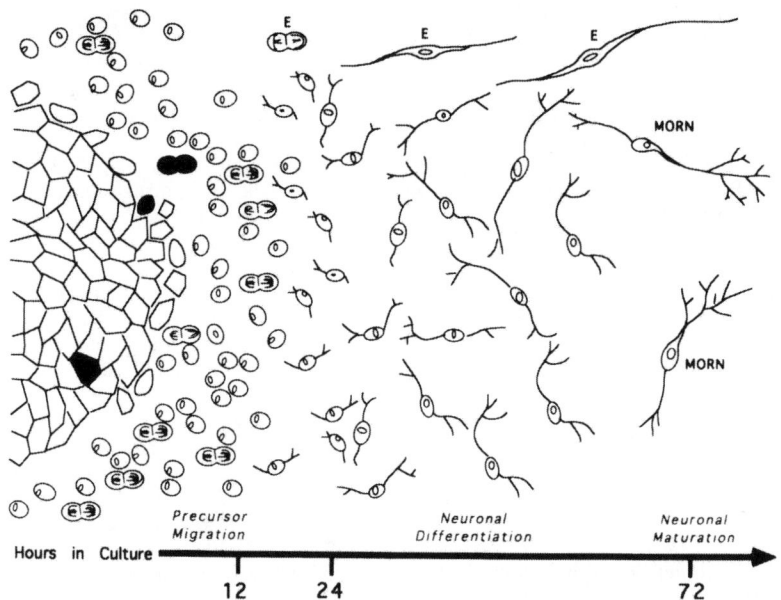

Hours in Culture ——————————————————————————————▶

| 12 | 24 | | 72 |

Precursor Migration   *Neuronal Differentiation*   *Neuronal Maturation*

Fig. 2. Timeline of cellular differentiation in olfactory epithelium cultures. Cellular transitions occurring in E14–E15 OE explant cultures are shown, from left to right, over the time course in which they occur. Shortly after explants are plated, olfactory neuron precursors (primarily INPs) sort out from basal epithelial cells and migrate extensively when plated on laminin-treated substrata. A subpopulation of these precursors (shown in *black*) are MASH1$^+$. Over the course of 2 days in culture, progeny of INPs differentiate into ORNs. When cultured under conditions that inhibit programmed cell death, ORNs mature biochemically (*MORN*, mature ORN), even in the absence of target tissue. *E*, progenitors of olfactory ensheathing cells, also derived from epithelial cells in OE cultures

of Thy 1.1-expressing ORNs surviving for 72 h in AT appears to be greater than the number that were present at the time of initial plating [20]. This suggests that AT treatment allows biochemical maturation of ORNs in the absence of their target tissue, suggesting that both initial differentiation [4] and subsequent biochemical maturation of ORNs are target-independent developmental processes.

Summary

Tissue cultures of embryonic mouse OE have been used to functionally test molecules regulating generation, differentiation, and maturation of ORNs. These in vitro experiments have also led to new insights into the cellular stages of olfactory neurogenesis. Our current concept of this process as it occurs in OE cultures is pictured in Fig. 2.

Acknowledgments. The authors thank Francois Guillemot for sharing information about MASH1 gene disruption experiments; Frank Margolis for OT-2 mice; and Miyuki Yamamoto, David Anderson, and William Matthew for gifts of hybridomas. This research was supported by grants to A.L.C. from the University of Iowa DERC (DK25295), and from the NSF (IBN-9253103), and to A.D.L. from the David and Lucile Packard Foundation.

References

1. Graziadei PPC, Monti Graziadei GA (1978) Continuous nerve cell renewal in the olfactory system. In: Jacobson M (ed) Handbook of Sensory Physiology, vol 9. Development of sensory systems. Springer, New York, pp 55–83
2. Graziadei PPC, MontiGraziadei GA (1979) Neurogenesis and neuron regeneration in the olfactory system of mammals. I: Morphological aspects of differentiation and structural organization of the olfactory sensory neurons. J Neurocytol 8:1–18
3. Verhaagen J, Oestreicher AB, Grillo M, Khew-Goodall YS, Gispen WH, Margolis FL (1990) Neuroplasticity in the olfactory system: differential effects of central and peripheral lesions of the primary olfactory pathway on the expression of B-50/GAP43 and the olfactory marker protein. J Neurosci Res 26:31–44
4. Calof AL, Chikaraishi DM (1989) Analysis of neurogenesis in a mammalian neuroepithelium: proliferation and differentiation of an olfactory neuron precursor in vitro. Neuron 3:115–127
5. Mackay-Sim A, Kittel P (1991) Cell dynamics in the adult mouse olfactory epithelium: a quantitative autoradiographic study. J Neurosci 11:979–984
6. Schwob JE, Szumowski KEM, Staskey AA (1992) Olfactory sensory neurons are trophically dependent on the olfactory bulb for their prolonged survival. J Neurosci 12:3896–3919
7. Calof AL, Lander AD (1991) Relationship between neuronal migration and cell-substratum adhesion: laminin and merosin promote olfactory neuronal migration but are anti-adhesive. J Cell Biol 115:779–794

8. Danciger E, Mettling C, Vidal M, Morris R, Margolis F (1989) Olfactory marker protein gene: its structure and olfactory neuron-specific expression in transgenic mice. Proc Natl Acad Sci USA 86:8565–8569

9. Burgess WH, Maciag T (1989) The heparin-binding (fibroblast) growth factor family of proteins. Annu Rev Biochem 58:575–606

10. Maisonpierre PC, Belluscio L, Friedman B, Alderson RF, Wiegand SJ, Furth ME, Lindsay RM, Yancopoulos GD (1990) NT-3, BDNF, and NGF in the developing rat nervous system: parallel as well as reciprocal patterns of expression. Neuron 5:501–509

11. Barde YA (1989) Trophic factors and neuronal survival. Neuron 2:1525–1534

12. Rohrer H (1990) The role of growth factors in the control of neurogenesis. Eur J Neurosci 2: 1005–1015

13. Calof AL, DeHamer M, Adusumalli M, Guevara J, Lopez A, Olwin B (1993) Generation of olfactory receptor neurons in vitro: characterization of neuronal progenitor cells and their mitogens. Soc Neurosci Abstr 19:1286

14. Lo L-C, Johnson JE, Wuenscehll CW, Saito T, Anderson DJ (1991) Mammalian achaetescute homolog 1 is transiently expressed by spatially restricted subsets of early neuroepithelial and neural crest cells. Genes Dev 5:1524–1537

15. Calof AL, Guevara JL (1993) Cell lines derived from retrovirus-mediated oncogene transduction into olfactory epithelium cultures. Neuroprotocols 3:222–231

16. Devon R, Doucette R (1992) Olfactory ensheathing cells myelinate dorsal root ganglion neurites. Brain Res 589:175–179

17. Calof A (1992) Sex, nose and genotype. Curr Biol 2:103–105

18. Calof AL, Yurchenco PD, O'Rear JJ, Lander AD (1992) Adhesion, anti-adhesion and migration of olfactory neurons and neuronal precursors are independently regulated by distinct molecular domains of laminin. Mol Biol Cell 3:74a

19. Mesner PW, Winters TR, Green SH (1992) Nerve growth factor withdrawal-induced cell death in neuronal PC12 cells resembles that in sympathetic neurons. J Cell Biol 119:1669–1680

20. Calof AL, Escandon E, Guevara JL, Mumm JS, Nikolics K, Whitehead SJ (1993) Regulation of olfactory receptor neuron maturation and survival in vitro. Mol Biol Cell 4:371a

Stem Cells of Olfactory Cells During Development

Yuko Suzuki and Masako Takeda[1]

Key words. Bromodeoxyuridine—Keratin—Neural cell adhesion molecule—Basal cells proper—Globose basal cells—Developing olfactory epithelium

Introduction

There are two types of basal cell in the basal region of the olfactory epithelium of mice: (1) The basal cells proper, which are in direct contact with the basement membrane, are positive for keratin as detected with anti-keratin antibodies [1–4]. (2) The globose basal cells, which lie between basal cells proper and olfactory cell nuclei, or often close to the basement membrane, are stained positively with antineural cell adhesion molecule (N-CAM) antibody [5,6] and are devoid of keratin [2]. An earlier study using [^{3}H]thymidine autoradiography revealed that basal cells proper divide and give rise to new olfactory cells via globose basal cells because olfactory cells are continuously replaced by stem cells [7]. We reported in a previous study combining immunohistochemistry of keratin with the bromodeoxyuridine (BrdU) method for labeling dividing cells that globose basal cells increase their mitotic activity after axotomy but that basal cells proper do not show any change in their mitotic rate, suggesting that globose basal cells are stem cells of olfactory cells [8]. These two types of basal cell have been shown to differentiate in the basal region of the olfactory epithelium at birth [2]; however, the changes in the mitotic activity of these cells during postnatal development have not been investigated.

Olfactory cells of mice appear at embryonic day 12 and increase in number during subsequent embryonic development [9]. Therefore the stem cells of the olfactory cells during the embryonic period are different from those observed postnatally, but ultrastructural and immunohistochemical characteristics of stem cells during the embryonic phase have not been fully investigated.

The aim of this study was to examine stem cells of olfactory cells during development. The developing olfactory epithelium of mice was examined using the combination of immunohistochemistry of keratin or N-CAM with BrdU and by conventional light and electron microscopy.

Materials and Methods

Dd mice of embryonic days 12, 14, 17, 19 and postnatal days 1, 3, 7, 14 were used. Unilateral bulbectomy was performed on some of the postnatal mice (P1–3) according to the surgical procedure reported by Graziadei and MontiGraziadei [10]. The bulbectomized mice were killed 5 days after bulbectomy. BrdU (Sigma) 50 mg/kg was injected intraperitoneally into normal and bulbectomized mice, and the animals were killed 1 and 24 h after injection of BrdU. The BrdU was detected immunohistochemically using anti-BrdU antibody.

For immunohistochemistry, mice were killed and their nasal cavities fresh-frozen or fixed in 4% paraformaldehyde. Sections 10 µm thick were prepared with a cryostat, and the incubation with antibodies was carried out as described previously [8]. Double immunostaining using antikeratin and anti-BrdU antibodies was performed on each postnatal day; and for the embryonic mouse specimens, antineural cell adhesion molecule antibody (N-CAM, Immunotech, Marseille, France) was used instead of antikeratin antibody. The number of BrdU-labeled cells was counted under a light microscope (10 × 40 magnification) equipped with an ocular micrometer along a 200 µm length of the olfactory epithelium. The specimens for electron microscopic study were prepared according to standard procedures.

Results

Embryonic Stage

In the 12-day-old embryonic mouse (E 12), the olfactory pits were already present, and their lateral parts showed secondary recesses. At 1 h after injection of BrdU, cells immunolabeled with anti-BrdU antibody were numerous throughout the olfactory epithelium, indicating active division. A few N-CAM-immunoreactive cells were found among he BrdU-labeled cells, and N-CAM-immunoreactive axons extended from the epithelium into the rostral forebrain. The secondary recesses of the lateral parts of the olfactory pits, consisting of thin epi-

[1] Department of Oral Anatomy, School of Dentistry, Higashi Nippon Gakuen University, Ishikari-Tobetsu, Hokkaido, 061-02 Japan

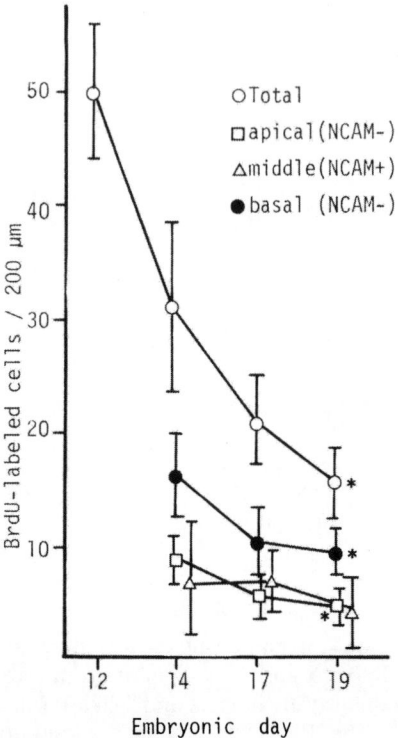

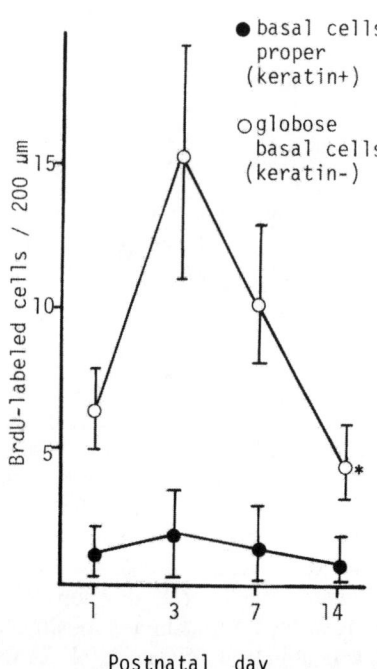

FIG. 1. BrdU-immunolabeled cells per 200 μm of olfactory epithelium on embryonic days 12, 14, 17, and 19. *Points* represents the mean and *vertical bars* the standard deviation. Significant decreases in the total number, the number in the basal region, and the number in the apical region occurred as development proceeded (*p < 0.01, t-test). No significant changes were observed in the middle region

FIG. 2. BrdU-immunolabeled cells per 200 μm of olfactory epithelium on postnatal days 1, 3, 7, and 14. *Points* represent the mean and *vertical bars* the standard deviation. Significant changes in the number of cells were seen among the globose basal cells at different ages (*p < 0.01, t-test). The peak number of labeled globose basal cells occurred at day 3. There were no significant differences in number of labeled basal cells proper between days 3 and 7

thelium, were immunolabeled with BrdU in their middle and basal regions, but N-CAM-immunoreactive cells did not appear in these regions. The BrdU-labeled cells were also numerous in the underlying mesenchyme. The development of these regions was chronologically behind that of the other regions of the olfactory pits, indicating that these regions were probably in the placodal stage. On E 14, N-CAM-immunoreactive cells, which were mostly negative for BrdU, increased in number and formed six or seven layers in the middle region of the olfactory epithelium. These cells corresponded to immature and mature olfactory cells. Most of the BrdU-labeled cells were located in the apical and basal regions, which were negative for N-CAM. There were mosaics of active or quiescent zones in the basal region where cells were labeled or unlabeled. Observation of embryos sacrificed 24 h after injection of BrdU, revealed many BrdU-labeled cells in the middle region of the olfactory epithelium and along the axons in the lamina propria. On E 17 to E 19, the BrdU-labeled cells in the apical and basal regions decreased in number. A few BrdU-labeled cells were present in the middle region of olfactory

epithelium. The apical and basal regions were still unstained with N-CAM at this stage. The changes in number and location of BrdU-labeled cells in the olfactory epithelium during embryonic development are summarized in Fig. 1.

Electron microscopy of E 12 specimens showed that a few olfactory cells and many columnar-shaped cells were present in the olfactory epithelium. In E 14 preparations, the columnar cells contained a small amount of rough endoplasmic reticulum (r-ER) and many free ribosomes, and they were located in the basal region of the olfactory epithelium, in direct contact with the basement membrane. During later development, the columnar cells decreased in number and round or pyramid-shaped cells appeared in the basal region. Near the border with the respiratory epithelium, the basal cells proper appeared as keratin-positive cells in the basal region of the olfactory epithelium of E 19 mice.

Postnatal Stage

The keratin-containing basal cells proper also appeared in all portions containing the border of olfactory epithelium on P 1 and were round or flat in shape. BrdU-labeling was mainly associated with nuclei of keratin-negative globose basal cells located above the basal cells proper. Several labeled cells in the apical and middle regions of the epithelium were supporting cells or the duct cells of Bowman's gland. The mature and immature olfactory cells in the middle region were not labeled with BrdU. On P 3, numerous globose basal cells became immunoreactive for BrdU, but the labeled globose basal cells gradually decreased in number as postnatal development proceeded. A few of the basal cells proper were labeled with BrdU on P 1 to P 7, but these cells rarely showed the labeling after P 14. The changes in numbers of the two types of labeled basal cells during postnatal development are summarized in Fig. 2.

The thickness of the olfactory epithelium was measured during postnatal development. It ranged from 60 to 100 μm on P 1 and increased 1.5-fold until P 7. No further change in thickness was observed thereafter, even at the adult stage.

In the mice that underwent bulbectomy on P 1 to P 3, the olfactory epithelium on the experimental side was thinner than that on the unoperated side. At 1 h after injection of BrdU, labeled basal cells proper and globose basal cells appeared to be more numerous in the olfactory epithelium of the bulbectomized animals. The clusters of cells, consisting of keratin-positive and keratin-negative cells, were out of the olfactory epithelium and located in the lamina propria of the experimental side.

Changes in the shape of basal cells were observed on P 3 to P 7 in both unoperated and operated mice. The basal cells proper contacting the BrdU-labeled globose basal cells were frequently pyramidal in shape. The globose basal cells were large and round and were located close to the basement membrane. After E 14, basal cells proper showed flat shapes, and globose basal cells lay above them. In electron micrographs of basal cells proper, mitochondria, r-ER, and tonofilaments were observed in the cytoplasm. The globose basal cells were devoid of tonofilaments and contained mitochondria, r-ER, and Golgi apparatus.

Discussion

Prior to this study, the origin of olfactory and supporting cells in the developing olfactory epithelium of the mouse was studied by light and electron microscopy. The first olfactory cells are found in the middle region of olfactory epithelium [5]. Mitotic figures are found only in the apical region of the developing olfactory epithelium up to E 13 and subsequently appear at the basal region and increase in number [11]. Mitotic figures in the apical region of the olfactory epithelium are thought to be related to

the differentiation of stem cells into supporting cells, whereas those in the basal region are considered to be stem cells of olfactory cells [5,11]. Our present study, combining immunohistochemistry of keratin or N-CAM with BrdU, identified the types of cell that were dividing in the developing olfactory epithelium. The columnar cells of the embryonic stage and globose basal cells of the postnatal stage were numerously labeled with BrdU, indicating active division. The labeling pattern suggests that the columnar cells, which are located throughout the olfactory epithelium on E 12 and then in the basal region after E 14, are stem cells of olfactory cells during the embryonic period, as labeled cells migrated into the olfacotory cell layer of the middle region 24 h after injection of BrdU. During the postnatal stage, active division of globose basal cells accompanied an increase in the thickness of the olfactory epithelium caused by an increase in the number of olfactory cells. On the other hand, there was a small number of labeled basal cells proper in the olfactory epithelium, and these cells did not correspond to the labeled globose basal cells. Therefore it is suggested that during the postnatal period olfactory cells originate from globose basal cells, not from basal cells proper. Moreover, at earlier stages (before embryonic day 14), our present data showed that cell divisions appeared in the middle and basal regions in the olfactory placode and then occurred throughout the epithelium of the olfactory pit on E 12, whereas mitotic figures had been found only at the apical region up to E 13 in an earlier anatomic study [11].

Bulbectomy-induced changes during early postnatal days are different from those observed after axotomy [8]. The mitotic activity of globose basal cells increased in the case of both operations, but the increase in the mitosis of basal cells proper was observed only after bulbectomy. Because anti-keratin antibody used in the present study reacted only with the basal cells proper and not with the supporting cells or Bowman's gland cells, the migrating keratin-positive cells are suggested to be basal cells proper. The cell types of keratin-negative cells that were migrating from the olfactory epithelium have been suggested elsewhere to be globose basal cells [12]. In the present study it is not clear whether dividing globose basal cells become olfactory cells or they migrate away from the olfactory epithelium.

Although columnar cells in the embryonic stage resemble globose basal cells in the postnatal stage morphologically, the immunostaining of N-CAM was not observed in columnar cells. In the vomeronasal organ (VNO), a previous study showed that columnar cells were present in the basal region of this organ at E 14 and expressed N-CAM in later embryonic days [13]. Thus columnar cells in the olfactory epithelium, which generate two types of basal cells after E 19 to P 1, may mature at a later stage than those in the VNO. On the other hand, Barber and Raisman [14] reported that the VNO of mice does not generate keratin-positive basal cells.

References

1. Key B, Akeson RA (1990) Immunohistochemical markers for the frog olfactory neuroepithelium. Dev Brain Res 57:103–117
2. Suzuki Y, Takeda M (1991a) Keratins in the developing olfactory epithelia. Dev Brain Res 59:171–178
3. Vollrath M, Altmannsberger K, Weber K, Osborn M (1985) An ultrastructural and immunohistochemical study of the rat olfactory epithelium: unique properties of olfactory cells. Differentiation 29:243–253
4. Yamagishi M, Hasegawa S, Takahashi S, Nakano Y, Iwanaga T (1989) Immunohistochemical analysis of the olfactory mucosa by use of antibodies to brain properties and cytokeratin. Ann Otol Rhinol Laryngol 98:384–388
5. Carr VM, Farbman AI, Lidow MS, Colletti LM, Hempstead JL, Morgan JI (1989) Developmental expression of reactivity to monoclonal antibodies generated against olfactory epithelia. J Neurosci 9: 1179–1198
6. Miragall F, Kadmon G, Husman M, Schachner M (1988) Expression of cells adhesion molecule in the olfactory system of the adult mouse: presence of the embryonic form of NCAM. Dev Biol 129:516–531
7. Graziadei PPC, Monti-Graziadei GA (1979) Neurogenesis and neuron regeneration in the olfactory system of mammals. 1. Morphological aspects of differentiation and structural organization of the olfactory sensory neurons. J Neurocytol 8:1–18
8. Suzuki Y, Takeda M (1991b) Basal cells in the mouse olfactory epithelium after axotomy: immunohistochemical and electron-microscopic studies. Cell Tissue Res 266:239–245
9. Cuscheri A, Bannister LH (1975) The development of the olfactory mucosa in the mouse: electron microscopy. J Anat 119:471–498
10. Graziadei PPC, Monti-Graziadei GA (1986) Neuronal changes in the forebrain of mice following penetration by regenerating olfactory axons. J Comp Neurol 247: 344–356
11. Smart I (1971) Location and orientation of mitotic figures in the developing mouse olfactory epithelium. J Anat 109:243–251
12. Monti-Graziadei AG (1992) Cell migration from the olfactory neuroepithelium of neonatal and adult rodents. Dev Brain Res 70:65–74
13. Suzuki Y, Takeda M (1993) Stem cells of olfactory cells during embryonic development. Higashi Nippon Dental J 12:175–184
14. Barber PC, Raisman G (1978) Cell division in the vomeronasal organ of the adult mouse. Brain Res 141:57–66

Growth Factor Regulation of Olfactory Cell Proliferation

ALBERT I. FARBMAN, JUDITH A. BUCHHOLZ, and RICHARD C. BRUCH[1]

Key words. Olfactory Epithelium—Mitosis—Neurogenesis—Growth factors—Cell division

Introduction

Vertebrate olfactory sensory cells are unique neurons in that they are continually replaced throughout the life of the animal [1–3]. Because neurogenesis is continuous, at any given time in an adult animal the olfactory sensory neuronal population is a mixture of cells of different ages. It was believed that the life span of the olfactory sensory cell was about 1 month [4,5], but it has since been shown that under some conditions cells can either live considerably longer [6,7] or die prematurely [8–10].

In recent years it has been shown that the *rate* of cell division in the olfactory neuron population can be increased or decreased experimentally. In adult rats, ablation of the olfactory bulb or severance of the olfactory nerves leads to an increase of more than twofold in the number of basal cells undergoing division [10–12], an effect that lasts for at least 7 weeks after surgery [10]. On the other hand, occlusion of one nostril in neonatal rats results, in 30 days, in a 40% reduction in the number of basal cells that incorporate tritiated thymidine ([³H]TdR), a marker of cell division [13].

These observations, that the mitotic rate in the olfactory sensory neuron population can be upregulated (by bulb removal) or downregulated (by naris occlusion), imply that regulatory factors modulate cell division in the basal cell population. We used an organ culture assay to explore the possibility that certain peptide growth factors are involved in this modulating process.

Material and Methods

Organ Culture

The nasal septa from 19-day fetal rats were explanted onto collagen-coated Millipore filter substrates resting on stainless steel grid platforms in humidified organ culture dishes [14,15]. Serum-free culture medium (Waymouth's MB/752 with supplements [16]) containing various doses of growth factors was used. Cultures were grown for 3 days; the medium was then replaced with medium containing the same dose of growth factor plus either $2\,\mu Ci/ml$ tritiated thymidine ([³H]TdR) or $10^{-4}\,M$ bromodeoxyuridine (BrdU) for 1 h. Explants were then fixed in 4% paraformaldehyde in 0.1 M phosphate buffer, pH 7.2, and prepared for autoradiography to reveal [³H]TdR uptake or for immunohistochemistry to reveal BrdU uptake. Counts were made of the numbers of cells that had incorporated the mitotic marker per 100-μm length of olfactory epithelium on 12 sections from each explant. From 6 to 14 explants were used for each dose of each growth factor.

Receptor Binding Assay

The method for the EGF receptor assay was modified from O'Keefe et al. [17]. The measurements were performed with samples of membrane preparations from olfactory mucosa of adult female rats (three animals in each of 10 experiments) containing 0.1 g of total protein in 0.1 ml Krebs-Ringer buffer containing 0.1% albumin. To these membranes was added either $50\,\mu l$ of buffer alone (total binding) or $50\,\mu l$ of buffer containing $2\,\mu g$ epidermal growth factor (EGF) (nonspecific binding); then 50 ng ^{125}I-EGF (in $50\,\mu l$) was added to each tube; the tubes were vortexed and incubated for 30 min at room temperature. Next, $800\,\mu l$ of ice-cold wash buffer (Krebs-Ringer) was added to each tube and the tubes placed on ice. The mixture was filtered through a 0.2-μm Millipore filter, followed by 10 ml of ice-cold wash buffer. Filters were placed in glass test tubes and counted in a gamma counter for 1 min. Specific binding was calculated as total binding minus nonspecific binding.

Immunohistochemistry

Rats were perfused with 0.1 M PBS to clear the vessels of blood, then with Bouin's fixative (15 parts 4% paraformaldehyde, 4 parts saturated picric acid, 1 part acetic acid). Tissues were decalcified and processed for paraffin embedding; 10-μm-thick sections were placed on subbed slides. Sections were deparaffinized to water and treated sequentially at 37°C with normal rabbit serum (20 min), antibody

[1] Department of Neurobiology & Physiology, Northwestern University, Evanston, IL 60208-3520, USA

EGF

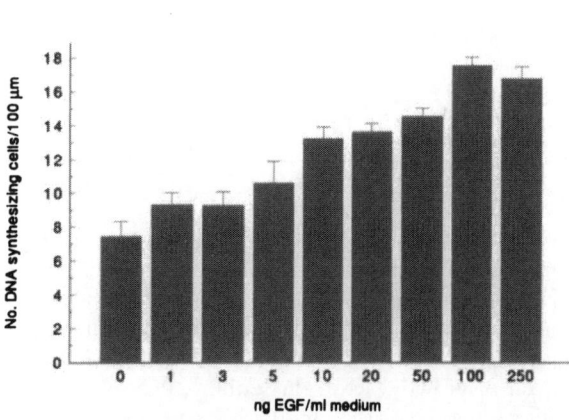

TGFα

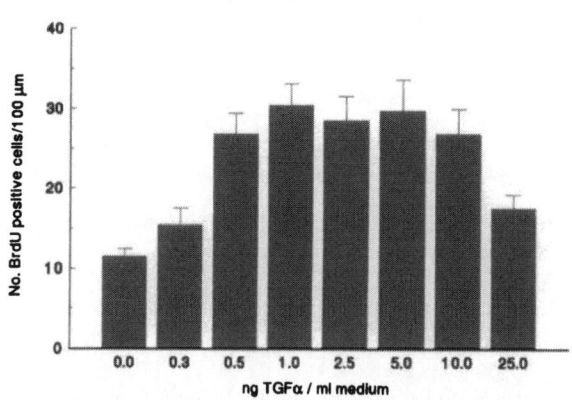

FIG. 1. Incorporation of DNA precursors by organ cultures of olfactory epithelium treated with various doses of EGF and TGF-α. Data from experiments using BrdU were combined with data from experiments in which [³H]TdR was used (*left*). Most EGF experiments used [³H]TdR as a DNA precursor but some used BrdU.

A [³H]TdR-labeled nucleus was counted when it had a minimum of five silver grains overlying it. The number of BrdU-positive nuclei in the control group (0 dose of growth factor) was about 20% higher than in control groups in which [³H]TdR was used. BrdU data were normalized to [³H]TdR values for the EGF graph

against EGF receptor (75 μg/ml for 60 min), the Vectastain kit for sheep antibodies (purchased from Vector Labs, Burlingame, CA) containing rabbit antisheep secondary antibody (30 min), and the ABC reagent (45 min). Each serum treatment was followed by three 5-min washes in 0.1 M phosphate buffer, pH 7.2. The specimens were then treated with H_2O_2 and diaminobenzidine to reveal EGF receptor immunoreactivity.

Results

EGF and transforming growth factor-alpha (TGF-α) increased the rate of cell division by two- to three-fold (Fig. 1) in a dose-dependent manner. TGF-α was 50- to 100-fold more potent as a mitogenic agent than EGF; the dose required to obtain a maximal response was 1 ng/ml, compared witn 50–100 ng/ml of EGF. Basic fibroblastic growth factor (bFGF) had virtually no effect (not shown).

The mitogenic activities of EGF and TGF-α, both members of the EGF family of growth factors, are mediated by their binding to the EGF receptor (170 kDa), a member of the tyrosine kinase group of receptors [18–20]. Consequently, if either or both of the two peptide growth factors are involved in regulation of olfactory neurogenesis, one can predict that the EGF receptor should be present. We made crude membrane preparations from homogenates of adult rat olfactory epithelium and assayed them for the presence of EGF receptor by electrophoresis and immunoblotting. A major band of 170 kDa was revealed in olfactory membranes by the antibody to EGF receptor (Fig. 2). A ligand-binding assay

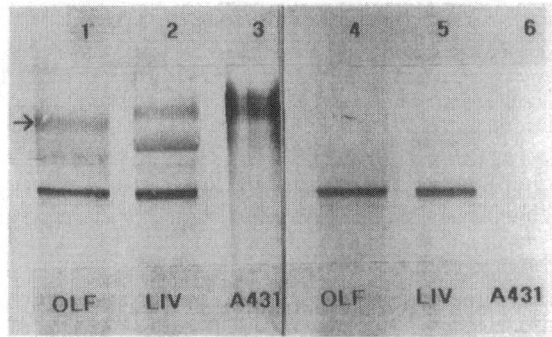

FIG. 2. Crude membrane preparations of adult rat olfactory mucosa (OLF: lanes 1, 4), liver (LIV: lanes 2, 5) and A431 cells (lanes 3, 6) were run on SDS-polyacrylamide gel electrophoresis, transferred to Immobilon P, then assayed for EGF receptor by immunochemistry. A major band at 170 kDa (*arrow*) and a minor band at 150 kDa were revealed in olfactory membrane homogenates by antibody to EGF receptor (10 μg/ml, lane 1). Liver (lane 2), known to bind EGF avidly, and A431 cells (lane 3), a mammary tumor cell line that overexpresses EGF receptor, were positive controls. Negative controls: phosphate-buffered saline (PBS) was substituted for EGF receptor antibody for olfactory (lane 4), liver (lane 5), and A431 cell membranes (lane 6). The intense band in lanes 1, 2, 4, and 5 is an artifact of staining by secondary antibody in rat tissues

on a membrane preparation of adult rat olfactory mucosa revealed that an average of 3.28 ng [¹²⁵I]EGF (SE, 0.6; *n* = 10) was specifically bound to the membranes.

If the EGF family of growth factors is involved in olfactory mitogenesis, one would predict that the receptor is localized on the cells that undergo mito-

FIG. 3. Histological section from olfactory mucosa of adult rat. Basal cells (*arrows*) are highly immuno-reactive; supporting cells are less so

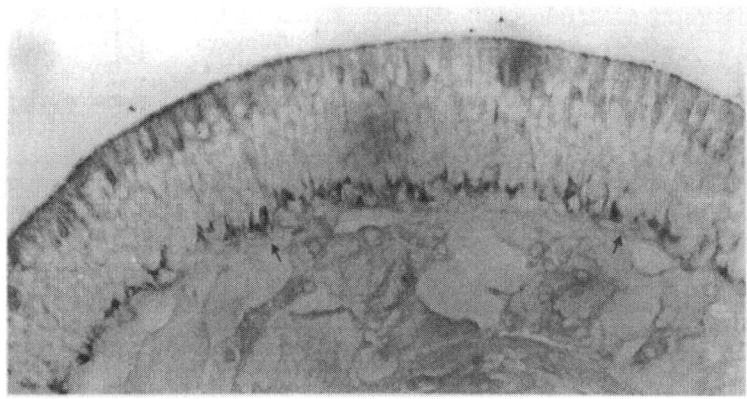

sis. An immunohistochemical analysis of olfactory mucosa using the antibody to EGF receptor showed that basal cells were indeed strongly immuno-reactive and that supporting cells were somewhat less immunoreactive (Fig. 3). Mature neurons did not appear to bind the antibody.

Discussion

These data, showing that EGF and TGF-α stimulate an increase in the number of dividing cells in olfactory epithelium, are consistent with observations in other epithelia, including epidermis, intestinal epithelium, liver, salivary glands, uterine epithelium, thyroid, thymus, lung, and prostate, all of which are stimulated to divide by EGF or TGF-α (reviewed in ref. [18]). Moreover, EGF is a known mitogen for neurons and astrocytes grown in vitro, whether the cells originate from young embryos [21–23] or from adult brains [24,25]. EGF is known to be produced by several tissues, especially epithelial glands, such as salivary [26], mammary [27], or sweat glands [28]. EGF is also found in nasal mucus [29], presumably a product of nasal glands or perhaps of the glands of Bowman found in olfactory mucosa.

Taken together, the mitotic responsiveness of olfactory epithelium to EGF and TGF-α in vitro, and the presence of significant amounts of EGF receptor on the mitotic cells of olfactory epithelium support the notion that EGF or TGF-α might be involved in regulation of cell division in olfactory epithelium in vivo. TGF-α is more potent in vitro and is considered more likely the active candidate in vivo. Given the multiple physiological actions of growth factors, however, it is probable that participation of the EGF family of peptides in regulation of olfactory mitosis is only part of a complex control system in which the regulation of cell death must also be a factor.

Acknowledgments. Research was supported by NIH Grant No. 1 P01 DC00347. The authors are grateful for the assistance of Shubhik DebBurman.

References

1. Thornhill RA (1970) Cell division in the olfactory epithelium of the lamprey, *Lampetra fluviatilis*. Z Zellforsch 109:147–157
2. Moulton DG, Celebi G, Fink RP (1970) Olfaction in mammals—two aspects: proliferation of cells in the olfactory epithelium and sensitivity to odours. In: Wolstenholme GEW, Knight J (eds) Ciba Foundation Symposium on Taste and Smell in Vertebrates. Churchill, London, pp 227–250
3. Graziadei PPC, Metcalf JF (1971) Autoradiographic and ultrastructural observations on the frog's olfactory mucosa. Z Zellforsch 116:305–318
4. Moulton DG (1975) Cell renewal in the olfactory epithelium of the mouse. In: Denton DA, Coghlan JP (eds) Olfaction and taste, V. Academic, New York, pp 111–114
5. Graziadei PPC, Monti Graziadei GA (1978) Continuous nerve cell renewal in the olfactory system. In: Jacobson M (ed) Handbook of sensory physiology, vol 9. Springer, Berlin, pp 55–82
6. Hinds JW, Hinds PL, McNelly NA (1984) An autoradiographic study of the mouse olfactory epithelium: evidence for long-lived receptors. Anat Rec 210:375–383
7. Mackay-Sim A, Kittel PW (1991) On the life span of olfactory receptor neurons. Eur J Neurosci 3:209–215
8. Breipohl W, Mackay-Sim A, Grandt D, Rehn B, Darrelmann C (1986) Neurogenesis in the vertebrate main olfactory epithelium. In: Breipohl W (ed) Ontogeny of olfaction. Springer, Berlin, pp 21–33
9. Schwob JE, Szumowski KEM, Stasky AA (1992) Olfactory sensory neurons are trophically dependent on the olfactory bulb for prolonged survival. J Neurosci 12:3896–3919
10. Carr VM, Farbman AI (1992) Ablation of the olfactory bulb up-regulates the rate of neurogenesis and induces precocious cell death in olfactory epithelium. Exp Neurol 115:55–59
11. Suzuki Y, Tateda M (1991) Basal cells in the mouse olfactory epithelium after axotomy: immunohistochemical and electron microscopic studies. Cell Tissue Res 266:239–245
12. Schwartz Levey M, Chikaraishi DM, Kauer JS (1991) Characterization of potential precursor populations in the mouse olfactory epithelium using immunocytochemistry and autoradiography. J Neurosci 11:3556–3564

13. Farbman AI, Brunjes PC, Rentfro L, Michas J, Ritz S (1988) The effect of unilateral naris occlusion on cell dynamics in developing rat olfactory epithhelium. J Neurosci 8:3290–3295

14. Farbman AI (1977) Differentiation of olfactory receptor cells in organ culture. Anat Rec 189:187–200

15. Chuah MI, Farbman AI (1983) Olfactory bulb increases marker protein in olfactory receptor cells. J Neurosci 3:2197–2205

16. Bottenstein JE, Skaper SD, Varon SS, Sato GH (1980) Selective survival of neurons from chick embryo sensory ganglionic dissociates utilizing serum-free supplemented medium. Exp Cell Res 125:183–190

17. O'Keefe E, Hollenberg MD, Cuatrecases P (1974) Epidermal growth factor. Characteristics of specific binding in membranes from liver, placenta, and other target tissues. Arch Biochem Biophys 164:518–526

18. Carpenter G, Wahl MI (1990) The epidermal growth factor family. In: Sporn MB, Roberts AB (eds) Peptide growth factors and their receptors I. Springer, Berlin Heidelberg New York, pp 69–172

19. Massagué J (1990) Transforming growth factor-α. A model for membrane-anchored growth factors. J Biol Chem 265:21393–21396

20. Pandiella A, Massagué J (1991) Transforming growth factor-α. Biochem Soc Trans 19:259–262

21. Anchan RM, Reh TA, Angello J, Balliet A, Walker M (1991) EGF and TGF-alpha stimulate retinal neuroepithelial cell proliferation in vitro. Neuron 6:923–936

22. Leutz A, Schachner M (1981) Epidermal growth factor stimulates DNA synthesis of astrocytes in primary cerebellar cultures. Cell Tissue Res 220:393–404

23. Simpson DL, Morrison R, de Vellis J, Herschman HR (1982) Epidermal growth factor binding and mitogenic activity of purified populations of cells from the central nervous system. J Neurosci Res 8:453–462

24. Reynolds BA, Weiss S (1992) Generation of neurons and astrocytes from isolated cells of the adult mammalian central nervous system. Science 255:1707–1710

25. Reynolds BA, Tetzlaff W, Weiss S (1992) A multipotent EGF-responsive striatal embyonic progenitor cell produces neurons and astrocytes. J Neuroscience 12:4565–4574

26. Cohen S (1965) The stimulation of epidermal proliferation by a specific protein (EGF). Dev Biol 12:394–407

27. Moran JR, Courtney ME, Orth DN, Vaughan R, Coy S, Mount CD, Sherrell BJ, Greene HL (1983) Epidermal growth factor in human milk: daily production and diurnal variation during early lactation in mothers delivering at term and at premature gestation. J Pediatr 103:402–405

28. Pesonen K, Viinikka L, Koskimies A, Banks AR, Nicolson M, Perheentupe J (1987) Size heterogeneity of epidermal growth factor in human body fluids. Life Sci 40:2489–2494

29. Tonnesen P, Thim E, Nexo E (1990) Epidermal growth factor and haptocorrin in nasal secretion. Scand J Clin Lab Invest 50:187–194

Lectin Histochemical Studies of the Development of Two Olfactory Receptors in *Xenopus laevis*

KEIKO SUZUKI, KAZUSHIGE OGAWA, and KAZUYUKI TANIGUCHI[1]

Key words. Amphibian olfaction—Lectin histochemistry

Lectin-binding patterns of the vomeronasal organ (VNO) in comparison with those of the olfactory epithelium (OE), were examined developmentally in larvae of *Xenopus laevis* to reveal the functional significance of the VNO in amphibians.

The larvae were examined soon after hatch to the end of metamorphosis. Stages of development were determined according to the description of Nieuwkoop and Faber [1]. Materials were fixed in Bouin's solution, embedded in paraffin, sectioned at 5 μm, and processed for lectin histochemical staining with 21 biotinylated lectins (Vector, Burlingame, Calif.).

Although the VNO and OE consist of equal numbers of receptor, supporting, and basal cells, we focused on receptor cells. Lectins showed diverse binding patterns in the VNO and OE, their binding patterns occasionally being changed during development. Changes in lectin-binding patters were roughly classified into four types: intense in early and late developmental stages, but weak in the middle stage (type 1); intense in the middle stage, but weak in early and late stages (type 2); intense in the early stage, becoming weak toward the late stage (type 3); and weak in the early stage, becoming intense toward the late stage (type 4).

On the free surface of the VNO, *Soranum tuberosum* lectin (STL), *Ricinus communis* agglutinin I (RCA I), and *Phaseolus vulgaris* agglutinin-E (PHA-E) showed the type 1 binding pattern, and *Pisum sativum* agglutinin (PSA) and *Lens culinaris* agglutinin (LCA) type 3, while peanut agglutinin (PNA) and *Ulex europaeus* agglutinin I (UEA I) showed faint staining throughout development. In the OE, however, PNA showed the type 1 pattern, and UEA I, PSA, LCA, and PHA-E the type 4, while STL and RCA I were intense throughout development. Wheat germ agglutinin (WGA), *Lycopersicon esculentum* lectin (LEL), *Dolichos biflorus* agglutinin (DBA), and *Bandeiraea simplici-folia* lectin I (BSL I) showed intense staining in both the VNO and the OE throughout development.

In the dendrites of the VNO, PHA-E showed the type 3 pattern, LCA showed type 1, while BSL II, BSL I, PNA, UEA I, and PSA showed weak staining throughout development. In the dendrites of the OE, however, BSL I, BSL II, PSA, and LCA showed the type 3 pattern, PNA, type 4; and UEA I, type 2, while PHA-E was intense throughout development. In common in the VNO and OE, LEL, RCA I, and *Erythrina cristagalli* lectin (ECL) showed type 1 and PHA-L type 3; WGA, DBA, and *Vicia villosa* agglutinin (VVA) were intense throughout development.

In the soma of the VNO, STL and UEA I showed the type 4 pattern, and RCA I and PHA-L type 2, while DBA, BSL I, VVA, and PSA showed intense staining throughout development. In the soma of the OE, however, PSA showed type 3, RCA I type 1, and UEA I type 2, while STL and PHA-L were moderate to intense throughout development. In addition, DBA, BSL I, and VVA showed various binding patterns to individual receptor cells of the OE throughout development. In both the VNO and the OE, PNA and ECL showed the type 4 pattern, and BSL II type 2; WGA, LEL, LCA, and PHA-E were intense throughout development.

Further studies will be necessary to determine the binding patterns of the other lectins, succinylated WGA, *Datura stramonium* lectin, soybean agglutinin, *Sophora japonica* agglutinin, jacalin, and concanavalin A to the VNO and OE during development.

In conclusion, the VNO and OE showed different binding patterns to several lectins during development. These results may reflect some kind of difference in olfactory function between the VNO and OE.

Reference

1. Nieuwkoop PD, Faber J (1967) Normal table of *Xenopus laevis* (Daudin): A systematical and chronological survey of the development from the fertilized egg till the end of metamorphosis. North-Holland Publishing, Amsterdam, pp 1–252

[1] Department of Veterinary Anatomy, Faculty of Agricultures, Iwate University, 3-18-8 Ueda, Morioka, 020 Japan

Effects of p-Chloroamphetamine on Recovery of Olfactory Function Following Olfactory Nerve Severance in Mice

Takaki Miwa[1], Tetsuji Moriizumi[2], Hideo Sakashita[1], Toshiaki Tsukatani[1], Yasuyuki Kimura[1], and Mitsuru Furukawa[1]

Key words. p-Chloroamphetamine—Olfactory-mediated behavior

Serotonin (5-HT) can influence neurite outgrowth and neuronal differentiation, including synaptogenesis. P-chloroamphetamine (p-CA) is a commonly used drug which has been proven to cause 5-HT depletion in the central nervous system. The purpose of this investigation was to evaluate the effects of p-CA on the recovery of axotomy-induced olfactory disturbance in adult mice.

Fourteen male ddy-conventional mice were used. Training for discrimination between 0.01% cycloheximide solution (CH) and distilled water (DW) was performed twice before olfactory nerve section. When a mouse drank DW, the response was interpreted as correct. When the mouse drank CH, it was interpreted as an error. The correct rate was calculated by dividing the number of correct responses by the number of trials (ten). The behavioral task was performed 7, 14, 21, 28, 35, and 42 days after surgery, in the same manner. Following training, bilateral olfactory nerve section was performed intracranially. A single dose of p-CA (10 mg/kg, 0.2 ml) was injected intraperitoneally to the p-CA group ($n = 7$), twice per week following surgery, while the control group ($n = 7$) received saline injections. Following the final test, 5-HT immunoreactivity in the olfactory bulb was assessed at the light microscope level.

Olfactory-mediated behavior was manifested as follows: All mice drank CH at least one time at the first trial, but they refused to drink it because of its specific odor and unpleasant taste at the second trial. Following surgery, the ratio of correct responses was decreased in both groups. Thereafter, it improved and all mice in the control group refused to drink CH 42 days after surgery, while recovery was delayed in the p-CA group. On day 42, the ratio of correct responses was 85.7%. Our immunohistochemical investigation detected 5-HT immunoreactivity in the glomerular layer in the control group; 5-HT immunoreactivity was weaker in the p-CA group.

We concluded that 5-HT may play an important role in the restoration of axotomy-induced olfactory disturbance.

Acknowledgments. This work was supported by the Fund for Medical Treatment of the Elderly (School of Medicine, Kanazawa University).

Departments of [1]Otorhinolaryngology and [2]Anatomy, School of Medicine, Kanazawa University, 13-1 Takara-cho, Kanazawa, 920 Japan

Neuronal and Glial Markers in Immortalized Olfactory Cell Lines

Kelli P.A. MacDonald[1], Gillian R. Bushell[1], Perry F. Barlett[2], and Alan Mackay-Sim[1]

Key words. Olfactory neurogenesis—Gene expression

Olfactory neurons are replaced in the adult olfactory epithelium by division and differentiation of cells close to the basement membrane of the olfactory epithelium. Among the basal cells is a stem cell which undergoes an asymmetric division, giving rise to a new stem cell and a precursor cell. The precursor cell divides again symmetrically and there may be additional symmetrical divisions before the progeny leave the cell cycle to differentiate [1]. We report here the characterization of cell lines created by retroviral insertion of the *n-myc* proto-oncogene into dividing cells of the adult mouse olfactory epithelium [2]. We have approximately 60 cell lines generated by this procedure, many of them cloned by single cell micromanipulation. These clones are heterogeneous in morphology but they fall into three broad categories: class 1 are mainly flat, epithelial-like cells; class 2 are a mixture of epithelial-like cells, round cells, and bipolar cells with short processes; and class 3 are a mixture of very large epithelial-like cells and bipolar cells with very long processes. Due to the slow growth of the class 3 cell lines and the difficulty of cloning them, we have concentrated our analysis on the class 1 and class 2 cells. Expression of neural and non-neural proteins was studied with a combination of immunocytochemistry, Western analysis, and reverse transcriptase polymerase chain reaction (RT-PCR). All cell lines express vimentin and keratin. Class 1 cell lines failed to express any neural cell markers when grown under our standard growth conditions (10% fetal calf serum, Dulbecco's modified Eagle's medium (DMEM), 5% CO_2). Class 2 cell lines expressed a variety of neuronal proteins (neurofilament, β-tubulin, MAP-2, NCAM), although not all cell lines expressed all markers. Some clonal cell lines expressed the intermediate filament protein of glial cells (GFAP) in addition to neurofilament. Due to the cross-reactivity of our GFAP and neurofilament antibodies it is not possible to determine whether individual cells express both markers. The morphology of class 2 cells but not class 1 cells could be changed by manipulating the growth conditions. Similar conditions often altered the expression of neuronal or non-neuronal markers. For example, one class 1 cell line up-regulated the expression of GFAP (but not any neuronal markers) when grown in low-serum conditions. Class 2 cell lines responded variably to the same growth conditions: some cell lines up-regulated expression of GFAP, whereas others up-regulated neuronal markers. Although the incorporation of the *n-myc* gene may cause cellular effects other than immortalization, the heterogeneity of these immortalized cell lines suggests that they derive from a heterogeneous population of dividing cells in the olfactory epithelium. The expression of both neuronal and glial proteins suggests that the basal cells of the olfactory epithelium could be multi-potent, potentially able to give rise to both neurons and glia. The induction of different protein expression by altered growth conditions suggests that the cell lineage of the basal cells in the olfactory epithelium is determined by locally released growth factors.

References

1. Mackay-Sim A, Kittel P (1991) Cell dynamics in the adult mouse olfactory epithelium: A quantitative autoradiographic study. J Neurosci 11:979–984
2. MacDonald K, Micozzi F, Bushell G, Mackay-Sim A, Bernard O, Bartlett PF (1989) Immortalisation of neuronal precursor cells from adult mouse olfactory epithelium. Neurosci Lett [Suppl] 34:S114

[1] Faculty of Science and Technology, Griffith University, Brisbane, Q 4111, Australia
[2] Walter and Eliza Hall Institute for Medical Research, Royal Parade, Melbourne, VIC 3052, Australia

Development of the Rat Olfactory Epithelium

Hideo Sakashita[1], Tetsuji Moriizumi[2], Yasuyuki Kimura[3], and Mitsuru Furukawa[3]

Key words. Olfactory epithelium—Postnatal development

Maturation of the rat olfactory epithelium was studied during development. As a developmental index of the olfactory epithelium, quantification of the thickness and the number of cell layers of the epithelium was performed. To clarify the number of mature receptor cells with their axons connecting with the olfactory bulb, we estimated the percentage of receptor cell labeling (number of labeled cells/ total number of receptor cells) in different developmental stages, with the aid of a retrograde fluorescent tracer (Fluoro-Gold) injected into the olfactory bulb. We also estimated the percentage of degenerated receptor cells after bulbectomy in different developmental stages.

The studies were performed on embryos, pups, and adult (body weight, 250–350 g) Sprague-Dawley rats. E0 (embryonic day 0) was the day that dams were sperm-positive. P0 (postnatal day 0) was the day of birth. The mean thickness of the olfactory epithelium at different stages was 84 μm on embryonic day 19 (E19), 94 μm on postnatal day 0 (P0), 98 μm on postnatal day 10 (P10), and 57 μm in adults. The average number of cell layers was seven (E19), eight (P0), ten (P10), and seven (adult). That is, the olfactory epithelium progressively increased in thickness between P0–P10, and later decreased in thickness and in the number of cell layers toward the adult stage.

Injection of the retrograde fluorescent tracer (Fluoro-Gold) into the olfactory bulb was performed 2 days before the perfusion. On (P3), labeled cells (one to two cell layers) were restricted to the superficial zone of the receptor cell layer. The percentage of receptor cell labeling was 34%. On P10, labeled cells (two to four cell layers) were present in the superficial zone of the receptor cell layer. The percentage of receptor cell labeling was 50%. In adults, the entire receptor cell layer was labeled. The percentage of receptor cell labeling was 78%.

Right bulbectomy was performed 7 days before the perfusion. On P8, the right epithelium was slightly decreased in thickness compared with the left epithelium. The degeneration rate was 31%. In adults, the right epithelium was markedly decreased in thickness compared with the left epithelium and the degeneration rate was 76%. These values were quite consistent with the percentages of receptor cell labeling shown by retrograde fluorescent tracer (Fluoro-Gold). This study revealed the exact percentages of mature receptor cells at different developmental stages and the presence of considerable amounts of immature receptor cells in the neonatal olfactory epithelium.

[1] Department of Otorhinolaryngology, Kouseiren Takaoka Hospital, 5-10 Eiraku-cho, Takaoka, Toyama, 933 Japan
Departments of [2] Anatomy and [3] Otorhinolaryngology, Kanazawa University, 13-1 Takara-machi, Kanazawa, 920 Japan

Acknowledgments. This work was supported by the Fund for Medical Treatment of the Elderly (School of Medicine, Kanazawa University).

Ontogeny of the Septal Olfactory Organ of Masera in the Rat

Harumi Saito, Kazushige Ogawa, and Kazuyuki Taniguchi[1]

Key words. Protein gene product 9.5—Septal olfactory organ of Masera

In mammals, olfactory receptor cells appear in the olfactory epithelium (OE), the vomeronasal organ (VNO), and the septal olfactory organ of Masera (MO). These cells are considered to be derived from the olfactory placode, to undergo some modification during development, and finally to differentiate, becoming responsible for different areas of olfaction [1,2]. However, there is little information on the development and function of the MO. Recently, protein gene product 9.5 (PGP), which has attracted much attention as a novel neuron-specific marker, has been detected in olfactory receptor cells [3]. In the present study, therefore, we examined the development of the MO immunohistochemically in rats, comparing it with that of the OE and VNO, using anti-PGP serum.

We examined fetuses at 13, 14, 15, 16, 17, 18, and 20 days of gestation, and newborn and postnatal animals at 3 and 7 days after birth. They were decapitated and the heads were immersed in Bouin's solution, embedded in paraffin, sectioned 4 μm and processed for immunostaining by the avidin-biotin complex (ABC) method.

At 13 days of gestation, PGP-immunopositive olfactory cells were rather numerous in the OE, but were small in number and restricted to a small part of the VNO. At 14 days of gestation, the immunopositive cells had increased in number in the OE, and were scattered throughout the VNO. Although the MO was not yet identified histologically, we observed a small number of immunopositive cells in the epithelium on the ventrocaudal side of the nasal septum at 13 and 14 days of gestation. At 15 days of gestation, the MO was first observed on the side described above as a thickening of the epithelium. PGP-immunopositive cells were numerous in the OE and VNO, and detectable in the MO. Thereafter, as development progressed, the immunopositive cells increased in number in the OE, in the epithelium lining the medial wall of the VNO, and in the MO, while they decreased in number in the epithelium lining the lateral wall of the VNO and in the epithelium lining the nasal septum between the OE and MO. These epithelia decreased in height and became pseudostratified at 20 and 17 days of gestation, respectively. Finally, these epithelia virtually lost PGP immunoreactivity, becoming the respiratory epithelium of the VNO and nasal septum.

To summarize, PGP immunoreactivity appeared earliest in the OE, next in the VNO, and last in the MO; these findings suggest that the functional development of the MO occurs later than that of the OE and VNO. Although the respiratory epithelium of the VNO and nasal septum are derived from the olfactory placode, it appears that they lose the characteristics of sensory epithelia during development.

References

1. Taniguchi K, Arai T, Ogawa K (1993) Fine structure of the septal olfactory organ of Masera and its associated gland in the golden hamster. J Vet Med Sci 55:107–116
2. Halpern M (1987) The organization and function of the vomeronasal system. Ann Rev Neurosci 10:325–362
3. Taniguchi K, Saito H, Okamura M, Ogawa K (1993) Immunohistochemical demonstration of protein gene product 9.5 (PGP 9.5) in the primary olfactory system of the rat. Neurosci Lett 156:24–26

[1] Department of Veterinary Anatomy, Faculty of Agriculture, Iwate University, 3-18-8 Ueda, Morioka, 020 Japan

Immunocytochemical Study of the Vomeronasal Epithelium of the Infant Rat

Junko Yoshida[1], Toshiya Osada[2], Yuji Mori[1], and Masumi Ichikawa[3]

Key words. Vomeronasal organ—Development

The vomeronasal organ (VNO) is thought to be a detector of chemical signals, i.e., pheromones [1,2]. The primordium of the organ is first observed on embryonic day 11 and then gradually increases in size with growth [3]. The shape of the VNO in the neonatal rat is similar to that in the adult, and its histological maturation is assumed to be completed by the 3rd postnatal week [4]. Past immunohistochemical studies indicate that the VNO has unique characteristics which distinguish it from the olfactory organ [5–7]. Its function, however, remains to be clarified. Recently, monoclonal antibodies (mAbs) VOM2, VOBM1, and VOBM2 have been generated against rat vomeronasal epithelium by Osada et al. [8]. The reactivities of these mAbs were observed only in the sensory epithelium, and VOM2 immunoreactivity (IR) was especially limited on the luminal surface of the sensory epithelium. Here we report the developmental changes in the IR of these three MAbs for the sensory epithelium of the VNO.

Postnatal day (P) 7, 11, 14, 17, 21, 24, 28, 35, 42, 49, and 56, and adult rats (Sprague-Dawley) were examined for VOM2, VOBM1, and VOBM2 IR. The rats were deeply anesthetized with pentobarbital and perfused with 0.9% saline, and then with 4% paraformaldehyde dissolved in 0.1 M phosphate buffer (pH 7.4). The VNOs were removed and kept in the same fixative for 24 h at 4°C. They were cut at a thickness of 20 μm in a cryostat, and the sections were mounted on gelatin-coated slides. The sections were incubated overnight at 4°C with these mAbs, diluted 1:400 with phosphate-buffere saline (PBS) containing 0.5% bovine serum albumin. Following a 30-min wash in PBS, the sections were incubated with anti-mouse Ig biotinylated antibody for 1 h at 37°C, washed in PBS for 30 min, and then incubated with streptavidin labeled with fluorescence dye. The sections were then again washed in PBS and coverslipped with water-based mountant, after which they were observed with a fluorescence microscope.

VOM2 IR was visible first at P 11. At this age, the luminal surface of the vomeronasal sensory epithelium showed a punctate IR. The IR was recognized as a discontinuous thin layer band on the surface of the sensory epithelium. No reactivity could be observed in the cell body of the sensory epithelium. VOM2 IR gradually increased with age, and by P 21, was observed as a continuous thin layer band on the surface of the epithelium. At P 35. VOM2 IR on the epithelium reacted at the same level as in the adult. VOBM1 IR was detectable on the luminal surface and in the cell body of the sensory epithelium at P 7 in some animals, with considerable individual variation being seen. After P 17, VOBM1 IR gradually increased in intensity. VOBM2 IR was already observed at P 7 on the luminal surface and in the sensory epithelium. The VOBM2 IR increased in intensity with age. At P 35, VOBM2 IR had reached the adult level.

The results of this study can be summarized as follows: (1) The three mAbs reacted specifically to the luminal surface of the sensory epithelium of the VNO. In addition, mAb VOBM1 and VOBM2 were specific to cell bodies of the vomeronasal sensory epithelium. (2) The IRs of these three mAbs were age-dependent. The first expression of the VOM2 IR and VOBM1 IR was around P 11, whereas VOBM2 IR was already observed at P 7. After P 28, the IRs of these three mAbs reached the adult level.

These results suggest that molecules recognized by these mAbs (VOM2, VOBM1 and VOBM2) could play a physiological role during the functional maturation of the VNO in rats. Immunoelectronmicroscopic study is necessary to elucidate the developmental changes of the ulrastructure of the mAb reaction sites in the VNO.

---

[1] Laboratory of Veterinary Ethology, The University of Tokyo, 1-1-1 Yayoi, Bunkyo-ku, Tokyo, 113 Japan
[2] Department of Biological Sciences, Tokyo Institute of Technology, 4259 Nagatsuta, Midori-ku, Yokohama, 227 Japan
[3] Department of Anatomy and Embryology, Tokyo Metropolitan Institute for Neuroscience, 2-6 Musashidai, Fuchu, Tokyo, 183 Japan

References

1. Wysocki CJ (1979) Neurobehavioral evidence for the involvment of the vomeronasal system in mammalian reproduction. Neurosci Biobehav Rev 3(4):301–341
2. Halpern M (1987) The organization and function of the vomeronasal system. Am Rev Neurosci 10:325–362
3. Yoshida J, Kimura J, Tsukise A, Okano M (1993)

Developmental study on the vomeronasal organ in the rat fetus. J Reprod Dev 39:47–54

4. Garrosa M, Iñiguez C, Fernandez JM, Gayoso MJ (1992) Developmental stages of the vomeronasal organ in the rat: A light and electron microscopic study. J Hirnforsch 33(2):123–132

5. Mori K (1987) Monoclonal antibodies (2C5 and 4C9) against lactoseries carbohydrates identify subsets of olfactory and vomeronasal receptor cell and their axons in the rabbit. Brain Res 408:215–221

6. Schwarting GA, Deutsch G, Gattey DM, Crandall JE (1992) Glycoconjugates are stage- and position-specific cell surface molecules in the developing olfactory system, 1: The CC1 immunoreactive glycolipid defines a rostrocaudal gradient in the rat vomeronasal system. J Neurobiol 23(2):120–129

7. Ichikawa M, Osada T, Ikai A (1992) *Bandeiraea simplicifolia* lectin and *Vicia villosa* agglutinin bind specifically to the vomeronasal axons in the accessory olfactory bulb of the rat. Neurosci Res 13:73–79

8. Osada T, Kito K, Ookata K, Graziadei PPC, Ikai A, Ichikawa M (1994) Monoclonal antibody (VOM2) specific for the luminal surface of the rat vomeronasal sensory epithelium. Neurosci Lett (in press)

Characterization of the Mucomicrovillar Complex in the Vomeronasal Epithelium with Lectinohistochemistry and Confocal Laser Scanning Microscopy

SHIGERU TAKAMI[1], MARILYN L. GETCHELL[2,3], and THOMAS V. GETCHELL[1,2,3]

Key words. Vomeronasal receptor neurons—Glyco-conjugates

Lectinohistochemistry and confocal laser scanning microscopy (CLSM) were utilized to resolve the mucomicrovillar complex (MMC) [1] of the vomeronasal sensory epithelium (VE) into mucoid and sensory components, and to identify glycoconjugates in these components. Six-week-old Sprague-Dawley rats of both sexes were anesthetized deeply with Nembutal and transcardially perfused with Zamboni's fixative. The vomeronasal organ (VNO), olfactory mucosa (OM), and the septal organ of Masera (SO) were sectioned serially in the transverse plane, using a cryostat. Binding sites for the isolectin *Bandeiraea simplicifolia*-I B_4 (BS-I-B_4), which specifically recognizes terminal α-D-galactose (α-Gal) sugar residues [2], were localized on tissue sections with the standard fluorescence technique for fluorescein isothiocyanate (FITC)-BS-I-B_4, and the avidinbiotin-peroxidase complex (ABC-peroxidase) technique for biotin-conjugated BS-I-B_4. Some sections were double-labeled with the antibody against olfactory marker protein (OMP) using the ABC-fluorochrome (Texas Red) technique and with FITC-BS-I-B_4 using the fluorescence technique. These sections were analyzed with a confocal laser scanning microscope (MultiProbe 2001; Molecular Dynamics, Sunnyvale, Calif.), using a filter set for simultaneous detection, that is, dual scanning, of the optimal emission spectrum of FITC (peak, 520 nm) and Texas Red (peak, 620 nm).

Intense fluorescence for BS-I-B_4 was observed throughout the entire MMC of the VE, whereas only a patchy distribution of BS-I-B_4 was observed in the apical surface of the non-sensory epithelium, which faces the VE through the fluid-filled lumen of the VNO [1]. In contrast, fluorescence for BS-I-B_4 was not observed in the mucociliary complex of either the OM or SO, indicating a characteristic expression of α-Gal sugar residues in the MMC of the VE. When the ABC-peroxidase technique was used, vomeronasal receptor neurons (VRNs) in the VE showed a characteristic punctate pattern of reactivity for BS-I-B_4 that was not observed in olfactory or septal receptor neurons. Intense fluorescence for BS-I-B_4 was observed in secretory granules of acinar cells in the vomeronasal glands (VNG) and posterior glands of nasal septum (PGNS) that secrete into the lumen of the VNO, whereas BS-I-B_4 fluorescence was not observed in the secretory granules of Bowman's glands in the OM or in the glands of the SO.

Omission of BS-I-B_4 from the labeling protocol resulted in no specific fluorescence. Binding of BS-I-B_4 was completely inhibited by 50 mM of the competing sugars Gal, methyl-α-galactopyranoside, and methyl-β-galactopyranoside. At lower concentrations (1 and 5 mM), methyl-α-galactopyranoside inhibited the binding of BS-I-B_4 to a greater extent than did methyl-β-galactopyranoside, indicating that BS-I-B_4 preferentially bound to α anomers of Gal.

Optical sectioning, using dual scanning CLSM, demonstrated that terminal α-Gal-containing glycoconjugates were present in the sensory and mucoid components of the MMC. OMP-immunoreactive dendritic terminals of VRNs contained intracellular glycoconjugates with terminal α-Gal sugar residues. Extracellular glycoconjugates with terminal α-Gal were localized around the dendritic terminals in areas clearly devoid of OMP immunoreactivity. Since the VNG and PGNS contained secretory glycoconjugates with terminal α-Gal sugar residues, the extracellular glycoconjugates identified in the MMC are probably derived from the VNG and PGNS. The terminal α-Gal-containing extracellular glycoconjugates in the VE of rats are similar to vomeromodulin, which is expressed in the rat VNO [3] and olfactomedin, which is expressed in the frog OM [4], in that all three glycoconjugates are localized in the mucosensory complexes of the chemosensory epithelia and secretory granules of the associated glands. The extracellular glycoconjugates with terminal α-Gal sugar residues, which were identified in the MMC of VE, contribute to the microchemical heterogeneity [5] in vomeronasal mucus and may play a role in perireceptor events [6] associated with vomeronasal transduction.

[1] Department of Physiology and Biophysics, [2] Sanders-Brown Center on Aging, [3] Division of Otolaryngology—Head and Neck Surgery, Department of Surgery, University of Kentucky College of Medicine, Lexington, KY 40536-0084, USA

Acknowledgments. We thank Dr. Nikk Katzman for technical assistance. This study was supported by NIH grants NIDCD-00159 (TVG) and NIDCD-01715 (MLG).

References

1. Rama Krishna NS, Getchell ML, Getchell TV (1992) Differential distribution of γ-glutamyl cycle molecules in the vomeronasal organ of rats. NeuroReport 3:551–554

2. Murphy LA, Goldstein IJ (1977) Five α-D-galactosyl-binding isolectins from *Bandeiraea simplicifolia* seeds. J Biol Chem 252:4739–4742

3. Khew-Goodall Y, Grillo M, Getchell ML, Danho W, Getchell TV, Margolis FL (1991) Vomeromodulin, a putative pheromone transporter: Cloning, characterization, and cellular localization of a novel glycoprotein of lateral nasal gland. FASEB J 5:2976–2982

4. Snyder DA, Rivers AM, Yokoe H, Menco BPh, Anholt RRH (1991) Olfactomedin: Purification, characterization, and localization of a novel olfactory glycoprotein. Biochemistry 30:9143–9153

5. Getchell TV, Su Z, Getchell ML (1993) Mucous domains: Microchemical heterogeneity in the muco-ciliary complex of the olfactory epithelium. In: Chardwick D, Marsh J, Goode J (eds) The molecular basis of smell and taste transduction, Ciba Found Symp 179, Wiley, Chichester, pp 27–50

6. Getchell TV, Margolis FL, Getchell ML (1984) Peri-receptor and receptor events in vertebrate olfaction. Prog Neurobiol 23:317–345

2. Taste Transduction

2.1 Biochemical Events

Gustducin and Transducin Are Present in Taste Cells

Susan K. McLaughlin[1], Peter J. McKinnon[1], Nancy Spickofsky[1], and Robert F. Margolskee[1]

Key words. G-proteins—Molecular cloning—RNA—cDNA—PCR

Introduction

The molecular mechanisms underlying taste transduction have been only partially elucidated. Each taste submodality has its own specific mechanism: salty taste results from Na^+ flux through apical Na^+ channels; sour is mediated by H^+ ion blockade of K^+ or Na^+ channels; bitter and sweet are mediated by guanine nucleotide binding protein- (G-protein-) dependent mechanisms. Sweet compounds cause a guanosine triphosphate-(GTP-) dependent generation of cyclic AMP (cAMP) in rat tongue membranes [1,2]. External application or microinjection of cAMP inactivates K^+ channels in vertebrate taste cells and leads to their depolarization [3–5]. It has been known for quite some time that taste tissue contains very high levels of adenylyl cyclase [6–8]. These observations suggest that sweet transduction involves sweet receptor activation of G_s or a G_s-like G protein, which leads to adenylyl cyclase-mediated elevation of cAMP. Bitter, at least in part, is also transduced via a G-protein-coupled receptor: the bitter compound denatonium leads to Ca^{2+} release from internal stores [9], presumably mediated by G-protein activation of phospholipase C to generate inositol trisphosphate (IP_3).

We set out to identify and molecularly clone proteins involved in vertebrate taste transduction: our initial focus was on G-protein α-subunits. These proteins are highly conserved, making it possible to design specific primers that can be used in the polymerase chain reaction (PCR) to amplify and then clone their DNAs. The identification of G protein α subunits with elevated or specific expression in gustatory tissue could provide insight into the mechanisms of the taste pathways. We have cloned a novel G-protein α-subunit, α-gustducin, and demonstrated that it is specifically expressed in taste cells [10]. In addition, we have cloned rod α-transducin ($α_{t\text{-rod}}$) and cone α-transducin ($α_{t\text{-cone}}$) from taste cells, and demonstrated that $α_{t\text{-rod}}$ mRNA and protein are specifically expressed in taste cells. The α-transducins previously had only been found in photoreceptor cells of the retina. The close relatedness of α-gustducin to the α-transducins and the presence of α-gustducin and $α_{t\text{-rod}}$ in taste cells suggest that they may play similar roles in taste transduction.

Materials and Methods

The circumvallate and foliate papillae from 100 Sprague-Dawley rats were harvested and immediately frozen in 100% ETOH at $-70°C$ [10]. An equivalent amount of nontaste lingual epithelium (devoid of taste buds) was likewise harvested. PolyA$^+$ mRNA was harvested from taste and nontaste tissue using the Quick Prep kit from Pharmacia (Uppsala, Sweden). The BRL pSPORT vector and the BRL superscript kit were used to make two cDNA libraries from 1 μg of taste and 1 μg of nontaste mRNA. The taste library contained 2.6×10^6 independent clones, and the nontaste library contained 4.8×10^6 independent clones. RNA for RNase protection assays was generated in vitro from the taste or nontaste cDNA library. ^{32}P-labeled antisense probes for RNase protection were generated in vitro by T3 or T7 polymerase transcription of linearized subclone DNAs. RNase protection assays used a kit from Ambion (Austin, Texas).

PCR primers corresponding to conserved amino acids of G proteins were made on an Applied Biosystems DNA synthesizer (Applied Biosystems, Foster City, Calif.). The following 5′ primers were used: DVGGQR; KWIHCF; HLFNSIC; TIVKQM; the following 3′ primers were used: FLNKKD; FLNKQD; VFDAVTD. The PCR conditions, cloning, transformation, and DNA sequencing were as described [10].

Results

To identify G proteins present in taste tissue, degenerate oligonucleotide primers from conserved regions of the G protein α-subunits were used in the PCR with taste tissue cDNA as the template. In this way, eight different types of α-subunit clones were isolated from the taste cDNA library (Table 1). Five

[1] Roche Research Center, Roche Institute of Molecular Biology, Nutley, NJ 07110, USA

TABLE 1. Isolates of α-subunit clones from PCR-amplified rat taste tissue cDNA[a].

	5'KWIHCF 3'FLNKKD	5'DVGGQR 3'FLNKKD	5'HLFNSIC 3'VFDAVTD	5'TIVKQM 3'FLNKQD
α_{i-2}	4	—	—	—
α_{i-3}	5	—	—	—
α_{14}	—	1	—	—
α_{12}	—	1	—	—
α_s	—	—	—	1
α_{gust}	5	—	4	1
$\alpha_{t\text{-rod}}$	3	—	2	1
$\alpha_{t\text{-cone}}$	—	—	2	—

[a] Degenerate PCR primers corresponding to conserved amino acids of G proteins were made and used pairwise in the polymerase chain reaction (PCR) to amplify DNA corresponding to G proteins. G-Protein α-subunit isolates cloned from PCR-amplified rat taste tissue cDNA are listed in left-hand column. The heading above each column lists upstream (5') and downstream (3') primers. Numbers in each column represent independent clonal isolates from the particular PCR amplification.

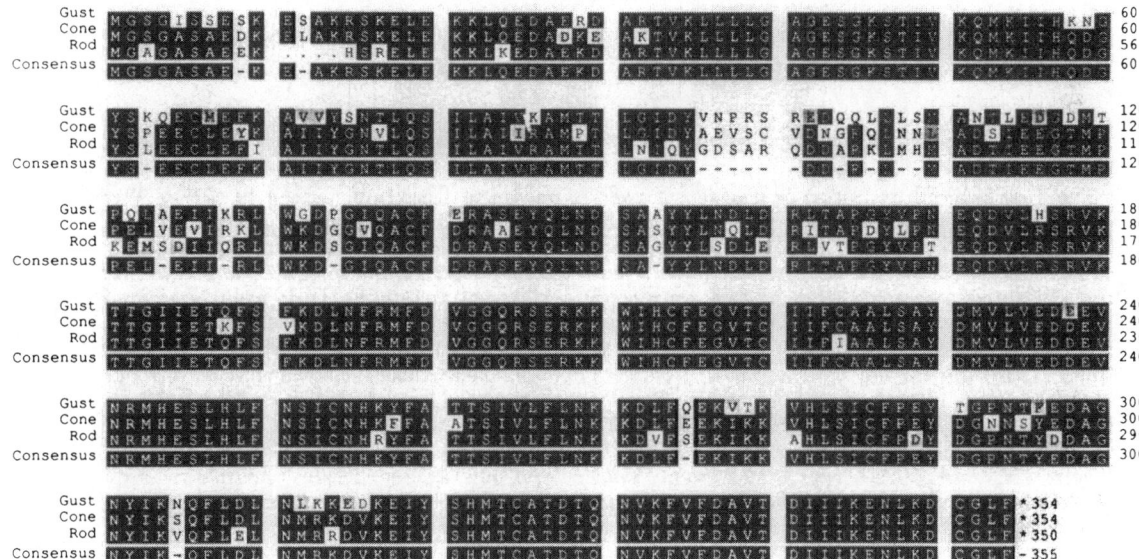

FIG. 1. Alignment of amino acid sequences of α-subunits of rat gustducin, rat rod transducin (rod), and bovine cone transducin (cone). Consensus sequence matches (i.e., at least two of three proteins match) are denoted by *white letters within black boxes* (reverse video). Conservative changes are denoted by *black letters within shaded boxes*. Nonconservative changes are depicted by *black letters in unboxed regions*. The consensus line shows positions where two or three of three proteins match; *dashes* in consensus sequence correspond to nonconserved positions. *Dots* in rod transducin sequence correspond to gaps to align it with the other sequences

of the α-subunit clones (α_{i-2}, α_{i-3}, α_{12}, α_{14}, α_s) have been previously identified and are expressed in several tissues other than lingual epithelium; two clones ($\alpha_{t\text{-rod}}$ and $\alpha_{t\text{-cone}}$) have been previously identified only in retina [11–14]. In addition to these previously known α-subunits, we isolated a novel clone, α-gustducin, which is taste-cell-specific and closely related to the transducins.

The entire coding sequences of α-gustducin and $\alpha_{t\text{-rod}}$ were obtained by PCR and screening of a rat taste tissue cDNA library. The α-gustducin cDNA encodes a protein of 354 amino acids; the $\alpha_{t\text{-rod}}$ cDNA encodes a protein of 351 amino acids. An alignment of rat α-gustducin, rat $\alpha_{t\text{-rod}}$, and bovine $\alpha_{t\text{-cone}}$ is presented in Fig. 1. These three proteins are highly homologous (79%–80% amino acid identity; 89%–90% similarity), suggesting that α-gustducin and the α-transducins are evolutionarily related members of a subgroup of G-protein α-subunits.

RNase protection demonstrates the presence of α-gustducin mRNA only in taste tissue-enriched preparations (Fig. 2A). No α-gustducin mRNA was detected in the other tissues examined: nontaste

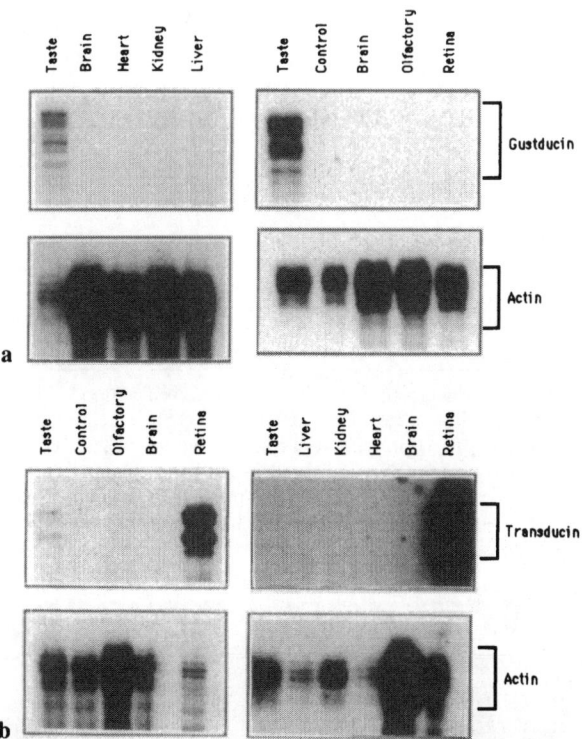

FIG. 2. **a** Tissue-specific expression of α-gustducin transcripts as assayed by RNase protection. *Right panels*, cDNA libraries from various tissues were prepared, linearized with restriction endonuclease, and in vitro runoff transcripts were generated. RNase protection was done simultaneously with α-gustducin (*upper panels*) and actin probes (*lower panels*) to normalize for expression. α-Gustducin transcripts are only present in RNA from taste cell-enriched library (Taste), and are absent from libraries derived from nontaste lingual epithelium (Control), retina, brain, and olfactory epithelium. *Left panels*, RNase protection with total RNA (15 μg) from various tissues demonstrates that α-gustducin is absent from brain, heart, kidney, and liver. **b** Tissue-specific expression of rod α-

transducin transcripts assayed by RNase protection of RNA generated in vitro (*left panels*). RNase protection was done simultaneously with rod α-transducin (*upper panels*) and actin probes (*lower panels*) to normalize for expression. α-Transducin transcripts are only present in RNA from the taste cell-enriched library (Taste) and the retinal library (Retina); α-transducin transcripts are absent from nontaste lingual epithelium (Control), olfactory epithelium, and brain. Note that expression of α-transducin is several hundredfold higher in retina than in taste tissue. *Right panels*, RNase protection with total RNA (15 μg) from various tissues demonstrates that rod α-transducin is absent from liver, kidney, heart, and brain but present in retina and taste tissue RNA

lingual tissue, olfactory epithelium, retina, brain, liver, heart, or kidney. RNase protection demonstrates the presence of $\alpha_{t\text{-rod}}$ mRNA only in retina and taste tissue (Fig. 2B). $\alpha_{t\text{-rod}}$ mRNA was not found in the other tissues examined: nontaste lingual tissue, olfactory epithelium, brain, liver, heart, or kidney. This is consistent with previous northern data, which demonstrated that $\alpha_{t\text{-rod}}$ is not expressed in heart, liver, and brain [12]. However, previous workers had not looked for the expression of transducin in taste tissue. Further, there is significantly more $\alpha_{t\text{-rod}}$ mRNA in retina than in taste bud (about several hundredfold more).

To determine directly if α-gustducin and $\alpha_{t\text{-rod}}$ proteins are expressed in the taste buds, immunohistochemistry with specific antisera was done on tongue sections containing taste papillae (Fig. 3A–G). These studies demonstrated that $\alpha_{t\text{-rod}}$ protein (Fig. 3A) and α-gustducin protein (Fig. 3D)

are present in the taste buds of the circumvallate papilla. α-Gustducin and rod α-transducin proteins are not present elsewhere in the nonsensory portions of the taste tissue. Preincubation of the rod α-transducin antiserum with the cognate α-transducin peptide blocks immunoreactivity in taste buds (Fig. 3B); preincubation of the α-transducin antiserum with the corresponding α-gustducin peptide has no effect (Fig. 3C) (i.e., the immunoreactivity is α-transducin specific). Preincubation of the α-gustducin antiserum with the cognate α-gustducin peptide blocks the immunoreactivity in taste buds (Fig. 3E); preincubation of the α-gustducin antiserum with the corresponding transducin peptide has no effect (Fig. 3F). These antisera specifically distinguish between α-gustducin and $\alpha_{t\text{-rod}}$, and demonstrate that both proteins are present in taste buds.

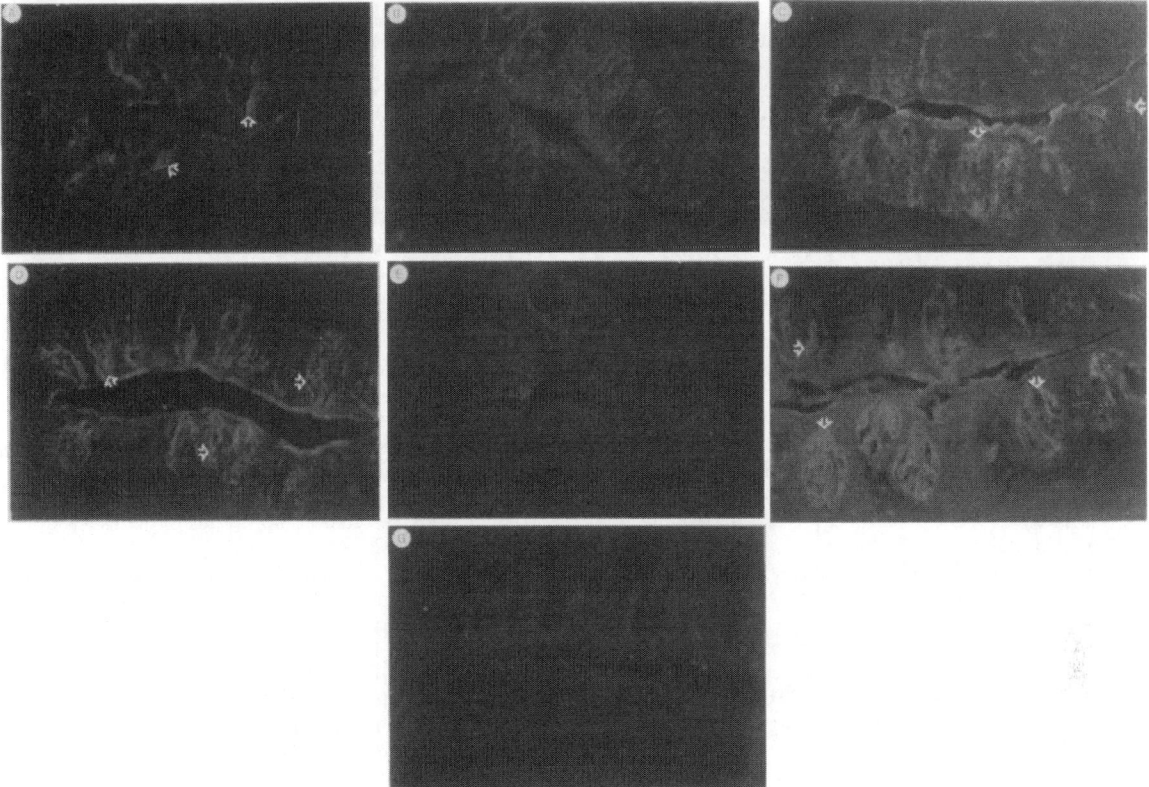

FIG. 3. Immunofluorescence photomicrographs (×400) of frozen sections of rat taste papillae treated with α-gustducin or $\alpha_{\text{t-rod}}$-specific antibodies and rhodamine conjugated secondary antibody. **A** Circumvallate papilla reacted with antitransducin antiserum. **B** Circumvallate papilla, antitransducin antiserum preblocked by adsorption to rod α-transducin peptide. **C** Circumvallate papilla, antitransducin antibody adsorbed to α-gustducin peptide (note that this peptide does not block transducin immunoreactivity). **D** Circumvallate papilla, antigustducin antibody. **E** Circumvallate papillae, antigustducin antibody preblocked by adsorption to α-gustducin peptide. **F** Circumvallate papilla, antigustducin antibody adsorbed to rod α-transducin peptide (note that this peptide does not block gustducin immunoreactivity). **G** Circumvallate papilla, no primary antibody. *Open arrows* indicate taste receptor cells expressing rod α-transducin (**A,C**) or α-gustducin (**D,F**)

Discussion

We have cloned α-gustducin and rod α-transducin from rat taste cells. α-Gustducin is only expressed within the taste buds of the tongue: its mRNA is not expressed in retina, olfactory epithelium, brain, liver, kidney, muscle, or heart. Rod α-transducin had previously been found only in retina: we have shown in this study that $\alpha_{\text{t-rod}}$ mRNA and protein are expressed in rat taste cells.

In the retina, the transducins relay activation of a receptor (rhodopsin or the color opsins) into activation of a cytoplasmic effector enzyme (cGMP phosphodiesterase (cGMP PDE)). Transducin disinhibits cGMP PDE by binding to the inhibitory γ-subunits, this leads to breakdown of cGMP and causes closure of cyclic nucleotide gated channels. We propose that α-gustducin and α-transducin transduce taste receptor activation into activation of taste cell PDEs. Consistent with this proposal, extremely high levels of cAMP PDE are present in taste tissue [7,8,15].

We propose that α-gustducin or α-transducin binds to an inhibitory subunit of a taste cell cAMP PDE to activate this enzyme and thereby decrease cAMP levels.

Several biochemical and electrophysiological studies argue for sweet pathway involvement of G_s (or a G_s-like protein) that activates adenylyl cyclase to raise taste cell cAMP levels. α-Transducin does not activate adenylyl cyclase, and α-gustducin by homology is unlikely to do so. α-Gustducin and α-transducin may inhibit the sweet response (PDE activation would decrease sweet-induced cAMP), or they may play a role in bitter transduction. In support of these proposals, previous workers have correlated bitter compounds with PDE activation [16], and several PDE inhibitors have been shown to enhance sweet perception [17,18]. The biochemical and molecular biological characterization of taste transduction has previously been hindered because taste receptor cells comprise only a small inaccessible fraction of the lingual epithelium. Using PCR we have cloned α-gustducin and α-transducins from rat taste cells. In like fashion, it should be possible to clone other components of the taste transduction pathways.

References

1. Striem BJ, Pace U, Zehavi U, Naim M, Lancet D (1989) Sweet tastants stimulate adenylate cyclase coupled to GTP binding protein in rat tongue membranes. Biochem J 260:121–126
2. Striem BJ, Yamamoto T, Naim M, Lancet D, Jakinovich W, Zehavi Y (1990) The sweet taste inhibitor methyl 4,6-dichloro-4,6-dideoxy-α-D-galactopyranoside inhibits sucrose stimulation of the chorda tympani nerve and of the adenylate cyclase in anterior lingual membranes of rats. Chem Senses 15:529–536
3. Avenet P, Lindemann B (1987) Patch-clamp study of isolated taste receptor cells of frog. J Membr Biol 97:223–239
4. Tonosaki K, Funakoshi M (1988) Cyclic nucleotides may mediate taste transduction. Nature (Lond) 331: 354–356
5. Avenet P, Hofmann F, Lindemann B (1988) Transduction in taste receptor cells requires cAMP-dependent protein kinase. Nature (Lond) 331:351–354
6. Kurihara K, Koyama N (1972) High activity of adenyl cyclase in olfactory and gustatory organs. Biophys Res Commun 48:30–34
7. Nomura H (1978) Histochemical localization of adenylate cyclase and phosphodiesterase activities in the foliate papillae of the rabbit I. Chem Senses Flavor 3:319–374
8. Asanuma N, Nomura H (1982) Histochemical localization of adenylate cyclase and phosphodiesterase in the foliate papillae of the rabbit II. Chem Senses Flavor 7:1–9
9. Akabas MH, Dodd J, Al-Awqati Q (1988) A bitter substance induces a rise in intracellular calcium in a subpopulation of rat taste cells. Science 242:1047–1050
10. McLaughlin SK, McKinnon PJ, Margolskee RF (1992) Gustducin is a taste-cell-specific G protein closely related to the transducins. Nature (Lond) 357: 563–569
11. Lochrie MA, Hurley JB, Simon MI (1985) Sequence of the α subunit of photoreceptor G protein: homologies between transducin, ras, and elongation factors. Science 228:96–99
12. Medynski DC, Sullivan K, Smith D, VanDop C, Chang F-H, Jung BK-K, Seeburg PH, Bourne HR (1985) Amino acid sequence of the α subunit of transducin deduced from the cDNA sequence. Proc Natl Acad Sci USA 82:4311–4315
13. Tanabe T, Nukada T, Nishikawa Y, Katsunori S, Suzuki H, Takahashi H, Noda M, Itaga T, Ichiyama A, Kangawa K, Minamino N, Matsuo H, Numa S (1985) Primary structure of the α-subunits of transducin and its relationship to ras proteins. Nature (Lond) 315:242–245
14. Yatsunami K, Khorana HG (1985) GTPase of bovine rod outer segments: the amino acid sequence of the α subunit as derived from the cDNA sequence. Proc Natl Acad Sci USA 82:4316–4320
15. Law JS, Henkin RI (1982) Taste bud adenosine-3'-5'-monophosphate phosphodiesterase: activity, subcellular distribution and kinetic parameters. Res Commun Chem Pathol Pharmacol 38:439–452
16. Schiffman SS, Gill JM, Diaz C (1985) Methylxanthines enhance taste: evidence for modulation of taste by adenosine receptor. Pharmacol Biochem Behav 22: 195–203
17. Price S (1973) Phosphodiesterase in tongue epithelium: activation by bitter taste stimuli. Nature (Lond) 241:54–55
18. Schiffman SS, Diaz C, Beeker TG (1986) Caffeine intensifies taste of certain sweeteners: role of adenosine receptor. Pharmacol Biochem Behav 24:429–432

Seven Transmembrane Receptors in Tongue Epithelia

KEIKO ABE and SOICHI ARAI[1]

Key words. Tongue epithelia—G-protein coupled receptor—G-protein—cDNA—Taste receptor

Taste transduction triggered by a variety of taste stimuli such as sweetness and bitterness is supposedly mediated by taste receptor cells in the taste buds of the tongue [1–3]. Several lines of physiologic evidence have revealed that taste-signaling may be related to other types of sensory signaling, such as phototransduction [4–6] and olfactory transduction [7,8] in terms of the neurotransmitters [9,10] and second messengers including cyclic nucleotides (cAMP and cGMP) [11,12] and calcium ion [13]. However, the molecular mechanism of taste transduction remains obscure, except for the recent identification of a novel taste-cell-specific G-protein (α-gustducin) [14].

We have conducted a cloning study to identify some of the molecules possibly involved in taste transduction in rat. Here we describe the characterization of a gene family comprised of more than 60 members encoding G-protein-coupled receptors that are expressed specifically on the surface of lingual epithelia, particularly in taste bud cells of fungiform papillae. These receptors show, to a greater or lesser extent, structural similarities to olfactory [8] and optic [15] receptors, suggesting the existence of a similar cell-signaling system for various sensory functions including gustation.

To identify G-protein-coupled receptors in lingual epithelia, several oligonucleotide primers corresponding to the amino acid sequences of olfactory receptor molecules [8] around transmembrane domains II, VI, and VII were prepared and subjected to a reverse transcription-polymerase chain reaction (RT-PCR) using mRNA obtained from apical epithelial cells of rat tongue tip. Screening was carried out with an internal oligonucleotide probe corresponding to the amino acid sequence of transmembrane domain III generally conserved in the G-protein-coupled receptor superfamily [16]. Nucleotide sequencing of the screened clones yielded more than 60 RT-PCR clones encoding closely related but partly different proteins, all of which contain several transmembrane domain motifs commonly observed in the various G-protein-coupled receptors [16]. By comparison of the deduced amino acid sequences, these 60 clones were classified into six distinct groups, each comprising 3 to 15 clones. The sequences of the RT-PCR clones show a typical feature common among other receptors in sensory systems [17].

The overall structure of the lingual epithelium receptors was determined using a cDNA clone (GUST27), which was obtained from a cDNA library constructed from mRNA of rat tongue tissue by hybridization with the mixed probes consisting of the six RT-PCR clones [17]. The deduced amino acid sequence of GUST27 (Fig. 1), which resembles PTE33 in particular [17], shows an overall structural similarity to various sensory receptors with seven-transmembrane domains [18]. Sequence homology was highest (56%) between the GUST27 protein and OLFF3, an olfactory receptor [8], and significant (33%) between the GUST27 protein and HGPO7J, a germ cell receptor [19]. A certain degree of homology was observed by comparison with rhodopsin [15] and other receptors [20–22] in terms of a few transmembrane domains. Thus a large number of G-protein-coupled receptors with similar structural characteristics are expressed in the epithelial cells of the tongue. To obtain further information about the gene family, genomic Southern analyses were conducted using the six RT-PCR clones [17] and GUST27 (Fig. 1) as probes. Under stringent conditions for hybridization, the DNA probes each gave a simple pattern, indicating that the genes are essentially single copy genes. However, under reduced stringency conditions, a probe (PTE45) hybridized several bands with various intensities, showing that the lingual epithelium receptor genes constitute a large gene family, probably made up of more than 60 members, as shown by the RT-PCR cloning.

From the above molecular studies, a large gene family of G-protein-coupled receptors is suggested to be involved in transmembrane signaling possibly for taste stimulus transduction in lingual epithelia. We next examined whether the expression of these receptors is tongue-specific. The results of the Northern blot analysis clearly demonstrate that the GUST27 mRNA is specifically expressed in epithelial cells of the tongue in the form of a single mRNA species about 2 kb in length; no expression was detected in other organs. Also, the signal was scarcely detected in whole tongue, suggesting that

---

[1] Department of Agricultural Chemistry, The University of Tokyo, Bunkyo-ku, Tokyo, 113 Japan

```
-132                                                    AGGGTCATTTTGTGATATGTCCGGTGCTATTATTTGGAAACT
 -90 AAACTTAAATTTTCTTTAACAGATACAAAAGTCATATGGAAACAGAGAACCACACAATGAGAACAGAATTTCACATCCTGGGTCTCTCAG

   1 ATGATCCTGAACTGCAACCCATTCTCTGGACTGTTCCTGTCCATGTATCTGGTCACAGTGCTTGGGAACTTGCTCATCATCCTGGCTGTC
   1 MetIleLeuAsnCysAsnProPheSerGlyLeuPheLeuSerMetTyrLeuValThrValLeuGlyAsnLeuLeuIleIleLeuAlaVal

  91 AGCTCTAATTCACATCTCCACAACCTCATGTATTTCTTCCTCTCCAATCTGTCCTTTGTTGACATCTGTTTCATCTCAACCACAATACCA
  31 SerSerAsnSerHisLeuHisAsnLeuMetTyrPhePheLeuSerAsnLeuSerPheValAspIleCysPheIleSerThrThrIlePro

 181 AAAATGCTAGTGAACATACATTCACAGACAAAAGACATCTCCTACATAGAATGCCTTTCACAGGTATATTTTTTAACTACTTTTGGTGGA
  61 LysMetLeuValAsnIleHisSerGlnThrLysAsdIleSerTyrIleGluCysLeuSerGlnValTyrPheLeuThrThrPheGlyGly

 271 ATGGATAATTTTTTACTCACTTTAATGGCCTGTGATCGCTATGTAGCCATCTGCCACCCCCTCAACTACACTGTAATCATGAACCTTCAG
  91 MetAspAsnPheLeuLeuThrLeuMetAlaCysAspArgTyrValAlaIleCysHisProLeuAsnTyrThrValIleMetAsnLeuGln

 361 CTGTGTGCCCTTCTGATTCTGATGTTTTGGTTAATCATGTTCTGTGTCTCCCTGATTCATGTTCTATTGATGAATGAATTGAACTTCTCC
 121 LeuCysAlaLeuLeuIleLeuMetPheTrpLeuIleMetPheCysValSerLeuIleHisValLeuLeuMetAsnGluLeuAsnPheSer

 451 AGAGGCACAGAAATTCCACATTTCTTCTGTGAACTGGCTCAAGTTCTTAAGGTAGCCAATTCTGACACTCATATCAATAATGTCTTCATG
 151 ArgGlyThrGluIlePheHisPhePheCysGluLeuAlaGlnValLeuLysValAlaAsnSerAspThrHisIleAsnAsnValPheMet

 541 TATGTGGTGACTTCCCTACTAGGACTGATCCCTATGACAGGAATACTTATGTCTTACTCACAGATTGCTTCATCCTTATTAAAGATGTCT
 181 TryValValThrSerLeuLeuGlyLeuIlePheMetThrGlyIleLeuMetSerTyrSerGlnIleAlaSerSerLeuLeuLysMetSer

 631 TCCTCTGTGAGTAAGTACAAGGCCTTTTCCACCTGTGGATCTCACCTCTGTGTGGTCTCTTTATTCTATGGGTCAGCAACTATAGTTTAC
 211 SerSerValSerLysTyrLysAlaPheSerThrCysGlySerHisLeuCysValValSerLeuPheTyrGlySerAlaThrIleValTyr

 721 TTCTGCTCTTCTGTGCTCCATTCTACACACAAGAAAATGATTGCTTCATTGATGTACACTGTAATCAGCCCCATGCTGAACCCCTTTATC
 241 PheCysSerSerValLeuHisSerThrHisLysLysMetIleAlaSerLeuMetTyrThrValIleSerProMetLeuAsnProPheIle

 811 TATAGCCTGAGAAACAAGGATGTAAAGGGTGCCCTTGGAAAACTTTTCATCCGAGTTGCCTCTTGCCCATTGTGGAGCAAAGACTTTAGA
 271 TyrSerLeuArgAsnLysAspValLysGlyAlaLeuGlyLysLeuPheIleArgValAlaSerCysProLeuTrpSerLysAspPheArg

 901 CCTAAATTCATACTAAAACCTGAAAGGCAAAGTTTATAAACAAACCTCTCCTGGGTCATTTGTATCATAAAATATATGCCTAATTTACAC
 301 ProLysPheIleLeuLysProGluArgGlnSerLeu*

 991 TATTCTAAAAGTATATATAGCTTGTCATTTGTGTACTTTCTACAAAAAATATTTTAATTCCCTATGCATATTGTTTAAAATTTGCAATTC
1081 TTGTTATGTC
```

FIG. 1. Nucleotide and deduced amino acid sequences of the cDNA clone GUST27. Nucleotides are numbered in the 5′ to 3′ direction starting with the first methionine of the longest open reading frame. The nucleotide sequence has been deposited in the EMBL Data Bank (accession no. D-12820)

expression may be restricted to the epithelium. Similar Northern analysis profiles were obtained when the six RT-PCR clones were used as probes (data not shown).

To obtain more precise information about the expression of the receptor genes, in situ hybridization experiments were carried out using digoxygenin-labeled antisense RNA probes to detect positive expressions in brownish color [18]. The GUST27 mRNA is specifically expressed in epithelial cells on the surface of the rat tongue; no positive expression was detected in the reverse side or in the muscle layer, horny layer, or connective tissue of the tongue. In some experiments in which the hybridization conditions were made relatively stringent, the signals predominated in the taste buds of the fungiform papillae, although in other experiments signals were observed in almost all epithelial cells including taste bud cells. The GUST27 mRNA is also expressed in the circumvallate papillae. The use of antisense RNA probes for the six RT-PCR clones gave essentially the same results, indicating that most, if not all, of the identified receptor genes are specifically expressed in the epithelium of the tongue, particularly in the taste bud cells. Such expression was not observed in other parts of tongue or in other organs.

We have shown here that multiple G-protein-coupled receptor genes, closely related to one another in terms of structure, are specifically expressed in epithelial cells. These receptors may be involved in taste signaling that starts with transmembrane events connected to intracellular signaling, leading to neural signaling of the neighboring axons of the sensory neurons. With such intracellular signaling, taste transduction is characterized by at least three events according to the kind of taste stimuli: changes in ion channel gating by salty and sour tastes [23,24], enhancement of cAMP concentration by sweet and bitter tastes [12,25], and Ca^{2+} influx by bitter taste [26]. Among them, the latter two at least may be mediated by transmembrane G-protein-coupled receptors located on the taste cell surface, to which the newly identified receptors described here may be related. The possibility thus exists that these lingual epithelium receptors may function by coupling with α-gustducin, reported to be a novel taste-cell-specific Gα protein possibly involved in taste stimulus transduction [14]. There is also the possibility that the lingual receptors couple with the sweet-taste-specific G_s protein, whose occurrence has been predicted by biochemical studies using rat tongue tip rich in fungiform papillae [12]. Taken together, the multiple species of seven-transmembrane proteins identified here are potent candidates for the taste signal receptors. Further cellular and molecular analyses will provide more direct evidence toward understanding the entire molecular mechanism of taste transduction.

Acknowledgments. We express our thanks to Yuko Kusakabe and Kentaro Tanemura, Faculty of Agriculture, The University of Tokyo, and Yasufumi Emori, Faculty of Science, The University of Tokyo, who have contributed to this work.

References

1. Ozeki M, Sato M (1972) Responses of gustatory cells in the tongue of rat to stimuli representing four taste qualities. Comp Biochem Physiol [A] 41:391–407
2. Ropers S (1983) Regenerative impulses in taste cells. Science 220:1311–1312
3. Arvidson K (1980) Human taste: response and taste bud number in fungiform papillae. Science 209:807–808
4. Berridge MJ, Irvine RF (1984) Inositole triphosphate, a novel second messenger in cellular signal transduction. Nature 312:315–321
5. Fesenko EE, Kolesnikov SS, Lyubarsky AL (1985) Induction by cyclic GMP of cationic conductance in plasma membrane of retinal rod outer segment. Nature 313:310–313
6. Nathans J, Thomas D, Hogness DS (1986) Molecular genetics of human color vision: the gene encoding blue, green, and red pigment. Science 232:193–202
7. Lowe G, Nakamura T, Gold GH (1989) Adenylate cyclase mediates olfactory transduction for a wide variety of odorants. Proc Natl Acad Sci USA 86:5641–5645
8. Buck L, Axel R (1991) A novel multigene family may encode odorant receptors: a molecular basis for ordor recognition. Cell 65:175–187
9. Deham R, Graziader PPC (1971) Functional anatomy of frog's taste organs. Experientia 27:823–826
10. Nagahama S, Kurihara K (1985) Norepinephrine as a possible transmission in frog taste organs and Ca dependence of its release. J Gen Physiol 85:431–442
11. Tonosaki K, Funakoshi M (1988) Cyclic nucleotides may mediate taste transduction. Nature 331:354–355
12. Striem BJ, Pace U, Zehavi U, Naim M, Lancet D (1989) Sweet tastant stimulate adenylate cyclase coupled to GTP-binding protein in rat tongue membranes. Biochem J 260:121–126
13. Kashiwayanagi M, Miyake M, Kurihara K (1983) Voltage-dependent Ca^{2+} channel and Na^+ channel in frog taste cells. Am J Physiol 24:C82–C88
14. McLaughlin SK, Mckinnon PJ, Margolskee RF (1992) Gustducin is a taste-cell-specific G protein closely related to the transducins. Nature 357:563–569
15. Nathans J, Hogness DS (1983) Isolation, sequence analysis, and intron-exon arrangement of the gene encoding bovine rhodopsin. Cell 34:807–814
16. Libert F, Parmentie M, Lefort A, Dinsart C, Sande J, Maenhaut C, Simons M, Dumont JE, Vassart G (1989) Selective amplification and cloning of four new member of the G protein-coupled receptor family. Science 244:569–572
17. Abe K, Kusakabe Y, Tanemura K, Emori Y, Arai S (1993) Multiple genes for G protein-coupled receptors and their expression in lingual epithelia. FEBS Lett 316:253–256
18. Abe K, Kusakabe Y, Tanemura K, Emori Y, Arai S (1993) Primary structure and cell-type specific expression of a gustatory G-protein-coupled receptor related to olfactory receptors. J Biol Chem 268:12033–12039
19. Parmentier M, Livert F, Schurmans S, Schiffmann S, Lefot A, Eggerickx D, Ledent C, Mollereau C, Gerard C, Perret J, Grootegoed A, Vassart G (1992) Expression of member of the putative olfactory receptor gene family in mammalian germ cells. Nature 355:453–455
20. Hirata M, Hayashi Y, Ushikubi F, Yokota Y, Kageyama R, Nakanishi S, Narumiya S (1991) Cloning and expression of cDNA for a human thromboxane A_2 receptor. Nature 349:617–620
21. Shapiro RA, Schere NM, Habecker BA, Subers EM, Nathanson NM (1988) Isolation, sequence, and functional expression of the mouse M1 muscarinic acetylcholine receptor gene. J Biol Chem 263:18397–18403
22. Masu Y, Nakayama K, Tamaki H, Harada Y, Kuno M, Nakanishi S (1987) cDNA cloning of bovine substance-K receptor through oocyte expression system. Nature 329:836–838
23. Kinnamon SC, Roper SD (1988) Membrane properties of isolated mudpuppy taste cells. J Gen Physiol 91:351–371
24. Kinnamon SC, Dionne VE (1988) Apical localization of K^+ channels in taste cells provide the basis for sour taste transduction. Proc Natl Acad Sci USA 85:7023–7027
25. Strien BJ, Yamamoto T, Naim M, Lancet D, Jakinovich Jr. W, Zwhavi U (1990) Chem Senses 15:529–536
26. Akabas MH, Dodd J, Al-Awqoti Q (1988) A bitter substance induces a rise in intracellular calcium in a subpopulation of rat taste cells. Science 242:1047–1050

Cloning of G-Protein-Coupled Receptors from Bovine Taste Tissue

Ichiro Matsuoka, Tetsuya Mori, Junko Aoki, and Taiji Sato[1]

Key words. G-protein-coupled receptor—Taste receptor—Peptide receptor—PCR

Introduction

In vertebrate taste systems, varieties of taste stimuli are received by taste receptor cells in taste buds located on the tongue epithelium. Several observations have suggested the involvement of a G-protein-mediated signaling system in taste transduction, especially for sweet and bitter stimuli [1,2]. Therefore, it is highly plausible that some of the receptors for taste stimuli belong to the superfamily of G-protein-coupled receptors (GCR). It is also plausible that GCRs are involved in modulating taste sensitivity or regeneration of taste cells. GCRs are characterized by the presence of seven transmembrane (TM) domains that share conserved amino acid residues [3,4]. Recent development of polymerase chain reaction (PCR) using degenerate primers deduced from these domains enabled the identification of new members of the GCR superfamily [5,6]. Making use of the PCR technology, we have identified novel members of two different subfamilies of the GCR superfamily expressed in bovine taste tissues that show significant similarities with the putative odorant receptor subfamily and neuropeptide receptor subfamily [7].

Materials and Methods

Total RNAs prepared from the bovine tongue taste papillae (circumvallate and fungiform papillae) and other bovine tissues were treated with DNase I, then converted to single-stranded cDNA with random hexamer and MMLV reverse transcriptase. The cDNAs were then subjected to a series of PCR experiments using degenerate oligonucleotide primers corresponding to the TM domains of the GCRs as shown: C1 (odorant receptor TM3), 5'-TT (T/C) (T/C) TI (T/C) TIGTIG-CIATIGCIT (A/T) (T/C) GA (T/C) (A/C) GITA-3'; D1 (odorant receptor TM6), 5'-ACIACI (G/C) (A/T) IA (A/G) (A/G) TGI (G/C)-(A/T) ICC-(A/G) CAIGTIGA (A/G) AA-3'; A1 (rhodopsin TM2), 5'-AA (T/C) T (G/A) (G/C) ATI (C/A) TI-(G/C) TIAA (T/C) (C/T) TIGCIGTIGCIGA-3'; B4 (thyrotropin receptor TM7), 5'-GC (C/T)-TTIGT (A/G) AAIATIGC (A/G) TAIAG (G/A)-AAIGG(G/A)TT-3'. PCR was performed under low stringent conditions [7]. Amplified products were analyzed by agarose gel electrophoresis, and bands of interest were cloned into a pCR1000 plasmid vector.

For Northern blot analysis, poly(A)$^+$ RNAs from various bovine tissues were electrophoresed on agarose gels and transferred to nylon membranes. The membranes were hybridized with ^{32}P-labeled cRNA probe transcribed from the plasmid containing PCR-amplified cDNA fragments [7].

Results and Discussion

To identify GCRs expressed in mammalian taste tissue, reverse transcription-PCR was performed with total RNA prepared from bovine taste papillae and various degenerate primer pairs corresponding to the transmembrane domains (TM2, 3, 6, and 7) of known subfamilies of GCR including the putative rat odorant receptors. Consequently, two types of cDNA fragments (type A, 408 bp; type B, 690 bp) were amplified, then cloned into plasmid vector and analyzed further.

Type A Clones, Which Are Homologous to Putative Rat Odorant Receptors

Sequence analysis of more than 50 clones of type A cDNA amplified with C1/D1 primers from bovine taste tissue identified more than 10 independent clones with high homology among each other (40%–90%). These clones encoded fragments of GCR between TM3 and TM6 that are very similar to the putative rat odorant receptors. Deduced amino acid sequences of 6 representative clones (TAS3, 4, 7, 8, 22, and 38) are shown in Fig. 1.

Northern blot analysis was performed to examine the tissue distribution of type A receptor expression (Fig. 2A). With cRNA probe synthesized from a mixture of TAS3- and TAS4-cDNA, a hybridization

[1] Faculty of Pharmaceutical Sciences, Hokkaido University, Sapporo, 060 Japan

```
TASTE TISSUE
       TM III_____    _____TM IV___                                                       _____TM V_____             _____TM VI___
TAS3   FLLVANAFDRYVAIYNPLLYSVSMSPRVYMPLINASYVAGILHATIHTVATFSLSFCGANEIRRVFCDIPPLLAISYSDTHTNQLLLFYFVCAIELVTILIVLISYGLILLAILKMYSAEGRRKVFSTCGSHLT-V    135
TAS22  FLLVANAYDRYVAICKPLLYSIIMTNRLIIRLKVLSFVGGLIHSIIHIGFLFRLTFCKSNIIHHFYCDIMPLFKISCTDPSINVLMVFIFSGSIQVFTILTVLISYILVLFTILKNKSIQGIRKAFSTCGTHFT-V     135
TAS4   FLLVANAYDRYNAIGNPLLYGSKMSRVACIRLITFTYIYGFLTSLAATLWTYGLYFCGKIEINHFYCADPPLIKMACAGTFVKEYTMIILAGINFTYSLTVIIISYLFILIAILRMHSAEGROKAFSTCGTHFSS-    135
TAS8   FLLVANAFDRYTVIGNPLPYGSKMSRDVCIRLITFPYIYGFLTSLTATLWTYGLYFCGKIEINHFYCADPPLIKMACAGTFVKEYTMLILAGINFTYSLTVIIISYLFILIALLGMRSAEGRORAFSTCGTHLSVV   136
TAS7   FLLVANAICNPLLYGSRMSKSVCSFLITVPYAYGALTGLMETMWTYNLAFCGPNEINHFYCADPPLIKLACSDTYNKELSNFYVAGWNLSFSLFIICISYLYIFPAILKIRSTEGRORGFSTCGSHLT-V      135
TAS38  FLLVANAFDRYVAICNPLRYPEVMCKAVYVPMAAGSWVAGSTSAMVOTSLAMRLPFCGDNVINHFTCEILAVLKLACADISINIISMAMTNVIFLGIPVLFIFFSYVFIIATILKIPSAERRKKAFSTCGNHLT-V   135
       ********* *** *.  ** *      * *                     *  *.       .* *       . .            *  **      ***     *  *** ****.**  *
```

```
OLFACTORY EPITHELIUM
       TM III_____    _____TM IV___                                                       _____TM V_____             _____TM VI___
OLF1   FLLVANAFDRYVAICKPLLYSVIMTNRLCIRLSVLSFVGGLIHSIIHIGFLFRLTFCKSNIIHHFYCDIMPLFKISCTDPSINVLMVFIFSGSIQVFTILTVLISYILVLFTILKNKSIQGIRKAFSICGTHFTV    135
OLF32  FLLVANAYDRYVAICKPLLYPVIMTNKLCIOLLVSSFVAGLIHPMIHVGALFRLTFCKSNIIHHFYCDIMPLFRISCIDPTINVLMVFISSGSIOVCSIMTVLVSYTFVLHTVLKNKSVOGIRKAFSICGTHFTV   135
OLF4   FLLVANAYDRYVAIYNPLLYSVSMAPRVYMPLIIAPYVOGIVHATVHTVATFSLSFCASNEIRHVFCDIPPLLAISCSDTHTNOLLLFYLVGSIEIVTVLIVLISYGFILLAILRMHSAEGRRKVFSICGTHFTV   135
OLF11  FLLVANAYDRYVAIYDPLLYGSKMSPRVYVPLIIASYVOGIVHATVHTVATFSLSFCASNEIIHVFCDIPPLLAISCSDTHTNOLLLFYLVGSIEIVTVLIVLISYSFILLAVLRMHSAEGRRKVFSTCGSHLTV   135
OLF22  FLLVANAFDRYLAICRPLHYPNIMTGHLCTKLVLICWVYGFLWFLVPIVLISOMPFCGPNIIDHVVCDPGPLFALACVSAPKTOLLCYSLSSIVIFGNFLFILGSYTLVLTAVLHMPSATGROKAFSICGTHFTV   135
OLF2   FLLVANAYDRYVATCNPLRYPVVMNKVACRPMAIGSWAAGITNSVVQTSOAVRLPFCGDNVINHFTCEILAVLKLACADISINVVSMVVANVIFLGVPVLFIFVSYVFIIATILRIPSGEGRRKAFSICGTHLSV   135
OLF36  FLLVANAICNPLRYPVVMNKVACRPMAIGSWAAGITNSVVQTSOAVRLPFCGDNVINHFTCEILAVLKLACADISINVVSMVVANVIFLGVPVLFIFVSYVFIIATILRIPSGEGRRKAFSTCGTHFSV   135
OLF31  FLLVANAICNPLRYPVAIGSWAAGITNSVVQTSOAVRLPFCGDNVINHFTCEILAVLKLACADISINVVSMVVANVIFLGVPVLFIFVSYVFIIATILRIPSGEGRRKAFSICGTHLTV   135
OLF12  FLLVANAFDRYVAICFPLHYTTVMSPRLCLFLVVLPWVLTTFHAMLHTLLMARLHFCEDNVIPHFFCDLSALLKLSCSDTRVNELVIFFVGGLIVIPFLLIIMSYARIVSSILKVPSAKGICKAFSTCGTHLCV   135
OLF30  FLLVANAICFPLHYTTIMSPRLCLFLVLLPWILTTFHAMLHTLLMARLHFCEDNVIPHFFCDSSALLKLSCSDTRVNELVIFFVGGLIIIPFLLIIMSYAOIVSSILKVPSAKGICKAFSICGTHLTV   135
OLF27  FLLVANAFDRYVAICYPLHYAALSSLKRCALLVVTPWVVCNLLSVLHIGLLGLLNFCDOREIPHFFCDLEPILRLACSDTOINDLTILVIGGAVIFIPPTFICVSYFLIGCTVLRIPSAKGKWKTFSTCGTHLSV   135
OLF16  FLLVANAIDRYIAICHPLRYSAIMSPRLCGLVILLTLFIIIMDALLHTLMLLOLTFCTDLEIPLFFCEVVOVIKLACSDTFINNILIYLATSIFOGIPVCGIIFSYIOIVSSVLRMPSASGKYRAFSICGTHLTV   135
OLF18  FLLVANAFDRYVAICNPLRYPLIMTNQLCGXILAAGCWFCGLMTAMIKIVFIAXLYYCGTPRINHYFCDISPLLNISCEDSSOAELVDFFLALMVIAVPLCVVVASYTTILTTVLRIPSSOGLHKAFSICGTHFTV   135
OLF19  FLLVANAICYPLHYTVIILSWGLSTRISAGTWACGFFFSLIHTFFTMRLPYCGPNMVWHYFCEGPSVRNLACMDTHVIEMVDLVISVFMVVAPLLIVASYIRIAQAILKLSMOARCKAFSTCGTHFTV   135
OLF26  FLLVANALDRYNAIGNPLLYGSKMSRVACIRLITFPYIYGFLTSLAATLWTYGIXYFCGKIEINHFYCADPPLIKMACAGTFVKEYTMIILAGINFTYSLTVIIISYLFILIAILRMHSAEGROKAFSICGSHLTV   135
       ********* *** *.  ** *      * *                     *  *.       .* *       . .           * *  **     *    * ** ** *.* *
```

```
KIDNEY
       TM III_____    _____TM IV___                                                       _____TM V_____             _____TM VI___
KID1   FLLVANAFDRYVAXCFPLHYTTIMSPRLCSFLVVLSWVLTMVPAMLYTLLMARLCFCEDNVITHFFCDMSAXLKLSCSDTRVNETW-CIFVVGGLIIVLPFLVIIMSYARIIFSILKVPSVKGICKAFSICGSHFTV   136
KID16  FLLVANAFDRYVAICFPLHYTTIMSPRLCLFLVVLSWVLTMXHAMLVILMSRLHFCEDNVIAHFFCDMSALLKLSCSDTRVNEL---VIFITGGLILVIPFLLIITSYAOVVSSTLKVPSAKGICKAFSICGTHLTV   135
KID6   FLLVANAFDRYVAIYNPLLYSVSMAPRVYMPLIIASYVOGIVHXTVHTVATFSLSFCASNEIRHVFRDIPPLLAISCSDTXTNOLL-LFYLVGS-TEIVTVLIVLISYGFILLAILRMHSAEGRRKVFSTCGTHFTV   135
KID28  FLLVANAFDRYVAIYNPLLYSVSMAPRVYMPLIIASYVOGIVHATVHTVATFSLSFCASNEIRHVFRDIPPLLAISCSDTHTNOLL-LFYLVGS-IEIVTVLIVLISYGFILLAILRMHSAEGRRKVFSICGTHFTV   135
KID15  FLLVANAFDRYVAIYNPLLYSASMAPRVYVPLIIASYVOGIVHATVHTVATFSLSFCVSNEIRHVFRDIPPLLAISCSDTHTNOLL-LFYLVGS-IEIVTVLIVLISYGFILLAILRMHSAEGRRKVFSICGTHFTV   135
KID29  FLLVANAYDRYLAICSPLHYPTIMTPKLCAWLTLGCCVCGFATPLPEIAWISTLLFCGSNHLEHIFCDFLPVLRLACTDTOAILMIOVVDVVHAVEIITAVMLIFMSYIGIVAVILRIRSTEGRRKAFSTRGSHLTV   137
       ********* *** *.  ** *      * *
```

Fig. 1. Amino acid sequences of clones of type A GCR cDNA fragments amplified from bovine taste tissue (circumvallate and fungiform papillae), olfactory epithelium, and kidney. Amino acid residues conserved in each tissue are indicated by *asterisks*. Amino acid residues conserved in most of clones and varied within amino acid residues of similar physicochemical properties are indicated by *periods*. Putative transmembrane domains are *overlined*

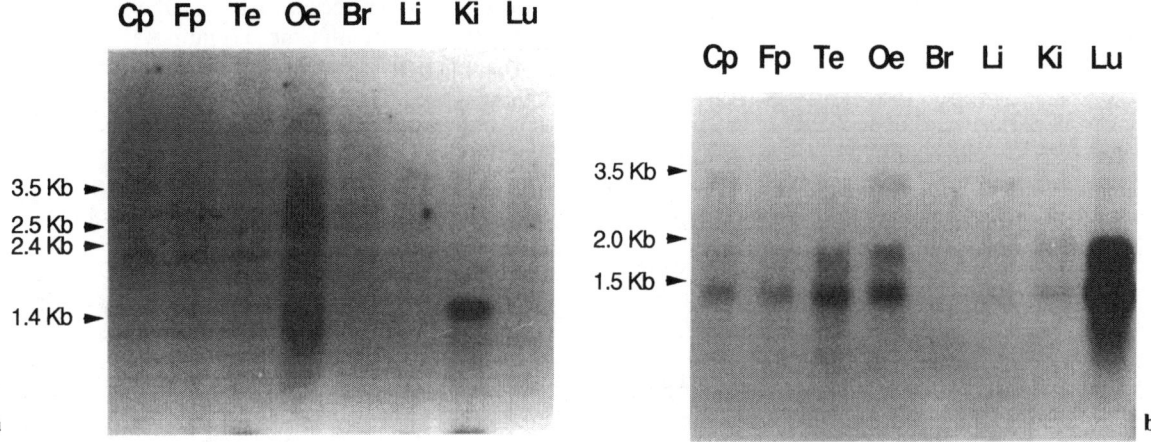

Fig. 2. Expression of type A and type B GCR mRNA in various bovine tissues. Northern blot analysis was performed with poly(A)$^+$-selected RNA from various bovine tissues (10 μg) as described in "Materials and Methods". Blotted membrane was subjected to multiple hybridizations with TAS3-4 cRNA probe (**a**) and PPR1-cRNA probe (**b**). Tissues used: circumvallate papillae (*Cp*), fungiform papillae (*Fp*), tongue epithelium bearing no taste papillae (*Te*), olfactory epithelium (*Oe*), brain (*Br*), liver (*Li*), kidney (*Ki*), lung (*Lu*)

band at 2.4 kb was detected in mRNA from circumvallate and fungiform papillae. A hybridization band of the same size was also detected in tongue epithelium devoid of taste papillae at a lesser strength than in taste papillae. Interestingly, mRNA from olfactory epithelium hybridized intensely with TAS3-4 probe at 3.5-4 and 1.5 kb in addition to 2.5 kb. If we assume that taste cells in circumvallate or fungiform papillae express type A receptors as taste receptors, the relatively low level of expression in the papillae compared to that in the olfactory epithelium is probably determined by low abundance of taste cells in the papillae. While mRNAs from most other tissues, including brain, liver, and lung, did not hybridize with TAS3-4 probe, kidney mRNA showed a hybridization band at 1.5 kb (Fig. 2A). When TAS38 cRNA was used as a probe, a hybridization pattern very similar to that with the TAS3-4 cRNA probe was observed (data not shown). We also examined the type A receptor expression by RT-PCR with C1/D1 primers and total RNAs from various bovine tissues. The ampli-

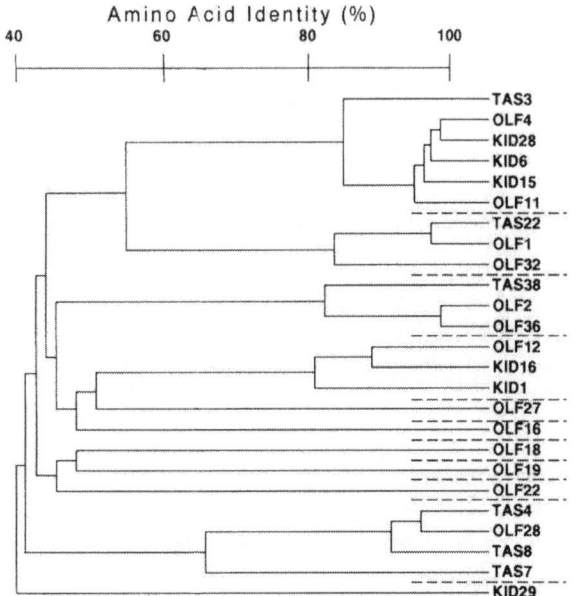

Amino Acid Identity (%)

FIG. 3. Dendrogram of type A cDNA clones. Phylogenic tree of bovine type A clones (Fig. 1) was made with CLUSTAL software. Scale on top indicates approximate percent amino acid identity between different clones at branching points. *TAS*, taste; *OLF*, olfactory; *KID*, kidney

fication pattern of the RT-PCR was very similar to the hybridization pattern of the Northern blot analysis (data not shown).

To examine the identity of the type A receptors expressed in various tissues, we cloned the RT-PCR products from bovine olfactory epithelium and kidney. Among more than 40 clones from olfactory epithelium and more than 30 clones from kidney, 13 independent clones from olfactory epithelium and 6 independent clones from kidney were identified (see Fig. 1). The independent clones from three tissues (taste, olfactory, and kidney) were subjected to sequence alignment analysis using CLUSTAL software, and a dendrogram was made (Fig. 3). Using the cutoff of about 60% amino acid identity, these receptors clones were divided into five subgroups and other individual clones. As a result, every subgroup contained clones from multiple tissues, and no one subgroup was specific to anyone particular tissue (Fig. 3). In other words, the type A receptor clones amplified from different tissues mix completely each other and constitute the putative bovine odorant receptor family [6].

In the mechanism of taste transduction, the roles of GCR and second messenger systems have been implicated in the reception of sweet and bitter substances [1,2]. In the reception of sweet substances (sugars), it was suggested that activation of adenylate cyclase in an G-protein-dependent manner is involved [1]. Although different sugar compounds induce a similar sweet taste, it was suggested that there are multiple types of receptors for sugar compounds [8,9].

On the other hand, transduction of bitter substances was suggested to be coupled to phospholipase C to produce inositol triphosphate (IP_3) [2]. The presence of multiple types of receptors for bitter substances also has been suggested [2]. It is interesting to note that olfactory responses are evoked also by bitter substances, which are usually hydrophobic compounds like odorants (Shoji and Kurihara, unpublished observation). Alternatively, gustatory responses are evoked by odorants [10]. On the other hand, coupling of the rat putative odorant receptors to phospholipase C to produce IP_3 has been confirmed by Raming et al. [11]. The length of the third cytoplasmic loop is constant among the subfamily of the putative rat odorant receptors and the bovine type A receptors, suggesting that they all interact with a similar type of G protein, such as Gq coupling to the IP_3 pathway. These notions suggest that some members of the type A receptors expressed in taste tissue function as receptors for bitter substances to produce IP_3. Further confirmation of type A receptors as taste receptors awaits histological localization of the receptors in taste buds and functional reconstitution of the receptors.

The functions of the type A receptors expressed in kidney also remain to be elucidated. In kidney, various hydrophobic molecules are transported from blood to urine by an as yet unidentified mechanism. The type A receptors in kidney may function as recognition molecules for such hydrophobic substances to be transported specifically. Alternatively, the type A receptors expressed on the cell surface of various tissues, including the taste tissue, may function as recognition molecules in a broad sense to define the "personality" of individual cells. In this context it is interesting to note that Parmentier et al. [12] have reported the presence of the odorant receptor transcripts in sperm cells. Members of the putative odorant receptor (type A receptor) family may play as yet unidentified roles in various tissues and cell types.

Type B Clone, Which Is Homologous to Peptide Receptors

Sequencing of type B cDNA fragments amplified with A1/B4 primers identified a single cDNA clone (AL11). ^{32}P-labeled AL11 cDNA was used as a probe to screen lamdaZAP cDNA library (6×10^5 independent clones) prepared from bovine taste papillae. A single positive clone (PPR1) containing a 2-kb insert was isolated. Sequence analysis of PPR1 identified a translational open reading frame encoding a protein of 350 amino acids (Fig. 4). The hydropathy profile of PPR1 is consistent with presence of seven stretches of hydrophobic TM domains.

FIG. 4. Alignment of amino acid sequence of PPR1 cDNA with peptide receptors. Amino acid residues conserved in all receptors are indicated by *asterisks*; Amino acid residues conserved in most receptors and varied within amino acid residues of similar physicochemical properties are indicated by *periods*. Putative transmembrane domains are *overlined*. Clones and their percent amino acid identity to PPR1: *IL8A HUMAN*, human high-affinity interleukin 8 receptor A (34.2%); *IL8B HUMAN*, human high-affinity interleukin 8 receptor B (34.2%); *NY3R BOVIN*, bovine neuropeptide Y receptor (35.2%); *VIPR HUMAN*, human VIP receptor (30.3%); *BRB2 RAT*, rat bradykinin receptor (26.0%); *AG2R BOVIN*, bovine type 1 angiotensin II receptor (31.5%); *C5AR HUMAN*, human anaphylatoxin C5a receptor (26.8%). Total number of amino acid residues are indicated at C-terminus

There are a number of conserved motifs typical for the GCR superfamily, including GN**V (TM1), LA*AD (TM2), I**DRY (TM3), and NP**Y (TM7). There are two potential N-glycosylation sites (N-X-S/T) in the extracellular N-terminal end, typical for the GCR superfamily. Several consensus acceptor sites for phosphorylation by ser/thr-protein kinases are found in the carboxy-terminus region.

Search of the SWISS-PROT protein databases revealed that PPR1 protein has not been previously identified, but has significant similarity with other known members of the GCR superfamily especially in the peptide receptor branch, for example, interleukin-8 receptor, neuropeptide-Y receptor, vasoactive intestinal peptide (VIP) receptor, and type 1 angiotensin II receptor (see Fig. 4). Most of the similarity between these receptors lies within the TM domains. The third cytoplasmic loops in most of these peptide receptors as well as PPR1 are short, suggesting the same type of G protein such as Gq is coupled. From these observations, it is highly plausible that PPR1 protein is also a receptor for peptide ligands.

Northern blot analysis was performed to determine the tissue distribution of PPR1 mRNA expression (see Fig. 2B). Multiple species of PPR1 mRNA (major, 1.5 kb; minor, 2.0 and 3.5 kb) were widely but unevenly detected in various adult bovine tissues. Circumvallate and fungiform papillae and olfactory epithelium expressed relatively high levels of PPR1 mRNA. Interestingly, lung expressed the most prominent level of PPR1 mRNA. Certain levels of PPR1 mRNA expression were also observed in liver, kidney, and tongue epithelium bearing no taste papillae, while the expression in brain was very low.

Using immunohistochemical technique, a variety of neuropeptides have been detected in vertebrate taste buds [13]. Much of the attention has focused on the roles of neuropeptides in taste tissue, including modulation of the sensitivity of taste cells and trophic action in maintenance and regeneration of taste cells [13]. Identification of the authentic ligand for the PPR1 receptor would certainly facilitate the understanding roles of neuropeptides in taste tissue.

Acknowledgments. This work was supported by a grant from Human Frontier Science Program. The authors are greatly indebted to Dr. K. Kurihara for his continuous encouragement and support.

References

1. Striem B, Pace U, Zehavi U, Naim M, Lancet D (1989) Sweet testants stimulate adenylate cyclase coupled to GTP-binding protein in rat tongue membranes. Biochem J 260:121–126
2. Hwang PM, Verma A, Bredt DS, Snyder SH (1990) Localization of phosphatidylinositol signaling components in rat taste cells: role in bitter taste transduction. Proc Natl Acad Sci USA 87:7395–7399
3. Dohlman HG, Thorner J, Caron MG, Lefkowitz RJ (1991) Model systems for the study of seven-transmembrane-segment receptors. Annu Rev Biochem 60:653–688
4. Probst WC, Snyder LA, Schuster DI, Brosius J, Sealfon SC (1992) Sequence alignment of the G-protein coupled receptor superfamily. DNA Cell Biol 11:1–20
5. Libert F, Parmentier M, Lefort A, Dinsart C, Jacqueline VS, Maehaut C, Simons MJ, Dumont JE, Vassart G (1989) Selective amplification and cloning of four new members of the G-protein-coupled receptor family. Science 244:569–572
6. Buck L, Axel R (1991) A novel multigene family may encode odorant receptors: a molecular basis for odor recognition. Cell 65:175–187
7. Matsuoka I, Mori T, Aoki J, Sato T, Kurihara K (1993) Identification of novel members of G-protein coupled receptor superfamily expressed in bovine taste tissue. Biochem Biophys Res Commun 194:504–511
8. Beidler LM, Tonosaki K (1985) Multiple sweet receptor sites and taste theory. In: Pfaff DW (ed) Taste, olfaction and the central nervous system. Rockefeller University Press, New York, pp 47–64
9. Shimada I, Shiraishi A, Kijima H, Morita H (1974) Separation of two receptor sites in a single labellar sugar receptor of the freshfly by treatment with p-chloromercuribenzoate. J Insect Physiol 20:605–621
10. Kashiwagura T, Kamo N, Kurihara K, Kobatake Y (1977) Responses of the frog gustatory receptors to various odorants. Comp Biochem Physiol C 56:105–108
11. Raming K, Krieger J, Strotmann J, Boekhoff I, Baumstark C, Breer H (1993) Cloning and expression of odorant receptors. Nature (Lond) 361:353–356
12. Parmentier M, Libert F, Schurmans S, Schiffmann S, Lefort A, Eggerickx D, Ledent C, Mollereau C, Gerard C, Perret J, Grootegoed A, Vassart G (1992) Expression of members of the putative olfactory receptor gene family in mammalian germ cells. Nature (Lond) 355:453–455
13. Roper SD (1992) The microphysiology of peripheral taste organs. J Neurosci 12:1127–1134

Strategies for Isolation of Taste Receptor Proteins

D. Lynn Kalinoski[1], Andrew I. Spielman[1,2], John H. Teeter[1,3], Isabella Andreini[1], and Joseph G. Brand[1,3,4]

Key words. Gustatory—Purification—Transduction — Chromatography — Lectins — Immunohistochemistry

Introduction

For almost two decades the channel catfish, *Ictalurus punctatus*, has proved a useful model for the study of the vertebrate gustatory system. During this period a combination of behavioral [1], neurophysiological [2–4], and biochemical [5–7] techniques have been directed at identifying specific taste receptors and characterizing the molecular mechanisms underlying sensory transduction mediated by these receptors. Neurophysiological experiments have established the presence of independent receptor sites for L-alanine (and possibly other short-chain neutral amino acids) [2,3], L-arginine [2,3], and L-proline [2–4]. In addition, receptors sites for some D-isomers of amino acids, in particular D-alanine and D-arginine, have been inferred from non-reciprocal cross-adaptation obtained in electrophysiological studies [3]. Receptor-binding methodologies utilizing both equilibrium and kinetic techniques have been employed to determine the apparent dissociation concentrations for the L-alanine ($1.5\,\mu M$) [5,6] and L-arginine ($18\,nM$ and $1.3\,\mu M$) [7] binding sites.

Competition binding studies have examined the specificity of these binding sites and established that neither L-alanine nor L-arginine is an effective competitor for the other L-amino acid binding site, suggesting in accordance with the electrophysiological studies that these amino acids interact with separate receptor proteins. Further, the D-enantiomers of these amino acids are competitive inhibitors for the respective L-amino acid binding sites; however, binding studies have also established the presence of an additional taste receptor specific for D-alanine [5]. Because results from electrophysiological studies suggest that a low affinity is likely for the L-proline binding site, making binding studies difficult when [^{3}H]-L-proline is used as a radioligand, the specificity of this binding site has not been studied using biochemical techniques.

In studies similar to those detailing the nature and specificity of the taste receptor sites in the channel catfish, the mechanisms underlying taste transduction have been examined using a multidisciplinary approach. These studies indicate that the major classes of amino acid taste receptors are coupled to activation of the receptor cell by at least two different mechanisms. Biochemical studies have demonstrated that L-alanine receptors produce an increase in the production of the second messengers inositol 1,4,5-trisphosphate (IP_3) and cyclic AMP (cAMP) via G-protein-mediated mechanisms [8]. In contrast, L-arginine and L-proline have no effect on the production of either second messenger.

Biophysical studies on membrane vesicles from catfish taste epithelia incorporated into lipid bilayers indicate that L-arginine and L-proline directly activate ion channels. While both of these amino acids increased bilayer conductance, they did so by activating different classes of cation channels [9,10]. The receptor–channel for L-arginine had a threshold near $0.5\,\mu M$ [9,10]. L-Arginine showed a half-maximal response at about $15\,\mu M$ and was saturated between 100 and $200\,\mu M$ [9,10]. D-Arginine acted as a potent antagonist to L-arginine stimulation. In agreement with data obtained from single unit recordings, L-proline channels displayed a relatively low affinity in comparison to sites for L-alanine or L-arginine [9,10]. Conductance measurements indicated a threshold of about $50\,\mu M$ and saturation above $2000\,\mu M$. Similar to the L-arginine receptor–channel, the D-enantiomer of proline was an antagonist to the responses produced by L-proline [10]. The L-proline channel was not activated by L-alanine or L-arginine. Similarly, L-arginine-activated receptor channels did not respond to L-proline.

These studies provide information establishing the identities and characteristics of specific amino acid taste receptors of the channel catfish. However, to further our knowledge of the properties, characteristics, and regulation of taste receptors, isolation of the proteins is essential. Specific taste receptor proteins may be isolated by cloning the gene that codes for the specific receptor followed by expression in *Xenopus* oocytes or a cell line, or, alternatively, traditional biochemical approaches may serve to purify the protein to homogeneity.

[1] Monell Chemical Senses Center, Philadelphia, PA, USA
[2] Division of Basic Sciences, College of Dentistry New York University, New York, NY, USA
[3] University of Pennsylvania, Philadelphia, PA, USA
[4] Veterans Affairs Medical Center, Philadelphia, PA, USA

Approaches to Isolation of Receptor Proteins

Affinity Chromatography

Several approaches exist for the purification of proteins. One of the most powerful of these is affinity
chromatography; a number of neurotransmitter
and hormone receptors have been isolated by this
method of separation [11]. This method takes
advantage of the specific reversible interactions
between biomolecules such as receptors and their
immobilized agonists or antagonists to obtain
purification from the heterogenous population of
proteins present in a cell membrane preparation.
Production of an affinity resin requires the covalent
attachment of the selected affinity reagent to a gel,
often by coupling to a spacer arm. These chemical
modifications often affect the ability of the chosen
ligand to interact with the protein to be isolated,
sometimes reducing the utility of the ligand for
isolation procedures.

Structure–activity studies on the L-alanine and
L-arginine receptors of the catfish taste system have
defined some of the structural requirements for
receptor activation and also delineated the effects
of structural modifications involved in coupling of
amino acids to affinity matrixes on ligand affinity.
Results of such studies on the L-arginine receptor
demonstrated that of all the guanidinium-containing
compounds tested, only L-arginine, L-α-amino-β-
guanidino propionic acid, and L-arginine methyl
ester were strong stimuli [12]. Similarly, structure–
activity studies of the L-alanine receptor-binding
properties demonstrated that of those derivatives
with side-chain modifications, L-serine, glycine, β-
chloro-L-alanine, and 1-amino-cyclopropane-1-
carboxylic acid were potent analogues with IC_{50}s
and neural responses similar to L-alanine [13].
Results of these studies indicate that functional
group substitutions and modifications required for
coupling of either L-arginine- or L-alanine-based
analogues to affinity matrixes generally resulted in
loss of stimulatory ability or affinity. As would be
predicted from these results, affinity purification of
L-amino acid taste receptors has proved difficult,
and to date there has been no report of a successful
isolation of amino acid taste receptor proteins using
this technique.

Lectin Chromatography

While isolation of a receptor protein by affinity
chromatography is among the most powerful techniques, purification by methods that depend on
other properties of proteins such as isoelectric point,
pH, or chemical modifications have also been employed to isolate membrane proteins [11]. Because
of the lack of success with affinity techniques, our
laboratory has decided to pursue some of these

methods to begin isolation of catfish amino acid taste
receptors.

Lectins, proteins of nonimmune origin that bind
carbohydrates and can cause precipitation of glycoproteins and agglutination of cells, have frequently
been used to characterize and partially purify membrane proteins [14,15]. Because many extracellular
facing membrane proteins are glycosylated or contain carbohydrate residues, our laboratory decided
to explore the practicality of using lectins in procedures to isolate catfish amino acid taste receptors.
In an initial study that examined the glycoprotein
nature of the membrane proteins from chemosensory tissues of the catfish, it was shown that the taste
membranes contained numerous glycoproteins when
probed with the lectin from concanavalin A (Con A)
[16]. However, the results of this study argued
against a role for carbohydrate subunits of receptor
proteins in stimulus recognition in the channel catfish because the three lectins, Con A, wheat germ
agglutinin (WGA), and peanut agglutinin (PNA),
inhibited binding of L-alanine and L-arginine to
olfactory cilia of the catfish but had little effect on
the binding of these ligands to taste epithelial
membranes. However, this study used only three of
the many available lectins, and in a subsequent study
we screened five additional lectins of varying carbohydrate specificities in the amino acid taste system of
the channel catfish.

This subsequent study demonstrated that the
receptor sites for L-alanine and L-arginine were distinguishable by their sensitivity to lectins [17]. Like
Con A and WGA, the lectin from *Pisum sativum*
(PSA) had little effect on the binding of either
L-alanine or L-arginine to taste membrane fractions.
However, the binding of L-arginine was selectively
inhibited by preincubation in presence of the lectin
from *Phaseolus vulgaris* (PHA) while the lectin from
Dolichos biflorus (DBA) selectively inhibited binding of L-alanine [17]. Binding of both amino acids
was almost fully inhibited by preincubation with the
lectin from *Ricinus communis* (RCA I) and was
partially inhibited by jacalin. Additionally, when the
biotinylated lectins were used to identify glycoprotein components of the taste plasma membranes
after sodium dodecyl sulfate polyacrylamide gel
electrophoresis (SDS-PAGE), only a few proteins
were labeled by PHA, DBA, and RCA I. This
differential labeling of the solubilized proteins of
taste membranes in combination with the differential inhibition of receptor binding suggested that
lectins would prove useful tools in the purification of
taste proteins.

Although both L-alanine and L-arginine binding
was selectively inhibited by lectins, a decision was
made to pursue isolation of the L-arginine receptor
because it would be possible to subsequently assay
purportedly purified fractions for both receptor
binding and physiological function (using reconstitu-

tion techniques). Because amino acid taste receptors generally possess low affinity for ligands when compared to many neurotransmitter and hormone receptors, and because this relatively low affinity binding makes it difficult to use many of the techniques to measure binding of soluble receptors, the availability of a second, highly sensitive method to test for the presence of receptor proteins was a decided advantage. Assays of L-alanine receptor function could be expected to prove difficult because is has been demonstrated that second messenger-activating receptors frequently dissociate from G-proteins during the manipulations required to extract them from their membrane environments, thus uncoupling the receptor from its ability to stimulate second messenger-producing enzymes.

Having made the decision to pursue the purification of the L-arginine receptor, we subjected detergent-extracted taste epithelial membrane preparations to lectin-affinity chromatography on either PHA E or RCA I coupled to agarose. Biophysical studies on phospholipid bilayers into which PHA E- or RCA I-purified proteins had been incorporated frequently displayed reversible increases in cation conductance in the presence of micromolar concentrations of L-arginine [18]. Similar to studies on bilayers into which taste membrane fractions had been incorporated, D-arginine was a potent inhibitor of L-arginine-activated conductances. SDS-PAGE followed by silver staining identified three proteins in PHA-purified material and four proteins in RCA-I-eluted fractions. Two of these proteins, a doublet of approximately 35000 MW, appeared in both RCA I- and PHA-purified fractions.

Further support for the hypothesis that these lectin-isolated glycoproteins were taste tissue specific was provided by a 1992 study by Böttger and Finger [19]. Using histochemical techniques, these investigators demonstrated that when mandibular barbels of the channel catfish were exposed to biotinylated PHA-E followed by fixation in 4% paraformaldehyde, lectin-binding sites appeared as round patches at the apex of each taste bud along the length of the barbel [19]. These patches appeared coincident with the taste pore region. When taste buds were viewed in profile, small reactive extensions could be seen extending from the surrounding taste pore area.

Results from the biophysical and histochemical studies support the hypothesis that the proteins isolated by lectin-affinity chromatography include the L-arginine receptor protein. To further test this hypothesis and to provide additional tools for the identification, purification, and characterization of the L-arginine receptor, SDS-PAGE-separated proteins from lectin columns were isolated by electroelution and used for the production of protein-specific antibodies [20].

Immunological Procedures

Antibodies are another useful tool in receptor purification. However, before starting the process of antibody production a number of important facts about the nature of the antigen, as well as the choice of monoclonal versus polyclonal antibody, need to be considered [21]. Discussion of these factors is beyond the scope of this article; however, a major consideration is the time and expense required for monoclonal antibody development and production. Additionally, because polyclonal antibodies recognize multiple antigenic determinants they generally act as precipitating antibodies, making them applicable to a large variety of methods.

Our laboratory has begun to produce and characterize polyclonal antibodies raised against the four proteins isolated from taste epithelial membranes applied to RCA-I affinity columns [22]. Although we have not completely characterized all these antibody populations, at least one of them, raised against the top band of the 35000-MW doublet (anti-R1), satisfies some of the criteria one would expect of an anti-L-arginine receptor antibody. Specifically, the anti-R1 antibody recognizes taste epithelial membrane fraction P2 in enzyme-linked immunosorbent assays (ELISAs), it localizes to the apical area of the taste pore in immunohistochemical studies, and in preliminary experiments anti-R1 blocks L-arginine-activated cation conductances of taste epithelial membrane fractions incorporated into lipid bilayers [22]. Control experiments with either nonimmune serum or control guinea pig IgG fractions had no effect in any of these experimental paradigms. Thus, the initial results of experiments using the R1 antibody suggest that the R1 antigen is a component of the L-arginine receptor–channel, and we are beginning to use this immunological tool to isolate the R1 protein.

Conclusion

Purification to homogeneity of the L-arginine taste receptor allows characterization of properties such as amino acid sequence, sites of chemical modification, and similarities and homologies to other families of proteins that should further our understanding of the function and regulation of specific receptor proteins in the taste system. While the advent of molecular biological techniques to the chemical senses has accelerated the processs of identifying, cloning, and characterizing chemosensory receptors, a traditional biochemical approach to receptor isolation can still provide useful information. This is particularly true for taste receptors such as the L-arginine receptor/channel that may not be members of gene families for which sequence information has already been established.

The studies outlined here provide the background for the isolation and characterization of the L-arginine receptor, and the next several years should see the successful purification and characterization of the properties of this taste receptor–channel.

References

1. Caprio J, Brand JG, Teeter JH, Valentinicic T, Kalinoski DL, Kohbara J, Kumazawa T, Wegert S (1993) The taste system of the channel catfish: from biophysics to behavior. TINS (Trends Neurosci) 16: 192–197

2. Caprio J (1978) Olfaction and taste in the channel catfish: an electrophysiological study of the responses to amino acids and derivativess. J Comp Physiol 123: 357–371

3. Wegert S, Caprio J (1991) Receptor site for amino acids in the facial taste system of the channel catfish. J Comp Physiol A Sens Neural Behav Physiol 168: 201–211

4. Kanwal JS, Caprio J (1983) An electrophysiological investigation of the oropharyngeal (IX–X) taste system in the channel catfish, *Ictalurus punctatus*. J Comp Physiol A Sens Neural Behav Physiol 150: 345–357

5. Brand JG, Bryand BP, Cagan RH, Kalinoski DL (1987) Biochemical studies of taste sensation. XIII. Enantiomeric specificity of alanine receptor sites in the catfish, *Ictalurus punctatus*. Brain Res 416:119–128

6. Kreuger JM, Cagan RH (1976) Binding of L-[³H]alanine to a sedimentable fraction from catfish barbel epithelium. J Biol Chem 251:88–97

7. Kalinoski DL, Bryant BP, Shaulsky G, Brand JG, Harpaz S (1989) Specific L-arginine taste receptor in the catfish, *Ictalurus punctatus*: biochemical and neurophysiological characterization. Brain Res 488: 163–173

8. Kalinoski DL, Huque T, LaMorte VJ, Brand JG (1989) Second messenger events in taste. In: Brand JG, Teeter JH, Kare MR, Cagan RH (eds) Receptor events and transduction in taste and olfaction. Dekker, New York, pp 85–102

9. Teeter JH, Brand JG, Kumazawa T (1990) A stimulus-activated conductance in isolated taste epithelial membranes. Biophys J 58:253–259

10. Teeter JH, Kumazawa T, Brand JG, Kalinoski DL, Honda E (1992) Amino acid receptor channels in taste cells. In: Corey D, Roper SD (eds) Sensory transduction. Rockefeller University Press, New York, pp 291–306

11. Deutscher MP (ed) (1990) Methods in enzymology. Volume 182, Guide to protein purification

12. Bryant BP, Harpaz S, Brand JG (1989) Structure/activity relationships in the arginine taste pathway of the channel catfish. Chem Senses 14:805–815

13. Bryant BP, Leftheris K, Quinn JV, Brand JG (1993) Molecular structural requirements for binding and activation of L-alanine taste receptors. Amino Acids 4:73–88

14. Hughes RC (1983) Glycoproteins. Chapman and Hall, London

15. Abbott WM, Strange PG (1985) Partial purification of dopamine D2 receptors using lectin affinity columns. Biosci Rep 5:303–308

16. Kalinoski DL, Bruch RC, Brand JG (1987) Differential interactions of lectins with chemosensory receptors. Brain Res 488:167–173

17. Kalinoski DL, Johnson LC, Bryant BP, Brand JG (1992) Selective interactions of lectins with amino acid taste receptor sites in the channel catfish. Chem Senses 17:381–390

18. Kalinoski DL, Teeter JH, Brand JG (1992) Partial purification of an L-arginine receptor from catfish taste epithelium (abstract). AChemS XIV

19. Böttger B, Finger TE (1992) Localization of PHA-E lectin binding (arginine receptors?) to taste buds on catfish barbels (abstract). AChemS XIV

20. Spielman AI, Bennick A (1989) Purification and characterization of six proteins from rabbit parotid saliva belonging to a unique family of proline-rich proteins. Arch Oral Biol 34:117–130

21. Harlowe E, Kane D (eds) (1988) Antibodies: a laboratory manual. Cold Spring Harbor Laboratory, Cold Spring Harbor, NY

22. Kalinoski DL, Teeter JH, Spielman AI, Brand JG (1992) Partial purification of an L-arginine receptor from catfish taste epithelium (abstract). European Chemoreception Research Organization Congress X

Molecular Mechanisms
of Taste Receptor Cell Signal Transduction

PAUL M. HWANG, SETH BLACKSHAW, XIAO J. LI, and SOLOMON H. SNYDER[1]

Key words. Bitter taste—Signal transduction—Denatonium—Potassium channel—Immunohistochemistry—Inositol triphosphate

Introduction

Some mechanisms by which taste signals are transduced have been identified [1]. Acids appear to mediate sour taste by blocking voltage-dependent K^+ channels localized on the apical surface of taste cells [2,3], and amiloride-sensitive sodium channels may mediate salty taste [4,5]. Sweet tastants stimulate adenylyl cyclase and the cyclic adenosine monophosphate (cAMP) produced may be the second messenger mediating sweet taste transduction [6]. At least two mechanisms have been proposed for bitter taste transduction, one involving direct blockade of K^+ channels by bitter substances [7] and the other involving receptor-mediated mobilization of intracellular Ca^{2+} stores [8].

The molecular mechanism of bitter taste transduction is unclear, but the existence of potently bitter substances suggests high affinity interactions of bitter substances with specific receptors. To study the role of second messengers in bitter transduction, Akabas et al. [8] loaded dissociated taste cells from rat circumvallate papilla with the Ca^{2+}-sensitive dye fura-2 to measure changes in intracellular Ca^{2+} concentration ($[Ca^{2+}]_i$). Upon stimulation by $1\,\mu M$ denatonium transient increases in $[Ca^{2+}]_i$ were produced in a subpopulation of taste cells. The $[Ca^{2+}]_i$ spikes were not elicited by $1\,mM$ saccharin and were not dependent on extracellular Ca^{2+}. These data indicate that taste cells may mobilize intracellular Ca^{2+} stores during bitter taste transduction and suggest the involvement of phosphoinositides (PI) as second messengers [9].

Delayed-rectifier K^+ currents are present in mammalian taste cells and may play important roles in taste transduction. Protonation and subsequent closure of voltage-activated K^+ channels may mediate sour taste [2]. cAMP and cyclic guanosine monophosphate (cGMP) modulate K^+ currents in

[1] Department of Neuroscience, Johns Hopkins University School of Medicine, 725 N. Wolfe Street, Baltimore, MD 21205, USA

taste cells [10,11]. Denatonium at millimolar concentrations appears to block taste-cell delayed-rectifier K^+ currents. The molecular identity of the taste cell K^+ channels remains unknown, and characterizing such channels may yield important clues to understanding the initial steps of taste transduction.

Methods

Male Sprague-Dawley rats ($150–200\,g$) were decapitated and the circumvallate papillae dissected and rapidly frozen in O.C.T. medium (Tissue-Tek). Cryostat sections ($14\,\mu m$) were cut and stored desiccated at $-20°C$. The glossopharyngeal nerve was bilaterally transected as described elsewhere [12,13]. Circumvallate papillae taste buds were allowed to degenerate a minimum of 10 days before denervated rats were used.

A novel method for the selective localization of intracellular Ca^{2+} stores in cryostat tissue sections was used [14,15]. Briefly, circumvallate papilla cryostat sections were permeabilized then transferred into uptake buffer containing K_2oxalate, K_2ATP, and $^{45}CaCl_2$ for 60 to 90 min at $25°C$. Then the sections were washed in buffer containing EGTA to remove nonsequestered $^{45}Ca^{2+}$, dried under a gentle stream of air, dipped in Kodak NTB2 emulsion, exposed 1 to 2 weeks, and developed in Kodak D19.

$[^3H]InsP_3$ autoradiography was performed as previously described [14] on fresh-frozen sections of rat circumvallate papillae, and PI turnover in circumvallate papillae was monitored by two previously described techniques, utilizing either $[^3H]$inositol [16] or $[^3H]$cytidine [17] as precursor. $InsP_3$ mass level measurements were performed by the method of Bredt et al. [18].

The primary antibodies were raised in rabbits against synthetic peptides of two rat K^+ channels as follows: CDRK, amino acids 774–788 [19]; and DRK1, amino acids 839–853 [20]. These antibodies were then affinity purified [21]. Slides of rat tongue tissue were prepared as previously described [21]. Standard blocking and incubation conditions were used as previously described for anti-CDRK ($1:250$) and anti-DRK1 ($1:250$) antibodies [21].

Results

Intracellular Calcium Sequestration and Release in Taste Receptor Cells

We developed a technique permitting autoradiographic localization of endoplasmic reticulum stores of calcium [14] and have used it to label these calcium pools in taste cells (Fig. 1A,B). Autoradiographic grains are localized to taste buds along the crypt of the circumvallate papillae and are most highly concentrated in apical areas. Virtually no $^{45}Ca^{2+}$ accumulation is detected in epithelial layers of the tongue not associated with taste transduction. The only other substantial $^{45}Ca^{2+}$ uptake occurred in von Ebner's glands, which is consistent with a

major role for the PI system and calcium in glandular function. The accumulated calcium is markedly depleted in sections treated with 100 nM InsP$_3$ (Fig. 1C,D), presumably by opening Ca^{2+} channels that release accumulated $^{45}Ca^{2+}$. Heparin (100 µg/ml), a potent inhibitor of InsP$_3$ receptor binding [22], reverses the effect of InsP$_3$ on $^{45}Ca^{2+}$ accumulation in the taste buds. To ensure that the observed $^{45}Ca^{2+}$ accumulation involves taste bud cells, we bilaterally transected the glossopharyngeal nerve (Fig. 1E,F). Following denervation, accumulation of $^{45}Ca^{2+}$ by apparent taste buds is abolished coincident with the disappearance of taste buds from the circumvallate papillae.

The ability of InsP$_3$ to release $^{45}Ca^{2+}$ from taste buds implies the existence of specific InsP$_3$ receptors, which we verified by autoradiography with [^{3}H]InsP$_3$ (Fig. 2). Autoratiographic grains are selectively associated with taste buds of circumvallate papillae. As with accumulated $^{45}Ca^{2+}$, [^{3}H]InsP$_3$ binding sites are most concentrated in the apical region of taste buds (Fig. 2C). Except for taste buds of the circumvallate papillae, no selective concentrations of [^{3}H]InsP$_3$ associated grains are seen in the tongue. Addition of 10 µM unlabeled InsP$_3$ completely abolishes specific labeling of taste buds by [^{3}H]InsP$_3$ (Fig. 2F).

PI Turnover in Taste Receptor Cells

To determine if tastants act via the PI cycle, we examined the influence of the bitter tastant denatonium on PI turnover measured in dissected portions of rat tongue enriched in circumvallate papillae (Table 1). PI turnover was monitored by two techniques, both of which require the use of lithium to block Ins(1)P$_1$ phosphatase to permit the accumulation, respectively, of [^{3}H]inositol phosphates [16] or ^{3}[H]CDP-DAG [17]. By either technique, 10 µM denatonium does not appear to alter PI turnover (Table 1).

Stimulation of PI turnover by neurotransmitters, hormones, or other regulatory substances results from activation of phospholipase C, producing increased mass levels of InsP$_3$. Accordingly, we measured InsP$_3$ mass levels in dissected circumvallate papillae using a sensitive radioreceptor assay [18]. Denatonium produces a rapid enhancement of InsP$_3$ mass levels, which is detected 15 to 30s after treatment with the tastant (Table 1) but is no longer apparent after 1 min (data not shown). This very rapid time course resembles the influence of neurotransmitters on InsP$_3$ mass levels in brain slices [23].

To examine whether the influence of denatonium is taste-specific, we examined its effects on InsP$_3$ levels in epithelium adjacent to circumvallate papillae, which lacks taste buds. No effect on InsP$_3$ mass levels is observed in these specimens (Table 1). We have also examined the effects of sucrose and

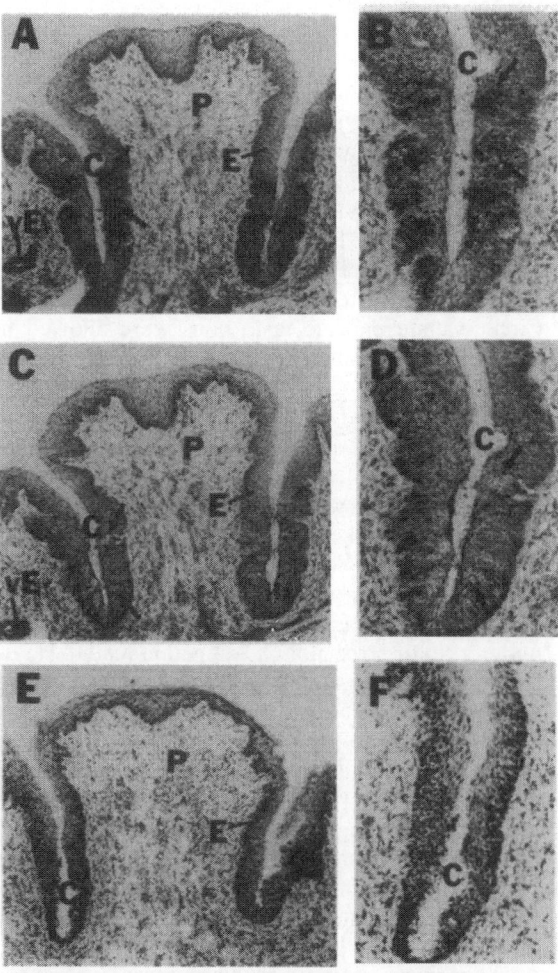

FIG. 1. Localization of $^{45}Ca^{2+}$ in the rat circumvallate papilla. **A,B** Uptake of $^{45}Ca^{2+}$ visualized by autoradiographic emulsion silver grains in light-field (dense, black silver grains) and dark-field (white silver grains), respectively. **C,D** Reduction in the net accumulation of $^{45}Ca^{2+}$ by 1 µM InsP$_3$. **E,F** No uptake of $^{45}Ca^{2+}$ in denervated circumvallate papilla due to taste bud degeneration. *P*, circumvallate papilla; *C*, crypt; *E*, epidermal layer; *vE*, von Ebner's gland; *arrows* indicate taste buds. **A–F**, ×100

Fig. 2. Localization of InsP$_3$ receptors in the circumvallate papilla. **A–C** High density of specific [^{3}H]InsP$_3$ binding sites in taste buds is revealed by dark-field silver grains selectively localized over taste buds. **D–F** Nonspecific binding with 10 µM unlabeled InsP$_3$. *P*, circumvallate papilla; *C*, crypt; *arrows* indicate taste buds. **A,D**, ×100; **B,C,E,F**, ×200

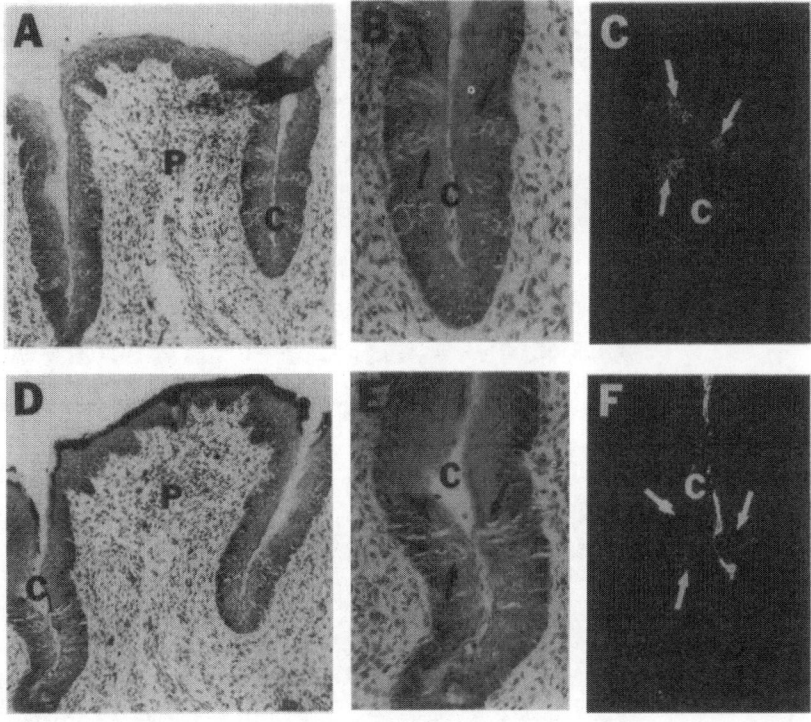

TABLE 1. Phosphoinositides in circumvallate papillae.

Measurement	No.	Data
A. PI turnover in circumvallate papillae		
[^{3}H]CDP-DAG accumulation		
Li$^+$ + carbachol	4	1.32 ± 0.06[a]
Li$^+$ + denatonium	3	1.01 ± 0.05
[^{3}H] Ins-P accumulation		
Li$^+$ + denatonium	3	0.99 ± 0.03
B. InsP$_3$ mass levels in response to denatonium		
Circumvalate papillae		
Control	29	42.9 ± 1.4
10 µM	16	50.0 ± 2.9[b]
1 mM	12	53.9 ± 3.8[c]
Nontaste epithelium		
Control	9	24.7 ± 2.2
1 mM	9	24.4 ± 2.1

Part A: Data are expressed as mean ratio of stimulated over control ± SEM. Concentrations used: LiCl 5 mM; carbachol 1 mM; denatonium 10 µM. No., number of experiments, each in duplicate. Mean control values: [^{3}H]CDP-DAG, 524 cpm per circumvallate papilla; [^{3}H]Ins-P, 4340 cpm per circumvallate papilla. Differs from Li$^+$ samples in absence of carbachol.
Part B: [InsP$_3$] concentrations are expressed as picomoles per milligiam of protein ± SEM. Denatonium concentrations are as indicated. No., total number of samples in three to five experiments, each in triplicate. Differs from control samples in absence of denatonium by two-tailed *t*-test.
[a] $p < 0.01$.
[b] $p < 0.02$.
[c] $p < 0.002$.

quinine at 1 mM concentration and have not observed a change in InsP$_3$ mass levels.

Immunolocalization of CDRK and DRK1 K$^+$ Channels in the Rat Circumvallate Papilla

To produce specific antibodies against CDRK or DRK1 K$^+$ channel, we synthesized peptides to portions of the channel proteins that had completely distinct sequences. Western blots with the anti-CDRK or anti-DRK1 sera reveal correct molecular mass protein bands specifically eliminated by preadsorption with the respective synthetic peptide [21]. In rat circumvallate papillae, DRK1 protein occurs exclusively in taste bud-containing receptor cells (Fig. 3B). The immunoreactivity is specifically blocked by preadsorption of the antiserum with excess DRK1 synthetic peptide (Fig. 3D). In contrast, no CDRK immunoreactivity occurs in taste buds (Fig. 3A). Instead, intense immunoreactivity is apparent in the basal cells of the epithelial layer of the circumvallate papillae. Additionally, some immunoreactivity is evident in cells lining the ducts of von Ebner's glands. Preadsorption with synthetic peptide antigen blocks CDRK immunoreactivity in the basal cells and the ducts of von Ebner's gland (Fig. 3C). Because in taste transduction three of the four major modalities (bitter, sour, sweet) provoke closure of potassium channels in the sensory cells [24], DRK1 localized to taste receptor cells may serve one or more of these modalities.

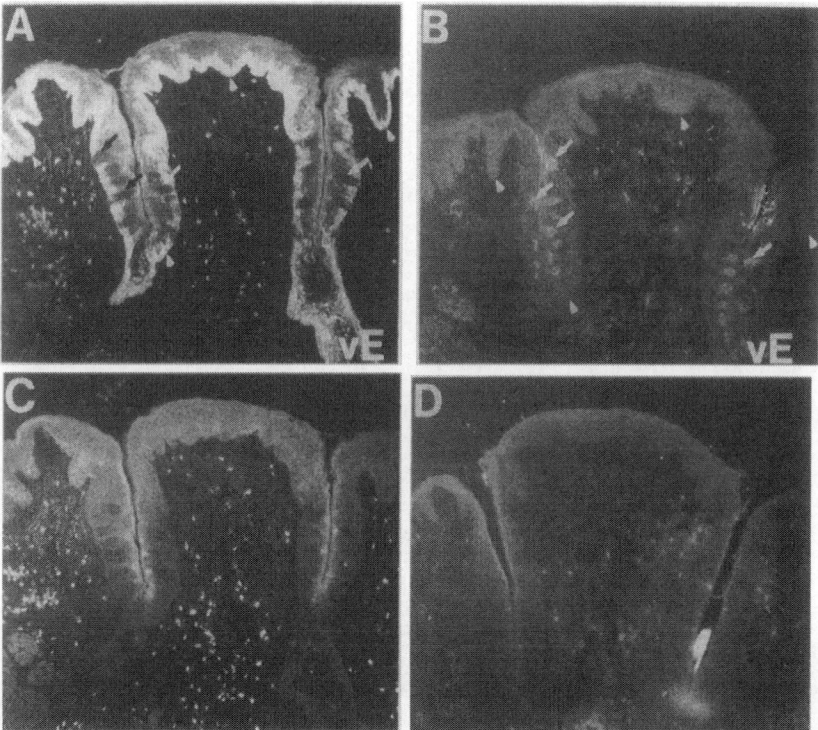

Fig. 3. Immunofluorescent labeling of CDRK and DRK1 K$^+$ channel proteins in the epithelial layer of the rat circumvallate papilla. **A** Anti-CDRK antibody intensely labels the basal cells between the epidermal and dermal layers (*arrowheads*). *Arrows* (black and white) indicate taste buds that are devoid of immunoreactivity in the epidermal layer. The epithelial cells lining von Ebner's gland ducts (*vE*) are also labeled. **B** Anti-DRK1 antibody labels the taste buds (*arrows*) containing taste receptor cells in the circumvallate papilla. Note the absence of labeling of basal cells at the epidermal-dermal junction (*arrowheads*). Preadsorption of anti-CDRK (**C**) and anti-DRK1 (**D**) antibodies with excess specific peptide abolishes immunoreactivity. **A–D**, ×10

Discussion

In our present studies we have demonstrated that the PI system components are highly enriched in the rat circumvallate papilla. This striking association of PI elements to the circumvallate papillae, as well as the elevation of InsP$_3$ mass levels in response to the potently bitter tastant denatonium, suggests a role for InsP$_3$ in bitter taste transduction. The transient enhancement of InsP$_3$ levels in response to denatonium resembles the transient rise in intracellular calcium with denatonium, which is detectable as early as 5 s but which has substantially subsided by 30 to 60 sec [14].

What may be the role of increased InsP$_3$ levels in taste receptor cells? InsP$_3$ mobilizes intracellular stores of Ca^{2+}, and the elevated [Ca^{2+}]$_i$ may play an important role in the control of neurotransmitter release by taste receptor cells, as in neuronal synapses in the brain. Another possible role of increased [Ca^{2+}]$_i$ is in the release of apical secretory vesicles on tastant stimulation, which has been well described [25].

DRK1 is expressed only in taste bud cells, and CDRK is highly expressed by the basal cells in the taste epithelium [26]. The DRK1 channel may be responsible for some of the K$^+$ currents previously described in taste receptor cells, a topic that deserves further study. CDRK appears to be associated with dividing and differentiating cells (unpublished data); hence it may have an important role in cell proliferation.

References

1. Kinnamon S (1988) Trends Neurosci 11:491–496
2. Kinnamon SC, Roper SD (1988a) Chemical Senses 13:115–121
3. McBride D, Roper S (1988) Biophys J 51:11a
4. Schiffman SS, Lockhead E, Maes FW (1983) Proc Natl Acad Sci USA 80:6136–6140
5. Heck GL, Mierson S, DeSimon JA (1984) Science 223:403–405
6. Striem BJ, Pace U, Zehavi U, Naim M, Lancet D (1989) Biochem J 260:121–126
7. Spielman AI, Mody I, Brand JG, Whitney G, MacDonald JF, Salter MW (1989) Brain Res 503:326–329
8. Akabas MH, Dodd J, Al-Awqati Q (1988) Science 242:1047–1050
9. Berridge MJ (1987) Annu Rev Biochem 56:159

10. Tonosaki K, Funakoshi M (1988) Nature 331:354–356
11. Avenet P, Hofmann F, Lindemann B (1988) Nature 331:351–354
12. Kennedy JG (1972) Arch Oral Biol 17:1197–1207
13. Guth L (1957) Anat Rec 128:715–731
14. Verma A, Ross CA, Verma D, Supattapone S, Snyder SH (1990) Cell Regulation 1:781–790
15. Hwang PM, Verma A, Bredt DS, Snyder SH (1990) Proc Natl Acad Sci USA 87:7395–7399
16. Berridge MJ, Downes CP, Hanley MR (1982) Biochem J 206:587–595
17. Godfrey PP (1989) Biochem J 258:621–624
18. Bredt DS, Mourey RJ, Snyder SH (1989) Biochem Biophys Res Commun 159:976–982
19. Hwang PM, Glatt CE, Breddt DS, Yellen G, Snyder SH (1992) Neuron 8:473–481
20. Frech GC, VanDongen AMJ, Schuster G, Brown AM, Joho RH (1989) Nature 340:642–645
21. Hwang PM, Fotuhi M, Bredt DS, Cunningham AM, Snyder SH (1992) J Neurosci (in press)
22. Worley PF, Baraban JM, Supattapone S, Wilson VS, Snyder SH (1987) J Biol Chem 262:12132–12136
23. Bredt DS, Snyder SH (1989) Soc Neurosci Abstr 15:1004
24. Avenet P, Kinnamon SC (1991) Curr Opin Neurobiol 1:198–203
25. Roper SD (1992) J Neurosci 12(4):1127–1134
26. Hwang PM, Cunningham AM, Peng YW, Snyder SH (1992) Neuroscience (in press)

Immunohistochemical Localization of G-Proteins in Mouse Taste Buds

ANDREW I. SPIELMAN[1,2], STEVEN DEMYER[1], GLORIA TURNER[1], and JOSEPH G. BRAND[2]

Key words. Immunohistochemistry—G-proteins

Taste is a specialized sensory system that provides us with an important screening ability to choose between useful or harmful compounds. Of particular importance is bitter taste for its low detection threshold and importance in screening potentially toxic compounds. Bitter taste is thought to be transduced by several mechanisms, including mediation through specific receptors for certain compounds [1]. One such substance is the bitter stimulant sucrose octaacetate (SOA) that interacts with a putative SOA-receptor in mouse taste tissue. This receptor is thought to be coupled to a GTP-binding regulatory protein. Receptor-mediated signal transduction in a variety of cells is facilitated by such G-proteins. Some G-proteins, in turn, stimulate a membrane-associated enzyme, phospholipase C, generating the second messenger inositol-1,4,5-trisphosphate (IP_3) [2]. At least 20 different G-proteins have been identified so far, including those specific to olfaction (Golf) and taste (gustducin) [3,4]. This diversity demonstrates the importance of unique G-proteins in signal processing. Sweet and some bitter compounds have also been proposed to be G-protein mediated [1,5]. However, the cellular location and specific function of these and other G-proteins in mouse taste tissue are not known. The purpose of this investigation was to identify the cellular localization and relative abundance of various G-protein alpha subunits in mouse taste tissue.

Perfused mouse tongues were frozen or paraffin-embedded, sectioned at 10 µm, and incubated with antisera to the α subunit of Gi1–2, Gi3, Go, Gs, gustducin, and Gq at dilutions ranging from 1:20 to 1:500. Immunohistochemistry demonstrated that only Gαi1–2 and Gαq showed significant immunoreactivity in the tested strains. Immunoreactivity was seen either in the apical region or over the entire surface of the taste bud. Antisera to Gαs and Gαo showed very weak immunoreactivity in both strains of mice tested (B6.SW and C57Bl/6J), while gustducin and Gαi3 displayed no immunoreactivity in B6.SW mice. Western blotting membrane preparations from both strains, and using the same α subunit-specific anti-sera, confirmed the abundance of Gαi1–2 and the absence of Gαi3 in mouse taste tissue.

The absence of gustducin in mouse (B6.SW) taste tissue is surprising, considering its abundance in rat taste tissue. Neither paraffin-embedded nor frozen sections showed immunoreactivity under the current conditions.

Gαi1 or Gαi2 is a pertussis toxin-sensitive, cholera toxin-insensitive G-protein. Recent evidence suggests that the signal transduction of sucrose octaacetate may utilize a similar G protein [1]. With the exception of sweet taste, no other taste quality has been associated with G-proteins. Although no functional correlation has been shown between Gi1–2 and SOA taste, the abundance of Gαi1–2 in mouse taste tissue suggests an important local function, perhaps one associated with bitter taste.

Acknowledgments. We thank Dr. G. Whitney for the B6.SW mice and Dr. R. Margolskee for anti-serum to gustducin. This study was supported by funds from VA, NIH, and NYU.

References

1. Spielman AI, Huque TI, Whitney G, Brand JG (1992) The diversity of bitter taste signal transduction mechanisms. In: Cory D, Roper S (eds) Sensory transduction. The Rockefeller University Press, New York, pp 307–324
2. Spielman AI, Huque T, Breer H, Boekhoff I, Whitney G, Brand JG (1993) Rapid kinetic measurements of bitter stimulus-induced IP3 in mouse taste tissue. Chem Senses 18:632–633
3. Jones DT, Reed RR (1987) Golf: An olfactory neuron-specific G-protein involved in odorant signal transduction. Science 244:790–795
4. McLaughlin SK, McKinnon PJ, Margolskee RF (1992) Gustducin: A taste cell-specific G-protein closely related to the transducins. Nature 357:563–569
5. Striem BJ, Pace U, Zehavi U, Naim M, Lancet D (1989) Sweet tastants stimulate adenylate cyclase coupled to GTP-binding proteins in rat tongue membranes. Biochem J 260:121–126

[1] Basic Science Division, New York University College of Dentistry, 345E 24th Street, New York, NY 10010, USA
[2] Monell Chemical Senses Center, 3500 Market Street, Philadelphia, PA 19104, USA

Kinetics of Second Messenger Formation in Taste Transduction

Hajimu Uebayashi, Kousei Miwa, and Keiichi Tonosaki[1]

Key words. Taste cell—Radioimmunoassay

Taste signal transduction occurs in the receptor membrane of taste sensory neurons [1]. Several lines of electrical and biochemical evidence indicate the involvement of adenylate cyclase or guanylate cyclase in vertebrate taste transduction [2–7]. The findings of some studies indicate that phospholipase C may play a role in vertebrate bitter taste transduction [6,7]. One important critierion for a candidate second messenger of taste transduction is that its formation must be accompanied by the onset of the taste stimulus. To investigate the mechanism of taste transduction, we developed a preparation of mouse taste cells that can be studied with techniques designed for radioimmunoassay measurement. This study demonstrated that taste stimuli can activate alternative second messenger pathways in taste cells. Here we report that a sweet taste stimulus raised the intracellular cyclic guanosine monphosphate (c-GMP) concentration in taste cells, and a bitter one increased the inositol 1, 4, 5 trisphosphate (IP_3) concentration. The sensation of taste produced by taste stimuli is produced by a chemoelectrical signal transduction process, which, upon specific interaction of the taste stimulus with the receptor cell, leads to membrane depolarization. Although there is much electrical and biochemical evidence for the involvement of second messengers in sensory receptor cells, little is known about the physiology of the cells or the molecular mechanism involved in the sensory transduction process. To address the problem of whether the type of second messenger system that is activated can be correlated to taste qualities we investigated distilled water (DW), 0.5 M sucrose, 0.01 M HCl, 0.02 M quinine, and 0.1 M NaCl to determine their potential to activate the cyclic adenosine monophosphate (c-AMP), the c-GMP, and the IP_3 pathways.

Adult male mice (Slc:ICR) were deeply anesthetized with pentobarbital sodium and tracheotomized. The tongue was continuously stimulated with a taste stimulus solution, the anterior 3- to 4-mm area was the immediately removed and immediately frozen. Each taste stimulus induced changes in second messenger concentrations in the tongue. We examined levels of c-GMP, c-AMP, and IP_3 with radioimmunoassay kits; that for c-GMP being supplied by Yamasa (Chiba, Japan), and those for c-AMP and IP_3 being supplied by Amersham (Arlington Heights, Ill.) (c-AMP[^{125}I]; code RPA 509, and IP_3[^{3}H]; code TRK 1000). The measurements were made according to the manufacturers' guidelines. Distilled water (DW) was always used as a control taste stimulus (c-GMP: 1.7 ± 0.6 fmol/mg, $n = 7$, $K_d = 54.3 \pm 8.7$ fmol; c-AMP: 350.4 ± 50.0 fmol/mg, $n = 4$, $K_d = 5.3 \pm 0.3$ pmol; IP_3: 774.6 ± 110.0 fmol/mg, n = 6, $K_d = 3.0 \pm 0.3$ pmol; values are mean $\pm$ SD). Upon application of sucrose, the concentration of c-GMP increased (2.87 ± 1.0 fmol/mg). In contrast, when DW (1.7 ± 0.6 fmol/mg), HCl (1.6 ± 0.5 fmol/mg), quinine (1.7 ± 0.4 fmol/mg), and NaCl (1.7 ± 0.5 fmol/mg), were applied, there were no significant differences in c-GMP concentrations among these agents. Upon application of quinine, the concentration of IP_3 increased (1705.4 ± 330.0 fmol/mg). The concentration of IP_3 also increased when sucrose (1136.0 ± 237.5 fmol/mg), HCl (1428.0 ± 337.5 fmol/mg), NaCl (1215.6 ± 300.0 fmol/mg), and distilled water (774.6 ± 110.0 fmol/mg) were applied to the tongue. However, these changes were always smaller than those produced by quinine. Changes in the concentration of c-AMP produced by the application of distilled water and the four individual taste stimuli were not significantly different for the different stimuli (DW: 350.4 ± 50.0 fmol/mg; sucrose: 379.7 ± 110.0 fmol/mg; HCl: 343.6 ± 70.0 fmol/mg; quinine: 317.6 ± 35.0 fmol/mg; NaCl: 321.9 ± 70.0 fmol/mg).

The tongue preparation contained large amounts of muscle, nerves and other tissues. However, muscles and nerves have an abundant supply of c-AMP and IP_3. To overcome this problem, we removed single fungiform papillae from the tongues (50–80 fungiform papillae from two to three mice), using special hand-made forceps. The fungiform papilla preparation thus obtained contained taste cells and little other tissue. The tongue was continuously stimulated with a taste stimulus solution throughout the operation. When a papilla was iso-

[1] Department of Oral Physiology, School of Dentistry, Asahi University, 1851 Hozumi, Hozumi-cho, Motosu, Gifu, 501-02 Japan

lated, it was immediately frozen. We were always careful to ensure that the receptor cells in the isolated papillae did not touch blood or other liquids, except for the taste solution, before they were frozen. The concentration of c-GMP was significantly increased with sucrose (DW: 46.9 ± 8.1 fmol/mg; sucrose: 69.1 ± 21.9 fmol/mg; HCl: 52.5 ± 10.6 fmol/mg; quinine: 52.5 ± 10.6 fmol/mg; NaCl: 46.9 ± 5.0 fmol/mg ($n = 5$, $K_d = 65.5 \pm 2.5$ fmol)). The concentration of IP$_3$ was markedly increased by quinine (DW: 1033.7 ± 225.0 fmol/mg; sucrose: 982.2 ± 237.5 fmol/mg; HCl: 1177.9 ± 155.0 fmol/mg; quinine: 1797.8 ± 245.0 fmol/mg; NaCl: 1093.8 ± 112.5 fmol/mg ($n = 5$, $K_d = 2.4 \pm 0.7$ pmol)). In contrast, the concentration of c-AMP was decreased by HCl and NaCl (DW: 24.8 ± 3.0 fmol/mg; sucrose: 31.3 ± 6.0 fmol/mg; HCl: 13.4 ± 1.3 fmol/mg; quinine: 20.9 ± 5.5 fmol/mg; NaCl: 15.2 ± 4.8 fmol/mg ($n = 6$, $K_d = 24.6 \times 3.4$ fmol)). These results indicate that c-GMP mediates sweet taste transduction, and that IP$_3$ mediates bitter taste transduction. Salt and sour taste transduction may be mediated by anti-c-AMP system(s) or by other transduction mechanism(s). The different magnitudes of c-GMP, c-AMP, and IP$_3$ formation induced by various tastes cannot readily be explained; however, it appears that taste receptor cells display a certain taste specificity and that the number of cells stimulated by a particular taste is different. This problem can be solved based on the data available here. We found that the concentrations of c-GMP were not different in the tongue and fungiform papilla preparations.

The weight of a tongue preparation was 41.66 mg ($n = 48$); this preparation had about 40 fungiform papillae that were visually countable under binocular observation. There are about 20 taste cells and almost no other tissues in a fungiform papilla. We compared the concentration ratios of each second messenger in tongue and fungiform papilla preparations. There was a good correlation between the two experimental values for c-GMP concentrations, whereas the concentration ratios for c-AMP and IP$_3$ were much larger than 1.0. This finding suggests that c-GMP is concentrated in the fungiform papilla (taste cell) and that IP$_3$ is mainly distributed in the papilla. In contrast, c-AMP is distributed not only in the papilla but also in other tissues. As stated above, the tongue preparation contained a large amount of muscle, nerves, and other tissues. Since there are high concentrations of c-AMP and IP$_3$ in the muscles and nerves, these results indicate that, when we study the concentrations of c-AMP or IP$_3$ in a receptor cell, we must pay attention either to the organ itself or to its surrounding tissues.

This study provides evidence for the notion that different second messengers may be involved in taste transduction in the mouse. The electrophysiological findings show that the cyclic nucleotide (c-GMP) may mediate sweet taste transduction [5] and that IP$_3$ may mediate bitter taste transduction [6,7]. Hitherto, specific changes in the concentration of putative intracellular messenger molecules have not been demonstrated in response to taste. The findings here indicate that sucrose was the only agent to increase the concentration of c-GMP, in the fungiform papilla and that quinine was the only one to increase the concentration of IP$_3$. Thus, at least two second messenger systems appear to be involved in taste transduction.

Acknowledgments. This work was supported in part by a Grant-in-Aid for Scientific Research from the Ministry of Education, Science, and Culture of Japan, and a Grant-in-Aid for Scientific Research from Miyata.

References

1. Beidler LM (1954) A theory of taste stimulation. J Gen Physiol 38:133–139
2. Beidler LM, Tonosaki K (1985) Multiple sweet receptor sites and taste theory. In: Pfaff DW (ed) Taste, olfaction, and the central nervous system. Rockefeller University Press, New York, pp 47–64
3. Roper SD (1989) The cell biology of vertebrate taste receptors. Ann Rev Neurosci 12:329–353
4. Avenet P, Hofmann F, Lindemann B (1988) Transduction in taste receptor cells requires cAMP-dependent protein kinase. Nature 331:351–354
5. Tonosaki K, Funakoshi M (1988) Cyclic nucleotides may mediate taste transduction. Nature 331:354–356
6. Tonosaki K (1990) Taste transduction mechanism. Neuroscience Res [Suppl] 12:s63–s72
7. Akabas MH, Dodd J, Al-Awqati Q (1988) Mechanism of transduction of bitter taste in rat taste bud cells. Science 242:1047–1050

Selective Suppression of Bitter Taste Responses by Lipoprotein

Yoshihisa Katsuragi[1] and Kenzo Kurihara[2]

Key words. Inhibitor of bitter taste—Lipoprotein

The development of a method to mask bitter taste is required in the fields of food and pharmaceutical sciences, and would be welcomed in the field of receptor physiology. A specific inhibitor for bitter taste would be most useful for this purpose, but, thus for, no such inhibitor has been available. In this study, we found that a lipoprotein, PA-LG, made of phosphatidic acid (PA) and β-lactoglobulin (LG) completely suppressed taste responses in the frog and taste sensation in humans to all bitter substances examined, without affecting responses to other taste stimuli [1]. Both PA and LG originate from foods, being prepared from soybeans and milk, respectively. Hence PA-LG can be safely used to mask bitter taste in foods and drugs.

We prepared the PA-LG lipoprotein by homogenizing a suspension of PA and LG in water and by freeze-drying the homogenate. The powder was easily dissolved in water, while PA alone was not soluble in water. First we examined the effects of PA-LG on taste responses by measuring the activity of the glossopharyngeal nerve. To exclude the possibility that PA-LG directly interacts with bitter substances, the tongue was pretreated with 0.3% PA-LG for 10 min and then stimulated by bitter substances dissolved in water. The responses to all eight bitter substances examined were greatly suppressed by PA-LG, while the responses to NaCl, acetic acid, L-alanine, and galactose were not suppressed. We also examined the effects of lipoproteins made of LG and other lipids, such as phosphatidylcholine, triacylglycerol, and diacylglycerol, on responses to bitter substances. These complexes did not suppress responses to the bitter substances.

We plotted the relative magnitude of frog taste nerve responses to quinine hydrochloride, L-leucine, caffeine, papaverine hydrochloride, and NaCl as a function of PA-LG concentration. Responses to the bitter substances were decreased with increased concentrations (percentages) of PA-LG, reaching a spontaneous level at 1%, while the response to NaCl was unaffected by all concentrations of PA-LG examined. These results indicated that PA-LG at high concentrations (e.g., 1%) completely suppressed the bitter responses by fully occupying the receptor sites. The data obtained with four bitter substances roughly followed a single curve, suggesting that if there are multiple receptor sites for bitter substances, PA-LG binds uniformly to these receptor sites.

In the above experiments, the tongue was pretreated with a PA-LG solution and then stimulated with a solution containing a bitter substance and no PA-LG. In this method, PA-LG bound to the receptor sites is detached during stimulation with a bitter substance, since there is no PA-LG in the stimulating solution. To obtain more complete suppression of the bitter responses, the tongue was treated with a solution containing a bitter substance for 30 s, and was subsequently stimulated with a solution containing a bitter substance and PA-LG. This method brought about more complete suppression of the bitter responses at lower concentrations of PA-LG.

We also examined the effects of PA-LG on human taste sensation to bitter substances by a psychophysical method. For usefulness in practical application, the tongue was not pretreated with PA-LG, but subjects tasted bitter substances dissolved in solutions containing PA-LG of varying concentrations. Standard solutions containing different concentrations of bitter substances and no PA-LG were prepared and the subjects were asked to choose a standard solution whose bitterness was equivalent to that of a test solution. While the salty taste of NaCl was unaffected, the bitter tastes of quinine hydrochloride, papaverine hydrochloride, and isoleucine were greatly suppressed by PA-LG, the degree of suppression of the bitter taste varying with the species of bitter substance. We emphasize that PA-LG is useful not only for masking the bitter taste of drugs and foods but also for exploring the characteristics of receptor sites for bitter taste.

[1] Kashima Research Laboratories, Kao Co., 20 Higashi-fukashiba, Kamisu-machi, Kashima-gun, Ibaraki, 314-02 Japan
[2] Faculty of Pharmaceutical Sciences, Hokkaido University, N-12 W-6, Kita-ku, Sapporo, 060 Japan

References

1. Katsuragi K, Kurihara K (1993) Specific inhibitor for bitter taste. Nature 365:213–214

Where and What Is the Target of the Sweet Taste Inhibitory Peptide, Gurmarin?

Akiko Miyasaka[1], Toshiaki Imoto[1], and Yuzo Ninomiya[2]

Key words. Sweet taste—Gurmarin

Substances that strongly inhibit the response to sweet taste stimuli could be potential tools to investigate the mechanism of sweet taste reception. Gymnemic acids (GA) a mixture of triterpene glucuronides found in the leaves of the Indian plant *Gymnema sylvestre*, have been thought to be suitable for that purpose. GA, however, does not have any significant effect on the taste responses of mammals such as rats, rabbits, and pigs [1]. We have recently found in the same plant a new sweet inhibitory peptide, gurmarin; this consists of 35 amino acids with a molecular weight of 4209 [2]. In contrast to GA, gurmarin strongly suppresses sweet responses in the rat chorda tympani, while it does not affect taste sensation in humans [3].

In this study, we performed electrophysiological and biochemical experiments to investigate the specificity of the effect of gurmarin on various taste stimuli in more depth; we also attempted to identify the proteins in the taste tissue that interact with gurmarin.

The response to 0.5 M sucrose of the rat chorda tympani was suppressed to about 20% of the original response by immersing the tongue in a solution of 20 µg/ml of gurmarin for 10 min, and the effect was still apparent even at 2 µg/ml (5×10^{-7} M). Once the sweet responses had been inhibited, they required at least 3 h for complete recovery; this was longer than the time required for other sweet suppressing substances, such as gymnemic acid, whose effect in humans disappeared within 1 h, and zizyphin, whose effect disappeared within 30 min.

Gurmarin suppressed not only the responses of various sugars, such as sucrose, glucose, fructose, and maltose but also those of sweet amino acids, such as glycine, alanine, D-phenylalanine, D-tryptophan, and the artificial sweetener, saccharin. On the other hand, the responses to L-histidine, L-lysine, and L-isoleusine, which elicit a rather bitter taste in humans, were barely affected by gurmarin. Thus, the inhibitory action of gurmarin is quite specific to the sweet taste response, suggesting a strong interaction of this agent with the sweet taste receptor or proteins that are closely related to the mechanism responsible for the transduction of sweet taste.

In order to examine whether there are specific proteins in the taste tissue that interact with gurmarin, we applied a protein mixture obtained from the homogenate of the tongue epithelium onto a gurmarin-linked Sepharose column. The bound fraction eluted from the column with 0.1 M glycine-HCl buffer solution (pH 2.2) was then analyzed by sodium dodecylsulfate-polyacrylamide gel electrophoresis (SDS-PAGE), followed by Western blotting onto a nitrocellulose membrane. After successive incubation of the membrane with biotinylated gurmarin and horseradish peroxidase (HRP)-linked streptavidin, the enzymic activity of bound HRP was detected by utilizing high sensitivity chemoluminescence.

Several proteins with molecular weights of around 70 to 80 kDa were detected in the dorsal epithelium of the rat tongue, whereas they were rarely seen in other tissues, such as tongue muscle, small intestine, liver, and salivary gland. Moreover, they seemed to be localized in the anterior part of the tongue, suggesting that they arose from the fungiform papilla. Further characterization of these proteins, together with a histochemical study, is now in progress to clarify the target of gurmarin.

References

1. Glaser D, Hellekant G, Brouwer JN, van del Wel H (1984) Effects of gymnemic acid on sweet taste perception in primates. Chem Senses 8:367–374
2. Kamei K, Takano R, Miyasaka A, Imoto T, Hara S (1992) Amino acid sequence of sweet taste-suppressing peptide (gurmarin) from the leaves of *Gymnema sylvestre*. J Biochem 111:109–112
3. Imoto T, Miyasaka A, Ishima R, Akasaka K (1991) A novel peptide isolated from the leaves of *Gymnema sylvestre*. I. Characterization and its suppressive effect on the neural responses to sweet taste stimuli in the rat. Comp Biochem Physiol A 100:309–314

[1] Department of Physiology, Faculty of Medicine, Tottori University, Nishimachi, Yonago, Tottori, 683 Japan
[2] Department of Oral Physiology, Asahi University School of Dentistry, Hozumi, Motosu, Gifu, 501-02 Japan

Effects of Various Liposomes on Frog Taste Nerve Responses to Odorants and Taste Stimuli

TAKASHI KUMAZAWA, NOBUHIKO HAYANARI, JUN TANAKA, and HARUMICHI SIMIZU[1]

Key words. Liposomes—Frog taste nerve response

In general, it is considered that olfactory responses to various odor stimuli and taste responses for amino acids, sugars, and bitter substances are induced by the binding of these stimuli to specific receptor proteins. It has also been reported that odorant and bitter substances elicit depolarization in liposomes [1,2]. Moreover, odorants elicit these responses not only in the olfactory but also in the taste system [3]. These results suggest that the interaction of these stimuli with lipid layers of the receptor membranes may also play an important role in the reception of odor and taste. However, whether lipids actually play a role in odor and taste reception in the olfactory and taste systems is still unknown. In this study, we measured the effects of liposomes on frog taste nerve responses to odorants and taste stimuli; this was done in adult bullfrogs, *Rana catesbeiana*. The glossopharyngeal nerve impulses were amplified and integrated with a time constant of 0.3 s. Stimulating solutions were dissolved in deionized water. Azolectin, phosphatidylserine, and phosphatidylethanolamine were used in this study; liposomes were prepared essentially as described previously [2]. The dried lipid in the flask was dispersed in 100 mM KCl solution containing 5 mM Hepes-NaOH buffer (pH 7.3). The lipid suspension was sonicated, and then centrifuged. The supernatant was used for the experiments. The frog tongue was bathed in Ringer's solution (110 mM NaCl, 5.4 mM KCl, 1.0 mM $CaCl_2$, 1.8 mM $MgCl_2$, and 5 mM Hepes-NaOH, pH 7.3) containing liposomes (2 mg lipids/ml) for 10 min, and was then washed out with deionized water for 1 min. After the response was adapted to the spontaneous level, the response of the taste nerve to the stimulus was measured.

For taste stimuli, we found that although treatment of the tongue with azolectin liposomes had little or no effect on the taste nerve responses to salts and sucrose, responses to quinine and L-arginine were enhanced about two to three times. The enhancement of responses to quinine and L-arginine was decreased by treatment with phosphatidylserine or phosphatidylethanolamine added to the azolectin liposomes. Saturation levels of responses to quinine before and after treatment with azolectin liposomes seemed to be different.

For odor stimuli, we found that the response to 0.2 mM *n*-nonanol was enhanced by treatment with azolectin liposomes, this response was greatly enhanced, about 13-fold, by treatment with phosphatidylserine added to the azolectin liposomes. Saturation levels of responses to *n*-nonanol before and after treatment with liposomes consisting of azolectin and phosphatidylserine (10:1) also seemed to be different. The effects of the changed lipid composition of the liposomes on the taste nerve response to various odorant stimuli were greater than these effects on taste stimuli and varied greatly among odorants.

These results suggest that some interaction actually occurs between receptors and lipids in the reception of odor and taste stimuli. The mechanism responsible for these effects is unknown; however, it is conceivable that these effects of lipids may be related to hydrophobic substances such as odorants and bitter substances.

References

1. Nomura T, Kurihara K (1987) Liposomes as a model for olfactory cells: Changes in membrane potential in response to various odorants. Biochemistry 26:6135–6140
2. Kumazawa T, Nomura T, Kurihara K (1988) Liposomes as a model for taste cells: Receptor sites for bitter substances and mechanism of membrane potential changes. Biochemistry 27:1239–1244
3. Kashiwagura T, Kamo N, Kurihara K, Kobatake Y (1977) Responses of the gustatory receptors to various odorants. Comp Biochem Physiol 56C:105–108

[1] Department of Environmental Engineering, Saitama Institute of Technology, 1690 Fusaiji, Okabe, Saitama, 369-02 Japan

Molecular Modelling Approach to the Structure of Taste Receptor Sites

NICOLAS FROLOFF, ANNICK FAURION, and PATRICK MAC LEOD[1]

Key words. Structure-taste relationships—Ligand similarity searching

The peripheral taste system (PTS) is a powerful chemical sense to discriminate a wide variety of molecular stimuli. About ten dimensions are needed to give a convenient description of the human taste space built from intensity estimate psychophysical data (Faurion, see elsewhere in this volume), setting a lower limit to the number of independent information channels used by the PTS to probe its chemical environment. We suggested the existence of multiple low affinity and low specificity receptor sites to account for the observed PTS discrimination function [1]. Within this hypothesis, two molecules are likely to bind to at least one common receptor site when biologically correlated. We exemplified this hypothesis on 14 rigid or conformationally restricted tastants: picric, 3-aminobenzoic, 3-nitrobenzenesulfonic, 2- and 3-nitrobenzoic acids, saccharin, cyclamate, perillartine, dulcin, 1-propoxy-2-amino-4-nitrobenzene, glycine, L-threonine, caffeine, and theophylline. Using molecular modelling, we developed a blind-searching methodology to extract common physicochemical patterns which could be complementary of taste receptor sites.

Geometries were retrieved from the Cambridge Crystallographic Database [2] or optimized in the MM2 force field [3]. Potential H-bonding atoms (HBAs) were specifically pointed to. Connolly solvent accessible surfaces were computed [4], and hydrophobic regions were defined as the sets of Connolly surface dots derived from nonpolar atoms. We developed an algorithm to split molecular models into many easily comparable quasi-planar fragments. The 14 molecular structures gave rise to 240 fragments. Most fragments contained 0–4 HBAs, and their hydrophobic patch area averaged $30 \text{\AA}^2$. The degree to which two fragments shared physicochemical properties was expressed by a correspondence index accounting for hydrogen bonding and hydrophobic interaction properties. Unrealistic or intramolecular redundant fragments were removed. The pairwise comparison of the 75 remaining fragments resulted in a list of 2775 fragment pair correspondence indices. From this list, we disclosed 12 independent clusters of fragments displaying similar HBA arrangements within $1 \text{\AA}$ resolution. These fragment clusters were considered to represent 12 potential taste stimulating patterns. A combinatorial analysis of any number among these 12 patterns was performed to determine which pattern subset best represented the structural intermolecular similarities compared to biological ones. For that purpose, a psychophysical taste distance matrix between molecules was used as a reference (Faurion et al., 1994, manuscript in preparation). An optimal subset of seven patterns significantly accounted for psychophysical supraliminal data. The hierarchical clustering of structural distances well reproduced the clustering of molecules found in biology, except for a weaker discrimination of molecules by the model, due to a non-exhaustive sampling of tastants. We generalized Kier's proposal [5] to a system of seven patterns having approximately three physicochemical features, satisfying both chiral discrimination properties and physical requirements of low affinity and low specificity of the putative taste receptor sites.

The seven patterns characterized and validated in our study are expected to give useful information for the direct three-dimensional modelling of taste receptors. They may also serve as a guide to the rational design of tastants with desired sensory properties. A similar methodology could also be applied in olfaction.

References

1. Faurion A, Saito S, Mac Leod P (1980) Sweet taste involves several distinct receptor mechanisms. Chem Senses 5:107–121
2. The Cambridge Crystallographic Database, University of Cambridge, U.K.
3. Allinger NL (1977) Conformational analysis. 130. MM2. A hydrocarbon force field utilizing V_1 and V_2 torsional terms. J Am Chem Soc 99:8127–8134
4. Connolly ML (1983) Analytical molecular surface calculation. J Appl Cryst 16:548–558
5. Kier LB (1972) A molecular theory of sweet taste. J Pharm Sci 61:1394–1397

[1] Laboratoire de Neurobiologie sensorielle, Ecole Pratique des Hautes Etudes, 1, Avenue des Olympiades, F-91305 Massy, France

Discrimination of Taste Between Polysaccharides and Common Sugars in Rats and Mice

Noritaka Sako, Takashi Kikuchi, Tsuyoshi Shimura, and Takashi Yamamoto[1]

Key words. Carbohydrate receptors—Sweet taste suppressants

Starch is one of the most widely available natural energy sources for animals. However, starch-derived polysaccharides are described as being tasteless, a common assumption. However, recent studies have suggested that rats can discriminate between the taste of polysaccharides, such as Polycose (Ross Lab., Columbus, OH) and common sugars [1–4]. The present study aimed to examine whether or not mice, as well as rats, could discriminate between the taste, of the two types of carbohydrates.

Adult male Wistar rats, A/J-CrSlc mice, BALB/cCrSlc mice, C3H/He-CrSlc mice, C57BL/6CrSlc mice, and DBA/2CrSlc mice were used. In behavioral experiments, the two-bottle preference method and conditioned taste aversion (CTA) paradigm were used. In 48-h two-bottle preference tests, all five strains of mice showed a typical preference for Polycose at concentrations of 1% and more, suggesting that all five strains of mice, as well as rats, have sensitivity to Polycose and common sugars.

In CTA tests, Wistar rats conditioned to 0.5 M sucrose clearly avoided common sugars such as sucrose, fructose, glucose, and maltose, but did not reject other taste stimuli, including 20% Polycose. The Wistar rats conditioned to 20% Polycose rejected 20% Polycose, but not the other taste stimuli, including 0.5 M sucrose. This suggests that, in Wistar rats, the taste of Polycose is different from the tastes of common sugars.

Mice showed strain differences in CTA tests. BALB/c, C57BL/6, and DBA/2 mice conditioned to 20% Polycose did not avoid 0.5 M sucrose, whereas A/J and C3H/He mice avoided it. However, the reverse was not true, i.e., BALB/c, C3H/He, and DBA/2 mice conditioned to 0.5 M sucrose avoided 20% Polycose, but C57BL/6 mice did not. Some A/J mice avoided it. Probably, mice have more complex carbohydrate receptors than rats.

In electrophysiological experiments, the effects of sweet taste suppressant (2% pronase E [5], 200 mg/ml gurmarin [6]) treatments on the magnitude of taste responses of the chorda tympani were analyzed. In C3H/He and C57BL/6 mice, the relative magnitude of the response ($NH_4Cl = 1.0$) to 20% Polycose was smaller than that to 0.5 M sucrose. However, in other strains of mice and in the Wistar rats, no significant difference was detected. Treatment of the tongue with pronase E suppressed responses to all the carbohydrate solutions, including Polycose, suggesting that the Polycose receptor consists of a protein. Treatment with gurmarin suppressed responses to 0.5 M sucrose in both rats and mice, except for BALB/c and DBA/2 mice, but did not suppress Polycose responses in these two species.

The present study shows that the taste of Polycose and common sugars is not so clearly discriminated in mice as in rats, although there are remarkable strain differences.

Acknowledgments. The authors wish to express their thanks to Professor A. Sclafani, New York State University, for his gift of Polycose, and to Dr. Imoto, Tottori University, for his gift of gurmarin. Supported by grants from the Japanese Ministry of Education (0330402, 04454465, 05267101).

References

1. Nissenbaum JW, Sclafani A (1987) Qualitative differences in polysaccharide and sugar tastes in the rat: A two-carbohydrate taste model. Neurosci Biobehav Rev 11:187–196
2. Sclafani A (1987) Carbohydrate taste, appetite, and obesity: An overview. Neurosci Biobehav Rev 11:131–153
3. Sclafani A, Hertwig H, Vigorito M (1987) Influence of saccharide length on polysaccharide appetite in the rat. Neurosci Biobehav Rev 11:197–200
4. Sclafani A, Nissenbaum WJ, Vigorito M (1987) Starch preference in rats. Neurosci Biobehav Rev 11:253–262
5. Hiji Y (1975) Selective elimination of taste responses to sugars by proteolytic enzymes. Nature 256:427–429
6. Imoto T, Miyasaka A, Ishima R, Akasaka K (1991) A novel peptide isolated from the leaves of *Gymnema sylvestre*—I. Characterization and its suppressive effect on the neural responses to sweet taste. Comp Biochem Physiol A 100(2):309–314

[1] Department of Behavioral Physiology, Faculty of Human Sciences, Osaka University, 1-2 Yamadaoka, Suita, Osaka, 565 Japan

Gustatory Responses of the Greater Superficial Petrosal Nerve to L- and D-Amino Acids Applied on the Soft Palate in the Rat

SHUITSU HARADA, SHOJI ENOMOTO, and YASUO KASAHARA[1]

Key words. Taste—Soft palate

Numerous taste buds are distributed on the soft palate in mammals, and these structures are innervated by the greater superficial petrosal nerve (GSP). Although the function of these taste buds remains unclear because of the operational difficulties in accessing this nerve, it has been shown that responses to sweet substances recorded from the GSP in rats were specifically robust [1], which suggests that such substances produce a "sweet" sensation in the animal. Also, in the hamster, we have revealed that sugars elicited two to three times larger responses in the GSP than in the chorda tympani (CT) nerve [2]. However, many amino acids also produce a sweet sensation in humans, and preferable in mammals. Accordingly, in this experiment, to determine the gustatory responses of the GSP to various L- and D-amino acids in the rat, we recorded these and compared them with the responses of the CT.

The experimental procedures were essentially as described before [2]. Relative response magnitudes were calculated, with the response to 0.1 M NaCl being taken as 100. The stimuli were concentration series of NaCl, HCl, quinine HCl, sucrose, 0.1 M L- and D-amino acids, and 0.01 M saccharin Na.

The magnitude of the total response [2] for each of the four basic taste stimuli was compared in the GSP and CT; we found that both the phasic and tonic responses to sucrose were larger in the GSP than in the CT. In the CT, basic amino acid HCl salts (L-Arg-HCl, L-Lys-HCl, and L-His-HCl), at 0.1 M, produced robust phasic and succeeding tonic responses that were quite similar to those for 0.1 M NaCl [3]. The responses to the D-basic amino acid HCl salts were similar to each of the L-form amino acids, and differences between the two forms were not significant. However, in the GSP, D-His-HCl produced a significantly larger response than L-His-HCl. The free base D-His was also more stimulatory than L-His. Multiple unit responses of fine strands of the GSP showed that the units that responded to

sucrose responded specifically to free base His. All eleven L-neutral amino acids (Asn, Ser, Gln, Ala, Met, Thr, Phe, Leu, Val, Pro, at 0.1 M, and Trp at 0.05 M), produced only 30%–40% of relative response magnitudes in both the CT and GSP, and there was strongly significant correlation ($P < 0.01$) between the two nerves ($r = 0.7290; n = 11$). In the CT, each response to the D-neutral amino acids (Asn, Ser, Gln, Ala, Met, Thr) was significantly smaller than that to each L-amino acid, only D-Trp producing a significantly larger response than that to L-Trp in this nerve. In the GSP, in contrast, most of the D-neutral amino acids produced significantly larger responses than the L-neutral amino acids ($P < 0.01$ for Asn, Ala, Phe, Leu, Val, and Trp; $P < 0.05$ for Met), only L-Pro producing a larger response than the D-form.

The effects of gurmarin, which is extracted from *Gymnema sylvestre*, and which specifically suppresses sweet tastes in the rat [4] were investigated in the CT and GSP after 15-min treatment of the tongue or the soft palate with 20 µg/ml gurmarin. In the CT, responses to sucrose decreased to 50% of control, while in the GSP, the suppression was so strong that responses to 0.5 M sucrose or 0.01 M saccharin decreased to 20%–30% of control; most of the responses to neutral amino acids of both forms were affected, with some exceptions, such as D-Trp.

These results suggest that the strong stimulatory effectiveness induced by D-amino acids in the rat GSP depends on strong responsiveness to sweet substances, and, further that the receptor sites or transduction mechanisms for amino acids in the rat might be different in the taste buds on the soft palate and on the tongue.

Acknowledgments. The authors wish to thank Dr. Toshiaki Imoto of Tottori University School of Medicine for providing gurmarin for this experiment.

References

1. Nejad MS (1986) The neural activities of the greater superficial petrosal nerve of the rat in response to chemical stimulation of the palate. Chem Senses 11: 283–293

[1] Department of Oral Physiology, Kagoshima University Dental School, 8-35-1 Sakuragaoka, Kagoshima, 890 Japan

2. Harada S, Smith DV (1992) Gustatory sensitivities of the hamster's soft palate. Chem Senses 17:37–51

3. Harada S (1987) Neural responses to amino acids in mice and rats. In: Roper SD, Atema J (eds) Olfaction and Taste IX. The New York Academy of Science, New York, pp 345–346

4. Imoto T, Miyasaka A, Ishima R, Akasaka K (1991) A novel peptide isolated from the leaves of *Gymnema sylvestre*-I. Characterization and its suppressive effect on neural responses to sweet taste stimuli in the rat. Comp Biochem Physiol 100A:309–314

The Effect of Topical Treatment with a Carbonic Anhydrase Inhibitor, MK-927, on the Response of the Chorda Tympani Nerve to Carbonated Water

Michio Komai[1,2], Bruce Bryant[2], Tomohiko Takeda[1], Hitoshi Suzuki[3], and Shuichi Kimura[1]

Key words. Carbonic anhydrase—Taste nerve sensation

Carbon dioxide is a potent irritant of the oral and nasal mucosa. In the mouth, carbonated solutions produce sensations of tingling, prickling, and burning. While much of this can be accounted for by the contribution of trigeminal innervation to these areas, gustatory nerve input may also be an important component of the total sensation. We have shown that carbonic anhydrase (CA) activity is required for the lingual trigeminal nerve responses to CO_2 [1]. In the chorda tympani nerve, however, the acute inhibition of CA produced by an intravenous injection of acetazolamide was not observed, as was the case in the lingual trigeminal nerve. MK-927 (Merck Sharp and Dohme, West Point, PA) is a topically applied carbonic anhydrase inhibitor that decreases intraocular pressure in rabbits and humans [2]. We tested the ability of MK-927 and two CA inhibitor analogs to affect the responses of the chorda tympani nerve to dissolved CO_2 stimuli on the tongue.

Multiunit (1–5) neural recordings were obtained from bundles of the chorda tympani nerve in anesthetized female Sprague-Dawley rats (sodium pentobarbital, 50 mg/kg; urethane, 150 mg/kg). Responses to dissolved CO_2 stimuli were measured both as (a) peak values of integrated multiunit responses and as (b) action potential activity responses. Two hundred μl of each of the CA inhibitors, MK-927 (50 mM, pH 3.8), acetazolamide (50 mM, pH 8.8), and methazolamide (48 mM, pH 8.8), and the same volume of control solvent for each inhibitor was applied to the tongue surface for 2 min. Neural activity before and after the application was recorded with a DAT recorder and later analyzed by MacLab/4 interface with Macintosh IIsi computer.

We discuss our results below:

1. Integrated responses to carbonated water (around 6000 ppm CO_2) after the topical application of MK-927 to the tongue decreased by 62% ($P < 0.001$, $n = 14$). The effect of MK-927 was dose-related, and sensitivity to CO_2 was recovered after 5–10 min of rinsing with water.
2. The recordings of spike activity especially showed that the initial part of the neural response to CO_2 was greatly decreased by the topical application of MK-927 to the tongue.
3. The low permeability CA inhibitor analog (acetazolamide and methazolamide) caused no inhibition. Therefore, it is intracellular, and not extracellular, CA that is sensitive to MK-927, and the inhibition of intracellular CA by MK-927 causes decreased gustatory nerve responses to CO_2. Our preliminary experiments suggest that neural responses to HCl and NaCl are not affected by MK-927.
4. This experiment showed, for the first time, that the topical application of a membrane-permeable CA inhibitor modified the neural response of the chorda tympani nerve to CO_2. This has important implications for the function of CA in the gustatory sensation of CO_2, as well as other gustatory stimuli [3,4].

Acknowledgments. This study was supported by Kirin Brewery Co. Ltd.

[1] Laboratory of Nutrition, Faculty of Agriculture, Tohoku University, 1-1 Tsutsumidori-Amamiyamachi, Aoba-ku, Sendai, 981 Japan
[2] Monell Chemical Senses Center, 3500 Market St., Philadelphia, PA 19104, USA
[3] Department of Biotechnology, Ishinomaki-Senshu University, Minamisakai, Ishinomaki, 986 Japan

References

1. Komai M, Bryant B (1993) Acetazolamide specifically inhibits lingual trigeminal nerve responses to carbon dioxide. Brain Res 612:122–129
2. Maren TH, Bar-Ilan A, Conroy CW, Brechue WF (1990) Chemical and pharmacological properties of MK-927, a sulfonamide carbonic anhydrase inhibitor that lowers intraocular pressure by the topical route. Exp Eye Res 50:27–36
3. Hansson HPJ (1961) On the effect of carbonic anhydrase inhibition on the sense of taste; an unusual side effect. Nord Med 65:566–567
4. Miller LG, Miller SM (1990) Altered taste secondary to acetazolamide therapy. J Fam Practice 31:199–200

2.2 Ionic Mechanisms

L-Arginine-Regulated Conductances in Catfish Taste Cells

John H. Teeter[1,2], Takenori Miyamoto[3], Diego Restrepo[1,2], Menekhem Zviman[1], and Joseph G. Brand[1,2,4]

Key words. Taste transduction—Catfish—Calcium —Amino acid

Introduction

Several distinct mechanisms have been implicated in mediating the responses of taste receptor cells to different classes of taste stimuli ([1–3], for reviews). Salts, acids, and some bitter substances have been shown to have direct effects on passive or voltage-dependent ion channels in the receptive membranes of the cells [4–8]. In contrast, sweeteners, amino acids, and other bitter compounds appear to act at specific receptor proteins located in the apical membranes of taste cells, which are coupled either directly or indirectly to changes in membrane conductance [7,9–14].

In the catfish taste system, several independent receptor sites for amino acids have been characterized on the basis of biochemical binding and second messenger assays, neural recordings, and membrane reconstitution techniques ([14,15], for recent reviews). Receptors activated by L-alanine (L-Ala), and apparently other short-chain neutral amino acids (L-Ser, L-Thr, and Gly), are coupled via G-proteins to the formation of cAMP and inositol 1,4,5-triphosphate (IP_3) [10]. How these messengers are coupled to the subsequent cellular response is not known; however, kinase-mediated phosphorylation and direct second messenger effects on ion channels are possibilities.

Two additional classes of amino acid receptors in the catfish, a high-affinity site(s) for L-arginine (L-Arg) and a low-affinity site for L-proline (L-Pro), have been shown, on the basis of reconstitution of taste epithelial membranes into lipid bilayers, to contain integral nonselective cation channels directly activated by binding of stimulus to the receptor ([13–15]; Kumazawa et al., manuscript in preparation).

The responses of catfish taste cells to amino acid stimuli are, however, only poorly characterized. We have used intracellular recordings from taste cells in vivo, and whole-cell recordings and fura-2 measurements of $[Ca_i]$ from isolated taste cells to further define amino acid-regulated conductances in catfish taste cells. In addition to responses consistent with direct activation of cation channels by L-Arg, we find evidence for a second, independent response in some cells, probably mediated by a direct effect of L-Arg on one or more classes of K conductances.

Materials and Methods

Intracellular recordings were made from cells in taste buds on the maxillary barbels of brown bullhead catfish (*Ictalurus nebulosus*). Taste cells were impaled under visual control with ultrafine micropipettes filled with KCl or potassium acetate. Cells with resting potentials more depolarized than $-25 \, mV$ were not analyzed further. Constant current pulses applied through the pipett via a bridge circuit were used to monitor membrane resistance. Conventional whole-cell recording techniques were used to record from taste cells isolated from the barbels of channel catfish (*I. punctatus*) by either enzymatic treatment or exposure to low pH. Measurements of $[Ca_i]$ were made in enzymatically dissociated cells loaded with the fluorescent Ca^{2+} indicator fura-2, using a Quantimet 570 image analysis workstation [16].

Results

Intracellular recordings from bullhead catfish taste cells revealed two types of responses to L-Arg (Miyamoto and Teeter, manuscript in preparation). In about 50% of the 15 responding cells, L-Arg at concentrations greater than $100 \, \mu M$ (applied concentration) elicited transient depolarizations associated with decreases in membrane resistance (Fig. 1A). The reversal potentials of these responses were variable, ranging from -20 to $+18 \, mV$, but were always positive to the resting potential of the cell. In the remaining cells, L-Arg induced a transient hyperpolarization, also associated with a decrease in

[1] Monell Chemical Senses Center, Philadelphia, PA 19104, USA
[2] University of Pennsylvania, Philadelphia, PA 19104, USA
[3] Nagasaki University School of Dentistry, Nagasaki, 852 Japan
[4] Veterans Affairs Medical Center, Philadelphia, PA 19194, USA

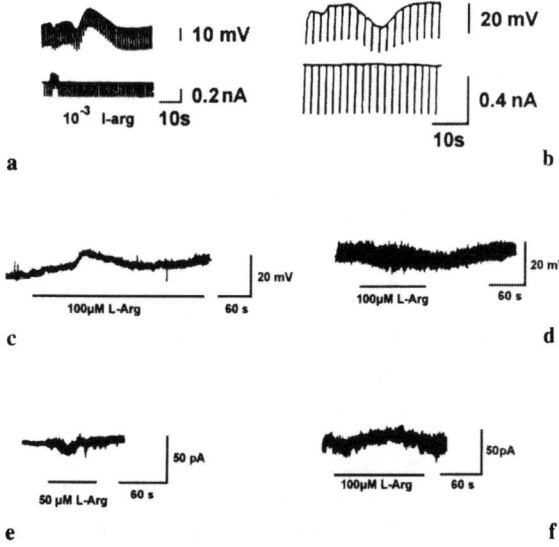

FIG. 1. L-Arginine-mediated responses in catfish taste cells. **a** Depolarizing and **b** hyperpolarizing changes in membrane potential elicited by L-arginine in taste cells on barbels of brown bullhead catfish. **c** Depolarizing and **d** hyperpolarizing changes in membrane potential produced by L-arginine in taste cells isolated from taste buds on barbels of channel catfish, under whole-cell current clamp. **e** Inward and **f** outward currents evoked by L-Arg in isolated channel catfish taste cells under whole-cell voltage clamp

membrane resistance (Fig. 1B). The reversal potentials for the hyperpolarizing responses were always negative of the resting potential, averaging −48 mV in 3 cells.

Taste cells isolated from the barbels of channel catfish also displayed two types of responses to L-Arg (Miyamoto and Teeter, manuscript in preparation). In about 40% of the cells that responded ($n = 16$), L-Arg (50–100 μM) elicited a transient depolarization under current clamp (inward currents under voltage clamp) (Fig. 1C,E). The remaining cells responded to L-Arg with membrane hyperpolarization under current clamp and enhanced outward currents under voltage clamp (Fig. 1D,F).

Preliminary results indicate that isolated channel catfish taste cells also display two general types of changes in [Ca$_i$] when stimulated with L-Arg. The cells maintained basal levels of Ca$_i$ in the range of 20–100 nM. In two cells, application of 1 mM L-Arg to the bath elicited transient increases in [Ca$_i$]. D-Arg alone had no effect on [Ca$_i$]. However, the increase in [Ca$_i$] produced by L-Arg was blocked when 1 mM D-Arg was present in the bath. In two additional cells, L-Arg produced decreases in [Ca$_i$] that were followed by a transient increase in [Ca$_i$] above the resting level on removal of L-Arg from the bath. In one of the cells, D-Arg also elicited a decrease in [Ca$_i$], followed by an increase when the D-Arg was removed.

Discussion

The results of both the electrophysiological and Ca-imaging experiments indicate that L-Arg elicits two fundamentally different types of responses in catfish taste cells. The depolarizing responses elicited by L-Arg in both in situ and isolated taste cells are consistent with direct stimulus activation of the non-selective cation channels previously characterized in reconstituted taste epithelial membranes [13]. The current through these channels would be inward at rest, even with 15 mM NaCl and 0.5 mM CaCl$_2$ in the external solution (Kumazawa et al., manuscript in preparation), which is consistent with the inward currents recorded in these cells under voltage clamp. The hyperpolarizing responses and potentiated outward currents observed in some taste cells are consistent with activation of one or more classes of K (or Cl) channels. Channel activation could result either from a direct effect of L-Arg on channels in the apical membrane of the cell or from activation of an indirect pathway, perhaps involving second messengers and phosphorylation of channels in the apical or basolateral membranes.

The transient increase in [Ca$_i$] elicited in some taste cells by L-Arg is consistent with activation of L-Arg-gated cation channels in the apical membrane, which are permeable to Ca^{2+}, desensitize quite rapidly when continuously stimulated, and are blocked by D-Arg ([13]; Kumazawa et al., manuscript in preparation). Although the basis of the decrease in [Ca$_i$] apparently elicited by either L-Arg or D-Arg is not known, the processes involved are clearly different from those mediating L-Arg-induced increases in [Ca$_i$].

Acknowledgments. This work was supported in part by grants DC-00327, DC-00356 and DC-01838 from the National Institutes of Health, BNS-8910042, BNS-9114153 and IBN-9209743 from the National Science Foundation and a grant from the Veterans Affairs Department, USA.

References

1. Brand JG, Teeter JH, Cagan RH, Kare MR (1989) Chemical senses. Volume 1, Receptor events and transduction in taste and olfaction. Dekker, New York

2. Roper SD (1989) The cell biology of vertebrate taste receptors. Annu Rev Neurosci 12:329–353

3. Avenet P, Kinnamon SC (1991) Cellular basis of taste reception. Curr Opin Neurobiol 1:198–203

4. Desimone JA, Heck GL, Persaud KC, Mierson S (1989) Stimulus-evoked transepithelial lingual currents and the gustatory neural response. In: Brand JG, Teeter JH, Cagan RH, Kare MR (eds) Chemical senses. Volume 1, Receptor events and transduction in taste and olfaction. Dekker, New York, pp 13–34

5. Roper SD, McBride DW Jr (1989) Distribution of ion channels on taste cells and its relationship to chemosensory transduction. J Membr Biol 109:29–39
6. Avenet P (1992) Role of amiloride-sensitive sodium channels in taste. In: Corey DP, Roper SD (eds) Sensory transduction. Rockefeller University Press, New York, pp 271–279
7. Kinnamon SC (1992) Role of K channels in taste transduction. In: Corey DP, Roper SD (eds) Sensory transduction. Rockefeller University Press, New York, pp 261–270
8. Gilbertson TA, Avenet P, Kinnamon SC, Roper SD (1992) Proton currents through amiloride-sensitive Na channels in hamster taste cells. J Gen Physiol 100:803–824
9. Striem BJ, Pace U, Zehavi U, Naim M, Lancet D (1989) Sweet tastants stimulate adenylate cyclase coupled to GTP-binding protein in rat tongue membranes. Biochem J 260:121–126
10. Kalinoski DL, Huque T, LaMorte VJ, Brand JG (1989) Second messenger events in taste. In: Brand JG, Teeter JH, Cagan RH, Kare MR (eds) Chemical senses. Volume 1, Receptor events and transduction in taste and olfaction. Dekker, New York, pp 85–101
11. Hwang PM, Verma A, Bredt DS, Snyder SH (1990) Localization of phosphatidylinositol signalling components in rat taste cells: role in bitter taste transduction. Proc Natl Acad Sci USA 87:7395–7399
12. Spielman AI, Huque T, Whitney G, Brand JG (1992) The diversity of bitter taste signal transduction mechanisms. In: Corey DP, Roper SD (eds) Sensory transduction. Rockefeller University Press, New York, pp 307–324
13. Teeter JH, Brand JG, Kumazawa T (1990) A stimulus-activated conductance in isolated taste epithelial membranes. Biophys J 58:253–259
14. Teeter JH, Kumazawa T, Brand JG, Kalinoski DL, Honda E, Smutzer G (1992) Amino acid receptor channels in taste cells. In: Corey DP, Roper SD (eds) Sensory transduction. Rockefeller University Press, New York, pp 291–306
15. Caprio J, Brand JG, Teeter JH, Valentincic T, Kalinoski DL, Kohbara J, Kumazawa T, Wegert S (1993) The taste system of the channel catfish: from biophysics to behavior. Trends Neurosci 16:192–197
16. Restrepo D, Okada Y, Teeter JH, Lowry LD, Cowart B, Brand JG (1993) Human olfactory neurons respond to odor stimuli with an increase in cytoplasmic Ca^{2+}. Biophys J 64:1961–1966

Ion Channels in Gustatory Transduction of Frog Taste Cells

Toshihide Sato, Yukio Okada, and Takenori Miyamoto[1]

Key words. Frog taste cell—Ion channel—Transduction—Four basic taste stimuli—Receptive and basolateral membranes

Introduction

Taste cells monitor a variety of chemical substances entering the mouth. After the binding of a gustatory substance to the receptor of the receptive membrane in a taste cell, the receptor potential is induced, which triggers an initiation of gustatory neural impulses. Molecular mechanisms underlying transduction of gustatory stimuli into electrical activities in a taste cell are not yet well understood. The purpose of this study was to elucidate ionic mechanisms of chemoelectrical transductions in frog taste cells in response to the four basic taste substances and to water.

Materials and Methods

Adult bullfrogs (*Rana catesbeiana*) were used. Most experiments were carried out with in situ taste cells in the fungiform papillae of the tongue using an intracellular microelectrode technique. Details of the experimental method and procedure are described elsewhere [1–3]. Some experiments were performed with whole-cell, cell-attached, inside-out, and outside-out configurations of the patch clamp technique. The channel currents were recorded with a patch pipette filled with various saline solutions such as 100 mM KCl, NaCl, CsCl, etc. For the four basic taste stimuli, NaCl, HCl, galactose, and quinine-HCl were usually used. All experiments were carried out at 20°–25°C (room temperature).

Results and Discussion

Salt Signal Transduction [1,2]

The taste cell membrane is mostly divided into the apical receptive membrane exposed to the oral cavity and the basolateral membrane. The former is bathed in the superficial fluid (SF) and the latter in the interstitial fluid (ISF). Amplitudes of the receptor potentials induced by salt stimuli were decreased by 32%–60% when interstitial Na^+ and Ca^{2+} were replaced with choline$^+$, tetramethylammonium$^+$, and tetraethylammonium$^+$. Addition of 5 mM Co^{2+} and 3 μM tetrodotoxin (TTX) to ISF did not affect the receptor potentials.

After the normal ionic composition of SF and ISF of the frog tongue had been changed with low-concentration Na^+ saline, the relationships between membrane potentials and receptor potentials in a frog taste cell evoked by various concentrations of NaCl and various types of salts were analyzed to examine the permeability of the taste-receptive membrane to cations and anions (Fig. 1). In this situation, the mean reversal potentials for depolarizing potentials of a taste cell in response to 0.05, 0.2, and 0.5 M NaCl were −40.0, 6.4, and 28.8 mV, respectively. When adding an anion channel blocker, SITS (4-acetamido-4'-isothiocyanostilbene-2, 2'-disulfonic acid), to a NaCl stimulus, the reversal potential for receptor potential with NaCl plus SITS became about twice larger than that with NaCl alone. This result indicates that Na^+ and Cl^- of the NaCl stimulus permeate the apical receptive membrane.

Reversal potentials for 0.2 M NaCl, LiCl, KCl, and NaSCN were 6.4, 25.4, −1.0, and −7.8 mV, respectively, indicating that permeability of the apical taste receptive membrane to cations of the Cl^- salts is of the order of $Li^+ > Na^+ > K^+$ and that the permeability to anions of the Na^+ salts is $SCN^- > Cl^-$. These results indicate that the NaCl stimulus-induced receptor potential results from an inflow of Na^+ and Cl^- across cation and anion channels on the taste-receptive membrane, as well as an inflow of interstitial Na^+ across cation channels on the basolateral membrane.

Patch pipette studies with isolated taste cells and excised patch membranes indicated that there are K^+ channels, nonselective cation channels, and Cl^- channels of various conductances, which are concerned with salt signal transduction, in the apical receptive membrane.

[1] Department of Physiology, Nagasaki University School of Dentistry, Nagasaki, 852 Japan

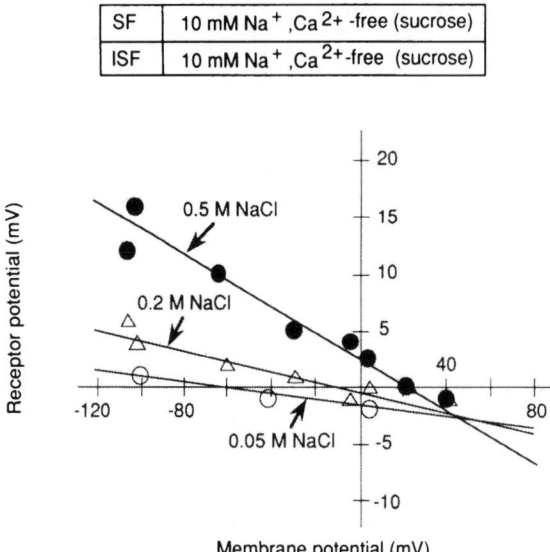

SF	10 mM Na$^+$, Ca^{2+}-free (sucrose)
ISF	10 mM Na$^+$, Ca^{2+}-free (sucrose)

FIG. 1. Relationship between membrane potentials and receptor potentials induced by NaCl stimuli in a frog taste cell. Superficial fluid (*SF*) and interstitial fluid (*ISF*) used are shown above graph. (From [2], with permission)

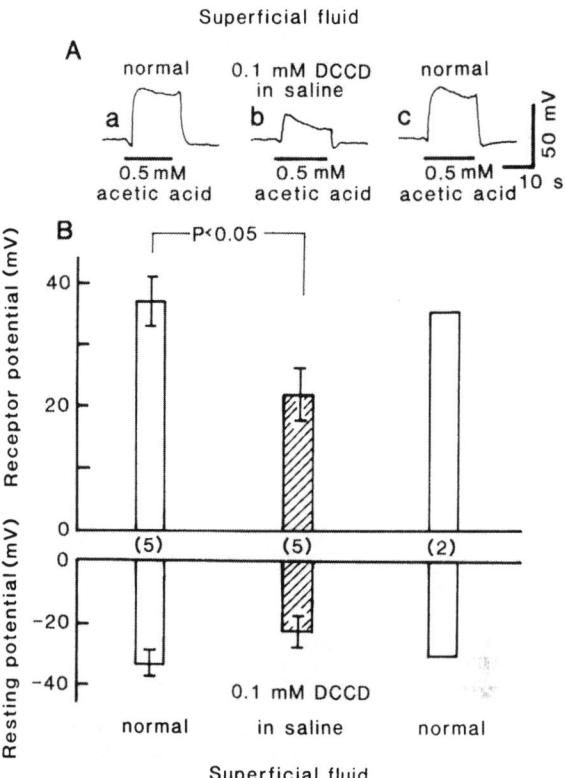

FIG. 2. Effect of superficial 0.1 mM DCCD in normal saline on acid-induced receptor potential. **A** Actual records: **a** control under normal saline; **b** test response 4.7 min after DCCD; **c** after control. **B** Acid-induced response (mean ± SE) under normal saline and DCCD. Numbers in parentheses are numbers of taste cells examined. (From [5], with permission)

Acid Signal Transduction [3–5]

The removal of Na$^+$, Ca^{2+}, and Cl$^-$ from the normal ISF did not affect the receptor potential induced by acid stimuli such as acetic acid and HCl. Interstitial 100 mM K$^+$ saline also did not affect the acid response. The receptor potential was reduced greatly when Ca^{2+} was removed from the superficial normal saline, but was increased when the Ca^{2+} concentration was elevated. The removal of superficial Cl$^-$ did not affect the receptor potential. The receptor potential elicited by an acid stimulus under superficial Ca^{2+}-free saline was partly caused by Na$^+$. Li$^+$, K$^+$, NH$_4^+$, or choline$^+$ substituted for Na$^+$ in producing the receptor potential. The amiloride-sensitive Na$^+$ channel on the receptive membrane did not contribute to the receptor potential. The receptor potential was unaffected by superficial TTX, but was blocked by superficial Ca^{2+} antagonists such as Co^{2+} and Cd^{2+}. Sr^{2+} and Ba^{2+} substituted for Ca^{2+} in generating the receptor potential. The receptor potentials observed under various concentrations of superficial Ca^{2+} became smaller when Na$^+$ was present in the SF, indicating a competition between Ca^{2+} and Na$^+$ passing through a Ca^{2+}-permeable conductance in the apical receptive membrane.

These findings indicate that a large portion of the receptor potential induced by acid stimuli is concerned with proton-gated Ca^{2+} channels on the taste-receptive membrane. Both divalent (Ca^{2+}, Sr^{2+}) and monovalent (Na$^+$, Li$^+$, K$^+$, NH$_4^+$, choline$^+$) cations can pass through the Ca^{2+} channel. Even after the tongue surface had been adapted to pure water, the amplitude of acid-induced response in a taste cell remained as large as 35% of the control. After 0.1 mM DCCD (N, N'-dicyclohexyl-carbodiimide), a proton pump inhibitor, was added to SF, the acid response was greatly suppressed, indicating a contribution of proton pump on the receptive membrane to the acid-induced receptor potential (Fig. 2).

The receptor current from a dissociated whole taste cell was recorded with a patch pipette filled with 100 mM CsCl. Application of an acetic acid stimulus containing 80 mM BaCl$_2$ to the cell initiated an inward current of about −50 pA at the holding potential of −40 mV. After the taste-receptive membrane alone was damaged, the receptor current induced by acetic acid stimulus containing the BaCl$_2$ greatly decreased, indicating that the inward receptor current was induced by Ba^{2+} passing across proton-gated Ca^{2+} channels on the apical receptive membrane. Cation permeability of the proton-gated Ca^{2+} channel was of the order of Ca^{2+} > Ba^{2+} > Sr^{2+} ≒ Na$^+$ ≒ Cs$^+$.

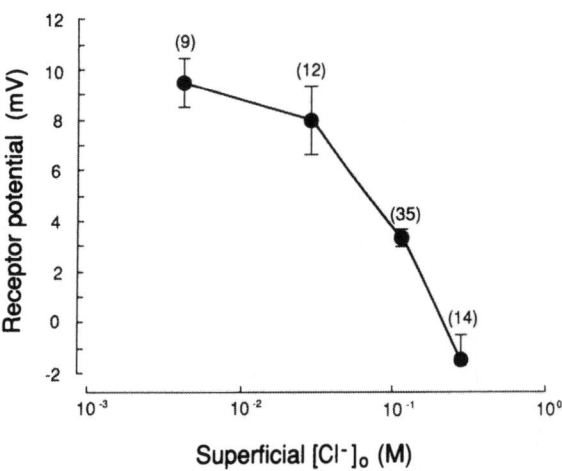

FIG. 3. Relationship between concentration of superficial Cl⁻ and amplitude of Q-HCl-induced responses. Numbers in parentheses are numbers of taste cells sampled. (From [6], with permission)

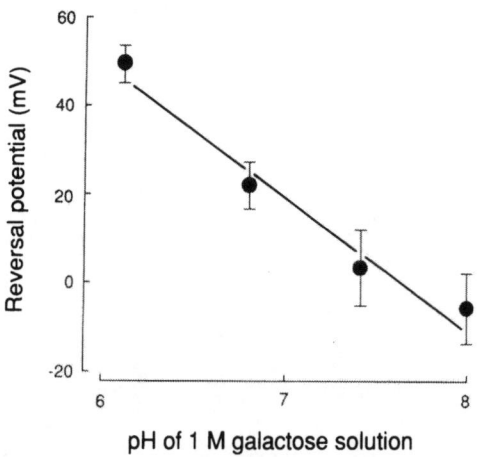

FIG. 4. Relationship between pH of 1M galactose and reversal potential for receptor potentials. *Points* are means from three or four taste cells; *bars* are SE. (From [7], with permission)

It is concluded that most of the acid-induced response in a frog taste cell is generated by a current carried through the proton-gated Ca^{2+} channel of the apical receptive membrane, and that the remaining portion of the acid response is generated by current carried through a DCCD-sensitive proton transporter of the receptive membrane.

Bitter Signal Transduction [6]

The ionic mechanism of the receptor potential elicited by quinine-HCl (Q-HCl) was studied. The frog taste cells whose receptive membranes were adapted to normal saline and deionized water generated depolarizing receptor potentials at Q-HCl concentrations higher than 2 and 0.01 mM, respectively. The input resistance of the taste cell during Q-HCl stimulation scarcely changed. The receptor potential did not change even when the membrane potential level was greatly changed. The magnitude of the receptor potential was increased by reducing the concentration of superficial Cl⁻ on the taste-receptive membrane (Fig. 3), but was independent of the concentration of superficial Na⁺.

Injection of Cl⁻ into a taste cell greatly increased the receptor potential. The magnitude of the receptor potential was greatly decreased by removing interstitial Na⁺ or Cl⁻, or both, surrounding the basolateral membrane of taste cell. Furosemide (1mM) added to the ISF decreased the receptor potential to 15%, while interstitial ouabain (0.1 mM) and superficial SITS (0.1 mM) did not influence it. From these results, we conclude: (1) an electro-neutral Na⁺/Cl⁻ cotransport occurs through the basolateral membrane of a taste cell in the resting state, so that Cl⁻ accumulates inside the cell. (2) Q-HCl stimulation induces the active secretion of

Cl⁻ across the taste receptive membrane, resulting in a depolarizing receptor potential.

Sweet Signal Transduction [7]

The frog taste cell generated a depolarizing receptor potential accompanying a remarkable reduction of input resistance in response to stimulation with galactose and sucrose. The magnitude of the receptor potential in response to a galactose solution increased linearly with decreasing pH in the pH range 6–8, but remained constant above pH 8. The reversal potential was increased by only 29 mV by a 10-fold increase in the H⁺ concentration of the stimulus, suggesting that there are pH-dependent and pH-independent components in the mechanism generating the receptor potential (Fig. 4). Na⁺-free, Ca^{2+}-free, and K⁺-free ISF did not affect the receptor potential, but the elimination of Cl⁻ from the ISF largely abolished it. Interstitial 0.1 mM DCCD completely inhibited the receptor potential, and interstitial 0.1 mM N-ethylmaleimide decreased the potential to 40% of the control value. Lowering the pH of ISF from 7.2 to 6.3 greatly decreased the receptor potential. It is concluded that part of the receptor potential in frog taste cells induced by sugar stimuli may be produced by an inflow of H⁺ through the taste-receptive membrane. The intracellular pH of the taste cell may be regulated by a Cl⁻-dependent H⁺ pump in the basolateral membrane.

Water Signal Transduction [8,9]

The frog taste cell located in the proximal portion of the tongue generated a depolarizing receptor potential that averaged 10 mV in response to stimulation with deionized water. Water-sensitive taste cells were classified into two types: Cl^- dependent and Cl^- independent. In Cl^--dependent cells whose input resistance was decreased or unchanged by deionized water, the magnitude of the water-induced depolarization decreased with an increase in concentration of superficial Cl^- in contact with the receptive membrane and with addition of blockers of anion channels (0.1 mM SITS and 0.1 mM DIDS) to deionized water. The reversal potential for the depolarization in this type shifted according to the concentration of superficial Cl^-. These properties of the responses were consistent with those of the glossopharyngeal nerve, which innervates the taste disc. In Cl^--independent cells whose input resistance was increased by deionized water, the reversal potential was approximately equal to the equilibrium potential for K^+ at the basolateral membrane. The water-induced response of the glossopharyngeal nerve was decreased to about 60% of the control value by addition of interstitial 2 mM Ba^{2+}. It is concluded that the water-induced receptor potential is produced by Cl^- secretion through the taste-receptive membrane in about 70% of water-sensitive frog taste cells, while it is generated by an inhibition of the resting K^+ conductance of the basolateral membrane in the remaining 30% of the cells.

Acknowledgments. These studies were supported in part by Grants-in Aid for Special Project Research, Co-operative Research, Developmental Scientific Research, and Scientific Research from the Ministry of Education, Science and Culture of Japan, and by Grants-in Aid from the Human Frontier Science Program Organization.

References

1. Miyamoto T, Okada Y, Sato T (1989) Ionic basis of salt-induced receptor potential in frog taste cells. Comp Biochem Physiol 94A:591–595
2. Miyamoto T, Okada Y, Sato T (1993) Cationic and anionic channels of apical receptive membrane in a taste cell contribute to generation of salt-induced receptor potential. Comp Biochem Physiol 106A:489–493
3. Miyamoto, T, Okada Y, Sato T (1988) Ionic basis of receptor potential of frog taste cells induced by acid stimuli. J Physiol 405:699–711
4. Okada Y, Miyamoto Y, Sato T (1992) Transduction of acid stimuli into the receptor potential in frog taste cells. J Physiol 446:387P
5. Okada Y, Miyamoto T, Sato T (1993) Contribution of proton transporter to acid-induced receptor potential in frog taste cell. Comp Biochem Physiol 105A:725–728
6. Okada Y, Miyamoto T, Sato T (1988) Ionic mechanism of generation of receptor potential in response to quinine in frog taste cell. Brain Res 450:295–302
7. Okada Y, Miyamoto T, Sato T (1992) The ionic basis of the receptor potential of frog taste cells induced by sugar stimuli. J Exp Biol 162:23–36
8. Okada Y, Miyamoto T, Sato T (1993) The ionic basis of the receptor potential of frog taste cells induced by water stimuli. J Exp Biol 174:1–17
9. Sato T, Ohkusa M, Okada Y, Sasaki M (1983) Topographical difference in taste organ density and its sensitivity of frog tongue. Comp Biochem Physiol 76A:233–239

Transcellular and Paracellular Pathways in Taste Buds Mediate Salt Chemoreception: The Location of Na$^+$ Transducing Channels

JOHN A. DeSIMONE, GERARD L. HECK, and QING YE[1]

Key words. Taste—Transduction—Channels—Sodium—Anions—Paracellular shunt

Introduction

Fugita has compared taste bud chemoreceptors and those of the gut [1]. All the duodenal chemoreceptor cells have apical and basolateral domains, as do most (but not all) taste cells. These domains are delineated by a continuous ring of fibrous protein linking the cells, the tight junctions [2]. This is a misnomer because it implies an impermeable barrier, which is only justified in some cases (e.g., rabbit urinary bladder) [3]. Other epithelia such as the gut [4] and the lingual epithelium [5–7] are leaky, i.e., paracellular shunts are major ion transport pathways. This distinction is more important for taste chemoreceptors which, unlike duodenal chemoreceptors, exist in cell clusters.

The ion permeability of tight junctions suggests that some ionic species traverse the junction and stimulate cells in the taste bud that do not have an apical process, e.g., basal cells [8]. Basal cells might be stimulated directly by a tight-junction-permeable ionic stimulus, or in some cases secondarily in a serial chain of receptor cell stimulation. Evidence also suggests paracrine interactions among taste bud cells [9]. These observations challenge the traditional view of taste cells as noninteracting entities, and indicate that multicellular groupings within a taste bud may be involved in transduction and quality coding.

Intravascular taste studies [10] and more recent work [11–13] indicate that transduction sites are not confined to the apical regions of taste cells. The apical membrane sensing elements for sodium are specific ion channels. These fail to develop in sodium-deprived rats ([14] and unpublished work). An independent sodium-sensing system is accessed through paracellular shunts [11,12,15,16]. We are using an in vivo voltage clamp method to dissect these pathways functionally.

[1] Department of Physiology, Virginia Commonwealth University, Richmond, VA 23298-0551, USA

Materials and Methods

The method of stimulus application and current and voltage clamp has been described [12]. Following anesthetization of the rat, a stimulation chamber was affixed to one side of the tongue. The chamber permitted solutions to contact a surface 7 mm in diameter containing about 25 fungiform papillae [17]. One of each pair of current-passing and voltage-sensing electrodes were housed in the flow chamber while two others were placed noninvasively beneath the tongue. The electrodes were connected to a voltage current clamp amplifier. To apply solutions, 3-ml aliquots were injected at a rate of 1 ml/sec. The rinse solution was 0.01 M KHCO$_3$. A 0.2 M NH$_4$Cl control stimulus was applied at the beginning and at the end of each experiment under zero current clamp.

Rats were anesthetized by intraperitoneal (IP) injection of sodium pentobarbital at a dose of 60 mg/kg. Additional IP injections (12 mg/kg) were administered as needed. Standard procedures were used to expose and prepare the chorda tympani (CT) for recording. The neural signal, recorded at the current clamp or voltage clamp, was amplified and stored on tape. The integrated CT response was obtained by full-wave rectification and integration using a time constant of 1.0 sec. The numerical value of a CT response was obtained as the area under the CT response curve for the first 30 sec. Sodium-restricted rats were obtained and maintained as described by Hill [14].

Results

Anion Effects

Anion influences on sodium salt responses are well known [18]. The mechanism involves passive ion transport through the paracellular shunts [5,11,12,15,16]. Ye et al. [11,12] have distinguished two anion effects. The first of these involves the field potential arising from the cation exchanger properties of tight junctions [5,6,11]. Voltage clamping the field potential removes this first anion influence for nonchloride salts and for NaCl up to about 50 mM

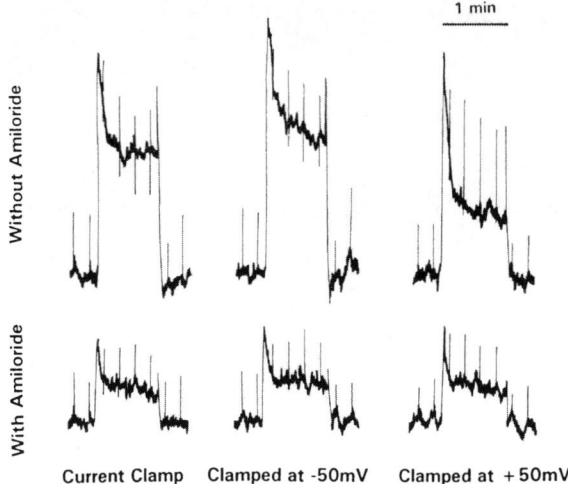

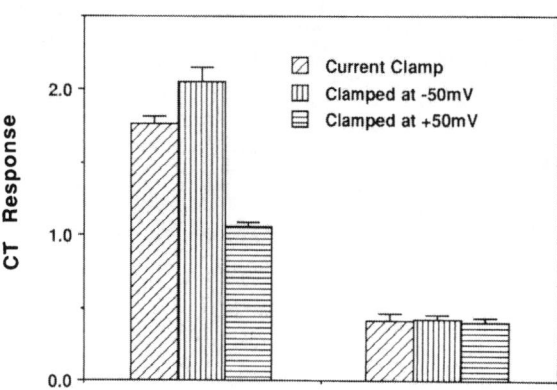

FIG. 1. Chorda tympani (CT) responses to 0.2 M NaCl. *Top row*, control responses. Relative to zero current clamp, response at −50 mV (mucosa ground) is enhanced while response at +50 mV is suppressed. *Bottom row*, effect of 0.1 mM amiloride. Current clamp response is reduced to 23% of control value and is now voltage-independent. Amiloride has eliminated the apical Na$^+$ channel response, which is voltage-dependent, unmasking a residual response that is mediated by the paracellular pathway and therefore voltage-independent

FIG. 2. Mean CT responses to 0.2 M NaCl. Numerical value of response was calculated as area under integrated response curve during first 30 sec. Each case represents the mean ± SEM ($n = 3$). In control cases (without amiloride), the current clamp response approaches the maximum possible response (saturation of available apical and submucosal channels). At −50 mV voltage clamp, maximum is achieved. At +50 mV, depolarizing Na$^+$ current into apical channels is reduced and CT response is consequently reduced. Blockage of apical channels with amiloride leaves only the voltage-independent (paracellular) component

[12]. At higher NaCl concentrations, a second anion effect emerges because the relatively high chloride permeability permits significant influx of NaCl through the tight junctions and subsequent stimulation of receptor cells below the tight junctions.

Voltage Perturbation of the Apical Na Channels

Because sodium receptor sites include apical ion channels, response intensity varies with both concentration and voltage difference across the channels. The actual stimulus intensity is the electrochemical concentration [12]. Figure 1 (top row) illustrates the effect of the voltage clamp in the response to NaCl in control rats. The first recording shows the response to 0.2 M NaCl at zero current clamp. The second record shows the enhanced response at −50 mV voltage clamp, while the third shows the suppression from +50 mV. At a concentration of 0.2 M, the second anion effect is in evidence; i.e., there will be a net influx of sodium through the tight junctions. The existence of this separate sodium-sensing pathway can be shown by blocking the apical channel pathway with amiloride. In this case, Fig. 1 shows that all responses are reduced to the same voltage-insensitive level. This is the Cl$^-$-mediated response through the paracellular pathway. Figure 2 illustrates both components. The voltage-insensitive (paracellular) component represents about 23% of the total current clamp response.

Development of the Sodium Taste Systems

In collaboration with D.L. Hill and R.E. Stewart, we are applying the voltage perturbation method to follow the development of the sodium taste system (unpublished work). Figure 3 shows dose–response relationships for rats at age 25 days (points connected by broken lines) and 50 days (points connected by solid lines). For each age, the dose–response curves were obtained at current clamp and at +50 mV and −50 mV voltage clamp. We observe that at +50 mV voltage clamp (open circles), both age groups give nearly the same suppressed responses. At current clamp and −50 mV voltage clamp (closed circles), the 50-day-old rats give larger responses than the 25-day-old rats. This result suggests that the apical channel (voltage-sensitive) component is far more developed in the older animals. An important limiting case is the sodium-restricted rat [14]. Figure 4 shows that it has virtually no voltage sensitivity, i.e., the responses are similar to normal rats treated with amiloride (cf. Figs. 1 and 2). Sodium-restricted rats lack functional apical membrane sodium channels in their taste cells (unpublished work).

Future Directions: Voltage-Sensitive Dye Recording

Thus far we have recorded only whole-nerve responses under voltage clamp. While much remains unexplored with that technique, it would also be

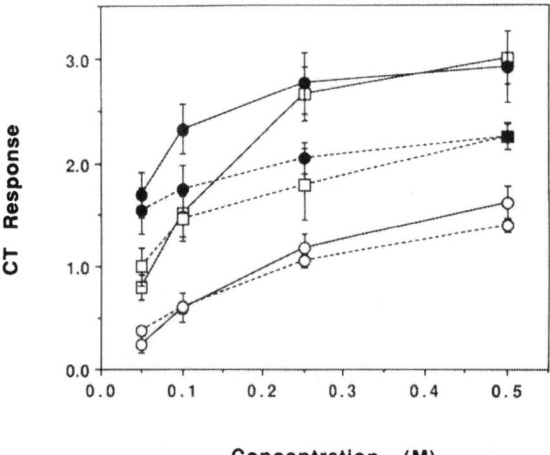

FIG. 3. Chorda tympani (CT) response as a function of NaCl concentration for 25-day-old rats (*broken lines*) and 50-day-old rats (*solid lines*). Parametric conditions include zero current clamp (*open squares*), −50 mV (*closed circles*), and +50 mV (*open circles*). At +50 mV there is no significant difference between 25- and 50-day-old rats. Increased current clamp response at 50 days relative to 25 days can be accounted for by addition of channel component (seen at maximum at −50 mv). Values are mean ± SEM ($n = 5$)

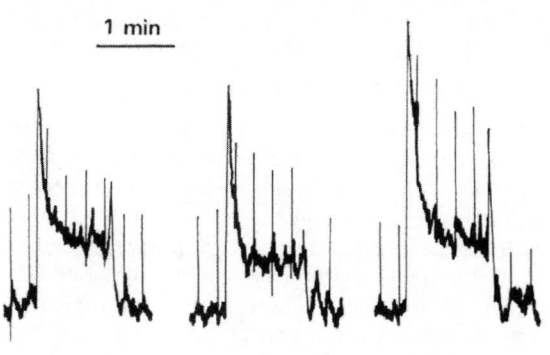

FIG. 4. Sodium-restricted rats give no voltage-dependent responses to NaCl. The apical channel component to the NaCl taste response is nonfunctional

desirable to obtain single fiber responses under voltage clamp. We are also exploring the use of voltage-sensitive dyes to record single- and multiunit data from the CT. In exploratory work, we have observed fluorescent light emitted from single CT fibers. Figure 5 illustrates the fluorescence from axons within the CT nerve of the rat. An excised segment of nerve was incubated for 5 min in a Krebs–Henseleit buffer solution containing 10 μM of the dye RH-795. This dye (Molecular Probes Inc., P.O. Box 22010, Eugene, OR) localizes in the outer leaflet of axonal membranes and undergoes a spectral shift in fluorescence with changes in transmembrane potential [19]. A small (1%–10%)

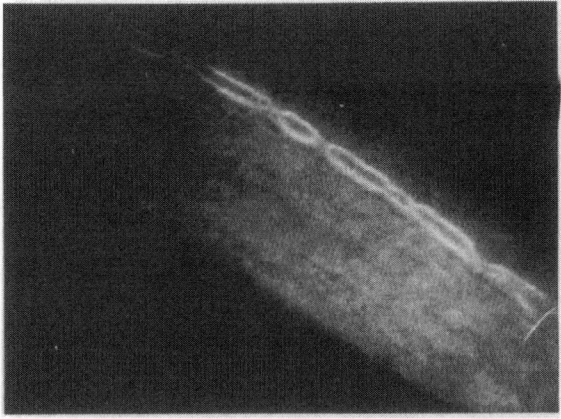

FIG. 5. Photomicrograph of excised chorda tympani nerve stained with RH-795 styryl dye. Brightly fluorescent threads are presumably axons in a common fascicle

voltage-sensitive change occurs during propagation of action potentials. As seen in Fig. 5, signal-to-noise ratios can be high for the in vitro fluorescence. Similar signal-to-noise ratios in vivo may allow the simultaneous measurement of a neural response from a number of isolated axons or isolated axon bundles. Once perfected, this technique may reveal cross-fiber patterns.

Discussion

Earlier views regarded tight junctions in taste buds as ion-impermeable barriers, but sodium is transported across the lingual epithelium, which has a resistance that places it in the "leaky" range [5,6]. For salts, receptor channels are located in the cell apical membranes, or on the submucosal side accessible through tight junctions. However, even leaky tight junctions should not be regarded as unregulated breaches in epithelial self-defense. Tight junctional permeability is physiologically regulated [20]. Our preliminary work suggests that chemical stimuli may regulate lingual epithelial tight junctions. Stimulation with HCl is accompanied by higher resistances than the same concentration of NaCl (unpublished observation). Whether this plays a role in proton taste transduction is unknown; however, the conductance of the tight junctions for acids is reduced, which must assist the cells in maintaining pH.

Data suggest that voltage sensitivity in the rat sodium system develops with age. The activation of apical channels depends on dietary sodium, because sodium-restricted rats do not develop the voltage-sensitive transducer. Salt taste transduction occurs over two topologically distinct pathways. Taste coding in general may involve a taste bud-residing topological component. Topological constraints may extend to the multi taste bud level. If so, a means of recording simultaneously from many units is desirable. Our preliminary results suggest that voltage-sensitive dyes in conjunction with the voltage perturbation method may provide a means of accomplishing this technique.

Acknowledgments. This work was supported by the National Institute on Deafness and other Communication Disorders (DC-00122) and the Human Frontiers of Science Program.

References

1. Fujita T (1991) Taste cells in the gut and on the tongue. Their common, paraneuronal features. Physiol Behav 49:883–885
2. Anderson JM, Stevenson BR (1991) The molecular structure of the tight junction. In: Cereijido M (ed) Tight junctions. CRC Press, Boca Raton, pp 77–90
3. Palmer LG (1986) The epithelial sodium channel. In: Poste G, Crooke ST (eds) New insights into cell and membrane transport processes. Plenum, New York, pp 327–344
4. Reuss L (1991) Tight junction permeability to ions and water. In: Cereijido M (ed) Tight junctions. CRC Press, Boca Raton, FL, pp 49–66
5. DeSimone JA, Heck GL, Mierson S, DeSimone SK (1984) The active ion transport properties of canine lingual epithelia in vitro: implications for gustatory transduction. J Gen Physiol 83:633–656
6. Simon SA, Garvin JL (1985) Salt and acid studies on canine lingual epithelium. Am J Physiol 249:398–C408
7. Holland VF, Zampighi GA, Simon SA (1989) Morphology of fungiform papillae in canine lingual epithelium: location of intercellular junctions in the epithelium. J Comp Neurol 279:13–27
8. Roper SD (1992) The microphysiology of peripheral taste organs. J Neurosci 12:1127–1134
9. Yoshie S, Wakasugi C, Teraki Y, Kanazawa H, Iwanaga T, Fujita T (1991) Response of the taste receptor cell to the umami-substance stimulus. An electron-microscopic study. Physiol Behav 49:887–889
10. Bradley RM (1973) Electrophysiological investigations of intravascular taste using perfused rat tongue. Am J Physiol 224:300–304
11. Ye Q, Heck GL, DeSimone JA (1991) The anion paradox in sodium taste reception: resolution by voltage-clamp studies. Science 254:724–726
12. Ye Q, Heck GL, DeSimone JA (1993) Voltage dependence of the rat chorda tympani response to Na$^+$ salts: implications for the functional organization of taste receptor cells. J Neurophysiol 70:167–178
13. Rehnberg BG, MacKinnon BI, Hettinger TP, Frank ME (1993) Anion modulation of taste responses in sodium-sensitive neurons of the hamster chorda tympani nerve. J Gen Physiol 101:453–465
14. Hill DL (1987) Susceptibility of the developing rat gustatory system to the physiological effects of dietary sodium deprivation. J Physiol (Lond) 393:413–424
15. Harper HW (1987) A diffusion potential model of salt taste receptors. Ann NY Acad Sci 510:349–351
16. Elliott EJ, Simon SA (1990) The anion in salt taste: a possible role for paracellular pathways. Brain Res 535:9–17
17. Miller IJ Jr (1976) Taste bud distribution and regional responsiveness on the anterior tongue of the rat. Physiol Behav 16:439–444
18. Beidler LM (1954) A theory of taste stimulation. J Gen Physiol 38:133–139
19. Grinvald A, Frostig E, Hildesheim R (1988) Optical imaging of neuronal activity. Physiol Res 68:1285–1365
20. Madara JL (1991) Relationships between the tight junction and the cytoskeleton. In: Cereijido M (ed) Tight junctions. CRC Press, Boca Raton, pp 105–119

Effect of Novobiocin on Cation Channel Formation and Enhancement of Salt Taste

ALEXANDER M. FEIGIN[1], JOSEPH G. BRAND[1,2,3], YUZO NINOMIYA[4], BRUCE P. BRYANT[1], SERGEY M. BEZRUKOV[5,6], PAUL A. MOORE[1], IGOR VODYANOY[5,7], and JOHN H. TEETER[1,2]

Key words. Salt—Taste—Novobiocin—Chorda tympani—Lipid bilayers—Enhancement

Introduction

One of the major mechanisms of salt taste transduction involves sodium entry into the taste receptor cell through epithelial sodium channels [e.g., 1–5]. These channels show pharmacologic similarities to other epithelial sodium channels located in other transporting epithelia [6]. These pharmacologic similarities make salt taste enhancement through channel modification possible [7].

We have been investigating the ability of the sodium transport-enhancing agent, novobiocin, to modify sodium taste. Novobiocin is an antibiotic that, when applied to the mucosal side of frog skin, brings about an amiloride-sensitive increase in short circuit current and an increase in transepithelial potential and conductance [8]. We have found that novobiocin also enhances the sodium-stimulated response of rat chorda tympani fibers that are amiloride-sensitive (N-fibers) but has no effect on the sodium response to fibers that are amiloride-insensitive (E-fibers). Preincubation of the lingual epithelium with amiloride prevents this enhancement by novobiocin. We have also found that novobiocin can form cation-selective ion channels in pure lipid bilayers, an observation that may make novobiocin's action on taste receptor cells (and on other transporting epithelia) more complex.

Materials and Methods

Adult Sprague-Dawley rats (250–300 g) were anesthetized with an initial dose of sodium pentobarbital

[1] Monell Chemical Senses Center, 3500 Market Street, Philadelphia, PA 19104-3308, USA
[2] University of Pennsylvania, Philadelphia, PA, USA
[3] Veterans Affairs Medical Center, Philadelphia, PA, USA
[4] Department of Oral Physiology, Asahi University, Hozumi, Motosu, Gifu, Japan
[5] National Institutes of Health, Bethesda, MD, USA
[6] St. Petersburg Nuclear Physics Institute, St. Petersburg, Russia
[7] Office of Naval Research, Arlington, VA, USA

(50–65 mg/kg) and urethane (150 mg/kg) and with supplemental doses of pentobarbital as needed. Access to the right chorda tympani nerve was obtained using a ventral approach. The chorda tympani nerve was sectioned approximately 1 cm central to the lingual nerve. For single-fiber recordings, the perineurium was freed from the distal portion of the chorda tympani nerve and the nerve divided into small strands, each of which was placed over a Pt/Ir wire electrode. An indifferent electrode was placed at the edge of the wound. Extracellular potentials were differentially amplified 10 000 to 20 000 times, monitored on an oscilloscope and audio monitor, and recorded on magnetic tape. Single units were discriminated and counted digitally off-line (RC Electronics, Goleta, CA).

Only fibers that responded to salts ("N-fibers" responded selectively to NaCl; "E-fibers" responded about equally to KCl and NaCl [4,9–11]) were used for this study. Of the 25 fibers surveyed, 18 fibers were classified as either N-fibers of E-fibers. E-fibers responded also to HCl, but neither E-fibers nor N-fibers responded to test concentrations of quinine HCl (1 mM) or sucrose (0.5 M).

Once a particular fiber was classified as either an N- or an E-fiber, the effect of novobiocin on the response characteristics of that fiber was assessed. For the studies detailed in Fig. 1 (see below), the tongue was pretreated by incubation for 3 min in 2 mM novobiocin. Thereafter solutions of 100 mM NaCl or 100 mM KCl, each in 2 mM novobiocin, were applied to the tongue. Between test solutions the tongue was rinsed with 2 mM novobiocin. Subsequently, the tongue was rinsed with water for 3 to 5 min, and solutions of NaCl and KCl in water were retested. The final test was the characterization of each fiber's sensitivity to amiloride to verify the fiber type. The tongue was incubated in amiloride (100 μM for 3 min) followed by stimulation with 100 mM NaCl and 100 mM KCl prepared in 100 μM amiloride.

"Solvent-free" membranes were prepared as described by Montal and Mueller [12]. The chamber [13] was made from Teflon. Two symmetric halves with volumes of 1 cm^3 were divided by a 15 μm thick Teflon partition containing a round aperture of about 100 μm diameter. Hexadecane in n-pentane (1:10, v/v) was used for aperture pretreatment. Detailed descriptions of methods used for membrane preparation have been published [13].

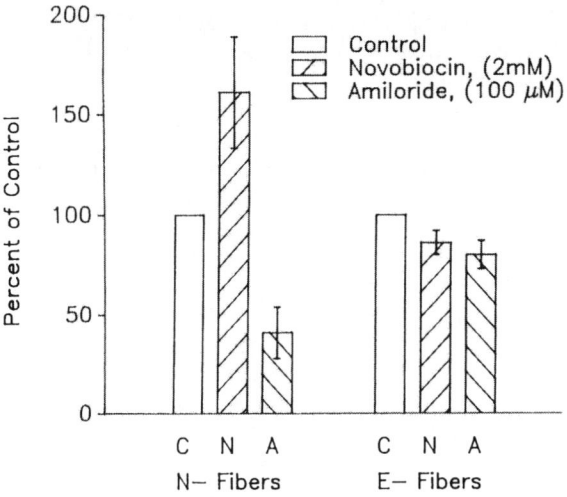

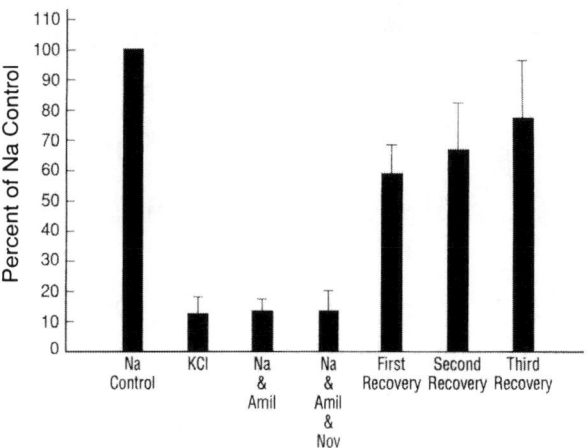

FIG. 1. Responses of N-type and E-type fibers (mean ± SEM) to 100 mM NaCl expressed as the number of spikes during the first 5 s (y-axis). *C*, control responses of fibers before application of any pharmacologic agents; *N*, responses during application of 2.0 mM novobiocin; *A*, responses following rinsing and subsequent application of 100 μM amiloride. For N-type fibers, *n* = 7; for E-type fibers *n* = 5

FIG. 2. Number of spikes during the first 5 s after stimulation, expressed as percent of the control number in response to 100 mM NaCl *vs* response to various pharmacologic manipulations. The initial firing rates in response to 100 mM NaCl and 100 mM KCl were determined. Next, the number of spikes in response to 100 mM NaCl + 100 μM amiloride was determined (*Na & Amil*) following a 3-min rinse with 100 μM amiloride. Following another 3-min rinse with 100 μM amiloride, the tongue was stimulated with 100 mM NaCl + 100 μM amiloride + 2 mM novobiocin (*Na & Amil & Nov*). Subsequent rinsing of the tongue with water for 3, 6, and 9 min allowed eventual recovery of the response to 100 mM NaCl. These results represent the responses (mean ± SEM) of six N-type fibers

Amiloride and the sodium salt of novobiocin were purchased from Sigma Chemical (St. Louis, MO). Synthetic lipids, dioleoylphosphatidylcholine (DOPC) and dioleoylphosphatidylethanolamine (DOPE) were obtained from Avanti Polar Lipids (Birmingham, AL). All other chemicals were reagent grade.

Results and Discussion

The mean responses of seven N-type and five E-type fibers (expressed as the mean number of spikes during the first 5 s) to 100 mM NaCl before application of pharmacologic agents (control), during application of 2 mM novobiocin, and (after rinsing) during application of 100 μM amiloride are presented in Fig. 1. Application of 2 mM novobiocin alone or 100 μM amiloride alone had no effect on the spontaneous firing rate of either N- or E-type fibers.

Novobiocin significantly ($P < 0.05$, Wilcoxon test) enhanced the response to NaCl of all (seven) N-type fibers. The enhancement varied between 15% and 230% (mean 55%) above control level.

The enhanced response of N-type fibers to NaCl (Fig. 1) was reversed and decreased to below control levels by rinsing with water and by the addition of 100 μM amiloride in rinse and subsequent stimulus solutions. After rinsing with water and amiloride, the response to 100 mM NaCl was 23% (mean) that of the novobiocin-induced (enhanced) response. The proportion of this residual response observed under amiloride that could be attributed to novo-

biocin enhancement cannot be precisely determined because in many cases responses to NaCl did not return completely to control levels during the water rinse phase.

To determine whether the enhancing effect of novobiocin on N-type fibers was sensitive to the presence of amiloride, a separate set of experiments was performed using a protocol that placed amiloride on the tongue prior to application of novobiocin. The results are shown in Fig. 2. A single fiber was first evaluated for its responses to 100 mM NaCl and 100 mM KCl, as before. If the fiber responded well to NaCl but poorly to KCl, it was chosen as a test fiber because it filled a criterion as an N-fiber. As shown in Fig. 2, the mean (± SEM) response to KCl was 11% ± 5% that of the response to NaCl. The tongue was then rinsed with water, subjected to application of 100 μM amiloride for 3 min, followed by application of 100 μM amiloride plus 100 mM NaCl. The fiber responded with a decrease in firing rate to a level of 13% ± 3% that of the initial control (Fig. 2). Following this test, the tongue was again rinsed with 100 μM amiloride for 3 min followed by rinsing with 100 μM amiloride plus 2 mM novobiocin for 3 min. A test application of 100 mM NaCl plus 100 μM amiloride plus 2 mM novobiocin was then evaluated. As shown in Fig. 2, the response to this test was a firing rate that was 14% ± 5% that of the

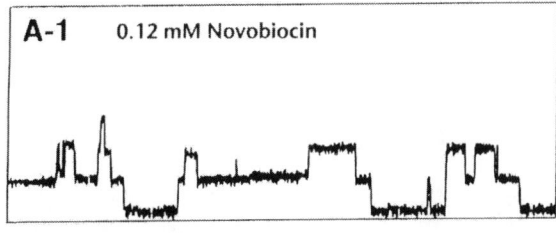

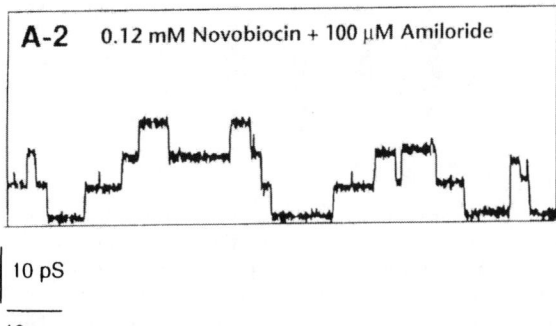

10 pS

----
10 sec

FIG. 3. Single-channel activity of a planar lipid bilayer (DOPC/DOPE = 1:1; mole/mole) after addition of 0.12 mM novobiocin at the *cis* side of the membrane (+100 mV at *cis* side). Electrolyte solutions at both sides were the same: 110 mM NaCl, 10 mM MOPS, pH 6.9. Signal was filtered by a low-pass Bessel filter with a 3-Hz corner frequency. **A-1** Prior to addition of amiloride. **A-2** Following addition of 100 μM amiloride (*cis* side)

initial control response. Following this test application, the tongue was rinsed with water for 3, 6, and 9 min, with a test application of 100 mM NaCl at the end of each 3-, 6-, and 9-min period. Figure 2 shows that the response of the fibers to 100 mM NaCl recovered over the course of the rinsing, reaching 59% ± 8%, 66% ± 16%, and 78% ± 19% that of the control value at 3, 6, and 9 min, respectively. The final value of 78% ± 19% was not significantly different from the initial control value (post hoc, Tukey-Kramer test).

Amiloride significantly inhibited the responses of these N-fibers to 100 mM NaCl and prevented the enhancing effect of 2 mM novobiocin seen in the absence of amiloride pretreatment ($P < 0.01$; post hoc, Tukey-Kramer test) (Fig. 2). These effects were completely reversible with extensive rinsing (Fig. 2).

In contrast to novobiocin's effect on N-type fibers, the responses of E-type fibers to NaCl (Fig. 1) were not significantly affected ($P > 0.05$, Wilcoxon test) by the application of novobiocin. As expected, amiloride also did not significantly influence the responses of E-type fibers to NaCl. The responses of E-type fibers to 100 mM KCl and the small responses of N-type fibers to 100 mM KCl were also not affected by novobiocin or by amiloride (data not shown).

The selective enhancement of the response to NaCl by novobiocin and the absence of enhancement in the presence of amiloride, as seen on N-type fibers, suggested that novobiocin increased the influx of Na^+ through amiloride-sensitive sodium channels. On the other hand, the possibility that novobiocin formed new channels in the apical membranes could not be discounted, even though the structure of novobiocin is dissimilar to that of other channel-forming drugs.

The addition of novobiocin (0.05–0.20 mM) to one side of a membrane formed of DOPE/DOPC = 1:1 (mole/mole) resulted in the appearance of cation-selective single channels with a conductance of 7.1 ± 0.5 pS (110 mM NaCl) and a mean open time of several seconds (Fig. 3,A-1). These parameters are similar to those of the well documented amiloride-sensitive sodium channel from apical membranes of epithelia [6,15,16]. However, the addition of amiloride (100 μM) at the same side of the bilayer did not influence the conductance and mean open time of single channels formed by novobiocin (Fig. 3,A-2).

The results with lipid bilayers suggest an additional possible mechanism of the stimulatory effect of novobiocin on the transport of sodium and on the enhancement of response of N-type fibers of rat chorda tympani nerve to NaCl stimulation: the formation of additional ion channels in apical membranes of epithelial cells. Yet considering that (1) in the presence of amiloride the enhancement of the NaCl response seen with novobiocin was not observed (Fig. 2) even though amiloride did not inhibit the activity of the novobiocin channels in lipid bilayers (Fig. 3), and that (2) novobiocin did not significantly alter the responses of amiloride insensitive E-fibers to either NaCl (Fig. 1) or KCl, it seems more likely that novobiocin on the rat tongue acts primarily as an activator of amiloride-blockable sodium channels. Of course, the ability of novobiocin to act as an enhancer of amiloride-sensitive epithelial channels does not preclude the possibility that it may also act as a channel former.

Acknowledgments. This work was supported in part by NIH grants DC-00327 and BRSG 507RR05825 and by a grant from the Veterans Affairs Department.

References

1. Heck GL, Mierson S, DeSimone JA (1984) Salt taste transduction occurs through an amiloride-sensitive sodium transport pathway. Science 223:403–405
2. Brand JG, Teeter JH, Silver WL (1985) Inhibition by amiloride of chorda tympani responses evoked by monovalent salts. Brain Res 334:207–214
3. Hettinger TP, Frank ME (1990) Specificity of amiloride inhibition of hamster taste responses. Brain Res 513:24–34

4. Ninomiya Y, Funakoshi M (1988) Amiloride inhibition of responses of rat single chorda tympani fibers to chemical and electrical tongue stimulations. Brain Res 451:319–325

5. Schiffman SS, Lockhead E, Mayes FW (1983) Amiloride reduces the taste intensity of Na and Li salts and sweeteners. Proc Natl Acad Sci USA 80:6136–6140

6. Garty H, Benos DJ (1988) Characteristics and regulatory mechanisms of the amiloride-blockable Na^+ channel. Physiol Rev 68:309–373

7. Schiffman SS, Simon SA, Gill JM, Beeker TG (1986) Bretylium tosylate enhances salt taste. Physiol Behav 36:1129–1137

8. Rick R, Dörge A, Sesselmann E (1988) Na transport stimulation by novobiocin: transepithelial parameters and evaluation of E_{Na}. Pfluegers Arch Eur J Physiol 411:243–251

9. Frank ME, Bieber SL, Smith DV (1988) The organization of taste sensibilities in hamster chorda tympani nerve fibers. J Gen Physiol 91:861–896

10. Ninomiya Y, Mizukoshi T, Higashi T, Funakoshi M (1984) Differential responsiveness of two groups of rat chorda tympani fibers to ionic chemicals and electrical tongue stimulation. Proc Jpn Symp Taste Smell 18:145–148

11. Ninomiya Y, Mizukoshi T, Nishikawa T, Funakoshi M (1987) Ion specificity of rat chorda tympani fibers to chemical and electrical tongue stimulation. Brain Res 404:350–354

12. Montal M, Mueller P (1972) Formation of bimolecular membranes from lipid monolayers and study of their properties. Proc Natl Acad Sci USA 65:3561–3566

13. Vodyanoy I, Hall JE, Balasurbramanian TM (1983) Alamethicin-induced current-voltage asymmetry in lipid bilayers. Biophys J 42:71–82

14. Bezrukov SM, Vodyanoy I (1993) Probing alamethicin channels with water-soluble polymers: effect on conductance of channel states. Biophys J 64:16–25

15. Hamilton KL, Eaton DC (1985) Single-channel recording from amiloride-sensitive epithelial sodium channel. Am J Physiol 249:C200–C207

16. Sariban-Sohraby S, Latorre R, Burg M, Olans L, Benos D (1984) Amiloride-sensitive epithelial Na^+ channels reconstituted into planar lipid bilayer membranes. Nature 308:80–81

Ion Channels Contributing to the Generation of Salt-Induced Response in Isolated Bullfrog Taste Cells

Takenori Miyamoto, Rie Fujiyama, Yukio Okada, and Toshihide Sato[1]

Key words. Salt-induced response—Ion channel

The contribution of the transduction pathway on the basolateral [1] and the receptive membranes [2] to the mechanism underlying the generation of the receptor potential induced by salt stimuli has been suggested individually by different groups. In previous studies, we reexamined the transduction mechanism of salt stimuli by changing the superficial and interstitial fluid of the frog tongue separately, using the lingual artery perfusion method [3,4]. Three important results were obtained: (1) the magnitude of the receptor potentials decreased by 50% after the removal of Na^+ and Ca^{2+} from interstitial fluid; (2) the reversal potential of salt-induced responses altered depending on the types and concentrations of salt stimuli after removal of Na^+ and Ca^{2+} from both the superficial and interstitial fluid; and (3) substitution of an organic anion for Cl^- and application of a Cl^- channel blocker changed the reversal potential of salt response. These results led us to the conclusion that at least three types of ion channels, non-selective cation channels, K^+ channels, and Cl^- channels, on the receptive membrane, and a Na^+ conductive pathway activated by an unknown second messenger on the basolateral membrane contribute to salt-induced responses in frog taste cells.

We carried out this experiment to investigate whether all these channels actually exist on the taste cell membrane; to this end we examined enzymatically isolated frog taste cells by the patch clamp method. The conductance of the receptive membrane increased greatly with a shift of the reversal potential after the pipette solution was replaced with 0.5 M NaCl. The permeability ratio of Na^+ to K^+ for this conductance was estimated to be approximately 0.1. Several K^+ channels, which opened spontaneously at the resting potential level, were frequently observed at the receptive membrane. These K^+ channels, which had a large conductance, were blocked by the NaCl stimulus, whereas the activity of K^+ channels with small conductance still remained. These findings suggest that K^+ channels, particularly small K^+ channels, play some role in the generation not only of the KCl response but also of the NaCl response. The open probability of the small K^+ channel obtained in the inside-out patch excised from the receptive membrane increased with increased NaCl concentration, indicating the possibility that the K^+ channel has a specific binding site to cations and a gating mechanism dependent on a higher concentration of cations. Even when 10 mM Ba^{2+} was added to the NaCl stimulus to block K^+ channels, some channel activities were still observed at both the receptive and basolateral membranes.

The reversal potentials and drug sensitivity of these channel activities indicated that they were derived from non-selective cation channels and Cl^- channels. The cation channels were also observed at the basolateral and receptive membrane. Taken together, the results here suggest that the transduction of salt stimuli in frog taste cells is mediated by multiple channel pathways.

Acknowledgments. This work was supported by Grants-in-Aid for Scientific Research (No. 01480433) and Developmental Scientific Research (No. 04557080) from the Ministry of Education, Science, and Culture of Japan.

References

1. Sato T, Beidler LM (1975) Membrane resistance change of the frog taste cells in response to water and NaCl. J Gen Physiol 66:735–763
2. Akaike N, Sato M (1976) Role of anions and cations in frog taste cell stimulation. Comp Biochem Physiol 55A:383–391
3. Miyamoto T, Okada Y, Sato T (1989) Ionic basis of salt-induced receptor potential in frog taste cells. Comp Biochem Physiol 94A:591–595
4. Miyamoto T, Okada Y, Sato T (1993) Cationic and anionic channels of apical receptive membrane in a taste cell contribute to generation of salt-induced receptor potential. Comp Biochem Physiol 106A:489–493

[1] Department of Physiology, Nagasaki University School of Dentistry, 1-7-1 Sakamoto, Nagasaki, 852 Japan

Distribution of Ion Channels on the Bullfrog Taste Cell Membrane

RIE FUJIYAMA, TAKENORI MIYAMOTO, and TOSHIHIDE SATO[1]

Key words. Distribution of ion channels—Frog taste cell

Recent studies of the gustatory transduction mechanism have been performed on isolated taste cells using the patch clamp method. Studies of ion channels on the taste cell membrane in mud puppies indicate that K^+ channel density is much greater on the receptive membrane than on the basolateral membrane [1]. We carried out the present experiments to examine the characteristics and distribution density of various channels on the receptive and basolateral membranes of bullfrog taste cells. Single channel currents were recorded using the inside-out configuration of the patch clamp technique.

We found that $40\,pS$ K^+ channels, $80\,pS$ K^+ channels, $20\,pS$ Cl^- channels, and $30\,pS$ cation channels were present in the bullfrog taste cell membrane. Their characteristics were: (1) $40\,pS$ K^+ channels were blocked by $5\,\mu M$ cAMP in the presence of $2\,mM$ ATP, but were not affected by Ca^{2+} concentration. (2) $80\,pS$ K^+ channels; the open probability of these channels increased with increasing Ca^{2+} concentration on the internal face of the cell membrane. (3) $20\,pS$ Cl^- channels were often observed with extracellular solution containing Ba^{2+} and intracellular solution containing Cs^+. The reversal potential for the Cl^- current obtained from the I-V relation was close to the equilibrium potential of Cl^-, and the current was suppressed by 4-acetamido-4'-isothiocyanostilbene-2, 2'-disulfonic acid · 2Na salt (SITS) as a Cl^- channel blocker. (4) $30\,pS$ cation channels; the permeability ratio of monovalent cations to Cs^+ was about 1, indicating that these were nonselective cation channels. The taste cell membrane was divided by a tight junction into the receptive membrane exposed to the superficial fluid and the basolateral membrane bathed in the interstitial fluid. Morphologically, the basolateral membrane was composed of dendrite, cell body, and proximal process membranes [2,3].

The $40\,pS$ K^+ channels were distributed over the entire taste cell membrane, but a high density was seen on the receptive and dendritic membranes. The $80\,pS$ K^+ channels were concentrated on the receptive and dendritic membranes. The $20\,pS$ Cl^- channels were distributed uniformly over all parts of the taste cell membrane; however, their density was lower than that of the K^+ channels. The $30\,pS$ cation channels were distributed throughout the taste cell membrane. It has been reported that after single taste cells were isolated by removing the tight junctions, membrane molecules invaded from the receptive membrane to the basolateral membrane [4]. However, in a study in which the frog taste receptive membrane was labeled alone with a lectin, no movement of the lectin was observed for $12\,h$ after labeling (R. Fujiyama, personal communication with T. Miyamoto). Since we performed all experiments in this study within $4\,h$ after cell isolation, our results should accurately reflect the in situ channel distribution in the frog taste cell membrane. Although the function of each channel in gustatory transduction will have to be clarified in future experiments, the present results suggest that each channel may play a different role in generating taste receptor potential.

References

1. Cummings T, Kinnamon SC (1992) Apical K^+ channels in *Necturus* taste cells. J Gen Physiol 99:591–613
2. DeHan RS, Graziadei PPC (1971) Functional anatomy of frog's taste organs. Experientia 27:823–826
3. Graziadei PPC, DeHan RS (1971) The ultrastructure of frog's taste organs. Acta Anat 80:563–603
4. Richter HP, Avenet P, Mestres P, Lindemann B (1988) Gustatory receptors and neighbouring cells in the surface layer of an amphibian taste disc. Cell Tiss Res 254: 83–96

[1] Department of Physiology, Nagasaki University School of Dentistry, 1-7-1 Sakamoto, Nagasaki, 852 Japan

Effects of Anions on the Responses to Ca, Mg, and Na in Single Water Fibers of the Frog Glossopharyngeal Nerve

Yasuyuki Kitada[1]

Key words. Salt taste response—Anion

Single water fibers of the frog glossopharyngeal nerve respond not only to water but also to salts such as those of Ca, Mg, and Na. The salt taste responses in the water fibers are strongly influenced by anions. In this study, we investigated the role of anions in salt taste reception in single water fibers of the frog glossopharyngeal nerve.

Isolated tongues from bullfrogs (*Rana catesbeiana*) were used. Antidromic nerve impulses, produced by chemical stimuli consisting of the application of various concentrations of salts dissolved in distilled water were recorded with a suction electrode. A solution of 50 mM NaCl was used as the adapting and rinsing solution. Only unitary discharges from single water fibers were analyzed. The number of impulses elicited during the period from 5 to 30 s after the onset of stimulation was counted with a spike counter.

Threshold concentrations for Ca, Mg, and Na salts differ appreciably and are dependent on the cation species, suggesting that cations stimulate taste cells. Anions modulate the responses produced by cations. The order of effectiveness of salts with different anions in producing a response was: 5 mM Ca^{2+} stimulation, $SO_4^{2-} > Cl^- = Br^- = NO_3^- > SCN^-$; 200 mM Mg^{2+} stimulation, $SO_4^{2-} < Cl^- = Br^- \leqslant NO_3^- < SCN^-$; and 500 mM Na^+ stimulation, $SO_4^{2-} = F^- < Cl^- < Br^- < NO_3^- \leqslant I^- < SCN^-$.

Thus, the anion effects vary with the cation species. The order of anion effects is similar to the sequence in a lyotropic series of anions. Since, in previous studies [1–4] it was found that SO_4^{2-} does not affect the response to Ca^{2+}, and that Cl^- does not affect the responses to Mg^{2+} and Na^+, the above results indicate that similar anions can both inhibit and enhance responses to cations. For example,

SCN^- has an inhibitory effect on the response to Ca^{2+}, whereas it has an enhancing effect on the responses to Mg^{2+} and Na^+. Concentrations for the appearance of the SCN^- effect varied with cation species. The results obtained here are consistent with a previous finding [1–3] that there are at least three distinct receptor sites for cations in the water fibers of the frog. Kinetic analysis of the data obtained for Na^+ stimulation demonstrated that SCN^- behaved as a non-competitive synergist in the response to Na^+, without altering affinity for Na^+.

It appears that each receptor site (cation-discriminative element) interacts with the respective anion-binding element that is affected by anion. Binding of an anion to the anion-binding element in the receptor domain modulates the permeability of the cation through the apical membrane, or the intrinsic activity of the cation (agonist)-receptor complex, without altering the affinity of the receptor for the respective cation. Since similar anions have different effects on responses to different cations, the mechanism of anion effect cannot be interpreted in terms of the anion-selective paracellular pathways between taste cells that have been proposed in the mammalian taste system. It is likely that the mechanism underlying the effects of anions on the salt taste response in the frog is different from that in mammals.

References

1. Kitada Y (1989) Taste responses to divalent cations in the frog glossopharyngeal nerve: competitive inhibition of the Mg^{2+} response by Ca^{2+}. Chem Senses 14:487–502
2. Kitada Y (1990) Taste responses to electrolytes in the frog glossopharyngeal nerve: initial process of taste reception. Brain Res 535:305–312
3. Kitada Y (1991) Competitive inhibition of the response to Na^+ by Ca^{2+} in water fibers of the frog glossopharyngeal nerve. Chem Senses 16:95–104
4. Kitada Y, Shimada K (1980) A quantitative study of the inhibitory effect of Na^+ and Mg^{2+} on the Ca^{2+} response of water fibers in the frog tongue. Jpn J Physiol 30:219–230

[1] Department of Physiology, Okayama University Dental School, 2-5-1 Shikata-cho, Okayama, 700 Japan

Electrophysiological Properties and Chemically-induced Responses of Mammalian Taste Bud Cells

Kumiko Sugimoto[1]

Key words. Mouse taste bud cells—Patch-clamp study

To elucidate the transduction mechanisms for sweet and umami tastes, it is necessary to use mammals for experiments, since they are more responsive to these tastes than amphibians, which have been used for numbers of patch-clamp studies. In the present study, therefore, we used isolated mouse taste bud cells and recorded voltage-gated and chemically-induced responses, using patch-clamp techniques.

Male ddY mice, 5–8 weeks of age, were used for the experiments. The epithelium, containing fungiform and circumvallate papillae, was peeled off 30 min after collagenase was injected into subepithelial parts of the tongue. Taste bud cells were finally dissociated by pipetting, following 10- to 30-min treatment of the isolated tongue epithelium with ethylenediamminetetraacetic acid (EDTA) and trypsin. Taste bud cells were identified by their long cell process and spindle or flask shape. Using whole-cell voltage clamp recording, we measured the currents in response to depolarizing voltage-steps and chemical stimulation in the isolated taste bud cells. Single channel activity of the basolateral membrane was also measured by on-cell and outside-out patch recordings. The taste solutions were ejected by air pressure from a micropipette situated close to the apical cell process. The standard external solution contained (in mM): 150 NaCl, 4.7 KCl, 3.3 CaCl$_2$, 0.1 MgCl$_2$, 2 N-2-hydroxyethylpiperazine-N'-2-ethanesulfonic acid (HEPES) and 7.8 glucose, and the standard pipette solution contained (in mM): 153 KCl, 2.5 MgCl$_2$, 0.5 CaCl$_2$, 10 HEPES, and 2 ethyleneglycol-bis-(β-aminoethyl ether) N,N,N',N'-tetraacetic acid (EGTA). Bath solutions were equilibrated with 100% O$_2$, and all solutions were adjusted to pH 7.4.

In the whole-cell configuration, the resting potentials of the isolated taste bud cells, measured at zero current, ranged from −40 to −95 mV. The cells were classified into three types, based on the currents in response to depolarizing voltage steps applied from a holding potential of −80 mV: (1) cells that demonstrated both initial transient inward currents and delayed outward currents, (2) cells which demonstrated only outward currents that were activated at −40 to −50 mV, and (3) cells which demonstrated only outward currents that were activated above 0 mV. It is an interesting question whether these three types correspond to three morphologically distinct types of spindle-shaped taste bud cells.

Some isolated taste bud cells elicited inward whole-cell currents at a −80 mV holding potential in response to 100 mM monosodium L-glutamate, which replaced NaCl in the standard external solution, while others elicited either outward currents or no current. This inward current was reversed at about −40 mV, indicating that the current may have passed through cation channels relatively permeable to K$^+$.

In outside-out patch recording, a single channel of 64 pS (at 25–40 mV) was blocked by 300 mM NaCl by extending the longer closed time component. This channel corresponded to high-threshold outward whole-cell current (type 3).

D-Phenylalanine (100 mM), a sweet stimulus, inhibited single channel activity recorded in the cell-attached configuration by shortening the mean open time and extending the mean closed time, indicating that intracellular events are involved in this effect. The relationship between the outward (from cell to pipette) channel current and pipette potential changed between −80 and −20 mV with the standard pipette solution, showing that the conductance of this channel was 33 pS and that the extrapolated reversal potential was about +40 mV.

Future pharmacological and ion-substituting experiments are needed to fully characterize the voltage-gated currents and the chemically-induced responses recorded from the isolated mouse taste bud cells in this experiment.

[1] Department of Neurobiology, Graduate School of Dentistry, Tokyo Medical and Dental University, 1-5-45 Yushima, Bunkyo-ku, Tokyo, 113 Japan

Patch Clamping of Single Mammalian Taste Receptor Cell

Yoshinori Uchida, Takenori Miyamoto, and Toshihide Sato[1]

Key words. Sweet stimuli—Gerbil taste receptor cells

This experiment was carried out to examine the mechanism for the transduction of sweet tastes in mammalian taste receptor cells. We used gerbil taste receptor cells, as the sensitivity to various sweeteners in these animals has been established neurophysiologically and behaviorally [1,2]. We made both whole cell and cell-attached patch recordings from the single taste cells. After the animal was killed by dislocating the cervical vertebrae, the tongue was removed and 1 mg/ml of elastase was injected into the muscle underlying its dorsal epithelium; the epithelium alone was then peeled off the underlying muscle. When taste receptor cells isolated from the fungiform papillae were incubated in 0.25% trypsin for 20 min at room temperature, a giga ohm seal of the cell membrane patch was easily established. Whole cell recordings, with the pipette containing a standard KCl solution, showed voltage-activated outward currents and inward currents in the taste cells. Outward K^+ currents were induced by depolarizing voltage steps from a holding potential of -60 mV, the threshold of activation being -20 mV. When K^+ in the pipette solution was replaced with Cs^+, inward rectifier K^+ currents were induced by hyperpolarizing voltage steps from a holding potential of -40 mV. The inward rectifier K^+ currents were marked at potentials more negative than -120 mV and were suppressed by 10 mM Ba^{2+} added to the bath. When outward currents were blocked by the addition of 10 mM Ba^{2+} to the bathing solution and by the replacement of K^+ in the pipette solution with Cs^+, sustained inward Ca^{2+} currents were observed. The Ca^{2+} currents were activated at -40 mV and reached a peak at $+20$ mV. In whole cell recordings, outward K^+ currents were suppressed reversibly by the application of 20 mM Na-saccharin to the bath. In cell-attached patch recordings, a single K^+ channel of 85 pS was blocked by 20 mM Na-saccharin, indicating that the blocking of the K^+ channel was mediated by a second messenger. In whole cell recordings, 5 mM cpt-cAMP applied to the bath also blocked outward K^+ currents. That cAMP is the second messenger has already been proposed by Béhé et al. [3] in their experiments in rat taste cells. In the present experiment, we clearly observed Na-saccharin-induced blocking of K^+ currents in both whole cell and single channel recordings. Our results support the hypothesis that the signal transduction of sweet tastes in mammalian taste receptor cells is mediated by cAMP-dependent blocking of K^+ channels.

References

1. Jakinovich W Jr, Oakley B (1975) Comparative gustatory responses in four species of gerbilline rodents. J Comp Physiol 99:89–101
2. Jakinovich W Jr (1981) Stimulation of the gerbil's gustatory receptors by artificial sweeteners. Brain Res 210:69–81
3. Béhé P, DeSimone JA, Avenet P, Lindemann B (1990) Membrane currents in taste cells of the rat fungiform papilla. J Gen Physiol 96:1061–1084

[1] Department of Physiology, Nagasaki University School of Dentistry, 1-7-1 Sakamoto, Nagasaki, 852 Japan

Electrical Responses of Taste Cells in Peeled Epithelium of Mouse Tongue

Masaya Etoh, Hidemasa Furue, and Kiyonori Yoshii[1]

Key words. Voltage-clamp—Salt responses

Several hypotheses have been advanced to explain the generation of receptor potentials in taste cells, intracellular messenger-mediated conductance changes being an important hypothesis in this area [1]. That is, intracellular messengers change the conductance of either the receptor or the basal membrane of taste cells.

We have previously suggested that the contribution of conductance on the receptor membrane to the generation of receptor potentials for amino acids is negligible [2]. However, more direct evidence is needed to define the taste transduction mechanism of where and how receptor potentials occur.

Isolated taste cells have provided much understanding of taste transduction mechanisms. However, these cells are not suitable for examining the role of the receptor and basal membranes, due to the difficulty of perfusing both membranes with different solutions under voltage clamp conditions. Typically, separation is important for investigating the receptor mechanisms for salts, since salts applied to the receptor membrane diffuse to the basal membrane.

In the present experiments, we made a recording chamber that separated the receptor membranes from the basal membranes of mice. Mouse tongue epithelium was set on the chamber. The receptor membrane was irrigated with deionized water and then stimulated with 100 mM NaCl, KCl, CaCl$_2$, or glycine while the basal membrane was perfused with a saline solution.

Cells in taste buds were voltage clamped with a glass electrode filled with 140 mM CsF and 300 μg/ml nystatin. The resistance of the electrode was about 10 MΩ. A computer provided command voltages and stored voltage clamp currents. It was difficult to obtain gigaseals of more than 1 GΩ.

We obtained the following results: (a) Zero current potentials were between −10 and −30 mV, probably due to poor gigaohm seals or gap junctions between the taste cells. (b) Depolarizing voltage steps produced transient inward currents followed by outward currents. (c) The outward currents were blocked by Cs added into the patch electrode. (d) Responses to salts were recorded from a few cells. At a holding potential of −50 mV, application of either 100 mM NaCl, KCl, or CaCl$_2$ solutions on the apical membrane adapted to deionized water elicited slow developing inward currents. (e) NaCl and CaCl$_2$ increased voltage-independent conductance, their reversal potentials being about at −20 mV. (f) KCl also produced a linear current-voltage relation with a reversal potential of about −20 mV. In addition, KCl increased slow developing conductance on hyperpolarization, apparently activated at −50 mV.

The equilibrium potentials of voltage-independent conductance elicited by the salts were similar, irrespective of the stimuli. Although further experiments are needed, it is suggested that the conductance occurs on the basal membrane because the ionic environment surrounding the basal membrane whereas that on the receptor membrane depends on the stimulating solutions.

It remains unknown where the slow developing conductance elicited by KCl occurs. This inward rectification may result from the asymmetric ionic composition of the solutions. Alternatively, an inward rectifier potassium channel may be responsible for the inward currents.

The results here indicate that this method is useful for the investigation of receptor mechanisms not only for salts but also for other taste substances. Although the probability of obtaining good gigaohm seals was low thus far, it will be worth while to improve this method.

Acknowledgments. This research was supported by the Salt Science Research Foundation of Japan.

References

1. McLaughin SK, McKinnon PJ, Margolskee RF (1992) Gustducin is a taste-cell-specific G-protein closely related to the transductions. Nature 357:563–569
2. Yoshii K, Kurihara K (1983) Ion dependence of the eel taste response to amino acids. Brain Res 280:63–67

[1] Department of Biochemical Engineering and Science, Kyushu Institute of Technology, 680-4 Kawazu, Iizuka, Fukuoka, 820 Japan

Role of Salivary Ions in Maintenance of Taste Sensitivity in the Chorda Tympani Nerve of Rats

RYUJI MATSUO and TOSHIFUMI MORIMOTO[1]

Key words. Chorda tympani nerve—Salivary gland

Previous behavioral [1] and neurophysiological [2] studies have shown that removal of the submandibular and sublingual salivary glands decreases taste sensitivity in rats. We examined the time course of decreases in sensitivity to the four standard taste stimuli (NaCl, sucrose, HCl, and quinine hydrochloride) and cold water (about 4°C), by comparing the activity of the chorda tympani nerve in normal and desalivated (2–28 days after removal of the submandibular-sublingual complexes) rats while the animals were deeply anesthetized with chlorpromazine and ketamine. We also examined the role of salivary ions and water in the maintenance of taste sensitivity. Distilled water or the ionic constituents of saliva (30 mM; NaHCO$_3$, and KCl) were supplied continuously (0.5 ml/h for 7 days) into the oral cavity of the desalivated rats, through the duct of the submandibular gland, and the taste and cold sensitivity of the chorda tympani nerve was examined while the animals were anesthetized.

We measured the initial peak and prolonged tonic (10 s after stimulus application) responses from the summated chorda tympani discharges. Since it is not valid to compare absolute response magnitudes between animals or groups, the ratios for responses to all stimuli were calculated relative to the amplitude of the baseline water-adapted activity [3]. As described below, alterations in the responsiveness of single chorda tympani fibers were similar to those of the whole nerve, after desalivation.

The supra-threshold responses to the four taste stimuli were sharply decreased by 7 days after desalivation, and then decreased slightly. By day 7, both the peak and tonic responses to NaCl, sucrose, HCl, quinine, and cold were decreased to about 55%, 50%, 40%, 40%, and 90%, respectively, of those of normal rats. By day 7, single fiber analyses revealed that the average discharge rates ($n = 55$) in response to NaCl, sucrose, HCl, and quinine were decreased to about 60%, 50%, 35%, and 40%, respectively. These attenuated discharge rates were seen overall in taste-responsive fibers. Discharge rates for cold stimulation were not altered. Elevation of the taste thresholds for the four taste stimuli was seen for 14 days after desalivation. The response to cold water gradually decreased to 80% by 28 days.

By day 7, the distilled water-supplied desalivated rats showed about 20% increases in NaCl peak and tonic responses and about 25% decreases in sucrose peak and tonic responses, whereas HCl, quinine, and cold responses were similar to those in normal rats. The thresholds for the four taste stimuli were not changed. On the other hand, the salivary ion-supplied rats showed similar magnitudes and thresholds of taste and cold responses to those of normal rats. These results suggest that salivary ions and water are important in maintaining normal taste sensitivity.

References

1. Brosvic GM, Hoey NE (1990) Taste detection and discrimination performance of rats following selective desalivation. Physiol Behav 48:617–623
2. Catalanotto FA, Frank ME, Contreras RJ (1986) Animal models of taste alteration. In: Meiselman M, Rifkin R (eds) Clinical measurement of taste and smell. Macmillan, New York, pp 429–442
3. Contreras RJ, Frank M (1979) Sodium deprivation alters neural responses to gustatory stimuli. J Gen Physiol 73:569–594

[1] Department of Oral Physiology, Faculty of Dentistry, Osaka University, 1-8 Yamadaoka, Suita, Osaka, 565 Japan

Mechanism of the Electric Response of Lipid Bilayers to Bitter Substances

MASAYOSHI NAITO[1], NAOYUKI SASAKI[2], and TAKESHI KAMBARA[3]

Key words. Bitter taste—Lipid bilayer

There is a marked similarity between the responses of taste cells and of lipid bilayers to bitter substances. The membrane potential of lipid bilayers is depolarized by the application of various bitter substances [1]. There is a good correlation between the minimum concentrations of various bitter substances required to depolarize lipid bilayers and those required to elicite bitter taste in humans [1]. As well as channel and receptor proteins, the lipid bilayer also seems to play an essential role in bitter taste transduction. It is quite important to investigate the characteristics of lipid bilayers to obtain a comprehensive understanding of the mechanism responsible for bitter taste transduction.

The lipid bilayer itself has interesting properties. It has been found that the responses of lipid bilayers to bitter stimuli have the following unique properties which cannot be explained by the usual model for membrane potential: (1) The response of the membrane potential appears even when there is no ion gradient across the membrane [1], (2) the response remains even when the salt in the stimulating solution is replaced with an impermeable cation salt [1], and (3) the direction of polarization is not reversed even when the ion gradient across the lipid multibilayer is reversed [2]. The membrane potential is usually described by ion permeation, as in the Goldman-Hodgkin-Katz model [3,4] for the potential induced by ion channels. All the above properties contradict predictions based on the channel model.

Here we report a theorectical study of the mechanism whereby the above unique properties of lipid bilayer responses to bitter stimuli are realized. Bitter substances may be adsorbed either on the surface region or on the deep inner region of the membrane. We assume that the adsorption of these substances on the surface region induces a change in the magnitude of the partition coefficient of ions between the membrane and the stimulation solution, and a change in the orientation of the electric dipoles of the membrane surface at the side where the stimuli are applied. We further assume that the adsorption of the substances on the inner region changes the magnitude of the diffusion constants of ions in the membrane.

We have found, based on a comparison of calculated and experimental results, that the response of lipid bilayers to bitter stimuli arises mainly from a changes in the partition coefficients. Protons play an essential role in the potential variation, due to changes in their partition coefficients. An increase in the proton partition coefficient induces depolarization of the membrane potential, as follows: An increase in the coefficient at the membrane surface on the side where the stimuli are applied produces an increase in the influx of protons, because the number of protons incorporated in the membrane from the side is increased. This increase in proton influx depolarizes the membrane, irrespective of the direction of the concentration gradient of the salt. This explains the above three properties (1), (2) and (3) of the bitter response of lipid bilayers.

References

1. Kumazawa T, Noma T, Kurihara K (1988) Liposomes as a model for taste cells: Receptor sites for bitter substances including N—C=S substances and mechanism of membrane potential changes. Biochemistry 27:1239–1244
2. Okahata Y, En-na G (1987) Electric responses of bilayer immobilized films as models of a chemoreceptive membrane. J Chem Soc Chem Commun 1987: 1365–1367
3. Goldman DE (1943) Potential, impedance and rectification in membranes. J Gen Physiol 27:37–60
4. Hodgkin AL, Katz B (1949) The effect of sodium ions on the electrical activity of the giant axon of the squid. J Physiol (Lond) 108:37–77

[1] Advanced Research Laboratories, Hitachi Ltd., Hatoyama, Saitama, 350-03 Japan
[2] Applied Mathematics Laboratory, The Nippon Dental University, Fujimi 1-9-20, Chiyoda-ku, Tokyo, 102 Japan
[3] Department of Applied Physics and Chemistry, The University of Electro-Communications, 1-5-1 Chofugaoka, Chofu, Tokyo, 182 Japan

A Theoretical Study of the Synergetic Effects of Mixed Taste Stimuli on the Responses of Taste Cells and Nerves

Yoshiki Kashimori[1], Akira Tuboi[1], Masayoshi Naito[2], and Takeshi Kambara[1]

Key words. Synergetic effects—Mixed taste

The understanding of the mechanism responsible for the recognition of mixed taste is quite important, since taste stimuli actually encountered in natural feeding are complex mixtures of substances with various kinds of taste. Many psychophysical studies of recognition of mixed taste and some physiological studies of the sensory effects of taste stimuli on nerve fibers have been carried out to investigate the effects of mixtures on the response to each component taste. These studies detected both synergetic and antagonistic responses, depending on the combinations of the two components tested and their concentrations. However, very few studies of the mixture effect have been made in regard to the response of taste receptor cells [1].

To clarify the response specificity intrinsic to receptor cell classes and to clarity taste nerve sensitivity, we carried out a theoretical investigation of the mechanism of transduction induced by single and mixed taste stimuli, using a well-defined realistic model of the taste receptor and afferent nerve systems. The receptor cells contained various combinations of four kinds of receptive proteins: (1) amiloride-sensitive Na^+ channels for NaCl, (2) H^+-blocking K^+ channels for HCl, (3) sugar-sensitive Na^+ channels, and (4) sugar-receptive protein + cAMP-blocking K^+ channels. We obtained various kinds of response patterns which had been observed experimentally for single taste stimuli, and we showed how the properties of taste cells and tight junctions affected these patterns. Whether the effects of taste mixing on the potential variation are synergetic, antagonistic, or nonexistent depend essentially not only on the properties of the taste cell

itself but also on the content of the interstitial space and the features of the tight junction. The mixture effects obtained from our present calculations can be briefly summarized as: (1) A mixture of NaCl and HCl solutions induced synergetic effects in those cells that had Na^+-sensitive sites for water adaptation, but had no effect in cells not showing this feature. (2) The effect of mixing of sugar with NaCl was synergetic in many kinds of taste cells, including those with sugar-sensitive Na^+ channels(S-N) or sugar-sensitive protein(S-K); however, this effect was antagonistic in some cells hyperpolarized by NaCl. (3) The mixture effect of sugar and HCl was antagonistic in many kinds of taste cells with S-N and in some cells with S-K hyperpolarized by HCl, but was synergetic in some cells with S-K depolarized by HCl. Based on the depolarization mechanism of apical and basolateral membranes, we considered how changes in the sensory effectiveness of the receptor site, due to the stimulation and the number of taste-insensitive ion channels, influenced the mixture effects.

We also studied nerve response in a model that consisted of a taste cell, synaptic connection, and afferent nerve. The calculated response patterns of the afferent nerve depended very closely on the rising of the depolarization curve of the taste cell. When the rising was rapid, the interspike interval histograms of impulse discharge resulted in an exponential distribution. As the rising became slower, the histograms showed a bimodal to γ-distribution. These distributions have been experimentally observed [2].

References

1. Ozeki M, Sato M (1972) Response of gustatory cells in the tongue of rat to stimuli representing four taste quantities. Com Biochem Physiol 41A:391–407
2. Nagai T, Ueda K (1981) Stochastic properties of gustatory impulse discharges in rat chorda tympani fibers. J Neuropysiol 45:574–593

[1] Department of Applied Physics and Chemistry, The University of Electro-Communications, 1-5-1, Choufugaoka, Choufu, Tokyo, 182 Japan
[2] Advanced Research Laboratory, Hitachi, Limited, 2520 Akanuma, Hatoyama-cho, Hiki-gun, Saitama, Japan

Response Properties of the Rat Lingual Trigeminal Nerve to Acidic Stimuli

B. Bryant and P.A. Moore[1]

Key words. Nociception—Irritation

The reception of oral irritants is mediated by free endings of the trigeminal nerve (V) that are located in the oral mucosa. These free nerve endings do not have direct access to the oral cavity but instead are located beneath the epithelial tight junctions. Thus, the reception and coding of chemical irritants are affected not only by the intrinsic response properties of the trigeminal neurons but also by the characteristics of the epithelium through which the stimuli must pass. This study addresses two questions regarding the reception of irritants within the oral cavity. First, we wanted to determine how potentially irritating acidic stimuli are encoded by single trigeminal neurons, and second, we wanted to determine the different possible pathways through which irritating stimuli gain access to free nerve endings. To this end, we tested a series of fatty acids (Cl to n-C8), as well as other acids (L-malic, L-tartaric, citric, oxalic, quinic, and CO_2), that come into contact with the tongue during ingestion. Neurons were also characterized by their responses to thermal stimuli. All of the neurons found in this study ($n = 51$) responded to a thermal stimulus. Of these, only 12% (six) of the neurons were sensitive to hot (>45°C) stimuli. Three of these neurons were hot nociceptors, not responding to any other stimuli, two also responded to citric but no other acid, and one neuron, a putative polymodal nociceptor, was sensitive to hot, cold, acidic, and mechanical stimuli. These latter two types of cells may mediate acid nociception. Forty-five neurons (88%) were sensitive to a decrease in temperature. The majority of these neurons (25) were not sensitive to any of the acids tested, while 20 neurons were sensitive to both cool or cold water and some type of acid. These results indicate that acidic and cool (16–20°C)/cold (4–8°C) stimuli are encoded by some of the same neurons and may therefore exhibit sensory interactions. The results also suggest that neurons which respond to innocuous stimuli may also be involved in the coding of noxious stimuli.

Physicochemical characteristics of acidic stimuli played an important role in determining the magnitude of response and latency to respond in acid-sensitive fibers. The mean response amplitude of single neurons to acid stimuli increased with increasing carbon chain length for straight chain fatty acids. The response to pentanoic acid was typically three to ten times greater than the response to formic acid for any single neuron. Moreover, the latency to respond decreased with increasing carbon chain length [responses to n-hexanoic acid (n-C6) were typically 3–8 s earlier than to formic acid (C1)]. Of the physicochemical properties we studied, only the octanol:H_2O partition coefficient, and not pK_a or pH, changed appreciably from formic acid to pentanoic acid. Thus, response amplitude and latency are both strongly correlated with the lipophilicity of this series of organic acids. Additional studies utilizing 10 mM $LaCl_3$ reversibly inhibited neural responses to 2.5 M KCl, but had no effect on response to either 100 mM pentanoic acid or to cool (17°C) water. This suggests that the access of fatty acids to free nerve endings through tight junctions is not as important as access through a lipid compartment in the epithelium. Together, these studies support the hypothesis that lipid solubility determines, to a great extent, the access and, therefore, the efficacy of fatty acid irritants in stimulating the lingual nerve endings.

[1] Monell Chemical Senses Center, 3500 Market St. Philadelphia, PA 19104, USA

Acknowledgments. Supported by NIH Grant DC-01921 and Kirin Brewery Co., Ltd.

The Effect of Gustatory Nerve Transections on Sucrose Responsiveness in the Rat

ALAN C. SPECTOR[1]

Key words. Taste—Ingestive behavior

Transection of the glossopharyngeal (GL) and/or the chorda tympani (CT) nerve in rats has relatively little effect on sucrose preference measured in long-term two-bottle tests [1]. Such tests, however, are vulnerable to the postingestive effects of the stimulus, and thus, the behavior might not be completely under gustatory control. In addition, no explicit attention was paid, in these prior studies, to the greater superficial petrosal nerve (GSP), a branch of the facial nerve which innervates palatal taste receptors, and which has since been shown electrophysiologically to be very responsive to sucrose applied to its receptive field [2]. Recent nerve-cut studies, employing behavioral methodology that is based on brief access exposures to taste stimuli, have demonstrated a role for palatal taste receptors in sucrose sensibility [3,4]. The present experiment was conducted to evaluate the relative contribution of the collective gustatory inputs of the 7th and 9th cranial nerves to the rat's unconditioned behavioral responsiveness to sucrose.

Rats were water deprived and then trained in a specially designed gustometer to lick a drinking spout to receive water. Next, rats were tested in a *water-deprived* state for their licking responses to 10-s presentations of water and a range of six sucrose concentrations (0.01–1.0 M). Water bottles were then replaced and the rats were retested 3 days later in a *nondeprived* state. This testing continued for a total of three sessions.

At the completion of testing before surgery, the rats were screened based on their performance. Only rats that had had at least five trials per stimulus over the 3-day non-deprived testing period were chosen to continue in the experiment. Three additional rats were excluded because they had excessively high lick scores to water compared with 1.0 M sucrose.

The rats were then deeply anesthetized and received bilateral transection of either: the CT (CTX, $n = 8$), the GL (GLX, $n = 9$), the GL and CT (GLX + CTX, $n = 6$), the GSP and CT (7thX, $n = 7$), or the GSP, CT, and GSP (7thX + 9thX, $n = 5$). An additional group served as a surgical control (CON, $n = 7$) and another group had the sublingual and submaxillary salivary glands removed (DESAL, $n = 6$). The sample sizes listed reflect those rats that survived surgery and had histologically confirmed complete nerve sections.

After recovery, the rats were retested for their responses to the stimulus array in a non-deprived state. At the end of the experiment, the water bottles were removed from the cages and the next day the rats received continuous access to water for 30 min in the apparatus. This was done to examine the motor competence of the rats.

A two-way analysis of variance (ANOVA; surgery × concentration) revealed that surgery significantly decreased sucrose responsiveness (non-deprived state) in the GLX + CTX, 7thX, 7thX + 9thX, and DESAL groups (all $Ps < 0.05$). The effect in the GLX + CTX group was relatively minor. An analysis of the change in the area under the concentration-response curves before vs after surgery indicated significant differences from 0 in the 7thX (-43.2%), DESAL (-65.1%), and 7thX + 9thX (-78.0%) groups (all $Ps < 0.05$). Although there was a significant increase in interlick interval following surgery in the 7thX and 7thX + 9thX groups, the magnitude of the apparent motor effect was not sufficient to completely account for the decreases in sucrose responsiveness observed as a function of the respective treatments.

These findings demonstrate that the proportion of taste buds eliminated by section of a given gustatory nerve is not related to the rat's unconditioned responsiveness to sucrose in any simple quantitative way. Moreover, the collective gustatory input from the branches of the 7th cranial nerve (CT + GSP) is necessary, and, most likely, sufficient to maintain normal unconditioned responsiveness to sucrose. In contrast, afferents of the 9th cranial nerve are neither necessary nor sufficient to maintain normal unconditioned responsiveness to sucrose. Finally, normal sucrose responsiveness depends, to a large extent, on the presence of the sublingual and submaxillary salivary glands.

[1]Department of Psychology, University of Florida, Gainesville, FL 32611, USA

Acknowledgments. I would like to thank Mircea Garcea, Steven St. John, Rachel Redman, Ray

Cauthon, and Perrin Klumpp for their technical assistance in this experiment.

References

1. Vance WB (1966) Water intake of partially ageusic rats. Life Sci 5:2017–2021
2. Nejad MS (1986) The neural activities of the greater superficial petrosal nerve of the rat in response to chemical stimulation of the palate. Chem Senses 11: 283–293
3. Krimm RF, Nejad MS, Smith JC, Miller IJ Jr, Beidler LM (1987) The effect of bilateral sectioning of the chorda tympani and the greater superficial petrosal nerves on the sweet taste in the rat. Physiol Behav 41:495–501
4. Spector AC, Travers SP, Norgren R (1993) Taste receptors on the anterior tongue and nasoincisor ducts of rats contribute synergistically to behavioral responses to sucrose. Behav Neurosci 107:694–702

3. Olfactory Transduction

Olfactory Reception: From Signal Modulation to Human Genome Mapping

Nissim Ben-Arie[1], Michael North[2], Miriam Khen[1], Ruth Gross-Isseroff[1], Naomi Walker[1], Shirley Horn-Saban[1], Uri Gat[1], Michael Natochin[1], Hans Lehrach[2], and Doron Lancet[1]

Key words. Human olfactory receptors—Olfactory threshold—Gene clusters—Receptor diversity—Genetic polymorphism—Molecular evolution

Introduction

Olfactory science has evolved into a comprehensive molecular definition brought about by the realization that broad odorant spectra may still be compatible with receptor genes encoding stereospecific proteins [1–3]. Evidence for the existence of specific molecular reception pathways arose with the discovery of G-protein-coupled second messenger transduction cascades in olfactory ciliary membranes [4–6]. Subsequently, a family of hundreds of G-protein-coupled receptors, first proposed based on the available biochemical data [7–9], became a reality with the cloning of olfactory receptors (ORs) in several vertebrate species [10–12]. It is proposed that when this molecular understanding is combined with human psychophysics and genome analysis, a better picture of human olfactory function will emerge.

Open Questions

The olfactory system is unique in that it responds to millions of chemicals [7,13,14], in contrast to most other reception devices, which have a much narrower scope. Olfactory sensory neurons undergo membrane depolarization when exposed to almost any volatile organic chemical [15]. Still, the pathway as a whole is at least as sensitive and selective as any other neuronal reception device. A long-time open question has been how discrimination is maintained despite the broad tuning of the individual sensory cells, seen in single neuronal electrophysiological recordings [16]. A favored answer has been that the brain performs pattern analysis, whereby the exact combination of activities across the different sensory cells provides a definition of odor quality and concentration [7,17–19]. Still, the ill-defined cellular response spectra have suggested to some researchers that specific receptors could not be at work. Rather, nonspecific odorant-induced mechanisms were proposed, such as membrane puncturing, lipid modulation, or surface potential effects [20,21]. Alternatively, it has been argued that each sensory cell might bear numerous protein receptor types—hence the broad response spectra. Current opinions and evidence tend more toward the notion that each sensory cell might have one or few receptor types (see below).

Another point of debate has been how many kinds of receptor are necessary for a functional olfactory system. An argument has been put forward that a few receptor types would suffice to code all odorants. A contrasting point of view, based on a probablistic molecular recognition model [22,23], claimed that 100 to 10000 receptors would be necessary [7]. These questions remained unsettled because continuous efforts failed to discover OR proteins and their genes.

An important open question is how the diversity of ORs is generated. Evolutionarily, the selective pressure appears to have been toward generating a large enough repertoire to be able to cope with practically all chemical configurations. It could have arisen solely by germ-line point mutations, or it could involve somatic mutations, recombination, and gene conversion mechanisms as well. Whatever the mechanism, a link must have been established between the identity of the receptor expressed in a given neuron and the central nervous system (CNS) target of that sensory neuron. A molecular understanding of OR genes should shed light on these and related questions.

Genetic Basis of Human Olfactory Thresholds

Evidence has been reported for a genetic basis of olfactory threshold variations [2,24–27]. The

[1] Department of Membrane Research and Biophysics, The Weizmann Institute of Science, Rehovot 76100, Israel
[2] Department of Genome Analysis, Imperial Cancer Research Fund, Lincoln's Inn Fields, London WC2A 3PX, U.K.

most likely molecular explanation is that threshold variations have to do with different individuals having different OR genes. A more quantitative statement is the *threshold hypothesis* [14], suggesting that an individual's olfactory threshold for a given odorant is determined by the receptor with the highest affinity toward this odorant. Specific anosmia (a decrement in the sensitivity for a particular odorant) may be explained by the absence of the gene for the highest affinity receptor, with the "taking over" of the receptor gene with the next highest affinity. The latter then becomes threshold-determining. Specific hyperosmia might likewise be explained as the occurrence of an atypically high affinity receptor whose gene is present in only a few individuals. This simple model predicts that specific anosmia will behave as a recessive trait (as indeed has been reported [2]). Likewise, specific hyperosmia is expected to behave as a dominant trait.

Cloning of Human OR Gene Clusters

Comparisons of ORs at the cDNA and genomic levels indicated that OR coding regions are intronless, 900 to 950 bases long [10,28]. This finding suggested that it would be relatively straightforward to carry out genomic cloning and mapping of such genes. As part of an effort to understand the genetics and molecular genetics of human olfaction, we have identified a cluster of 16 OR genes within a contiguous stretch of approximately 350 kb of DNA at the telomeric end of the p arm of human chromosome 17 [29,39]. It was afforded through the screening of a single chromosome cosmid library [30] with a PCR-generated OR probe, resulting in the isolation of 70 OR-positive cosmids. The cloned OR

genes on chromosome 17 appear to be clustered with a density of about five receptors per 100 kb, and several OR coding regions are often found on one cosmid clone. Other OR gene clusters exist on chromosome 19 [28] and possibly several others. Altogether, there may be several hundred OR genes in the human "olfactory subgenome," perhaps occupying a few megabases of DNA.

Patterns of OR Variation: A Putative OR Hypervariable Region

Comparing the sequences of OR genes encoded within the human chromosome 17 cluster, we were able to show that the highest degree of intersequence variability occurs in the fourth and fifth putative transmembrane segments (TM4 and TM5). Interestingly, one report [31] suggested that the extracellular loop connecting these two transmembrane segments (E2) may be part of an odorant binding domain. It is based on the high prevalence of mutations that alter the polarity of the coded amino acid in this loop. Taken together, the results suggest that the entire contiguous segment of the OR polypeptide, between the beginning of TM4 and the end of TM5, including the loop that connects them, may be important for odorant binding and could be regarded as the olfactory receptor *hypervariable region* (Fig. 1).

OR Polymorphism

Specific anosmia and hyperosmia are highly prevalent: Practically any individual tested displays

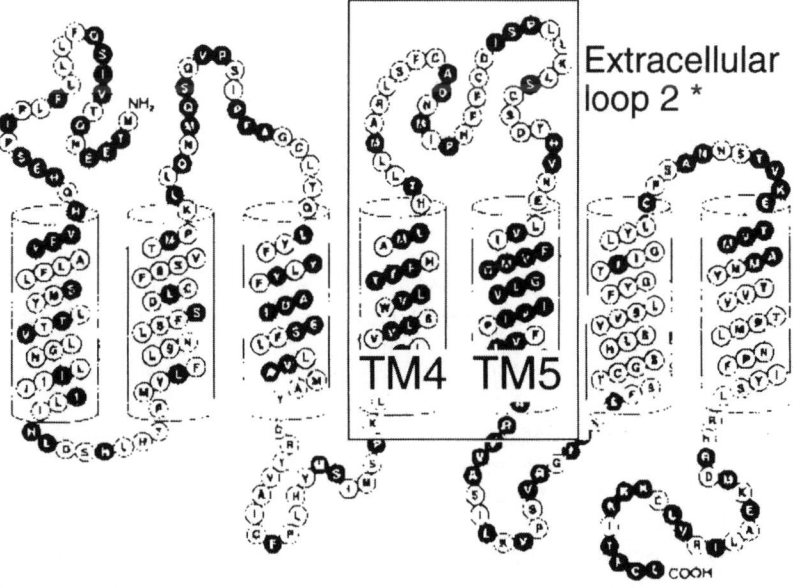

FIG. 1. Olfactory receptors' hypervariable region. The region spanning TM4, extracellular loop 2, and TM5, which is the largest variable region between ORs, is marked by a *box*. *Open circles* represent positions of conservation between rat ORs [10]

extreme olfactory threshold values for one of a few dozen odorants tested [24]. An explanation of such widespread individual differences in olfactory thresholds is that each human individual has only part of the entire OR gene repertoire of the species. In other words, the number of alleles in the population may much exceed the number of gene loci in an individual. This scenario holds for major histocompatibility complex genes (HLA in humans) [32] and manifests as a high level of heterozygosity: At each locus of an individual there is a high probability of having two different alleles. The evolutionary regimen that brings about this phenomenon is known as "balancing selection" or "overdominant selection" [33]. The alleles maintained by balancing selection are often different at multiple positions and display a high ratio of nonsynonymous to synonymous substitutions.

Individual variation in the ability to smell specific odorants is analogous to color vision abnormalities. Most of the individual variations in color sensitivity are related to the green and red pigments, a situation related to the occurrence of these two cone opsin genes in a small cluster that usually contains one red opsin and one to four copies of the green opsin [34]. Variations arise owing to unequal crossing over, which results in an increase or decrease of the number of green opsin genes, in the loss of the green or red opsin (or both), or in a red plus green opsin recombination. It is likely that similar phenomena occur within OR gene clusters. The number of variants that may arise in the latter case is much larger owing to the high gene count. As a result, it is possible that no two human individuals (except monozygotic twins) are identical in their OR repertoire.

Evolution of OR Genes

Olfactory receptors appear to be more related to photoreceptor opsins than to the other seven transmembrane domain receptors. This similarity is seen mainly in the minimal intracellular and extracellular loop structures of both sensory receptor classes. For example, the third cytoplasmic loop (connecting transmembrane domains 5 and 6) is about 20 amino acid long in both ORs and photoreceptor opsins but is usually larger in other superfamily members, up to 250 residues long in some of the muscarinic acetylcholine receptors. Likewise, both the amino and carboxy termini of all sensory receptors are about 25 residues long, whereas in many other superfamily members one or both termini are much longer. This minimal structure raises the possibility that the sensory receptors represent ancestral proteins in the seven transmembrane domain receptor superfamily.

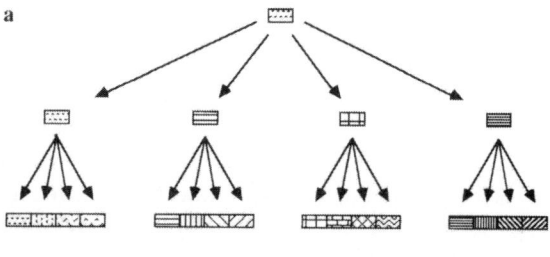

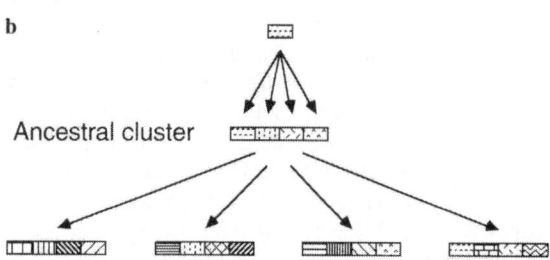

FIG. 2. Alternative scenarios for the evolution of the olfactory receptor gene repertoire. **a** Original single OR gene is duplicated in separated genomic regions, perhaps on different chromosomes. Each of these distant single genes then duplicates to form a cluster. The homology within each cluster is larger than that among members of different clusters. **b** Original OR gene first forms a local cluster by duplication, which is termed the ancestral cluster. This cluster is then duplicated as a whole in different genomic regions, including on different chromosomes. In this case ORs within a cluster are not particularly similar to each other, and in some cases it is perhaps possible to define homology groups across clusters (i.e., find ORs in two clusters that are more similar to each other than to other members of their respective clusters)

Macrosmatic mammals (e.g., dog and rat) may have many more OR genes than microsmatic ones (e.g., humans); and organisms with even less broadly disposed olfactory systems (e.g., fish and many invertebrates) could have even fewer ORs. A possible scenario is that early in evolution the number of OR genes of a species was small (10–30), as appears to be the case for fish [35,36], *C. elegans* and *Drosophila*. In some mammals, at later evolutionary stages, the number appears to have grown to several hundred [10,39]. Interestingly, this increase in OR diversity seems to have occurred solely at the germline level, without resort to ontogenetic DNA rearrangement, as is the case with immune receptors [37,38]. Human evolution may have witnessed a recent decrease in receptor gene number, as vision replaced olfaction as the dominant sense.

The original gradual increase in OR gene count may have occurred by gene duplication and diversification, as proposed for photoreceptor opsins [34]. Two alternative scenarios may be envisioned (Fig. 2): (1) OR genes were first individually scattered on different chromosomes, subsequently undergoing gene duplication. (2) First, a small single OR gene cluster formed, then the cluster as a whole duplicated several times. The homology relations among the OR genes on human chromosome 17 and their comparison to OR genes in other species suggest that the second possibility may be the correct one. It is because the OR genes in the chromosome 17 cluster show mutual differences nearly as large as those observed for randomly cloned OR genes from other species [39,40].

Future Directions

In the long run, correlating polymorphic genotypes to threshold variation phenotype could help identify olfactory receptors with known odorant specificity. Molecular modeling based on known OR sequences and assisted by the predicted structures of other seven transmembrane domain receptors could help elucidate the structure of the OR odorant binding site and its ligand pharmacology. This information could eventually provide a molecular-genetic basis for threshold polymorphisms, including specific anosmia and hyperosmia.

References

1. Amoore JE (1974) Evidence for the chemical olfactory code in man. Ann NY Acad Sci 237:137–43
2. Whissell-Buechy D, Amoore JE (1973) Odour-blindness to musk: simple recessive inheritance. Nature 242:271–273
3. Beets J (1971) Olfactory response and molecular structure. In: Beidler LM (ed) Handbook of sensory physiology, vol 4, Springer, Berlin, pp 257–321
4. Pace U, Hanski E, Salomon Y, Lancet D (1985) Odorant-sensitive adenylate cyclase may mediate olfactory reception. Nature 316:255–258
5. Sklar PB, Anholt RR, Snyder SH (1986) The odorant-sensitive adenylate cyclase of olfactory receptor cells: differential stimulation by distinct classes of odorants. J Biol Chem 261:15538–15543
6. Breer H, Boekhoff I, Tareilus E (1990) Rapid kinetics of second messenger formation in olfactory transduction. Nature 345:65–68
7. Lancet D (1986) Vertebrate olfactory reception. Annu Rev Neurosci 9:329–355
8. Lancet D, Pace U (1987) The molecular basis of odor recognition. Trends Biochem Sci 12:63–66
9. Reed RR (1990) How does the nose know? Cell 60:1–2
10. Buck L, Axel R (1991) A novel multigene family may encode odorant receptors: a molecular basis for odor recognition. Cell 65:175–187
11. Parmentier M, Libert F, Schurmans S, Schiffman S, Lefort A, Eggerickx D, Ledent C, Mollereau C, Gerard C, Perret J, et al. (1991, in press) Expression of members of the putative olfactory receptor gene family in mammalian germ cells. Nature
12. Raming K, Krieger J, Strotmann J, Boekhoff I, Kubick S, Baumstark C, Breer H (1993) Cloning and expression of odorant receptors. Nature 361:353–356

13. Shepherd GM (1991) Sensory transduction: entering the mainstream of membrane signaling. Cell 67: 845–851
14. Lancet D (1992) Olfactory reception: from transduction to human genetics. In: Corey DP, Roper SD (eds) Sensory transduction. Rockefeller University, New York, pp 73–91
15. Getchell TV (1986) Functional properties of vertebrate olfactory receptor neurons. Physiol Rev 66:772–818
16. Sicard G, Holley A (1984) Receptor cell responses to odorants: similarities and differences among odorants. Brain Res 292:283–296
17. Shepherd GM (1985) The olfactory system: the uses of neural space for a non-spatial modality. Prog Clin Biol Res 176:99–114
18. Kauer JS (1991) Contributions of topography and partial processing to odor coding in the vertebrate olfactory pathway. Trends Neurosci 14:79–85
19. Schild D (1988) Principles of odor coding and a neural network for odor discrimination. Biophys J 54:1011–1011
20. Davies JT (1971) Olfactory theories. In: Beidler LM (ed) Handbook of Sensory Physiology, vol 4. Springer, Berlin, pp 322–350
21. Nomura T, Kurihara K (1987) Liposomes as a model for olfactory cells: changes in membrane potential in response to various odorants. Biochemistry 26: 6135–6140
22. Lancet D (1983) What determines the size of the immune receptor repertoire? A probabilistic analysis of molecular interactions 1-A169. Jerusalem
23. Lancet D, Sadovsky E, Seidemann E (1993) Probability model for molecular recognition in repertoires of biological receptors: significance to the olfactory system. Proc Natl Acad Sci USA 90:3715–3719
24. Amoore JE, Steinle S (1991) A graphic history of specific anosmia. In: Wysocki CJ, Kare MR (eds) Chemical senses, vol 3. Dekker, New York, pp 331–352
25. Wysocki CJ, Beauchamp GK (1984) Ability to smell androstenone is genetically determined. Proc Natl Acad Sci USA 81:4899–4902
26. Wysocki CJ, Beauchamp GK (1991) Individual differences in human olfaction. In: Wysocki CJ, Kare MR (eds) Chemical senses, vol 3. Dekker, New York, pp 353–373
27. Gross-Isseroff R, Ophir D, Bartana A, Voet H, Lancet D (1992) Evidence for genetic determination in human twins of olfactory thresholds for a standard odorant. Neurosci Lett 141:115–118
28. Levy NS, Bakalyar HA, Reed RR (1991) Signal transduction in olfactory neurons. J Steroid Biochem Mol Biol 39:633–637
29. Ben-Arie N, North M, Khen M, Margalit T, Lehrach H, Lancet D (1992) Mapping the olfactory receptor "sub-genome": implications to human sensory polymorphisms. In: Myers R, Porteous D, Roberts R (eds) Genome Mapping and Sequencing. Cold Spring Harbor Laboratory, Cold Spring Harbor, pp 280–280
30. Nizetic D, Zehetner G, Monaco P, Gellen L, Young BD, Lehrach (1991) Construction, arraying and high-density screening of large insert libraries of human chromosome X and 21: their potential use as reference libraries. Proc Natl Acad Sci USA 88:3233–3237
31. Hughes LA, Hughes MK (1993) Adaptive evolution in the rat olfactory receptor gene family. J Mol Evol 36:249–254

32. Klein J, Figueroa F, Nagy ZA (1983) Genetics of the major histocompatibility complex: the final act. Annu Rev Immunol 1:119–142

33. Altukhov YP (1991) The role of balancing selection and overdominance in maintaining allozyme polymorphism. Genetica 85:79–90

34. Nathans J, Piantanida TP, Eddy RL, Shows TB, Hogness DS (1986) Molecular genetics of inherited variation in human color vision. Science 232:203–210

35. Ngai J, Dowling MM, Buck L, Axel R, Chess A (1993) The family of genes encoding odorant receptors in the channel catfish. Cell 72:657–666

36. Ngai J, Chess A, Dowling MM, Necles N, Macagno ER, Axel R (1993) Coding of olfactory information—topography of odorant receptor expression in the catfish olfactory epithelium. Cell 72:667–680

37. Berek C, Milstein C (1988) The dynamic nature of the antibody repertoire. Immunol Rev 105:5–26

38. Kronenberg M, Siu G, Hood LE, Shastri N (1986) The molecular genetics of the T-cell antigen receptor and T-cell antigen recogntion. Annu Rev Immunol 4:529–591

39. Lancet D, Ben-Arie N (1993) Olfactory receptors. Current Biol 3(10):668–674

40. Ben-Arie N, Lancet D, Taylor C, Khen M, Walker N, Ledbetter DH, Carrozzo R, Patel K, Sheer D, Lehrach H, North MA (in press) Olfactory receptor gene cluster on human chromosome 17: possible duplication of an ancestral receptor repertoire. Human Mol Genetics 1994

Olfactory Receptor Family: Diversity and Spatial Patterning

Susan L. Sullivan, Kerry J. Ressler, and Linda B. Buck[1]

Key words. Olfactory neuron—Odorant receptors—Olfactory receptors—Spatial patterning—Expression zones—Zonal organization

Introduction

To gain insight into the mechanisms underlying olfactory perception in mammals, we have focused on the primary receptive elements of the olfactory system, the odorant receptors. In our initial experiments, we identified a large, diverse multigene family that appears to encode hundreds of odorant receptors that are expressed on olfactory sensory neurons [1]. To determine how the information provided by such a large collection of diverse receptors might be organized, we have examined the patterns of expression of various odorant receptor genes in the olfactory epithelium. We have observed distinct topographic patterns of odorant receptor RNAs that indicate that the olfactory epithelium is divided into a series of expression zones [2]. These zones are likely to provide a broad organization of sensory information in the nasal cavity, which is maintained in the axonal projection to the olfactory bulb.

Materials and Methods

The materials and methods employed have been described previously [1,2].

Results and Discussion

Olfactory Multigene Family

We initially sought to identify putative odorant receptors, the premise being that the isolation and characterization of the these molecules would provide a means of investigating the molecular mechanisms underlying olfactory perception. In these experiments, we identified and characterized a novel multigene family that appeared to encode odorant receptors on rat olfactory sensory neurons [1]. Our

data indicated that this family is expressed in the olfactory epithelium but not in a number of other tissues, and that within the olfactory epithelium it may be exclusively expressed by olfactory sensory neurons. Consistent with previous observations implicating heterotrimeric G-proteins in olfactory signal transduction [3,4], the olfactory multigene family encodes proteins that are members of the large superfamily of receptors that transduce signals via interactions with G-proteins. Southern blotting experiments and genomic library screens indicated that the olfactory multigene family is composed of hundreds of genes.

On the basis of these observations, we concluded that the multigene family was likely to encode a large variety of odorant receptor proteins that are expressed on olfactory sensory neurons in the rat. This conclusion is supported by functional analyses of members of this family that are expressed by cell lines following transfection of odorant receptor cDNAs [5].

Families of genes homologous to the rat olfactory multigene family have now been identified in a variety of species, including humans [6–9], the dog [6], mouse [2,10], salamander (A. Jesurum, D. Chicaraishi, and J. Kauer, ACHEMS abstract, 1993), catfish [11], and zebrafish (S. Korshing and H. Baier, ACHEMS abstract, 1992). In at least three of these species (rat, mouse, catfish), the expression of the odorant receptor genes has been localized to olfactory sensory neurons in the olfactory epithelium [2,10–13; R. Reed et al., unpublished]. We have also observed expression of members of the olfactory multigene family in the mouse vomeronasal organ by in situ hybridization (E. Liman and L. Buck, unpublished).

Patterns of Diversity and Subfamilies in the Odorant Receptor Family

An alignment of the odorant receptors reveals several intriguing features of the olfactory receptor family that may be important to its role in olfactory perception [1]. First, as shown in Fig. 1, there is extensive sequence diversity in this receptor family. Interestingly, such diversity is seen in several of the transmembrane domains implicated in ligand binding in other G-protein-coupled receptors [14,15].

[1] Department of Neurobiology, Harvard Medical School, 220 Longwood Avenue, Boston, MA 02115, USA

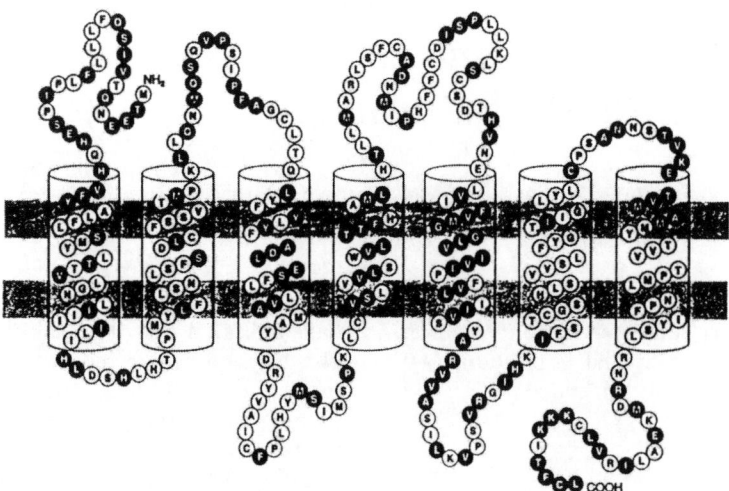

FIG. 1. Sequence diversity in the odorant receptor family. The protein encoded by rat odorant receptor cDNA clone I15 is shown traversing the plasma membrane seven times with its *N*-terminus located extracellularly and its *C*-terminus intracellularly. The vertical cylinders delineate the seven putative α helices spanning the membrane. Positions at which extensive diversity is observed in a series of rat odorant receptors are shown as black balls. The high degree of sequence variability, particularly that evident in transmembrane domains 3, 4, and 5, may enable the odorant receptor family to recognize a variety of structurally diverse odorous ligands. (From Buck and Axel [1], with permission)

This diversity in potential ligand binding regions is consistent with the ability of the olfactory system to recognize a variety of structurally diverse odorants.

Another striking feature is the presence of subfamilies within the family [1]. Although the cDNAs we sequenced encode a diverse family of receptors, some of these receptors are almost identical to each other. Southern blotting experiments with individual cDNAs revealed that, in fact, members of the olfactory multigene family belong to subfamilies of 1 to about 20 highly related genes. We propose that different subfamilies recognize different structural classes of odorants, and individual members of a subfamily detect subtle differences between structurally related odorants.

Topographic Patterning of Odorant Receptors in the Olfactory Epithelium

The remarkable size and diversity of the odorant receptor multigene family indicate that odor discrimination relies on the differential binding properties of hundreds of odorant receptors. How does the olfactory system organize the information provided by such a large collection of diverse receptors? One possibility is that the olfactory system, like other sensory systems, uses physical space within the nervous system to organize information [16–19]. If so, "spatial codes" or "topographic maps" for odors might exist within the olfactory epithelium or bulb. The molecular basis for these maps might be the selective expression of different receptor genes in spatially segregated sets of olfactory neurons in the olfactory epithelium or the selective formation of synapses by such neuronal sets at distinct sites within the olfactory bulb.

To investigate this possibility, we analyzed the patterns of expression of different odorant receptor genes in the olfactory epithelium of the mouse [2]. In preliminary experiments, we amplified members of the mouse olfactory multigene family from genomic DNA by polymerase chain reaction (PCR), cloned the individual gene segments, and analyzed a number of clones using nucleotide sequencing and Southern blotting.

We next performed in situ hybridization studies in which we hybridized ^{35}S-labeled antisense RNAs prepared from individual clones to coronal and horizontal sections spanning the mouse olfactory epithelium. In-depth analyses were performed with three receptor clones (K4, K7, K18), and limited studies were carried out with a large number of additional receptor clones.

We found that each of the odorant receptor subfamily probes hybridizes to cells that are centrally located within the olfactory epithelium, confirming that the odorant receptor genes are expressed in olfactory neurons but not in olfactory stem cells or supporting cells. Each hybridized neuron is surrounded by many neurons that have not hybridized to the probe, arguing against the hypothesis that spatial codes for odors consist of tightly clustered neurons that recognize the same odors.

Each receptor probe hybridized to neurons in many, but not all, areas of the nasal cavity. Moreover, the patterns of hybridized neurons in the two

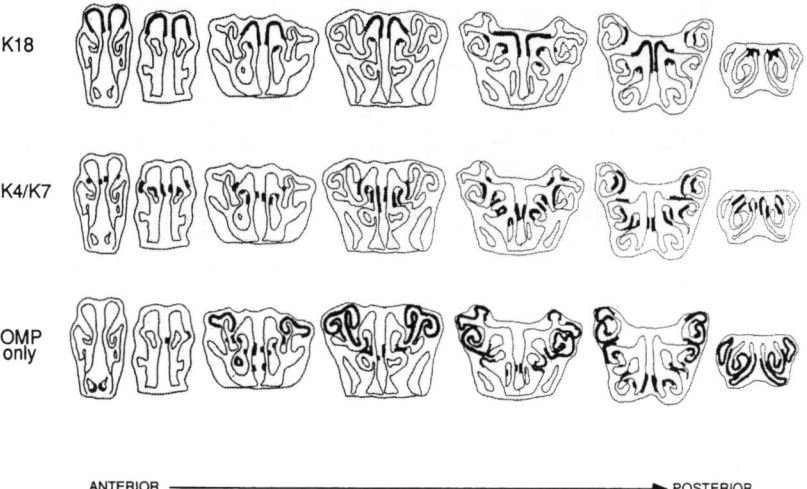

ANTERIOR ━━━━━━━━━━━━━━━━━━━━━━━━━━━━━━━━▶ POSTERIOR

FIG. 2. Mouse nasal cavity showing regional patterns of odorant receptor gene expression. Tracings of coronal sections from regions spaced approximately 300 μm apart along the anteroposterior axis of the nasal cavity are shown. The shaded areas of the olfactory epithelium indicate regions that hybridize to the K18 receptor probe, the K4 and K7 receptor probes, or the OMP probe, but not to the K4, K7, or K18 probe. (From Ressler et al. [2], with permission)

nasal cavities appeared to be bilaterally symmetric. To define the hybridization patterns more precisely, we made tracings from projected photographs of a large number of coronal sections that had been hybridized to the K4, K7, or K18 receptor subfamily probe or to the OMP probe, which is expressed in all mature olfactory neurons [20]. The diagrams in Fig. 2 were constructed on the basis of these data. From the coronal diagrams, we prepared diagrams of predicted medial and lateral views of the hybridization patterns (Fig. 3).

Our observations indicate that odorant receptor genes are expressed in highly specified topographic patterns in the olfactory epithelium [2]. The expression of each receptor subfamily is limited to distinct regions of the olfactory epithelium. These regions exhibit bilateral symmetry in the two nasal cavities and are virtually identical in different individuals. We found that the K4 and K7 subfamilies are always expressed in the same regions, whereas the K18 subfamily is always expressed in different regions. In fact, the K4/K7 and K18 regions appear to be mutually exclusive except at shared boundaries, where a small degree of overlap is observed. About 93% to 99% of cells that hybridize to each probe fall within the boundaries indicated in Fig. 3.

Each receptor subfamily is expressed in several bands located in different regions of the nasal cavity (e.g., the septum and individual turbinates). These bands extend along the anteroposterior axis of the nasal cavity (Figs. 2 and 3). Interestingly, the K4/K7 and K18 expression zones show a consistent dorsoventral or mediolateral relation to one another. The K18 hybridizing regions are always located dorsal or medial to the K4/K7 regions. Interestingly,

regions that hybridize to the OMP probe, but not to K4, K7, or K18, are always located ventral or lateral to the K4/K7 regions.

Our measurements indicate that the K4/K7 and K18 expression zones each cover about one-quarter of the olfactory epithelium. If the olfactory epithelium is divided into nonoverlapping expression zones of approximately equal size, there could be as few as four such zones. To investigate this possibility, we performed limited in situ hybridization studies with a large number of additional odorant receptor probes that recognize different subfamilies. Many of these probes show patterns of hybridization characteristic of those observed with K4 and K7 or with K18, including some that appear to recognize only a single gene in Southern blots. However, others hybridize to regions that hybridize to the OMP probe but not to K4, K7, or K18 probes. The patterns of hybridization we have observed thus far indicate that there are at least four expression zones that are organized along the dorsoventral and mediolateral axes.

Although the expression of each receptor subfamily is confined to neurons in a single zone, we have not observed clustering or any other spatial ordering of hybridized neurons within an expression zone. Quantitations of the densities of hybridizing neurons indicate that there is a similar density of cells that hybridize to each probe throughout the expression zone. Within their respective zones, the K4 and K18 subfamilies are expressed in about 0.7%, and the K4 subfamily in about 0.3%, of the olfactory neuron population [2].

Our experiments indicate that the zonal assignment of each odorant receptor gene is strictly

LATERAL

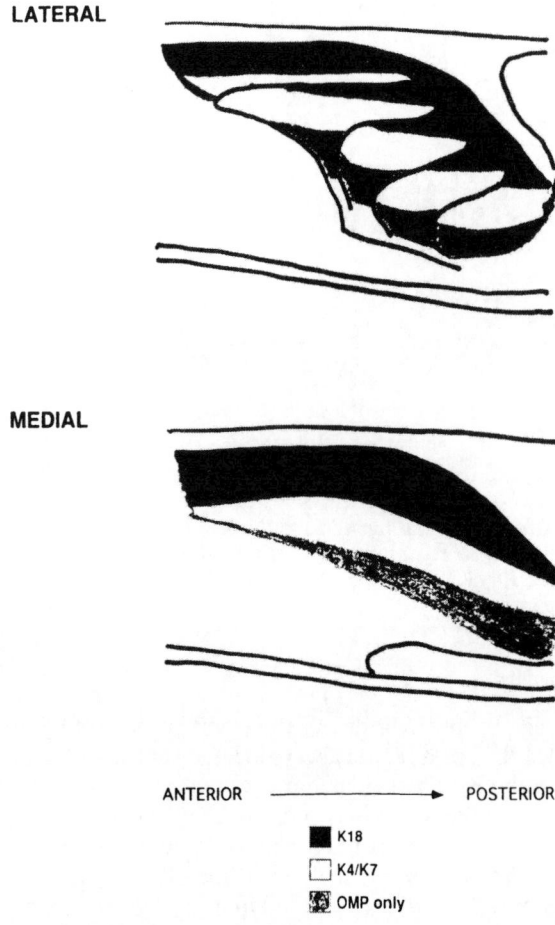

MEDIAL

ANTERIOR ————————————→ POSTERIOR

▉ K18
☐ K4/K7
▨ OMP only

FIG. 3. Zonal expression of odorant receptor genes in medial and lateral views of the nasal cavity. Predicted views of the medial and lateral nasal cavity based on reconstructions of coronal sections are shown. A lateral view of the nasal cavity shows that cells that hybridize to the K18 probe are located on the roof, the dorsal lateral wall, and the dorsal surface of each turbinate (II–IV). Cells that hybridize to the K4 and K7 probes lie just ventral to the K18⁺ regions on each of these structures. Regions that hybridize to the OMP probe but not to the receptor probes are ventral to the K4/K7⁺ regions. The medial view shows bands of expression extending from anterior to posterior along the nasal septum

regulated. However, within each zone neurons that express a particular receptor gene are broadly distributed. It is therefore likely that when an olfactory neuron or its progenitor chooses which odorant receptor gene(s) to express it is restricted to a single zonal gene set but may choose a receptor gene (or set of genes) to express from among the members of the set via a stochastic mechanism [2].

Expression Zones: Functional Implications and Derivation

Our studies indicate that neurons that express a given odorant receptor gene—and therefore must

recognize the same odorants—are confined to a single expression zone. The restriction of multiple members of the same subfamily to the same zone further indicates that neurons that express related odorant receptor genes that are likely to recognize related odorants are also located in the same zone. These findings suggest that the expression zones might provide for a broad organization of incoming sensory information in the olfactory epithelium.

Numerous studies of the axonal projection from the olfactory epithelium to the olfactory bulb indicate that this projection, like the expression zones, is organized along the dorsoventral and mediolateral axes [21–25]. In some studies, retrograde tracers have been placed in different bulbar regions and the patterns of back-labeled neurons in the epithelium examined [21,25,26; T. Schoenfeld, personal communication]. Comparisons of our zonal patterns with the patterns of labeled neurons seen in those studies suggest that neurons that belong to the same odorant receptor expression zone project axons to the same broad region along the dorsoventral axis of the bulb. Thus the zonal organization of sensory information in the epithelium appears to be maintained during the transmission of this information to the bulb. It is possible that, within each bulbar region, a further refinement of this patterning occurs such that neurons that express the same odorant receptor gene synapse within the same set of glomeruli. By analogy with the visual system [27], such a sharpening of the axonal projection might conceivably arise via ligand-dependent, activity-dependent mechanisms [19].

Theoretically, odorant receptor expression zones could be an intrinsic characteristic of the olfactory epithelium or could arise indirectly from a programming of the epithelial to bulb projection via retrograde influences imposed by the olfactory bulb. However, we have found that zonal patterns are in existence in the olfactory epithelium prior to the development of the olfactory bulb (S. Sullivan, K. Ressler, and L. Buck, unpublished data). This finding suggests that zonal specification occurs early during development in the absence of the olfactory bulb. We have proposed that the zones might derive from patterns laid down along the dorsoventral axis of the olfactory placode prior to the development of the nasal cavity [2].

References

1. Buck L, Axel R (1991) A novel multigene family may encode odorant receptors: a molecular basis for odor recognition. Cell 65:175–187
2. Ressler KJ, Sullivan SL, Buck LB (1993) A zonal organization of odorant receptor gene expression in the olfactory epithelium. Cell 73:597–609
3. Reed RR (1992) Signaling pathways in odorant detection. Neuron 8:205–209

4. Breer H, Boekhoff I, Krieger J, Raming K, Strotmann J, Tareilus E (1992) Molecular mechanisms of olfactory signal transduction. In: Sensory transduction. Rockefeller University Press, New York

5. Raming K, Krieger J, Strotmann J, Bockhoff I, Kubick S, Baumstark C, Breer H (1993) Cloning and expression of odorant receptors. Nature 361:353–356

6. Parmentier M, Libert F, Schurmans F, Schiffmann S, Lefort A, Eggerickx D, Ledent C, Mollereau C, Gerard C, Perret J, Grootegoid A, Vassart G (1992) Expression of members of the putative olfactory receptor gene family in mammalian germ cells. Nature 355:453–455

7. Reed RR (1993, in press) CIBA Foundation Symposium 179. CIBA Foundation, London

8. Selbie LA, Townsend-Nicholson A, Iismaa TP, Shine J (1992) Novel G protein-coupled receptors: a gene family of putative human olfactory receptor sequences. Brain Res 13:159–163

9. Lancet D (1993, in press) CIBA Foundation Symposium 179. CIBA Foundation, London

10. Nef P, Hermans-Borgmeyer I, Artieres-Pin H, Beasley L, Dionne VE, Heinemann SF (1992) Spatial pattern of receptor expression in the olfactory epithelium. Proc Natl Acad Sci USA 89:8948–8952

11. Ngai J, Dowling MM, Buck L, Axel R, Chess A (1993). The family of genes encoding odorant receptor in the channel catfish. Cell 72:657–666

12. Koshimoto H, Katoh K, Yoshihara Y, Mori K (1992) Distribution of putative odour receptor proteins in olfactory epithelium. Neuroreport 3:521–523

13. Strotmann J, Wanner I, Krieger J, Raming K, Breer H (1992) Expression of odorant receptors in spatially restricted subsets of chemosensory neurones. Neuroreport 3:1053–1056

14. Strader CD, Sigal IS, Dixon RAF (1989) Structural basis of beta-adrenergic receptor function. FASEB 3:1825–1832

15. Dohlman HG, Thorner J, Caron MG, Lefkowitz RJ (1991) Model systems for the study of seven-transmembrane-segment receptors. Annu Rev Biochem 60:653–688

16. Kauer JS (1991) Contributions of topography and parallel processing to odor coding in the vertebrate olfactory pathway. Trends Neurosci 14:79–85

17. Stewart WB, Kauer JS, Shepherd GM (1979) Functional organization of rat olfactory bulb, analyzed by the 2-deoxyglucose method. J Comp Neurol 185:715–734

18. Adrian ED (1956) The action of the mammalian olfactory organ. J Laryngol Otol 70:1–14

19. Buck LB (1992) The olfactory multigene family. Curr Opin Gen Dev 2:467–473

20. Danciger E, Mettling C, Vidal M, Morris R, Margolis F (1989) Olfactory marker protein gene: its structure and olfactory neuron-specific expression in transgenic mice. Proc Natl Acad Sci USA 86:8565–8569

21. Clancy AN, Schoenfeld TA, Macrides F (1985) Topographic organization of peripheral input to the hamster olfactory bulb. Chem Senses 10:399–400

22. Saucier D, Astic L (1986) Analysis of the topographical organization of olfactory epithelium projections in the rat. Brain Res Bull 16:455–462

23. Schwarting GA, Crandall JE (1991) Subsets of olfactory and vomeronasal sensory epithelial cells and axons revealed by monoclonal antibodies to carbohydrate antigens. Brain Res 547:239–248

24. Schwob JE, Gottlieb DI (1986) The primary olfactory projection has two chemically distinct zones. J Neurosci 6:3393–3404

25. Astic L, Saucier D, Holley A (1987) Topographical relationships between olfactory receptor cells and glomerular foci in the rat olfactory bulb. Brain Res 424:144–152

26. Astic L, Saucier D (1986) Anatomical mapping of the neuroepithelial projection to the olfactory bulb in the rat. Brain Res Bull 16:445–454

27. Shatz CJ (1990) Impulse activity and the patterning of connections during CNS development. Neuron 5:745–756

Signal Recognition and Transduction in Olfaction

HEINZ BREER[1]

Key words. Odorants—Receptors—Transduction—Second messengers—Kinases

Introduction

The sense of smell depends fundamentally on the detection of specific molecules and its conversion into a form readable by nerve cells. Olfactory receptor cells use specialized forms of ubiquitous biochemical pathways mediating signalling across cell membranes. Interaction of odorants with the chemosensory membrane of olfactory neurons triggers an amplified cellular response which deflects the membrane potential within 10–100 ms, thereby encoding the strength, duration, and quality of odorant stimuli into distinct patterns of afferent neuronal signals which travel along the cell's axon towards the brain. Our understanding of the molecular mechanisms mediating the primary events of odor detection has lagged behind the knowledge about other sensory modalities.

Results

Second Messenger Signalling

Questions concerning the time course of odorant-induced second messenger responses have been addressed using a rapid kinetic methodology which allows the monitoring of odorant-induced formation of second messengers in olfactory preparations on a subsecond time scale [1]. Mixing a suspension of isolated olfactory cilia from rat with odorants, such as citralva or isomenthone, elicited a rapid and transient elevation of the cyclic adenosine monophosphate (cAMP) level. The odorant-induced "pulse" of cyclic nucleotide preceded the electrical response of olfactory receptor cells, and the buildup of cAMP-levels may account for the latency of electrical responses (Fig. 1). Surveying the second messenger response of rat cilia to a variety of different odorous compounds demonstrated that only certain odorants induced a rapid rise in cAMP; inactive odors, such as pyrazine, induced a rapid and transient change of the inositol trisphosphate (IP$_3$)

levels [2]. The functional importance of odor-induced rapid IP$_3$-responses is emphasized by the recent identification of IP$_3$-gated ion channels in the plasma membrane of olfactory cells [3,4] (Fig. 1). Coexistence of two olfactory signalling pathways in the same cell would allow an efficient interaction between both systems, thereby providing the poten-

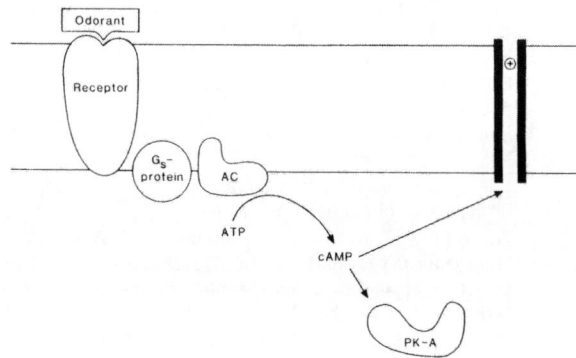

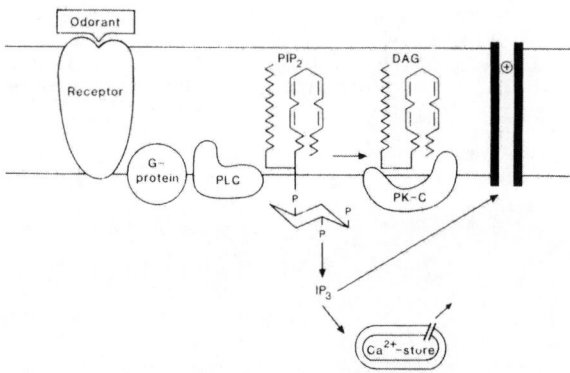

FIG. 1. A schematic representation of the alternative chemoelectrical transduction pathways in olfactory neurons. Interaction of odorants with specific receptor proteins triggers via characteristic G-proteins the activation of an intracellular reaction cascade generating second messengers. The resulting transiently elevated levels of second messengers (cAMP; IP$_3$ + DAG) have a dual function: firstly, activation of second messenger gated cation channels, thus eliciting the generator potential; secondly, stimulation of protein kinases which uncouple the reaction cascade via phosphorylation of activated receptors (signal termination). *IP$_3$*, inositol trisphosphate; *cAMP*, cyclic adenosine monophosphate; *DAG*, diacylglycerol; *AC*, adenylate cyclase; *PK-A(c)*, protein kinase A(c); *PLC*, phospholipase C; *PIP$_2$*, phosphatidyl inositol bisphosphate

[1] University of Stuttgart-Hohenheim, Institute of Zoophysiology, 70593 Stuttgart, Germany

tial for considerable positive or negative "cross talks". Thus, convergent integration of different odor stimuli may already begin at the level of primary reactions in olfactory receptor cells.

One of the characteristic features of the odorant-induced second messenger response is the transient nature of this molecular signal. Pulsatile changes in second messenger levels are essential for sensory receptor neurons responding precisely to iterative stimulation. According to the desensitization of many hormone and neurotransmitter systems [5], the "turn off" reaction is blocked in the presence of the kinase inhibitors [6]. The odorant-induced second messenger response is terminated by the kinase controlled by the second messenger which is generated in that particular cascade; the cAMP pathway by kinase A and the IP_3 pathway by kinase C (Fig. 1). These observations suggest that odorant-induced second messenger signalling is turned off via a negative feedback reaction uncoupling the reaction cascade by phosphorylating an element of the transduction apparatus. Recent evidence indicates that another kinase, a receptor-specific kinase, is involved in "turning off" the olfactory cascade. In fact, it was found that a characteristic subtype of the β-adrenergic receptor kinases (βARK)-2-like kinase, is preferentially expressed in the olfactory neuroepithelium and this kinase is specifically involved in the rapid desensitization of olfactory signalling [7]. Since desensitization phenomena have been attributed to phosphorylation of the receptor proteins, it was conceivable that in the olfactory system, receptors for odorants may be phosphorylated. Recent immunoprecipitation experiments employing receptor-specific antibodies have indeed confirmed that receptor proteins are phosphorylated upon odorant stimulation.

Receptors for Odorants

Exploring the nature and diversity of receptors for odorants has long been considered as the key for understanding the molecular basis of olfaction. The recent discovery of a novel gene family which is supposed to encode odorant receptors [8] has now opened the way to explore this critical element of the transduction apparatus. In order to prove whether the encoded proteins in fact bind specific odorants and interact with second messenger systems, identified cDNAs have been expressed in nonneuronal surrogate cells. Sf9 cells infected with baculovirus harbouring the receptor-encoding cDNA were assayed for odorant-induced second messenger formation. A large panel of odorous compounds, including representatives of different odor classes and compounds representing different chemical classes, was analyzed. Some odorous aldehydes were found to induce a significant generation of IP_3 in a dose-dependent manner [9]. Thus, candidate

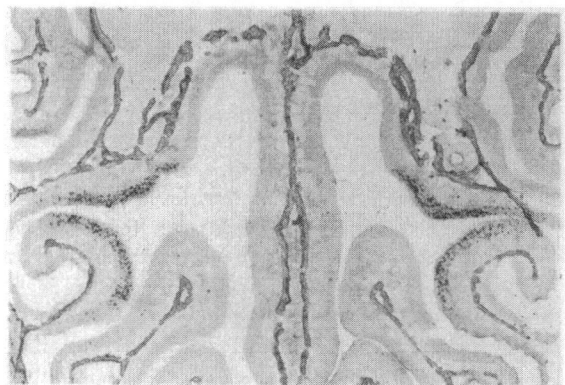

Fig. 2. Spatial distribution of olfactory receptor neurons expressing a particular odorant receptor subtype (OR37). Coronal sections of the rat nasal neuroepithelium were probed with digoxigenin-labelled antisense RNA riboprobe. In sections of the olfactory epithelium, OR37 transcripts can be visualized in a subset of chemosensory neurons segregated in restricted areas on endoturbinate II and ectoturbinate 3. The clusters of reactive cells are located symmetrically on both sides of the septum in the nasal cavities

odorant receptors can be expressed in surrogate cells and the second messenger response elicited by odorants can be used to monitor the affinity and specificity of ligand/receptor interaction.

The identification of cDNA clones encoding odorant receptors allowed the exploration of the spatial segregation of olfactory neurons expressing particular receptor genes using in situ hybridization approaches. Under high stringency conditions a punctated pattern of labelling was observed in certain regions of the sections; large areas were nonreactive. These observations indicate that cells expressing a particular receptor type are not randomly distributed throughout the neuroepithelium, but rather segregated in a defined, yet rather wide area. Some of the subtypes were found within the same regions, others in complementary zones. One of the receptors was found to be expressed only in a spatially very restricted subset of cells (Fig. 2) [10]. A detailed mapping of the nasal neuroepithelium for the spatial expression pattern of related and divergent odorant receptor subtypes may contribute to the unravelling of the chemotopic organization of the olfactory system.

Discussion

Recent advances in the physiology, biochemistry, and molecular biology of olfaction support the concept that second messengers provide the critical link between the initial odor recognition by specific receptor proteins and the elicitation of generator currents in olfactory receptor cells.

Second messenger signalling is terminated by uncoupling the reaction cascades via kinases catalyzing the phosphorylation of activated odorant receptors. The status of odorant receptors controlled by kinase/phosphatase systems may determine the sensitivity of olfactory receptor cells.

The expression of odorant receptors in surrogate cells is a first step in reconstituting the olfactory cascade, which may contribute to the exploration of important aspects of olfactory signalling, such as the specificity of odorant/receptor and receptor/G-protein interaction.

Acknowledgments. The work from this laboratory was supported by the Deutsche Forschungsgemeinschaft.

References

1. Breer H, Boekhoff I, Tareilus E (1990) Rapid kinetics of second messenger formation in olfactory transduction. Nature 345:65–68
2. Boekhoff I, Tareilus E, Strotmann J, Breer H (1990) Rapid activation of alternative second messenger pathways in olfactory cilia from rats by different odorants. EMBO J 9:2453–2458
3. Restrepo D, Miyamoto T, Bryant BP, Teeter JH (1990) Odor stimuli trigger influx of calcium into olfactory neurons of the channel catfish. Science 249: 1166–1168
4. Fadool DA, Ache BW (1992) Plasma membrane inositol 1,4,5-trisphosphate-activated channels mediate signal transduction in lobster olfactory neurons. Neuron 9:907–918
5. Lefkowitz RJ, Hausdorf WP, Caron MG (1990) Role of phosphorylation in desensitization of the β-adrenoceptor. TIPS 11:190–194
6. Boekhoff I, Breer H (1992) Termination of second messenger signaling in olfaction. Proc Natl Acad Sci USA 89:471–474
7. Schleicher S, Boekhoff I, Arriza J, Lefkowitz RJ, Breer H (1993) A β-adrenergic receptor kinase-like enzyme is involved in olfactory signal termination. Proc Natl Acad Sci USA 90:1420–1424
8. Buck L, Axel R (1991) A novel multigene family may encode odorant receptors: a molecular basis for odor recognition. Cell 66:175–187
9. Raming K, Krieger J, Strotmann J, Boekhoff I, Kubick S, Baumstark C, Breer H (1993) Cloning and expression of odorant receptors. Nature 361:353–356
10. Strotmann J, Wanner I, Krieger J, Raming K, Breer H (1992) Expression of odorant receptors in spatially restricted subsets of chemosensory neurons. NeuroReport 3:1053–1056

Role of Inositol Triphosphate (IP$_3$) in Olfactory Transduction

Lisa FitzGerald[1], Yukio Okada[1,4], D. Lynn Kalinoski[1], Christian DellaCorte[1], Joseph G. Brand[1,3], John H. Teeter[1,2], and Diego Restrepo[1,2]

Key words. Olfactory transduction—Inositol-1,4,5-triphosphate—Adenosine 3'5'-cyclic monophosphate—Stopped flow—Calcium

Introduction

Olfactory neurons respond to stimulation with odorants by increasing the frequency of action potential discharge. The sequence of biochemical reactions that culminates in action potential discharge is believed to start with the interaction of odor stimuli with specific G-protein-coupled receptors located on the plasma membrane of cilia that extend into the mucous layer. Families of putative odorant receptors have been identified in several species [1–3], and a putative receptor clone has been shown to cause stimulation of second messenger formation when expressed in a heterologous system [4].

Some odorants stimulate G-protein-mediated formation of adenosine 3',5'-cyclic monophosphate (cAMP), and in amphibian neurons stimulation with certain odorants causes opening of cAMP-gated channels leading to membrane depolarization and action potential firing (reviewed in [5–7]). However, some potent odorants, such as isovaleric acid in air breathers and certain amino acid odorants (e.g., L-cysteine) in fish, do not stimulate cAMP formation as measured in isolated olfactory cilia preparations [8,9]. Work in our laboratory has been directed towards studying the possible role of the second messenger inositol-1,4,5-triphosphate (IP$_3$) in olfactory transduction. In other biological systems, IP$_3$ causes release of Ca^{2+} from internal stores via efflux of Ca^{2+} from the endoplasmic reticulum through IP$_3$-gated channels, which is followed by stimulation of Ca^{2+} influx through the plasma membrane by yet unknown mechanisms [10]. In this manuscript, we summarize biochemical, biophysical, and electrophysiological studies which indicate that IP$_3$ plays a mediatory role in olfactory transduction acting in an unorthodox manner involving influx of extracellular Ca^{2+} without release of Ca^{2+} from internal stores.

Materials and Methods

Catfish olfactory cilia were isolated using a Ca^{2+} shock procedure described previously [9]. To study second messenger formation, the cilia were stimulated with amino acid odorants or guanosine 5'-O-(3-thiotriphosphate) (GTPγS) and the reaction was stopped quickly (10–500 ms) in a quenched flow apparatus [9]. Rat olfactory neurons were isolated by treatment with papain in divalent cation-free Ringer solution [11]. The neurons were loaded with the fluorescent Ca^{2+} indicator fura-2, and intracellular calcium was measured employing a Quantimet 570 image analysis workstation by procedures described in detail in [11].

Results

The olfactory system of the channel catfish can detect amino acids at nanomolar concentrations [12,13]. It was recognized early on that certain potent amino acid stimuli such as L-cysteine did not stimulate cAMP formation [14], but rather, appeared to elicit hydrolysis of phosphatidylinositol-4,5-diphosphate to diacylglycerol and inositol-1,4,5-triphosphate [15]. On the basis of these observations, Huque and Bruch [15] proposed that IP$_3$ could play a mediatory role in olfactory transduction for certain odorants. More recently, we have measured second messenger formation in isolated catfish olfactory cilia in the subsecond time range, and we find that the potent amino acid odorants L-alanine and L-cysteine cause G-protein mediated increases in IP$_3$ formation within 50 ms after stimulation [9]. Significant increases in IP$_3$ formation were measured with 10 nM odorant. In contrast, these odorants did not stimulate cAMP or cGMP formation within this time frame, even at concentrations as high as 1 mM.

In the rat, Breer and coworkers have shown that some odorants stimulate cAMP formation while

[1] Monell Chemical Senses Center, Philadelphia, PA 19014, USA
[2] Department of Physiology, School of Medicine, [3] Department of Biochemistry, School of Dental Medicine, University of Pennsylvania, Philadelphia, PA 19104, USA
[4] Department of Physiology, Nagasaki University School of Dentistry, 1-7-1 Sakamoto, Nagasaki, 852 Japan

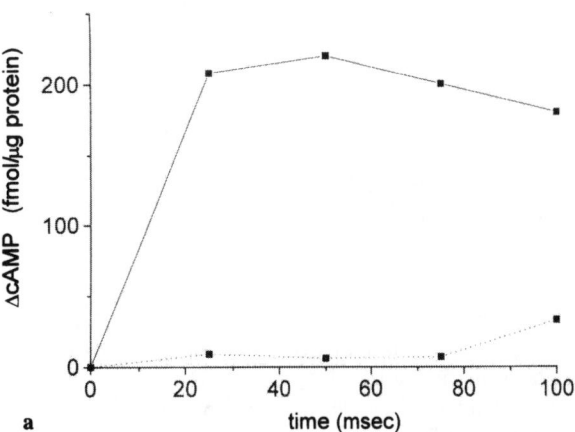
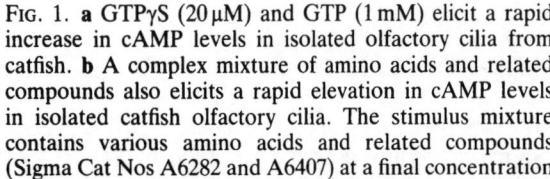

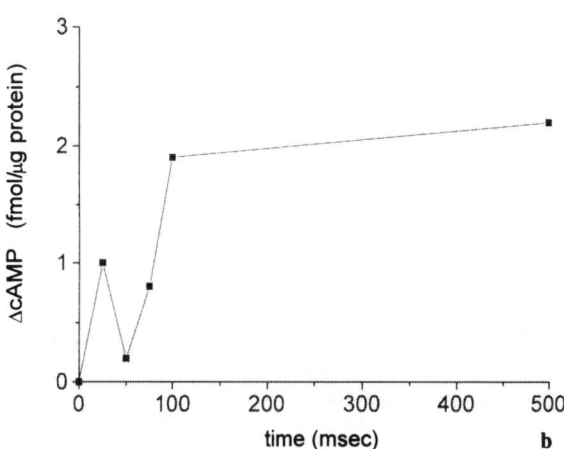

Fig. 1. **a** GTPγS (20 μM) and GTP (1 mM) elicit a rapid increase in cAMP levels in isolated olfactory cilia from catfish. **b** A complex mixture of amino acids and related compounds also elicits a rapid elevation in cAMP levels in isolated catfish olfactory cilia. The stimulus mixture contains various amino acids and related compounds (Sigma Cat Nos A6282 and A6407) at a final concentration of 100 μM each, except for L-cysteine which is present at 50 μM. cAMP levels are plotted as the difference from basal ($\Delta cAMP$). The basal level of cAMP in the experiment in **a** was 45 fmol/μg protein and it averaged 3.2 ± 0.7 (mean ± SD, $n = 6$) for the experiment in **b**. Results are representative of three experiments for **a** and two experiments for **b**

others stimulate formation of IP$_3$ [8]. Although, as discussed already, in catfish some potent amino acid stimuli elicit IP$_3$ formation only, other stimuli could elicit cAMP formation. To explore this issue we measured cAMP formation in catfish olfactory cilia using guanine nucleotides and a complex mixture of amino acids as stimuli. Stimulation of catfish olfactory cilia with GTPγS causes a rapid increase in cAMP formation (Fig. 1a), and a complex mixture of amino acids also causes a rapid elevation in cAMP levels (Fig. 1b). We do not know yet whether this increase in cAMP is triggered by one or several of the compounds included in the odorant mixture used in Fig. 1b, or by the relatively large total odorant concentration. However, these experiments are the first to show that the olfactory system of the catfish can respond to odorants with increases in either cAMP or IP$_3$ in a manner analogous to the olfactory system of mammals [8]. Presumably, the increase in cAMP causes membrane depolarization in catfish by direct gating of an olfactory-specific cAMP-gated channel which was recently cloned from catfish olfactory epithelium [16].

These studies of IP$_3$ formation in catfish and rat show that certain odorants stimulate production of IP$_3$ in the absence of cAMP formation. The action of IP$_3$ in other cell systems is typically mediated by release of Ca^{2+} from internal stores [10]. To explore the mechanism of action of IP$_3$ in the olfactory system we measured intracellular calcium ([Ca$_i$]) in isolated rat and catfish olfactory neurons using the fluorescent Ca^{2+} indicator fura-2 ([17] and [25]). Rat olfactory neurons responded to stimulation with odorants known to elicit IP$_3$ formation in rat olfactory cilia with an increase in [Ca$_i$] (Fig. 2a), and the increase in [Ca$_i$] did not take place in the absence of extracellular Ca^{2+} (Fig. 2b). These results are similar to those obtained previously by us in isolated catfish olfactory neurons stimulated with amino acid odorants [17]. These experiments indicate that IP$_3$ acts in olfactory neurons in an unusual manner which does not involve release of Ca^{2+} from internal stores. In fact, in the rat many olfactory neurons appear to accumulate little Ca^{2+} in internal stores as evidenced by the fact that thapsigargin, which releases Ca^{2+} from internal stores causing an increase in [Ca$_i$] in other cells [10], does not affect the levels of [Ca$_i$] in rat olfactory neurons (Fig. 2c). Consistent with the notion that these neurons do not accumulate large amounts of Ca^{2+} in internal stores, the addition of ionomycin in the absence of extracellular Ca^{2+} causes no change in the measured [Ca$_i$] (not shown).

Discussion

The studies of second messenger formation and [Ca$_i$] regulation described here indicate that IP$_3$ may act in a novel manner in olfactory neurons not involving release of Ca^{2+} from internal stores. The way in which IP$_3$ may exert its action in the olfactory system is suggested by electrophysiological and biochemical experiments in rat and catfish: biochemical studies indicate that olfactory cilia from rat [18] and catfish [19] possess an IP$_3$ receptor with a specificity for inositol polyphosphates that is clearly different from the IP$_3$ receptor located in the endoplasmic reticulum in brain and other peripheral tissues. The

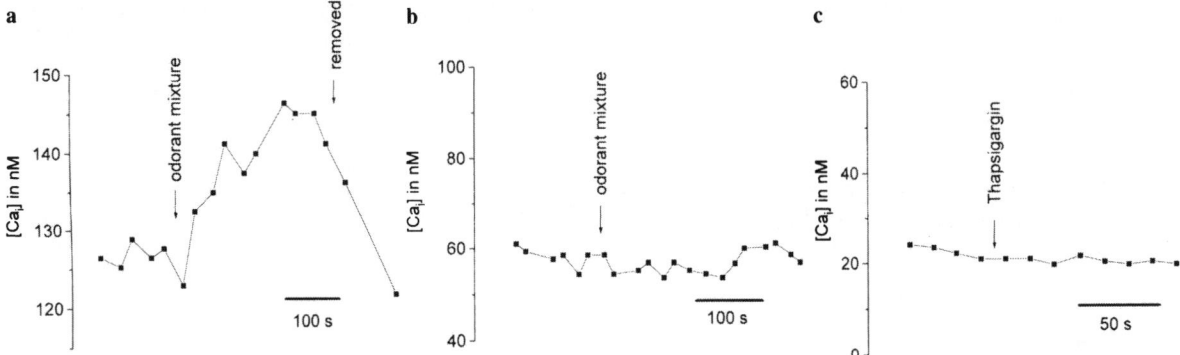

FIG. 2. Measurements of intracellular Ca^{2+} in isolated rat olfactory neurons. **a** A mixture of odorants that is known to stimulate inositol triphosphate (IP$_3$) formation in rat olfactory cilia [8] elicits a transient increase in [Ca$_i$]. The odorant mixture contained 100 µM each of lilial, lyral, ethyl vanillin, isovaleric acid, trimethyl amine, and phenyl ethyl amine. **b** The same odorant mixture does not affect [Ca$_i$] when applied in the absence of extracellular Ca^{2+}. Ca$_0$ was removed by adding 1 mM EGTA to Ringer's solution without added CaCl$_2$. **c** Thapsigargin (2 µM) does not elicit an increase in [Ca$_i$] in an isolated rat olfactory neuron. Representative of three experiments

olfactory IP$_3$ receptor is recognized by an antibody raised against the carboxyl terminus of the brain type I IP$_3$ receptor, and this antibody distinctly labels the surface of the olfactory epithelium ([20] and C. DellaCorte, D. Restrepo, and D.L. Kalinoski, manuscript in preparation). Furthermore, when olfactory cilia membranes from rat [18] or catfish [17] are incorporated into artificial phospholipid bilayers at the tip of a patch pipette, IP$_3$ elicits opening of a cationic conductance which is slightly selective for divalent cations. The IP$_3$-gated channel in both species is inhibited by the hexavalent cation ruthenium red. These studies indicate that isolated olfactory cilia membranes possess an IP$_3$-receptor/channel protein.

In addition, in experiments in which IP$_3$ is dialyzed into olfactory neurons from either rat [26] or catfish [17,21], IP$_3$ elicits a depolarization under current clamp, and an inward current at negative holding potentials under voltage clamp. The current–voltage curves are consistent with a cation channel slightly selective for divalent cations, and the responses to intracellular IP$_3$ are inhibited by ruthenium red. The whole cell response does not appear to be due to an increase in [Ca$_i$] as evidenced by the fact that addition of ionomycin causes development of an outward K$^+$ current clearly different from the IP$_3$-induced current.

On the basis of these experiments we have proposed that IP$_3$ may act in the olfactory system by directly opening a ciliary (IP$_3$-gated) cation channel [17]. This hypothesis is consistent with results from experiments in other laboratories in which the action of IP$_3$ was studied in the olfactory systems of lobster [22], rat [20], and frog [23]. Some studies, however, appear to be inconsistent with a direct role for IP$_3$ in olfactory transduction [24]. Further work is necessary to determine whether this hypothesis is viable, to establish the extent of interaction of different second messenger systems, and to clarify the role of the olfactory IP$_3$-gated channel(s).

Acknowledgments. This work was supported by National Institutes of Health grant DC-00566, by a grant to J.G. Brand from the Veterans Affairs Department, and by cooperative agreement No 14-16-0009-91-930 between the US Fish and Wildlife Service and the Monell Chemical Senses Center.

References

1. Buck L, Axel R (1991) A novel multigene family may encode odorant receptors: a molecular basis for odor recognition. Cell 65:175–187
2. Ngai J, Dowling MM, Buck L, Axel R, Chess A (1993) The family of genes encoding odorant receptors in the channel catfish. Cell 72:657–666
3. Ressler KJ, Sullivan SL, Buck L (1993) A zonal organization of odorant receptor gene expression in the olfactory epithelium. Cell 73:597–609
4. Raming K, Krieger J, Strotmann J, Boekhoff I, Kubick S, Baumstark C, Breer H (1993) Cloning and expression of odorant receptors. Nature 361:353–356
5. Reed R (1992) Signaling pathways in odorant detection. Neuron 8:205–209
6. Anholt RRH (1993) Molecular neurobiology of olfaction. Crit Rev Neurobiol 7:1–22
7. Ronnett GV, Snyder SH (1992) Molecular messengers of olfaction. Trends Neurosci 15:508–513
8. Breer H, Boekhoff I (1991) Odorants of the same odor class activate different second messenger pathways. Chem Senses 16:19–29
9. Restrepo D, Boekhoff I, Breer H (1993) Rapid kinetic measurements of second messenger formation in olfactory cilia from channel catfish. Am J Physiol (Cell Physiol) 264:C906–C911
10. Irvine RF (1992) Inositol lipids in cell signalling. Curr Opinions Cell Biol 4:212–219
11. Restrepo D, Okada Y, Teeter JH, Lowry LD, Cowart B, Brand JG (1993) Human olfactory neurons respond

to odor stimuli with an increase in cytoplasmic Ca^{2+}. Biophys J 64:1961–1966

12. Caprio J (1978) Olfaction and taste in the channel catfish: an electrophysiological study of the responses to amino acids and derivatives. J Comp Physiol 123: 357–371

13. Bruch RC, Rulli RD (1988) Ligand binding specificity of neutral L-amino acid olfactory receptor. Comp Biochem Physiol (B Comp Biochem) 91:535–540

14. Bruch RC, Teeter JH (1990) Cyclic AMP links amino acid chemoreceptors to ion channels in olfactory cilia. Chem Senses 15:419–430

15. Huque TH, Bruch RC (1986) Odorant- and guanine nucleotide-stimulated phosphoinositide turnover in olfactory cilia. Biochem Biophys Res Comm 137: 36–42

16. Goulding EH, Ngai J, Kramer R, Colicos S, Axel R, Siegelbaum S, Chess A (1992) Molecular cloning and single-channel properties of the cyclic nucleotide-gated channel from catfish olfactory neurons. Neuron 8: 45–58

17. Restrepo D, Miyamoto T, Bryant BP, Teeter JH (1990) Odor stimuli trigger influx of calcium into olfactory neurons of the channel catfish. Science 249: 1166–1168

18. Restrepo D, Teeter JH, Honda E, Boyle AG, Marecek JF, Prestwich GD, Kalinoski DL (1992) Evidence for an InsP₃-gated channel protein in isolated rat olfactory cilia. Am J Physiol (Cell Physiol) 263:C667–C673

19. Kalinoski DL, Aldinger S, Boyle AG, Huque T, Marecek JF, Prestwich GD, Restrepo D (1992) Characterization of an inositol-1,4,5-trisphosphate receptor in isolated olfactory cilia. Biochem J 281:449–456

20. Wood SF, Ronnett GV, Snyder SH (1990) Activation of inositol-phosphate metabolism in primary olfactory cell cultures (abstract). Chem Senses 15:655

21. Miyamoto T, Restrepo D, Cragoe EJ Jr, Teeter JH (1992) IP₃- and cAMP-induced responses in isolated olfactory receptor neurons from the channel catfish. J Membr Biol 127:173–183

22. Fadool DA, Ache BW (1992) Plasma membrane inositol-1,4,5-trisphosphate-activated channels mediate olfactory transduction in lobster olfactory receptor neurons. Neuron 9:1–20

23. Suzuki N (1992) IP₃-activated ion channel activity in frog olfactory receptor cells (abstract). Chem Senses 17:87

24. Lowe G, Gold GH (1993) Contribution of the ciliary cyclic nucleotide-gated conductance to olfactory transduction in the salamander. J Physiol 462:175–196

25. Restrepo D, Okada Y, Teeter JH (1993) Odorant-regulated Ca^{2+} gradients in rat olfactory neurons. J Gen Physiol 102:907–924

26. Okada Y, Teeter JH, Restrepo D (1994) Inositol 1,4,5-trisphosphate-gated conductance in isolated rat olfactory neurons. J Neurophysiol (in press)

β-Adrenergic Receptor Kinase, β-Arrestin, and cGMP as Mediators of Olfactory Desensitization

G.V. Ronnett, T.M. Dawson, D.E. Jaworsky, A.J. Roskams, R.E. Bakin, H.H. Cho, and R.J. Lefkowitz[1]

Key words. Olfaction—Second messengers—Desensitization—Kinase—Cyclic GMP—Cell culture

Introduction

Desensitization of signal transduction may occur through a number of processes, including receptor internalization, receptor uncoupling mediated by receptor phosphorylation, and receptor downregulation. Heterologous desensitization (nonagonist mediated) may proceed via protein kinase A (PKA), protein kinase C (PKC), or other kinases. Homologous or agonist-induced desensitization of the β_2-adrenergic receptor occurs through agonist-activated specific receptor phosphorylation that is catalyzed by a specific receptor kinase called the β-adrenergic receptor kinase (β-ARK). Complete quenching of the signal requires the action of β-arrestin (β-ARR). As the putative odorant receptors are members of the G-protein-coupled receptor family, we investigated the possibility that these proteins were involved in olfactory desensitization.

Additionally, abundant evidence indicates that nitric oxide (NO) is a physiologic messenger molecule. NO is a short-lived free radical gas that can activate the soluble guanylyl cyclase; it is synthesized by nitric oxide synthase (NOS). The cloning of NOS reveals it has close homology with only one other mammalian protein, cytochrome P-450 reductase. This enzyme also serves as the electron donor for heme oxygenase, which degrades heme to biliverdin and releases carbon monoxide (CO) in the process. We investigated the localization of heme oxygenase-2 (HO-2) and the ability of HO to regulate cGMP levels in olfactory neuronal cultures. The role of cGMP in signal transduction or desensitization was then investigated.

Methods

Rat olfactory cilia are prepared as previously described [1]. Primary cultures of rat olfactory neurons are prepared as described by Ronnett et al. [2]. Assays of adenylyl cyclase (AC) and phospholipase C (PLC) activities are performed as previously described [3]. Rapid stop-flow experiments and immunohistochemistry is performed as described by Dawson et al. [4].

Results

β-ARK and β-ARR isoforms were identified in the olfactory neuroepithelium with affinity-purified rabbit antibodies raised against glutathione-S-transferase fusion proteins containing the carboxyl terminus of rat β-ARK-1, β-ARK-2, β-ARR-1, or β-ARR-2. β-ARK-2 immunoreactivity is evident in the olfactory neuroepithelium, whereas β-ARK-1 immunoreactivity is completely absent. This immunoreactivity localizes to the dendritic knobs in cilia. Unilateral bulbectomy confirms localization of β-ARK-2 to the distal processes of the olfactory receptor neuron (ORN). Similar staining was performed for both isoforms of arrestin, and results demonstrate that β-ARR-2 is localized within the apical dendritic knobs of the ORNs. Bulbectomy markedly attenuated the staining. These antibodies were shown to be specific for the appropriate isoforms by immunoblot analysis. To further examine the roles of these proteins in signal desensitization, we performed rapid stop-flow analyses of the effects of neutralizing antibodies to these two isoforms on odorant-induced cAMP responses in isolated rat cilia. As previously shown, the odorant citralva causes a rapid elevation of cAMP levels within rat olfactory cilia isolated by Ca^{2+} shock, with levels descending rapidly thereafter, suggesting desensitization. A 30-min pretreatment of cilia with either anti-β-ARK-2 or anti-β-ARR-2 antibodies resulted in an elevation of the peak cAMP levels in the presence of citralva, whereas use of the other isoforms did not. For both β-ARK-2 and β-ARR-2, levels are maintained at a plateau, presumably due to loss of desensitization. Preincubation with both

[1] Departments of Neuroscience and Neurology, Johns Hopkins University School of Medicine, Baltimore, MD 21205, USA

antibodies together resulted in additive effects. Thus, the regulation of odorant signaling may involve the same components involved with the regulation of neurotransmitter and hormonal signaling elsewhere.

Cyclic GMP is an intracellular signaling molecule that regulated ion channels, cAMP concentrations (via its actions on selective phosphodiesterases), and protein kinases (PKG), to alter cellular processes. cGMP is synthesized from GTP by two forms of guanylyl cyclase (GC), one of which is a cell surface receptor family and one of which is a family of cytoplasmic receptors. This latter group of GC's appears to exist as a heterodimer containing heme as a prosthetic group and is activated by, among other things, NO. Using primary cultures of ORNs, we have demonstrated that these cells contain high levels of cGMP, in the range of 7–10 pmol/mg protein. We initially investigated whether NOS may be present in olfactory neurons to account for the production of cGMP. By immunoblot analysis, enzyme activity (as measured by the conversion of arginine to citrulline), and immunohistochemistry, we could find no evidence for NOS in the olfactory neuroepithelium or in cultured cells. In addition to NO, CO may act as a physiologic second messenger which can similarly stimulate GC. We have demonstrated by enzyme assay that ORNs contain high levels of HO-2, the form of HO found in the brain, and that inhibition of HO by zinc protoporphyrin-9 (ZN PP-9), which inhibits HO activity, results in depletion of cGMP in these cultures. By in situ hybridization we have previously demonstrated that HO-2 is abundant in the ORNs. Exposure of ORNs in culture to iso-butylmethoxypyrazine (IBMP) results in the several-fold elevation of cGMP levels, but with a slower time course than that seen for second messenger cAMP and inositol 1,4,5-trisphosphate. This slower time course suggests that cGMP may function in desensitization. To begin to address this, we have measured the effect of perturbation of cGMP levels on cGMP levels. Preliminary studies suggest that cGMP may function to regulate the levels of cGMP and ORNs.

References

1. Sklar PB, Anholt RRH, Snyder SH (1986) The odorant-sensitive adenylate cyclase of olfactory receptor neurons. J Biol Chem 261:15538–15543
2. Ronnett GV, Hester LD, Snyder SH (1991) Primary culture of neonatal rat olfactory neurons. J Neurosci 11:1243–1255
3. Ronnett GV, Cho H, Hester LD, Wood SF, Snyder SH (1993) Odorants differentially enhance phospho-inositide turnover and adenylyl cyclase in olfactory receptor neuronal cultures. J Neurosci 13(4):1751–1758
4. Dawson TM, Arriza JL, Lefkowitz RJ, Jaworsky D, Ronnett GV (1993) β-Adrenergic receptor kinase-2 and β-arrestin-2: Mediators of odorant-induced desensitization. Science 259:825–829

G-Proteins and Type III Adenylyl Cyclase During Rat Olfactory Epithelium Ontogeny

Bert P.M. Menco[1], Francesca D. Tekula[2], Albert I. Farbman[2], and Waleed Danho[2]

Key words. G-proteins—Adenylyl cyclase—Olfactory epithelium—Development—Electron microscopic immunocytochemistry—Light microscopic immunocytochemistry

Introduction

Results from biochemistry and physiology suggest that the sites of olfactory signal-transduction are located in knob-shaped dendritic endings of olfactory receptor cells, especially in cilia sprouting from these [1,2]. In part olfactory signal-transduction involves G-proteins and type III adenylyl cyclase (AC) [3]. Cytochemistry helps to understand this process and its onset at a morphological level. So far it has uncovered the (sub)cellular sites of G-proteins and AC [4–8]. This study continues these efforts, especially in relationship to development [7,9].

Materials and Methods

Light Microscopy

Cryostat sections of 4% paraformaldehyde-immersion-fixed rat embryos were preincubated in phosphate buffered saline supplemented with 0.1% sodium dodecyl sulfate (SDS) and then with normal goat serum. They were next reacted with polyclonal primary antibodies. Binding of these was visualized with secondary antibodies conjugated to biotin, followed by an avidin-biotin-peroxidase reaction using rabbit ABC kits (Vector Labs., Burlingame, CA); diaminobenzidine served as chromophore [8].

Electron Microscopy

Nasal epithelia of 4% paraformaldehyde-immersion-fixed rat embryos were cryoprotected in glycerol/PBS, rapidly frozen in liquid propane, and freeze-substituted in methanol: 0.1% uranyl acetate (after

[10]). Other specimens, especially those of older embryos, were not fixed with aldehydes, but immediately after dissection were rapidly frozen by slamming on a liquid nitrogen-cooled copper block and then freeze-substituted in acetone rather than methanol [8]. The specimens were embedded at low temperature in Lowicryl K11M (Chemische Werke Lowi, Waldkraiburg, Germany). Acetylated BSA (Aurion, Wageningen, The Netherlands) [11] was used for blocking. Antibody binding was visualized with secondary goat–antirabbit antibodies conjugated to 10-nm gold particles.

Results

At E15 (E1, sperm-positive; E23 = P1, day of birth), some primary and secondary cilia [12] of olfactory receptor cells label with antibodies to AC (Figs. 2 and 5) and $G_{s\alpha}$ (Fig. 4) but they do not yet bind antibodies to $G_{olf\alpha}$ (Figs. 1 and 3). At E16 all three antibodies label some olfactory cilia. As development progesses, the numbers of receptor cells with labeled cilia increase (Figs. 6 and 7), but $G_{olf\alpha}$ and AC are later in labeling continuous regions of labeled cells than $G_{s\alpha}$.

Dendritic knobs and ciliary necklaces (the basal 0.2- to 0.3-µm area of area of the cilia) label little or

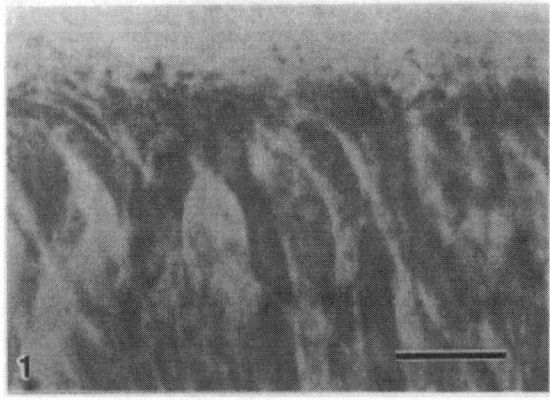

Fig. 1. Light micrograph of E15 rat embryonic olfactory epithelium incubated with antibodies to $G_{olf\alpha}$ (dilution, 1:10), which does not label any epithelial structure at this embryonic age. Bar: 10 µm

[1] Department of Neurobiology and Physiology, O.T. Hogan Hall, Northwestern University, Evanston, IL 60208, USA
[2] Peptide Research Department, Hoffmann-La Roche, Inc., 340 Kingsland Street, Nutley, NJ 07110, USA

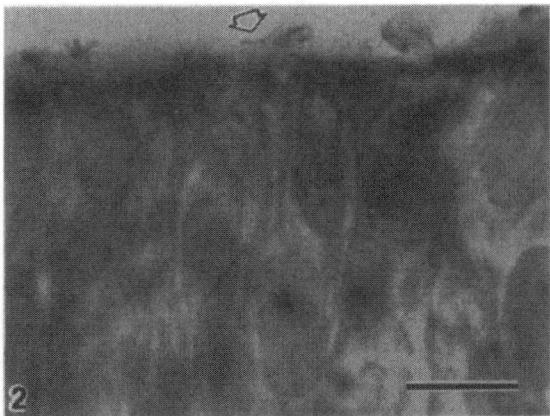

FIG. 2. As in Fig. 1; E15 olfactory epithelial surface labeled with antibodies to AC instead of $G_{olf\alpha}$ (dilution, 1:100). Cilia of some knobs (*arrow*) bind antibodies at this embryonic age. Bar: 10 μm

◄─────────────────────────────

Discussion

This study showed that ontogeny of olfactory signal transduction proteins coincides with ciliogenesis at the distal ends of olfactory receptor cells. The α-subunits of the stimulatory G-proteins G_{olf} and G_s, as well as AC, are expressed in the same cellular compartments, the olfactory cilia. These cilia are the cellular compartments likely to encounter odorous molecules first [4,7,8]. Ciliary necklaces may be responsible for the compartmentalization between cilia and dendritic knobs. Our findings confirm and

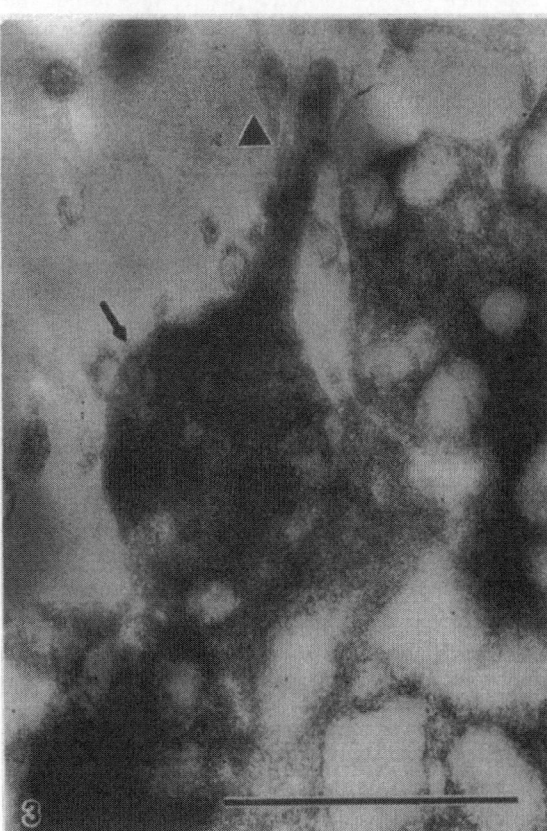

FIG. 3. E15 as in Fig. 1; at ultrastructural level (fixed). Antibodies to $G_{olf\alpha}$ (dilution, 1:4) do not label olfactory cilia (*arrowhead*) or dendritic knobs (*arrow*). Bar: 1 μm

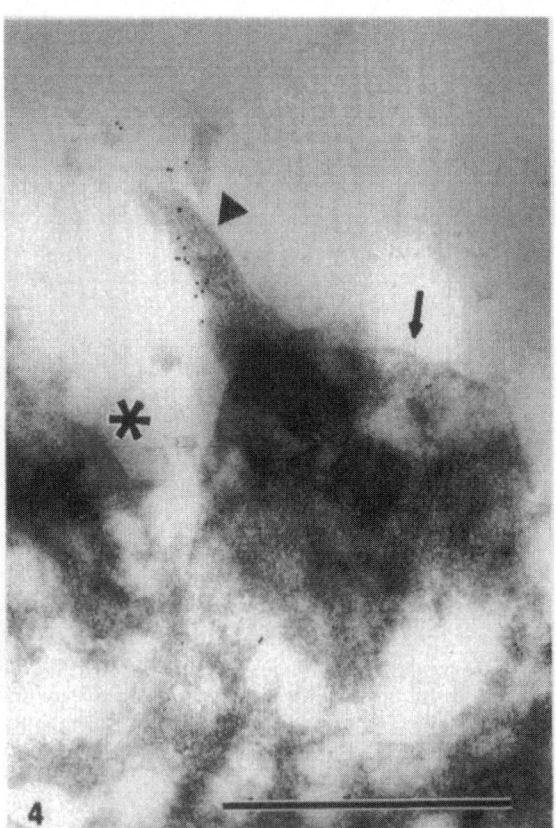

FIG. 4. E15 as in Fig. 3; with antibodies to $G_{s\alpha}$ (dilution, 1:25) rather than $G_{olf\alpha}$. Cilia, also primary ones (*arrowhead*), bind antibodies; dendritic knobs (*arrow*) and supporting cell microvilli (*asterisk*) do not. Bar: 1 μm

─────────────────────────────►

FIG. 6. $G_{olf\alpha}$ as in Fig. 3; olfactory epithelial surface of E19 embryo (unfixed). Proximal (*arrowhead*) and especially distal parts (*large arrow*) of olfactory cilia bind antibodies; dendritic knobs (*small arrow*) and supporting cell microvilli (*asterisk*) do not. Bar: 1 μm

not at all (Figs. 3–8). Dendritic knobs also do not label when they do not yet have cilia (Fig. 8). New cilia label along their lengths (Figs. 3–7); on maturation most of the label is on their long distal processes [8]. Cilia of individual cells bind the antibodies with variable intensity (Fig. 7). Olfactory receptor cell bodies and axons show some immunoreactivity to $G_{s\alpha}$, but not to $G_{olf\alpha}$ and AC.

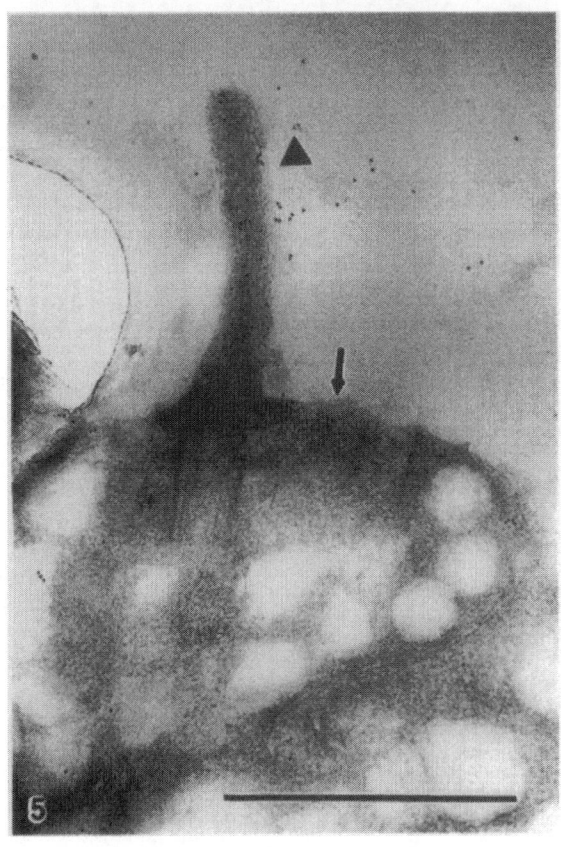

FIG. 5. E15 as in Figs. 3 and 4; with antibodies to AC (dilution, 1:8). These antibodies give the same pattern of labeling (*arrow*) as antibodies to $G_{s\alpha}$ (see Fig. 4). Bar: 1 μm

extend those of Margalit and Lancet [13], who showed that the mRNA for AC begins to appear at E15 whereas that for G_{olf} begins to appear at E16.

That the developmental expression of $G_{olf\alpha}$ lags behind that of $G_{s\alpha}$ and AC could mean that other, yet unknown, G-proteins [6,14] that bind antibodies to $G_{s\alpha}$ are involved in olfactory transduction. There may also be a molecular switch replacing G_s with G_{olf}. This could account for the electrophysiological onset of olfactory specificity with embryonic development [15]. Finally, $G_{s\alpha}$ of olfactory receptor cell bodies and axons may be transported to axon terminals where it may participate in transduction of signals between target and sensory cells.

Acknowledgements. Dr. R.R. Reed (Johns Hopkins, Baltimore, MD) is thanked for antibodies to $G_{olf\alpha}$ and AC. We appreciated the help of Dr. R.C. Bruch and Mr. E.W. Minner with affinity purification of the antibodies to $G_{olf\alpha}$ and photography, respectively. Supported by NSF (IBN-9109851 to BPhMM) and NIH (DC-00347 to AIF).

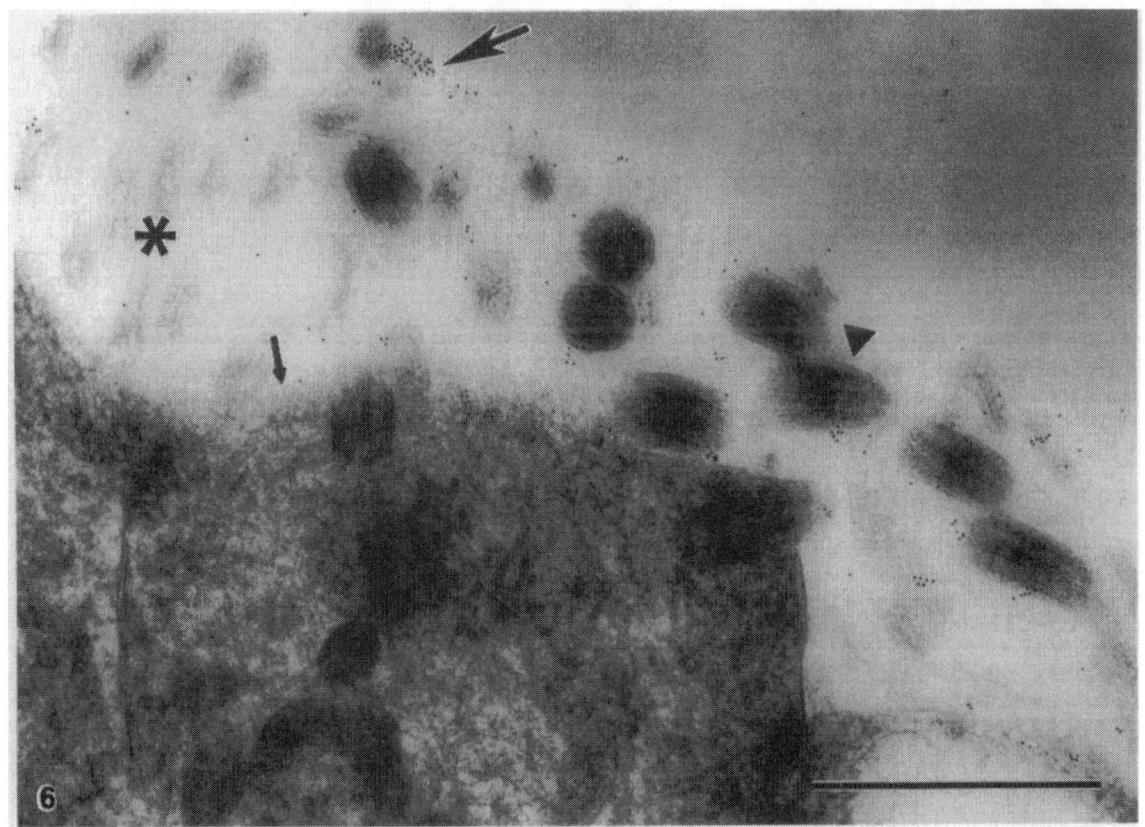

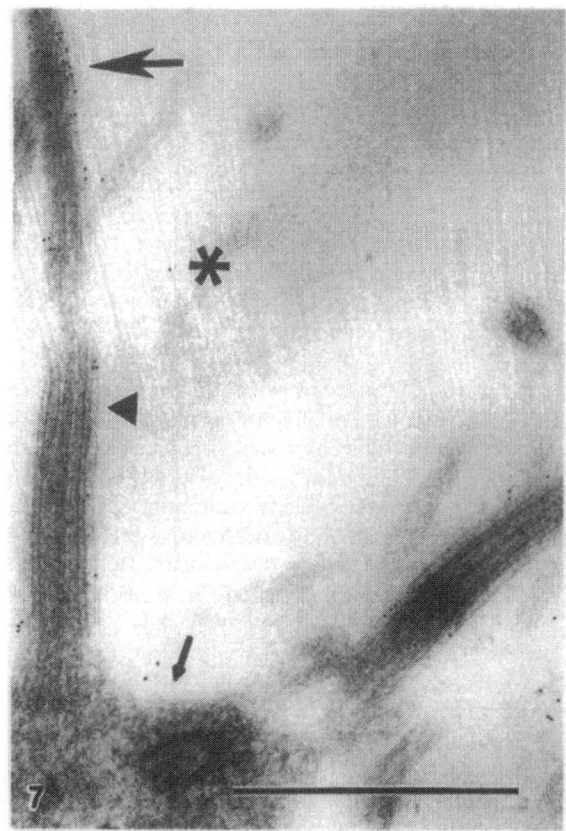

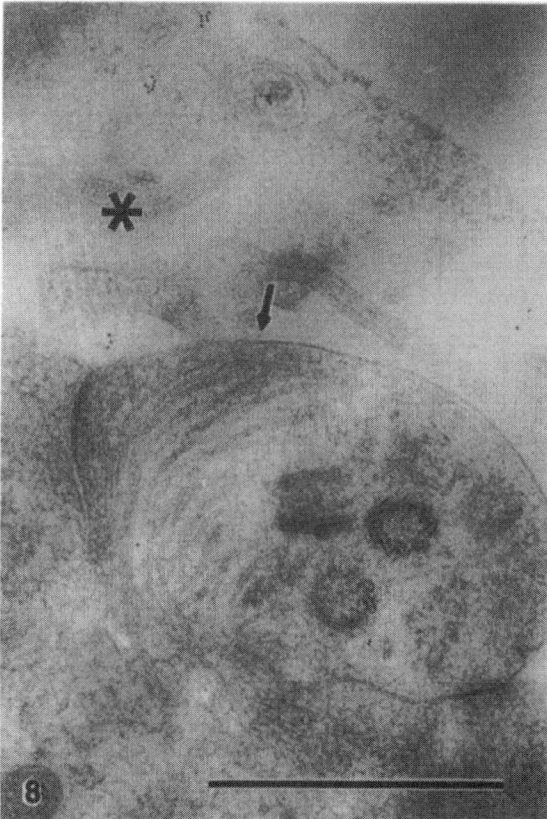

FIG. 7. AC as in Fig. 5; epithelial surface of E19 embryo. Pattern of labeling resembles that of G$_{olfa}$ at this embryonic age. *Large arrow*, distal part olfactory cilium; *arrowhead*, proximal part olfactory cilium; *small arrow*, dendritic knob; *asterisk*, supporting cell microvilli. Bar: 1 µm

FIG. 8. AC at E19 as in Fig. 7; dendritic knob with basal bodies that did not yet develop olfactory cilia (*arrow*). Labeling is absent from this knob and surrounding supporting cell microvilli (*asterisk*). Bar: 1 µm

References

1. Corey DP, Roper SD (eds) (1992) Sensory transduction. Rockefeller University Press, New York
2. Farbman AI (1992) Cell biology of olfaction. Cambridge University Press, Cambridge, UK
3. Reed RR, Bakalyar HA, Cunningham AM, Levy NS (1992) The molecular basis of signal transduction in olfactory sensory neurons. In: Corey DP, Roper SD (eds) Sensory transduction. Rockefeller University Press, New York, pp 53–60
4. Asanuma N, Nomura H (1991) Cytochemical localization of adenylate cyclase activity in rat olfactory receptor cell. Histochem J 23:83–90
5. Bakalyar HA, Reed RR (1990) Identification of a specialized adenylyl cyclase that may mediate odorant detection. Science 250:1403–1406
6. Jones DT, Reed RR (1989) G$_{olf}$: an olfactory neuron-specific G protein involved in odorant signal transduction. Science 244:790–795
7. Mania-Farnell B, Farbman AI (1990) Immunohistochemical localization of guanine nucleotide-binding proteins in rat olfactory epithelium during development. Dev Brain Res 51:103–112
8. Menco BPM, Bruch RC, Dau B, Danho W (1992) Ultrastructural localization of olfactory transduction components: the G protein subunit G$_{olfa}$ and type III adenylyl cyclase. Neuron 8:441–453
9. Dau B, Menco BPM, Bruch RC, Danho W, Farbman AI (1991) Appearance of the transduction proteins G$_{sa}$, G$_{olfa}$ and adenylate cyclase in the olfactory epithelium of rats occurs on different prenatal days. Chem Senses 16:511–512 (abstr 36)
10. Van Lookeren Campagne MB, Oestreicher B, van der Krift TP, Gispen WH, Verkleij AJ (1991) Freeze-substitution and Lowicryl HM20 embedding of fixed rat brain: suitability for immunogold ultrastructural localization of neural antigens. J Histochem Cytochem 39:1267–1279
11. Leunissen JLM (1990) Background suppression using Aurion BAS-C and/or Tween-20®. Aurion Newsletter 1, Wageningen, The Netherlands
12. Menco BPM, Farbman AI (1985) Genesis of cilia and microvilli of rat nasal epithelia during prenatal development. II. Olfactory epithelium, a morphometric analysis. J Cell Sci 78:311–336
13. Margalit T, Lancet D (1993) Expression of olfactory receptor and transduction genes during rat development Dev Brain Res 73:7–16
14. Jones DT, Reed RR (1987) Molecular cloning of five GTP-binding protein cDNA species from rat olfactory neuroepithelium. J Biol Chem 29:14241–14249
15. Gesteland RC, Yancey RA, Farbman AI (1982) Development of olfactory receptor neuron selectivity in the rat fetus. Neuroscience 7:3127–3136

Differential Expression of Vomeromodulin mRNA in the Nasal Mucosa of Rats During Ontogeny

N.S. Rama Krishna[1], Marilyn L. Getchell[2,3], Frank L. Margolis[4], and Thomas V. Getchell[1,2,3]

Key words. Vomeromodulin—In situ hybridization —Lateral nasal glands—Vomeronasal organ— Vomeronasal glands—Posterior glands of the nasal septum—Perireceptor events

Introduction

Vomeromodulin is a 70-kDa glycoprotein that has been identified and cloned from the nasal mucosa of rats [1]. Its primary protein structure and partial amino acid sequence have been determined. The 2.2-kb mRNA encodes the primary protein of 60 kDa that is posttranslationally modified by N-glycosylation to form the 70-kDa glycoprotein. Immunocytochemical studies with antibodies raised against vomeromodulin showed that this protein is localized primarily in the acinar cells of the lateral nasal glands (LNG) located in the lateral wall of the nasal cavity and, to a lesser extent, in the glandular acini of vomeronasal and posterior glands of the nasal septum. Further, these studies also demonstrated that vomeromodulin is abundantly present in the mucus covering the sensory and nonsensory epithelia of the vomeronasal organ. These observations led to the suggestion that this glycoprotein may function as a molecular transporter for pheromones to vomeronasal receptor neurons; hence, it was called a putative pheromone transporter, vomeromodulin [1]. We have reported the expression of vomeromodulin mRNA in glands of the nasal mucosae of postnatal rats [2]. To obtain further insight into its functional properties, we have used in situ hybridization to investigate the expression of the vomeromodulin gene in the nasal mucosa during ontogeny to correlate its expression with that of other molecular markers signifying functional maturation of the nasal chemosensory systems.

Materials and Methods

Embryos were taken from timed pregnant Sprague Dawley rats (Harlan Sprague Dawley, Indianapolis, IN, USA). Nasal tissues from rats of different developmental ages were collected as described [3]. Zamboni's-fluid-fixed nasal mucosae of embryonic day (E) 14, 16, and 19, postnatal day (P) 2, 6, 11, 17, and 27, and 6-week-old rats were sectioned at $10\,\mu m$, thaw-mounted onto Vectabond-treated slides, and stored at $-20°C$ until use.

For in situ hybridization, radiolabeled antisense and sense cRNA transcripts were synthesized from the vomeromodulin cDNA subcloned into pSP64 (antisense) and pSP65 (sense) vectors. The plasmids were linearized with *Eco*RI/*Bam*HI and in vitro transcribed with SP6 polymerase using α-[^{35}S]UTP as the radioactive nucleotide.

The sections were rinsed in phosphate buffer containing 1 mM glycine, digested with $0.5\,\mu g/ml$ proteinase K in Tris buffer (pH 8.0), and acetylated with 0.25% acetic anhydride (Sigma Chemical, ST. Louis, MO) in 0.1 M triethanolamine. Following rinses in SSC ($1\times$ SSC = 0.15 M NaCl plus 0.015 M sodium citrate), the sections were dehydrated in ethanol, delipidated in chloroform, air dried, and hybridized in a mixture containing 50% formamide, 10% dextran sulphate, $4\times$ SSC, Denhardt's solution, salmon sperm DNA ($300\,\mu g/ml$), yeast tRNA ($150\,\mu g/ml$), 40 mM dithiothreitol (DTT), and $0.7 \times 10^6\,cpm/100\,\mu l$ radiolabeled vomeromodulin probe. The hybridization was performed at 45°C for 16 h. Following hybridization, the slides were washed with $4\times$ SSC, and treated with RNase ($2.5\,\mu g/ml$) in Tris-NaCl buffer (pH 8.0) at 45°C for 30 min. Slides were then washed in $2\times$ SSC at room temperature, then $0.5\times$ SSC at 60°C for 30 min, dehydrated in ethanol, and exposed to hyperfilm β-max (Amersham, Arlington Heights, IL, USA) for 36 h. The slides were then coated with NTB-2 nuclear track emulsion (Eastman Kodak, Rochester, NY) and exposed in lighttight boxes for 5–10 days. The autoradiograms were developed in Kodak D-19 developer, rinsed in water, and fixed in Kodak fixer. Sections were counterstained with cresyl violet, dehydrated in ethanol, cleared in xylene, and coverslipped with Permount.

[1] Department of Physiology and Biophysics, [2] Division of Otolaryngology—Head and Neck Surgery, Department of Surgery, [3] Sanders–Brown Center on Aging, University of Kentucky College of Medicine, Lexington, KY 40536, USA
[4] Roche Institute of Molecular Biology, Roche Research Center, Nutley, NJ 07110, USA

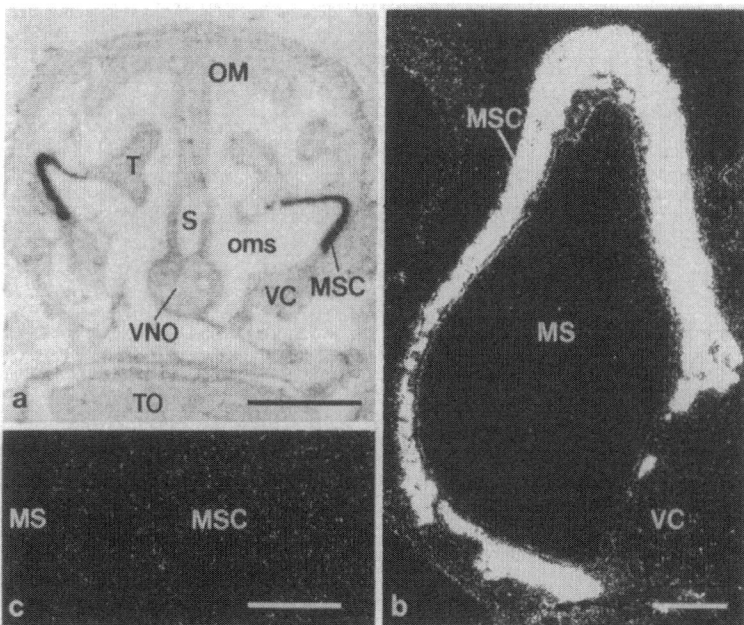

FIG. 1. Expression of vomeromodulin mRNA in P6 rat nasal mucosae. **a** X-Ray film autoradiogram shows intense hybridization signal in maxillary sinus component (*MSC*), but not ventral component (*VC*) of lateral nasal glands. Bar: 2mm. **b** Emulsion autoradiogram of lateral wall of caudal nasal cavity shows most of maxillary sinus encircled by MSC except a small patch of VC. Note that high density of silver grains is associated with glandular acini of MSC but not VC of LNG. Bar: 250μm. **c** Control section hybridized with vomeromodulin sense cRNA probe shows even, low density of silver grains over section, indicating no specific hybridization signal. Bar: 250μm. *MS*, maxillary sinus; *OM*, olfactory mucosa; *oms*, ostium of the maxillary sinus; *S*, nasal septum; *T*, turbinate; *TO*, Tongue; *VNO*, vomeronasal organ

To determine the specificity of the probes used, the following control experiments were performed: (a) omission of the probe from the hybridization mixture; (b) prior treatment of the tissue sections with 100μg/ml RNase (Boehringer Mannheim, Indianapolis, IN, USA); or (c) substitution of antisense probe with labeled sense probe in the hybridization mixture. Finally, parallel immunocytochemistry was performed to localize the glycoprotein using polyclonal antibody to vomeromodulin, as previously described [1].

Results

The nasal mucosae of rats consist of three chemosensory systems: the olfactory mucosa, the vomeronasal organ (VNO), and the septal organ of Masera; and a nonsensory respiratory mucosa. The lamina propria of these mucosae contains Bowman's glands in the olfactory mucosa, vomeronasal glands (VNG) in the VNO, glands of the septal organ, and respiratory glands including the anterior and posterior glands of the nasal septum in the respiratory mucosa. The lateral nasal glands (LNG) comprise the major glandular complex in the lateral wall of the nasal cavity. The LNG of rats can be divided into the maxillary sinus component (MSC) and the ventral component (VC) on the basis of their cellular, histochemical, biochemical, and molecular properties. At the rostral level, the glandular acini of the MSC is primarily associated with the maxillary sinus, while the VC surrounds the tooth root of the upper incisor. At the level of the nasal cavity where the ostium of the maxillary sinus opens into the nasal cavity, both MSC as well as VC surrounds the maxillary sinus (Fig. 1). At the caudal level, the MSC of LNG mostly encircles the maxillary sinus.

Expression of Vomeromodulin mRNA and Protein in Adult Rats

In adult rats, hybridization with vomeromodulin antisense cRNA probe revealed the expression of mRNA in three locations: the MSC of LNG, VNG, and in posterior glands of the nasal septum (PGNS). At the rostral level of the nasal cavity, the most intense hybridization signal was observed in the glandular acini of MSC; no signal was evident in the neighboring acinar cells of the VC. In addition, the expression of vomeromodulin mRNA was also detected in the acini of VNG located in the junctional region of the sensory and nonsensory epithelia of the VNO. In the nasal septum, expression of

TABLE 1. Expression and relative intensities of vomeromodulin mRNA and protein in nasal mucosa of rats.

	mRNA[a]	Protein[a]
Olfactory mucosa		
Mucociliary complex (MC)	−	+
Olfactory epithelium (OE)	−	−
Bowman's glands (BG)	−	−
Vomeronasal mucosa		
Mucomicrovillar complex (MMC)	−	+ +
Mucociliary complex (MC)	−	+ +
Vomeronasal glands (VNG)	+ +	+ +
Respiratory mucosa		
Mucociliary complex (MC)	−	+ + +
Respiratory epithelium (RE)	−	−
Respiratory glands (RG)	−	−
Nasal septum		
Anterior glands of nasal septum (AGNS)	−	−
Posterior glands of nasal septum (PGNS)	+ +	+ +
Lateral nasal glands (LNG)		
Maxillary sinus component (MSC)	+ + + +	+ + + +
Ventral component (VC)	−	−

[a] −, negative; +, weak; + +, moderate; + + +, intense; + + + +, very intense.

vomeromodulin mRNA was localized in the acinar cells of PGNS but not in the adjacent anterior glands of the nasal septum. In parallel immunocytochemical studies, the pattern of vomeromodulin immunoreactivity indicating the presence of the glycoprotein was similar to that of the expression of mRNA in the acinar cells of the MSC of LNG, VNG and PGNS (Table 1). In the extracellular compartments, intense vomeromodulin immunoreactivity was observed in the mucus covering the sensory and nonsensory epithelia of the vomeronasal organ, respiratory epithelium, maxillary sinus epithelium, and also in the mucus covering the nasopharyngeal region. In addition, olfactory mucus also exhibited small patches of immunoreactivity (Table 1).

Developmental Expression of Vomeromodulin mRNA

During ontogeny, the first hybridization signal for vomeromodulin mRNA was detected in the LNG of E16 rats. At this stage, a weak hybridization signal was localized in the glandular acini located in the lateral wall of the nasal cavity. During development, the intensity of the vomeromodulin mRNA signal and the size of the LNG increased progressively until birth. Shortly after birth, the expression of vomeromodulin mRNA in the LNG greatly increased. A high density of silver grains was associated exclusively with the MSC but not the VC of LNG (Fig. 1a,b). At P27 and later, in addition to the MSC of LNG, the acinar cells of vomeronasal glands and PGNS also expressed vomeromodulin mRNA.

Controls

The highly restricted patterns of the hybridization signal and the congruent immunoreactivity confirmed the specificity of the hybridization signal. In control experiments, omission of the probe in the hybridization mixture or treatment of the tissue sections with 100 μg/ml of RNase before hybridization with the antisense probe did not reveal any specific hybridization signal. Similarly, hybridization with the sense probe resulted in a uniform low density of silver grains over the entire section (Fig. 1c), demonstrating a lack of specific expression signal.

Discussion

The results of this study demonstrate that the onset of vomeromodulin gene transcription is different at three locations of the nasal mucosa during development. These include the acinar cells of the MSC of the LNG, VNG, and PGNS. These observations are complemented by the coincident pattern of immunoreactivity for vomeromodulin protein in the glandular acini, confirming these glands as the site of synthesis of this glycoprotein. During ontogeny, vomeromodulin mRNA first occurs about day E16. The MSC of the LNG represents the primary site for vomeromodulin gene expression observed early in the nasal mucosa during development. The VNG and PGNS first express vomeromodulin in the late postnatal period (that is, about P27).

The LNG is among the largest glandular complexes in the nasal mucosa [4]. Within the LNG, vomeromodulin is localized exclusively in the MSC, but not the VC. In the past, different terminologies have been used for the same glands in the lateral wall. Although the MSC and VC of the present study have been called dorsal and ventral glands [5], LNG and serous glands [4], maxillary and lateral glands [6], maxillary sinus gland and LNG [7], and dorsal and lateral parts of the lateral glandular tissue [8], respectively, these glandular elements were generally considered as components of the LNG. In our study also, in view of their location in the lateral wall, relationship to the maxillary sinus, and biochemical and molecular properties, we consider them to be two components of the lateral nasal glandular complex. Clone RYF3 has also been localized in the lateral nasal glands of rats [9]. A comparison of the data presented by Dear et al. [9] in their Fig. 2g with our in situ hybridization data suggests that the clone RYF3 is localized in the MSC of the LNG. The RYF3 clone has substantial nucleotide homology (to 98.6%) with the vomeromodulin clone, and both are localized at similar sites in the LNG, suggesting that they represent independent isolates of the same gene transcript.

In adult rats, the vomeromodulin mRNA and immunoreactivity were localized primarily in the acinar cells of the MSC of LNG and also, to some extent, in the VNG and PGNS. The secretions of the MSC of LNG are presumably delivered through short ducts into the lumen of the maxillary sinus and eventually into the main nasal cavity through the ostium of the maxillary sinus [6,8] (Fig. 1a). Within the nasal cavity, intense vomeromodulin immunoreactivity was localized at the surface of the maxillary sinus mucosa, as well as in the mucociliary complex of respiratory and olfactory mucosae. The products of the VNG and PGNS are known to be secreted into the lumen of the VNO [7,10]. The expression of vomeromodulin mRNA in the VNG and PGNS, and its protein product in the glands as well as in vomeronasal mucus, may support its putative role as a transporter for pheromones to the vomeronasal receptor neurons in young adults [1].

During development, olfactory receptor neurons first show odorant-evoked neural activity at E14 [11] and begin to express olfactory marker protein (OMP) [12–14] in utero. The expression of vomeromodulin mRNA was observed in acinar cells of the LNG at E16. The timing of these events suggests that vomeromodulin may serve as transporter for stimuli associated with intrauterine olfactory function. In contrast, the expression of vomeromodulin in VNG and PGNS is not detected until shortly after weaning at P27, which is much later than the time at which the vomeronasal receptor neurons begin to express OMP at P2 (our unpublished data). This further suggests that vomeromodulin expression in the VNG and PGNS may be induced by extrinsic sensory cues or intrinsic hormonal changes.

In conclusion, this study demonstrates that vomeromodulin mRNA and its translational product are synthesized in LNG, VNG, and PGNS. It also suggests that the expression of a single protein may be regulated by different signals (extrinsic versus intrinsic) at different developmental stages in different glands. The occurrence of vomeromodulin mRNA and protein in the VNG and PGNS and its protein product in vomeronasal mucus confirms and extends our earlier results [1]. In addition, its temporal expression during development and its occurrence in olfactory mucus may indicate that vomeromodulin may also have a broader role in perireceptor events [15] such as odorant transport.

Acknowledgments. The authors thank Dr. Nikk Katzman for technical assistance. This work was supported by NIH grants DC-00159 (TVG), and DC-01715 (MLG).

References

1. Khew-Goodall Y, Grillo M, Getchell ML, Danho W, Getchell TV, Margolis FL (1991) Vomeromodulin, a putative pheromone transporter: cloning characterization, and cellular localization of a novel glycoprotein of lateral nasal gland. FASEB J 5:2976–2982
2. Rama Krishna NS, Getchell TV, Margolis FL, Getchell ML (1993) Vomeromodulin gene expression in the nasal mucosa of rats during postnatal development. Chem Senses 18:617
3. Rama Krishna NS, Getchell TV, Getchell ML (1994) Alpha, Mu, and Pi classes of glutathione S-transferases are differentially expressed in the nasal mucosae of rats during development. Cell Tissue Res 275:435–450
4. Bojsen-Moller F (1964) Topography of the nasal glands in rats and some other mammals. Anat Rec 150:11–24
5. Warshawsky H (1963) Investigations on the lateral nasal gland of the rat. Anat Rec 47:443–455
6. Vidic B, Greditzer HG (1971) The histochemical and microscopical differentiation of the respiratory glands around the maxillary sinus of the rat. Am J Anat 132:491–514
7. Cuschieri A, Bannister LH (1974) Some histochemical observations on the mucosubstances of the nasal glands of the mouse. Histochem J 6:543–558
8. Klaassen ABM, Kuijpers WK, Denuce JM (1981) Morphological and histochemical aspects of the nasal glands in the rat. Anat Anz 149:51–63
9. Dear TN, Boehm T, Keverne EB, Rabbitts TH (1991) Novel gene for potential ligand-binding proteins in sub regions of the olfactory mucosa. EMBO J 10:2813–2819
10. Rama Krishna NS, Getchell ML, Getchell TV (1992) Differential distribution of γ-glutamyl cycle molecules in the vomeronasal organ of rats. NeuroReport 3:551–554
11. Gesteland RC, Yancy RA, Farbman AI (1982) Development of olfactory receptor neuron selectivity in the rat fetus. Neuroscience 7:3127–3136
12. Farbman AI, Margolis FL (1980) Olfactory marker protein during ontogeny: immunohistochemical localization. Dev Biol 74:205–215

13. Monti Graziadei GA, Stanley RS, Graziadei PPC (1980) The olfactory marker protein in the olfactory system of the mouse during development. Neuroscience 5:1239–1252

14. Farbman AI (1991) Developmental neurobiology of the olfactory system. In: Getchell TV, Bartoshuk LM, Doty RL, Snow JB Jr (eds) Smell and taste in health and disease. Raven, New York, pp 19–33

15. Getchell TV, Margolis FL, Getchell ML (1984) Perireceptor and receptor events in vertebrate olfaction. Prog Neurobiol 23:317–345

A Temperature Increase Abolishes Ability of Turtle Olfactory Receptors to Discriminate Odorants with Close Structures

Takuya Hanada, Makoto Kashiwayanagi, and Kenzo Kurihara[1]

The olfactory system can recognize and discriminate multifarious odorant molecules. It has been assumed that responses to odorants are induced by the binding of odorants with specific receptor proteins on olfactory receptor cells [1]. Receptor proteins in biological membranes are surrounded by lipid molecules and, hence, the perturbation of the lipid layers may affect the olfactory responses. If this is true, a temperature change in olfactory cells, which will induce a membrane fluidity change, may affect olfactory responses. In this study, we changed the temperature of the turtle olfactory epithelium by perfusing it with Ringer solutions of different temperatures and examined the effects of the temperature change on the ability of the turtle olfactory system to discriminate odors by applying a cross-adaptation method to the olfactory bulbar responses [2].

A decrease of temperature to 5°C or an increase to 40°C brought about no irreversible effect on the olfactory responses. The magnitudes of the olfactory responses to odorants reached a maximal level around 30°C. The cross-adaptation experiments indicated that the olfactory system well discriminated all eight pairs of odorants examined at 5°C and 18°C. However, the ability of the olfactory receptors to discriminate pairs of odorants with similar odors (e.g., trans-3-hexenol and cis-3-hexenol; n-amyl acetate and isoamyl acetate; d-carvone and l-carvone) was abolished by increasing the temperature to 40°C, whereas discrimination of odorants (e.g., anisole and cineole; l-limonene and cineole) was much less affected.

In order to explore the relation between the abolition of the odor-discriminating ability and fluidity changes in the olfactory receptor membranes, we used a fluorescence dye to measure the temperature dependence of the membrane fluidity of cells isolated from turtle olfactory epithelia and that of liposomes made of lipids extracted from turtle olfactory epithelia. The results indicated that the temperature dependence of the membrane fluidity changes was closely related to that of the odor-discriminating ability, suggesting that increases in membrane fluidity are related to the abolition of odor-discriminating ability. One possible explanation for these results is that the temperature change induces a conformational change in a specific receptor protein for an odorant, via fluidity changes in the lipid layers, which change leads to a change in the specificity of the protein to the odorant. It is well known that olfactory and gustatory systems carry specific receptor proteins for amino acids. Hence, we examined the effects of temperature changes on the cross-adaptation of the responses to amino acids by recording the olfactory bulbar responses in the carp. The results indicated that the ability of the carp olfactory receptors to discriminate L-alanine and D-alanine was unaffected by an increase of temperature. Thus, these results do not support the possibility that a temperature increase leads to a change in the specificity of the receptor protein to the odorant. The present results favor the mechanism that an interaction of odorants with lipid layers of the receptor membranes is involved in the odor-discriminating mechanism [3]. At a lower temperature, the lipid structure is rather rigid and different odorants are adsorbed on different domains. At a higher temperature, the fluidity of the lipid layers is increased, and the domains for the odorants then become flexible. At 40°C, the receptor pocket for cis-3-hexenol, for example, accepts trans-3-hexenol and, hence, is desensitized by the previous application of trans-3-hexenol. We emphasize that domains composed of receptor proteins and lipids are concerned with odor reception.

The present results also address the mechanism of desensitization of the olfactory response. It is possible that desensitization occurs at the cellular level. This mechanism cannot, however, explain the finding that the response to cis-3-hexenol, for example, did not appear after trans-3-hexenol at 40°C; there is no reason why desensitization of one cell should induce desensitization of an other cell. The present results are more reasonably explained as follows: one cell has receptor sites for both trans-3-hexenol and cis-3-hexenol, and desensitization occurs at each receptor site level.

---

[1] Faculty of Pharmaceutical Sciences, Hokkaido University, N12 W6 Kita-ku, Sapporo, 060 Japan

150

References

1. Buck L, Axel R (1991) A novel multigene family may encode odorant receptors: A molecular basis for odor recognition. Cell 65:175–187
2. Hanada T, Kashiwayanagi M, Kurihara K (1994) A temperature increase abolishes ability of turtle olfactory receptors to discriminate odorants having similar odors. Am J Physiol (in press)
3. Nomura T, Kurihara K (1987) Liposomes as a model for olfactory cells: Changes in membrane potential in response to various odorants. Biochemistry 26:6135–6140

Enhancement of Turtle Olfactory Responses to Fatty Acids by Treatment of Olfactory Epithelium with Phosphatidylserine

MUTSUO TANIGUCHI, SHUICHI ENOMOTO, and KENZO KURIHARA[1]

Key words. Turtle olfactory bulbar response—Phosphatidylserine

It is generally considered that an olfactory response is induced by the binding of an odorant to specific receptor proteins in olfactory receptor membranes. On the other hand, it has been pointed out that odorants are, in general, hydrophobic, and that interaction of odorants with the lipid layers of the receptor membranes may also play an important role in odor reception [1]. In a previous study, we found that the addition of phosphatidylserine (PS) to phosphatidylcholine (PC) liposomes greatly increased the magnitude of the membrane potential changes of the liposomes, especially in response to fatty acids such as *n*-valeric, isovaleric and *n*-butyric acids [2]. In order to examine whether PS enhanced the response in the *in vivo* olfactory system, we treated turtle olfactory epithelium with suspensions of various lipids and examined the effects on the olfactory responses by measuring the olfactory bulbar responses [3].

PS treatment enhanced responses to *n*-valeric, isovaleric, and *n*-butyric acids, by a factor of four to five. The responses to 14 other odorants, including lilial, geraniol, lyral, β-ionone, *l*-citronellal, and citralva, were little increased or practically unchanged by PS treatment. The findings that the olfactory responses to the fatty acids were greatly enhanced by PS treatment closely resemble those observed with liposomes. The enhanced responses to the fatty acids induced by PS treatment returned to the original level about 10 h after the treatment. The threshold concentration for *n*-valeric acid was about 5×10^{-4} M before the treatment. PS treatment greatly lowered the threshold of the response to *n*-valeric acid, to about 10^{-6} M, and enhanced the magnitude of the responses at all concentration ranges examined.

The olfactory epithelium was also treated with cardiolipin (CL) or phosphatidic acid (PA). CL treatment did not significantly affect the responses to the odorants examined. The responses to *n*-butyric and *n*-valeric acids and lyral were reduced, but not significantly, and the response to cineole was little increased by the treatment. Similarly to CL treatment, PA treatment did not affect or only slightly reduced the responses to the odorants examined. The results described above indicate that the effect on the olfactory responses varies greatly with the species of lipid used for the treatment of the olfactory epithelium.

To examine whether the PS used for the treatment was incorporated into the olfactory epithelium, we incubated the epithelium with PS suspension containing [^{14}C]PS for 1 h, and then washed it out thoroughly with Ringer solution for 15 min. This condition was nearly the same as that of the experiment described above. The results obtained indicated that 0.28% of the PS used for the treatment was incorporated into the olfactory epithelium. It is not clear into which part of the epithelium the PS was incorporated, but it is possible that it was incorporated into the olfactory receptor membranes.

It is known that PS activates adenylate cyclase [4]. Hence, it is possible that PS is incorporated into the olfactory receptor membranes and activates the second messenger system, which leads to enhancement of the olfactory responses. The olfactory response to isovaleric acid, which does not increase cyclic adenosine monophosphate (cAMP), but increases inositol 1,4,5-trisphosphate IP$_3$ [4], was greatly increased by the PS treatment. The olfactory responses to geraniol, β-ionone, and citronellal, which increase cAMP [5], were virtually not enhanced by the PS treatment. Hence, it is unlikely that this treatment enhanced the responses to the odorants via the second messenger system. A simple explanation for the present results is that PS is incorporated into the olfactory receptor membranes and modifies the receptor site for the fatty acids, so that the affinity of the receptor membranes to the fatty acids is increased. We conclude that lipids, as well as proteins, in the receptor membranes play an important role in odor reception, at least for the fatty acids.

[1] Faculty of Pharmaceutical Sciences, Hokkaido University, N12 W6, Kita-ku, Sapporo, 060, Japan

References

1. Enomoto S, Kashiwayanagi M, Kurihara K (1991) Liposomes having high sensitivity to odorants. Biochim Biophys Acta 1062:7–12
2. Enomoto S, Kawashima S, Yoshimura A, Kurihara K (1992) Effects of changed lipid cmposition and addition of proteins on odor specificity of liposomes. Sensors and Materials 4:153–164
3. Taniguchi M, Kashiwayanagi M, Kurihara K (1994) Enhancement of the turtle olfactory responses to fatty acids by treatment of olfactory epithelium with phosphatidylserine. Brain Res (in press)
4. Levey GS (1971) Restoration of glucagon responsiveness of solubilized myocardial adenyl cyclase by phosphatidylserine. Biochem Biophys Res Commun 43:108–113
5. Breer H, Boekhoff I (1991) Odorants of the same odor class activate different second messenger pathways. Chem Senses 16:19–29

Olfactory Transduction Studied with Caged Cyclic Nucleotides

Graeme Lowe and Geoffrey H. Gold[1]

Key words. Olfaction—Receptor cell—Cyclic AMP —Transduction

Introduction

Although olfaction has captured the attention of scientists and philosophers alike for most of recorded history, it is only during the past twenty years that our present understanding of odor detection (transduction) has begun to emerge. In 1972, Kurihara and Koyama [1] demonstrated that the olfactory epithelium contains a high concentration of adenylyl cyclase, the enzyme that synthesizes cyclic AMP (adenosine monophosphate). Based on this observation, they proposed that the intracellular messenger cyclic AMP plays an important role in olfactory transduction. In 1973 Minor and Sakina [2] reported that dibutyryl cyclic AMP induced an electroolfactogram (EOG) response similar to that evoked by odorants, thus providing the first electrophysiological evidence that cyclic AMP is an excitatory messenger in olfactory transduction. Similar experiments were reported by Menevse et al. in 1977 [3]. The first biochemical evidence in support of the excitatory hypothesis was provided in 1985 by Pace et al. [4], who showed that the olfactory adenylyl cyclase can be stimulated by odorants. Perhaps more importantly, they showed that adenylyl cyclase activation was GTP-dependent, suggesting that olfactory receptor cells contain a G-protein-linked enzymatic cascade similar to those that mediate signal transduction in a variety of other cell types. In 1987, Nakamura and Gold [5] discovered a likely target for cyclic AMP, an ion channel that was activated directly by this intracellular messenger. This ion channel is quite similar to the cyclic guanosine monophosphate- (GMP-) gated channel that mediates visual transduction in photoreceptors [6], further indicating a close relationship between olfactory transduction and other signal transduction mechanisms. The final link between olfactory and other signal transduction mechanisms was provided by the cloning of a very large family of olfactory receptor proteins by Buck and Axel in 1991, based on sequence homology with other G-protein-linked receptor proteins [7].

This paper summarizes our recent experiments [8] using caged cyclic AMP to test the hypothesis that cyclic AMP mediates olfactory transduction by a direct effect on the cyclic nucleotide-gated channels. The use of caged cyclic AMP allowed controlled and repeatable production of cyclic AMP within the receptor cells, enabling a number observations to be made that were not possible in earlier studies [9–13].

Materials and Methods

Membrane currents were recorded in the whole-cell configuration from salamander olfactory receptor cells, isolated as described previously [8]. Caged cyclic nucleotides were introduced by extracellular application or pipette perfusion, also as described previously [8].

Results

Time Course of Responses to Photolysis and Odorants

Brief illumination of solitary olfactory receptor cells loaded with $100 \mu M$ caged cyclic AMP caused a transient inward current under whole-cell voltage clamp at $-50 mV$. The photolysis response had a short latency (12 ms), time to peak (108 ms), and duration (210 ms to decay to half-peak amplitude), compared with those of an odorant response recorded from the same cell (latency, 170 ms, time to peak, 360 ms, and duration, 510 ms). These data provide evidence that cyclic AMP acts directly on the cyclic nucleotide-gated conductance and demonstrate that the kinetics of the transduction current are not limited by the gating kinetics of the conductance, as suggested recently by Zufall et al. [14].

Voltage Dependence of Responses to Photolysis and Odorants

Photolysis flashes and odorant pulses were repeated at a series of holding potentials to compare the

[1] Monell Chemical Senses Center, 3500 Market St., Philadelphia, PA 19104, USA

voltage dependencies of the currents induced by odorants and by cyclic AMP. The two voltage dependences were nearly identical, indicating that the transduction current is mediated by the same conductance or conductances that are activated by cyclic AMP.

Cyclic Nucleotide Specificity of Photolysis Responses

Currents induced by the photolysis of 100 μM caged cyclic GMP were similar to those induced by photolysis of 100 μM caged cyclic AMP; that is, they exhibited a similar amplitude and latency (mean peak current ± SD, 256 ± 95 pA; mean latency ± SD, 7.5 ± 2.0 ms, for six cells). This is consistent with the similar affinities of the cyclic nucleotide-gated conductance for cyclic AMP and cyclic GMP [5].

Localization of the Cyclic Nucleotide-Gated Conductance

The spatial distribution of the cyclic nucleotide-gated conductance was investigated by recording the whole-cell current in response to local photolysis of caged cyclic AMP. Local photolysis was accomplished by using a diaphragm to limit the portion of the cell that was illuminated; the cell was moved relative to the image of the edge of the diaphragm to vary the region illuminated. When illumination was limited to the distal portions of the cilia, the peak amplitudes of the photolysis responses increased in proportion to the sums of the lengths of the ciliary segments that were illuminated. In contrast, the latencies of the responses were constant at about 8–10 ms. This indicates that the cyclic nucleotide-gated channels are approximately uniformly distributed throughout the cilia.

On the other hand, when illumination was limited to the basal end of the cell, the photolysis response exhibited a long latency (>1 s) that decreased in proportion to the square of the distance from the bases of the cilia. This provides evidence that the channels are localized to the cilia, consistent with data from excised patches [15]. Also, simultaneous measurements of the somatic and whole-cell currents during dialysis of cells with cyclic AMP provided an estimate of the ratio of channel densities in the ciliary and somatic membranes of 800:1. Ciliary localization supports the role of these channels in transduction, because we have shown that the transduction current is also localized to the cilia [16].

Summation of Odorant and Photolysis Responses

Summation of odorant and photolysis responses was characterized using paired stimuli, consisting of an odorant stimulus followed by a light flash timed to occur at the peak of the odorant response. Summation was found to be nonlinear: a small odorant-induced current caused an increase in the amplitude of the photolysis-induced current relative to that in the absence of an odorant stimulus, and a large odorant-induced current caused attenuation of the photolysis-induced current. These effects were expected if the odorant stimulus and the photolytic production of cyclic AMP both induced a current by acting on a common cyclic nucleotide-gated conductance.

The data were analyzed quantitatively by fitting the data to an equation derived using the Hill equation and assuming additivity of the cyclic AMP concentrations caused by the two stimuli. Although the data were fitted well by this equation, this required a Hill coefficient higher than that observed in excised patches. We propose that the greater cooperativity observed in an intact cell is caused by the activation of the Ca-dependent Cl current that has been described [17–19].

Discussion

We have shown that the voltage dependence, spatial localization, and summation of odorant and cyclic AMP-induced currents are consistent with olfactory transduction being mediated by the ciliary cyclic nucleotide-gated conductance. Our results confirm and extend previous studies [9–13], which have also concluded that cyclic AMP mediates the transduction current. This evidence that cyclic AMP serves as an excitatory messenger in olfactory transduction supports further the hypothesized roles for the olfactory (type III) adenylyl cyclase [20], G_{olf} [21], and the putative olfactory receptor proteins [7] in odorant detection.

The great variety of odorous compounds, however, raises the question of whether this is the only transduction mechanism in vertebrates or whether other transduction mechanisms exist. For example, it is possible that odorants may elicit or modulate olfactory responses by nonspecific effects, as proposed by Kurihara's group (e.g., ref. [22]). In addition, it is possible that other intracellular messengers, such as IP_3, play a role in olfactory transduction [23]. Sklar et al. [24] observed that many odorants produce low, or undetectable, cyclic AMP production, and Breer and coworkers [25] have shown that the same odorants induce rapid increases in IP_3 concentration in olfactory cilia. These data have been interpreted as evidence that cyclic AMP and IP_3 both serve as excitatory messengers, but for different odorants. Although this hypothesis is attractive on biochemical grouds, it has not yet received much electrophysiological support. For example, Restrepo and coworkers [26] reported IP_3-induced depolarization and small IP_3-induced currents during whole-cell dialysis, but several groups have failed to confirm these observations [8,11,27]. IP_3-induced single-channel currents in somatic patches have been described at this meeting [28], but because of the low density of these channels it cannot yet be concluded that they contribute to the transduction current, which is localized to the cilia [16].

Furthermore, Lowe et al. [29] showed that the amplitudes of EOG responses to individual odorants are correlated with cyclic AMP production caused by the same odorants. This result supports the hypothesis that cyclic AMP mediates transduction for a wide variety of odorants, even those that produce small or undetectable cyclic AMP production by the assays used. This interpretation also predicts that cells that respond to the weak adenylyl cyclase-stimulating odorants are relatively rare in the olfactory epithelium. Thus, the small adenylyl cyclase activities caused by many odorants argues against, rather than for, the existence of another transduction mechanism.

In summary, although the existence of the IP_3 pathway in olfactory cilia strongly suggests a role in transduction, additional data are needed before that role can be satisfactorily established.

References

1. Kurihara K, Koyama N (1972) High activity of adenyl cyclase in olfactory and gustatory organs. Biochem Biophys Res Commun 48:30–34
2. Minor AV, Sakina NL (1973) Role of cyclic adenosine-3′,5′-monophosphate in olfactory reception. Neirofiziologiya 5:415–422
3. Menevse A, Dodd G, Poynder TM (1977) Evidence for the specific involvement of cyclic AMP in the olfactory transduction mechanism. Biochem Biophys Res Commun 77:671–677
4. Pace U, Hanski E, Salomon Y, Lancet D (1985) Odorant-sensitive adenylate cyclase may mediate olfactory reception. Nature (Lond) 316:255–258
5. Nakamura T, Gold GH (1987) A cyclic nucleotide-gated conductance in olfactory receptor cilia. Nature (Lond) 325:442–444
6. Fesenko EE, Kolesnikov SS, Lyubarsky AL (1985) Induction by cyclic GMP of cationic conductance in plasma membrane of retinal rod outer segment. Nature (Lond) 313:310–313
7. Buck L, Axel R (1991) A novel multigene family may encode odorant receptors: a molecular basis for odor recognition. Cell 65:175–187
8. Lowe G, Gold GH (1993) Contribution of the ciliary cyclic nucleotide-gated conductance to olfactory transduction in the salamander. J Physiol (Lond) 462:175–196
9. Suzuki N (1989) Voltage- and cyclic nucleotide-gated currents in isolated olfactory receptor cells. In: Brand JG, Teeter JH, Cagan RH, Kare MR (eds) Chemical Senses, vol 1: Receptor events and transduction in taste and olfaction. Dekker, New York, pp 469–493
10. Kurahshi T (1990) The response induced by intracellular cyclic AMP in isolated olfactory receptor cells of the newt. J Physiol (Lond) 430:355–371
11. Firestein S, Darrow B, Shepherd GM (1991) Activation of the sensory current in salamander olfactory receptor neurons depends on a G protein-mediated cAMP second messenger system. Neuron 6:825–835
12. Firestein S, Zufall F, Shepherd GM (1991) Single odor-sensitive channels olfactory receptor neurons are also gated by cyclic nucleotides. J Neurosci 11:3565–3572
13. Frings S, Lindemann B (1991) Current recording from sensory cilia of olfactory receptor cells in situ. I. The neuronal response to cyclic nucleotides. J Gen Physiol 97:1–16
14. Zufall F, Hatt H, Shepherd GM, Firestein S (1993) Rapid application and removal of second messengers reveals integrative properties of olfactory signal transduction. In: 15th Annual Meeting of the Assocuation for Chemoreception Science abstr 51
15. Kurahashi T, Kaneko A (1991) High density cAMP-gated channels at the ciliary membrane in the olfactory receptor cell. Neuroreport 2:5–8
16. Lowe G, Gold GH (1991) The spatial distributions of odorant sensitivity and odorant-induced currents in salamander olfactory receptor cells. J Physiol (Lond) 442:147–168
17. Kleene S, Gesteland RC (1991) Calcium-activated chloride conductance in frog olfactory cilia. J Neurosci 11:3624–3629
18. Kleene SJ (1993) Origin of the chloride current in olfactory transduction. Neuron 11:123–132
19. Kurahashi T, Yau K-W (1993) Co-existence of cationic and chloride components in odorant-induced current of vertebrate olfactory receptor cells. Nature (Lond) 363:71–74
20. Bakalyar HA, Reed RR (1990) Identification of a specialized adenylyl cyclase that may mediate odorant detection. Science 250:1403–1406
21. Jones DT, Reed RR (1989) G_{olf}: an olfactory neuron-specific G protein involved in odorant signal transduction. Science 244:790–795
22. Nomura T, Kurihara K (1987) Liposomes as a model for olfactory cells: changes in membrane potential in response to various odorants. Biochemistry 26:6135–6140
23. Boyle AG, Park YS, Huque T, Bruch RC (1987) Properties of phospholipase C in isolated olfactory cilia from the channel catfish (Ictalurus punctatus). Comp Biochem Physiol B 88:767–775
24. Sklar PB, Anholt RRH, Snyder SH (1986) The odorant-sensitive adenylate cyclase of olfactory receptor cells. Differential stimulation by distinct classes of odorants. J Biol Chem 261:15538–15543
25. Breer H, Boekhoff I (1991) Odorants of the same odor class activate different second messenger pathways. Chem Senses 6:19–29
26. Restrepo D, Miyamoto T, Bryant B, Teeter JH (1990) Odor stimuli trigger influx of calcium into olfactory receptor neurons. Science 249:1166–1168
27. Miyamoto S, Tsuru K, Nakamura T (1993) Transmigration of calcium ions in the olfactory receptor cell. In: Abstracts, 11th International Symposium on Olfaction and Taste, p 124
28. Suzuki N (1993) IP3-activated ion channel activities in olfactory receptor cells from different vertebrate species. In: Abstracts, 11th International Symposium on Olfaction and Taste, p 137
29. Lowe G, Nakamura T, Gold GH (1989) Adenylate cyclase mediates olfactory transduction for a wide variety of odorants. Proc Natl Acad Sci USA 86:5641–5645

cAMP-Gated Channels from Olfactory Neurons of the Rat: Chemical Gating, Inhibition by Ca Ions, Noise Analysis

BERND LINDEMANN[1] and JOSEPH W. LYNCH[2]

Key words. Olfaction—Transduction—cAMP—Chemical gating—Calcium—Noise analysis—Dose-response curves

Introduction

One odorant transduction pathway of olfactory receptor cells causes generation of cAMP. Binding of this messenger to cation channels present in the ciliary membranes induces opening of the channels, inflow of Na ions, and thereby membrane depolarization [1]. Termination of the inward current involves inactivation of several members of the transduction chain and depends in part on the presence of Ca ions. We excised membrane patches from olfactory knobs and induced currents by exposure to cAMP. Based on (a) voltage-dependent Ca-blocking ratios, (b) steady state noise analysis, and (c) cAMP dose-response curves, we describe three inhibitory effects which Ca ions exert on the channel. They occur at local Ca concentrations >100 μM.

Material and Methods

Rat olfactory mucosa was dissociated into single cells with trypsin, and inside-out membrane patches were excised from apical knobs and exposed to 160 mM Na^+ on both sides, nominally low Ca^{2+} on the outside and varying concentrations of cAMP and Ca^{2+} on the inside [2]. Current–voltage curves, I(V), were recorded from the patches with slow voltage sweeps, and blocking ratios at different Ca^{2+} concentrations as well cAMP dose-response curves were constructed from them as described before [2]. In blocking-ratio experiments, patches were exposed to 3 mM Mg^{2+} for 1 min prior to recording the effects of cytosolic Ca^{2+}, in order to disable the "desensitization" described below (see "Dose-response curves"). For noise analysis, patch currents were sampled at 5 kHz (anti-aliasing filter at 1.2 kHz) and blocks of 4096 data points subjected to Fourier transformation [3]. Up to 256 steady state power density spectra (PDS) were averaged and condensed to yield one of the spectra used for integration. Current variance (σ^2_I) was obtained by integrating the PDS with respect to frequency. Membrane voltage is given as inside minus outside.

Results

Ca^{2+}-Blocking Ratios

Inside-out patches were pulled from apical knobs of isolated olfactory receptor cells of the rat. They contained hundreds of cAMP-gated channels. I(V) curves were generated in the range ±100 mV. A saturating cAMP concentration (100 μM) was used on the inside of the patch. The cAMP-activated current was calculated by subtracting the current obtained in the absence of cAMP. These cAMP-activated currents are essentially Na currents observed with symmetrical high [Na] on both sides of the membrane. When the Ca^{2+} concentration on the inside of the patch was increased above 100 μM, Na currents became smaller especially at positive membrane voltage (V).

Blocking ratios (BR) of currents ($I_{at\ Ca>0}/I_{at\ Ca=0}$) were calculated and plotted against V. Figure 1 shows as a dashed line the BR(V) for the case that cytosolic Ca^{2+} can occupy the cAMP-gated channel, but cannot transgress it completely. The data followed this curve in the voltage range −50 to 0 mV, indicating that access of Ca^{2+} to a blocking site in the channel was favored at less negative voltages, as expected for an open-channel block. However, the BR(V) found deviated from the dashed curve in two important respects. First, at positive V the block was less than expected. This is explicable by escape of the blocking Ca ion from the channel to the trans-side (the outside). The escape would be facilitated at positive voltages, lowering the blocking efficiency. Second, at negative voltages the block was stronger than expected. This deviation indicated an inhibitory effect of Ca ions at voltages where entrance of

[1] Department of Physiology, Universität des Saarlandes, 66421 Homburg/Saar, Germany
[2] Present address: The Garvan Institute of Medical Research, St. Vincents Hospital, Darlinghurst, Sydney, NSW 2010, Australia

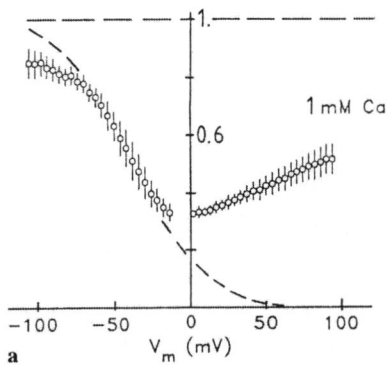
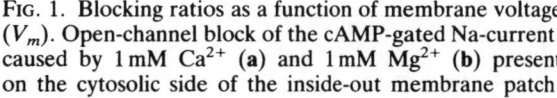
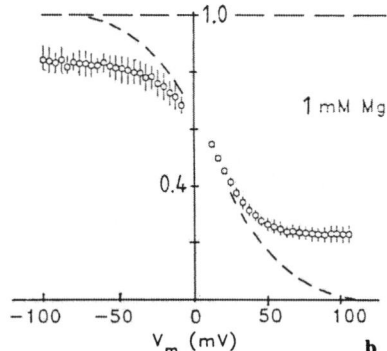

FIG. 1. Blocking ratios as a function of membrane voltage (V_m). Open-channel block of the cAMP-gated Na-current, caused by 1 mM Ca^{2+} (a) and 1 mM Mg^{2+} (b) present on the cytosolic side of the inside-out membrane patch.

Dashed sigmoid line, curve expected for a divalent cation blocking the channel by occupation from the cytosolic side, but unable to leave the channel towards the extracellular side

Ca^{2+} into the channel was disfavored by voltage. Similar results were obtained with Mg^{2+}, Mn^{2+}, Sr^{2+}, and Ba^{2+}. Mg^{2+} at a concentration of 1 mM had a BR of 0.8 at −100 mV and of 0.25 at +100 mV.

Using outside-out patches, we attempted to obtain blocking ratios for extracellular Ca^{2+}. This was possible in one experiment, which showed at 1 mM Ca^{2+} a BR of 0.75 at +100 mV and of 0.1 in the range 0 to −100 mV. A voltage-independent block component was seen at +100 mV in this case.

Noise Analysis

The fluctuations of cAMP-induced Na current were transformed to power density spectra. One dominating Lorentzian of corner frequency near 10 Hz was found. Its plateau power changed with [cAMP], having a maximum at the K_m of the cAMP dose/response of 3 μM (gating Lorentzian). A further increase of [cAMP] up to 100 μM caused the low-frequency power (Lorentzian plateau) to decrease. In contrast, the high-frequency power (above 100 Hz) increased monotonously with [cAMP]. Most of the high-frequency noise results from the rapid flickering of operative (successfully gated) channels [4]. This fast open-close flickering occurs in so-called bursts. The total burst time, and therefore the high-frequency power, increases with [cAMP]. In contradistinction, the low-frequency Lorentzian is related to the cAMP-gating process, which determines whether a channel is closed or bursting. Hereafter, the variance of the gating noise will be called σ^2_g and that of the intra-burst flicker σ^2_b.

Variance (σ^2) was computed by integrating power spectra with respect to frequency. Variance vs current plots [5] were parabolic, σ^2_t becoming small at low and at high [cAMP] (Fig. 2a,b). The gating variance σ^2_g was extracted from σ^2_t and was plotted against the mean cAMP-induced current (Fig. 2c,d). From such plots, the number of channels (N), the

time-averaged single channel current within a burst (i_b), and the probability (P_b) to be in the burst or "gate open" state (i.e., fully cAMP-ligated) were estimated by curve fitting. In the absence of divalent cations, i_b was 0.5 ± 0.1 (SD) pA, corresponding to an apparent single channel conductance of 16 pS. N varied from 40 to 700 per patch, and P_b was 0.5 at 2–3 μM cAMP (Fig. 2c).

When raising the inside Ca^{2+} concentration to 0.2 mM, i_b decreased to 0.28 ± 0.1 pA, as expected from a fast open-channel block. In addition, the parameter P_b decreased strongly (Fig. 2d), requiring about 40 μM cAMP to reach the value 0.5. Thus, the presence of Ca ions increased the apparent K_m value of cAMP-gating. Plots of the corner frequency of the gating Lorentzian vs cAMP concentration were linear in the concentration range used, indicating for the gating process an on-rate constant of 18 s^{-1} $μM^{-1}$ and an off-rate constant of 28 s^{-1}. In the presence of 0.2 mM inside Ca^{2+}, the on-rate decreased 30-fold and the off-rate increased 3-fold. The increased K_m of cAMP, induced by Ca^{2+} at concentrations >100 μM, suggests that Ca^{2+} acts not only as an open-channel blocker but, more importantly, decreases the sensitivity of the cAMP-gated channels to cAMP.

Dose-Response Curves

From I(V) curves obtained at different [cAMP], as just described, dose-response relationships for the activation of patch current by cAMP were constructed. In the absence of divalent cations, the K_m was 4 μM at −50 mV and 2.5 μM at +50 mV. Addition of Ca^{2+} to the solution on the cytosolic side of the membrane in concentrations of up to 100 μM had no effect. However, addition of 0.2 or 0.5 mM Ca^{2+} shifted the K_m of cAMP reversibly to higher cAMP concentrations (35 or 90 μM respectively, measured at −50 mV), as shown in Fig. 3.

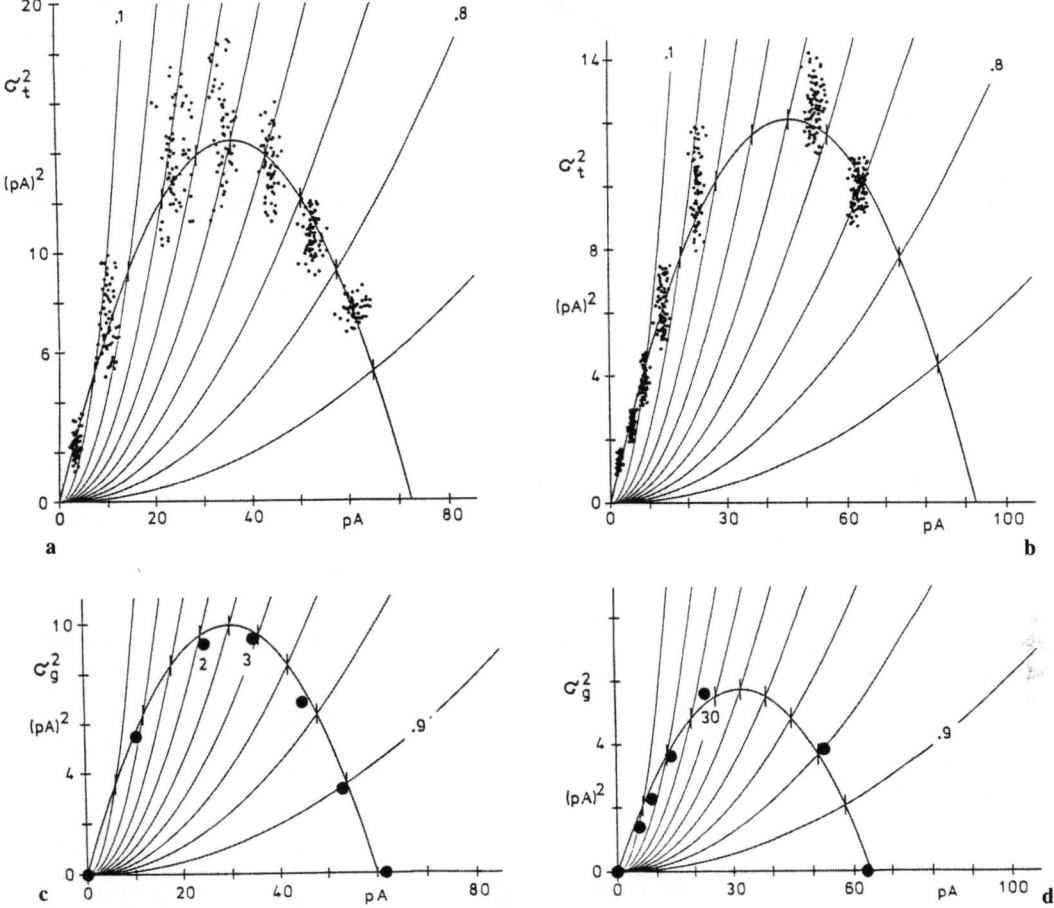

Fig. 2. Variance of cAMP-gated patch current, plotted against the mean current (I). Numbers at the data points indicate concentration of cAMP used. *Upward-open parabolic curves* are lines of equal probability. For **c** and **d**, they are lines of equal burst probability P_b. The K_m of the cAMP dose response is found at $P_b = 0.5$. **c,d** *Downward-open parabolas* were obtained by fitting the data points with $\sigma^2_g = i_b I - I^2/N$, where i_b is the mean single channel current in a burst. **b,d** Obtained after adding $0.2\,mM$ Ca^{2+} to the cytosolic side of the patch

Among divalent cations, the series of effectiveness of this "desensitization" was $Ca^{2+} > Sr^{2+} > Mn^{2+} > Ba^{2+} > Mg^{2+}$, of which Mg^{2+} (up to $3\,mM$) did not shift the K_m at all. The sequence is compatible with a calmodulin-mediated effect. Brief transient exposure to $3\,mM$ Mg^{2+} (or Ca^{2+} or Ba^{2+}) had a protective effect in that, following washout of Mg^{2+}, subsequent exposure to lower $[Ca^{2+}]$ no longer caused desensitization. Thus, binding of divalent cations controls the sensitivity of the cAMP-gated channels to cAMP in multiple ways.

Discussion

For cAMP-gated channels of the rat our data did not show inhibitory effects of $[Ca^{2+}]$ below $100\,\mu M$, even though such effects are known for the salamander [6] and the catfish [7]. Above $100\,\mu M$, Ca^{2+} on the cytosolic side of the patch exerted

(a) a strongly V-dependent block
(b) a poorly V-dependent block
(c) desensitization with respect to cAMP

Effect (a) appears to be due to Ca^{2+} occupying the channel (rapid open-channel block), suggested by the fact that the apparent single channel current i_b was lowered and that all divalent cations tested, including Mg^{2+}, had the potency to exert this type of block. (b) is explicable by an additional binding site for divalent cations near the channel entrance. The site may be reached from the cytosolic bulk solution in a diffusional step which is little affected by membrane voltage. Its occupancy by a divalent cation would raise the barrier for Na passage through the channel. A similar explanation was put forward by Menini [8], based on data obtained with the cGMP-gated channel of photoreceptor cells.

The cytosol will have a $[Mg^{2+}]$ in the order of $1\,mM$ and the extracellular space a $[Ca^{2+}]$ of similar

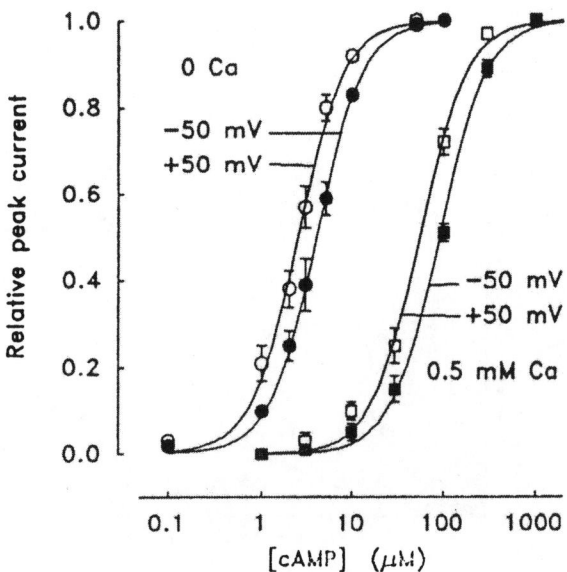

FIG. 3. cAMP dose response curves at two membrane voltages. The desensitization (shift to the *right*) caused by 0.5 mM Ca^{2+} present on the cytosolic side of the patch, was reversible. In the desensitization capacity, Mg^{2+} did not substitute for Ca^{2+}

magnitude. For such a distribution of divalent cations, our BR(V) curves indicate that at -80 mV the cAMP-gated channels are blocked to more than 90% by effects (a) and (b). The block lowers i_b, presumably providing for a less noisy receptor potential.

Effect (c) is a Ca^{2+}-induced shift of the cAMP dose-response to higher [cAMP]. Typically, this inhibition can be overcome by raising the [cAMP]. Based on the change in Lorentzian corner frequencies, more than one rate constant of the gating process was affected by Ca^{2+}. Another characteristic feature was that Mg^{2+} would not substitute for Ca^{2+}. On the contrary, brief exposure to Mg^{2+} locked the channel in the sensitive state, protecting it from future "desensitizations" by Ca^{2+}. Effect (c) may be related to the Ca^{2+}-inhibitions described for the salamander by Zufall et al. [6] and for the catfish by Kramer and Siegelbaum [7]. However, in the rat

and with the excised patches used in our experiments, $[Ca^{2+}] > 100\,\mu M$ were required. While the local cytosolic $[Ca^{2+}]$ may transiently reach $100\,\mu M$ in some systems [9,10], it remains to be shown which $[Ca^{2+}]$ are reached at the cytosolic opening of ciliary cAMP-gated channels during odorant stimulation.

Acknowledgments. Supported by the Deutsche Forschungsgemeinschaft through SFB 246, project A10 and C1, and by the CSIRO Sensory Research Center, NSW, Australia.

References

1. Nakamura T, Gold GH (1987) A cyclic nucleotide-gated conductance in olfactory receptor cilia. Nature 325:442–444
2. Frings S, Lynch JW, Lindemann B (1992) Properties of cyclic nucleotide-gated channels mediating olfactory transduction: activation, selectivity, and blockage. J Gen Physiol 100:45–67
3. Avenet P, Lindemann B (1990) Fluctuation analysis of amiloride-blockable currents in membrane patches excised from salt-taste receptor cells. J Basic Clin Physiol Pharmacol 1:383–391
4. Matthews G, Watanabe SI (1988) Activation of single ion channels from toad retinal rod inner segments by cyclic GMP: concentration dependence. J Physiol 403:389–405
5. Sigworth FJ (1980) The variance of sodium current fluctuations at the node of Ranvier. J Physiol 307: 97–129
6. Zufall F, Shepherd GM, Firestein S (1991) Inhibition of the olfactory cyclic-nucleotide gated ion channel by intracellular calcium. Proc Royal Soc Lond B 246:225–230
7. Kramer RH, Siegelbaum SA (1992) Intracellular Ca^{2+} regulates the sensitivity of cyclic uncleotide-gated channels in olfactory receptor neurons. Neuron 9: 897–906
8. Menini A (1990) Currents carried by monovalent cations through cyclic GMP-activated channels in excised patches from salamander rods. J Physiol 424:167–185
9. Augustine GJ, Neher E (1992) Calcium requirements for secretion in bovine chromaffin cells. J Physiol 450:247–271
10. Llinas R, Sugimoro M, Silver RB (1992) Microdomains of high Ca concentration in a presynaptic terminal. Science 256:677–679

Transmigration of Calcium Ions in the Olfactory Receptor Cell

SIO MIYAMOTO, KAZUNOBU TSURU, and TADASHI NAKAMURA[1]

Key words. Olfaction—Transduction—Ca^{2+}—Fura-2—Membrane current—Second messengers

Introduction

It is now widely believed that olfactory transduction is carried out in the receptor cell by the cascadal reactions in which cAMP is a second messenger for most of the odorants [1–4]. However, some reports have proposed that there is an alternative or additional transduction pathway(s) in which inositol 1,4,5,-triphosphate (IP_3) is a second messenger instead of cAMP [5–10]. On the other hand, several studies showed that IP_3 may not be a direct second messenger in the amphibian [2,11,12].

Thus, the behavior of IP_3 or Ca^{2+} in the olfactory transduction has yet to be studied. We have started to measure the intracellular concentration of Ca^{2+} in the isolated olfactory cell using the Ca^{2+}-sensitive fluorescent dye Fura2.

In this study, we investigated if those second messenger candidates could induce the increase of intracellular Ca^{2+} and the membrane current.

Methods

Olfactory epithelium was dissected from the newt and incubated for 5 min with 0.1% collagenase. Solitary olfactory cells were prepared by trituration of the epithelium with a Pasteur pipette. Fura2 was loaded into the cell by incubation with 5 µM Fura-2-AM for about 1 h at room temperature.

The low Ca^{2+} (pCa > 7.5) pseudo-intracellular solution was prepared by use of the ion exchanger (Chelex-100) column. The cell was irradiated by an epi-irradiating system at 340 or 380 nm, and the fluorescence emitted from intracellular Fura-2 was detected by a Silicone Intensified Target camera fixed on the microscope through a 500- to 530-nm bandpass filter. The image was recorded on a video cassette tape in S-VHS format and given further processing with an image analyzer (Argus 100, Hamamatsu, Japan).

Using the conventional patch electrode and the patch amplifier (Axopatch 1C, Axon, USA), the whole-cell membrane current was recorded with a video cassette recorder equipped with a PCM data processor (DP-16, Shosin EM, Japan) or with a personal computer (PC9801FX, NEC, Japan) at the clamped voltage.

When whole-cell mode was achieved at the cell body with the patch electrode that contained cAMP or IP_3 dissolved in the pseudo-intracellular solution, these chemicals were dialyzed into the cell. During dialysis, the membrane current and the "Ca^{2+} image" were recorded simultaneously.

Results

Responses to cAMP

When 0.5–1 mM cAMP was dialyzed through the patch pipette into the cell at the membrane potential of −50 mV, transient inward current was recorded as reported elsewhere [13–15]. This current is generally believed to result from the cation influx through the cAMP-gated channels from the extracellular side.

We then checked whether this cation influx contained Ca^{2+}. While the cAMP-gated current was recorded by dialysis into the cell that contained Fura-2, the fluorescence of the cell under the 340-nm excitation light increased significantly. Similar results were obtained in more than 10 cells. These observations suggested that Ca^{2+} influxed together with Na^+ from the extracellular side of the cell through the cAMP-gated channels.

Responses to IP_3

Using the same method as for cAMP, 0.1–1 mM IP_3 was dialyzed into the receptor cell. The fluorescence change occured mainly at the cell body. However, current was not significantly induced except the fast, small baseline shift on the achievement of whole-cell mode that had also appeared in the preceding part of cAMP-induced current. Similar results were obtained in more than five cells.

Thus, IP_3 appeared to increase $[Ca^{2+}]$ in the olfactory cell. However, the result indicates that this

[1] Department of Applied Physics and Chemistry, The University of Electro-Communications, 1-5-1 Chofugaoka, Chofu, Tokyo, 182 Japan

162   S. Miyamoto et al.

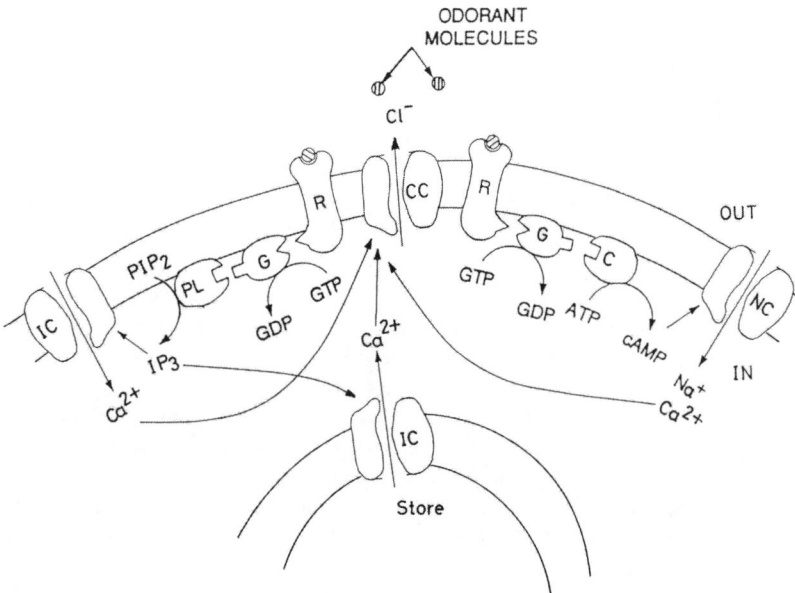

ODORANT
MOLECULES

Fig. 1. Diagram of putative pathways for olfactory transduction in Amphibia. Three different channels may contribute to the olfactory transduction. *R*, receptor; *G*, G-protein; *C*, adenylate cyclase; *PL*, phospholipase C; *NC*, cyclic nucleotide-gated channel; *CC*, calcium-gated chloride channel; *IC*, IP₃-gated calcium channel

Ca^{2+} increase is not related to the transductory membrane current.

Discussion

As mentioned previously, cAMP applied into the intracellular side of the newt olfactory receptor cell induced both the inward current and a Ca^{2+} increase. This result indicates that Ca^{2+} influxed through the cAMP-gated channels.

On the other hand, IP₃ did not induce significant membrane current, but induced the increase of Ca^{2+} concentration mainly at the cell body. Failure to detect the IP₃-induced currents has been reported by other laboratories [12,15], and is consistent with our report that IP₃-gated conductance was barely detectable in the frog cilia [18]. This result indicates that the contribution of the IP₃-gated channel on the plasma membrane to the increase of Ca^{2+} is not significant in the case of the newt.

However, our experiments do not slight the role of Ca^{2+} in the transduction. It is possible that the population of the cell bearing the IP₃ pathway in the olfactory epithelium of the newt is too low to detect by our method. On the other hand, recent biochemical work [16] has shown that all odorants stimulate both the cAMP and the IP₃ systems, suggesting that IP₃ may contribute to the cAMP cascade by such a mechanism as modulation of the enzymatic sensitivity. It is well known that Ca^{2+} contributes to the adaptation process in other sensory systems such as photoreceptors. Moreover, recent discovery of the contribution of the Ca^{2+}-gated Cl⁻ channel to the receptor potential [17] increased the importance of Ca^{2+}. In fact, we have confirmed that

Ca^{2+}-gated Cl⁻ conductance and cAMP-gated conductance together comprise the major conductance of the amphibian olfactory cilia [18]. Thus, we can draw the putative transduction pathways in the olfactory cell as shown in Fig. 1. It is important to investigate whether the Ca^{2+} increase induced by IP₃ contributes to those processes in the olfactory receptor.

Acknowledgments. We thank Ms. Yukimi Fukase for her experimental assistance. This work was partly supported by the special grant for promoting research and education in the University of Electro-Communications.

References

1. Nakamura T, Gold GH (1987) A cyclic nucleotide-gated conductance in olfactory receptor cilia. Nature (Lond) 325:442–444
2. Lowe G, Nakamura T, Gold GH (1989) Adenylate cyclase mediates olfactory transduction for a wide variety of odorants. Proc Natl Acad Sci USA 86: 5641–5645
3. Anholt RRH (1992) Molecular aspects of olfaction. In: Serby MJ, Chobor KL (eds) Science of olfaction. Springer, Berlin Heidelberg New York Tokyo, pp 51–79
4. Pace U, Hanski E, Salomon Y, Lancet D (1985) Odorant-sensitive adenylate cyclase may mediate olfactory reception. Nature (Lond) 316:255–258
5. Huque T, Bruch RC (1986) Odorant and guanine nucleotide-stimulated phosphoinositide turnover in olfactory cilia. Biochem Biophys Res Commun 137: 36–42
6. Boekhoff I, Tareilus E, Strotmann J, Breer H (1990) Rapid activation of alternative second messenger

pathways in olfactory cilia from rats by different odorants. EMBO J 9:2453–2458

7. Miyamoto T, Restrepo D, Cragoe EJ Jr, Teeter JH (1992) IP$_3$- and cAMP-induced responses in isolated olfactory receptor neurons from the channel catfish. J Membr Biol 127:173–183

8. Restrepo D, Miyamoto T, Bryant BP, Teeter JH (1990) Odor stimuli trigger influx of calcium into olfactory neurons of the channel catfish. Science 249: 1166–1168

9. Restrepo D, Teeter JH, Honda E, Boyle AG, Marecek JF, Prestwich GD, Kalinoski DL (1992) Evidence for an InsP$_3$-gated channel protein in isolated rat olfactory cilia. Am J Physiol 263:C667–C673

10. Sato T, Hirono J, Tonoike M, Takebayashi M (1991) Two Types of increases in free Ca^{2+} evoked by odor in isolated frog olfactory receptor neurons. Neuroreport 2:229–232

11. Firestein S, Darrow B, Shepherd GM (1991) Activation of the sensory current in salamander olfactory receptor neurons depends on a G protein-mediated cAMP second messenger system. Neuron 6:825–835

12. Lowe G, Gold GH (1993) Contribution of the ciliary cyclic nucleotide-gated conductance to olfactory transduction in the salamander. J Physiol 462:175–196

13. Kurahashi T (1990) The response induced by intracellular cyclic AMP in isolated olfactory receptor cells of the newt. J Physiol 430:355–371

14. Suzuki N (1989) Voltage- and cyclic nucleotide-gated currents in isolated olfactory receptor cells. In: Brand JG, Teeter JH, Cagan RH, Kare MR (eds) Chemical senses, vol 1, Receptor events and transduction in taste and olfaction. Dekker, New York, pp 469–493

15. Firestein F, Darrow B, Shephered GM (1991) Activation of the sensory current in salamander olfactory receptor neurons depends on a G protein-mediated cAMP second messenger system. Neuron 6:825–835

16. Ronnett GV, Cho H, Hester LD, Wood SF, Snyder SH (1993) Odorants differentially enhance phosphoinositide turnover and adenylyl cyclase in olfactory receptor neuronal cultures. J Neurosci 13:1751–1758

17. Kurahashi T, Yau KW (1993) Co-existence of cationic and chloride components in odorant-induced current of vertebrate olfactory receptor cells. Nature (Lond) 363:71–74

18. Nakamura T, Satoh T, Miyamoto S (in press) Ligand-gated ionic conductances in the olfactory cilia. In: Proceedings, International Symposium on Mechanisms of Sensory Chemoreception in Vertebrates

Cyclic AMP-Gated Channels and Ca-Activated Cl⁻ Channels in Vertebrate Olfactory Receptor Cells

Takashi Kurahashi and King-Wai Yau[1]

Key words. Olfaction—Receptor cell—cAMP—Ca—Signal transduction—Ion channel

Introduction

Odor stimulation induces a depolarization of olfactory receptor cells. It is now generally accepted that this signal transduction process is mediated by second messenger cascades, cAMP and/or IP$_3$ pathway [1–6]. Because both cAMP- and IP$_3$-gated channels are known to be cation selective, it is natural to think that odorant-induced current is carried solely by cations. In this study, however, we show that the odorant-induced current also contains an inward chloride component triggered by calcium influx through the cation-selective channel. The cationic component has characteristics very similar to the current induced by an injection of cyclic AMP (cAMP), suggesting that the cationic conductance is identical to cAMP-activated channels found in olfactory cilia [1]. Because the Cl component is dependent on the presence of external Ca, it seems likely that the anionic channel is identical to the Ca-activated Cl channel also found in cilia [7].

The cationic and anionic components carry similar amounts of current during response. The co-presence of cationic and chloride components in the odorant response, possibly unique among sensory transduction mechanisms, may serve to reduce variations in the transduction current resulting from changes in external ionic concentrations around the olfactory cilia. Our findings can solve the long-standing puzzle that removal of most mucosal cations still does not diminish the amplitude of the olfactory receptor cell response [8–10]. Part of the current study has been published elsewhere [11–17].

Materials and Methods

Olfactory cells isolated from newts (*Cynops pyrrhogaster*) were used mainly for these experiments.

Similar experiments on the adult tiger salamander (*Ambystoma tigrinum*) olfactory neurons gave identical results. The animal was pithed and the dorsal bone covering the olfactory cavity was removed. The olfactory epithelia were excised and then dissociated as previously described [11]. In short, the epithelia were cut into pieces of 4 to 9 mm^2 and treated with 0.1% collagenase (Sigma Type IA) for 5 min at 37°C in Ca- and Mg-free saline. After being rinsed several times with normal saline, the epithelial pieces were dissociated by trituration with a Pasteur pipette. The dissociated cells were plated onto ConA-coated culture dishes and stored under normal saline at 4°C for as long as 10 h until use.

Membrane currents were recorded with the whole-cell voltage clamp method. In the current study, cells were, unless otherwise indicated, bathed in physiological saline solution (120 mM NaCl, 5 mM KCl, 2 mM CaCl$_2$, 1 mM MgCl$_2$, 10 mM NaHEPES, and 10 mM glucose, pH, 7.4). The whole-cell recording pipette contained 120 mM CsCl, 5 mM NaEGTA, and 10 mM NaHEPES; pH, 7.4. The patch pipette had a tip opening of about 0.5-μm in diameter and a resistance of about 10 megaohms (MΩ). Seal resistance on establishment of a membrane patch was about 10 gigaohm. The odorant used for stimulation consisted of a mixture of *n*-amylacetate (1 μl/ml), iso-amylacetate (1 μl/ml), and limonene (0.2 μl/ml) dissolved in normal saline solution. It was ejected by pressure from a glass pipette (0.5 μm opening) with its tip brought within 20 μm from the cilia of the recorded cell. All experiments were done at room temperature.

For perforated-patch recordings, the pipette solution contained 120 mM K-gluconate, 10 mM NaHEPES, 5 mM NaEGTA, and 200 μg/ml nystatin (diluted from a stock solution of 50 mg/ml in DMSO). After formation of the gigaseal, the conductance of the membrane patch gradually increased as indicated by the development of an outward potassium current at 0 mV, reflecting the incorporation of the ionophore into the membrane. Odorant stimulation was applied after the outward membrane current became stable, typically about 20 s after establishment of the gigaseal.

[1] Howard Hughes Medical Institute, and Department of Neuroscience, The Johns Hopkins University School of Medicine, Baltimore, MD 21205, USA

Results

Odorant-Induced Responses
in Isolated Olfactory Receptor Cells

Isolated receptor cells responded to odorant with a monophasic inward current when cells were voltage clamped at their resting membrane potential (about $-50\,mV$). The response amplitude was reduced when membrane potential was shifted to positive and become undetectable at $+3\,mV$ [11]. Further shift of membrane potential reversed current polarity to outward.

The odorant-induced current observed in isolated preparations was confirmed to be a real response, based on following properties. (1) The cell sensitivity was maximum when the cilia or the apical part of the dendrite was locally stimulated. (2) They showed strong adaptation during the long-lasting stimulation (9 s); this adaptation has been shown to be abolished when external Ca was removed [12]. (3) Responsiveness to two different odorants (amylacetate and limonene) varied from cell to cell. The type a cell responded to only amylacetate, the type b cell responded only to limonene, and the type c cell responded to both, but the type d cell did not respond to either. This variation is thought to have a cellular basis of odor discrimination [13].

Reversal Potential of Odorant-Induced Current

In the normal saline, the reversal potential was quite insensitive to the change in $[Na^+]_0$ even at a very low concentration of Na (i.e., 10 mM). This result is consistent with previous reports that removal of external Na does not diminish the olfactory response in situ [8–10]. This phenomenon can be explained by the fact that significant fraction of transduction current is also carried by Cl, which is triggered by the influx of Ca through cationic conductance (see below).

Odorant-Induced Cationic Conductance
Observed in the Absence of Calcium

To isolate the cationic component, odor responses were recorded under a zero Ca condition [11]. In the absence of external Ca^{2+}, the reversal potential of odorant-activated conductance depended strongly on external Na^+. The reversal potential shifted by 57 mV per 10-fold change in $[Na^+]_0$, consistent with the opening of a cation-selective conductance. The permeability ratios of cationic conductance to alkali metal ions were

$$P_{Li} : P_{Na} : P_K : P_{Rb} : P_{Cs} = 1.25 : 1 : 0.98 : 0.84 : 0.80$$

The permeability to Ca was also examined under Ca/Cs bi-ionic condition in which all Cl was replaced

with glutamate [12]. The permeability ratio between Ca and Na was

$$P_{Ca} : P_{Na} = 6.5 : 1$$

cAMP-Induced Current and
cAMP-Gated Channels

Isolated receptor cells also showed a large response to introduced cAMP from a whole-cell recording pipette [14]. The maximum current amplitude was 300–600 pA when cells were voltage clamped at $-50\,mV$. This value is almost identical to the maximum inward current observed in cells stimulated with odorant. Further, the properties of the cAMP-induced responses were very similar to the odorant-induced current; (1) cAMP responses showed a decay with a time course similar to that of odorant responses; the adaptation was dependent on the influx of Ca. (2) The permeability ratio of the cAMP-induced conductance was almost identical to the odorant-induced conductance, as follows:

$$P_{Li} : P_{Na} : P_K : P_{Rb} : P_{Cs} = 0.93 : 1 : 0.93 : 0.91 : 0.72$$

(3) The maximum sensitivity to cytoplasmic cAMP was also found to be localized at the ciliary membrane. Latency of response was the shortest when cAMP was introduced from the apical dendrite, but latency was longer when cAMP was introduced from the cell body [14]. The channel density measured with inside-out patch preparations was $1000/\mu m^2$ at the ciliary membrane but $2/\mu m^2$ at the dendrosomatic membrane [15,16].

Ca-Dependent Cl Component
in Transduction Current

In the normal Ringer's solution containing Ca^{2+} the reversal potential became quite insensitive to $[Na^+]_0$ and remained near 0 mV even with very low external Na^+ [17]. This result suggests that another major current component is present in the olfactory response. This component cannot be carried by Ca^{2+}, because Ca^{2+} is not a major current carrier through the odorant-induced cation conductance. Instead, it must be triggered by a rise in intracellular Ca^{2+}, most probably a Ca^{2+}-activated Cl^- current; a Ca^{2+}-activated K^+ current would have been blocked by internal Cs^+ used in the experiments or otherwise would have given a reversal potential very negative from 0 mV. Finally, the existence of a Ca^{2+}-activated Cl^- conductance has been reported in frog olfactory cilia [7].

To check the involvement of chloride, we reduced the external Cl^- concentration, $[Cl^-]_0$, to 34 mM by replacement with gluconate. In this case, the odorant-induced current became biphasic at $+24\,mV$, consisting of an outward current (cationic current) followed by an inward component (anionic current).

All cells tested ($n = 10$) showed similar results, suggesting that the Cl$^-$ current is probably present in all receptor cells.

The presence of a chloride component in the olfactory response was further confirmed by the chloride channel blocker SITS. Application of this drug selectively removed the inward current in the biphasic odorant response elicited in low external Cl$^-$ and at a positive membrane potential. The effect of SITS was rapid and reversible. This drug had little effect on the pure cationic current observed in the absence of external Ca^{2+}, indicating that it specifically blocked the Cl$^-$ component.

Fraction of Cationic and Anionic Components

To measure the fraction of Cl component under physiological conditions, we have also performed perforated-patch recording with a nystatin-containing pipette solution [18] that considerably slowed Cl$^-$ exchange between the cell cytoplasm and the pipette interior. The odorant-activated current initially still had a reversal potential near 0 mV. Further, even with a pipette containing no Cl$^-$, there was still an SITS-sensitive component in the inward transduction current (measured at -65 mV). Thus, an inward Cl$^-$ current in the olfactory response appears to be a physiological phenomenon. Put in another way, the normal chloride concentration within the olfactory cilia must already be very high, most probably because of a chloride uptake mechanism. From a total of seven perforated-patch experiments (four from newt and three from salamander), the Cl$^-$ component accounts for $36 \pm 25\%$ in peak current amplitude and $60 \pm 22\%$ in total charge influx of the olfactory response. The Cl$^-$ contribution is very substantial, in particular for the total charge influx, which is the more important parameter in determing excitation of a receptor cell when the odorant stimulus lasts for more than a couple of seconds.

Discussion

In this study we have demonstrated that both cationic and anionic conductances are responsible for the olfactory transduction. The Cl current is triggered by Ca^{2+} influx through the cationic channel which is most likely to be a cAMP-gated channel [13] (Fig. 1). The presence of a Ca-activated Cl channel has been demonstrated in olfactory receptor cilia [7]. Further, Ca^{2+} influx causes a negative feedback on this system, presumably via phosphodiesterase activity or modulation of cAMP-gated channel sensitivity, which explains an olfactory adaptation [13] (see Fig. 1).

The olfactory receptor cells are unusual among sensory neurons in that their transduction machinery is in direct contact with the external environment.

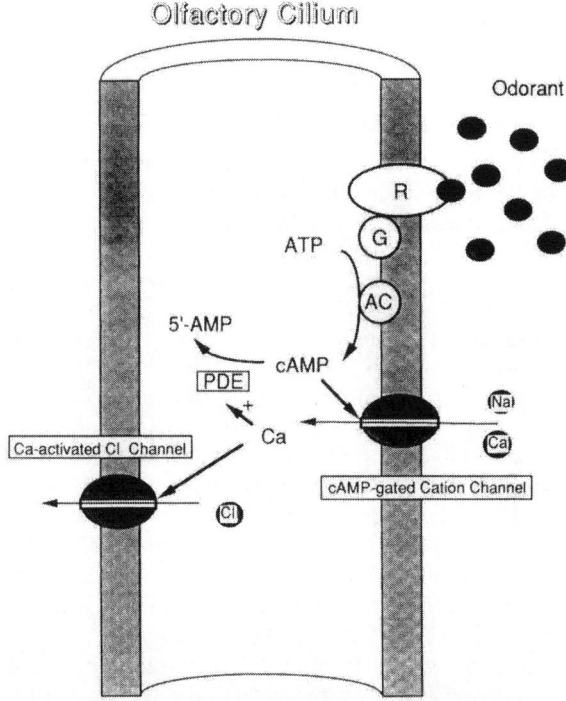

FIG. 1. Scheme of olfactory transduction mechanism. R, receptor molecule; G, GTP-binding protein; AC, adenylyl cyclase

While generally bathed in secreted mucus, the olfactory cilia may not always be exposed to constant ionic concentrations. A mucus dilution by, say, fresh water would reduce the transduction current if cations were the sole current carriers. With the copresence of a Cl$^-$ component, on the other hand, such a reduction in inward cationic current would be compensated by an increase in the inward anionic current, thus maintaining the effectiveness of the transduction mechanism. Indeed, it has been shown that the olfactory receptor responses were not diminished when all external Na ions were removed [8–10]. The experiments presented here therefore provide a simple explanation for this intriguing observation.

Acknowledgments. T.K. was supported by a fellowship from the Uehara Memorial Foundation.

References

1. Gold GH, Nakamura T (1987) Cyclic nucleotide-gated conductance: a new class of ion channels mediates visual and olfactory transduction. Trends Biochem Sci 8:312–316
2. Bakalyar HA, Reed RR (1991) The second messenger in olfactory receptor neurons. Curr Opin Neurobiol 1:204–208
3. Breer H, Boekhoff I (1992) Second messenger signalling in olfaction. Curr Opin Neurobiol 2:439–443

4. Firestein S (1992) Electrical signals in olfactory transduction. Curr Opin Neurobiol 2:444–448

5. Ronnet GV, Snyder SH (1992) Molecular messengers of olfaction. Trends Neurol Sci 15:508–513

6. Anholt RRH (1993) Molecular neurobiology of olfaction. Crit Rev Neurobiol 7:1–22

7. Kleene SJ, Gesteland RC (1991) Calcium-activated chloride conductance in frog olfactory cilia. J Neurosci 11:3624–3629

8. Tucker D, Shibuya T (1965) A physiologic and pharmacologic study of olfactory receptors. Cold Spring Harbor Symp Quant Biol 30:207–215

9. Suzuki N (1978) Effect of different ionic environments on the responses of single olfactory receptors in the lamprey. Comp Biochem Physiol A61:461–467

10. Yoshii K, Kurihara K (1983) Role of cations in olfactory reception. Brain Res 274:239–248

11. Kurahashi T (1989) Activation by odorants of action-selective conductance in the olfactory receptor cell isolated from the newt. J Physiol (Lond) 419:177–192

12. Kurahashi T, Shibuya T (1990) Ca^{2+}-dependent adaptive properties in the solitary olfactory receptor cell of the newt. Brain Res 515:261–268

13. Kurahashi T, Kaneko A, Shibuya T (1990) Ionic mechanisms of the olfactory transduction studied on isolated receptor cells of the newt. Neurosci Res 12:S85–S96

14. Kurahashi T (1990) The response induced by intracellular cyclic AMP in isolated olfactory receptor cells of the newt. J Physiol (Lond) 430:355–371

15. Kurahashi T, Kaneko A (1991) High density cAMP-gated channels at the ciliary membrane in the olfactory receptor cell. NeuroReport 2:5–8

16. Kurahashi T, Kaneko A (1993) Gating properties of the cAMP-gated channel in toad olfcatory receptor cells. J Physiol (Lond) 466:287–302

17. Kurahashi T, Yau K-W (1993) Co-existence of cationic and chloride components in odorant-induced current of vertebrate olfactory receptor cells. Nature (Lond) 363:71–74

18. Horn R, Marty A (1988) Muscarinic activation of ionic currents measured by a new whole-cell recording method. J Gen Physiol 92:145–159

Patch-Clamp Measurements on Mammalian Olfactory Receptor Neurons

PETER H. BARRY[1], JOSEPH W. LYNCH[2], and SUNDRAN RAJENDRA[1]

Key words. Patch-clamp—Olfaction—Receptor—Neurons—Channels

Introduction

The olfactory system is an exquisitely sensitive system for detecting picomolar concentrations of odorant molecules [1]. Our investigations used the patch-clamp technique to characterize the properties of the ion-selective channels and ion currents in olfactory receptor neurons so that we could provide information of value in understanding the process of olfactory transduction. Although the cells are smaller and more difficult to work with, we chose dissociated cells from adult rats because of their more probable similarity to human olfactory cells. These cells had a number of properties that seemed to be particularly appropriate for controlling the extremely sensitive response to an olfactory stimulus.

Material and Methods

The cell dissociation techniques are discussed in detail in [2]. Briefly, the olfactory epithelium from adult female Wistar rats (killed by CO_2 inhalation) was enzymatically dissociated using 0.022% trypsin in divalent cation-free Dulbecco's Modified Eagles Medium and incubated at 37°C for 30–35 min. Cells were then triturated and placed in a flow-through recording chamber situated on the microscope and superfused with a general mammalian Ringer's (GMR) solution (containing NaCl, 140 mmol/l; KCl, 5 mmol/l; MgCl, 1 mmol/l; CaCl$_2$, 2 mmol/l; glucose, 10 mmol/l; Na-HEPES, 10 mmol/l; pH, 7.2). The olfactory receptor neurons were identified by their bipolar, ciliated structure and the presence of large transient Na^+ currents. Standard single-channel and whole-cell recording techniques [3] were used. The pipette solutions typically contained KCl,

145 mmol/l; K-HEPES, 10 mmol/l; CaCl$_2$, 2 mmol/l; MgCl$_2$, 1 mmol/l; NaHEPES, 10 mmol/l; glucose, 10 mmol/l; pH, 7.4, for intact patches, and KF, 125 mmol/l; KCl, 15 mmol/ll; K-HEPES, 10 mmol/l; EGTA, 11 mmol/l; pH 7.2, for whole-cell measurements. Replacement of K^+ by Cs^+ in the pipette was also used to block K^+ currents in Na^+ current measurements. Changes in $[K]_0$ or the addition of various pharmacological agents were compensated for by decreasing $[Na]_0$ to maintain osmolality. Data generally were filtered at 1 kHz (and digitized at 2.5 kHz). All membrane potentials were expressed as the potential of the internal with respect to the external membrane surfaces and were corrected for liquid junction potentials (as outlined in ref [4]). Current measurements were made by using an EPC-7 amplifier (List, Darmstadt, Germany) and recorded on videotape. Analysis was done on an IBM-AT computer using PNSCROLL software [5], SigmaPlot (Jandel, Corte Madera, California) and other software developed in our laboratory.

Results

In some early intact patch measurements it became clear that these neurons must have an extremely high input impedance because a channel opening within the patch could cause a current spike across it. This was shown to result from an action potential being generated elsewhere in the cell [6]. In fact, in one cell in particular, the opening of a certain channel (with a pipette potential greater than about 20 mV) invariably initiated such a current spike response (Fig. 1a). It was seen that when the channel opened, the current waveform was not rectangular but seemed to initially "droop", drifting slowly back toward the baseline, then reached a threshold level and displayed a fast and somewhat biphasic spike. Reducing the pipette potential increased the time to reach this threshold level and initiate the spike response (Fig. 1b). A complete mathematical analysis of the full electrical model of the cell and patch clearly showed that such a response was to be expected for a cell with a very high input impedance [4,6]. The opening of the channel in the patch causes current to flow through it and across the rest of the cell. Because of its high

[1] School of Physiology and Pharmacology, University of New South Wales, PO Box 1, Kensington, Sydney, NSW 2033, Australia

[2] Current address: Neurobiology Division, Garvan Institute of Medical Research, 384 Victoria St., Darlinghurst, NSW 2010, Australia

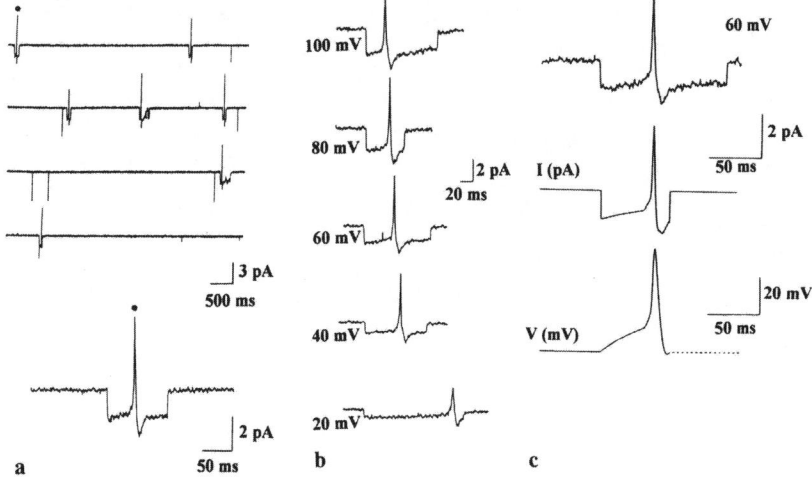

FIG. 1. Action potential spikes recorded across intact patch of mammalian olfactory receptor neuron initiated by opening of a single ion channel in that patch. **a** Top four traces show a continuous record in which a spike is generated each time that channel opens; bottom trace is expanded version of the first event ($V_p = +70$ mV; from Fig. 3 of [11]; see also Fig. 1 of [6]). **b** Effect of pipette potential on changes in delay of spike initiation and magnitude of "droop" of the current (from part of Fig. 4 of [6]). **c** Change in cell membrane potential following a channel opening predicted from a full circuit analysis of the cell and patch and resultant predicted current waveform compared with a somewhat typical experimental record at $V_p = +60$ mV (from part of Fig. 6 of [6]). **a** from [11], with permission; **b** and **c** from [6], with permission of the Rockefeller University Press

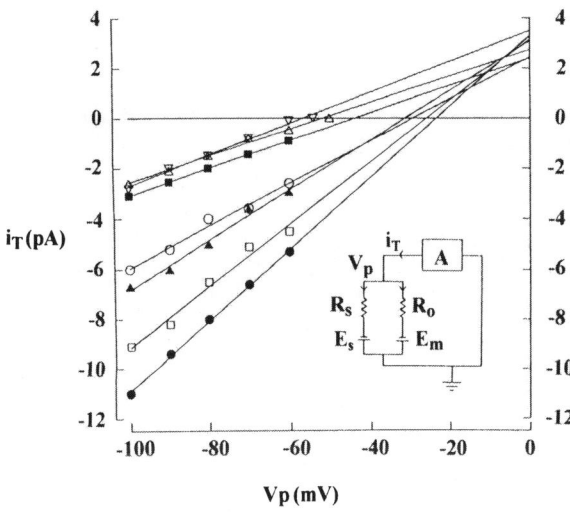

FIG. 2. Relationship between (whole-cell) total current, i_T, recorded by the patch-clamp amplifier (**A**), and pipette potential, V_p, for a small olfactory receptor cell in which seal resistance, R_s, is comparable to cell input resistance, R_o. In equivalent circuit shown, E_s and E_m refer to potential difference across seal and zero current resting membrane potential, respectively. Results are shown for seven different cells. (Modified from parts of Fig. 1 of [2] and reproduced from [2], with permission of the Rockefeller University Press)

input impedance, the cell then becomes more and more depolarised (see first curved part of bottom trace of Fig. 1c), the driving force across the channel being concomitantly reduced (the current therefore "droops"; see middle trace of Fig. 1c and current

records in Fig. 1a–c) until, if the patch potential is high enough, the threshold for an action potential is reached (perhaps in the initial segment of the cell).

By fitting the time-to-reach-threshold, the amount of current "droop" observed, and the amount of charge passing through the channel before the spike, each as a function of pipette potential, it was possible to uniquely determine whole-cell resistance, R_o, channel conductance, γ_c, and its reversal potential, E_o. The value of R_o was about 40 GΩ, giving a specific membrane resistance of about $100 \, k\Omega \cdot cm^2$, comparable to data obtained for the much bigger amphibian receptor cells [7]; the specific membrane capacitance was close to $1 \, \mu F/cm^2$, the channel conductance was about 29 pS (similar to measurements in excised patches, but very different from the apparent conductance without corrections for small cell effects), and the channel reversal potential was about -44 mV. Using these values and a somewhat typical simulated action potential waveform (the fast transient section of the bottom trace of Fig. 1c), analysis predicted the current response across the patch (middle trace; 60% capacitative and remainder ionic), in good agreement with the experimental record.

Whole-cell measurements also confirmed the extremely high value of R_o for these cells. By considering the circuit shown in the inset of Fig. 2, it can be seen [2] that the total current, i_T, will be proportional to the pipette potential, V_p. If the seal resistance is very nonselective, so that $E_s \approx 0$, and the membrane is very K^+ selective, so that $E_m \approx E_K$, then the input resistance will be simply given by

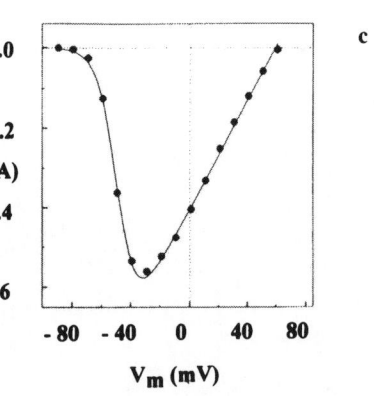

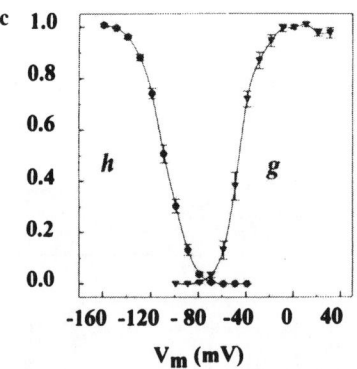

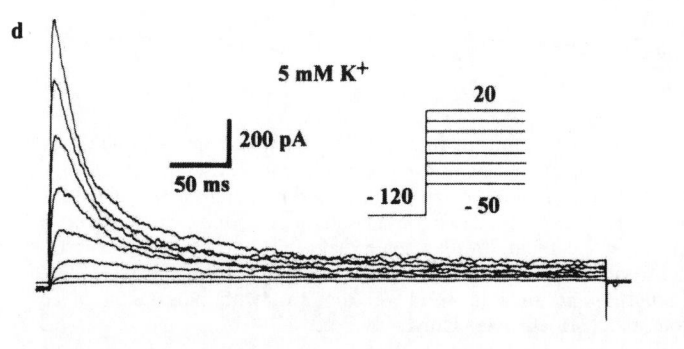

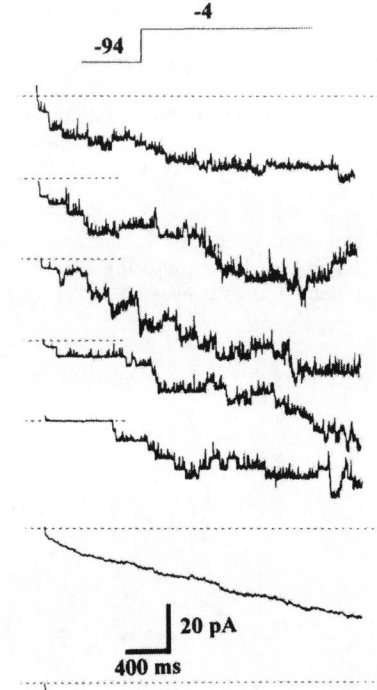

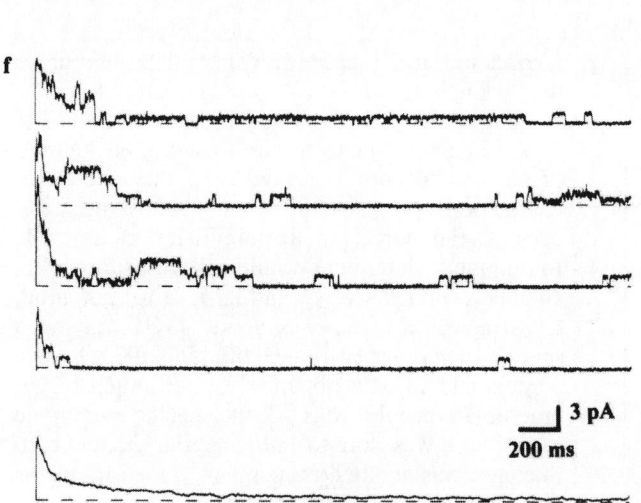

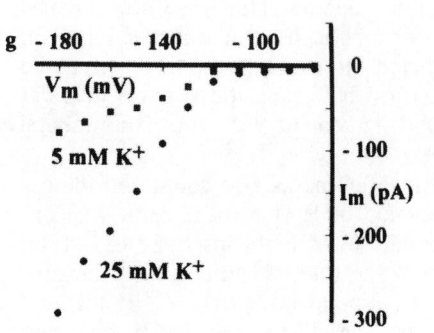

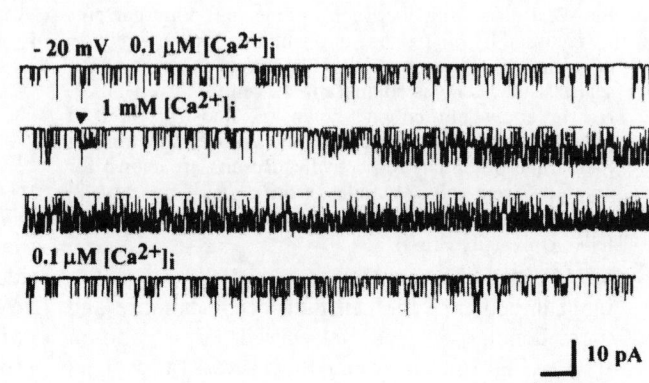

$R_o \approx -E_K/i_T^o$, where i_T^o is the total current with $V_p = 0$. Using this relationship to estimate R_o, data from seven cells indicated a value of about $26 \pm 2\,G\Omega$ (mean $\pm$ SEM, range $19\text{–}31\,G\Omega$) [2], comparable to the value determined by the action potential spike data already discussed.

An interesting range of different types of channels and currents was also observed in these olfactory neurons. The properties of some are depicted in Fig. 3. A voltage-activated Na^+ current was typically present in these neurons. It could be blocked completely by the addition of either $10\,\mu M$ tetrodotoxin (TTX) or $1\,mM\ Zn^{2+}$ to the external bathing solution (Fig. 3a); its reversal potential was, as expected, close to the Na^+ equilibrium potential (Fig. 3b). Of particular interest was that it had an especially negative voltage dependence of steady-state inactivation (Fig. 3c; complete at potentials more positive than $-70\,mV$, half complete at $-110\,mV$, and with a very slow recovery), which suggested that these cells had an unusually negative resting potential, probaly more negative than $-90\,mV$ [8].

Delayed rectifier K^+ currents were rather rare. In contrast, a transient K^+ current that had some interesting characteristics was often seen in these cells [2]. Figure 3d shows the typical transient response of this current to depolarizing voltage steps. It rapidly activated and then inactivated with a double exponential time course (with time constants of 22 and 143 ms). Steady-state inactivation was complete at $-20\,mV$ and completely removed at $-80\,mV$, and recovery was extremely slow (about 50 s at $-100\,mV$). The distinct feature of this current was that increasing $[K^+]_0$ from 5 to 25 mM partially reduced the inactivation without affecting the kinetics of activation. In addition, the underlying single-channel currents observed in excised patches (Fig. 3f) were activated in large clusters and had single-channel conductances of 17 and 26 pS and a permeability ratio of $P_{Na}/P_K \approx 0.1$. The channel was partially blocked by 1 mM tetraethylammonium (TEA) and slightly more effectively blocked by 4-aminopiridine (4-AP), which reduced both the amplitudes and open times of the underlying channels.

In a small proportion ($\leqslant 5\%$) of membrane patches, an unusual, extremely slowly activating K^+ current was observed [9] that activated very slowly with depolarization over a period of about 10 s. Excised patch records of such channel activation are shown in Fig. 3e. The channels were insensitive to Ca^{2+}, had a single-channel conductance of about 135 pS, and could be abolished by 20 mM TEA but were resistant to 5 mM 4-AP.

Figure 3h also displays some continuous calcium-activated K^+ channel records with the open-probability of the channels increasing as $[Ca^{2+}]_0$ is increased [10,11].

Finally, an inwardly rectifying cation-selective current was also observed [12] that was activated by hyperpolarization to potentials more negative than $-100\,mV$ (Fig. 3g). The currents were blocked by $5\,mM\ [Cs^+]_0$ but not by $20\,mM\ [TEA]_0$ or $5\,mM\ [4\text{-}AP]_0$.

Discussion

The above-mentioned electrophysiological properties of these olfactory neurons seem very appropriate for their exquisite odorant sensitivity [1]. The high input resistance of these mammalian cells indicated a specific membrane resistance of about $100\,k\Omega \cdot cm^2$, comparable in magnitude to the much bigger amphibian receptor cells [7] but two orders greater than the membranes of most other neuronal cells. In such olfactory cells, it is clear that the coupling of the odorant-activated channel activity in the cilia and dendritic knob, where the odorant receptors are predominantly found, to the soma and initial segment where the action potential spikes are generated would be greatly facilitated by such an extremely high resistance, electrically compact, neuron.

FIG. 3. Some examples of whole-cell and single-channel currents and their properties recorded in adult mammalian olfactory neurons. **a** Whole-cell Na^+ currents being blocked by Zn^{2+} (control-before, during and after bath washout; from part of Fig. 2b of [8]). **b** Peak whole-cell Na^+ current as a function of membrane potential (V_m) with a reversal potential of $+60\,mV$ close to the Na^+ equilibrium potential of $+67\,mV$ (from part of Fig. 3 of [8]). **c** Steady-state inactivation (*h*) and activation (*g*) as a function of V_m for same Na^+ currents (relative values from part of Fig. 3 of [8]). **d** Whole-cell transient K^+ currents recorded in 5 mM $[K^+]_0$ (from part of Fig. 2 of [2]). **e** Single-channel recordings of slowly-activating K^+ currents in an inside-out patch. Top five traces show separate responses to single depolarizations; bottom shows ensemble mean of many such responses (from Fig. 3 of [9]). **f** Single-channel recordings of transient K^+ currents in an inside-out patch. As in **e**, top traces show separate responses to single depolarizations, and bottom shows ensemble mean (from Fig. 11 of [2]). **g** Current–voltage relationship of the inwardly rectifying cation current in bath solutions of 5 mM and 25 mM K$^+$ (from part of Fig. 2 of [11]). **h** Continuous record of single Ca^{2+}-activated K^+ channels recorded in an inside-out patch as $[Ca^{2+}]_i$ is changed from 0.1 to 1.0 and back to $0.1\,\mu M$ (*dashed line*, closed state). **d** and **f** from [2], with permission of the Rockefeller University Press; **e** and **g** with permission of the Royal Society of London; **h** from [11], with permission of Today's Life Science

The negative voltage sensitivity of inactivation of the Na^+ currents seems more striking than in amphibians, although one study [13] did suggest a half-inactivation value of $-82\,mV$. It was suggested that axotomy may have removed a different population of channels with a more positive voltage dependence. In addition, in cultured rat neonatal neurons, an inactivation midpoint of $-63\,mV$ was observed [14] that may indicate either different properties of these immature cells or a different population of channels compared to those remaining in adult neurons after dissociation.

The other channels were broadly similar to those observed in other mammalian and non-mammalian studies with the exception of two newly reported channels. The inwardly rectifying cation channel, not seen in any other olfactory neuron, may be modulated by cAMP [15] and may therefore be involved in the control of spiking behavior during odorant stimulation. The slowly activating K^+ channel would be expected to help to terminate spike activity following any sustained burst of odorant excitation. In addition, the transient K^+ channel was seen to have its inactivation significantly reduced by an increase in $[K]_0$, as might be expected to happen during sustained odorant stimulation, and so could help to modulate the adaptation of frequency and the termination of a burst of spikes.

Acknowledgments. This work was supported by the Australian Research Council and the National Health and Medical Research Council of Australia.

References

1. Frings S, Lindemann B (1991) Current recording from sensory cilia of olfactory receptor cells in situ. I: The neural response to cyclic nucleotides. J Gen Physiol 97:1
2. Lynch JW, Barry PH (1991) Properties of transient K^+ currents and underlying single K^+ channels in rat olfactory receptor neurons. J Gen Physiol 97: 1043–1072
3. Hamill OP, Marty A, Neher E, Sakmann B, Sigworth FS (1981) Improved patch-clamp techniques for high resolution current recording from cells and cell-free membrane patches. Pfluegers Arch 391:85–100
4. Barry PH, Lynch J W (1991) A topical review. Liquid junction potentials and small cell effects in patch-clamp analysis. J Membr Biol 121:101–117
5. Barry PH, Quartararo N (1990) PNSCROLL, a software package for graphical interactive analysis of single channel patch clamp currents and other binary file records: under mouse control. Comp Biol Med 20:193–204
6. Lynch JW, Barry PH (1989) Action potentials initiated by single channels opening in a small neuron (rat olfactory receptor). Biophys J 55:755–768
7. Firestein S, Werblin FS (1987) Gated currents in isolated olfactory receptor neurons of the laval tiger salamander. Proc Natl Acad Sci USA 84:6292–6296
8. Rajendra S, Lynch JW, Barry PH (1992) Negative voltage-dependence and slow recovery of Na^+ currents in rat olfactory receptor neurons. Pfluegers Arch 420: 342–346
9. Lynch JW, Barry PH (1991) Slowly-activating K^+ channels in rat olfactory receptor neurons. Proc R Soc Lond Ser B Biol Sci 244:219–225
10. Lynch JW (1991) A patch clamp study of membrane ion channels in mammalian olfactory receptor neurons. Ph.D. thesis, The University of New South Wales, Sydney, Australia
11. Lynch JW, Barry PH (1992) Studying olfactory transduction using patch-clamping. Today's Life Sci 4: 26–42
12. Lynch JW, Barry PH (1991) Inward rectification in rat olfactory receptor neurons. Proc R Soc Lond Ser B Biol Sei 243:149–153
13. Pun RYK, Gesteland RC (1991) Somatic sodium channels of frog olfactory receptor neurons are inactivated at rest. Pfluegers Arch 418:504–551
14. Trombley PQ, Westbrook GL (1991) Voltage-gated currents in identified rat olfactory receptor neurons. J Neurosci 11:435–444
15. Pape H-C, McCormick DA (1989) Noradrenaline and serotonin selectively modulate thalamic burst firing by enhancing a hyperpolarisation-activated cation current. Nature (Lond) 340:715–718

IP₃-Activated Ion Channel Activities in Olfactory Receptor Neurons from Different Vertebrate Species

Noriyo Suzuki[1]

Key words. Olfactory transduction—Olfactory receptor neuron—Inositol phosphate—IP₃-activated channel—Cyclic nucleotide-activated channel—Vertebrate olfaction

Introduction

It is now widely accepted that vertebrate olfactory transduction is mediated through a cAMP pathway in which an odorant-occupied receptor molecule activates a G-protein-linked adenylate cyclase that produces a second messenger, adenosine $3',5'$-cyclic monophosphate (cAMP). cAMP then gates cyclic nucleotide-activated cation channels localized in the membrane of olfactory cilia [1–7]. In addition, recent experiment on fish and rats suggest that not only cAMP, but also inositol 1,4,5-trisphosphate (IP₃), serves as a second messenger in an additional transduction pathway in vertebrate olfaction.

For example, a rapid transient formation of either cAMP or IP₃ has been detected with the stopped-flow technique in rat olfactory cilia when stimulated by specific odorants [8]. IP₃-binding proteins that appear to be the effector channel molecules for IP₃ have been identified in the membrane fractions from catfish and rat olfactory epithelia, and their reconstitution into an artificial phospholipid bilayer displayed the ion channel activities or an increase in conductance on exposure to IP₃ [9,10].

Further, an intracellular injection of IP₃ into catfish olfactory neurons induced depolarization that was accompanied by Ca^{2+} entry through the plasma membrane of olfactory neurons [11,12]. These experiments strongly suggest that the effector ion channels for IP₃ are present in the plasma membrane of olfactory neurons. However, no recordings of the IP₃-activated channel activities were made directly from vertebrate olfactory neurons. In this study, the membrane responses to IP₃ were recorded using patch clamp techniques from olfactory neurons isolated enzymatically from bullfrogs, mice, and rats, and the properties of the IP₃-activated ion channels were examined to clarify their role as the effector ion channels in vertebrate olfactory transduction.

Material and Methods

Bullfrog olfactory receptor neurons were isolated enzymatically by methods described previously [6]. Receptor neurons from female mice (ddY, 9–12 weeks) and rats (Wistar, 9–15 weeks) were isolated using enzyme treatment (0.3% collagenase/dispase, Boehringer 269638, or 0.2% papain, Calibiochem 5125), divalent cation-free medium containing 2 mM EDTA, and mechanical disruption. The isolated receptor neurons from mice and rats had a characteristic morphology as reported previously [13] and 3–10 immotile cilia [14] varying in length from 5 to 17 μm. The isolated receptor neurons were plated on a con A-coated glass coverslip and were maintained before use in normal Ringer's solution for bullfrog receptor neurons: 110 mM NaCl, 2.5 mM KCl, 1 mM CaCl₂, 1.6 mM MgCl₂, 5 mM HEPES, 2.8 mM NaOH, pH 7.4; or in normal saline for mouse and rat receptor neurons: 140 mM NaCl, 5 mM KCl, 2 mM CaCl₂, 1 mM MgCl₂, 5 mM HEPES, 2.8 mM NaOH, pH 7.4.

Inside-out membrane patches [15] were excised mainly from olfactory soma membranes of bullfrog, mouse, and rat receptor neurons and were stimulated with IP₃ and cAMP using a microflow superfusion chamber, as described previously [6]. Membrane patches were also excised from olfactory vesicles of bullfrog receptor neurons to compare the success rate for recording IP₃-activated channel activities between different membrane regions. Patch pipettes with tip lumens of about 0.5 μm (bubble number, 4.5–5.0) were filled with Ba external solution: 86 mM BaCl₂ for frogs or 110 mM BaCl₂ for mice, 5 mM HEPES, 1.6 mM Ba(OH)₂, pH 7.4. Ba^{2+} was chosen as the current carrier because IP₃-activated channels were expected to resemble Ca^{2+} channels [11] and because Ba^{2+} has been shown to be more permeant through traditional Ca^{2+} channels [16].

Membrane patches were exposed to either K or KF internal solution buffered with Ca-EGTA to maintain free Ca^{2+} and Mg^{2+} concentrations at 10^{-8} M and 1.25 mM, respectively; K solution con-

[1] Animal Behavior and Intelligence, Division of Biological Sciences, Graduate School of Science, Hokkaido University, Sapporo 060, Japan

tained 100 mM KCl for frogs or 132 mM KCl for mice, 0.68 mM $CaCl_2$, 1.73 mM $MgCl_2$, 5 mM EGTA-2K, 5 mM HEPES, 3.8 mM KOH, at pH 7.4; for the KF solution, KCl was substituted by 30 mKF at an equimolar basis. IP_3 (Sigma I4009, D-myo-inositol-1,4,5-trisphosphate, potassium salt) and cAMP (Sigma A6885, sodium salt) were dissolved in these internal solutions and applied to the membrane patches using the air bubble–partition method [6] when the response latency was measured.

Calibration of this method indicated that a complete solution exchange at the membrane patch terminated within 20 msec. For comparisons of the response latency and the distribution of IP_3-activated channels with those of cyclic nuclotide-activated channels, membrane patches from bullfrog receptor somas were also exposed symmetrically to divalent cation-free solution: 116.2 mM NaCl, 0.1 mM EGTA, 0.1 mM EDTA, 5 mM HEPES, 2.4 mM NaOH, at pH 7.4. Theoretical values of the liquid junction potential derived from different combinations of the test solutions were calculated with JPCalc software [17] and were used to correct the holding potential.

Macroscopic currents induced by intracellular injections of IP_3 were recorded in the whole-cell configuration [6,15] from rat receptor neurons adapted in the normal saline. Patch pipettes were filled with K internal solution in which IP_3 or cAMP was dissolved at a concentration of 20, 50, or 100 μM.

Current signals were amplified with a conventional patch clamp amplifier at a 3-kHz cutoff frequency (Nihon Kohden CEZ2200, Nihon Kohden, Tokyo, Japan) and stored on magnetic tape of a PCM-recorder (DC, 13 kHz). Data were digitized and analyzed on an NEC PC9801VM or IBM-AT 286 compatible computer using BASIC software [6] or pClamp software (Axon Instruments).

Results

Characteristics of IP_3-Activated Channels

Applying either 1, 2.7, or 5.3 μM IP_3 to the internal face of 125 patches from bullfrog, mouse, and rat receptor neurons induced responses in 27 patches. Of the 27 patches, single-channel activities could be resolved in 10 patches (8 from bullfrog and 2 from mouse olfactory somas). Flickering current fluctuations that corresponded to the openings and closings of single channels were observed when 1 μM IP_3 was applied. As observed in the cyclic nucleotide-activated channels [6,7,18], the channels activated by IP_3 did not show desensitization even after prolonged stimulus application (>85 s). Removal of IP_3 quickly abolished the channel activity, indicating that the effect of IP_3 on the channels was direct and

reversible because the perfusing internal solution contained no other phosphates such as ATP or GTP. In addition, the IP_3-activated channels did not respond to cAMP even at a concentration of 10 μM, indicating that the IP_3-activated ion channels differ from the cyclic nucleotide-activated channels, which did not respond to 5.3 μM IP_3. Unitary currents, which occurred in bursts, were more remarkable in the patches when stimulated by 5.3 μM IP_3 (Fig. 1). The event histogram analyses for the unitary currents were carried out in the two different channel groups recorded from bullfrog receptor somas using K and KF internal solutions combined with the Ba external solution. The mean of the histogram fit was 1.48 to 2.14 pA (grand mean, 1.75 ± 0.48 pA, $n = 4$) at the corrected holding potential of −85.1 mV for the channels recorded with K internal solution, and was 1.25 to 1.98 pA (grand mean, 1.53 ± 0.32 pA, $n = 4$) at the corrected holding potential of −86.1 mV for the channels recorded with KF internal solution, giving the corresponding conductances of 20.5 and 17.8 pS, respectively.

Response Latency

To examine whether the IP_3-activated channels respond fast enough to be qualified as the effector channels for the transduction pathway, the latency of the IP_3-activated channel response to IP_3 was compared with that of the cyclic nucleotide-activated channel response to cAMP in the different patches from bullfrog receptor somas. The IP_3-activated channel response to 1 μM IP_3 appeared after a latency of 44–1420 ms (mean, 462.2 ± 582.4 ms, $n = 10$), while the cyclic nucleotide-activated channel response to 1 μM cAMP appeared after a latency of 25–1620 ms (mean, 252.5 ± 415.1 ms, $n = 15$).

Subtraction of the delay for solution exchange at the patch (20 ms) from these minimal values gave the minimal latency of 24 ms for the IP_3-activated channels and 5 ms for the cyclic nucleotide-activated channels, respectively. Thus, the minimal latency for the IP_3-activated channels were not greatly different from that for the cyclic nucleotide-activated channels at the same stimulus concentration. The IP_3-activated channel response to 5.3 μM IP_3 appeared after a latency of 30–530 ms (mean, 210.5 ± 235.5 ms, $n = 4$), giving a minimal latency of 10 ms. These minimal response latencies to the different concentrations of IP_3 were short compared with the latency measured in the salamander receptor neurons, where the odorant-induced whole-cell current appeared after a latency of 140–570 ms [19].

Comparison of the Success Rate for Recording

If it is assumed that the conditions for the recording of single-channel activity, such as the pipette dimensions and the procedure for obtaining patches or cell

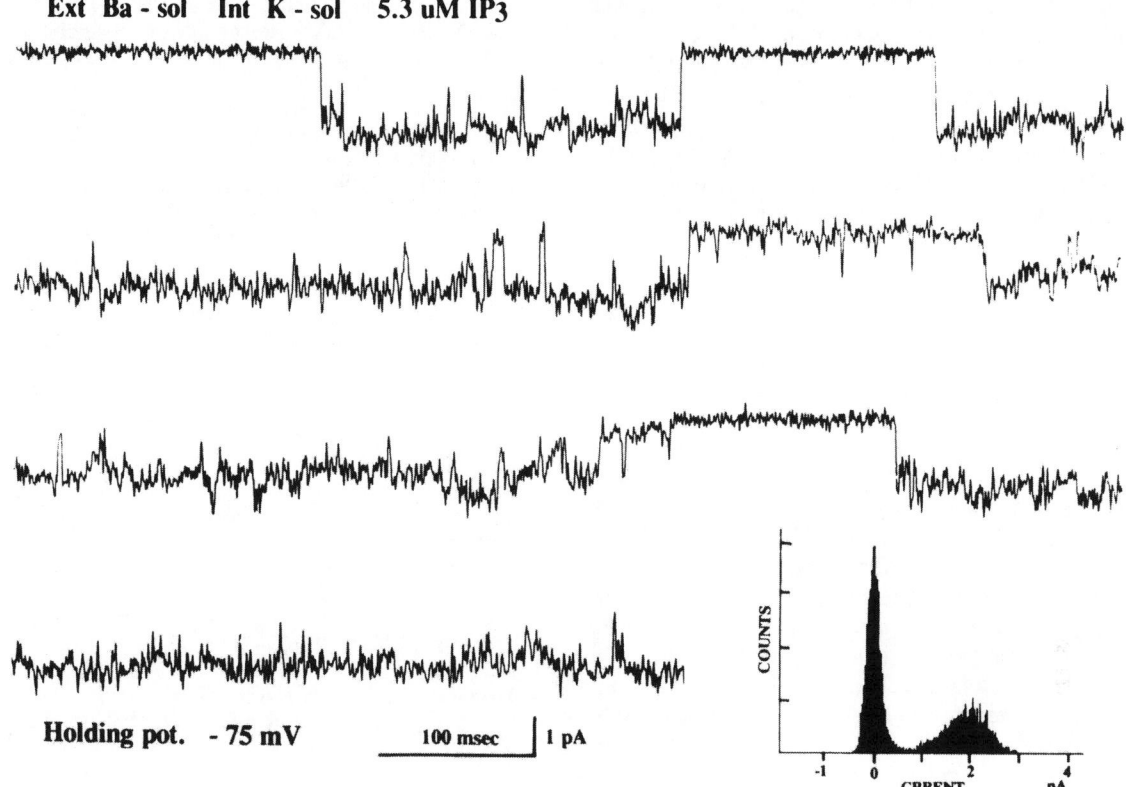

Ext Ba - sol   Int K - sol   5.3 uM IP3

Holding pot. - 75 mV   100 msec   1 pA

COUNTS

-1   0   2   4

CRRENT   pA

FIG. 1. Single-channel responses of inside-out patch from bullfrog olfactory receptor soma to 5.3 μM IP₃. Patch was held at −75 mV. Downward direction indicates channel opening. Patch pipette contained Ba external solution. IP₃ was dissolved in K internal solution and applied to patch from its cytoplasmic side. Record was low-pass-filtered at 1 KHz with a digital Gaussian filter after sampling at 5 KHz and was selected to illustrate various features associated with openings and closings of the IP₃-activated channel. Inset at *bottom right* shows total amplitude histogram of record; smaller peak at ~2 pA corresponds to channel open state

viability, do not differ greatly between the different experiments, then the success rate for the recording of single-channel activity would be an index for the estimation of both the channel density between different membrane regions of the receptor neuron and the difference in density between different species of ion channels. In bullfrog receptor neurons, the IP₃-activated channel responses were obtained in 7 of 12 patches (58.3%) from olfactory vesicles, while the responses were obtained in 18 of 80 patches (22.5%) from the receptor somas. The difference in the success rate between two different membrane regions was significant (Fisher's exact probability test, $p < 0.05$). These results suggest that the density of the IP₃-activated channel is higher in the vesicle membrane than in the soma membrane. By contrast, the cyclic nucleotide-activated channel responses were obtained in 47 of 76 patches (61.8%) from the receptor somas. This success rate for the recording of cyclic nucleotide-activated channel responses in the receptor somas differed significantly from that for the IP₃-activated channel responses in the receptor somas (Fisher's exact probability test, $p <$ 0.001), suggesting that the density of the IP₃-activated channels is much lower than that of the cyclic nucleotide-activated channels in the soma membrane.

Whole-Cell Response to IP₃ and cAMP

Of 32 rat receptor neurons tested in normal saline, IP₃ responses were obtained in 2 receptor neurons (6.3%). With a pipette containing 20 μM IP₃, a small inward current with a peak amplitude of about 20 pA at a holding potential of −60 mV was induced after the transition to whole-cell configuration. As observed previously in frog receptor neurons injected with cAMP [5] and in lobster olfactory receptor neurons injected with IP₃ [20], the inward currents induced by IP₃ in the rat receptor neurons were also composed of a fast rise and a slower decay with a time constant of about 1.2 s. To compare the success rate for the recording of the IP₃ response with that for the cAMP response in the rat receptor neurons, cAMP was injected into 32 additional receptor neurons. Of the 32 receptor neurons, similar inward current responses to those observed previously in frog receptor neurons [5] were obtained in 9 receptor neurons (28.1%) by using pipettes containing 50 μM cAMP. This larger success rate for the recording of cAMP responses than that for the IP_3 responses indicates that a smaller population of individual receptor neurons responded to IP_3.

Discussion

This study demonstrated that the channels activated directly by IP_3 are indeed present in the plasma membrane of olfactory receptor neurons of different vertebrate species. Although the permeability of the IP_3-activated channels for different cation species was not examined in the current study, the channels probably belong to a type of Ca^{2+} channels as observed in other cells [16] because the channels could be detected with Ba external solution. An examination of response latency confirmed that the IP_3-activated channels responded as quickly as the cyclic nucleotide-activated channels. In addition, the density of the IP_3-activated channels appeared to be higher in the vesicle membrane than in the soma membrane. A similar spatial different in the channel density for cyclic nucleotide-activated channels has been shown in isolated receptor neurons of frogs and newts [21]. Consequently, the properties of these IP_3-activated channels agree well with the criteria for the IP_3-activated channels to be the effector channels for the IP_3 pathway in vertebrate olfactory transduction.

In previous studies of catfish olfactory receptor neurons [11,12], both cAMP and IP_3 pathways were postulated to exist in individual receptor neurons; however, the low success rate for recording IP_3-activated single-channel and macroscopic current responses from individual receptor neurons in this study questions this possibility. The current results lead to the speculation that cAMP and IP_3 pathways operate independently in individual receptor neurons, and that a smaller population of olfactory receptor neurons utilize the IP_3 rather than the cAMP pathway for olfactory transduction.

Acknowledgments. This work was supported by a Grant-in-Aid 04640647 from the Japanese Ministry of Education, Science and Culture.

References

1. Pace U, Hanski E, Salomon Y, Lancet D (1985) Odorant-sensitive adenylate cyclase may mediate olfactory reception. Nature (Lond) 316:255–258

2. Sklar PB, Anholt RRH, Snyder SH (1986) The odorant-sensitive adenylate cyclase of olfactory receptor cells—differential stimulation by distinct classes of odorants. J Biol Chem 261:15538–15543

3. Nakamura T, Gold GH (1987) A cyclic nucleotide-gated conductance in olfactory receptor cilia. Nature (Lond) 325:442–444

4. Kurahashi T (1989) Activation by odorants of cation-selective conductance in the olfactory receptor cell isolated from the newt. J Physiol 419:177–192

5. Suzuki N (1989) Voltage- and cyclic nucleotide-gated currents in isolated olfactory receptor cells. In: Brand JG, Teeter JH, Cagan RH, Kare MR (eds) Chemical senses, vol 1, Receptor events and transduction in taste and olfaction. Dekker, New York, pp 469–493

6. Suzuki N (1990) Single cyclic nucleotide-activated ion channel activity in olfactory receptor cell soma membrane. Neurosci Res [Suppl] 12:S113–S126

7. Firestein S, Zufall F, Shepherd GM (1991) Single odor-sensitive channels in olfactory receptor neurons are also gated by cyclic nucleotides. J Neurosci 11:3565–3572

8. Boekhoff I, Tareilus E, Strotmann J, Breer H (1990) Rapid activation of alternative second messenger pathways in olfactory cilia from rats by different odorants. EMBO J 9:2453–2458

9. Kalinoski DL, Aldinger SB, Boyle AG, Huque T, Marecek JF, Prestwich GD, Restrepo D (1992) Characterization of a novel inositol 1,4,5-trisphosphate receptor in isolated olfactory cilia. Biochem J 281:449–456

10. Restrepo D, Teeter JH, Honda E, Boyle AG, Marecek JF, Prestwich GD, Kalinoski DL (1992) Evidence for an IP_3-gated channel protein in isolated rat olfactory cilia. Am J Physiol 263:c667–c673

11. Restrepo D, Miyamoto T, Bryant BP, Teeter JH (1990) Odor stimuli trigger influx of calcium into olfactory neurons of the channel catfish. Science 249:1166–1168

12. Miyamoto T, Restrepo D, Cragoe EJ Jr, Teeter JH (1992) IP_3- and cAMP-induced responses in isolated olfactory receptor neurons from the channel catfish. J Membr Biol 127:173–183

13. Maue RA, Dionne VE (1987) Preparation of isolated mouse olfactory receptor neurons. Pfluegers Arch 409:244–250

14. Lidow MS, Menco BPM (1984) Observations on axonemes and membranes of olfactory and respiratory cilia in frogs and rats using tannic acid-supplemented fixation and photographic rotation. J Ultrastruct Res 86:18–30

15. Hamill OP, Marty A, Sakmann B, Sigworth FJ (1981) Improved patch-clamp techniques for high-resolution current recording from cells and cell-free membrane patches. Pfluegers Arch 391:85–100

16. Kuno M, Gardner P (1987) Ion channels activated by inositol 1,4,5-trisphosphate in plasma membrane of human T-lymphocytes. Nature (Lond) 326:301–304

17. Barry PH, Lynch JW (1991) Liquid junction potentials and small cell effects in patch-clamp analysis. J Membr Biol 121:101–117

18. Zufall F, Firestein S, Shepherd GM (1991) Analysis of single cyclic nucleotide-gated channels in olfactory receptor cells. J Neurosci 11:3573–3580

19. Firestein S, Werblin F (1989) Odor-induced membrane currents in vertebrate-olfactory receptor neurons. Science 244:79–82

20. Fadool DA, Ache BW (1992) Plasma membrane inositol 1,4,5-trisphosphate-activated channels mediate signal transduction in lobster olfactory receptor neurons. Neuron 9:907–918

21. Kurahashi T, Kaneko A (1991) High density cAMP-gated channels at the ciliary membrane in the olfactory receptor cell. NeuroReport 2:5–8

Mediation of Opposing Transduction Cascades in Lobster Olfactory Receptor Neurons by cAMP and IP$_3$

D.A. Fadool and B.W. Ache[1]

Key words. Inositol—Transduction—Cyclic nucleotide—G protein—Culture—Second messenger

Cyclic 3,5-adenosine monophosphate (cAMP) and inositol 1,4,5-trisphosphate (IP$_3$) have been implicated as olfactory second messengers, although the role of having more than one transduction pathway in olfaction is unclear [1–6]. In the spiny lobster, *Panulirus argus*, odors inhibit as well as excite olfactory receptor neurons (ORNs) by generating opposing currents, a finding that implies the cells express at least two, opposing transduction pathways [7] (Fig. 1). We established the ORNs in primary cell culture [8], which permits characterizing the transduction pathways by patch-clamping the somata, rather than the thin outer dendrites that normally contain the transduction machinery in situ. The cultured ORNs mimic their counterparts in situ in terms of their basic biophysical properties as well as their specificity (tuning) and sensitivity (threshold) to odors independently of the length, number, or presence of processes [9]. There is evidence for G-protein-coupled second messenger cascades in both the excitatory and inhibitory pathways. Odors evoke unitary currents in cell-attached recordings [10]; and cells perfused with the nonhydrolyzable analogs of GTP and GDP, guanosine 5'-O-(3-thiotriphosphate) (GTPγS) and guanosine 5'-O-(2-thiodiphosphate) (GDPβS), respectively, show measurably changed odor-evoked macroscopic inward and outward currents (paired *t*-test, $p \leq 0.05$) [11,12].

To further explore the G proteins involved in olfactory transduction, we screened purified membrane preparations of lobster brain and olfactory tissue with antibodies directed against the G-protein subunits: $G_{i\beta}$, $G_{olf\alpha}$, $G_{o\alpha}$, $G_{\alpha 11}$ (E976), and G_q (E973). By Western blot analysis, a protein immunoreactive to anti-$G_{o\alpha}$ at 40.5 kDa was localized specifically to the outer dendrite and not to the soma. These antibodies, as well as an antibody directed against G_q (Z811-5DE), were then perfused into the cells to determine if any of the antibodies could functionally perturb odor-evoked macroscopic

inward or outward currents. Patch electrodes were tip-filled approximately 1 mm with patch solution and then back-filled with test antibody solution. Odor-evoked currents were monitored every min to minimize adaptation or desensitization for up to 30 min. None of the antibodies or nonimmune rabbit IgG or normal patch solution altered the magnitude of odor-evoked outward current. In contrast, anti-G_q (Z811-5DE) and anti-$G_{o\alpha}$ significantly or completely blocked the odor-evoked inward current within several minutes. The data therefore support G_o-, and possibly G_q-like G proteins in the excitatory cascade. The identity of the G protein(s) in the inhibitory cascade is as yet unknown.

Several lines of evidence from cultured ORNs suggest that IP$_3$ is the excitatory second messenger. Introducing IP$_3$ into the cells via the patch pipet evoked a sustained inward current in the absence of odorant stimulation [10] (Fig. 2a). Applying IP$_3$ to the inside face of cell-free patches of the plasma membrane directly gated one of two types of ion channel in 110 of 184 patches studied [10,13] (Fig. 2B). The ion channels differed in their average slope conductance (30 versus 74 pS) but had similar mean open times, concentration dependencies and sensitivity to blockade by ruthenium red and heparin [10]. Further characterization of the IP$_3$-gated channels demonstrate that both types are pH-sensitive, are modulated by Ca^{2+}, can exhibit mode behavior, and are permeable to Ca^{2+} and Na$^+$ (Fadool and Ache, this meeting). An antibody directed against the last 19 amino acids of the carboxyl-terminus of a clone of a mammalian IP$_3$ receptor (anti-IP$_3$-R) that identified a larger than 200-kDa protein, selectively increased the odor-evoked inward current over 400% when perfused into the cells [10]. Co-application of the antibody plus IP$_3$ to the inside face of cell-free patches of membrane increased the Pr$_{open}$ and second exponential of the mean open time (t$_{o2}$) of the IP$_3$-activated channels without affecting the conductance of the channels [10]. Finally, it was possible to insert a pipette with an inside-out patch containing an IP$_3$-activated channel taken from one cell into a second cell. "Spritzing" an excitatory odorant mixture on the second, recipient cell induced channel activity in the patch, presumably (though not necessarily) due to elevation of intracellular IP$_3$ [10] (Fig. 3).

[1] Whitney Laboratory and Departments of Zoology and Neuroscience, University of Florida, 9505 Ocean Shore Boulevard, St. Augustine, FL 32086, USA

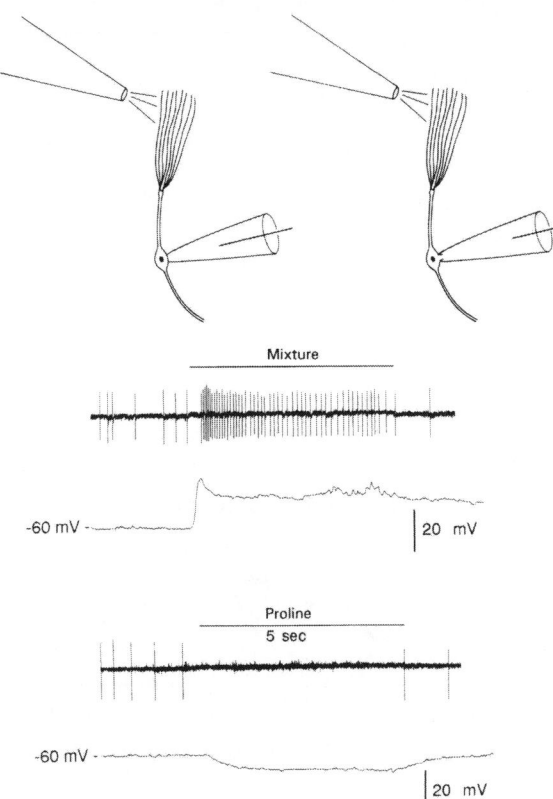

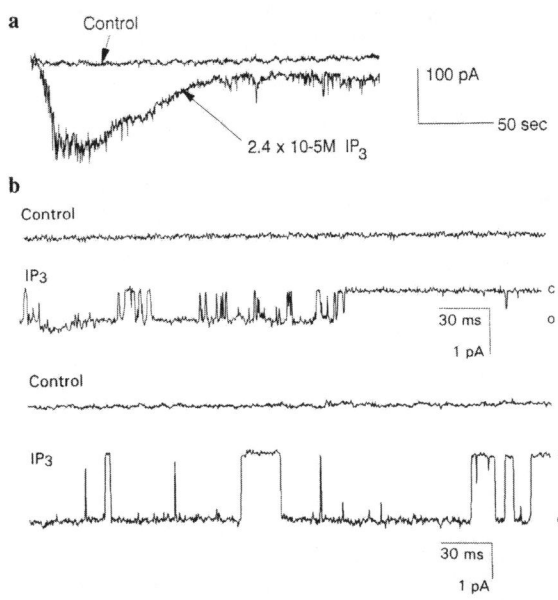

FIG. 1. Cell-attached extracellular (*left diagram* and *upper trace of each pair*) and subsequent whole-cell, current-clamp recordings (*right diagram* and *lower trace of each pair*) of a single lobster olfactory receptor neuron (ORN) in situ. Different odors excite and inhibit the same ORN. Odors that increase the spontaneous activity of the cell depolarize the membrane potential, and odors that decrease the spontaneous activity of the cell, hyperpolarize the membrane potential. (Modified from [15], with permission)

FIG. 2. **a** Whole-cell, voltage-clamp recording of a cultured lobster ORN that was sequentially patched, first with normal patch solution (control) and then with inositol 1,4,5-trisphosphate (IP$_3$) in the patch electrode. Holding potential was $-60\,$mV. **b** Unitary current recordings of a cell-free patch of membrane from a cultured ORN when patch solution (control) or IP$_3$ was applied to the inside face of the excised patch. Two types of IP$_3$-activated channel were observed with approximately equal frequency: one channel with a slope conductance of 30 pS (*upper traces*) and a second channel with a slope conductance of 74 pS (*lower traces*). Holding potential was $-60\,$mV. *c*, closed; *o*, open. Modified from [10], with permission

Several lines of evidence from the cells in situ provided by the work of William Michel, formerly in our laboratory, suggest that cAMP is the inhibitory second messenger [14]. Superfusing the outer dendrites, the presumed site of signal transduction, with forskolin, an activator of the cAMP-producing enzyme adenylyl cyclase (AC), evoked currents that mimicked the time course, magnitude, and polarity of odor-evoked outward currents in 19 of 27 cells tested. Similarly, the phosphodiesterase inhibitor IBMX evoked outward currents in 26 of 33 cells tested (Fig. 4a). The membrane permeant probes for cyclic nucleotides 8-bromo-cAMP and 8-bromo-cGMP selectively elicited macroscopic outward currents in 11 of 13 cells tested (Fig. 4b). The pharmacology and ion dependency of the odor-evoked outward current is consistent with it being carried by K$^+$ [15], in contrast to the cAMP-dependent inward current mediating excitation in

vertebrate ORNs, which is nonselective for cations [16–18].

Work from our laboratory done in collaboration with the laboratories of Hanns Hatt (Technical University, Munich, Germany) and Heinz Breer (U. Stuttgart-Höhenheim, Germany) confirmed that the IP$_3$-mediated pathway characterized in the soma of the cultured ORNs occurs together with the cAMP-mediated pathway in the outer dendrites [19]. The outer dendrites naturally vesiculate with time to form 2- to 5-μm diameter spheres that can be patched directly. Applying cAMP to the inside face of cell-free patches of outer dendritic membrane activated a 28 pS channel that strongly rectified at depolarizing potentials under the recording conditions. Applying IP$_3$ in the same configuration activated two types of channels that did not rectify and could therefore be distinguished from the cAMP-gated channels. Each of the two IP$_3$-gated channels had chord conductances within several picosiemens of those reported for the IP$_3$-activated channels in the cultured neurons [10]. Odors also rapidly and transiently stimulated the production of both IP$_3$ and cAMP in the outer dendrites [20].

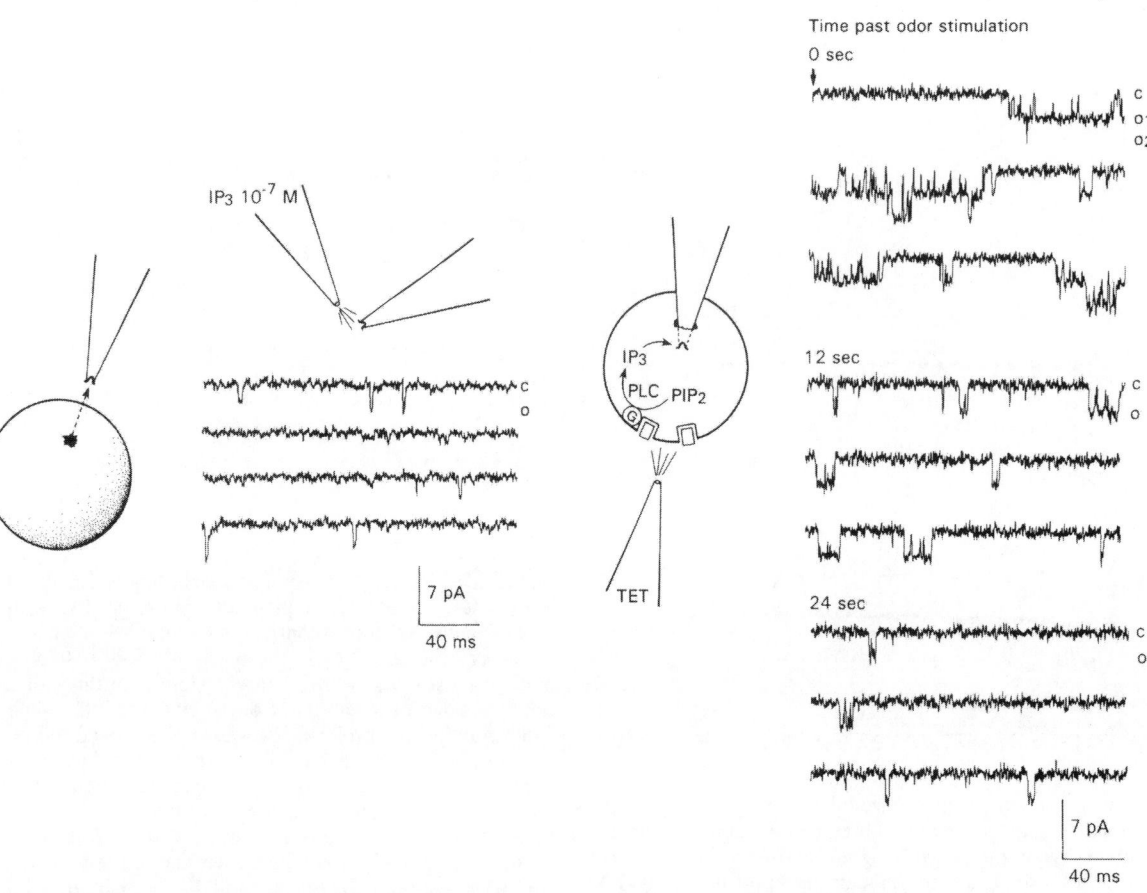

Fig. 3. Cell-free patch of membrane containing an IP₃-activated 74 pS channel from one cultured lobster ORN was inserted into a second, recipient cell that was stimulated with an excitatory odorant mixture (*TET*). Odor-evoked channel activity decreased with time after stimulation. Model depicts the hypothesized binding of an odor molecule to a G-protein-linked receptor to activate the PLC conversion of PIP₂ to IP₃, which presumably activates the channel activity. Holding potential was −60 mV. *c*, closed; *o₁* and *o₂*, open. (From [10], with permission)

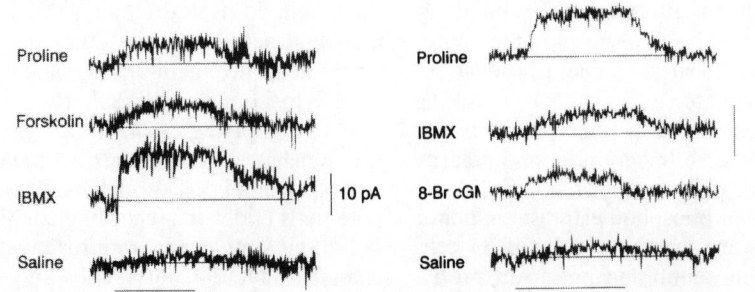

Fig. 4. Whole-cell, voltage-clamp recordings from two lobster ORNs in situ whose dendrites were superfused with either an odor or a membrane-permeant probe of the cyclic adenosine monophosphate (cAMP) second messenger cascade. **a** Application of an inhibitory odor (proline), an adenylyl cyclase agonist (forskolin), or a phosphodiesterase inhibitor (IBMX) each evoked an outward current. *Bar* denotes 5-s duration of the application. **b** Application of an inhibitory odor (proline), a phosphodiesterase inhibitor (IBMX), or a cyclic nucleotide analog (8-Br cGMP) to another cell had the same effect as in **a**. *Bar* denotes 4-s duration of the application. (From [10], with permission)

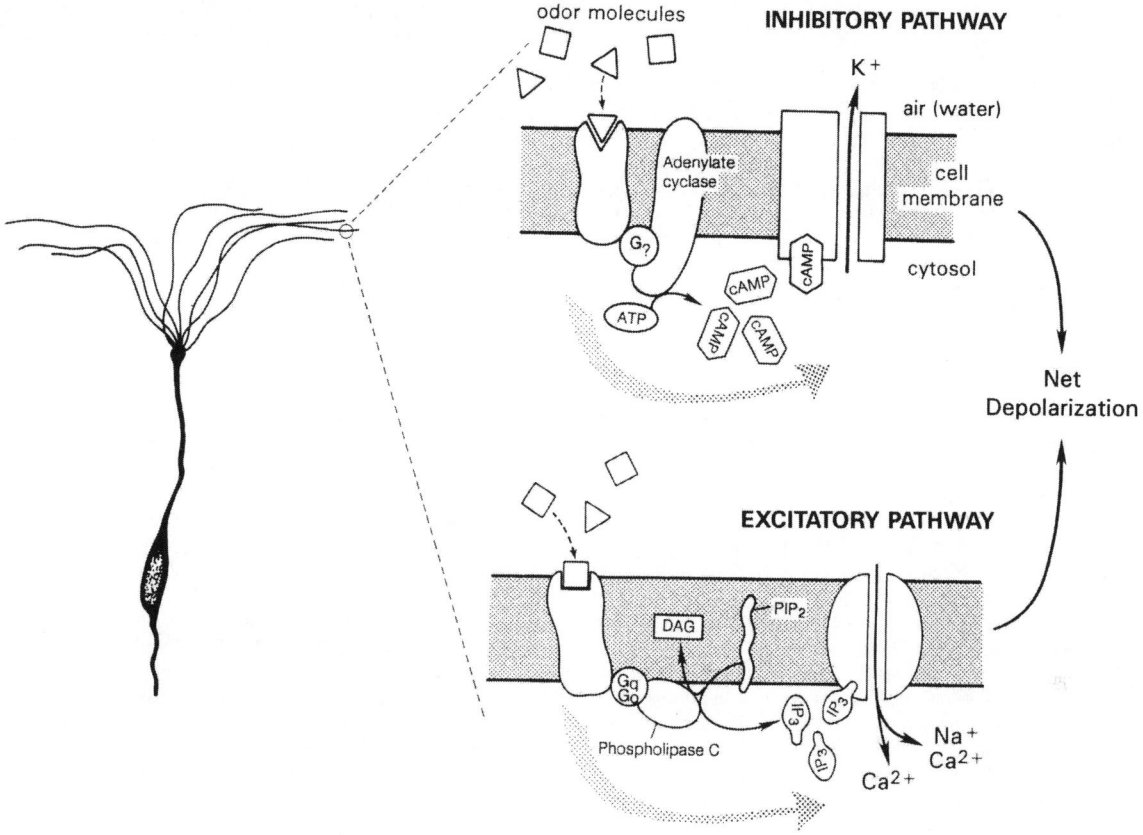

FIG. 5. Model of the dual, opposing second messenger cascades hypothesized to mediate olfactory transduction in lobster ORNs. Some odor molecules (*triangles*) bind to receptor proteins linked to an as yet undefined G protein. This G protein mediates adenylyl cyclase conversion of ATP to cAMP, which directly gates a potassium channel to hyperpolarize the cell. Other odor molecules (*squares*) bind to receptor proteins linked to a G$_o$- or G$_q$-like G protein. This G protein mediates phospholipase C conversion of PIP$_2$ to IP$_3$, which directly gates either a Ca^{2+} selective or nonselective cation channel to depolarize the cell. The output of the cell to the CNS depends on the net level of depolarization resulting from coactivation of the two transduction cascades

In summary, lobster ORNs can have two parallel input pathways, each mediated by a different second messenger and coupled to an opposing conductance. Our data are consistent with a model (Fig. 5) in which some components of naturally occurring odor mixtures act through a G-protein-coupled receptor(s) to activate adenylate cyclase and increase intracellular cAMP. cAMP then directly activates a potassium-selective channel to generate a hyperpolarizing, outward current. Other components, in contrast, act through distinct, G$_o$- or G$_q$-like protein coupled receptors to activate the conversion of PIP$_2$ to IP$_3$ by phospholipase C. IP$_3$ then directly activates a Ca^{2+}-selective or a nonselective cation channel to generate a depolarizing, inward current. The two opposing currents together set the rise time and magnitude of the receptor potential, which in turn drives the output of the cell to the central nervous system. The ability to encode inhibitory input, in this instance by tempering excitation, expands the overall potential for coding odor information at the level of the ORN. Consideration of the extent to which this model for the role of dual transduction pathways in olfaction generalizes to other animals is beyond the scope of this brief review but has been presented elsewhere [2].

Acknowledgments. The authors would like to thank Drs. R. Bruch, J. Exton, and P. Sternweiss for the generous gifts of G-protein antibodies and peptides; Dr. P. DeCamilli for the generous gift of the IP$_3$ receptor antibody; Ms. L. VanEkeris and Mr. Kelly Jenkins for technical assistance; Ms. L. Milstead for artistic assistance; and Mr. James Netherton for photographic assistance. This investigation was supported by ONR grant N0014-90-J-1566, NIH grants 1F31MH10124 and 1R01DC01655, and the Dan Charitable Trust Fund of Japan.

References

1. Breer H, Boekhoff I (1992) Second messenger signalling in olfaction. Curr Opin Neurobiol 2:439–443
2. Ache B (1994) Towards a common strategy for transducing olfactory information. Semin Cell Biol 5
3. Reed RR (1992) Signaling pathways in odorant detection Neuron 8:205–209
4. Firestein (1992) Electrical signals in olfactory transduction Curr Opin Neurobiol 2:444–448
5. Ronnett GV, Snyder SH (1992) Molecular messengers of olfaction TINS 15:508–513
6. Anholt RRH (1993) Molecular neurobiology of olfaction. Crit Rev Neurobiol 7:1–22
7. McClintock TS, Ache BW (1989) Hyperpolarizing receptor potentials in lobster olfactory receptor cells: implications for transduction and mixture suppression. Chem Senses 14:637–647
8. Fadool DA, Michel WC, Ache BW (1991) Sustained primary culture of lobster (*Panulirus argus*) olfactory receptor neurons. Tissue Cell 23:719–732
9. Fadool DA, Michel WC, Ache BW (1993) Odor sensitivity of cultured lobster olfactory receptor neurons is not dependent on process formation. J Exp Biol 174:215–233
10. Fadool DA, Ache BW (1992) Plasma membrane inositol 1,4,5-trisphosphate-activated channels mediate signal transduction in lobster olfactory receptor neurons. Neuron 9:907–918
11. Michel WC, Fadool DA, Ache BW (1992) cAMP mediates the odor-evoked inhibitory conductance in lobster olfactory receptor cells. Chem Senses 17:669
12. Fadool DA, Michel WC, Ache BW (1991) G-proteins and inositol-phospholipid metabolism implicated in odor response of cultured lobster olfactory neurons. Chem Senses 16:518–519
13. Fadool DA, Ache BW (1993) Ionic selectivity and ligand specificity of IP_3-gated channels mediating excitatory transduction in lobster olfactory receptor neurons. Chem Senses 18:553
14. Michel WC, Ache BW (1992) Cyclic nucleotides mediate an odor-evoked potassium conductance in lobster olfactory receptor cells. J Neurosci 12:3979–3984
15. Michel WC, McClintock TS, Ache BW (1991) Inhibition of lobster olfactory receptor cells by an odor-activated potassium conductance. J Neurophysiol 65:446–453
16. Kolesnikov SS, Zhainazarov AB, Kosolapov AV (1990) Cyclic nucleotide-activated channel in the frog olfactory receptor plasma membrane. FEBS 266:96–98
17. Firestein S, Zufall F, Shepherd GM (1991) Single odor-sensitive channels in olfactory receptor neurons are also gated by cyclic nucleotides. J Neurosci 11:3565–3572
18. Zufall F, Firestein S, Shepherd GM (1991) Analysis of single cyclic nucleotide-gated channels in olfactory receptor cells. J Neurosci 11:3573–3580
19. Ache BW, Hatt H, Breer H, Boekhoff I, Zufall F (1993) Biochemical and physiological evidence for dual transduction pathways in lobster olfactory receptor neurons. Chem Senses 18:523
20. Boekhoff I, Michel WC, Breer H, Ache BW (in press) Single odors differentially stimulate dual second messenger pathways in lobster olfactory receptor cells. J Neurosci

The Significant Contribution of the cAMP-Independent Pathway to Turtle Olfactory Transduction

Makoto Kashiwayanagi[1]

Key words. cAMP—Turtle—Independent pathway—Olfactory transduction—Forskolin—Ruthenium red—SITS

Introduction

Recent studies on the olfactory transduction have suggested that olfactory responses are induced by the following mechanisms. Binding of odorants to receptors leads to stimulation of the G-protein-coupled adenylate cyclase [1], and subsequently intracellular cAMP increases the membrane conductance via cAMP-gated cation channels [2]. As a result, the olfactory neurons are depolarized. It has also been pointed out that other transduction pathways exist. For example, some odorants increase inositol 1,4,5-trisphosphate (IP_3) concentration in the rat olfactory cilia [3], and there is an IP_3-gated channel in the cilia [4]. These results suggest that the IP_3-dependent pathway is also involved in the transduction mechanism. It is possible that a Ca-activated Cl channel is also involved in the transduction pathway [5]. In addition, unknown pathways may exist. In this study, we suggest that the cAMP-independent pathway, which is probably a novel pathway, greatly contributes to the turtle olfactory transduction.

Results and Discussion

Olfactory neurons were isolated from the turtle olfactory epithelium without enzymes, and cAMP-independent responses were measured by the whole-cell clamp technique. Application of $50\,\mu M$ forskolin, which activates adenylate cyclase, induced an inward current (Fig. 1a). The inward current was adapted immediately in the case of this neuron. Because the response to a high concentration of forskolin is adapted, the cAMP-dependent pathway must be fully desensitized. If odor responses are induced only through the cAMP-dependent pathway, it is expected that odor responses will not appear under this condition. However, a large inward cur-

rent in response to the odorant mixture is induced even after forskolin. The odorant mixture is composed of 0.2 mM each of citralva, hedione, eugenol, *l*-carvone, and cineole, which were reported to increase cAMP concentration in rat and bullfrog olfactory cilia preparations. Our results indicate that so-called cAMP-dependent odorants induce inward currents through the cAMP-independent pathway in the turtle.

To investigate the effect of forskolin on in situ odor response, olfactory nerve impulses were measured from the olfactory cilia according to Frings and Lindemann [6]. Figure 1b shows a typical recording from a single cilium. Forskolin is applied first, and after the response to forskolin is adapted to a spontaneous level, the odorant mixture is applied. As seen from Fig. 1, the odorant mixture increases impulse frequency, suggesting that the cAMP-independent pathway contributes to the generation of olfactory nerve impulses.

The olfactory information received at olfactory neurons is transmitted to the olfactory bulb. Whole-cell recording from isolated cells gives direct information on the transduction, but the properties of the isolated cells may not be identical with those of an in vivo system. To obtain information in vivo, we examined the effects of forskolin on odor responses recorded from the olfactory bulb.

Figures 2a and b show typical olfactory bulbar responses to 0.1 mM citralva alone (Fig. 2a) and 0.1 mM citralva solution containing $50\,\mu M$ forskolin after the response to $50\,\mu M$ forskolin is adapted to a spontaneous level (Fig. 2b). Figure 2c shows the plot of the peak heights of the responses as a function of forskolin concentration. The magnitude of response to 0.1 mM citralva is taken as unity. The response to forskolin is increased with an increase in forskolin concentration and reaches a plateau level at $10\,\mu M$. The response to citralva after forskolin is decreased with an increase in forskolin concentration to $5\,\mu M$; a further increase in concentration does not significantly affect the odor response. The magnitude of response to citralva after forskolin of $50\,\mu M$, at which the response to forskolin itself is sufficiently saturated, is 56% of that of citralva applied alone. These results indicate that the cAMP-independent component greatly contributes to olfactory bulbar responses.

[1] Faculty of Pharmaceutical Sciences, Hokkaido University, Sapporo, 060 Japan

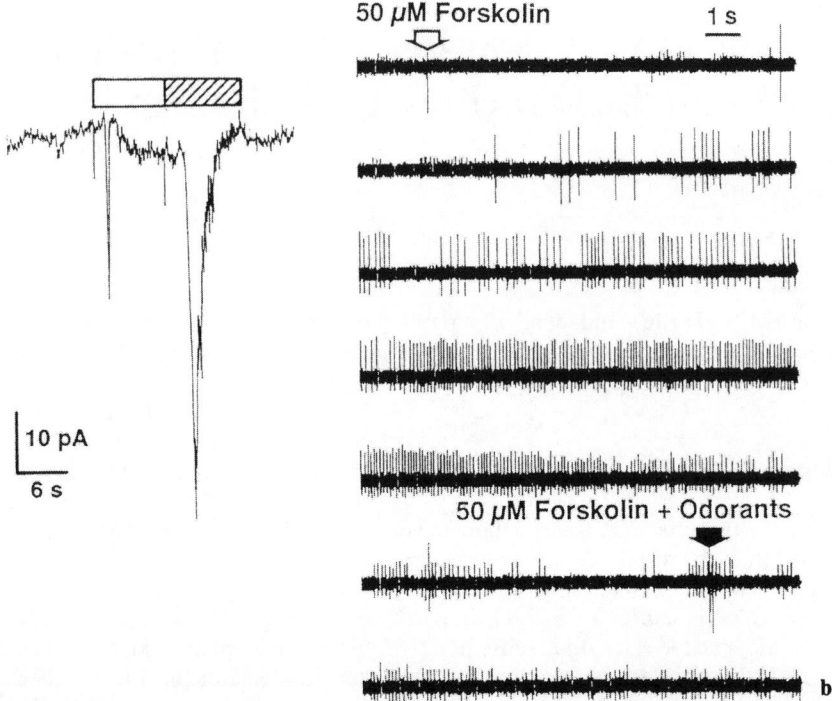

FIG. 1. **a** Inward current induced by odorant mixture after forskolin. **b** Increase in impulse frequency induced by odorant mixture after forskolin

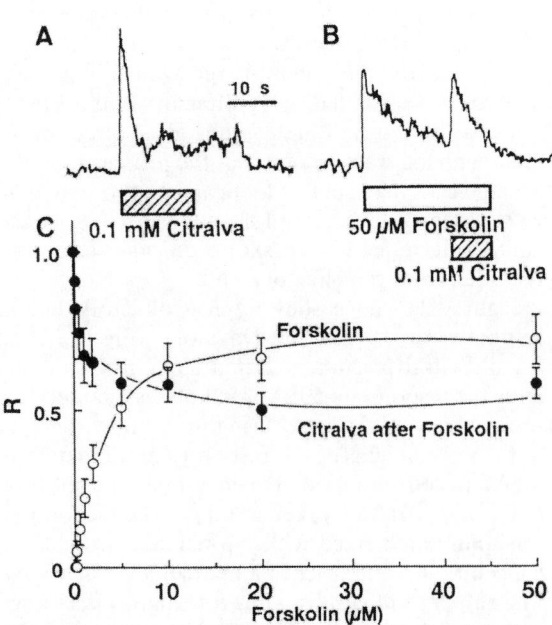

FIG. 2. Effects of forskolin on olfactory bulbar responses to citralva

In these experiments, relatively high concentrations of odorants were used. Therefore, one may consider that the forskolin-independent responses are induced only by high concentrations of odorants.

To test this possibility, the effects of forskolin on the responses to citralva of varying concentrations were examined. As shown in Fig. 3, citralva of low concentrations elicits appreciable responses even when applied after $50\,\mu M$ forskolin. For example, at $5\,\mu M$, the magnitude of the response to citralva after forskolin was 67% of that to citralva applied alone. This indicates that the cAMP-independent pathway contributes to the response to the odorant of low concentrations as well as that of high concentrations.

It is expected that application of 3-isobutyl-1-methylxanthine (IBMX) to the olfactory epithelium leads to an increase of cAMP level in the olfactory neurons by inhibition of phosphodiesterase. The response to IBMX itself reached a plateau level at $0.3-1\,mM$. The magnitude of the response to citralva after $1\,mM$ IBMX was 73% of that to citralva applied alone. To increase cAMP level in the olfactory cells to a maximal level, a mixture of $3\,mM$ IBMX and $50\,\mu M$ forskolin was applied to the olfactory epithelium. After the response to the mixture was adapted, $0.1\,mM$ citralva solution containing the mixture induced a large response. This suggests that an odor response is induced even at the maximal cAMP level.

The effects of 8-(4-chlorophenylthio)-adenosine 3'-5'-cyclic monophosphate (cpt-cAMP), a membrane-permeant cAMP analogue, on the responses to $0.1\,mM$ citralva were examined. The response to cpt-cAMP itself reached a plateau level

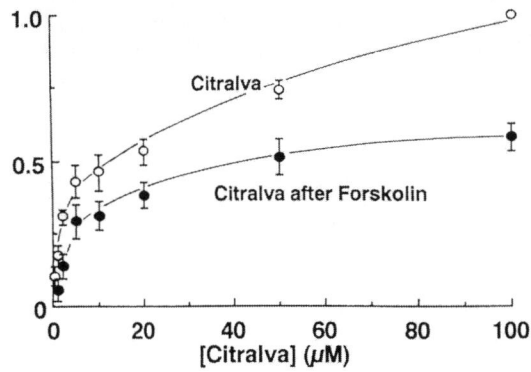

FIG. 3. Effects of 50 μM forskolin on dose–response relationship for citralva

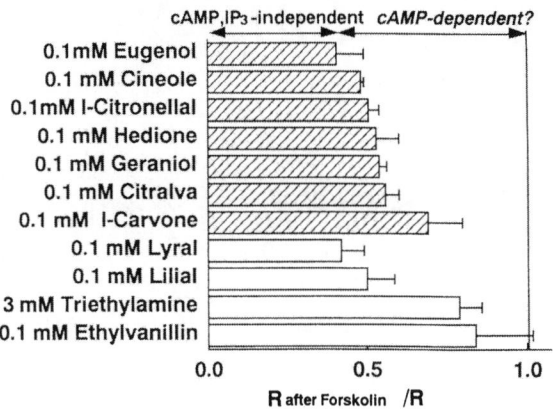

FIG. 4. Responses to various odorants after responses to 50 μM forskolin

at about 1 mM. The magnitude of response to citralva after 3 mM cpt-cAMP was 51% of that to citralva applied alone. The degree of cAMP-independent odor response after cpt-cAMP was similar to that after forskolin and IBMX plus forskolin.

Figure 4 summarizes effects of 50 μM forskolin on the responses to various odorants. The value in the figure represents relative magnitude of the response to an odorant after 50 μM forskolin. The magnitude of the response to the odorant alone is taken as unity. The magnitude of the odor response after forskolin varies from 45% to 80% of that when forskolin is applied alone. This suggests that the cAMP-independent pathway contributes to the odor responses by as much as 50% or more. In isolated rat cilia preparations, many odorants increased cAMP concentration, but some odorants such as lilial, lyral, triethylamine, and ethyl vanillin did not enhance cAMP accumulation with increasing IP_3 concentration [3]. As shown in Fig. 4, however, there is no essential difference in the degree of the reduction of the response to forskolin between the two groups. It is known that forskolin not only activates adenylate cyclase but also affects other functions, such as inhibiting glucose transport, modulating the voltage-dependent K channel and inhibiting the GABA-induced Cl flux. It is true that the cAMP pathway contributes to the forskolin-sensitive response, but the suppression observed in the present study may be brought about partly by nonspecific effects of forskolin.

Breer and Boekhoff reported that odorants which accumulate cAMP do not accumulate IP_3 [3]. This suggests that the cAMP-independent odor responses to so-called cAMP-dependent odorants are not IP_3-dependent. To test this possibility, the effect of ruthenium red, which is reported to be an IP_3-channel blocker in olfactory cells, on olfactory responses was examined. Ruthenium red also blocked the electrolfactogram (EOG) in response to amino acid in the catfish [7]. Ruthenium red brought about no significant effect on the responses to citralva, citralva

after 50 μM forskolin, and 50 μM forskolin (Fig. 5a). This result suggests that IP_3 does not play a role in the generation of olfactory responses to so-called cAMP-dependent odorants.

Finally, we tested to what extent the Ca-activated Cl channel contributes to turtle olfactory transduction. Figure 5b shows the effects of 2 mM SITS (4-acetamido-4'-isothiocyanostilbene-2, 2'-disulfonic acid), a Cl-channel blocker, on olfactory responses. The magnitude of response to 0.1 mM citralva is slightly reduced by application of SITS, but not significantly, suggesting that Cl channels contribute only partially to odor response in the turtle. The magnitude of the forskolin-insensitive response to citralva is practically unchanged by the application of SITS, suggesting that the Cl channel does not contribute to cAMP-independent odor responses.

Figure 5c illustrates the effects of removal of Cl^- on odor responses. There is no essential difference between the magnitude of odor responses in normal Ringer's solution and that in Cl^--free Ringer's solution. As shown in Fig. 5b, the degree of contribution of the Ca-activated Cl channel to odor responses is not as large, and thus we might not detect an increase in the magnitude of odor response when Cl^- was removed.

Results obtained in the current study suggest that the cAMP-independent pathway greatly contributes to turtle olfactory transduction. At the olfactory bulb level, the cAMP-independent odor information was more than 50%. In addition, IP_3- and Ca-activated Cl channels may not be involved in this pathway. Although the data are not shown, effects of db-cGMP on odor responses were small in the turtle, suggesting that the role of cGMP in odor transduction is not significant in the turtle [8]. Thus, our results suggest that there is a new transduction pathway that greatly contributes to turtle olfactory transduction. The mechanism of this previously unknown pathway is not clear at present. Margalit

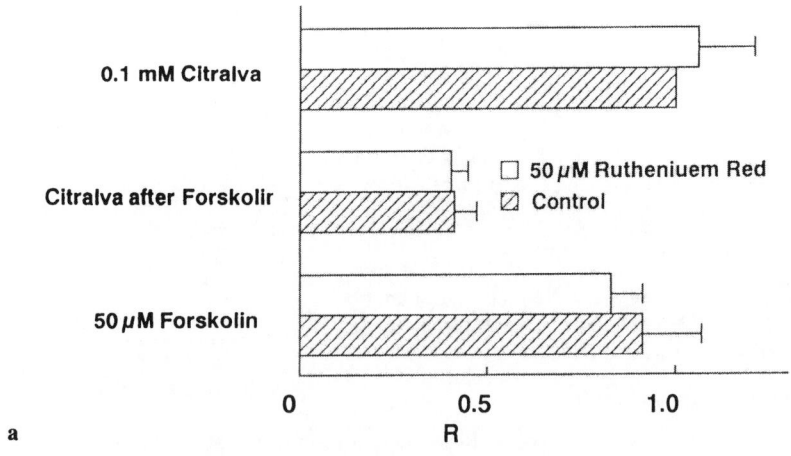

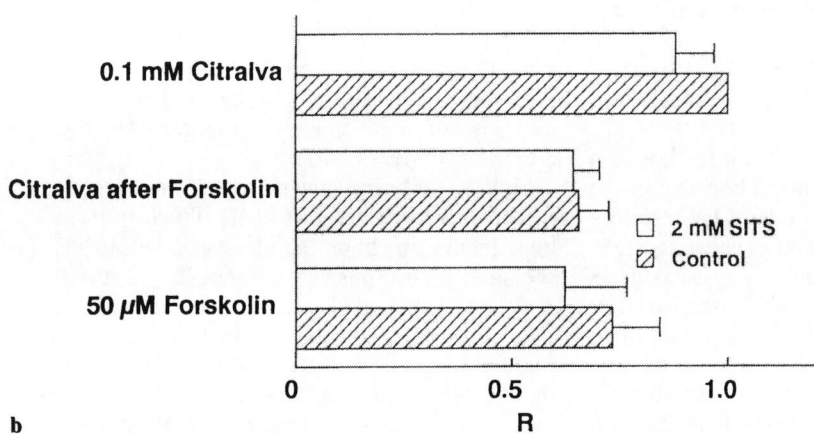

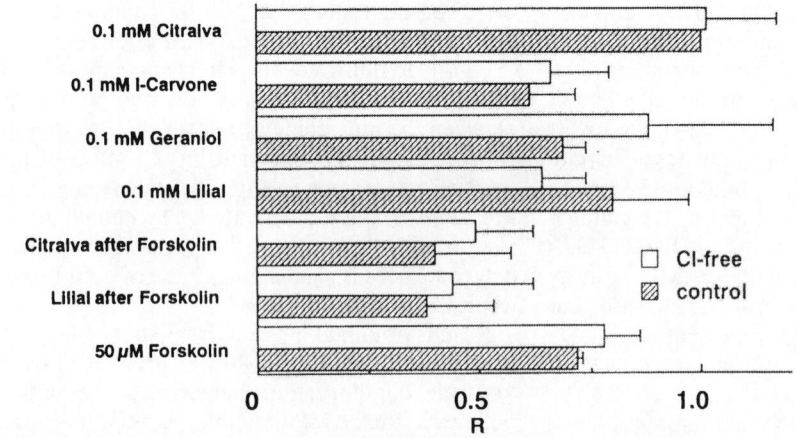

FIG. 5. Effects of ruthenium red (**a**), SITS (**b**), and removal of Cl⁻ on olfactory bulbar responses

and Lancet have measured the appearance of mRNA encoding receptors and cyclic nucleotide-gated channels during development of rat olfactory cells [9]. They reported that fetal olfactory responses were induced before the appearance of the olfactory receptors and cyclic nucleotide-gated channels, which suggests that odor response does not appear via the olfactory receptors and cAMP-gated channels in early stages of embryonic development. The new

transduction pathway observed in the current study may be related to the pathway observed in rat fetal olfactory cells.

Summary

(1) The effects of cAMP-inducing agents on turtle olfactory responses recorded from isolated olfactory

neurons, olfactory cilia, and the olfactory bulb were examined. The results demonstrated that large responses to various odorants appeared even after $50\,\mu M$ forskolin, $3\,mM$ IBMX plus $50\,\mu M$ forskolin, and $3\,mM$ cpt-cAMP were applied. (2) The cAMP-independent pathway contributed to responses to odorants; it was reported to increase cAMP concentration in the rat and bullfrog by as much as 50%. (3) The cAMP-independent odor responses were not affected by $50\,\mu M$ ruthenium red, $2\,mM$ SITS, or removal of Cl^-. (4) Results obtained in the current study suggested that the cAMP-independent pathway, which is probably IP$_3$- and Ca activated and Cl-channel independent, greatly contributes to olfactory transduction in the turtle.

Acknowledgements. I am grateful to Prof. Kenzo Kurihara for improvements to this manuscript and for his support. I also express may gratitude to Takasago International for supplying very pure odorants.

References

1. Pace U, Hasnski E, Salomon Y, Lancet D (1985) Odorant-sensitive adenylate cyclase may mediate olfactory reception. Nature (Lond) 316:255–258
2. Nakamura T, Gold GH (1987) A cyclic nucleotide-gated conductance in olfactory receptor cilia. Nature (Lond) 325:442–444
3. Breer H, Boekhoff I (1991) Odorants of the same odor class activate different second messenger pathways. Chem Senses 16:19–29
4. Restrepo D, Teeter JH, Honda E, Boyle AG, Marecek JF, Prestwich GD, Kalinoski DL (1992) Evidence for an InsP$_3$-gated channel protein in isolated rat olfactory cilia. Am J Physiol 263:C667–C673
5. Kleene SJ, Gesteland RC (1991) Calcium-activated chloride conductance in frog olfactory cilia. J Neurosci 11:3624–3629
6. Frings S, Lindemann B (1990) Current recording from sensory cilia of olfactory receptor cells *in situ*. I: The neuronal response to cyclic nucleotides. J Gen Physiol 97:1–16
7. Miyamoto T, Restrepo D, Teeter JH (1992) Voltage-dependent and odorant-regulated currents in isolated olfactory receptor neurons of the channel catfish. J Gen Physiol 99:505–530
8. Kashiwayanagi M, Kawahara H, Hanada T, Kuribara K (1994) A large contribution of a cyclic AMP-independent pathway to turtle olfactory transduction. J Gen Physiol (in press)
9. Margalit T, Lancet D (1993) Expression of olfactory receptor and transduction genes during rat development. Dev Brain Res 73:7–16

Functional Properties of Frog Vomeronasal Receptor Cells

DIDIER TROTIER[1], KJELL B. DØVING[2], and JEAN-FRANÇOIS ROSIN[1]

Key words. Olfaction—Vomeronasal organ—Microvillar receptor cells—Patch-clamp—Voltage-dependent currents

Introduction

Many animals rely on their chemical senses for the major life functions such as reproduction, feeding, and predator avoidance. Most terrestrial vertebrates have developed two olfactory chemosensory systems: the main olfactory system and the vomeronasal, or "accessory," olfactory system. The main olfactory system is exposed to volatile odorant molecules brought to the sensory epithelium by the respiratory airstream.

The architecture and functional properties of the vomeronasal organ (VNO), or Jacobson's organ, in nonamphibian vertebrates suggest that this organ is primarily accessible to nonvolatile substances excreted in biological fluids [1]. We have recently studied the functional properties of the VNO in the frog and demonstrated that it fits this description [2]. In this article, we extend our observations on the stimulus access to the VNO in frog.

The other aspect considered here is the properties of the receptor cells of the VNO. The ciliated receptor cell in the main olfactory organ has been intensively studied, whereas there are few studies on the microvillar receptor cells found in the VNO. We have described some of their electrophysiological properties [3].

Methods

Experiments were performed on the frog *Rana esculenta*. Details on the methods used in the morphological studies and the patch-clamp procedures have already been published [2,3]. All recordings reported here were obtained from freshly isolated cells, after dissociation of the vomeronasal epithelium with papain [3]. All presented a long dendrite (65–110 µm)

[1] Laboratoire de Neurobiologie Sensorielle, E.P.H.E., 1 avenue des Olympiades, 91305 Massy, France
[2] Department of Biology, University of Oslo, P.O. Box 1051, 0316 Oslo 3, Norway

and a piece of axon. Numerous microvilli (more than 10) were observable under Nomarski illumination (Zeiss, ×400) (Fig. 1). The recording electrode was sealed onto the somatic membrane. To prevent modification of the intracellular content, recordings were done using the cell-attached Nystatin-perforated patch configuration [3]. The pipette contained K-aspartate, 100 mM; NaCl, 24 mM; HEPES, 10 mM; pH 7.1. The bath contained NaCl, 136.8 mM; KCl, 2.7 mM; $CaCl_2$, 0.9 mM; $MgCl_2$, 0.5 mM; Na_2HPO_4, 8.1 mM; KH_2PO_4, 1.5 mM; glucose, 10 mM; pH 7.1.

Results

Stimulus Access

When freely moving frogs were kept in about 3 cm of water containing a fluorescent dye (cresyl violet), histological observations showed that only the VNO and the communicating cavities had taken the dye, not the main olfactory organ, indicating that the VNO is designed to sample water-borne substances.

The vomeronasal organ in the frog is anatomically and functionally separated from the main olfactory organ [4]. The vomeronasal sensory epithelium covers a cul-de-sac at the most anteromedial part of the inferior cavity. This inferior cavity communicates with the naris through the middle cavity and the vestibule of the superior cavity by narrow channels or slits. The nasolacrymal duct empties into the posterolateral corner of the middle cavity. The vestibule of the superior cavity is located just beneath the external naris. The ventral part of the inferior cavity communicates with the lateral part of the nonsensory epithelium, which surrounds the *eminentia olfactoria* in the principal nasal cavity. The cavities and the narrow slits between them are lined with kinocilia transporting water and mucus from the slits of the vestibule through the VNO, and then out toward the choana.

As the inferior cavity is surrounded by a set of rigid cartilages, and because there is no cavernous tissue [5], an active pumping of fluids to the sensory epithelium is improbable. As the frog spends most of the time with the head just over the interface between water and air, the frequent openings of the naris certainly allow water to enter the vestibule and

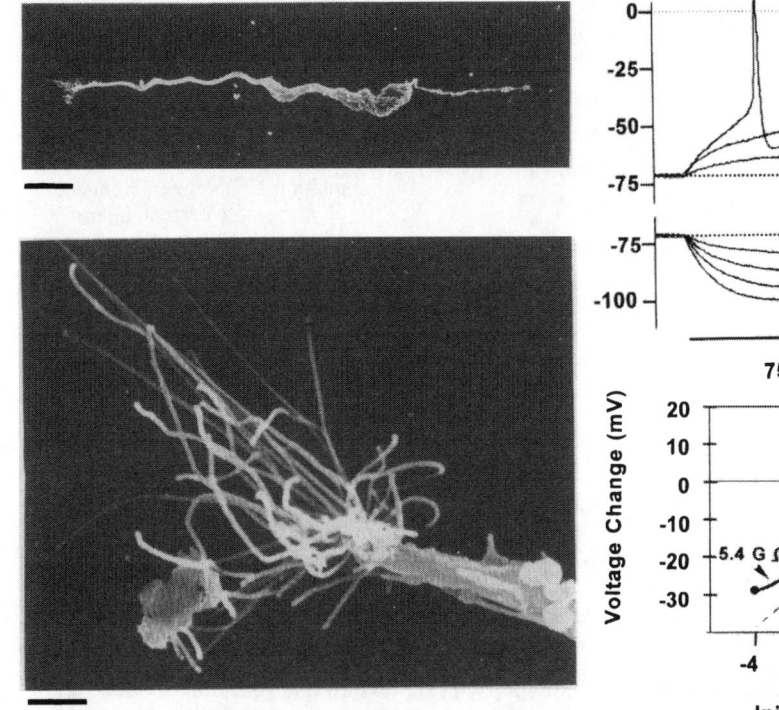

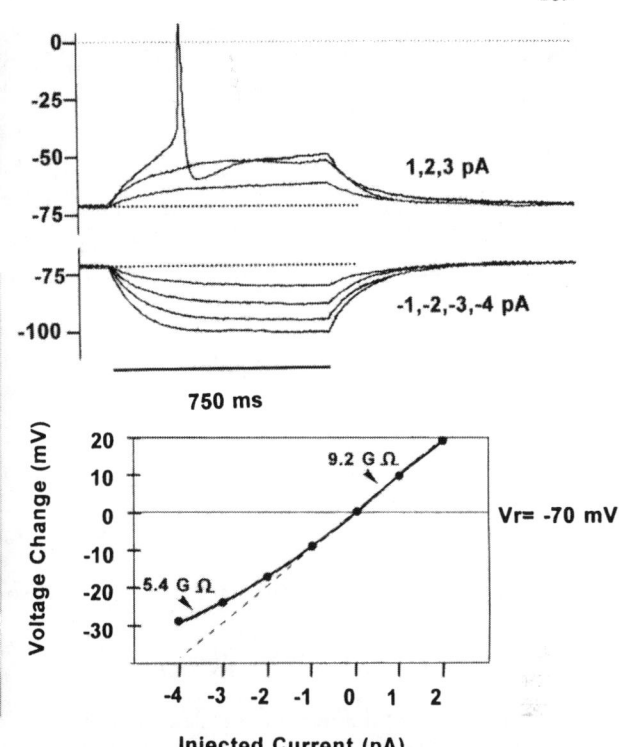

FIG. 1. Scanning electron micrographs of general morphology of isolated frog vomeronasal receptor cell (*top*; scale = 10 μm) and detailed view of microvilli at apex of dendrite (*bottom*; scale = 1 μm). Note characteristic enlargement of proximal part of dendrite near cell body. (See [2] for methods)

FIG. 3. Vomeronasal receptor cells are very sensitive to small current intensities. *Upper traces*, Voltage responses to depolarizing and hyperpolarizing current pulses injected through recording pipette during 750 ms. Note that depolarizing current of 3 pA was large enough to bring potential to firing threshold and elicit overshooting action potential. *Lower graph*, Voltage–current relationship of this recording measured at end of pulse. Slope of curve indicates input resistance of 9.2 GΩ near the resting potential. The resistance decreases to 5.4 GΩ on hyperpolarization because of activation of inward rectifier conductance (From [3] with permission of Oxford University Press)

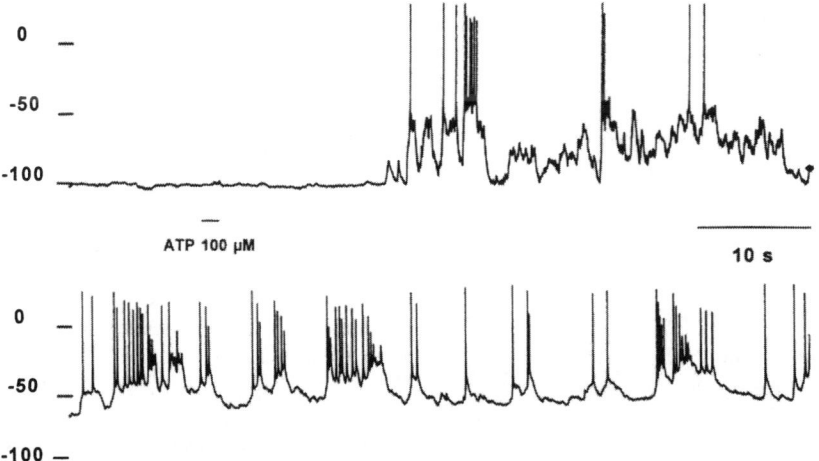

FIG. 2. Responses of vomeronasal receptor cells to 100 μM ATP applied externally during 2 s. *Upper trace*, membrane potential of this cell was initially very stable near −100 mV. After application of 100 μM ATP, membrane potential showed large fluctuations; some of them reached the firing

threshold and triggered action potentials. *Lower trace*, another cell responded similarly, but presented long-lasting rythmic fluctuations of membrane potential together with periodic firing of action potentials. ATP was applied about 2 min before beginning of trace

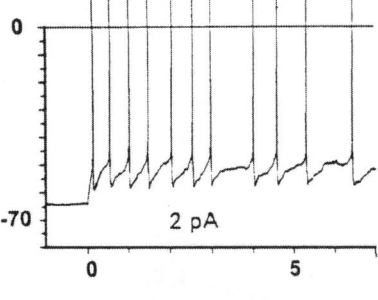
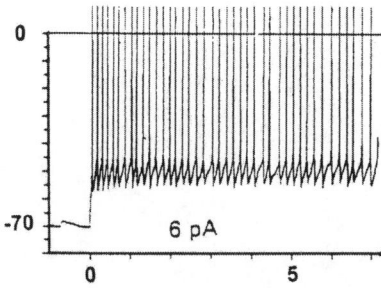
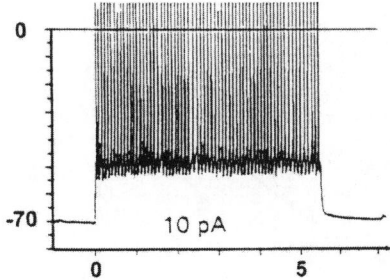
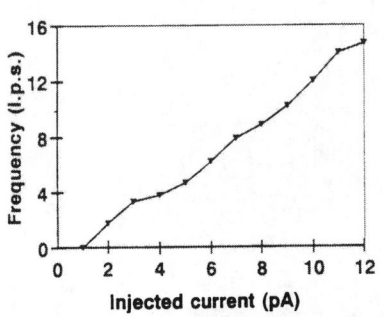

FIG. 4. Responses of vomeronasal receptor cell to long-lasting injections of depolarizing current through recording electrode. Indicated current intensities were applied at time 0. Graph indicates nearly linear relationship between frequency of firing and current intensity. Time scale in seconds; voltage scale in millivolts.

flow, probably with the help of lateral glands secretions, onto the vomeronasal epithelium. The water–air interface would be an interesting structure for chemoreceptive signals. The frog, as other animals, could produce a lipid secretion, from skin or specialized glands, which would be present at a high concentration at the interface.

Microvillar Vomeronasal Receptor Cells

Sensory neurons in the VNO are bipolar receptors, with a long dendrite reaching the surface, and a long axon that sends the sensory information to the accessory olfactory bulb. Morphologically, vomeronasal microvillar receptor cells differ from ciliated receptor cells in the main olfactory system by at least two features (see Fig. 1): the presence of numerous microvilli (7–10 μm in length, about 100 nm in diameter) at the apex of the dendrite; and a typical enlargement of the proximal part of the dendrite, near the cell body, which results from the presence of an hypertrophy of the smooth endoplasmic reticulum [6].

Electrophysiological Properties

Most cells had a stable resting potential ranging from about −60 to −90 mV, with spontaneous fluctuations of a few millivolts, and virtually no spontaneous action potential. Other cells had less stable resting potentials, with fluctuations of sometimes a few tens of millivolts, in the range of −50 to −120 mV. Even in this less stable state, the spontaneous firing frequency was generally very low. Interestingly, some stable cells entered the fluctuating state in

response to a brief external application of adenosine triphosphate (ATP), a nonvolatile substance (Fig. 2, top). In a few cases, long-lasting, often rhythmic, fluctuations of the membrane potential, together with a somewhat periodic firing of action potentials, was observed (Fig. 2, bottom).

These cells have a high input resistance and are very sensitive to small depolarizing currents (Fig. 3). Overshooting action potentials were elicited with injected depolarizing currents in the range of 1–10 pA [3]. Above a threshold, these receptor cells fired repetitively during long-lasting current injections, showing no accommodation (Fig. 4). The frequency of firing was linearly related to the current intensity (Fig. 4). They were also able to fire in response to slow ramps of depolarizing current [3].

The firing threshold corresponds to the activation of a fast and transient inward current, which could be blocked by tetrodotoxin. Other voltage-dependent currents have been characterized: I_H, I_{Ca}, and various I_K [3]. The properties of these currents are similar to those already described in ciliated olfactory receptors. Injections of cAMP in the microvillar receptor cells failed to elicit a membrane current similar to the cAMP-induced current observed in ciliated receptor cells, suggesting a distinct mechanism of transduction, which remains to be investigated.

Acknowledgments. D.T. was supported by a Human Frontier Science Program Grant "Molecular Recognition and Transduction in Chemosensory Systems." K.B.D. was supported by Elf Aquitaine Norge A.S. through the University of Oslo "Support Program" and by the Norwegian Research Council for Science and the Humanities.

References

1. Halpern M (1987) The organization and function of the vomeronasal system. Annu Rev Neurosci 10:325–362
2. Døving KB, Trotier D, Rosin J-F, Holley A (1993) Functional architecture of the vomeronasal organ of the frog. Acta Zool (Stockh) 74:173–180
3. Døving KB, Trotier D, Rosin J-F (1993) Voltage-dependent currents in microvillar receptor cells of the frog vomeronasal organ. Eur J Neurosci 5:995–1002
4. Scalia F (1976) Structure of the olfactory and accessory olfactory systems. In: Llinas R, Precht W (eds) Frog neurobiology. Springer, Berlin, pp 213–233
5. Franceschini F, Sbarbati A, Zancanaro C (1991) The vomeronasal organ in the frog, *Rana esculenta*. An electron microscopy study. J Submicrosc Cytol Pathol 23(2):221–231
6. Bannister LH, Dodson HC (1992) Endocytotic pathways in the olfactory and vomeronasal epithelia of the mouse: ultrastructure and uptake of tracers. Microsc Res Technol 23:128–141

Experimental Study of Olfactory Desensitization in Electro-Olfactogram

Tadashi Ishimaru, Yasuyuki Kimura, Takaki Miwa, and Mitsuru Furukawa[1]

Key words. Desensitization—Electro-olfactogram

Introduction

The sensory system has a desensitization mechanism which contributes to the expansion of the sensory dynamic range and to the protection of the sensory organ from overstimulation.

In both vision [1] and olfaction [2], desensitization seems to be related to Ca^{2+}. It is known that inward Ca^{2+} flows elicit desensitization in a single olfactory cell. To our knowledge, no studies have addressed the relationship between desensitization and Ca^{2+} in the olfactory mucosa. Hence we carried out an electro-olfactographic study of olfactory mucosae superfused with and without Ca^{2+}.

Material and Methods

Bullfrogs (*Rana catesbiana*) were killed by decapitation. The dorsal olfactory mucosa was removed as a single sheet. The ciliated surface was continuously superfused (at 1 ml/s) with solutions containing Ca^{2+} (standard solution) or without Ca^{2+} (Ca^{2+}-free solution). The composition of the standard solution was (mM): NaCl 100, KCl 2.5, $NaHCO_3$ 10, $CaCl_2$ 1, $MgSO_4$ 1, and glucose 10. The composition of the Ca^{2+}-free solution was (mM): Nacl 100, KCl 2.5, $NaHCO_3$ 10, $MgSO_4$ 1, and glucose 13. These solutions were saturated with 98.5% O_2/1.5% CO_2 at pH 7.6. *n*-Amyl acetate (1 mM) was dissolved in solutions identical with the superfusates, and these serred as stimulants. Electro-olfactograms (EOG) represented the differences in transepithelial voltage responses without and with an odorant.

Results

The odorant stimulus was administered for a long period (50 s). The olfactory responses slowly declined in the solutions with and without Ca^{2+}. The amplitude of the responses decayed to $50 \pm 7.8\%$ (mean $\pm$ SE, three mucosae) and $68 \pm 20\%$ (three different mucosae) in the standard and Ca^{2+}-free solutions, respectively, 40 s after their peaks. The difference was not significant (*t*-test, $P > 0.05$). This observation indicates that desensitization occurred at an almost identical speed, and to an identical extent, regardless of the presence of Ca^{2+} on the ciliated surface.

Discussion

While the single olfactory cell did not exhibit desensitization in the external cell solution without Ca^{2+}, the olfactory mucosa exhibited desensitization in the Ca^{2+}-free superfusion. The difference in these results may be explained as follows:

1. The difference was probably due to the incomplete removal of Ca^{2+} from the micromilieu surrounding the cilia. In our experiments, the olfactory mucosa was not free of mucus, which contains Ca^{2+} secreted from Bowman's glands and/or supporting cells.

2. Inflows of Ca^{2+} occur not only via ionic channels located in the plasma membrane of olfactory cilia but also in the somatic plasma membrane of the cell under the tight junction between the supporting cell and the olfactory cell. A single olfactory cell therefore cannot have Ca^{2+} influx in the external cell solution without Ca^{2+}. In the olfactory mucosa, the olfactory cell, with the supporting cells, has an external cell solution containing Ca^{2+} below the tight junction, while the ciliated surface is superfused with Ca^{2+}-free solution.

At the Congress of the Japanese Symposium of Taste and Smell, 1991, we reported that a reduction in the Ca^{2+} level of the superfusate on the ciliated surface of olfactory mucosa augmented EOG amplitude. This result suggests that the conductance of cyclic adenosine monophosphate (c-AMP) activated channels in cilia was modulated by the Ca^{2+} level of the superfusate. The removal of Ca^{2+} from the micromilieu surrounding the cilia, therfore, seems to be sufficient to influence desensitization. We believe that explanation (2) above is a more likely one than explanation (1).

[1] Department of Otorhinolaryngology, School of Medicine, Kanazawa University, 13-1 Takara-machi, Kanazawa, 920 Japan

References

1. Nakatani K, Yau K-W (1988) Calcium and light adaptation in retinal rods and cones. Nature 334:69–71

2. Kurahashi T, Shibuya T (1990) Ca^{2+}-dependent adaptive properties in the solitary olfactory receptor cell of the newt. Brain Res 515:261–268

Survey of Ligand-Gated Conductance in Olfactory Receptor Cilia

Tomo-Oki Satoh*, Hirokatsu Kobayashi, and Tadashi Nakamura[1]

Key words. Second messenger—Transductory conductance

A series of biochemical [1], electrophysiological [2], and molecular biological [3] experiments has proven that the olfactory receptor cells of vertebrates transduce chemical signals to electrical responses via cyclic adenosine monophosphate (cAMP)-gated channels in the cilia [4].

However, more recently, it was suggested that the mechanism is more complex, with two other channels. The first one is an inositol-1,4,5-triphosphate (IP_3)-gated Ca^{2+} channel that has been proposed to serve for the transduction [5], and the second one is a Ca^{2+}-gated Cl^- channel. In the present study, we surveyed these ligand-gated conductances in isolated cilia, using the ciliary patch procedure described by Kleen and Gesteland [6]. This procedure is suitable for detecting small conductances with low channel density or low unit conductance. The membrane conductance was measured by recording current-voltage relationships (I–V curve) during bath-application of these substances into the cilia. In all the cilia tested, we observed a large membrane conductance with $200\,\mu M$ cAMP and a moderate level of conductance with $1\,mM$ Ca^{2+}. When we exchanged most of the Cl^- in the bath solution to gluconage$^-$, in order to identify the charge carrier of the Ca^{2+}-gated conductance, there was a large shift in the I–V curve, giving minus reversal potential. This shift in the I–V curve was due to the gradient of Cl^- across the membrane, which indicates that this conductance is specific to Cl^-. These results are consistent with those reported by Kleen and Gesteland [7]. On the other hand, the I–V curve was little changed by $50-100\,\mu M$ IP_3 applied into the cilia by the same method ($n = 15$). This result dose not support the idea that IP_3 is the second messenger to generate the receptor potential, even if it exists in the cell. In most of the cilia, the Ca^{2+}-gated Cl^- conductance was more than 10% of that gated by $200\,\mu M$ cAMP in the absence of divalent cations. This substantial value suggests a large contribution of this conductance to the receptor potential, because, to a large extent, the cAMP-gated conductance at the normal concentration of divalent cations is blocked.

Very recently, Kurahashi and Yau [8] reported that the odor-induced current in the olfactory receptor cells included a Cl^--component that was regulated by Ca^{2+}. Our present observation is consistent with their conclusion.

References

1. Pace U, Hanski E, Salomon Y, Lancet D (1985) Odorant-sensitive adenylate cyclase may mediate olfactory reception. Nature 316:255–258
2. Nakamura T, Gold GH (1987) A cyclic nucleotide-gated conductance in olfactory receptor cilia. Nature 325:442–444
3. Reed RR, Heather Bakalyar AH, Cunningham AM, Levy NS (1992) The molecular basis of signal transduction in olfactory sensory neurons. In: Corey DP, Roper SD (eds) Society of General Physiologists Series, 47, Sensory Transduction. The Rockefeller University Press, New York, pp 53–59
4. Firestein S, Darrow B, Shepherd GM (1991) Activation of the sensory current in salamander olfactory receptor neurons depends on a G-protein-mediated cAMP second messenger system. Neuron 6:825–835
5. Miyamoto T, Restrepo D, Gragoe EJ Jr, Teeter JH (1992) IP_3- and cAMP-induced responses in isolated olfactory receptor neurons from the channel catfish. J Membr Biol 127:173–183
6. Kleen SJ, Gesteland RC (1991) Currents in frog olfactory cilia. J Membr Biol 120:75–81
7. Kleen SJ, Gesteland RC (1991) Calcium-activated chloride conductance in frog olfactory cilia. J Neurosci 11:3624–3629
8. Kurahashi T, Yau KW (1993) Coexistence of cationic and chloride components in odorant-induced current of vertebrate olfactory receptor cells. Nature 363:71–74

[1] Department of Applied Physics and Chemistry, University of Electro-Communications, 1-5-1 Chohfugaoka, Chohfu, Tokyo, 182 Japan
* Present address: Yamanouchi Pharmaceutical Co., Ltd., 21 Miyukigaoka, Tsukuba, Ibaraki, 305 Japan

IP$_3$-Gated Current in Isolated Rat Olfactory Neurons

YUKIO OKADA*, JOHN H. TEETER, and DIEGO RESTREPO[1]

Key words. IP$_3$-gated current—Rat olfactory neuron

Odorants interact with receptor proteins located in the cilia that extend from the dendrite of olfactory neurons, triggering a biochemical cascade that results in membrane depolarization and discharge of action potentials. Stopped-flow experiments have shown that stimulation of isolated olfactory cilia with odorants results in increases in the levels of adenosine 3',5'-cyclic monophosphate (cAMP) or inositol 1,4,5-trisphosphate (IP$_3$) [1]. Odorants that stimulate cAMP formation are believed to induce the opening of a cAMP-gated cation channel in the ciliary membrane [2]. In catfish olfactory neurons, the increase in IP$_3$ concentration may activate a cationic conductance which has been shown to be different from that activated by cAMP [3]. In contrast, it is unclear how an odorant-induced increase in IP$_3$ concentration elicits a depolarization in mammalian olfactory neurons.

In the present study, the effect of intracellular application of 1,4,5-IP$_3$ from the patch pipette was analyzed in isolated rat olfactory neurons under whole-cell patch clamp. Intracellular dialysis of 10 μM 1,4,5-IP$_3$ in K$^+$-internal solution induced a sustained depolarization of 35.8 ± 10.5 mV (mean ± SD, $n = 16$). The IP$_3$-induced response was observed in 75% of the cells dialyzed with IP$_3$, but not when 10 μM ruthenium red was also included in the pipette solution (four cells). Lower concentrations (50–100 nM) of 2,4,5-IP$_3$ induced similar responses in 5 of 8 olfactory neurons. Steady-state I-V relationships of IP$_3$-gated currents with K$^+$-internal solution were classified into two types: slightly outwardly rectifying and N-shaped. In Cs$^+$-internal solution linear and slightly outwardly rectifying patterns were observed. The IP$_3$-induced currents were inhibited by external Cd^{2+} (1 mM). The reversal potentials of the Cd^{2+}-inhibitable currents of the slightly outwardly rectifying and N-shaped types were −16.1 mV ($n = 2$) and −29.0 ± 7.1 mV ($n = 3$), respectively, in K$^+$-internal solution. The reversal potential was −5.9 ± 6.8 mV ($n = 5$) in the Cs$^+$-internal solution. In contrast, the Ca^{2+}-ionophore, ionomycin (5 μM) hyperpolarized the olfactory neurons and greatly potentiated the outward currents at positive holding potentials. It is concluded that an increase in intracellular IP$_3$ depolarizes rat olfactory neurons without mediation by intracellular Ca^{2+}. This is consistent with previous work indicating that odorant-induced increases in IP$_3$ directly gate a class of cation-selective channels in the plasma membrane of olfactory receptor neurons.

References

1. Breer H, Boekhoff I (1991) Odorants of the same odor class activate different second messenger pathways. Chem Senses 16:19–29
2. Nakamura T, Gold GH (1987) A cyclic nucleotide-gated conductance in olfactory receptor cilia. Nature 325:442–444
3. Miyamoto T, Restrepo D, Cragoe EJ, Teeter JH (1992) IP$_3$- and cAMP-induced responses in isolated olfactory receptor neurons from the channel catfish. J Membr Biol 127:173–183

[1] Monell Chemical Senses Center, 3500 Market Street, Philadelphia, PA 19104, USA
*Present address: Department of Physiology, Nagasaki University School of Dentistry, 1-7-1 Sakamoto, Nagasaki, 852 Japan

Ligand and Ion Specificity of IP_3-Gated Channels in Lobster Olfactory Receptor Cells

D.A. Fadool and B.W. Ache[1]

Key words. Inositol—Transduction

Previously, we have demonstrated that inositol 1,4,5-trisphosphate (IP_3) directly activates two types of ion channels (30 pS and 74 pS) in the plasma membrane of cultured lobster olfactory receptor cells [1]. We have now further characterized these channels to better understand the role of each or both in mediating the odor-evoked inward current.

Clusters of the olfactory neurons were isolated from the lateral filament of the antennule (olfactory organ) of intermolt specimens of *Panulirus argus* and placed into primary cell culture, as previously described [2]. Recordings were made by patch-clamping the cells at a holding potential of -60 mV in the whole-cell configuration or from cell-free patches of membrane in the inside-out configuration. Odorants, drugs, or inositol phospholipids were "spritzed" on cells or patches from a multibarrel glass micropipette.

The Pr_{open} of both channels increased two- to tenfold as the pH of the IP_3 solution was increased from pH 8 or 9 to pH 10. Increasing the Ca^{2+} concentration on the inside face of the patch ($[Ca^{2+}]_i$) from 0.001 to 30 mM increased the Pr_{open} and decreased the unitary conductance of the channels. The slope conductance of each channel was reduced by as much as 50% and, in some instances, the channels inwardly rectified at depolarizing potentials in high Ca^{2+}. Modes occurred for a small percentage of the channels; both the 30 and 74 pS channel could enter a long maintained open state followed by a long closed state with high flicker activity, before falling back into their most observed state of $Pr_{open} \approx 0.33$.

The action of ruthenium red, Co/Cd, and tetrodotoxin (TTX) on the odor-evoked macroscopic inward current mimicked these agents' effects on both IP_3-activated channels. Tetraethylammonium (TEA), Cs, and Mn were without effect on either the macroscopic or unitary currents. Thapsigargin, a known inhibitor of cytoplasmic Ca-pumps, had no effect on the magnitude of the odor-evoked macroscopic inward currents (t'statistic, $P \leq 0.05$).

The 30 pS and 74 pS channels showed different patterns of ion selectivity. The 30 pS IP_3-activated channels were largely nonselective for cations, reversed near 0 mV; independently of E_{Na}, E_{Ca}, or E_K (21 of 32 trials). The 74 pS IP_3-activated channels were Ca^{2+} selective, reversing at or near E_{Ca} in 8 of 11 trials.

None of six products of inositol phospholipid metabolish (inositol, IP(4), IP_2(1,4), cyclic IP(1,2), IP_4(1,3,4,5), and IP_6) directly gated the IP_3-activated channels. However, IP_4 activated a novel unitary current in 25 of 42 patches. The channel had a larger slope conductance (191 ± 12.8 pS), a more dense distribution ($0.60-10.8$ channels/μm^2), and a shorter mean open time (5.03 ± 0.97 ms) than did either of the channels gated by IP_3. The IP_3- and IP_4-gated channels interacted when co-localized in the same patch. The Pr_{open} of the 191 pS IP_4-gated channel was greater upon activation of a neighboring IP_3 channel (paired t-test, $P \leq 0.05$, $n = 16$). Conversely, the Pr_{open} of the 30 pS or 74 pS IP_3-gated channel was suppressed upon activation of a neighboring IP_4 channel (paired t-test, $P \leq 0.05$, $n = 6$).

We propose that odors excite lobster olfactory receptor neurons (ORNs) by activating two different types of IP_3-gated ion channels in the plasma membrane. The action of IP_3 on these channels can be regulated by intracellular pH and Ca^{2+}, and by interaction with a novel IP_4-gated plasma membrane receptor, whose direct role in olfactory transduction, if any, is as yet, unknown.

Acknowledgments. Supported by: ONR N0014-90-J-1566, NIH NRSA 1-F31MH10124-01A1 and 1R01DC01655-01, and the Dan Charitable Trust Fund of Japan.

References

1. Fadool DA, Ache BW (1992) Plasma membrane inositol 1,4,5-trisphosphate-activated channels mediate signal transduction in lobster olfactory receptor neurons. Neuron 9:907–918
2. Fadool DA, Michel WC, Ache BW (1991) Sustained primary culture of lobster (*Panulirus argus*) olfactory receptor neurons. Tissue Cell 23:719–732

[1] Whitney Laboratory and Departments of Zoology and Neuroscience, University of Florida, 9505 Ocean Shore Blvd., St. Augustine, FL 32086, USA

Origin of the Chloride Component of Olfactory Receptor Current

Steven J. Kleene[1]

Key words. Olfactory transduction—Chloride current

In the olfactory cilia of amphibians, odorants increase the activity of adenylate cyclase [1]. The cyclic adenosine monophosphate (cAMP) formed directly gates channels in the ciliary membrane [2], leading to a depolarizing inward current carried by cations. Recently it was shown [3] that the odorant response in amphibians consists not only of a cationic current but also of a Cl^- current that depends on external Ca^{2+}. It seemed likely [3] that Ca^{2+} entering the cilium through the cAMP-gated channels might, in turn, activate Ca^{2+}-activated Cl^+ channels known to exist in the ciliary membrane [4]. This model allows the following predictions:

1. cAMP should activate a ciliary Cl^--dependent current.
2. This activation should also require external Ca^{2+}.
3. The Cl^- current should appear sooner and/or be larger if intraciliary Ca^{2+} buffering is reduced.
4. The Cl^- current should require both the cAMP-gated cationic channels and the Ca^{2+}-activated Cl^- channels.

I have tested and confirmed each of these predictions. Single cilia were excised from frog olfactory receptor neurons and studied with voltage-clamp recording [5]. The ciliary membrane potential was held at 0 mV and then jumped to various other potentials for 4 s. In the presence of 100 μM cytoplasmic cAMP, voltage jumps to negative potentials resulted in a biphasic inward current. The first phase was due to the influx of cations through the cAMP-gated channels. The second phase was carried by Cl^- and was eliminated when cytoplasmic Cl^- was replaced by methanesulfonate$^-$. The second phase was also blocked by Cl^--channel inhibitors (300 μM niflumic acid or 100 μM 3',5-dichlorodiphenylamine-2-carboxylate; DCDPC). These inhibitors were found to have no effect on the cAMP-activated cationic current. As predicted by the model, activation of the Cl^--dependent current by cAMP also required external Ca^{2+}. In the presence of 100 μM cAMP and 1 mM external Ca^{2+}, the Cl^--dependent current at −80 mV averaged −297 ± 25 pA (n =

49). In other cilia tested with cAMP but no external Ca^{2+}, the Cl^--dependent current was just −4 ± 4 pA (n = 8). Reduction of cytoplasmic Ca^{2+} buffering decreased the latency and/or increased the amplitude of the second phase. With 100 μM cytoplasmic cAMP present, the amount of secondary Cl^- current, on average, corresponded to an intraciliary Ca^{2+} concentration >10μM, which saturates the Cl^- channels. Inhibitors of the cAMP-gated channels (3',4'-dichlorobenzamil or *l-cis*-diltiazem, each at 300 μM) eliminated both phases of the cAMP-activated current. The cAMP-activated cationic current was blocked directly. This prevented the Ca^{2+} influx needed to activate the secondary Cl^- current. Neither of these inhibitors directly inhibited the Ca^{2+}-activated Cl^- conductance. Finally, it was found that very small cationic currents through the cAMP-gated channels can produce large secondary Cl^- currents. With reduced cytoplasmic Ca^{2+} buffering, 0.3 μM cAMP produced a cationic current at −80 mV of −3.6 ± 2.5 pA and a secondary Cl^--dependent current of −79 ± 13 pA (n = 5). Thus, in some neurons, the secondary Cl^- current could serve to amplify the depolarizing receptor current [3]. In other neurons, internal Cl^- may be low [6], and activation of the Cl^- current could help to repolarize the neuron and terminate the odorant response.

Acknowledgments. This work was supported by the National Institutes of Health.

References

1. Pace U, Hanski E, Salomon Y, Lancet D (1985) Odorant-sensitive adenylate cyclase may mediate olfactory reception. Nature 316:255–258
2. Nakamura T, Gold GH (1987) A cyclic nucleotide-gated conductance in olfactory receptor cilia. Nature 325:442–444
3. Kurahashi T, Yau K-W (1993) Co-existence of cationic and chloride components in odorant-induced current of vertebrate olfactory receptor cells. Nature 363:71–74
4. Kleene SJ, Gesteland RC (1991) Calcium-activated chloride conductance in frog olfactory cilia. J Neurosci 11:3624–3629
5. Kleene SJ, Gesteland RC (1991) Transmembrane currents in frog olfactory cilia. J Membr Biol 120:75–81
6. Dubin AE, Dionne VE (1993) Conductances in the ciliated dendrites of mudpuppy olfactory receptor neurons (abstract). Chem Senses 18:548

[1] Department of Anatomy and Cell Biology, University of Cincinnati, Cincinnati, OH 45267-0521, USA

Study of Olfactory Transduction by Analysis of Dynamics of Odor-Induced [Ca^{2+}]$_i$ Increases in Receptor Neurons

TAKAAKI SATO, JUNZO HIRONO, MITSUO TONOIKE, and MASAMINE TAKEBAYASHI[1]

Key words. Cytoplasmic calcium—Olfactory transduction

The regulation of odor-induced changes in cytosolic free Ca^{2+} ([Ca^{2+}]$_i$) and odor-responsiveness was studied by analyzing the dynamics of the fluorescence intensity of intracellular fura-2 in the knobs and/or somata of isolated olfactory receptor neurons. Isolated olfactory receptor cells were enzymatically obtained from the olfactory epithelia of wild bullfrog or newt anesthetized with MS-222, or from an adult female mouse anesthetized with Ketalar 50, using the tissue-printing method. Fura-2 was loaded to isolated cells by incubation with fura-2/AM solution. The measurement of [Ca^{2+}]$_i$ was carried out at seven positions, as described in our previous reports [1,2]. Odorants dissolved in normal Ringer's solution (NR) were applied to the cells by perfusion of the bath solution.

Simultaneous patch-clamp and [Ca^{2+}]$_i$ recordings were performed to clarify the relationship between the dynamics of odor-induced increases in [Ca^{2+}]$_i$ (OIC) and receptor potentials. The rising phase of the OIC corresponded with the duration of the odor-induced inward current, indicating that the rising phase of the OIC almost overlapped the duration of the receptor potential [3]. To examine the odor-induced Ca^{2+}-release in mice, OICs were measured under extracellular Ca^{2+}-free (CF) conditions. In four of six cells responsive to citralva, two components of OICs occurred in the somata or olfactory knobs. The early component of OICs, which occurred under CF conditions, with some delays from the normal OIC, indicated that odorants evoked the Ca^{2+}-release both in the soma and in the knob. The late OICs occurred after the recovery of the extracellular ionic condition and were sustained. In 70% of the tested cells, the little quenching by manganese 10 s after the onset of the late component of OICs suggested that the sustained OIC in CF was mediated by the inactivation of a Ca^{2+}-exclusive apparatus. The time course of OICs showed that the duration of activation of the Ca^{2+}-release channel was the shortest among Ca^{2+}-permeable channels such as the voltage-dependent Ca^{2+}-channel and the cAMP-gated channel. The normalized increases in [Ca^{2+}]$_i$ in the soma and knob indicated an excess Ca^{2+} pathway in the knob, which, perhaps, was mediated by ciliary cyclic adenosine monophosphate (cAMP)-gated channels.

The response specificity of mouse receptor neurons for n-fatty acids and n-aliphatic alcohol was examined. One-third of the responsive cells were selective to a subtype of n-fatty acids whose hydrocarbon chain length were similar to that responded to by mitral/tufted cells [4]. The other cells responded to both n-fatty acids and *n*-aliphatic alcohol, the highest sensitivities to individual series of *n*-fatty acids and n-aliphatic alcohol being obtained in odorants with hydrocarbon chains of similar length. These results suggested that receptor specificity was determined by the affinity of odorants to receptors, this being dependent on fine stereochemical structures.

References

1. Sato T, Hirono J, Tonoike M, Takebayashi M (1991) Two types of increases in free Ca^{2+} evoked by odor in isolated frog olfactory receptor neurons. Neuroreport 2:229–232
2. Hirono J, Sato T, Tonoike M, Takebayashi, M (1992) Simultaneous recording of [Ca^{2+}]$_i$ increases in isolated olfactory receptor neurons retaining their original spatial relationship in intact tissue. J Neurosci Methods 42:185–194
3. Sato T, Hirono J, Tonoike M (1992) Molecular mechanism of olfactory transduction: as viewed from the dynamics in cytoplasmic free calcium. Sensors and Materials 4:11–20
4. Mori K, Imamura K, Mataga N (1992) Differential specificities of single mitral cells in rabbit olfactory bulb for a homologous series of fatty acid odor molecules. J Neurophysiol 67:786–789

[1] Life Electronics Research Center, Electrotechnical Laboratory, 3-11-46 Nakoji, Amagasaki, Hyogo, 661 Japan

Optical Recordings of Calcium Responses in Neighboring Olfactory Receptors Suggest that the Distribution of Subtypes of Receptors is Heterogeneous

Junzo Hirono, Takaaki Sato, Mitsuo Tonoike, and Masamine Takebayashi[1]

Key words. Odor responsivity—Local distribution of subtypes

The cellular distribution of odor responsiveness in the olfactory epithelium is unclear. To obtain evidence for this distribution, we examined the odorant responsiveness of neighboring olfactory receptor neurons in mouse and frog. Using our new isolation method, "tissue printing" [1], we prepared samples in which many neurons were isolated while their original spatial relationship in intact tissue was retained. The enzymes used in the isolation procedure were papain (1 mg/ml, 5–10 min, for frog) and trypsin (0.025%, 12 min, for mouse). We made measurements of cytoplasmic free calcium concentration ($[Ca^{2+}]_i$) in isolated neurons, using the Ca^{2+} indicator dye, fura-2. The fluorescence intensity of fura-2 was recorded in the somata of several cells within the optical field of view ($210 \times 235 \,\mu m^2$) at $\frac{1}{3}$-s intervals. Each fluorescence intensity was the integrated value of 5×5 pixels ($2 \times 2.4 \,\mu m^2$ on a cell) in eight video frames ($\frac{1}{30}$-s intervals). The odor stimulus solutions used were: citralva (CT; 100 μM), isoamyl acetate (AM; 10 mM), pyrazine (PY; 1 mM), and several fatty acids in normal Ringer's solution. Cell viability and physiological intactness were confirmed by among others, determining the responses to high concentration potassium solution and forskolin, determining the stable resting level of $[Ca^{2+}]_i$, and investigating cell morphology.

The odorant-induced responses, determined as increases in $[Ca^{2+}]_i$, were measured in 43 mouse samples (5.5 cells per sample on average), with CT, AM, and PY stimuli. CT and AM are related to the cyclic adenosine monophosphate (cAMP)-mediated transduction system and PY is related to the inositol 1,4,5 trisphosphate (IP_3) system. Eighteen of 235 cells responded to CT only, this being the major population (7.7%) in regard to response to these odorants; this population was also the major type for the frog. Nine cells responded only to AM, and one cell responded only to PY. There were also generalist types, that is, seven cells responded to both CT and AM, and one cell responded to all three odorants. As we found in our previous studies [1,2], almost cells responsive to CT or AM also responded to the adenylate cyclase activator, forskolin. The fact that the number of cells responsive to CT or AM was larger than that of PY-selective cells indicated that olfactory transduction in the majority of receptor cells was mediated by a cAMP second messenger, and that the cells with the IP_3-mediated system might be distributed at a low density.

The findings were classified according to the number of odorant responsive cells in each preparation and the degree of heterogeneity of their responsiveness. In three preparations, three cells responded to some odor stimuli. In one of the preparations, the selectivity of the three cells differed. The odor selectivity in the other two preparations was two of types. In one preparation, there were two cells whose knobs were contiguous, i.e., they were doublets, and both of them responded to CT. In the three of six preparations where two cells showed odor responses, their odor-selectivity differed, while in the other three preparations, the selectivity was the same. In 15 preparations, only one cell responded to the tested odorants. Many cells did not respond to the three odorants, but it is possible that they might respond to other odorants. Our results here suggested that several response-types might coexist within each small region in the olfactory epithelium and, that, in some regions, cells of a similar subtype might be distributed sparsely. The findings for the frog and the findings obtained with the fatty acid odorants also supported this idea of heterogeneous distribution.

References

1. Hirono J, Sato T, Tonoike M, Takebayashi M (1992) Simultaneous recording of $[Ca^{2+}]_i$ increases in isolated olfactory receptor neurons retaining their original spatial relationship in intact tissue. J Neurosci Methods 42:185–194
2. Sato T, Hirono J, Tonoike M, Takebayashi M (1991) Two types of increases in free Ca^{2+} evoked by odor in isolated frog olfactory receptor neurons. Neuroreport 2:229–232

[1] Life Electronics Research Center, Electrotechnical Laboratory, 3-11-46 Nakoji, Amagasaki, Hyogo, 661 Japan

Olfactory Receptor Responses to Chemical Stimulation

Keiichi Tonosaki, Kousei Miwa, and Hajimu Uebayashi[1]

Key words. Odor response—Olfactory nerve twig

Although there have been several investigations of odor-evoked neural responses in vertebrate olfactory receptor cells, the role of the olfactory receptor cells in the coding of olfactory odor information is poorly understood. It is impossible to isolate and record from a single nerve fiber, since the axon of the olfactory receptor cell is a nonmyelinated nerve fiber. The olfactory nerve twig method can be used to record the discharge of the olfactory receptor cells over a long period [1]. The nerve twig contains many olfactory receptor cell axons, and this technique records multiunit neural acitivity in response to odor stimulation in teased fine strands of olfactory nerves. Electrophysiological studies of olfactory receptor cells in the box turtle (a terrestrial animal) have shown that a response was generated in olfactory nerve twigs when odorous air and aqueous odorous solution were applied to the mucosa. In this study, we investigated air and aqueous odor responses in the olfactory receptor cells of the box turtle as revealed by olfactory nerve twig recordings. The air odorous stimuli from an air olfactometer were delivered to a nasal breathing chamber with a continuous flow of air over the nares at 100 cc/s. Concentrations were varied over the range 10^{-3} to unity vapor saturation at 20°C [1]. The aqueous odorous stimuli from the air olfactometer were introduced into the aqueous flow, with a gas-to-liquid odorant exchanger, just before the flow entered the nose. The device used was constructed from the body of a 50-ml volumetric pipette. This method was developed by Dr. D. Tucker [2]. Aliphatic n-fatty acids (C3-C10) were used as olfactory stimuli and n-amyl acetate was used as the standard for comparison. The box turtle's magnitude of response to n-amyl acetate odor stimuli increased with increasing odor concentration, results which were observed with both the air and aqueous odor stimulation experiments. The magnitude of response to aliphatic n-fatty acids also increased with increasing odor concentration, results which were also observed in both air and aqueous odor stimulation experiments. However, there was a significant difference between the experimental results for air and aqueous solution odor stimulation. In the air stimulation experiments, the relative response magnitude plotted against the carbon chain length of aliphatic n-fatty acids increased with an increase in carbon chain length up to C7 and then decreased; the response magnitude increased again slightly at C9, and then decreased with increased carbon chain length. A similar response profile has been reported in various other species [1]. In contrast, in the aqueous odor solution experiments, the relative response magnitude increased with increases carbon chain length up to C7, and then decreased. It is reasonable to assume that an increase in molecular size would be correlated with increased adsorption losses in water. Indeed, as the carbon chain length increases, the adsorption energy of the air-water surface decreases; this indicates that more energy is required to make a hole in water as the length of the carbon chain increases [3]. It has been reported that the newt, which lives on land, responds well to air odor stimulation, but does not respond to aqueous odor stimulation [4]. These phenomena depend on the characteristics of the olfactory cell.

Acknowledgments. This work was supported, in part, by a Grant-in-Aid for Research from Miyata.

References

1. Tonosaki K, Tucker D (1980) Olfactory receptor cell responses of dog and box turtle to aliphatic n-acetate and aliphatic n-fatty acids. Behav Neur Biol 35:187–199
2. Tucker D, Shibuya T (1965) Physiologic and pharmacologic study of olfactory receptors. Cold Spring Harbor Symp Quant Biol 30:207–215
3. Brink F, Posternack JM (1948) Thermodynamic analysis of the effectiveness of narcotics. J Cell Comp Physiol 32:211–233
4. Shibuya T, Takagi SF (1963) Electrical response and growth of olfactory cilia of the olfactory epithelium of the newt in water and on land. J Gen Physiol 47:71–82

[1] Department of Oral Physiology, School of Dentistry, Asahi University, 1851 Hozumi, Hozumi-cho, Motosu, Gifu, 501-02 Japan

Whole Tissue Voltage Clamp of Frog Olfactory Mucosa Using a Modified Ussing Chamber

Anna Maria Pisanelli and Krishna C. Persaud[1]

Key words. Voltage-clamp—Frog olfactory mucosa—Furosemide—Amiloride—K^+ ion currents

Introduction

In the olfactory receptor neuron, at least five membrane currents have been identified using patch-clamp techniques. These include two inward sodium currents, and three distinct outward potassium currents, some of which may be calcium modulated [1–4]. Odorant-induced currents in isolated cells may pass through nonselective monovalent cation channels, which may be blocked at high concentrations by amiloride, a drug whose action is normally to block sodium channels [5,6]. The interaction of the various ion channels in the homeostasis of the intect olfactory mucosa is still only poorly understood. The understanding of what are now termed "perireceptor events" in olfaction [7], how they interact with olfactory transduction events, the feedback mechanisms, and the molecular interactions between different cell types in the olfactory mucosa, are open questions.

Heck et al. [8] applied whole tissue voltage-clamp techniques to the olfactory epithelium of the bullfrog, and demonstrated that this was a viable technique for investigating the intact olfactory system, since drugs and ion channel blockers could be applied to both sides of the olfactory mucosa. Because of the vast increase in detailed knowledge of the molecular events in isolated cells of the olfactory mucosa, it is now an opportune time to re-apply whole tissue voltage-clamp methods to study the interaction of the various molecular events in the intact tissue.

Here, we present the design of a modified Ussing chamber adapted for the maintenance of olfactory mucosa, and for the delivery of a defined bolus of odorant solution to the ciliated side of the mucosa. This system has been used to investigate the effects of the drugs furosemide and amiloride on basal and odorant-stimulated olfactory mucosa, in order to distinguish the contributions of the various ion currents to the intact system.

Materials and Methods

Furosemide, amiloride, and the constituents of amphibian Ringer's solution were obtained from Sigma (St. Louis, Mo.) *Rana temporaria* were used for these experiments, and the optimum amphibian Ringer's solution was found to consist of 112 mM NaCl, 2.5 mM KCl, 2.5 mM $CaCl_2$, 5 mM $MgCl_2$, 1.1 mM Na_2HPO_4, and 0.4 mM NaH_2PO_4, pH 7.3, at 22°C, when aerated. This is a slight modification of Ringer's solution previously used with the American bullfrog *Rana catesbiana*. For some experiments, KCl was substituted with CsCl, and for other experiments a calcium-free Ringers solution was used, the osmolarity of the solution being maintained by a corresponding increase in NaCl concentration. Ringer's solution bathing the submucosal side of the mucosa also contained 10 mM glucose. 1,8-Cineole (10^{-4} M) was dissolved in Ringer's solution and this was presented as a bolus to the ciliated side of the mucosa in order to record the odorant-induced short-circuit currents. Experimental procedures were as described in [9,10].

Results and Discussion

Design of an Adapted Ussing Chamber

The ciliated side of the mucosa needed to be bathed with a continuous flow of solution, into which a bolus of odorant solution could be introduced as required. Several practical problems were considered, and as a result, a vertical design was produced, illustrated in Fig. 1. The top chamber consisted of a conically tapered flow chamber with a volume of 0.5 ml. Omnifit (Cambridge, UK) fast protein liquid chromatography (FPLC) type connectors were threaded into both chambers for the easy insertion of both agar bridge voltage sensing electrodes, current passing Ag/AgCl electrodes, and the solution inlet and outlet connectors, which consisted of Pharmacia (Pharmacia-Uppsala, Sweden) FPLC fittings that allowed a leak-proof system to be

[1] Department of Instrumentation and Analytical Science, UMIST, PO Box 88, Sackville Street, Manchester M60 1QD, England, UK

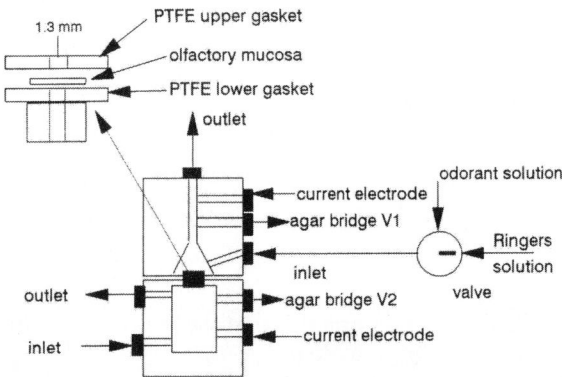

FIG. 1. Modified Ussing chamber designed for intact olfactory mucosa. *PTFE*, Polytetrafluroethylene

created. Special emphasis was placed on the inlet to the system, where a tapered capillary insert allowed the flow of solution to be directed onto the olfactory mucosa with no dead-volume dilution of incoming odorant. The bottom chamber was less critical, and consisted of a chamber with a volume of 2 ml, with solution input and output ports placed so that solution entered through the lower port and exited through the upper port, while bathing the lower surface of the olfactory mucosa. The excised olfactory mucosa was sandwiched between two polytetrafluoroethylene (PTFE) gaskets shown in Fig. 1 (inset), the working diameter being 1.3 mm. In assembling the system, the olfactory mucosa was centred on to the lower PTFE gasket, a silicone medical spray adhesive then being used to attach the lips of the upper gasket to the lower gasket, thus clamping the olfactory mucosa non-destructively. A cotton-wool wick inserted into the lower gasket below the mounted tissue prevented accumulation of air bubbles on the submucosal side of the tissue.

An high performance liquid chromatography (HPLC) pump was used to deliver a continuous,

non-pulsatile flow of 1.0 ml min^{-1} to the ciliated side of the olfactory mucosa. The submucosal side of the tissue was bathed in oxygenated Ringer's solution containing glucose. Odorant solution was switched into the upper chamber via a sample loop (volume 120 μl) via an FPLC sample-introduction valve made of PTFE.

The short-circuit current of the olfactory mucosa was measured as described previously [9,10]. This was digitized to a resolution of 16 bits via an integrating type analog-to-digital converter, a sampling interval of 1 s being adequate for the signals being recorded.

Effect of Furosemide

Previous work has indicated that furosemide has a profound reversible influence on the resting short-circuit current. We have investigated this effect more closely. In studying transporting epithelia with more than one cell type, it is difficult to define precisely the contributions of individual cell components.

The response to 10^{-4} M cineole was measured until reproducible responses were obtained. Then 120 μl of 10^{-4} M furosemide was introduced. A large outward current was measured (shown in Fig. 2), which recovered to baseline conditions after about 40 min, without affecting the amplitude of subsequent odorant responses. The response to odorant was also measured at the point where the maximum change in current was observed with furosemide; there was little effect observed on the odorant response. This would indicate that the ion transport affected by furosemide is independent of the olfactory transduction system. The change in short-circuit current observed was highly reproducible, and reversible, and at least three or four such experimental cycles could be performed with a single preparation. The effect of furosemide in many tissues is to block Na/Cl cotransport.

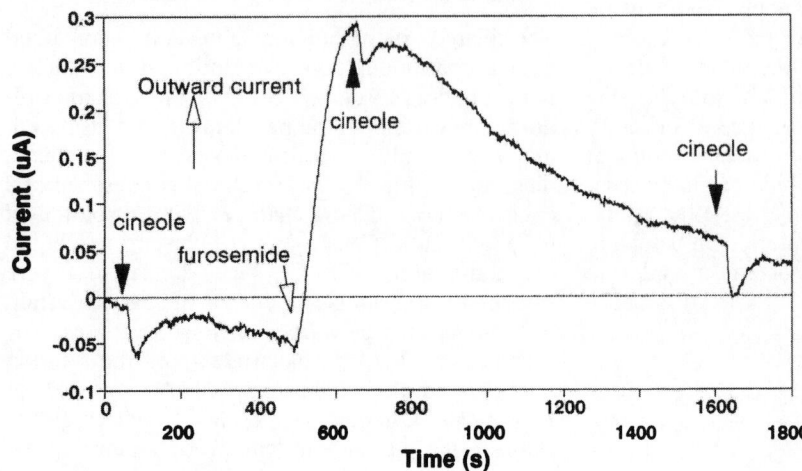

FIG. 2. Effect of furosemide on the basal and odorant stimulated currents. The figure shows the responses to the odorant 1,8-cineole (1×10^{-4} M), used as a control, and the response to a bolus of furosemide (1×10^{-4} M). An inward current is observed with odorant stimulation and an outward current with furosemide

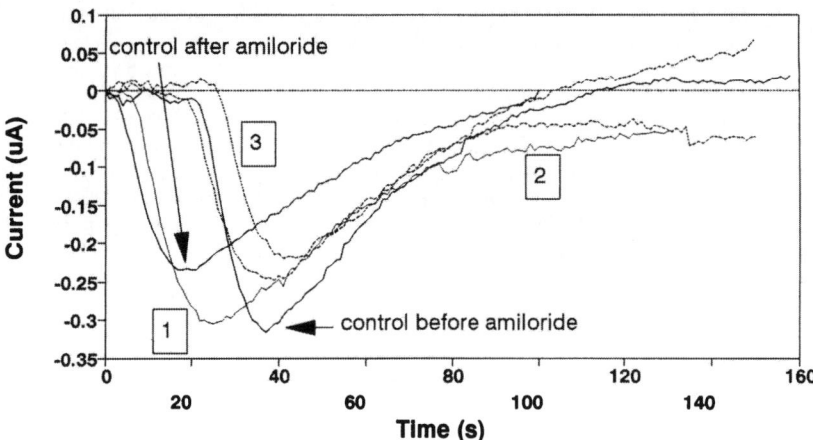

FIG. 3. Effect of amiloride on the odorant stimulated current. The figure shows the responses to the control odorant, 1,8-cineole, then to three consecutive applications of a mixture of amiloride and cineole (both 1×10^4 M) labelled *1–3*, followed by a control of cineole only. There is progressive diminution of the odorant response with amiloride followed by partial recovery

There may be some analogies here to the frog gastric mucosa, where the surface epithelial cells play a role in non-acidic chloride secretion, chloride being accumulated in the surface cells against its electrochemical potential difference at the serosal membrane [11]. The ready effect of furosemide applied to the apical side of the olfactory mucosa indicates that the reduction of the electrochemical gradient for chloride occurs on this side rather than on the submucosal side, and that this is independent of the odorant-stimulated ion transport events.

Effect of Amiloride

Patch-clamp experiments on isolated olfactory cells now indicate that the cation transport events mediated by odorants may allow more than one type of cation to pass through, and that these events are rather nonspecific in nature. Evidence also exists that they can be inhibited by amiloride at high concentrations [5]. Previous work on whole tissue voltage clamp of the olfactory mucosa had indeed indicated that an amiloride-blockable site existed, but was unable to assign this to a single channel type or a certain population of odorant-modulated channels. Amiloride seems to have little influence on the resting short-circuit current of the tissue. The present work indicates that the odorant-mediated channels are partially inhibited by concentrations greater than 10^{-4} M amiloride, but that 100% inhibition is difficult to achieve. Figure 3 shows odorant responses to 10^{-4} M cineole (control responses) and to mixtures of the same odorant concentration and amiloride (10^{-4} M) presented consecutively. Repeated presentations of amiloride together with the odorant caused a continuing diminution of the odorant response; this effect was reversible with time, as the subsequent control cineole response was of larger amplitude than the previous response to amiloride mixed with odorant. The presentation of amiloride at 2×10^{-4} M mixed with cineole, caused a much greater

diminution of the response, to less than 50% of the original amplitude, but this effect may not be completely reversible. The inhibition by amiloride may indicate that the ion channels do have a preferential specificity to sodium ions. However, the high concentration of amiloride necessary to achieve inhibition indicates low affinity of amiloride for these sites, and may corroborate the patch-clamp studies indicating low specificity of the odorant-gated ion channels.

Conclusion

We have optimized the design of a suitable apparatus for voltage-clamp experiments of the olfactory mucosa. The system is capable of maintaining stable responses from the frog olfactory mucosa for 3–5 h. We have applied the apparatus to experiments on *Rana temporaria*, a frog species which has not previously been used for whole tissue voltage clamp. The results of furosemide-blockable basal currents and amiloride-blockable odorant-stimulated currents are broadly consistent with results observed from former studies on *Rana catesbeiana*. The interdependence of different ion transport systems in the olfactory mucosa can now be studied using the techniques developed.

Acknowledgments. Dr. A.M. Pisanelli was supported by a grant from the Science and Engineering Research Council, UK.

References

1. Firestein S, Werblin FS (1987) Gated currents in isolated olfactory receptor neurons of the larval tiger salamander. Proc Natl Acad Sci USA 84:6292–6296
2. Trotier D (1986) A patch-clamp analysis of membrane currents in salamander olfactory receptor cells. Pfluegers Arch 407:589–595

3. Maue RA, Dionne VE (1987) Patch-clamp studies of isolated mouse olfactory receptor cilia. J Gen Physiol 90:95–126

4. Kurahashi T (1989) Activation by odorants of cation-selective conductance in the olfactory receptor cell isolated from the newt. J Physiol 419:177–192

5. Frings S, Lindemann B (1988) Odorant response of isolated olfactory receptor cells is blocked by amiloride. J Membr Biol 105:233–243

6. Li JH, de Sousa RC (1979) Inhibitory and stimulatory effects of amiloride analogues on sodium transport in frog skin. J Membr Biol 46:155–169

7. Getchell ML, Getchell TV (1984) β-Adrenergic regulation of the secretory granule content of acinar cells in olfactory glands of the salamander. J Comp Physiol A 155A:435–443

8. Heck GL, DeSimone JA, Getchell TV (1984) Evidence for electrogenic active ion transport across the frog olfactory mucosa in vitro. Chem Senses 9:273–283

9. Persaud KC, DeSimone JA, Getchell ML, Heck GL, Getchell TV (1987) Ion transport across the frog olfactory mucosa: The basal and odorant stimulated states. Biochim Biophys Acta 902:65–79

10. Persaud KC, Heck GL, DeSimone SK, Getchell, TV, DeSimone JA (1988) Ion transport across the frog olfactory mucosa: The action of cyclic nucleotides on the basal and odorant-stimulated states. Biochim Biophys Acta 944:49–62.

11. Curci S, Schettino T (1984) Effect of external sodium on intracellular chloride activity in the surface cells of frog gastric mucosa: Microelectrode studies. Pfluegers Arch 401:152–159

Mucosal Activity Patterns and Odorant Quality Perception

PAUL F. KENT[1,3], STEVEN L. YOUNGENTOB[2,3], and DAVID E. HORNUNG[2,3]

Key words. Electrophysiology—Behavior

One goal of sensory physiology has been to explore the relationship between the neurophysiologically recorded activity of sensory receptors and the perception of the stimuli they engender. That is, are various parameters of neural activity at this primary level of the nervous system (e.g., spatial pattern, temporal pattern, and magnitude) predictive of psychophysical data or is further processing through complex neural networks required? To examine this question for the olfactory system, the present study evaluated the relationship between odorant-induced inherent activity patterns [1] and odorant quality identification in the rat [2,3].

A cross modal association paradigm and an odorant identification confusion matrix task [2,3] were used to train five male Long-Evans Hooded rats to differentially report (i.e., identify) the odorants propanol, carvone, citral, propyl acetate, and ethylacetoacetate. Following acquisition training, the animals were tested 40 times, using a randomized blocks design. Stimuli were presented in blocks of five trials, with each block consisting of the five different odorants. The order of presentation within each block was randomized, and testing continued for 20 consecutive blocks of trials. The response to each of the 100 presentations of odorant (five different odorants × twenty blocks) were entered into a standard 5 × 5 confusion matrix. Data were recorded in the matrix such that the rows represented the stimuli presented and the columns represented the same odorants as alternative response choices. The entries in the cells of the matrix were the frequency with which the presentation of a particular odorant was responded to with each response alternative. Since the mathematical properties of confusion matrices are such that they may be analyzed using multidimensional scaling techniques (MDS), the results of the behavioral tests were subjected to an MDS analysis which established a two-dimensional perceptual odor space for the rats.

At the completion of testing, each animal was sacrificed and their odorant-induced mucosal inherent activity patterns were recorded, using a voltage-sensitive dye technique [4]. Using the dye, di-4-ANEPPS, we monitored fluorescence changes at 100 contiguous sites in a 10 × 10 photodiode array on the olfactory mucosa of each rat's septum and turbinate in response to the same five odorants used in behavioral training. Each odorant was humidified and puffed uniformly upon the entire olfactory mucosa for a duration of 500 ms, at a flow rate of 600 cc/min. The odorants were presented in randomized sequence at two concentrations, in a 3:1 ratio. For both the septum and the turbinates, although the entire mucosa responded to all of the five odorants tested, each stimulus produced a unique spatial pattern of activity with differing regions of maximal response. The physical location, in two dimensions, of a particular odorant's "hot spot" on the mucosa was determined by calculating the centroid (center of mass) of the mucosal inherent activity pattern.

The relative position of the five electrophysiologically determined "hot spots", in two dimensions, were then compared to their relative position in the behaviorally determined two-dimensional perceptual odor space. The results of these qualitative comparisons were striking in their apparent similarity. That is, the distances between the "hot spots" in the electrophysiological space paralleled those in the perceptual odor space, thereby suggesting a possible predictive relationship. To evaluate this provocative possibility, a regression analysis was performed in order to formally examine whether the observed electrophysiological data were predictive of the observed MDS data. The results of this analysis demonstrated a highly significant relationship ($P = $ Nil) between the relative position of an odorant's mucosal "hot spot" and the relative position of the same odorant in the perceptual odor space. These data suggest that the mucosal inherent activity patterns are preserved through further neural processing and that they, in turn, serve as the substrate for the perception of odorant quality.

Acknowledgments. This work was supported by National Institutes of Health grants DC-00220 and DC-00072.

Departments of [1] Neurology, [2] Physiology, and [3] The Clinical Olfactory Research Center, SUNY Health Science Center, Syracuse, NY 13210, USA

References

1. Kubie JL, MacKay-Sim A, Moulton DG (1980) Inherent spatial patterning of response to odorants in the salamander epithelium. In: H. Van der Starre (ed) Olfaction and taste VII. IRL, London, pp 163–166
2. Youngentob SL, Markert LM, Mozell MM, Hornung DE (1990) A method for establishing a five-odorant identification confusion matrix in rats. Physiol Behav 47:1053–1059
3. Youngentob SL, Markert LM, Hill TW, Matyas EP, Mozell MM (1991) Odorant identification in rats: An update. Physiol Behav 49:1293–1296
4. Kent PF, Mozell MM (1992) The recording of odorant-induced mucosal activity patterns with a voltage sensitive dye. J Neurophysiol 68:1804–1819

Olfactory Response to an Inorganic Metal Compound

Noriaki Kishi, Katsuyoshi Toriyama[1], Mitsuo Tonoike, and Takaaki Sato[2]

Key words. Olfactory event related potentials (ERPs)—Inorganic metal compound

Introduction

Inorganic metal compounds are generally thought of as odorless since they do not meet the criteria of odorific substances, which are those that have vapor pressure and dissolve in water or lipids. In this study, we measured the changes in cytosolic free Ca^{2+} concentration in isolated murine olfactory cells and the changes in olfactory event related potentials (ERPs) in humans, and analyzed these to determine whether an inorganic metal compound, e.g., aluminum hydroxide, $Al(OH)_3$, could be physiologically recognized as having an odor.

Response of Olfactory Cell

We used fura-2 to monitor the changes in the cytosolic free Ca^{2+} concentration ($[Ca^{2+}]_i$) in an olfactory cell isolated from a mouse using the tissue printing method. We conducted image analysis of the changes in the fluorescence intensity of the cell arising when $[Ca^{2+}]_i$ was stimulated. The measurement system consisted of an inverted microscope, an incident-light fluorescence device (DM400 dynamic mirror), and an electrically-driven filter switching device, all manufactured by Nikon (Tokyo, Japan), and a 150 W Xenon lamp, a SIT camera, and an image processor (ARGUS-100), all manufactured by Hamamatsu Photonics (Hamamatsu, Japan). We used $Al(OH)_3$ (22 μM) as the inorganic metal compound which served as the stimulus. In the isolated olfactory receptor neuron, there were no changes when only Ca-free Ringer's solution was used, whereas, when $Al(OH)_3$ was applied, we observed transient increases in $[Ca^{2+}]_i$ in the soma (10 of 85 cells).

Olfactory Event Related Potentials (ERPs)

The subject was a healthy 30-year-old man with a normal olfactory sense. The experiments were conducted while he was in an odorless room with his eyes open and in a tranquil pose. We supplied the stimuli by the blast method in synchronization with the subject's respiration cycles. The stimuli, each being supplied for 100 ms, were presented randomly in two to nine respirations. Each wave of ERPs summed 20 times was recorded using electrodes placed at 16 locations, based on the international electroencephalograph (EEG) 10–20 system. For the measurement we used an EEG-1A97 electroencephalograph manufactured by NEC San-ei (Tokyo, Japan) and an analysis processor, EP-WORKS, manufactured by Kissei-Comtech (Matsumoto, Japan). The stimuli were given under four conditions: humidified air ventilated through the inorganic metal compound, dry air ventilated through the inorganic metal compound, humidified air only, and dry air only.

When the inorganic metal compound was added to the stimulus, we observed an electropositive fluctuation with an amplitude of 10 μV from the baseline 568 ms after the stimulation. When no inorganic metal compound was added, we did not observe a similar amplitude.

Discussion

The increase in $[Ca^{2+}]_i$ indicates a response to the stimulus. From this, we consider that the increase in $[Ca^{2+}]_i$ observed when the inorganic metal compound was applied directly to the olfactory cell indicates that the cell perceived this compound as an odor and responded accordingly. The waveform fluctuation obtained from the olfactory ERPs in the human subject is the cognitive response element, P 300, which appears when a given stimulus is recognized physiologically. P 300 was detected only when humidified air was ventilated through the inorganic metal compound, probably because some material was ionized, due to the moisture, before reaching the olfactory epithelium. The subject did not notice any odor in the stimulus containing no inorganic metal compound, but reported a dusty odor in the humidified air ventilated through the inorganic metal compound. These results support our proposal that even an inorganic metal compound is recognized as an odor if it reaches the olfactory cell. In time we will investigate responses to other inorganic metal compounds.

[1] System Development Engineering Department, Nippondenso Co., Ltd., 1-1 Showa-machi, Kariya, Aichi, 448, Japan
[2] Bioelectronic Interface Section, Life Electronics Research Center, Electrotechnical Laboratory, 3-11-46, Nakoji, Amagasaki, Hyogo, 661 Japan

4. Genetics and Molecular Neurobiology of Chemoreception

Ligand Recognition Mechanism of Bacterial Chemoreceptors Revealed by Site-Specific Sulfhydryl Modification

SUMIKO GOMI, LAN LEE, TOMONORI IWAMA, YASUO IMAE, and IKURO KAWAGISHI[1]

Key words. Aspartate receptor—Chemoreception—*Escherichia coli*—Ligand recognition—Site-directed mutagenesis—Sulfhydryl modification

Introduction

All organisms are forced to live in an ever-changing environment and have specialized systems for responding to it. Even bacteria can sense various chemical stimuli and can migrate in a more "favorable" direction, a phenomenon known as chemotaxis [1,2].

In *Escherichia coli* such stimuli are detected by four homologous transmembrane chemoreceptors: Tsr for serine, Tar for aspartate and maltose, Trg for ribose and galactose, and Tap for dipeptides. They share a common structure: The *N*-terminal periplasmic domain is responsible for ligand-binding, whereas the highly homologous *C*-terminal cytoplasmic domain interacts with the cytoplasmic chemotaxis proteins, thereby eliciting signals [3,4]. We investigated how Tar and Tsr recognize their specific ligands.

Materials and Methods

Bacterial Strains, Phages, Plasmids

The *E. coli* strain HCB339 [5] and the plasmids carrying *tar* [6] and *tsr* [7], as well as other strains and vectors, were described elsewhere [8–10].

Site-Directed Mutagenesis

Site-directed mutagenesis was carried out according to Kunkel et al. [11].

Sulfhydryl Modification and Chemotaxis Assays

The procedures for in vivo modification with 5,5'-dithiobis-2-nitrobenzoic acid (DTNB) and *N*-

ethylmaleimide (NEM) [8] and the temporal stimulation assay [10] were described previously. The magnitude of the aspartate response was defined as the percent increase in smooth-swimming cells after the addition of aspartate.

Results

Mutated Chemoreceptors with Altered Recognition of Specific Ligands

Escherichia coli tsr mutants with altered serine sensing were isolated by Hedblom and Adler [12]. We established that the amino acid substitutions occurred at two sites, Arg-64 and Thr-156 [9]. Various substitutions caused various levels of reduction in serine sensing, whereas repellent responses were not affected (Table 1).

Despite their ligand specificities, Tsr and Tar have significant homology: both Arg-64 and Thr-156 in Tsr are conserved in Tar as Arg-64 and Thr-154, respectively. The Arg-64 to Cys mutation in Tar was reported to cause about five-order reduction in aspartate sensing but to cause only a slight reduction in maltose sensing [13–15]. We confirmed it by showing a reduced aspartate response (Table 1) and normal repellent response. Various substitutions of Thr-154 [10] also caused various levels of reductions in aspartate sensing without changing the repellent response (Table 1). It is therefore concluded that the two common residues in the periplasmic domain of Tar and Tsr are specifically involved in the specific recognition of these amino acid attractants.

Site-directed Sulfhydryl Modification of the Ligand-Binding Sites

Because Tar contains no Cys, it may be possible to specifically modify an introduced Cys with sulfhydryl (SH)-modifying reagents such as NEM and DTNB. Therefore we replaced Arg-64 and Thr-154 as well as Arg-69, which is also implicated in aspartate recognition [16], with Cys in order to modify the resultant residues.

Address correspondence to: I. Kawagishi
[1] Department of Molecular Biology, Faculty of Science, Nagoya University, Chikusa-ku, Nagoya, 464-01 Japan

TABLE 1. Mutated chemoreceptors with altered recognition for a specific ligand.

Receptor	Residue	Replacement	Apparent affinity for serine and aspartate[a]
Tsr	Arg-64	None	1×10^{-6} M for serine
		Cys	1×10^{-1} M
		His	1×10^{-2} M
	Thr-156	None	1×10^{-6} M for serine
		Ile	1×10^{-4} M
		Pro	1×10^{-1} M
Tar	Arg-64	None	2×10^{-7} M for aspartate
		Cys	5×10^{-3} M
	Thr-154	None	2×10^{-7} M for aspartate
		Ser	1×10^{-6} M
		Cys	5×10^{-6} M
		Ile	1×10^{-4} M
		Pro	1×10^{-1} M

[a]Chemoresponse was measured using HCB339 cells with wild-type or various mutant receptors. The value indicates the concentration of serine or aspartate required to induce a response in about 50% of the cells.

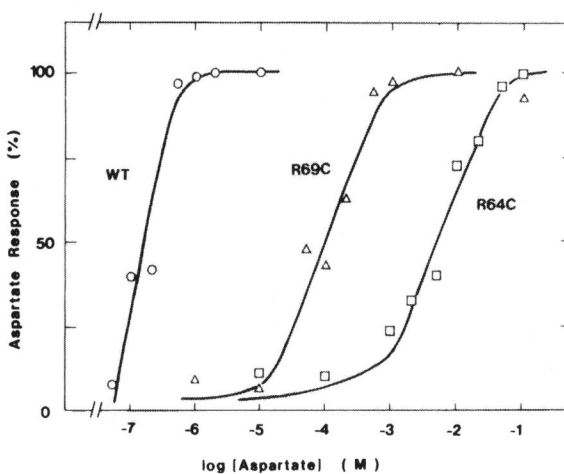

FIG. 1. Aspartate-sensing abilities of the mutant Tars. HCB339 cells with Tar (○), Tar-R64C (□), and Tar-R69C (△), pretreated with 1 M glycerol, were stimulated with various concentrations of aspartate. The change in the fraction of smooth-swimming cells after 30 s was measured at 25°C

Arginine-64 and Arginine-69

Mutant Tars with Cys substitutions at Arg-64 and Arg-69 were designated Tar-R64C and Tar-R69C, respectively. These receptors were expressed in the receptorless strain HCB339 [5] and were analyzed by temporal stimulation assays (Fig. 1). All the transformant cells responded normally to 1 M glycerol, indicating that the mutant receptors retain the ability to generate chemotactic signals. The cells with Tar-R64C and Tar-R69C required about 5 mM

and 0.1 mM aspartate for attractant response in 50% of the population, respectively, whereas 0.2 μM aspartate was enough to induce 50% response for the cells with the wild-type Tar. The dose-response curves for the mutant Tars, however, are almost parallel with that for the wild-type Tar, suggesting that only the affinity for aspartate is altered by the Cys substitution.

The cells with Tar-R64C or Tar-R64C were subjected to SH modification. Their aspartate-sensing ability was quickly lost by 50 μM NEM or 500 μM DTNB, whereas normal ability was retained in the cells with the wild-type Tar even after 60 min of treatment (Fig. 2). In the case of DTNB treatment, addition of 1,4-dithiothreitol (DTT) completely restored aspartate-sensing ability, as expected. Furthermore, repellent response to 1 M glycerol was normal even after the aspartate response was mostly lost (Fig. 2). Thus, covalent modification of the SH group of Cys-64 or Cys-69 results in a specific defect in aspartate sensing.

When the cells with Tar-R64C or Tar-R69C were preincubated with 0.1 M aspartate, inactivation of aspartate sensing by NEM was drastically delayed (Fig. 3). 0.1 M serine, however, had no effect. These results indicate that aspartate specifically protects both Cys-64 and Cys-69 from NEM modification. The level of protection is dependent on the concentration of aspartate (Fig. 4). The concentration required for 50% protection (at least 10 mM) was much higher than that for 50% chemotactic response. Thus it is plausible that the protection occurs as a result of the competition between NEM and aspartate at Cys-64 or Cys-69.

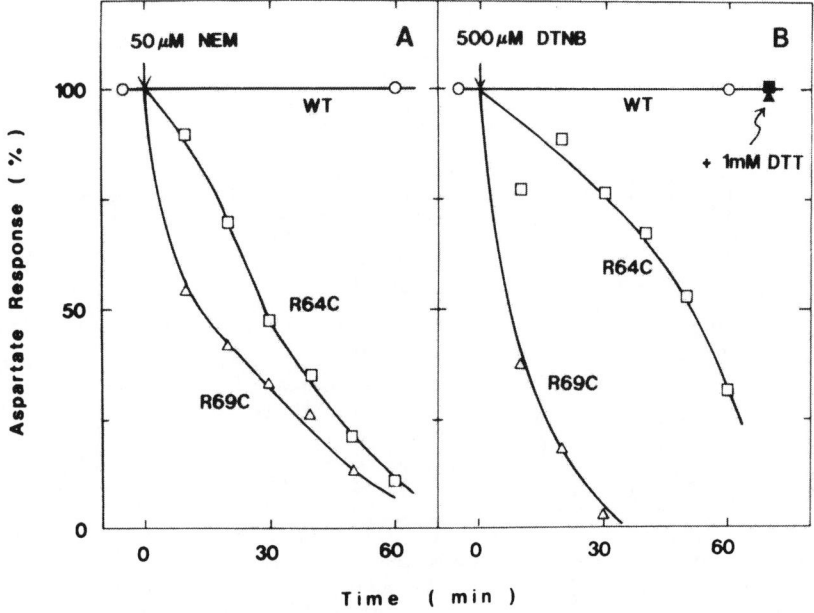

Fig. 2. Inactivation by NEM and DTNB of aspartate sensing of the Cys-replaced mutant Tar. At the time indicated, $50\,\mu M$ NEM (A) or $500\,\mu M$ DTNB (B) was added to HCB339 cells with Tar ($\bigcirc$), Tar-R64C ($\square$), and Tar-R69 ($\triangle$) suspended in motility medium (pH 7.6). At intervals a small aliquot was taken and pretreated with 1 M glycerol, and the response to 1 mM aspartate was measured. After 60 min of DTNB treatment, 1 mM DTT was added and the aspartate response was measured (*closed symbols*)

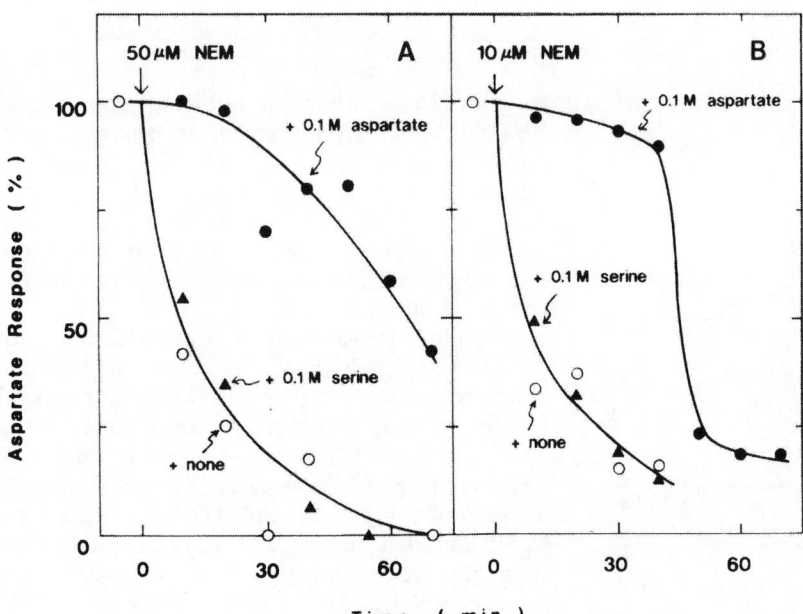

Fig. 3. Protective effect of aspartate on the NEM modification of Cys-64 and Cys-69. At the time indicated, either $50\,\mu M$ (A) or $10\,\mu M$ NEM (B) was added to HCB339 cells with Tar-R64C (A) or Tar-R69C (B) in motility medium (pH 7.6) supplemented with (or not) 10 mM aspartate or 10 mM serine. Aspartate response was measured after washing out aspartate or serine. $\bigcirc$, none; $\bullet$, 10 mM aspartate; $\blacktriangle$, 10 mM serine

Threonine-154

Similar results were obtained for the Thr-154 to Cys mutant Tar, Tar-T154C [8]. The mutation caused only about a two-order reduction in aspartate response (Table 1). On a semisolid agar plate containing 0.1 mM aspartate, the cells with Tar-T154C produced a comparable size of swarm to the cells with the wild-type Tar. Even in the presence of $50\,\mu M$ DTNB, the cells with the wild-type Tar swarmed normally. The cells with Tar-T154C, however, produced no swarm. Incubation of the cell suspension with $25\,\mu M$ DTNB resulted in specific

loss of aspartate response within 20 min without affecting repellent responses to 0.5 M glycerol and 0.5 mM Ni^{2+}. Aspartate response was completely restored by 0.2 mM DTT.

10 mM aspartate, but not 10 mM serine, drastically slowed the inhibition, indicating that aspartate protected Cys-154 from DTNB modification. While $2\,\mu M$ aspartate induced a 50% chemotactic response, a much higher concentration of aspartate (about 1 mM) was required for 50% protection. Thus the protection occurs as a result of the competition between DTNB and aspartate at Cys-154.

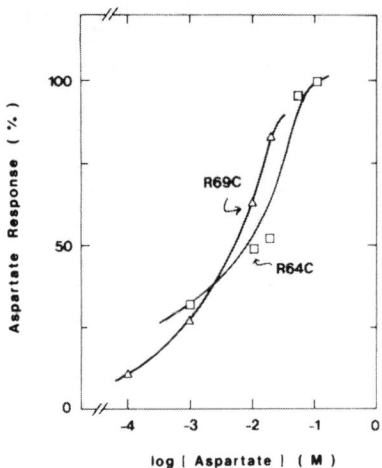

FIG. 4. Concentration dependence of aspartate on the protection of Cys-64 and Cys-69 from the NEM modification. HCB339 cells with Tar-R64C (□) or Tar-R69C (△), suspended in motility medium (pH 7.6) supplemented with various concentrations of aspartate, were incubated with 50 μM (for Tar-R64C) or 10 μM NEM (for Tar-R69C) for 20 min. After washing, the aspartate response was measured

Discussion

Our results indicate that Cys-64, Cys-69, and Cys-154 are located in the hydrophilic region to which the SH reagents have easy access, and that their modification does not induce marked distortion in the overall conformation. Cys-64 and Cys-69 required higher concentrations of DTNB for modification than Cys-154. Therefore the former two could be positioned a little inside the molecule.

The protective effect of aspartate indicates that all of these Cys residues—hence Arg-64, Arg-69, and Thr-154 in Tar—constitute the aspartate-binding site. Because Arg has a positive charge and Thr has a hydroxyl group, it is reasonable to assume that the α-carboxyl group and the amino group of aspartate are recognized by Arg and Thr, respectively.

Milburn et al. [17] and Yeh et al. [18] determined the three-dimensional structure of the periplasmic portion of *Salmonella typhimurium* Tar, which is highly homologous with *E. coli* Tar. The results indicate that Arg-64/Arg-69 and Thr-154 interact with the α-carboxyl group and the amino group of aspartate, respectively. As for the location and function of these residues, our in vivo results with *E. coli* Tar are consistent with their crystallographic data of the *N*-terminal fragment of *S. typhimurium* Tar.

The crystallographic data show that Arg-69 also interacts with the β-carboxyl group of aspartate. Beause serine has a β-hydroxyl group instead of a β-carboxyl group, Arg-69 in Tsr may function differently. The mechanism underlying the precise

discrimination of aspartate and serine remains to be elucidated.

Summary

Escherichia coli detects various stimuli with four closely related transmembrane chemoreceptors. Two conserved residues, Arg-64 and Thr-154 (Thr-156 in the serine receptor), in the aspartate and serine chemoreceptors were found to be specifically involved in recognition of the amino acid attractants. The roles of these residues and Arg-69 were further analyzed by substituting them with Cys and treating the mutated receptors with SH-modifying reagents. In vivo modification of the introduced Cys residues caused specific defects in aspartate recognition. An excess of aspartate specifically protected them from SH modification. It is concluded that the three residues are located in the ligand-binding site.

Acknowledgments. We thank Drs. J. Adler, H.C. Berg, J. Gebert, and M.I. Simon for providing us with bacterial strains and plasmids; Drs. N. Kamo and M. Homma for critically reading the manuscript; Mr. T. Nara for helpful discussions; and Ms. N. Nishioka for drawing.

This article is dedicated, with deep sorrow, respect, and affection, to the late Professor Y. Imae, who died of a cerebral hemorrhage on July 2, 1993.

References

1. Bourret RB, Borkovich KA, Simon MI (1991) Signal transduction pathways involving protein phosphorylation in prokaryotes. Annu Rev Biochem 60:401–441
2. Stock JB, Lukat GS, Stock AM (1991) Bacterial chemotaxis and the molecular logic of intracellular signal transduction network. Annu Rev Biophys Chem 20:109–136
3. Hazelbauer GL, Yaghmai R, Burrows GG, et al. (1990) Transducers: transmembrane receptor proteins involved in bacterial chemotaxis. In: Armitage JP, Lackie JM (eds) Society for General Microbiology Symposium, vol 46. Biology of the chemotactic response. Cambridge University Press, Cambridge, pp 107–134
4. Hazelbauer GL (1992) Bacterial chemoreceptors. Curr Opin Struct Biol 2:505–510
5. Wolff AJ, Conley MP, Kramer TJ, Berg HC (1987) Reconstitution of signaling in bacterial chemotaxis. J Bacteriol 169:1878–1885
6. Krikos A, Conley MP, Boyd A, Berg HC, Simon MI (1985) Chimeric chemosensory transducers of Escherichia coli. Proc Natl Acad Sci USA 82:1326–1330
7. Manoil C, Bechwith J (1986) A genetic approach to analyzing membrane protein topology. Science 233:1403–1408
8. Gomi S, Lee L, Iwama T, Imae Y (1993) Inhibition of aspartate chemotaxis of Escherichia coli by site-

directed sulfhydryl modification of the receptor. J Biochem 133:208–213

9. Lee L, Mizuno T, Imae Y (1988) Thermosensing properties of Escherichia coli tsr mutants defective in serine chemoreception. J Bacteriol 170:4769–4774

10. Lee L, Imae Y (1990) Role of threonine residue 154 in ligand recognition of the Tar chemoreceptor in Escherichia coli. J Bacteriol 172:377–382

11. Kunkel TA, Robert JD, Zakour RA (1987) Rapid and efficient site-specific mutagenesis without phenotype selection. Methods Enzymol 154:367–382

12. Hedblom ML, Adler J (1980) Genetic and biochemical properties of Escherichia coli mutants with defects in serine chemotaxis. J Bacteriol 144:1048–1060

13. Wolff C, Parkinson JS (1988) Aspartate taxis mutants of the Escherichia coli Tar chemoreceptor. J Bacteriol 170:4509–4515

14. Kossmann M, Wolff C, Manson MD (1988) Maltose chemoreceptor of Escherichia coli: interaction of maltose-binding protein and the Tar signal transducer. J Bacteriol 170:4516–4521

15. Mowbray ML, Koshland DE Jr (1990) Mutations in the aspartate receptor of Escherichia coli which affect aspartate binding. J Biol Chem 265:15638–15643

16. Gardina P, Conway C, Kossmann M, Manson M (1992) Aspartate and maltose-binding protein interact with adjacent sites in the Tar chemotactic signal transducer of Escherichia coli. J Bacteriol 174:1528–1536

17. Milburn MV, Prive GG, Milligan DL, Scott WG, Yeh J, Jancarik J, Koshland DE Jr, Kim SH (1991) Three-dimensional structures of the ligand-binding domain of the bacterial aspartate receptor with and without a ligand. Science 254:1342–1347

18. Yeh JI, Biemann H-P, Pandit J, Koshland DE Jr, Kim S-H (1993) The three-dimensional structure of the ligand-binding domain of a wild-type bacterial chemotaxis receptor: structural comparison to the cross-linked mutant form. J Biol Chem 268:9787–9792

Genetics of Taste in *Drosophila*

Veronica Rodrigues, Maneesha Inamdar, Krishanu Ray, and K. VijayRaghavan[1]

Key words. Chemosensory—*Drosophila*—Taste mutants—*Scalloped* gene—Taste sensilla

Introduction

An organism's response to the environment requires a neural circuit accurately programmed to detect, integrate, and respond to a stimulus. The fruitfly *Drosophila melanogaster* exhibits highly sensitive and stereotyped responses to a variety of taste stimuli. Subtle defects at any level of the pathway are likely to be amplified and hence detectable as a behavioral abnormality [1]. Mutant animals with defects in taste behavior not only allow analysis of the taste pathway per se but may provide a handle to studies on the nervous system in general. Chemosensory assays most often identify partial loss of function or hypomorphic mutations in these genes. The phenotype resulting from the complete loss of function can be studied by generating alleles at the locus and examining the phenotypes in stages and tissues where the gene is expressed.

The taste organs of *Drosophila* are sensilla located on the proboscis, legs, and wings. There are about 36 labellar chemosensory bristles located at defined positions on the proboscis of *D. melanogaster*. The structure [2] and function [3] of individual sensilla have been described. A typical labellar bristle is innervated by four chemosensory and one mechanosensory neuron. Each of the chemosensory neurons is functionally distinct, one responding to sugars, one to water, and two to salts [4]. Each of these neurons project to different regions in the subesophageal ganglion in the brain [5]. The mechanisms that determine the specificity of these neurons is not understood.

In this chapter we describe strategies to identify genes involved in the development and functioning of the taste pathway in *Drosophila*. We use behavioral assays to isolate and characterize mutants generated by insertional mutagenesis. The "enhancer-trap" approach [6] allows rapid facility for isolating cDNAs that represent mRNA coding regions in the vicinity of the insertion of the P-element. The β-galactosidase reporter in the P-element and mRNA localization by in situ hybridizations allow detection of the pattern of expression of the gene of interest. Sequencing of the cDNA and production of antibodies against the native molecule allow us to follow the spatial and temporal expression of the molecule. Newly described mosaic methods allow us to examine the role of the gene product specifically in the chemosensory system, all other parts of the fly being normal [7]. We draw examples from our work on the gene *scalloped* (*sd*) to illustrate our approach.

Materials and Methods

Feeding Preference Test

The feeding preference test was designed by Tanimura et al. [8]. The stimulus to be tested is dissolved in 1% agar and placed in alternate wells of a microtiter plate. When the response to an attractant is to be assayed, the remaining wells contained 0.2% of a food dye, carmoisine red, in 1% agar. For repellents the stimulus is placed in the same wells as the food dye, and the remaining wells contained agar solution alone. Flies, 2 to 4 days old, starved in humid chambers for 18 h are introduced into each test plate and left undisturbed in the dark for 1 h. Flies are subsequently immobilized by cooling, and the color of their abdomens is scored under a dissection microscope. The acceptance of a stimulus is measured by the fraction of flies with uncolored abdomens. The tolerance response to a repellent is given by the fraction of colored flies. Wild-type and mutant flies are tested in parallel under identical conditions. Appropriate controls showed that it is the taste stimulus and not the agar or the food color itself that determines the result [9].

Proboscis Extension Test

Flies 2 to 4 days old, starved in moist chambers for 15 h prior to the proboscis extension test, are immobilized by cooling on ice and fixed ventral side up on a microscope slide using myristic acid wax (m.p. = 58.5°C). After a recovery period of 3 hours, flies are allowed to drink deionized water until satiated. The tarsus of the first leg is touched with a drop of the stimulus from a microsysringe needle

---

[1] Molecular Biology Unit, Tata Institute of Fundamental Research, Homi Bhabha Road, Bombay 400005, India

and the extension of the proboscis scored [10]. Each fly is given five trials and taken as a responder if it extended its proboscis during at least three trials. When the response of mutants is being evaluated, the wild type was tested in parallel under identical conditions.

Strains

The Canton-S strain was used as a wild-type strain in most experiments. The allele used in this study $sd^{(Pry+ETX4)}$ had an P element insertion in the sd locus [11].

Results

Behavioral Analysis

The feeding preference assay and the proboscis extension test allow analysis of the gustatory behavior of populations of flies to a variety of chemical stimuli. Flies are attracted to sugars and low-concentration sodium chloride and are repelled by most other salts at high concentrations. Both tests yielded qualitatively similar results. The behavioral tests we used allow the detection of strains with altered thresholds of responses or changes in sensitivity (Fig. 1A,B). Figure 1 shows the response of a mutant in the sd locus [11]. Adults of this strain showed altered responses to sugars and salts (Fig. 1). It is possible to measure taste responses of larvae in assays where animals are presented a choice between control agar and agar containing the stimulus. Wild-type larvae are attracted by sugars and low-concentration sodium chloride and are repelled by high concentrations of sodium chloride and potassium chloride [10]. The larvae of sd mutants showed reduced sensitivities to sodium chloride (data not shown).

Behavioral defects to taste stimuli could, in principle, arise by a lesion at any step of the pathway, from stimulus detection to integration and the motor response. It is possible to record the electrophysiological activity of the sensory neurons [3]. Mutants that show normal peripheral physiology but defects in behavior are likely to be affected at a more central level of the taste pathway. Electrophysiological recordings from sd mutants have shown that the peripheral neurons have normal responses to stimuli.

Role of sd in the Development of the Taste Pathway

The sd mRNAs are expressed in a dynamic fashion in the developing *Drosophila* nervous system, suggesting that sd plays a role in the development of this system [12]. Several alleles at the sd locus have been identified, some of which mutate to lethality and show defects in the development of the peripheral

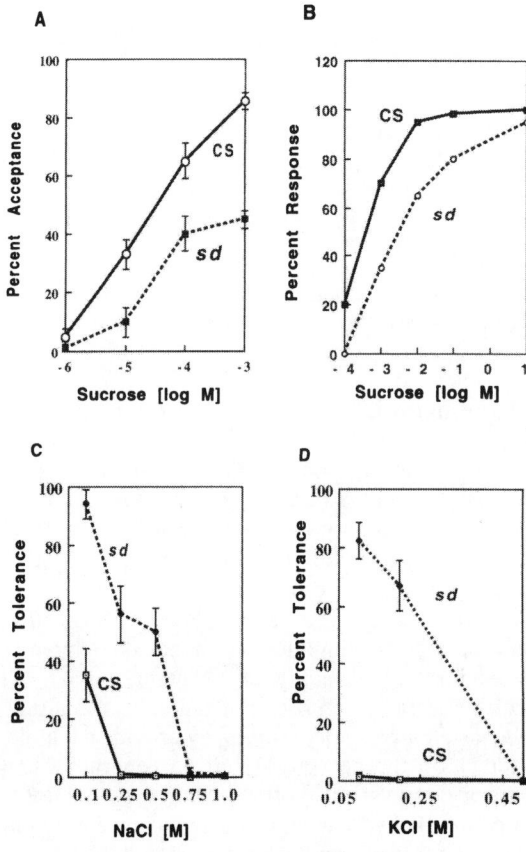

FIG. 1. Behavioral responses of wild-type and sd adults to taste stimuli. **A** Dose-response curves of sd and CS flies, to sucrose. **B** Proboscis extension response of wild-type (CS) and sd to sucrose. **C,D** Tolerance of CS and sd flies to NaCl (C) and KCl (D). **A,C,D** Points represent the means and standard deviations for 10 assays for each genotype at each concentration. **B** Each point represents the average response of 50 flies

as well as central nervous system of the embryo. The defects in chemosensory behavior in larvae and adults in hypomorphic alleles could reflect developmental defects in the taste pathways. Anatomical inspection of the nervous system of hypomorphic alleles did not reveal significant defects. The effects of nervous system clones null for sd gene product in the adult brain needs to be examined.

The sd gene has been cloned, and the sequence of the putative *scalloped* (SD) protein shows a high degree of homology to the human transcription factor TEF-1 [12]. This conserved motif defines a novel DNA binding domain denoted TEA [12]. If sd is indeed a transcription factor, its expression in the context of the developing taste circuits would give important insights into its function.

The development of the gustatory pathways in *Drosophila* and other dipterans is only just beginning to be investigated [13–16]. The adult labellar sensilla arise from a pair of labial discs during pupation. We have followed the neurogenesis

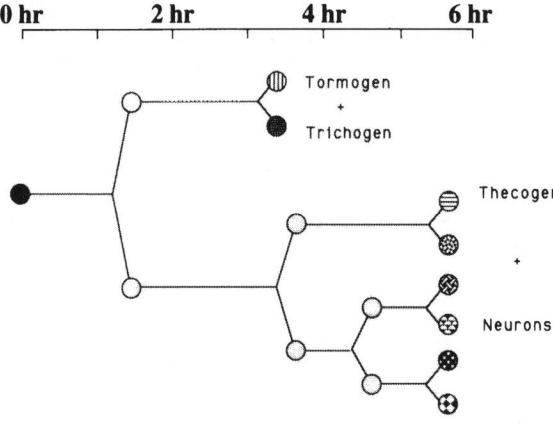

FIG. 2. Lineage of a labellar taste sensillum. Staged pupae were injected with brdU. Animals were allowed to develop until 26 h into pupation, and the pupal proboscis were then stained with an antibody to brdU. The identity of the cells was confirmed by counterstaining preparations with the neuron-specific marker mab22C10. The diagram represents data from 50 pupa injected at various times after pupation

in these discs using neuron-specific markers: cell-marking techniques using enhancer-trap lines and bromodeoxyuridine (brdU) uptake.

The cells that form a sense organ are derived from a single precursor cell called a sensory mother cell (SMC) [17,18]. The SMCs for the labellar sensilla arise in three successive waves of specification [16]. This kind of specification is likely to serve as a mechanism for spacing of the sensilla on the surface of the epidermis [19]. A single SMC divides to produce the five neurons and three support cells comprising a single sensillum. Data from brdU experiments are summarized in Fig. 2. The mechanisms by which cells differentiate to achieve distinct functional roles is not understood.

Drawing parallels from work on the visual system in *Drosophila*, we suggest that the differentiation of specific cell types depends on the interaction of cells with each other and with their environment. Several of the genes involved in these processes have been shown to encode surface ligands and receptors, other molecules involved in intracellular signaling, and finally transcriptional factors that are the last step in the pathway that responds to the external signal [20,21].

Discussion

The chemosensory system performs perceptual tasks that are in many ways distinct from any other type of sensory system. The chemical senses require a high degree of sensitivity and the ability to discriminate between similar chemicals [22]. Studies on how these neural systems are built and how they function are likely to reveal novel developmental mechanisms. The genetic approach to the analysis of the taste system involves the isolation of mutants in behavioral screens, cloning the genes, and studying the role of the gene products in the developing taste pathway. The cells in a single taste sensillum are related by lineage but acquire distinct functional specificities through the action of different genes. The genes that specify the early events in specification of the SMC are shared with the other sensory sensilla elsewhere on the body of the adult and in the embryo. They include members of the proneural genes: *daughterless*, the *achaete-scute* complex, and the neurogenic genes *Notch* and *Delta* [19]. The genes that act in the terminal differentiation of the chemosensory cells are yet to be identified. Hypomorphic alleles of genes specifying these steps are likely to exhibit taste defects.

Most of the taste mutants isolated so far are not affected in the physiology of the peripheral neurons and by that criterion are speculated to act at a more central level in the processing [23,24]. The products specified by at least some of these genes are likely to play important roles in the development of the taste pathways. Studies on the phenotypes themselves have in some cases provided clues to how information is processed in the taste pathway. The genetic approach combined with behavioral and physiological analysis provides a means to study the taste pathway at the cellular and molecular level and to understand the complex chemosensory behavior of the animal.

References

1. VijayRaghavan K, Kaur J, Paranjape J, Rodrigues, V (1992) The *east* gene of *Drosophila melanogaster* is expressed in the developing nervous system and is required for normal olfactory and gustatory responses of the adult. Dev Biol 154:23–36
2. Nayak SV, Singh RN (1983) Sensilla on the tarsal segments and mouthparts of adult *Drosophila melanogaster* Meigen (Diptera: Drosophilidae). Int J Insect Morphol Embryol 12:115–129
3. Fujishiro N, Kijima H, Morita H (1984) Impulse frequency and action potential amplitude in labellar chemosensory neurons of *Drosophila melanogaster*. J Ins Physiol 30:317–325
4. Rodrigues V, Siddiqi O (1978) Genetic analysis of chemosensory pathway. Proc Ind Acad Sci [B] 87: 147–160
5. Shanbhag S, Singh RN (1992) Functional implications of the projections of neurons from individual labellar sensillum of *Drosophila melanogaster* as revealed by the neuronal-marker horseradish peroxidase. Cell Tissue Res 262:273–282
6. O'Kane C, Gehring WJ (1987) Detection in-situ of genomic regulatory elements in *Drosophila*. Proc Natl Acad Sci USA 80:7641–7645
7. Xu T, Rubin G (1993) Analysis of genetic mosaics in developing and adult *Drosophila* tissues. Development 117:1223–1237

8. Tanimura T, Isono K, Shimada I (1992) Genetic dimorphism in taste sensitivity to trehalose in *Drosophila melanogaster*. J Comp Physiol 147:433–437

9. Balakrishnan R, Rodrigues V (1991) The *Shaker* and *shaking-B* genes specify elements in the processing of gustatory information in *Drosophila melanogaster*. J Exp Biol 157:161–181

10. Rodrigues V, Sathe S, Balakrishnan R, Pinto L, Siddiqi O (1991) Closely linked lesions in a region of the X-chromosome affect central and peripheral steps in gustatory processing in *Drosophila*. Mol Gen Genet 226:265–276

11. Inamdar M, VijayRaghavan K, Rodrigues V (1993, in press) The *Drosophila* homolog of the human transcription factor TEF-1, *scalloped* is essential for normal taste behavior. J Neurogenet

12. Campbell SD, Inamdar M, Rodrigues V, et al. (1992) The *scalloped* gene encodes a novel evolutionarily conserved transcription factor required for sensory organ differentiation in *Drosophila*. Genes Dev 6: 367–379

13. Tompkins L (1979) Developmental analysis of two mutations affecting chemotactic behavior in *Drosophila melanogaster*. Dev Biol 73:174–177

14. Possidente DR, Murphey RK (1989) Genetic control of sexually dimorphic axon morphology in *Drosophila* sensory neurons. Dev Biol 132:448–457

15. Lakes R, Pollack GS (1990) The development of the sensory organs of the legs in the blowfly, *Phormia regina*. Cell Tissue Res 259:93–104

16. Ray K, Hartenstein V, Rodrigues V (1993) Development of the taste bristles on the labellum of *Drosophila*. Dev Biol 155:26–37

17. Bate CM (1978) Development of sensory system in arthropods. In: Jacobson M (ed) Handbook of sensory physiology, vol 9. Springer, Berlin, pp 1–53

18. Hartenstein V, Posakony JW (1989) Development of adult sensilla on the wing and notum of *Drosophila melanogaster*. Development 107:389–405

19. Ray K, Rodrigues V (1993, in press) The function of proneural genes achaete and scute in the spatiotemporal patterning of the adult bristles of *Drosophila melanogaster*. Roux Arch Dev Biol

20. Banerjee U, Zipursky SL (1990) The role of cell-cell interaction in the development of the *Drosophil* visual system. Neuron 4:177–187

21. Rubin G (1989) Development of the *Drosophila* retina: inductive events studied at single cell resolution. Cell 57:519–520

22. Dethier V (1976) *The hungry fly*. Harvard University Press, Cambridge, MA.

23. Siddiqi O, Joshi S, Arora K, Rodrigues V (1989) Genetic investigation of salt perception in *Drosophila*. Genome 31:646–651

24. Arora K, Rodrigues V, Joshi S, Shanbhag S, Siddiqi O (1987) A gene affecting the specificity of chemosensory neurons in *Drosophila*. Nature 330:62–63

Genetic Dissection of Taste Transduction Mechanisms in *Drosophila*

Teiichi Tanimura[1]

Key words. *Drosophila*—Insect—Chemosensory transduction—Mutant—IP$_3$

Introduction

Physiologic and molecular studies have given us a detail view of the molecular mechanisms of the visual transduction pathway in the eyes. It has been established in vertebrates that a cyclic nucleotide (cGMP) is the second messenger mediating the light signal from a receptor protein to open ionic channels. However, in invertebrates several lines of evidence suggest that inositol-1,4,5-triphosphate (IP$_3$) acts as a second messenger. In the olfactory transduction pathways in both invertebrates and vertebrates, several electrophysiologic and biochemical experiments have suggested that both cyclic adenosine monophosphate (cAMP) and IP$_3$ are involved as second messengers [1].

The molecular mechanisms involved in taste transduction remain unknown, though the cAMP and IP$_3$ are suggested to be involved [2]. In *Drosophila*, a large number of visual mutants have been isolated, some of which have been studied at a molecular level. One of the mutants, *norpA* (no-receptor-potential A) shows reduced or no receptor potential on light stimuli depending on the allele but has normal visual pigment [3,4]. Biochemical studies suggested that phosphatidylinosito (PI)-specific phospholipase C (PI-PLC) activity is reduced in *norpA* [5]. Cloning of the *norpA* gene was undertaken and proved that the *norpA* gene encodes a phosphatidylinositol-specific phospholipase C [6]. These studies support the view that the phototransduction cascade in the fly is mediated by the second messenger IP$_3$.

If there involved a common second messenger molecule, for both vision and chemoreception, *norpA* flies should show an abnormality in taste or olfactory responses. To examine this possibility we studied the feeding responses to sugars in the *norpA* mutants utilizing food color as a maker of intake. The results obtained showed that IP$_3$ is involved in the chemosensory transduction in the fly.

[1] Biological Laboratory, Kyushu University, Ropponmatsu, Fukuoka, 810 Japan

Materials and Methods

Fly Stocks

Canton-S was used as a wild-type strain. Visual mutants examined were *norpA*, receptor-degeneration A (*rdgA*), receptor-degeneration B (*rdgB*), and transient-receptor-potential (*trp*). The mutant strains used were described in Lindsley and Zimm [7].

Behavioral Test

The two-choice preference tests were carried out as described previously [8] using microtest plates with 60 wells. The concentrations of the food dye were 0.125 mg/ml for the blue dye and 0.5 mg/ml for the red dye. Flies were allowed to feed on the plate for 2 h in the dark. To measure the amount of intake, the blue dye was used at the concentration of 0.5 mg/ml. A sugar-agar solution was filled into the petri dish, and flies were freely allowed to feed for 2 h. Then flies were homogenized in 50% ethanol in phosphate buffer pH 7.0 and centrifuged; the absorbance of the supernatant was then measured at 630 nm.

Results and Discussion

To examine the taste response to sugars, we first measured the amount of intake of sugar solution. Flies were allowed to feed on sugar-agar solution colored with blue food dye for 2 h. At a low concentration of sugar the amount of intake was proportional to the feeding time. Flies were homogenized in the buffer, and the amount of intake was calculated after measuring the absorbance of the homogenate. Visual mutants, *norpA*, *rdgA*, *rdgB*, and *trp* were examined. The amount of intake of sucrose, glucose, and trehalose were markedly reduced to about 20% in *norpA* mutants compared with that of the wild-type flies. The amount of fructose intake, however, was not changed in the mutant. We then examined several *norpA* alleles (EE5, KO50, and JM11) and found that the EE5 allele showed the most severe defect. This result coincides with results on ERG phenotype. In

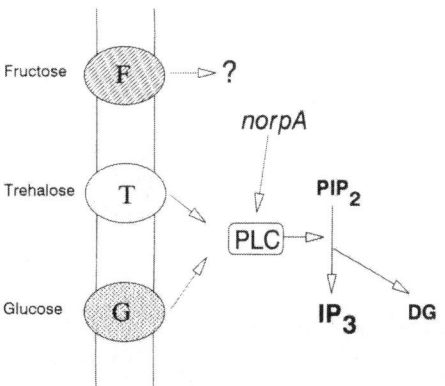

FIG. 1. Taste transduction pathways through multiple sugar receptor sites in *Drosophila*. Three kinds of receptor site are postulated for sugar responses: *G*, glucose, sucrose, and glycopyranosides: *F*, furanose; *T*, trehalose. The responses through G and T sites are mediated by IP_3. The second messenger for the F site remains unknown. Because not all responses to sugars are eliminated in the *norpA* mutant, the G and F sites may transmit signals to the unknown pathway

norpA[EE5] the ERG response is nearly eliminated. The reduced intake to sugars is not due to a defect in feeding behavior itself (i.e., muscle-nerve systems of ingestive behavior), as the fructose intake is normal.

To confirm this point, we behaviorally measured the taste sensitivity to sugars using the two-choice preference test. The details of this test have been described elsewhere; in brief, flies were allowed to choose between two kinds of sugar-agar solution, each colored with red and blue food dye. Knowing that fructose response is unaffected in the *norpA*[EE5] mutants, the two-choice preference test was performed between fructose and glucose of different concentrations. The concentration of glucose that gives 50% preference index value is the one with equal stimulative effectiveness to 20 mM fructose. The value was 30 mM in *Canton-S* and 250 mM in *norpA*. Using deletion chromosomes near the *norpA*[EE5], the taste sensitivity phenotype was mapped on the *norpA* gene itself. Previously we showed the presence of three sugar receptor sites in *Drosophila*: T-site, G-site, and F-site [9,10] (Fig. 1). In the *norpA* mutants, the response at G- and T-sites are markedly affected, whereas the response to fructose remain unaffected.

These behavioral results were confirmed by electrophysiologic recordings from the labeler chemosensilla. These results demonstrate that IP_3 is one of the possible second messengers in taste transduction in *Drosophila*. It is important to note that the responses to glucose, sucrose, and trehalose are not completely eliminated in the mutants. It remains unknown from what mechanisms these residual responses come. It is possible that another second messenger system may account for the residual taste responses, and another isoform of PLC encoded by a separate gene may be involved.

The next question is to know whether the olfactory response is also affected in the *norpA* mutant. We have recorded electroannenograms (EAGs) from the *norpA* flies and found that the responses to several odorants are markedly reduced (T. Tanimura and M. Yokohari, unpublished results). These results proved that the PI signaling is involved in visual, taste, and olfactory transduction pathways. Previous studies showed that PLC is exclusively localized in the retina [5,6], but our studies suggest aht PLC must be present in the chemosensory cells. Immunohistochemical studies using antibody to PLC would help to clarify this point.

If IP_3 is involved in chemosensory transduction, there should be a molecule that bind IP_3 and then transmits the signal to ionic channels. IP_3 receptor is known to act as a calcium release channel. To reveal the presence of IP_3 binding proteins in *Drosophila*, a binding assay was performed using membrane preparation. Membrane proteins were prepared from various developmental stages and from defined parts of adult flies. Using tritiated IP_3, the binding activity was detected in embryos, larvae, and pupae. In adult flies, activity was low in the abdomen and high in the head. If such a receptor protein is involved phototransduction, it should be abundant in the retina. In fact, we found that the binding activity is reduced in the *eyes-absent* mutant flies with no compound eyes. The highest binding activity was found in the appendage fraction containing legs, antennae, and sensilla.

The IP_3 receptor gene has been cloned from mouse [11]. Using mouse cDNA as a probe, we have screened a *Drosophila* cDNA library and obtained a homolog of the mouse IP_3 receptor [12]. The *Drosophila* IP_3 receptor deduced from the cDNA sequence is composed of 2833 amino acids with a molecular weight of 319 kDa. The overall amino acid sequence shares 57% identity between the *Drosophila* and the mouse receptor. Hydropathy plot revealed the presence of six membrane-spanning segments. The last two membrane-spanning sequences are highly homologous between *Drosophila* and the mouse receptor, showing about 90% homology. Another significant homology was found near the *N*-terminal region, a region shown to form the IP_3-binding domain. Northern blot analyses indicated that the transcript is abundant in the fraction containing antennae, legs, and chemosensory bristles. The *Drosophila* IP_3 receptor gene is cytologically mapped on the third chromosome. Unfortunately, no mutants have been isolated within this interval. So far we have been unable to detect transcripts on the tissue sections of the fly by in situ hybridization.

Further molecular studies should clarify the role of IP_3 receptor in the chemosensory transduction pathways. Studies suggest that there is an IP_3 receptor on the plasma membrane of olfactory epithelia [13,14]. It should be determined whether the cloned *Drosophila* IP_3 receptor is located on the plasma membrane of the chemosensory cells.

Acknowledgments. Part of the work presented here was done in collaboration with Drs. F. Yokohari, S. Yoshikawa, A. Miyawaki, T. Furuichi, and K. Mikoshiba. It was supported by grants from the Ministry of Education, Science, and Culture of Japan.

References

1. Reed RR (1992) Signaling pathways in odorant detection. Neuron 8:205–209
2. Hwang PM, Verma A, Bredt DS, Snyder SH (1990) Localization of phosphatidylinositol signaling components in rat taste cells: role in bitter taste transduction. Proc Natl Acad Sci USA 87:7395–7399
3. Pak WL, Grossfield J, Arnold K (1970) Mutants of the visual pathway of *Drosophila melanogaster*. Nature 227:518–520
4. Hotta Y, Benzer S (1970) Genetic dissection of the *Drosophila* nervous system by means of mosaics. Proc Natl Acad Sci USA 67:1156–1163
5. Inoue H, Yoshioka, T, Hotta Y (1985) A genetic study of inositol triphosphate involvement in phototransduction using *Drosophila* mutants. Biochem Biophys Res Commun 132:513–519
6. Bloomquist BT, Shortridge RD, Schneuwly S, et al. (1988) Isolation of putative phospholipase C gene of *Drosophila*, *norpA*, and its role in phototransduction. Cell 54:723–733
7. Lindsley DL, Zimm GG (1992) The Genome of *Drosophila melanogaster*. Academic, London
8. Tanimura T, Isono K, Takamura T, Shimada I (1982) Genetic dimorphism in the taste sensitivity to trehalose in *Drosophila melanogaster*. J Comp Physiol 147: 433–437
9. Tanimura T, Isono K, Yamamoto MT (1988) Taste sensitivity to trehalose and its alteration by gene dosage in *Drosophila melanogaster*. Genetics 119: 399–406
10. Tanimura T (1991) Genetic alteration of the multiple taste receptor sites for sugars in *Drosophila*. In: Wysocki CJ, Kare MR (eds) Chemical senses, vol 3. Genetics of perception and communication. Dekker, New York, pp 125–135
11. Furuichi T, Yoshikawa S, Miyawaki A, et al. (1989) Primary structure and functional expression of the inositol 1,4,5-triphosphate-binding protein P_{400}. Nature 342:32–38
12. Yoshikawa S, Tanimura T, Miyawaki A, et al. (1992) Molecular cloning and characterization of the inositol 1,4,5-triphosphate receptor in *Drosophila melanogaster*. J Biol Chem 267:16613–16619
13. Restrepo D, Miyamoto T, Bryant BP, Teeter JH (1990) Odor stimuli trigger influx of calcium into olfactory neurons of the channel catfish. Science 249: 1166–1168
14. Fadool DA, Ache BW (1992) Plasma membrane inositol 1,4,5-triphosphate-activated channels mediate signal transduction in lobster olfactory receptor neurons. Neuron 9:907–918

Maxillary Palp: A Second Olfactory Organ of *Drosophila*

Juan Riesgo-Escovar, Debasish Raha, and John R. Carlson[1]

Key words. *Drosophila*—Olfaction—Antenna—Maxillary palp—Mutant—Electroantennogram

Introduction

A wide variety of animals, including many amphibians, reptiles, and mammals, have an olfactory organ and another chemosensory organ, the vomeronasal organ [1–3]. This second organ is associated with pheromone detection and the modulation of sexual behavior. In humans it has been reported to appear during embryonic development but to be absent or rudimentary in adults. In many species the vomeronasal organ is located near the olfactory organ and sends projections to an accessory olfactory bulb, rather than to the main olfactory bulb.

This chapter concerns olfactory organs of the fruit fly *Drosophila melanogaster*, which responds to an extensive array of volatile chemicals [4,5]. The antenna has long been known to serve as an olfactory organ, initially from behavioral experiments and subsequently from physiologic experiments. We have provided physiologic evidence that a second organ, the maxillary palp, functions as an olfactory organ in *Drosophila* [6], as it does in certain other insects.

The existence of two organs serving the same sensory modality in *Drosophila* raises questions about the functional, developmental, and evolutionary relations between them. Mechanistic questions are also raised: for example, to what extent do the two organs share the same genetic and molecular underpinnings? Some of these issues are of broad interest in the context of sensory system biology in that there are examples of other sensory modalities served by two organs. For example, *Drosophila* has both a compound eye and a simple eye, or ocellus.

In the case of the antenna and the maxillary palp, previous studies have shed light on some of these questions. Anatomic studies have documented morphologic similarities between the two organs. Both contain sensory hairs: sensilla basiconica, trichodea, and coeloconica cover the surface of the third antennal segment, the olfactory segment [7,8]; sensilla basiconica and trichodea cover the maxillary palp [9,10]. In the case of each organ, a zone rich in basiconica and one rich in trichodea can be distinguished. Both organs send projections to the antennal lobes of the brain, which consist of approximately 35 glomeruli [9,11]. Interestingly, the glomeruli innervated by the maxillary palp are apparently distinct from those innervated by the third antennal segment [11] (although an earlier study had reported overlap [9]). The antenna and the maxillary palp have a common developmental origin in that both derive from the eye-antennal imaginal disc.

There is evidence that the maxillary palp plays a role in sensing certain pheromones. Wild-type males show low levels of courtship behavior toward mated females, an effect believed to be mediated by inhibitory pheromones, or "antiaphrodisiacs" released from mated females [12,13]. Surgical removal of maxillary palps from male flies caused elevated levels of courtship toward mated females [14]. It was thus suggested that the maxillary palp senses one or more inhibitory pheromones. There is also behavioral evidence that the maxillary palp responds to short-chain alcohols and organic acids [15], as does the antenna.

The issue of genetic and molecular relations between the two olfactory organs is an issue for which *Drosophila* is particularly well suited as an experimental organism. It is the focus of the remainder of this chapter.

Results

The maxillary palp, shown in Fig. 1, has been described in some detail [9,10]. One maxillary palp projects from each side of the basiproboscis but does not make contact with food sources while the animal is feeding. Each maxillary palp contains about 60 sensilla basiconica, approximately 8 to 10 μm long. The ultrastructure of the basiconica is characteristic of olfactory sensilla: each is perforated by about 500 pores 0.05 μm in diameter. The maxillary palp also contains about 20 sensilla trichodea (whose ultrastructure is characteristic of mechanosensory sensilla) and a large number of uninnervated spinules.

Figure 2 shows the response of the maxillary palp to ethyl acetate vapor. The recording is analogous to the electroantennogram (EAG), and we thus refer to it as an electropalpogram (EPG). It is an extracellular recording that likely measures the summed receptor potentials of neurons in the vicinity of

[1] Department of Biology, Yale University, PO Box 6666, New Haven, CT 06511-8112, USA

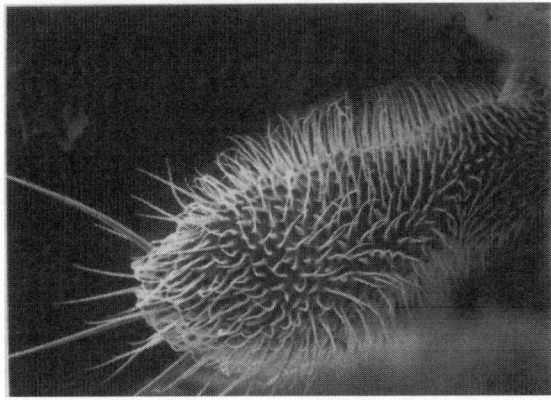

FIG. 1. Maxillary palp of *Drosophila*. Scanning electron micrograph, ×500

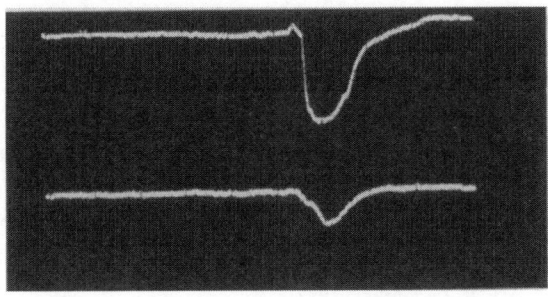

FIG. 2. Recording of the maxillary palp response to stimulation with ethyl acetate vapor. *Top*, control wild-type (Canton-S 5); *bottom*, *acj6*. Air is injected over a small volume of ethyl acetate, diluted 10^{-2} in paraffin oil, and placed into an airstream directed toward the fly. The recording electrode was placed in the dorsal aspect of the distal tip of the palp

the recording electrode [6]. The top trace shows a typical response, whose peak amplitude is about 10 mV, obtained from our wild-type strain, Canton-S 5 [16].

Does olfactory physiology in the maxillary palp rely on the same genes as the antenna? We have used three approaches to identify olfactory genes of *Drosophila*: behavioral genetics, an enhancer trap screen, and substractive screens. The *abnormal chemosensory jump 6* (*acj6*) gene was identified in a screen for mutants defective in olfactory-driven behavior [17]. The *acj6* mutant was found to have a defective EAG [6,18]. The mutation has been mapped to position 49.4 on the X chromosome, is recessive, and does not affect the electroretinogram (ERG). The bottom trace in Fig. 2 illustrates the EPG, which is defective. The peak amplitude in this recording is only about 3.5 mV, or one-third of the wild-type value. Thus the *acj6* gene is apparently required for normal response in both the antenna and the maxillary palp.

We have carried out a large-scale enhancer trap screen for olfactory genes of *Drosophila* [19]. Specifically, we screened 6400 lines of *Drosophila*, each carrying an independent insertion of an enhancer trap element containing a *lacZ* reporter gene coupled to a weak promoter and no enhancer. Those insertions that reside near the enhancer of an olfactory gene are expected to show *lacZ* expression in the olfactory system. We screened the 6400 lines initially at the larval stage and isolated 120 lines, which showed staining associated with the larval olfactory organ but little staining elsewhere. These 120 lines were then stained at the adult stage, and 12 lines were recovered that showed staining of the third antennal segment but relatively little staining elsewhere. Thus these 12 lines showed staining in two morphologically and developmentally distinct olfactory organs.

Interestingly, 5 of the 12 lines show reporter gene expression in the maxillary palps. An example is line 7502, which shows uniform expression throughout the third antennal segment, maxillary palp, labellum (which contains taste hairs), and anterior wing margin (which contains a row of contact chemosensory hairs).

We have also isolated a number of olfactory genes directly by subtractive hybridization. Specifically, we constructed an antennal cDNA library and screened it for olfactory-specific genes by one of two methods. With one method the antennal cDNA library was screened with a subtracted probe; with the other method the library was used to construct a subtracted antennal cDNA library, and we then performed a differential screen with labeled antennal, leg, body, and head cDNA probes.

From this screening we identified five genes, which are expressed in the antenna but not elsewhere in the adult head—except that one gene, OS9, is expressed in the maxillary palp. Figure 3 shows an in situ hybridization of OS9 to a tissue section of the maxillary palp. The figure shows hybridization near the periphery of the palp, where cell bodies are located. OS9 encodes a predicted protein of 18 kDa. We have examined its expression in the maxillary palp and have found evidence that it is expressed in neuronal nuclei.

Discussion

We have reviewed (1) morphologic, physiologic, and behavioral evidence that the maxillary palp has olfactory function in *Drosophila*; (2) genetic evidence that at least one gene required for odorant response in the antenna is also required in the maxillary palp; and (3) molecular evidence that some, but not all, antennal genes are also expressed in the maxillary palp.

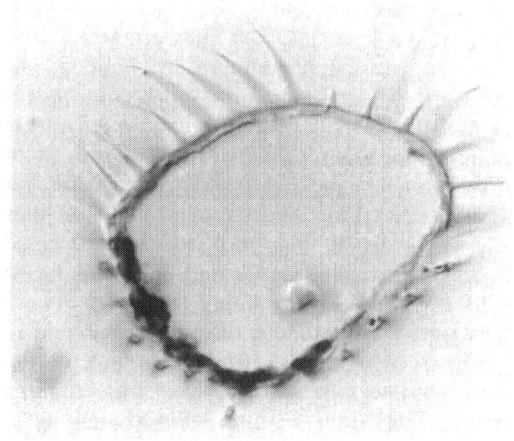

FIG. 3. In situ hybridization to RNA in a tissue section of a maxillary palp. The probe is digoxygenin-labeled OS9 antisense RNA

There are intriguing parallels between the maxillary palp and the vomeronasal organ in vertebrates. Each has been associated with pheromone response and the modulation of sexual behavior. The vomeronasal organ sends projections to an accessory olfactory bulb, which does not overlap with the main olfactory bulb. Likewise, the sensilla basiconica of the maxillary palp have been reported to send projections to glomeruli in the antennal lobe that do not overlap those innervated by antennal sensilla [11; cf. 9]. Interestingly, the two organs may differ in terms of the stimuli that gain access to them. There is evidence that large, nonvolatile molecules have access to the vomeronasal organ [20]; the ultrastructure of the maxillary palp sensilla suggests that they function primarily in sensing volatile, airborne molecules (and mechanosensory stimuli).

It is striking that during the course of *Drosophila* evolution two distinct adult olfactory organs have arisen, both developing from the same imaginal disc. It will be interesting to determine to what extent their functions differ. One means of addressing this issue is to determine by physiologic recording whether there are differences in the sensitivities of the two organs to any of a variety of olfactory stimuli, including cuticular extracts likely to contain pheromones.

We note finally that functional differences between the antenna and the maxillary palp could be engendered by virtue of their different positions on the head and by the different orientations of their olfactory hairs. The maxillary palps are located on the basiproboscis, which plays a role in both feeding and courtship behavior (the male licks the female genitalia). It is possible that the maxillary palp sensilla may become exposed to higher local concentrations of some odorants than antennal sensilla,

a factor that could be especially significant for molecules of low volatility. Moreover, the animal grooms itself with its legs, which may carry pheromones or food sources into closer proximity with one organ than the other.

Acknowledgments. This work was supported by National Institutes of Health grants GM36862 and GM39813 and by a McKnight Scholars Award.

References

1. Wysocki C, Meredith M (1987) The vomeronasal system. In: Finger T, Silver W (eds) Neurobiology of taste and smell. Wiley, New York, pp 125–150
2. Farbman A (1992) Cell Biology of Olfaction. Cambridge University Press, Cambridge
3. Tucker D (1971) Nonolfactory responses from the nasal cavity: Jacobson's organ and the trigeminal system. In: Beidler L (ed) Handbook of sensory physiology, vol IV. Chemical senses 1. Springer, Berlin, pp 151–181
4. Siddiqi O (1987) Neurogenetics of olfaction in *Drosophila melanogaster*. Trends Genet 3:137–142
5. Carlson J (1991) Olfaction in *Drosophila*: genetic and molecular analysis. Trends Neurosci 14:520–524
6. Ayer RK, Carlson J (1992) Olfactory physiology in the *Drosophila* antenna and maxillary palp: acj6 distinguishes two classes of odorant pathways. J Neurobiol 23:965–982
7. Venkatesh S, Singh R (1984) Sensilla on the third antennal segment of *Drosophila melanogaster* meigen. Int J Insect Morphol Embryol 13:51–63
8. Stocker RF, Gendre N (1988) Peripheral and central nervous effects of lozenge 3: a *Drosophila* mutant lacking basiconic antennal sensilla. Dev Biol 127: 12–24
9. Singh RN, Nayak SV (1985) Fine structure and primary sensory projections of sensilla on the maxillary palp of *Drosophila melanogaster* meigen (Diptera: Drosophilidae). Int J Insect Morphol Embryol 14: 291–306
10. Harris WA (1972) The maxillae of *Drosophila melanogaster* as revealed by scanning electron microscopy. J Morphol 138:451–456
11. Stocker RF, Lienhard MC, Borst A, Fischbach KF (1990) Neuronal architecture of the antennal lobe in *Drosophila melanogaster*. Cell Tissue Res 262:9–34
12. Jallon JM (1984) A few chemical words exchanged by *Drosophila* during courtship and mating. Behav Gen 14:441–478
13. Tompkins L (1984) Genetic analysis of sex appeal in *Drosophila melanogaster*. Behav Genet 14:411–440
14. Stocker RF, Gendre N (1989) Courtship behavior of *Drosophila* genetically or surgically deprived of basiconic sensilla. Behav Genet 19:371–385
15. Venard R, Stocker RF (1991) Behavioral and electroantennogram analysis of olfactory stimulation in lozenge: a *Drosophila* mutant lacking antennal basiconic sensilla (Diptera: Drosophilidae). J Insect Behav 4:683–705
16. Woodard C, Huang T, Sun H, Helfand S, Carlson J (1989) Genetic analysis of olfactory behavior in *Drosophila*: a new screen yields the ota mutants. Genetics 123:315–326

17. McKenna M, Monte P, Helfand S, Woodard C, Carlson J (1989) A simple chemosensory response in *Drosophila* and the isolation of acj mutants in which it is affected. Proc Natl Acad Sci USA 86:8118–8122
18. Ayer RK, Carlson J (1991) acj6: a gene affecting olfactory physiology and behavior in *Drosophila*. Proc Natl Acad Sci USA 88:5467–5471
19. Riesgo-Escovar J, Woodard C, Gaines P, Carlson J (1992) Development and organization of the *Drosophila* olfactory system: an analysis using enhancer traps. J Neurobiol 23:947–964
20. Wysocki C, Wellington J, Beauchamp G (1980) Access of urinary nonvolatiles to the mammalian vomeronasal organ. Science 207:781–783

Genetics of the Gurmarin-Sensitive Sweet Receptor in Mice

Yuzo Ninomiya[1], Toshiaki Imoto[2], Akiko Miyasaka[2], and Kazumichi Mochizuki[1]

Key words. Sweet taste receptor—Gurmarin—Congenic strain—Genetics—Chorda tympani nerve—Mouse

Introduction

Taste receptor mechanisms for sweet substances have been investigated in a variety of animals with neurophysiologic, biochemical, behavioral, and other techniques by a number of researchers. However, in mammals it is not fully established whether the sweet receptor site is single or multiple [1,2]. This lack of information is at least in part due to possible limitations of each methodology and the lack of strong and specific inhibitors of sweet taste responses of the available experimental animals, such as rats, hamsters, and mice.

As a possible way to overcome these problems, we chose the genetic approach, because if taste variants that lack a particular sweet receptor site can be found the genetic approach using such variants might lead to the determination of the existence of multiple sweet receptor sites. In this area we have successfully isolated mouse taste variants for a sweet-tasting amino acid, D-phenylalanine (D-Phe; nonsweet tasters for D-Phease BALB and C3H mice), among inbred strains (sweet tasters for D-Phease C57BL mice) using both neurophysiologic and behavioral taste testing [3–5]. Subsequent genetic analysis using C57BL, BALB, and their F1, F2, and B1 hybrids have demonstrated that taste sensitivity to D-Phe is controlled by a single gene (designated dpa) on chromosome 4, and the site of action of the dpa gene is at the sweet receptor site on the taste cell membrane [6–9]. Thus our previous genetic analysis strongly suggested the existence of multiple sweet receptor sites in mice.

However, because of the possible influence of the genetic background other than the taste gene, the specificity of the effect of the proposed dpa gene on sweet taste responses might not be fully established.

Therefore to further evaluate the dpa gene, the development of a congenic strain was needed whose genetic background is identical to that of the non-sweet-taster BALB strain, except with a gene segment containing the dpa locus derived from the sweet-taster C57BL strain.

Imoto et al. [10] have isolated a peptide, gurmarin, from the leaves of Gymnema sylvestre and found that this peptide specifically inhibits chorda tympani nerve responses to sweet substances in rats and hardly affects responses to other taste stimuli. In their neurophysiologic studies, they also found some individual differences among Wistar closed-colony rats (not inbred strain) in terms of gurmarin inhibition on sweet responses. This finding suggested to us that we employ gurmarin in our mouse genetic studies on the sweet receptor mechanism.

Therefore, in the present study we developed a dpa sweet-taster congenic strain using conventional techniques and compared chorda tympani responses of congenic strain, its donor C57BL strain, and inbred partner BALB strain to various taste stimuli with and without gurmarin treatment.

Materials and Methods

Production of a D-Phe Sweet-Taster Congenic Strain

A congenic strain was produced by a succession of back-crosses [11]. Crosses were begun between male and female C57BL/6CrSlc (D-Phe sweet-taster) and BALB/cCrSlc (D-Phe non-sweet-taster) strains. C57BL strain is a donor strain that donates a chromosomal segment containing the dpa locus (dominant sweet-taster allele); BALB strain is the inbred partner that donates the genetic background. The procedure for producing a congenic strain is as follows (Fig. 1).

First, a C57BL donor animal is mated to a BALB inbred partner. All the N1 progeny are expected to be heterozygotes at the dpa locus and to be uniformly sweet-tasters. Phenotypes of all offspring are tested, and then one N1 donor is back-crossed to a BALB inbred partner. The N2 progeny of this back-cross are expected to be of two classes. One-

---

[1] Department of Oral Physiology, Asahi University School of Dentistry, Hozumi, Motosu, Gifu, 501-02 Japan
[2] Department of Physiology, Medical School, Tottori University, Yonago, Tottori 683 Japan

FIG. 1. Back-cross protocol for a *dpa* congenic strain development involving transfer of the dominant sweet-taster allele from the C57BL (Dpa/Dpa) donor strain onto the BALB(dpa/dpa) inbred partner genomic background. Only phenotypic taster (Dpa/dpa) from generations N1 to N11 are selected for back-crossing to the BALB mice. Sweet-tasters from generation N12 are sibling intercrossed

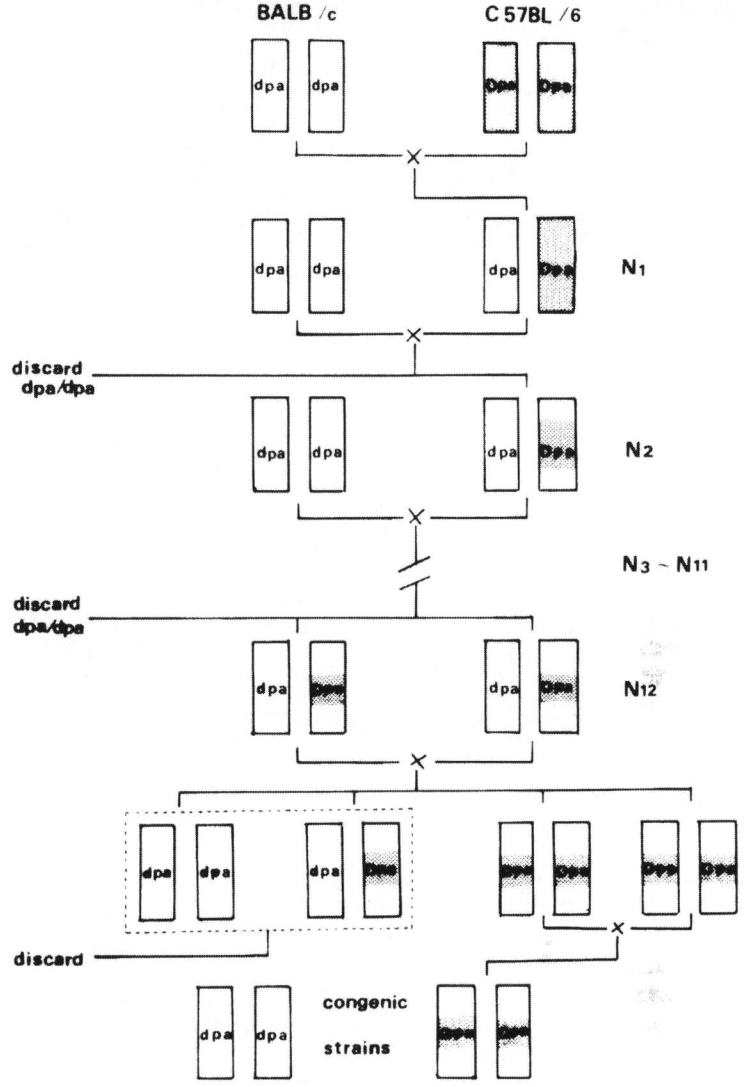

half should be heterozygous sweet-tasters and half homozygous non-sweet-tasters. After testing the phenotype, one sweet-taster donor is again back-crossed to a BALB inbred mouse. This cycle of behavioral taste testing and selective back-crossing continues across generations. At each back-cross generation, one-half of the remaining genetic material from the C57 donor strain is expected to be lost, except for material closely linked on the chromosome bearing the differential *dpa* locus. After 12 generations of back-crossing, a heterozygous sweet-taster brother and sister are intercrossed, and progeny from this last cross, which have the homozygous sweet-taster genotype, are selected.

To measure behavioral responses, we used a conditioned taste aversion paradigm and sometimes a two-bottle preference test as comparison. Based mainly on whether the aversion conditioning to D-Phe would generalize to sucrose (sweet-taster) or not (non-sweet-taster), we classified mice into sweet-

and non-sweet-tasters. The details of the experimental procedures were described in previous reports [4,5,8]. The conditioning stimulus was 0.1 M D-Phe. For aversion conditioning, animals were given access to D-Phe and then given an intraperitoneal injection of LiCl (230 mg/kg) to induce gastrointestinal malaise. Test stimuli mainly used were 0.1 M D-Phe, 0.1 M L-phenylalanine, 0.03 M D-tryptophan, 0.1 M NaCl, 0.1 to 0.5 M sucrose, 1 mM HCl, and 0.1 mM quinine HCl.

Chorda Tympani Nerve Responses to Various Taste Stimuli in C57, BALB, and Congenic Strains

Subjects were male and female mice of C57BL, BALB, and congenic strains and ranged in weight from 20 to 35 g. Procedures of dissection of the chorda tympani nerve, recording of the whole nerve responses to various taste stimuli, and stimulation of the tongue with taste solutions were described in previous reports [3,5].

Solutions used for taste stimuli were 0.001 to 1.000 M sucrose, 0.5 M fructose, 0.5 M glucose, 0.5 M maltose, 0.02 M sodium saccharin, 0.1 M NH₄Cl, 0.1 M NaCl, a mixture of 0.1 M NaCl and 0.1 mM amiloride, 0.1 M KCl, 0.1 M D-Phe, 0.1 M L-alanine, 0.3 M L-proline, 0.03 M D-tryptophan, 0.1 M D-histidine, 0.01 M HCl, 0.02 M quinine HCl, and mixtures of 0.1 mM sodium saccharin with 0.1 M D-Phe or 0.03 M D-tryptophan. These solutions were made up in distilled water at about 20°C.

To examine gurmarin inhibition of sweet-taste responses, the tongue was treated with gurmarin 1.0 to 100.0 μg/ml dissolved in 5 mM acetate or phosphate buffer (pH 5.0–7.0) for 10 min in the same manner as described by Imoto et al. [10].

For data analysis, the magnitude of the integrated response at 10 s after stimulus onset was measured. The relative response magnitude for each stimulus was calculated, with the response magnitude to 0.1 M NH₄Cl was taken as an unity (1.0), and this figure was used for statistical analysis.

Results

Production of a D-Phe Sweet-Taster Congenic Strain

Using the phenotypic classification based on behavioral taste testing mentioned above, sweet-tasters and non-sweet-tasters were segregated in each back-cross generation. In all back-cross generations from N2 to N12, the ratios of sweet- and non-sweet-tasters were not significantly different from the expected 1:1 ratio for the single locus model. Thus using a back-cross protocol for congenic strain development, we could established a D-Phe sweet-taster congenic strain whose genetic background is identical to that of non-sweet-taster BALB strain, except for a gene segment containing the *dpa* locus that is derived from the sweet-taster C57BL strain. Behavioral responses of this congenic strain to D-Phe as well as other sweeteners were not different from those of the C57BL donor strain.

Chorda Tympani Responses to Various Taste Stimuli in C57BL, BALB, and Congenic Strains

Strain Differences in Response to Sweeteners and Salts

Relative responses to various sweeteners of congenic mice were not significantly different from those of C57BL mice. For example, in both strains, responses to 0.1 M D-Phe and 0.3 M L-proline were inhibited by the proteolytic enzyme pronase, and D-Phe responses were enhanced by saccharin; these response characteristics were not observed in BALB

mice. The relative response magnitudes of these strains to 0.5 M sucrose and 0.02 M saccharin were about double those in BALB mice. Also, responses of C57BL and congenic mice to other sweet-tasting amino acids and sugars were slightly larger than those of BALB mice. However, congenic mice are similar to BALB mice when tested with salts. Unlike the C57BL mice, these two strains showed no inhibition of NaCl responses by the sodium channel blocker amiloride.

These results suggest that the transferred gene segment from C57BL mice to congenic mice contain the gene responsible for responses not only to D-Phe but to the other sweeteners as well, whereas the transferred gene segment does not contain the gene responsible for salt taste responses.

Gurmarin Inhibition of Sweet-Taste Responses

We found a prominent strain difference in the effect of gurmarin on sweet-taste responses among the three strains. In C57BL and congenic mice, responses to sucrose at concentrations of 0.1 M or more were significantly reduced after gurmarin treatment, whereas no such inhibition of sucrose responses was observed in BALB mice. After gurmarin the magnitude of sucrose responses in sweet-taster strains decreased to about 50% of control. The reduction was maximum when tested with the concentration of gurmarin at 20 μg/ml or more. The strains difference in the magnitude of relative responses to sucrose between then almost disappeared.

Responses to all sweeteners employed were more or less suppressed by gurmarin in both C57BL and congenic mice, whereas no significant reduction of responses was found in BALB mice. In C57BL and congenic mice, the magnitude of the response after gurmarin varied considerably among sweeteners and ranged from about 50% for sucrose, saccharin, and a mixture of D-Phe, saccharin, and D-histidine to 75% for fructose. Note that in the sweet-taster strains the enhancing effect of saccharin on D-Phe responses disappeared after gurmarin, whereas that on D-tryptophan responses remained, as in the BALB mice. Again, strain differences in the response to all sweeteners almost disappeared after gurmarin. The important points of our results are summarized in Table 1. The results strongly suggest that the strain difference in sweet-taste responses between C57BL and BALB mice corresponds to its gurmarin-sensitive component.

Discussion

The present results strongly suggest that there are at least two types of sweet-taste receptor in mice: One is gurmarin-sensitive and the other gurmarin-

TABLE 1. Chorda tympani responses to various taste stimuli in C57BL, congenic, and BALB strains.

Response	C57BL(Dpa/Dpa)	Congenic (Dpa/—)	BALB (dpa/dpa)
Sucrose response (0.5 M)			
Relative magnitude (0.1 M NH$_4$Cl = 1.0)	1.02[a]	1.01	0.53
After gurmarin	0.50	0.52	0.51
(Gurmarin inhibition)	(+)	(+)	(−)
D-Phenylalanine response			
Pronase inhibition	+	+	−
Enhancement by saccharin	+	+	−
Gurmarin inhibition	+	+	−
Enhancement by saccharin after gurmarin	−	−	−
D-Tryptophan response			
Pronase inhibition	+	+	+
Enhancement by saccharin	+	+	+
Gurmarin inhibition	+	+	−
Enhancement by saccharin after gurmarin	+	+	+
NaCl response			
Amiloride inhibition	+	−	−

+ positive; − negative.

[a] Relative magnitude of response when the response to 0.1 M NH$_4$Cl was taken as unity (1.0).

insensitive. The sweet-taster mice (e.g., C57BL and congenic strains) probably possess both types of sweet receptor, whereas the non-sweet-taster BALB mice lack the gurmarin-sensitive type but possess the gurmarin-insensitive type. The strain difference in taste responsiveness to a sweet-tasting amino acid, D-Phe, therefore would be due to the difference in whether the mouse has the gurmarin-sensitive type of sweet receptor, which probably contains the receptor site for D-Phe.

Ninomiya and Kajiura [12] have demonstrated that saccharin selectively enhances the response to D-amino acids with a ring structure, such as D-Phe, D-tryptophan, and D-histidine in C57BL mice and D-tryptophan and D-histidine in BALB mice, but saccharin did not affect responses to other sweet-tasting D- and L-amino acids and sugars. These results suggest that the receptor site for saccharin and the above mentioned D-amino acids would be different from that for sugars and other sweet-tasting amino acids; and probably the receptor site for saccharin is different from that for the D-amino acids.

The receptor sites for saccharin and D-amino acids may be able to cooperate allosterically with each other on the taste cell membrane for enhancement. In the present study, after gurmarin administration the enhancing effect of saccharin on D-Phe responses disappeared in C57BL and congenic mice, indicating that only the gurmarin-sensitive type would contain the receptor site for D-Phe, which could interact with that for saccharin. In C57BL and congenic mice, the enhancing effect of saccharin on D-tryptophan responses remained even after gurmarin, as was shown in BALB mice before and after gurmarin. This finding suggests that both gurmarin-sensitive and gurmarin-insensitive types would contain distinct receptor sites for saccharin and D-tryptophan in addition to the receptor sites for sugars and other sweet-tasting amino acids.

Concerning the site of action of the *dpa* gene, there are two possibilities. One is that the *dpa* gene may regulate the synthesis of only the receptor site for D-Phe. In this case, there may exist many sweet-taste genes on the transferred chromosomal segment in the congenic strain, each of which regulates the synthesis of each gurmarin-sensitive sweet receptor site. The other possibility is that the *dpa* gene would control the synthesis of the whole gurmarin-sensitive receptor molecule. If this is the case, in our genetic studies we might have classified the phenotypes of mice on the basis of the D-Phe response as representative of their gurmarin-sensitive sweet sensitivity because the D-Phe response could be the most differential component of the gurmarin-sensitive sweet responses. In other words, the D-Phe site might be the most salient part of the gurmarin-sensitive receptor that could be differentiated from the gurmarin-insensitive receptor.

However, to examine these possibilities, we may need to differentiate sweet-taste receptor sites genetically by producing further back-cross generations in the continuous congenic development.

Acknowledgments. This study was supported in part by grants-in-aid for scientific research (no. 05671560) from the Ministry of Education, Science, and Culture of Japan. Authors wish to express thanks to Drs. N. Sako (Osaka University), H. Katsukawa, and H. Kajiura for their help in developing the congenic strain.

References

1. Sato M (1985) Sweet taste receptor mechanisms. Jpn J Physiol 35:875–885

2. Beidler LM, Tonosaki K (1985) Multiple sweet receptor sites and taste theory. In: Pfaff DW (ed) Taste, olfaction and Central Nervous System. Rockefeller University Press, New York, pp 47–64

3. Ninomiya Y, Mizukoshi T, Higashi T, Katsukawa H, Funakoshi M (1984) Gustatory neural responses in three different strains of mice. Brain Res 302:305–314

4. Ninomiya Y, Higashi T, Katsukawa H, Mizukoshi T, Funakoshi M (1984) Qualitative discrimination of gustatory stimuli in three different strains of mice. Brain Res 322:83–92

5. Ninomiya Y, Nomura T, Katsukawa H (1992) Genetically variable taste sensitivity to D-amino acids in mice. Brain Res 596:349–352

6. Funakoshi M, Tanimura T, Ninomiya Y (1987) Genetic approaches to the taste receptor mechanisms. Chem Sens 12:285–294

7. Ninomiya Y, Higashi T, Mizukoshi T, Funakoshi M (1987) Genetics of the ability to perceive sweetness of D-phenylalanine in mice. Ann NY Acad Sci 510: 527–529

8. Ninomiya Y, Sako N, Katsukawa H, Funakoshi M (1991) Taste receptor mechanisms influenced by a gene on chromosome 4 in mice. Chem Senses 3: 267–278

9. Ninomiya Y, Funakoshi M (1993) Genetic and neurobehavioral approaches to the taste receptor mechanism in mammals. In: Simon SA, Roper SD (eds) Mechanisms of taste transduction. CRC Press, Boca Raton, FL, pp 253–272

10. Imoto T, Miyasaka A, Ishima R, Akasaka K (1991) A novel peptide isolated from the leaves of Gymnema sylvestre. 1. Characterization and its suppressive effect on the neural responses to sweet taste stimuli in the rat. Comp Biochem Physiol 100A:309–314

11. Flaherty L (1981) Congenic strains. In: Foster HL, Smith JD, Fox JG (eds) The mouse in biomedical research vol, 1 Academic, New York, pp 215–222

12. Ninomiya Y, Kajiura H (1993) Enhancement of murine gustatory neural responses to D-amino acids by saccharin. Brain Res 626:287–294

Molecular-Genetic Studies of Mouse Proline-Rich Protein Genes and Bitter Taste

EDWIN A. AZEN and LANG ZHUO[1]

Key words. Proline-rich proteins—Bitter

Proline-rich proteins (PRPs) represent an abundant and functionally poorly understood class of proteins that are produced by salivary glands and von Ebner's gland. According to earlier studies, *Prp* genes are either the same or closely linked to gene(s) for quinine and raffinose acetate bitter taste [1] and are localized to distal mouse chromosome 6 [2]. Monogenic inheritance in mice of the *Soa* (sucrose octaacetate) gene for bitter taste was previously well confirmed by studies of Lush and Whitney during the 1980s. More recent genetic linkage and association studies conducted in Whitney's laboratory using recombinant, congenic, and CFW mouse strains could not distinguish *Prp* and *Soa* genes [3–5]. There is also close correspondence between Soa-avoidance phenotypes and PRP genomic DNA fragment patterns [5]. Also PRPs are expressed in von Ebner's gland [6], whose secretions may play a role in taste stimulation. These experiments, although suggesting a possible role for PRPs in bitter taste, do not establish this proposed function.

We chose to study the possible role of *Prp* genes in bitter Soa taste, as the behavioral test for the *Soa* gene was shown by other investigators to be reliable and reproducible. Furthermore, the Soa^a taster allele is dominant over the Soa^b nontaster allele, and the expression of the Soa^a taster allele should be detectable in a nontaster Soa^b genetic background. Therefore transgenic experiments are in progress to test the possible function of PRPs in bitter tasting. The transgenic mice were prepared and housed in the laboratory of Messing. Thus *Prp* genes from the Soa taster mouse have been introduced into Soa nontasting BXJ recipient mice in an attempt to convert the Soa nontaster recipient to the Soa-taster phenotype and thus establish the biologic role of PRPs in bitter taste.

In previous work from Carlson's laboratory, two *Prp* genes, MP-2 and M-14, were isolated from the CD-1 mouse and characterized [7,8]. The mouse *Prp* genes are relatively small (3–5 kbp) and contain 3 exons, the most 3' exon being noncoding. We have isolated two *Prp* genes ("MP-2-like" and "M-14-like") from the Soa-tasting SWR mouse using *Prp* probes obtained from Carlson. Thus a partial SWR *SauIIIAl* genomic library in λ-phage Charon 40 was screened with the rat PRP33 cDNA probe and an M-14-like gene was isolated. The library was re-screened with an upstream probe from the CD-1 MP-2 gene, and an MP-2-like gene was also isolated. Four SWR *Prp* transgenes have been constructed: (1) a 17-kbp DNA fragment containing an MP-2-like gene; (2) a promotor construct from the MP-2-like gene to test by histochemistry the upstream promotor function (8.2 kb upstream promotor region fused to the *LacZ* reporter gene); (3) a 13-kbp DNA fragment containing an M-14-like gene with SV40/CAT DNA substituted for the 3' untranslated terminal exon to provide a marker for detection, as SV40 and CAT sequences are not present in mice; (4) a promotor construct from the M-14-like gene (7.8 kb upstream promotor region fused to the *LacZ* reporter gene). The first three SWR transgenes have been introduced into BXJ recipients yielding 5, 7, and 2 founders, respectively. Some offspring of these founders inherit the transgenes, and the offspring are being tested by behavioral taste assays (two-bottle choice and conditioned aversion) in the laboratory of Hellekant.

References

1. Azen EA, Lush IE, Taylor BA (1986) Close linkage of mouse genes for salivary proline-rich proteins (PRPs) and taste. Trends Genet 2:199–200
2. Azen EA, Davisson MT, Cherry M, Taylor BA (1989) Prp (proline-rich protein) genes linked to markers Es-12 (esterase-12), Ea-10 (erythrocyte alloantigen) and loci on distal mouse chromosome 6. Genomics 5: 415–422
3. Capeless CG, Whitney G, Gannon KS, et al. (1990) The sucrose octaacetate taste gene (Soa) is on distal mouse chromosome 6 and is closely linked (or identical) to salivary proline-rich protein genes (Prp). Chem Senses 15:559
4. Capeless, CG, Whitney G, Azen EA (1992) Chromosome mapping of Soa, a gene influencing gustatory sensitivity to sucrose-octaacetate in mice. Behav Genet 22:655–663

[1] Departments of Medicine and Medical Genetics, University of Wisconsin, 1300 University Avenue, Madison, MI 53706, USA

5. Harder DB, Capeless CG, Maggio JC, et al. (1992) Intermediate sucrose octaacetate sensitivity suggests a third allele at mouse bitter taste locus Soa and Soa-Rua identity. Chem Senses 17:391–401

6. Azen EA, Hellekant G, Sabatini LM, Warner TF (1990) mRNAs for PRPs, statherin and histatins in von Ebner's gland tissues. J Dent Res 69:1724–1730

7. Ann DK, Carlson DM (1985) The structure and organization of a proline-rich protein gene of a mouse multigene family. J Biol Chem 260:15863–15872

8. Ann DK, Smith MK, Carlson DM (1988) Molecular evolution of the mouse proline-rich protein multigene family. J Biol Chem 263:10887–10893

Odortypes Determined by the Major Histocompatibility Complex in Mice

Kunio Yamazaki[1], Gary K. Beauchamp[1], Yoshihisa Imai[1], Maryanne Curran[1], Judith Bard[2], and Edward A. Boyse[2]

Key words. MHC (Major histocompatibility complex)—Mouse—Odortype—Pups—Fetuses—Urine

Introduction

The Major Histocompatibility Complex (MHC), a string of at least 50 or more genes, has been subject to a vast amount of cellular and molecular analysis because of its critical role in governing the immune response [1]. These genes, the products of which are found on most cell surfaces, are so polymorphic that they uniquely code every individual, save identical twins. These genes have also been shown to provision mice with an individual olfactory signal or odortype [2–6], found most strongly in urine and apparently serving to regulate mating preferences [7–9] and maintenance of pregnancy [10,11]. MHC-selective mating depends on familial imprinting, as shown by the altered mating choices of appropriately fostered males [12]. In the context of familial MHC imprinting, and in other reproductive contexts such as maternal identification of progeny in relation to nursing, it is necessary to know whether MHC odortypes are expressed in early life, not only to elucidate further the reproductive significance of MHC odortypes but also because data on the early expression of MHC odortypes are essential to a number of potential studies on the generation of odortypes.

General Methods

The Y-Maze

As detailed elsewhere [3], air is conducted through two odor chambers, containing urine samples exposed in Petri dishes, or perforated containers housing the infant mice, to the arms of the maze.

[1] Monell Chemical Senses Center, 3500 Market Street, Philadelphia, PA 19104, USA
[2] Department of Microbiology and Immunology, University of Arizona, College of Medicine, Tucson, AZ 85724, USA

Gates are raised and lowered in timed sequence in training or testing sessions of up to 48 consecutive trials. As before [3], the samples were assigned to the left or right odor boxes of the Y-maze according to a series of random numbers. The reward for a correct response is a drop of water, the trained mouse having been deprived of water for 23 h.

Training

Preliminary training in the present study progressed from gross to fine distinctions in stages. Mice were first trained to discriminate between urine donors of two unrelated strains [C57BL/6(B6) vs AKR]. When this was successfully completed, adult panels of congenic mice differing only in the MHC (B6 vs B6-H-2^k) next served as urine donors.

Experiment 1

The purpose of this experiment was to determine the age of onset of MHC-determined odortypes. We tested the ability of mice trained in a Y-maze to discriminate odors of pups differing only at the MHC.

Methods

Source of Odors

To ensure that any odor difference found would be due to the infants and not some odor derived from the mothers, litters of B6 and B6-H-2^k mice, which would form the odor-donor panels, were removed from their mothers within 16 h of birth and fostered onto lactating BALB/c females.

Only males were used to provide urine. When picked up by the skin of the back, the pups emitted drops of urine that were drawn up into a test tube. Usually five or six infant mice provided sufficient urine (0.2–0.3 ml) to cover the bottom of a 3.5-cm diameter Petri dish, but sometimes more mice were needed. Urine samples were frozen at $-20°C$ until needed. Freshly defrosted samples at room temperature from different donors were provided for each day's testing.

233

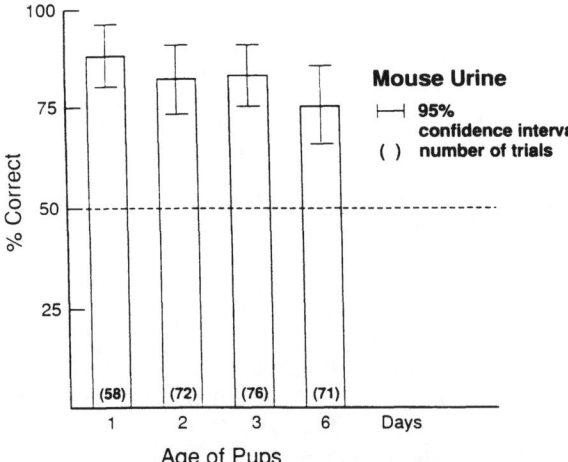

FIG. 1. Percent correct choices for 4 adult mice (2 females, 2 males) trained to differentiate urine odors of B6 vs B6-H-2^k pups in the Y-maze. Trained animals were each given 16–36 trials at each pup age. Testing proceeded from older to younger pup urines

Training

When preliminary training was successful, a series of four trials was begun with adult urines followed by four trials with infant mice or their urines, with all correct responses being reinforced.

Results and Discussion

When whole pups were used as odor sources, MHC-determined odors were not discriminated by trained mice until the pups reached 11–14 days of age. However, mice were successfully trained in the Y-maze to distinguish urine of B6 (H-2^b) mice of age 6 days, 3 days, 2 days, or 1 day from urines of B6-H-2^k congenic mice of the same age (Fig. 1). The results clearly demonstrated that urine of pups as young as one day of age express MHC-determined odor [13]. MHC-determined odortypes may play a prominent role in the ability of the mother mouse to discriminate her own pups from alien pups beginning at least as early as day 1 of age.

The fact that infant mice manifest MHC-determined odortypes, and the finding that urine appears to be the main or sole source, are crucial not only to broader appreciation of the contexts in which odortype recognition affects behavior but also to improved understanding of how these odortypes are composed.

A probable reason why urinary H-2 odortypes are not apparent when infant mice are tested in the Y-maze can be found in the fact that infant mice characteristically do not urinate spontaneously but are stimulated to do so when the mother licks them [14], and thereby no doubt ingests the odortype-

expressing urine which serves to distinguish her own pups from others [15]. Since urine serves as a chemical signal between pups and mother [16], it is likely that MHC-regulated odortypes provide a basis for recognition.

The apparent absence of odortype from the cleaned non-urinating infant mouse affords further evidence that the urinary H-2 odortype is the most potent source of these odors.

Experiment 2

The purpose of this experiment was to determine whether H-2 odortypes may even be expressed by the fetus, and whether a fetal odortype might appear in urine of the pregnant female, thereby altering the female odortype to a mixture of female plus fetus odortype.

Methods

Source of Odors

B6 females were divided into two groups; one group was mated with B6 males yielding bb fetuses, whereas the other group of B6 females was mated with congenic B6-H-2^k males, yielding bk fetuses. Thus, these pregnant females were identical but carried fetuses of differing H-2 types. Urine of each separated pregnant female of each group was collected overnight in a metabolic cage and stored frozen at −20°C until after the litter was born following 19–21 days of gestation. Samples contaminated with feces were discarded. Fetal age is taken as the gestation period, expressed as days after observation of a vaginal plug.

Training

We tested the ability of the trained mice to discriminate between urine collected from these two groups. Each day's training and testing in the Y-maze employed freshly-thawed urine samples maintained at room temperature. For training each day, 6–8 urine donors were randomly selected from panels of pregnant females.

Generalization

The purpose of this procedure is to introduce duplicate urine-donor panels, of the same paired genotypes, and test them without reward, thereby excluding the possibility that incidental or genetically unrelated cues are being learned and responded to (without reward, there can be no learning of new adventitious cues). Generalization is conducted blind with coded samples, which is possible because no reward is called for. To maintain responsiveness,

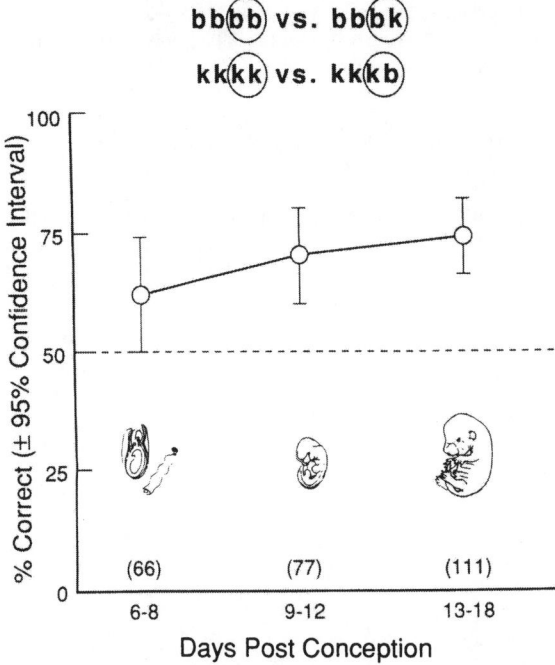

bb(bb) vs. bb(bk)

kk(kk) vs. kk(kb)

FIG. 2. By the ninth to twelfth day of gestation, the maternal urinary H-2 odortype is modified by the presence of a fetus with a nonmaternal H-2 haplotype. Five mice previously trained to distinguish the urinary H-2 odortype of B6 (**bb**) adult males (reward) from that of (B6×B6-H-2^k) F_1 congenic hybrid (**bk**) adult males (no reward) were next trained to select the odor of **bb**/*bb* pregnancy urine from that of **bb**/*bk* pregnancy urine. Similarly, two mice previously trained to choose H-2 odortype **kk** over odortype **kb** of adult male urine were trained to select the odor of **kk**/*kk* pregnancy urine from **kk**/*kb* pregnancy urine. Concordance data for these two test sets did not significantly differ and so have been combined in this figure. Generalization trials, conducted with urines of females that were never encountered during training, included urines from 44 females with 13–18-day-old fetuses, urines from 29 females with 9–12-day-old fetuses, and urines from 20 females with 6–8-day-old fetuses

however, the unrewarded blind samples from new panels for generalization are interspersed with concurrent rewarded and unrewarded tests of the familiar panels. Significant generalization is final proof that the trained mice are discriminating the odor sources on the basis of class to which the donors belong.

Results and Discussion

Mice were successfully trained to distinguish maternal urinary odortypes of these pregnancies at 13–18 days of gestation, and their ability to do so extended to duplicate groups of 13–18 days pregnant females, whose urine had not previously been presented, in interspersed unrewarded trials (generalization). In further generalization tests, the same trained mice

also distinguished these same two maternal urinary odortypes at 9–12 days of gestation but not at 6–8 days (Fig. 2). In further experiments [17], it was shown that the normal H-2-determined maternal urinary odortype is a compound of maternal and fetal H-2 odortypes. Thus, fetuses do express their own H-2 odortype as early as 9–12 days of gestation. This is the earliest time at which H-2 genes become demonstrably expressed in the fetus [18,19]. This striking correlation between the first evidence for production of class I products and the appearance of H-2 odortypes suggests that odor production is a fundamental aspect of H-2 genes. This is the first evidence in vertebrates that the genetic identity of the fetus can be signalled to the outside world via its smell.

Acknowledgments. We thank S. Corisdeo and M. DePamphilis for excellent technical assistance. This work was supported in part by grants from the National Institutes of Health (GM-32096 and CA-39827) and the Richard Lounsbery Foundation. EAB is an American Cancer Society Research Professor.

References

1. Klein J (1986) Natural history of the major histocompatibility complex. Wiley, New York
2. Yamazaki K, Yamaguchi M, Baranoski L, Bard J, Boyse EA, Thomas L (1979) Recognition among mice: Evidence from the use of a Y-maze differentially scented by congenic mice of different major histocompatibility types. J Exp Med 150:755–760
3. Yamaguchi M, Yamazaki K, Beauchamp GK, Bard J, Thomas L, Boyse EA (1981) Distinctive urinary odors governed by the major histocompatibility locus of the mouse. Proc Natl Acad Sci USA 78:5817–5820
4. Boyse EA, Beauchamp GK, Yamazaki K (1987) The genetics of body scent. Trends in Genetics 3:97–102
5. Boyse EA, Beauchamp GK, Bard J, Yamazaki K (1991) Behavior and the major histocompatibility complex (MHC), H-2, of the mouse. In: Ader R, Felter DL, Cohen N (eds) Psychoneuroimmunology-II, Academic, Orlando, pp 831–846
6. Singh PB, Brown RE, Roser B (1987) MHC antigens in urine as olfactory recognition cues. Nature 327: 161–164
7. Yamazaki K, Boyse EA, Mike V, Thaler HT, Mathieson BJ, Abbott J, Boyse J, Zayas ZA, Thomas L (1976) Control of mating preferences in mice by genes in the major histocompatibility complex. J Exp Med 144:1324–1335
8. Potts WK, Manning JL, Wakeland EK (1991) MHC genotype influences mating patterns in semi-natural populations of Mus. Nature 352:619–621
9. Egid K, Brown JL (1989) The major histocompatibility complex and female mating preferences in mice. Anim Behav 38(3):548–550
10. Yamazaki K, Beauchamp GK, Wysocki CJ, Bard J, Thomas L, Boyse EA (1983) Recognition of H-2 types in relation to the blocking of pregnancy in mice. Science 221:186–188

11. Yamazaki K, Beauchamp GK, Matsuzaki O, Kupniewski D, Bard J, Thomas L, Boyse EA (1986) Influence of a genetic difference confined to mutation of H-2K on the incidence of pregnancy block in mice. Proc Natl Acad Sci USA 83:740–741

12. Yamazaki K, Beauchamp GK, Kupniewski D, Bard J, Thomas L, Boyse EA (1988) Familial imprinting determines H-2 selective mating preferences. Science 240:1331–1332

13. Yamazaki K, Beauchamp GK, Imai Y, Bard J, Boyse EA (1992) Expression of H-2 odortypes by infant mice. Proc Natl Acad Sci USA 89:2756–2758

14. Gubernick DJ, Alberts JR (1985) Maternal licking by virgin and lactating rats: Water transfer from pups. Physiol Behav 34:501–506

15. Chantrey DF, Jenkins BAB (1982) Sensory processes in the discrimination of pups by female mice (Mus musculus). Anim Behav 30:881–885

16. Londei T, Segala P, Leone VG (1989) Mouse pups' urine as an infant signal. Physiol Behav 45:579–583

17. Beauchamp GK, Yamazaki K, Curran M, Bard J, Boyse EA (1994) Fetal H-2 odortypes are evident in the urine of pregnant female mice. Immunogenetics 39:109–113

18. Ozato K, Wan Y, Orrison BM (1985) Mouse major histocompatibility class I gene expression begins at midsomite stage and is inducible in earlier-stage embryos by interferon. Proc Natl Acad Sci USA 82:2427–2431

19. Jaffe L, Robertson EJ, Bikoff EK (1991) Distinct patterns of expression of MHC class I and β_2-microglobulin transcripts at early stages of mouse development. J Immunol 147:2740–2750

Isolation of Variants for Fructose or Glucose Taste from a Natural Population of Hawaiian *Drosophila adiastola*

LINDA M. KENNEDY and JASON E. POSKANZER[1]

Key words. Sweet—Receptor

Strains variant in behavioral and receptor cell responses to trehalose, sucrose, glucose, and/or pyranose sugars, but not to fructose or furanose sugars, are known in the small fruitfly, *Drosophila melanogaster*. Genes on the X or autosomal chromosomes are involved [1–5]. Here we report variants for fructose, as well as glucose, in the larger species, *D. adiastola*.

In studies of the population, behavioral and receptor cell thresholds and parameters of response functions were similar for sucrose (s) and fructose (f), but different for glucose (g). Rising-phase slopes for f and g were significantly different (analysis of covariance, $P < 0.001$). For each of the three sugars, behavioral and receptor cell response parameters corresponded, and rising-phase slopes were well-correlated ($r^2 \geqslant 0.79$). These data suggested separate mechanisms for f and g taste in or before the receptor cells [6–8].

By serial two-choice tests (f vs g at concentrations which are just-suprathreshold for the natural population [6,7]), 937 flies were screened to select "wild-type" (WT) and two variant types: "fructose nontaster" (FN) and "glucose nontaster" (GN). The frequencies ($\pm$SE) were 5.7 ($\pm$0.8) % FN and 4.9 ($\pm$0.9) % GN. Since there were no significant differences in male vs female frequencies for FNs or GNs (χ^2 test, $P > 0.05$), it appears that the variant traits are autosomal. Mendelian crosses are planned for confirmation. Neurophysiological studies to confirm location of the defects in or before the receptor

cells and the breeding of an FN strain for molecular genetic analyses are in progress.

Acknowledgments. Supported by NSF BNS-9118858, NIH DC01563, and Clark University Faculty Development Awards to LMK. We thank L. Rudnick and E. Pinunsky for assistance, K. Kaneshiro for flies, and L. Tompkins, T. Lyerla, and D. Thurlow for discussion and suggestions.

References

1. Isono K, Kikuchi T (1974) Autosomal recessive mutation in sugar response of *Drosophila*. Nature 248: 243–244
2. Rodrigues V, Siddiqi O (1981) A gustatory mutant of *Drosophila* defective in pyranose receptors. Mol Gen Genet 181:406–408
3. Tanimura T, Isono K, Takamura T, Shimada I (1982) Genetic dimorphism in the taste sensitivity to trehalose in *Drosophila melanogaster*. J Comp Physiol 147: 433–437
4. Tanimura T, Isono K, Yamamoto M-T (1988) Taste sensitivity to trehalose and its alteration by gene dosage in *Drosophila melanogaster*. Genetics 119:399–406
5. Tanimura T (1991) Genetic alteration of the multiple taste receptor sites for sugars in *Drosophila*. In: Wysocki CJ, Kare MR (eds) Chemical senses, vol 3, genetics of perception and communication. Marcel Dekker, New York, pp 125–136
6. Her C, Kennedy LM (1991) Behavioral and neurophysiological responses to taste stimuli in Hawaiian fruitflies. Chem Senses 16:533
7. Poskanzer JE, Rudnick L, Her C, Kennedy LM (1992) Isolation of Hawaiian *Drosophila* variants which prefer glucose to fructose at equipotent suprathreshold concentrations. Chem Senses 17:684
8. Poskanzer JE, Kennedy LM (1993) Fructose and glucose nontasters in the Hawaiian, *Drosophila adiastola*. Chem Senses 18:615

[1] Department of Biology and Neuroscience Program, Clark University, Worcester, MA 01610, USA

Molecular Population Genetics of Olfactory Systems in *Drosophila melanogaster* Complex: Cloning of Putative Olfactory Receptor Genes

ATSUKO TANABE[1], HUI SUN[1], MAKI KANEKO[2], RYU UEDA[3], DAISUKE YAMAMOTO[3], ETSUKO TAKANASHI MATSUURA[1], and SADAO ISHIWA CHIGUSA[1]

Key words. *Drosophila*—Olfactory receptor genes

A large number of different odorant molecules are discriminated by the olfactory system. The putative olfactory receptor genes of the rat [1], catfish [2], and a few other organisms have recently been cloned. Interestingly, the following sequence analyses suggest that the genetic variation of these gene subfamilies is maintained by a certain type of Darwinian selection [2,3]. Fruit flies seem to have a relatively simple, olfactory system and can be studied by molecular, genetic and behavioral methods [4]. However, no nucleotide sequence data of any *Drosophila* olfactory receptor gene has yet been reported. *Drosophila* should be very useful as material for studies of molecular population genetics and molecular evolution. To elucidate the genetic basis of olfaction and the mechanisms that maintain genetic diversity, we have begun to clone the olfactory receptor genes of *Drosophila*, based on three kinds of methods; the polymerase chain reaction (PCR), cDNA library screening, and the enhancer trap assay:

1. In PCR amplification we have constructed six series of PCR primers, corresponding to the highly conserved regions between the putative olfactory receptor genes of the rat [1] and the catfish [2]. Many PCR products were amplified and sequenced, but unfortunately, we found no clone showing significant homology to the olfactory receptor genes published. At present, we are examining other DNA products, using the primers 5′ CGGAGCTCGA (CT) (AC)GITA(CT)GTIGCIAT(ACT)TG and 3′ GCGGATCCTA(AGT)AT(AG)AAIGG(AG) TTIA(AG)CAT.

2. cDNA library screening plays an important role in our experiment. We are constructing libraries from the whole body and the head and we have prepared a series of oligonucleotides for the use of library screening. After performing Southern blot hybridization experiments, we found two appropriate oligonucleotides as a probe.

3. We are planning to isolate genome and cDNA clones by means of the enhancer trap assay. Following jump starter methods for the construction of insertional mutant lines with the P-lwB vector, we (R.U. and D.Y.) have obtained nearly 1000 homozygous fertile lines which may contain a single P-1wB insert in the second or third chromosome. After staining for β-galactosidase expression, we carefully observed the third antennal segment and maxillary palpus, since these organs are believed to have olfactory function [5]. Thus far, from this assay, we have found several lines with interesting staining patterns. We will be continuing with all the experiments described above. After we have obtained the DNA sequence data, which we should have in the near future, we will discuss *Drosophila* olfactory systems from the molecular evolutionary viewpoint.

Acknowledgments. This study was supported by Grants in Aid from the Ministry of Education, Science, and Culture of Japan.

References

1. Buck L, Axel R (1991) A novel multigene family may encode odorant receptors: A molecular basis for odor recognition. Cell 65:175–187
2. Ngai J, Dowling MM, Buck L, Axel R, Chess A (1993) The family of genes encoding odorant receptors in the channel catfish. Cell 72:657–666
3. Carlson J (1991) Olfaction in *Drosophila*. Genetic and molecular analysis. Trends Neurosci 14:520–524
4. Hughes AL, Hughes MK (1993) Adaptive evolution in the rat olfactory receptor gene family. J Mol Evol 36:249–254
5. Ayer RK, Carlson J (1992) Olfactory physiology in the *Drosophila* antenna and maxillary palp: *acj6* distinguishes two classes of odorant pathways. J Neurobiol 23:965–982

[1] Department of Biology, Ochanomizu University, 2-1-1 Ohtsuka, Bunkyo-ku, Tokyo, 112 Japan
[2] Department of Biology, Brandeis University, Bassine 235 Waltham, MA 02254-9110, USA
[3] Mitsubishi Kasei Institute of Life Sciences, 11 Minamiooya, Machida, Tokyo, 194 Japan

Chemical Characterization of Odortypes in Mice

Tatsuhiko Yajima[1], Alan G. Singer[2], Gary K. Beauchamp[2], and Kunio Yamazaki[2]

Key words. MHC-determined odortypes—Mouse urine

The odors of individual mice are specified by the genes of the major histocompatibility complex (MHC), which are the same genes that help regulate the immune response. Specifically, mice can discriminate urine samples obtained from strains of inbred mice that are genetically identical except for genes of the MHC [1]. The odors specified by the MHC genes have been called odortypes. These odors are involved in mate selection, which is strongly biased toward preference for a mate of a histocompatibility type different from that of the individual's parents. This tendency for outbreeding appears to be important to maintain the genetic diversity of MHC alleles and serves to help avoid inbreeding.

Our research goal is to understand how the genes of the MHC specify the odortype. Identifying the chemicals that make up the odortypes in mice is the most important initial step in attaining this goal. This study was carried out to identify the chemical basis of MHC odortypes in mice. Urine samples from two inbred congenic strains of mice differing in many genes throughout the MHC complex (C57BL/6, B6 and C57BL/6-H-2^k, B6-H-2^k) were used for these experiments.

Urine was ultrafiltered with an Amicon Centricon-10 to remove proteins larger than 10 000 molecular weight. Ultrafiltrates containing no proteins were readily discriminated by mice [1]. In this study, MHC-determined odorants in ultrafiltered urine samples were extracted by the solid phase extraction method. The sorbents used were silica-based materials chemically modified with functional groups possessing non-polar (C18, octadecyl; C8, octyl; CH, cyclohexyl; PH, phenyl), polar (CN, cyanopropyl; 2OH, diol; SI, silica; NH$_2$, aminopropyl; PSA, ethylenediaminepropyl), ion exchange (SCX, propylbenzenesulfonyl; SAX, trimethylamino-

propyl), or carbohydrate-selective (PBA, propylphenylboronic acid) characteristics. Ultrafiltered B6 and B6-H-2^k urine samples were separately passed through different tubes containing identical sorbents. The olfactory differences between effluents were bioassayed by the Y-maze discrimination technique [1].

The five mice participating in these tests were trained as previously reported [2]. Three of them were reinforced for choosing the urine odor of B6 male mice and two were reinforced for choosing B6-H-2^k urine odors. Repeated screening tests demonstrated that NH$_2$ and PSA were better at retaining odortype constituents than any other sorbents. The strong anion exchanger, SAX, did not alter odortypes. These observations suggest that MHC-determined odorants consist of very polar molecules and interact with amino groups by polar interaction rather than by ionic interaction mechanisms. PBA did not retain odortype constituents; this result indicates that the odorants have no vicinal diol moieties in their molecular structures. MHC odorants in urine samples were also found to be effectively extracted by the solvent isopropanol-hexane (1:1) when the urine was saturated with sodium sulfate and acidified with phosphoric acid. The less polar isopropanol-hexane solvent matrix improved the extraction efficiency of the NH$_2$ sorbent for odortype constituents. MHC-regulated odorants, therefore, may have carboxylic groups. This idea was further supported by the esterification reaction with 1-pyrenyldiazomethane, which abolished the discriminability of extracts from B6 and B6-H-2^k urine samples. From these experiments, we obtained a basis for the isolation and chemical characterization of MHC-determined odors in mice.

References

1. Singer AG, Tsuchiya H, Wellington JL, Beauchamp GK, Yamazaki K (1993) Chemistry of odortypes in mice: Fractionation and bioassay. J Chem Ecol 19: 569–579
2. Yamazaki K, Beauchamp GK, Imai Y, Bard J, Phelan SP, Thomas L, Boyse EA (1990) Odortypes determined by the major histocompatibility complex in germ-free mice. Proc Natl Acad Sci USA 87:8413–8416

[1] Saitama Institute of Technology, 1690 Fusaiji, Okabe, Saitama, 369-02 Japan
[2] Monell Chemical Senses Center, Philadelphia, PA 19104-3308, USA

Induction of Olfactory Receptor Sensitivity in Mice

Hai-Wei Wang, Charles J. Wysocki, and Geoffrey H. Gold[1]

Key words. Olfaction—Receptor cell—Plasticity—Specific anosmia

Introduction

In 1989, Wysocki, Dorries, and Beauchamp [1] reported that exposure to the odorant androstenone induced olfactory sensitivity to that odorant in individuals that could not perceive the odor of androstenone previously. This striking observation challenged the existing view that specific anosmia is a genetic defect, analogous to color blindness. To investigate the cellular and molecular mechanisms underlying olfactory induction, we attempted to observe induction using electro-olfactogram (EOG) recordings from mice. If successful, this would establish an animal model for studying induction and would also demonstrate that induction occurs, at least in part, at the receptor cell level. We also attempted to observe induction with two odorants, androstenone and isovaleric acid, to determine if induction is a general phenomenon. A detailed description of our results has appeared recently [2].

Materials and Methods

Mice were exposed to an odorant for around 4 weeks (16 h daily) and were sacrificed for EOG recording. Recordings were made from an exposed and an unexposed (control) animal each day to minimize procedural variability, and the recordings and data analysis were done without knowledge of which animal was control or exposed. The mice used were of inbred strains that had relatively high or low behavioral sensitivity to the exposure odorant, to determine whether induction was correlated with low olfactory sensitivity, as suggested by psychophysical data [1].

Results

In androstenone-insensitive mice (NZB/B1NJ strain), exposure to androstenone caused up to a 3.6-fold increase in the EOG response amplitude to androstenone, but did not significantly affect responses to isoamyl acetate. However, in androstenone-sensitive mice (CBA/J strain) androstenone exposure had no effect on response amplitudes to either odorant. These findings demonstrate that induction of androstenone sensitivity: (1) can occur in mice, (2) occurs, at least in part, at the receptor cell level, (3) occurs only in a strain that initially has low sensitivity to androstenone, and (4) increases sensitivity specifically for androstenone.

To test the generality of this phenomenon, induction of sensitivity to isovaleric acid was attempted in strains which are relatively insensitive (C57BL/6J) and sensitive (AKR/J) to isovaleric acid. The results paralleled those with androstenone, i.e., isovaleric acid exposure increased isovaleric acid response amplitude (3.4-fold) only in the C57BL/6J strain and had no effect on responses to isoamyl acetate in either strain.

Discussion

Our findings demonstrate that induction of olfactory sensitivity is observable in mice and that this phenomenon closely resembles the induction of olfactory sensitivity that has been demonstrated psychophysically in humans [1], i.e., the sensitivity increase is specific for the exposure odorant and induction occurs only in animals that initially have low sensitivity to that odorant. Therefore, demonstrating induction in mice provides an animal model for studying this phenomenon at the cellular and molecular levels. The fact that induction can occur with two unrelated odorants, i.e., androstenone and isovaleric acid, provides evidence that induction may be a general property of specific anosmias. Because a large fraction of the human population exhibits specific anosmia to at least one odorant (50% are anosmic to androstenone alone), our findings suggest that a large fraction of the human population may experience changes in olfactory sensitivity and perception as a result of olfactory experience. This phenomenon may need to be considered in research that requires stable measurements of olfactory sensitivity.

References

1. Wysocki CJ, Dorries K, Beauchamp GK (1989) Proc Natl Acad Sci USA 86:7976–7978
2. Wang H-W, Wysocki CJ, Gold GH (1993) Induction of olfactory receptor sensitivity in mice. Science 260:998–1000

[1] Monell Chemical Senses Center, 3500 Market Street, Philadelphia, PA 19104-3308, USA

5. Structure and Function of Flavor Compounds

Structures and Functions of Antisweetness Substances, Sweetness-Inducing Substances, and Sweet Proteins

Yoshie Kurihara[1]

Key words. Gymnemic acid—Zizipin—Mabinlin—Curculin—Miraculin—Strogin

This chapter reviews the progress in exploring the structures and functions of antisweet substances (gymnemic acid, ziziphin), sweet proteins (mabinlin, curculin), and sweetness-inducing substances (strogin, miraculin, curculin). Table 1 summarizes the sources and functions of these substances.

Antisweet Glycosides

It has been known that the leaves of *Gymnema sylvestre* contain the active principle gymnemic acid, function of which is to suppress sweet taste. During the late 1960s, Reichstein and coworkers purified gymnemic acid and determined its chemical structure [1]. The sample they isolated, however, contained many homologs in at the high-pressure liquid chromatography (HPLC) level. We purified two of the homologs and determined their structures to be glucuronide of triterpene whose OH group at C-21 in the genin is esterified with 2-methylbutyric acid or 2-methylcrotonic acid [2]. Hydrolysis of the acyl group led to compete loss of activity.

A number of gymnemic acid homologs were isolated by Arihara's group in Tokushima-Bunri University [3]. The structural difference between these homologs is attributed to the number and species of acyl groups in gymnemagenin, which has six hydroxy groups. In general, the antisweet activity of gymnemic acid homologs decreases with decreasing number of acyl groups in genin.

The leaves of the plant *Ziziphus jujuba* contain a sweetness-inhibiting substance named ziziphin. We purified ziziphin and determined its structure [4]. It was a glycoside of triterpene, jujubogenin. Antisweet activity of ziziphin suppressed the taste of all the sweeteners examined (D-glucose, D-fructose, stevioside, sodium saccharin, and aspartame), but showed no suppressive effect on the salty taste of NaCl, the sour taste of HCl, or the bitter taste of quinine.

Sweet Protein

Plants of *Capparis masaikai* Lévl. grow in the south of Yunnan in China and bear fruit the size of tennis balls. The seeds of the mature fruit have a sweet taste, and the sweet principle was named mabinlin. In collaboration with Prof. Hu of the Kunming Institute of Botany, we purified five homologs of mabinlin [5]. Among five homologs, mabinlin II is most heat-stable. Its sweet activity was unchanged by incubating a solution of it at 80°C over 48h. On the other hand, the sweet activity of mabinlin I-1 was abolished by 1h of incubation at 80°C. The circular dichroism spectrum of mabinlin II was unchanged by a 1-h incubation at 80°C, whereas that of mabinlin I-1 was greatly changed. Hence the α-helix of mabinlin II was unchanged by the 80°C incubation, whereas the helix of mabinlin I-1 was completely destroyed by it.

We determined the amino acid sequence of mabinlin II and mabinlin I-1. Both proteins are composed of an A-chain and a B-chain. There is a high similarity between the two proteins. Both proteins have eight cysteine residues and the same disulfide structure [6]. A-chains and B-chains are connected by two disulfide bridges, and there are two intrachain disulfide bonds in the B-chain (Fig. 1). The difference in heat stability of the proteins seems to derive from the difference in some part of the amino acid sequence of the proteins.

Sweetness-Inducing Proteins and Glycoside

Miracle fruit has unusual properties. For example, lemon elicits a sweet taste after chewing it. Dr. Beidler kindly supplied the seeds of miracle fruits. Prof. Asoh of our University germinated them and cultured the plants in the green house in Yokohama National University. A number of investigators have tried to purify the active principle miraculin, but none has succeeded in obtaining pure miraculin. We developed a new extraction method using 0.5M NaCl solution and succeeded in purifying miraculin [7]. Miraculin in the fruit exists as a homodimer of polypeptide with a molecular weight of 24 600 daltons. In the pure state, miraculin dimer aggregates into a tetramer. Both dimer and tetramer have taste-

[1] Department of Chemistry, Faculty of Education, Yokohama National University, Yokohama, 240 Japan

TABLE 1. Sources and functions of taste modifiers and sweet proteins.

Substance	Source	Growing district	Activity
Gymnemic acid	Leaves of *Gymnema sylvestre*	India	Suppression of sweetness
Ziziphin	Leaves of *Ziziphus jujuba*	North China	Suppression of sweetness
Strogin	Leaves of *Staurogyne merguensis*	Malaysia	Sweet taste
			Water → sweet taste
Mabinlin	Seeds of *Capparis masaikai*	South China	Sweet taste
Miraculin	Fruit of *Richadella dulcifica*	West Africa (Ghana)	Sour taste → sweet taste
Curculin	Fruit of *Curculigo latifolia*	Malaysia (Pennang)	Sweet taste
			Sour taste → sweet taste
			Water → sweet taste

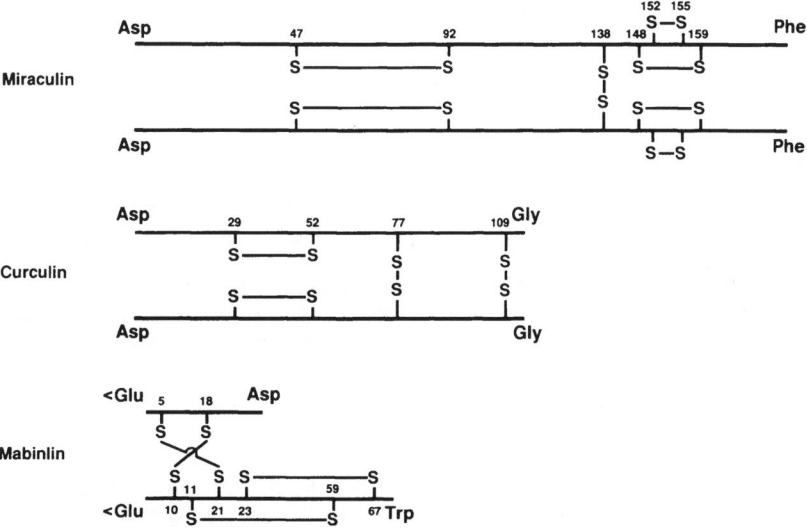

FIG. 1. Structures of mabinlin, miraculin, and curculin

modifying activity [8]. We then determined the primary structure of miraculin. Miraculin monomer is composed of 191 amino acid residues [9], and there are three intrachain disulfide bridges. The 138th cysteine residue forms an interchain disulfide bridge to form the dimer [8]. There are two sugar chains, which are connected to the 42nd and 186th asparagine residues. There are five species of sugar chains that distribute to the chains connected to the 42nd and 186th asparagine residues in different ratios. Among the five sugar chains, new types, not seen in other glycoproteins, were found [10]. Surprisingly, the amino acid sequence of miraculin has high homology with that of soybean trypsin inhibitor. The *N*-terminal sequence has 36.4% homology, and the *C*-terminal sequence has 51.1% homology. It is unknown why miraculin has high homology with trypsin inhibitor.

On the basis of the established amino acid sequence of miraculin, we constructed a DNA duplex using an automated solid-phase phosphite approach. The 601-basepair DNA duplex was inserted into the expression vector. Western blot-

ting analysis showed that miraculin is induced in *Escherichia coli*. This miraculin has shown no sweetness-inducing activity.

Curculigo latiforlia is grown under gum trees in Malaysia and at the root bears fruit with a sweet taste. In addition, after chewing the fruit, water elicits a sweet taste. For example, black tea elicits a sweet taste withiout sugar. Sour substances such as citric acid or ascorbic acid induce strong sweetness similar to that of miraculin. We purified the active principle of the fruits by a method similar to that employed for purification of miraculin, and are named it curculin [11]. Using the purified sample, we determined the amino acid sequence of curculin. Curculin is a homodimer of a polypeptide whose molecular weight is 27000 daltons. Curculin monomer is composed of 114 amino acid residues. It has four cysteine residues, which form two interchain disulfide bridges and one intrachain disulfide bridge (Fig. 1). We have succeeded in crystallizing curculin by the hanging-drop vapor diffusion method. The size of the crystal is sufficiently large for X-ray diffraction studies.

TABLE 2. Properties of mabinlin, miraculin, and curculin.

Property	Mabinlin	Miraculin	Curculin
Molecular weight			
Monomer	12 400	24 600	12 400
Dimer	—	43 000	27 800
Carbohydrate content (%)	None	13.9	None
Isoelectric point	11.3	9.1	7.1
Taste of protein	Sweet	No taste	Sweet
Taste of acids after protein	—	Sweet	Sweet
Taste of water after protein	—	—	Sweet

TABLE 3. Properties of antisweet and taste-modifying glycosides.

Property	Gymnemic acid	Ziziphin	Strogin
Aglycon	Triterpene	Triterpene	Triterpene
Carbohydrate	Glucuronic acid	Rhamnose (2 mol)	Rhamnose (1 mol)
		Arabinose (1 mol)	Xylose (2 mol)
Taste	Bitter	Bitter	Sweet
Effect on sweetness	Suppression	Suppression	—
Effect on water	—	—	Sweet

The maximum sweetness of curculin is equivalent to that of 0.3 M sucrose. The sweetness of curculin is abolished after it is held in the mouth, but, application of water to the mouth elicits a sweet taste. The maximum sweetness induced by water was equivalent to that of 0.3 M sucrose. Curculin can modify a sour taste into a sweet taste. The maximum sweetness induced by 0.1 M citric acid was equivalent to that of 0.3 M sucrose. As described earlier, the sweet taste of curculin disappeared a few minutes after holding it in the mouth. We examined the components in saliva that suppress the sweetness and found that 1 mM Ca^{2+} or Mg^{2+} completely suppressed the sweetness.

The taste-modifying action of curculin is explained as follows. The active site of curculin binds to the sweet receptor site, which induces sweet taste. The divalent cations in saliva suppress the stimulation of curculin on the sweet receptors. That is, the active site does not stimulate the sweet receptor site in the presence of the divalent cations, and hence the sweetness is abolished. Elimination of saliva by tasting water leads to the generation of sweet taste again. Similar to miraculin, curculin elicits a sweet taste in response to acids. In the presence of acids, divalent cations do not suppress the sweetness induced by curculin, probably because divalent cations do not bind to the receptor site at acidic pH. Therefore the sweetness lasts a longer time in the presence of acids.

To determine the active site of curculin, we carried out experiments using molecular cloning of curculin protein. We extracted RNA by the phenol-SDS method and used it to synthesize cDNA. Two clones with longer cDNA inserts were subjected to nucleotide sequencing analysis. The encoded prepro-curculin is composed of 158 amino acid residues, including a signal sequence of 22 residues and a carboxy-terminal extension peptide of 22 residues [12]. Curculin expressed in E. coli had no sweet activity at present.

Curculin antibody crossed with miraculin and vice versa [13], suggesting that there is a structure common to both proteins. However, there is no major similarity in amino acid sequence in the two proteins. In addition, there is no marked similarity between mabinlin, monellin, thaumatin, miraculin, and curculin. That these proteins have a sweet activity seems to have no biologic significance and to be accidental in the development. It is the reason why there is no marked amino acid sequence similarity among species of sweet proteins and sweetness-inducing proteins.

The leaves of Staurogyne mergensis Kuntze grown in Malaysia contain a principle that elicits sweet taste in response to water similar to curculin. We have purified the active principle and named it strogin. Strogin has a sweet taste. After strogin is held in the mouth, cold water elicits a sweet taste. The structure of strogin was determined as a glycoside of triterpene.

Tables 2 and 3 summarize properties of the proteins and glycosides discussed herein.

Acknowledgments. I started this line of study under the guidance of Prof. L.M. Beidler of the Florida State University. I wish to express my heartfelt gratitude to Prof. Beidler. I also express to my thanks to Prof. Asoh who cultured the plants used in the present study and to my collaborators: Profs. Y. Nakamura and K. Nakaya of Showa University; Prof. N. Takahashi of Nagoya City University; Profs.

Y. Arata and S. Arai, and Dr. K. Abe of University of Tokyo; Prof. T. Nishino of Yokohama City University; Prof. Z. Hu and Mr. X. Liu of Kunming Institute of Botany; Dr. S. Theerasilp of Ayudhya Teachers College; Mrs. H. Yamashita and T. Akabane of Asahi Denka Kogyo Co.; and Drs A. Hasegawa and K. Sakamoto of Tonen Co.

References

1. Stocklin W, Weise E, Reichstein T (1968) Gymnemasäure: das antisaccharine Prinzip von Gymnema sylvestre R. Br Helv Chim Acta 50:474–490
2. Maeda M, Iwashita T, Kurihara Y (1989) Studies on taste modifiers. II. Purification and determination of gymnemic acids, antisweet active principle from Gymnema sylvestre leaves. Tetrahedron Lett 30:1547–1550
3. Yoshikawa K, Amimoto K, Arihara S, Matsuura K (1989) Structure studies of new antisweet constituents from Gymnema sylvestre. Tetrahedron Lett 30:1103–1106
4. Kurihara Y, Ookubo K, Tasaki H, et al. (1988) Studies on the taste modifiers. I. Purification and structure determination of sweetness inhibiting substance in leaves of Ziziphus jujuba. Tetrahedron 44:61–66
5. Liu X, Maeda S, Hu Z, et al (1993) Purification, compete amino acid sequence and structural characterization of the heat-stable sweet protein, mabinlin II. Eur J Biochem 211:281–287
6. Nirasawa S, Liu X, Nishino T, Kurihara Y (1993) Disulfide bridge structure of the heat-stable sweet protein, mabinlin II. Biochim Biophys Acta 1202:277–280
7. Theerasilp S, Kurihara Y (1988) Compete purification and characterization of the taste-modifying protein, miraculin from miracle fruit. J Biol Chem 263:11536–11539
8. Igeta H, Tamura Y, Nakaya K, Nakamura Y, Kurihara Y (1991) Determination of disulfide array and subunit structure of taste-modifying protein, miraculin. Biochim Biophys Acta 1079:303–307
9. Theerasilp S, Hitotsuya H, Nakajo S, et al. (1989) Compete amino acid sequence and structure characterization of the taste-modifying protein, miraculin. J Biol Chem 264:6655–6659
10. Takahashi N, Hitotsuya H, Hanzawa H, Arata Y, Kurihara Y (1990) Structural study of asparagine-linked oligosaccharide moiety of taste-modifying protein, miraculin. J Biol Chem 265:7793–7798
11. Yamashita H, Theerasilp S, Aiuchi T, et al. (1990) Purification and compete amino acid sequence of new type of sweet protein with taste-modifying activity, curculin. J Biol Chem 265:15770–15775
12. Abe K, Yamashita H, Arai S, Kurihara Y (1992) Molecular cloning of curculin, a novel taste-modifying protein with sweet taste. Biochim Biophys Acta 1130:232–234
13. Nakajo S, Akabane T, Nakaya K, Nakamura Y, Kurihara Y (1992) An enzyme immunoassay and immunoblot analysis for curculin, a new type of taste-modifying protein: cross-reactivity of curculin and miraculin to both antibodies. Biochim Biophys Acta 1118:293–297

Further Studies of L-Ornithyltaurine HCl, A New Salty Peptide

HIDEO OKAI[1]

Key words. L-Ornithyltaurine—Salty peptide—Sodium ion dietary compounds—Enhancing effect of saltiness—Ornithyl-β-alanine—Delicious peptide

In 1984 our research group found that a series of basic dipeptides, such as H-L-Orn-Tau-OH·HCl [1] and others, possessed a salty state.

$$
\begin{bmatrix}
NH_3^+ \\
| \\
CH_2 \\
| \\
CH_2 \\
| \\
CH_2 \\
| \\
NH_3^+-CH-CONH-CH_2-CH_2-SO_3^-
\end{bmatrix} Cl^-
$$

These salty peptides are expected to have good effects on various adult diseases, such as hypertension, gestosis, diabetes mellitus, and so on because they contain no sodium ion. Our study began when we found that ammonium chloride (NH$_4$Cl) can be converted to other amino compounds that retain its saltiness.

We have focused our attention on the research and development of sodium ion dietary compounds since then. In this presentation, we report the latest information gained from our investigation.

Molecular Design of Salty Peptides and Saltiness: Chemical Structure Relations

The synthetic route of Orn-Tau is shown below:

In the original method, Z-Orn(Z)-OH and Tau were condensed by the HOSu method. The reaction mixture was acidified and then extracted with ethylacetate (AcOEt) and dried over anhydrous sodium sulfate for 3 hours. Z-Orn(Z)-Tau was obtained in poor yield because Z-Orn(Z)-Tau has high solubility in water. Today we can obtain it as a barium salt, and the barium salt is easily removed by H$_2$SO$_4$ in a good yield.

[1]Department of Fermentation Technology, Faculty of Engineering, Hiroshima University, 1-4-1 Kagamiyama, Higashihiroshima, Hiroshima, 724 Japan

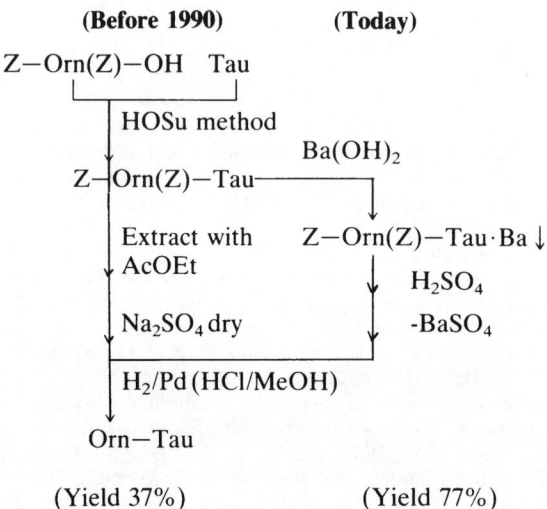

(Before 1990)   (Today)

(Yield 37%)   (Yield 77%)

The relation of salty and umami tastes are summarized in Fig. 1. It is suggested that exhibition of the umami taste requires that the strict length of the dipeptide skeleton be maintained (m = 1), as shown in Fig. 1. It is not, however, affected by the side chain length of the basic amino acid residue. The salty taste requires just the opposite.

Enhancing Effect of Saltiness of L-Orn-Tau

The saltiness of L-Orn-Tau and of ornithyl-β-alanine (OBA) [2] depends on the pH of the solution. Solutions of 30 mM L-Orn-Tau containing 0 to 1.3 equivalents of HCl were prepared by adding HCl to an HCl-free L-Orn-Tau solution. The result of the sensory analysis is shown in Table 1. When L-Orn-Tau contained HCl in less than 0.67 equivalents, it did not produce the salty taste. L-Orn-Tau containing 0.67 equivalents of HCl had increased saltiness, but the sourness originating from HCl became evident when more than 1.00 equivalent of HCl was added. We obtained the best saltiness by adding 1.2 equivalents of HCl to the L-Orn-Tau solution, which produced a weakly sour taste.

We prepared solutions of L-Orn-Tau·1.2 HCl in several concentrations and NaCl solutions that produced the same saltiness in each concentration, as shown Table 2. L-Orn-Tau needs about 0.9 times the weight percentage (or 22% molar concentration)

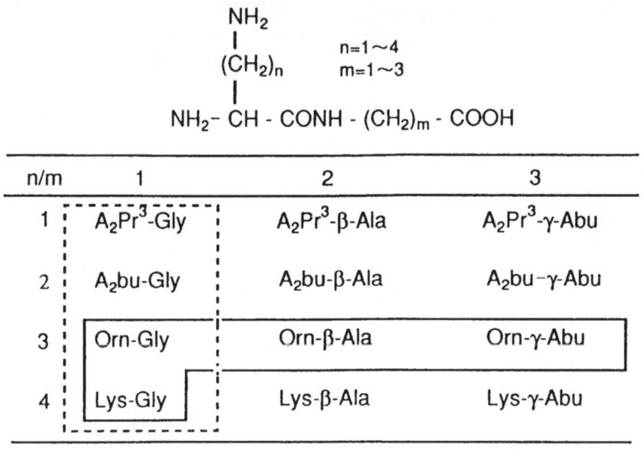

$$NH_2$$
$$|$$
$$(CH_2)_n \qquad n=1\sim4$$
$$| \qquad\qquad m=1\sim3$$
$$NH_2-CH-CONH-(CH_2)_m-COOH$$

n/m	1	2	3
1	A$_2$Pr^3-Gly	A$_2$Pr^3-β-Ala	A$_2$Pr^3-γ-Abu
2	A$_2$bu-Gly	A$_2$bu-β-Ala	A$_2$bu-γ-Abu
3	Orn-Gly	Orn-β-Ala	Orn-γ-Abu
4	Lys-Gly	Lys-β-Ala	Lys-γ-Abu

FIG. 1. Relation of salty and umami tastes. *Solid lines* indicate the salty peptides and *dashed lines* the umami peptides

TABLE 1. Influence of HCl on the salty taste of Orn-Tau.[a]

HCl (eq.)	pH	Score	Sourness[b]
0	8.9	0	
0.11	8.0	0	
0.67	7.0	1	
1.00	6.4	2	±
1.10	**6.1**	**2.5**	+
1.20	5.9	3	+
1.30	3.4	3.5	++

[a] Boldface entries indicate the point at which the best saltiness was obtained.
[b] Measured in 30 mM Orn-Tau concentration.
[c] Scoring: 5 = 0.5% NaCl; 3 = 0.25% NaCl; 1 = 0.1% NaCl.

TABLE 4. Sodium ion enhancing effect.

Substance	Mol conc. base (%)
Orn-Tau	85
OBA	41
Gly-OEt·HCl	41
Lys·HCl	19

to produce the same saltiness as NaCl. The taste behaviour is similar to the exhibition of OBA saltiness investigated by Seki et al. [2] except for the minimum equivalent of HCl for saltiness production and HCl equivalents for excellent saltiness production (in the case of OBA 1.3 HCl).

We then prepared solutions of L-Orn-Tau·1.2 HCl and NaCl and mixed them in several ratios, as shown in Table 3. All of the solutions in each of these series had the same degree of saltiness. Hence the saltiness of L-Orn-Tau·1.2 HCl and NaCl was produced independently. In the case of 60 mM L-Orn-Tau·1.2 HCl and 0.063% NaCl, the sourness originating from HCl had almost disappeared even though the pH was still acidic.

We have reported elsewhere on the enhancing effect on saltiness by several compounds, such as OBA. Gly-OEt·HCl [3,4], and basic amino acids [3–5]. These effects are shown in Table 4. It is reported that the average sodium chloride intake of Japanese is 12 g/day. The aim of a restricted sodium chloride diet is to decrease the intake to 6 g/day for individuals whose disease is not serious and to 3 g/day for those with serious conditions. In other words,

TABLE 2. Strength of the salty taste of Orn-Tau·1.2 HCl.[a]

Orn-Tau·1.2 HCl aq. (mM)	Score	NaCl aq. (mM)	Score
120 (2.88%, pH 5.1)	+6.0	107.0 (0.625%)	+6
100 (2.40%, pH 5.2)	+5.0	85.6 (0.5%)	+5
60 (1.44%, pH 5.3)	+4.0	64.2 (0.375%)	+4
45 (1.08%, pH 5.4)	+3.5	—	—
30 (0.72%, pH 5.3)	+3.0	42.8 (0.25%)	+3
15 (0.36%, pH 5.4)	+1.0	10.7 (0.063%)	+1

[a] Threshold value of Orn-Tau·1.2 HCl is 7.59 mM.

TABLE 3. Enhancement of saltiness of Orn-Tau·1.2 HCl by adding NaCl.[a]

Orn-Tau · HCl + NaCl		pH	Saltiness
100 mM (2.40%)	0% (0 mM)	5.2	+5 (contains weak sourness)
60 mM (1.44%)	**0.063% (10.7 mM)**	**5.2**	**+5** (contains very weak sourness)
30 mM (0.72%)	0.125% (21.4 mM)	5.3	+5
15 mM (0.36%)	0.375% (64.2 mM)	5.3	+5
0 mM (0%)	0.500% (85.6 mM)		+5

[a] Boldface entries indicate the point at which the sourness of Orn-Tan·HCl almost completely disappeared.

TABLE 5. Taste of "fragment-shifted" delicious peptides and analogs.

Compound	Taste	Threshold value (mM)[a]
Ser-Leu-Ala-Asp-Glu-Glu-Lys-Gly	Umami/sour	1.25
Ser-Leu-Ala-Lys-Gly-Asp-Glu-Glu	Sour/umami	1.50
Lys-Gly-Ser-Leu-Ala-Asp-Glu-Glu	Sour/umami/sweet	0.78
Lys-Gly-Asp-Glu-Glu	Sour/salty/sweet	2.50
Glu-Glu-Asp-Gly-Lys	Sour/umami/sweet/astringency	1.25
Lys-Gly-Asp-Glu-Glu-Ser-Leu-Ala	Umami/sour	1.41

[a] For a umami or salty taste.

the purpose of a sodium chloride diet is to reduce the amount of sodium chloride, respectively, to one-half and one-fourth that usually consumed.

Enhancing Effects Trials

Production of Low-Sodium Soy Sauce

In 1991 Muramatsu et al. reported an improved method for manufacturing soy sauce. By their method, especially during the aging process, the NaCl concentration could be reduced to one-third that in conventional soy sauce. The amino acid content of the product is similar to that of natural soy sauce, the only difference being in the NaCl concentration. We studied this preparation of artificial soy sauce by mixing the reduced NaCl product with salty peptides or substances that enhance saltiness (e.g., Gly-OEt·HCl and Lys·HCl) developed in our laboratory. We confirmed that the samples prepared by this method had effects similar to those seen with the model systems. That is, when we added salty peptides (ratio of peptides/NaCl = 2:1), the resulting samples had degrees of saltiness similar to that of natural soy sauce.

Delicious Peptide from Beef Soup

In 1978 Yamasaki and Maekawa isolated an octapeptide from beef soup [6]. Its primary structure was proposed as Lys-Gly-Asp-Glu-Glu-Ser-Leu-Ala. In 1987 our research group synthesized an octapeptide that possessed the simple umami taste and that had a threshold value approximately the same as that of monosodium glutamate and with a slightly sour taste [7]. It was found that the taste in produced by three components: Lys-Gly (salty and umami), Asp-Glu-Glu (sour), and Ser-Leu-Ala (bitter). We then synthesized several fragment-shifted analogs. As shown in Table 5, The tastes of these analogs are almost the same. This situation is completely different from the case of bitterness and sweetness. Probably mechanisms that produce sweetness and bitterness are different from those that produce other tastes.

Conclusion

L-Orn-Tau·HCl is the best "salty peptide" for use as an effective dietary substitute. Other salty peptides are also effective, but they are not expected to be used as sodium chloride substitutes in the near future because of the difficulty of their synthesis and their high cost. In addition, the saltiness of these other peptides is in the range of that of sodium chloride. The problem in the next stage of development is how to prepare extremely salty substances. Our plan is to search for new salty peptides that contain both cations and anions in a molecule such as:

Orn-Pro-Tau   cyclo-(Orn-Asp)

References

1. Tada M, Shinoda I, Okai H (1984) L-Ornithyltaurine, a new salty peptide. J Agric Food Chem 32:992–996
2. Seki T, Kawasaki Y, Tamura M, Tada M, Okai H (1990) Further study on the salty peptide ornithyl-β-alanine: some effects of pH additive ions on the saltiness. J Agric Food Chem 38:25–29
3. Kawasaki Y, Seki T, Tamura M, et al. (1988) Glycine methyl or ethyl ester hydrochloride as the simplest examples of salty peptides and their derivatives. Agric Biol Chem 52:2679–2681
4. Tamura M, Seki T, Kawasaki Y, et al. (1989) An enhancing effect on the saltiness of sodium chloride of added amino acids and esters. Agric Biol Chem 53: 1625–1633
5. Tamura M, Kawasaki Y, Seki T, Tada M, Okai H (1989) Enhanced effects of saltiness of sodium chloride using amino acids and their esters. In: Ueki M (ed) Peptide chemistry 1988. Protein Research Foundation, Osaka, pp 35–38
6. Yamasaki Y, Maekawa K (1978) A peptide with delicious taste. Agric Biol Chem 42:1761
7. Tamura M, Nakatsuka T, Tada M, et al. (1989) The relationship between taste and primary structure of "delicious peptide" (Lys-Gly-Asp-Glu-Glu-Ser-Leu-Ala) from beef soup. Agric Biol Chem 53:319–325

Behavioral Evidence for the Primacy of the Chorda Tympani Nerve in Sodium Chloride Taste

ALAN C. SPECTOR[1] and HARVEY J. GRILL[2]

Key words. Salt—Ingestive behavior—Glossopharyngeal nerve—Nerve transection—Taste coding

Introduction

A conceptually rich depiction of sensory function can result from the comparison of behavioral and electrophysiologic data. Until recently, however, the correspondence between the electrophysiology of the peripheral gustatory system and the rat's behavioral responsiveness to NaCl remained obscure. This chapter briefly reviews the interpretive limitations of the early behavioral work and confirms the value of animal psychophysics in revealing the functional significance of the chorda tympani nerve (CT) with respect to NaCl taste sensibility.

It has long been known that the rat CT, which innervates the taste receptors on the front of the tongue, is highly responsive to NaCl [e.g., 1]. With the advent of single-fiber recording techniques, it became clear that many taste fibers respond to more than one chemical stimulus. There are some fibers, however, that show narrow tuning.

A population of CT fibers that respond with relative specificity to Na and Li salts have been identified [2]. These are referred to as N units and are contrasted with the H units of the chorda tympani that respond nonselectively to sodium and non-sodium salts, HCl, and quinine. There are sodium-responsive fibers in the glossopharyngeal nerve (GL), but, unlike the N units, these fibers do not selectively respond to sodium [3]. Boudreau and coworkers independently confirmed these findings by recording from single units in the respective ganglia of these two nerves [4,5].

Based on these data, one might expect that the CT nerve plays a significant role in the peripheral coding of sodium taste. Surprisingly, evaluation of the behavioral effects of CT section uniformly revealed normal NaCl intake and preference [e.g., 6,7]. One interpretation was that the remaining taste buds maintained the behavior; after all, the CT nerve innervates only 15% of the total oropharyngeal taste bud population. Yet when more extensive denervation of the oral cavity was examined, dramatic effects were not seen. Thus the correspondence between the electrophysiology and the behavior remained to be elucidated.

An erroneous inference from these studies was that the CT nerve does not provide a unique input that characterizes NaCl taste. The basis for this misleading inference involves the limitations and assumptions associated with the long-term intake test methodology used to assess sensory function. There are at least two weaknesses associated with such tests. First, they rely on the inherent hedonic characteristics of the stimuli to drive behavior. This reliance could be problematic, especially at low concentrations or with taste stimuli that are neither preferred nor avoided. Also, changes in the discriminative properties of a taste stimulus (e.g., taste quality) as the result of a given manipulation can potentially go unnoticed because the motivational properties of the stimulus have not been significantly altered. Second, nongustatory factors, such as postingestive stimuli, can influence ingestion.

We sought to overcome the shortcomings associated with the intake test methodology and more directly challenge the gustatory capacities of intact and nerve-sectioned rats with respect to NaCl. We have conducted, along with our associates, several experiments, all of which lead to the same conclusion. Namely, that the afferent input provided by the CT nerve plays a significant role in the coding of sodium taste. The remainder of this chapter focuses on three experiments that support this conclusion.

Experiments

We began by incorporating several methodological features into our behavioral tests. First, some of our experiments employed instrumental conditioning procedures in which taste was used as a signal for some other event (e.g., reinforcement, punishment). Consequently, gustatory responsiveness was assessed, independent of hedonic processes. Second, small volumes of taste stimuli were presented. Third,

[1] Department of Psychology, University of Florida, Gainesville, FL 32611, USA
[2] Department of Psychology, University of Pennsylvania, Philadelphia, PA 19104, USA

immediate behavioral responses to the taste stimulus were quantified. The latter two features raised the confidence that the behavior was under oral sensory control. The experiments were conducted in specially designed computer-operated gustometers that afforded precise control of stimulus delivery and automatic monitoring of licking behavior [8,9].

NaCl Detection Thresholds

In the first experiment detection thresholds were assessed in a taste-signaled shock avoidance paradigm [10]. The rats' sensitivity to low concentrations of NaCl and sucrose was measured before and after bilateral CT section. In this procedure, 5-s stimulus trials were intermittently presented, and the rats were trained to *maintain* spout licking when the stimulus was water and to *suppress* spout licking when the stimulus was taste. Threshold determination was accomplished by lowering the limit of the concentration range of stimuli across sessions until it included perithreshold concentrations. The results indicated that CT section increased the NaCl detection threshold by about two orders of magnitude. In contrast, the sucrose detection threshold, measured in the same rats during the same session, was only marginally affected by the CT section. These results have since been replicated using a different operant conditioning procedure [11]. It is also worth noting that in the few rats we have tested to date, GL section has had no consistent effect on NaCl detection thresholds.

Although it is clear that CT section substantially raises the detection threshold for NaCl, these rats can still detect higher concentrations of NaCl. This finding leads to the question: does deafferentation of the taste receptors on the anterior tongue merely

reduce the intensity of the NaCl signal, or does it also alter the signal's qualitative signature? The electrophysiologic data that were described previously can be interpreted to suggest that the CT provides a unique contribution to the peripheral signal representing NaCl. The second experiment we describe provides compelling psychophysical evidence supporting this hypothesis.

NaCl Versus KCl Discrimination

In this experiment a modified version of the taste-signaled shock avoidance procedure was used to test the rats' performance on an NaCl *vs* KCl discrimination task, before and after selective gustatory denervation [12]. The main difference between this procedure and the one used to measure taste thresholds was that instead of pitting taste against water, we pitted two tastes against each other. Half of the rats were trained to maintain licking for KCl and to suppress licking to NaCl, and the other half were trained with the opposite contingencies. Three concentrations (0.05, 0.10, 0.20 M) of each stimulus were presented in an effort to make intensity an irrelevant cue. The design favored discrimination based on quality, not intensity.

Figure 1 depicts the performance of two representative rats on the four sessions before and after bilateral CT section, collapsed across concentration. The response distributions represent licks during the last 3 s (avoidance period) of the 5-s taste trial for KCl and NaCl. Potassium chloride signaled shock for rat 1A and NaCl signaled shock for rat 3A. Note that before CT section the NaCl and KCl distributions are clearly separate; after CT section there is a substantial breakdown in discrimination performance as indicated by the extensive overlap in .

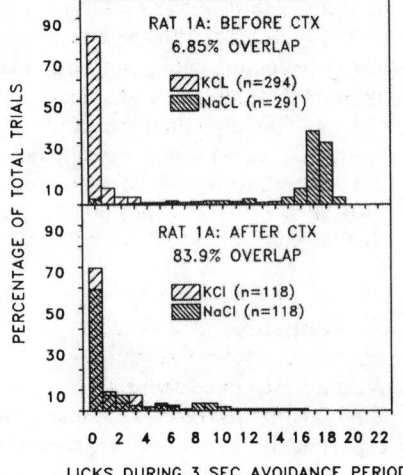

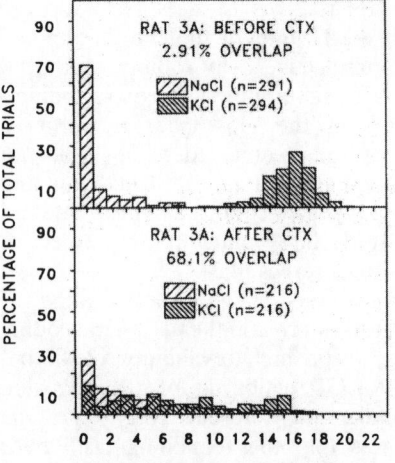

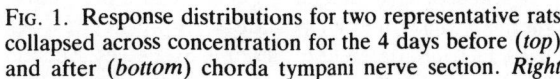

FIG. 1. Response distributions for two representative rats collapsed across concentration for the 4 days before (*top*) and after (*bottom*) chorda tympani nerve section. *Right*

hatched bars: stimulus signals shock; *left hatched bars*: stimulus signals the absence of shock. (From Spector and Grill [12], with permission)

the areas of the two distributions. Similar effects have been obtained in all of the rats that we have tested to date.

There is apparently some neuro- and chemo-specificity involved in this effect. When the GL is sectioned, removing four times as many taste buds, rats show no impairment in their salt discrimination performance. Section of the CT does not appear to compromise the rats' performance in a sucrose *vs* quinine discrimination task.

These results tell us that the discriminability between NaCl and KCl is impaired when the afferent input of the CT nerve is removed. There are, however, some caveats worthy of consideration.

First, we do not wish to claim that rats with CT sections are entirely unable to discriminate between these two salts. With further training, such rats might be able to use discriminable elements that potentially remain in the signals to guide their performance. Moreover, these results pertain to the particular concentrations used in this experiment.

Second, it is important to recognize that in addition to afferent axons, the chorda tympani nerve contains preganglionic parasympathetic efferents supplying glandular and vascular structures associated with the oral cavity. Most notably, the CT provides a portion of the innervation of the sublingual and submaxillary salivary glands. Perhaps the effects of CT section, although taste-related in origin, are attributable to the interruption of these efferents, which in turn detrimentally affects the function of all oropharyngeal taste buds.

Although our experimental results do not explicitly discount this possibility, Brosvic and Hoey [13] have reported that removal of the sublingual and submaxillary glands in rats affects neither NaCl detectability nor performance on an NaCl *vs* KCl discrimination task. Moreover, the chemospecificity associated with these effects argue against a general debilitation of taste bud function.

A final consideration is that these findings do not inherently tell us which signal is affected by the nerve section. It could be NaCl, KCl, or both. Nevertheless, it is most parsimonious to assume that the afferents that show the greatest differential response to the two salts are the principal components. Accordingly, the so-called N-units [2] appear to be the best candidates. This hypothesis gains support from behavioral [e.g., 14] and electrophysiologic [e.g., 15] data involving the epithelial sodium transport blocker amiloride.

Depletion-Induced Sodium Appetite

The final behavioral experiment we would like to discuss attacks the problem from a more ecologic perspective. When rats are depleted of sodium, they enhance their intake of NaCl even at concentrations that are normally avoided [e.g., 16]. There

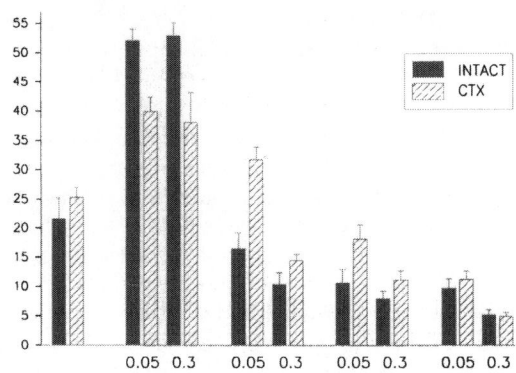

FIG. 2. Mean ($\pm$ SE) licks to the various taste stimuli during 10-s trials in intact rats (*solid bars*) and rats with bilateral chorda tympani section (*hatched bars*). All rats were tested in a sodium-depleted state. (From Breslin et al. [17], with permission)

is evidence suggesting that this adaptive response, commonly referred to as sodium appetite, is innate, taste-guided, and specific to the sodium cation [16]. Accordingly, we, along with Breslin, hypothesized that if the information in the CT nerve was critical to the rat's identification of NaCl, then section of this nerve should compromise the rats performance in a sodium appetite paradigm.

In this experiment [17], intact rats and rats with bilateral CT sections were depleted of sodium by administration of furosemide and given a sodium-deficient diet. These rats were otherwise not deprived. The next day they were tested for their unconditioned licking responses to randomly delivered 10-s trials of water, NaCl, and various non-sodium chloride salts. The salts were presented at two concentrations, 0.05 M and 0.30 M.

The mean number of licks to the various stimuli are illustrated in Fig. 2. Note the sharpness of the response profile for sodium-depleted intact rats. They responded best to the NaCl solutions and did not respond much to the other nonsodium salts. In contrast, the response profile was broadened in the rats with bilateral CT section. There was a decreased response to sodium chloride and increased response to some of the other salts, especially the low concentrations of KCl and NH_4Cl. Nevertheless, the sodium appetite was apparently not completely eliminated by CT section.

Interestingly, the sodium-depleted rats with CT section did discriminate between 0.3 M KCl and both concentrations of NaCl. This result is seemingly inconsistent with the effect of CT section on salt discrimination performance in the shock avoidance paradigm. There are several methodologic factors that may contribute to this disparity. First, the highest concentration tested was different in the two experiments. As mentioned previously, the effect of CT section on salt discrimination may be concentration-dependent. Second, discrimination behavior was measured under different physiologic states in the two experiments: water deprivation *vs* sodium depletion. It is known that challenges to sodium homeostasis can have effects on the gustatory system [e.g., 18]. Third, these tasks differ in the circumstances driving the behavior. In the case of sodium appetite the response is presumably based on the comparison of the taste signal with an innate template or a search image for sodium [e.g., 16]. With the shock avoidance task, the response is based (at least initially) on the comparison of the taste signal with preoperatively established contingencies. Despite these differences, the findings from the two studies, with the above-noted exception, are largely in agreement.

Hence it appears as though the signal that remains after CT section is sufficient to allow the affective neural circuits that are potentiated by the combination of sodium taste and sodium depletion to show partial competence, but the specificity and robustness of the behavior are clearly compromised.

Conclusion

The three experiments discussed here collectively demonstrate the primacy of the chorda tympani nerve with respect to NaCl taste sensibility in the rat. As shown above, the search for correspondences between single-neuron electrophysiologic data and the results from appropriate behavioral techniques has been effective in revealing part of the functional organization of the peripheral gustatory system with respect to sodium sensibility in the rat.

Acknowledgments. This research was suported by USPHS grants DC-00161 to A.C.S. and MH-43787 to H.J.G. We would like to thank Steven St. John for reading an earlier version of this manuscript.

References

1. Beidler LM (1953) Properties of chemoreceptors of tongue of rat. J Neurophysiol 16:595–607
2. Frank ME, Contreras RJ, Hettinger TP (1983) Nerve fibers sensitive to ionic taste stimuli in chorda tympani of the rat. J Neurophysiol 50:941–960
3. Frank ME (1991) Taste-responsive neurons of the glossopharyngeal nerve of the rat. J Neurophysiol 65:1452–1463
4. Boudreau JC, Hoang NK, Oravec J, Do LT (1983) Rat Neurophysiological taste responses to salt solutions. Chem Senses 8:131–150
5. Boudreau JC, Do LT, Sivakumar L, Oravec J, Rodriguez CA (1985) Taste systems of the petrosal ganglion of the rat glossopharyngeal nerve. Chem Senses 12:89–127
6. Pfaffmann C (1952) Taste preference and aversion following lingual denervation. J Comp Physiol Psychol 45:393–400
7. Vance WB (1967) Hypogeusia and taste preference behavior in the rat. Life Sci 6:743–748
8. Spector AC, Grill HJ (1988) Differences in the taste quality of maltose and sucrose in rats: issues involving the generalization of conditioned taste aversions. Chem Senses 13:95–113
9. Spector AC, Andrews-Labenski J, Letterio FC (1990) A new gustometer for psychophysical taste testing in the rat. Physiol Behav 47:795–803
10. Spector AC, Schwartz GJ, Grill HJ (1990) Chemospecific deficits in taste detection after selective gustatory deafferentation in rats. Am J Physiol 258:R820–R826
11. Slotnick BM, Sheelar S, Rentmeister-Bryant H (1991) Transection of the chorda tympani and insertion of ear pins for stereotaxic surgery: equivalent effects on taste sensitivity. Physiol Behav 50:1123–1127
12. Spector AC, Grill HJ (1992) Salt taste discrimination after bilateral section of the chorda tympani or glossopharyngeal nerves. Am J Physiol Regul Integr Comp Physiol 263:R169–R176
13. Brosvic GM, Hoey NE (1990) Taste detection and discrimination performance of rats following selective desalivation. Physiol Behav 48:617–623
14. Hill DL, Formaker BK, White KS (1990) Perceptual characteristics of the amiloride-suppressed sodium chloride taste response in the rat. Behav Neurosci 104:734–741
15. Ninomiya Y, Funakoshi M (1988) Amiloride inhibition of responses of rat single chorda tympani fibers to chemical and electrical tongue stimulations. Brain Res 451:319–325
16. Wolf G (1969) Innate mechanisms for regulation of sodium intake. In: Pfaffman C (ed) Olfaction and taste (vol 3). Rockefeller University Press, New York, pp 548–553
17. Breslin PAS, Spector AC, Grill HJ (1993) Chorda tympani section decreases the cation specificity of depletion-induced sodium appetite in rats. Am J Physiol 264:R319–R323
18. Contreras RJ (1989) Gustatory mechanisms of a specific appetite. In: Cagan RH (ed) Neural Mechanisms in taste. CRC Press, Boca Raton, FL, pp 119–146

Effect of Gurmarin on Glucoreceptor-Originated Autonomic Reflex

Akira Niijima[1]

Key words. Gurmarin—Sweet taste receptor—Glucoreceptor—Vagal pancreatic efferent—Vagal gastric efferent—Adrenal efferent

Introduction

It is assumed that sweetness is a marker for a carbohydrate-rich diet. It has been reported that oral stimulation by sweet taste solution elicits insulin secretion accompanied by enhancement of vagal pancreatic activity [1,2].

In 1978 the presence of vagal intestinal glucoreceptors was demonstrated by recroding vagal units in the nodose ganglion of the cat [3]. Neurons from the intestinal glucoreceptors were activated by perfusing the small intestine with glucose solutions, and, stimulation of these intestinal glucoreceptors produced enhancement of insulin release via the vagus nerve [4].

However, the reflex regulation of the visceral functions is not limited to sensory units involving taste receptors and intestinal glucoreceptors. Glucose-sensitive afferents have also been observed in the hepatoportal system. An increase in glucose concentration in the portal venous blood induced by intraportal injection of isotonic glucose solution resulted in a decrease in afferent discharge rate in the hepatic branch of the vagus nerve, and the mean discharge rate of glucose-sensitive units was inversely related to the concentration of glucose in the portal blood. Reflex inhibition of the adrenal nerve activity and reflex activation of vagal pancreatic efferents due to glucose injection into the portal vein have also been reported [5]. It can be assumed that these responses cause a decrease in catecholeamine secretion from the adrenal medulla and an increase in insulin secretion from the pancreas.

In relation to oral, intestinal, and hepatoportal sugar sensors, a new substance, gurmarin, which strongly suppresses the sweet taste responses of the rat chorda tympani, was produced [6]. This substance is a peptide purified from the leaves of Gymnema silvestre and consists of 35 amino acids with a molecular weight of 4000 daltons.

This chapter deals with the effect of gurmarin on the reflex change in autonomic outflows to the stomach, pancreas, and adrenal medulla due to sweet taste stimulation or stimulation of intestinal glucoreceptors and hepatoportal glucose sensors.

Method

Wistar rats weighing about 300 g were used. Animals had free access to water but were deprived of food for 12 h before the experiment. Rats were anesthetized with urethane (1 g/kg IP).

Small nerve bundles were dissected from the pancreatic and gastric branch of the vagus nerve and adrenal branches of the splanchnic nerve. Under the dissecting microscope, nerve filaments were isolated from the central cut end of these nerve bundles to record the efferent activity with a pair of silver wire electrodes. Nerve activity was amplified by means of a conventional differential amplifier and displayed on an oscilloscope and stored on magnetic tape. All analysis of nerve activity was performed after conversion of raw data to standard pulses using a window discriminator that distinguished nerve activity from the background noise. A rate meter with a reset time of 5 s was used with a pen recorder. The effect of stimulation on efferent activity was investigated by comparing the mean number of spikes per 5 s over 50 s (i.e., mean value of 10 successive measured samples). Significance was determined by ANOVA test ($p < 0.05$).

For taste stimulation, a 10% sucrose solution was used. For intestinal infusion and intraportal injection, a 5% glucose solution was used. Before use gurmarin was dissolved in saline in a concentration of 20 µg/ml. The solutions were kept at 32° to 35°C. Animal body temperature was maintained by means of a heating pad.

Results

Figure 1 shows the effects of gurmarin on the sweet taste receptor-originated autonomic reflex. The effects of sweet taste stimulation with 10% sucrose

[1] Department of Physiology, Niigata University School of Medicine, Niigata, 951 Japan

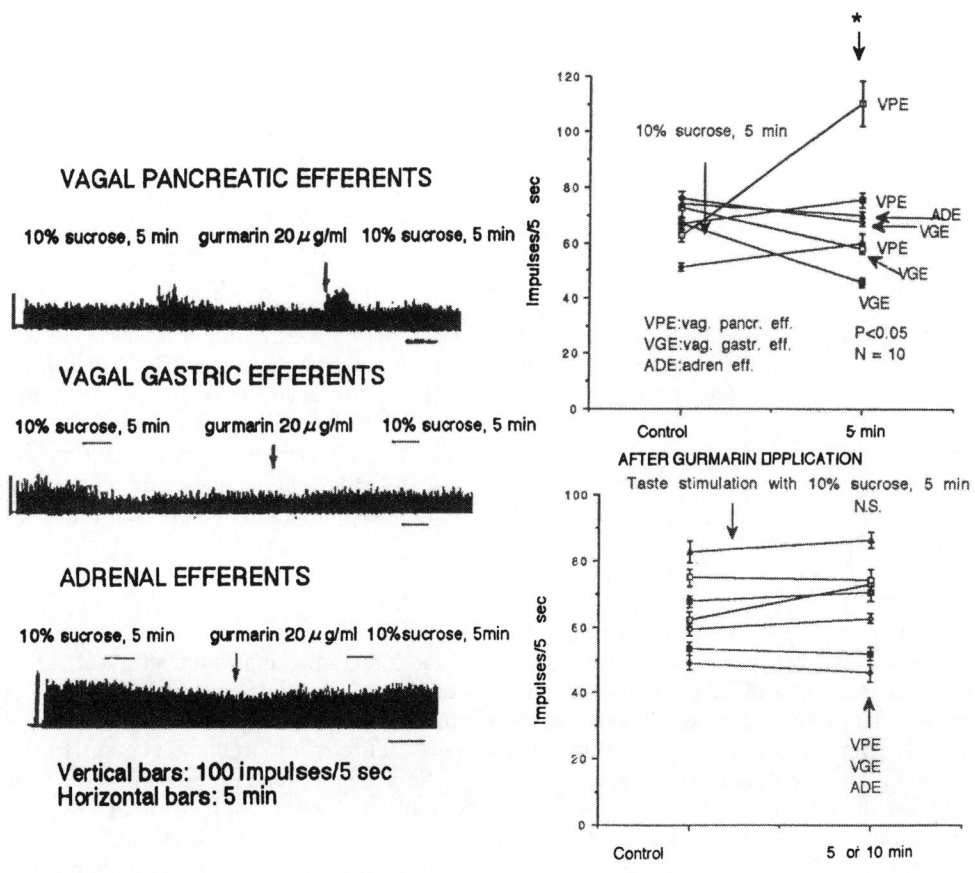

FIG. 1. Effect of gurmarin on the sweet taste receptor-originated autonomic reflex. (*Right panels*) Nerve activities just before stimulation (control) and 5 min after sweet taste stimulation

solution for 5 min on the efferent activity of the vagal pancreatic, vagal gastric, and adrenal nerves were compared before and after application of gurmarin solution. For sweet taste stimulation, sucrose solution was applied to the tongue surface with a syringe for 5 min, and the tongue was flushed with distilled water at the end of each taste stimulation. As shown in the top trace of Fig. 1, before gurmarin application sweet taste stimulation enhanced efferent activity of the vagal pancreatic efferents; however, 10 min after the application of gurmarin (20 µg/ml) sweet taste stimulation caused no remarkable change in efferent activity. In the middle trace, sweet taste stimulation before gurmarin application resulted in a suppression of the discharge rate of the vagal gastric efferents. No significant change in discharge rate was observed with sweet taste stimulation after gurmarin administration. In the bottom trace, gurmarin blocked the suppressive effect of sweet taste stimulation on the efferent activity of the adrenal nerve. The upper right panel of Fig. 1 shows the facilitatory effect of sweet taste stimulation on the vagal pancreatic efferents (three examples) and the inhibitory effect on the vagal gastric efferents (three examples) and

the adrenal efferents (one example). In all seven of these observations the facilitatory or inhibitory effects were significant ($p < 0.05$). After gurmarin application sweet taste stimulation was without effect on vagal pancreatic activity (three examples), vagal gastric activity (three examples) and adrenal nerve activity (one example), as shown in the lower panel of Fig. 1.

Figure 2 shows the effect of gurmarin on the intestinal glucoreceptor-originated autonomic reflex. A segment of a proximal part of the small intestine including the duodenum and jejunum (about 15 cm in length) was used for intestinal infusion. After ligation at the pyloric region and the jejunum, cannulation with small catheters was accomplished both proximally and distally. For stimulation of intestinal glucoreceptors 5% glucose solution (5 ml) was slowly infused into the duodenal area of the intestine for 5 min. At the end of each stimulation the intestine was flushed with saline. As shown in the top traces in the left panels of Fig. 2, the glucose solution infusion clearly enhanced efferent activity of the vagal pancreatic nerve, but no clear increase in efferent activity was observed after gurmarin administration. The middle and bottom

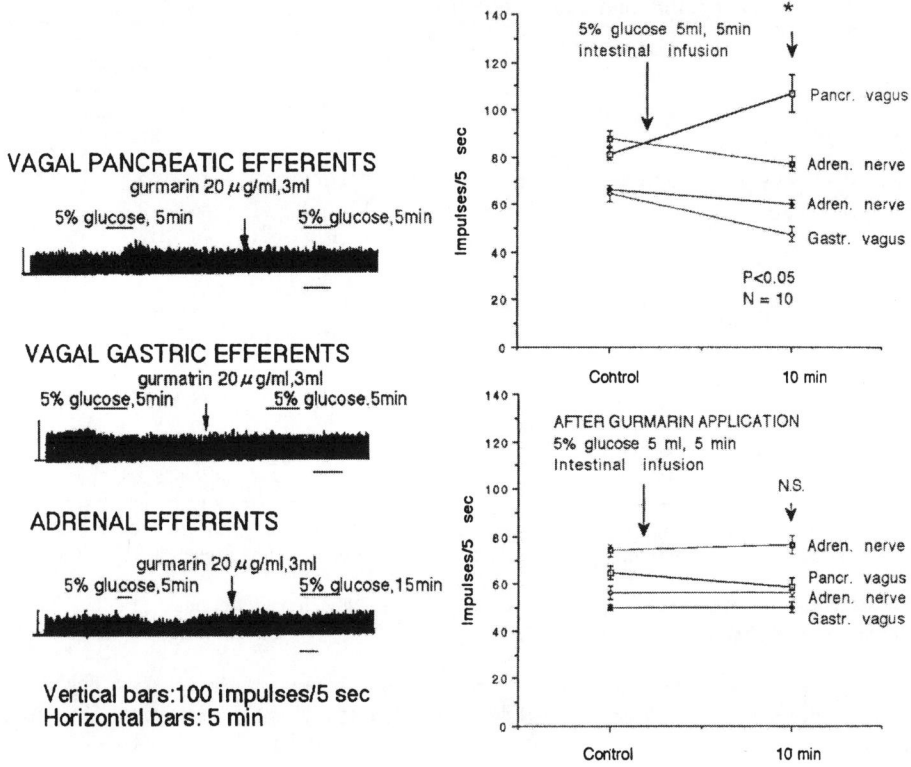

FIG. 2. Effect of gurmarin on the intestinal glucoreceptor-originated autonomic reflex. (*Right panels*) Nerve activities just before stimulation (control) and 10 min after stimulation

traces show the inhibitory effect of intestinal gluco-receptor stimulation on the vagal gastric and adrenal efferents, which was blocked after gurmarin application. The right panels shown the effect of intestinal glucoreceptor stimulation on vagal pancreatic efferents (one example), vagal gastric efferents (one example), and adrenal efferents (two examples). All four observations demonstrate significant reflex facilitation or inhibition (upper panel), which were blocked after administration of gurmarin (lower panel). The reflex effects of sweet taste stimulation and glucose infusion into the intestine before and after gurmarin application are summarized in Table 1.

As mentioned, a decrease in afferent discharge rate in the hepatic branch of the vagus nerve following intraportal injection of glucose solution was usually observed. Even after gurmarin administration into the portal vein (20 µg/ml, 0.2 ml), intraportal injection of glucose solution resulted in the same suppressive response. It is likely that gurmarin does not block the sensory function of hepatoportal glucose sensors. This observation suggests the existence of autonomic reflex originated from hepatoportal glucose sensors even after administration of gurmarin. A reflex increase in efferent activity of the vagal pancretic nerve [7] and similar reflex facilitation was observed on the activity of vagal pancreatic

efferents even after administration of gurmarin. Furthermore, reflex inhibition due to glucose infusion into the portal vein was observed in vagal gastric nerve and adrenal nerve activities after gurmarin administration (Fig. 3).

Discussion

The experimental results indicate that sweet taste stimulation with 10% sucrose and infusion of 5% glucose into the intestine and portal vein accelerated efferent activity of the pancreatic branch of the vagus nerve and inhibited efferent activities of the gastric branch of the vagus nerve and the adrenal branch of the splanchnic nerve. The change in autonomic outflows due to sweet taste stimulation and glucose infusion into the intestine was blocked by prior administration of gurmarin (Figs. 1 and 2).

It has already been reported that sweet taste stimulation facilitates vagal pancreatic activity [2] and suppresses adrenal nerve activity [8], and that stimulation of intestinal glucoreceptors enhances vagal pancreatic nerve activity [9]. It has also been reported that an infusion of glucose into the portal vein causes an increase in pancreatic vagal activity and a decrease in adrenal nerve activity [5]. It was further observed that the hyperglycemic situation

TABLE 1. Nerve activity before and after gurmarin.

Experiment	Efferent	Nerve activity (impulses/5 seconds)		Nerve activity (impulses/5 seconds)	
		Control	Before gurmarin[a]	Control	After gurmarin (20 μg/ml)
Taste stimulation					
1	VPE	62.5 ± 2.3	110.5 ± 8.1	62.3 ± 2.2	73.1 ± 2.2
2	VPE	51.1 ± 1.5	60.0 ± 3.2	49.1 ± 2.3	46.0 ± 2.8
3	VPE	87.1 ± 2.2	75.5 ± 2.5	67.8 ± 1.8	70.6 ± 2.3
4	VGE	76.1 ± 2.2	67.5 ± 1.5	59.4 ± 1.8	62.8 ± 1.5
5	VGE	66.9 ± 2.0	45.7 ± 1.6	53.4 ± 2.0	51.8 ± 2.0
6	VGE	72.8 ± 3.9	57.8 ± 2.1	75.0 ± 2.7	74.4 ± 3.0
7	ADE	74.4 ± 1.8	70.1 ± 1.2	82.7 ± 3.3	86.5 ± 2.4
Intestinal infusion (5% glucose, 5 minutes)					
8	VPE	81.1 ± 2.6	106.9 ± 7.8	64.5 ± 2.9	58.3 ± 4.0
9	VGE	66.4 ± 1.9	50.0 ± 1.7	50.0 ± 0.9	50.1 ± 2.0
10	ADE	64.5 ± 3.4	47.4 ± 3.0	56.2 ± 2.8	56.2 ± 1.0
11	ADE	87.5 ± 3.4	77.3 ± 3.2	74.0 ± 2.6	76.5 ± 3.8

Estimations of nerve activity (mean ± SEM; $n = 10$) were done at 10 min after stimulation for experiments 9 and 11; the others were done at 5 min after stimulation. Taste stimulation consisted in exposure to 10% sucrose for 5 min.
VPE, vagal pancreatic efferents; VGE, vagal gastric efferents; ADE, adrenal efferents.
[a] Significant increase ($P < 0.05$).

	Sweet taste st. (10% sucrose)	Intestinal inf. (5% glucose)	Portal inject. (5% glucose)
BEFORE GURUMARIN ADMINISTRATION			
Vagal pancreatic eff.	↑	↑	↑
Vagal gastric eff.	↓	↓	↓
Adrenal eff.	↓	↓	↓
AFTER GURMARIN ADMINISTRATION			
Vagal pancreatic eff.	→	→	↑
Vagal gastric eff.	→	→	↓
Adrenal eff.	→	→	↓

FIG. 3. Effect of gurmarin on the autonomic reflex originated from the sweet taste receptors, intestinal glucoreceptors, and hepatoportal glucose sensors (with permission from [7])

following intravenous administration of glucose resulted in an acceleration of vagal pancreatic nerve activity [10] and suppression of gastric vagal nerve [11] and adrenal nerve [10] activities. It is interesting that the hyperglycemic situation in the brain [12], sweet taste stimulation, and stimulation of intestinal glucoreceptors or hepatoportal glucose sensors (all glucose-related stimulations) result in activation of vagal pancreatic nerve activity and inhibition of vagal gastric and adrenal nerve activities. Glucose-related sensors in the tongue, intestine, hepatoportal region, and brain may play a common role in glucose homeostasis and regulation in visceral functions.

The experimental results deomonstrated that gurmarin blocks the reflex change in autonomic outflows originated from sweet taste receptors and from intestinal glucoreceptors, but not from glucose sensors in the hepatoportal system. It might be because of the difference in the sensory mechanism between sweet taste receptors, intestinal glucoreceptors, and hepatoportal glucose sensors. Further study is expected to explain the difference in glucoreceptor function between taste receptors, intestinal glucoreceptors, and hepatoportal glucose sensors in relation to the effect of gurmarin.

Acknowledgments. This work was supported by a grant from the Umami Manufacturers Association of Japan. The author expresses his thanks to Drs. Y. Hiji and T. Imoto, First Department of Physiology, Tottori University School of Medicine, for kindly supplying the gurmarin.

References

1. Niijima A, Togiyama T, Adachi A (1990) Cephalic phase insulin release induced by taste stimulus of monosodium glutamate (umami taste). Physiol Behav 48:905–908
2. Niijima A (1991) Effect of taste stimulation on the efferent activity of the pancreatic vagus nerve in the rat. Brain Res Bull 26:161–164
3. Mei N (1978) Vagal glucoreceptors in the small intestine of the cat. J Physiol (Lond) 282:485–506
4. Mei N, Arlhac A, Boyer A (1981) Nervous regulation of insulin release by the intestinal vagal glucoreceptors. J Auton Nerv Syst 41:351–363
5. Niijima A (1980) Glucose sensitive afferent nerve fibers in the liver and regulation of blood glucose. Brain Res Bull 5(suppl 4):175–179
6. Imoto T, Miyasaka A, Ishima R, Akasaka K (1991) A novel peptide isolated from the leaves of *Gymnema sylvestre*. 1. Characterization and its suppressive effect on the neural responses to sweet taste stimuli in the rat. Comp Bioshem Physiol [A] 100:309–314

7. Nagase H, Inoue S, Tanaka K, Takamura Y, Niijima A (1993) Hepatic glucose-sensitive unit regulation of glucose-induced insulin secretion in rats. Physiol Behav 53:139–143
8. Niijima A (1991) Effects of taste stimulation on the efferent activity of the autonomic nerves in the rat. Brain Res Bull 26:165–167
9. Niijima A, Mei N (1987) Glucose sensors in viscera and control of blood glucose level. News Physiol Sci 2:164–167
10. Niijima A (1975) Studies on the nervous regulatory mechanism of blood sugar levels. Pharmacol Biochem Behav 3(suppl):139–143
11. Hirano T, Niijima A (1980) Effects of 2-deoxy-D-glucose, glucose and insulin on efferent activity in gastric vagus nerve. Experientia 36:1197–1198
12. Niijima A (1975) The effect of glucose on the activity of the adrenal nerve and pancreatic branch of the vagus nerve in the rabbit. Neurosci Lett 1:159–162

Chirality in Tea Aroma Compounds

Akio Kobayashi and Kikue Kubota[1]

Key words. Tea—Linalool—Methyljasmonate—Aroma character—Gas chromatography—Enantiodifferentiation

Introduction

The analytic enantiomer separation by gas chromatography (GC) has been applied to chemical research on natural and synthetic products. However the resolution of chiral compound in aroma constituents has been limited because of the microquantity and complexity of the aroma compounds. Since the introduction of chemically bonded chiral liquid phase to high-resolution gas chromatography coupled with multidimensional operation, and the GC mass spectrometry (GC-MS) system [1], it has become possible to separate quantitatively the volatile enantiomers originated from fruit, vegetables, and other food materials [2] without derivatization or other analytic pretreatments.

With the development of such separation techniques, it became clear that there are many aromatic compounds that have different aroma characteristics and strengths, even between enantiomers; therefore the enantioseparation becomes indispensable for correlating flavor quality and gas chromatographic analysis as well as for estimating the biosynthetic route of the corresponding compound and analyzing the linkage of the physiologically active molecule to the receptor site of the human sensory organ. Based on this historical background of the flavor research, we describe the application of chiral analysis to tea aroma.

Tea is the most popular beverage in the world and is classified into three types according to the manufacturing processes: (1) fermented tea, or black tea type; (2) semifermented tea, or oolong tea type; and (3) nonfermented tea, or green tea type. "Fermented" does not refer to microbiologic fermentation but to the enzymatic activity in fresh tea leaves, therefore black tea has the largest amount of essential oil, and oolong tea shows different aroma constituents because of the different type of fermentation.

More than 300 compounds have been identified as the constituents of tea aroma [3], among which linalool and its derivatives are the main components of black tea aroma.

First we discuss the enantioseparation of these compound in black tea and then of other chiral aroma compounds in oolong tea. Finally, the R/S ratios of these compounds in the different types of tea are compared to discuss the formation and characterization of tea aroma.

Materials and Methods

Tea Sample

Black tea was manufactured at the Tea Research Institute of Sri Lanka in 1993, made from Clone DT-1 cultivar (DT-1). Tie Guan Yin was purchased from China as a typical oolong tea sample. Other commercially available teas—Wen Shan and Tai Cha-12 made in Taiwan, Huang Jin Gui and Shui Xian made in China, Dimbula in Sri Lanka and Darjeeling made in India—were purchased from the respective countries.

Preparation of Aroma Concentrate

Each tea sample (100 g) was powdered and steam-distilled under reduced pressure of 20 mm Hg. Aroma compounds was extracted from the distillate and the residual oil after distilling off the solvent was treated as aroma concentrate. The detailed extraction process is shown in Fig. 1.

Gas Chromatography

Analytic separation were performed on the following two gas chromatographs.

1. *GC condition 1, for general separation*: A shimadzu GC-9A gas chromatograph equipped with flame ionization detector was used. An FS-WCOT column was coated with PEG 20 M (50 m × 0.25 mm i.d.). Nitrogen gas was used as carrier gas with flow rate of 1.2 ml/min. Oven temperature was programmed from 60°C (4-minute hold) to 180°C at the rate of 2°C/min. The injection port was held at 170°C to avoid thermal isomerization.

2. *GC condition 2, for chiral separation*: A CP-Cyclodextrin β-236M capillary column (50 m ×

[1] Laboratory of Food Chemistry, Department of Nutrition and Food Science, Ochanomizu University, 2-1-1 Ohtuka, Bunkyo-ku, Tokyo, 112 Japan

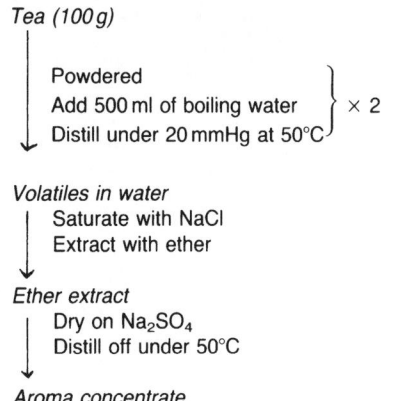

Tea (100 g)

Powdered
Add 500 ml of boiling water   } × 2
Distill under 20 mmHg at 50°C

Volatiles in water
    Saturate with NaCl
    Extract with ether

Ether extract
    Dry on Na$_2$SO$_4$
    Distill off under 50°C

Aroma concentrate

FIG. 1. Preparation of aroma concentrate by steam distillation under reduced pressure

0.25 mm i.d.) was used. Oven temperature starts from 60°C (8-minute hold) and increases to 200°C at a rate of 1°C/min. Other conditions are same as those described for condition.

GC-MS

Gas chromatographic conditions are same as for GC analysis except for using helium as the carrier gas. A JEOL DX-300 mass spectrometer was directly coupled to a Hewlett-Packard 5790A GC with a JWA-DA5000 data processing system.

Results and Discussion

As shown in Fig. 2, 11 peaks on a gas chromatogram of black tea aroma concentrate were identified as those of chiral compounds, among which main six were linalool and its derivatives. In contrast, oolong tea showed a more complex aroma pattern, with relatively small amounts of linalool and its derivatives plus another three compounds that were newly identified as chiral compounds; that is, 14 chiral compounds were separated on the tea aroma gas chromatogram. All identified chiral compounds were listed in the Fig. 2 legend. Peak nos. 14 and 15 were methyljasmonate and methylepijasmonate, respectively, which were separated on this GC and are discussed in a later section.

The sum of the peak area of linalool, four linalool oxides (LOs), and 3,7-dimethyl-1,5,7-octatrien-3-ol (trienol) occupies 61% of the total peak area on the gas chromatogram, whereas that of oolong tea is only 9%. S-(−)-Linalool has a sweet, petitgrain-like aroma and R-(+)-linalool has a more woody and citrus note; therefore enantiodifferentiation of linalool in the tea aroma concentrate should be required to evaluate the contribution of linalool to black tea aroma. Moreover, LOs were known to be

formed from linalool by air oxidation without an enzymatic reaction. If there were any difference in the R/S ratio between linalool and LOs, it would suggest that some enzymatic interaction is ongoing during the formation of LOs from linalool.

Four diasteromes of LOs are synthesized from an enantiomer of linalool, and four racemic pairs of LOs can be prepared from racemic linalool. Askari and Mosandl [4] have already shown these stereoisomers can be separated effectively on a cyclodextrin GC column and established the elution order of these compounds. We also synthesized respective LOs from R-linalool and (R,S)-linalool following the known method [5]. (R)-Trienol and (R,S)-trienol were also derived from R- and (R,S)-linalool, respectively [6]. The identification of each enantiomer and calculation of the R/S ratio were performed to compare retention indices (KI value) of those standard samples to the indices on the gas chromatogram of the tea aroma concentrate. In the case of an enantiomer peak overlapping with another peak, mass chromatographic technique was effective for calculating the concentration because each enantiomer has its own mass spectrographic pattern.

The percents of R and S epimers in two pouchong teas, three oolong teas, and three black teas are shown in Fig. 3. Almost all semifermented tea has more S-enantiomer and black tea has more R-enantiomer. The R/S ratios of linalool, linalool oxides, and trienol were calculated as the mean value from three oolong and black teas and are summarized in Table 1. The R/S values of LOs were different from their corresponding linalool enantiomer. R-Trienol is not present in black tea and much less than any other R-enantiomers in oolong tea. These variable ratios of linalool and its derivatives seems to affect the aroma charactor of each tea.

Among the chiral compounds in oolong teas, it has been claimed that methyljasmonate is one of the flavors because of its jasmine-like odor and its low threshold value. However, Acree et al. [7] reported that (+)-methylepijasmonate (1R,2S) shows a strong odor (TH value 3 ng/ml), although its epimer is odoless and epimerization of the former to trans-methyljasmonate (1R,2R) results in decreasing the odor strength. We [8] found in an earlier study that epijasmonate is easily transformed to jasmonate at higher temperature; however, chiral separation of these interesting compounds in tea aroma has not been accomplished. Under the same chiral GC condition, methyljasmonate was separated clearly with an approximately 1:1 ratio of R and S enantiomers. Unfortunately, enantiomers of methylepijasmonate showed nearly KI values, and it is difficult to differentiate each peak on the same GC (Fig. 4). However, there is a like amount of methyljasmonate epimers, suggesting that they are transformed from epijasmonate by heating.

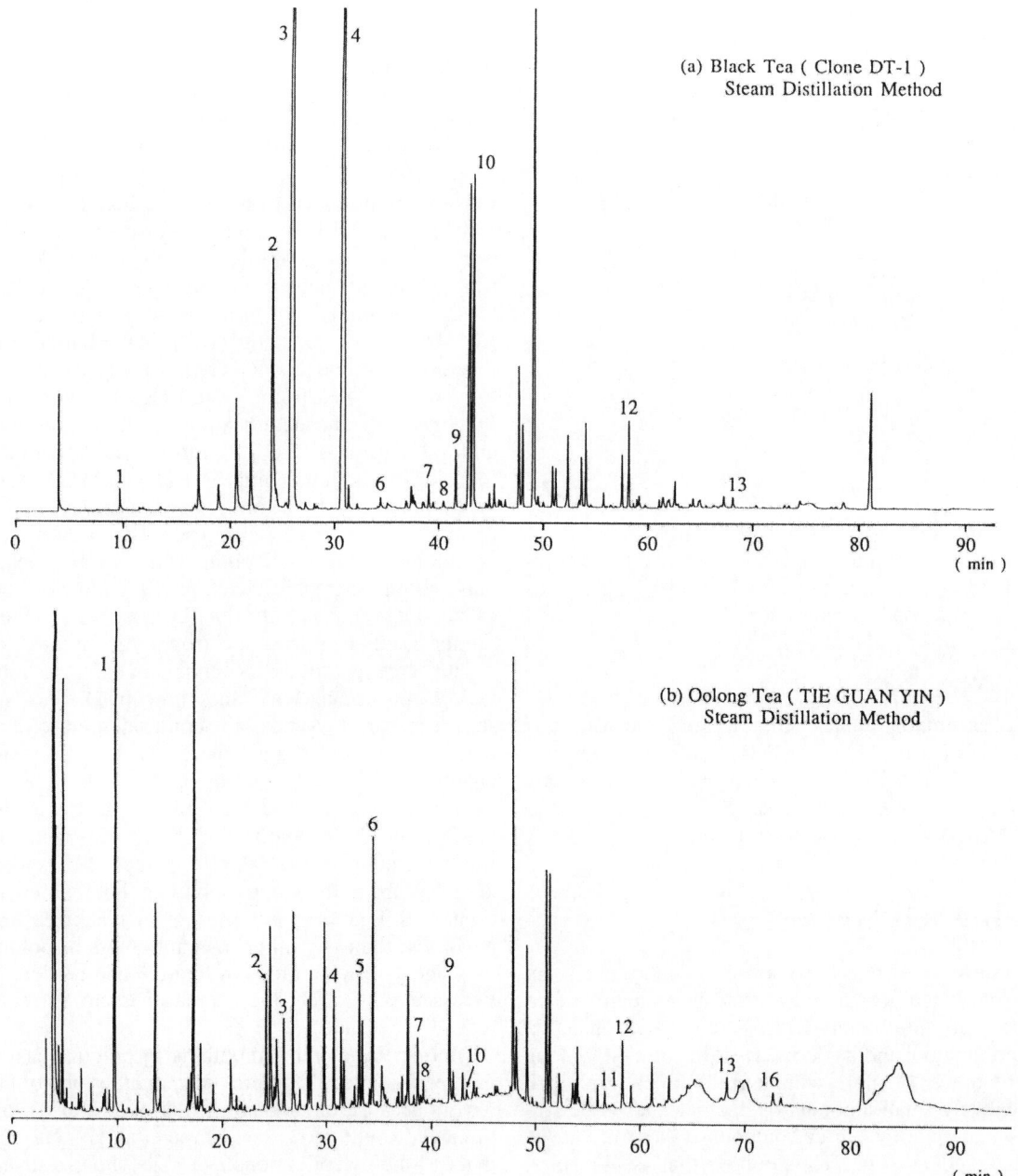

FIG. 2. Gas chromatograms of aroma concentrates from black and oolong teas. The numbered peaks are those of chiral compounds: *1* = 1-penten-3-ol; *2* = linalool oxide I; *3* = linalool oxide II; *4* = linalool; *5* = 2,6,6-trimethyl-2- hydroxycyclohexanone; *6* = 3,7-dimethyl-1,5,7-octatrien-3-ol; *7* = α-terpineol; *8* = 4-hexanolide; *9* = linalool oxide III; *10* = linalool oxide IV; *11* = 4-nonanolide; *12* = nerolidol; *13* = jasmine lactone; *16* = dihydroactinidiolide

In Fig. 4, the highest peak of jasmine lactone also showed a single peak, but it was not established whether the single peak consists of an enantiomer or a racemate because this chiral column has often shown the deficiency of stereodifferentiation for six lactones in the same family. Peaks 1, 8, 9, 12, and 16 have a similar problem. Optimal GC conditions are required for the resolution of these compounds.

Conclusion

The main chiral aroma compounds in tea were separated by direct injection onto GC equipped with a chirospecific capillary column. Linalool and its derivatives showed variable R/S ratios in various tea samples. The enantiodifferentiation of these compounds is important when considering the aroma

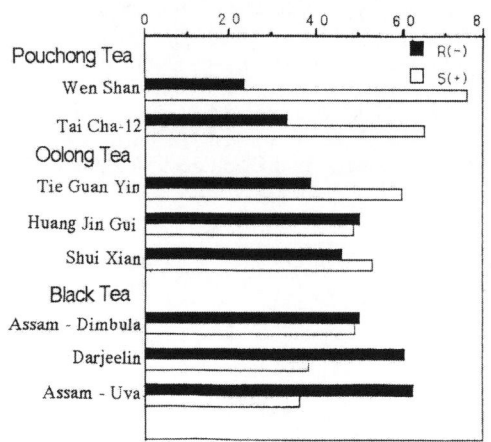

FIG. 3. Percent of R and S epimers present in various teas

TABLE 1. Enantiomer ratio^a of linalool and its derivatives in various teas.

Compound	KI value		R/S ratio	
	3-R	3-S	Oolong	Black
Linalool	1252	1254	0.83	1.39
Linalool oxide I^b	1222	1231	2.35	0.21
Linalool oxide II	1201	1217	0.38	0.10
Linalool oxide III	1377	1370	0.54	0.24
Linalool oxide IV	1364	1360	1.27	0.63
3,7-Dimethyl-1,5,7-octatrien-3-ol	1243	1278	0.07	s-100%

^{a} Enantiomer ratio (R/S) is the mean value of the respective three oolong teas and three black teas.
^{b} I = *cis*-furanoid; II = *trans*-furanoid; III = *cis*-pyranoid; IV = *trans*-pyranoid.

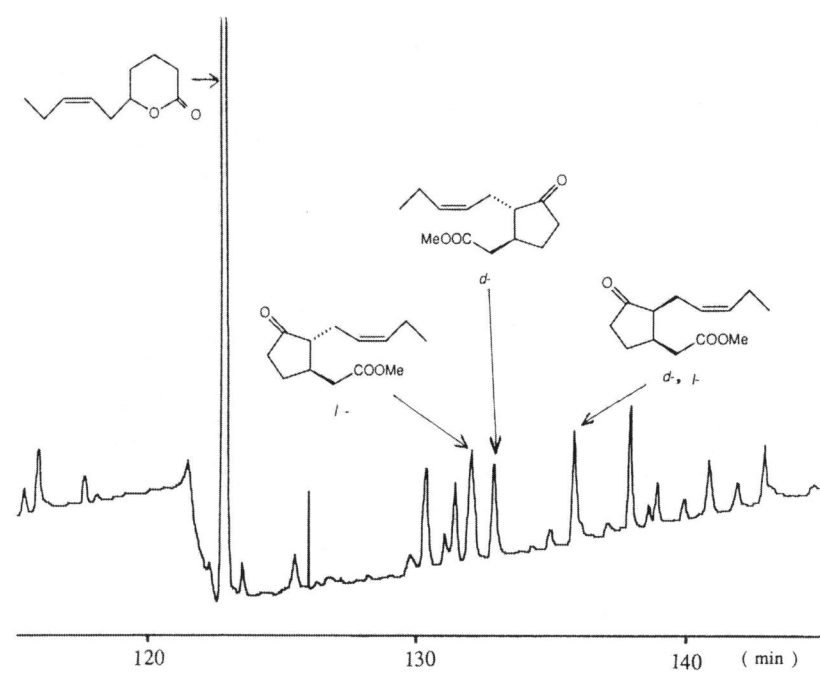

FIG. 4. Separation of methyl-jasmonate and methylepijasmonate on a chiral GC column

character. These results also suggest that linalool oxides and dehydrolinalool (trienol) were not derived from linalool without enzymatic activity. Methylepijasmonate was not separated clearly into enantiomers, but it was changed to racemic methyl-jasmonate nonenzymatically with decreasing of aroma. As described above, chiral differentiation is valuable not only for correlating the chemical structure with the aroma character but also when considering the formation process of the aroma compound in food materials.

References

1. König WA (1984) New development in enantiomer separation by capillary gas chromatography. In: Schreier P (ed) Analysis of volatiles. de Gruyter, Berlin, pp 72–92
2. Schurig V (1993) Enantiomer separation by GLC SFC and CZE on high-resolution capillary columns coated with cyclodextrin-derivatives. In: Schreier P, Winterhalter P (eds) Progress in flavour precursor studies. Allured, Carol Stream, IL, pp 64–75
3. Flament I (1989) Coffee, cocoa and tea. Food Rev Int 5:325–361
4. Askari C, Mosandl A (1991) Stereo-isomeric flavour compounds LII: separation and structure elucidation of

the furanoid linalool oxide stereoisomers using chirospecific capillary gas chromatography and nuclear magnetic resonance spectroscopy. Phytochem Anal 2:211–214

5. Felix D, Melera M, Seibl J, Kovats E (1963) Zur Konstitution des Linalooloxids. Helv Chim Acta 46: 1513–1536

6. Nakatani Y, Sato S, Yamanishi T (1969) 3S(+)-3,

7-Dimethyl-1,5,7-octatrien-3-ol in the essential oil of black tea. Agric Biol Chem 52:967–968

7. Acree TE, Nishida R, Fukami H (1985) Odor thresholds of the stereoisomers of methyl jasmonate. J Agric Food Chem 33:425–427

8. Kobayashi A, Kawamura M, Yamamoto Y, Kubota K, Yamanishi Y (1988) Methyl jasmonate in the essential oil of tea. Agric Biol Chem 52:2299–2303

Important Odorants in Roasted White and Black Sesame Seeds

PETER SCHIEBERLE[1]

Key words. Aroma extract dilution analysis—Stable isotope dilution analysis—Sesame seeds—Sesame flavor—Flavor precursors—Flavor formation

Introduction

Because of their unique odor, roasted sesame seeds (*Sesamum indicum* L.) are widely used as a food flavoring. The overall flavor note generated by the roasting process depends on either the roasting conditions [1,2] or the variety used [3,4] and is often characterized as burnt, nutty, roasted, or meat-like.

The volatile fractions of different roasted sesame varieties have been studied in several investigations, and more than 220 compounds have been identified [4–8]. However, the sensory contribution of single volatiles to the overall sesame flavor has not yet been evaluated. Furthermore, compounds responsible for the flavor differences in the odors of different varieties are still unknown.

An aroma extract dilution analysis (AEDA) ranks volatiles of a food extract according to their odor activities. [reviewed in 9]. By applying this technique to an extract of roasted white sesame seeds we detected 46 odor-active compounds among which 2-furfurylthiol, 2-methoxyphenol, and 4-hydroxy-2,5-dimethyl-3(2H)-furanone showed the highest odor activities [10].

The purpose of the following investigating was to identify the most odor-active compounds in a black sesame variety and to compare them on the basis of their odor activity values (OAVs: ratio of concentration/odor threshold) with the most important odorants identified in the white variety. We also studied the influence of the roasting temperature on the sesame flavor.

Material and Methods

Black sesame seeds (Egypt) were heated in small glass vessels (1 cm internal diameter) for 30 min in a metal block maintained at 180°C. The roasted material (50 g) was immediately frozen in liquid

nitrogen, ground in a commercial blender, and then extracted with diethylether (three times, total volume 300 ml) at 20°C. The extract was concentrated to about 150 ml and the volatiles and solvent isolated by sublimation in vacuo [11]. The distillate was concentrated to 200 µl and evaluated by aroma extract dilution analysis (AEDA) on a 30 m × 0.32 mm fused silica column [DB-5 (SE-54); Fisons Instruments, Mainz, Germany] as described elsewhere [12]. The most important odorants were subsequently identified. Their quantification was performed by stable isotope dilution assays as reported for 4-hydroxy-2,5-dimethyl-3(2H)-furanone [13]. The syntheses of the further labeled internal standards are reported elsewhere [14; P. Schieberle, manuscript in preparation]. As an example, the synthetic route developed for the preparation of d_2-2-phenylethylthiol is given in Fig. 1.

Results and Discussion

Samples of black and white sesame seeds were roasted under the same conditions (180°C, 30 min). A sensory evaluation of the crushed kernels revealed that both materials exhibited a typical sesame-like odor, but whereas in the white seeds a roasty, burnt note predominated, in the black seeds a distinct oily, fatty note was perceivable.

Important Odorants in Roasted Black Sesame Seeds

To reveal the most important odorants, the volatile fraction from the roasted black seeds was isolated and the odor-active compounds evaluated by aroma extract dilution analysis (AEDA). The flavor-dilution (FD) chromatogram obtained as a result of the AEDA is displayed in Fig. 2.

Among the 34 odor-active compounds detected in the FD factor range 2^4 to 2^{12} (16–4096), compound no. 31 with a fatty odor, no. 20 (burnt), and no. 27 (fatty, tallowy) followed by no. 8 (coffee-like), no. 11 (sulfurous), no. 18 (caramel-like), and no. 19 (fried potato) showed the highest FD factors. Of the 20 most odor-active compounds (FD factor $> 2^7$), 17 could be identified. The results of the identification experiments (Table 1) in combination with the FD

[1] Universität/GH Wuppertal, Lebensmittelchemie (FB 9), Gauß-Straße 20, D-42119 Wuppertal, Germany

263

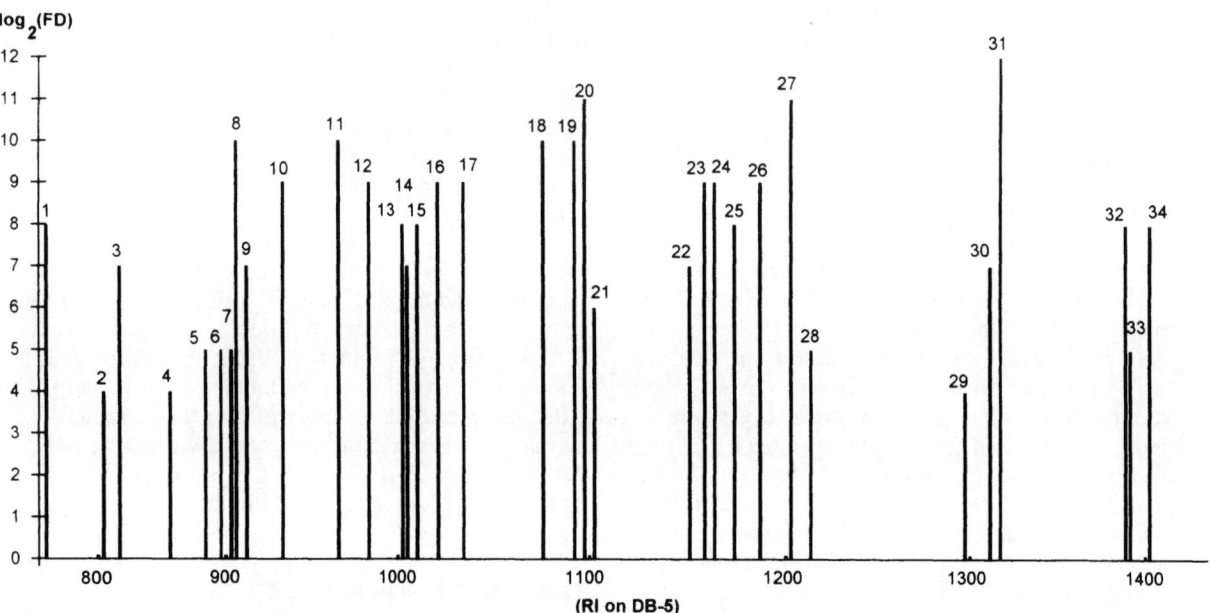

Fig. 1. Synthetic route used for the preparation of deuterium-labeled 2-phenylethylthiol

Fig. 2. Flavor dilution (FD) chromatogram of roasted black sesame seeds

factors (Fig. 2) revealed (E,E)-2,4-decadienal (no. 31), 2-methoxyphenol (no. 20), and 2-pentylpyridine (no. 27) as the most important odorants in the roasted black seeds. In addition, 2-furfurylthiol (no. 8), 4-hydroxy-2,5-dimethyl-3(2H)-furanone (no. 18), 2-ethyl-3,5-dimethylpyrazine (no. 19), and an unknown compound showing a sulfurous note (no. 11) contributed with high FD factors to the flavor of the black seeds.

Differences in the Most Important Odorants of Black and White Seeds

Seven of the odorants showing the highest FD factors in the black seeds are contrasted in Table 2 with those identified recently by us as the most important odorants in roasted white seeds [10]. The most significant difference was found for (E,E)-2,4-decadienal. Its FD factor was by a factor of 32 higher in the black than in the white seeds. In addition, the FD factors of 2-pentylpyridine and 2-ethyl-3,5-dimethylpyrazine were higher in the black variety.

On the other hand, the FD factors of 2-furfurylthiol and 2-phenylethylthiol were comparatively lower than in the white seeds.

The AEDA has been developed as a screening method to detect the most odor-active compounds in food extracts. Because, for example, the FD factors are based on the odor thresholds in air, the sensory significance of an odorant identified in a food extract for the flavor of the food itself must be established on the basis of quantitative data and odor thresholds in the food matrix.

Seven of the key sesame odorants were selected for quantification by stable isotope dilution assays. Either carbon 13- or deuterium-labeled internal standards were synthesized. The internal standards were added to the material immediately after roasting; and after enrichment by column chromatography, the concentrations of the odorants were determined by mass chromatography.

The amounts of seven key odorants present in roasted black and white sesame seeds are compared in Table 3. The data reveal that especially the

TABLE 1. Important odorants (FD factor > $2^7 = 128$) in roasted (180°C, 30 min) black sesame seeds.

No.[a]	Odorant[b]	Odor quality	RI on DB-5 (SE-54)
1	3-Methylbutanal	Malty	650
8	2-Furfurylthiol	Coffee-like	912
10	Unknown	Cooked cabbage	—
11	Unknown	Sulfurous	—
12	1-Octen-3-one	Mushroom-like	981
13	2-Ethyl-3(and 5)-methylpyrazine + trimethylpyrazine	Fruity, potato-like	1003/1004
15	Unknown	Fruity	—
16	2-Acetylpyrazine	Roasty	1024
17	Unknown	Sulfurous	—
18	4-Hydroxy-2,5-dimethyl-3(2H)-furanone	Caramel-like	1078
19	2-Ethyl-3,5-dimethylpyrazine	Fried potato	1084
20	2-Methoxyphenol	Burnt	1091
23	2,3-Diethyl-5-methylpyrazine	Fried potato	1156
24	(E)-2-Nonenal	Green, tallowy	1162
25	2-Phenylethylthiol + 2-methyl-3-[methyldithio]-furane	Rubbery, meat-like	1178
26	Unknown	Leaf-like, green	—
27	2-Pentylpyridine	Fatty, tallowy	1192
31	(E, E)-2,4-Decadienal	Fatty, waxy	1318
32	4,5-Epoxy-(E)-2-decenal	Metallic	1382
34	Vanillin	Vanilla-like	1402

[a] Numbering refers to the FD chromatogram in Figure 2.
[b] The compounds were identified by comparing their mass spectra (MS/EI and MS/CI), retention indices (two columns of different polarity), and odor qualities and thresholds (perceived at the sniffing port) with data obtained from reference substances.

TABLE 2. Comparison of the most important odorants (FD $\geq$ 1024 = 2^{10}) in roasted (180°C, 30 min) black and white sesame seeds.

Odorant	FD factor	
	Black seeds	White seeds[a]
(E,E)-2,4-Decadienal	4096	128
2-Methoxyphenol	2048	2048
2-Pentylpyridine	2048	256
2-Furfurylthiol	1024	4096
2-Ethyl-3,5-dimethylpyrazine	1024	256
4-Hydroxy-2,5-dimethyl-3(2H)-furanone	1024	512
Unknow (sulfurous)	1024	512
2-Phenylethylthiol	256	1024

[a] The FD factors published recently [10] were divided by 5 because in the previous experiment 250 g of roasted material was used to prepare the extract for AEDA.

TABLE 3. Concentrations of the seven most important odorants in roasted (180°C, 30 min) black and white sesame seeds.

Odorant	Concentration (µg/kg)[a]	
	Black seeds	White seeds
(E,E)-2,4-Decadienal	1 103	212
2-Methoxyphenol	2 652	4974
2-Pentylpyridine	904	255
2-Furfurylthiol	673	2461
2-Ethyl-3,5-dimethylpyrazine	394	238
4-Hydroxy-2,5-dimethyl-3(2H)-furanone	11 685	9155
2-Phenylethylthiol	12	44

[a] Calculated on the basis of the amount of unroasted material used.

TABLE 4. Odor thresholds and odor activity values of seven important odorants in black and white sesame seeds.

Odorant	Odor threshold[a] (μg/kg sunflower oil)	Odor activity value[b]	
		Black seeds	White seeds
(E,E)-2,4-Decadienal	180[c]	6	1
2-Methoxyphenol	19	139	262
2-Pentylpyridine	5	181	51
2-Furfurylthiol	0.4	1682	6152
2-Ethyl-3,5-dimethylpyrazine	3	131	79
4-Hydroxy-2,5-dimethyl-3(2H)-furanone	50	234	183
2-Phenylethylthiol	0.05	240	880

[a] Odor thresholds were determined by the triangle test. Values detected by at least eight of ten panelists are reported.
[b] Odor activity values were calculated by dividing the concentrations (Table 3) by the odor thresholds.
[c] Adapted from [14].

concentrations of (E,E)-2,4-decadienal and 2-pentylpyridine were significantly higher in the black variety, whereas the amounts of 2-furfurylthiol and 2-phenylethylthiol were comparatively lower than in the white seeds.

Sesame seeds contain more than 50% oil. Therefore the sensory relevance of the seven odorants for the sesame flavor was established by calculating their OAVs on the basis of odor thresholds in oil. The data (Table 4) revealed that because of their extremely low odor thresholds 2-furfurylthiol and 2-phenylethylthiol are the most important odorants in both varieties. However, although in the white seeds the OAV of the most important odorant 2-furfurylthiol was higher than that of 2-pentylpyridine by a factor of 120, its OAV was higher in the black seeds only by a factor of 9. Because of its relatively high odor threshold in sunflower oil the flavor contribution of (E,E)-2,4-decadienal was low in both seed types (Table 4). These data implied that especially the increase in the OAV of the fatty-tallowy smelling 2-pentylpyridine compared with the OAV of the coffee-like smelling 2-furfurylthiol is mainly responsible for the fatty, oily note prominent in the black seeds.

2-Pentylpyridine has been shown [10] to be formed in significant amounts by a reaction of (E,E)-2,4-decadienal with ammonia. Because decadienal is known as a degradation product from the peroxidation of linoleic acid [15] it may be assumed that in the black seeds the fatty acid degradation is favored, probably due to a lower concentration of antioxidants in these seeds.

Influence of the Roasting Temperature

In addition to the influence of the sesame variety, the roasting time has been shown to influence significantly the amounts of odorants formed during roasting [10]. In a further experiment, the influence of the roasting temperature was studied. Four samples of white sesame seeds were roasted for 20 min at 180°, 200°, 220°, and 240°C and the overall odors of the roasted, crushed materials sensorially evaluated. In the samples that had been roasted at 180°C and 200°C the typical roasted, sweet-sulfur note of sesame predominated, whereas in the samples roasted at higher temperatures (220° and 240°C) the burnt, rubbery, and fatty odor notes were enhanced.

To gain insight into these flavor changes, the concentrations of four of the key sesame odorants were determined in the four samples. The data summarized in Table 5 revealed that each odorant increased with the increase in roasting temperature, but to a different extent. Whereas an increase in the temperature from 180°C to 240°C enhanced the amounts of 2-furfurylthiol by a factor of about 4, the concentrations of 2-pentylpyridine, 2-methoxyphenol, and 2-phenylethylthiol were increased by factors of about 30.

A calculation of the odor activity values reflected the differences in the overall flavors of the four samples (Table 6). In the seeds roasted at 180°C, the coffee-like smelling 2-furfurylthiol showed the highest OAV, whereas in the sample roasted at 240°C the 2-phenylethythiol (exhibiting a rubbery note) became the most important odorant among the four compounds analyzed.

Conclusion

The data reveal that 2-furfurylthiol, 2-phenylethylthiol, 2-methoxyphenol, 4-hydroxy-2,5-dimethyl-3(2H)-furanone, 2-pentylpyridine, 2-ethyl-3, 5-dimethylpyrazine, and an unknown compound with a sulfurous note are the key odorants in roasted black seeds as well as in roasted white sesame seeds. A tallowy, fatty odor note, most prominent in the roasted black seeds, was shown to be mainly due to a comparatively higher odor activity value (OAV) of the tallow-smelling 2-pentylpyridine and a lower OAV of especially the coffee-smelling 2-furfuryl-

TABLE 5. Influence of roasting temperature on the amounts of four key odorants formed during roasting (20 min) of white sesame seeds.

Odorant	(μg/kg seeds)			
	180°C	200°C	220°C	240°C
2-Furfurylthiol	490	1266	2032	2130
2-Methoxyphenol	836	6451	25347	27711
2-Pentylpyridine	89	368	1385	2627
2-Phenylethylthiol	18	108	448	540

TABLE 6. Influence of the roasting temperature on the odor activity values (OAV) of four key odorants in white sesame seeds.

Odorant	OAV[a] in sample			
	180°C	200°C	220°C	240°C
2-Furfurylthiol	1225	3165	5080	5235
2-Methoxyphenol	44	340	1334	1458
2-Pentylpyridine	18	74	277	525
2-Phenylethylthiol	360	2160	8960	10800

[a] The odor activity values (OAV) were calculated by dividing the amounts present in the roasted material (Table 5) by the odor threshold given in Table 4.

thiol. Furthermore, the appearance of tallowy, rubbery odor notes induced by roasting white seeds above a temperature of 200°C was well correlated with comparatively higher odor activities of 2-pentylpyridine as well as 2-phenylethylthiol (rubbery) and 2-methoxyphenol (burnt). These results imply that 2-furfurylthiol, 2-pentylpyridine, 2-phenylethylthiol, and 2-methoxyphenol could be useful as indicator odorants for the objective assessment of roasted sesame flavor.

Summary

Application of an aroma extract dilution analysis on an extract of roasted black sesame seeds revealed (E,E)-2,4-decadienal, 2-methoxyphenol, and 2-pentylpyridine following by 2-furfurylthiol, 4-hydroxy-2,5-dimethyl-3(2H)-furanone, 2-ethyl-3,5-dimethylpyrazine, and an unknown compound with a sulfurous note as the most important odorants in the overall flavor. A comparison with the key odorants recently identified in roasted white seeds on the basis of odor activity values (OAVs: ratio of concentration/odor threshold) indicated that the fatty, tallowy note predominant in the black seeds was mainly due to a comparatively higher OAV of the tallow-smelling 2-pentylpyridine and a lower OAV of the coffee-smelling 2-furfurylthiol in the black seeds. Furthermore, unpleasant odor notes induced by roasting white seeds at temperatures above 200°C were well correlated with the high OAVs of 2-phenylethylthiol, 2-methoxyphenol, and 2-pentylpyridine.

Acknowledgments. I thank Mrs. Dagmar Karrais and Mrs. Jana Kubickova for skillful technical assistance and Mrs. Rita Berger for typing the manuscript.

References

1. Yen GC, Shyu SL, Lin T-C (1986) Studies on improving the processing of sesame oil. I. Optimum processing conditions. Shi Plin O Husuch (Taipei) 13:198–211
2. Yen GC (1990) Influence of seed roasting process on the changes in composition and quality of sesame oil. J Sci Food Agric 50:563–570
3. Yen GC, Lai SH (1987) Comparison of sesame oil from different varieties of sesame. Shi pin Kexue (Taipei) 184:190
4. El-Sawy AA, Osman F, Fadel AM (1989) Volatile components of roasted black sesame seeds. Seifen Öle Fette Wachse 115:677–679
5. Takei Y, Nakatani Y, Kobayashi A, Yamanishi T (1969) Studies on the aroma of sesame oil. Nippon Nogei Kagaku Kaishi 43:667–674
6. Manley CH, Vallon PP, Erickson RE (1974) Some aroma components of roasted sesame seeds. J Food Sci 39:73–76
7. Soliman MM, Kinoshita S, Yamanishi T (1975) Aroma of roasted sesame seeds. Agric Biol Chem 39:973–977
8. Nakamura S, Nishimura O, Masuda H, Mihara S (1989) Identification of volatile flavor components of the oil from roasted sesame seeds. Agric Biol Chem 53:1891–1899
9. Grosch W (1993) Detection of potent odorants in foods by aroma extract dilution analysis. Trends Food Sci Technol 4:68–73
10. Schieberle P (1993) Studies on the flavour of roasted white sesame seeds. In: Schreier P, Winterhalter P (eds) Progress in flavour precursor studies. Allured, Carol Stream, IL, pp 343–360
11. Guth H, Grosch W (1989) 3-Methylnonane-2,4-dione—an intense odour compound formed during flavour reversion of soya-bean oil. Fat Sci Technol 6:225–230
12. Schieberle P (1991) Primary odorants in popcorn. J Agric Food Chem 39:1141–1144
13. Sen A, Schieberle P, Grosch W (1991) Quantitative determination of 4-hydroxy-2,5-dimethyl-3(2H)-furanone and its methyl ether using a stable isotope dilution assay. Lebensm Wiss Technol 24:364–369
14. Guth H, Grosch W (1990) Deterioration of soya-bean oil: quantification of primary flavour compounds using a stable isotope dilution assay. Lebensm Wiss Technol 23:513–522
15. Grosch W (1987) Reactions of hydroperoxides—products of low molecular weight. In: Chan HWS (ed) Autoxidation of unsaturated Lipids. Academic, London, pp 95–139

Flavor Development in the Beans of *Vanilla planifolia*

Tsuneyoshi Kanisawa, Kazuhiko Tokoro, and Seiichi Kawahara[1]

Key words. *Vanilla planifolia*—Vanillin—Glucoside—Enzyme—Vanillin formation—Pathway

Vanilla planifolia is in the family Orchidaceae, native to Central America. The mature green vanilla beans possess no characteristic flavor, although the subsequent curing process after harvest generates a strong flavor with vanillin as the main component. This flavor is extracted with alcohol and water or other solvents and becomes the vanilla extract that has been widely used in food manufacturing from ancient times. Thus the vanilla bean is an important source of flavor. Yet the chemical and biologic mechanism of flavor formation is not fully understood. The present study was initiated to clarify the mechanism of formation of vanillin and other related chemical components.

Glucosides in Green Vanilla Beans

It is generally known that phenolic glucosides are the important precursors of vanilla flavor [1–3]. We initiated a study to analyze glucosides present in the green vanilla beans. Part of this study was presented at the Sixth Weurman Symposium in Geneva in 1990 [4]. In this study the green beans, stems, and leaves of *Vanilla planifolia* were harvested each month after pollination in Indonesia. The glucosides from these samples were extracted by methanol and fractionated by Amberlite XAD-2 and silica gel columns. Four major glucosides were isolated by high performance liquid chromatography (HPLC) equipped with a C-18 ODS column; and they were identified by ^{1}H-NMR, ^{13}C-NMR, CI-MS, EI-MS, FAB-Ms, IR, and optical rotation. The major glucosides were quantitatively estimated by HPLC with an ultraviolet (254 nm) detector with external standards. Minor glucosides were identified through their aglycones by HRGC-MS. The identification of some of the aglycones were confirmed with the HPLC retention times compared with synthetic glucosides.

The four major glucosides present were found to be glucovanillin; *p*-hydroxybenzyl-β-D-glucopyranoside (glu-HBalc); bis[4-(β-D-glucopyrano-syloxy)-benzyl] 2-isopropyl tartrate (glucoside A); and bis (β-D-glucopyranosyloxy)benzyl]2-(1-methylpropyl) tartrate (glucoside B). Glucosides A and B have not been reported in the scientific literature to be found in vanilla beans. Both have a chemical structure to that of loroglossins found in various Orchidaceae [5]. The identified aglycones were vanillin, vanillic acid, vanillyl alcohol, 4-vinylguaiacol, acetovanillon, caffeic acid, ferulic acid, methyl-3,4-dihydroxy cinnamate, 2-methoxy-4-cresol, homovanillyl alcohol, 3,4-dihydroxybenzoic acid, ethyl-4-hydroxy-3-methoxyphenylacetate, *p*-cresol, 4-vinylphenol, methylsalicylate, *p*-hydroxybenzylmethyl ether, *p*-hydroxybenzaldehyde, *p*-hydroxybenzyl alcohol, *p*-hydroxycinnamic acid, cinnamic alcohol, cinnamic acid, phenethyl alcohol, 3-phenylpropanol.

Changes in the Content of Glucosides During Maturation of Green Vanilla Beans

Changes in the major four glucosides during maturation are shown in Fig. 1. The amounts of glucosides A and B rapidly increase until the third month after pollination and then gradually decrease thereafter. On the other hand, glucovanilln appeared around the fourth month and significantly increased, finally reaching a level of 1.4 mmol at the eighth month. Glu-HBalc appeared before glucosides A and B, gradually increased until the third month, and decreased thereafter. It then again increased until the sixth month, decreasing again, suggesting that glu-HBalc was generated vigorously during the early stage of maturation and accumulated along with glucosides A and B. Accompanying the decrease in glucoside A and B, glu-HBalc may regenerate and be immediately oxidized to *p*-hydroxybenzaldehyde (*p*-HBald), which we suggest changes to vanillin and other minor glucosides. This subject is discussed in detail later. Glucosides A and B are also detected in the leaves and stems, although in small amounts.

Enzymes Related to the Metabolism of Phenolic Compounds in Green Vanilla Beans

To establish the biosynthetic pathway for vanillin, we tried to detect various enzymes believed to be present in the green beans [6]. A supernatant of the

[1]Central Research Laboratory, Takasago International Corporation, 1-4-11 Nishiyawata, Hiratsuka, Kanagawa, 254 Japan

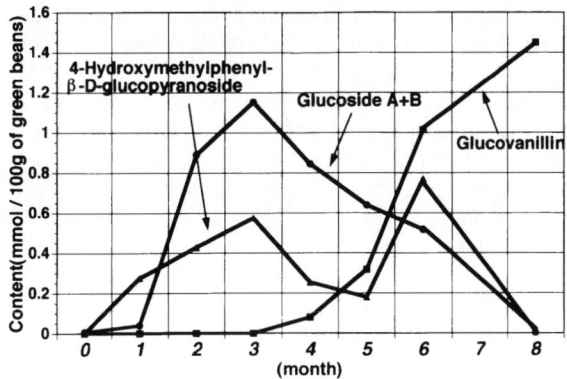

FIG. 1. Changes of abundant glucosides during maturation of green vanilla beans

and separated by cation-exchange chromatography. Two activity peaks for β-glucosidase (I and II) were observed on the chromatogram of the green beans, showing that the green beans have two isozymes of β-glucosidase. On the other hand, only one peak (β-glucosidase II) was detected for the green leaves. β-Glucosidase I had high activity for glucovanillin and glu-HBald. β-Glucosidase II has been found to have relatively broad substrate activity. The measurement of glucosidase activity during maturation of the green beans showed that maximum activity was detected during the fifth month after pollination, suggesting that there is active turnover of glucosides around the fifth month after pollination. At the eighth month the activity of glucosidase might be enough to cleave the glucosides into aglycones and glucose during the curing process.

Another enzyme that was detected in the green beans was 3-O-methyltransferase. This crude enzyme had the highest activity to protocathecuic aldehyde. The activity to caffeic acid was about half that to protocathecuic aldehyde. In contrast, it was

macerated green beans was shown to have strong β-glucosidase activity. This enzymatic activity was also detected in the leaves. The crude enzyme was partially purified by ammonium sulfide precipitation

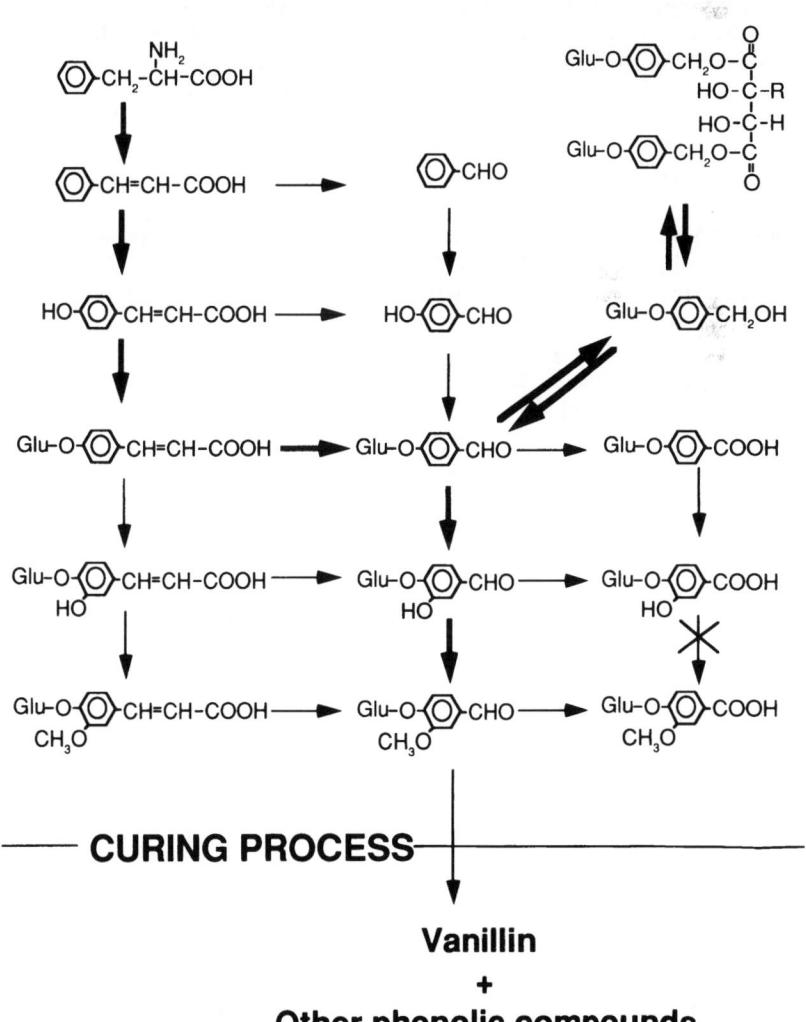

FIG. 2. Proposed pathway for the formation of vanillin and other phenolic compounds in the beans of *Vanilla planifolia*

found to have no activity to protocathecuic acid. Such substrate activity suggests that side chain cleavage occurred before the 3-methylation. This hypothesis coincides with *p*-HBalc accumulation in glucosides A and B during the early stage of maturation. *p*-HBalc is a possible intermediate to vanillin.

Pathway for Vanillin Formation in *V. planifolia*

Various pathways for vanillin formation have been proposed. Zenk proposed that ferulic acid generated in vanilla beans might be cleaved into vanillin through the β-oxidation pathway [2]. Anwar suggested that vanillin formation occurred via enzyme hydrolysis and oxidation of coniferin [1]. Our study identified two new glucosides, A and B, as intermediate products and suggested a new pathway for vanillin formation, as indicated in Fig. 2. The pathway from phenylalanine to coumaric acid is generally accepted to proceed via the lignin biosynthesis route. Tyrosine may be considered a precursor of coumaric acid. Side-chain cleavage is believed to be performed by β-oxidation, as mentioned above; but it has yet to be verified. Yazaki et al. reported that coumaric acid was cleaved to *p*-HBald by lyase in the cell-free extract of the *Lithospermum erythrizon* cell culture [7]. This enzyme may also be active in *V. planifolia*. In our study glu-HBald was detected in a relatively large amount.

Considering the accumulation of glucosides A and B during the early stages of maturation, *p*-HBald may be reduced to *p*-HBalc, which may be acumulated in glucosides A and B. After the third month following pollination, glucosides A and B decrease gradually as glucovanillin increases. Hence glucosides A and B might be cleaved and give rise to *p*-HBalc, which is enzymatically oxidized to *p*-HBald. The *p*-HBald is imediately coverted to protocatecuic aldehyde by 3-monooxygenase and may then be converted to vanillin by 3-O-methyltransferase, which is present in high levels. The heavy arrows in Fig. 2 indicate the most likely main pathway for vanillin formation. All other glucosides in the figure are detected in the green vanilla beans, and the pathway indicated by the thin arrows is also possible. Various phenolic glucosides containing vanillin as the main component formed in the green vanilla beans give various phenolic compounds during the curing process through the cleavage activity of β-glucosidase.

Vanillin contributes the most important flavor character of vanilla beans. Many enzymes are present in green vanilla beans, and other flavor components are generated by these enzymes during the curing process. Heat generated during the curing process may also contribute to the formation of flavor components. Vanilla beans have a complex profile, and controlling the curing process is the key to developing high quality flavor.

Acknowledgments. We thank Dr. M.J.C. Rhodes, Dr. R.J. Robins, and Dr. A. Narbat of the Institute of Food Research in England for their collaboration in the enzyme studies.

References

1. Anwar MH (1963) Paper chromatography of monohydroxyphenols in vanilla extract. Anal Chem 12:1974–1976
2. Zenk MH (1965) Biosynthese van vanillin in Vanilla planifolia Andr. Z Pflanzenphysiol 53:404–414
3. Leong G, Archavlis A, Derbsy M (1989) Research on the glucoside fractin of the vanilla beans. J Ess Oil Res 1:33–41
4. Tokoro K, Kawahara S, Amano A, Kanisawa T, Indo M (1990) Glucosides in vanilla beans and changes of their contents during maturation. In: Bessiere Y, Thomas AF (eds) Flavour science and technology. Wiley, Chichester, pp 73–76
5. Glay RW, Guggisberg A, Segebarth KP, Hesse M, Schmid T (1977) Die Konstitution des Loroglossins. Helv Chem Acta 60:1304–1307
6. Narbad A, Robins RJ, Rhodes MJC, Tokoro K, Kanisawa T (1993, in preparation) Planta
7. Yazaki K, Heide L, Tabata M (1991) Formation of p-hydroxybenzoic acid from p-coumaric acid by cell free extract of Lithospermum erythrorhizon cell cultures. Phytochemistry 30:2233–2236

Odor Thresholds of Cyclic Esters

Gary R. Takeoka, Roy Teranishi, and Ron G. Buttery[1]

Key words. Cyclic esters—Odor thresholds—Mass spectra—Polycyclic esters

Introduction

We have previously determined the odor thresholds of various cyclic and branched esters and found them to be surprisingly low [1]. In this chapter we examine the effect of substitution of different heteroatoms in the ring system on odor thresholds. In addition, we determine the odor thresholds of various polycyclic esters.

Materials and Methods

Reference Compounds

Reference standards were obtained commercially or synthesized according to established procedures. The carboxylic acids were obtained from Aldrich Chemical Company (Milwaukee, WI) or TCI America (Portland, OR). The mass spectra for the substances used are as follows.

Ethyl cyclopentaneacetate: 156(M^+,2), 141(1), 128(6), 113(8), 111(28), 99(4), 89(23), 88(100), 83(32), 81(8), 70(25), 67(23), 61(22), 60(26), 55(26), 41(41); Kovats index (DB-1) 1109

Ethyl-2-tetrahydrofurancarboxylate: 116(8), 88(4), 72(4), 71(100), 43(45), 41(27), 39(16); Kovats index (DB-1) 1049

Ethyl 2-tetrahydrofuranacetate (TCI America): 130(35), 115(14), 113(12), 111(6), 88(13), 87(16), 84(17), 83(10), 71(100), 55(13), 43(52), 41(24); Kovats index (DB-1) 1128

Methyl cyclohexanecarboxylate (TCI America): 142(M^+,32), 127(9), 113(23), 111(17), 110(21), 101(14), 87(72), 83(54), 82(24), 81(16), 74(29), 67(20), 59(15), 55(100), 41(45), 39(30); Kovats index (DB-1) 1040

Ethyl 2-pyridinecarboxylate (Aldrich): 151(M^+,3), 107(26), 106(17), 79(100), 78(51), 52(15), 51(25), 50(8), 39(3); Kovats index (DB-1) 1209

Ethyl 2-piperidinecarboxylate (Aldrich): 157(M^+, 2), 84(100), 85(6), 82(3), 56(15), 55(7), 42(4), 41(4); Kovats index (DB-1) 1169

Ethyl 1,3-dithiolane-2-carboxylate (Aldrich): 180(3), 179(3), 178(M^+,33), 107(9), 106(5), 105(100), 77(4), 61(11), 59(5), 45(20); Kovats index (DB-1) 1334

Ethyl 1,3-dithiane-2-carboxylate (Aldrich): 194(4), 193(3), 192(M^+,30), 121(9), 120(6), 119(100), 91(4), 85(3), 75(5), 73(5), 59(3), 45(18); Kovats index (DB-1) 1415

Methyl 1-adamantanecarboxylate: 194(M^+,16), 136(11), 135(100), 107(11), 93(22), 91(10), 79(22), 77(9), 67(6), 65(4), 41(7), 39(8); Kovats index (DB-1) 1425

Ethyl 1-adamantanecarboxylate (TCI America): 208(M^+,9), 180(7), 136(12), 135(100), 107(12), 93(18), 91(10), 79(21), 77(9), 67(6), 53(4), 41(6); Kovats index (DB-1) 1490

Propyl 1-adamantanecarboxylate: 222(M^+,<1), 182(5), 181(41), 136(12), 135(100), 107(10), 93(19), 91(10), 79(20), 77(9), 67(5), 41(10); Kovats index (DB-1) 1594

Ethyl 1-adamantaneacetate: 222(M^+,5), 177(4), 148(8), 136(12), 135(100), 107(10), 93(16), 91(17), 79(16), 77(9), 67(5), 41(8); Kovats index (DB-1) 1584

Odor Threshold Determinations

Thresholds were determined (with samples purified by preparative gas chromatography) as described previously [2] using a panel of 16 to 22 members. Solutions were presented in odor-free Teflon squeeze bottles equipped with Teflon tubing at the top.

Results and Discussion

Earlier studies [1] have shown that ethyl cyclopentanecarboxylate has a low threshold of 0.001 ppb. The addition of a CH_2 group in front of the cyclopentane ring had little effect, as the odor threshold of ethyl cyclopentaneacetate was 0.002 ppb (Fig. 1). Substitution of two sulfur atoms into the cyclopentane ring results in a 400-fold loss in odor potency, and the substitution of an oxygen atom into the ring causes an 80 000-fold reduction in odor potency. The substitution of a CH_2 group between the tetrahydrofuran ring and carbonyl group results in a

[1] Western Regional Research Center, U.S. Department of Agriculture—Agricultural Research Service, 800 Buchanan Street, Albany, CA 94710, USA

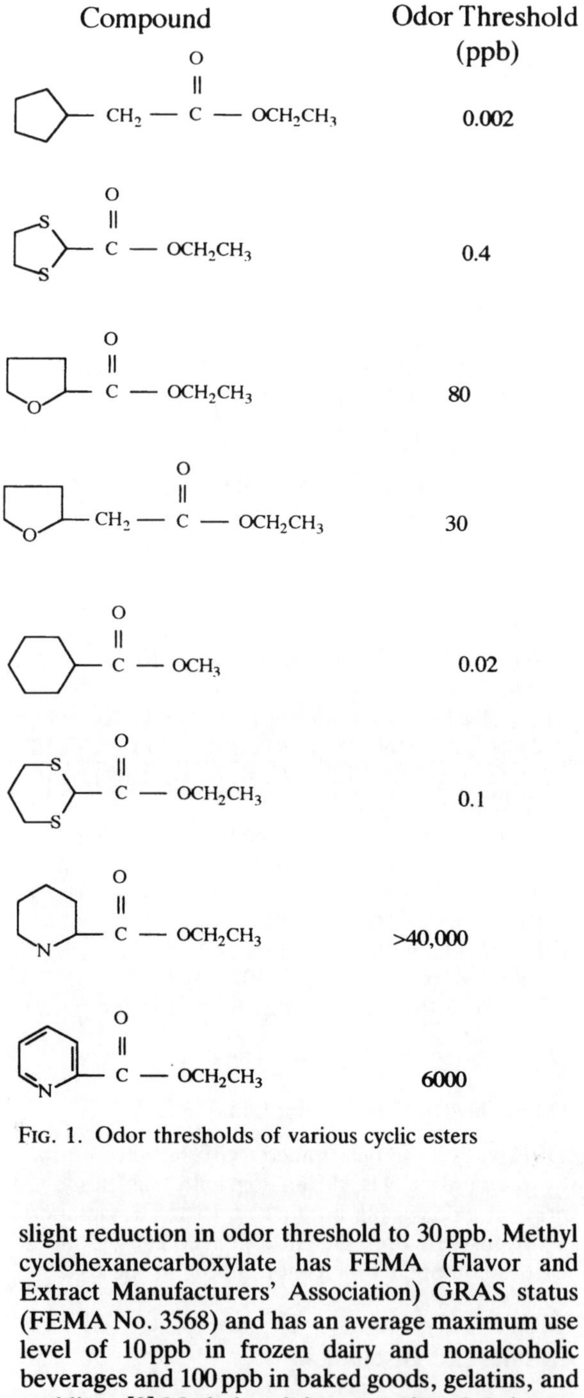

Compound	Odor Threshold (ppb)

cyclopentane—CH$_2$—C(=O)—OCH$_2$CH$_3$ 0.002

1,3-dithiolane-2-yl—C(=O)—OCH$_2$CH$_3$ 0.4

tetrahydrofuran-2-yl—C(=O)—OCH$_2$CH$_3$ 80

tetrahydrofuran-2-yl—CH$_2$—C(=O)—OCH$_2$CH$_3$ 30

cyclohexane—C(=O)—OCH$_3$ 0.02

1,3-dithiane-2-yl—C(=O)—OCH$_2$CH$_3$ 0.1

piperidine-2-yl—C(=O)—OCH$_2$CH$_3$ >40,000

pyridine-2-yl—C(=O)—OCH$_2$CH$_3$ 6000

FIG. 1. Odor thresholds of various cyclic esters

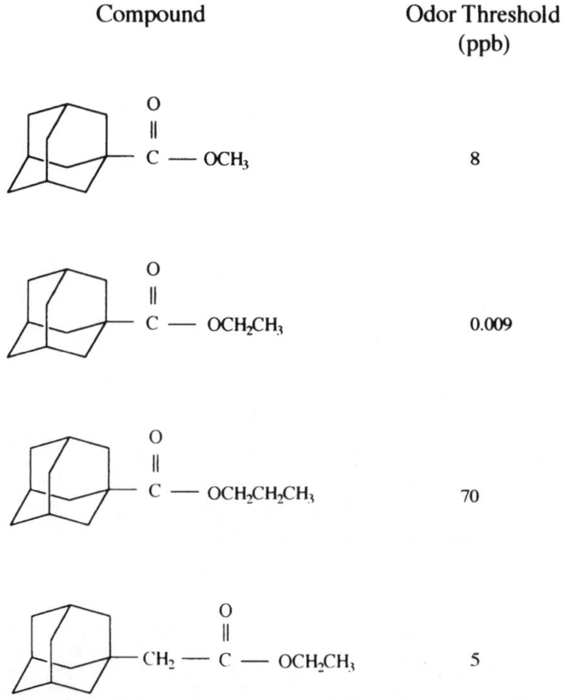

Compound	Odor Threshold (ppb)

adamantane—C(=O)—OCH$_3$ 8

adamantane—C(=O)—OCH$_2$CH$_3$ 0.009

adamantane—C(=O)—OCH$_2$CH$_2$CH$_3$ 70

adamantane—CH$_2$—C(=O)—OCH$_2$CH$_3$ 5

FIG. 2. Odor thresholds of some polycyclic esters

slight reduction in odor threshold to 30 ppb. Methyl cyclohexanecarboxylate has FEMA (Flavor and Extract Manufacturers' Association) GRAS status (FEMA No. 3568) and has an average maximum use level of 10 ppb in frozen dairy and nonalcoholic beverages and 100 ppb in baked goods, gelatins, and puddings [3] Methyl cyclohexanecarboxylate has an odor threshold of 0.02 ppb, which is 20 times higher than that of the corresponding ethyl ester [1]. Guth and Grosch [4] found ethyl cyclohexanecarboxylate in virgin olive oils. They confirmed the potent nature of this ester, reporting an odor threshold in air of 7.5 pg/L. Using their aroma extract dilution analysis method, they found that this compound was an important contributor to the fruity note in certain virgin olive oils. Substitution of two sulfur atoms into the cyclohexane ring leads to a 100-fold reduc-

tion in odor potency, and substitution of a nitrogen atom into the ring results in a dramatic 40 million-fold loss in odor strength. Ethyl benzoate has an odor threshold of 60 ppb [1]. Substitution of a nitrogen atom to give the pyridine ring causes a 100-fold increase in odor threshold to 6000 ppb.

The effect of the alcohol group on the odor threshold was also evident for the polycyclic esters (Fig. 2). Ethyl 1-adamantanecarboxylate is covered by a Dutch patent [5] for use as a perfume constituent. This ester has an odor threshold of 0.009 ppb. The corresponding methyl and propyl esters had thresholds of 8 and 70 ppb, respectively. The addition of a CH$_2$ group between the polycyclic ring and carbonyl group (ethyl 1-adamantaneacetate) increased the threshold to 5 ppb.

Acknowledgments. We thank Jean G. Turnbaugh for her help in the odor threshold determinations.

References

1. Takeoka GR, Buttery RG, Turnbaugh JG, Teranishi R (1991) Cyclic esters: compounds possessing remarkably low odor thresholds. Lebensm Wiss Technol 24:569–570
2. Guadagni DG, Buttery RG (1978) Odor threshold of 2,3,6-trichloroanisole in water. J Food Sci 43:1346–1347

3. Oser BL, Ford RA (1978) Recent progress in the consideration of flavoring ingredients under the food additives amendment. 11. GRAS substances. Food Technol 32:60–70

4. Guth G, Grosch W (1991) A comparative study of the potent odorants of different virgin olive oils. Fat Sci Technol 93:335–339

5. N.V. Philips' Gloeilampenfabricken Neth. Appl. 67 15,903 (Cl. C 07c), 28 May 1969, Appl. 23 Nov 1967; 6 pp. Chem Abstr 1969, 71, 94695w

Synthesis of "Delicious Peptide (Lys-Gly-Asp-Glu-Glu-Ser-Leu-Ala)" Analogs; Investigation of Their Tastes

Takashi Nakata, Masaru Nakatani, Masatoshi Takahashi, and Hideo Okai[1]

Key words. Delicious peptide—Umami

In 1978, Yamasaki and Maekawa isolated a peptide from extracts of beef treated with papain. They proposed that the primary structure of the isolated peptide was Lys-Gly-Asp-Glu-Glu-Ser-Leu-Ala, and they reported that this octapeptide produced a delicious taste, just like beef soup [1,2].

In 1989, our research group re-synthesized this octapeptide independently. Sensory analysis of the synthetic "delicious peptide" supported the finding that this peptide produced a simple umami taste, the threshold level of which was almost the same as that of monosodium glutamate (MSG), as well as a slightly sour taste. In order to investigate the taste-active sites of this "delicious peptide" in detail, we synthesized several kinds of peptide fragments and organoleptically evaluated their tastes. We found that this taste consisted of that of three fragments; Lys-Gly as a satly/umami taste, Asp-Glu-Glu as a sour taste, and Ser-Leu-Ala as a bitter taste. When these equimolar taste-active fragments were mixed, a umami taste was reformed at the same level as that of the original "delicious peptide". We proposed that the cation of the basic portion and the anion of the acidic portion in the molecule must be located in tight apposition to produce the salty and umami taste [3,4].

In this study, our research group synthesized "fragment-shifted delicious peptides" in which the fragments of the original "delicious peptide" [5] and its analogs (Lys-Gly-Asp-Glu-Glu and Glu-Glu-Asp-Gly-Lys) were shifted in order to investigate the effect of the sequence or spatial structure of these peptides on their tastes and to elucidate the umami mechanism. Sensory analysis of the synthetic analogs showed that the umami taste was produced at almost the same level as that of the original "delicious peptide", those analogs which possessed acidic amino acids in the C-terminal portion in particular produced a strong sour taste. The sequence

of the synthetic "delicious peptide" analogs did not always appear to be important for the production of a umami taste, and the presence of the umami taste in the "delicious peptide" had nothing to do with the spatial structure of these peptides.

It is well-known that a umami taste is exhibited at concentrations of the order of 10^{-3} M, whereas sweet and bitter tastes are produced at concentrations of the order of 10^{-8} M. A umami taste was produced more broadly than a sweet or bitter taste. For instance, the threshold value was remarkably variable when only one methylene group of a sweet or bitter peptide was substituted.

In the "delicious peptide" analogs, both the kinds of taste and the threshold values of these analogs were slightly different, but they exhibited essentially the same taste. It seems that the umami mechanism is quite different from that of a sweet or bitter taste. We propose that the umami taste of the "delicious peptide" and its derivatives is exhibited as the result of a stimulatory transmission between ionic groups (the cation and anion) and the surface of the cell membrane on the tongue.

References

1. Yamasaki Y, Maekawa K (1978) A peptide with delicious taste. Agric Biol Chem [Biosci Biotech Biochem] 42:1761–1765
2. Yamasaki Y, Maekawa K (1980) Synthesis of a peptide with delicious taste. Agric Biol Chem 44:93–97
3. Tamura M, Nakatsuka T, Tada M, Kawasaki Y, Kikuchi E, Okai H (1989) The relationship between taste and primary structure of "delicious peptide (Lys-Gly-Asp-Glu-Glu-Ser-Leu-Ala)" from beef soup. Agric Biol Chem 53(2):319–325
4. Tamura M, Seki T, Miyoshi T, Okai H (1990) The relationship between taste and primary Structure of "delicious peptide (Lys-Gly-Asp-Glu-Glu-Ser-Leu-Ala)". In: Yanaihara N (ed) Peptide chemistry 1989. Protein Research Foundation, Osaka, pp 195–198
5. Nakata T, Takahashi M, Nakatani M, Kuramitsu R, Tamura M, Okai H (1993) Synthesis of "Delicious peptide (Lys-Gly-Asp-Glu-Glu-Ser-Leu-Ala)" and its derivatives (Part II). In: Yanaihara N (ed) Peptide chemistry 1992, pp 431–433

[1] Department of Fermentation Technology, Faculty of Engineering, Hiroshima University, 1-4-1 Kagamiyama, Higashihiroshima, Hiroshima, 724 Japan

The Tastes of *O*-Aminoacyl Sugars

Kozo Nakamura, Masahiro Tamura, and Hideo Okai[1]

Key words. *O*-Aminoacyl sugar—Chemicale

In 1982, our research group reported a series of *O*-aminoacyl sugars [1]. Although these sugars are composed of very simple elements (carbohydrates, amino acids or peptides, and an ester linkage), they are rarely found in natural products. To study the relationship between the taste and structure of these sugars, were prepared a number of them containing neutral, hydrophobic, and basic amino acids and we carried out a sensory analysis, using their aqueous solutions.

O-Aminoacyl sugars in which hydrophobic amino acids were introduced into the 2-*O*- and 3-*O*-positions of methyl-α-D-glucopyranoside produced bitterness. For example, methyl 2,3-di-*O*-(L-phenylalanyl)-α-D-glucopyranoside produced strong bitterness, 20 times stronger than that of caffeine [2]. By the systematic synthesis of bitter peptides, we known that strong bitterness is produced when two bitterness-producing groups, which are a hydrophobic group and a hydrophobic or basic group, are located close to each other [3,4]. In the case of this compound, two phenylalanyl groups might be close enough to produce the strong bitterness.

O-Aminoacyl sugars containing neutral amino acids with low hydrophobicity at the 2-*O*- and 3-*O*-positions of methyl-α-D-glucopyranoside produced a sweetness. Methyl 2,3-di-*O*-(L-α-aminobutyryl)-α-D-glucopyranoside produced a sweet taste 50 times stronger than that of sucrose [5]. *O*-Aminoacyl sugars containing glycyl, alanyl, and valyl residues at these positions also produce sweetness. The intensity of the sweetness became stronger as the hydrophobicity of the amino acid increased. *O*-Aminoacyl sugars containing neutral amino acids with higher hydrophobicity than valyl residues produced bitterness. The hydrophobicity of the amino acid moiety was important for producing sweetness, although higher hydrophobicity makes the compounds bitter.

Recently, we employed sucrose as the sugar skeleton and synthesized *O*-aminoacyl sugars containing triphenylmethyl groups and several neutral amino acids (glycine, β-alanine, γ-amino-*n*-butyric acid, 5-aminovaleric acid, and 6-amino-*n*-propionic acid). All these compounds produced bitterness, the bitter intensity of which became stronger as the length of the amino acid chain increased. The 6-amino-*n*-caproyl derivative produced the strongest bitterness. The bitter potency of this compound (threshold value, 0.0032 mM) was equal to that of strychnine [6].

O-Aminoacyl sugars containing basic amino acids (diaminopropionic acid, diaminobutyric acid, ornithine, and lysine) produced sourness, umami, and saltiness regardless of the positions at which they were introduced and regardless of the sugar moieties. The taste seemed to depend on the type of the basic amino acid. Methyl-2,3,4-tri-*O*-(L-α,γ-diaminobutyryl)-α-D-glucopyranoside produced saltiness and had an excellent dietary effect on sodium ion intake, reducing this intake by 10% [7].

The results described above indicate that the taste quality of *O*-aminoacyl sugars is dependent on the properties of the amino acids introduced into the sugars.

References

1. Kinomura K, Tada M, Okai H (1983) *O*-Aminoacyl sugars. A new utility of amino acid residues as the protecting groups of sugar hydroxyls (in Japanese). In: Sakakibara S (ed) Peptide chemistry 1982. Protein Research Foundation, Osaka, pp 35–40
2. Tamura M, Nakatuka T, Kinomura K, Okai H, Fukui S (1985) Methyl 2,3-di-*O*-(L-α-aminobutyryl)-α-D-glucopyranoside, a new sweet substance. Agric Biol Chem 49:891–893
3. Ishibashi N, Kouge K, Shinoda I, Kanehisa K, Okai H (1988) A mechanism for bitter taste sensibility in peptides. Agric Biol Chem 52:819–827
4. Tamura M, Miyoshi T, Mori N, Kinomura K, Kawaguchi M, Ishibashi N (1990) Mechanism for the bitterness tasting potency of peptides using *O*-aminoacyl sugars as model compounds. Agric Biol Chem 54:1401–1409
5. Tamura M, Shoji M, Nakatuka T, Kinomura K, Okai H, Fukui S (1985) Methyl 2,3-di-*O*-(L-α-aminobutyryl)-α-D-glucopyranoside, a new sweet substance, and tastes of related compounds of neutral amino acids and D-glucose derivatives. Agric Biol Chem 49:2579–2586
6. Nakamura K, Segawa D, Kinomura K, Tamura M, Okai H (1993) Synthesis of saccharoide (complex containing saccharide and peptide). In: Yanaihara N (ed) Peptide chemistry 1992
7. Tamura M, Nakamura K, Kinomura K, Okai H (1993) Relationship between taste and structure of *O*-aminoacyl sugars containing basic amino acids. Biosci Biotech Biochem 57:20–23

[1] Department of Fermentation Technology, Faculty of Engineering, Hiroshima University, 1-4-1 Kagamiyama, Higashihiroshima, Hiroshima, 724 Japan

Role of Arginine and Lysine Residues in Sweetness Expression of Thaumatin

NAOFUMI KITABATAKE and MITSUAKI KUSUNOKI[1]

Key words. Thaumatin—Sweet protein

Thaumatin I, one of the sweet-tasting proteins from the fruit of *Thaumatococcus Daniellii Benth*, is about 100 000 times sweeter than sucrose. Although its three-dimensional structure has been reported [1], the active site of its sweetness is not known. However, there have been various reports concerning the chemical modification of arginine [2] and lysine residues in this regard [3,4]. In this study, selective modification of the arginine residue(s) that comprise the active site of sweetness was carried out by controlling the modification conditions.

Thaumatin I, purified by ion-exchange chromatography on SP-Sephadex C-25, gave a single band on native polyacrylamide gel electrophoresis (PAGE). Arginine residues were modified with 1,2-cyclohexanedione in borate buffer at pH 9. The modified thaumatin reverted to the original thaumatin I following treatement with hydroxylamine. Lysine residues were modified with pyridoxal phosphate. A taste assay was performed by sensory analysis, using 10 µg/ml of either thaumatin I or modified sample solution.

Modification of one or two arginine residues of thaumatin I, as confirmed by PAGE analysis, resulted in the loss of the sweet taste. When modified thaumatin I, which has no sweet taste, was treated with hydroxylamine, it reverted to the original thaumatin I, as shown by PAGE, and sweetness was restored. This result indicates that one or two specific arginine residues are related to the expression of sweetness.

Thaumatin I was modified with pyridoxal phosphate until no bands corresponding to those of the original thaumatin I were observed on PAGE. This modified thaumatin I showed only a slight decrease in sweetness. When arginine-modified thaumatin I, which retains some sweetness after 1 h of reaction, was similarly modified with pyridoxal phosphate, its sweetness completely disappeared. These results indicate that lysine and arginine residue(s) may constitute part of the active site for the expression of sweetness and that both types of residues are close together.

References

1. De Vos AM, Hatada M, Van der Wel H, Krabbendam H, Peerdeman AF, Kim SH (1985) Three-dimensional structure of thaumatin I, an intensely sweet protein. Proc Natl Acad Sci USA 82:1406–1409
2. Van der Wel H (1977) Structural investigation of thaumatin I, a sweet tasting protein from *Thaumatococcus Daniellii Benth*. In: Magnen J Le, MacLeod P (eds) Proceedings of the sixth international symposium on olfaction and taste. Information Retrieval, London, p 72
3. Van der Wel H, Bel WJ (1976) Effect of acetylation and methylation on the sweetness intensity of thaumatin I. Chem Senses Flavor 2:211–218
4. Shamil S, Beynon RJ (1990) A structure-activity study of thaumatin using pyridoxal 5'-phosphate (PLP) as a probe. Chem Senses 15:457–469

[1] Research Institute for Food Science, Kyoto University, Gokasho, Uji, Kyoto, 611 Japan

Purification and Structural Characterization of Homologues of the Heat-Stable Sweet Protein, Mabinlin

Satoru Nirasawa[1], Xiaozhu Liu[2], Zhong Hu[2], Tomoko Nishino[3], Masato Katahira[4], Seiichi Uesugi[4], and Yoshie Kurihara[1]

Key words. Heat-stable protein—Sweet protein

The plants, *Capparis masaikai* Lévl. (local name mabinlang), which grows in the subtropical region of Yunnan province in China, bears fruit of tennis-ball size. The mature seed is used as traditional Chinese medicine and elicits a sweet taste. The seeds of *C. masaikai* contain the heat-stable sweet protein named mabinlin.

We established a method for the isolation of mabinlin and purified five homologues (I-1, I-2, II, III and IV) [1]. The seeds of *C. masaikai* were defatted with petroleum ether and extracted with 0.5 M NaCl solution. After ammonium sulfate fractionation, the sample was applied to a CM-Sepharose CL-6B column (Pharmacia LKB Biotechnology Inc., Uppsala, Sweden). Mabinlin homologues were eluted from the column by a linear gradient of NaCl (0.25–1.0 M). Each peak was further pruified with a Bio gel P-30 column (Bio-Rad Lab., Richmond, CA) (gel filtration). The sample thus obtained was lyophilized as pure mabinlin.

Since mabinlin II and I-1 are abundant, we focused on these homologues for experiments on the relationship between heat stability and structure.

The sweetness of mabinlin II was equivalent to that of mabinlin I-1. On the other hand, the heat stability of both proteins was quite different. That is, while the sweet activity of mabinlin II was unchanged by incubation at 80°C for 48 h, the sweet activity of mabinlin I-1 was gradually reduced by 80°C incubation and was completely abolished by 1-h incubation. The circular dichroism spectrum of mabinlin I-1 was changed by incubation at 80°C, in parallel with the decrease of the sweet activity, while that of mabinlin II was unchanged by 48-h incubation. We concluded that changes of the α-helix of mabinlin I-1 into a random coil with the temperature increase lead to the abolition of its activity.

The complete amino acid sequences of mabinlin II and I-1 were determined by automated Edman degradation. Mabinlin II was composed of an A-chain with 33 amino acid residues and a B-chain with 72 amino acid residues, and mabinlin I-1 was composed of an A-chain with 32 amino acid residues and a B-chain with 72 amino acid residues. Both proteins had a relatively high content of glutamine and arginine. The homology match of the amino acid sequence between mabinlin II and I-1 was 66.7% for the A-chain and 68.1% for the B-chain. The difference in the amino acid sequence seemed to contribute to the difference in heat stability without influencing the sweetness of mabinlin II and I-1.

Both mabinlin II and I-1 contain two cysteine residues in the A-chain and six cysteine residues in the B-chain, and the position of all cysteine residues and disulfide bridges of mabinlin II and I-1 were coincident. There were two interchain disulfide bridges, and two intrachain disulfide bridges in the B-chain [2]. Therefore, it seems that the disulfide bridges are independent of the difference in heat stability between mabinlin II and I-1.

Although mabinlin II is very heat-stable, mabinlin I-1 in also heat-stable compared with most other proteins. Four disulfide bridges of relatively small molecular mass (12.4 kDa) in mabinlin seem to fix the molecular conformation peptides in this protein, which contributes to its heat stability. This, together with the finding that there is a difference in heat stability between mabinlin I-1 and mabinlin II, suggests that not only disulfide bridges but also other factors (e.g., the hydrophobic bond between peptides) contribute to the heat stability of this protein.

[1] Department of Chemistry, Facutly of Education, Yokohama National University, Hodogaya-ku, Yokohama, 240 Japan
[2] Kunming Institute of Botany, The Academy of Sciences of China, Yunnan, Kunming, China
[3] Department of Biochemistry, Yokohama City University School of Medicine, Kanazawa-ku, Yokohama, 236 Japan
[4] Department of Bioengineering, Faculty of Engineering, Yokohama National University, Hodogaya-ku, Yokohama, 240 Japan

References

1. Liu X, Maeda S, Hu Z, Aiuchi T, Nakaya K, Kurihara Y (1993) Purification, complete amino acid sequence, and structural characterization of the heat-stable sweet protein, mabinlin II. Eur J Biochem 211:281–287
2. Nirasawa S, Liu X, Nishino T, Kurihara Y (1993) Disulfide bridge structure of the heat-stable sweet protein, mabinlin II. Biochim Biophys Acta 1202:277–280

Sensory Identification of Effective Components for Masking Bitterness of Arginine in Synthetic Extract of Scallop

KYOKO MICHIKAWA and SHOJI KONOSU[1]

Key words. Arginine—Bitterness

The adductor muscle of the scallop, *Patinopecten yessoensis*, is a tasty seafood. One of the authors (SK) and collaborators [1] previously reported that a simplified synthetic extract of scallop muscle containing only eight components, glycine (Gly), alanine (Ala), glutamic acid (Glu), arginine (Arg), adenosine 5'-monophosphate (AMP), Na^+, K^+, and Cl^-, satisfactorily reproduced the taste of the natural extract. They also pointed out that neither the synthetic nor the natural extract were bitter, despite their having a content of more than 300 mg of Arg/100 ml, the threshold concentration of Arg being 50 mg/100 ml [2]. This finding indicates that one or more of the above components are effective for reducing or masking the bitterness of Arg. In order to make a sensory identification of the effective component(s), we conducted the following tests.

First, the taste quality of 0.3% Arg solution was examined, this concentration being approximately that of the natural and synthetic extracts. The triangle difference test with the Arg solution and deionized water at 25°C was performed by a panel consisting of one man and nine women. They were also asked to evaluate the magnitude of difference in five basic tastes between the two samples on a four-point rating scale (0–+3). The results were statistically analyzed by the Student's *t*-test. *P* values less than 0.05 were regarded as significant.

The values obtained clearly demonstrated that the Arg solution was perceived as strongly bitter ($P < 0.001$), with weak sweetness ($P < 0.01$). The tests on two dilute Arg solutions, 0.15% and 0.075%, also gave similar results.

For identifying the component(s) effective for decreasing or masking the bitterness of Arg, we carried out two kinds of addition test, i.e., (1) addition of Gly, Glu, AMP, NaCl, KCl, or Glu+AMP to Arg, and (2) the addition of Arg to each of these six components. Ala was excluded from the tests, as it had been found to be ineffective in a preliminary test. The concentrations of these components were the same as those in the natural and synthetic extracts (mg/100 ml): Arg, 323; Gly, 1925; Glu, 140; AMP, 172; Na^+, 73; K^+, 218; Cl^-, 95. The pH of all test solutions was adjusted to 6.1, the same value as that of the natural extract, and the temperature was 25°C. The method for sensory tests was essentially the same as that described above, except that seven to ten panelists were employed and the magnitude of difference in bitterness was evaluated on a seven-point rating scale (−3–+3).

In the first addition test, Glu ($P < 0.001$), AMP ($P < 0.05$), NaCl ($P < 0.001$), and Glu+AMP ($P < 0.01$) were found to decrease the strong bitterness due to Arg. Some of the panelists stated that the addition of Gly reduced the bitterness, but the result was not significant ($P > 0.05$). Among the components tested, NaCl was found to be the most effective for reducing the bitterness; most of the panelists stated that the bitterness of Arg was greatly lowered by the addition of NaCl.

The effectiveness of NaCl was confirmed by the second addition test, in which the panel could not make significant discriminations between NaCl and NaCl+Arg solutions in terms of bitterness ($P > 0.05$), while the addition of Arg to Gly ($P < 0.001$), Glu ($P < 0.001$), AMP ($P < 0.05$), and Glu+AMP ($P < 0.01$) elicited bitterness to some extent.

In conclusion, Glu, AMP, and NaCl reduced the bitterness of Arg at concentrations found in the scallop muscle extract, with NaCl being the most effective in this regard.

References

1. Watanabe K, Lan H, Yamaguchi K, Konosu S (1990) Role of extractive components of scallop in its characteristic taste development (in Japanese). Nippon Shokuhin Kogyo Gakkaishi (J Japan Soc Food Sci Technol) 37:439–445
2. Yoshida M, Ninomiya T, Ikeda S, Yamaguchi S, Yoshikawa T, Ohara M (1966) Studies on the taste of amino acids. Part I. Determination of threshold values of various amino acids (in Japanese). Nippon Nogei Kagaku Kaishi (J Agric Chem Soc Japan) 40:295–299

[1] Faculty of Home Economics, Kyoritsu Women's University, 2-2-1 Hitotsubashi, Chiyoda-ku, Tokyo, 101 Japan

Comparison of Extracts from Species of the Mackerel Family

Yuko Murata, Hikaru Henmi, and Fujio Nishioka[1]

Key words. Extracts—Taste of skipjack

Skipjack is a very tasty fish and, in Japan, is processed as dried flakes (bonito) and as canned skipjack tuna; it is also served raw as "sashimi". However, its slight astringency and sourness are sometimes disliked. In the mackerel family, skipjack and frigate mackerel are sourer and more astringent than other members such as tunas and mackerel. In this study, we examined the characteristics of extracts of the mackerel family and the peculiar taste of skipjack.

Ten species were used: skipjack, frigate mackerel (*Auxis thazard* and *A. tapeinosoma*), yaito, yellowfin, albacore, southern bluefin, bluefin tuna, mackerel, and Spanish mackerel. Hot water (HW) and perchloric acid (PCA) extracts were prepared from the minced dorsal muscle of each sample.

Twenty ml of HW extract was prepared from 20 g homogenate by successive deproteinization (heated at 90°C for 10 min) and removal of fat (washed with diethyl ether three times). PCA (10%) was used instead of hot water to prepare the PCA extracts.

Free amino acids (FAA) and organic acids (ORA) in the HW extracts were analyzed with an automated amino acid analyzer and by conducting high-performance liquid chromatography (HPLC), respectively. Nucleotide (NUC) analysis of the extracts was conducted by HPLC.

Total FAA exceeded 1000 mg/100 g in skipjack and frigate mackerel, but was only 300–400 mg/100 g in Spanish mackerel.

Imidazole compounds (IMI), i.e., histidine (His) and related compounds, constituted 82%–96% of all ninhydrine-positive substances in skipjack, frigate mackerel, and tunas, 76%–86% in mackerel, and only 56%–68% in Spanish mackerel. Suyama and Yoshizawa [1] found that the skeletal muscles of migratory fish contained high levels of IMI. In mackerel, the ratio of His to IMI was higher than that in other species, and anserine could not be detected.

Skipjack and frigate mackerel had a sour taste and a low pH, possibly due to the presence of organic acids. Both species were richer in lactic and malic acids than the other species tested.

Total NUC content in all species was 230–500 mg/100 g, and the total content of inosine monophosphate (IMP), the major NUC, was 150–410 mg/100 g. Ehira and Uchiyama [2] noted three types of species based on considerable levels of inosine (HxR), hypoxanthine (Hx), or both, during adenosine triphosphate (ATP) degradation: HxR-forming, Hx-forming, and intermediate. Most of the mackerel family were found to be HxR-forming species. Yaito was an Hx-forming species and bluefin tuna, an intermediate species.

Based on the present findings, it appears that the astringency, sourness, and strong taste of skipjack and frigate mackerel may be due to the high content of FAA, ORA, and NUC, and to the presence of various taste-active components in the extracts. The HPLC chromatogram for ORA and NUC separation indicated many small unidentifiable peaks in skipjack and frigate mackerel, suggesting the presence of many components in the muscle.

Acknowledgments. The authors thank Mr. T. Murase of Hagoromo Foods Co. and Mr. I. Shibata of Kanagawa Prefecture Fisheries Experimental Station for providing the fish samples.

References

1. Suyama M, Yoshizawa Y (1973) Free amino acid composition of the skeletal muscle of migratory fish (in Japanese). Bull Jpn Soc Sci Fish 39:1339–1343
2. Ehira S, Uchiyama H (1973) Formation of inosine and hypoxanthine in fish muscle during ice storage (in Japanese). Bull Tokai Reg Fish Res Lab 75:63–73

[1] Marine Biochemistry Division, National Research Institute of Fisheries Science, 2-12-4 Fukuura, Kanazawa-ku, Yokohama, 236 Japan

Odor of (R)-Ethyl Citronellyl Oxalate and Its Conformations

Fumiko Yoshii, Shuichi Hirono, and Ikuo Moriguchi[1]

Key words. Complex odor—Conformations of odorants

It is quite difficult to classify odor qualities, one reason being that many odorants have a complex odor, i.e., a few odor facets. Why does a single odorant have a complex odor? In 1971, Beets had already pointed out that odorants possessing conformational freedom have several odor facets [1]. It is probable that a few stable conformers of a single odorant fit different odor proteins and cause complex odor. Experimental evidence concerning odorant-receptor interaction has not yet been shown by X-ray or nuclear magnetic resonance (NMR), however. In this work, we tried to analyze and study stable conformers of (R)-ethyl citronellyl oxalate (ECO), the same odorant noted by Beets in 1971. ECO was synthesized and its odor quality was evaluated as "mainly musk-like odor with partially fresh and sweet rose-like odor" at the laboraories of the Takasago International Corporation (Hiratsuka, Japan).

The following calculations and comparisons were done: (1) conformational analysis of ECO using CAChe (Ver. 3, CAChe Scientific, Beaverton, Ore.), (2) structure optimization of selected conformers of ECO, using PM3 (MOPAC Ver. 6, QCPE#455, Indiana University, Bloomington, Ind.), and (3) comparison of conformers of ECO, with (R)-muscone, a three-dimensional structure model for benzenoid musks, and with a conformer of (S)-citronellol, using MidasPlus (Ver. 1.5, UCSF San Francisco, Calif.). For conformational analysis, multiple passes sequential search was done, using the conjugate gradient as the optimization method with Allinger's MM2 parameters as force field parameters. Ten single bonds and one double bond were rotated for conformational analysis. The three-dimensional structure model for benzenoid musks was previously constructed in 1992 [2]. One conformer of (S)-citronellol was previously selected by comparison with the conformers of phenylethyl alcohol and diphenyloxide, which are rose odor compounds. The hardware used was Apple Macintosh Quadra 700 (Apple, Cupertino, Calif.) and SiliconGraphics IRIS 4D/420GTX (SiliconGraphics, Mountain View, Calif.).

By conformational analysis, 37 stable conformers (20.9–22.9 kcal/mol) were chosen. The resemblance of the structures of 37 conformers was judged by root mean square (RMS) error values calculated by the least squares fit method. When the two conformers were considered similar, the conformer with a higher energy was deleted. Fifteen conformers were thus selected. The structures of these 15 conformers were then optimized by PM3, and the optimized structures were divided into three groups. Five rather stretched conformers constituted group 1; group 3 contained two bent conformers; their shapes were like the letter "C". The rest of the conformers were in group 2. On comparison of graphic images of the two conformers (Conformer A and B) in group 3 and (R)-muscone, it was seen that the whole molecule and one oxygen atom of both odorants occupied the common space. Also, one conformer (Conformer A) in group 3 fitted well to the three-dimensional structure model for benzenoid musks. One conformer (Conformer C) in group 2 partially fitted the conformer of (S)-citronellol. Furthermore, a good correspondence in the area with positive electrostatic potential and the area that was negative was seen between Conformer A of ECO and (R)-muscone, and between Conformer C and (S)-citronellol.

These calculations, showed that Conformer A, a stable conformer of ECO, resembled both (R)-muscone and three-dimensional structure model for benzenoid musks, and Conformer C, another stable conformer of ECO, partially resembled (S)-citronellol. Therefore, the complex odor of ECO could be explained by these stable conformers. These findings also showed that conformational analysis, using a computer, was useful to predict conformers that probably fit receptor proteins.

Acknowledgments. We thank Mr. Takeshi Yamamoto for the synthesis of (R)-ethyl citronellyl oxalate and for information about it, and the chief perfumer, Takasago International Corporation, Mr. Takemi Kato, for the sensory evaluation of (R)-ethyl citronellyl oxalate.

[1] Department of Physical Pharmacy, School of Pharmaceutical Sciences, Kitasato University, 5-9-1 Shirokane, Minato-ku, Tokyo, 108 Japan

References

1. Beets MGJ (1982) Odor and stimulant structure. In: Theimer ET (ed) Fragrance chemistry, the science of the sense of smell. Academic, Orlando, pp 77–122

2. Yoshii F, Hirono S, Moriguchi I (1992) Three-dimensional structure model for benzenoid musks expressed by computer graphics. Chem Senses 17(5): 573–582

Statistical Analysis of Flavor Components and Sensory Data for Tea

Asako Tamura[1], Masashi Omori[1], Miyuki Kato[2], Toshiko Onoue[3], Naohisa Koremura[4], Shuuichi Fukatsu[5], and Ryoyasu Saijo[5]

Key words. Multivariate analysis—Gas-chromatographic data

In this study, we formed a data base by collecting and analyzing preference terms. Important words relating to tea were selected in accordance with the method of Ruhn et al. [1]. We carried out a preference survey by adding terms for examining teas to these important words; correlations between the results of the preference survey and the analytical data for tea components were analyzed with plural variables.

Black, oolong, and green tea belong, respectively, to the categories fermented, semi-fermented, and unfermented tea. We found that the amino acid content increased with the progress of fermentation. That is to say, the amino acid content in oolong tea was higher than that of green tea but lower than that of black tea. Regarding each amino acid, however, green tea contained aspartic acid and threonine in larger amounts than black tea. Each of these teas contained glutamic acid and theanine, the main amino acids of tea leaves, at high levels. We also found that the content of flavor components increased as the fermentation progressed. Black tea contained the largest amounts of these flavor components.

Nine samples in all, namely, three samples each of green, black, and oolong tea, were subjected to statistical treatment by quantification system III (Shakai Joho Service, Multi Tokei System). The x axis was referred to as axis 1, and the y axis as axis 2. A higher score on the x axis means words for more preferable images, such as "suitable smell," "brisk smell," or "delicious smell," while a lower score means words for less preferable images, such as "greenish," "smoky," or "heavy." A higher score on the y axis means "pungent," "burnt," or "baked smell," while a lower score means "tainted" or "sweaty".

Next, we determined the correlations among the scores (axes 1 and 2) obtained by quantification system III and peaks in the gas chromatograms obtained by analyzing the flavor components. In green tea, axis 1 was highly correlated to β-phenyl ethanol, while axis 2 was highly correlated to linalool oxide I. Green tea contained the largest amount of β-phenyl ethanol and was highly correlated to the words "brisk" and "delicious" in the positive direction of axis 1.

In conclusion, we found that a difference in the preference pattern of each tea was correlated to changes in flavor components.

[1] Department of Food Science, Otsuma Women's University, 12 Sanban-cho, Chiyoda-ku, Tokyo, 102 Japan
[2] Department of Education, Kagawa University, 1-1 Saiwai-cho, Takamatsu, Kagawa, 760 Japan
[3] Mejiro Gakuen Women's College, 4-31-1 Nakaochiai, Shinjuku-ku, Tokyo, 161 Japan
[4] Tokyo University Agric, 1-1-1 Sakuragaoka, Setagaya-ku, Tokyo, 156 Japan
[5] National Research Institute, Vegetables, Ornamental Plants, and Tea, 2769 Kanaya, Kanaya-cho, Haibara-gun, Shizuoka, 428 Japan

References

1. Ruhn HP (1985) The automatic creation of literature abstracts. IBM J Res Devel 4(159)

6. Psychophysics of Taste and Olfaction

Behavior Manifestations Indicative of Hedonics and Intensity in Chemosensory Experience

JACOB E. STEINER[1]

Adequate stimulation of a sensory system initiates a cascade of neural events. This results in the arousal of a "sensation", or "feeling". Stimulating events and their neural sequels occur in the public domain, and therefore are subject to direct measurement. Sensations in contrast, are psychological events, occurring in the private domain, and are therefore not subject to direct measurement.

Assessment of sensory functions, for clinical or other purposes, requires quantitative measures. Biology, psychology, and other disciplines have developed testing methods that use two major approaches: (1) the subjective, psychophysical approach, based on cognitive processes, mainly to assess detection and identification of stimuli in conscious, collaborative, verbal human subjects, and (2) objective tests, based on recording of biological events that occur in fixed, causal, and temporal relationship to the delivered stimulus. For these, bioelectrical manifestations, that is, motor or secretory reactions, are critical measures. Subjective and objective methodolgies alike only indicate sensations but cannot directly observe them. For both testing methods, stimuli should be delivered under well-defined criteria with controls for gender, age, developmental and physiological states and other variables of the subjects.

Gustatory and olfactory stimuli, probably more than others, have a special pleasure aspect (*hedonics*). When senses of taste or smell are studied, information regarding subject affect or motivation toward stimuli should also be obtained. Evidently, human verbal reports may sensitively mirror these aspects. We presumed that some somatic manifestations are equally indicative of the pleasure aspects of chemical sensations. Although Darwin, in his classic book [1], had emphasized that emotions, moods, and feelings are expressed by behaviors in humans and animals, these aspects attracted little attention in chemosensory studies. Throughout the past three decades, efforts have been made to gain better insights into the determination of behavior by chemical stimuli.

Our aim was to show that stimulus-dependent motion stereotypes of the orofacial musculature and changes in autonomous reactions, for example, in heartrate (HR) or in respiratory rate (RR), are comparably sensitive indicators of taste and odor hedonics, like psychophysical tests. Our first studies evinced that the full-term, healthy, perinatal human infant displays distinct, differential facial expressions to different gustatory stimuli before the first food intake. These responses of "acceptance", "indifference", and "aversion", were named the gusto-facial reflex (GFR) [2]. GFR was found to be a rich, easily evaluable repertoire of nonverbal communicational signs of wide cross-cultural appeal. Water always induces the response of "indifference", sweet and umami-taste, that of "acceptance", and sour and bitter taste triggers "aversion". Examples of GFR are shown in Fig. 1.

Neonates who were born with severe forebrain defects were found to display identical reactions. From this, it could be inferred that GFR is not only innate and probably inherited, but primarily controlled by the brain stem [2]. Experiments using an animal model [3] strongly supported this assumption. An analysis of GFR using a special notational system showed that it can be used as a quantitative indicator of both hedonic taste and taste intensity [4]. GFR also served as an adequate tool to demonstrate classical conditioning in perinatal human infants [5]. Similar sets of responses induced by odor stimuli were also described and named the nasofacial reflex (NFR) [6,7]. Because GFR and NFR could be used effectively as diagnostic tools [8–10], and analogous reactions in many animal species were also reported [11–17], we have suggested the routine use of this behavioral testing whenever nonverbal human subjects are to be tested or cognitive tests are inappropriate [18]. Other experiments indicated that taste hedonics are sensitively reflected also by differential HR acceleration [19,20]. When hedonic estimates of adult human testees were first simultaneously recorded with ratings of the same testees' videotaped facial behaviors, these two measures were found to be in close correlation [21]. A current, large-scale, multidisciplinary study aims at further investigation of interrelationships between cognitive and behavioral measures of taste and odor hedonics. Data from

---

[1] Department of Oral Biology, The Hebrew University, Hadassah School of Dental Medicine, Jerusalem, Israel

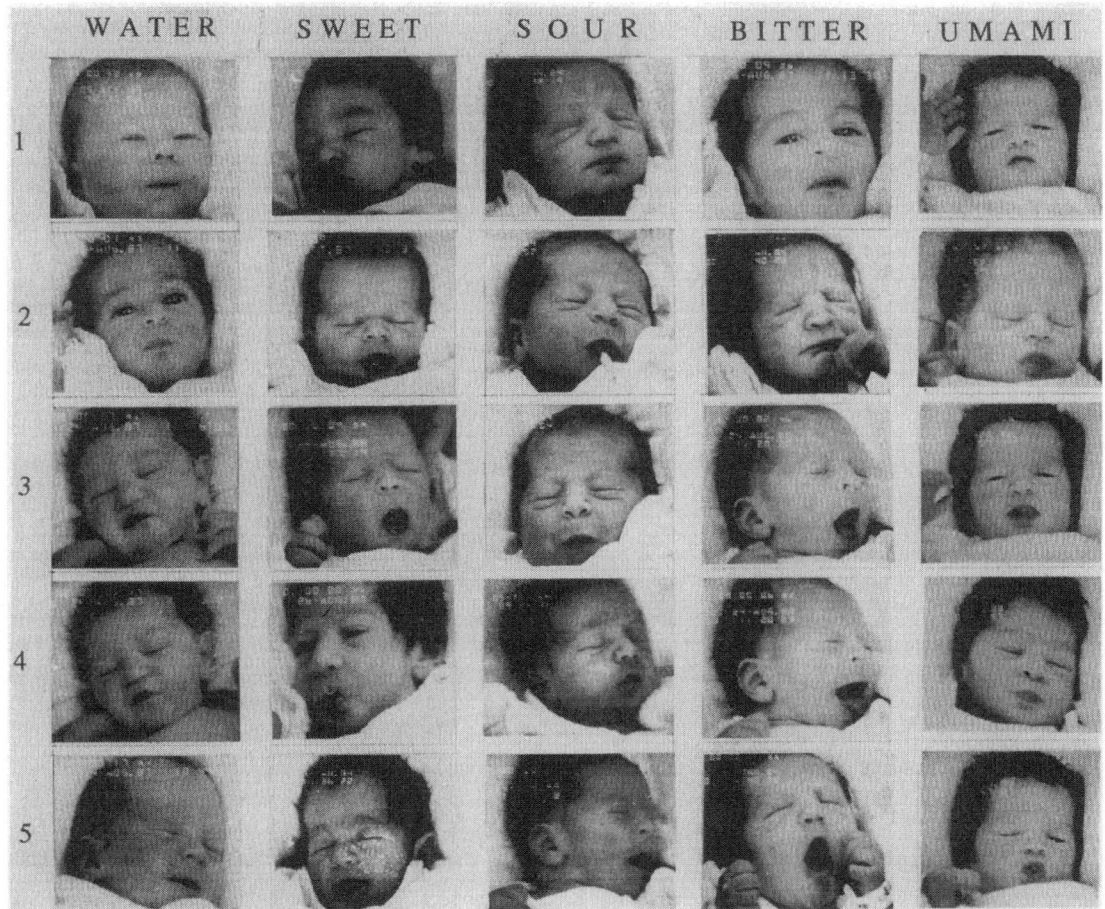

FIG. 1. Synoptic tableau of neonate infant facial reactions (GFR) to intraorally presented water, sweet, sour, bitter, and umami stimuli. All infants were tested 2–8 h after birth. (Stillframes from videotaped sequences; original photographs of author. Reproduced with permission from Marcel Dekker, Inc.)

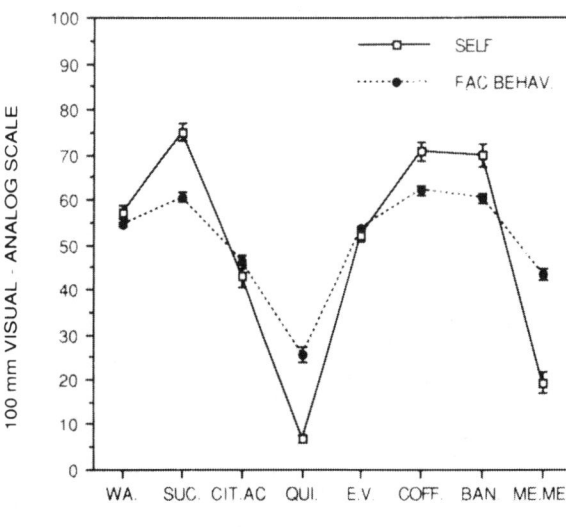

FIG. 2. Graph of mean hedonic self-estimates of 80 testees (SELF) and of mean hedonic ratings of their stimulus-induced facial behaviors (FAC. BEHAV.), given by evaluators, for intraoral and nasal stimuli. WA., water; SUC., 0.3 M sucrose solution; CIT.AC., 0.2 M citric acid solution; QUI., 0.0007 M quinine HCl solution; E.V., odorless empty vial; COFF., fresh ground coffee; BAN., amyl-acetate (banana-flavor); ME.ME., methymercaptane

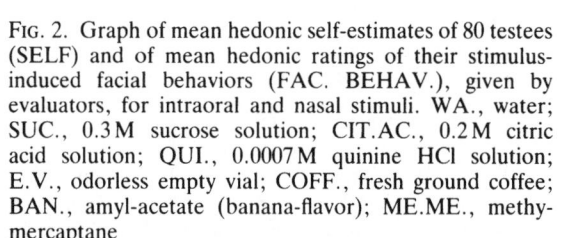

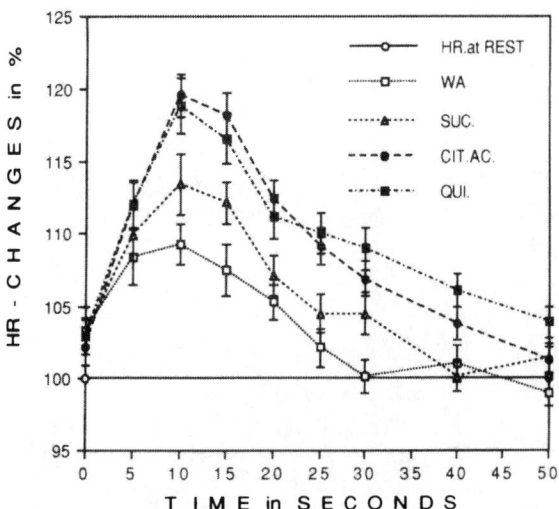

FIG. 3. Graph of gradual differential changes in heart rate in 80 volunteer subjects relative to HR at REST (= 100%) induced by swallowed stimulants: WA., water; SUC., 0.3 M sucrose solution; CIT.AC., 0.2 M citric acid; QUI., 0.0007 M quinine HCl solution

testing 80 young, healthy volunteers show that: (1) mean hedonic ratings of videotaped facial expressions given by a trained and a naive evaluator (in a double-blind setting) yield a high interobserver correlation ($r > 0.75$); (2) mean hedonic estimates given by testees and mean ratings of their facial behaviors, given by the evaluators, also show as impressive correlation for all stimuli ($r > 0.8$) (Fig. 2); and (3) all swallowed, intraoral, gustatory and sniffed olfactory stimuli induce a transient acceleration in both RR and HR. Results revealed that taste-induced HR acceleratory responses were found to be gradually differential: water caused the least response, sucrose somewhat more, and sour and bitter induced the highest peaks of HR increment (Fig. 3).

Further analysis aims to assess similar findings for gradually differential duration of facial responses as well as HR and RR acceleratory responses because raw data indicate that these values also reflect on hedonic polarity of stimuli.

In summary, stimulus-dependent facial expressions as well as HR and RR acceleration were found to mirror, sensitively and reliably, hedonics of perceived tastes and odors comparable to the indication of pleasure by common psychophysical tests. These somatic indicators can therefore be used as alternatives to psychophysical tests.

References

1. Darwin C (1872) The expression of the emotions in man and animals. [Reprinted 1965] University of Chicago Press, Chicago-London

2. Steiner JE (1973) The gustofacial response: observation on normal and anencephlaic newborn infants. In: Bosma JF (ed) Symposium on oral sensation and perception/IV. Bethesda, MD, NIH, U.S. Department of Health, Education, and Welfare, pp 254–278

3. Grill HI, Norgren R (1978) The taste reactivity test. II: Mimetic response to gustatory stimuli in chronic thalamic and chronic decerebrate rats. Brain Res 143: 281–297

4. Ganchrow JR, Steiner JE, Daher M (1983) Neonatal facial expressions in response to different qualities and intensities of gustatory stimuli. Infant Behav Dev 6:189–200

5. Blass EM, Ganchrow JR, Steiner JE (1984) Classical conditioning in newborn humans 2–48 hours of age. Infant Behav Dev 7:223–235

6. Steiner JE (1977) Facial expressions of the neonate infant, indicating the hedonics of food-related chemical stimuli. In: Weiffenbach JM (ed) Taste and development, the genesis of sweet preference. Bethesda, MD, NIH, U.S. Department of Health, Education, and Welfare, pp 173–188

7. Steiner JE (1979) Human facial expressions in response to taste and smell stimuli. In: Reese H, Lipsitt LP (eds) Advances in child development and Behavior, Volume 13. Academic, New York, pp 257–293

8. Kalmus H (1976) PTC-tasting in infants. Ann Hum Genet 169:139–140

9. Steiner JE, Abraham FE (1978) Gustatory and olfactory functions in patients affected by Usher's syndrome. Chem Senses and Flavor 3:93–98

10. Steiner JE, Hattab JY, Efrat J, Meir D (1993, in press) Taste- and odor-reactivity in autistic and psychotic children and adolescents. J Am Acad Child Adolesc Psychiatry

11. Steiner JE, Glaser D (1984) Differential behavioral responses to taste stimuli in nonhuman primates. J Hum Evol 13:709–723

12. Steiner JE, Glaser D (1985) Orofacial motor behavior-patterns induced by gustatory stimuli in apes. Chem Senses 10:375–367

13. Grill HI, Norgren R (1978) The taste reactivity test. I: Mimetic response to gustatory stimuli in neurologically normal rats. Brain Res 143:263–269

14. Ganchrow JR, Oppenheimer M, Steiner JE (1979) Behavioral display to gustatory stimuli in newborn rabbit pups. Chem Senses and Flavor 4:49–61

15. Ganchrow JR, Steiner JE, Canetto S (1986) Behavioral display to gustatory stimuli in newborn rat pups. Dev Psychobiol 19:163–174

16. Ganchrow JR, Steiner JE, Bartana A (1990) Behavioral reactions to gustatory stimuli in young chicks (Gallus gallus domesticus). Dev Psychobiol 23: 103–117

17. Steiner JE, Harpaz S (1987) Behavioral stereotypes of food-acceptance in the freshwater prawn: Macrobrachium rosenbergii. Chem Senses 12:89–97

18. Steiner JE (1990) Reeflectory behavioral responses induced by chemical stimuli as an objective testing-method in sensory evaluation. In: Doving K (ed) ISOT-X (Proceedings of 10th International Symposium on Olfaction and Taste), Oslo, Norway, July 16–20, 1989, GCS A/S, Oslo, pp 362

19. Steiner JE, Reuveni J, Beja Y (1982) Simultaneous multidisciplinary measures of taste-hedonics. In: Steiner JE, Ganchrow JR (eds) Determination of be-

havior by chemical stimuli. IRS Press, London, pp 149–160

20. Steiner JE, Steinberger A (1987) Multidisciplinary measures of taste-hedonics. Chem Senses 12:558

21. Perl E, Shay U, Hamburger R, Steiner JE (1992) Taste- and odor-reactivity in elderly demented patients Chem Senses 17:779–794

Semantics of Astringency

Harry T. Lawless and Carol J. Corrigan[1]

Key words. Astringency—Sensory evaluation—Texture—Taste—Touch

Introduction

Astringency denotes a class of tactile sensations induced by chemical stimulation of the oral cavity. Astringent materials include many polyphenols, acids, salts of multivalent cations, and dehydrating agents such as ethanol [1]. Polyphenols are ubiquitous in plant materials, and human foods are no exception. Rinsing the mouth with acids (even those considered prototypical "sour" tastants such as citric acid) induces astringency [2–4].

The theory of astringency most frequently cited is that of Bate-Smith [5,6]. In his view, polyphenolic compounds such as tannins form complexes with salivary proteins or mucopolysaccharides, either precipitating them or causing sufficient conformational changes that they lose their lubricating power, thus making the mouth feel rough, puckery, and dry. From this perspective, astringency may be viewed as a kind of "temporarily induced xerostomia." Guinard et al. [7] have suggested that after saliva and mucus layers covering the epithelium are stripped away by astringent agents, reactions with epithelial proteins themselves might take place.

There is disagreement over the definition of astringency. The American Society for Testing and Materials Committee E-18 on Sensory Evaluation has defined astringency as "the complex of sensations due to shrinking, drawing or puckering of the epithelium as a result of exposure to substances such as alums or tannins" [8]. Some workers have stressed the drying aspects of astringents [9]. In some historical classifications, astringency was considered a taste [10]. However, other workers [11] have classified astringency as a chemically induced tactile sensation, because it can be evoked from oral areas devoid of taste receptors.

We propose that astringency is not one simple sensory experience, but is made up of at least three distinct cues. Drying of the oral cavity, sensations of increased roughness of oral tissues, and a puckering or drawing sensation felt in the buccal musculature may all contribute to the impression of astringency. Drying is felt as a lack of lubrication, or friction between surfaces in the mouth. Roughing is felt as a physical bumpiness of the tissues, not unlike coarse sandpaper. Finally, puckering is a tightening or drawing sensation that can be felt in the cheeks and muscles of the face. Taken at face value, it would seem that panelists rating astringency could attend to any or all these cues. Evidence for this position and for some alternative models is presented here.

Note that two of the astringent subqualities, drying and roughing, require mouth movements for clear perception. This implies that astringency requires an active perceiver. It would be difficult to study astringency in an immobilized subject, such as an animal held in a stereotaxic device. Thus, the study of astringency is well suited to human psychophysical methods.

Techniques for Studying Relationships Among Sensory Qualities

Understanding the sensory language used to describe experience entails a certain amount of qualitative observation. That is, one cannot decide on the proper language for "carving nature at its joints" just by consulting a dictionary. Consumer meanings change over time, and people may disagree about how to best describe their sensations. For this reason, we have always included consumer interviews with samples to taste and discuss in the focus group format popular in marketing research [12,13].

Quantitative techniques have also been popular, using correlations to assess the relatedness of terms and analyses such as factor analysis or principal components analysis [14]. Redundant terms may group together on a single factor, identifying the main sensory qualities in a domain such as food texture [15]. Examples of such techniques can be found in Yoshikawa et al. [16–18]; they aptly describe these as a "mathematical phenomenology." Major factors were uncovered for moisture, elasticity, temperature-related terms, and hardness. The correlational approach is not without some pitfalls, however. It is important to have a wide and repre-

[1] Department of Food Science, Cornell University, Ithaca, NY, USA

sentative sample of foods or stimuli in sufficient numbers. Appropriate words must be chosen for inclusion. Also, deviations from Gaussian data may influence the usefulness of correlational measures.

Such approaches may give insights into how terms covary in foods. It may also give insights into how terms are related in consumers' minds. However, correlations may overemphasize the commonalities of different words, and they are not evidence that words have equivalent meanings or may be substituted in psychophysical analysis. An alternative approach that we call "decorrelation analysis" has as its goal discovering how perceptual ratings of different terms do *not* change in a parallel manner when stimuli or physical conditions are changed.

An example of decorrelation used in texture analysis training refers to the difference between hardness and denseness. These two terms covary in many foods. However, some foods are hard, but contain sufficient air to be not dense at all; an example is the American confection "malted milk balls." Conversely, cream cheese is dense, but not hard. Giving panelists such examples helps them to differentiate the terms [19].

Separation of complex terms into their component parts has value in applied sensory evaluations of foods and consumer products. Many words used by consumers are integrated impressions of several sensory experiences, and are poorly suited for use by sensory panels in descriptive analysis. An example is the term "creamy", usually a positive consumer description in American speech. Impressions of creaminess may be driven by more basic sensations, such as viscosity (thickness), lubricity (slippery mouthfeel), and even dairy-type aromas. If sensory panelists are asked to rate the creaminess of a product, they may weight these different contributing factors to different extents, and therefore have disagreement and increased error variance. It is usually more reliable to have descriptive analysis panelists focus on the basic components of a complex sensory experience. Of course, for some consumer tests the global impression is also needed.

Decorrelation in Astringency Ratings

In the experiments described here, category scale ratings were made of oral sensations following 15-s mouth rinses with various concentrations of astringent materials. Ratings were made repeatedly for several minutes so that time–intensity curves could be constructed. Details of the experimental procedure may be found in Lee [20] and Lee and Lawless [21]. Ratings for astringency, drying, roughing, puckery/drawing sensations, sourness, and bitterness were collected.

Aluminum ammonium sulfate (alum) was more drying and roughing than tannins and acids in

its astringent subqualities. Conversely, some acids are higher in puckery/drawing sensations. Figure 1 shows an example for alum and tartaric acid, replotted from Lee and Lawless [21]. Concentrations were adjusted to produce approximately equivalent overall sensory impact and approximately equal overall astringency at 1.36 g/l alum and 0.53 g/l tartaric acid. The alum and tartaric acid samples were judged about equal in the puckery/drawing sensations they produced. However, the alum was judged to be much more drying and more roughing than the tartaric acid.

A similar pattern is seen in Fig. 2 for a comparison of alum (0.5 g/l) and gallic acid (3.0 g/l), replotted from [22]. The overall astringency is about the same, but gallic acid is initially more puckery while alum is

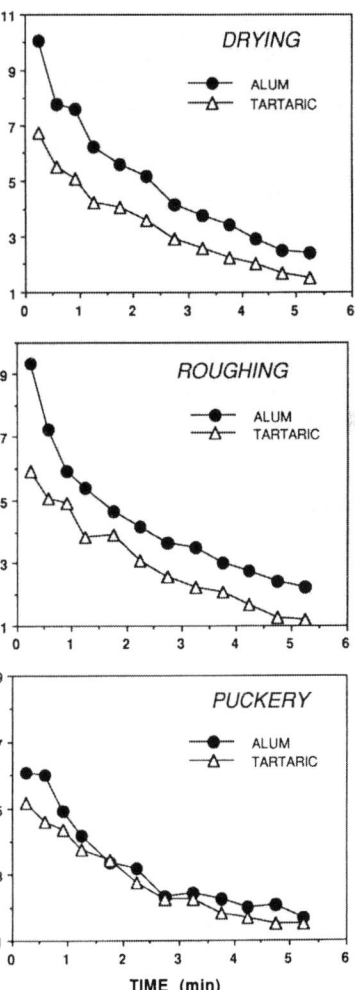

FIG. 1. Ratings for perceived drying, roughness of oral tissues, and puckery/drawing sensations in the mouth after 15-s mouth rinses with 1.36 g/l alum and 0.53 g/l tartaric acid. Mean intensity over time was higher for alum on drying and roughing scales ($p < 0.01$) but not significantly different on the scale for puckery. (From [21], with permission)

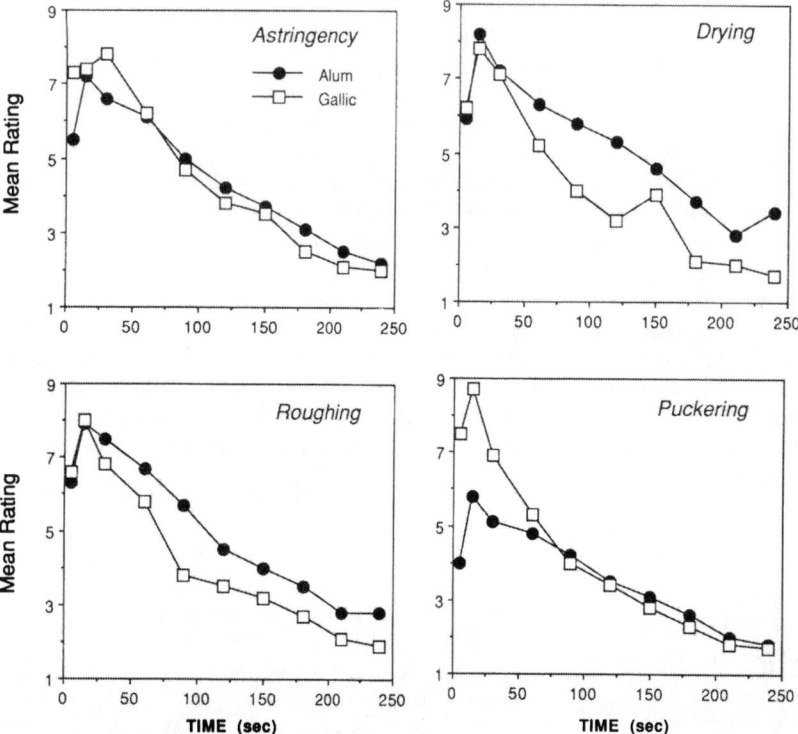

FIG. 2. Ratings for perceived astringency, drying, roughness of oral tissues, and puckery/drawing sensations in the mouth after 15-s mouth rinses with 0.5 g/l alum and 3.0 g/l gallic acid. (From [26], with permission)

more roughing and drying. We conclude from such results that these terms are different in the minds of subjects and that they are not redundant. Because different compounds produce different effects in the three subqualities, they are not interchangeable, and may each be contributing additional information about the qualitative pattern of astringent sensations.

One additional effect is seen in Fig. 2. Subjects were collecting whole mouth saliva [23], and expectorating or clearing the mouth at about 150 s. This can be seen as a bump in the curve for rated dryness after gallic acid after 150 s. Other astringent stimuli also showed an elevation in the drying attribute [20,22] when the mouth was cleared, but an elevation was not seen for the other subqualities. We interpret this as additional evidence that drying is not totally redundant with roughing, but is affected by other physical processes in the mouth. As an acid, gallic acid may enhance saliva flow [20] creating a baseline from which it is possible to see this rise.

Semantic Models

Figure 3A shows a model we developed based on the conjecture of astringent subqualities. A Venn diagram is a useful communication tool for this purpose. We interpret overlapping circles to mean strong correlations between attributes, while lack of overlap implies independence. Other data in-

fluenced us to question this model, however. Multidimensional scaling (MDS) was conducted with a variety of terms describing oral sensations [23]. Groupings usually separated taste, pain, tactile, and astringent attributes while drying, puckering, and astringent usually plotted together. However, this pattern was not the same for all subjects. MDS output for one group of females differed from the output for males. In this case, puckery fell near sour in the "taste" cluster and astringent was plotted near tactile words. This led us to revise the model as shown in Fig. 3B. We conjectured that because puckery sensations and sourness were both common results of stimulation with acids, they might be related in subjects' minds.

A series of experiments was then conducted to assess the possible relatedness of sourness and puckery sensations using acids [25]. Many common acids have astringent properties [2,26]. Ratings were collected as in the experiments described previously, with astringent subqualities, sourness, and bitterness evaluated over time after 15-s mouth rinses. Acids were fairly similar in their astringent profiles, although inorganic acids (hydrochloric and phosphoric) were higher than organic acids in astringency relative to sourness.

Ratings for puckering paralleled the astringency ratings more than the sourness differences. A correlation matrix and factor loadings for one experiment with acids [25; experiment 3] is shown in Table 1. Puckering was most highly correlated with overall astringency, while sourness was less highly correlated

A

First Model: Lee and Lawless (1991)

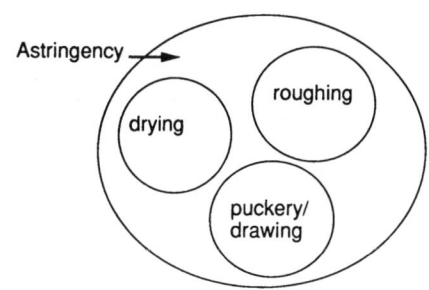

B

Potential modification:

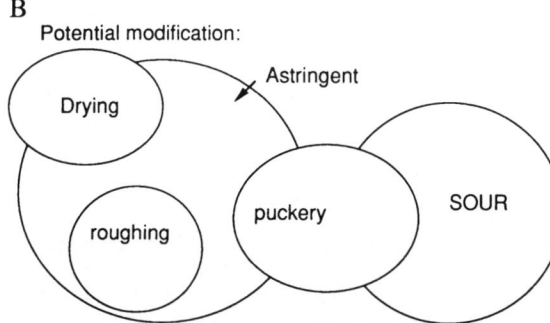

FIG. 3. Pictorial representations of relationships among astringent terms. **A** Original model from [21]. **B** Revised model. Overlap of circles implies degree of correlation across all hypothetical experiments

(some positive correlation is expected because concentrations were also varied). In the principal components analysis, sourness had a much higher loading on factor II than any other attribute. Drying, roughing, and sourness appeared on separate factors, with only astringency and puckering loading together. Thus the revised model (Fig. 3B) received less support than the original model (Fig. 3A).

Discussion: An Unanswered Question

Stimulation with astringent materials produces complex changes in the oral cavity. Semitrained subjects are able to differentiate several sensations and would seem to attend to them separately. An alternative interpretation is possible, however. Stimulation with chemicals results in a variety of sensory experiences. Stimulation with strong salts, for example, produces irritation [27]. However, this does not imply that irritation is part of saltiness or vice versa. Stimulation with irritants such as capsaicin produces a number of sensations and reflexes, such as increased salivation [28]. Because salivary output parallels the perceptual ratings for irritation, it is possible for subjects to use salivary flow as an additional cue for the intensity of irritation. However, we would be unwise to call saliva flow a sensory subquality of irritation; it is rather a *concomitant reaction* that accompanies the perception of oral irritation.

Are the astringent subqualities simply concomitant reactions? The tightening/drawing sensations we describe as puckery seem most closely associated with astringency. In contrast, drying is also evoked by simply expectorating or by putting a hygroscopic agent such as corn starch in the mouth. However, the drying feeling from starch would rarely be described as astringent. At this time, there is no clear criterion for assigning the status of subqualities to some sensations and concomitant reactions to others. This idea awaits a new theory.

Acknowledgments. The authors thank Christopher B. Lee, Sandy Glatter, Mary Johnston, Rich Tucciarone, Mary Lou McGiff, and Renee Vaia for assistance. Supported by NIH grant DC 00902.

TABLE 1. Correlation matrix from [25], experiment 3.

	Sourness	Drying	Roughing	Puckering
Drying	0.51			
Roughing	0.47	0.84		
Puckering	0.65	0.67	0.74	
Astringent	0.65	0.77	0.85	0.91

Factor loadings from principal component analysis (values < |0.4| omitted)

	Factor I	Factor II	Factor III	Factor IV
Astringent	—	—	0.69	—
Puckering	—	—	0.86	—
Roughing	0.54	—	—	0.72
Drying	0.87	—	—	—
Sourness	—	0.93	—	—
Variance explained (%)	27	24	31	16

References

1. Deshpande SS, Cheryan M, Salunhke DK (1985) Tannin analysis of food products. CRC Crit Rev Food Sci Nutr 24:401–449
2. Straub A (1992) Power function determination of sourness and time-intensity measurements of sourness and astringency for selected acids. M. Sc. thesis, Oregon State University, Corvallis, OR
3. Rubico SM, McDaniel MR (1992) Sensory evaluation of acids by free-choice profiling. Chem Senses 17:273–289
4. Corrigan CJ, Lawless HT (1992) Sensory evaluation terminology and the semantics of astringency. In: 17th Annual Meeting, American Society for Enology and Viticulture, Eastern Section, Corning, NY
5. Bate-Smith EC (1954) Astringency in foods. Food Proc Pack 23:124–127
6. Bate-Smith EC (1973) Haemanalysis of tannins: the concept of relative astringency. Phytochemistry (Oxf) 12:907–912
7. Guinard J-X, Pangborn RM, Lewis MJ (1986) The time course of astringency in wine upon repeated ingestion. Am J Enol Vitic 37:184–189
8. Anonymous (1989) E-253: Standard definitions of terms relating to sensory evaluation of materials and products. Volume 15.07, End use products. American Society for Testing and Materials, Philadelphia, PA, p 2
9. Lyman B, Green BG (1990) Oral astringency: effects of repeated exposure and interactions with sweeteners. Chem Senses 15:151–164
10. Bartoshuk LM (1978) History of taste research. In: Carterette EC, Friedman MP (eds) Handbook of perception, Volume VIA, Tasting and smelling. Academic, New York, pp 3–18
11. Breslin PAS, Gilmore MM, Beauchamp GK, Green BG (1993, in press) Psychophysical evidence that oral astringency is a tactile sensation. Chem Senses
12. Chambers E, Smith EA (1991) The uses of qualitative research in product research and development. In: Lawless HT, Klein BP (eds) Sensory science theory and applications in foods. Dekker, New York, pp 395–412
13. Marlow P (1987) Qualitative research as a tool for product development. Food Technol 41:74, 76, 78
14. Powers JJ (1984) Current practices and application of descriptive methods. In: Piggott JR (ed) Sensory analysis of foods. Elsevier, New York, pp 179–242
15. Szczesniak AS, Loew BJ, Skinner EZ (1975) Consumer texture profile technique. J Food Sci 40:1254–1256
16. Yoshikawa S, Nishimaru S, Tashiro T, Yoshida M (1970) Collection and classification of words for description of food texture. III: classification by multivariate analysis. J Texture Stud 1:452–463
17. Yoshikawa S, Nishimaru S, Tashiro T, Yoshida M (1970) Collection and classification of words for description of food texture. I: collection of words. J Texture Stud 1:437–442
18. Yoshikawa S, Nishimaru S, Tashiro T, Yoshida M (1970) Collection and classification of words for description of food texture. II: texture profiles. J Text Stud 1:443–451
19. Civille GV, Lawless HT (1986) The importance of language in describing perceptions. J Sens Stud 1:259–274
20. Lee CB (1992) Sensory evaluation of astringent compounds. Master's thesis, Cornell University, Ithaca, NY
21. Lee CB, Lawless HT (1991) Time-course of astringent materials. Chem Senses 16:225–238
22. Lawless HT, Lee CB, Corrigan CJ (1994, in press) Mixtures of astringent substances. Chem Senses
23. Navazesh M, Christensen CM (1982) A comparison of whole mouth resting and stimulated salivary measurement procedures. J Dent Res 61:1158–1162
24. Bertino M, Lawless HT (1993) Understanding mouthfeel attributes: a multidimensional scaling approach. J Sens Stud 8:101–114
25. Corrigan CJ (1993) Time-intensity measurements of astringent subqualities in selected organic and inorganic acids. Master's thesis, Cornell University, Ithaca, NY
26. Rubico SM, McDaniel MR (1992) Sensory evaluation of acids by free-choice profiling. Chem Senses 17:273–289
27. Green BG, Gelhard B (1989) Salt as an oral irritant. Chem Senses 14:259–271
28. Lawless HT (1984) Oral chemical irritation: psychophysical properties. Chem Senses 9:143–155

Central and Peripheral Aspects of Spatial Summation and Mixture Suppression in Taste

Jan H.A. Kroeze, Miriam R.I. Linschoten, and Corinne A. Ossebaard[1]

Key words. Mixture suppression—Spatial summation in taste—Central *vs* peripheral—Taste mixtures—Threshold summation—Enhancement in taste

Introduction

Peripheral and central spatial summation and mixture suppression in taste were investigated by the "split-tongue" technique. In this method, stimuli are applied to one or both sides of the tongue in different combinations. Interactions between tongue sides must be central, whereas comparison with unilateral results provides information about peripheral *vs* central contributions [1].

Material and Methods

Spatial Summation Experiments

Circular ($63.62 \, mm^2$; diameter, 9 mm) and half-circle filter papers were used. After rinsing and drying the tongue, a stimulus-dipped paper of $35.7°C$ was applied. Subjects received a subset of NaCl concentrations from a larger range and indicated which of two successive stimuli differed from water (temporal 2-AFC method). The papers were removed by a circular glass filter connected to a $5.3 \, N/cm^2$ vacuum, which also removed the residual stimulus liquid [2]. Intertrial interval was 50 s.

First Experiment

Five levels of concentration bracketing the individual threshold were selected from a range of 11 concentrations: the highest was 0.067 M; each dilution step was 0.125 log-unit. One tongue side was used and 12 subjects judged half- and full-circle papers 32 times per concentration with the intrapair blank–stimulus order counterbalanced.

Second Experiment

Thirteen subjects; both tongue sides were used.

[1] Psychological Laboratory, Utrecht University, Heidelberglaan 2, 3584 CS Utrecht, The Netherlands

Third Experiment

Nine subjects. Only half-circle papers, either alone or combined, were used with identical or different NaCl concentrations. Thus, doubling the area provided two equal or different detection probabilities. Four levels bracketing individual detection thresholds were used, selected from six concentrations (0.007–0.053 M); each higher step was 1.5 fold stronger.

Mixture Suppression Experiments

Corresponding parts of the tongue, enclosed in a tongue chamber [1] at either side of the midline, were repeatedly exposed to a stimulus flow ($34.5°C$; flow rate, 3 ml/s; duration, 2 s). Stimuli were sucrose (S: 0.10, 0.15, 0.22, 0.34, 0.50 M), NaCl (N: 0.10, 0.17, 0.30, 0.52, 0.90 M), and citric acid (A: 0.0031, 0.0063, 0.0125, 0.0250, 0.0500 M). Four conditions were constructed: NA-sourness (NaCl-citric acid mixtures with the instruction to estimate sourness), NA-saltiness, NS-sweetness, and NS-saltiness.

All 25 combinations of the five intensities within each of the four mixture-judgment conditions occurred. To each condition, five unmixed stimuli per compound were added. Stimulus presentations proceeded under three split-tongue conditions: (1) unilateral unmixed: an unmixed stimulus on one tongue side and distilled water on the contralateral side; (2) unilateral: a mixture of two concentrations, one per compound, applied to one tongue side, and distilled water applied contralaterally; and (3) bilateral: one mixture component applied to one tongue side, the other component contralaterally. Each stimulus-instruction combination was repeated 10 times in 23 student subjects who responded by magnitude estimation [3].

Results

Summation at Threshold

First Experiment

Detection thresholds (half-circle, 0.029 M/l; full circel, 0.0196 M/l) showed a ratio of 1.46. In a hyperbolic function [4,5], this ratio fits $IA^{1/2} = c$, equivalent to Piper's law in vision (I = intensity; A

= area, c = constant). This equals probability summation, predicting $P_{pf} = 2P_{oh} - (P_{oh})^2$ where P_{pf} = predicted probability for the full circle and P_{oh} = observed probability for the averaged half-circle. The observed detection probabilities for the full circle and the probabilities predicted on the basis of the half circles were similar at all five stimulus levels.

Second Experiment

The combined detection probabilities of half-circle stimuli originating from the same tongue side equaled the combined probabilities of different tongue sides, and either combination predicted accurately the detection probabilities for all stimulus levels of the full-circle threshold function, according to probability summation.

Third Experiment

The probability summation model was appropriate when equal detection probabilities were combined, and the data were similar for ipsilateral and bilateral combinations (Table 1). This was also true for unequal probabilities in the bilateral condition. In the ipsilateral condition, however, an exception occurred where two half-circle stimuli were combined on the same tongue side: with stronger combinations of different intensities, probability summation overpredicted detection probability of twice the stimulus area.

Mixture Suppression

In Fig. 1, open circles, denoted "theoretical," are the reference values of 0% suppression. Filled (solid) triangles denote the bilateral condition; that is, any difference with the theoretical condition is of central

origin. Open triangles represent the unilateral condition, and the deviation from the theoretical condition in this case reflects a combined central and peripheral contribution. All differences with the theoretical condition include a correction for side tastes [6], revealing that enhancement was always peripheral. Suppression may be central as well as peripheral. Table 2 summarizes the most significant results.

Discussion

Spatial Summation

The most parsimonious explanation of the spatial summation findings for combinations of stimuli with equal detection probabilities is that central probability summation accurs at threshold.

However, with ipsilateral stimulation and unequal detection probabilities, the probability summation model overpredicts. We speculate that with two equally intense stimuli applied within the same receptive field, the inhibitory effects will be symmetrical, because both stimuli recruit equal numbers of cells; however, with stimuli of unequal intensity, recruiting different numbers of cells, the inhibition effects will be asymmetrical and the probability summation model progressively overpredicts detection as stimulus intensity increases.

This explanation assumes that the net result of stimulation is a compromise between spatial inhibition and spatial summation. It also predicts that when the two filter papers are placed more and more spatially apart, then the spatial inhibition disappears. At a certain distance the sites no longer influence each other and independent combination of detection probabilities occurs, as in the bilateral case.

Mixtures

The mixture results suggest that mixture enhancement has no central origin. On the other hand, suppression may be central as well as peripheral. Peripheral suppression of two qualitatively different taste substances does not necessarily rely directly on receptor activity. Lawless [7], studying NaCl-sucrose mixtures, showed that differences in flow rate (4 ml/s compared to 0.8 ml/s) may account for peripheral suppression differences, the higher flow rate producing a small but significant amount of peripheral sweetness suppression as compared to the lower flow rate. In the mixture experiment reported here, the flow rate was 3 ml/s, but instead of sweetness suppression, as in Lawless' experiment, peripheral sweetness enhancement was found. With a much lower flow rate (0.8 ml/s) we found neither peripheral sweetness suppression nor peripheral saltiness suppression [8,9]. For now we can only conclude

TABLE 1. Mean predicted and observed detection probabilities for equal and unequal pairs in ipsi- and bilateral stimulation conditions.

Pair	Ipsilateral		Bilateral	
	Pred	Obs	Pred	Obs
1 1	0.20	0.22	0.24	0.22
2 2	0.40	0.40	0.43	0.40
3 3	0.64	0.60	0.65	0.63
4 4	0.84	0.78	0.83	0.82
1 2	0.30	0.36	0.34	0.37
1 3	0.47	0.49	0.49	0.47
1 4	0.65	0.59	0.64	0.59
2 3	0.54	0.39	0.55	0.50
2 4	0.70	0.60	0.69	0.65
3 4	0.77	0.65	0.76	0.74

Pred, predicted; obs, observed.

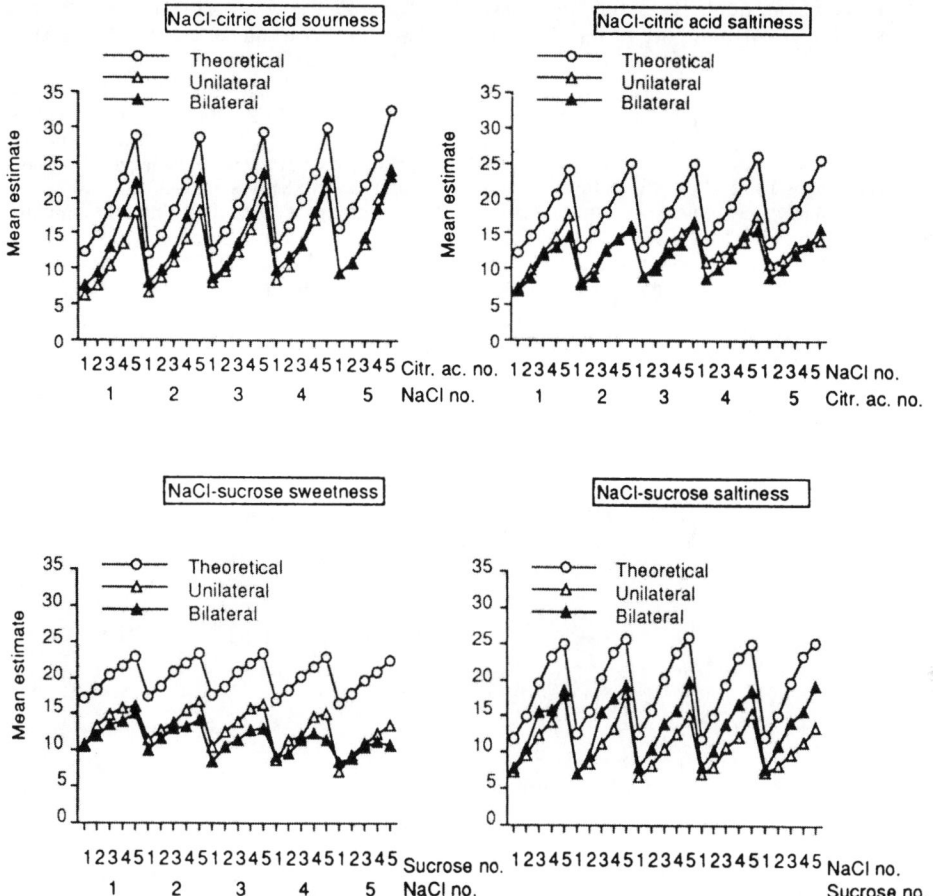

FIG. 1. Mixture results: conditions indicated by *rectangles*: NaCl-citric acid sourness means that in this mixture sourness was estimated. All 25 intensity combinations per condition are shown. "Theoretical", estimate of unmixed substance; unilateral, both components on same side of tongue; bilateral, each component on different side of tongue

TABLE 2. Summary of the results of the mixture experiment with (C) and without side tastes taken into account.[a]

	NA sourness	NA saltiness	NS saltiness	NS sweetness
Central		enh. + suppr.	suppr.	suppr.
Peripheral		enh.	suppr.	enh.
Central (C)	suppr.	suppr.	suppr.	suppr.
Peripheral (C)	suppr.		suppr.	enh.

[a] NA sourness means that sourness of NaCl-citric acid mixtures was estimated. The other headings are analogous. With side tastes taken into account, suppression is without exception centrally located.
Enh., enhancement; suppr., suppression.

that the conditions of peripheral suppression are not clear.

It is not surprising that receptor-theories are generally more compatible with results from taste-cell and first-order neuron recordings than with behavioral indicators. This is particularly the case with complex stimuli, causing central neural inter-actions. Knowledge of these interactions is required for a full understanding of taste responses.

Acknowledgments. This research was partially supported by a grant from the Netherlands Organization for Scientific Research, awarded to the first author (NWO-PSYCHON grant no. 560-262-025). The authors are grateful to students Ilse Polet and Gerko Blok, who contributed substantially to the project.

References

1. Kroeze JHA, Bartoshuk LM (1985) Bitterness suppression as revealed by split-tongue taste stimulation in humans. Physiol Behav 35:779–783
2. Ganzevles PGJ, Kroeze JHA (1986) The use of filter paper as stimulus carrier in taste research. Chem Senses 11:383–387
3. Stevens SS (1969) Sensory scales of taste intensity. Percept Psychophys 6:302–308
4. Bujas Z, Ostojcic A (1941) La sensibilité gustative en fonction de la surface excitée. Acta Inst Psychol Univ Zagrebiensis 2:3–12
5. McBurney DH (1969) A note on the relation between area and intensity in taste. Percept Psychophys 6:250
6. Kroeze JHA (1982) The relationship between side tastes of masking stimuli and masking in binary mixtures. Chem Senses 7:23–37
7. Lawless HT (1982) Adapting efficiency of salt-sucrose mixtures. Percept Psychophys 32:419–422
8. Kroeze JHA (1978) The taste of sodium chloride: masking and adaptation. Chem Senses Flavor 3:443–449
9. Kroeze JHA (1979) Masking and adaptation of sugar sweetness intensity. Physiol Behav 22:347–351

Temporal Patterns of Perceived Tastes Differ from Liquid Flow at the Tongue

Bruce P. Halpern[1]

Key words. Time-intensity—Taste—Reaction time
—Caffeine—Amiloride—NaCl

Introduction

The time at which a predetermined behavior (a response) begins after the onset of a specified and more or less controlled change in the environment (a stimulus) can be investigated during psychophysical analysis of a sensory system. Subjects are instructed to respond only when a particular perceptual state, for example, a change in perceived taste, occurs. The resultant time interval between stimulus and response is called a reaction time. This one measure allows the temporal difference between stimulus arrival and the start of a taste perception to be indexed [1]. By itself a taste reaction time does not permit the physically determined total time course of a taste stimulus to be compared with that of a perceptual response, because only the disparity between one physicochemical measurement at the tongue and one response is timed.

The observed separations of at least several hundred milliseconds between arrival of a taste stimulus at the tongue and the fastest reports of any perceived change in taste [1,2] demonstrate that physical events at the human tongue and taste perception do not fully correspond in time. This difference could denote only an offset, with taste stimuli and taste perceptions having equal durations. On the other hand, it may connote an overall lack of temporal correspondence between taste perception and taste stimuli, with perceived tastes perhaps continuing long after stimulation at the tongue has ended. The answer can be discovered only if one measures the time at which the taste perception ends as well as taste reaction time.

Changes in perceived taste descriptions or taste intensity during or after a stimulus are also important. The usual psychophysical measures are designed to normally obtain a single value for the intensity or other characteristics of each stimulus presented [3]. These measures of taste perception can be taken repeatedly at intervals of 5–10 s. This has been done to track taste intensity over time (see [4]). However, human drinking and taste-dependent decisions occur over a much faster time course. Descriptions of taste stimulus quality have reaction times of 1 sec or less [5], while taste intensity reaction times can be less than 2 s. Taste-dependent decisions to reject or swallow liquids require less than 4 s [6]. It is apparent that a measurement every 5 s cannot capture the perceptual bases of these gustatory behaviors [4].

High-temporal-resolution tracking of perceived taste permits changes in taste quality descriptions and taste intensity to be followed over a time course that corresponds to human taste-dependent behavior [4]. Time-quality tracking of taste with high temporal resolution has been previously reported [7]. This paper reports the results of high-temporal-resolution taste time-intensity tracking studies.

Material and Methods

The tongue was stimulated using an automated closed-flow taste stimulus delivery apparatus [8,9] that was calibrated by a flow-through conductivity cell in the liquid delivery tube [8]. Solution duration and waveform were monitored during data collection trials by a conductivity cell in the liquid discard channel. Each trial consisted of a 10-s purified water (polished reverse osmosis, pH, ~6; conductivity, <1.3 µS; refractive index, 1.3330) flow through the liquid delivery tube and across a 39.3-mm² area of the anterodorsal tip region of the tongue, followed immediately by a 4-s stimulus solution presentation and then a 5-s purified water (H_2O) flow. Trials were separated from each other by 60 s.

Each subject participated in 11 sessions; the first 3 sessions were practice. Each of the solutions used was presented 8 times during the 8 data collection sessions, in random order. Solutions, all prepared in H_2O, were 100 mM, 250 mM, and 500 mM NaCl (A.C.S.); 10 µM and 100 µM amiloride (Sigma Chemical, St. Louis, MO); and 33 µM, 50 µM, 100 µM, 8.3 mM, and 12.5 mM caffeine (99%, Aldrich Chemical, Milwaukee, WI). In addition, mixtures consisting of the three concentrations of

[1] Department of Psychology and Section of Neurobiology and Behavior, Cornell University, Ithaca, NY 14853-7601, USA

NaCl and the amiloride or caffeine concentrations were used, for a total of 19 different solutions for each subject; however, none of the mixture results are included in this paper (see [5,6]).

A nonlinguistic, high-resolution, visually guided taste intensity-tracking procedure provided continuous measures of taste intensity with 100-ms resolution, as previously described [4]. Briefly, the six subjects (age: range, 18–22; median, 19.5; mean, 19.7; five were female) indicated taste intensity by using a single-axis joystick to control the vertical position of a moving computer display that they viewed on a monochrome video monitor located in front of them. Solution arrival at the tongue was simultaneous with the beginning of the display. Subjects were asked "What does it taste like?" after every stimulus solution trial, but taste quality results are not included in this paper (see refs. [5,6]).

Results

Tracked taste intensity values were above zero for at least 1000 ms after the stimulus solution had been removed from the tongue. Figure 1a illustrates this for one subject; the median ± standard error of the median tracked taste intensity for 100 mM NaCl reached zero at 1500 ms after solution removal from the tongue. In Figs. 1 and 2, the time required for NaCl concentration at the tongue to decrease to 100 μM NaCl after the post-NaCl H_2O flow began [4,8] is schematically illustrated. Median results for 100 mM NaCl for six subjects, including the subject

of Fig. 1a, are shown in Fig. 1b. Median results for 250 mM NaCl are shown in Fig. 2a and, for 500 mM NaCl, in Fig. 2b. Across these concentrations of NaCl, median tracked taste intensity remained above zero for 1000 ms to more than 4000 ms after NaCl solutions were removed from the tongue. Figure 3 shows tracked taste intensity for 8.3 mM and 12.5 mM caffeine, which individual subjects had selected as similar to the taste of 100 μM amiloride; Fig. 4 shows tracked taste intensity for 100 μM amiloride. For both caffeine and amiloride, tracked taste intensity remained well above zero for 1000 to more than 4000 ms after the post-solution H_2O flow began.

Discussion

Differences between the time course of stimulus events and human perceptions are regularly observed when the stimulus events are very brief [10]. For brief taste stimuli, differences have been reported not only for reaction times [2,11] for also for time-intensity tracking responses [4]. The current data demonstrate that for taste stimuli with durations similar to those found during the taste-dependent decision to swallow or spit out ingested liquids [12,13], perceived taste intensities have durations much longer than the time during which a stimulus solution is in contact with the tongue. Because physicochemical measurements at the tongue indicated rapid removal of the stimulus solution when a closed-flow taste stimulus delivery apparatus [8,9] was used, these differences between durations of

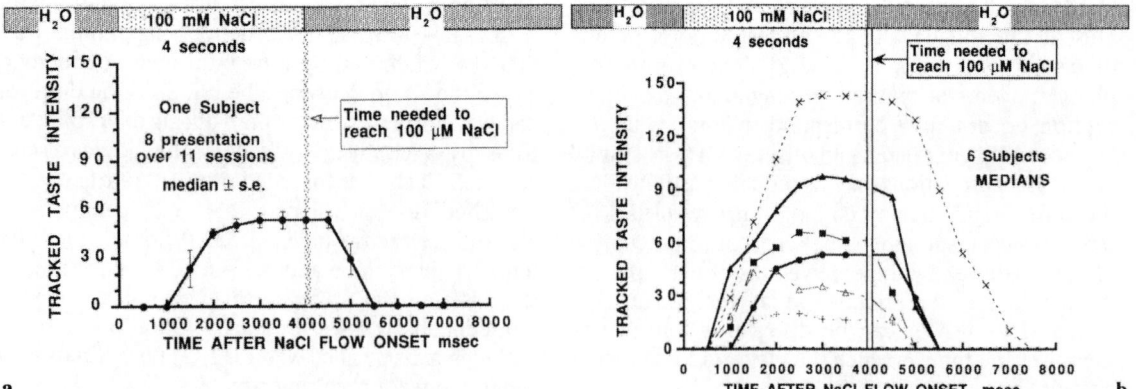

a

b

Fig. 1. Vertical axis for all figures is median, visually guided, high-temporal-resolution, time-intensity tracking of 4-s duration (calibrated at the tongue) stimulus solution presentations; horizontal axis is time in milliseconds after solution reaches tongue. For all experiments, angular position of single-axis joy stick controlled by subject determined vertical location of moving display on computer monitor every 100 ms; display started at time of solution arrival at the tongue. For all figures, rectangular bar above graph indicates 10-s purified water (H_2O) prestimulus rinse (gray bar) immediately followed by a 4-s duration contin-

uous flow of aqueous solution (patterned bar) that was immediately followed by a 5-s H_2O stimulus removal rinse (gray bar). In this figure, dotted vertical column (a) or open bar (b) starting at 4000 ms represents time required for H_2O stimulus-removal rinse (2nd H_2O) to lower 100 mM NaCl to concentration of 100 μM at the tongue. a Median ± standard error of median time-intensity tracking by one subject of eight 4-s presentations of 100 mM NaCl. b Median tracked taste intensities of 100 mM NaCl for each of six subjects

Fig. 2. Median tracked taste intensities of 250 mM NaCl (**a**) or 500 mM NaCl (**b**) for each of six subjects. *Vertical bar* starting at 4000 ms represents time required for H_2O stimulus-removal rinse (2nd H_2O) to lower NaCl to concentration of 100 µM at the tongue

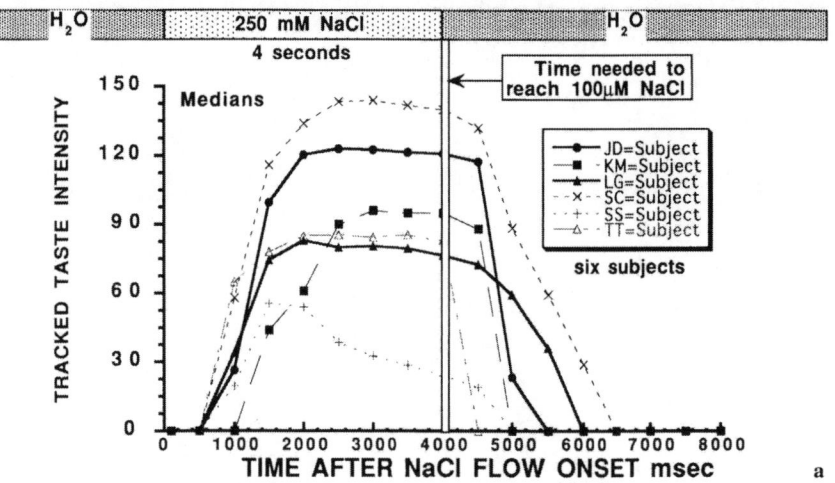

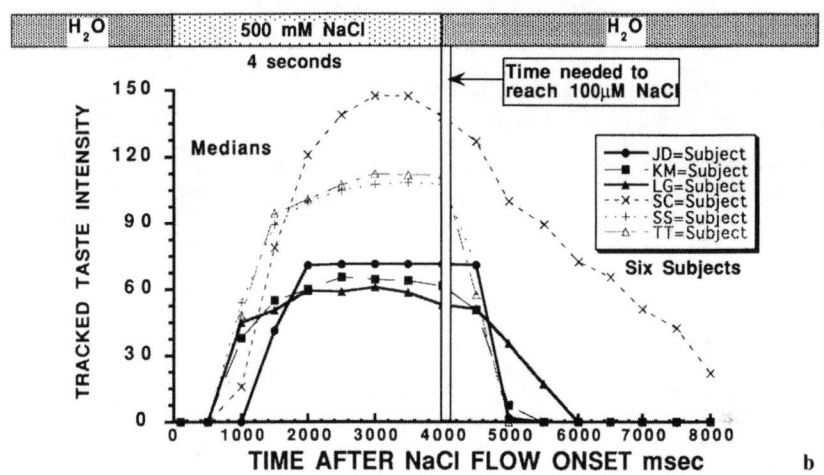

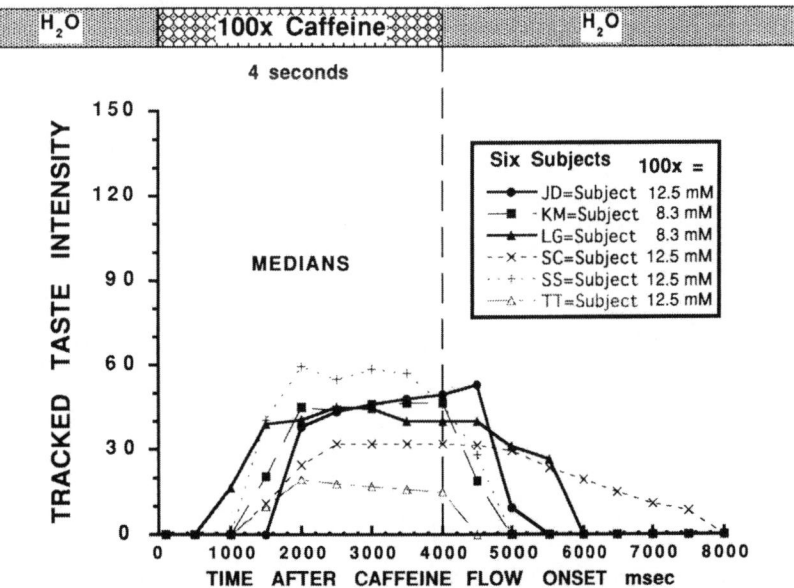

Fig. 3. Median tracked taste intensities of 8.3 mM or 12.5 mM caffeine for each of six subjects. Subjects selected these concentrations of caffeine to match overall bitterness of 100 µM amiloride. *Vertical broken line* at 4000 ms indicates start of H_2O stimulus removal rinse

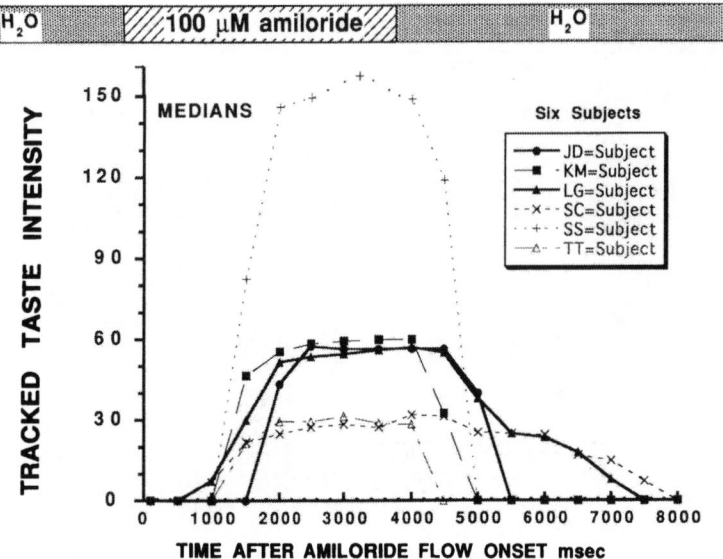

FIG. 4. Median tracked taste intensities of 100 µM amiloride for six subjects

chemical stimuli and perceived taste intensity can be attributed to events in the sensory system.

Acknowledgments. Support was received from Campbell Research and Development during the collection of these data.

References

1. Halpern BP (1986) Constraints imposed on taste physiology by human taste reaction time data. Neurosci Biobehav Rev 10:135–151
2. Kelling ST, Halpern BP (1987) Taste judgments and gustatory stimulus duration: simple taste reaction times. Chem Senses 12:543–562
3. Luce RD, Krumhansl CL (1988) Measurement, scaling, and psychophysics. In: Atkinson RC, Herrnstein RJ, Lindzey G, Luce D (eds) Stevens' handbook of experimental psychology, 2nd ed, vol 1: Perception and Motivation. J Wiley, New York, pp 3–74
4. Halpern BP (1991) More than meets the tongue: temporal characteristics of taste intensity and quality. In: Lawless HT, Klein BP (eds) Sensory science theory and applications in foods. Dekker, New York, pp 37–105
5. Halpern BP, Kelling ST, Davis J, Dorries KM, Haq A, Meltzer JS (1992) Effects of amiloride on human taste responses to NaCl: time-intensity and taste quality descriptor measures. Chem Senses 17:637 (abstr)
6. Halpern BP, Meltzer JS, Darlington RD (1993, in press) Effects of amiloride on tracked taste intensity and on taste quality descriptions of NaCl: individual differences and dose-response effects. Chem Senses 18:566 (abstr)
7. Zwillinger SA, Halpern BP (1991) Time-quality tracking of monosodium glutamate, sodium saccharin, and a citric acid-saccharin mixture. Physiol Behav 49:855–862
8. Kelling ST, Halpern BP (1986) Physical characteristics of open flow and closed flow taste delivery apparatus. Chem Senses 11:89–104
9. Kelling ST, Halpern BP (1987) Comparisons between human gustatory judgments of square wave trains and continuous pulse presentations of aqueous stimuli: reaction times and magnitude estimates. Ann NY Acad Sci 510:406–408 (ISOT IX)
10. Halpern BP, Kelling ST (1983) Taste flashes: reaction times, intensity, and quality. Science 291:412–414
11. Kelling ST, Halpern BP (1988) Taste judgments and gustatory stimulus duration: taste quality, taste intensity, and reaction time. Chem Senses 13:559–586
12. Delconte JD, Kelling ST, Halpern BP (1992) Speed and consistency of human decisions to swallow or spit sweet and sour solutions. Experientia (Basil) 48: 1106–1109.
13. Halpern BP (1985) Time as a factor in gustation: temporal patterns of taste stimulation and response. In: Pfaff DW (ed) Taste, olfaction, and the central nervous system. Rockefeller University Press, New York, pp 181–209

Structure and Dimension of the Taste Sensory Space: Central and Peripheral Data

ANNICK FAURION[1]

Key words. Hamster whole chorda tympani—Single units—Human threshold—Iso-intense concentrations—Receptor sites—Semantics

Introduction

The theory of four tastes emerged during the nineteenth century, preceding any experimental work on the matter [1]. Based on semantics and introspective psychology, this theory led to studying taste physiology using only four standard stimuli, which I have previously called an epistemological mistake. This is because for more than 30 years, from 1940 to the 1970s, electrophysiologists were locked within this dogma, exploring the taste function with a set of four stimuli. In 1916, however, the psychologist Henning [2] did describe taste perception as a tridimensional sensory continuum in which the four descriptors appear as mere landmarks. In 1974, Pfaffmann [3] questionned whether the taste space structure would be best represented by a unique continuum or four independent subcontinua centered on each semantic category. A taste space continuous aspect was further confirmed [4,5], on the basis of qualitative similarity measurements between paired stimuli.

We showed, in 1977, that sweet taste itself is a continuum. After measuring the individual sensitivities of human subjects to each compound of a series of sweeteners, we could build an argument explaining that sweet taste depends on multiple receptor sites: several sites cooperating to the sensitivity for one and the same stimulus, and each receptor site accepting different ligands, with different affinities. Subjects' sensitivities should depend on their genetic characteristics, which code for different quantities of each receptor type [6].

In this paper we depict the whole taste space, its structure and its dimension, on a much wider series of compounds. As the same set of gustatory fibers has been shown to encode quality as well as intensity [7], we decided to measure the stimulatory efficiency of tastants on the human or animal taste system by intensity measurements, recording the response amplitude in the hamster gustatory nerves or the intensity produced in human subjects by a series of different stimuli.

Methods

A threshold experiment in the human gathered data on 37 stimuli. An electrophysiological study at threshold response level on single units recorded in the hamster papillae gathered 18 stimuli. At suprathreshold level, two psychophysical experiments dealt, respectively, with 21 and 26 stimuli; two electrophysiological experiments considered, respectively, 38 and 51 stimuli.

In the human, we measured a supraliminal parameter, namely iso-intense concentrations, on 124 subjects. Each subject participated in a 1-hour session per day, 4 days per week for one or several semesters, 14 to 16 weeks per semester (they received 32.50FF per hour). Subjects compared 60–100 pairs of stimuli per session, answering the question "which is stronger?", from an NaCl reference (1.7 g/l) or the stimulus, the concentration of which varied according to the Up and Down technique of Dixon [8]. The whole experiment was automatized, solutions being selected, diluted, and distributed by the computer, in an order that looked random to the subject because of intermingling several tests, although each response given to the computer determined the following stimulation within a given test. Thresholds were measured on 90 subjects in the same way except that the reference solution was water (Faurion et al., manuscript in preparation).

Solutions were made sterile; the temperature was controlled at $0.1°–0.2°C$ between paired elements. Each subject used a plaster mold of his own nostrils, which admitted an airflow of about 200 l/h (depending on the subject's need) while tasting, to thoroughly suppress retronasal olfaction. Long-term reproducibility was tested for each stimulus and each subject. A training period was respected for each subject and each molecule, which means data of the first sessions were often discarded for each stimulus.

In animals, we first measured the response frequency of action potentials recorded in the papillae

---

[1] Laboratoire de Neurobiologie Sensorielle de l'E.P.H.E., 91305 Massy, France

[9] and the amplitude of integrated chorda tympani nervous activity in response to application of stimuli (two experimental sets). Each set of data gathered about 60 nerves or units. The signal was obtained through digital on-line data acquisition as well as analogical summation. Every stimulus was repeated at least twice to assess reproducibility. Stimuli were made sterile every month and kept frozen in small vials.

For data analysis, in each case we obtained a rectangular matrix of data in which each value represented the sensitivity of one individual to a given stimulus. Results enabled us to compare the sensitivity spectra of every subject (line) to a series of stimuli (columns) or the sensitivity spectra of all subjects for selected stimuli. We could do the same with hamster nerves and hamster papillae single units.

The intrasubject variance was found to be about 10 fold less than the interindividual variance. The method we chose consisted in studying the covariance between either subject profiles or stimulus profiles. Correlation coefficients have been calculated on paired stimulus profiles (across subjects). Factor analyses, factor analysis of correspondences (FAC), and principal component analysis (PCA), as well as hierarchical clustering, were performed on all data matrices.

Results

At threshold level in the human, we found no transmission of learning from one compound to another, which indicates that the central nervous system performs a task of form recognition of the elicited sensory image even for detecting a stimulus. At threshold and suprathreshold level, we found a high discriminating power, indicating that the system codes a different image for each different stimulus irrespective of the semantics:

1. Every subject, whether human or a hamster, presented a different sensitivity spectrum across all stimuli tested: all subjects are different. Interindividual differences of sensitivities are large, significant, and stable over time; no "standard" or "average" observer can represent a population of individuals.

2. No stimulus is found similar to any other one; the system discriminates every compound, and we could not find a grouping into four taste qualities. Most correlation coefficients between responses elicited by all pairs of stimuli are near zero (48 of 485 are above 0.7 in the human, and 78 of 2095 are above 0.6 in the hamster at supraliminal level), which means that all stimuli are well discriminated: the peripheral acceptors code for sensory images all of which are different. Only a very few stimuli

form covarying groups: in the human, for example, cyclamate, acesulfam, fructose, aspartame, on the one hand, and orthonitrobenzoic acid, aminobenzoic acid, propoxyaminonitro-benzoic acid, on the other hand, which are two different groups of sweet-tasting compounds; meta-nitrobenzoic acid, nitrosulfobenzoic acid, on the one hand, and pentobarbital, theophylline, niacinamide, on the other hand, which are two groups of bitter-tasting compounds. Sodium salts also form a group. We find again the famous so-called four standard qualities, but as landmarks in a field of many compounds, the relative distances between them represent all degrees of similarity.

3. Thus, taste space appears as a true unique continuum at threshold (Faurion et al., manuscript in preparation) and supraliminal [10] levels in both human and hamster experiments, which means that peripheral taste is a system oriented toward a discriminative function and not a categorizing function. This follows the intuition of Henning. It also means that one and the same set of multiple receptors cooperates in coding any sensation. Contrary to what was believed until now, there are no four groups of receptors coding for each of four qualities.

We can find several examples of correlations between molecules belonging to different qualities, for example, between a sweet-tasting and a bitter-tasting compound. This indicates that the set of sapid receptors may not be divided into subgroups devoted to specific qualities at the periphery. We must also take into account the numerous molecules that taste neither sweet nor bitter and are all distinguishable from one another: each one triggers a particular set of receptors selected among all of them.

4. This continuum needs a certain number of dimensions to be conveniently represented: unlike Henning's description and unlike color vision, which are tridimensional, the taste space needs about 10 factors to conveniently represent a minimum of 75% of relevant information. This means that at least such a number of independent information channels (receptors) are needed at the interface between internal and external media to encode taste information.

5. It seems that the dimension of the taste space is limited: increasing the number of molecules from 4 to 20 rapidly increases the number of factors, but increasing the number of molecules to 50 tends to an asymptote. We conclude that most receptors are sampled with a few tens of tastants.

6. We suggest that highly correlated stimuli share a maximal proportion of common receptors whereas poorly correlated ones share perhaps only a few receptors or even no receptor. After some preliminary attempts [11], molecular modeling is in progress to propose some receptor site features (see Froloff et al., this volume).

Discussion

Taste is essentially a discriminative system, including many more different qualities than the four semantically described categories, and comparable to olfaction with which it is so intimately associated. We would need as many names as stimuli for describing taste. It is very clear that our four descriptors are not sufficient. Fifty subjects describing the taste of pure compounds with four descriptors (sweet, bitter, salty, acid) may lead, for some stimuli (such as D-threonine) to 25% of each descriptor cited. This does not mean the compound is a quarter sweet, a quarter bitter, etc., but rather that this set of descriptors is not convenient to describe this compound and that all four categories have been randomly used. Semantic categorization is probably a process involving structures which are more central than the sensory chain itself. For two similar peripheral sensory images or neuronal patterns, duplicate information, that is, similarity and difference, seems to be always available. One can focus on only one part of the information, extracting a subset of the activated pattern and scotomizing the other one, and decide whether they are similar or different.

We can understand that, as for olfaction, the lack of semantic descriptor results from the lack of perceptual consensus among individuals. We can experimentally demonstrate that intensity of perceptions are all different from one subject to another, as are quality descriptions and qualitative similarities: the neuronal pattern elicited by one and the same stimulus is not the same from one subject to another one. No word can emerge if no common sensation occurs among individuals. Taste appears as a personal experience independent from any reference to other individuals, and as such is not communicable. In the case of sweet, we can communicate thanks to a unique chemically pure reference: sugar. Replace sucrose with sweeteners, and the lack of agreement between several subjects' quality descriptions tells you that we probably do not experience the same perception even for sucrose. Semantic descriptors have changed throughout history, from 2, sweet opposed to bitter for Aristotle [12], to 12 with Linneus, and since olfaction was properly distinguished from taste by Chevreul [13] in 1824, only 4 have been retained: *sweet, sour, bitter, salty*. Ikeda [14] introduced a fifth one: *umami*, for glutamate and ribonucleotide taste. Looking for the relativity of these terms all over the world, in remote ethnic groups, for example [15], we find confusion everywhere between the different sensory modalities falsely confused within one: taste. Taste (in the everyday life acceptance) is a multimodal sensation including somesthesis (tactile, thermal sense), chemical trigeminal, and olfactory perceptions as well as taste. Semantic description is not related to "physiological taste" and is not relevant to the study of taste chemoreception. Semantics, as a product of daily culture, is not a good guide to studying the taste quality concept. Our culture has good reasons to disagree with the physiologist.

Conversely, where do these four descriptors come from? The taste of salt is salty by definition; salt is a crystallized pure compound, not actually a description of the perception. The same is true for sweet, and this is particularly obvious in the French language, where sweet is said "sucré" (sugared) to tell the taste quality and "doux" to tell the somesthetic sensation. Most probably, salty, sweet (in French, sugared), and, perhaps, acid, are pure compound names: as we daily refer to those compounds kept in the kitchen cupboard, whatever be our perception it will be possible to name it by reference to a constant and objective reference. Bitter is usually used for everything that is hedonically negative, even confusing bitter odors and tastes. Bitter is also confused with acid in 50% of cases by untrained subjects, indicating a very poor definition of these words in our daily culture. Acid is well defined in terms of chemistry but contaminated by various odors in the kitchen (citric, acetic acids) and is moreover confused with the unpleasant trigeminal tingling sensation elicited by acids at pH lower than 2.4. It is very probably that the origin of bitter and sweet (in its English meaning) is the hedonic opposition between bad and good, as in the monodimensional taste continuum of Aristotle.

In conclusion, there are no four "qualities" but an unlimited number of qualitatively different taste sensations. Each tastant, whether pure or complex, elicits a different taste. There are no four actual descriptors but names of compounds used in daily life and descriptors deriving from hedonics, which is not surprising: describing is a difficult mental task to achieve, as it relates to abstraction and conceptual processing, whereas telling whether one likes or does not like something is expressed without needing to think about it.

Acknowledgments. Part of this material is original and was presented as an overview in [10]. The author is indebted to research assistants T. Lardier, L. David, V. Dirand, C. Courchay, and A-M. Pillias for data collection.

References

1. Erickson RP (1984) Ohrwall, Henning, von Skramlik: the foundation of the four primary position in taste. Neurosci Behav Rev 8:105–127
2. Henning H (1916) The quality range of taste. Z Psychol 74:203–219
3. Pfaffmann C (1974) Specificity of the sweet receptors of the squirrel monkey. Chem Senses Flavor 1:61–67
4. Yoshida M, Saito S (1969) Multidimensional scaling of the taste of aminoacids. Jpn Psychol Res 11:149–166

5. Schiffman SS, Erickson RP (1980) The issue of primary taste versus a taste continuum. Neurosci Behav Rev 4:109–117

6. Faurion A, Saito S, Mac Leod P (1980) Sweet taste involves several distinct receptor mechanisms. Chem Senses 5:107–120

7. Yamamoto T, Kawamura Y (1972) A model of neural code for taste quality. Physiol Behav 9:559–563

8. Dixon WJ, Nassey J (1960) Sensitivity experiments. In Introduction to statistical analysis. McGraw Hill, New York, pp 377–394

9. Faurion A, Courchay C (1990) Taste as a highly discriminative system. Brain Res 512:317–332

10. Faurion A (1993) The physiology of sweet taste and molecular receptors. In: Mathlouti M, Kanters JA, Birch GG (eds) Sweet taste chemoreception. Elsevier, Barking, pp 291–316

11. Faurion A, MacLeod P (1982) Sweet taste receptor mechanisms. In: Birch GG, Parker KJ (eds) Nutritive sweeteners. Applied Science, London, pp 247–273

12. Aristotle De anima. Livre II, Chapitre 10,422b, pp 10–15

13. Chevreul ME (1824) Des différentes manières dont les corps agissent sur l'organe du goût. J Physiol Exp Pathol 4:117–129

14. Ikeda K (1912) 8th International Congress of Applied Chemistry, Washington, D.C. and New York

15. Chamberlain AF (1903) Primitive taste words. Am J Psychol 14:146–153

Increased Taste Discrimination Ability by Flowing Stimuli over the Tongue

RYUICHI NONAKA[1] and MICHAEL O'MAHONY[2]

Key words. Taste psychophysics—Dorsal flow

Flow presentation procedures have been used by several researchers for controlled taste stimulation to stabilize the environment of the taste receptors and to control adaptation effects. O'Mahony [1] has shown flow presentation to elicit remarkable taste sensitivity in subjects; they were able to discriminate between once and twice distilled water. O'Mahony et al. [2] later demonstrated that subjects showed more sensitivity for taste discrimination with the flow presentation than with a sip-and-spit presentation. This study was carried out to determine the reason for this phenomenon; we also used the flow presentation technique for discrimination testing with foodstuffs.

A flow delivery system was built to deliver taste stimuli flowise over the anterior dorsal surface of the tongue. The onset of the stimulus was controlled by the subject, but the sequence of stimuli delivered was pre-programmed. The flow rate was 301–309 ml/min. Taste discrimination was measured for stimuli delivered flowise to the tongue and was compared to discrimination for the same stimuli tasted by traditional sip-and-spit procedures. For both procedures, the subjects had a warm-up [3] before the session began. All experiments were performed at constant room temperature (20°–22°C). The psychophysical procedure was an A, Not-A test, with added sureness judgments to counter response bias. Twenty signals and 20 noises were given to compute R-index values (here equivalent to signal detection P(A) values) as measures of taste discrimination. The session length required was usually about 30 min to complete both flow and sip-and-spit procedures.

For taste discrimination between purified water and low concentration NaCl (1.0 mM), a significantly higher mean R-index ($n = 18$) was obtained for the flow procedure than for the sip-and-spit procedure. Measures of the physical signal strength for both procedures and the stimulus concentration change in the mouth upon tasting, analyzed by atomic absorption spectrophotometry, indicated that the superior performance for the flow procedure was due to a reduction in variance.

To apply the flow procedure to a foodstuff, we investigated discrimination between two brands of mineral water. In this experiment, the sample that contained more mineral was selected as the signal. Nose clips were used for half of each session to ensure that the subject discriminated only by taste. Thus, two sessions were required to complete 20 signal and noise presentations both with and without noseclips. Again, higher mean R-indices ($n = 22$) were obtained both with and without noseclips for the stimulus flow condition.

The ability to discriminate between 1–3 mM caffeine and purified water, was different for trained subjects ($n = 13$) and untrained subjects ($n = 9$). The flow procedure gave higher R-indices than the sip-and-spit procedure, provided judges were trained to recognize the appropriate time during adaptation at which the stimulus should be presented and the appropriate area of the tongue with which to detect the signal.

In conclusion, the flow procedure, if it was conducted appropriately, produced higher discrimination ability than the sip-and-spit procedure. The results of the physical signal strength measurements indicated that the flow procedure produced this higher discrimination ability by reducing stimulus variance.

References

1. O'Mahony M (1972) Purity effects and distilled water taste. Nature 240:489
2. O'Mahony M, Gardner L, Long D, Heintz C, Thompson B, Davies M (1979) Salt taste detection: An R-index approach to signal detection measurements. Perception 8:497–506
3. O'Mahony M, Thieme U, Goldstein LR (1988) The warm-up effect as a means of increasing the discriminability of sensory difference tests. J Food Sci 53:1848–1850

[1] Sensory Evaluation Group, Research and Development Division, Kirin Brewery Co., Ltd., 26-1 Jingumae 6-chome, Shibuya-ku, Tokyo, 150 Japan
[2] Department of Food Science and Technology, University of California, Davis, CA 95616, USA

The Effects of Tasting Foods on the Flow Rate of Whole Saliva

Shigeru Watanabe, Kaori Imai, Mineko Ohnishi, and Seiji Igarashi[1]

Key words. Gustatory stimulation—Salivary flow rate

In a previous report [1], we evaluated the effects of tasting and chewing foods on the flow rate of whole saliva. The flow rate in response to tasting, as a percentage of that in response to chewing, was 86.7 ± 12.2 (rhubarb pie), 79.3 ± 17.9 (rice), and 73.4 ± 15.4 (carrot), respectively. The objective of this study was to determine the effects of tasting foods on the flow rate of whole saliva in children.

The subjects were 14 boys and 16 girls, all 5 years of age, who were all in good health and had complete primary dentition. The studies were carried out about noon, before they had consumed lunch and at least 1 h after their last meal. The foods selected were boiled rice, apples, cookies, and pickled radish. In a preliminary study, the weight of food per mouthful and the chewing time for one mouthful, for each of the four foods, were calculated. The same total weights of the same four foods as those in the preliminary study were each divided into small portions with weights equivalent to one normal mouthful, as determined for each subject in the preliminary study. The foods were weighed to the nearest 10 mg on a top-loading balance. For each food, one series of portions (a) was not treated further, while the other (b) was either finely chopped or mashed to reduce the particle size to approximately that achieved during chewing. For food of type (a), the subjects chewed the individual portions on both sides of their mouths in their normal way for the length of time determined during the preliminary study. However, instead of swallowing, they spat out the food into a weighed container. The volume of saliva secreted was determined by subtracting the initial weight of the food from the weight of the food/saliva mixture. Foods of type (b) were simply held between the dorsum of the tongue and the palate, with slight movement of the tongue to ensure continued gustatory stimulation, but with no masticatory movements. At the end of the time that would normally have been taken to chew the food, the food/saliva mixture was spat into a weighed container, as above. Calculation of food swallowed or retained was done by the method of Watanabe and Dawes [1].

The smallest weight of food per mouthful was for the cookies. Boiled rice and pickled radish required the longest chewing time per mouthful. In the main study, there were no significant differences in the flow rate response of boys and girls. With all four foods, chewing was a significantly more effective secretory stimulus than tasting alone ($P < 0.001$). The flow rate with cookies was significantly higher ($P < 0.05$) than with rice. The response to tasting as a percentage of that to chewing was highest with cookies (95.1%) and lowest with rice (71.2%), and this difference was statistically significant ($P < 0.01$). The mean percentage of food which was either inadvertently swallowed or remained stuck to the teeth or oral mocosa was calculated to be 12.4 ± 2.5 for chewing and 12.1 ± 3.1 for tasting; these two values were not significantly different.

Our major finding is that, in humans, the gustatory stimulation of food appears to be much more important in eliciting salivary flow than the mechanical stimulation of chewing. With the four different foods, gustatory stimulation alone elicited flow rates that ranged from 71.2% to 95.1% of those produced when food was chewed. This average percentage was very similar to that found in adults, evaluated in a previous study [1].

[1] Department of Pediatric Dentistry, School of Dentistry, Health Sciences University of Hokkaido, Ishikari-Tobetsu, Hokkaido, 061-02 Japan

References

1. Watanabe S, Dawes C (1988) The effects of different foods and concentrations of citric acid on the flow rate of whole saliva in man. Arch Oral Biol 33:1–5

The Effect of Taste Adaptation on Salivary Flow Rate in Children

Shigeru Watanabe[1], Yuji Ishizuka[2], Junko Amasaka[1], Seiji Igarashi[1], and Tokuro Ichida[2]

Key words. Taste adaptation—Salivary flow rate

Dawes and Watanabe [1] measured changes in the flow rate of parotid saliva while a tastant was infused continuously. They concluded that the half-time for adaptation of flow rate was independent of the nature or concentration of the stimulus, and averaged 11.3 s. There is little information about salivary secretion in children. The objective of this study was to determine how salivary flow rate changes during constant gustatory stimulation in children.

Parotid saliva from one gland was collected with Lashley cannulae. The subjects were four boys and six girls, all 5 years of age. Saliva was allowed to run from the tubing of the cannula into a beaker on a balance. The balance was connected to a Sartorius integrator (1202 MP; Sartorius, Göttingen, Germany), which responded to increments of 0.01 ml of saliva, and the integrator was connected to a chart recorder (VP-6541A, National, Tokyo, Japan), whose paper moved at a speed of 5 cm/min. The taste solutions were sodium chloride (15 and 5 g/dl) and citric acid (0.5 and 0.1 g/dl). Initially, water was infused into the mouth until the parotid saliva flow rate had stabilized. Then, one of the taste solutions was infused for about 3 min, followed again by water. The flow rates of parotid saliva were determined from tangents drawn to the curves on the chart recorder. To calculate the half-time for salivary flow rate adaptation (the time for the flow rate to decrease by half), we expressed all flow rates with each tastant as a percentage of the maximum flow rate with that tastant. Whole saliva was collected from ten children (the same subjects as those whose parotid saliva was examined) and ten adults (five females and five males). A lemon drop was placed into each of 20 cups that had been weighed previously. The subjects tasted the lemon drop for 20 s and then spat saliva and the drop into the cup. This procedure was repeated 20 times continuously. The volume of whole saliva secreted during the tasting of the lemon drop was determined by subtracting the initial weight of the lemon drop and cup from the total weight.

All subjects showed adaptation of salivary flow rate in response to the constant gustatory stimulation. Plots of the logarithms of the parotid salivary flow rates, expressed as a percentage of the maximum flow rate, against time, were essentially linear. The mean value of r^2 from the regression analyses was 0.94, with a standard deviation of 0.05. The unstimulated flow rates of parotid saliva (when water was being infused) averaged 0.034 ± 0.010 (SD) ml/min. Comparisons of the maximum flow rates achieved with the higher and lower concentrations of each of the two stimuli showed that the higher concentrations elicited significantly higher flow rates for both types of stimulus ($P < 0.01$). Analysis of variance revealed that none of the half-times for the four different tastants was significantly different from the others. The mean half-time for parotid salivary flow rate adaptation after maximum flow rate had been achieved was 12.68 ± 1.22 (SD) s.

The unstimulated flow rates (children, 0.22 ± 0.13 ml/min; adults, 0.45 ± 0.19 ml/min) and the maximum flow rates of whole saliva (children, 4.46 ± 0.98 ml/min; adults: 5.64 ± 1.77 ml/min) for children were lower than those for adults ($P < 0.001$, $P < 0.01$, respectively). The mean half-time for whole salivary flow rate in children (278.4 ± 73.8 s) was significantly higher than that in adults (391.4 ± 79.7 s) ($P < 0.05$).

Shannon [2] reported no decrease in whole salivary flow rate with time during 3 h of gustatory stimulation with sour candies. However, in our study, whole salivary flow rate decreased during the continuous tasting of the lemon drop (the mean half-time was about 280 s). Probably this difference depended on the movements of the tongue.

References

1. Dawes C, Watanabe S (1987) The effect of taste adaptation on salivary flow rate and oral sugar clearance. J Dent Res 66:740–744
2. Shannon IL (1958) Sodium and potassium levels of saliva and salivary glands following prolonged stimulation. J Dent Res 37:981

Departments of [1]Pediatric Dentistry and [2]Oral Biochemistry, School of Dentistry, Health Sciences University of Hokkaido, Ishikari-Tobetsu, Hokkaido, 061-02 Japan

Relation Between Human Taste and Ion Concentration in Saliva

Ichiro Ohsawa[1]

Key words. Saliva—Gustatory test

I have already reported the relation between human taste and ion concentration in saliva in ISOT X [1]. At that time, I used semiquantitative papers for the measurement of ions in saliva, and concluded that human taste was related to anions or some kinds of protein in saliva. However, since there are many kinds of ions in human saliva, both anions and cations, there still seemed to be problems. In this study, I used ion-selective electrodes with the normal solution-dropping method for the measurement of ion concentration. Because human taste sensibility seemed to be affected by salivary pH and ion concentration, I took the mental or physical condition of subjects into consideration.

I measured the concentration of an anionic surface active agent and of the following ions: F^-, Cl^-, Ag^+, Pb^{2+}, Cu^{2-}, Ca^{2+}, Na^+ and S^{2+} in mixed saliva. I collected a mixture of saliva from ten subjects who held a piece of paraffin wax in their mouths for 15 min and determined the pH of the saliva with BTB (brom-thymol-blue) paper (Toyoroshi, Tokyo, Japan). Before measuring the ions, I collected 1 ml of saliva from the saliva collected over the 15 min and mixed this with the regulating solution of ion concentration to increase the measured ion concentration. I selected the regulating solution: 20% CH_3COONa and 10% KNO_3 (adjusted to pH 6 with CH_3COOH) for Ag^+, Pb^{2+}, Cu^{2+}, and Ca^{2+}, 10% triorsaminomethane (adjusted to pH 10 with CH_3COOH) for Na^+, 10 M NaOH for S^{2+}, and CH_3COONa (adjusted to pH 4 with CH_3COOH) for the anionic surface active agent. For F^-, Cl^-, Ag^+, Pb^{2+}, Cu^{2+}, Ca^{2+}, Na^+, and the anionic surface active agent, I mixed the saliva sample with the same volume of regulating solution; for S^{2+}, the saliva was mixed with a tenfold volume of regulating solution. The ion concentration of these samples was measured with an ionmeter (model IOL-50; DKK, Tokyo, Japan) in a 37°C thermostat. The volume of normal dropping solution was 0.1 ml.

Ag^+, Pb^{2+}, and Cu^{2+} were not measured, and the anionic surface active agent was not stable. F^-, Cl^-, Ca^{2+}, Na^+, and S^{2+} were measured. I compared these ion concentrations and the results of the gustatory test. For F^-, Cl^-, Ca^{2+}, and Na^+, there was an adequate correlation, but for S^{2+}, there was no correlation with the gustatory test.

Human taste sensitivity is affected by many factors, for example, systemic disease and medication, the effects being direct or indirect; taste sensitivity is also affected by human mentality. Of course, taste sensitivity in subjects with many carious teeth, or with periodontal diseases, is lower than in those who are healthy. Moreover, salivation activity is reduced in conditions of too little sleep. The function of saliva in taste sensitivity is to carry the taste substances to the taste bud. Molecular activity seems to influence the speed at which the taste substance permeates the taste bud.

In this experiment, there was an adequate correlation for F^-, Cl^-, Ca^{2+}, and Na^+, in regard to taste sensitivity. Fluoride and calcium ions were related to dental caries, and chloride and sodium ions were related to clinical dental inflammation. Systemically, chlorine and sodium ions were related to kidney function.

I concluded that, although saliva ion concentration was important for human taste, saliva protein was much more important, in the light of this and previous experiments [2–7]. However, in the dental clinic, saliva ion measurement was effective for determining human taste and the treatment of taste defects.

Acknowledgments. Nihon University School of Dentistry, Sato grant 1991.

References

1. Ohsawa I (1989) A study about human taste and saliva hydrogen ion concentration. In: Døving KB (ed) Proceedings of the Tenth International Syposium on Olfaction and Taste. Oslo, Norway, p 351
2. Ohsawa I, Yoshida M, Ishikawa S, Gotoh M (1988) A study about the relation between human taste and saliva (in Japanese). In: Morita H (ed) Proceedings of the 22nd JASTS. Fukuoka, Japan, pp 65–68

---

[1] Department of Oral Diagnosis, Nihon University School of Dentistry, 1-8-13, Kanda-Surugadai, Chiyoda, Tokyo, 101 Japan

3. Ohsawa I, Gotoh M, Sugiura M (1988) Occurrence of carious teeth in autonomic nerve system sites (in Japanese). J Dent Health 38:552–553
4. Ohsawa I, Hagihara Y, Yoshida M, Ishikawa S, Gotoh M (1990) A clinical report about taste lesion patients (in Japanese). In: Hidaka I (ed) Proceedings of the 24th JASTS. Tsu, Japan, pp 199–202
5. Ohsawa I (1990) A study about the saliva viscosity in human taste. In: Kroeze J (ed) Abstracts of the 9th ECRO. Noordwijkerhout, Netherlands
6. Ohsawa I, Watanabe M, Hara K, Hagihara Y, Yoshida M, Ishikawa S, Gotoh M (1992) A study of the whole mouth method in gustatory tests (in Japanese). In: Arai I (ed) Proceedings of the 26th JASTS. Tokyo, Japan, pp 185–188
7. Ohsawa I (1992) A study of saliva physical condition in human taste. In: Hatt H (ed) Abstracts of the 10th ECRO. Munich, Germany

Approach to Tastiness Evaluation: Electrophysiological Evaluation of Effect of Mood State on Perception of Taste

MASASHI NAKAGAWA, HAJIME NAGAI, TAKAKO INUI, and MIYUKI NAKAMURA[1]

Key words. Tastiness—Electrophysiological

Many qualitative and quantitative studies have been performed to determine emotional states in humans. Psychophysiological methods such as component analysis of electroencephalograms (EEGs), heart rate, blood pressure, galvanic skin reflex (GSR), and pupillary light reflex, have been used. However, few studies have addressed the psychophysiological evaluation of smell and taste, in particular, we know of no reports of the evaluation of tastiness. We carried out a study of the role played by odor compounds in mood states in humans in 1989 [1], evaluating the mood state quantitatively with a medical electronic instrument. In this study, we developed two basic methods of approach to the evaluation of tastiness. This first approach was to evaluate emotional changes produced by a refreshing drink, and the second was to evaluate the feeling in the mouth, depending on the temperature differences of drinks. Pupillary light reflex and EEGs served as a physiological index, and the profile of mood state (POMS) [2] as a psychological index. These determinations were made before and after mental arithmetic tasks in the experiment with the refreshing drink. Subjects performed mental calculations, two figure addition and subtraction, for 60 min according to a computer program. The calculation problems increased in difficulty in a step-by-step fashion to prevent habituation due to practice. Carbonated drinks with mint flavor and carbonated drinks without mint flavor (control) were given. The recovery ratio and tolerance to stress when subjects performed 20-min arithmetic tasks were compared for the two drinks and after that the subjects continued the 60-min performance again. There were improvements due to the test drink in the correct answers for arthmetic tasks ($F[1,6] = 3.12$, $P = 0.021$), but there was no effect on the total number of trials in both conditions. There was no significant difference in amount of alpha-EEGs between conditions during arithmetic tasks. However, alpha-wave activity was increased gradually toward the end of the tasks by the test drink, while alpha-wave activity decreased in the last session with control drinks. There was no recovery effect on the constriction ratio of the pupillary light reflex for either drink, but the irritation of the mint flavor in the test drink reduced the pupillary constriction just after subjects drank it. The POMS showed that the test drink prevented decreased activity, increased fatigue, and increased confusion, compared with control, during the arithmetic tasks. In the second experiment, the feeling in the mouth of different temperature drinks was evaluated by topographic analysis of EEGs and by questionnaire. The samples used were whisky and water at three different temperatures (8.5% alcohol; 0°C, 10°C, and 20°C). There were correlations for total sense of irritation and chilliness of drinks ($r = 0.695$), and total sense of irritation and amount of alpha-wave ($r = -0.374$), and richness of aroma and localization of alpha-wave (covariance) in the cortex. There were some changes in physiological index, for example, diminution of pupillary constriction and reduction of alpha-EEG, produced by irritation i.e., the mint flavor or chilliness. We assume that such irritation in the mouth influences the mind. If these irritations are related to good experiences, we will retain them as good memories, and we may sense that such drinks are "tasty".

References

1. Nakagawa M, Nagai H, Nakamura M, Fujii W, Inui T (1992) Influence of odors on human mental stress and fatigue. In: Doty RL, Schwarze DM (eds) Chemical signals in vertebrates, VI. Plenum, New York, pp 518–585
2. McNair DM, Lorr M (1964) An analysis of mood in neurotics. J Abnorm Soc Psychol 69:620

[1] Suntory Ltd. Institute for Fundamental Research, Shimamoto-cho, Mishima, Osaka, 618 Japan

Free-Choice Profiling vs Conventional Descriptive Analysis in Characterizing the Sensory Properties of Acids on an Equi-Sour Basis

Sonia M. Rubico and Mina R. McDaniel[1]

Key words. Acids—Free-choice profiling—Descriptive analysis

Previous studies on the tastes of acids have focused on threshold, equi-sour levels, and time intensity values of sourness. Studies on characterizing the sensory properties of organic and inorganic acids have been very limited. Recently, the characterization of the sensory properties of 15 acid samples, using free-choice profiling, was reported by Rubico and McDaniel [1]. Our results, analyzed by general Procrustes analysis (GPA), indicate that on a weight (0.08% w/w or v/v) basis, acids differ in their sensory profiles. Panelists found the concentration of hydrochloric, phosphoric, and succinic acids to be very strong in sensory characteristics other than sourness. Another study on a molar basis (0.005 M) still found significant differences in astringency, bitterness, and sourness, although similarities of the fruit acids were more evident [2].

This study was designed to determine the sensory profile of selected acids (adipic, citric, fumaric, glucono-delta-lactone, hydrochloric, lactic, malic, phosphoric, quinic, succinic, tartaric, and citric: fumaric, citric: malic and fumaric: malic combinations) on an equi-sour basis while comparing two sensory methods, free-choice profiling (FCP) and conventional descriptive analysis (CDA). In FCP, panelists independently develop individual vocabularies, while in CDA, a consensus on descriptors is obtained from the panel. The former method was analyzed by Generalized Procrustes Analysis and the latter with Principal Component Analysis with the principal axes scores analyzed by analysis of variance and least significant difference. The relationship between chemical measures and sensory data was also reported.

An equi-sour concentration of each of the different acids was prepared, based on the methods used by Rubico [3]. Five molar concentrations of each acid were evaluated by 16 panelists by comparing each concentration to 0.005 M citric acid. The equi-sour concentrations were estimated based on their power

functions. The same group of four male and eight female panelists were used first for FCP and and then for CDA. The panelists evaluated the acid samples when they were in the mouth and after expectoration.

For FCP, the number of descriptors used varied from 3 to 15, while for CDA, the lists of descriptors were consolidated into a consensus vocabulary of 8 terms for sensations perceived when the samples were in the mouth (overall intensity, sourness, astringency, bitterness, sweetness, saltiness, monosodium glutamate (MSG), metallic) and 6 terms for the samples after expectoration (astringency, sourness, bitterness, metallic, chalky, and persistent). Then, these attributes were rated, using a 16-point scale (0, none; 7, moderate; 15, extreme).

The final result of a GPA is a consensus configuration of the samples for the different principal axis (PA). The first two dimensions accounted for 31% and 25% of the original variance. The first PA showed a contrast between bitter/aspirin, MSG/salty, and "yucky"/sick/spoiled/musty on the positive side, with astringent and sour on the negative side. The negative axis was also described as sharp, takes off enamel, abrasive, mouth-puckering, and chalky. All the characteristics on the positive axis referred to succinic acid, which was found to be significantly different from the rest of the acids. The inorganic acids, with lactic and quinic, had the more astringent characteristics.

For CDA, the first two principal components (PC) accounted for 72.31% of the explained variation, with the first PC described as MSG, bitterness, persistent, and metallic aftertaste as the most influencing attributes. This axis was mainly a contrast between succinic and the rest of the acids. For the second PC, the inorganic acids, HCl and phosphoric, were significantly more astringent than the organic acids. The sourness aftertaste was the key attribute for the second PC, rather than the sourness in the mouth.

The two sensory methods gave the same pattern of information regarding the acid samples. The acids differed in their flavor and taste dynamics even at their equi-sour concentrations, with succinic acid having a very intense bitterness and MSG taste. Hydrochloric and phosphoric acids were more

[1] Sensory Science Laboratory, Oregon State University, Corvallis, OR 97331-6602, USA

astringent than sour. While the similarities of several organic acids and their mixtures were very evident, no chemical measure could explain the differences well. It was concluded that acids had other sensory properties aside from sourness that must be considered in a given food application.

References

1. Rubico S, McDaniel M (1992) Sensory evaluation of acids by free-choice profiling. Chem Senses 17:273–289
2. Rubico S, McDaniel M (1991) Sensory evaluation of acids (weight and molar basis) by free-choice profiling. Paper No. P233, presented at the 8th World Congress of Food Science and Technology, Toronto, Canada, Sept. 29–Oct. 4
3. Rubico S (1993) Perceptual characteristics of selected acidulants by different sensory and multivariate methods. PhD thesis. Oregon State University

Phenylthiocarbamide (PTC) Taste Test in Various Ethnic Groups in China

TOSHIHIDE MATSUMOTO, KAZUAKI NONAKA, TETSURO OGATA, and MINORU NAKATA[1]

Key words. Phenylthiocarbamide (PTC)—Ethnic groups in China

We carried out this investigation to determine taste sensitivity of phenylthiocarbamide (PTC) in various Chinese ethnic groups. While the Han ethnic group accounts for more than 90% of the population in China, there are 55 other ethnic groups, distinguished from each other by their historical origin, customs, language, and religion. Taste sensitivity to PTC is controlled by a gene, and differences in this sensitivity have been investigated in various populations.

The subjects of this investigation were three populations from the Xinjiang Uigur Autonomous Region [the Uigur ($n = 272$; age 10.9 ± 2.1 years; from Urumqi), the Kazak ($n = 244$; age 10.7 ± 2.7 years; from Ining), and the Xibo ($n = 403$; age 11.2 ± 2.2 years; from Ining)], and one population from the Inner Mongolia Autonomous Region [the Mongolians ($n = 373$; age 11.7 ± 2.6 years; from Damauqi)]. The controls were three groups of Han [from Beijing ($n = 355$; age 11.2 ± 2.6 years), from Leshan ($n = 359$; age 11.3 ± 2.8 years), and from Hong Kong ($n = 233$; age 11.5 ± 2.4 years)], and one group of Japanese [from Fukuoka ($n = 250$; age 11.0 ± 2.4 years)]. (All ages are expressed as mean $\pm$ SD.) Commercial PTC tasting test paper (Advantec, Tokyo, Japan) was used. A sheet of paper was placed on the tongue of each subject and allowed to remain for 5 s, after which the subjects were divided into PTC nontasters or tasters.

Because there were no significant sex differences in the frequency of PTC nontasters in these populations ($P < 0.05$), the data for both sexes were pooled. The percentage of PTC nontasters in each population was 34.6% in the Uigur, 23.8% in the Kazak, 7.0% in the Xibo, 9.9% in the Mongolians, 7.3% in the Han (Beijing), 7.0% in the Han (Leshan), 7.7% in the Han (Hong Kong), and 14.0% in the Japanese. The interpopulation differences were tested by the Chi-square test. There was no difference between the Uigur and the Kazak, but these two populations differed significantly from the other populations ($P < 0.05$). The gene frequency of the PTC gene is explained by the Hardy-Weinberg law, whereas the genotype frequency follows Mendel's laws. We thus estimated the gene frequency of the PTC gene. The frequency of the PTC recessive gene (t) in the Uigur was greater than that of the PTC dominant gene (T), while the frequency of t in the Xibo, the Mongolians, the Han, and the Japanese was smaller than that of T. The results of this study were compared with the data reported previously in various populations [1–3]. It was concluded that the Uigur and the Kazak peoples are closer to the Caucasoid than the Xibo, the Mongolians, the Han, or the Japanese.

References

1. Parr LW (1934) Taste blindness and race. J Hered 25:187–190
2. Allison AC, Blumberg BS (1959) Ability to taste phenylthiocarbamide among Alaskan Eskimos and other populations. Hum Biol 31:352–359
3. Grunwald P, Herman C (1962) Distribution of taste sensitivity for phenylthiocarbamide in Yugoslavia. Nature 194:95

[1] Department of Pediatric Dentistry, Kyushu University Faculty of Dentistry, Maidashi 3-1-1, Higashi-ku, Fukuoka, 812 Japan

Cross-Cultural Studies of Japanese and Australian Taste Preferences

JOHN PRESCOTT and GRAHAM BELL[1]

Key words. Cross-cultural—Tastes

Despite major differences between the diets of Japan and western countries such as Australia, the extent to which the taste perceptions and preferences of Japanese and Australians differ is uncertain. Collaborative research between the Commonwealth Scientific and Industrial Research Organisation (CSIRO) Sensory Research Centre (Australia) and Chuo University (Japan) has investigated cross-cultural taste perceptions and preferences, using a range of tastes and food types.

To determine if there were differences in taste sensitivity between Japanese and Australians, we examined sensitivity to variations in four tastes: sweet (sucrose), sour (citric acid), salty (sodium chloride), and bitter (caffeine). For both strong and weak intensities, panels of Japanese and Australian adults were required to taste a series of pairs of solutions: a standard plus another of either identical strength, or one of a series of slightly more intense solutions. The subjects judged whether the solutions were of the same or different intensity. Japanese and Australian subjects were found to be no more nor less sensitive than one another for all four tastes, and for both strong and weak concentrations of the tastes [1].

Further research on perception of tastes compared Japanese and Australian ratings of sweetness or saltiness intensity in 36 sweet and 36 salty foods (unpublished data). The foods were Japanese and Australian products in the following categories: chocolates, jams, beverages, fruit drinks, biscuits, and breakfast cereals (sweet); nuts, crackers, snacks, fish products, peanut butter, and soups (salty). With very few exceptions, Japanese and Australians agreed on the sweetness or saltiness intensity of these foods. In subsequent studies using foods (unpublished data), we manipulated the following key tastants: (a) sweetness in orange juice, cornflakes, and ice cream; (b) saltiness in breakfast cereal; (c) sourness in orange juice and salad dressing; and

(d) bitterness in grapefruit juice. For each food, Japanese and Australian subjects rated the intensity of the key taste within four versions of the food, each with a different tastant level. Again, in most cases, Japanese and Australians were in close agreement regarding the actual intensity of the key tastes.

These findings suggests that Japanese and Australians have few, if any, differences in the way they perceive or discriminate taste intensity. This suggests that any differences that are found in terms of food preference are most likely due to dietary experience and other cultural factors.

To investigate Japanese and Australian taste preferences, we initially compared hedonic responses to tastes in solution using six different concentrations of sweet (sucrose), sour (citric acid), salty (sodium chloride), bitter (caffeine), and *umami* (monosodium glutamate [MSG]; inosine monophosphate [IMP]; guanosine monophosphate [GMP]) tastes in solution [2]. For sourness and the *umami* tastes at high concentrations, the Japanese showed a greater preference than the Australians. The greater preference for *umami* tastes is perhaps because of greater Japanese familiarity with such tastes. The increased preference shown by Japanese for very sour tastes may reflect the greater use of sour foods in Japan as compared to Australia. Given the differences in diet of the two groups, it is interesting that the Japanese and Australians were *not* different in the degree of liking for sweet, salty, and bitter tastes.

Preferences for sweetness and saltiness were determined using the same range of foods mentioned above. In contrast to perceptions of intensity, or preferences for tastes in solution, preference for these taste qualities in foods showed considerable differences between Japanese and Australians. It is not possible to make general statements about the responses of either group to sweetness or saltiness overall, as the preferences were dependent on the particular food context. Thus, while the Japanese liked the saltiness of Australian nuts, they disliked the saltiness of Australian seafood, which was considered too salty. This context dependence is shown as a low correlation across the range of foods between ratings of intensity and preference, i.e., taste intensity does not predict liking, for either Japanese or Australians.

[1] Sensory Research Centre, CSIRO Division of Food Science and Technology, PO Box 52, North Ryde, NSW 2113, Australia

References

1. Laing DG, Prescott J, Bell GA, Gillmore R, James C, Best DJ, Allen S, Yoshida M, Yamazaki K (1993) A cross-cultural study of taste discrimination with Australians and Japanese. Chem Senses 18(2):161–168

2. Prescott J, Laing DG, Bell GA, Yoshida M, Gillmore R, Allen S, Yamazaki K, Ishii R (1992) Hedonic responses to taste solutions: A cross-cultural study of Japanese and Australians. Chem Senses 17(6):801–809

A Basic Study of the Initial Development of Food Preference: Feeding Experiments with Diet Containing Capsaicin

Tadao Kurata and Kazue Nakamura[1]

Key words. Food preference—Capsaicin

It is well known that the feeding pattern of mothers is the most important and decisive factor in the formation of food preferences in child after weaning. However, the details of the initial development of food preferences are still not clear.

In a preliminary study, we fed rats a diet containing capsaicin (CAP), a substance usually not liked by animals because of its pungency. We found that a preference for pungent taste could be developed when intake was repeated from infancy.

In this study, we carried out similar feeding experiments with a diet containing CAP to investigate the influence of the eating behavior of mother rats on the formation of food preference in their pups.

In experiment 1, a mother rat was fed a diet containing 200 ppm CAP and 50% sucrose during lactation. The mother and pups were kept together in a cage for 5 weeks after birth. During this period, the pups appeared to be completely weaned, and food consumption increased, suggesting that they might have shared the food.

After separation from the mother, the pups were also fed a diet containing 200 ppm CAP. When they were fed with this diet without sucrose, their food intake did not decrease, and they showed continuous growth. We carried out a second experiment to more precisely identify the relationship between the

presence of the mother and the food preference of the pups. The mother rat in the test group was fed a diet containing 20 ppm CAP during pregnancy, and throughout the experimental period. After birth, the pups (7 males and 5 females) were kept with the mother for 3 weeks; they were then divided into two groups: group A (4 males and 2 females) remained with the mother for a further 2 weeks, while group B (3 males and 3 females) was separated from the mother. Both groups were fed the diet containing 20 ppm CAP. A control group of pups (2 males and 4 females) was kept with the mother for 5 weeks after birth, as for group A; they were fed a control diet containing no CAP. In the preference test, group A ate more of the CAP-containing diet than group B and the control group when CAP levels were 200 ppm, 20 ppm, and 10 ppm.

In conclusion, the acceptance of food seemed to be easier for the pups when they were with their mother, suggesting that the eating behavior of the mother affects that of the pups, a factor which might be essential for the formation of food preference in mammals. The duration specific of the period after weaning might also be important for progeny to perceive and accept new foods other than breast milk.

In these experiments, we observed the differences between the sexes in food preference; in general, the female rats ate more of the CAP-containing diet than the males. It appeared that the female rats might adapt to a pungent diet more easily than the males, irrespective of the diet they had been fed previously.

[1] Institute of Environmental Science for Human Life, Ochanomizu University, Otsuka 2-1-1, Bunkyo-ku, Tokyo, 112 Japan

Why Do You Like Bitter Taste?

K. Mizuma, M. Nakagawa, and K. Namba[1]

Key words. Bitter—Stress

We generally reject foods with a bitter taste, this taste being a signal for the rejection. Nevertheless, many adults often crave bitter foods, such as beer, coffee, chocolate, and so on. Why do they prefer bitter foods like this? One reason is thought to be that the intake of bitter food and drink is related to the level of stress in today's society. We carried out a study to determine the relationship between bitter taste and liking, and bitter taste, stress, and acceptability in 1991 [1]. Here, we report two experiments carried out to investigate the effects on acceptability of bitter taste in humans who were exposed to conditions of mental stress (experiment 1) and physical stress (experiment 2).

In experiment 1, the subjects were 18 healthy volunteers between the ages of 24 and 46 years. A bitter-tasting sample, quinine sulfate solution (1.36 mg/100 ml), was prepared. Subjects performed mental tasks for 40 min on a personal computer. Before and after these sessions, the duration of the aftertaste and the intensity of the bitter sensation were recorded by time-intensity (TI) evaluation [2], and, in addition, psychological mood state was evaluated with a Japanese version of POMS (profile of mood state) [3]. TI evaluation is one of the most convenient methods for determining sensation intensity in real time scaling. The subjects retained 10 ml of the bitter sample in their mouths for 10 s, after which the sample was expectorated, and the intensity of bitterness was rated every 5 s. These findings were compared with those in control conditions in which subjects relaxed by watching an environmental video program on TV for 40 min. TI evaluation, showed that, after the mental task period, the duration of the bitter sensation was shortened. Total bitterness amount (before/after analysis of variance (ANOVA) $F[1,17] = 42.81$, $P <$ 0.001) and maximum intensity (before/after ANOVA $F[1,17] = 22.61$, $P < 0.001$) showed significant reductions, whereas there was no change of TI patterns for before/after tasks in the relaxed condition. Psychological mood state scores (tension, impatience) increased after computer tasks, but after subjects watched the video program there were no changes in these indexes.

Eleven healthy subjects between the ages of 23 and 27 participated in experiment 2. The samples and evaluations were the same as those used in experiment 1. Subjects were given physical workloads with an ergometer for 10 min; before and after the workloads TI and psychological mood state evaluations were carried out. The data for physical stress were compared with those obtained in the control condition, i.e., relaxing by watching an environmental video program on TV for 10 min. In this experiment, there were no significant differences in TI patterns between before and after physical workloads on between before and after the relaxed condition. In conclusion, mental stress, caused by tension and impatience, appeared to have a strong influence on the acceptability of bitter taste. In states of high tension and impatience when subjects were exposed to mental stress, the level of acceptability of bitter taste was reduced. Therefore, under such stressful conditions, we may prefer more bitter tastes such as beer, coffee, and so on.

References

1. Mizuma K, Namba K, Nakagawa M (1992) Taste preference of bitterness (II): Mental stress and receivability of bitterness (in Japanese with English abstract). Proceedings of the 26th Japanese Symposium on taste and smell, pp 149–152
2. Takagi M, Asakura Y (1984) Taste time-intensity measurements using a new computerized system (in Japanese). Proceedings of the 18th Japanese Symposium on taste and smell, pp 105–108
3. Mcnair DM, Lorr M (1964) An analysis of mood in neurotics. J Abnorm Soc Psychol 69:620

[1] Suntory Ltd. Institute for Fundamental Research, Shimamoto-cho, Mishima, Osaka, 618 Japan

Production of Low Sodium Chloride Soy Sauce

Rie Kuramitsu[1], Masatoshi Takahashi[2], Daisuke Segawa[2], Kozo Nakamura[2], and Hideo Okai[2]

Key words. Low sodium chloride soy sauce—Saltiness-enhancing substance

Excessive intake of NaCl is implicated in hypertension, toxemia of pregnancy, and other diseases in adults. Soy sauce, the most representative seasoning in Japan, contains a high concentration of NaCl, on average, 16%. We have studied salty substances that can be substituted for NaCl and have successfully developed some saltiness-enhancing substances [1]: Orn-β-Ala, Gly-OEt·HCl, Lys·HCl, and Orn·HCl. In 1991, Muramatsu et al. [2] succeeded in improving the soy sauce manufacturing process, decreasing the NaCl concentration to a maximum of 4.59%. We designated this preparation, pre-soy sauce. Using this as the starting material, and adding the saltiness-enhancing substances, Gly-OEt · HCl, Lys · HCl, and KCl, we designed the experimental manufacture of a new low-salt soy sauce. Amino acid derivatives are relatively easy to use in practice in terms of metabolic safety and economic efficiency in the manufacturing process, and this metal chloride has long been known as a salty substance. The preparations thus obtained were compared with commercial soy sauce to determine the most desirable manufacturing conditions for the production of low-salt soy sauce.

The low-salt soy sauce designated here is defined as a preparation made by adding NaCl and additives to pre-soy sauce to adjust the saltiness to the equivalent of 16.2% NaCl. The amount of the additive as a proportion of the entire salty components was expressed as a percentage and was termed the adding rate. The rates used for the experiments were 72%, 67%, 50%, and 33%. Samples were diluted to NaCl concentrations of 0.063%–2%, and evaluated in sensory tests by several panelists. The results of the sensory tests were: (1) For 72% adding rate:

Except for gly-OEt·HCl, which gave a sufficiently salty taste close to that of soy sauce, the saltiness of the other additives was not close to that of soy sauce. Each substance produced a characteristic taste. (2) For 67% adding rate: Samples with added Lys·HCl or KCl were improved in saltiness. However, in regard to taste constitution, some improvement was obtained at low added concentration while the taste characteristic of each substance was prevalent at higher added concentration. (3) For 50% adding rate: Gly-OEt·HCl produced a saltiness comparable to that of soy sauce. Other additives also produced a similar salty taste. In regard to quality, pre-soy sauce was evaluated to be comparable to soy sauce in low concentration (0.063%–0.25%) but in high concentration (0.5%–1.0%) its umami taste was more prominent, indicating that it was clearly inferior to say sauce in taste quality. (4) For 33% adding rate: The samples with added Lys·HCl and KCl had a saltiness almost equivalent to that of soy sauce. Quality and taste constitution were also improved by the addition of these substances at this adding rate, and depending on the concentration, the samples with added Gly-OEt·HCl and Lys·HCl were comparable to soy sauce in quality.

The experimental results revealed that a 33% adding rate of the salty components produced preparations not inferior to commercial soy sauce, and preparations with an adding rate of 50% were also suggested to be of use in some selected fields. In the present studies we also suggest that, for manufacturing soy sauce of good quality, a balance of salty and umami tastes is very important, not to mention the contribution of bitterness, acidity, and other contaminant tastes to the complete taste of foods.

[1] Department of Chemistry, Faculty of General Education, Akashi College of Technology, Uozumi, Akashi, Hyogo, 674 Japan
[2] Department of Fermentation Technology, Faculty of Engineering, Hiroshima University, 1-4-1 Kagamiyama, Higashihiroshima, Hiroshima, 724 Japan

References

1. Tada M, Shinoda I, Okai H (1984) L-Ornithyltaurine, a new salty peptide. J Agric Food Chem 33:792–795
2. Muramatsu S, Sano Y, Takeda T, Uzuka Y (1991) The study of digestive conditions of soy sauce *koji* at high temperatures (in Japanese). J Brew Soc Jpn 86: 610–615

Cognitive and Physiological Factors That Affect the Perception of Odor Mixtures

DAVID G. LAING[1]

Key words. Odor mixtures—Human perception—Temporal processing—Suppression—Stimulus fusion—Odor identification

Introduction

Perception of odor mixtures by humans is characterized by two major phenomena, odor suppression and odor blending or fusion. Both tend to reduce the ability of subjects to perceive mixture constituents. Although little is understood of the underlying mechanisms of these phenomena, the evidence available suggests that suppression can occur at the periphery [1,2] and at central olfactory structures [3]. Likely peripheral mechanisms for odor suppression include competition between odorants for common receptor sites, or for cells if they operate through different transduction pathways. In this regard it has been reported [4] that the patterns of receptor cells normally activated by different odorants are altered and become unrecognizable when odorants are presented as mixtures. Such alterations can result in the partial or complete suppression of an odorant and the loss of information about its quality and identity. In other words, only part of the information about an odorant may remain when it has been processed as a mixture constituent by the receptor cells, and this part may not be sufficient for identification. Thus, although an olfactory sensation may be perceived in response to a mixture of odorants, much of the information about individual odorants can be lost, giving the impression that many of the odorants may have blended or fused, with only a few of the components being perceived. The studies described here provide psychophysical evidence for these mechanisms of suppression or fusion.

The Capacity of Humans to Identify Odorants in Mixtures

This study aimed to determine how readily odorants could be identified in mixtures. A limited capacity could indicate the extent to which suppression and blending affect the perception of odorants in mixtures.

Methods

An eight-channel olfactometer was used to deliver stimuli containing one to eight odorants [5]. In each of the three test procedures, subjects were instructed by a computer screen when to sample a single odor outlet and to identify which odorants were perceived during the 50-sec sampling period in each trial. They responded by touching odor labels on a digitizing pad with a light-sensitive pen. The test conditions involved:

a. different levels of experience and training with the test odorants using untrained, trained, and expert (perfumer and flavorist) subjects;
b. different types of odorants that were classified by perfumers as good and poor blending substances; and
c. different test methods; selective attention (only one odorant identified per trial) and multiple identification (all odorants in a mixture to be identified per trial) methods

Results and Discussion

The results from each of the test conditions are shown in Fig. 1. In each case the outcome is clear; subjects had great difficulty in identifying more than three or four odorants in a mixture regardless of their training and experience, type of odorant, or test method. Such an outcome cannot be caused by cognitive factors; rather, it must be due to a physiologically imposed limit. But what of the mechanisms?

As discussed, loss of information most likely occurred as a result of competition for receptor sites and cells, resulting in the loss of identity of some or most of a mixture's constituents. Although not demonstrated in the current experiments, it is possible that odor sensations other than those identified were perceived as a result of blending. Whatever the reasons, the results show that massive loss of information about the identity of individual odorants increases with increasing numbers of mixture constituents.

[1] Centre for Advanced Food Research, University of Western Sydney, Richmond, NSW, Australia 2753

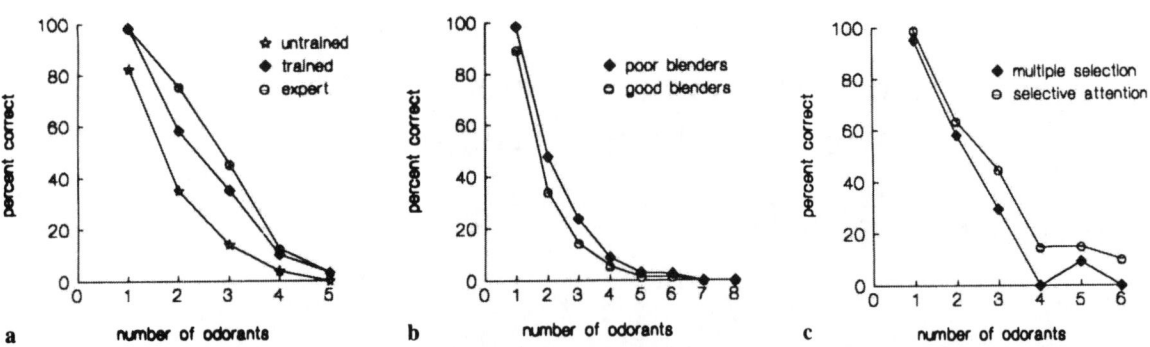

FIG. 1. Percent correct identification of odorants in mixtures **a** by expert, trained, and untrained subjects; **b** using good and poor blending odors; **c** with selective attention or multiple identification procedures

Processing of Mixtures Consisting of Complex Odorants

It is apparent from the first set of results that, despite increasing loss of information about odorants as mixture complexity grew, it was always possible to identify several of the constituents. Thus, sufficient information remained intact after processing at the periphery to allow identification of up to four constituents. The second question asked concerns the mode of processing of complex odorants such as chocolate. If four complex odorants can be identified in a mixture of, say, eight complex odorants, this would suggest that complex odorants are processed or encoded similarly to single chemical odorants, that is, as a gestalt. On the other hand, if a mixture of eight complex odorants produced a nondescript olfactory sensation, it would suggest that the mechanisms of encoding were different.

Methods

The methods used were identical to those described in the first part.

Results and Discussion

Figure 2 clearly demonstrates that the outcome of processing mixtures of complex odors is similar to that for single chemicals, with only three or four complex odorants being identified by each subject. This is not a surprising result when considered together with physiological data. Evidence from 2-deoxyglucose studies [7] with single odorants and complex mixtures, for example, indicates that the patterns of activated cells in the olfactory bulb are of similar complexity and contain similar numbers of cells. The current result, therefore, is in agreement with the latter data. Thus, complex odorants such as chocolate appear to be processed as a single entity, a gestalt. Alteration in encoding patterns as a result of cell suppression appears to be the primary mechanism underlying the present result.

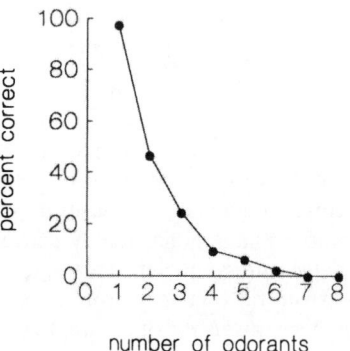

FIG. 2. Percent correct identification of mixtures consisting of as many as eight complex odorants

Loss of Information Through Temporal Filtering

The information loss observed in experiments already described was attributed to alterations in activity patterns and has been aptly labeled spatial filtering [4]. A recent physiological study [8] indicates that loss of information could also occur via another mechanism, namely temporal filtering. The latter study showed that different odorants require different times to stimulate receptor cells, with differences of the order of hundreds of milliseconds being reported. Accordingly, it was proposed [9] that if the time differences recorded at the periphery are maintained during processing at other olfactory centers, it is possible that the constituents of mixtures are perceived in series. Odorants with very different stimulating times may be more readily identified than those with similar times, so that with mixtures consisting of many components it could be difficult to perceive each odor because of the minute time differences in stimulation. In addition, "fast" odorants may be more competitive, reaching receptor sites first, or blocking other odorants, with the result that they may be a suppressor in mixtures.

This study aimed to establish whether odorants in mixtures are perceived in series, that is, are temporally coded, and whether "fast" odorants, being first to stimulate receptor cells, suppress "slow" odorants.

Methods

A 6-channel olfactometer was used to deliver two odorants as mixtures or in series separated by ±25, 50, 100, 200, or 400 msec. Subjects were instructed by a computer and were only required to indicate (using a mouse) which odorant was perceived first. The time difference required for both odorants to be perceived first on 50% of trials was proposed as being equivalent to the critical time difference occurring at the periphery. A total of four odor pairs were studied. Previous studies showed that only one odor in each pair was suppressed at the specific concentrations used.

Results and Discussion

Table 1 shows the time differences found with each of the odor pairs, and demonstrates that odorants in mixtures are perceived in series. The magnitude of the time differences are of the same order as reported earlier [8]. Further, the "slow" odorant of each pair was the suppressed odorant, indicating that "fast" odorants are first to stimulate receptor cells or first to block reception of the slow odorant via antagonistic mechanisms. Whatever the mechanism, fast odorants appear to have a distinct advantage over slow odorants, and perception of these is affected less than that of their slower counterparts. Temporal filtering, therefore, provides yet another mechanism that results in loss of information about mixture constituents.

Studies underway are also investigating the role of the two transduction pathways identified for olfaction in mixture perception [10]. Indirect evidence [1] suggests that an odorant operating via the adenylate cyclase pathway (carvone) will dominate in mixtures

with odorants that stimulate cells via the inositol phosphate pathway (propionic acid). This proposal requires that both transduction pathways operate within a single cell, an occurrence that was recently described [11].

In summary, our current studies have provided evidence that odorants in mixtures are subject to suppression and fusion, both of which result in loss of information about the identity of the constituents. Two mechanisms, namely spatial and temporal filtering, have been proposed to account for the effects seen. In addition, evidence for the temporal coding of mixtures was presented.

References

1. Bell GA, Laing DG, Panhuber H (1987) Odour mixture suppression: evidence for a peripheral mechanism in human and rat. Brain Res 426:8–18
2. Laing DG, Willcox ME (1987) An investigation of the mechanisms of odour suppression using physical and dichorhinic mixtures. Behav Brain Res 26:79–87
3. Derby CD, Ache BW, Kennel EW (1985) Mixture suppression in olfaction: electrophysiological evaluation of the contribution of peripheral and central neural components. Chem Senses 10:301–316
4. Daniel PC, Derby CD (1991) Mixture suppression in behaviour: the antennular flick response in the spiny lobster towards binary odorant mixtures. Physiol Behav 49:591–601
5. Laing DG, Francis GW (1989) The capacity of humans to identify odours in mixtures. Physiol Behav 46:809–814
6. Laing DG, Eddy A, Francis GW, Stephens L (1992) The role of temporal coding in the perception of odour mixtures. In: 14th Annual Meeting of the Association for Chemoreception Sciences, Sarasota, FL, abstract 116
7. Stewart WB, Kauer JS, Shepherd GM (1979) Functional organisation of rat olfactory bulb analysed by the 2-deoxyglucose technique. J Comp Neurol 185:715–734
8. Getchell TV, Margolis FL, Getchell ML (1984) Perireceptor and receptor events in vertebrate olfaction. Prog Neurobiol (Oxf) 23:317–345
9. Laing DG (1987) Coding of chemosensory stimulus mixtures. Ann NY Acad Sci 510:61–66
10. Ronnett GV, Cho H, Hester LD, Wood SF, Snyder SH (1993) Odorants differentially enhance phosphoinositide turnover and adenylyl cyclase in olfactory receptor neuronal cultures. J Neurosci 13:1751–1758
11. Ache BW, Hatt H, Breer H, Boekhoff I, Zufall F (1993) Biochemical and physiological evidence for dual transduction pathways in lobster olfactory receptor neurons. In: 15th Annual Meeting of the Association for Chemoreception Sciences, Sarasota, FL, abstract 48

TABLE 1. Stimulation time differences.

Odor pair	Time difference (ms)
Carvone—Limonene[a]	92
Carvone—Propionic Acid[a]	159.4
Limonene—Propionic Acid[a]	311.9
Carvone—Benzaldehyde[a]	579.6

[a] Required a time advantage and was the suppressed odorant.

Nasal Trigeminal Chemoreceptors May Have Affector and Effector Functions

WAYNE L. SILVER[1], THOMAS E. FINGER[2], and BÄRBEL BÖTTGER[2]

Key words. Ethmoid nerve, Neuropeptide

The nasal cavity embraces a wide variety of chemoreceptors, including those of the olfactory, vomeronasal, and trigeminal systems. Chemical stimulation of nasal trigeminal receptors is usually associated with pain or irritation. Accordingly, trigeminal fibers sensitive to chemical irritants have been considered part of the common chemical sense [1], although, more recently, this trigeminal "chemosense" has been called chemesthesis [2,3]. The nasal cavity is innervated by the ethmoid and nasopalatine branches of the trigeminal nerve. Ethmoid nerve endings in the nasal cavity respond to a wide variety of chemical stimuli [3]. Ethmoid nerve fibers which respond to chemical stimuli appear to be capsaicin-sensitive, containing calcitonin gene-related peptide (CGRP) and substance P [4]. For the present report, experiments were performed on rats injected with capsaicin (50 mg/kg) on the 2nd day postnatally and examined at least 40 days later. Systemic capsaicin treatment in neonatal animals produces a long-lasting insensitivity to noxious chemicals (for review, see [5]). In our experiments, neonatal capsaicin administration eliminated ethmoid nerve responses to chemical stimuli, while leaving responses to mechanical stimuli intact. In addition, neonatal capsaicin administration decreased the number of CGRP-containing fibers in the ethmoid nerve by 90% or more. These were primarily unmyelinated, or fine myelinated fibers. Such immunoreactive fibers reach to within 1 μm of the surface of the nasal epithelium [6]. Capsaicin-sensitive, CGRP-immunoreactive fibers course alongside olfactory receptor axons within the olfactory nerve to terminate within glomeruli of the olfactory bulb. Removal of the trigeminal ganglion eliminated such CGRP-immunoreactive fibers from the olfactory nerve and bulb. Double-label experiments, in which different fluorescent retrograde tracers were injected into the olfactory bulb and nasal epithelium, revealed that single trigeminal ganglion cells have a sensory arborization with branches in both the nasal epithelium and olfactory bulb. We suggest that trigeminal nerve fibers in the nasal cavity have both an affector and effector function. Capsaicin-sensitive, peptidergic sensory neurons in other systems are known to release, via an axon reflex, neuropeptide mediators which have an effector function on surrounding tissues [7]. Accordingly, nasal trigeminal sensory fibers may not only transmit sensory information centripetally, but may act via an axon reflex to release neuroactive peptides into the nasal mucosa and olfactory bulbar neuropil, thereby affecting processing of simultaneous or subsequent olfactory stimuli. Thus, trigeminal chemoreception should not be viewed as a simple sensory system that can be studied in isolation, but as a potential modulator of olfactory processes.

References

1. Moncrieff RW (1951) The chemical senses. Leonard Hill, London, p 760
2. Green BG, Mason JR, Kare MR (1990) Preface. In: Green BG, Mason JR, Kare MR (eds) Chemical senses, vol 2, Irritation. Dekker, New York, pp v–vii
3. Silver WL, Finger TE (1991) The trigeminal system. In: Getchell TV, Doty RL, Bartoshuk LM, Snow JB Jr (eds) Smell and taste in health and disease. Raven, New York, pp 97–108
4. Silver WL, Farley LG, Finger TE (1991) The effects of neonatal capsaicin administration on trigeminal nerve chemoreceptors in the rat nasal cavity. Brain Res 561: 212–216
5. Buck SH, Burks TF (1986) The neuropharmacology of capsaicin: Review of some recent observations. Pharm Rev 38:179–226
6. Finger TE, St Joer V, Kinnamon JC, Silver WL (1990) Ultrastructure of substance P- and CGRP-immunoreactive nerve fibers in the nasal epithelium of rodents. J Comp Neurol 294:293–305
7. Maggi CA, Meli A (1988) The sensory-efferent function of capsaicin-sensitive neurons. Gen Pharmacol 19:1–43

[1] Department of Biology, Wake Forest University, Winston-Salem, NC 27109, USA
[2] Department of Cellular and Structural Biology, University of Colorado Health Science Center, Denver, CO 80262, USA

Effects of Fragrance of Whiskey: An Electrophysiological Study

YOSHIKAZU SHUTARA[1], YOSHIHIKO KOGA[1], KENICHI FUJITA[1], MASAHIKO MOCHIDA[1], KENICHI TAKEMASA[1], KEN NAGATA[2], IWAO KANNO[3], and HIDEAKI FUJITA[3]

Key words. Positron emission tomography (PET)—Event-related potentials (ERPs)

We have already shown [1] that the alpha wave power at a frequency of 10 Hz was greater for the fragrance of whiskey than for the fragrances of distilled water, ethanol, and nonmature whiskey. Therefore, it was assumed that the fragrance of mature whiskey can bring about a moderate resting state, i.e., a relaxed condition without drowsiness, in those who like it. We have also demonstrated that the fragrance of whiskey can influence the rate of information processing in the brain. In the present study, we measured the regional cerebral blood flow (rCBF) of the limbic system, including the olfactory and the auditory systems, in ten subjects while they performed an auditory oddball task in the presence of an ambient fragrance. Event-related potentials (ERPs) were also recorded during the same task.

The subjects were ten right-handed normal male adults with a mean age of 29.3 years. For each subject, the r-CBF was measured by positron emission tomography (PET) while they performed an auditory oddball task. During the task, a glass with a fragrant sample was set below his nose. The subject was instructed to count the target stimuli while he breathed in the fragrance. The stimuli used for the oddball task consisted of two tone bursts of different frequencies (1000 Hz and 2000 Hz). The 2000 Hz tone was the target, with a probability of 0.2. Three different fragrances whiskey, ethanol, and distilled water were presented during the task.

Regional CBF was measured by $H_2^{15}O$ bolus; rCBF values were obtained from 14 regions. ERPs were elicited by the oddball task from 16 scalp sites referred to the linked ears.

The rCBF value was highest for the fragrance of whiskey in every region of interest (ROI) in the

right hemisphere. In particular, in the right nucleus accumbens septi, the rCBF value for the fragrance of whiskey was significantly higher than that for the fragrance of ethanol. It was also highest in the pyriform cortex, orbitofrontal cortex, nucleus accumbens septi, primary auditory cortex, and angular gyrus of the left hemisphere. Each subject assigned the three fragrances in order of preference: the favorite, second most favorite, and least favorite. Eight of the ten subjects designated the fragrance of whiskey as the favorite one. The rCBF value for the favorite fragrance was highest in every ROI in the right hemisphere. The rCBF value for the favorite fragrance was significantly higher than those of the others in the right nucleus accumbens septi. There was no significant difference in the P300 peak latency at Pz among distilled water, whiskey, and ethanol. With respect to the peak amplitude of the P300, it was largest for the fragrance of whiskey; the P300 peak amplitude at Cz for the fragrance of whiskey was significantly higher than those for the other two fragrances. Moreover, the peak amplitude of the P300 for the favorite fragrance was significantly higher than that for the second most favorite fragrance at Cz and P4. No high correlations were obtained between rCBF value in each ROI and P300 latency and peak amplitude at four sites.

The results of PET showed that the fragrance of whiskey increased the rCBF in the majority of the ROI. The regions in which blood flow increased belong to the limbic system, including the olfactory system, suggesting that the fragrance of whiskey and the favorite fragrance enhanced the function of the limbic system through activating the olfactory system. As the regions which showed marked increase of rCBF are located in the right hemisphere, which is thought to be closely related to emotion, it appears that the fragrance of whiskey may predominantly influence the emotional function of the limbic system. The result that the subjects' preference for fragrance influenced the rCBF of the right hemispheric regions supports this hypothesis. The peak amplitude of the P300 reflects the amount of allocated information processing resources. Thus, our ERP findings show that the fragrance of whiskey and favorable fragrances activate the function of auditory information processing in the brain. How-

[1] Department of Neuropsychiatry, Kyorin University School of Medicine, 6-20-2 Shinkawa, Mitaka, Tokyo, 181 Japan
[2] Department of Neurology, Research Institute for Brain and Blood Vessels, Akita, 010 Japan
[3] Department of Radiology and Nuclear Medicine, Research Institute for Brain and Blood Vessels, Akita, 010 Japan

ever, there was no relationship between rCBF and
ERPs. Thus, the enhancement of the limbic system,
shown by the increase in rCBF, might not directly
affect the stage of processing of auditory input reflected by P300. However, the other stages may
relate more closely to the olfactory input, as indicated by the findings that the rCBF values in the
transverse temporal gyri of both hemispheres were
increased by the fragrance of whiskey and the favorable fragrances.

Reference

1. Shutara Y, Fujita K, Takeuchi H, Koga Y, Takemasa
   K (1992) The electrophysiological study on the effect of
   fragrance of whisky. Proceedings of the 26th Japanese
   Symposium on Taste and Smell, pp 349–352

A Study of Fragrance Preferences

Tamaki Mitsuno and Kazuo Ueda[1]

Key words. Preference—Fragrance

Preferences for fragrances are affected by many factors, e.g., sex, age, and environmental scents. Here we describe the ranking of preference for nine fragrances, iso-amyl acetate, cinnamic aldehyde, orange oil (from Florida), L-carvone, vanillin, milk, β-phenyl-ethyl alcohol, 5α-androst-16-en-3α-ol, and galaxolide; all were tested in 1%–10% dipropylene glycol solutions. We thank Takasago Int. Corp. for supplies of fragrances. Our subjects were young Japanese women assigned to four experimental groups: junior high school students, 14–15 years old ($n = 11$); senior high school students, 15–18 years old ($n = 35$); undergraduates, 18–19 years old ($n = 30$); and graduates, 23–25 years old ($n = 18$). One third of the latter two groups were perfume users; majority of the former two were not.

Fragrances to be tested were placed into bottles (10 ml volume and 10-mm caliber). Subjects opened the bottles in random order and inhaled the fragrances for ranking their preferred fragrances (one session). We examined the results of four sessions carried out on different days and at different times. Ranking correlation coefficients of Spearman ($\alpha \leq 0.05$) were calculated between four sessions. Correlations among eight parameters (the best and worst rankings of fragrances, age frequency in use of perfume/deodorant, habit of drinking/smoking, and number of perfume users in their family) were examined.

The preferred ranking of fragrances for junior and senior high school students varied from session to session, showing that their preference for fragrances had not yet been established. Of the perfume nonusers, both undergraduates and graduates liked orange or vanillin best, while habitual users of perfume liked vanillin or 5α-androst-16-en-3α-ol (the undergraduates' preference) and orange or galaxolide (the graduates' preference) best. The coefficient of variation in the preferred ranking of fragrances decreased with age. Two major factors affecting the preference for fragrances were suggested to be ages of subjects and frequencies in use of perfumes.

[1] Biological Laboratory, Faculty of Home Economics, Kyoritsu Women's University, 1-710 Motohachioji, Hachioji, Tokyo, 193 Japan

A Study of Perfume Preferences

TAMAKI MITSUNO and KAZUO UEDA[1]

Key words. Preference—Perfume

Preferences for perfumes are affected by many factors, e.g., sex, age, and environmental scents. Here we describe the ranking of preference for nine perfumes (imitations of Portugal, Lauren, L'Air du Temps, Chanel No. 5, Poison, Obsession, Mitsouko, Cabochard, and Alliage; all 20% alcohol solutions). We thank Takasago Int. Corp. for supplies of perfumes. Our subjects were young Japanese women assigned to four experimental groups: junior high school students, 14–15 years old ($n = 11$); senior high school students, 15–18 years old ($n = 35$); undergraduates, 18–19 years old ($n = 30$); and graduates, 23–25 years old ($n = 18$). One-third of the latter two groups were perfume users; majority of the former two were not.

The subjects were asked to smell the top note and the middle/last note of an individual perfume (one session). We examined the results of four sessions carried out on different days and at different times, changing the order of presentation of the perfumes. Ranking correlation coefficients of Spearman ($\alpha \leq 0.05$) were calculated between four sessions. Correlations among ten parameters (the best and worst rankings of perfumes in two respective notes, age, frequency in use of perfume/deodorant, habit of drinking/smoking, and number of perfume users in their family) were examined.

The preferred ranking of perfumes for junior and senior high school students varied from session to session, showing that their preference for perfumes had not yet been established. Of the habitual users of perfume, both undergraduates and graduates liked L'Air du Temps and Poison best in either notes, while perfume nonusers of them and high school students, whether users or nonusers, liked Portugal best. The same perfumes that had been ranked first (and last) in the top note were chosen as best (and worst) in the middle/last note ($\alpha \leq 0.01$). The coefficient of variation in the preferred ranking of perfumes decreased with age. Two major factors affecting the preference for perfumes were suggested to be ages of subjects and frequencies in use of perfumes.

[1] Biological Laboratory, Faculty of Home Economics, Kyoritsu Women's University, 1-710 Motohachioji, Hachioji, Tokyo, 193 Japan

Factor Structure in the Imagery of Fragrance: Study of Differences Between the Sexes

M. Kodama[1], M. Fujiwara[1], K. Kaneko[2], and T. Asakoshi[2]

Key words. Imagery—Fragrance

Fragrances influence humans both physiologically and psychologically, and they have also been used to reduce stress responses, as in aromatherapy. The physiological response to fragrance has been determined bioelectronically in many studies [1,2], but there have been few studies of the psychological evaluation of fragrance by imagery. Takeuchi et al. [3], in their study of imagery and fragrance, noted that five factors: freshness, nature, passion, activity, and femininity were associated with the images of 19 kinds of fragrances. They also found that the scores for each factor differed in the 19 kinds of fragrances tested and that there were sex differences in the factor scores.

In this study, we examined sex differences in the psychological evaluation of imagery for six fragrant oils: lemon, sandalwood, lavender, rose, peppermint, and eucalyptus.

The imagery for each fragrant oil was based on 20 semantic differentials: dynamic-static, hard-soft, familiar-unfamiliar, plain-showy, old-new, warm-cool, passionate-intellectual, light-heavy, coarse-polished, sweet-bitter, calm-violent, complicated-simple, rustic-sophisticated, weak-strong, mature-immature, shallow-deep, vulgar-elegant, disgusting-nice, passive-active, and old-young. Each fragrant oil sample was presented to subjects on a separate piece of paper (0.5 × 15 cm). The subjects were 66 male undergraduate students and 47 female undergraduates; they were asked to score each pair of adjectives on a 7-point intensity scale to assess their association of imagery with each of the oils.

The imagery scores were factor-analyzed, using the principle factor method with varimax rotation; interpretable factors emerged in different order for each fragrant oil as follows: (1) lemon—popularity, activity, maturity, coolness; (2) sandalwood—activity, popularity, maturity, newness, coolness; (3) lavender—popularity, activity, newness, immaturity, coolness; (4) rose—popularity, activity, warmth, newness, maturity; (5) peppermint—popularity, activity, unpopularity, maturity; (6) eucalyptus—popularity, activity, warmth, maturity.

For all fragrant oils, the popularity factor (e.g., elegant, sophisticated, familiar) and activity factor (e.g., violent, strong, active) were denoted factor 1 and factor 2, respectively. The popularity factor included the dignity factor and familiarity factor. Lemon and sandalwood oils had similar factor structures (coolness) while rose and eucalyptus oils had the opposite factor, warmth. The latter two oils are thought to have physiologically sedative effects; the image of warmth might influence these effects. The other two oils, lavender and peppermint also had implication of coolness and were similar to the former two oils. But lavender had immaturity factor and peppermint had unpopularity, exceptionally.

One-factor ANOVA showed that females had higher scores than males for factor 1 (popularity) of rose oil. It was suggested that females were familiar with the rose fragrance because of their use of perfumes. However, there were no significant sex-related differences for the other factors of the fragrant oils. The results were not consistent with those of Takeuchi et al. [3], who found sex-related differences in factor scores. However, compared with that study, our population samples were larger and there were fewer kinds of fragrances, that is, our study was more reliable. Hama and Nishimura [4] reported that of five fragrance oils, only for mint were significant differences found between the sexes. We concluded that, in young people, there were no differences between the sexes regarding the imagery of fragrant oils. It has generally been held that women, especially young women, are more sensitive to the perception and discrimination of fragrances than men. The singular physiological factor, the menstrual cycle [5], and differences in cultural roles were thought to explain this. However, it would appear that, since images of fragrances are formed according to experience and memories of the past, a rich foundation of common experience, rather than cultural differences between the sexes, led to similar images in our male and female subjects.

References

1. Kubota M, Ikemoto T, Yamashita Y, Koori T (1989) Effective components in lavender oil and jasmin absolute on contingent negative variation (in Japanese).

[1] Department of Human Sciences, Waseda University, 2-579-15 Mikajima, Tokorozawa, Saitama, 359 Japan
[2] Hasegawa Corporation, 4-4-14 Honcho, Nihonbashi, Chuo-ku, Tokyo, 103 Japan

Proceedings of the 23rd Japanese Symposium on Taste and Smell, pp 317–320

2. Yoshida T, Kanamura S (1989) The effect of odors on frequency fluctuation of brain waves—the 2nd report (in Japanese). Proceedings of the 23rd Japanese Symposium on Taste and Smell, pp 293–296

3. Takeuchi S, Miyazaki Y, Yatagai M, Ide M, Kobayashi S (1990) Organic functional (sensuous) evaluation of essential oils (in Japanese). Proceedings of the 24th Japanese Symposium on Taste and Smell, pp 191–194

4. Hama H, Nishimura N (1992) Difference of preference for fragrance among generations: by perfume (in Japanese). Proc Jpn Assoc Health Psychol 5:42–43

5. Shiihara Y, Asakoshi R, Kaneko K, Kodama M (1992) Influence of menstrual cycle on image of fragrance (in Japanese). Proc Jpn Assoc Health Psychol 5:44–45

Factor Structure in the Imagery of Fragrance: Study of Generational Differences

M. Fujiwara[1], M. Kodama[1], K. Kaneko[2], and R. Asakoshi[2]

Key words. Imagery—Fragrance

The psychological and physiological effects of fragrances on humans have been addressed in a number of studies [1–3]. However, there have been few reports about the evaluation of fragrance by imagery [1]. In this study, we examined generational differences in the evaluation of imagery for six fragrant oils: lemon, sandalwood, lavender, rose, peppermint, and eucalyptus.

Each fragrant oil was presented to subjects on a separate piece of paper (0.5×15 cm). Imagery was evaluated using a scale of 20 semantic differentials: dynamic-static, hard-soft, familiar-unfamiliar, plain-showy, old-new, warm-cool, passionate-intellectual, light-heavy, coarse-polished, sweet-bitter, calm-violent, complicated-simple, rustic-sophisticated, weak-strong, mature-immature, shallow-deep, vulgar-elegant, disgusting-nice, passive-active, and old-young. Our subjects, 47 young females (undergraduates, aged 18–21), 37 middle-aged females (40–59 years), and 61 elderly females (60–89 years), scored each pair of adjectives on a 7-point intensity scale to assess their association of imagery with each fragrant oil.

The imagery scores were factor-analyzed, using the principle factor method with varimax rotation. The results of one-factor ANOVA showed definite effects in factors of all fragrant oils. Furthermore, as one of the methods of multiple comparison, Tukey's test was used.

Five factors emerged in different order for each oil as follows: (1) lemon—popularity, activity, dignity, maturity, coolness; (2) sandalwood—popularity, activity, dignity, coolness, maturity; (3) lavender—popularity, softness, newness, maturity, activity; (4) rose—popularity, softness, newness, immaturity, coolness; (5) peppermint—popularity, inactivity, coolness, maturity, newness; (6) eucalyptus—popularity, activity, newness, coolness, immaturity.

For lemon oil, the young subjects scored higher than the middle-aged and elderly for both factor 1 (popularity) and factor 3 (dignity). Conversely, for factor 4 (maturity), young females scored lower than middle-aged females. For sandalwood oil, young females scored lower than middle-aged females on factor 2 (activity). For lavender oil, young and middle-aged females scored higher than the elderly on factor 2 (softness), and middle-aged females scored higher than the young and elderly on factor 3 (newness). Young females scored higher than the middle-aged and elderly on factor 5 (activity). For rose oil, young and middle-aged females scored higher than the elderly on factor 2 (softness). For peppermint oil, the young and middle-aged females scored higher than the elderly on factor 1 (popularity). Conversely, on factor 4 (maturity), young and middle-aged females scored lower than the elderly, and the middle-aged females scored higher than the elderly on factor 2 (inactivity). On factor 3 (coolness) young females scored higher than the middle-aged and elderly females. For eucalyptus oil, the middle-aged females scored higher than the elderly on factor 1 (popularity) but lower on factor 2 (activity). On factor 4 (coolness), young females scored higher than the middle-aged and elderly. Young and middle-aged females scored higher than the elderly on factor 5 (immaturity).

The results showed that the older the subjects, the lower were the scores. One exception was noted: on the factor of maturity (for both lemon and peppermint oil) the older the subjects, the higher were their scores. For both lemon and peppermint oil, on the factor of popularity, younger females scored higher than older females, these results suggesting that for these oils, maturity and popularity had opposite images for the generations.

Kamiyama et al. [4] reported that middle-aged females evaluated the fragrances of wood, grass, and moss as calm, more than other generations. This finding was similar to the results of the present study for eucalyptus oil, in that middle-aged females scored higher than the elderly on the factor of popularity and lower on the factor of activity. However, for sandalwood oil, middle-aged females scored higher than the elderly on the factor of activity. This was inconsistent with the results of Kamiyama et al. [4], who also reported that young females evaluated floral fragrances as calm. In our study, similarly, the young females had higher scores on the factor of

[1] Department of Human Sciences, Waseda University 2-579-15 Mikajima, Tokorozawa, Saitama, 359 Japan
[2] Hasegawa Corporation, 4-4-14 Honcho, Nihonbashi, Chuo-ku, Tokyo, 103 Japan

softness for rose oil. However, for lavender oil, young females had higher scores on both softness and activity and this result was inconsistent with the results of Kamiyama et al. [4].

We concluded from this study that there were generational differences in females with regard to the imagery of fragrance. However, since the results of this study are somewhat contradictory to results in other studies, we suggest that further data are necessary to establish conclusive results.

References

1. Nagai N, Nakagawa M, Miyake S, Asakura Y (1989) Effects of odorants on physiological and subjective responses during the Kraepelin work load (I). Proceedings of the 23rd Japanese Symposium on Taste and Smell, pp 55–58
2. Nakagawa M, Nagai H, Miyake S, Asakura Y (1989) Effects of odorants on physiological and subjective responses during the Kraepelin work load (II). Proceedings of the 23rd Japanese Symposium on Taste and Smell, pp 305–308
3. Yoshida T, Kanamura S (1989) The effect of odors on frequency fluctuation of brain waves—the 2nd report. Proceedings of the 23rd Japanese Symposium on Taste and Smell, pp 293–296
4. Kamiyama K, Suzuki F (1986) Aromatherapy and phytoncide (in Japanese). Jpn J Fragrance 77:26–31

Effects of Fragrances on Work Efficiency

Y. Sakuta[1], M. Kodama[1], M. Fujiwara[1], K. Kaneko[2], and T. Asakosi[2]

Key words. Fragrance—Monotonous work

Aromatherapy, which has been used for a long time, has shown that fragrances have not only psychological but also physiological influence in humans. The effects of fragrances on recovery from fatigue brought about by a mental workload have been reported [1–3]; however, to our knowledge, there are few studies of the effects of fragrances on psychological stress induced by monotonous work. We carried out this study to examine such effects.

The subjects were eight healthy undergraduate female students who made less than 50% errors in an olfactory test. For this test, the subjects had to select the designated fragrance when presented with three different fragrances. We chose female subjects because females have a keener sense of smell than males. Two fragrant oils, lemon oil (Lemon) and sandalwood oil (Sandalwood), and one blank oil for control (Blank) were used. The subjects filled out the questionnaire for the Maudsley Personality Inventory (MPI), and were assigned a monotonous task on a pegboard. One trial of the task was counted when the subject finished moving all pegs (48 pieces) to a symmetrical part of the board. The work was repeated in three sessions, one session per day, each session being limited to 20 min. Within the limit, the task was continued over and over until the subjects reached the limit of their patience. In each of the three sessions, a different fragrance, Lemon, Sandalwood, or Blank, was sprayed for 1 s per min onto the subjects from behind. The order of presenting the fragrances was randomized. While the subjects were doing the task, two measures of autonomic nervous system activity were carried out: skin potential level (SPL) and skin temperature (ST).

The SPL and ST data were divided into four blocks. Each block consisted of two trials of the task, except for the first two trials: the first block had the third and fourth trials, the second block had the fifth and sixth trials, and so on.

For determining the changes of SPL in each fragrance condition, two factors, three fragrances and four blocks of changes in SPL were subjected to analysis of variance. The sedative effect of fragrances (Lemon > Sandalwood > Blank) was shown to be significant. However the changes of ST in each fragrance condition were not significant.

To determine changes in work speed in each fragrance condition, two factors—three fragrances and four blocks of work speed—were subjected to analysis of variance; there was no effect of fragrances on work speed.

In the last analysis, subjects were divided into two groups according to their MPI scores: a neurotic group and non-neurotic. To determine the mean SPL in each of these groups under three fragrance conditions, two factors—the two groups and three fragrances—were subjected to analysis of variance. The sedative effect of Lemon on both groups and the same effect of Sandalwood on the neurotic group were thus shown. In the same way, an analysis of variance for the mean ST in each group under three fragrance conditions was done. The ST in the neurotic group was higher than that in the non-neurotic group.

The experiment showed that fragrances had no effect on work speed (work efficiency). However, fragrances did have a sedative effect on SPL. Thus, it can be concluded that such a sedative effect would lead to the promotion of work efficiency. On the other hand, the experiment also showed that effects of fragrances differed according to degree of neuroticism Sandalwood having the strongest sedative effect on those who were neurotic. This result suggests that fragrances could be effective for the treatment of neurosis.

References

1. Nagai G, Nakagawa T, Miyake S, Asakura Y (1989) The effect of fragrances on workload in Kraepelin's test (part 1) (in Japanese). Proceedings of the 23rd Japanese symposium on taste and smell, pp 55–58
2. Nagai G, Nakagawa T, Miyake S, Asakura Y (1989) The effect of fragrances on workload in Kraepelin's test (part 2) (in Japanese). Proceedings of the 23rd Japanese symposium on taste and smell, pp 305–309
3. Sugano H (1989) Mental operation of fragrance (in Japanese). Proceedings of the 23rd Japanese symposium on taste and smell, pp 301–304

[1] Department of Human Sciences, Waseda University, 2-579-15 Mikajima, Tokorozawa, Saitama, 359 Japan
[2] Hasegawa Corporation, 4-4-14 Honcho, Nihonbashi, Chuo-ku, 103 Japan

Evaluation of the Effect of Plant-derived Odors Based on Fatigue Investigation

Kazunori Shimagami[1]

Key words. Plant odor—Fatigue investigation

Generally, the effects of odors on humans are evaluated by focusing on odor-exposure methods, adjusting the concentration of the odorant while stabilizing environmental conditions in an artificial weather chamber, and measuring various functions of autonomic nerves and brain waves in the subjects. However, many of these methods are often carried out under conditions that are very different from the actual living environment, and the experiments themselves may be a great burden to the subject. Accordingly, it is necessary to provide an evaluation method in which the mental and physical burden of the experimental investigation is small, and in which experimental results follow practical life more closely, with respect to a group living in an environment where behaviors are controlled to some extent. In line with these ideas, this study was conducted to examine the effects of several plant-derived odors on humans. The method used was based on the investigation of fatigue. The usefulness of this investigation method was evaluated.

Degree of fatigue was determined in two ways: by measuring flicker value (an indicator of objective fatigue) and by assessing subjective symptoms. For this assessment, a questionnaire, the "Subjective Symptoms Inquiry", was used. Three categories of fatigue were assessed: category I (sleepiness and weariness) category II (difficulty in concentrating), and category III (localized physical problems). The total complaint rate (as determined by the questionnaire) was calculated to evaluate the degree of fatigue. On the first day of the investigation, all subjects underwent Cornel Medical index (CMI) and fatigue measurement. Based on these results, they were divided into two groups, an odor-exposure

group and a control group. On and after the 2nd day, the daytime fatigue in the two groups was compared.

The subjects of this study were 355 female office workers and sales clerks (aged 18–38 years). They lived according to a given schedule for the duration of the study (5–7 days). All the subjects had the same study and mealtimes during the daytime. At the end of the day, until the next morning, each group was divided into subgroups of about five subjects each, to stay in their respective bedrooms. For the odor exposure group, the bedrooms were scented with an odor at night. No information on the odor was given to the subjects. The concentration of the odorant was adjusted to a level so that it could be slightly smelled. Fatigue during the day was measured in the morning (before the beginning of the morning study) and in the evening (immediately after the afternoon study).

Four investigations were conducted. Odors used for evaluation were alpha-pinene, white cedar leaf oil, bornyl acetate, jasmine oil, and eucalyptol.

Our results can be summarized thus:

1. In the alpha-pinene, bornyl acetate, and white cedar leaf oil groups, there was a relief from fatigue that was not observed in the eucalyptol and jasmine oil groups.
2. Time sequence changes in relief from fatigue in the alpha-pinene, bornyl acetate, and white cedar leaf oil groups showed a decrease in subjective symptoms compared to the control group at a relatively early stage after the beginning of odor exposure (on days 2–3) and, at the same time or a little later, there was a retention of flicker values, which were generally lower in the control group.

These results indicate that fatigue investigation is a suitable method for evaluating the effects of different odors on humans; the method is not complex and the experimental conditions are quite close to actual living conditions.

[1] Beauty Culture Laboratory, Kanebo Ltd., 3-1-7 Kamiosaki, Shinagawa-ku, Tokyo, 141 Japan

A Case Study of Olfactory Fatigue to an Odorant (Tertiary-Butyl-Mercaptan) in Town Gas in the Matsumoto City Area of Japan

Yasuyuki Hoshika[1]

Key words. Olfactory fatigue—Tertiary-butyl-mercaptan

It is important that town gas possess a smell, so that accidental carbon monoxide poisoning can be avoided as far as possible. Tetrahydrothiophene (THT) has been used for this purpose. In the United States, three types of odorant have been used: mercaptans, aliphatic sulfides, and THT, which is chemically stable, economical, and easily added to town gas. In the Matsumoto City area in Japan, tertiary-butyl-mercaptan is added as an odorant to the town gas; however, there is little practical information regarding the constituents and the olfactory fatigue of the odorant in the town gas.

In this experiment, the olfactory fatigue and recovery to an odorant were investigated to clarify safe levels of odorant in the town gas in the Matsumoto City area. The odorant was tertiary-butyl-mercaptan, at a density of around $6.2\,mg/m^3$ per detector tube (Gas tec No. 75L); the odor recognition threshold value in town gas in the Matsumoto City area was $0.00033\,mg/m^3$ (carbon monoxide, about 3.6% per detector tube). The subjects were ten (seven male and three female) medical students with normal olfactory sensitivity; they passed the T and T olfactometer test. Their ages ranged from 18 to 30 years. A dynamic flow test for olfactory fatigue was carried out by passing the gas through an air pump (flow rate $3\,l/min$) with a three-way Teflon cock (odor-free air line and another line) from a 100-l Tedlar bag (filled with diluted town gas, 1/1000). The subjects evaluated odor perception signals (presence, absence, intensity, and other). In the continuous flow of the diluted town gas, the odor was detected until the town gas odor was not detected. After rest intervals of 5, 10, and 15s with odor-free air (recovery stage), olfactory fatigue (in the continuous flow of the diluted town gas) was also determined. The average recovery time from olfactory fatigue was 8.3s, which was shorter than primary fatigue.

[1] Department of Hygiene, Shinshu University School of Medicine, 3-1-1 Asahi, Matsumoto, Nagano, 390 Japan

Changes in Detection Thresholds of Androstenol and Pentalide During the Menstrual Cycle

Saho Ayabe-Kanamura[1], Yasuhiro Takashima[2], and Sachiko Saito[3]

Key words. Olfactory sensitivity—Menstrual cycle

Fluctuations in women's sensory sensitivity, feelings, and behavior during the menstrual cycle have been suggested. Most studies of olfactory sensitivity have suggested increased sensitivity during the ovulatory phase [1,2,3]. However, several reports have noted a secondary peak during the luteal phase [1] and some have reported no changes during the menstrual cycle [3]. The physiological mechanisms for these responses related to each phase are not fully understood. Some authors have suggested that changes in olfactory sensitivity might be due to peripheral mechanisms that limit the access of odorant molecules to olfactory receptors [2], while others have indicated that these changes might be directly related to the central nervous system, which is affected by sexual hormones [3]. The possible involvement of human pheromones is very interesting. Androstenol and androstenone, which have musk-like odors, occur both in urine and as a product of the apocrine glands in humans and are secreted more by men than by women. Although these pheromones are implicated in the sexual behavior of the boar, their effects on humans are not well known. Some studies [3] have suggested that there is no relationship between the menstrual cycle and olfactory sensitivity for non-human odors, but that sensitivity does change for particular human odors, such as that of androstenone.

In this study, we investigated changes in the detection thresholds of pentalide and androstenol. Pentalide (oxacyclohexadecan-2-one) is a synthetic compound with a musk-like odor. Androstenol (5a-androst-16-en-3a-ol) was also synthesized. Both odorants were diluted with liquid paraffin. In the preliminary experiment, six Japanese women (mean age, 41 years), who were married and multiparous,

were tested in the morning on 3 days per week through three menstrual cycles. Each subject measured her basal body temperature before getting up every day. They were not naive to the olfactory experiment. Detection thresholds were measured by the triangle test method on an ascending series, using filter paper strips. The thresholds of all three subjects dropped until the end of the second menstrual cycle. During the period of the three menstrual cycles, the cycles of only half of the subjects were always stable. These three subjects, whose menstrual cycles were 24–28 days, participated in the main experiment, carried out over two menstrual cycles. The ages of these women were 39, 41, and 42 years. The procedure was the same as in the preliminary experiment. The phases of the menstrual cycle were identified according to the data for basal body temperature and by taking into consideration the lifespan of the corpus luteum, i.e., 2 weeks $\pm$ 2 days. The luteal phase lasted for 12 days before menstruation; the ovulatory phase for 5 days before the luteal phase; and the follicular phase for the remaining days, including menstruation.

We found that: (1) The threshold for Pentalide was more variable than that for androstenol throughout two menstrual cycles. (2) The threshold for pentalide was lower in the luteal than in the follicular phase in all three subjects. (3) The threshold for androstenol was lower in the ovulatory than in the follicular phase in all subjects and was significantly lower in the luteal than in the follicular phase in one subject. Although there were changes in the detection thresholds of pentalide and androstenol during the menstrual cycle in this experiment, patterns of fluctuation related to the menstrual cycle were not the same for the subjects or odorants. The results also indicated that there were cyclic changes in the thresholds of both human-related and non-human-related odors, suggesting that the likelihood of sexual hormones directly affecting olfactory sensitivity was low. Although it could be assumed that increased arousal in the luteal phase influenced the increased sensitivity in this phase, it will be necessary in future research to examine sensitivity to other chemicals and to measure the threshold during many cycles, before the factors underlying changes in olfactory sensitivity during the menstrual cycle can be established.

[1] Institute of Psychology, University of Tsukuba, Tennoudai 1-1-1 Tsukuba, Ibaraki, 305 Japan
[2] Takasago International Corporation, Nishiyawata 1-4-11, Hiratsuka, Kanagawa, 254 Japan
[3] National Institute of Bioscience and Human Technology, Higashi 1-1, Tsukuba, Ibaraki, 305 Japan

References

1. Doty RL, Snyder PJ, Huggins GR, Lowry LD (1981) Endocrine, cardiovascular, and psychological correlates of olfactory sensitivity changes during the human menstrual cycle. J Comp Physiol Psychol 95:45–60

2. Mair RG, Bouffard JA, Engen T, Morton TH (1978) Olfactory sensitivity during the menstrual cycle. Sensory Processes 2:90–98

3. Parlee MB (1983) Menstrual rhythms in sensory processes: A review of fluctuations in vision, olfaction, audition, taste, and touch. Psychol Bull 93:539–548

A Study of Aromatherapy for Pregnant Women

Izumi Saitoh[1]

Key words. Pregnant women—Aromatherapy

Introduction

From ancient times, human beings have enjoyed fragrances and made attempts to use them for therapy, against which background "aromatherapy" has been established. As a nurse and midwife, I had planned to use aromatherapy in nursing.

Purpose

In this study, aromatherapy was employed for pregnant women in the latter half of pregnancy, to allow them to spend this period in a more comfortable physical and mental condition; the effects were evaluated.

Method

Subjects

The subjects were seven pregnant women who visited the department of Obstetrics of Ehime Prefectural Central Hospital; all had an uneventful course of pregnancy and were supervised from the mid-term of pregnancy, by students majoring in a nurse midwifery course at Ehime College of Medical Technology.

Materials

Natural lemon essential oil and natural lavender essential oil were used (provided by Takasago International Corporation, Japan).

Concentration

Both essential oils were diluted with sterile distilled water to make a 10% solution.

Method of Application

One ml of the 10% solution of lemon or 1 ml of the lavender solution was injected into a cotton ball (standard No. 25) with a syringe. The cotton ball was placed at the center of a gauze folded into four, and tied with a ribbon to make a round scent bag.

Experimental Method

Students visited the homes of the subjects during the latter half of pregnancy, learned about the life-styles of the women in each region, performed physical examinations, and gave appropriate health instructions, which included private instruction on a respiratory method to aid the course of effective delivery. Aromatherapy was explained to the subjects when these instructions were given, and the patients' informed consent was obtained.

Blood pressure, heart rate per min, respiratory rate per min, fetal heart rate per min, and body temperature were determined at rest, and then immediately after application of the respiratory method without aromatics, immediately after the respiratory method was employed while the subject smelled the lavender scent bag for 1 min, and immediately after application of the respiratory method while the subject smelled the lemon scent bag for 1 min. The respiratory exercises were performed at appropriate intervals.

In addition, the subjects were interviewed during home visits 2–4 weeks after delivery, and asked for their opinion on the delivery, as well as their impressions of the aromatherapy and its effects, to determine the psychological and mental effects of aromatherapy.

Results and Discussion

Physical Information

Blood pressure, respiratory rate, heart rate, body temperature, and fetal heart rate did not change sufficiently before or after inhalation of aromatics to affect the subjects or fetuses. That is, the physical effects peculiar to lemon and lavender described in the literature were not observed in this study.

In other words, the use of the aromatics did not affect the physical condition of the pregnant women

[1] Master's Course of Medical Science, Tsukuba University, Takezono 1-6-2, 905-203, Tsukuba, Ibaraki, 305 Japan

in terms of affecting circulation and respiration or fetal circulation.

Investigation via Interviews After Delivery

When asked their impressions of aromatherapy, none of the subjects had a negative reaction to the sedative and relaxing effects of the therapy. Only one subject said she had no positive feelings, including that of feeling, refreshed, which feelings are useful for the respiratory method, good for diverting the mind, and for feeling comfortable. None of the subjects stated that the therapy had caused discomfort.

All seven subjects a stated that the therapy had a positive psychological effect, and approximately half of them felt that it had a positive physical effect. Six of the seven subjects stated that the therapy had a positive effect on pregnancy and delivery, and all subjects agreed that this was an interesting trial.

Regarding the use of the aromatherapy during delivery, in the early first stage of delivery, some wished to try the aroma in the delivery room and others did not wish to do so.

Most subjects were positive about the aromatherapy experience, more positive than we had expected, i.e., they stated a desire to try other aromatics and methods under other conditions. None of the items in the questionnaire were analyzed in depth in this study. However, as no physical changes were observed and the therapy had a favorable psychological effect, we believe that further study is required, with more stringent experimental conditions and a larger study population.

It can be concluded that aromatherapy has no effect on the physical condition of pregnant women, but has a marked psychological effect.

References

1. Thisland R (1991) Aromatherapy. Takayama R (translated), Fragrance Journal
2. Takayama R (1992) Introduction to aromatherapy. Aromatopia 1:36–39
3. Torii S (1991) Effect of aroma on brain wave. Nioi-no-Kagaku, pp 201–207
4. Davis P (1991) Dictionary of aromatherapy. Takayama R (translated), pp 325–327
5. Davis P (1991) Dictionary of aromatherapy. Takayama R (translated), pp 315–319
6. Ojima N, Takemura K, Amamiya Y (1991) Foundation and application of new Lamaze method, pp 140–141
7. Takeuchi H (1992) Application of aromatherapy to mental disorders. Aromatopia 1:9–15

Looking-Time in the Presence of Congruent and Incongruent Odors

SUSAN C. KNASKO[1]

Key words. Ambient odors—Approach behavior

Environmental stimuli are often used to influence an individual's emotional responses of pleasure and arousal. In turn, these responses are said to mediate a variety of approach-avoidance behaviors [1]. Approach behavior includes movement toward a stimulus, liking, and exploration; avoidance is the opposite. Theoretically, the more pleasant the environment, the greater the approach behavior.

In a previous study, people lingered longer in sections of a store when the areas were scented with pleasant odors compared to when they were not scented [2]. The objective of the present study was to explore how ambient pleasant odors influence approach behavior when they are congruent or incongruent with the setting. It was hypothesized that the presence of pleasant odors would lead to more approach behavior (i.e., increased looking-time and improved evaluations) when the odors were congruent with the slides shown compared to when they were incongruent.

Ninety subjects between the ages of 18–35 were randomly assigned to one of three odor conditions (15 women and 15 men in each group). The odor conditions included: no experimental odor, baby powder odor, and chocolate odor (IFF odors no. 3367-HS and no. 3372-HS respectively). The testing room was scented with perfumed blotters (four in the chocolate condition, eight in the baby powder condition), each containing approximately 0.12 g of the odorant.

In the first part of the study, subjects viewed the 24 slides at their own pace while, unknown to them, the computer timed how long they looked at each slide. Six slides were of babies, 6 were of chocolate items, and 12 were control slides (e.g., pine trees). In pilot testing, the chocolate slides were rated as being congruent with chocolate odor but incongruent with baby powder odor while the baby slides were rated the opposite. The control slides were rated as being incongruent with both odors. In the second part of the study, subjects viewed the slides again, this time evaluating the slides (e.g., on how interesting they found each one). Subjects then answered a questionnaire concerning their mood (pleasure and arousal) and a health questionnaire (checking whether or not they had any of six symptoms and rating symptom severity).

Subjects exposed to a pleasant room scent looked for a significantly longer time at the 24 slides compared to subjects exposed to no experimental odor ($F[1.88] = 5.9$, $P = 0.02$; odor $= 152 \pm 9$, no odor $= 120 \pm 9$); the congruency of the odors did not affect looking time. The odor condition did not influence the rating of how interesting the slides were or how easy it was to imagine oneself in the setting of the slides.

Subjects exposed to a pleasant odor were in a significantly more pleasant mood than subjects exposed to no odor ($F[1.88] = 4.6$, $P = 0.04$; odor $= 10 \pm 1$, no odor $= 6 \pm 2$). Arousal was affected by the type of odor present. Chocolate odor resulted in reports of greater arousal compared to no odor ($F[2.87] = 4.0$, $P = 0.02$; chocolate $= -0.07 \pm 1$, no odor $= -4.2 \pm 1$, baby powder $= -3.3 \pm 1$).

The number of health symptoms reported varied among the three odor conditions. Subjects in the baby powder condition reported fewer symptoms than subjects in the no odor condition ($F[2.87] = 3.0$, $P = 0.05$; baby powder $= 1.7 \pm 0.3$, no odor $= 2.7 \pm 0.4$, chocolate $= 2.3 \pm 0.3$).

Congruency did not play a role in the findings. Pleasant odors had certain effects compared to no odor (i.e., longer looking-time and better mood). Other effects were related to specific odors (arousal was highest in the chocolate condition and there were fewer health symptoms in the baby powder condition). This suggests that pleasant odors may have some general effect due to their hedonic value, while other effects may be due to particular odor associations.

Acknowledgments. The author would like to thank Connie Papaziekos for running the subjects, Dr. Michael Tordoff for developing the computer slide program, and International Flavors and Fragrances Inc. (Union Beach, N.J.) for supplying the odorants. This research was funded, in part, by a grant from the Olfactory Research Fund.

References

1. Mehrabian A, Russell JA (1974) An approach to environmental psychology. MIT Press, Cambridge, Mass.
2. Knasko SC (1989) Ambient odor and shopping behavior. Chem Senses 14:718, A94

[1] Monell Chemical Senses Center, 3500 Market Street, Philadelphia, PA 19104-3308, USA

Cross-Adaptation Effects Among Five Odorants of the T and T Olfactometer

Yutaka Kurioka, Shunshiro Ohnishi, and Yasunori Okabayashi[1]

Key words. Cross adaptation—T and T olfactometer

The T and T olfactometer (Daiichi Yakuhin Kogyo Co., Ltd., Tokyo, Japan) is composed of five odorants; beta-phenyl ethyl alcohol (A), methyl cyclopentenolone (B), isovaleric acid (C), gamma-undecalactone (D), and skatole (E). After a high concentration level of one of the odorants was sniffed by a subject for 1 min, the influence on the thresholds of the four other odorants (cross-adaptation effects) was checked (filter paper method). In a previous report, the cross-adaptation (CA) effects of odor A influenced by C and odor A by D were detected for three subjects and odor C by A for one subject [1].

In this study, the subject's nose was first stimulated by pulses of odorant X of duration $\tau 1$ and period T. The number of pulses (Nps) which the subject recognized was counted with the number of stimulations (Ns). The nose was then stimulated alternately with pulses of two odorants, X and Y, at durations $\tau 1$ and $\tau 2$, respectively. The number of pulses counted by the subject was designated Npa. The number for Npa would be smaller than Nps if the CA effect existed between two odorants. The decreasing rate (DR) = ([Nps-Npa]/Nps) × 100(%) expresses the magnitude of the CA effect.

The experiment was performed for 20 combinations of 5 odorants, whose concentrations were 128 times the threshold level, and the Nps and Npa were counted for Ns = 50. Durations $\tau 1$ and $\tau 2$ were 1 s each, and period T was 20 s. The subject was a 58-year-old male.

For 20 combinations, only the DR of odor A influenced by C was large enough (18%), and the

CA effect was detected. The others were small (less than 4%), and the CA effect was not detected. The result was somewhat different from that for the filter paper method.

Combinations of the three odorants A, C, and D, among which the CA effects were detected in the filter paper method, were examined with increased duration $\tau 2$. Here, the ratio of pulse durations, $R = \tau 2/\tau 1$, was used, which expresses the magnitude of the influence of odorant Y. In this experiment, R equals 1. For $\tau 1 = 1$ s and $\tau 2 = 2$ s (R = 2), the DRs of odor A influenced by C, A by D, and D by A were 20%, 11%, and 16%, respectively [2]. Others, C by A, C by D and D by C, were less than 6%. For $\tau 1 = 0.4$ s and $\tau 2 = 2$ s (DR = 5), the DRs of odor A influenced by C, C by A, A by D, and D by A all exceeded 10%; the others, C by D and D by C, were less than 8% [2].

We concluded that: (1) the CA effect becomes larger with increases of duration $\tau 2$. (2) By changing the ratio of pulse durations, $R = \tau 2/\tau 1$, the results obtained with the filter paper method were confirmed.

Acknowledgment. The authors express their thanks to Takasago International Co. Hiratsuka, Japan for financial support.

References

1. Ohnishi S, Kurioka Y (1991) Cross adaptation effect of human subjects to the five standard test odorants of T and T olfactometer (in Japanese). Jpn J Ergonomics 27-3:135–142
2. Ohnishi S, Kurioka Y, Okabayashi Y, Indo M, Kawasaki M, Takashima Y (1992) Detection of cross adaptation effect using alternative stimulation of two kinds of odor pulses (part 3: Influence of pulse duration). In: Arai S (ed) Proc of the 26th symposium on taste and smell. Institute of Biological Sciences University of Tsukaba, Ibaraki, Japan, pp 337–340

[1] Kinki University, Faculty of Science and Technology, Department of Electrical Engineering, 3-4-1 Kowakae, Higashi-Osaka, Osaka, 577 Japan

T&T Olfactometer for Standardized Olfactory Test and Its Uses

Katsunori Saiki[1], Osamu Fukazawa[1], Hideyo Asaka[2], and Sadayuki Takagi[3]

Key words. Olfactory test—Deodorizer

Today, most countries have standardized methods for testing vision and hearing. However, there are no set rules anywhere in the world for testing the olfactory sense, despite a growing need. To answer this need, an olfactory sense test has been developed, which uses a testing device called the T&T Olfactometer (Daiichi Yakuhin Sangyo, Tokyo, Japan). In 1975, a study group was set up with funding from the Ministry of Education and Cultural Affairs of Japan to formulate standards for testing the sense of smell. A further study group, under Professor Sadayuki Takagi of Gunma University, funded by the Ministry of Health and Welfare, was set up in the same year to research treatment for the impairment of the olfactory sense.

As a result of research spanning several years, the first standard odors for testing the sense of smell have been created. The feature of this olfactory test with the T&T Olfactometer is that the test simulates the action of sniffing in everyday life. Therefore, this olfactory test should be useful in testing olfactory sensitivity and in selecting persons with normal sense who can act as examiners. The T&T Olfactometer consists of five different types of odors, labelled A, B, C, D, and E, that have different chemical compositions. These are provided in eight different concentrations. The concentration "0" represents the detection threshold level for a person with a normal sense of smell.

Each concentration is ten times stronger than the one below it. The weakest level is −2, and the strongest is 5. The test should be carried out in an odorless room. The examiner dips a test paper into the standard odor liquid to a depth of about 1 cm and passes it to the subject. To smell the paper, the subject holds the tip of the paper about 1 cm away from the nose.

For each of the five different odors, the subject must state when the odor is first detected. That concentration is the detection threshold value for that particular odor. The subject then continues to smell the five odors at increasing concentrations until the type of smell can actually be recognized.

This is the subject's recognition threshold value for that particular odor. An olfactogram form is a plot of a subject's detection and recognition threshold values for each of the five test odors. An average is taken of the five detection threshold values, or of the five recognition threshold values, from the subject's olfactogram. Of the two, the average recognition threshold value should normally be used, as it more accurately represents the use of the sense of smell in everyday life.

From these results, a judgment can then be made on the degree of impairment of a subject's sense of smell. Olfactory tests done in clinics should be carried out without olfactory environmental contamination. Therefore, it is desirable to use a deodorizer such as the T&T Olfactometer. This olfactory test has how been adopted at 514 medical facilities in Japan.

[1] Daiichi Yakuhin Sangyo Co., Ltd., Japan, 2-14-4 Nihonbashi, Chuo-ku, Tokyo, 103 Japan
[2] Department of Otorhinolaryngology, Takatsu-Chuo Hospital, 211 Mizonokuchi, Takatsu-ku, Kawasaki, Kanagawa, 213 Japan
[3] Gunma University, 3-39-22 Showa-cho, Maebashi, 371 Japan

Offensive Odor Control Measures in Japan

Manami Fujikura[1]

Key words. Offensive odor control law—Olfactory sensory test

Annually, in Japan, the second greatest number of complaints about pollution, following those about noise arc complaints regarding offensive odors. In financial year 1991, 10 616 complaints of offensive odor were registered.

The emission of offensive odors resulting from business activities is regulated by the Offensive Odor Control Law enacted in 1971. Based on this law, the Environment Agency (EA) has designated "offensive odor substances" and their criteria. Unlike criteria for other pollutants, values are not designated as a single value, but as ranges of concentration.

The following are the offensive odor substances presently designated, with the lower limits of the ranges shown in parenthesis (ppm): ammonia (1), methyl mercaptan (0.002), hydrogen sulfide (0.02), dimethyl sulfide (0.01), dimethyl disulfide (0.009), trimethylamine (0.005), acetaldehyde (0.05), propionaldehyde (0.05), butyraldehyde (0.009), iso-butyraldehyde (0.02), valeraldehyde (0.009), isovaleraldehyde (0.003), isobutyl alcohol (0.9), ethyl acetate (3), methyl isobutyl ketone (1), toluene (10), styrene (0.4), xylene (1), propionic acid (0.03), butyric acid (0.001), valeric acid (0.0009), and isovaleric acid (0.001).

These substances are common causes of the offensive odors emitted from the major sources, such as livestock farms, fish meal processing plants, and paint manufacturers. The concentration ranges are derived from the relationship between the concentration and the odor intensity, this relationship being obtained by carrying out an odor-free-chamber test.

The emission of offensive odors in regulated areas is controlled. These areas are designated by the prefectural governor. At the time of a designation the governor must establish the regulation standards which are within the designated range. If emissions from industrial or business activity exceed the standards, the mayor of the municipality can recommend or order which measures are to be taken.

The EA has also been studying the possible use of a triangular odor bag method as an olfactory sensory test. In January 1993, the EA authorized an examination for the qualification of operators of the test to enhance its reliability. The authorized operators are now required to have the ability to distinguish five odorants in the T and T olfactometer (Daiichi Yakuhin Kogyo Co., Ltd., Tokyo, Japan).

[1] Special Pollution Division, Environment Agency, Government of Japan, 1-2-2 Kasumigaseki, Chiyoda-ku, Tokyo, 100 Japan

Typical Olfactory Properties
of the Aromas of Green, Oolong, and Black Teas

Tadakazu Takeo[1]

Key words. Tea aroma—Composition of tea aroma

Green, oolong, and black teas, the major commercial teas used throughout the world, can be made from the same tea leaves by changing the manufacturing process. However, such teas have typical aroma profiles that affect our olfactory sense.

These unique aroma profiles are related to the characteristics of the volatile compounds that constitute the aroma of the tea. Tea aroma reflects four components that express a green left-like smell (C-1) felt strongly with green tea, a light and fresh flowery smell (C-2) which is important in the black tea aroma, a heavy flowery smell (C-3), which is typical of the oolong tea aroma, and a roasted smell (C-4), particular to oolong tea. It has recently been shown that tea aroma compounds are the products of nonvolatile precursors that are accumulated in tea leaves by biochemical and chemical reactions during the tea manufacturing process [1,2].

The green-leaf aroma (C-1) consists of leaf aldehyde and leaf alcohol and its esters, which are oxidative-degradation products of linolenic acid liberated from leaf lipid. The bright flowery aroma (C-2) consists of terpene-alcohols liberated by tea leaf glucosidase from the glucosides in tea leaves. These glucosides have been identified in tea leaves. The heavy flowery aroma (C-3) is related to several volatiles, such as benzylalcohol,2-phenyl-ethanol, jasmine-lactone, and indole, which are also produced from nonvolatile compounds in tea leaves by biochemical reactions during the manufacturing process. The routes of synthesis for these compounds, except for benzylalcohol, which is made from its glucoside, have not yet been elucidated. The roasted aroma (C-

4) is produced from aminoacids and sugars in the tea leaf by heat reaction.

These typical aroma components are found in different concentrations in green, oolong, and black tea, respectively. The specific aroma profile of each tea accepted by our olfactory sense is related to the characteristic composition of these four aroma components.

The aroma of green tea contains the three components C-1, 2, and 3, in about the same proportions. The intensity of the green tea aroma is 100 times lower than that of oolong and black tea. The aroma of oolong tea consists of four components, the specific property of the oolong tea aroma being that the ratio of the C-3 component is comparatively higher than that of the C-1 and C-2 fractions and that there is a considerable content of the C-4 component.

The aroma pattern of black tea consists of three components and is characterized by high ratios of the C-1 and C-2 components and a fairly low ratio of the C-3.

These different ratios of the four aroma components are related to the characteristic aroma profile of the three kinds of tea.

It appears that the specific aroma profile of each tea accepted by our olfactory sense is based on the characteristic composition of the four aroma components.

References

1. Saijo R, Takeo T (1975) Increases of cis hexen 1 ol content in tea leaves following mechanicae injury. Photochemistry 14:181–182
2. Takeo T (1981) Production of linalol and geraniol by hydrolytic breakdown of bound forms in disrupted tea shoots. Phytochemistry 20:2145–2147

[1] Ito-En Co. Ltd., Central Research Institute, Megami, Sagara-cho, Haibara, Shizuoka, 421-05 Japan

Excitatory and Sedative Effects of Body Odor-Related Chemicals on Contingent Negative Variation

Y. Okazaki, Y. Takashima, and T. Kanisawa[1]

Key words. Contingent negative variation—Body odor-related chemicals

Human body odor and its related chemicals (especially Δ16-androstens) can affect human behavior, mood, attitude, menstrual cycle length, and so on [1]. Recently some unsaturated aliphatic acids (7-octenoic and *cis-* and *trans-*3-methyl-2-hexenoic acids) have been identified in axillary sweat in a Japanese population [2].

This study was carried out to investigate the effects of the odor of these chemicals on the arousal level of males and females and on osmatics and anosmatics. Three acids, androstenol, androstenone, and androsterone were used. The odor samples were diluted to a concentration of 1% in diethyl phthalate. Contingent negative variation (CNV) was measured from the frontal region (Fz) to detect variations in arousal level between blank and odor. There were some subjects who were anosmatic to each of the Δ16-androstens, but all subjects were osmatic to each aliphatic acid. In the Δ16-steroid conditions, the averaged amplitude of the early component (400–1000 ms after S1) of CNV from both anosmatic

males and anosmatic females showed no significance between blank and odor. Osmatic males showed a significant decrease in androstenol and androstenone and a significant increase in androsterone. In contrast, osmatic females showed opposite and significant changes. These results for the Δ16-androstens suggest that odorous substances affect human arousal levels psychologically rather than pharmacologically. Both males and females showed a significant decrease in *cis-* and *trans-* 3M2H acids, no sex differences were observed. In the 7-octenoic acid condition, the distributions of each individual were not found to be normal, and were clearly bimodal. Half of the individuals showed significant increases, while the others showed a significant decrease, although there were no differences in hedonics and the perceived intensity of the odor. We are unable to explain this distributions; however, certain uncontrolled personal factors may influence or cause interactions with this odor.

References

1. Gower DB, Nixon A, Mallet AI (1988) The significance of odorous steroids in axillary odour. In: van Toller S, Dodd GH (eds) Chapman and Hall, London
2. Zen XN, Leyden JJ, Lawley HJ, Sawano K, Nohara I, Preti G (1991) Analysis of characteristic odors from human male axillae. J Chem Ecol 17:1469–1492

[1] Central Research Laboratories, Takasago International Corporation, 1-4-11 Nishi-Yawata, Hiratsuka, Kanagawa, 254 Japan

The Effect of Video Images on Psychophysiological Changes Concomitant with Exposure to Odors

Tomoyuki Yoshida[1], Yoshiaki Misumi[2], and Wataru Kameda[2]

Key words. Odor-image—EEG

An odor evokes an emotional image in one's mind. Though most images of good odors usually evoke a positive mood, they may sometimes evoke a negative mood, because they may be connected with images of bad odors that evoke negative feelings to a certain degree. The purpose of this study was to examine whether a video program that was pleasant or unpleasant could change the emotional image of an odor, psychophysiological measurements were used for evaluation. Twenty subjects were divided into two groups. Each subject was seated in a chair in a shielded soundproof room. In front of each subject, there was a video monitor and an odor-presenting tube. The first group (ten subjects) was exposed to a good odor and the second group to a bad one, for 5 min, after which they assessed both their own moods and feelings of arousal, using a self-rating questionnaire (scale from −100 to +100). Next, half of the subjects in each group watched a natural scene on the video program for 5 min while the other half watched a toilet scene; they then evaluated their own condition. After that, the subjects smelled the same odor as before for 5 min again and gave a final evaluation of their condition.

An electroencephalogram (EEG) was performed, using eight scalp locations (F3,F4,C3,C4,P3,P4,O1, O2), during both the odor presentation and video watching, using a telemetric style amplifier, and this was recorded on a pulse cord modulation (PCM) data recorder. The EEG was analyzed using the filtering and zero-crossing method. The recorded EEG was passed through an eight-channel analog band-pass filter (−24 dB/oct) and the frequency component in each channel to be analyzed (8–13 Hz) was extracted. Next, by serially generating a pulse at the midpoint in the up-going slope of the filtered wave, using an electric circuit, the onset times of the pulses were recorded on a floppy disk. By this method, we transformed the alpha wave cycles into a time series of pulses. The interval between pulses was sequentially calculated using these data. The fluctuation of frequency in alpha waves was expressed by frequency of the time series (inverse of interval). A power spectrum of frequency fluctuation of alpha waves for 5 min was calculated by fast Fourier transform (FFT) and transformed into logarithmic values, and the slope of the envelope of the spectrum in the low frequency range (below 2.0 Hz), which reflected the degree of frequency fluctuation of alpha waves, was calculated, using a simple regression analysis.

In previous repeated tests, we have realized that a slope coefficient correlates extremely well with mood and feeling of arousal. That is, when a subject feels a negative mood and is highly aroused, the slope becomes lower (the coefficient is near 0 point, that is, the subject's rhythm of frequency fluctuation of alpha waves is lost), whereas the slope is near −1 (1/f) if the subject feels in a positive mood and is feeling lower arousal. The result of this study agrees well with the previous results, indicating that the video program changed the emotional image of odors. A positive mood and lower feeling of arousal before watching the video in the good odor group changed into a negative mood and higher feeling of arousal after they watched the toilet scene only, even though the subjects smelled the same odor. This psychological change also correlated well with the change in the slope of the spectrum. That is, a higher slope became lower after subjects watched the toilet scene. However, the bad odor group always had a negative mood and high feeling of arousal, neither of which depended on what was shown on the video program. These results lead us to conclude that a positive mood evoked by good odors tends to change to a negative one more easily than a negative mood evoked by bad odors is changed to a positive one, this being because we have more images regarding good odors than bad ones.

[1] Human Informatics Department, National Institute of Bioscience and Human-Technology, 1-1 Higashi, Tsukuba, Ibaraki, 305 Japan
[2] Okayama Laboratory, Ogawa & Co., Ltd., 1-2 Taiheidai, Shoocho, Katsuta-gun, Okayama, 709-43 Japan

Effects of Fluctuating Odor on Odor Intensity and Annoyance

S. Saito[1], K. Iio[1], T. Yoshida[1], S. Ayabe-Kanamura[2], and T. Sadoyama[1]

Key words. Fluctuating odor—Unpleasantness

Introduction

The influence of fluctuation of an odor as a pollutant on the perception is not clear, since there has been little research on the effect of fluctuating odors on humans. Hence, the present study was undertaken to clarify the effect of the time-pattern of fluctuating odors on overall perceived intensity and feelings of annoyance.

Method

A computer-controlled olfactometer was made to continuously emit odor stimuli of varied concentrations and duration. The odor stimulus was presented for a certain period to a subject in a small ventilated odorless room, and the perceived intensity (or unpleasantness) was rated in real time by the subject sliding a small lever attached to a slide potentiometer. These data were sent to a computer for monitoring and analysis. The physical characteristics of the odor stimuli given to a subject were continuously measured by a semiconductor gas sensor and the concentration of odor stimuli was analyzed by the gas chromatography method. The response of the gas sensor was highly correlated with the logarithmic concentration of the odor ($r = 0.991$). The subject's olfactory sensitivity was confirmed with a T&T olfactometer (Tokyo) and by using the smell test of AIST, University of Tsukuba, and Takasago Int. Corp (Tokyo) (STAUTT). This is a smell test that has been developed for a Japanese population. In this experiment, 5 single odors (same as the T&T olfactometer) and 13 compound odors mostly recalled by Japanese (orange, Indian ink, gardenia, perfume, rose, cigarette, milk, feces, face powder, sweat, sake, wood, grass) were microcapsulated and printed on a paper card (1.6 × 2.8

in.), and the odor was presented to the subject by breaking the capsules.

Experiment 1

The effect of intermittent and lasting odors on overall perceived intensity and unpleasantness [1] was examined. The overall intensity and unpleasantness of the intermittent odor were significantly higher than those of the lasting odor (2.3 ppm) in the case of the same volume of triethylamine, which was often reported by subjects as having an offensive, rotten or excretal odor. The overall unpleasantness of triethylamine was significantly correlated with the sum of the real time-intensities, this being more than intensity which corresponds to the recognition threshold.

Experiment 2

The effects of the fluctuations in concentration of the fluctuating odor on the perception of overall intensity and unpleasantness were examined. Sixteen different fluctuating patterns of triethylamine, in eight different conditions, were presented during a 12-min period to a subject; there were two concentration levels (35.7 and 58.0 ppm) and four levels for the coefficient of variation of concentration. Thirteen subjects participated in this experiment and rated the perceived intensity in real time; they then rated the overall perceived intensity and unpleasantness after exposure to the odor. We found that the overall intensity and unpleasantness increased with the coefficient of variation in both concentration levels, except for one stimulus, the lasting odor (smallest coefficient of variation) of high concentration. This result showed that adaptation was difficult at a high concentration level. Overall unpleasantness was significantly correlated with the sum of real time-intensities more than the intensity which corresponds to the recognition threshold ($r = 0.933$). In conclusion, the overall unpleasantness of the fluctuating odor of triethylamine in the same exposure volume, an odor that was offensive to all subjects, was higher for intermittent odor than when the odor was lasting. Further, the overall unpleasantness became higher at similar concentration levels as the coefficient of variation of the fluctuating odor concentration increased.

[1] Human Informatics Department, National Institute of Bioscience and Human-Technology, AIST, MITI, Higashi 1-1, Tsukuba, Ibaraki, 305 Japan
[2] Institute of Psychology, University of Tsukuba, Tennoudai 1-1-1, Tsukuba, Ibaraki, 305 Japan

References

1. Saito S, Iio K, Yoshida T, Sadoyama T, Ayabe-Kanamura S, Hayano Y (1992) A comparison of unpleasantness between lasting odor and intermittent odor presented by a time controlled olfactometer. Proceedings of the 26th Symposium on Taste and Smell, pp 321–324

7. Umami: Physiology, Nutrition and Food Science

Perception of Monosodium Glutamate in Water and Food by Young and Elderly Subjects

Susan S. Schiffman and Elizabeth A. Sattely-Miller[1]

Key words. Monosodium glutamate—Taste—Elderly—Thresholds—Preference—Foods

Introduction

During aging, decrements in taste perception occur at both threshold and suprathreshold levels [1,2]. Many studies have demonstrated elevated taste thresholds in the elderly for solutions of sodium salts, sweeteners, acids, bitter compounds, and amino acids including monosodium glutamate [1,3]. Magnitude estimation and identification tests of suprathreshold taste sensitivity in aging populations also reveal decrements in intensity and in identification of taste solutions.

This study was designed to expand on previous research [3] by incorporating food products into threshold experiments. Both detection and recognition thresholds were determined for monosodium glutamate (MSG) and MSG plus 0.5 mM inosine-5'-monophosphate (IMP) in a variety of food substances for young and elderly subjects. The concentration levels at which the MSG was preferred were determined as well.

Materials and Methods

Subjects

Fourteen students or employees at Duke University with a mean age of 23.8 ± 5.8 years were selected for the young group; 44% were male and 56% female. Ten female residents (mean age 87.5 ± 3.3 years) of the Methodist Retirement Home in Durham, NC served as the elderly subjects. Mean thresholds and preference scores were obtained from each group.

Stimuli

Subjects were offered nine foods (Kroger frozen peas, Kroger frozen corn, fresh carrots, Perdue ground turkey, Perdue ground chicken, Kroger cubed steak, Campbell's low sodium tomato soup, Campbell's low sodium chicken broth, and Lipton's dried onion soup mix) with MSG alone or with 0.5 mM IMP added. The frozen vegetables were steamed, the raw carrots were shredded, the meats were microwaved, and the soups were prepared according to package directions. Foods were presented at room temperature (72°F).

The experimental design was as follows: thresholds were determined for all foods with MSG alone in young subjects and with MSG + IMP in young and elderly subjects. Preferences were established for all foods with MSG alone and MSG + IMP in young subjects and for corn, carrots, chicken broth, and cubed steak in elderly subjects.

Procedure

Threshold Tests

Thresholds were determined for all foods with MSG alone and MSG + 0.5 mM IMP in young subjects and for MSG + 0.5 mM IMP in elderly subjects, as tests with young subjects had revealed no differences between thresholds for MSG alone or with IMP in foods. Two days were used to familiarize the subjects with threshold procedures and with "umami." One food per day was tested with either MSG alone or MSG + IMP added, and the order of foods was randomized. Five-gram samples of each food mixed with 1.5 ml of the MSG solution (MSG alone or with IMP) or with 1.5 ml of deionized water (control) were offered in 30-ml plastic containers. Concentrations of MSG for both conditions in young subjects ranged from 0.0034% (w/w) MSG to 3.45% (w/w) MSG; and for elderly subjects in the MSG + IMP condition the range was from 0.0270% (w/w) MSG to 6.91% (w/w) MSG.

The forced choice ascending method of limits was employed to establish thresholds. Young subjects tasted 11 rows of two samples each at one tasting session, and elderly subjects tasted 9 rows of two samples each. A given concentration of MSG solution was added to one sample, and deionized water was added to the other. The weakest concentration of the MSG solution was in the first row, followed by progressively stronger concentrations in subsequent rows. A dilution factor of 2 separated the concen-

[1] Department of Psychology, Duke University, Durham, NC 27706, USA

trations. Each sample was masticated for 10 s and then expectorated. A half-min delay separated each cup within a row, and there was a delay of 1 min between rows. During the delays, subjects used deionized water to rinse their mouths.

Correct identification of the stronger cup in a pair at least five rows in succession (i.e., at five increasing concentrations) established the detection threshold. An umami/MSG recognition threshold was similarly established.

Preference Tests

Young subjects were tested for preference ratings in all nine foods with six concentrations of MSG alone and MSG + 0.5 mM IMP, and the elderly subjects were tested for four foods with five concentrations of MSG alone and MSG + 0.5 mM IMP. Before food samples were offered, each subject was given an umami taste reference, a 1.105% (w/w) solution of MSG.

Each day a single food was randomly selected, either with the MSG alone or with MSG plus IMP, to be tested. Five-gram samples of each food, with either 1.5 ml of each concentration of the MSG or the MSG + IMP solutions added, were offered in 30-ml plastic containers. A dilution factor of 4 separated the concentrations. The range of MSG concentrations tested in young subjects was from 0.004% (w/w) MSG to 4.421% (w/w) MSG; in the elderly subjects, it was from 0.017% (w/w) MSG to 4.421% (w/w) MSG.

A tray with the six concentrations randomly arranged was presented to each young subject; elderly subjects were given five concentrations in random order. Each subject was requested to rate each sample on a scale of 1 to 10 as follows: dislike/like, too weak/too strong, not sweet/too sweet, not sour/

too sour, not salty/too salty, not bitter/too bitter, not umami/too umami.

Again, each sample was masticated 10 s and then expectorated. A 2-minute delay intervened between expectoration of one sample and tasting the next. Subjects rinsed their mouths with deionized water and gave their responses during the delay.

Results

Threshold Tests

Mean detection and recognition thresholds for young and elderly subjects in the MSG + IMP condition are shown in Table 1 and Fig. 1, respectively. Ratios are given in Table 1 for each threshold (elderly threshold/young threshold). Detection thresholds for MSG + IMP of a given food which differ significantly between young and elderly subjects are indicated. Although detection thresholds were found for all subjects, some did not demonstrate recognition thresholds for umami.

An analysis of variance performed on young subjects' detection thresholds revealed no significant main effects of condition (MSG alone or MSG + IMP) or of food, or a significant food–condition interaction. For recognition thresholds, analysis of variance could not be performed because so many young subjects did not get umami recognition thresholds within the range of concentrations tested.

An analysis of variance of the detection thresholds for young and elderly subjects in the MSG + IMP condition revealed significant main effects of food and of age, as well as a significant food–age interaction. Individual F-tests revealed that young and elderly subjects had significant differences in detection threshold scores for peas, corn, carrots,

TABLE 1. Mean detection and recognition thresholds for umami for young and elderly subjects in the MSG + IMP condition.

Food	Detection threshold			Recognition threshold		
	Young	Elderly	E/Y ratio	Young	Elderly	E/Y ratio
Carrots	0.104	0.343	3.30	0.529	1.813	3.43
Corn	0.209	0.880	4.21	0.389	3.562	9.16
Peas	0.296	1.036	3.50[a]	0.739	2.267	3.07
Chicken broth	0.062	0.151	2.44	0.255	1.527	5.99
Onion soup	0.733	0.680	0.93	1.512	1.093	0.72
Tomato soup	0.746	3.621	4.85[a]	0.949	0.864	0.91
Cubed steak	1.323	0.637	0.48	1.073	2.027	1.89
Ground chicken	0.819	0.645	0.79	1.082	2.704	2.50
Ground turkey	0.234	1.133	4.84[a]	0.855	2.159	2.53
Average			2.82			3.36

Results are given as percent of MSG (w/w).
E/Y = elderly threshold/young threshold.
[a] Significantly different detection thresholds between elderly and young subjects.

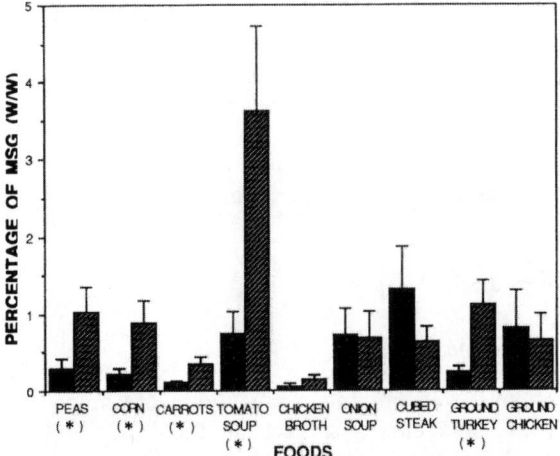

FIG. 1. Mean detection thresholds for young and elderly subjects in the MSG + IMP (0.5 mM) condition. *Solid bars*, young subjects; *hatched bars*, elderly subjects. *Asterisks* indicate significantly different detection thresholds between elderly and young subjects

tomato soup, and ground turkey. Because many of the young and elderly subjects did not achieve recognition thresholds for umami within the range of concentrations tested for the MSG + IMP condition in all nine foods, an analysis of variance could not be performed.

Preference Tests

Mean umami and preference ratings for young subjects for corn with MSG + IMP, as a representative, are compared in Fig. 2. Preference values are based on "dislike/like" ratings, and umami ratings are based on "not umami/too umami" ratings. Comparisons of Fig. 2 with Table 1 show that the umami ratings increased sharply at concentrations slightly higher than the detection threshold values in foods. Young subjects' preference ratings were constant or slightly increasing until detection threshold concentrations; they then tended to decrease.

Umami and preference ratings for elderly subjects are also given in Fig. 2 for carrots with MSG + IMP, as a representative. Similar to the umami ratings of young subjects, the umami ratings for elderly subjects increased sharply at concentration levels slightly above detection threshold values. Elderly subjects' preference ratings varied with the foods. Preference ratings for carrots paralleled the pattern in young subjects, increasing slightly until detection threshold concentrations and then decreasing.

A significant main effect of concentration was revealed by an analysis of variance performed on the preference scores of the young subjects. However, there was not a significant main effect of condition or a significant concentration–condition interaction. No significant effects of concentration or condition or a significant concentration–condition interaction were revealed for elderly subjects.

An analysis of variance was performed on the preference scores of young and elderly subjects for the four foods they tested in common. For carrots, the ANOVA indicated significant main effects of concentration of MSG and of age, as well as a significant concentration–age interaction. There was no significant main effect of condition for carrots or a significant concentration–condition interaction.

Analysis of variance performed on the preference scores of young and elderly subjects for corn in-

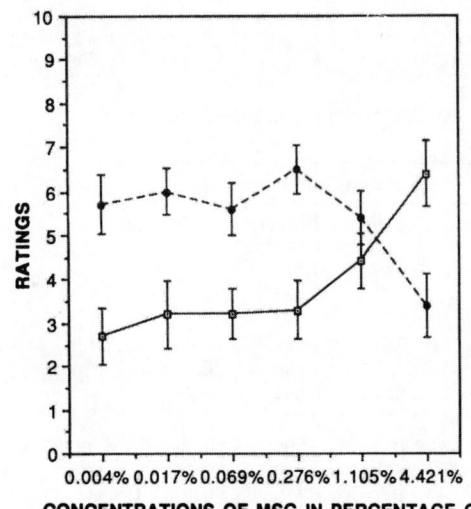

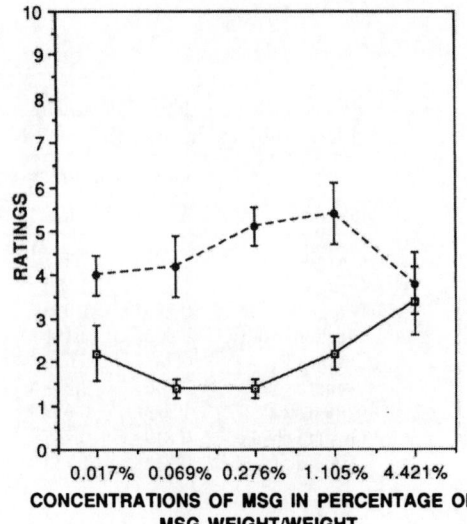

FIG. 2. Mean preference and umami scores for MSG + IMP **a** in corn for young subjects and **b** in carrots for elderly subjects. *Solid circles*, preference; *open squares with dots*, umami

dicated a significant main effect of the concentration of MSG and a significant concentration–age interaction. There were no significant main effects of age for corn or of condition, and there was no significant concentration–condition interaction.

An analysis of variance performed on the preference scores of young and elderly subjects for cubed steak revealed significant main effects of MSG concentration and age and a significant concentration–age interaction. Neither significant main effect of condition nor significant concentration–condition interaction was seen for cubed steak.

An analysis of variance performed on the preference scores of young and elderly subjects for chicken broth revealed no significant main effects of MSG concentration, age, or condition, nor was there a significant concentration–age or concentration–condition interaction.

Discussion

Detection Thresholds

The main finding of this study was that detection thresholds for MSG + IMP in foods were 2.82 times higher on average in elderly subjects than in young subjects. Significant differences were demonstrated in detection threshold scores for young and elderly subjects for MSG + IMP in peas, corn, carrots, tomato soup, and ground turkey.

In addition, detection thresholds in foods were higher than in water. For young subjects, the mean detection threshold for MSG in water is between 0.012% and 0.015% (w/v) [3,4], and for MSG + 0.1 mM IMP in water it is 0.002% (w/v) [3]. However, the mean detection threshold for MSG in foods for young subjects in this study ranged from 0.089% (w/w) in carrots to 0.819% (w/w) in cubed steak; for MSG with 0.5 mM IMP in foods, it ranged from 0.062% (w/w) in chicken broth to 1.323% (w/w) in cubed steak.

The mean detection threshold for MSG in water for elderly subjects is 0.089% (w/v) [3]; for MSG + 0.1 mM IMP in water it is 0.031% (w/v) [3]. However, in the present investigation, the mean detection thresholds for MSG with 0.5 mM IMP in foods for elderly subjects ranged from 0.151% (w/w) in chicken broth to 3.621% (w/w) in tomato soup.

Variability in Detection and Recognition Thresholds

The variability in the ability to detect MSG (with or without IMP) in foods far exceeded the variability previously demonstrated in water, especially for young subjects. In this study, the ratio of the highest/lowest detection threshold for MSG in chicken broth

was 2032 which is dramatically higher than the same ratio for MSG in water, which was 16 [3]. This variability of detection thresholds in foods may result from several attributes of MSG. First, the taste of glutamate may synthesize with other tastes in foods to become a new quality [5,6]. Fuke and Konosu [6] demonstrated that synthesis of glutamate with other food ingredients, including amino acids and nucleotides, created the characteristic tastes of seafood, meat, and tomatoes. Another possibility is that Americans have no word for the MSG/umami flavor and hence may have more difficulty detecting and discriminating it in foods [7]. The analogy is posed here that the difficulty discriminating umami in foods by American subjects resembles the linguistic difficulty experienced by native speakers of Japanese in distinguishing the spoken [r] from the spoken [l] in speech [8].

Effect of IMP on MSG Thresholds and Preferences

In threshold and preference protocols, adding 0.5 mM IMP to the MSG failed to affect food ratings significantly, nor did mean detection thresholds in the MSG alone and the MSG + IMP conditions in foods differ statistically for young American subjects. Schiffman et al. [3], however, found a significant difference between the detection thresholds for MSG and MSG + IMP for American subjects in water solutions. The different effect of IMP in foods and water might not have occurred if the concentration of IMP used in foods in this study had been higher.

Preference for MSG

The concentrations of MSG most preferred in foods were those below the detection or recognition levels in foods but above the level detected in water for Americans. A possible reason has been suggested earlier: MSG synthesizes with other chemicals in the food, producing a new taste. In concentrations above the point of recognition in foods, it may be that MSG no longer synthesizes with the other tastes and is therefore no longer preferred. Another reason may be that the taste cells are partially depolarized by food chemicals and that complete depolarization occurs in the presence of MSG.

The preference ratings of foods in both the MSG and MSG + IMP conditions tended to be lower, especially for carrots and cubed steak, in elderly subjects than in young subjects. This finding supports earlier results reported by Schiffman [9] and Schiffman and Warwick [10], who demonstrated that elderly subjects perceive foods as weaker in taste and smell and less easily identifiable when compared with the perceptions of young subjects.

Acknowledgments. Supported in part by a grant from the International Glutamate Committee and by NIA grant AG00443.

References

1. Schiffman SS (1991) Taste and smell perception in elderly persons. In: Fielding JE, Frier HI (eds) Nutrition research: future directions and applications. Raven, New York, pp 61–73
2. Schiffman SS, Gatlin CA (1993) Clinical physiology of taste and smell. Annu Rev Nutr 13:405–436
3. Schiffman SS, Frey AE, Luboski JA, Foster MA, Erickson RP (1991) Taste of glutamate salts in young and elderly subjects: role of inosine 5'-monophosphate and ions. Physiol Behav 49:843–854
4. Yamaguchi S (1991) Basic properties of umami and effects on humans. Physiol Behav 49:833–841
5. Schiffman SS, Erickson RP (1993, in press) Psycho-physics: insights into transduction mechanisms and neural coding. In: Simon SA, Roper S (eds) Taste. CRC Press, Cleveland
6. Fuke S, Konosu S (1991) Taste-active components in some foods: a review of Japanese research. Physiol Behav 49:863–868
7. Ishii R, O'Mahony M (1987) Taste sorting and naming: can taste concepts be misrepresented by traditional psychophysical labelling systems? Chem Senses 12:37–51
8. Miyawaki K, Strange W, Verbrugge R, et al. (1975) An effect of linguistic experience: the discrimination of [r] and [l] by native speakers of Japanese and English. Percept Psychophys 18:331–340
9. Schiffman SS (1977) Food recognition by the elderly. J Gerontol 32:586–592
10. Schiffman SS, Warwick ZS (1991) Changes in taste and smell over the lifespan: effects on appetite and nutrition in the elderly. In: Friedman MI, Tordoff MG, Kare MR (eds) Chemical Senses, vol 4. Appetite and Nutrition. Dekker, New York, pp 341–365

Humans and Appreciation of the Umami Taste

Shizuko Yamaguchi and Ikuko Kobori[1]

Key words. Umami—MSG—IMP—Adaptation—Flow system—Successive time intensity

Introduction

Umami is the main factor in controlling the taste of natural foods and contributes greatly to the palatability of foods. We have been studying the effects and qualities of umami for a long time. During this period, we have published many papers discussing the basic properties of umami and its effects on humans [1–3]. Here we discuss some recent findings regarding the special aspects of umami.

Sensitivity to Umami: Low in Flow Stimulation but High in Natural Oral Movement

Sensitivity to umami of monosodium glutamate (MSG) or inosine-5'-monophosphate (IMP) as compared with NaCl was measured using the flow method. In this experiment subjects extended their tongues, and continuous streams of solutions at 34°C were delivered over the anterior tongue surface using a simple flow system. Pure water was delivered for 10s followed by a test solution for 10s at a rate of 5 ml/s. Subjects were asked to indicate the kind of taste they experienced while the test solutions were being delivered by pointing to one of the five taste groups or by indicating "no taste" or "uncertain" on a board in front of them while their tongues were still protruding. Following this stage pure water was again delivered over the tongue to remove the residue of taste before the subjects retracted their tongues. Thus the subjects never tasted the samples in their mouths. Some subjects tested different concentrations of MSG and NaCl and the other subjects IMP and NaCl. The order of sample presentation was semirandomized, in ascending concentrations. As for MSG and IMP, even with concentrations as high as 160 mM, fewer than 50% of subjects tasted umami. Meanwhile, NaCl was tasted as definitely more salty, even at a lower concentration (20 mM).

However, sensitivity to umami increased greatly when the whole mouth method or the lick method was employed. Figure 1 shows the comparison of recognition sensitivities to the basic tastes, as measured by the whole mouth method and the lick method. For the whole mouth method, subjects were asked to hold 10 ml of sample in their mouths for 10s. For the lick method, 0.01 ml of solution was placed on a small plastic spoon using a micropipet, and subjects were asked to lick it carefully and indicate the kind of taste they experienced. The certainty of the taste was also scored on a 4-point scale, from absolutely (3) to no taste (0). In both experiments the concentrations of the samples were scaled by the power of 2. The series of test solutions were given in ascending order to avoid prejudice regarding the taste quality. The data obtained are summarized in Fig. 1.

In the case of sucrose and NaCl, the data for the whole mouth and the lick methods were not largely different. In the case of tartaric acid, however, the two methods gave a markedly different result. Using lick stimulation, the recognition threshold was increased markedly, probably due to the change in pH of the samples. The differences between the whole mouth and the lick methods for MSG and IMP were not largely different, particularly for IMP, which suggests that a small volume of sample seems to be related to diminishing adaptation and dilution of saliva (discussed later).

Figure 1 also compares the detection thresholds measured by the whole mouth method and the lick method. This comparison was arrived at by means of triangular tests using whole mouth stimulation and 0.01 ml lick simulation. For these experiments, a series of test samples were given in decreasing concentrations. The descending order helped the subjects focus their attention on the target stimuli successively. The results of these tests show that recognition and detection thresholds have similar tendencies, although detection thresholds are slightly lower than recognition thresholds.

Adaptation to Umami in Aqueous Solution and Foods

Next, adaptation to MSG and IMP was examined. For this experiment, subjects were given a series of 10 samples that had the same substance and con-

[1] Food Research & Development Laboratories, Ajinomoto Co., Inc., 1-1 Suzuki-cho, Kawasaki-ku, Kawasaki, Kanagawa, 210 Japan

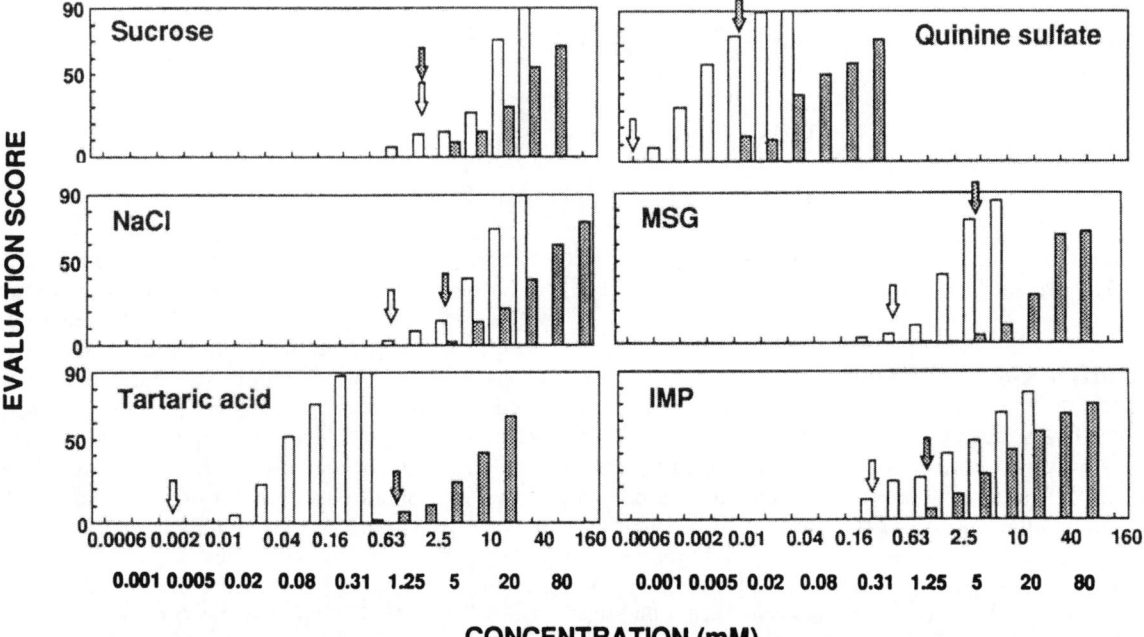

FIG. 1. Comparisons of recognition thresholds of the basic tastes, as measured by the whole mouth method and the lick method (*n* = 30). The evaluation score is the product sum of the evaluation points (0-no taste; 1-unsure; 2-perhaps; 3-absolutely) and the number of people. *Open and dotted arrows* represent detection thresholds for the whole mouth method (10 ml) and the lick method (0.01 ml), respectively

centration, but the subjects were not aware the samples were the same. They were asked to hold 10 ml of the first sample in their mouths for 5 s and regard its umami intensity as 10. After they expectorated the sample they rested for 10 s without rinsing their mouths and then tested the next sample in the same manner, scoring the umami intensity with reference to the first sample. They repeated this action 10 times. Adaptation to both MSG and IMP took place to some extent, but adaptation to IMP occurred more rapidly than to MSG. However, when a small amount of MSG (1.5 ppm—the salivary level of glutamate) was added to IMP, the adaptation was suppressed. Adaptation to IMP and MSG was diminished in a solution of a flavor composed of NaCl and volatile components extracted from a beef stew. This mixture gave the impression of a solution similar to actual food.

Umami of IMP:
Enhancement by Glutamate

We tried to compare the subjective umami intensity of MSG and IMP [4]. These substances cannot be compared directly because of their strong synergism, nor can they be tested repeatedly without rinsing owing to subjects' taste adaptability. Therefore we had to measure their taste separately by means of an absolute judgment. Each subject was given a series of test samples. The samples were presented one by

one, and the subjects were not informed of how many samples were going to be presented. Subjects tasted 10 ml of each sample in a mouthful and scored the umami intensity on a 10-point scale. After rinsing their mouths thoroughly, they rested for 30 s to avoid adaptation, and then they tested the next sample. When the samples were presented in increasing concentration, the ratings were affected by the intensity of the initial sample, concentration interval, and so on. However, when the order of presentation was randomized, the ratings were not affected by such conditions and comparable dose-response curves for MSG and IMP were obtained. The slope of the curve for IMP was less steep than that of MSG. However, when a small amount of MSG (even the normal salivary level) was added to IMP, the IMP curve was definitiely elevated.

These experiments suggest two possibilities. One is that IMP may have a small amount of umami present itself. The other is that there is no umami in IMP. In this case the synergistic effect between IMP and salivary glutamate creates an umami taste. However, at this stage of research, no definite conclusions can be made.

Umami: Strong Aftertaste
and Synergistic Actions

Figure 2 shows the successive time-intensity (T-I) curves for MSG and IMP, together with NaCl and

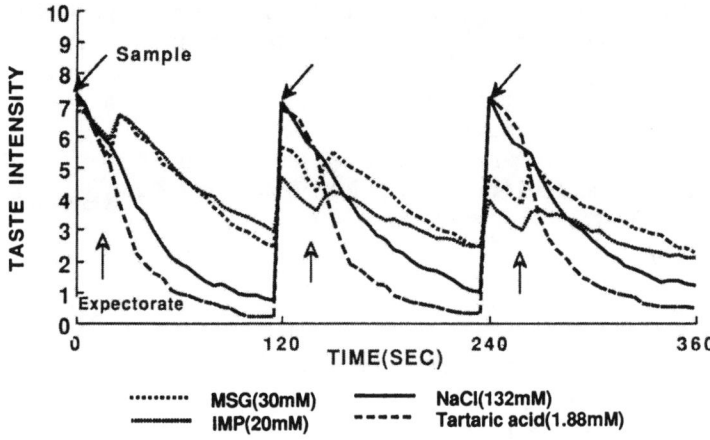

FIG. 2. Successive time-intensity curves in response to the umami of MSG and IMP, the saltiness of NaCl, and the sourness of tartaric acid ($n = 30$)

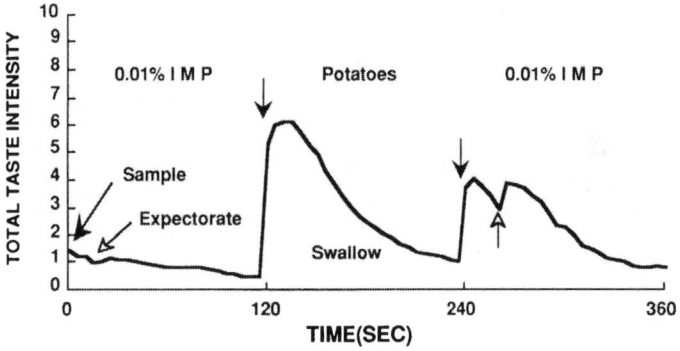

FIG. 3. Successive time-intensity curve in response to the total taste intensity of 0.01% IMP when three cubes of potatoes were eaten in the interim ($n = 20$)

tartaric acid. The subjects were given three identical samples. They then placed 10 ml of the first sample in their mouths and recorded the perceived umami intensity every 5 s, following the directions of a sound signal delivered by a microcomputer. After 20 s (Fig. 2), the solution was expectorated, while the subjects continued to record the perceived umami intensity for another 100 s. This process was repeated with all three samples. The sourness of tartaric acid decreased continuously and quickly. The saltiness of NaCl also decreased quickly but at a slightly slower rate than the tartaric acid. The T-I patterns of the three replications were almost the same for these two cases. On the other hand, the T-I curves for MSG and IMP were different. After the samples were expectorated, the umami taste became strong again, and the T-I curves showed two peaks, representing a strong aftertaste and continuity. For the three replications, the heights of the peaks decreased gradually, showing some adaptation. The MSG and IMP curves showed almost the same pattern in the first sample, but in the following samples the IMP curve showed more adaptation. The adaptation of MSG was also suppressed in actual food systems. For example, in a beef extract solution, adaptation was hardly observed. Adaptation to IMP was also suppressed in the beef extract

solution, although slightly more adaptation was observed.

The successive T-I pattern for MSG and IMP when tasted successively was also measured. The concentration level was 0.03%, only slightly higher than taste thresholds. The second sample was tasted 100 s after the first sample was expectorated. Although the umami intensities of the two samples were almost the same, and weak by themselves, the T-I curve was remarkably elevated after the second sample was tested. This experiment shows that the synergistic effect of MSG and IMP occurs not only when these substances are tasted simultaneously, but also successively. If the order of presentation is changed, we should be able to obtain almost the same result.

In the last experiment, 0.01% IMP, which is less than the taste threshold, was tasted first, and then three cubes of boiled potatoes, measuring $1 \, cm^3$, were eaten at a normal speed, and completely swallowed. IMP 0.01% was then tasted again. The successive T-I curve obtained is shown in Fig. 3. The second peak in the curve is thought to be the result of the glutamate contained in potatoes naturally.

In our diet, we enjoy various kinds of foods that range from salty or sweet to sour or bitter. To enjoy a meal, it is necessary that not only each dish have a pleasant taste, but that the foods we choose complement each other, because the aftertaste of one food can affect the taste of the next dish we eat. For example, the sourness of pickles diminishes fast and makes us feel refreshed. On the other hand, the umami taste in soup or steak continues for a long time, which should increase the depth or intensity of foods that are eaten successively.

Conclusion

These findings suggest several ways umami can be used in experiments and in our daily lives. The effects of umami in an actual diet and in a natural oral movement are far different from those in simple aqueous solutions or under unrealistic experimental conditions. Umami is far more complicated and delicate than the other basic tastes. Thus in order to understand the real value of umami for humans, we should take into consideration normal oral movements including saliva secretion. We should appreciate umami not only in individual foods but also in complete menus of a well-rounded diet for humans.

References

1. Yamaguchi S (1979) The umami taste. In: Boudreau JC (ed) Food taste chemistry. American Chemical Society, Washington, DC, pp 33–51
2. Yamaguchi S (1987) Fundamental properties of umami in human taste sensation. In: Kawamura Y, Kare MR (eds) Umami: a basic taste. Dekker, New York, pp 41–73
3. Yamaguchi S (1991) Basic properties of umami and effects on humans. Physiol Behav 49:833–841
4. Kobori I, Yamaguchi S (1992) A study on the causing mechanism of umami of IMP. In: Proceedings of the 26th Japanese Symposium on Taste and Smell, Japanese Association for the Study of Taste and Smell. Gakushi Kaikan, Kanda, Tokyo, pp 157–160

Enhancing Effect of Nucleotides on Sweetness of Heated Prawn Muscle

Shinya Fuke[1], Katsuko Watanabe[2], and Shoji Konosu[3]

Key words. Free amino acids—Nucleotides—Prawn—Sensory test—Sweetness—Taste

Introduction

Various kinds of prawns, lobsters, and shrimp are eaten raw or heated in Japan. However, most of the studies related to their taste and taste components focused on the raw muscle. Hujita et al. [1], who analyzed free amino acids in the raw muscles of nine kinds of prawns and lobsters, reported that their palatability is closely related to the total amount of sweet amino acids, such as glycine (Gly), proline, serine (Ser), and alanine (Ala), especially the amount of Gly. Take et al. [2] also examined the taste of aqueous extracts of dried shrimp (*Sergestes lucens*) and fresh prawn (*Pandalus borealis*). They treated the extracts with either glutamate decarboxylase or nucleotidase and noted that the disappearance of glutamate resulted in a slight decrease in umami, but that adenosine 5′-monophosphate (AMP) was not responsible for the taste. The taste components of prawns and lobsters have been reviewed by Konosu [3].

According to Matsumoto et al. [4], adenosine 5′-triphosphate (ATP) quickly decomposed mostly to AMP when the tail muscle of decapitated live Kuruma prawn (*Penaeus japonicus*) was heated in water at 48°C to 50°C for 30 s. Even in the live prawn, a large amount of AMP was produced from ATP in the muscle when it was kept under anaerobic conditions for 48 h in sawdust [5]. The decreased ATP was easily recovered by putting the prawn into seawater for 15 min.

It is known that the sweetness of heated Kuruma prawn is stronger than that of the raw prawn. The objective of this study was to confirm sensorially that AMP formed from ATP is responsible for the increase in sweetness.

[1] Faculty of Education, Tokyo Gakugei University, 4-1-1 Nukui-kitamachi, Koganei, Tokyo, 184 Japan
[2] Faculty of Agriculture, The University of Tokyo, 1-1-1 Yayoi, Bunkyo-ku, Tokyo, 113 Japan
[3] Faculty of Home Economics, Kyoritsu Women's University, 2-2-1 Hitotsubashi, Chiyoda-ku, Tokyo, 101 Japan

Materials and Methods

Sample

Live Kuruma prawn (*P. japonicus*) was purchased from the Tokyo Central Wholesale Fish Market. Twenty prawns with an average body weight of 20 g were decapitated and the tail muscle quickly dissected into small pieces. Ten grams each of the muscle was wrapped in polychlorovinylidene film (Asahikasei, Tokyo, Japan), placed in water, and heated for 1, 3, 5, 10, 15, or 20 min. Before and after heating, the muscles were extracted with 5% trichloroacetic acid (TCA) as described by Konosu et al. [6]. At the same time, the temperatures of water and muscle were measured.

Analytic Methods

The TCA extracts were analyzed for free amino acids, nucleotides, and sodium and potassium ions. Free amino acids were determined on a Hitachi 835 type automatic amino acid analyzer (Hitachi, Tokyo, Japan). Nucleotides were estimated by high-performance liquid chromatography (HPLC) after Harmsen et al. [7]. Sodium and potassium ions were analyzed with an atomic absorption spectrophotometer (Seiko, SAS-760, Tokyo, Japan) according to the usual method.

Synthetic Extracts

To elucidate the taste effect of 5′-nucleotides, synthetic extracts containing free amino acids, NaCl, and KH_2PO_4 were prepared on the basis of the analytic data. The amounts of NaCl and KH_2PO_4 were calculated from the contents of the sodium and potassium ions, respectively. Then either AMP or a mixture of AMP plus inosine 5′-monophosphate (IMP) was added to the extracts, varying the amount of AMP. The pH of the extracts were adjusted to 6.5.

Sensory Tests

Two tests were done.

1. *Effect of AMP level on the taste of synthetic extract without IMP*. Varying amounts of AMP were

357

added to 100 ml of synthetic extracts as shown in Fig. 2, and their tastes were compared at room temperature using the triangle difference test with the combination of A and B, B and C, and C and D. The same test was repeated twice by six panelists and 12 responses were combined and analyzed statistically.

The panel was also asked to evaluate the magnitude of difference in five basic tastes—sweetness, sourness, saltiness, bitterness, umami—between the two synthetic extracts on a 5-point rating scale. The scores given by the panelists who answered correctly in the triangle difference test were combined and analyzed statistically.

2. *Effect of AMP level on the taste of synthetic extract with IMP.* Varying amounts of AMP and 4 mg of IMP were added to 100 ml of synthetic extracts as shown in Fig. 2. Their tastes were then compared by the same method as above with the combination of A and E, E and F, F and G, and G and H.

Results and Discussion

Table 1 shows changes in free amino acids in the muscle of Kuruma prawn during heating. Gly decreased from 816 mg (hereafter expressed in terms of milligrams or millimoles per 100 g of muscle) in raw muscle to 596 mg in muscle heated for 20 min. Arginine also decreased from 629 mg in raw muscle to 580 mg after 20 min of heating. The decrease in glutamic acid (Glu), which has been reported to be

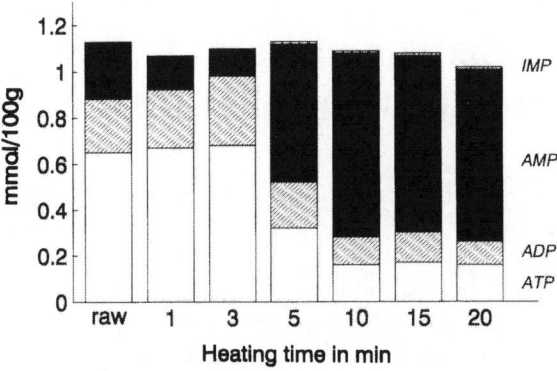

FIG. 1. Changes in nucleotides in the muscle of Kuruma prawn during heating

one of the key components producing characteristic taste in snow crab [8], scallop [9], dried skipjack [10], and short-necked clam [10], was remarkable after 20 min of heating.

Changes in nucleotides in the muscle during heating are illustrated in Fig. 1. The amount of ATP in raw muscle (0.65 mmol) was retained up to 3 min, but declined rapidly to 0.32 mmol after 5 min. After 10 min, the level was maintained at about 0.16 mmol up to 20 min. Adenosine 5'-diphosphate (ADP) increased slightly from 0.23 mmol in raw muscle to 0.30 mmol in muscle heated for 5 min; it then decreased gradually. The amount of AMP in raw muscle was found to be 0.25 mmol, which decreased to 0.12 mmol after 3 min. However, it increased to 0.59 mmol after 5 min and reached the maximum level of 0.83 mmol after 10 min, then decreasing slightly.

The temperatures of the muscle and water bath, respectively, during heating were as follows: 13° and 18°C before heating; 18°–19°C and 30°C after 1 min; 30°–32°C and 52°C after 3 min; 47°–49°C and 68°C after 5 min; 64°–66°C and 84°C after 7 min; and 86°–87°C and 95°C after 10 min; and 95°C and 95°C after 20 min. The quantity of AMP produced during 7 min after 3 min was 0.71 mmol, two-thirds of which was produced during the first 2 min. In other words, most of the AMP produced by heating occurred during about 5 min. The muscle temperature was about 48°C at that time. These results are consistent with those obtained by Matsumoto et al. [4]. A small amount of IMP (ca. 0.01 mmol) was detected after 5 min of heating. The total amounts of these purine nucleotides were about 1.1 mmol for all samples analyzed.

The contents of sodium and potassium ions were 268 and 804 mg, 355 and 920 mg, and 275 and 776 mg for raw muscles and for muscles after 10 and 20 min of heating, respectively.

In the triangle difference test, two kinds of synthetic extract were discriminated from each other significantly in all the combinations mentioned above

TABLE 1. Changes in free amino acids in the muscle of Kuruma prawn during heating.

Amino acid	Amino acid conc. after heating (mg/100 g)		
	0 min	10 min	20 min
Tau	87	114	80
Asp	3	3	2
Thr	10	14	10
Ser	25	26	17
Asn	19	21	20
Glu	115	113	49
Pro	456	388	426
Gly	816	785	596
Ala	66	77	77
Val	20	23	20
Met	11	12	10
Cth*	10	11	7
Ile	8	9	8
Leu	16	15	15
Tyr	15	15	15
Phe	5	5	6
Orn	1	1	1
Lys	15	23	25
His	15	15	15
Arg	629	609	580

* Cystathionine

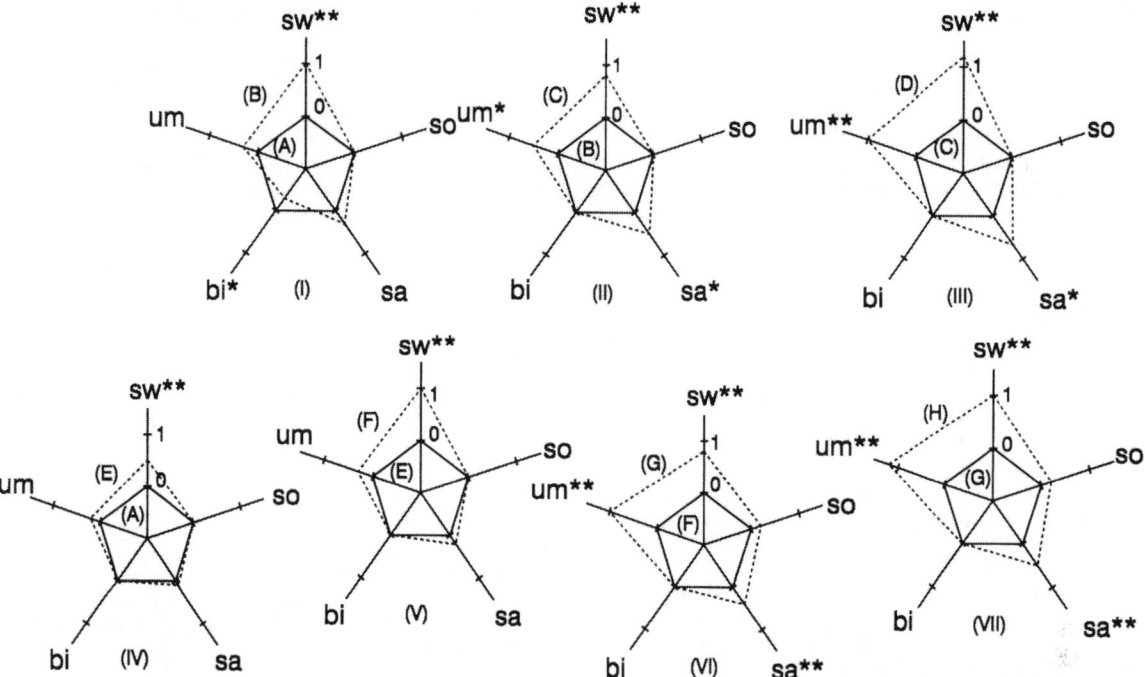

FIG. 2. Effect of 5′-nucleotides on the taste of synthetic extracts. (**I–VII**) Results of a comparison of five basic tastes of synthetic extracts ($A–H$). The concentrations of AMP in A, B, C, and D were 0, 50, 100, and 200 mg/dl, respectively. E, F, G, and H were added with 4 mg of IMP to A, B, C, and D, respectively. sw, sweetness; so, sourness; sa, saltiness; bi, bitterness; um, umami. $*P < 0.05$; $**P < 0.01$

($P < 0.01$). Results of sensory evaluation of the difference in the five basic tastes of synthetic extracts are demonstrated in Fig. 2.

1. Effect of AMP level on the taste of synthetic extract without IMP. Synthetic extract containing no nucleotide (A) was sweet with slight saltiness and bitterness. By adding 50 mg of AMP to 100 ml of synthetic extract A (B), the sweetness significantly increased and bitterness disappeared. Sweetness and saltiness were stronger in (C), which contained 100 mg of AMP, than in (B). Umami was detected in (C). The elevation of AMP to 200 mg (D) caused a further increase in sweetness, saltiness, and umami. Thus it can be stated that sweetness is increased by the increase of AMP. When the AMP level was elevated to more than 100 mg, it imparted umami as well. As the increase of added AMP lowered the pH of synthetic extracts, a larger amount of NaOH was required to adjust the pH to 6.5. The increase in saltiness detected in the extracts containing the larger amount of AMP might be attributed to the increased amount of sodium ion. In the synthetic extract of the short-necked clam, omission of AMP led to a decrease in saltiness [10]. Ugawa et al. [11] reported that the sweetness of sweet amino acids such as Gly, Ser, and Ala was enhanced significantly by a sub-threshold amount of NaCl and slightly by Na phosphate. The same effect might have occurred in our synthetic extracts of Kuruma prawn.

2. Effect of AMP level on the taste of synthetic extract with IMP. The addition of 4 mg of IMP to 100 ml of synthetic extract without AMP (E) induced a slightly stronger sweetness than in (A). When the added AMP was increased, sweetness was increased, as observed in experiment 1, above. Weak umami was detected in (F), and it was strengthened by an increase in AMP. Weak bitterness was detected in (E) to (H).

The role of nucleotides in producing characteristic taste was studied in several types of seafood. In abalone [12], the omission of AMP (90 mg) from the synthetic extract did not affect the sweetness, though umami and the characteristic flavor were weakened remarkably. In sea urchin [13], AMP (only about 10 mg) did not play an important role in the taste. In show crab [8] and short-necked clam [10], AMP and Glu contributed to the sweet taste.

Results obtained in experiments 1 and 2 led us to postulate that AMP chiefly enhances sweetness, and IMP potentiates umami, which is strengthened by AMP in Kuruma prawn. The enhancement of sweetness by AMP was consistent with that reported for the crab [8], scallop [9], and short-necked clam [10]. However, some of the sweetness increased by AMP in our synthetic extracts might be due to the sodium ion added to adjust pH, as pointed out by Ugawa et al. [11]. Further study is required to clarify this problem.

References

1. Hujita M, Endo K, Shimidu W (1972) Studies on muscle of aquatic animals. XXXXVI. Free amino acids, trimethylamine oxide and betaine in shrimp muscle. Mem Fac Agric Kinki Univ 5:60–67
2. Take T, Honda R, Otsuka H (1964) Studies on the tasty substances of various foods. Part 3. On the tasty substances of prawn and shrimp. J Jpn Soc Food Nutr 17:268–274
3. Konosu S (1979) The taste of fish and shellfish. In: Boudreau JC (ed) ACS Symposium Series. No. 115. Food and Taste Chemistry. American Chemical Society, Washington, DC, pp 185–203
4. Matsumoto M, Yamanaka H, Hatae K (1991) Effect of "arai" treatment on the biochemical changes in the Kuruma prawn muscle. Nippon Suisan Gakkaishi. Bull Jpn Soc Sci Fish 57:1383–1387
5. Furusho S, Umezawa Y, Honda A (1988) Changes in the concentration of ATP-related compounds and lactic acid in muscles of live prawn *Penaeus japonicus*. Nippon Suisan Gakkaishi. Bull Jpn Soc Sci Fish 54:1209–1212
6. Konosu S, Watanabe K, Shimizu T (1974) Distribution of nitrogenous constituents in the muscle extracts of eight species of fish. Nippon Suisan Gakkaishi. Bull Jpn Soc Sci Fish 40:909–915
7. Harmsen E, De Tombe PPh, De Jong JW (1982) Simultaneous determination of myocardial adenine nucleotides and creatine phosphate by high-performance chromatography. J Chromatogr 230:131–136
8. Hayashi T, Yamaguchi K, Konosu S (1981) Sensory evaluation of taste-active components in the extract of snow crab meat. J Food Sci 46:479–483, 493
9. Watanabe K, Lan HL, Yamaguchi K, Konosu S (1990) Role of extractive components of scallop in its characteristic taste development. Nippon Shokuhin Kogyo Gakkaishi. J Jpn Soc Food Sci Technol 37:439–445
10. Fuke S, Konosu S (1991) Taste-active components in some foods: a review of Japanese research. Physiol Behav 49:863–868
11. Ugawa T, Konosu S, Kurihara K (1992) Enhancing effect of NaCl and Na phosphate on human gustatory responses to amino acids. Chem Senses 17:811–815
12. Konosu S (1973) Taste of fish and shellfish with special reference to taste-producing substances. Nippon Shokuhin Kogyo Gakkaishi. J Jpn Soc Food Sci Technol 20:432–439
13. Komata Y (1964) Studies on the extractive components of "uni"-IV: taste of each component in the extract. Nippon Suisan Gakkaishi. Bull Jpn Soc Sci Fish 28:749–756

Enhancing Effects of Salts on Canine Taste Responses to Amino Acids and Umami Substances

Tohru Ugawa and Kenzo Kurihara[1]

Key words. Umami taste—Canine taste response—Enhancing effect of salts—Amino acids—Anion dependence—Cation dependence

Introduction

The receptor mechanism of umami substances has been studied mainly with the rat. In the rat, synergism between monosodium glutamate (MSG) and nucleotides was observed [1,2], but the synergism was not as remarkable as that seen in humans. Both MSG and the nucleotides are Na salts, and hence Na ion in the umami substances stimulates salt receptors of the rat. Single fibers of rat chorda tympani nerve that were sensitive to MSG were also sensitive to NaCl [1]. Hence it has been considered that the umami substances stimulate only the salt receptors.

Large Synergism Between MSG and Nucleotides in the Dog

We have recorded the canine chorda tympani nerve responses to umami substances and found that the synergism between MSG and the nucleotides is marked in the dog [3]. The magnitude of the tonic response to MSG in the absence and presence of 0.5 mM guanosine monophosphate (GMP) is plotted in Figure 1a as a function of MSG concentrations. The synergism is remarkable at low concentrations of MSG and becomes small at high concentrations of MSG. These results are explained by an allosteric model. The umami receptor is postulated to have two binding sites: one for MSG (M-site) and another for nucleotides (N-site). Binding of GMP to the N-site increases affinity of the M-site to MSG. Binding of MSG to the M-site increases affinity of the N-site to GMP.

Umami Response Independent of Salt Response

As described above, GMP is a Na salt, and there is a possibility that GMP stimulates salt receptors. In the canine chorda tympani response, the threshold concentration of GMP is around 0.2 mM and that in the presence of 100 mM NaCl is 0.01 mM (Fig. 1b). The threshold for NaCl is much higher than that of GMP, and hence it is unlikely that the response to GMP is induced by the Na ion.

To examine the contribution of the Na component to the umami response in more detail, we examined the effects of amiloride, an inhibitor of the salt response [4], on the responses to umami substances [5]. The response to MSG was mostly suppressed by amiloride, but the response to GMP alone or that induced by the synergism between MSG and GMP was not inhibited by amiloride. These results suggested that the response to GMP alone and that induced by the synergism are composed of a pure umami response.

Responses to Umami Substances in Various Animals

There are a number of differences between the synergism in the rat and the dog. (1) The synergism between MSG and the nucleotides was much greater than that in the rat. (2) In the dog the synergism was observed only between MSG and the nucleotides, whereas in the rat the synergism was observed between all amino acids examined and the nucleotides.

The synergism in dogs and humans was compared as follows. (1) In both humans and the dog, the large synergism occurs between MSG and IMP or GMP. (2) The synergism in the dog does not occur between the nucleotides and amino acids other than MSG, similar to that in humans. (3) In humans, MSG alone elicits a distinct umami taste, and the umami taste induced by GMP or IMP is much weaker than that induced by MSG. Hence it seems that MSG acts as a main agonist, and GMP acts mainly as a modulator in humans. On the other hand, GMP and IMP elicit a large response at low concentrations in the dog. Hence nucleotides are the main agonists in the dog, and MSG is a modulator.

[1] Faculty of Pharmaceutical Sciences, Hokkaido University, Sapporo, 060 Japan

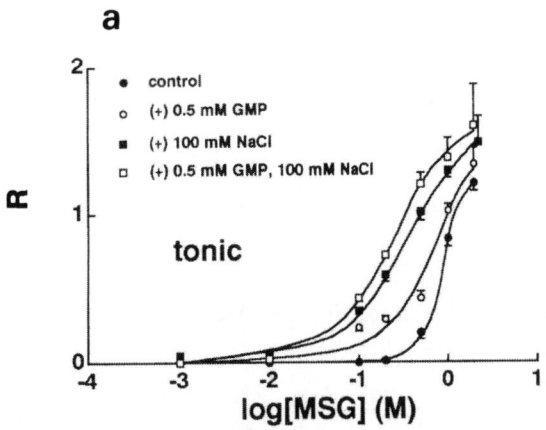

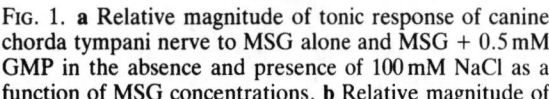

FIG. 1. **a** Relative magnitude of tonic response of canine chorda tympani nerve to MSG alone and MSG + 0.5 mM GMP in the absence and presence of 100 mM NaCl as a function of MSG concentrations. **b** Relative magnitude of tonic response of canine chorda tympani nerve to GMP of varying concentrations in the absence and presence of 100 mM NaCl

As described above, the responses to the umami substances in the rat are much different from those of humans. Hence the data on the responses to the umami substances observed with the rat cannot simply be applied to humans. Among various animals, a dog is the most suitable experimental animal as a model for humans as far as the umami taste is concerned.

Effects of Salts on the Responses to Amino Acids

Konosu et al. [6] applied the omission test to an extract of snow crab meat and explored whether a mixture of glycine, alanine, arginine, MSG, inosine-5'-monophosphate (IMP), NaCl, and K_2HPO_4 can produce the characteristic taste of the crab meat. In the absence of NaCl, a mixture of amino acids, IMP, and K_2HPO_4 elicited only weak taste. These results suggest that the presence of NaCl enhances tastes of amino acids, umami substances, or both.

First we examined the effects of NaCl and Na phosphate on the taste of single amino acids such as glycine, alanine, and serine by the psychophysical method [7]. There was no difference in the quality of sweetness between the amino acids in the presence and those of absence of salts, as low concentrations of NaCl and Na phosphate having no salty taste were used. The sweetness of the amino acids was greatly increased with an increase in NaCl concentration. The sweetness of 100 mM amino acids in the presence of 30 mM NaCl was equivalent to that of 500 to 600 mM amino acids containing no salt. On the other hand, Na phosphate little affected the sweetness of the amino acids. Figure 2a shows the enhancing effects of NaCl and Na phosphate on the sweetness of alanine.

The effects of salts on the taste responses to amino acids were also examined by recording the activity of the canine chorda tympani nerve [8]. The responses to most amino acids examined were significantly enhanced by the presence of various salts. The degree of the enhancement varies with species of amino acid and salt. For example, the responses to glutamine and asparagine were not enhanced by the salts, but responses to most other amino acids were more or less enhanced. The degree of the enhancement also varied with both species of cation and anion of the salts. For example, the responses to most amino acids were enhanced by sodium, potassium, and calcium salts, but not enhanced by magnesium salts. Figure 2b shows the effects of NaCl and Na phosphate on the response to alanine as a function of Na concentration. The response is increased with an increase of NaCl concentration up to 100 mM and is decreased with a further increase in NaCl concentration. Na phosphate also shows an enhancing effect, but the Na concentration required to give a peak response is much higher than that for NaCl; 300 mM Na is needed to give a peak response for Na phosphate, whereas 100 mM Na is needed for NaCl. At a concentration below 100 mM, NaCl has a much larger enhancing effect than Na phosphate at equimolar Na ion, which is consistent with the results obtained by the psychophysical method (Fig. 2a).

Effects of Salts on Responses to Umami Substances

The effects of salts on canine taste responses to umami substances were examined by recording the activity of the chorda tympani nerve. In Fig. 1a the

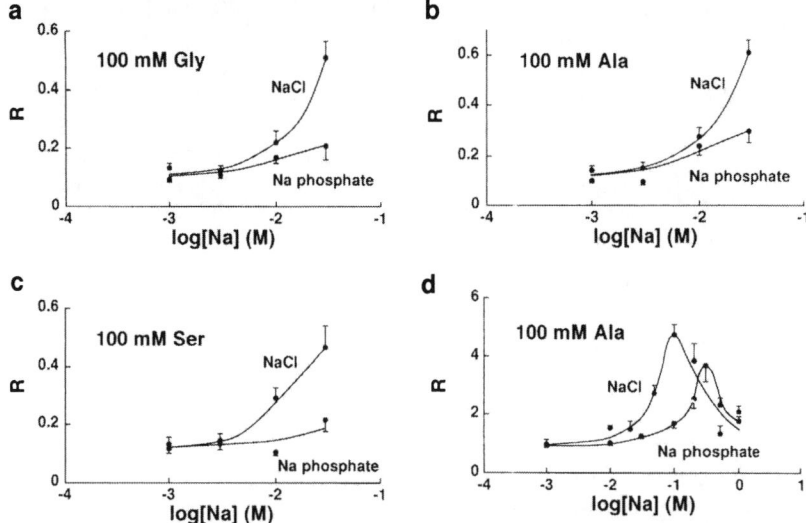

FIG. 2. Effects of NaCl and Na phosphate on sweetness of 100 mM glycine (**a**), alanine (**b**), and serine (**c**) as a function of logarithmic Na concentration evaluated by psychophysical method. R in the ordinate represents molar concentration of standard alanine solution containing no salt whose sweetness is equivalent to the sweetness of an alanine solution containing the salts. **d** Effects of NaCl and Na phosphate on canine chorda tympani nerve response to 100 mM alanine as a function of logarithmic Na concentration. Plotted response (R) was calculated relative to the magnitude of the tonic response to 100 mM alanine alone

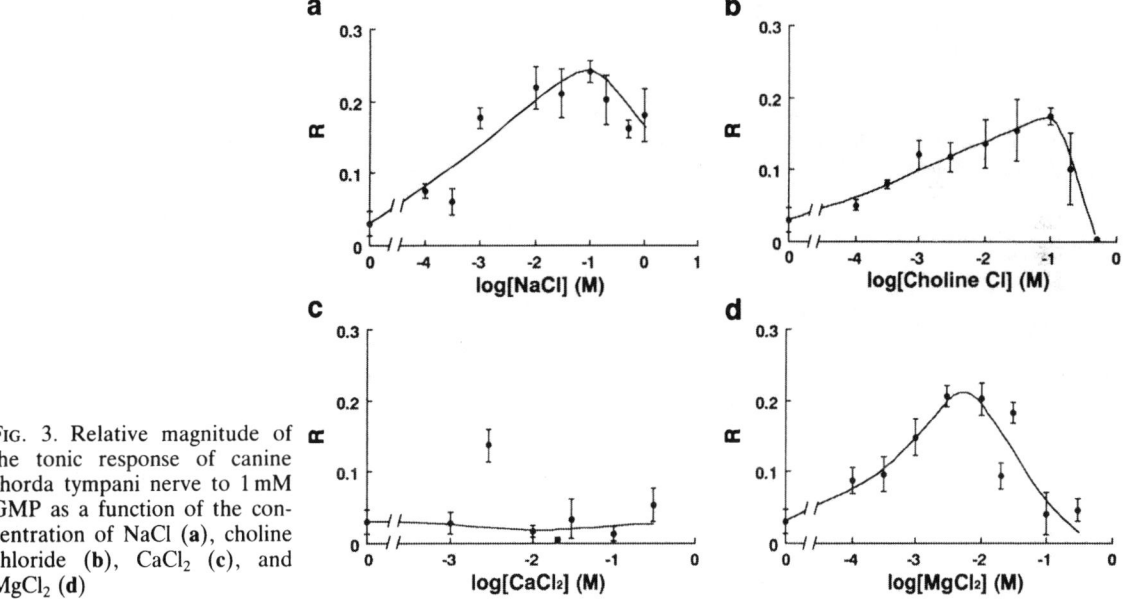

FIG. 3. Relative magnitude of the tonic response of canine chorda tympani nerve to 1 mM GMP as a function of the concentration of NaCl (**a**), choline chloride (**b**), $CaCl_2$ (**c**), and $MgCl_2$ (**d**)

magnitude of the tonic responses to MSG and MSG + 0.5 mM GMP in the absence and presence of 100 mM NaCl is plotted as a function of MSG concentration. The presence of 100 mM NaCl greatly enhances the response to MSG at all concentrations examined. The presence of NaCl also enhances the response induced by the synergism between MSG and GMP. In Fig. 1b, the effects of 100 mM NaCl on the response to GMP are shown as a function of GMP concentration. As seen from the figure, the

response to GMP is also greatly enhanced by the presence of NaCl.

Figure 3 shows the effects of various salts on the response to 1 mM GMP as a function of the salt concentration. The response to GMP starts to increase at 0.1 mM NaCl and reaches a peak at 100 mM NaCl. Any further increase in NaCl concentration decreases the enhanced response. The figure also shows that the response to GMP is enhanced by the presence of choline and $MgCl_2$, but it is not en-

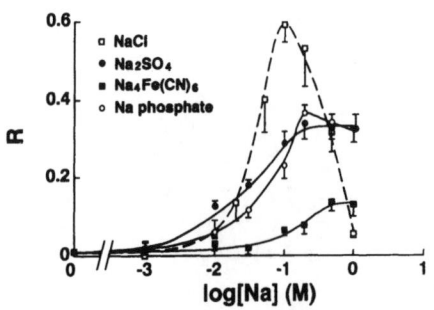

Fig. 4. Relative magnitude of the tonic response of canine chorda tympani nerve to 200 mM MSG as a function of the NaCl, Na$_2$SO$_4$, Na$_4$Fe(CN)$_6$, and Na phosphate concentrations

hanced by CaCl$_2$. The response to GMP was also enhanced by salts carrying organic cations such as Tris, choline, N-methyl-D-glucamine, and bis-tris propane.

The degree of the enhancement depended not only on species of cation but also on the anion. Figure 4 shows the effects of sodium salts carrying different anions on the response to 200 mM MSG as a function of the Na concentration. As seen from the figure, the enhancing effect greatly varies with species of anion; that is, 100 mM NaCl exhibits a large enhancing effect, whereas the enhancing effects of Na phosphate and Na$_2$SO$_4$ are much smaller than that of NaCl at equimolar Na ion. Na$_4$Fe(CN)$_6$ is much smaller than that of other salts at equimolar Na ion.

Mechanism of Enhancing Effects of Salts on the Response to Amino Acids and Umami Substances

As described above, the enhancing effects of salts on the responses to amino acids and umami substances depend on both species of cation and anion. Salts of impermeable cations showed large enhancing effects. This point was also true about the enhancing effects of salts on canine chorda tympani nerve response to sugars [10]. Hence the enhancing effects cannot be explained in terms of ion permeability at apical membranes of taste cells, but they can be explained in terms of binding of salts on the receptor membranes. It is postulated that both cation and anion bind to the receptor membranes, which

induces a conformational change in the receptor membranes. The conformational change affects the affinity of the ligands to the receptors. The changes also lead to exposure of the receptor sites on the membranes or the burying of sites in the membranes. The above mechanism is highly speculative, however, and further study is needed to clarify the mechanism of the enhancing effect of salts on the umami responses.

Acknowledgments. This work was supported by a Grant-in-Aid for Science Research from the Ministry of Education, Science and Culture of Japan and Human Frontier Science Program.

References

1. Sato M, Yamashita S, Ogawa H (1967) Patterns of impulses produced by MSG and 5′-ribonucleotides in taste units of the rat. In: Hayashi T (ed) Olfaction and Taste, vol, II. Pergamon Press, Oxford, pp 399–410
2. Yoshii K, Yokouchi C, Kurihara K (1986) Synergistic effects of 5′-nucleotides on rat taste responses to various amino acids. Brain Res 367:45–51
3. Kumazawa T, Kurihara K (1990) Large synergism between monosodium glutamate and 5′-nucleotides in canine taste nerve responses. Am J Physiol 259:R420–R426
4. Nakamura M, Kurihara K (1990) Non-specific inhibition by amiloride of canine chorda tympani nerve responses to various salts: do Na$^+$-specific channels exist in canine taste receptor membranes? Brain Res 524:42–48
5. Nakamura M, Kurihara K (1991) Canine taste nerve responses to monosodium glutamate and disodium guanylate: differentiation between umami and salt components with amiloride. Brain Res 541:21–286
6. Konosu S, Hayashi T, Yamaguchi K (1987) Role of extractive components of boiled crab in producing the characteristic flavor. In: Kawamura Y, Kare MR (eds) Umami: A Basic Taste. Marcel Dekker, New York, pp 235–253
7. Ugawa T, Konosu S, Kurihara K (1992) Enhancing effects of NaCl and Na phosphate on human gustatory responses to amino acids. Chem Senses 17:811–815
8. Ugawa T, Kurihara K (1993) Large enhancement of canine taste responses to amino acids by salts. Am J Physiol 246:R1071–R1076
9. Ugawa T, Kurihara K (1994, in press) Enhancement of canine taste responses to umami substances by salts. Am J Physiol
10. Kumazawa T, Kurihara K (1990) Large enhancement of canine taste responses to sugars by salts. J Gen Physiol 95:1007–1018

Umami Taste in the Chimpanzee (*Pan troglodytes*) and Rhesus Monkey (*Macaca mulatta*)

Göran Hellekant[1] and Yuzo Ninomiya[2]

Key words. Umami—Chimpanzee—Monkey—Chorda tympani—Single fiber

Introduction

The similarity between the sense of taste in humans and that in monkeys and chimpanzees makes data from the monkey and chimpanzee especially interesting. This similarity applies also to the umami taste. As is well known, the umami taste is elicited by monosodium glutamate (MSG) and a number of 5'-ribeonucleotides which increase the palatability of certain foods in humans. The umami taste has been characterized by some authors as a separate taste quality [1], although this opinion is not shared by everyone [2].

We earlier presented summated and single fiber recordings from the chorda tympani proper nerve of chimpanzees [3] to MSG and guanosine 5'-monophosphate, disodium salt (GMP) and to mixtures of the two. The summated recordings showed no or slight synergism between MSG and GMP. Synergism is here defined as a stronger responses to the mixture of MSG and GMP than the sum of the responses to each of them. In single fibers we found that 10 mM MSG with and without 0.3 mM GMP elicited moderate response and that the responses were largest in sweet fibers.

In humans the taste of umami has been reported to become more salty as the concentrations of the umami compounds increase. Therefore we decided to study the responses to two concentrations of MSG, low (10 mM) and high (70 mM), with and without 0.3 mM GMP in our next chimpanzee study. Here we present results from additional 20 single chorda tympani nerve fibers in chimpanzees (*Pantroglodytes troglodytes*). To shed further light on the umami taste in primates, we also present single-fiber

data from the Old World monkey (*Macaca mulatta*) to 70 mM MSG with and without 0.3 mM GMP.

Methods

Animals

The data presented here were obtained from three adult female chimpanzees (*Pan troglodytes*), 35 to 45 kg, housed at the Laboratory for Experimental Medicine and Surgery in Primates (LEMSIP), New York Medical Center, Tuxedo, NY, and five male and one female Old World monkeys (*M. mulatta*), 2 to 6 years old weighing 3 to 8 kg, housed at Wisconsin Regional Primate Center, Madison, WI.

Surgery

The animals were anesthetized with an intramuscular injection of ketamine 400 mg/animal and atropine 0.5 mg/animal. After intubation the chimpanzees were maintained on a mixture of oxygen, halothane, and N_2O_2, and the monkeys were kept on a mixture of oxygen and halothane. Fluid was replenished with intravenous 5% dextrose and lactated Ringer's solution.

The right chorda tympani proper (CT) nerve was exposed through an incision along the mandibular angle between the rostral lobes of the parotid gland and the mandibular bone. First the tissue attached to the mandibular angle was sectioned. Then blunt dissection was used to follow the caudomedial side of the pterygoid muscle to its origin at the pterygoid plate of the skull and to the CT nerve. The nerve forms one bundle and is surrounded by small veins. There are fewer veins surrounding the CT nerve in the chimpanzee than in the monkey. After the experiment deeper layers were resutured with adsorbable suture, and the skin was closed with nylon. The nylon sutures were removed after 7 to 10 days.

Apparatus

The nerve impulse activity was recorded with a PAR 113 amplifier, monitored over a loudspeaker and an oscilloscope, and fed into a recorder (Gould ES

---
[1] University of Wisconsin and Wisconsin Regional Primate Center, 1655 Linden Drive, Madison, WI 53706, USA
[2] Asahi University, Gifu, Japan

1000, Gould Stathan, Oxnard, CA) and an IBM PC-AT computer via a DAS-Keithley interface.

Stimulation

A taste stimulation system was used that delivers 32 different solutions at given intervals and under conditions of constant flow and temperature (33°C) on the tongue. Each stimulation lasted 8 s. Between the stimulations, the tongue was rinsed for 30 s with artificial saliva.

Test Substances and Procedure

In the chimpanzees we recorded the responses to tongue stimulation with 0.07 M NaCl, 0.3 M sucrose, 0.04 M citric acid, 0.005 M quinine hydrochloride, and 10 and 70 mM MSG with and without 0.3 mM GMP. In the monkeys we used the same compounds and concentrations, except for MSG, which was used at 70 mM concentration with and without 0.3 mM GMP. All compounds but quinine, which for solubility reasons was dissolved in distilled water, were dissolved in artificial saliva.

Results

The addition of GMP to 70 mM MSG had no synergistic effect on the summated nerve response in either the chimpanzees or the monkeys. An example from the monkeys is shown in Fig. 1.

The single fibers were classified in salt (salt-best), bitter, sour, and sweet fibers based on their response to a much larger array of compounds. This array has been presented elsewhere [3]. The single fibers from both species could readily be classified into salt, bitter, sour, and sweet fibers.

Chimpanzees

In the chimpanzees, 10 mM MSG elicited a response in 30% of sweet fibers and 46% of the salt fibers. In general the response was small. Synergistic effects of GMP on 10 mM MSG responses were observed in five of the eight sweet fibers that responded. Synergism is here defined as a larger response to the mixture of MSG and GMP than the sum of the responses to each of them. Two of the six salt fibers responding to MSG showed synergism, but the synergism was less prominent than in the sweet fibers.

When the concentration of MSG was increased from 10 mM to 70 mM, salt fibers were predominantly stimulated. Sweet fibers that responded to the lower concentration of MSG responded more to 70 mM, but the increase was less than the one observed in salt fibers. Synergism between MSG and GMP was

Rhesus-92 summated responses

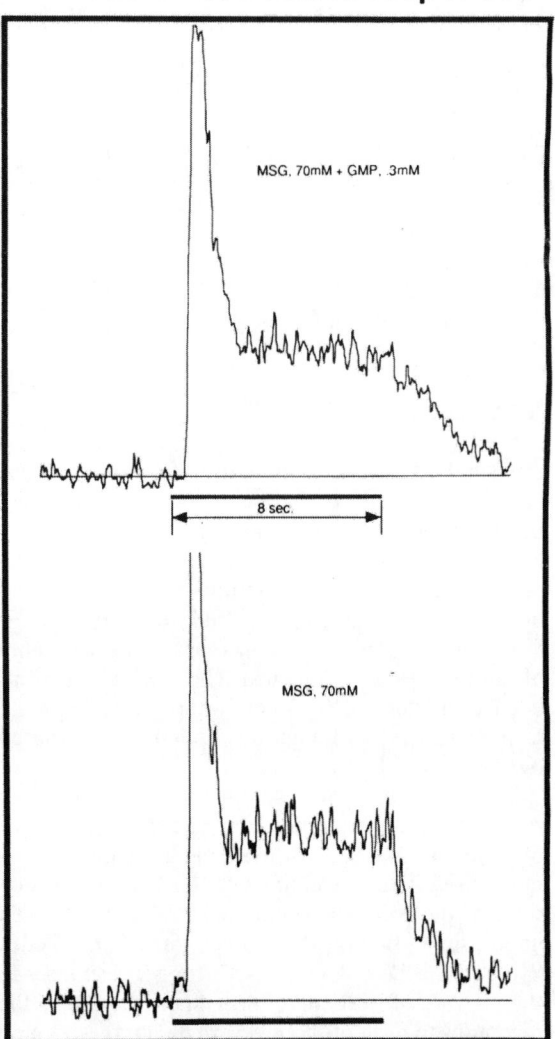

MSG, 70 mM + GMP, .3 mM

8 sec.

MSG, 70 mM

FIG. 1. Summated monkey chorda tympani nerve responses to 0.3 mM GMP mixed with 70 mM MSG (*top*) and 70 mM MSG (*bottom*). The time for stimulation (8 s) is illustrated by the *lower horizontal line*. The *straight horizontal line* that continues into the summated trace shows the average spontaneous nerve activity before stimulation

observed in one sweet fiber but not in any salt fibers. No bitter or sour fibers responded.

In four taste fibers, 70 mM MSG with and without GMP evoked a larger response than NaCl, quinine, citric acid, or sucrose (Fig. 2). Figure 2 suggests that fibers responding to sour and bitter stimuli do not respond to MSG. This finding indicates that the taste of bitter and sour compounds are not relayed by the same fiber types as the ones mediating the taste of MSG and GMP.

Monkeys

In the monkeys, 70 mM MSG with and without GMP stimulated all 11 salt fibers and 6/15 (40%)

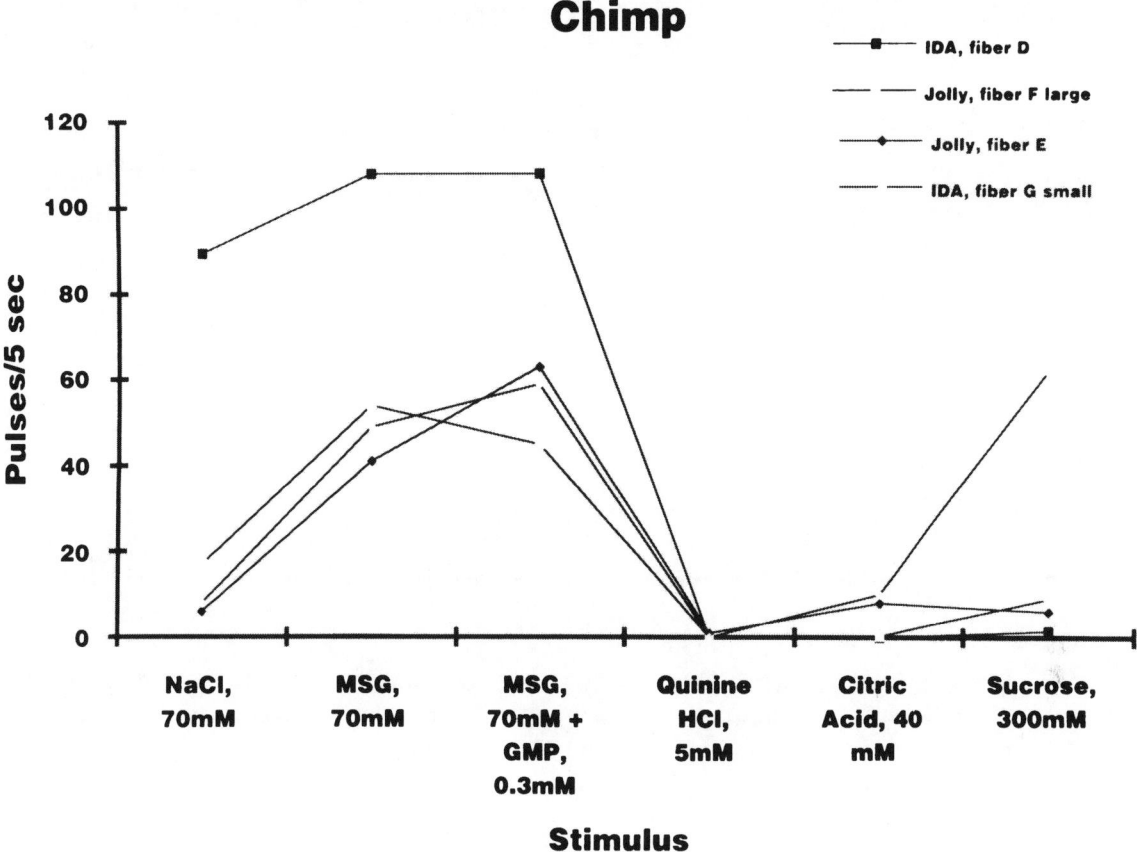

Fig. 2. Response in chimpanzee single fibers, which responded best to MSG or MSG plus GMP. Tastants used during stimulation are shown below the x-axis

of the sweet fibers. As in the chimpanzees, the strongest responses were observed in salt fibers. No response was recorded in bitter fibers.

Figure 3 shows the response spectrum of the five fibers in which we recorded a larger response to the umami compounds than to any other tastant. The main difference between the response spectrum of the chimpanzee fibers, in Fig. 2, and that of the monkeys, in Fig. 3, is that the fibers from the monkeys responded to the acid as well.

Discussion

This study shows that in both monkeys and chimpanzees, the higher (70 mM) concentration of MSG stimulated mostly salt fibers. In the chimpanzees, in which we recorded the effect of both 10 and 70 mM MSG, the results may be interpreted as a shift in the taste quality of MSG, when the concentrations increased, from sweet/salty to a more salty taste.

This interpretation is based on the assumption that our earlier conclusion [3] is correct: that the identity of the taste fiber carries the information on the taste quality. We base this assumption on the

finding that the classification of a single taste fiber is independent of stimulus concentrations but dependent on the taste quality of the stimulus. For example, increasing the concentration of the NaCl does not shift the response spectrum of the fibers. As a matter of fact, we have tested this response, although it is not shown here, by recording the response to 70 and 300 mM NaCl as well as three different concentrations of sucrose and aspartame. The data we obtained did not change the classification of the taste fibers. Thus, for example, aspartame never, used within physiologic ranges, stimulates nonsweet fibers. In short, changing the concentrations of compounds representing the four taste qualities does not change the classification of the taste fibers in the monkey or the chimpanzee.

This finding refutes the suggestions of Erickson [4] that were presented for the first time at ISOT I and then in a series of studies. He suggested that "*there are many fiber types representing gustatory quality* as for pitch discrimination, rather than a few fiber types as in color vision."

If our conclusions above are correct, the shift of fiber type responding when the concentration of MSG was increased from 10 mM to 70 mM indicates

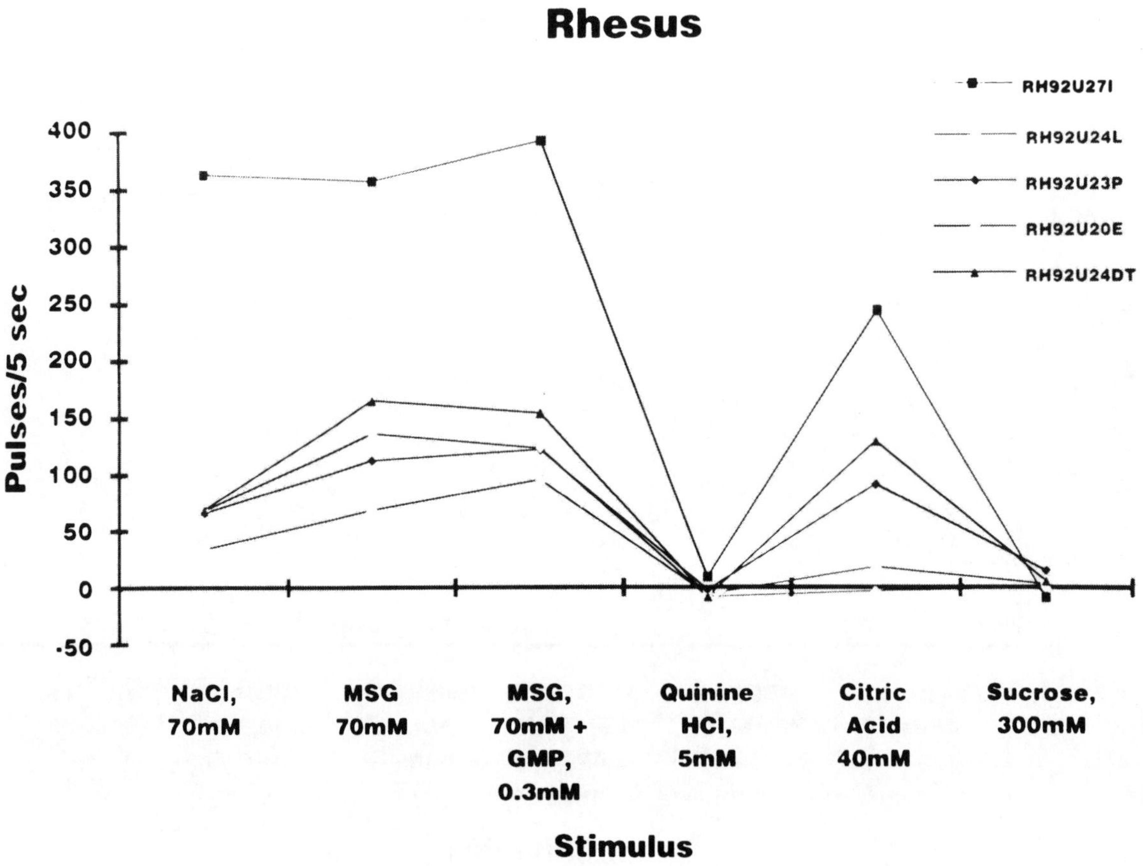

Rhesus

FIG. 3. Response in the five monkey single fibers, which responded best to MSG or MSG plus GMP. Tastants used during stimulation are shown below the x-axis

that there is no umami taste quality in the sense that there exist specific umami taste fibers in the chorda tympani nerve in the monkey or chimpanzee. However, because we have not recorded from other taste nerves, it is possible that this conclusion applies only to fibers in the chorda tympani nerve. It may not apply to glossopharyngeal nerve fibers. For example, Ninomiya and Funakoshi [5] described fibers specifically sensitive to MSG in the glossopharyngeal nerve of mice. Furthermore, because there are differences between the sense of taste in the monkey and the chimpanzee [6] and perhaps also between humans and chimpanzees, and fiber specificity is higher in the chimpanzee than in the monkey judged by the breadth of tuning, it is possible there are specific umami fibers in humans. On the other hand, our findings do explain the varying results of attempts to describe or categorize the umami taste in humans [2].

Acknowledgments. The data were collected with financial support from the Umami Manufacturers Association of Japan; NutraSweet R&D, Mt. Prospect, IL, USA; and grant AA09391 from the NIH.

We thank Drs. J. Moor-Jankowski, E. Muchmore, and W.W. Socha for permission to study the chimpanzees at the Laboratory for Experimental Medicine and Surgery in Primates (LEMSIP), New York Medical Center, Tuxedo, NY, USA.

References

1. Yamaguchi S (1991) Basic properties of umami and effects on humans. Physiol Behav 49:833–841
2. O'Mahony M, Ishii R (1987) The umami taste concept: implications for the dogma of four basic tastes. In: Kawamura Y, Kare MR (eds) Umami: A Basic Taste. Dekker, New York, pp 75–93
3. Hellekant G, Ninomiya Y (1991) On the taste of umami. Physiol Behav 49:922–934
4. Erickson RP (1963) Sensory nerural patterns and gustation. In: Zotterman Y (ed) Olfaction and taste. Pergamon Press, Oxford, pp 205–214
5. Ninomiya Y, Funakoshi M (1987) Qualitative discrimination among "umami" and the four basic taste substances in mice. In: Kawamura Y, Kare MR (eds) Umami: A Basic Taste. Dekker, New York, pp 365–385
6. Hellekant G, van der Wel H (1989) Taste modifiers and sweet proteins. In: Cagan R (ed) Neural Mechanisms of taste. CRC Press, Boca Raton, FL, pp 85–96

Umami Taste in the Forebrain of the Alert Macaque

Thomas R. Scott[1], Carlos R. Plata-Salaman[2], Edmund T. Rolls[3], Zoltan Karadi[4], and Yutaka Oomura[5]

Key words. Umami—Taste—Macaque—Monkey—Electrophysiology

Introduction

"Umami" is the term proposed by Ikeda in 1909 to represent the taste of glutamate, the salt of the amino acid he had isolated from flavor-enhancing sea tangle. It has since been shown to be transduced by specific protein receptor molecules in a variety of animals [1]. Peripheral and central taste neurons are responsive to monosodium glutamate (MSG), and some show exclusive sensitivity to it. In multidimensional taste spaces generated from either human psychophysical or rat electrophysiologic data, MSG lies beyond the tetrahedron created by connecting the positions of the four traditional basic taste stimuli [2,3]. This fact implies that the taste of MSG cannot be accounted for by any combination of these basic stimuli. MSG is typically situated closer to NaCl than it is to sucrose, HCl, or quinine, suggesting a salty component to its taste. Yet when sodium transduction was disrupted in the rat by lingual application of amiloride, the impact on MSG's response was minimal, and its neural code remained unaltered [4]. Therefore MSG presumably elicits a taste quality that transcends saltiness—one whose identity is sustained even as saltiness is largely blocked.

The taste of MSG cannot be generalized to those of the other basic stimuli in behavioral tests using rats; nor does the addition of glutamate to solutions of salts, acids, sugars, or quinine enhance the neural effects of these stimuli [5]. Thus, evidence from studies of receptor processes, through sensory coding and extending to behavior, indicates that umami is a taste quality of independent standing.

We have studied the effects of umami compounds on the taste systems of macaque monkeys. MSG served as a taste stimulus during recordings from the cortex [6–8], amygdala [9], and hypothalamus [10,11]; and inosine monophosphate (IMP) was used in the cortex. Here we characterize the electrophysiologic responses of the macaque to these stimuli.

Responses to Umami Compounds

Monosodium glutamate is an effective stimulus throughout the monkey's taste system. A few cortical neurons responded to MSG at concentrations as low as 0.001 M, but the dynamic middle range of intensities was 0.009 to 0.300 M, and these concentrations were used in our studies.

At each of the four synaptic levels studied—primary (insular-opercular) taste cortex, secondary (orbitofrontal) taste cortex, amygdala, and lateral hypothalamic area—about one-third of taste-responsive neurons were activated by MSG. The mean net discharge rate to MSG was 3.5 spikes/s, about equal to those elicited by sodium-lithium salts at the same concentrations. Many of the same neurons that were activated by MSG responded well to IMP at concentrations as low as 0.1 mM.

Taste Quality of Monosodium Glutamate

It has been proposed that activity in the primary taste cortex is the purveyor of quality-intensity discriminations in the macaque and human [12]. This hypothesis derives from the finding that similarities among neural patterns evoked by tastants in the primary cortex bore a close resemblance to human reports of the relative similarities among taste qualities [7,8,13]. At higher-order gustatory synapses of the orbitofrontal cortex, amygdala, and hypothalamus, however, this association was degraded. Taste-responsive cells at these relays could not account for the fine discriminative capacity reported by humans, but they assumed another role: analysts of the hedonic character of a gustatory stimulus. Neurons in the orbitofrontal cortex and ventral forebrain exhibited discharge rates proportional to the degree of appeal a tastant held for the subject [14,15].

[1] Department of Psychology, [2] Department of Biology, University of Delaware, Newark, DE 19716, USA
[3] Department of Experimental Psychology, University of Oxford, Oxford OX1 3UD, UK
[4] Institute of Physiology, University Medical School, Pecs 7643, Hungary
[5] Institute of Bioactive Science, Nippon Zoki Pharmaceutical Co., Kato-Gun, Hyogo 673-14, Japan

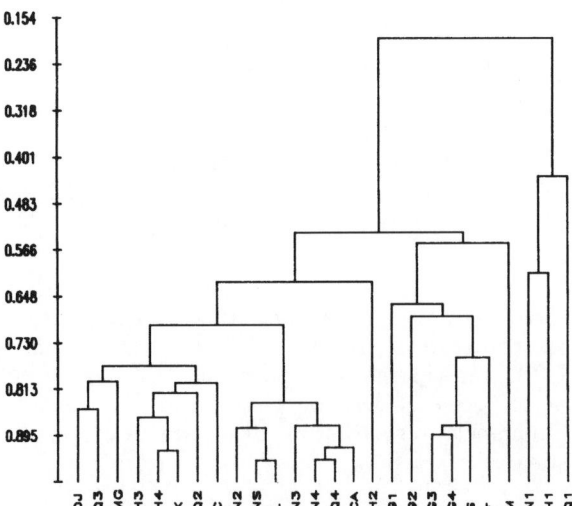

FIG. 1. "Dendrogram" showing the relative similarity among 26 taste stimuli as seen in the amygdala of the macaque monkey. The measure of similarity between any pair of stimuli was the correlation coefficient between the response profiles they evoked. The matrix of coefficients between all possible pairs of stimuli ($26 \times 25/2 = 325$) served as the basis for linking individual pairs and eventually groups until all 26 chemicals were joined. The level of correlation at which any stimulus or group is paired with another is indicated by the *horizontal line* between them. The scale of coefficients to which these lines may be referred is shown on the ordinate. The length of the *vertical line* leading to any chemical or group is a measure of its isolation from other stimuli. The three chemicals on the *far right* (N1, H1, Q1) were applied at concentrations below threshold and should not be considered taste stimuli. The remaining 23 tastants split at a correlational level of +0.54 into MSG (M) and sugars [glucose (G1, G2, G3, G4), sucrose (S), fructose (F)] toward the *right*, *vs* salts, acids, and alkaloids toward the *left* (all abbreviations are not listed). MSG established its independence from the sugars almost immediately thereafter, at a level of +0.56. Therefore the taste quality of MSG is at least as distinct from those of the other basic tastes as they are from one another

Monosodium glutamate evoked a response profile that bore little resemblance to those elicited by sucrose, HCl, or quinine at any synaptic level tested. However, in the primary taste cortex the correlation between the response patterns generated by MSG and NaCl was high: +0.71 [7]. Accordingly, MSG has a pronounced salty component for humans [16], a perception likely to derive from responses at this neural level.

At successive synapses, the relation between MSG and NaCl deteriorated. In the amygdala a cluster analysis of similarity among taste stimuli (Fig. 1) implied that MSG had become more closely affiliated with sugars than with sodium salts, a manifestation of one of the translations of umami: "deliciousness."

In sum, the relations between the neural profile generated by MSG and those of the four basic taste stimuli is increasingly distant and labile as one progresses through the system. At higher-order gustatory relays, MSG is no more similar to the recognized basic tastes than they are to each other. These neurophysiologic results reinforce the data from receptor and behavioral studies in arguing for the independent status of MSG as a basic taste stimulus serving as the prototype for the umami quality.

Neural Response to MSG as a Function of Satiety

In primary (insular-opercular) taste cortex, neurons responded to chemicals as a function of their quality and intensity. At the succeeding gustatory synapse in the orbitofrontal cortex, activity was based on the hedonic value of the stimulus to the monkey at the moment. Thus, as the monkey continued to consume a hedonically rewarding stimulus, responsiveness to that solution was reduced, approaching the spontaneous rate as satiety was achieved [15]. This neural concomitant of the phenomenon of sensory-specific satiety, restricted heretofore to sugars, has now been demonstrated for MSG. Evoked responses to MSG declined in both the visual and gustatory modalities as monkeys consumed a 0.1 M solution to satiety.

Conclusion

Monosodium glutamate and inosine monophosphate are effective stimuli in the forebrain of the macaque. Although closely associated with NaCl in the primary gustatory cortex, MSG achieves independence at higher-order synapses. In the secondary taste cortex and in gustatory regions of the ventral forebrain, the response profile evoked by MSG is disparate from those of NaCl, sucrose, HCl, and quinine, implying independence of the umami taste from the four recognized basic qualities. Because activity in these neural areas is thought to be related to the hedonic value of a stimulus, MSG presumably elicits a hedonic tone distinct from those of the other basic tastants.

Acknowledgments. Research grant BNS-9001213 from the National Science Foundation supported much of the research reported here.

References

1. Kurihara K (1987) Recent progress in taste receptor mechanisms. In: Kawamura Y, Kare MR (eds) Umami: A basic taste. Dekker, New York, pp 3–39
2. Pritchard TC, Scott TR (1982) The taste of amino acids. II. Quality and neural coding. Brain Res 253: 93–104

3. Schiffman SS, Erickson RP (1971) A psychophysical model for gustatory quality. Physiol Behav 7:617–633
4. Giza BK, Scott TR (1991) The effect of amiloride on taste-evoked activity in the nucleus tractus solitarius of the rat. Brain Res 550:247–256
5. Yoshii K, Yokouchi C, Kurihara K (1986) Synergistic effects of 5′ nucleotides on rat taste responses to various amino acids. Brain Res 367:45–51
6. Baylis LL, Rolls ET (1991) Responses of neurons in the primate taste cortex to glutamate. Physiol Behav 49:973–979
7. Plata-Salaman CR, Scott TR, Smith-Swintosky VL (1992) Gustatory neural coding in the monkey cortex: L-amino acids. J Neurophysiol 67:1552–1561
8. Smith-Swintosky VL, Plata-Salaman CR, Scott TR (1991) Gustatory neural coding in the monkey cortex: stimulus quality. J Neurophysiol 66:1156–1165
9. Scott TR, Karadi Z, Oomura Y, et al. (1993, in press) Gustatory neural coding in the amygdala of the alert macaque monkey. J Neurophysiol 69:1810–1820
10. Karadi Z, Oomura Y, Nishino H, et al. (1992) Responses of lateral hypothalamic glucose-sensitive and glucose-insensitive neurons to chemical stimuli in behaving rhesus monkeys. J Neurophysiol 67:389–400
11. Oomura Y, Nishino H, Karadi Z, Aou S, Scott TR (1991) Taste and olfactory modulation of feeding related neurons in behaving monkey. Physiol Behav 49:943–950
12. Scott TR (1992) Taste, feeding and pleasure. Progr Psychobiol Physiol Psychol 15:231–291
13. Plata-Salaman CR, Scott TR, Smith-Swintosky VL (1993) Gustatory neural coding in the monkey cortex: the quality of sweetness. J Neurophysiol 69:482–493
14. Burton MJ, Rolls ET, Mora F (1976) Effects of hunger on the response of neurons in the lateral hypothalamus to the sight and taste of food. Exp Neurol 51:668–677
15. Rolls ET, Sienkiewicz ZJ, Yaxley S (1989) Hunger modulates the responses to gustatory stimuli of single neurons in the caudolateral orbitofrontal cortex of the macaque monkey. Eur J Neurosci 1:53–60
16. O'Mahoney M, Ishii R (1987) The umami taste concept: implications for the dogma of four basic tastes. In: Kawamura Y, Kare MR (eds) Umami: A basic taste. Dekker, New York, pp 75–93

Central Function in Preference for Amino Acids

Taketoshi Ono[1], Kunio Torii[2], Eiichi Tabuchi[1], Takashi Kondoh[1], and Teruko Uwano[1]

Key words. Preference—Rat—Amino acid—LHA—Lysine deficiency—Ingestion

Introduction

Food intake in animals is influenced by the composition of amino acids in the diet. Preference for monosodium L-glutamate (MSG), also known as umami, is evident when dietary proteins are sufficient, whereas preference for NaCl is induced during marginal deficiency of some dietary proteins [1]. Deficiency of an essential amino acid in the diet causes a change of the taste preference to induce selective behavior that favors the deficient amino acid [1,2]. Thus animals skillfully and precisely change their preference for food or liquid when a change of their diet results in a deficiency of some component(s).

In the lateral hypothalamic area (LHA), we found MSG-specific neurons in a control diet situation and lysine-specific neurons during lysine deficiency [3]. In combined neuronal recording and behavior studies, significantly more of these neurons responded specifically to the solution that contained the substance that was deficient; MSG was the preferred amino acid in the control diet condition, and lysine was most preferred during lysine deficiency. The data strongly suggest the plasticity of LHA neuron responses to stimulation by specific tastes and their integration of taste with the body's physiologic need of that substance. The concentration of an essential amino acid in the plasma and the brain decreased a few hours after rats were exposed to a diet deficient in the essential amino acid [1,2]. The data indicate that the central nervous system (CNS), especially the LHA, might influence amino acids and protein metabolism through regulation of amino acid ingestion.

Three studies were undertaken to investigate central functions regarding preference for amino acids and their dependence on the blood level. In experiment 1 LHA neuronal responses to ingestion of amino acid solutions and NaCl were recorded in lysine-deficient rats after they learned selective ingestion in their respective conditions. In Experiment 2 LHA neuronal responses to iontophoretic application of amino acids were recorded. In experiment 3 the effects of intraperitoneal (IP) injection of lysine or a lysine-degrading agent on selective ingestion of lysine solution were examined.

Methods and Results

The control diet consisted of starch, wheat gluten, L-amino acid mixture, minerals, and vitamins. The lysine-deficient diet consisted of the control diet with a level of lysine too low to maintain appropriate physiologic processes and functions (Lysine is an essential amino acid for rats). The lysine-deficient and control diets were isocaloric and isonitrogenic.

Experiment 1: LHA Neuronal Responses to Ingestion of Amino Acid Solutions Under Different Nutritional Conditions

Method

Five rats of each group were housed in a wooden cage and fed a control diet or a lysine-deficient diet and six solutions (0.2 M L-lysine HCl, 0.15 M MSG, 0.05 M L-arginine, 0.5 M glycine, 0.15 M NaCl, and distilled water) ad libitum. We then confirmed that rats fed the control diet selectively ingested MSG and arginine solutions, and that rats fed the lysine-deficient diet learned to ingest lysine solution selectively. To relate neuronal activity to amino acid ingestion, a receptacle mounted surgically on the head of each rat was specially designed and constructed to accept modified ear bars. This device restrained an unanesthetized rat painlessly in a stereotaxic instrument for neuronal recording [3].

After the stereotaxic surgical procedure, the rat was placed in a stereotaxic apparatus and trained to lick a drop of one of the six solutions presented through a spout. The spout was automatically extended close to its mouth for 2 s following a 2-s cue tone stimulus (1000–8750 Hz), which had a different frequency for each solution. Single LHA neuron activity was recorded through glass microelectrodes (4 M NaCl, 2–3 MΩ) during ingestion of taste solu-

[1] Department of Physiology, Faculty of Medicine, Toyama Medical and Pharmaceutical University, Toyama, 930-01 Japan
[2] Ajinomoto Co., Inc., Central Research Laboratories, Yokohama, 244 Japan

tions and the conditioned cue tones. Between recording sessions, each rat was fed either the control
or the lysine-deficient diet.

Results

During the cue tone and subsequent ingestion of
amino acid solutions and saline, 379 LHA neurons
(176 in control and 203 during lysine deficiency)
were analyzed. For the control situation 77 neurons
(44%) responded to any cue tone, and 92 (52%)
responded to ingestion of any solution. During lysine
deficiency, 94 neurons (46%) responded to any cue
tone, and 117 (58%) responded to ingestion of any
solution. Of these neurons, 17 (9.6%) in the control
situation and 17 (8.4%) during lysine deficiency
differentiated the cue tone, the ingestion of the taste
solutions, or both. During lysine deficiency, more
neurons responded to the cue tone or ingestion
of lysine and NaCl solution, and fewer neurons
responded to ingestion of arginine solution than in
the control situation (Fig. 1A). Typically, during
lysine deficiency neurons that responded to the cue
tone or ingestion of the lysine or MSG solution also
responded to the cue tone or ingestion of the NaCl
solution (Fig. 1B), but this relation was not evident
in the control environment. Some neurons in the
control situation responded specifically to ingestion
of MSG solution; and during lysine deficiency some
responded specifically to ingestion of lysine solution.
The proportion of neurons responsive to MSG and
arginine solutions tended to be higher in the control
environment, and more were responsive to lysine
solution and saline during lysine deficiency. These
data were consistent with preference in the behavioral study.

Experiment 2: LHA Neuronal Response
to Application of Amino Acids
Under Different Nutritional Conditions

Method

After recovery from the stereotaxic operation, the
rat was allowed to eat the lysine-deficient diet or the
control diet ad libitum with free drinking of water.
Activity of LHA neurons was then recorded during
iontophoretic application of various amino acids.
Applied substances and their concentrations were
0.05 M MSG, 0.05 M glycine, 0.05 M threonine,
0.05 M lysine HCl, 0.05 M arginine, and 0.15 M NaCl.
These six solutions were administered from a seven-
barreled glass micropipet that had an attached glass
recording microelectrode. The tip of the recording
electrode projected about 10 μm beyond the tip of
the micropipet. Neuronal activity was recorded
during iontophoretic application of amino acids and
NaCl with strict criteria to avoid artifacts.

Results

The activity of 250 LHA neurons (122 in control
situation, 128 during lysine deficiency) was recorded
during application of amino acids. Of 250 neurons,
163 (73 in control situation, 90 during lysine deficiency) were analyzed and compared in two nutritive conditions; 87 (49 in control situation, 38 during
lysine deficiency) were omitted according to our
criteria for incorrect responses to iontophoresis. Of
163 neurons, 112 (49 in control situation, 63 during
lysine deficiency) responded during the application
of any amino acid, and 51 (24 in control situation, 27
during lysine deficiency) did not respond. With the
control diet situation, neurons responsive only to
glutamate were often observed (19/49, 39%), and
most of those responsive to several amino acids responded to glutamate (37/49, 76%). Some neurons
responded selectively to glutamate and threonine
(7/49, 14%). During lysine deficiency, neurons
responsive only to glutamate were also observed
(23/63, 36%), and those responsive to any amino
acid usually responded to glutamate (47/63, 75%).
Some neurons responded specifically to lysine or
lysine plus glutamate (7/63, 11%) (Fig. 2), and some
responded selectively to glutamate and threonine, as
in the control situation (7/63, 11%).

Experiment 3: Effects of IP Injection of Lysine or
a Lysine-Degrading Agent on Selective Ingestion

Method

Five rats of each group were housed together in an
equilateral octagonal cage. They were continuously
fed one diet, either a control or a lysine-deficient
diet, from a food container in the center of the floor,
and they could freely drink any of eight solutions.
The rats were usually exposed to light from 9:00
a.m. to 9:00 p.m. and to darkness from 9:00 p.m.
to 9:00 a.m. Each cylinder contained one of eight
solutions: 0.2 M lysine HCl, 0.15 M MSG, 0.05 M
arginine, 0.5 M glycine, 0.4 M threonine, 0.05 M
histidine, 0.15 M NaCl, and distilled water. The
concentration of each amino acid solution used was
that reported to be most preferred by rats [1]. In the
control situation and during lysine deficiency, the
volume consumed and the pattern of consumption of
each solution were measured in each of six groups;
three were administered a lysine-degrading agent
(lysine α-oxidase, EC 1.4.3.14, 100 IU/kg IP), and
three were given lysine (5 mmol/kg IP q12h) for 5
consecutive days to maintain the blood level of
lysine constantly low or high.

Results

Control rats much preferred MSG and arginine solutions to lysine solution. Lysine intake by rats fed the
control diet was less even when the blood level

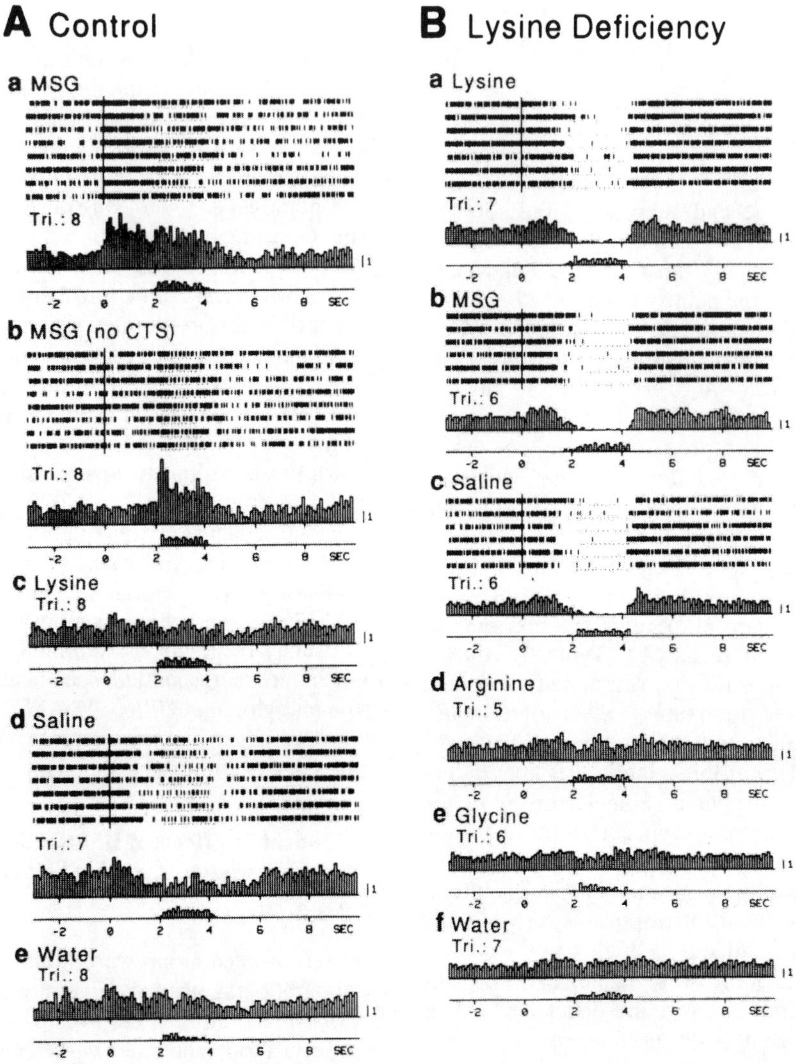

A Control

a MSG

Tri.: 8

b MSG (no CTS)

Tri.: 8

c Lysine

Tri.: 8

d Saline

Tri.: 7

e Water

Tri.: 8

B Lysine Deficiency

a Lysine

Tri.: 7

b MSG

Tri.: 6

c Saline

Tri.: 6

d Arginine

Tri.: 5

e Glycine

Tri.: 6

f Water

Tri.: 7

FIG. 1. Two examples of differential neurons in control situation (**A**) and during lysine deficiency (**B**). **A** Neuronal responses excite selectively during cue tone and ingestion of MSG solution and slightly inhibit during those of saline, but not for lysine solution and water. This neuron also responded to ingestion of MSG when its predicting cue tone was absent. **B** Neuronal responses inhibit selectively during ingestion of lysine solution, MSG solution, or saline, but not other solutions. During lysine deficiency LHA neurons often responded differentially to lysine solution and to saline. In each pair of averaged histograms: *top*, neuronal responses; *bottom*, lick signals; *Tri.*, trial number. *Dots* in each raster display, the time of licking. Calibration of one impulse is shown at the *right* of each *upper* histogram

of lysine was maintained at a low level by IP injection of lysine α-oxidase (Fig. 3). During lysine deficiency they preferred lysine solution and saline. The lysine intake during lysine deficiency increased even when the blood level of lysine was maintained at a high level by IP injection of lysine (Fig. 3).

Discussion

Experiments 1 and 2

Some LHA neurons responded specifically to ingestion of MSG solution under the control condition, and during lysine deficiency some of the neurons responded specifically to ingestion of lysine solution. The proportion of neurons responsive to MSG and arginine solutions was higher in the control situation and more were responsive to lysine solution and saline during lysine deficiency. During lysine deficiency LHA neurons responded frequently to applied lysine, whereas few responded to applied lysine in the nutritively complete condition. These data suggest that animals have the ability to adapt their sensors to external and internal stimuli under different nutritive conditions. This ability, which is probably involuntary, permits adjustment of the intake of nutrient substances necessary for their body.

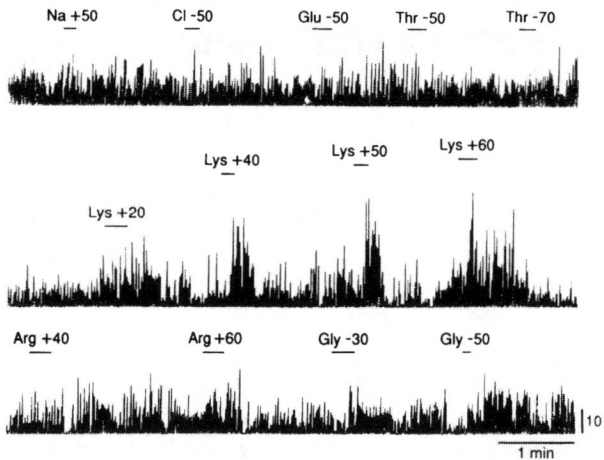

Na +50   Cl -50   Glu -50   Thr -50   Thr -70

Lys +40   Lys +50   Lys +60

Lys +20

Arg +40   Arg +60   Gly -30   Gly -50

|10

1 min

Fig. 2. Neuronal responses specific to iontophoretic application of lysine during lysine deficiency. This neuron responds selectively to lysine (Lys) applied by +20 nA and shows dose-dependent responses. During lysine deficiency LHA neurons frequently responded selectively to lysine or lysine plus glutamate (*Glu*). *Ordinate* shows impulses per second. *Lines* under applied substances and numbers indicate the current duration and the current intensity. Calibration of five impulses is shown at *bottom right*

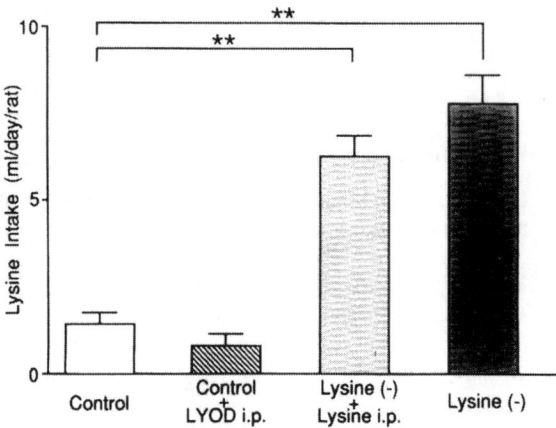

Fig. 3. Lysine intake in control situation, during lysine deficiency, and after IP injection of lysine α-oxidase or lysine. Lysine intake increased significantly above control during lysine deficiency ($p < 0.01$). Lysine intake in controls did not increase during IP injection of lysine α-oxidase. During lysine deficiency, lysine intake also increased during IP injection of lysine. *Ordinate* shows mean lysine intake of three groups on the fourth day after each diet or agent given

Experiment 3

The blood level of lysine was changed when the control diet or the lysine-deficient diet was offered. However, the IP injection did not affect the selective ingestion of lysine, which suggests that selective ingestion of lysine depends on more than the blood level of lysine alone. Experiments are presently in progress to verify this hypothesis and to establish additional cause and effect relations.

Conclusion

The LHA might be involved in intrinsic regulation of amino acids (i.e., the monitoring of amino acids) in the blood and body fluids. The blood lysine level, lowered by loss of lysine intake, could, along with some other factor(s), trigger LHA neurons directly or indirectly and elevate the responsivity of LHA neurons to lysine. This reaction could induce another system(s) to direct the selection of lysine ingestion. A reward response could promote selective ingestion of a deficient amino acid, if available. The present findings suggest that the LHA might help integrate intrinsic and extrinsic regulation of amino acids; moreover, preference for some substances that are deficient in the body might be caused by plastic changes of the peripheral nervous system (including alimentary tract sensors from the oral cavity to the intestines) and the CNS, including the LHA.

References

1. Mori M, Kawada T, Ono T, Torii K (1991) Taste preference and protein nutrition and L-amino acid homeostasis in male Sprague-Dawley rats. Physiol Behav 49:987–995
2. Torii K, Mimura T, Yugari Y (1987) Biochemical mechanism of umami taste perception and effect of dietary protein on the taste preference for amino acids and sodium chloride in rats. In: Kawamura Y, Kare MR (eds) Umami: a basic taste. Dekker, New York, pp 513–563
3. Tabuchi E, Ono T, Nishijo H, Torii K (1991) Amino acid and NaCl appetite, and LHA neuron responses of lysine-deficient rat. Physiol Behav 49:951–964

Effect of Umami Taste of Monosodium Glutamate on Early Humoral and Metabolic Changes in the Rat

R. Caulliez, C. Viarouge, and S. Nicolaidis[1]

Key words. Monosodium glutamate—Umami—Metabolism—Thermic effect of food—Indirect calorimetry—Anticipatory reflexes

Introduction

Nutritional substrates in the internal milieu act on their specific effectors to produce the appropriate endocrine responses and, as a result, their proper cellular utilization. Before reaching the internal milieu, nutrients have been in contact with the oral and gastrointestinal walls and their nutrient specific receptors. It was shown that specific neuronal messages from the gustatory and gastrointestinal system reach hypothalamic structures [1–3] and trigger endocrine and metabolic responses in an anticipatory way; that is, they produce an effect similar to the one they produce after having reached the systemic compartment. These anticipatory responses were shown to be sensory-specific [4].

Monosodium glutamate (MSG) associated with other nutrients enhances their hedonic impact and acceptability [5] and was suggested to be the sensory message alerting the organism to the arrival of amino acids [6]. Because amino acids elicit a double α and β islet activation, we asked, in a first experiment [7], if the oral sensing of MSG could be followed by changes in circulating glucose, insulin, and glucagon concentrations similar to those seen after protein ingestion. Are these possible changes, if any, specifically due to the sensory effect of MSG, or are they also produced by systemic action of MSG? In the second [7] and the third [8] experiments, we investigated the action of MSG at the oral level on energy metabolism, paired to either a non-nutritious or a standard nutritious meal.

Methods

Experiment 1

Two groups of Wistar rats were implanted with an intrajugular catheter [9], allowing blood sampling and intravenous MSG administration in one group of animals. The other group of rats received, in addition, a chronic intraoral cannula [10] in order to produce the oral stimulation with MSG. The animals were deprived of food for 4 h and a blood sample was obtained before the MSG injections. Then either MSG (0.5 ml, 0.05 M) or vehicle (distilled water) was injected via either the intravenous or oral catheter. Three more blood samples were withdrawn, at 5, 15, and 30 min posttreatment. Plasma glucose, insulin, and glucagon levels were assayed [7,11].

Experiment 2

Energy metabolism was measured using an open circuit calorimeter [12]. This experimental device consisted of a small cage mounted on three dynamic force transducers so the spontaneous activity produced by the animal could be quantitatively measured. The food cup was weighed independently, and the temperature in the cage was closely controlled by a powerful feedback-controlled thermoregulatory system. This metabolic device allowed continuous measurement of both the respiratory quotient ($RQ = CO_2/O_2$), total metabolism, energy cost of spontaneous activity, and hence the background metabolism (BM). BM represents the resting metabolism in the nonresting animal and was computed by subtracting the energy cost of spontaneous activity from the total non-resting metabolic rate calculated by the Lusk formula [13,14]. Thus BM is crucial for the appreciation of post-stimulation metabolism because, in free-moving animals, it varies considerably and so makes comparisons almost impossible. Placed in the metabolic device, rats which had been deprived of food for 4 h were given 6 g of control or test meal. Control meals consisted of a gel made of alimentary gelatin powder in tap water (40 g/l). Test meals were supplemented (0.05 M) with MSG. These meals provided

[1] Laboratoire de Neurobiologie des Régulations, Collège de France, 11 Place Marcelin Berthelot, 75231 Paris Cedex 05, France

0.16 kcal/g, which is a negligible amount of calories. The experiment lasted until the next day.

Experiment 3

Ten male Wistar rats implanted with an intraoral cannula and placed in the metabolic device were given, after one night of food deprivation, 3 g of powder chow. MSG solution 0.5 ml (0.01 or 0.15 M) in distilled water or its vehicle (distilled water) alone was injected through the oral cannula. The injection was performed so it could be stopped whenever the rat made a short pose. In this way the injection was always associated with the taste of the food. The animal's BM and RQ were monitored for 8 h after the beginning of the meal. Because a meal induces a normal increase in BM, referred to as diet-induced thermogenesis (DIT), the area between the increasing curve of BM induced by the meal and the horizontal line corresponding to the average premeal BM (expressed in joules) was calculated for 10-min intervals during the first 2 h. RQ changes were expressed as the difference between the higher value of RQ reached almost 1 h after the onset of the meal and the average level of RQ just before the meal. These variations were also expressed as percentages of the initial premeal value of the RQ.

Results

Experiment 1

Figures 1, 2, and 3 show that sensing of pure umami taste induced by an MSG water solution did not induce significant changes in circulating glucose, insulin, or glucagon. However, a delayed (30 min) and significant increase in plasma insulin ($p < 0.05$) was found in response to the intravenous injection of MSG.

Experiment 2

Figure 4 shows the pooled data on metabolic changes in response to the ingestion of a gelatin gel supplemented with MSG. The amount consumed was 1.23 ± 0.26 (SEM) g of MSG flavored and 0.90 ± 0.45 g of the nonflavored gel. The BM and RQ remained unchanged regardless of whether the gelatin meal was umami flavored.

Experiment 3

Quantitative variations of DIT are pooled in Fig. 5. A clear bimodal shape can be shown with two acrophases, one near 30 min and the other near 80 min after the onset of the meal in each case of injection. The injection of MSG induced a significantly larger increase of DIT compared to vehicle at

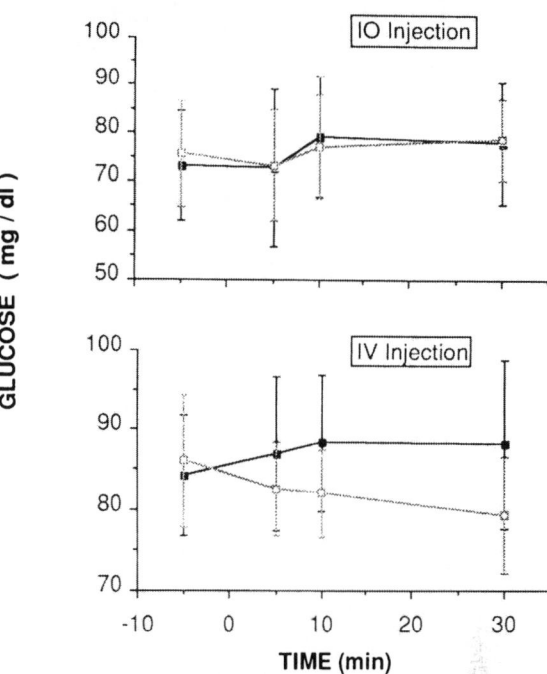

FIG. 1. Changes in blood glucose levels before and after intravenous (*IV*) or oral (*IO*) injection (time 0 min) of 0.5 ml MSG (0.05 M) (*filled squares*) or vehicle (*open squares*). Data are expressed as mean ± standard deviation (SD). (From [7], with permission)

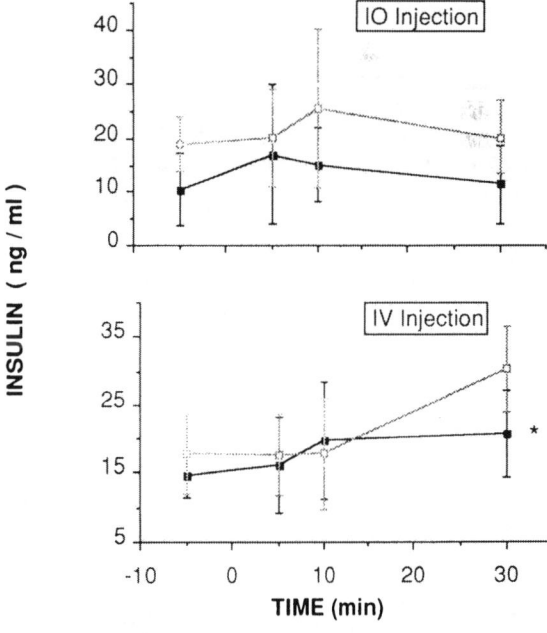

FIG. 2. Changes in blood insulin levels before and after intravenous or oral injection (time 0 min) of 0.5 ml MSG (0.05 M) (*filled squares*) or vehicle (*open squares*). Data are expressed as mean ± SD. *$p < 0.05$. (From [7], with permission)

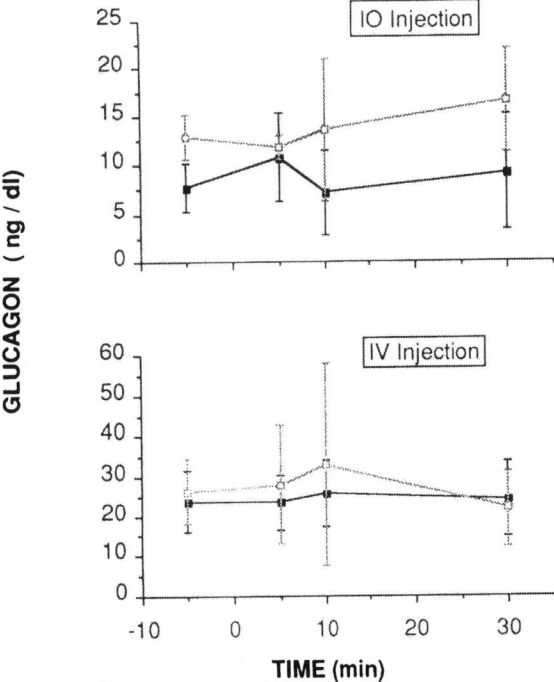

FIG. 3. Changes in blood glucagon levels before and after intravenous or oral injection (time 0 min) of 0.5 ml MSG (0.05 M) (*filled squares*) or vehicle (*open squares*). Data are expressed as mean ± SD. (From [7], with permission)

10 and 20 min after the onset of the meal. After 30 min, this effect seemed to be reversed, and a significantly smaller increase of DIT in the case of MSG injection was found at 50 min for MSG 0.01 M and at 80 and 100 min for MSG 0.15 M. No significant difference was found between the effect of the two concentrations of MSG used in this experiment. The rate of ingestion (as measured by the slope of the food-cup strain-gauge) did not differ regardless of whether the chow was supplemented with MSG.

The RQ (expressed as a percent of the premeal value) (Fig. 6) showed a sustained increase after the onset of the meal in both the MSG and control subjects. However, the MSG groups showed a significantly lesser increase. This RQ change was not different between the two MSG concentrations.

Discussion

Data from the two first experiments indicate that the umami taste provided by the MSG solution does not seem to produce, by itself, a significant effect on metabolic parameters. Neither plasma glucose nor plasma insulin and glucagon showed the hypothesized increase that could ultimately affect the metabolic rate and RQ. The absence of clear-cut changes

of circulating metabolic parameters is in agreement with the absence of changes in BM and RQ that vary minimally, only as a function of the amount consumed spontaneously by the rat regardless of whether MSG was added. Thus it could be concluded that, contrary to previous reports on sweet-taste-induced hypermetabolism [1], MSG by itself does not enhance metabolism. The results suggest that umami taste could actually act on metabolism under some ingestive conditions. Because the aqueous, noncaloric test meal by itself had no effect on metabolism, little could be expected by enhancing such a non-thermogenic meal. MSG could be (as in the case of taste) only an enhancer of thermogenesis, not a thermogenic agent per se. This idea is reinforced by results from the third experiment, which demonstrate that the taste of MSG added to a complete diet brings about significant changes in metabolism. DIT shows an earlier, more dramatic enhancement than in control meals, and this enhancement lasts approximately 20 min after the oral stimulation. MSG-supplemented meals show, as normal meals do [1,15–17], a bimodal shape in their DIT; but this bimodal shape is more pronounced than in control meals. Not only is the first thermogenic phase (anticipatory phase) larger and sharper but the metabolic depression between the first and the second phase (the latter probably corresponds to the postabsorptive metabolism of ingestants) is also deeper and lasts 30 to 100 min. This quantitative change is accompanied by a qualitative one, as RQ itself shows a lesser increase. The absence of a dose-dependent effect when a higher concentration of MSG was used is probably due to a ceiling effect at the low concentration.

These early metabolic responses correspond to the anticipatory reflexes previously described for sweet and for "fat" taste [1]. The umami-induced fake announcement of ingestion of a supplement of proteins reflexly results in what the ingestion of a real protein supplement would have induced, i.e., an extra increase in DIT and a switch of RQ changes toward figures that characterize protein (RQ = 0.83) rather than carbohydrate (RQ = 1.0) utilization. The MSG-induced enhancement of the early phase does not seem to be due to an increased palatability bound to the MSG addition because, according to Le Magnen [18], such an increase in palatability should have also increased the rate of ingestion, something that was not observed in the present experiment.

It therefore appears that the taste of MSG is metabolically efficient only when it is associated with an already caloric nutrient. According to Beauchamp [19], the pleasantness of umami may be manifest only in combination with other flavors. An explanation is that the chow, but not the gel solution, provides the nucleotides known to potentiate the enhancing properties of MSG [20].

FIG. 4. Metabolic and behavioral parameters preceding and 20 min after the onset of 12 control gelatin meals (*top*) vs 9 test gelatin + MSG meals (*bottom*). Values are expressed as means ± SEM. From the *upper* to the *lower* part of the figure: respiratory quotient (*RQ*); total and background metabolism (*TM & BM*); food intake (*FI*); locomotor activity (*Act*). (From [7], with permission)

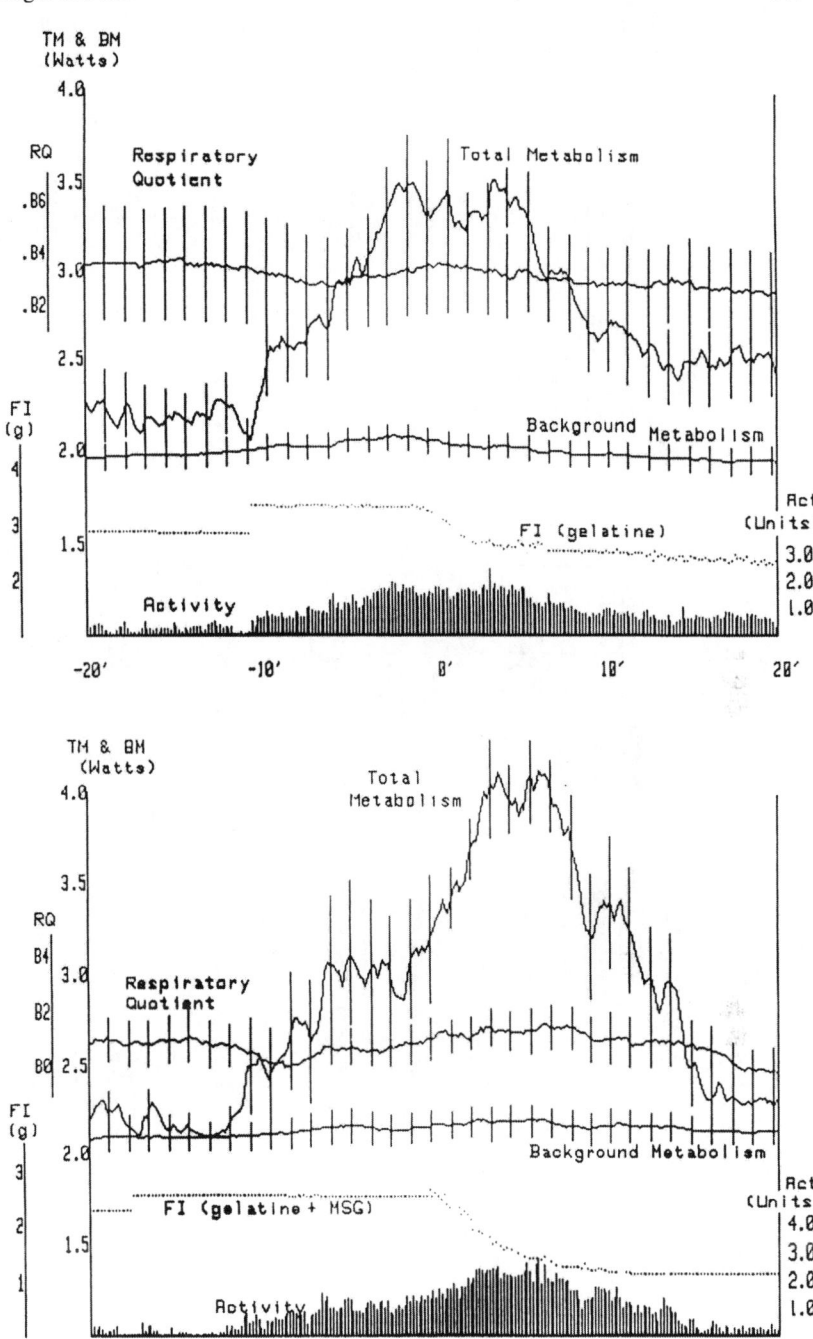

Conclusion

The umami taste of MSG reveals another macronutrient (protein)-specific anticipatory reflex leading to an early adaptation of the metabolic mechanism toward the gustatory announcement of the subsequent intestinal absorption and utilization of a perceived protein-rich diet.

Acknowledgments. We are grateful to Patrick Even for his contructive comments and to Laurent Poirier for his excellent technical assistance. This work was supported by grants from (180089013-00296 APE9311, MRT 88G-0523) and from the Fédération Française pour la Nutrition.

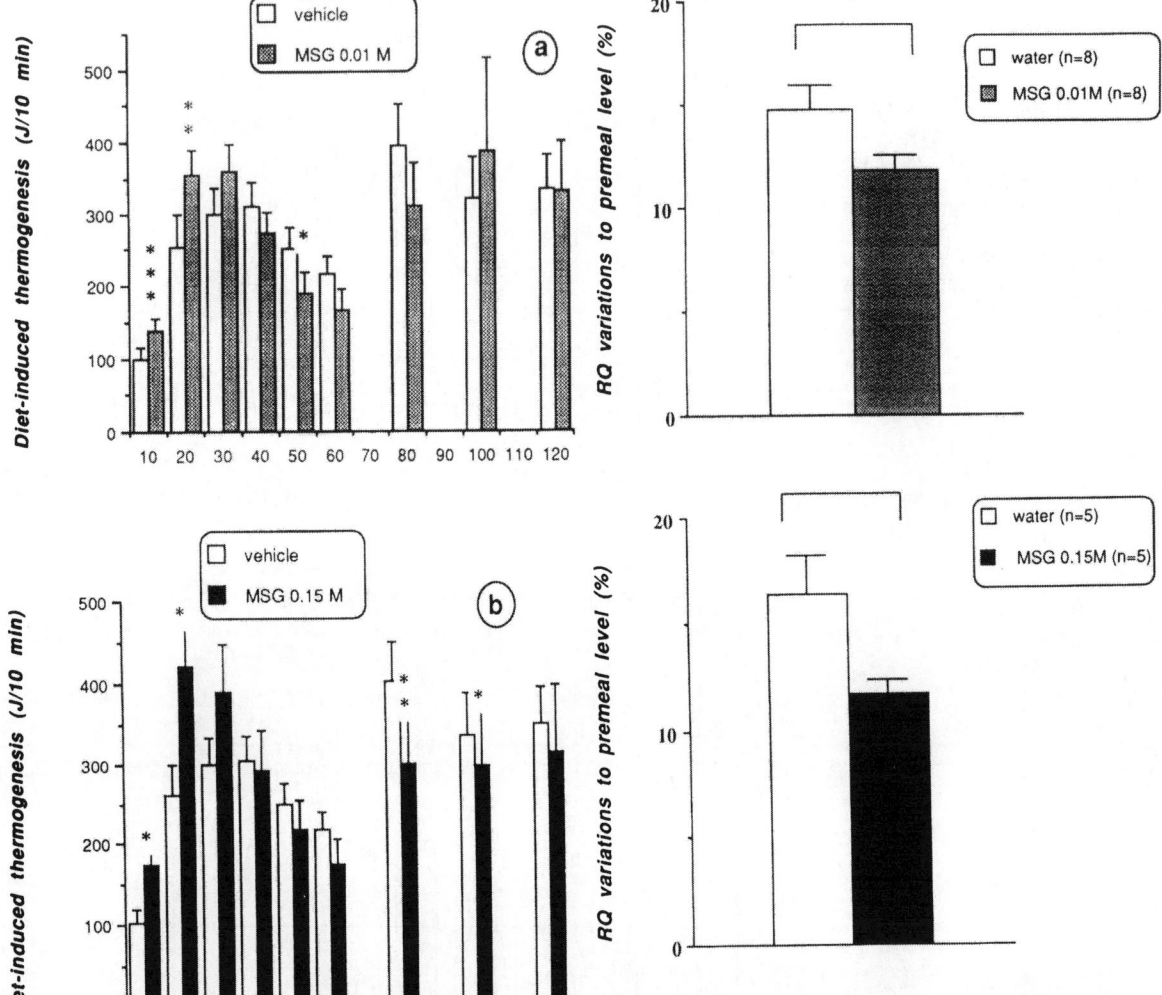

Time after the onset of meal (min)

FIG. 5. Comparison of the effect on diet-induced thermogenesis of an intraoral injection of vehicle, 0.01 M MSG ($n = 8$) (*top*), or 0.15 M MSG ($n = 5$) (*bottom*) during a meal of standard chow. $*p < 0.05; **p < 0.01; ***p < 0.001$. (From [8], with permission)

FIG. 6. Comparison of the effect of intraoral infusion of distilled water, 0.01 M MSG ($n = 8$) (*top*), or 0.15 M MSG ($n = 5$) (*bottom*) during a meal of standard chow on diet-induced increase of RQ as a percentage of the RQ premeal value. $*p < 0.05$. (From [8], with permission)

References

1. Nicolaïdis S (1969) Early systemic responses to orogastric stimulation in the regulation of food and water balance: functional and electrophysiological data. Ann NY Acad Sci 157:1176–1203
2. Nicolaidia S (1963) Effets sur la diurèse de la stimulation des afférences buccales et gastriques par l'eau et les solutions salines. J Physiol (Paris) 55:309–310
3. Norgren R, Leonard CM (1971) Taste pathways in rat brainstem. Science 173:1136–1139
4. Yin TH, Tsai WH, Barone FC, Wayner MJ (1979) Effect of continuous intramesentric infusion of glucose and amino acids on food intake in rats. Physiol Behav 22:1207–1210
5. Maga JA (1987) Organoleptic properties of umami substances. In: Kawamura Y, Kare MR (eds) Umami: A basic taste. Marcel, New York, pp 255–269
6. Steiner JE (1987) What the neonate can tell us about umami. In: Kawamura Y, Kare MR (eds) Umami: A basic taste. Dekker, New York, pp 97–123
7. Viarouge C, Even P, Pougeot C, Nicolaïdis S (1991) Effects on metabolic and hormonal parameters of monosodium glutamate (umami taste) ingestion in the rat. Physiol Behav 49:1013–1018
8. Viarouge C, Caulliez R, Nicolaïdis S (1992) Umami taste of monosodium glutamate enhances the thermic effect of food and affects the respiratory quotient in the rat. Physiol Behav 52:879–884
9. Nicolaidis S, Rowland N, Meile MJ, Marfaing-Jallat P, Pesez A (1974) A flexible technique for long term infusions in unrestrained rats. Pharmacol Biochem Behav 2:131–136

10. Phillips MJ, Norgren RE (1970) A rapid method for permanent implantation of an intraoral fistula in rats. Behav Res Methods Instrum 2:124
11. Fraker PJ, Speck JC (1978) Protein and cell membrane iodination with sparingly soluble chloroamid 1,3,4,6-tetrachlo-3α,6α-diphenylglycoluryl. Biochem Biophys Res Commun 80:849–857
12. Even P, Nicolaidis S (1984) Le metabolisme de fond: définition et dispositif de mesure. C R Acad Sci (Paris) 298:261–266
13. Lusk G (1928) The Elements of the Science of Nutrition. Saunders, Philadelphia
14. Even P, Perrier E, Aucouturier JL, Nicolaidis S (1991) Utilisation of the method of Kalman filtering for performing the on-line computation of background metabolism in the free-moving, free-feeding rat. Physiol Behav 49:177–187
15. Diamond P, Brondel L, Leblanc J (1985) Palatability and postprandial thermogenesis in dogs. Am J Physiol 248:E75–E79
16. Leblanc J, Brondel L (1985) Role of palatability on meal-induced thermogenesis in human subjects. Am J Physiol 248:E333–E336
17. Leblanc J, Cabanac M (1989) Cephalic postprandial thermogenesis in human subjects. Physiol Behav 46: 479–482
18. Le Magnen J (1984) Bases neurobiologiques du comportement alimentaire. In: Delacour J (ed) Neurobiologie des comportements. Hermann, Paris, pp 1–54
19. Beauchamp G (1990) Human development and umami taste. In: Proceedings of the Second International Symposium on Umami, Sicily
20. Yoshii K (1987) Synergistic effects of 5'-nucleotides on rat taste responses to various amino acids. In: Kawamura Y, Kare MR (eds) Umami: A basic taste. Dekker, New York, pp 219–232

The Molecular Biology of Glutamate Receptors in Rat Taste Buds

N. Chaudhari[1], C. Lamp[1], H. Yang[1], A. Porter[1], M. Minyard[1], and S. Roper[2,3]

Key words. Umami—PCR

Glutamate is an important taste stimulus thought to evoke a distinctive taste, *umami*. Glutamate may also be a neurotransmitter in taste buds [1]. Several glutamate receptors (GluRs) from the brain have been cloned and sequenced [2]. We postulate that GluRs in taste buds are closely related to those in brain. GluRs fall into four families based on sequence and function: α-amino-3-hydroxy-5-methyl-4-isoxazole propionic acid (AMPA), kainate, N-methyl-D-aspartic acid (NMDA), and metabotropic. We have used the reverse-transcriptase-polymerase chain reaction (RT-PCR) to test for the presence of mRNAs from these families in vallate papillae of rats.

AMPA Receptors

GluR1–4 are ion channels that are activated by glutamate and AMPA. We designed PCR primers that would recognize the shared region of this gene family. We carried out RT-PCR on cDNAs reverse transcribed from poly(A)RNAs from the following tissue sources: cerebellum (a known source of GluR1–4); vallate papillae (highly enriched in taste buds); lingual epithelium just anterior to vallate papillae (containing few or no taste buds); and skeletal muscle and liver (not believed to contain GluRs). Our data indicate that mRNAs for one or more AMPA-type GluRs are present in lingual tissues. Vallate papillae, as well as lingual epithelium containing few taste buds, expressed these mRNAs. The negative control tissues, liver and skeletal muscle, did not show an amplification product. We have subcloned the PCR product from vallate papillae. The DNA sequences for several clones were identical to brain GluR1.

We also designed specific PCR primers for each of the AMPA-type GluRs, based on sequences

near the 5′ ends. Our results indicated that only GluR1 was consistently found in vallate papillae. GluR4 was detected only rarely, and GluR2,3 not at all.

NMDA Receptors

These are also ion channels. They are believed to include one or more NR1 subunits, possibly in combination with related subunits, NR2a,b,c,d. Several sites of alternative splicing have been described for the NR1 gene. We designed PCR primers specific for NR1 that would be expected to amplify two products, corresponding to the splice variation near the 5′ end. Our data showed that mRNA for NR1 is present in vallate papillae, but not in skeletal muscle or liver. We also detected mRNA for NR1 in lingual epithelium containing few taste buds. The PCR products indicated that both splicing variants described in brain are also present in lingual tissues.

Metabotropic Glutamate Receptors

These are coupled to intracellular second messengers through G-proteins. The sequences of brain mGluRs (1–5) reveal seven membrane-spanning regions. We designed degenerate primers near the conserved transmembrane segments 1 and 3 of the mGluRs. Diagnostic restriction digests demonstrated that mGluR2–5 were readily amplified from brain mRNAs. RT-PCR indicated that mRNAs for mGluRs were present in vallate papillae. Most importantly, *no mRNAs for these receptors were detected in lingual epithelium containing few or no taste buds*. That is, mGluRs appear to be selectively expressed in taste buds or tissue closely associated with taste buds. No mRNAs for mGluRs were detected in skeletal muscle.

To analyze which members of the mGluR family were expressed in vallate papillae, we subcloned the above PCR product. All 21 clones analyzed by restriction analysis corresponded to mGluR4. This was also confirmed by sequencing one clone. PCR, using primers specific for mGluR1, indicated that this mRNA was absent in vallate papillae.

Departments of [1] Physiology and [2] Anatomy and Neurobiology, Colorado State University, Fort Collins, CO 80523, USA
[3] The Rocky Mt. Taste and Smell Center, University of Colorado Health Sciences Center, Denver, CO 80262, USA

In conclusion, PCR indicates that GluR1 (AMPA), NR1 (NMDA), and mGluR4 (metabotropic) glutamate receptors are present in rat lingual tissues. Of these, mGluR4 is found selectively in regions enriched in taste buds (vallate papillae). We are also testing for the presence of kainate GluRs in lingual tissues. Experiments are in progress to identify the cellular localization of glutamate receptor expression, using in situ hybridization, and to verify the PCR results with Northern blot and RNase protection assays. These experiments represent the first step in identifying the receptor for glutamate taste.

Acknowledgments. We acknowledge the support of NIH-BRSG, the Umami Manufacturers Association of Japan, Kraft General Foods, and the National Institutes of Health.

References

1. Jain SB, Roper SD (1991) Immunocytochemistry of GABA, glutamate, serotonin, and histamine in *Necturus* taste buds. J Comp Neurol 307:675–682
2. Nakanishi S (1992) Molecular diversity of glutamate receptors and implications for brain function. Science 258:597–603

Monosodium Glutamate Binding to Bovine Epithelial Cells Is Affected by the Ionic Composition of the Medium

YUKAKO HAYASHI, TAKASHI TSUNENARI, and TOMOHIKO MORI[1]

Key words. *Umami* taste—Ligand-receptor binding assay

Umami is a unique taste sensation, reflecting a synergistic effect between monosodium glutamate (MSG) and 5'-nucleotides. This synergistic effect of *umami* taste increases the perception of deliciousness. Humans, cows, mice, and many other mammals have *umami* taste perception. It is now conceivable that the MSG taste receptor is a ionotropic receptor [1]. The present investigation was designed to make clear how the surrounding medium of the taste cells is related to the perception of *umami* taste. We prepared a taste cell-rich fraction of circumvallate papillae obtained from bovine tongue immediately after slaughter and we examined the effects of the ionic composition of the medium on the binding activity of MSG to the plasma membrane.

The papillae were washed three times with Hanks' solution and were incubated in 5 mM ethylenediaminetetraacetic acid (EDTA) in divalent ion-free Hanks' solution for 1 h at room temperature, during which time the epithelial membrane was peeled off with fine forceps. Next, the scraped epithelium was treated with 0.2% collagenase/dispase, for 30 min at 25°C. After being rinsed three times with ice-cold Hanks' solution, the epithelium was agitated gently by aspirating in and out of a Pasteur pipette to loosen the cells. Under microscopic observation, the majority of cells in the supernatant were taste cells. It was likely that only taste cells were displaced from the epithelium and that these could be separated easily from the remaining epithelial cells. Using this taste cell-rich fraction, we used a ligand-receptor binding method to assay *umami* perception activity. Buffer (pH 7.4), [^{3}H]glutamic acid neutralized with sodium hydroxide, and 25 mM MSG were mixed. The cell suspension was added to the mixture, which was then incubated on ice. After 2 h, the reaction mixture was filtered rapidly through 0.45-μm Millipore filters under continuous suction. The filters were quickly rinsed with ice-cold buffer. The filtration and rinse steps took less than 30 s for each sample. After these steps, the radioactivity retained on the filter was measured in scintillation fluid. *Umami*-specific activity was distinguished by the synergistic effect in the presence of guanosine monophosphate (GMP).

The *umami* substance, monosodium glutamate, bound to the intact cells in Hanks' solution, and the binding activity was enhanced by guanosine-5'-monophosphate(5'-GMP, 2 mM). When the cells were incubated in Hanks' solution in which Na$^+$ was exchanged for K$^+$ (K-Hanks' solution—this makes a depolarized condition) or in divalent ion-free Hanks' solution (F-Hanks), no significant MSG binding activity was found, and 5'-GMP had no effect on the binding activity. These results indicate that two factors are needed for the binding of MSG to the receptor, the first being a divalent ion in the surrounding medium of the cells, and the second, polarization of the cells.

To examine the effect of depolarization and divalent ion-free medium on the binding of ligand (MSG) to the receptor, and its release, cells were washed with F-Hanks' solution or K-Hanks' solution after binding MSG to the cells in Hanks' solution. The MSG binding activity was decreased in a short time (within 30 s) in the divalent ion-free rinsing condition, while it remained in the depolarized rinsing conditions. This result suggests that the divalent ion is at least one important factor in keeping the glutamic acid bound to receptors. In natural conditions, it is unlikely that no divalent ion exists in the oral cavity, since saliva contains about 1–10 mM divalent ions. The results for taste cells also show that polarization is needed to bind glutamic acid to the receptors, but not to release it. However, in the natural condition in the epithelium of the tongue, the taste cells will always be polarized because of their ionic environment, and so will readily bind MSG. This implies the involvement of other indirect factors, such as kinase, for releasing glutamic acid from the receptor.

References

1. Teeter JH, Kumazawa T, Brand JG (1992) Monosodium glutamate-activated channels in mouse taste epithelial membranes (abstract). Chem Senses 17:707–708

[1] The Research Institute for Food Science, Kyoto University, Uji, Kyoto, 611 Japan

Role of Yeast in the Formation of Inosinic Acid as the Taste Compound in Leavened Bread

Kazue Fujisawa[1] and Masataka Yoshino[2]

Key words. Inosinic acid—Bread

Yeast cells contain high levels of adenosine monophosphate (AMP) deaminase, which catalyzes AMP to inosinic acid (IMP), the taste compound. This enzyme is also distributed in vertebrate muscles, but not in prokaryotes, and it has very low activity in higher plants. AMP deaminase acts as a key enzyme in the degradation of adenine nucleotides, and controls glycolysis in yeast. Stimulation of fermentation is thus expected to produce high IMP content in yeast. In this study, we analyzed the relationship between IMP content and the process of bread fermentation. The substantial role played by yeast in

this process was confirmed by noting the content of IMP in the leavened bread.

IMP content was enzymatically determined in bread made of different wheat flours. Dough made of hard flour contained considerably higher levels of IMP than that made of medium and soft flour. We further analyzed the role of the yeast *Saccharomyces* in the production of IMP during the fermentation of bread made of hard flour. After the dough was mixed with yeast, the IMP content increased with time of fermentation at 30°C; after fermentation for 13 h, the content had risen to a level 2.1-fold that originally determined in the dough. On the other hand, a non-fermented bread, chapati, made without yeast, did not show any increase in IMP at 30°C. In conclusion, baker's yeast with high AMP deaminase activity can contribute to increasing the content of IMP, the taste compound in bread, through the fermentation process.

[1] Department of Food and Nutrition, Chukyo Women's University, Ohbu, Aichi, 474 Japan
[2] Institute for Developmental Research, Aichi Prefectural Colony, Aichi, Japan

8. Central Coding Mechanisms of Taste and Olfaction

Neural Coding of Taste in Macaque Monkeys

MASAYASU SATO[1] and HISASHI OGAWA[2]

Key words. Neural coding—Gustatory information—Chorda tympani fibers—*Cynomolgus* monkeys—Taste qualities

Introduction

Almost 20 years ago Sato and his colleagues of the Kumamoto University Medical School undertook experiments to record responses to various gustatory stimuli in single chorda tympani (CT) nerve fibers of cynomolgus monkeys (*Macaca fascicularis*). They were examining the possibility that the monkeys, which are phylogenetically more developed than rodents, might possess taste nerve fibers narrowly tuned to either one or two of the four basic taste stimuli. They had also hoped to be able to understand better the neural coding mechanism of taste in humans by studying gustatory information in monkeys, which are close to humans. The results, reported by Sato et al. [1], indicated that when CT fibers were classified according to the best-stimulus criterion, the sucrose-best fibers were almost specifically responsive to sucrose and other sweeteners, whereas some of the NaCl-best fibers responded to HCl and many HCl-best fibers responded to NaCl. The quinine-HCl (QHCl)-best fibers, which are few in number in the CT of hamsters and rats, occupied one-sixth of the monkey CT nerve fibers sampled and were narrowly tuned to QHCl.

In this review we first present the results of the early experiments on CT fiber responses in cynomolgus monkeys in a more quantitative manner than before by reanalyzing them by using data-processing methods proposed by Smith and Travers [2] and Bieber and Smith [3]. We next compare them with responses of CT fibers and gustatory relay neurons in rodents and primates, and then with human taste perception elicited by anterior tongue stimulation.

In the paper published in 1975 [1] we presented as a measure of response to a taste stimulus in each fiber the number of spikes per 5 s without subtracting the spontaneous discharge rate. This method was used because the spontaneous rate in some fibers was large—its mean and standard deviation (SD) of 67 fibers sampled being 9.8 ± 8.8 spikes/5 s—and when this figure was subtracted from the evoked discharge rate, the adjusted rate sometimes became negative. We now subtract the spontaneous discharge rate from the evoked one in each fiber, and the negative response value is considered nonresponsive, or zero. One fiber of the 67 was omitted because it showed no positive response to 0.3 M sucrose, 0.3 M NaCl, 0.01 M HCl, or 0.003 M QHCl, which were the four standard stimuli employed in the experiments.

Response Properties of Monkey CT Fibers to the Four Prototypical Taste Stimuli

Response profiles of 66 single CT fibers for the four standard stimuli, freshly obtained, are presented in Fig. 1. The sucrose-best and QHCl-best fibers are almost specifically responsive to sucrose and QHCl, respectively. On the other hand, some NaCl-best fibers responded well to HCl, and some HCl-best fibers showed moderate responses to NaCl. Of 66 fibers, 35 (53%) responded to sucrose, and 16 of them (24% of the sample) were sucrose-best fibers; 53 (83%) responded to NaCl, and 28 of them (42%) were NaCl-best fibers. Forty (61%) responded to HCl, and 11 (17%) of them were HCl-best fibers. Thirty-one fibers (47%) responded to QHCl, and 11 of them (17%) were QHCl-best fibers.

The average response magnitude for each of the four standard stimuli in each fiber group is presented in Fig. 2. Among the sucrose-best fibers the response magnitude for sucrose is large compared with those for other stimuli, indicating that this fiber group is almost specifically responsive to sucrose. Similarly, the QHCl-best fibers are narrowly tuned to QHCl only. On the other hand, the magnitude of HCl response in the NaCl-best fibers is about 22.5% of that for NaCl, and that of NaCl response in the HCl-best fibers is 57% of that for HCl.

Breadth of Responsiveness

Of 66 fibers, 11 (17%) responded to only one stimulus: 3 sucrose-best, 7 NaCl-best, and 1 QHCl-best. Twenty fibers (30%), consisting of 10 NaCl-best, 5 sucrose-best, 2 HCl-best, and 3 QHCl-best,

[1] Brain Science Foundation, Honda Yaesu Bldg., 2-6-20 Yaesu, Chuo-ku, Tokyo, 104 Japan
[2] Department of Physiology, Kumamoto University Medical School, Kumamoto, Japan

FIG. 1. Response profiles of monkey chorda tympani fibers to the four standard gustatory stimuli (0.3 M sucrose, 0.3 M NaCl, 0.01 M HCl, 0.003 M QHCl). Fibers are grouped in the four best-stimulus categories and, within those categories, are arranged in the descending order of response magnitude to the best stimulus. The ordinate shows the adjusted number of spikes in 5 s

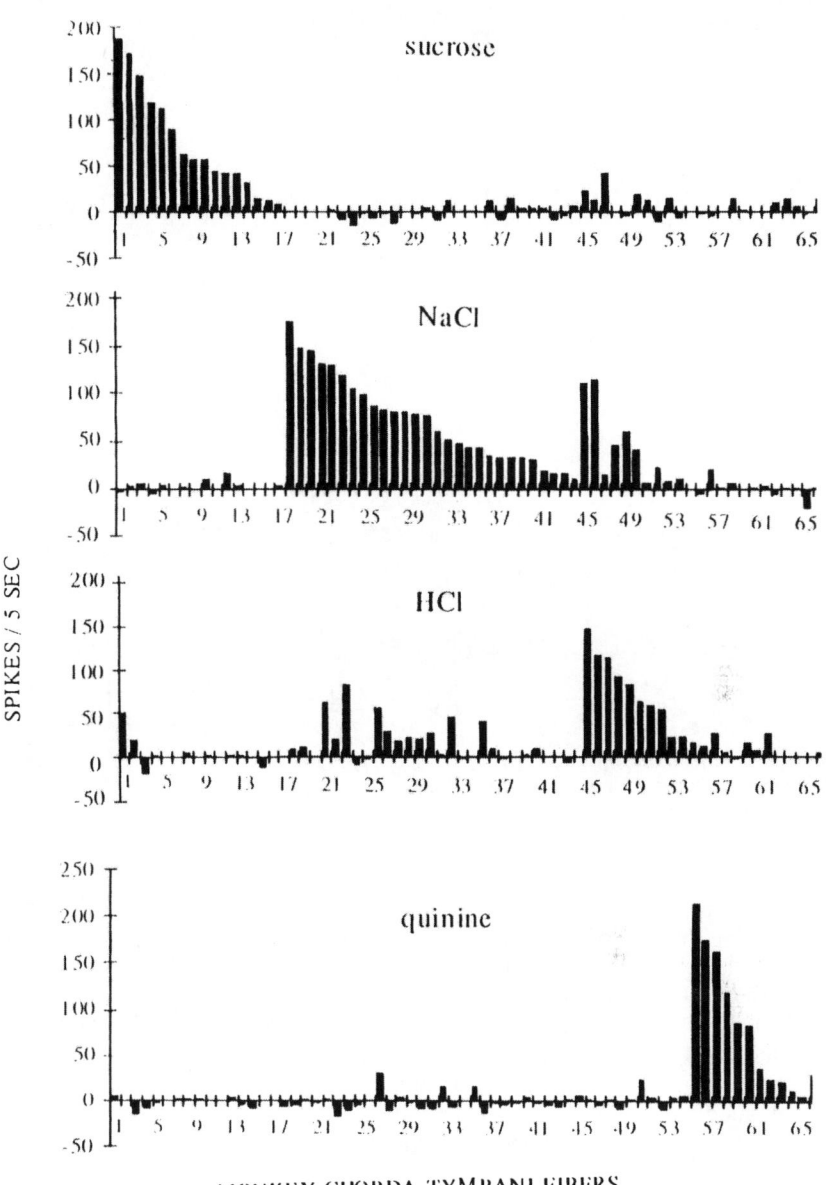

MONKEY CHORDA TYMPANI FIBERS

responded to two stimuli. Therefore 47% of 66 fibers were responsive to one or two stimuli, and the remaining 53% were composed of fibers responsive to three (25 fibers) and four (10 fibers) stimuli.

The entropy coefficient H, representing a measure of breadth of responsiveness, was calculated for each fiber. The value of H varies from 1.0, characterizing a unit that responds equivalently to all the four stimuli, to 0, characterizing a unit that responds exclusively to one stimulus [2]. The average value of H for all the fibers was 0.375. It was the smallest for sucrose-best fibers (0.276), followed by NaCl-best (0.348), QHCl-best (0.374), and HCl-best (0.634) fibers. Therefore the sucrose-best fibers are most narrowly tuned to the four stimuli, and the HCl-best fibers are most broadly tuned.

The mean entropy value of 0.375 for all the fibers is much smaller than that for CT fibers of rats and hamsters, which ranges from 0.54 to 0.61 [4–6], indicating that CT fibers of monkeys are more narrowly tuned to the four prototypical stimuli than those of rodents, but the H value of 0.375 is larger than that of chimpanzee CT fibers (0.217) [7].

Responses of CT Fibers to a Variety of Gustatory Stimuli

We recorded responses of 25 fibers to 14 stimuli, consisting of, in addition to the four standard stimuli, 0.01 M Na saccharin, 0.3 M Na saccharin, 0.3 M LiCl, 0.3 M KCl, 0.3 M NH₄Cl, 0.3 M CaCl₂, 0.3 M

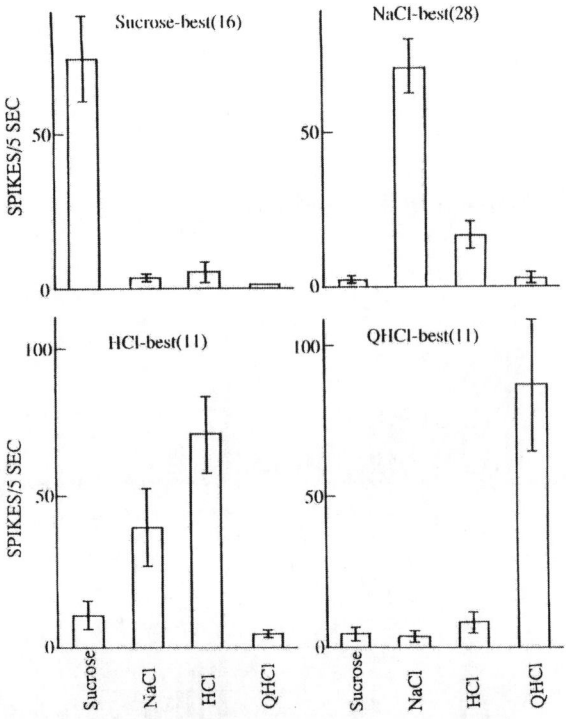

FIG. 2. Mean response profiles of chorda tympani fibers to the four standard gustatory stimuli in four best-stimulus categories. *Numbers in parentheses* indicate numbers of fibers. *Bars* represent ±1 SEM

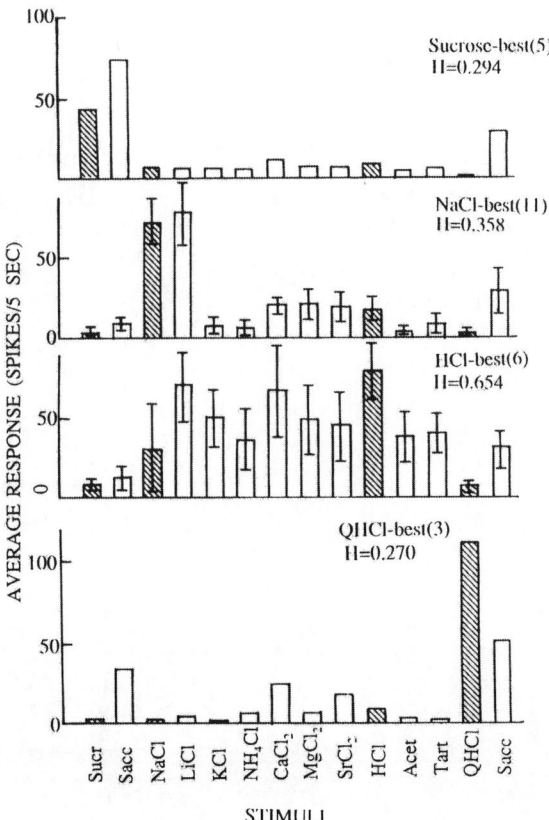

FIG. 3. Mean response profiles of chorda tympani fibers to 14 gustatory stimuli in four best-stimulus categories. The stimuli are (*from left*): 0.3 M sucrose, 0.01 M Na saccharin, 0.3 M NaCl, 0.3 M LiCl, 0.3 M KCl, 0.3 M NH$_4$Cl, 0.3 M CaCl$_2$, 0.3 M MgCl$_2$, 0.3 M SrCl$_2$, 0.01 M HCl, 0.02 M acetic acid, 0.01 M tartaric acid, 0.003 M QHCl, 0.3 M Na saccharin. *Bars* represent ±1 SE of the mean. *Shadowed blocks* are responses to the four standard stimuli. *Numbers in parentheses* indicate numbers of fibers

MgCl$_2$, 0.3 M SrCl$_2$, 0.02 M acetic acid, 0.01 M tartaric acid. The 25 fibers were classified into 5 sucrose-best, 11 NaCl-best, 6 HCl-best, and 3 QHCl-best fibers. The average response magnitudes for the 14 stimuli in the groups of fibers are presented in Fig. 3. The sucrose-best fibers responded predominantly to sucrose and Na saccharin. The NaCl-best fibers showed marked responses to NaCl and LiCl, although they responded moderately to HCl, three divalent salts, and 0.3 M Na saccharin. The response to Na saccharin may be due to Na ions in this salt. The HCl-best fibers responded well to all the salts and acids. The QHCl-best fibers showed a marked response to QHCl and a moderate response to Na saccharin. Thus each of the four groups of fibers showed a distinctly different response pattern to a variety of gustatory stimuli from others.

Hierarchical Cluster Analysis

Responses of 25 fibers to 14 stimuli were subjected to hierarchical cluster analysis [3]. The 14 stimuli were classified into three clusters and two single layers: *N* cluster consisting of NaCl and LiCl, *A* cluster consisting of acids and salts other than NaCl and LiCl, and *S* cluster consisting of sucrose and 0.01 M Na saccharin. QHCl was a single layer, and

0.3 M Na saccharin was situated between *S* cluster and QHCl. The *A* cluster was composed of two subclusters, one consisting of three divalent salts and the other of three acids, NH$_4$Cl, and KCl. Therefore 14 stimuli could be classified into four categories corresponding to each of the four basic tastes. The results reflect four distinct patterns of response (NaCl/LiCl, acid/nonsodium salt, sweet, and QHCl).

Information Transference from Peripheral Nerve to Brain

Information transference from the peripheral nerve fibers to the brainstem neurons and from there to thalamic and cortical neurons has been studied extensively in rats, hamsters, and monkeys by a number of investigators. In hamsters [4] and rats [5] it has been demonstrated that although neurons in the nucleus tractus solitarius (NTS) show a slightly higher discharge rate and a slightly broader responsiveness to the four basic taste stimuli than CT fibers, the relations found among the neural patterns in the CT are fairly well preserved in the NTS. It was suggested that there was probably a rather simple "straight through" connection between individual first- and second-order taste neurons [8]. Further studies using unanesthetized rats [9] have revealed that in rats the breadth of tuning within the gustatory system remained virtually unchanged from the periphery to the cortex.

Scott et al. [10] recorded taste responses from 48 neurons in the NTS of the alert cynomolgus monkeys. They found that the tuning of the neurons to one of the prototypical four gustatory stimuli was in most cases broad; 84% of the neurons showed some response to three or four prototypical stimuli. The breadth of tuning measure, the entropy coefficient H, was 0.87 for 48 neurons tested. This finding contrasts with the narrow tuning of CT fibers of the monkey. The mean evoked discharge rate reported for NTS neurons is much smaller than that for CT fibers. It was also reported that in NTS neurons the correlation between HCl and QHCl was the largest, and acid and nonacid was the most clearest distinction; whereas in CT fibers the largest correlation was obtained between NaCl and HCl, and the sucrose response was most easily differentiated from other responses [1].

Pritchard et al. [11] recorded activities from 50 single neurons in the ventromedial nucleus of the thalamus of alert rhesus monkeys during stimulation of the oral cavity. Of the 50 gustatory neurons, 40 responded to sucrose and 28 of them were sucrose-best; 22 neurons responded to NaCl, and 12 of them were classified NaCl-best. HCl-best and QHCl-best neurons comprised only 14% of the sample. The average breadth of responsiveness of the 50 thalamic neurons, as expressed by the entropy coefficient, was 0.73.

Comparing the data obtained on CT nerve fibers, NTS, and thalamic neurons, it is apparent that marked discrepancies exist among the three sets of data. The breadth of responsiveness of neurons to the four stimuli is narrow in the CT but broad at the NTS; and at the thalamus it becomes somewhat narrow again. Especially noteworthy is the marked discrepancy in the response patterns among neurons in the CT, NTS, and thalamus.

In the experiments on unanesthetized alert monkeys, gustatory stimuli were delivered into the oral cavity but not limited to the anterior tongue; and the regions in the NTS and thalamus, where recordings of single neuron activities were made, were not limited to the region that receives input from the CT. Moreover, concentrations of the four standard test stimuli were not the same in the three experiments, nor were the criteria for a significant neuronal response rate. Such differences in the experimental procedure and the data processing could markedly affect the response pattern and the breadth of responsiveness of neurons. It is thus premature at the present stage to conclude that the tuning of the gustatory neurons in the monkey becomes broad when gustatory information is transferred from the peripheral nerve to the NTS and becomes narrow again when it ascends further from the NTS to the thalamus.

Comparison of Monkey CT Fiber Responses with Human Taste Sensation Elicited by Anterior Tongue Stimulation

Sandick and Cardello [12] measured the magnitude of taste sensations elicited by stimulating a small area of the anterior tongue containing several fungiform papillae with four taste stimuli. They indicated that sucrose and QHCl elicited almost exclusively sweet and bitter taste, respectively, whereas NaCl and tartaric acid elicited both salty and sour tastes. Furthermore, they [13] demonstrated that salts such as NaCl, LiCl, and KCl elicited not only salty but also sour taste, and that acids such as hydrochloric, tartaric, and citric acids produced sour as well as salty taste. Thus human taste perception evoked by stimulation of the anterior tongue innervated by the CT is essentially similar to the response profile across the four CT fiber groups of the monkey, although in humans 0.25 and 0.4 M KCl elicited primarily salty sensation accompanied by sour and bitter tastes [13], whereas in the monkey CT 0.3 M KCl evoked the largest response in the HCl-best fibers and the second largest in the NaCl-best fibers (Fig. 3). The similarity of the human taste perception evoked by the anterior tongue stimulation to the response profile in the monkey CT fibers suggests that the peripheral gustatory nerve system in humans is organized similarly to that in the monkey. It may further suggest a possibility that gustatory information from the peripheral nerve in the monkey is transmitted to the cortex without significant modification of its pattern.

References

1. Sato M, Ogawa H, Yamashita S (1975) Response properties of macaque monkey chorda tympani fibers. J Gen Physiol 66:781–810.
2. Smith DV, Travers JB (1979) A metric for the breadth of tuning of gustatory neurons. Chem Sens Flav 4: 215–229
3. Bieber SL, Smith DV (1986) Multivariate analysis of sensory data: a comparison of methods. Chem Sens 11:19–47
4. Travers JB, Smith DV (1979) Gustatory sensitivities in neurons of the hamster nucleus tractus solitarius. Sens Process 3:1–26

5. Ogawa H, Imoto T, Hayama T (1984) Responsiveness of solitario-parabrachial relay neurons to taste and mechanical stimulation applied to the oral carity in rats. Exp Brain Res 54:349–358

6. Yamamoto T, Yuyama N, Kato T, Kawamura Y (1984) Gustatory responses of cortical neurons in rats. I: J Neurophysiol 51:616–635

7. Hellekant G, Ninomiya Y (1991) On the taste of umami in chimpanzee. Physiol Behav 49:927–934

8. Doetsch GS, Erickson RP (1970) Synaptic processing of taste-quality information in the nucleus tractus solitarius of the rat. J Neurophysiol 33:490–507

9. Nakamura K, Norgren R (1991) Gustatory responses of neurons in the nucleus of the solitary tract of behaving rats. J Neurophysiol 66:1232–1248

10. Scott TR, Yaxley S, Sienkiewicz ZJ, Rolls ET (1986) Gustatory responses in the nucleus tractus solitarius of the alert cynomolgus monkey. J Neurophysiol 55:182–200

11. Pritchard JC, Hamilton RB, Norgren R (1989) Neural coding of gustatory information in the thalamus of *Macaca mulatta*. J Neurophysiol 61:1–14

12. Sandick B, Cardello AV (1981) Taste profiles from single circumvallate papillae: comparison with fungiform papillae. Chem Sens 6:197–214

13. Sandick B, Cardello AV (1983) Tastes of salts and acids on circumvallate papillae and anterior tongue. Chem Sens 8:59–69

Role of Calcitonin-Gene-Related Peptide (CGRP) as a Chemical Marker for the Thalamocortical Visceral Sensory System

Clifford B. Saper[1]

Key words. Calcitonin gene-related peptide— Thalamocortical visceral sensory system—Taste

The earliest studies of the organization of the visceral sensory system focused on the special visceral sensory modality, taste. Lesion studies dating back to the early part of this century established, in a variety of mammalian species including primates and humans, that the rostral insular cortex contains a region that is critical for taste discrimination (see [1 and 2] for review). This gustatory cortex is near but distinct from the tongue somatosensory area. Subsequent electrophysiological studies demonstrated the presence of taste-responsive neurons in this field and in its thalamic relay cell group, the ventroposterior medial parvocellular nucleus.

Other studies demonstrated the pathways taken by gustatory afferents to the forebrain. As early as 1905, Herrick [3] had pointed out that taste afferents from the equivalent of the nucleus of the solitary tract in fishes synapse in a tertiary taste nucleus in the dorsolateral midbrain. However, it was not until the landmark work of Norgren and colleagues in the mid-1970s in the rat [1,4–6] that it was appreciated that taste afferents in mammals take a similar course, terminating in the parabrachial nucleus (PB). The PB in the rat then projects to the sites in the thalamus, hypothalamus, and basal forebrain that participate in taste appreciation and in feeding behavior.

Beginning in the late 1970s, parallel studies on the general visceral sensory system in the rat demonstrated that it is organized in a topographic fashion, with taste as the most anteriorly represented visceral sensory modality [7–9]. Hence, within the nucleus of the solitary tract, gastrointestinal afferents are topographically represented just caudal to taste, in the middle portion of the nucleus. Cardiovascular afferents end more caudally still, and respiratory afferents terminate caudally and laterally [10–13]. These different zones project to distinct regions of the PB, which in turn project to distinct terminal fields in the forebrain [14–16].

The presence of a general visceral sensory area in the cortex and the thalamus was presaged by early observations from studies employing evoked potentials, demonstrating that electrical stimulation of the proximal end of the cut cervical vagus nerve produced responses in the anterior insular cortex and the ventroposterior medial parvocellular thalamus, very near the gustatory areas [17–19]. More refined studies by Yamamoto and colleagues [20] placed the cortical areas responsive to glossopharyngeal and vagal stimulation just caudal to the gustatory region in the insular cortex of the rat.

The nature of the physiological responses of the neurons in these areas was studied by Cechetto and Saper [21], who demonstrated that the taste responses are predominantly found in the rostroventral dysgranular insular area (see also [22–24]). Neurons responding to gastric stretch were found just dorsocaudally to this region, in the granular insular area, although a few neurons responded to both modalities. Neurons that were sensitive to cardiovascular stimulation (atrial stretch or blood pressure elevation by a bolus of phenylephrine) were seen more caudally in the granular insular area, and these were admixed with neurons responding to respiratory stimuli (hypercarbia, hypoxia, or pulmonary stretch). Very few neurons responded to more than one class of stimulus, although a few reacted both to blood pressure and blood gas manipulation.

Injections of horseradish peroxidase at the sites of responsive neurons in the insular cortex retrogradely labeled the related thalamic relay nucleus. Neurons labeled from gustatory cortical sites were mainly found in the medial part of the ventroposterior medial parvocellular nucleus, whereas neurons labeled from gastrointestinal sites were found in the lateral part of the ventroposterior medial parvocellular nucleus. Neurons retrogradely labeled from cardiovascular and respiratory sites were found more lateral still, extending beyond the recognized borders of the ventroposterior medial parvocellular nucleus. These latter neurons were initially termed the ventroposterior lateral parvocellular nucleus, but

[1] Department of Neurology, Beth Israel Hospital, Harvard Medical School, 330 Brookline Avenue, Boston, MA 02215, USA

more recent studies [25,26] have found that the neurons of the visceral sensory relay group form a continuum, and the term ventroposterior parvocellular nucleus has been introduced to identify the entire visceral sensory strip in the thalamus.

Single unit recordings in the thalamus have verified the organotopic arrangement of the responses of neurons in the ventroposterior parvocellular nucleus (Cechetto and Saper, unpublished observations). Furthermore, injections of tracers into this cell group in the rat have demonstrated a topographically organized projection to it from the contralateral external medial PB nucleus [21]. Although many neurons are retrogradely labeled in the ipsilateral internal and central lateral, medial, and waist subnuclei of the PB following injections into the medial part of the ventroposterior parvocellular nucleus (which spread into the adjacent midline and intralaminar thalamic nuclei), anterograde transport studies using *Phaseolus vulgaris* leucoagglutinin have demonstrated that these PB subnuclei mainly project to the midline and intralaminar thalamic nuclei, rather than the visceral sensory relay nucleus (Cechetto and Saper, unpublished observations). Injections of retrograde tracers into the lateral part of the ventroposterior parvocellular nucleus retrogradely label predominantly neurons in the contralateral external medial PB subnucleus, as these injections are too far lateral to spread into the intralaminar nuclei.

These observations are supported by a parallel series of studies in which we have studied the participation in this system by neurons that are immunoreactive for calcitonin-gene-related peptide (CGRP). We confirmed the observation by Shimada and colleagues [27] that there is CGRP-like immunoreactive innervation of the insular visceral sensory cortex, and demonstrated in addition an intense CGRP-immunoreactive terminal field filling the ventroposterior parvocellular nucleus [25]. This was surrounded by CGRP-immunoreactive neurons in the posterior intralaminar thalamic complex, but none of these neurons appeared to be part of the visceral sensory relay nucleus. By combining retrograde transport of fluorescent dyes with immunofluorescence histochemistry, it was established that none of the neurons in the ventroposterior parvocellular nucleus that were retrogradely labeled from the insular cortex contained CGRP. On the other hand, CGRP-immunoreactive neurons in the ventral lateral PB subnucleus were retrogradely labeled, establishing the PB as the major source of insular CGRP innervation. Conversely, the CGRP-immunoreactive neurons in the posterior intralaminar complex were found to project to the perirhinal cortex (caudal to the insular visceral sensory area) and to the amygdala and striatum [26].

Injections of fluorescent dyes into the ventroposterior parvocellular nucleus retrogradely labeled CGRP-immunoreactive neurons in the external medial PB subnucleus, predominantly on the contralateral of the brain. These observations confirm that the ventroposterior parvocellular nucleus is indeed the thalamic target of the contralateral external medial PB, and indicate that CGRP is an excellent marker for this pathway.

A major controversy in the organization of the visceral sensory pathways in primates is the role played by the PB in relaying taste information to the thalamus. In their studies of the projection from the taste portion of the nucleus of the solitary tract, Beckstead et al. [28] reported that the most rostral part of the nucleus of the solitary tract projects directly to the ventroposterior parvocellular nucleus of the thalamus, bypassing the parabrachial nucleus (which is an obligate relay in rats). However, they did find projections to the PB from more caudal parts of the taste portion of the nucleus of the solitary tract, and there have been no studies published on the PB projections to the thalamus in primates.

To examine whether CGRP might serve as a marker that might help elucidate this system of pathways in primates, including humans (where tracer studies are not possible), we have recently investigated the distribution of CGRP-like immunoreactivity in the thalamus and parabrachial nucleus in humans and monkeys. Our preliminary data [29] indicate that this system may be organized in virtually an identical fashion in monkeys and in humans. We find that the neurons of the external medial nucleus are CGRP-like immunoreactive in both species, and that a CGRP-like immunoreactive terminal field fills the ventroposterior parvocellular nucleus. We hypothesize that this peptide marks a major PB projection to the ventroposterior parvocellular nucleus in primates as well as in rodents. Furthermore, the presence of this CGRP-immunoreactive projection system in primates strongly suggests the participation by the PB in processing taste information in primates, including humans.

Acknowledgments. The author thanks Quan Hue Ha for excellent technical assistance in the studies that are described. This work was supported in part by USPHS grant, NS22835.

References

1. Norgren R (1984) Central neural mechanisms of taste. In: Smith D (ed) Handbook of physiology, Sect I: the nervous system, Vol III, sensory processes. American Physiological Society, Bethesda, MD, pp 1087–1128
2. Cechetto DF, Saper CB (1990) Role of the cerebral cortex in autonomic function. In: Loewy AD, Spyer KM (eds) Central regulation of autonomic functions. Oxford University Press, New York, pp 208–223

3. Herrick CJ (1905) Central gustatory paths in the brains of bony fishes. J Comp Neurol 15:375–456

4. Norgren R, Leonard CM (1973) Ascending central gustatory connections. J Comp Neurol 150:217–238

5. Norgren R (1976) Taste pathways to hypothalamus and amygdala. J Comp Neurol 166:17–30

6. Norgren R (1993) The gustatory system. In: Paxinos G (ed) The rat nervous system. Academic Press, San Diego

7. Ricardo JA, Koh ET (1978) Anatomical evidence of direct projections from the nucleus of the solitary tract to the hypothalamus, amygdala, and other forebrain structures in the rat. Brain Res 153:1–26

8. Saper CB, Loewy AD (1980) Efferent connections of the parabrachial nucleus in the rat. Brain Res 197:291–317

9. Saper CB (1982) Convergence of autonomic and limbic connections in the insular cortex of the rat. J Comp Neurol 210:163–173

10. Seiders EP, Stuesse SL (1984) A horseradish peroxidase investigation of carotid sinus nerve components in the rat. Neurosci Lett 46:13–18

11. Housley GD, Martin-Body RL, Dawson NJ, Sinclair JD (1987) Brain stem projections of the glossopharyngeal nerve and its carotid sinus branch in the rat. Neuroscience 22:237–250

12. Altschuler SM, Bao X, Bieger D, Hopkins DA, Miselis RR (1989) Viscertopic representation of the upper alimentary tract in the rat: sensory ganglia and nuclei of the solitary and spinal trigeminal tracts. J Comp Neurol 283:248–268

13. Finley JCW, Katz DM (1992) The central organization of carotid body afferent projections to the brainstem of the rat. Brain Res 572:108–116

14. Herbert H, Moga MM, Saper CB (1990) Connections of the parabrachial nucleus with the nucleus of the solitary tract and the medullary reticular formation in the rat. J Comp Neurol 293:540–580

15. Moga MM, Herbert H, Hurley KM, Yasui Y, Gray TS, Saper CB (1990) Organization of cortical, basal forebrain, and hypothalamic afferents to the parabrachial nucleus in the rat. J Comp Neurol 295:624–661

16. Bernard JF, Alden M, Besson JM (1993) The organization of the efferent projections from the pontine parabrachial area to the amygdaloid complex. A *Phaseolus vulgaris* leucoagglutinin (PHA-L) study in the rat. J Comp Neurol 329:201–209

17. Bailey P, Bremer F (1938) A sensory cortical representation of the vagus nerve with a note on the effects of low blood pressure on the cortical electrogram. J Neurophysiol 1:405–412

18. Wall PD, Davis GD (1951) Three cerebral cortical systems affecting autonomic function. J Neurophysiol 14:507–517

19. Dell P, Olson R (1951) Projections thalamiques corticales et cerebelleuses des afferences viscerales vagales. C R Soc Seances Soc Biol Fil 145:1084–1088

20. Yamamoto T, Matsuo R, Kawamura Y (1980) Localization of cortical gustatory area in rats and its role in taste discrimination. J Neurophysiol 44:440–455

21. Cechetto DF, Saper CB (1987) Evidence for a viscerotopic sensory representation in the cortex and thalamus in the rat. J Comp Neurol 262:27–45

22. Kosar E, Grill HJ, Norgren R (1986) Gustatory cortex in the rat. I. Physiological properties and cytoarchitecture. Brain Res 379:329–341

23. Kosar E, Grill HJ, Norgren R (1986) Gustatory cortex in the rat. II. Thalamocortical projections. Brain Res 379:342–352

24. Ogawa H, Ito S, Murayama N, Hasegawa K (1990) Taste area in the granular and dysgranular insular cortices in the rat identified by stimulation of the entire oral cavity. Neurosci Res 9:196–201

25. Yasui Y, Saper CB, Cechetto DF (1989) Calcitonin gene-related peptide immunoreactivity in the visceral sensory cortex, thalamus, and related pathways in the rat. J Comp Neurol 290:487–501

26. Yasui Y, Saper CB, Cechetto DF (1991) Calcitonin gene-related peptide (CGRP) immunoreactive projections from the thalamus to the striatum and amygdala in the rat. J Comp Neurol 308:293–310

27. Shimada S, Shiosaka S, Hillyard CJ, Girgis SI, MacIntyre I, Emson PC, Tohyama M (1985) Calcitonin gene-related peptide projection from the ventromedial thalamic nucleus to the insular cortex: a combined retrograde transport and immunocytochemical study. Brain Res 344:200–203

28. Beckstead RM, Morse JR, Norgren R (1980) The nucleus of the solitary tract in the monkey: Projections to the thalamus and brain stem nuclei. J Comp Neurol 190:259–282

29. de Lacalle S, Saper CB (1992) Calcitonin gene-related immunoreactivity in visceral sensory pathways in the human brain. Soc Neurosci Abstr 18:805

Functional Organization of the Orally Responsive NST

Susan P. Travers, David C. Becker, Christopher B. Halsell, Marie I. Harrer, and Joseph B. Travers[1]

Key words. Nucleus of the solitary tract—Topographic organization—Taste—Tactile—Chemotopy—Biocytin—c-fos

Introduction

The nucleus of the solitary tract (NST) has long been recognized as a primary sensory relay with diverse afferent inputs. Until recently, however, the most emphasized functional organization was its single partition into rostral gustatory and caudal visceral regions. Studies have now underscored regularities inherent in the organization of the diverse inputs within each of these NST divisions [e.g., 1,2] and have begun to define morphologic [e.g., 3], efferent [e.g., 4], and neurochemical [e.g., 5] heterogeneities that also exhibit organizational regularities. The work in our laboratory has concentrated specifically on the organization of the heterogeneous oral afferent inputs reaching the rostral NST (rNST).

Materials and Methods

The results reported here summarize outcomes from several neurophysiologic and neuroanatomic experiments [6–10]. The subjects were adult male Sprague-Dawley rats. The neurophysiologic studies [8–10] used an acute preparation that made it possible to visualize and stimulate defined oral regions with gustatory or somatosensory stimuli [9]. One neuroanatomic experiment used this recording preparation, together with injections of biocytin [11], to trace efferent projections arising from functionally defined regions in the rNST [6]. The second neuroanatomic investigation used fluorescent, retrograde tracers in a double-labeling paradigm to define the relation between the two major efferent pathways arising from the rNST [7]. The final experiment [12] is currently using immunohistochemistry to detect the presence of the protein product of the

[1] Departments of Oral Biology and Psychology, College of Dentistry, Ohio State University, Columbus, Ohio 43210, USA

immediate-early gene, c-fos, in the NST following taste stimulation.

Results

Afferent Organization of the NST

Neuroanatomic studies of central projections arising from cranial nerves V, VII, and IX have demonstrated that the rNST receives somatosensory afferent projections from nearly all of the oral soft tissues and gustatory afferent projections from several spatially discrete taste bud subpopulations. These studies suggest an orderly arrangement of afferent information [e.g., 2], but intranuclear connections [6,13,14] and the lack of specific peripheral innervation have precluded a definitive description of rNST topographic and receptive field organization based only on anatomy. Our neurophysiologic investigations have been designed to clarify these issues.

Modality

Our studies of modality in the rNST have centered on gustatory and mechanical sensitivity. The results of one single-unit investigation [10] indicated the presence of two populations of orally responsive neurons based on modality. In this study, 93 rNST neurons were tested for responsiveness to whole-mouth gustatory stimulation, and approximately one-half responded to this stimulus. A subset of these gustatory neurons were also assessed for mechanical sensitivity, and most responded, albeit less so, to this modality too. Of the cells not responsive to gustatory stimulation, most ($n = 44/46$) responded to innocuous mechanical stimulation of the oral soft tissues.

These findings were corroborated by a study of multiunit responses to mechanical and gustatory stimulation of the foliate and circumvallate papillae of the posterior tongue [8]. The latter study also identified two main recording site types: one responsive to both gustatory and tactile stimulation of the posterior tongue ($n = 24$) and a second responsive only to tactile stimulation ($n = 48$).

The data from these two studies were combined with data from a third investigation [9] of gustatory-

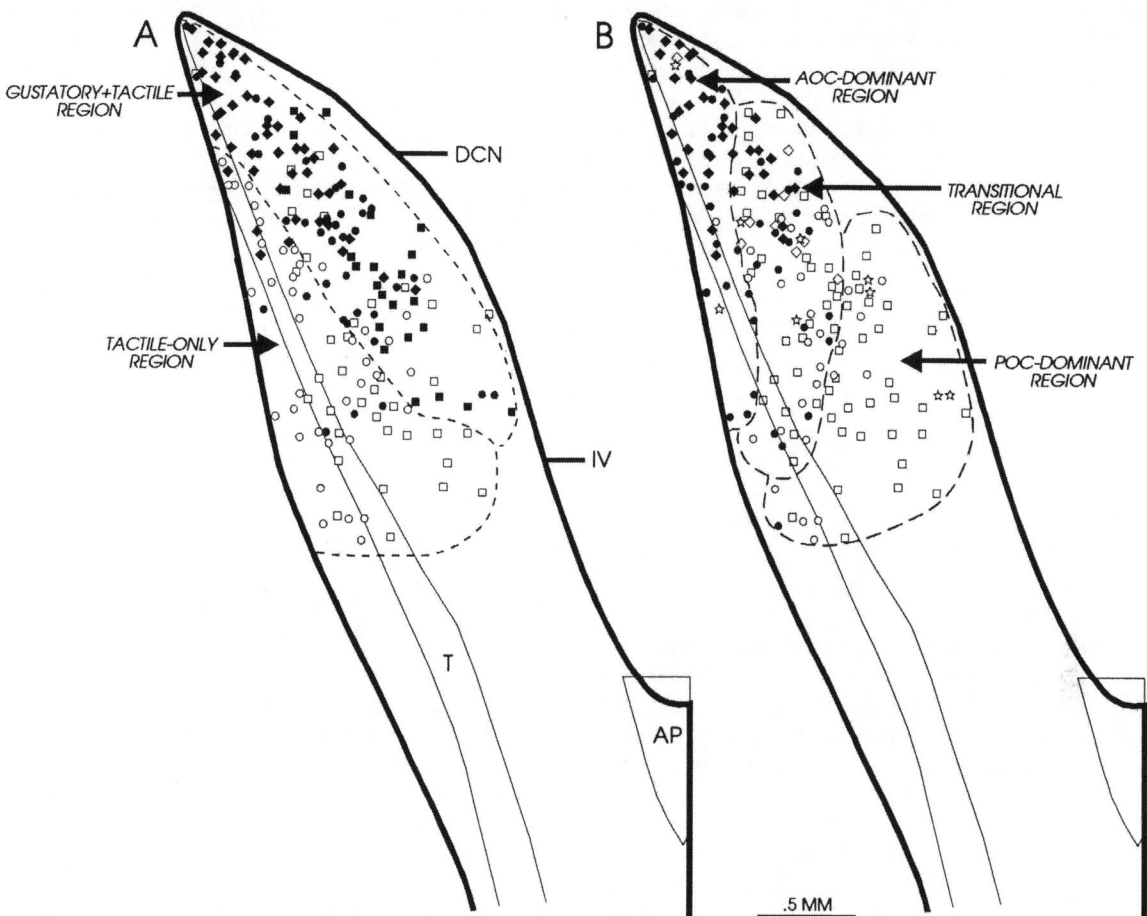

FIG. 1. **A** Horizontal diagram of the NST depicting topographic organization by modality. *Open symbols* depict locations of single neurons or multiunit recording sites responsive to oral mechanical but not gustatory stimulation. *Closed symbols* depict single- or multiunit recording sites responsive to gustatory (and sometimes also to mechanical) stimulation. The symbols identify the source of the data: *diamonds*, single-unit sites [10]) *circles*, single-unit sites [9]; *squares*, multiunit sites [8]. *AP*, area postrema; *T*, solitary tract; *DCN*, level of the NST coincident with the caudal border of the dorsal cochlear nucleus; *IV*, level where the NST abuts the fourth ventricle and border between the rostral and caudal divisions. *Top*, rostral; *right*, medial. **B** Horizontal diagram of the NST depicting topographic organization by receptive field. *Closed symbols* depict recording sites optimally responsive to gustatory or mechanical stimulation of the AOC. *Open symbols*, except *stars*, represent multiunit recording sites that exhibited robust responses to gustatory or tactile stimulation of the posterior tongue and single-unit sites optimally responsive to gustatory or mechanical POC stimulation. *Open stars* depict single neurons responsive to gustatory stimulation of the whole mouth but not individual receptor subpopulations. The various symbols depict the data source and are coded as in **A**

responsive rNST neurons in order to compare the anatomic distribution of gustatory and tactile responses within the rNST. The summary revealed a clear topographic organization of the rNST by modality (Fig. 1A; total recording sites = 183). A combined *gustatory + tactile* region circumscribed boundaries occupied by 98 sites that responded to sapid stimulation (and often tactile stimulation as well), but only 19 responsive exclusively to the tactile stimulus. This region extended to the rostral pole of the NST and caudally was medial to a *tactile-only* region. The tactile-only region extended slightly caudal to the gustatory region and circumscribed 57

sites responsive exclusively to oral tactile stimulation but only 9 responsive to gustatory stimulation.

Receptive Field

Several laboratories have reported that single gustatory neurons can respond to multiple taste receptor subpopulations [e.g., 15,16]. Our data concur and further suggest that convergence occurs systematically [9,10]. In two studies, we tested a total of 98 neurons for responsiveness to independent stimulation of several taste receptor subpopulations. Forty-six responded to only one stimulation, and

10 reponded only when the entire oral cavity was stimulated; the remaining 42 responded to independent stimulation of multiple receptor subpopulations. Most of these *convergent* neurons (26/42) responded *most vigorously* to stimulation of two receptor subpopulations in the anterior oral cavity (AOC)—the anterior tongue and nasoincisor ducts—whereas the next most frequent type ($n = 9$) received their two most robust inputs from receptor subpopulations in the posterior oral cavity (POC)—the foliate papillae, soft palate, or retromolar mucosa ($p < 0.0005$, Fisher's-exact test for "best" vs "second-best" receptive field). Thus convergent gustatory neurons in the rNST tend to receive their major inputs from apposing or adjacent receptor subpopulations within the AOC or POC, but not from both areas.

Topographic Organization

As described earlier, modality is one important determinant of rNST topographic organization. It is not, however, the sole determinant: An organization by receptive field location is also evident. This organization was summarized using the same 183 gustatory and oral mechanical recording sites used to summarize organization by modality (Fig. 1A). Multiunit recording sites were classified as POC-responsive because they responded to stimulation of the only oral location tested, the posterior tongue, whereas single-unit recording sites were classified according to whether their *optimal* receptive field was located in the AOC or POC. The resulting plot demonstrated clear orotopy in rNST (Fig. 1B). Three oblong regions, staggered from anterior and lateral to posterior and medial were identified: the most anterior and lateral *AOC-dominant region* was occupied mostly by AOC recording sites (45/50) and the most caudal and medial *POC-dominant region* nearly exclusively by POC sites (41/46); the area between them formed a *transitional region* (28 AOC and 55 POC).

Efferent Organization of the rNST

Efferent Organization of Topographic Regions

Because our neurophysiologic studies of the rNST suggested an orderly topographic afferent organization, we wished to determine whether this topography was related to the rNST efferent organization. Injections of the anterograde tracer biocytin were made into rNST sites following electrophysiologic characterization [6,17]. Punctate injections were made into seven sites in the rostral pole of NST that were primarily responsive to AOC gustatory stimulation, nine sites further caudal and medial responsive to mechanical and probably also gustatory POC stimulation, and nine sites caudal to the rostral pole but in the lateral one-third of the nucleus

that were optimally responsive to AOC mechanical stimulation.

The results indicated differences in efferent organization, particularly for the ascending projection to the parabrachial nucleus (PBN). Despite significant overlap, projections arising from the AOC and POC gustatory-responsive NST maintained some orotopy in the PBN, with AOC gustatory projections centered more laterally and caudally in the medial and ventral lateral PBN subnuclei. More strikingly, the intensities of the terminal field labeling varied: Injections into the AOC gustatory-responsive region produced dense PBN terminal labeling but injections into the POC gustatory-responsive regions produced sparser labeling that appeared to be composed of finer fibers. The most dramatic difference in ascending projections occurred for sites responsive to mechanical stimulation of the AOC, which produced barely detectable PBN labeling.

Ascending Versus Descending Pathways from rNST

A second neuroanatomic study [7] suggests a further differentiation of rNST efferent organization. Dual injections of different fluorescent retrograde tracers were made into the terminal fields of the two major efferent pathways arising from the rNST: the PBN and the medullary reticular formation. Within the NST, few double-labeled neurons were observed, and the distribution of the two populations was distinctive. Neurons contributing to the ascending pathway were focused more dorsally and medially in the area of the central subnucleus [3] compared to neurons contributing to the descending pathway, whose distribution also encompassed the lateral and ventral subdivisions [3]. These data suggest that separate neural populations in the rNST contribute to the two pathways. The separate populations do not appear related to rNST orotopic organization, but they may be partially related to the organization of gustatory vs mechanical sensitivity.

Chemosensitive Organization

Implications of Orotopy

The distinctive chemical sensitivities of different gustatory receptor subpopulations [e.g., 18,19], in concert with the orotopy we have observed, prompt an obvious question: Is the rNST chemotopically organized? Two of our studies bear on this question [8,9].

In one experiment [9] 15 neurons responsive to both anterior tongue and nasoincisor duct stimulation and 10 neurons responsive only to anterior tongue stimulation were located in the rostral pole of the NST. On average, these cells responded best to NaCl or sucrose, depending on whether the tongue or the palate was stimulated.

In a more recent study [8], chemosensitive profiles elicited by posterior tongue stimulation were determined for 24 multiunit recording sites located mainly posterior and medial to the AOC neurons and they responded optimally to HCl. These studies suggest a chemotopic organization of the rNST, with optimal responsiveness to the sweet and salty qualities located anterior to the sour quality. These results are inconclusive, however, because of the relatively small number of recording sites and because only the tongue was stimulated in the latter study. The conclusions are also tentative because none of our neurophysiologic investigations have yet revealed the location of robust rNST responses elicited by bitter-tasting stimuli.

A New Approach

We have begun to use a new technique to study rNST functional organization [12], in particular to address the question of chemotopy. This approach employs standard immunocytochemical techniques to detect the presence of the protein product of the immediate-early gene, c-*fos*, following gustatory stimulation in awake animals. Alternate deliveries of four 150-µl aliquots of sucrose (1.0 M) and three 100-µl aliquots of quinine HCl (0.03 M) resulted in strong expression of the c-*fos* gene in the rNST in four subjects. Robust Fos-like immunoreactivity (Fos-LI) was observed in the nuclei of many cells in the medial, gustatory + tactile region of the rNST, but not in the lateral, tactile-only region. This labeling extended into the caudal division [3] of the NST (i.e., caudal to where we have observed oral gustatory responses electrophysiologically) but dropped off sharply at the level of the area postrema.

The labeling pattern produced by gustatory stimulation was distinct from that which occurred when capsaicin was injected into the oral mucosa of urethane-anesthetized subjects. In these preparations ($n = 4$), robust labeling also occurred in the NST but was intense at the level of the area postrema and dropped off sharply at the caudal border of the rNST. Infusions of water through intraoral cannulas ($n = 1$) also produced minimal rNST labeling.

We have begun to explore the distribution of Fos-labeled nuclei in response to delivery of a single gustatory stimulus. Quinine stimulation produced pronounced labeling in the NST ($n = 2$) that was robust in a restricted medial zone in the caudal half of rNST, a region corresponding partially to the POC gustatory + mechanical region observed neurophysiologically (Fig. 2). Following sucrose delivery ($n = 4$), Fos-like immunoreactivity also was observed in the rNST but was not nearly as robust as that produced by quinine, and it was difficult to define its anatomic distribution. Thus at present the major difference observed between c-*fos* expression after sucrose and quinine stimulation is quantitative and, oddly, in direct opposition to what we have observed neurophysiologically.

Discussion

The principles by which orosensory information is organized in the rNST have emerged somewhat slowly. For example, there has been considerable disagreement about the deceptively straightforward but fundamental question of whether gustatory neurons can be divided into distinct categories based on their chemosensitive profiles [20–22]. Although this issue has not been resolved, there is growing evidence that most taste neurons fit into one of a relatively small number of categories whose response profiles exhibit a systematic organization [21–23].

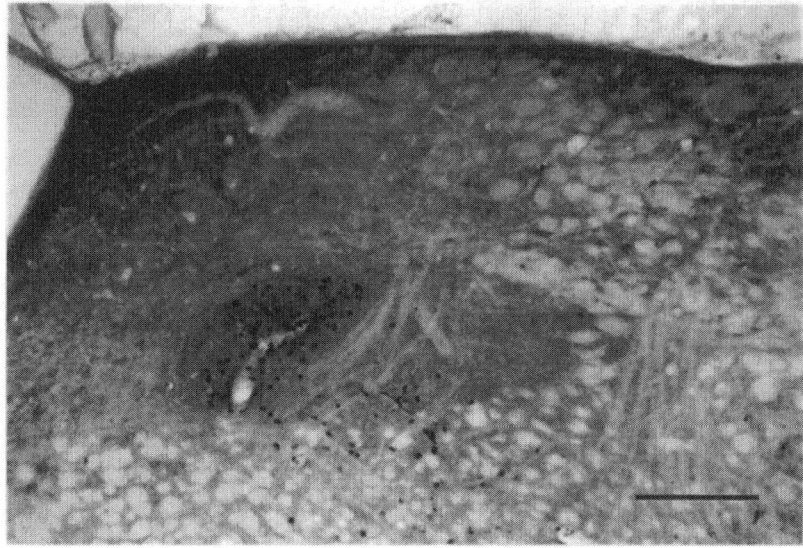

Fig. 2. Photomicrograph of a coronal section through the caudal half of the rNST, showing the distribution of Fos-LI immunoreactivity following stimulation with 0.03 M quinine HCl. Scale bar = 200 µM

Our work has also demonstrated regularity in the response properties of NST orosensory neurons but highlights variables in addition to chemosensitivity relevant to defining their functional identity: modality, receptive field, anatomic location, and efferent projections. The way in which all these variables interact to define rNST organization appears behaviorally meaningful. For example, Frank [24] initially demonstrated that the orderliness of chemosensitive response profiles is best revealed when responses are plotted against stimuli ordered on the basis of behavioral preference.

Our work on gustatory convergence within single neurons similarly suggests that the most integration occurs between taste receptor subpopulations that are located in the same half of the oral cavity and thus are most likely to be activated simultaneously in the rostral-to-caudal movement of ingesta during consummatory behavior. We demonstrated previously [9] that the chemosensitive convergence that is a consequence of this spatial convergence involves hedonically (although not qualitatively) similar pairs of stimuli and occurs in a manner that preserves orderly response profiles. When the organization of responsiveness *across* single NST neurons is considered, a clear orotopy is evident that based on a rostral to caudal oral gradient. Our data, as well as several other sources (reviewed in [23]) suggest that this orotopy may underlie a chemotopic organization in the rNST. The available data suggest that the chemotopy may be partially characterized by a scheme suggested by Nowlis [25]: Preferred stimuli are represented rostrally and the nonpreferred stimuli caudally. Such an organization would provide a discrete anatomic locus for the afferent limbs of the distinctive behavioural responses elicited by these classes of tastants.

Acknowledgments. This work was supported by NIH Grant DC00416 to S.P.T., NIH Grant DC00417 to J.B.T., and the Ohio State University College of Dentistry. The excellent technical support of Lisa Akey, Nicole Burkhardt, Tim Copelin, and Lisa McConnell is much appreciated. The neurophysiologic preparation was developed in the laboratory of Dr. Ralph Norgren during postdoctoral work by S.P.T.

References

1. Altschuler SM, Boa X, Bieger D, Hopkins DA, Miselis RR (1989) Viscerotopic representation of the upper alimentary tract in the rat: sensory ganglia and nuclei of the solitary and spinal trigeminal tracts. J Comp Neurol 283:248–268
2. Hamilton RB, Norgren R (1984) Central projections of gustatory nerves in the rat. J Comp Neurol 222: 560–577
3. Whitehead MC (1988) Neuronal architecture of the nucleus of the solitary tract in the hamster. J Comp Neurol 276:547–572
4. Cunningham E Jr, Sawchenko P (1989) A circumscribed projection from the nucleus of the solitary tract to the nucleus ambiguus in the rat: anatomical evidence for somatostatin-28-immunoreactive interneurons subserving reflex control of esophageal motility. J Neurosci 9:1668–1682
5. Herbert H, Saper C (1990) Cholecystokinin-, galanin-, and corticotropin-releasing factor-like immunoreactive projections from the nucleus of the solitary tract to the parabrachial nucleus in the rat. J Comp Neurol 293: 581–598
6. Becker DC (1992) Efferent projections of electrophysiologically identified regions of the rostral nucleus of the solitary tract in the rat. Ohio State University, Columbus, OH [Master's thesis]
7. Halsell CB, Travers SP, Travers JB (1993) Ascending and descending efferent neurons in the nucleus of the solitary tract comprise two distinct populations. Soc Neurosci Abstr 19:1282
8. Halsell CB, Travers JB, Travers SP (1993) Gustatory and tactile stimulation of the posterior tongue activate overlapping but distinctive regions within the nucleus of the solitary tract. Brain Res 632:161–173
9. Travers SP, Pfaffmann C, Norgren R (1986) Convergence of lingual and palatal gustatory neural activity in the nucleus of the solitary tract. Brain Res 365: 305–320
10. Travers SP, Norgren R (1988) Oral sensory responses in the nucleus of the solitary tract. Soc Neurosci Abstr 14:1185
11. King MA, Louis PM, Hunter BE, Walker DW (1989) Biocytin: a versatile anterograde neuroanatomical tract-tracing alternative. Brain Res 497:361–367
12. Harrer M, Dinkins M, Travers J, Travers S (1993) Gustatory-elicited expression of c-*fos* in brainstem nuclei. Soc Neurosci Abstr 19:1431
13. Davis BJ (1988) Computer-generated rotation analyses reveal a key three-dimensional feature of the nucleus of the solitary tract. Brain Res Bull 20:545–548
14. Travers JB (1988) Efferent projections from the anterior nucleus of the solitary tract of the hamster. Brain Res 457:1–11
15. Hayama T, Ito S, Ogawa H (1985) Responses of solitary tract nucleus neurons to taste and mechanical stimulations of the oral cavity in decerebrate rats. Exp Brain Res 60:235–242
16. Sweazey RD, Smith DV (1987) Convergence onto hamster medullary taste neurons. Brain Res 408: 173–185
17. Travers SP (1993) Orosensory processing in neural systems of the nucleus of the solitary tract. In: Simon S, Roper S (eds) Mechanisms of taste transduction. CRC Press, Boca Raton, FL, pp 339–394
18. Frank ME (1991) Taste-responsive neurons of the glossopharyngeal nerve of the rat. J Neurophysiol 65:1452–1463
19. Nejad MS (1986) The neural activities of the greater superficial petrosal nerve of the rat in response to chemical stimulation of the palate. Chem Senses 11: 283–293
20. Erickson RP (1985) Grouping in the chemical senses. Chem Senses 10:333–340
21. Scott TR, Plata-Salaman CR (1991) Coding of taste quality. In: Getchell TV, Bartoshuk LM, Doty RL,

Snow JB (eds) Smell and taste in health and disease. Raven, New York, pp 345–368

22. Smith DV, Van Buskirk RL, Travers JB, Bieber SL (1983) Gustatory neuron types in hamster brain stem. J Neurophysiol 50:522–540

23. Hettinger T, Frank M (1992) Information processing in mammalian gustatory systems. Curr Opin Neurobiol 2:469–478

24. Frank M (1973) An analysis of hamster afferent taste nerve response functions. J Gen Physiol 61:588–618

25. Nowlis GH (1977) From reflex to representation: taste-elicited tongue movements in human newborns. In: Weiffenbach JM (ed) Taste and Development. The Genesis of Sweet Preference. Department of Health and Human Services, NIH, Bethesda, MD, pp 190–204

Information Processing in the Parabrachial Nucleus of the Pons

PATRICIA M. DI LORENZO[1], SCOTT MONROE[2], and GERALD S. HECHT[1]

Key words. Taste—Parabrachial nucleus of the pons —Electrophysiology—Neural code—Rat—Nucleus of the solitary tract

Introduction

In traditional investigations of the taste system, electrophysiological responses to taste stimuli are recorded within a particular neural structure and the function of that structure in the perceptual process is deduced from analyses of those responses. The analysis of the responsive properties of a single neural structure necessarily views those responses out of the context of their input. This method has the disadvantage that, at higher levels of neural processing, the relationship of neural responses to activity evoked peripherally may become increasingly obtuse.

At the core of most analyses of the neural code for gustation is the response measure, which consists most commonly of the number of spikes that occur during some arbitrary interval of time when a tastant is on the tongue. Typical response measures are based on the idea that the system integrates activity over time without regard to the dynamic properties of taste responses. Because these temporal patterns of activity have been shown to be distinctive and reliable, it has been suggested that this aspect of the neural taste response conveys useful, if not essential, information [1–9]. Moreover, when activity is summed over time, the critical interval of measurement is problematic. Despite evidence that a taste stimulus can be identified within 200 ms of contact with the tongue [see 10], virtually all theories of taste coding are based on response measures taken over 3–5 s of response.

In our investigations of the parabrachial nucleus of the pons (PbN), the second relay in the central pathway for gustation [11,12], we have tried to avoid some of these potential pitfalls associated with traditional investigations. First, we have made no assumptions about the critical response interval. Instead, we have focused on the dynamic aspects of the neural response, not of individual neurons, but of the population of responsive neurons taken together, as reflected in our sample. By examining the changes in the neural code as it unfolds over time, we have been able to provide some detail about the role of the PbN in taste coding of individual taste stimuli. Second, we have been careful to form hypotheses about the function of the PbN strictly in the context of activity recorded in the nucleus of the solitary tract (NTS), the first relay in the neural pathway for taste and the major source of input to the PbN [11,12]. With this approach, we have gained some insight into what the PbN adds to the signal for taste as it ascends in the neural pathway.

Experiments

We began our investigation by recording the electrophysiological responses of PbN neurons to a range of concentrations of representatives of the four basic taste qualities: NaCl, HCl, quinine HCl, and sucrose. Fifteen male rats, anesthetized with urethane (1.5 g/kg), were tested. Using quantitative techniques which have been described previously [13], we analyzed the difference in across-neuron patterns (ANPs) as they were elaborated over the time course of the response. Briefly, the ANP of response to a given taste stimulus was represented as a vector in n-dimensional space, where n is the number of taste-responsive neurons. The coordinates of this vector were determined by the number of spikes that occurred during a specified interval during which the taste stimulus was bathed over the tongue. The length of each vector associated with a particular stimulus is a function of the overall magnitude of response. The angle between vectors represents an index of similarity; small angles imply similar ANPs of response and large angles imply just the opposite. By changing the interval over which the response was measured, it was possible to chart the time course of angular separation of ANPs as the response unfolded.

Changes in the ANPs of response over time were analyzed for 22 taste-responsive PbN cells. Initially,

---

[1] Department of Psychology, PO Box 6000, State University of New York at Binghamton, Binghamton, NY 13902-6000, USA
[2] Current address: Department of Anesthesiology, Southwest Medical Center, University of Texas, 5323 Harry Hines Boulevard, Dallas, TX 75235-9068, USA

402

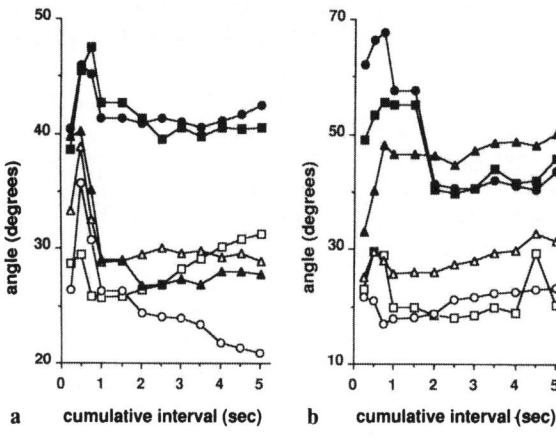

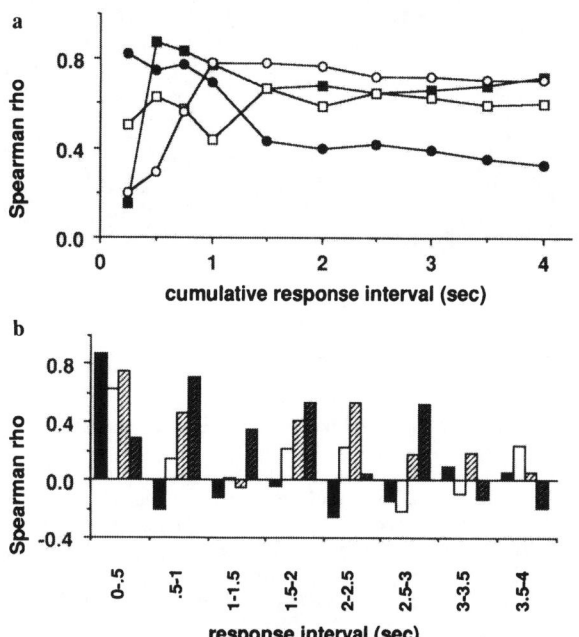

FIG. 1. Changes over time in the angular separation of across-neuron patterns (ANPs) associated with weak tastants (**a**) and with a variety of concentrations of NaCl (**b**). *N*, 0.001 *M* NaCl; *H*, 0.001 *M* HCl; *Q*, 0.001 *M* quinine HCl; *S*, 0.001 *M* sucrose; *N1*, 0.1 *M* NaCl; *N2*, 0.01 *M* NaCl; *N3*, 0.001 *M* NaCl; *N4*, 0.0001 *M* NaCl. **a** *Open squares*, N-H; *open circles*, N-Q; *closed squares*, N-S; *open triangles*, H-Q; *closed circles*, H-S; *closed triangles*, Q-S. **b** *Open squares*, N1-N2; *open triangles*, N1-N3; *closed squares*, N1-N4; *open circles*, N2-N3; *closed circles*, N2-N4; *closed triangles*, N3-N4

FIG. 2. Spearman correlations of average temporal patterns of activity in the nucleus of the solitary tract (NTS) and the parabrachial nucleus of the pons (PbN) in cumulative response intervals (**a**) and in successive 0.5-s intervals (**b**). **a** *Closed squares*, N; *open squares*, H; *closed circles*, S; *open circles*, Q. **b** *Closed bars*, N; *open bars*, H; *open hatched bars*, S; *shaded hatched bars*, Q

we examined the changes in the angles associated with ANPs for tastants presented at very low concentrations (all taste stimuli were presented at 0.001 *M*), where taste quality would presumably be just barely identifiable. For these stimuli, the ANPs of response are most different from each other during the first 0.5 s of the responses (Fig. 1a). Those pairwise comparisons involving sucrose produce the largest angles in the very earliest part of the response interval. These data imply that information about the identification of a taste quality may be contained in the very earliest portions of the PbN response, in agreement with the behavioral literature [see 10].

In contrast to information about taste quality, information about the intensity of a taste stimulus does not appear to be expressed by changes in the ANP of responses over time. Figure 1b shows the angles between across-unit patterns of responses for different concentrations of NaCl across time. It can be seen that, with the exception of comparisons between subthreshold (N4) and suprathreshold (N1–N3) stimuli, these patterns were relatively similar throughout the response interval. Data were similar for all four taste stimuli used. These results support the conclusion that taste intensity is not encoded by the dynamic properties of ANPs of response in the PbN.

Although these initial studies provided some insight into the possible significance of the dynamic aspects of taste responses in the PbN, there remained the issue of defining the specific role of the PbN in

the perceptual process. To accomplish this, taste responses in the PbN were examined in the context of activity from the NTS. To this end, we recorded taste responses simultaneously from pairs of neurons, one located in the NTS and the other in the PbN in 22 rats anesthetized with urethane (1.5 g/kg). Following identification and isolation of a pair of taste responsive cells, representatives of the four basic taste qualities were bathed over the tongue. Details of stimulus presentation have been described previously [13].

Electrophysiological responses to taste stimuli were recorded from 31 pairs of neurons in the NTS and PbN. The first question that we asked pertained to the extent to which the temporal patterns of activity that characterized NTS responses to taste were preserved at the level of the PbN. To answer this question, we averaged the temporal pattern of evoked response at both neural levels as an approximation of the population vector associated with the time course of the neural responses. Spearman correlations of this averaged activity were calculated over progressively longer intervals of time (Fig. 2a) and in successive 0.5 s epochs (Fig. 2b). Results of these analyses suggest that the temporal pattern of response of neurons in the PbN follows that of neurons in the NTS within different time frames for different taste stimuli. Neural activity in the PbN is

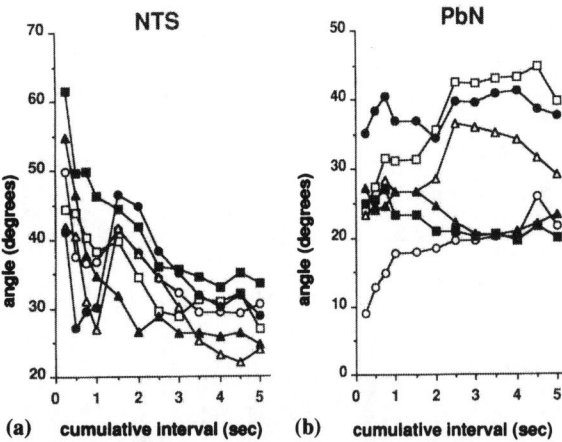

FIG. 3. Changes over time in the angular separation of ANPs associated with midrange concentrations in the NTS (a) and in the PbN (b). *N*, 0.1 *M* NaCl; *H*, 0.01 *M* HCl; *Q*, 0.01 *M* quinine HCl; *S*, 0.5 *M* sucrose. **a,b** *Open squares*, N-H; *open circles*, N-Q; *open triangles*, H-Q; *closed squares*, N-S; *closed circles*, H-S; *closed triangles*, Q-S

entrained to NTS activity within the first 0.25 s for sucrose, within 0.5 s for NaCl and HCl, and within 1 s for quinine. Correlations decrease rapidly for NaCl and HCl, but remain relatively high for sucrose and quinine for the first 2.5–3.0 s. After 3 s of response, the time course of PbN activity appears to be independent of the time course of NTS activity for all tastants tested.

Next, we examined the changes in the ANPs over time in both the NTS and the PbN as a way of discovering what the PbN was adding to the neural code for taste. Figure 3 shows the comparisons of ANPs of response among taste stimuli presented at midrange concentrations in the NTS (Fig. 3a) and the PbN (Fig. 3b). In the NTS the angles between ANPs generally decreased over the response interval, but in the PbN, these angles increased over time. If it is assumed that differences in the ANPs associated with taste stimuli imply that information is being conveyed, then these data imply that neural activity in the NTS contains the most information in the earliest portions of the response while the information conveyed in the PbN intensifies during the later period of the response.

Summary and Conclusions

As a result of the investigations outlined, some general hypotheses about information processing in the PbN emerge. First, the PbN appears to receive the most information from the NTS in the initial portion of the response. The time course of this transfer of information varies according to the particular stimulus involved; however, information about sucrose consistently appears to be the first conveyed.

Although it is clear from data from the presentation of very weak tastants that the activity in the PbN reflects this information in the very earliest parts of the responses, data from the presentation of midrange concentration tastants suggests that the activity in the PbN, unlike the activity in the NTS, continues to convey information in the later portions of the response. This additional information does not appear to be related to the intensity of a particular tastant, but instead may relate to the amplification of taste quality information.

Acknowledgments. This work was supported by a grant from the Whitehall Foundation to P.M.Di Lorenzo. This paper could not have been prepared without the help of Dr. S. Robinson, Dr. A.F. Smith, and Mr. L. Borowski. Dr. Jelle Atema graciously read an earlier version of this manuscript.

References

1. Bradley RM, Stedman HM, Mistretta CM (1983) Superior laryngeal nerve response patterns to chemical stimulation of sheep epiglottis. Brain Res 276:81–93
2. Covey E (1980) Temporal coding in gustation. Doctoral dissertation, Duke University, NC
3. Di Lorenzo PM, Schwartzbaum JS (1982) Coding of gustatory information in the pontine parabrachial nuclei of the rabbit: Temporal patterns of neural response. Brain Res 251:244–257
4. Fishman IY (1957) Single fiber gustatory impulses in rat and hamster. J Cell Comp Physiol 49:319–334
5. Funakoshi M, Ninomiya Y (1977) Neural code for taste quality in the thalamus of the dog. In: Katsuki Y, Sato M, Takagi SF, Oomura Y (eds) Food intake and the chemical senses. University Park Press, Tokyo, pp 223–232
6. Nagai T, Ueda K (1981) Stochastic properties of gustatory impulse discharges in rat chorda tympani fibers. J Neurophysiol 45:574–592
7. Ogawa H, Sato M, Yamashita S (1973) Variability in impulse discharges in rat chorda tympani fibers in response to repeated gustatory stimulations. Physiol Behav 11:469–479
8. Ogawa H, Yamashita S, Sato M (1974) Variation in gustatory nerve fiber discharge pattern with change in stimulus concentration and quality. J Neurophysiol 37:443–457
9. Travers SP, Norgren R (1989) The time course of solitary nucleus gustatory responses: influence of stimulus and site of application. Chem Senses 14:55–74
10. Halpern BP (1985) Time as a factor in gustation: Temporal patterns of stimulation and response. In: Pfaff DW (ed) Taste, olfaction and the central nervous system. Rockefeller University Press, New York, pp 181–209
11. Norgren R (1978) Projections from the nucleus of the solitary tract in the rat. Neurosci 3:207–218
12. Norgren R (1985) Taste and the autonomic nervous system. Chem Sens 10:143–161
13. DiLorenzo PM (1989) Across unit patterns in the neural response to taste: Vector space analysis. J Neurophysiol 62(4):823–833

Taste Information Processing in the Insular Cortex of Rats

H. Ogawa, K. Hasegawa, S. Otawa, and T. Nakamura[1]

Key words. Taste neurons—Cortex—Rats—Transmitters—Peptides—Iontophoresis

Introduction

Cortical taste area (CTA) is located in two cytoarchitectonically different areas of the cerebral cortex in rats: granular and dysgranular insular areas (areas GI and DI). In both areas, taste neurons have been found among mechanoreceptive neurons with receptive fields in the oral cavity [1]. Cortical taste neurons in rats are characterized by small response magnitudes, large receptive fields, and response profiles with two peaks [2–4]. Such response features are considered to be the results of convergence of afferents at various relay stations along the central gustatory pathway and in the CTA, together with the action of excitatory and inhibitory interneurons in these regions. In these synaptic processes, many substances operate as transmitters or modulators.

Several substances have been found as neurotransmitters or neuromodulators in the cerebral cortex. Excitatory amino acids (EAAs), e.g., glutamate, and gamma-aminobutyric acid (GABA) are identified as excitatory or inhibitory neurotransmitters in other sensory cortices, such as visual or somatosensory cortices [5–8] though their actions in the CTA are not known. Among neuromodulators, calcitonin gene-related peptide (CGRP) is rather rich in the insular cortex [9–11] in contrast to substance P (SP) which is rather sparse in the cerebral cortex including the CTA [12]. CGRP is reportedly involved in taste aversion learning [13]. The aim of the present paper is to understand the underlying mechanism of taste coding in the cortex in terms of neurotransmitters or modulators. We studied the effects of iontophoretic applications of these putative transmitters, glutamate and GABA, and neuromodulators, CGRP and SP, on the taste cortical neurons, and also studied antagonists of certain glutamate and GABA receptors by means of a multibarrel electrode.

Materials and Methods

Adult SD-strain albino rats, female, weighing 180–320 g, were anesthetized with urethane (1 g/kg body weight, i.p.). After cannulation of the femoral vein and trachea, the animal was mounted on a standard stereotaxic instrument with a pair of ear bars following the Paxinos and Watson method [14]. During the experiment, the rats were tilted by 45° with the left side up. The animals were immobilized with *d*-tubocurarine and artificially ventilated. End-tidal CO_2 concentration was maintained at 3.5%–4.5%. Urethane (100 mg/kg) was supplemented whenever necessary. Body temperature was kept at 37°C with a water heater. ECG was monitored throughout the experiment.

The left buccal wall was cut from the mouth corner to the anterior edge of the ramus of the mandibula, the mouth was opened at approximately 30°–40°, and the tongue was stretched out anteroventrally. A bone covering the left middle cerebral artery was resected to expose the left CTA, and a small opening was made to insert a recording electrode. Cut wounds and pressure points were infiltrated with 1% xylocaine.

Single unit activities were recorded extracellularly from the soma of the cortical neurons with a glass micropipette filled with a saturated solution of pontamine sky blue in 1 *M* Na acetate (8–15 *M*Ω). The recording micropipette was glued to a multibarrel micropipette which was used for iontophoretic application of drugs. The drugs employed are listed in Table 1. Retaining currents were 2–10 nA, and ejection currents were less than 100 nA.

The taste stimuli used were 0.1 *M* NaCl, 0.5 *M* sucrose, 0.01 *N* HCl, and 0.02 *M* quinine-HCl, and taste stimulation of the entire oral cavity was made in the following sequence: rinse with water (15 s), taste stimulus (10 s), and then water-rinse (15 s). Between taste stimulations, the oral cavity was repeatedly rinsed with water. Taste responses were identified when, during the 10 s of taste stimulation, there was a change in the discharge rate at least 1.0 s long and 2 SD above or below the prestimulus average [15]. The control of solenoid valves for taste stimulation and the collection of impulses during the stimulation period were performed using a 16-bit microcomputer (PC 9801RX, NEC, Tokyo, Japan).

[1] Department of Physiology, Kumamoto University School of Medicine, Honjo 2-2-1, Kumamoto, 860 Japan

TABLE 1. List of drugs used for iontophoresis.

Glutamate	(0.5 M, pH 8.0)
N-methyl-D-aspartate (NMDA)	(50 mM, pH 8.0)
Kainate (an agonist for the non-NMDA receptor)	(20 mM, pH 8.0)
D-amino-5-phosphonovalerate (AP-5; a specific antagonist for NMDA)	(5 mM, pH 8.0)
6-cyano-7-nitro-quinoxaline-2,3-dione (CNQX, a specific antagonist for non-NMDA)	(1 mM, pH 8.0)
Gamma-aminobutyric acid (GABA)	(0.5 M, pH 3.0)
Bicuculline methiodide (BMI)	(5 mM, pH 3.0)
Substance P (SP)	(0.2 mM, pH 8.0)
Calcitonin gene-related peptide (CGRP)	(0.02 mM, pH 6.0)
165 mM NaCl for balancing	

Dye was deposited at interesting points of the brain through the recording electrode. At the termination of the experiment, the animal was deeply anesthetized and intracardially perfused with buffered 10% formalin. The brain was frozen, serially sectioned at 50 μm, and stained with thionin. Recording sites were reconstructed by referring to the dye marks and micromanipulator readings of the recording sites.

Results

Excitatory Transmitters and Glutamate Receptors

In about 60% of the 38 taste neurons tested, glutamate of 50 nA increased spontaneous discharges, and decreased them in a few cases. Among the ionotrophic glutamate receptors, two main subtypes are known: N-methyl-D-aspartate (NMDA)-receptors and non-NMDA-receptors [16]. Agonists for these receptors, NMDA and kainate, excited most of the 49 taste neurons tested (78%). Both of the selective antagonists of these receptors, D-amino-5-phosphonovalerate (AP-5) for NMDA receptors and 6-cyano-7-nitro-quinoxaline-2,3-dione (CNQX) for non-NMDA receptors, antagonized taste responses. AP-5 also decreased spontaneous discharges in some neurons. A small number of neurons were selectively affected by one of these antagonists, and a few were not affected at all. In the distribution of taste neurons with NMDA and/or non-NMDA receptors, no difference was noted between areas GI and DI, or across cortical layers. This is not in agreement with previous histochemical and physiological results in the visual cortex in cats, where NMDA receptors or neurons with them were rare in layer V.

Inhibitory Transmitters and Effects of Inhibitory Interneurons in Taste Coding

In about 61% of the taste neurons tested ($n = 83$), GABA of 50 nA decreased spontaneous discharges. Bicuculline methiodide (BMI), a selective antagonist of the GABA$_A$ receptor, antagonized this inhibitory action of GABA. BMI itself increased spontaneous discharges when a relatively large amount of ejection current was used.

To examine the role of inhibitory interneurons on taste coding in the CTA, the effects of BMI on taste responses were tested. Ejection current of BMI was usually 4–8 nA, below the threshold of the drug to increase spontaneous discharges. In 34 of the 85 taste neurons tested, BMI increased taste responses. In several neurons, discharges were increased in the prestimulus period when water-rinsing was carried out. This was probably due to an increase in responses of the neurons to the mechanical stimulation of rinsing, because many taste neurons were receptive fields for mechanical stimulation [1]. When discharges in the prestimulus period were greatly increased, the net magnitudes of some taste responses were decreased in comparison with those in control states without BMI. It is probable that discharges due to taste and mechanical stimulations became nonlinear when both responses were very large: subtraction of the discharge rate in the prestimulus period (mechanical and thermal stimulation for adaptation) from that in the stimulus period (addition of taste stimulation to the prestimulus period) sometimes gave negative values in contrast to positive ones in the control.

In about 40% of the neurons, changes in response profiles were seen during BMI application: from single-peaked type to double-peaked type, or vice versa. Even among the neurons whose response type was not affected, the best stimulus did change in about 50%. The size of receptive fields was increased in almost all neurons (93% of the 15 tested), except one in which the taste responses were decreased by the drug but the receptive fields were not affected. Increase in the receptive fields was also noted in neurons without taste responses significantly affected by the drug.

Some taste neurons were discovered during BMI application. They were spontaneously silent and did not respond to taste stimulation without application of the drug. BMI disclosed taste responses and receptive fields to taste stimulation. The best stimulus was defined in these neurons in the presence of BMI.

Effects of Neuromodulators on Taste Responses

Neuromodulators, such as SP and CGRP, were tested on spontaneous discharges and taste responses of CTA neurons. SP and CGRP often increased spontaneous discharges, as seen in the somatosensory cortex [17]. The fraction of neurons affected by SP was larger in area DI than in area GI, but that for CGRP was larger in area GI. The present findings indicate that the distribution of neurons sensitive to SP and/or CGRP was different from that of afferents containing the peptides [9,10,12,15,18], with that of receptors studied with immunohistochemical methods [11,19].

Responses to SP were produced in a short latency, but those to CGRP in some neurons were sluggish and outlasted the period of drug application, which suggests that CGRP activates intracellular second messengers to depolarize cortical neurons, as reported for both peptides [20,21]. CTA neurons had a tendency to have receptors for both SP and CGRP or for both glutamate and SP.

Both peptides affected taste responses in a large fraction of neurons tested. SP affected taste responses in almost all neurons tested, even when it did not modulate spontaneous discharges, but CGRP changed taste responses and/or spontaneous discharges independently. The actions of both peptides on taste responses and spontaneous discharges were not always in the same direction.

Discussion

It is evident that CTA neurons have receptors for EAAs, GABA, and peptides (Fig. 1). Both MNDA- and non-NMDA receptors for EAAs are indicated to contribute to normal transmission of taste information to CTA neurons, as in the case of other sensory cortices [1]. The present findings indicate

that GABA-ergic inhibitory interneurons contribute to taste coding by shaping the response profiles and receptive fields of cortical neurons. The findings also indicate the possible presence of tonic inhibition to suppress activities of some taste neurons. The findings that SP and CGRP affected spontaneous discharges and/or taste responses suggest that peptides act at the presynaptic terminals as well as the postsynaptic cells. The peptides probably modulate the release of EAAs from the presynaptic terminals, or act at the postsynaptic cells to modulate ionic channels through NMDA receptors [22] or by activating intracellular second messengers [20,21]. Since CGRP has been indicated to increase with the formation of taste aversion [13], it must be clarified in future whether the peptides interact with NMDA receptors in the CTA in the formation of taste memory.

References

1. Ogawa H, Ito S, Murayama N, Hasegawa K (1991) Taste area in granular and dysgranular insular cortices in the rat identified by stimulation of the entire oral cavity. Neurosci Res 9:196–201
2. Ogawa H, Hasegawa K, Murayama N (1992) Difference in taste quality coding between two cortical taste areas, granular and dysgranular insular areas, in rats. Exp Brain Res 91:415–424
3. Ogawa H, Murayama N, Hasegawa K (1992) Difference in receptive field features of taste neurons in rat granular and dysgranular insular cortices. Exp Brain Res 91:408–414
4. Yamamoto T, Yuyama N, Kato T, Kawamura Y (1984) Gustatory responses of cortical neurons in rats. I: Response characteristics. J Neurophysiol 51:616–635
5. Sillito AM (1977) The effectiveness of bicuculline as an antagonist of GABA and visually evoked inhibition in the cat's striate cortex. J Physiol (Lond) 250:287–304
6. Dykes RW, Llandry P, Metherate R, Hicks TP (1984) Functional role of GABA in the cat primary somatosensory cortex: Shaping receptive fields of cortical neurons. J Neurophysiol 52:1066–1093
7. Tamura H, Hicks TP, Hata Y, Tsumoto T, Yamatodani A (1990) Release of glutamate and aspartate from visual cortex of the cat following activation of afferent pathways. Exp Brain Res 80:447–455
8. Hicks TP, Kaneko T, Metherate R, Oka JI, Start CA (1991) Amino acids as transmitters of synaptic excitation in neocortical sensory processes. Can J Physiol Pharmacol 69:1099–1114
9. Kruger L, Sternini C, Brecha NC, Rogers WT, Mantyh PW (1988) Distribution of calcitonin gene-related peptide immunoreactivity in relation to the rat central somatosensory projection. J Comp Neurol 273:149–162
10. Yasui Y, Saper C, Cechetto DF (1989) Calcitonin gene-related peptide immunoreactivity in the visceral sensory cortex, thalamus and related pathways in the rat. J Comp Neurol 290:487–501

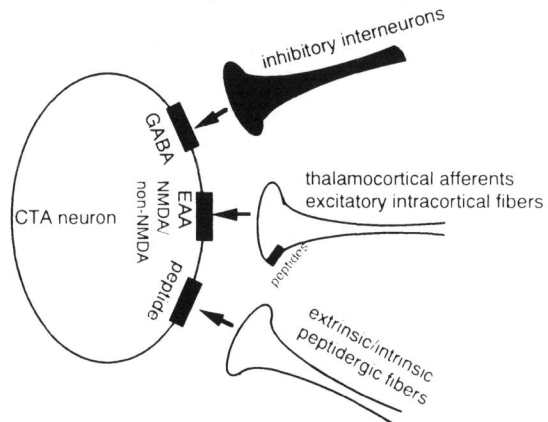

FIG. 1. Synaptic actions of various inputs on neurons in the cortical taste area (CTA). CTA neurons have receptors for gamma aminobutyric acid (GABA), excitatory amino acids (EAAs), and peptides, i.e., substance P (SP) and calcitonin-gene-related peptide (CGRP). Terminals of excitatory afferents are supposed to have receptors for peptides. NMDA, N-methyl-D-aspartate

11. Skofitsch G, Jacobositz D (1992) Calcitonin- and calcitonin gene-related peptide: receptor binding sites in the central nervous system. In: Bjorklund A, Hokfelt T, Kuhar MI (eds) Handbook of chemical neuroanatomy, vol 11, Elsevier, Amsterdam, pp 97–144

12. Cuello CA, Kanazawa I (1978) The distribution of substance P immunoreactive fibers in the rat central nervous system. J Comp Neurol 178:129–156

13. Yamamoto T, Matsuo R, Ichikawa H, Wakisaka S (1990) Aversive taste stimuli increase CGRP levels in the gustatory insular cortex of the rat. Neurosci Lett 112:167–172

14. Paxinos G, Watson C (1982) The rat brain in stereotaxic coordinates. Academic, New York

15. Ogawa H, Hayama T, Ito S (1984) Location and taste responses of parabrachiothalamic relay neurons in rats. Exp Neurol 83:507–517

16. Armstrong-James M, Welker E, Callahan CA (1993) The contribution of NMDA and non-NMDA receptors to fast and slow transmission of sensory information in the rat SI barrel cortex. J Neurosci 13:2149–2160

17. Lamour Y, Dutar P, Jobert A (1983) Effects of neuropeptides on rat cortical neurons: Laminar distribution and interaction with the effect of acetylcholine. Neurosci 10:107–117

18. Iritani S, Fujii M, Satoh K (1988) The distribution of substance P in the cerebral cortex and hippocampal formation: An immunohistochemical study in the monkey and rat. Brain Res Bull 22:205–303

19. Mantyh PW, Gates T, Mantyh CR, Maggio JE (1989) Autoradiographic localization and characterization of tachykinin receptor binding sites in the rat brain and peripheral tissues. J Comp Neurol 9:258–279

20. Watson SP, Downes CP (1983) Substance P induces hydrolysis of inositol phospholipids in guinea pig ileum and rat hypothalamus. Eur J Pharmacol 93:245–253

21. Van Valen F, Piechot G, Jurgens H (1990) Calcitonin gene-related peptide (CGRP) receptors are linked to cyclic adenosine monophoshate production in SK-N-MC human neuroblastoma cells. Neurosci Lett 119:195–198

22. Rusin KI, Ryu PD, Randic M (1992) Modulation of excitatory amino acid responses in rat dorsal horn neurons by tachykinins. J Neurophysiol 68:265–286

Optical Recording of Neural Signals in Rat Geniculate Ganglion

MICHIO NAKASHIMA, MAMIKO YANAURA, SATOSHI YAMADA, and SATORU SHIONO[1]

Key words. Optical recording—Geniculate ganglion

In the gustatory system, the geniculate ganglion (GG), located in the facial canal, contains taste sensory fibers innervating from the frontal tongue (chorda tympani nerve; CT) and from the soft palate (greater superficial petrosal nerve; GSP). Recent electrophysiological and behavioral experiments have shown that the GSP, in addition to the CT, plays an important role in taste perception [1].

To understand taste perception, it is important to measure spatio-temporal activity from many taste fibers. Multi-site optical recording, utilizing a voltage-sensitive dye, is a powerful means of overcoming the limitation of microelectrode techniques; it allows the simultaneous detection of activity from a large population of neurons [2]. We employed this optical recording with the rat GG in vitro to monitor neural signals. We observed optical signals responding to repetitive stimulation of electric pulses to the GSP nerve.

Sprague-Dawley strain rats were anesthetized with urethane (1 g/kg) and maintained at surgical level. The GG was carefully dissected out by an exposing technique (S. Harada, 1992, personal communication). The isolated preparation was attached to the silgard bed of an experimental chamber by pinning it with tungsten wires. The preparation was kept in a bathing solution of rat Ringer equilibrated with oxygen.

The cut end of the GSP nerve bundle was drawn into a suction electrode. The preparation was stained with a voltage-sensitive dye, RH155, for 15 min and the transmitted light was detected by a 24×24 photodiode matrix array with 448 active elements, using a 720 ± 25 nm interference filter. Objective magnification of $40\times$ was chosen to monitor neural signals from the GG. One pixel of each element covered 23×23 μm in the objective plane. The GSP nerve was orthodromically stimulated by applying depolarizing square-current pulses (200 μA, 0.3 ms, 4 Hz). When optical recordings were performed, the Ringer solution contained 1 mM tetraethylammonium to prolong the duration of action potentials.

Optical signals responding to the electric stimulation were obtained by averaging 40 trials. The signal to noise ratio (S/N) of the averaged trace was about 2, and the fractional change in optical intensity ($\Delta I/I$) of the optical signals was 2×10^{-4}. The optical signals were clearly different from the vibration noise observed in the marginal area of the GG. The area where the optical signals were observed was located in the dorsal left half layers of the GSP nerve adjacent to the GG. It extended to about 1 mm along the GSP nerve bundle. Each optical signal showed no significant difference in peak time within this area. The optical signals seemed to be due to the population activity of action potentials caused by the electric stimulation.

References

1. Harada S, Smith DV (1992) Gustatory sensitivity of the hamster's soft palate. Chem Senses 17:37–51
2. Nakashima M, Yamada S, Shiono S, Maeda M, Satoh F (1992) 448-Detector optical recording system: Development and application to *Aplysia* gill-withdrawal reflex. IEEE Trans Biomed Eng 39:26–36

[1] Embryonic Science and Technology Department, Central Research Laboratories, Mitsubishi Electric Corporation, 8-1-1 Tsukaguchi-Honmachi, Amagasaki, Hyogo, 661 Japan

Fos Protein Expression in Taste Neurons Following Electrical Stimulation of Afferent Nerves

Theresa A. Harrison and Nancy W. Miller[1]

Key words. c-fos—Taste pathway—Rostral NST organization—Taste nerve stimulation—Rat

The immediate early gene c-fos is turned on in neurons when they are activated synaptically, Ca^{2+} presumably triggering an intracellular signaling pathway [1]. The protein product of the c-fos gene has been used to functionally map afferent pathways in several sensory systems, including audition and nociception. In the present study, Fos protein immunocytochemistry was used in an attempt to localize discrete populations of neurons which are activated by input via different afferent nerves, through several stages of the gustatory pathway. Although it is well known that primary taste afferents from different regions of the oral cavity travel in different nerves and terminate in the nucleus of the solitary tract (NST) in more or less distinct regions, and that localization of taste responses within the nucleus depends also on stimulus location within the oral cavity, the distribution of taste-responsive neurons which are activated serially at successive levels through the pathway by input via the different nerves is not known.

In our experiments, either the lingual-tonsilar (IXth) nerve, or the chorda tympani (VIIth) nerve, was stimulated electrically in rats anesthetized with Nembutal (60 mg/kg, IP) or a Ketamine-Xylazine-Acepromazine cocktail (5 mg/kg-2.5 mg/kg-0.85 mg/kg, IM). High frequency (85 Hz) pulse trains lasting 500 ms were delivered once per s for 3 h. This frequency was selected to approximate response frequencies of IXth nerve afferent fibers during the initial 500 ms of their responses to tastants applied to the posterior tongue [2]. Sham-operated rats, treated identically to experimentals except that the electrical stimulus was not turned on, served as controls.

Frozen sections of paraformaldehyde-fixed brains, cut at 50 μm, were reacted with an affinity-purified rabbit polyclonal anti-c-Fos-peptide antibody (Santa Cruz Biotechnology, Inc.; Santa Cruz, California; c-fos [4]); 1:2000 overnight at 4°C). Fos immunoreactivity was visualized with the avidin-biotin-horseradish peroxidase (HRP) procedure.

In the caudal NST, experimental and control animals had similar, bilateral distributions of c-Fos-labeled neuronal nuclei. Significant numbers of cells were found scattered in the ventrolateral, dorsolateral, caudal central, and medial subdivisions of the nucleus [3]. In sham-operated animals, labeling did not extend into the rostral division of the NST.

After IXth nerve stimulation, in addition to this bilateral caudal distribution, labeled cells were located unilaterally in the rostral NST, extending more than 1 mm from 500 μm anterior to the obex to within 250 μm of the anterior pole of the nucleus. These cells were concentrated in the rostral central subdivision, but extended into the rostral lateral and ventral subdivisions to a limited degree. Chorda tympani-activated cells were also found unilaterally in the rostral central subdivision, but were largely anterior and lateral to the IXth nerve-activated neurons. In our analysis, very little overlap was seen in the two populations activated by the different nerves. However, the extent of overlap in these populations would be more conclusively established by comparing the results of stimulating both nerves, one right and one left, in the same animal: these experiments are currently underway.

In the parabrachial nucleus (PBN) of control rats, c-Fos-labeled neurons were consistently seen bilaterally in the dorsal lateral and external lateral subdivisions of the nucleus [4]. In experimental animals, an identical distribution of labeled neurons was seen. The extent of labeling was also similar in the two groups. Despite the presence of stimulation-induced c-Fos labeling within the NST, no labeled neurons were ever encountered in the medial or ventral lateral subdivisions of the PBN, the two regions in which taste responses have been recorded [5].

Given the lack of stimulation-specific c-Fos labeling in the PBN, it was not surprising that no labeling was seen in the gustatory zone of the thalamus. Presumably, either activation of higher level centers via NST output neurons was not sufficient to produce detectable levels of c-Fos, or c-fos is not the immediate early gene turned on in these cells when they are activated.

[1] Department of Cellular Biology and Anatomy, Medical College of Georgia, Augusta, GA 30912-2000, USA

Acknowledgments. This work was supported by NIH grant no. R29-DC00690 to Theresa A. Harrison.

References

1. Sheng M, Greenberg ME (1990) The regulation and function of *c-fos* and other immediate early genes in the nervous system. Neuron 4:477–485
2. Frank ME (1991) Taste responsive neurons of the glossopharyngeal nerve of the rat. J Neurophysiol 65: 1452–1463
3. Whitehead MC (1988) Neuronal architecture of the nucleus of the solitary tract in the hamster. J Comp Neurol 276:547–572
4. Fulwiler CE, Saper CB (1984) Subnuclear organization of the efferent connections of the parabrachial nucleus in the rat. Brain Res 319:229–259
5. Halsell CB, Frank ME (1991) Mapping study of the parabrachial taste-responsive area for the anterior tongue in the golden hamster. J Comp Neurol 306: 708–722

Coding of Binary Heterogeneous Taste Mixtures in the Hamster Parabrachial Nucleus

MARK B. VOGT and DAVID V. SMITH[1]

Key words. Taste mixtures—Neurophysiology

Mixtures are an integral feature of mammalian gustation: Taste experience naturally arises from a mixture of chemical stimuli bathing the receptors, and it has long been known that a gustatory stimulus may be perceived differently depending on whether it is presented in mixture with another stimulus or it is presented alone. However, most neurophysiological studies of the mammalian taste system have focused on responses to single chemical components. Recently [1,2] we have investigated the response frequencies of single third-order neurons in the hamster parabrachial nucleus (PbN) to anterior tongue stimulation with binary mixtures of heterogeneous taste stimuli; sucrose, NaCl, citric acid, and QHCl, each stimulus at a range of concentrations. Most mixture responses did not differ from the response to the *more effective component* (MEC) in the mixture presented alone, and very few were greater. However, about one-third of responses involved mixture suppression, or a mixture response that was significantly *less* than the response to the MEC. Robust mixture suppression was displayed primarily by sucrose-best neurons to mixtures that contained stronger concentrations of sucrose with citric acid or QHCl; mixtures of NaCl with citric acid evoked relatively weak suppression. Mixture suppression was apparent in measures of absolute response frequency, which are thought to be related to perceived stimulus intensity. The across-neuron patterns of activity, which may code taste quality, were similar for each mixture and its MEC, except for those mixtures that evoked robust suppression where the mixture pattern was more similar to that of the less effective component. Thus, mixture suppression is apparent in the responses of third-order neurons in the hamster gustatory system, and may be associated with pasychophysical mixture suppression, as observed in humans. Furthermore, our findings suggest that robust mixture suppression is associated with certain types of heterogeneous mixtures, specifically, with combinations of strongly appetitive and aversive stimuli. Additional studies of PbN neurons, employing other stimuli, will enable us to test the hypothesis that mixture suppression is related to the hedonic value of the mixture stimuli, and comparable studies of chorda tympani nerve fibers will enable us to identify the contribution of central and peripheral processes in mixture suppression.

Acknowledgments. Research supported by NIDCD Grant DC-00066 to DVS.

References

1. Vogt MB, Smith DV (1993) Responses of single hamster parabrachial neurons to binary taste mixtures: Mutual suppression between sucrose and QHCl. J Neurophysiol 69:658–668
2. Vogt MB, Smith DV (1993) Responses of single hamster parabrachial neurons to binary taste mixtures of citric acid with sucrose or NaCl. J Neurophysiol 70: 1350–1364

[1]Department of Otolaryngology—Head and Neck Surgery, ML 258, University of Cincinnati College of Medicine, Cincinnati, OH 45267, USA

Location of a Taste-Related Region in the Thalamic Reticular Nucleus in Rats

T. Hayama[1], K. Hashimoto[2], and H. Ogawa[2]

Key words. Thalamic reticular nucleus—Rat

The thalamic reticular nucleus (Rt) plays an important role in sensory information processing in the thalamo-cortical system [1]. Physiological studies have revealed visual-, auditory-, and somatosensory-related regions in the rat Rt [1]. It is not known, however, whether any part of the Rt plays a role in taste information processing. The present study anatomically explored a region in the Rt that has connections with the thalamic taste relay nucleus and the cortical taste area.

Sprague Dawley rats, anesthetized with amobarbital sodium (80 mg/kg) or urethane (1000 mg/kg), received injections of neuronal tracers, horseradish peroxidase (HRP), wheat germ agglutinin-HRP (WGA-HRP), and biocytin, in: (1) the thalamic taste relay, i.e., the parvicellular part of the thalamic posteromedial ventral nucleus (VPMpc), (2) the cortical taste area (CTA), and (3) the Rt. The histochemical reactions for HRP and WGA-HRP were carried out after the tetramethyl benzidine method. Biocytin was visualized with peroxidase-conjugated streptoavidin and 3,3'-diaminobenzidine.

Injections of WGA-HRP into the VPMpc labeled axons traveling ipsilaterally from the VPMpc through the medial lemniscus, the Rt, the internal capsule, and the basal ganglia, reaching the CTA. In the Rt, both anterograde and retrograde labels were observed in a limited portion located at the most ventromedial portion of the nucleus, at a level of 1.1–1.3 mm anterior to the rostral end of the VPMpc. In the CTA, anterograde labels were observed in layer IV, the deep portion of layer III, and the superficial portion of layer V, and labeled somata were observed in layer VI. Biocytin injections in the VPMpc labeled both somata and axon terminals in the same region of the Rt as found with the WGA-HRP injections.

Cortical injections of HRP or WGA-HRP in the CTA labeled axons traveling to the VPMpc by the same route, as found with the VPMpc injections. Anterograde labels were observed in the same region of the Rt as found with the thalamic injections. Both labeled somata and axon terminals were observed in the VPMpc. These results strongly suggest that the most ventromedial portion of the Rt is a taste-related region of the nucleus.

When WGA-HRP was injected into this taste-related region of the Rt, we observed labeled somata in layer VI in the CTA, in the VPMpc and its vicinity we found densely labeled axon terminals and a few lightly labeled somata.

These results indicate that the taste-related region is a very limited portion, different from other sensory portions in the Rt. This region is ventral to the somatosensory part, and anterior to the visual and auditory parts of the rat Rt [1]. The results also showed that the taste-related region in the Rt received inputs from the CTA and VPMpc and sent outputs to the VPMpc. This suggests that GABAergic neurons [2] in the Rt taste-related region have an inhibitory effect on the VPMpc taste neurons, as in other sensory systems [1], thereby inducing their inhibitory responses [3,4] to electrical stimulation of the CTA.

References

1. Shosaku A, Kayama Y, Sumitomo I, Sugitani M, Iwama K (1988) Analysis of recurrent inhibitory circuit in rat thalamus: Neurophysiology of the thalamic reticular nucleus. Prog Neurobiol 32:77–102
2. Howser CR, Vaughn JE, Barber RP, Roberts E (1980) GABA neurons are the major cell type of the nucleus reticularis thalami. Brain Res 200:341–354
3. Yamamoto T, Matsuo R, Kawamura Y (1980) Corticofugal effects on the activity of thalamic taste cells. Brain Res 193:258–262
4. Ogawa H, Nomura T (1988) Receptive field properties of thalamo-cortical taste relay neurons in the parvicellular part of the posteromedial ventral nucleus in rats. Exp Brain Res 73:364–370

[1] Department of Physiology, Kumamoto University School of Medicine, Honjo 2-2-1, Kumamoto, 860 Japan
[2] Department of Medical Technology, College of Medical Science, Kumamoto University, Kumamoto, 860 Japan

Convergence of Afferent Inputs from Tongue, Pharynx, and Larynx in the Insular Cortex of the Rat

T. Hanamori, T. Kunitake, K. Hirota, and H. Kannan[1]

Key words. Rat—Insular cortex—Convergence

It has been shown that gustatory afferent input projects to the anterior portion of the insular cortex in rats. Gustatory information from the taste buds distributing the oral and extraoral regions is conveyed by the chorda tympani (CT), greater superficial petrosal (GSP) nerve, the lingual-tonsilar branch of the glossopharyngeal (LT-IXth) nerve, the pharyngeal branch of the glossopharyngeal (PH-IXth) nerve, and the superior laryngeal (SL) nerves. How afferent information from these gustatory nerves is processed in the neurons of the insular cortex is still unknown. On the other hand, it has been demonstrated that visceral afferent information projects to the posterior portion of the insular cortex. There is a possibility that integration of the gustatory and visceral information may occur in the insular cortex. In the present experiment, we investigated the convergence of insular cortex neurons from all taste nerves except the GSP, as well as investigating changes in the spontaneous activity of the insular cortex neurons during chemo- and baroreceptor stimulation.

Anesthetized and paralyzed rats were used. The trachea was cannulated for artificial ventilation. The femoral artery and vein were cannulated for measurement of blood pressure (BP) and for administration of drugs, respectively. The heart rate was obtained from the recording of BP. Four nerves, the CT, LT-IXth, PH-IXth, and SL, were dissected free and transected; their central cut ends were placed on a pair of platinum wire electrodes for electrical stimulation. Extracellular unit responses of the neurons in the insular cortex were recorded with

glass microelectrodes filled with pontamine sky blue in sodium acetate. The recording sites were determined histologically following iontophoretic deposition of dye. For chemo- and baroreceptor stimulation, sodium nitroprusside, methoxamine hydrochloride, and sodium cyanide were intravenously injected.

Neurons which responded to electrical stimulation of at least one of the four nerves were located in the insular cortex between 2.9 mm anterior and 1.4 mm posterior to the anterior edge of the joining of the anterior commissure (AC) and between 0.1 and 2.9 mm dorsal to the rhinal sulcus. Most of the neurons were recorded from the rostral side to the anterior edge of the joining of the AC. The mean latency of the first unitary spikes of the insular cortex neurons elicited following electrical stimulation of the CT, LT-IXth, PH-IXth, and SL nerve was 21.0, 36.1, 38.3, and 36.0 ms, respectively. Of a total 50 neurons, 20% responded to one, 14% responded to two, 58% responded to three, and 8% responded to all four nerves stimulated. Of the specific neurons, 70% responded to the CT and 30% responded to the LT-IXth nerve. No specific neurons responding to the PH-IXth or the SL nerve were found. Most neurons (80%) had converged from more than two nerves. Among the converging neurons, 58% had converged from the CT, LT-IXth, and PH-IXth nerves. Of eight of the converging neurons, six showed a decrease or an increase of the spontaneous activity during chemo- and baroreceptor stimulation, while the remaining two showed no changes in spontaneous activity.

These results show that many insular cortex neurons have converged from the LT-IXth, PH-IXth, and SL nerves, and that the spontaneous activity in some of these neurons is changed by chemo- and baroreceptor stimulation.

[1] Miyazaki Medical College, 5200 Ooazakihara, Kiyotake-cho, Miyazaki-gun, Miyazaki, 889-16 Japan

NMDA and Non-NMDA Receptors Mediate Taste Afferent Inputs to Cortical Taste Neurons in Rats

Satoshi Otawa and Hisashi Ogawa[1]

Key words. Cortical taste area—Excitatory amino acid receptor

Ionotrophic receptors for excitatory amino acids are categorized into two main groups: N-methyl-D-aspartate (NMDA) and non-NMDA receptors. Our previous studies [1] revealed that many neurons in the cortical taste area (CTA) of rats were sensitive to glutamate. The aim of this study was to examine which of the receptor subtypes mediated the taste afferents to CTA neurons, and whether there was any difference in the distribution of glutamate receptor subtypes between the two CTAs, i.e., granular and dysgranular insular areas (GI and DI areas).

Forty adult female, Sprague-Dawley rats (250–375 g body weight) were used. After being anesthetized with an i.p. injection of urethane, the animal was mounted on a conventional stereotaxic instrument and the bone covering the left middle cerebral artery was removed. A small opening was made in the dura and a multibarrel pipette was inserted. Cortical taste neurons were recorded with a glass micropipette (5–10 MΩ). The recording pipette was glued to a five-barreled micropipette that was used for the iontophoretic application of drugs. The drugs employed were: NMDA (50 mM, pH 8.0), kainate (an agonist for the non-NMDA receptor; 20 mM, pH 8.0), D-amino-5-phosphono-valerate (AP-5; an antagonist for NMDA, 5 mM, pH 8.0), 6-cyano-7-nitro-quinoxaline-2,3-dione (CNQX; an antagonist for non-NMDA; 1 mM, pH 8.0), and 165 mM NaCl for balancing. Neuronal responses to NMDA and kainate were recorded on a pen-recorder through a pulse counter. The amount of the ejection current of the antagonists was determined in each neuron to block the specific but not the other subtype of glutamate receptor. During application of the antagonists, taste responses were collected and compared with control ones. To evoke taste responses in cortical neurons, the entire oral cavity was stimulated with four putative taste stimuli, 0.1 M NaCl, 0.5 M sucrose, 0.01 N HCl, and 0.02 M quinine HCl. Criteria to identify taste responses have been described elsewhere [2].

A total of 49 cortical taste neurons were recorded from 40 rats. Both AP-5 and CNQX suppressed taste responses in 37 neurons (GI, 18; DI, 19), AP-5 selectively blocked these responses in 4 neurons (GI, 1; DI, 3), and CNQX also selectively blocked these responses in 4 neurons (GI, 1; DI, 3). The remaining 4 neurons (GI, 2; DI, 2) were not affected by either AP-5 or CNQX. No difference was noted between the GI and DI areas in the fraction of neurons affected by these antagonists. However, most of the neurons affected by AP-5 only or CNQX only, and these not affected by either of these drugs, were located in the border between the GI and DI areas. A large fraction of neurons in all layers of the GI and DI areas were affected by both AP-5 and CNQX.

These results indicate that both NMDA and non-NMDA receptors mediate the synaptic transmission of taste afferents in both CTAs, the GI and DI areas. No difference was noted in the distribution of taste neurons with NMDA or non-NMDA receptors between the GI and DI areas, or across cortical layers.

References

1. Ogawa H, Hasegawa K, Ikeda I (1991) Sensitivity to glutamate and GABA of neurons in cortical gustatory area in rats. Neurosci Res [Suppl] 14:S25
2. Ogawa H, Hasegawa K, Murayama N (1992) Difference in taste quality coding between two cortical taste areas, granular and dysgranular insular areas, in rats. Exp Brain Res 91:415–424

---

[1] Department of Physiology, Kumamoto University School of Medicine, Honjo 2-2-1, Kumamoto, 860 Japan

Effects of Substance P and Calcitonin Gene-Related Peptides on Cortical Taste Neurons in Rats

K. Hasegawa, T. Nakamura, and H. Ogawa[1]

Key words. Substance P—Calcitonin gene-related peptides

The cortical taste area (CTA) in rats is located in the granular insular (GI) and dysgranular insular (DI) areas and contains neurons with receptors for glutamate [1]. Substance P (SP) and calcitonin gene-related peptide (CGRP)-like immunoreactivity, however, is found in the gustatory pathway from the tongue to the CTA, although there is a mismatch between nerve terminals and receptors for the peptides. CGRP-containing fibers [2], and SP receptors [3] are numerous in the DI area, whereas CGRP receptors [4] are numerous in the GI area and SP fibers [5] spread throughout the insular cortex. In this study, we examined the distribution of neurons sensitive to these peptides in the CTA, and the effects on taste responses.

We used adult female, Sprague Dawley rats anesthetized with urethane. While stimulating the entire oral cavity with four taste solutions, we recorded neural activity in the CTA with a glass microelectrode filled with pontamine sky blue in 1 M Na acetate, glued to a multibarrel electrode for drugs. Drugs employed were glutamate (0.5 M, pH 8), SP (0.2 mM, pH 8), CGRP (0.2 mM, pH 6), and NaCl (165 mM). Taste stimuli used were 0.1 M NaCl, 0.5 M sucrose, 0.01 N HCl and 0.02 M quinine HCl. At the termination of the experiment, dye marks were made along the tract electrode. Recording sites were reconstructed histologically.

We recorded a total of 747 neurons (386 in the GI area, 361 in the DI area), including 74 taste neurons (39 in the GI area, 35 in the DI area). By setting ejection currents at 30–50 nA, we surveyed neurons sensitive to each drug. All drugs facilitated spontaneous discharges in most cases. Neurons responded to SP with a short latency, but some produced sluggish responses to CGRP, which responses outlasted the period of drug application. In the whole sample, glutamate was effective in 44% of GI neurons and in 52% of DI neurons, SP in about 40% of GI and DI neurons, and CGRP in about 35% of GI and DI neurons. About 20% of neurons in both areas responded to glutamate only and 25% of neurons in either area responded to a given combination of two of the three drugs. Sensitivity to SP and glutamate or to SP and CGRP in both areas occurred concomitantly ($P < 0.05$, χ^2-test). No difference was noted in the distribution of neurons sensitive to the three drugs between the GI and DI areas, except that SP-sensitive taste neurons were numerous in layer II–III. SP affected taste responses in most of the neurons tested, and CGRP was effective in more than half the neurons tested. Facilitatory effects were seen in 50% of the neurons affected. In 65% of the neurons, both peptides were effective. Both peptides modulated taste responses, even in neurons in which the spontaneous discharges were not affected. It is suggested that peptide receptors are present in both pre- and post-synaptic membranes. SP or CGRP affected response profiles in some neurons.

It is evident that SP and CGRP contribute to taste information processing in the CTA by modulating cortical taste responses.

References

1. Ogawa H, Hasegawa K, Ikeda I (1991) Sensitivity to glutamate and GABA of neurons in cortical gustatory area in rats. Neurosci Res [Suppl] 14:S.25
2. Yasui Y, Saper C, Cechetto DF (1989) Calcitonin gene-related peptide immunoreactivity in the visceral sensory cortex, thalamus, and related pathways in the rat. J Comp Neurol 290:487–501
3. Mantyh PW, Gates T, Mantyh CR, Maggio JE (1989) Autoradiographic localization and characterization of tachykinin receptor binding sites in the rat brain and peripheral tissues. J Comp Neurol 9:258–279
4. Skofitsch G, Jacobositz D (1992) Calcitonin- and calcitonin gene-related peptides: Receptor binding sites in the central nervous system. In: Bjorklund A, Hokfelt T, Kuhar MI (eds) Handbook of chemical neuroanatomy, vol 11. Elsevier, Amsterdam, pp 97–144
5. Cuello AC, Kanazawa I (1978) The distribution of substance P immunoreactive fibers in the rat central nervous system. J Comp Neurol 178:129–156

[1] Department of Physiology, Kumamoto University School of Medicine, Honjo 2-2-1, Kumamoto, 860 Japan

Gustatory Responses of Cortical Neurons in the Awake Monkey

Y. Miyaoka[1] and T.C. Pritchard[2]

Key words. Cortical gustatory neuron—Alert monkey

Behavioral experiments conducted in our laboratory have shown that the taste preference of Old World monkeys for sucrose (Suc) is reduced when it is dissolved in 0.03 M monosodium glutamate (MSG) instead of distilled water (DW). No reduction in preference is observed, however, when the solvent is switched from DW to 0.03 M sodium chloride (NaCl). The present study examined this issue further by recording the activity of cortical gustatory neurons in alert monkeys as they consumed a variety of sapid stimuli.

Extracellular recording of neurons in the primary taste area was done in two alert female rhesus monkeys (*Macaca mulatta*). Data collection began after each monkey was acclimated to the primate chair used for electrophysiological recording. A 1.0-ml plastic syringe was used to deliver 18 sapid stimuli (vol, 0.3 ml). The stimuli, dissolved in DW were: 1.0 M Suc, 0.03 M and 0.1 M NaCl, 0.003 M hydrochloric acid (HCl), 0.001 M quinine hydrochloride (QHCl), 0.03 M MSG, 0.03 M polycose, 0.3 M glycine, 0.1 M proline, and 0.1 M malic acid. The stimuli, except for HCl and QHCl, were also dissolved in 0.03 M MSG, and sucrose was mixed with 0.03 M NaCl. The interstimulus interval was about 1 min. Spontaneous activity and the response evoked by DW and each sapid stimulus were measured for 3 s. The responses to sapid stimuli were corrected for the activity evoked by DW.

A total of 129 cortical gustatory neurons was isolated in the primary taste area. The spontaneous firing rate of these neurons was 1.9 ± 1.2 spikes/s (mean ± SE). Distilled water alone had a negligible effect, 2.2 ± 1.7 spikes/s, upon the firing rate. The cortical gustatory neurons were classified into six groups according to the best stimuli: sucrose ($n = 32$); NaCl ($n = 19$); HCl ($n = 19$); QHCl ($n = 17$); MSG ($n = 14$); other ($n = 28$). More than 70% of these gustatory neurons, except those in the other category, were responsive to only one of the five sapid stimuli. The responsiveness of 19 Suc-best neurons to 1.0 M Suc was reduced after changing the solvent from DW to 0.03 M MSG (4.0 to −0.2 spikes/s; t(19) = 6.864, $P < 0.001$). This reduction in the firing rate was consistent with behavioral data that have shown a weaker preference for Suc/MSG mixtures [1].

A similar reduction in the average response of 3.7 spikes/s was shown in 18 Suc-best neurons by changing the solvent from DW to 0.03 M NaCl (t(18) = 6.501, $P < 0.001$). This result conflicts with the preference behavior of Old World monkeys [1], as well as with previously published reports of human psychophysical studies. Human subjects reported suppression of sweetness in sucrose-salt mixtures only when the salt concentration used as the adulterant was much higher than the 0.03 M concentration used in the present experiment [2,3].

The neurophysiological findings in the present study were inconsistent with the preference behavior of Old World monkeys, suggesting that the primary gustatory cortex does not play a prominent role in taste-guided preference behavior.

Acknowledgment. This study was supported by PHS grant DC-00246.

References

1. Pritchard TC, Norgren R (1991) Physiol Behav 49: 1003–1007
2. Beebe-Center JG, Rogers MS, Atkinson WH, O'Connell DN (1959) J Exp Psychol 57:231–234
3. Kamen JM, Pilgrim FJ, Gutman NJ, Kroll BJ (1961) J Exp Psychol 62:348–356

[1] Department of Oral Physiology, Niigata University, 5274 Niban-cho, Gakkochodori, Niigata, 951 Japan
[2] Department of Neuroscience and Anatomy, Pennsylvania State University, Hershey, PA 17033, USA

Artificial Neural Networks Analyze Response Patterns of Cortical Taste Neurons in the Rat

H. Katayama[1], T. Nagai[2], K. Aihara[1], M. Adachi[1], and T. Yamamoto[3]

Key words. Coding—Neural network model

For many years the similarity of taste stimuli has been measured mainly by the determination of a correlation coefficient across the neurons activated by the stimuli. In a previous study, we developed a new method for analyzing gustatory neural activities, using artificial neural networks [1]. In the present study, we examined further the operation of these networks in regard to neural responses induced by a variety of taste stimuli, and the effect of "pruning" on the network output. Three-layer neural networks were trained by the back-propagation learning algorithm to classify the neural response patterns to the four basic taste qualities (1.0 M sucrose, 0.03 M HCl, 0.01 M quinine HCl, and 0.1 M NaCl). The networks had four output units representing the four basic taste qualities: sweet (S), sour (H), bitter (Q), and salt (N). The input units represented rat cortical neurons from which the response patterns (impulses/3 s) were recorded in a separate physiological experiment [2]. After training, the response patterns to test stimuli (0.02 M sodium saccharin, 0.3 M KCl, 0.3 M $MgCl_2$, 0.1 M $NaNO_3$, 0.01 M tartaric acid, 0.3 M $CaCl_2$, 0.1 M monosodium L-glutamate [MSG] and 0.1 M inosine 5'-monophosphate [IMP]) were presented to the input units. For $NaNO_3$, the networks produced large outputs (around 0.9) almost exclusively in the N unit, showing pure salt taste in the stimulus. Large and exclusive outputs were also produced in the H unit for KCl (around 0.7) and in the Q unit for tartaric acid (around 0.5). On the other hand, outputs suggesting mixed bitter and salt tastes were produced for $CaCl_2$ and $MgCl_2$. As to the similarity of the test stimuli to the four basic taste qualities, the neural networks presented a clearer and more definite result than the conventional correlation analysis. For MSG and IMP, the networks produced exclusive but small outputs (around 0.2) in the N unit, although some investigators classify these stimuli as an umami taste that is independent of the four basic taste qualities.

It is possible that the outputs in the N unit were produced by sodium dissociated in the taste solutions. We tentatively constructed the neural networks with five output units representing five basic taste qualities, but due to the involvement of the salt taste in MSG, the networks did not adequately classify the test stimuli. We evaluated the relative contribution of the input units to the taste discrimination of the networks by examining their connection weights to the hidden layer. The input units with weaker connection weights were "pruned" from the trained network. The response patterns to the four basic taste qualities were presented again to the pruned network. When the trained network with 32 input units was pruned of less than 7 units, the output values of the four output units stayed very close to those in the original network. However, when more units were pruned, the output values gradually decreased. We calculated across-neuron correlations among neural activities from which neurons corresponding to the pruned inputs were eliminated. These correlations were not different from those in the original set of neural activities. These results show that the pruned units did not positively contribute to the taste discrimination. On the contrary, when even a few input units with stronger connection weights were pruned, the output values of the network for the response patterns to the four basic taste qualities were rapidly reduced. This suggests that certain taste neurons are predominant in the coding of the four basic taste qualities. Pruning of the input units in artificial neural networks allows us to evaluate taste neurons in terms of their relative contribution to information processing.

Acknowledgments. This work was supported by Grants-in-Aid for Scientific Research on Priority Areas (02255107 to KA and 05267101 to TY) from the Ministry of Education, Science, and Culture of Japan.

References

1. Nagai T, Yamamoto T, Katayama H, Adachi M, Aihara K (1992) A novel method to analyze response patterns of taste neurons by artificial neural networks. Neuro Report 3:745–748
2. Yamamoto T, Yuyama N, Kato T, Kawamura Y (1985) Gustatory responses of cortical neurons in rats. III. Neural and behavioral measures compared. J Neurophysiol 53:1370–1386

[1] Department of Electronic Engineering, Faculty of Engineering, Tokyo Denki University, 2-2 Kandanishiki-cho, Chiyodu-ku, Tokyo, 101 Japan
[2] Department of Physiology, Teikyo University School of Medicine, 2-11-1 Kaga, Itabashi-ku, Tokyo, 173 Japan
[3] Department of Behavioral Physiology, Faculty of Human Sciences, Osaka University, 1-2 Yamadaoka, Suita, Osaka, 565 Japan

Amygdalar Neuronal Responses to Taste and Other Sensory Stimuli in Awake Rats

Hisao Nishijo[1], Teruko Uwano[1], Makoto Yonemori[2], and Taketoshi Ono[1]

Key words. Amygdala—Taste

Introduction

The amygdala (AM) is a focus of the sensory modalities in which information converges via the thalamus and sensory association cortices. The AM receives taste inputs from the brainstem taste area, and from the thalamic and cerebral taste areas. However, the basic characteristics of the AM taste neurons have not yet been systematically analyzed in awake rats. In the present study, neuronal activity was recorded in and around the AM of unanesthetized rats during presentation of gustatory and other sensory stimuli.

Methods

Each rat was prepared for the stereotaxic placement of a recording electrode by the formation of a receptacle of dental cement on its head to accept modified ear bars. The receptacle permitted fixing the skull painlessly in the appropriate stereotaxic planes. A 1200 Hz cue tone signaled availability of sucrose solution, and 4300 Hz signaled availability of intracerebral self stimulation (ICSS). A 2800 Hz tone was neutral (associated with neither reward nor aversion). Either light (visual) or an air puff (somatosensory) was similarly associated with either sucrose solution or ICSS, respectively, as in the auditory discrimination paradigm. Neuronal responses during ingestion were further analyzed in detail by intraoral infusion of various sapid solutions (0.1 M NaCl, 0.3 M sucrose, 0.01 M citric acid, 0.00003 M quinine HCl, 0.1 M monosodium glutamate, and 0.2 M lysine HCl). Each trial consisted of the delivery of 0.05 ml of water, a similar amount of a sapid stimulus, and then a water rinse.

Results

Ten rats were used. Recordings were made from 692 neurons in the AM and amygdalostriatal area. Of the 303 responsive neurons, 163 responded exclusively to one sensory stimulation (58, auditory [A]; 4, visual [V]; 51, oral sensory [OS]; 50, somatosensory [SS]), 115 responded to various combinations of the sensory stimuli, and 25 could not be classified.

When these responsive neurons were tested for responses to oral sensory stimulation, 88 responded. All neurons recorded were screened by intraoral infusion of a mixture of four basic sapid stimuli. If intraoral infusion of the mixture of sapid solutions or ingestion of sucrose solutions during licking in the task elicited a response significantly larger than the spontaneous firing rate, the neuron was further tested with each of the six sapid solutions. In 21 of the 88 neurons that responded, sapid stimuli elicited significantly larger responses than water. Of these 21 taste neurons, 7 also responded to other sensory stimulations. Based on the relative magnitudes of their responses to these four standard sapid stimuli, the taste neurons were classified as follows: 8 sucrose-best, 4 NaCl-best, 3 citric acid-best, and 6 quinine HCl-best. One neuron responded significantly to only lysine HCl. Of the taste neurons, 4 responded to only one sapid stimulus each. The mean entropy for the absolute value of the taste responses was 0.82; for the excitation alone, it was 0.74. Configuration by 21 taste neurons in hierarchical cluster analysis and multidimensional scaling was largely in accordance with best-stimulus classification.

Discussion

The present study demonstrated AM neuron responsiveness to unimodal or multimodal sensory stimuli in awake behaving rats. The characteristics of the AM neuron responses to sapid solutions differed from those in brainstem and cortical taste neurons in the following ways: (1) One-third of the AM taste neurons responded to other sensory stimuli, whereas brainstem and cortical taste neurons responded exclusively to taste stimuli. (2) The ratio of quinine HCl-best neurons among all of the taste

Departments of Physiology[1] and Dental and Oral Surgery[2], Faculty of Medicine, Toyama Medical and Pharmaceutical University, 2630 Sugitani, Toyama, 930-01 Japan

neurons in awake rats was higher than the proportion of brainstem taste neurons. (3) The mean entropy of the AM taste neurons was larger than that of brainstem taste neurons. (4) The response rate to taste stimuli was relatively small in comparison with brainstem taste neurons, but similar to that of cortical taste neurons. However, despite these differences, taste quality coding in the AM taste neurons was not very different from that in the brainstem taste neurons in the limited stimulus array studied.

Role of Taste in the Frog

HIROMICHI NOMURA[1]

Key words. Water receptor—Jaw reflex

The frog has two kinds of chemoreceptors, one kind being a water receptor [1] and the other a chemoreceptor sensitive to Quinine HCl and other compounds [2]. Excitation of these chemoreceptors elicits jaw and tongue reflexes [3,4]. To elucidate the roles of these chemoreceptors, we studied these reflexes more precisely in the frogs *Rana nigromaculata* and *Rana catesbeiana*. Reflex activity was studied by recording reflex contractions of the jaw and tongue muscles and by recording reflex neural discharges in the nerves innervating these muscles.

In the reflex elicited by excitation of the water receptor, the following phenomena were observed: (a) clenching, caused by a tonic contraction of the medial major masseter muscle; (b) nostril-closure, caused by tonic contractions of the intermandibular and medial major masseter muscles; (c) elevation of the buccal floor, caused by contractions of the intermandibular and geniohyoid muscles; (d) cession of gorge respiration; and (e) one or a few phasic contractions at the beginning of the reflex.

In the reflex elicited by the excitation of the other type of chemoreceptor, folding of the tongue, which had been stretched, was observed. In this reflex, efferent discharges were elicited in the nerves innervating the intrinsic tongue, genioglossal, and hyoglossal muscles, but not the sternohyoid muscle, indicating that no retraction of the tongue occurs. The role of this reflex has not yet been revealed.

The water receptor in the frog has been suggested proposed to play a role in: (a) the inhibition of respiratory movements, protecting the airways from being filled with water, and (b) the reduction of an obvious increase of water intake for stabilizing water balance [5]. The finding that the geniohyoid muscle contracted together with the intermandibular and medial major masseter muscles indicates that the role of the water receptor is the inhibition of respiratory movements, but not the reduction of an obvious increase of water intake, because the geniohyoid muscle does not play a role in the jaw and nostril closure which are necessary for the reduction of an obvious increase of water intake.

In the reflex elicited by excitation of the water receptor, tonic reflex discharges were elicited with relatively constant latencies, while phasic discharges had varied latencies. This finding suggests that the reflex arc of phasic reflex discharges is different from that of tonic reflex discharges.

References

1. Nomura H, Sakada S (1965) On the "water response" of frog's tongue. Jpn J Physiol 15:433–443
2. Nomura H, Kumai T (1982) A specific chemoreceptor to the linguo-hypoglossal chemoreflex of the frog. Jpn J Physiol 32:683–687
3. Nomura H, Kumai T (1981) Reflex discharge evoked by water stimulation on the frong tongue. Brain Res 221:198–201
4. Kumai T (1981) Reflex response of the hypoglossal nerve induced by gustatory stimulation of the frog tongue. Brain Res 208:432–435
5. Zotterman Y (1949) The response of the frog's taste fibres to the application of pure water. Acta Physiol Scand 18:181–189

[1] Department of Oral Physiology, Matsumoto Dental College, 1780 Hirooka-gobara, Shiojiri, Nagano, 399-07 Japan

Comparison of Hypoglossal and Reticular Formation Neural Activity During Gustatory-Evoked Ingestion and Rejection

J.B. Travers[1] and L.A. DiNardo[2]

Key words. Taste—Sensory-motor processing

Neurons in the reticular formation (RF) lateral to the hypoglossal nucleus (mXII) contribute to a distribution of brainstem neurons that project to the oral motor nuclei [1]. This region of the medullary RF, implicated as part of a central pattern generator (CPG) for swallowing [2], appears contiguous with a CPG for rhythmic masticatory function [3]. Other neuroanatomical studies indicate that this region of the RF may also receive projections from oro-sensory nuclei, including the nucleus of the solitary tract [4,5] and the trigeminal sensory complex [6], and is thus well situated to integrate oro-sensory signals with ingestive function. Acute neurophysiological studies, in fact, have shown that a small subset of neurons in the caudal RF lateral to mXII are active during both cortically induced fictive mastication and swallowing induced by superior laryngeal nerve stimulation [7]. This suggests that the CPGs for mastication/licking and swallowing share some premotor neurons. The purpose of the present study was to determine the relationship between neural activity during ingestion (licks and swallows) and rejection (gapes) of gustatory stimuli from neurons in the medullary RF compared to a population of mXII neurons in awake, freely-moving animals.

Under surgical anesthesia (Nembutal 60 mg/kg), rats ($n = 42$) were implanted with electromyographic (EMG) electrodes in a subset of the oro-pharyngeal musculature to monitor the occurrence of licks, swallows, and gapes. A chronic microdrive, containing a bundle of fine wire electrodes, was positioned within the brainstem to record single cells from either the medullary RF or mXII [8]. An intraoral cannula was implanted to deliver gustatory stimuli and water rinses. Rhythmic neurons within the RF and mXII were identified using the autocorrelation technique. For each neuron, the number of spikes per swallow (Ns: RF = 16; mXII = 26) and gape (Ns: RF = 23; mXII = 31) were statistically compared to the number of spikes/lick, using a *t*-test.

Of 31 mXII neurons that were rhythmically active during lick cycles, 17 (55%) were significantly altered during QHCl-induced gape cycles (12 were excited and 5 suppressed). In contrast, only 6/26 (23%) mXII neurons showed altered activity during a swallow (2 excitatory and 4 suppressed). Similar to observations in mXII, a high proportion of neurons in the RF were also significantly altered during gape cycles (17/23; 74%). Significantly more RF neurons, however, that were rhythmically active during lick cycles were modulated during a swallow (7/16; 44%: Chi-square, $P < 0.05$). In addition, unique types of cells observed in the RF that were not present in mXII included neurons that were specifically active during swallowing, and neurons with excitatory gustatory responses to NaCl and sucrose.

Although gape cycles and swallows are independent behaviors, both make use of the same oral musculature, and further evidence indicates that they share some of the same motoneurons. Thus, in the present study, 4/26 rhythmic neurons in mXII showed significant changes in firing activity during *both* gapes and swallows. Although ingestion (lick cycles and swallows) and rejection (gape cycles) have different behavioral outcomes, they are not mutually exclusive, i.e., swallows can occur *during* a bout of gape cycles. More neurons in the RF (7/16) showed significantly altered response magnitudes during both swallows and gape cycles compared to the population of mXII neurons. This suggests that some of the sensory-motor processing that drives these two behaviors is located in neural pools outside the motor nucleus. This neural substrate coordinates the two behaviors, allowing them to share some common final motoneurons, yet establishing a preemptive hierarchy such that a swallow can interrupt either a lick or gape cycle. The existence of gustatory responses in this region of the RF further suggests a substrate responsible for producing the lingual resequencing that differentiates a gape from a lick cycle.

Departments of Oral Biology[1] and Psychology[2], Ohio State University, 305 W. 12th Ave., Columbus, OH 43210, USA

Acknowledgments. This work was supported by PHS DC-00417.

References

1. Travers JB, Norgren R (1983) Afferent projections to the oral motor nuclei in the rat. J Comp Neurol 220: 280–298
2. Kessler JP, Jean A (1985) Identification of the medullary swallowing regions in the rat. Exp Brain Res 57:256–263
3. Chandler SH, Tal M (1987) The effects of brain stem transections on the neuronal networks responsible for rhythmical jaw muscle activity in the guinea pig. J Neurosci 6:1831–1842
4. Norgren R (1978) Projections from the nucleus of the solitary tract in the rat. Neuroscience 3:207–218
5. Travers JB (1988) Efferent projections from the anterior nucleus of the solitary tract of the hamster. Brain Res 457:1–11
6. Borke RC, Nau ME (1987) The ultrastructural morphology and distribution of trigemino-hypoglossal connections labeled with horseradish peroxidase. Brain Res 422:235–241
7. Amri M, Lamkadem M, Car A (1991) Effects of lingual nerve and chewing cortex stimulation upon activity of the swallowing neurons located in the region of the hypoglossal motor nucleus. Brain Res 548:149–155
8. Travers JB, Jackson LM (1992) Hypoglossal neural activity during licking and swallowing in the awake rat. J Neurophysiol 67:1171–1184

Oropharyngeal/Laryngeal Mechanisms for Vasopressin Release

Takao Akaishi and Shinji Homma[1]

Key words. Pharyngolaryngeal mucosa— Vasopressin

In humans, taking 20 min to drink a very small volume of water or hypertonic (0.30 M) saline (0.15–0.16 ml/kg body weight) results in a slight hypotonic diuresis or hypertonic antidiuresis, respectively. Isotonic saline has no effect. We have shown that there is a significant linear relationship between both changes in urine volume and osmotic pressure throughout the concentration range of sodium chloride in the solutions ingested [1,2]. These previous findings suggested the presence of neural afferent pathways from the oropharyngeal/laryngeal mucosa to the hypothalamic mechanism regulating vasopressin (AVP) release. In the present study, therefore, we recorded the single unit activity of the hypothalamic AVP-producing cell before and after the oropharyngeal application of water and 0.15 M and 0.30 M saline in the male rat.

All experiments were performed on male rats (Wistar, 250–360 g) anesthetized with urethane (1.2 g/kg). Tracheotomy and esophagectomy were performed to avoid effects originating from these mucosa. Double-barreled vinyl tubing was inserted into the oral cavity so that the tips of the tube reached the pharynx. A bipolar stimulating electrode, made from stainless wire (0.2 mm diameter), was implanted into the neurohypophysis to apply electrical stimulation (500 μs, 0.8 Hz) for antidromic identification of the unit discharge activity of the magnocellular neurosecretory cells. The antidromic identification of discharge activities of the AVP-producing cells was followed by the conventional criteria.

A total of 67 AVP cells were studied; their mean discharge rate was 6.7 ± 1.1 spikes/s. In the first step, the effects of the oropharyngeal/laryngeal application of distilled water and of 0.15 M and 0.30 M sodium chloride (0.15 ml/kg) on the spontaneous activity of the AVP cells were examined. Application of water ($n = 11$) and 0.30 M sodium chloride ($n = 13$) resulted in a decrease and increase, respectively, in the firing activity of the AVP cells examined. These changes reached a maximum within 1.5–2.0 min and maintained this level for several min. Isotonic saline (0.15 M NaCl) had no effect on their activity ($n = 10$). Throughout the molarity range (0, 0.15, and 0.30 M) of sodium chloride, there was a significant linear relationship between the percent change (-17.1 ± 4.5, -3.4 ± 2.5, and $41.7 \pm 26.0\%$; mean ± SEM) in total spike number generated during the initial 3 min following the application of each test solution and the molarity of the sodium chloride in the applied solution ($P < 0.01$, $\gamma = 0.956$).

In the next step of this study, we examined the effects of amiloride, which reduces the perceived intensity of sodium chloride at the level of the receptor site [3], on evoked responses in AVP cells. Each test solution was prepared in 0.05 mM amiloride. The application of 0.05 mM amiloride ($n = 10$) caused a significant ($P < 0.05$) decrease in firing activity during the initial 3 min. Interestingly, hypertonic saline (0.30 M, $n = 12$) in amiloride caused an insignificant change in the activity of the AVP cells, i.e., the evoked response was partially, but not entirely, inhibited. Isotonic saline ($n = 11$) in amiloride had no effect.

These findings indicate that there is a neural connection between the oropharynx/larynx and hypothalamic AVP cells, and that this mechanism seems to depend on the molarity of sodium chloride in the applied solution. Amiloride-sensitive components also appear to be involved in the oropharyngeal/laryngeal mechanism that regulates the release of AVP.

Acknowledgments. This study was supported by the Salt Science Research Foundation Grants Nos. 9137, 92051.

References

1. Akaishi T, Shingai T, Miyaoka Y, Homma S (1989) Hypotonic diuresis following oropharyngeal stimulation with water in humans. Neurosci Lett 107:70–74
2. Akaishi T, Shingai T, Miyaoka Y, Homma S (1991) Antidiuresis immediately caused by drinking a small volume of hypertonic saline in man. Chem Senses 16:277–281
3. Heck GL, Mierson S, DeSimone JA (1984) Salt taste transduction occurs through an amiloride-sensitive sodium transport pathway. Science 223:403–405

[1] Department of Physiology, Niigata University School of Medicine, Asahimachi-dori 1, Niigata, 951 Japan

Synaptic Circuitry of Olfactory Bulb Glomeruli

Charles A. Greer and Juan C. Bartolomei[1]

Key words. Synaptic organization—Local circuits—Periglomerular cells—Mitral cells—Tufted cells—Olfactory receptor cell axons

Introduction

The classic studies of Hirata [1], Pinching and Powell [2], White [3,4], and Hinds [5] established the fundamental synaptic organization of the mammalian olfactory bulb glomeruli. These studies showed that the circuitry of the glomerulus included primary afferent synapses from olfactory receptor cell axons onto the apical dendrites of projection neurons, mitral and tufted cells, and interneurons, peri- or juxtaglomerular cells. In addition, these works showed clearly the presence of reciprocal dendrodendritic local circuits between the projection neurons and interneurons. The latter circuit appears similar in morphology to that described previously by Rall et al. [6] for granule cell–mitral/tufted cell interactions in the external plexiform layer of the olfactory bulb. More recently, a number of laboratories have characterized the axonal and dendritic constituents of the glomerulus and shown unequivocally the presence of multiple subpopulations. For example, several reports have shown that the molecular properties of receptor cell axons can differ between, as well as within, glomeruli [7–11]. In a complementary manner, studies from the laboratories of Mori [12] and Scott [13] have characterized subpopulations of mitral cells as well as showing the differential granule cell innervation of mitral versus tufted cells. In addition, immunocytochemical techniques have revealed multiple subpopulations of peri/juxtaglomerular cells. Cells immunoreactive for tyrosine hydroxylase (TH) [14,15], γ-aminobutyric acid (GABA) [16], and calcium binding protein (CaBP) [14] have been described. Whereas colocalization of proteins is seen in some cells [16,17], other juxtaglomerular cells appear to express only one of the above [14].

It has been recognized that the distribution of primary afferents, olfactory receptor cell axons, within the glomerular neuropil is topologically restricted [18]. Upon entering the glomerulus, a typical receptor cell axon arborizes six times and exhibits approximately eight *en passant* or terminal varicosities. Of particular interest, a typical arbor appears to occupy less than 3% of the glomerular volume. These data suggest that subglomerular compartments may exist and, coupled with the evidence for subpopulations of intraglomerular dendritic constituents, present the possibility that the synaptic circuitry of the glomerulus is not homogeneous.

To explore intraglomerular organization more fully, we have employed immunocytochemical analyses of TH at the ultrastructural level. Our approach enables us to discriminate the intraglomerular synaptic organization of dendritic processes that are and are not immunoreactive for TH or other proteins. The data provide provocative evidence supporting the notion of parallel intraglomerular synaptic pathways.

Methods

Sprague-Dawley rats, 200 to 250 g, were perfused through the heart with 4% paraformaldehyde and 0.1% glutaraldehyde in 0.1 M phosphate buffer. Tissue sections not used for immunocytochemistry were routinely processed, sectioned at approximately 70 nm, and lead-stained for transmission electron microscopy. Tissue sections used for immunocytochemistry electron microscopy (IC-EM) were cut at 100 μm on a Vibratome and processed for TH (Eugene Tech, Ridgefield Park, N.J.) immunocytochemistry using ABC elite Kits (Vector Labs, Burlingame, Calif.). The sections were mounted in Epon between slides and coverslips previously coated with Liquid Release Agent (Electron Microscopy Sciences, Fort Washington, PA). During light microscopic assessment, glomeruli were excluded from further analysis if they showed evidence of innervation by TH immunoreactive cells in the external plexiform layer, most likely tufted cells. Thus at this point we focused attention on putative immunoreactive peri/juxtaglomerular cells. Glomeruli that exhibited good staining, including visible cell bodies and intraglomerular processes, were blocked from the Epon film, remounted on Epon blanks, and thin-sectioned as above. The IC-EM tissue sections were examined with and without lead staining.

[1] Sections of Neurosurgery and Neurobiology, Yale University School of Medicine, 333 Cedar Street, New Haven, CT 06510, USA

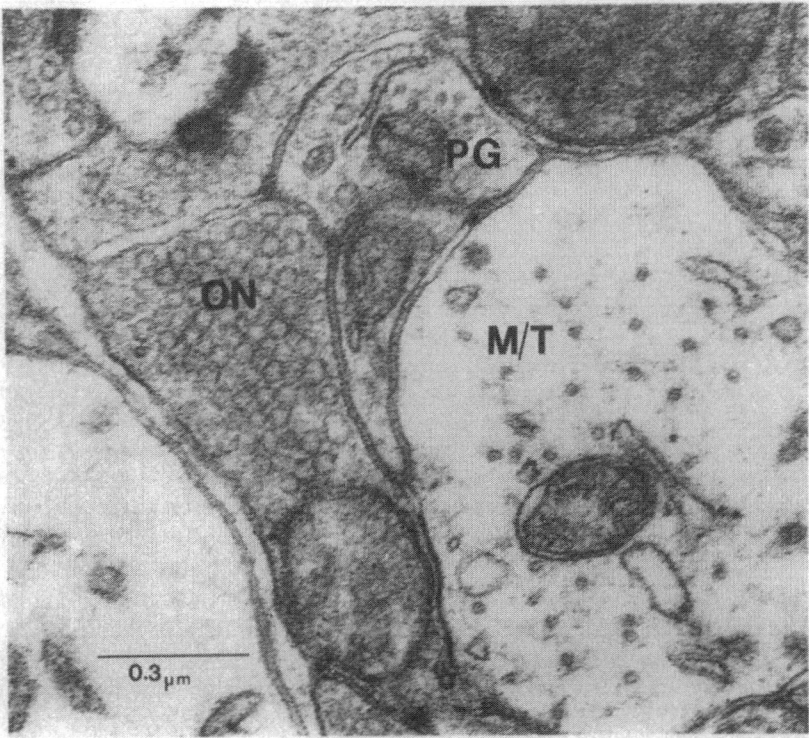

FIG. 1. Conventional transmission electron micrograph showing the principal profiles seen in an olfactory bulb glomerulus. *ON*, olfactory nerve terminal; *PG*, peri/juxtaglomerular cell dendrite; *M/T*, mitral/tufted cell dendrite

Results

The typical glomerular constituents are shown in Fig. 1. The olfactory nerve terminal (ON) is flocculent in appearance and densely packed with spherical vesicles. The peri/juxtaglomerular dendrite (PG) is also flocculent in appearance, though less so than the ON, with evidence of smooth endoplasmic reticulum, mitochondria, and pleomorphic vesicles. The mitral/tufted (M/T) dendrites are characteristically pale or electron-lucent in appearance with regular arrays of microtubules, smooth endoplasmic reticulum, mitochondria, and occasional clusters of 2 to 10 small spherical vesicles.

The synaptology of the glomerulus is shown, in part, in Fig. 2. A peri/juxtaglomerular dendrite is seen receiving multiple asymmetric synapses from ONs. An M/T dendrite receives an equivalent synapse from an ON. In addition, two M/T dendrites are seen receiving symmetric synapses from the peri/juxtaglomerular dendrite. Not seen in the figure is the reciprocal asymmetric synapse from the M/T dendrites back onto the peri/juxtaglomerular dendrites.

In Fig. 3 the fundamental distribution of ON synapses onto intraglomerular TH immunoreactive and nonimmunoreactive processes is seen. The TH immunoreactive processes are easily recognized owing to the deposition of the electron-dense reaction product. As is apparent, the ONs make asymmetric synapses with both dendritic processes. Of particular interest, however, is the observation that

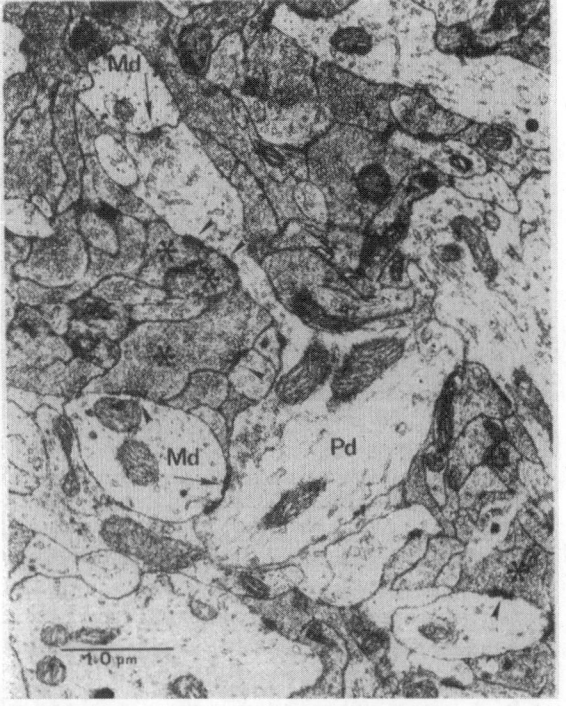

FIG. 2. Conventional transmission electron micrograph showing some of the typical synaptic contacts found in olfactory bulb glomeruli. *Md*, mitral/tufted cell dendrite; *Pd*, peri/juxtaglomerular cell dendrite; *asterisk*, olfactory nerve terminal. *Arrowheads* indicate asymmetric synapses from olfactory nerve terminals onto Mds/Pds; *arrows* indicate symmetric synapses from Pds onto Mds. (Adapted from [23], with permission)

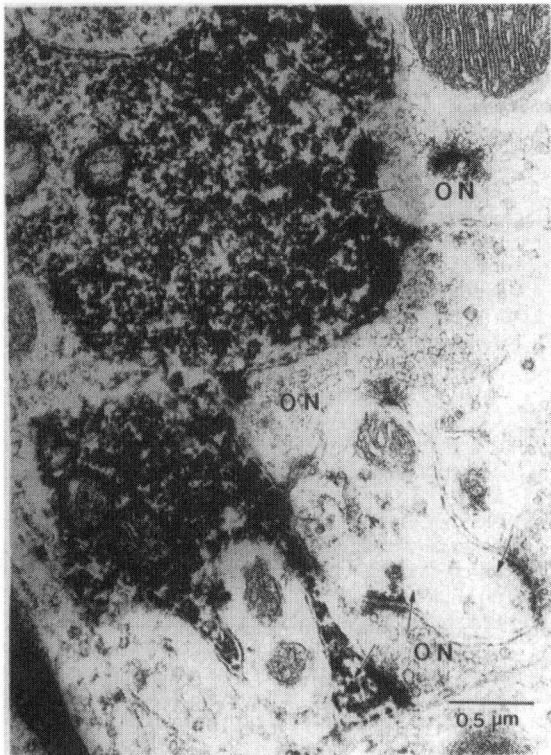

FIG. 3. Olfactory nerve (*ON*) synapses onto tyrosine hydroxylase (TH) immunoreactive profiles and unstained profiles. *Arrows* indicate polarity of the synapses. Note in the *lower right* that a single ON synapses with both an immunoreactive and an unstained process

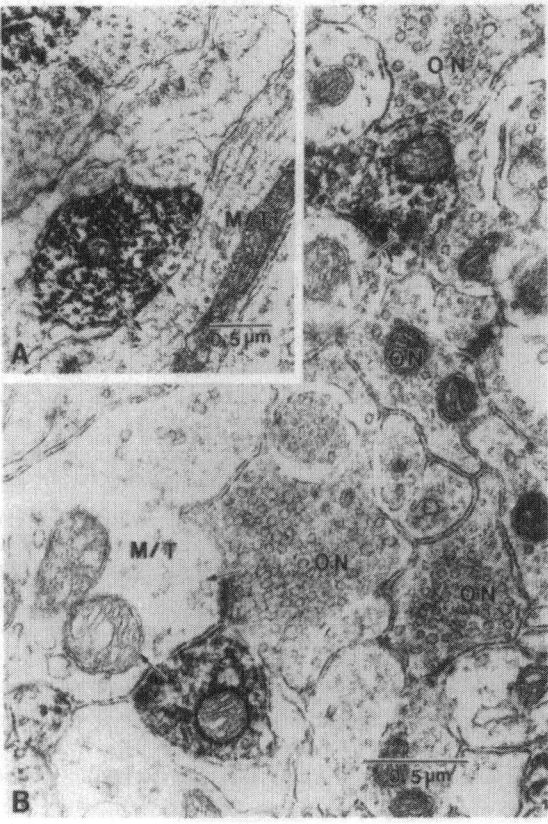

FIG. 4A,B. TH immunoreactive profiles making symmetric synapses (*arrows* indicate polarity) onto unstained mitral/tufted (*M/T*) processes. **A** An unstained process, most likely a peri/juxtaglomerular cell dendrite, is also seen making a symmetric synapse (*arrowhead* indicates polarity) with a probable mitral/tufted cell dendrite. *ON*, olfactory nerve terminals

in some cases a single ON can simultaneously make asymmetric synapses onto both immunoreactive and nonimmunoreactive processes, as seen in the lower right of Fig. 3.

In Fig. 4 TH immunoreactive processes are seen making symmetric synapses with probable mitral/tufted cell dendrites. In Fig. 4A, upper left, an unlabeled peri/juxtaglomerular-like profile is seen making a symmetric synapse with a transversely cut probable M/T profile. In Fig. 4B, an M/T profile receiving a symmetric synapse from a TH immunoreactive process is also seen receiving an asymmetric synapse from an adjacent ON.

Quantitative assessment of synapse distribution showed an equivalent distribution of synapses on both TH immunoreactive peri/juxtaglomerular processes and nonimmunoreactive peri/juxtaglomerular-like processes (6.18 ± 0.43 and 6.13 ± 0.39 synapses/$100\,\mu m^2$ of dendritic profile, respectively). In contrast, mitral/tufted processes had a higher density of synapses/unit area (8.35 ± 0.43 synapses/$100\,\mu m^2$ of dendritic profiles).

Discussion

The conventional transmission electron micrographs showed intraglomerular synaptic organization consistent with that previously reported [1–5,19]. ONs made asymmetric synapses onto mitral/tufted processes and peri/juxtaglomerular processes. In turn, mitral/tufted processes established dendrodendritic asymmetric synapses with peri/juxtaglomerular processes. Although not routinely seen in the same section, the peri/juxtaglomerular dendrites made reciprocal dendrodendritic symmetric synapses with the mitral/tufted processes.

The TH IC-EM showed that TH immunoreactive processes are present in each of the circuits summarized above. They receive asymmetric axodendritic input from ONs and dendrodendritic input from large electron-lucent dendrites that are likely mitral/tufted cells. In addition, they make symmetric dendrodendritic synapses with probable mitral/tufted cell dendrites. Coupled with a typical cell body diameter of approximately 8 µm, these observations suggest that the immunoreactive cell bodies and processes described are periglomerular neurons (see Shepherd and Greer [20] and Scott et al. [21] for review).

These results are consistent with those from prior studies of synaptic organization of TH processes [17] and suggest further that there is an equivalent distribution of synapses between TH and non-TH peri/juxtaglomerular processes. Although one example of an olfactory receptor cell axon simultaneously synapsing with both an immunoreactive and a non-immunoreactive process was observed, the results thus far seem to favor the notion that subpopulations of peri/juxtaglomerular neurons form parallel synaptic circuits. This interpretation is consistent with the prior suggestion that local circuits in the external plexiform layer of the olfactory bulb also form parallel mitral versus tufted cell pathways [12,13,22]. Thus analyses of glomerular and external plexiform circuits provide evidence that supports the hypothesis of parallel odor processing through the olfactory bulb.

Acknowledgments. The authors express their appreciation to Christine Kaliszewski and Kenny Chiu for their expert technical assistance. This work was supported in part by NIH grants DC00210 and NS10174.

References

1. Hirata Y (1964) Some observations on the fine structure of the synapses in the olfactory bulb of the mouse, with particular reference to the atypical synaptic configurations. Arch Histol Jpn 24:302–317
2. Pinching AJ, Powell TPS (1971) The neuropil of the glomeruli of the olfactory bulb. J Cell Sci 9:347–377
3. White EL (1972) Synaptic organization in the olfactory glomerulus of the mouse. Brain Res 37:69–80
4. White EL (1973) Synaptic organization of the mammalian olfactory glomerulus: new findings including an intraspecific variation. Brain Res 60:299–313
5. Hinds JW (1970) Reciprocal and serial dendrodendritic synapses in the glomerular layer of the rat olfactory bulb. Brain Res 17:530–534
6. Rall W, Shepherd GM, Reese TS, Brightman MW (1966) Dendrodendritic synaptic pathway for inhibition in the olfactory bulb. Exp Neurol 14:44–56
7. Mori K (1987) Monoclonal antibodies (2C5 and 4C9) against lactoseries carbohydrates identify subsets of olfactory and vomeronasal receptor cells and their axons in the rabbit. Brain Res 408:215–221
8. Fujita SC, Mori K, Imamura K, Obata K (1985) Subclasses of olfactory receptor cells and their segre-
gated central projections demonstrated by a monoclonal antibody. Brain Res 326:192–196
9. Schwob JE, Gottlieb DI (1986) The primary olfactory projection has two chemically distinct zones. J Neurosci 6:3393–3404
10. Mori K (1993) Molecular and cellular properties of mammalian primary olfactory axons. Microsc Res Technique 24:131–141
11. Schwarting GA, Crandall JE (1991) Subsets of olfactory and vomeronasal sensory epithelial cells and axons revealed by monoclonal antibodies to carbohydrate antigens. Brain Res 547:239–248
12. Mori K, Kishi K, Ojima H (1983) Distribution of dendrites of mitral, displaced mitral, tufted, and granule cells in the rabbit olfactory bulb. J Comp Neurol 219:339–355
13. Orona E, Rainer EC, Scott JW (1984) Dendritic and axonal organization of mitral and tufted cells in the rat olfactory bulb. J Comp Neurol 226:346–356
14. Halasz N, Hökfelt T, Norman AW, Goldstein M (1985) Tyrosine hydroxylase and 28k-vitamin-D-dependent calcium binding protein are localized in different subpopulations of periglomerular cells of the rat olfactory bulb. Neurosci Lett 61:103–107
15. Baker H (1988) Neurotransmitter plasticity in the juxtaglomerular cells of the olfactory bulb. In: Margolis FL, Getchell TV (eds) Molecular neurobiology of the olfactory system. Plenum, New York, pp 185–216
16. Kosaka T, Hataguchi Y, Hama K, Nagatsu I, Wu JY (1985) Coexistence of immunoreactivities for glutamate decarboxylase and tyrosine hydroxylase in some neurons in the periglomerular region of the rat main olfactory bulb: possible coexistence of gamma-aminobutyric acid (GABA) and dopamine. Brain Res 343:166–171
17. Gall CM, Hendry SHC, Seroogy KB, Jones EG, Haycock JW (1987) Evidence for co-existence of GABA and dopamine in neurons of the rat olfactory bulb. J Comp Neurol 266:307–318
18. Halasz N, Greer CA (1993) Terminal arborizations of olfactory nerve fibers in the glomeruli of the olfactory bulb. J Comp Neurol 337:307–316
19. Zheng LM, Jourdan F (1988) Atypical olfactory glomeruli contain original olfactory axon terminals: an ultrastructural horseradish peroxidase study in the rat. Neuroscience 26:367–378
20. Shepherd GM, Greer CA (1990) Olfactory bulb. In: Shepherd GM (ed) Synaptic organization of the brain (3rd edn). Oxford University Press, New York, pp 133–169
21. Scott JW, Wellis DP, Riggott MJ, Buonviso N (1993) Functional organization of the main olfactory bulb. Microsc Res Technique 24:142–156
22. Greer CA, Halasz N (1987) Plasticity of dendrodendritic microcircuits following mitral cell loss in the olfactory bulb of the murine mutant Purkinje cell degeneration. J Comp Neurol 256:284–298
23. Greer CA (1991) Structural organization of the olfactory system. In: Getchell T, Doty R, Bartoshuk L, Snow J (eds) Smell and taste in health and disease. Raven, New York, pp 65–81

Molecular Receptive Range Properties of Mitral/Tufted Cells in the Mammalian Main Olfactory Bulb

KENSAKU MORI[1]

Key words. Molecular conformation—Molecular recognition—Axonal convergence—Glomerulus—Single unit recording—Molecular receptive range

Introduction

Elucidation of sensory stimulus parameters that are essential in characterizing individual neurons gives an insight into understanding sensory information processing in the central nervous system. Because the olfactory image of an object is mediated by a number of different odor molecules, it has long been supposed that critical stimulus parameters in analyzing responses of individual neurons of the central olfactory system may be related to some specific properties of odor molecules. However, because of the existence of an immense number of different odor molecules, it has been technically difficult to elucidate the critical determinants that characterize response specificity of individual central neurons, in a manner similar to that used in the other sensory systems.

We have been studying response selectivity of individual mitral/tufted cells in the rabbit main olfactory bulb [1–3; Katoh et al., manuscript in preparation], and have noticed that by using a large number of different odor molecules having a systematic variation of molecular conformation, it is possible to obtain a clue to critical stimulus parameters for characterizing individual bulbar neurons. This paper summarizes the results of our recent studies on the tuning specificity of individual mitral/tufted cells tested with a carefully selected battery of odor molecules.

In the course of analyzing the tuning specificity of individual mitral/tufted cells, we thought it useful to use the term "molecular receptive range" of a bulbar neuron [2,4]. The molecular receptive range of a mitral/tufted cell has been defined as the range of odor molecules that either excite or inhibit the cell. The range of odor molecules that excite the cell is called the "excitatory molecular receptive range," while the "inhibitory molecular receptive range" of

the cell comprises odor molecules that inhibit discharges of the cell. Because different mitral/tufted cells typically show different molecular receptive ranges, individual mitral/tufted cells in the olfactory bulb can be characterized by the molecular receptive range property.

Despite the difference in their molecular receptive ranges, the different mitral/tufted cells showed the following property in common: odor molecules within the excitatory molecular receptive range of the cell share a similar molecular conformation. We discuss the possible neuronal circuitry underlying the molecular receptive range properties of mitral/tufted cells in correlation with the recently reported knowledge on the spatial distribution of different odor receptor proteins in the sensory sheet of the olfactory epithelium, and the axonal convergence of olfactory receptor neurons onto the glomeruli.

Response Specificity of Mitral/Tufted Cells to Odor Molecules

Our initial trials to correlate the tuning specificity of individual mitral/tufted cells to the structure of odor molecules were made with single unit recordings of spike discharges from neurons located in the dorsomedial region of the main olfactory bulb. Rabbits were anesthetized with urethane or a mixture of urethane and α-chloralose, and a double tracheal cannulation was performed for controlled inhalation of odor-containing air. Single unit discharges were recorded extracellularly from neurons in the mitral cell layer and external plexiform layer. Mitral cells were identified by their antidromic spike responses to the stimulation of the lateral olfactory tract.

When a homologous series of aliphatic compounds (n-fatty acids, n-aliphatic aldehydes, n-aliphatic alcohols, n-alkanes, esters, and ketones) was presented, a clear correlation was noted between the tuning specificity of individual mitral/tufted cells and the stereochemical structure of the aliphatic compounds with respect to: (1) length and/or structure of hydrocarbon chain; (2) difference in functional group; and (3) position of the functional group within the molecule. A tuning specificity of a mitral cell is exemplified in Fig. 1. It can be clearly seen that the

[1] Department of Neuroscience, Osaka Bioscience Institute, 6-2-4 Furuedai, Suita, Osaka, 565 Japan

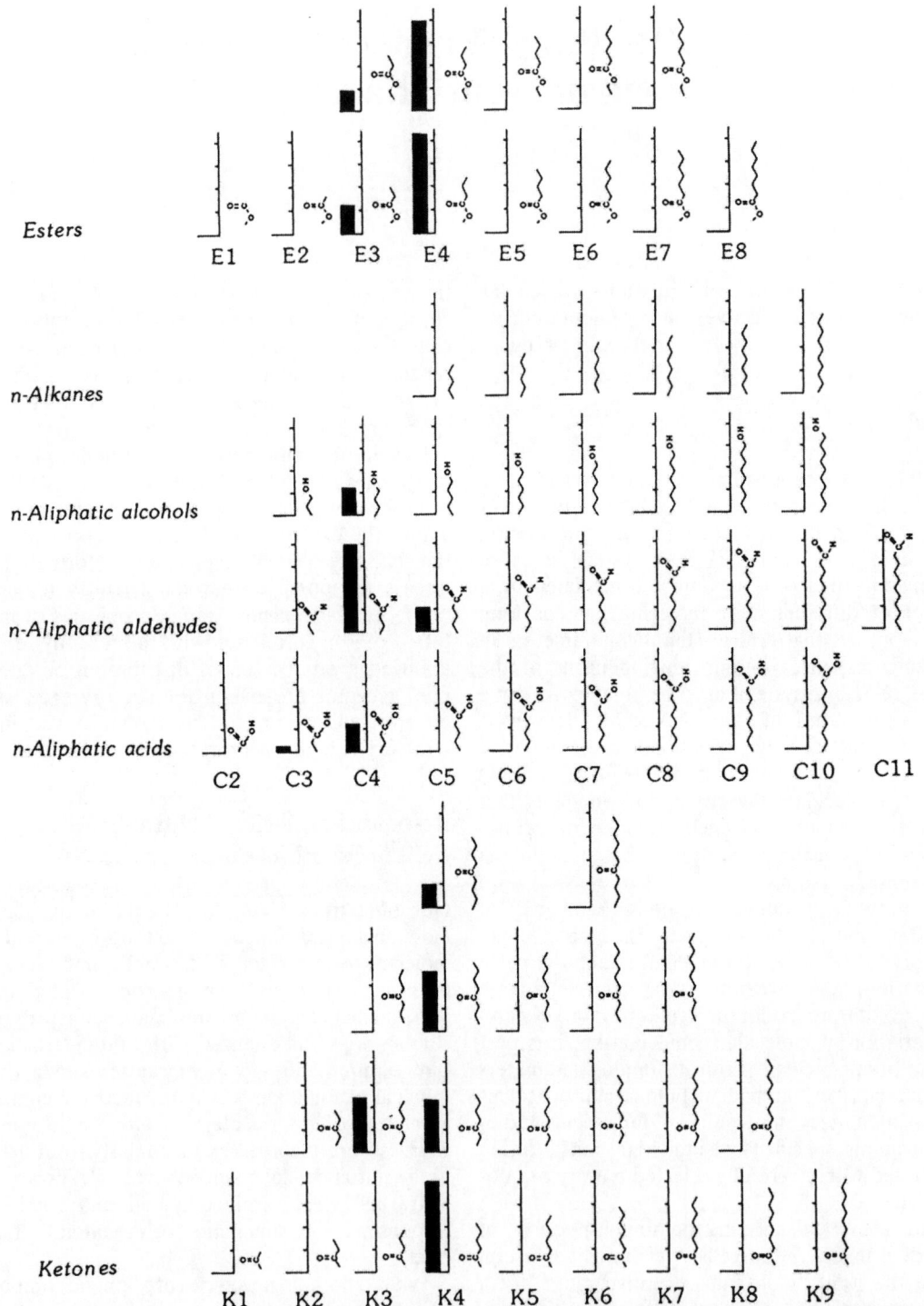

FIG. 1. Representative example of the molecular receptive range of a mitral cell tested with *n*-primary aliphatic compounds (*n*-alkanes, *n*-aliphatic alcohols, *n*-aliphatic aldehydes, and *n*-aliphatic acids) and nonprimary compounds (esters and ketones). *Filled bars*, mean number of spikes per 1-inhalation cycle elicited by stimulation with respective odor molecules; *no bar*, no facilitatory response. One digit of the scale indicates 10 impulses. Molecular formula of compounds are shown on the *right* side of each set of coordinates. C2–C11 indicate the number of carbon atoms in the side chain of the primary compounds. Esters and ketones were arranged according to the length of 1 side chain. (From [2], with permission)

recorded mitral cell was activated by a range of the aliphatic compounds having, at least in part, a similar stereochemical structure. For a number of the recorded cells in the dorsomedial region, it was possible to characterize individual neurons by their excitatory molecular receptive range properties. Odor molecules within the molecular receptive range of individual cells showed a characteristic stereochemical structure in common, and different neurons typically showed different molecular receptive ranges.

Because aliphatic compounds are flexible and assume a number of different conformations, we extended this study by including, as stimulus odor molecules, aromatic compounds which have relatively fixed molecular conformations. The results revealed that single mitral/tufted cells show excitatory impulse responses to a range of odor molecules having similar conformations (Katoh et al., manuscript in preparation). In other words, odor molecules within the excitatory molecular receptive range of individual mitral/tufted cells share a characteristic molecular conformation. In these experiments also, it was observed that different mitral/tufted cells typically showed different excitatory molecular receptive ranges in terms of characteristic molecular conformations.

Functional Convergence of Olfactory Axons onto Glomeruli

Mitral/tufted cells of the rabbit main olfactory bulb project a single primary dendrite to a single glomerulus. Thus, the excitatory molecular receptive range of a mitral/tufted cell reflects strongly the molecular receptive range of olfactory axons converging onto the glomerulus. The excitatory molecular receptive range properties just described suggest that the response specificity of individual mitral/tufted cells reflects strongly the response specificity of a single receptor protein to odor molecules bearing common conformations (or odotopes [4,5]).

Why does the response specificity of individual mitral/tufted cells reflect strongly the response specificity of a single receptor protein? Figure 2 shows a working model for explaining the tuning specificity of individual mitral/tufted cells. This model presupposes three major hypotheses: (1) a single receptor cell is assumed to express a single or a few different kind(s) of odor receptor protein(s); (2) a single receptor protein can be activated by a range of odor molecules having similar conformations; and (3) receptor cells that express the same or similar receptor protein(s) project their axons to a functionally specific glomerulus. According to this working model, the excitatory molecular receptive range of an individual mitral/tufted cell reflects strongly the molecular receptive range of receptor cells that pro-

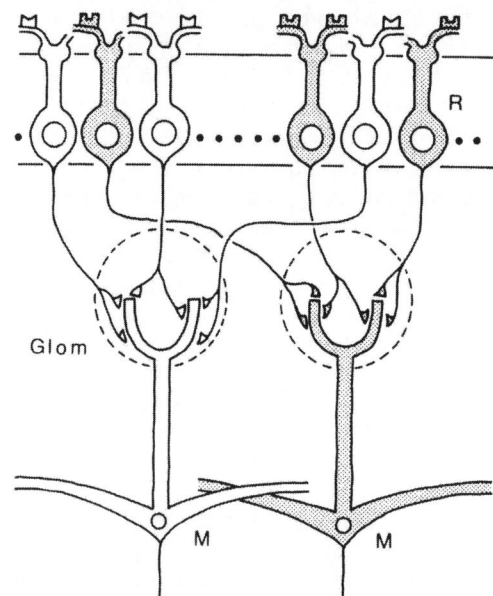

FIG. 2. Hypothetical scheme to explain the molecular receptive range properties of individual mitral/tufted cells in the rabbit main olfactory bulb. In this model, individual glomeruli receive converging axonal inputs from a number of olfactory receptor neurons expressing the same or similar receptor protein subtype(s). For simplicity, this scheme shows only 2 subsets of receptor neurons (*R*, *shaded* and *blank*), each subset expressing a specific subtype of receptor protein on the cilial membrane. *Glom*, glomerulus; *M*, mitral cell. (From [2], with permission)

ject their axons to that mitral/tufted cell. Because the receptor cells express one or a few common odor receptor protein(s), the molecular receptive range of these receptor cells may reflect strongly the tuning specificity of the receptor protein(s). Thus, the excitatory molecular receptive range of the mitral/tufted cell may reflect strongly the tuning specificity of the receptor protein(s) that are commonly expressed by these receptor cells.

Discussion

Recent advances in understanding the molecular mechanisms for the recognition of odor molecules by receptor proteins [6–10] provided insight into the critical stimulus parameters useful for analyzing the response specificity of individual central neurons in the olfactory system. The odor receptor proteins are members of the G-protein-coupled class of receptors with seven transmembrane domains. This suggests that the basic mechanism of molecular interaction between odor ligand molecules and receptor proteins can be deduced from knowledge of well-characterized members of the G-protein-coupled receptor family such as the β-adrenergic receptor and the photoreceptor. By analogy with the ligand-receptor protein interactions in photoreceptor and β-adrenergic receptor proteins, it can be assumed that an individual odor receptor protein may recognize a range of odor molecules that have conformations spatially complementary to the receptive site of the odor receptor protein. This argument thus supports hypothesis (2) and suggests that conformational features of odor molecules play important roles in sensory signal processing in the central olfactory system. Our experimental results indeed indicate that individual mitral/tufted cells of the main olfactory bulb are tuned to specific conformational features of odor molecules.

In catfish, Ngai et al. [7] provided evidence that supports hypothesis (1), that an individual receptor cell expresses only one or a few different kinds of receptor protein(s). In addition, it was demonstrated that receptor cells expressing different receptor proteins are intermingled with each other in the olfactory epithelium and form a mosaic-like sensory sheet. In the mammalian olfactory epithelium, three spatially segregated zones for expressing receptor proteins were demonstrated [8]. However, within each zone, a number of different odor receptor proteins are distributed like a mosaic.

According to these reports, it can be assumed that receptor cells expressing a particular receptor protein are distributed relatively widely in the olfactory epithelium and are intermingled with a number of different subsets of receptor cells. However, axons of the receptor cells expressing the same receptor protein may converge their axons and project to one or a few common glomeruli. Subsets of receptor cells projecting to specific glomeruli have been suggested by histochemical studies using monoclonal antibodies [11]. Thus, signals from different receptor proteins may be sorted out into different glomeruli and may be transmitted to different mitral/tufted cells in the mammalian olfactory bulb. Different mitral/tufted cells that project their primary dendrites to different glomeruli may thus show different

excitatory molecular receptive range properties [2,12].

References

1. Mori K, Imamura K, Onoda N (1990) Signal processing in the rabbit olfactory bulb. In: Døving KB (ed) ISOT X. GCS A S, Oslo, pp 134–141
2. Imamura K, Mataga N, Mori K (1992) Coding of odor molecules by mitral/tufted cells in rabbit olfactory bulb. I: Aliphatic compounds. J Neurophysiol 68: 1986–2002
3. Mori K, Imamura K, Mataga N (1991) Differential specificities of single mitral cells in rabbit olfactory bulb for a homologous series of fatty acid odor molecules. J Neurophysiol 67:786–789
4. Mori K, Shepherd GM (1994) Emerging principles of molecular signal processing by mitral/tufted cells in the olfactory bulb. Semin Cell Biol (in press)
5. Shepherd GM (1987) A molecular vocabulary for olfaction. In: Atema J (ed) Olfaction and Taste IX. New York Academy of Science, pp 98–103
6. Buck L, Axel R (1991) A novel multigene family may encode odorant receptors: a molecular basis for odor recognition. Cell 65:175–187
7. Ngai J, Chess A, Dowling MM, Necles N, Macagno ER, Axel R (1993) Coding of olfactory information: Topography of odorant receptor expression in the catfish olfactory epithelium. Cell 72:667–680
8. Ressler KJ, Sullivan SL, Buck LB (1993) A zonal organization of odorant receptor gene expression in the olfactory epithelium. Cell 73:597–609
9. Raming K, Klieger J, Strotmann J, Boekhoff I, Kubic S, Baumstark C, Breer H (1993) Cloning and expression of odorant receptors. Nature 361:353–356
10. Lancet D (1986) Vertebrate olfactory reception. Annu Rev Neurosci 9:329–355
11. Mori K (1993) Molecular and cellular properties of mammalian primary olfactory axons. Microscopic Res Tech 24:131–141
12. Buonviso N, Chaput MA (1990) Response similarity to odors in olfactory bulb output cells presumed to be connected to the same glomerulus: electrophysiological study using simultaneous single unit recordings. J Neurophysiol 63:447–454

Properties of Salamander Olfactory Bulb Circuits

John S. Kauer, Joel White, David P. Wellis, and Angel R. Cinelli[1]

Key words. Odor coding—Optical recording—Computer modeling—Olfactory bulb—Salamander—Patch clamp electrophysiology

Introduction

This laboratory is using electrophysiologic recording, optical recording, and computer modeling to study how odors are encoded by the olfactory epithelium (OE) and bulb (OB). These studies are carried out in the tiger salamander (*Ambystoma tigrinum*), which has the advantages of large neurons, a simplified nasal cavity permitting controlled odorant stimulation, the possibility of both in vivo and in vitro physiologic experiments, and expression of odor-guided behavior.

The hypothesis guiding this work holds that single, "monomolecular" odorant stimuli are encoded by activity distributed across many cells in the OE and OB, rather than by different, separate pathways, each dedicated to carrying information about only a particular compound. Such a distributed processing scheme does not imply lack of specificity in either individual cells or the ensemble circuit. Rather, it asserts that there are elemental odorant attributes that, taken together, characterize individual odorants [1,2] and that these attributes are encoded by parallel activation of distributed mechanisms [3–8]. Thus there is a combinatorial code in which multiple, elemental odorant attributes define odor compounds. At present the number and characteristics of these putative elemental odorant attributes are unknown. It has been suggested that there may be dynamic [9] or other kinds [7,10,11] of functional modules in the bulb that could relate to this encoding process, possibly on a one-to-one basis. This idea is attractive because it provides for the broad discriminative abilities and high sensitivity typical of the olfactory pathway as well as for its flexibility, fault tolerance [12], and redundancy. Such distributed mechanisms may also be pertinent for understanding other regions of the brain.

We have studied the physiologic properties of both single neurons and ensemble circuits in order to examine the degree to which these properties conform to this hypothesis. Previous experiments in this preparation include studies on: (1) properties of transduction mechanisms [13]; (2) single receptor neuron responses to odorants [14]; (3) odor-elicited activity distribution across the OE [15]; (4) projections of the OE onto the OB [16,17]; (5) responses of single OB cells to controlled odorant and electrical stimulation [18–20]; (6) patterns of voltage-sensitive dye (VSD) fluorescence in OB neurons after electrical [21] and odor stimulation [3]; and (6) properties of odor-guided behavior [22].

This chapter describes recently acquired information on analysis of single mitral/tufted (M/T) and granule cells using pharmacologic and whole-cell patch-clamp methods [23,24] and on VSD imaging of OB circuits after stimulation with homologous series of odorants with which the salamander has been behaviorally tested [22]. These and previously obtained data are incorporated into a schematic description of how odorants may be encoded.

Results

Mitral/Tufted and Granule Cell Properties Observed with Whole-Cell Patch-Clamp Recording

Figure 1 shows responses using whole cell patch clamp recording from an identified M/T (bottom left) and granule cell (bottom middle) after olfactory nerve (ON) electrical stimulation. The recordings were made in OBs with attached telencephalons that were removed from the animal and the dorsal bulbar layers exposed by a horizontal cut. Method details are described elsewhere [24]. Unequivocal cell identification was accomplished by recording in the appropriate layer under visual control and by intracellular staining with biocytin. An example of a filled granule cell is shown at the top of Fig. 1.

Records on the left and middle of Fig. 1 demonstrate voltage and current recordings from typical M/T and granule cells. The top row voltage records show that orthodromic stimulation evoked a brief depolarization and spike followed by long-lasting hyperpolarization in the M/T cell. This stimulation evoked a long depolarization and spike in the granule cell. These events were also examined in current records (second row). Here, in the M/T cell, an ON

[1] Section of Neuroscience, Departments of Neurosurgery and of Cell Biology and Anatomy, Tufts Medical School and New England Medical Center, 750 Washington Street, Boston, MA 02111, USA

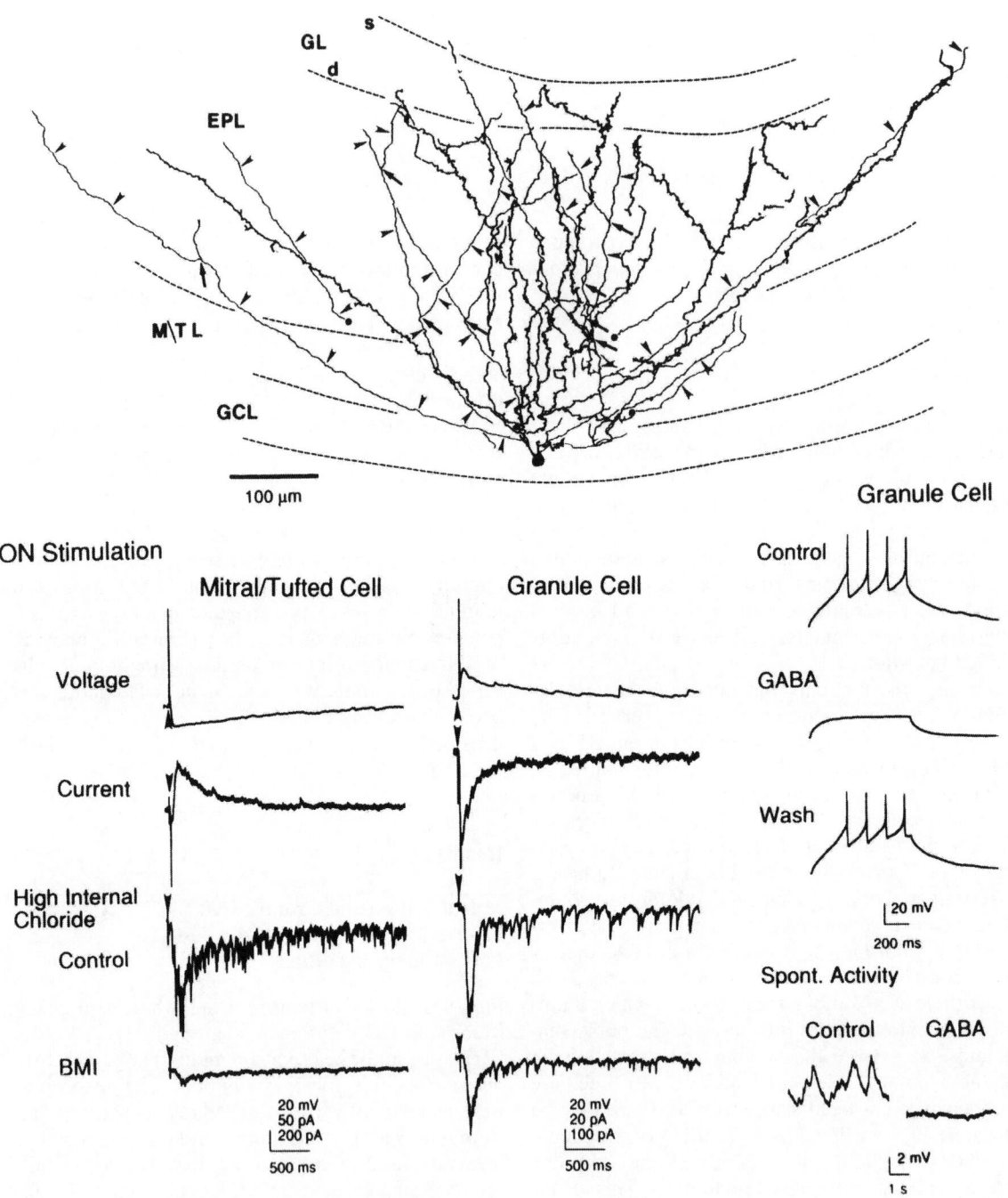

FIG. 1. **Top** Biocytin-filled spiny salamander granule cell. Note the axon-like processes indicated by arrowheads. *GL*, glomerular layer (*s*, superficial; *d*, deep); *EPL*, external plexiform layer; *M/TL*, mitral/tufted cell layer; *GCL*, granule cell layer. **Bottom** Records from identified M/T cell (*left column*) and granule cell (*middle columns*) showing voltage and current traces under normal electro-

lyte conditions (*top two traces*) and with high internal chloride (*bottom two traces*) after olfactory nerve stimulation. Higher sweep speed records in the right column show the sensitivity of a granule cell to GABA after current injection. The three vertical calibration values refer to voltage, current, and current for both high internal chloride traces, respectively

shock generated a brief inward current corresponding to excitatory depolarization followed by a longer-lasting outward inhibitory postsynaptic current corresponding to hyperpolarization generated

by γ-aminobutyric acid (GABA$_A$)-mediated inhibitory influences from granule cells. Further evidence that this outward current is mediated by GABA$_A$ receptor activation is seen in the bottom two records from the M/T cell showing inward currents (outward movement of Cl^-) under symmetric chloride conditions at -70 mV holding potential and by sensitivity to bicuculline (bottom trace).

A surprising finding is shown in the granule cell current records [23]. Here, under normal electrolyte conditions (second row), a large inward current related to the depolarization was seen. If, however, the same symmetrical chloride and holding potential conditions were applied as they were to the M/T cells (bottom two records), a chloride-mediated, bicuculline-sensitive current is also revealed *in the granule cells* suggesting that they, like M/Ts, receive GABAergic, inhibitory input. As shown in the rapid sweep records (right), GABA application hyperpolarized this granule cell, blocked spike activity, and shunted spontaneous currents.

An additional new finding was that each of the well stained, spiny granule cells showed the presence of thin, smooth processes that we interpret as axons (Fig. 1, top). These processes arose near the soma and projected into the external plexiform layer region occupied by the granule cell dendrites. These findings suggest, at least for the salamander, that granule cells may be more complex than originally thought.

VSD Imaging of OB Responses to Behaviorally Tested Odorants

Figure 2A illustrates six temporal series of VSD images obtained after stimulation of the ventral olfactory mucosa with two sets of homologous compounds for which there are behavioral data (see below); at left are responses to propyl, butyl, and amyl (3-, 4-, 5-carbon) alcohols; at right are responses to propyl, butyl, and amyl acetates. Odors were delivered at 10^{-1} of vapor saturation (as in the behavioral tests, see below) for 0.5 s. Details of the odor delivery [19], optical data acquisition [21], and behavioral [22] methods are given elsewhere. The stacked response profiles represent sequential frames showing depolarizing (clear profiles) or hyperpolarizing (shaded profiles) activity over time across the dorsal surface of the bulb (time proceeds left to right). Inset photographs show how representative VSD frames relate to black and white images of the OB from which they were obtained. At this plane in the OB, the dominant response to the alcohols was hyperpolarization whereas that to the acetates was initial depolarization followed by hyperpolarization.

The graph sets in Fig. 2B(a,b) are taken from a behavioral study [22] in which salamanders were tested for their ability to discriminate or generalize among the same homologous series of alcohols and acetates as were used for the VSD recordings. This study showed that the salamanders trained to recognize and avoid butyl alcohol also avoided propyl alcohol but not amyl alcohol. That is, the animals

generalized between the 3- and 4-carbon alcohols but discriminated between the 4- and 5-carbon alcohols.

The findings for the acetates were, however, different. With these odorants, animals trained to avoid butyl acetate also avoided amyl acetate but not propyl acetate. Thus in contrast to the alcohols, they generalized between the 4- and 5-carbon acetates but discriminated between the 4- and 3-carbon acetates. These data suggest that the animals discriminated odorants with different carbon chain lengths differently for these two homologous series.

These same odors were chosen for the VSD experiments (above) in order to examine how the OB activity patterns correlated with the behavior. As seen in Fig. 2A, the patterns evoked by propyl and butyl alcohol are more similar to one another than either pattern is to that evoked by amyl alcohol, consistent with the behavioral findings showing that the salamander generalized between the 3- and 4-carbon alcohols. Similarly, the patterns evoked by amyl and butyl acetate are more similar to one another than either is to the pattern evoked by propyl acetate, consistent with behavioral generalization between the 5- and 4-carbon acetates.

These findings suggest that it should be possible to carry out extensive comparisons between OB global response patterns and the behavioral abilities of the salamander. Such data should provide insight into the dimensionality of odor space and into how specific physical/chemical properties of odorants are encoded by the system.

Conclusion

Figure 3 displays a diagram incorporating information presently available on the salamander peripheral olfactory circuit schematizing how this system might encode odor information (see details in [4,9]). This model is based on: (1) responses observed in individual receptor cells; (2) response distribution across the receptor sheet; (3) connectivity between the receptor mucosa and the OB; (4) responsivity of single OB neurons to global and punctate odor and electrical stimulation; and (5) the distribution of activity in the OB.

Briefly, this hypothesized circuit represents a synthesis of anatomic and physiologic information suggesting that odorant substances are encoded by recognition of multiple chemical constituent components, which could be the building blocks of odorant recognition. Examples of such putative components are labeled I, II, and III on odors A and B in Fig. 3. Monomolecular odorant molecules thus interact with multiple molecular receptor site types residing on many, distributed receptor cells. Candidates have been found for such molecular receptors [25]. Although receptors of a single type have

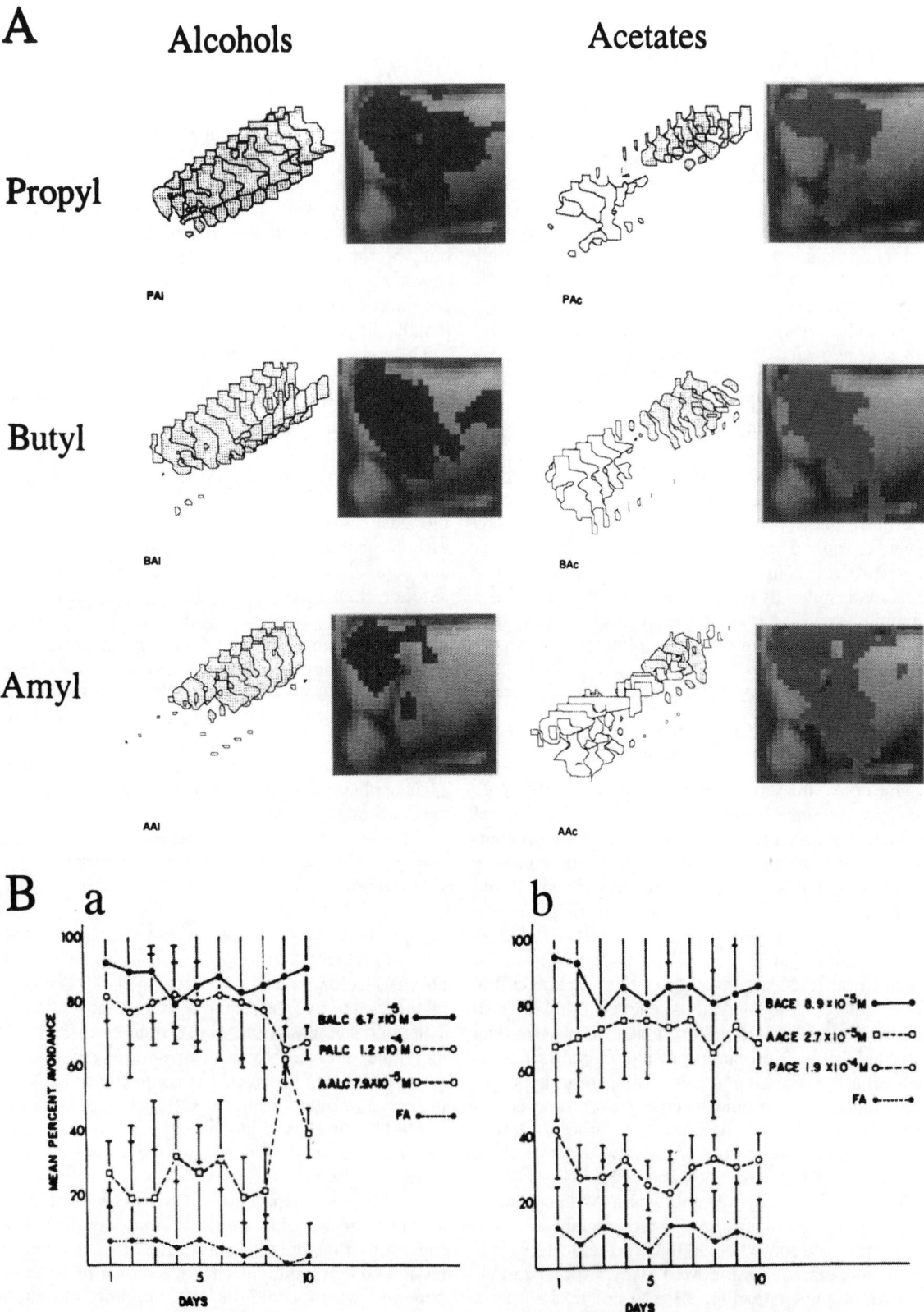

FIG. 2. **A** Sets of real-time images (taken by summing 16-ms frames into 133-ms bins) of voltage-sensitive dye fluorescence of activity in *in vivo* salamander olfactory bulb after stimulation with homologous series of propyl, butyl, and amyl alcohols (*left*) and acetates (*right*). *Shaded profiles* represent regions of hyperpolarization; *open profiles* represent depolarization; time goes *from left to right* into the plane of the page. Insets show detailed relations between the distribution of activity and the dorsal view of the bulb. **B,a** Results of behavioral testing of salamanders with propyl and amyl alcohols after training to butyl alcohol (from [22], with permission). *BALC*, butyl alcohol; *PALC*, propyl alcohol; *AALC*, amyl alcohol. **B,b** Behavioral results of testing with propyl and amyl acetates after training to butyl acetate. *BACE*, butyl acetate; *AACE*, amyl acetate; *PACE*, propyl acetate

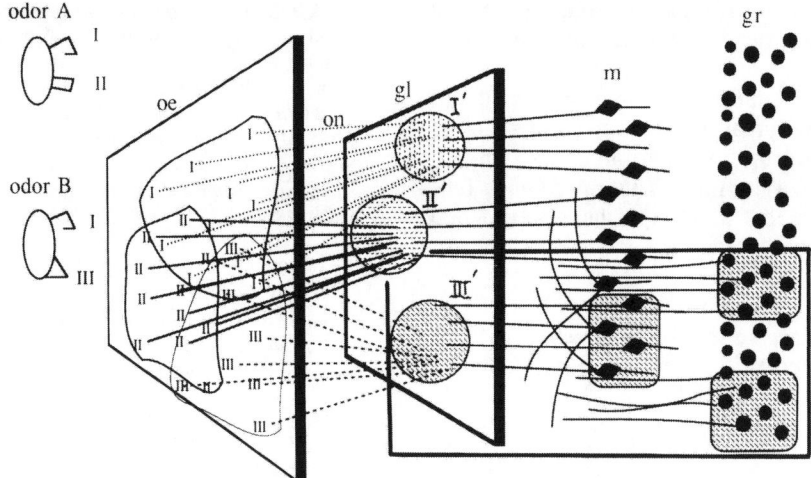

FIG. 3. Hypothesis of how modular activity might arise from distribution of responsivity across the olfactory epithelium (*oe*), from patterns of connectivity between the epithelium and bulb, and from specific aggregates of cells in each layer of the bulb participating in the response. I′, II′, and III′ on the oe illustrate the distribution of three putative receptor binding sites. Receptor sites of type I project to glomeruli of type I′, and so on. Groups of periglomerular, M/T (*m*), and granule (*gr*) cells are associated with each glomerulus (*gl*) or group of glomeruli. The *rectangle* delineates the region of activity we are defining as a module (for III′). For odorant A, which has odor properties I and II, modules I′ and II′ are activated. Odorant B activates a different set of modules but because it shares odor property II with odorant A the activation patterns in both oe and ob overlap. The *shaded elements* in module III are those activated by odorant attribute III. (Modified from [9], with permission)

been found in sensory cells distributed across wide mucosal regions [26,27], it is not yet known definitively whether single sensory cells express one or several such putative molecular receptors. Odorants interacting with a single molecular receptor type thus activate many, dispersed cells. This distribution is consistent with many physiologic observations [14,15,18,28] and is shown in Fig. 3 by the distribution of Roman numerals across the OE sheet. Highest regions of OE activity occur where the distribution fields of multiple molecular receptors interacting with one odorant, overlap.

Projections from the OE to the OB are complex [3,4,16,17,29] and do not appear to follow a simple point-to-point topography. The complex convergent/divergent connections are illustrated by the olfactory nerve lines drawn from the OE to the glomerular layer (gl) in Fig. 3. This complex OE/OB topography may represent convergence of axons arising from receptor cells having similar response profiles (or expressing similar molecular receptors) onto localized groups of one or more glomeruli [3,4,16]. This possibility is consistent with the long-held view that glomeruli may serve as "functional units" [9,30,31] similar to modules in other sensory systems.

The major cell groups in the OB include output M/T cells and periglomerular and granule interneurons, both generally considered to be inhibitory, although this explanation is clearly a gross oversimplification. In the canonical circuit presented in Fig. 3, M/T cells receive excitatory input from sensory receptor cell axons converging from distributed regions of the OE, receive feedforward and feedback inhibition from periglomerular cells, and receive feedback inhibition from granule cells [32].

The present hypothesis holds that this circuit is activated in a modular fashion in which the modules can be dynamic, changing over time and space with the time course and the attributes of the stimulus application. This could occur by: (1) activation of several distributed sets of different receptor types, each set looking at a different attribute of the odorant stimulus; (2) convergence of sensory cells carrying information about each particular odorant attribute onto one or several contiguous glomeruli, consistent with 2-deoxyglucose, VSD, and receptive field mapping observations [9,18,33]; and finally (3) activation of a number of M/T, periglomerular, granule cell modules (shaded regions on the right in Fig. 3), each responsive to a different odorant attribute, operating as a parallel distributed processor to encode the entire odorant stimulus [3,34–37]. Thus an odorant compound is encoded by activation of multiple modules each carrying information about a different odor attribute and together defining the overall properties of the compound. The complete encoded message is highly distributed; and only a relatively limited number, perhaps no more than hundreds, of different receptor types [25] are necessary to encode large numbers (thousands) of compounds, including those to which the system had never previously been exposed. This kind of highly distributed system has the requisite broad band responsivity, resistance to damage, and great flexibility.

This hypothesis is undoubtedly oversimplified, but it provides a conceptual framework within which to try to understand the available anatomic, physiologic, pharmacologic, and behavioral data. It provides a platform on which specific experimental predictions can be tested.

Acknowledgments. Supported by the USPHS, ONR, and the Department of Neurosurgery, Tufts New England Medical Center.

References

1. Beets MGJ (1970) The molecular parameters of olfactory response. Pharmacol Rev 22::1–34
2. Polak EH (1973) Multiple profile-multiple receptor site model for vertebrate olfaction. J Theoret Biol 40:469–484
3. Kauer JS (1991) Contributions of topography and parallel processing to odor coding in the vertebrate olfactory pathway. TINS 14:79–85
4. Kauer JS (1987) Coding in the olfactory system. In: Finger TE, Silver WL (eds) Neurobiology of taste and smell. Wiley, New York, pp 205–231
5. Holley A, Doving KB (1977) Receptor sensitivity, acceptor distribution, convergence and neural coding in the olfactory system. In: LeMagnen J, MacLeod P (eds) Olfaction and Taste VI. IRL Press, London, pp 113–123
6. Schild D (1988) Principles of odor coding and a neural network for odor discrimination. Biophys J 54:1011
7. Shepherd GM (1993) Principles of specificity and redundancy underlying the organization of the olfactory system. Microsc Res Techniques 24:106–112
8. Lancet D (1992) Olfactory reception: from transduction to human genetics. In: Corey DP, Roper SD (eds) Sensory transduction. Rockefeller University Press, New York, pp 74–91
9. Kauer JS, Cinelli AR (1993) Are there structural and functional modules in the vertebrate olfactory bulb? Microsc Res Techniques 24:157–167
10. Macrides F, Schoenfeld TA, Marchand JE, Clancy AN (1985) Evidence for morphologically, neurochemically and functionally heterogeneous classes of mitral and tufted cells in the olfactory bulb. Chem Senses 10: 175–202
11. Scott JW, Harrison TA (1987) The olfactory bulb: anatomy and physiology. In: Finger TE, Silver WL (eds) Neurobiology of taste and smell. Wiley, New York, pp 151–178
12. Risser JM, Slotnick BM (1987) Suckling behavior in rat pups with lesions which destroy the modified glomerular complex. Brain Res Bull 19:275–281
13. Firestein S, Werblin FS (1989) Odor-induced membrane currents in vertebrate olfactory receptor neurons. Science 244:79–82
14. Getchell TV, Shepherd GM (1978) Responses of olfactory receptor cells to step pulses of odour at different concentrations in the salamander. J Physiol (Lond) 282:521–540
15. Mackay-Sim A, Shaman P, Moulton DG (1982) Topographic coding of olfactory quality: odorant-specific patterns of epithelial responsivity in the salamander. J Neurophysiol 48:584–596
16. Kauer JS (1980) Some spatial characteristics of central information processing in the vertebrate olfactory pathway. In: van der Starre H (ed) Olfaction and Taste VII. IRL Press, London, pp 227–236
17. Mackay-Sim A, Nathan MH (1984) The projection from the olfactory epithelium to the olfactory bulb in the salamander, Ambystoma tigrinum. Anat Embryol (Berl) 170:93–97
18. Kauer JS, Moulton DG (1974) Responses of olfactory bulb neurones to odour stimulation of small nasal areas in the salamander. J Physiol (Lond) 243:717–737
19. Kauer JS, Shepherd GM (1975) Olfactory stimulation with controlled and monitored step pulses of odor. Brain Res 85:108–113
20. Hamilton KA, Kauer JS (1988) Responses of mitral/tufted cells to orthodromic and antidromic electrical stimulation in the olfactory bulb of the tiger salamander. J Neurophysiol 59:1736–1755
21. Kauer JS (1988) Real-time imaging of evoked activity in local circuits of the salamander olfactory bulb. Nature 331:166–168
22. Mason JR, Stevens DA (1981) Discrimination and generalization among reagent grade odorants by tiger salamanders (Ambystoma tigrinum). Physiol Behav 26:647–653
23. Wellis DP, Kauer JS (1994) GABAergic and glutamatergic synaptic input to identified granule cells in salamander olfactory bulb. J Physiol (Lond) 475:419–430
24. Wellis DP, Kauer JS (1993) $GABA_a$ and glutamate receptor involvement in dendrodendritic synaptic interactions from salamander olfactory bulb. J Physiol (Lond) 469:315–339
25. Buck L, Axel R (1991) A novel multigene family may encode odorant receptors: a molecular basis for odor recognition. Cell 65:175–187
26. Ressler KJ, Sullivan SL, Buck LB (1993) A zonal organization of odorant receptor gene expression in the olfactory epithelium. Cell 73:597–609
27. Nef P, Hermans-Borgmeyer I, Artieres-Pin H, et al. (1992) Spatial pattern of receptor expression in the olfactory epithelium. Proc Natl Acad Sci USA 89: 8948–8952
28. Revial MF, Sicard G, Duchamp A, Holley A (1982) New studies on odour discrimination in the frog's olfactory receptor cells. I: Experimental results. Chem Senses 7:175–191
29. Astic L, Saucier D, Holley A (1987) Topographical relationships between olfactory receptor cells and glomerular foci in the rat olfactory bulb. Brain Res 424:144–152
30. Leveteau J, MacLeod P (1966) Olfactory discrimination in the rabbit olfactory glomerulus. Science 153:175–176
31. Shepherd GM (1981) The olfactory glomerulus; its significance for sensory processing. In: Katsuki Y, Norgren R, Sto M (eds) Brain Mechanisms of Sensation. Wiley, New York, pp 209–223
32. Mori K (1987) Membrane and synaptic properties of identified neurons in the olfactory bulb. Prog Neurobiol 29:275–320
33. Stewart WB, Kauer JS, Shepherd GM (1979) Functional organization of rat olfactory bulb analysed by

the 2-deoxyglucose method. J Comp Neurol 185:715–734

34. Buonviso N, Chaput MA (1990) Response similarity to odors in olfactory bulb output cells presumed to be connected to the same glomerulus: electrophysiological study using simultaneous single-unit recordings. J Neurophysiol 63:447–454

35. Imamura K, Mataga N, Mori K (1992) Coding of odor molecules by mitral/tufted cells in rabbit olfactory bulb. I: Aliphatic compounds. J Neurophysiol 68: 1986–2002

36. Meredith M (1986) Patterned response to odor in mammalian olfactory bulb: the influence of intensity. J Neurophysiol 56:572–597

37. Wilson DA, Leon M (1987) Evidence of lateral synaptic interactions in olfactory bulb output cell responses to odors. Brain Res 417:175–180

Central Coding of Odor Information in the Brain of the Slug, *Limax marginatus*

Tetsuya Kimura, Atsushi Yamada, Haruhiko Suzuki, Euji Kono, Tatsuhiko Sekiguchi, and Yukihiro Sugiyama[1]

Key words. Oscillating neural network—Olfaction

In terrestrial molluscs, odor information is detected at the tips of two paired tentacles, and is transferred into the procerebral lobe (PCL) via tentacle ganglia located at the tentacle tips. The PCL, a brain area highly developed in terrestrial molluscs, is constituted of numerous interneurons with small cell bodies. Using a 2-deoxyglucose labeling technique in *Helix*, Chase [1] showed that the cell activity of the PCL interneuron increases during odor stimulation. Further, Gelperin and Tank [2] reported that the PCL interneurons always showed oscillatory electrical activity, and that odor stimuli applied to the posterior tentacle modulated the frequency. However, the role of the PCL in odor recognition remains obscure.

In the present study, we used slugs, *Limax marginatus*, which had been conditioned to food odors (carrot, cucumber) either aversively or appetitively. In aversive conditioning, the animals were exposed to carrot or cucumber odor for 10 s, and then 200 µl of saturated quinidine sulfate solution, a very bitter substance, noxious for slugs, was dropped on the anterior body surface, including the lip. In appetitive conditioning, the slugs were also exposed to the food odor, and then they were allowed to eat the odor source. From preliminary experiments, it was confirmed that these procedures changed the odor-induced food aversive behavior.

An isolated inferior nose-brain preparation obtained from a conditioned slug was placed into our experimental chamber, which was separated into three blocks, each of which were always perfused with saline; the local field potential (LFP) of the PCL was recorded via one or two glass pipets (tip diameter about 50 µm) placed on the PCL cell layer surface. When an odor stimulus was applied, the saline level of the nose block was decreased to expose the olfactory epithelium.

When the LFPs of two different regions were simultaneously recorded on the surface of the PCL cell layer with two electrodes, the observed oscillations were always sychronous, and there was no phase lag along the dorsal-ventral axis, this lag occurring along the medial-lateral axis. An oscillation recorded from any region always preceded that from a more medial region. Thus, oscillatory activity in the entire PCL could be regarded as periodic waves which traveled from the lateral to the medial margin. Furthermore, similar oscillatory activity was recorded from all small pieces (200-µm-thick slices) of the PCL examined. These observations indicate that: (1) the PCL consists of multiple oscillating elements, and (2) suggest that the traveling wave is produced by coordination of the activity of each element.

Application of both aversive learned carrot and cucumber odor modulated the oscillatory state of the PCL. That is, although the phase-lag between two regions always fluctuated in spontaneous activity, the odor stimuli decreased the fluctuation range. This indicates that the oscillatory activity of the recorded regions is extensively phase-locked. The value of the phase-lag during the phase-locked state induced by the carrot odor was not equal to that during the steady state induced by the cucumber odor.

We also examined the effects of conditioning on the frequency modulation. We found that the aversively learned carrot (or cucumber) odor usually decreased the frequency, while the appetitively learned carrot (or cucumber) odor increased it. This finding suggests that these types of frequency modulation induced by food odors reflect the learned preference of each animal, but not the quality of the applied odors.

In conclusion, odor information, including the learned preference, is expressed as an oscillating state in the PCL network, shown as phase-lag relation and frequency.

References

1. Chase R (1985) Responses to odors mapped in snail tentacle and brain by [14C]-2-deoxyglucose autoradiography. J Neurosci 5:2930–2939
2. Gelperin A, Tank DW (1990) Odour-modulated collective network oscillations of olfactory interneurons in a terrestrial mollusc. Nature 345:437–440

[1] Tsukuba Research Center, SANYO Electric Co. Ltd., 2-1 Koyadai, Ibaraki, 305 Japan

Differential Regulation of Neurotrophic Factor Expression in the Olfactory Bulb

Hiroshi Nagao, Ichiro Matsuoka, and Kenzo Kurihara[1]

Key words. Ciliary neurotrophic factor—Olfactory bulb

The olfactory system possesses unusual capacity to generate neurons throughout life. Although the neurogenesis of olfactory neurons seems to occur independently of trophic support from the olfactory bulb (OB), the subsequent maturation of these neurons is dependent on this support [1]. Recent studies have demonstrated the expression of several neurotrophic factors in the OB, such as the nerve growth factor (NGF) family and ciliary neurotrophic factor (CNTF) [2,3]. In the central nervous system the expression of CNTF is highly concentrated in the OB, suggesting that this expression is related to the specific function of OB.

In this study, we investigated the regulatory mechanism for the expression of CNTF-mRNA in cultured astrocytes obtained from rat OB. Cells were dissociated from the OB of 2-day-old Wistar rats, and were cultured for 1 week in DMEM supplemented with 10% FCS and antibiotics. Cells thus obtained consisted of >95% astrocytes, as assessed by immunostaining with anti-glial fibrillary acidic protein (GFAP) antibody. Northern blot analysis revealed that the cultured astrocytes expressed both CNTF-mRNA and NGF-mRNA. The basal level of the CNTF-mRNA was much higher than that of NGF-mRNA, similarly to the in vivo situation in the OB. Activation of adenylate cyclase by $20\,\mu M$ forskolin (FK) led to an increase of NGF-mRNA in cultured astrocytes, while, in contrast, CNTF-mRNA was abolished by FK. Four-h treatment with $20\,\mu M$ FK led to a decrease of CNTF-mRNA, to half the original level, and CNTF-mRNA was reduced to less than 20% of the original level after 12 h. The effect of FK was mimicked by the cAMP-linked agonists, vasoactive intestinal peptide (VIP) and isoproterenol. Of various transmitters, neuropeptides, and growth factors examined, VIP was the most effective agonist in suppressing CNTF-mRNA levels. Stimulation of cultured astrocytes with 100 nM VIP led to a 40% decrease of CNTF-mRNA at 10 h. The effect of VIP was reversed after 12 h, presumably by receptor desensitization. A nuclear run-off assay, using nuclear extracts from cultured astro-cytes, revealed that in vitro CNTF-mRNA transcription was reduced by 3-h FK treatment, to 30% of the initial level. Cycloheximide, an inhibitor of protein synthesis, did not prevent the FK-induced decrease of CNTF-mRNA. These findings suggest that the cAMP-induced decrease of CNTF-mRNA was predominantly due to the inhibition of CNTF-mRNA transcription via preexisting factor(s). On the other hand, CNTF-mRNA in cultured astrocytes without stimulation had a relatively long half-life, of 6 h; stimulation of the astrocytes with FK slightly shortened its half-life. The effects of FK treatment on the half-life of CNTF-mRNA were reversed by actinomycin D, an inhibitor of transcription. These results suggest that the destabilizing effect of FK on CNTF-mRNA occurred via the de novo synthesis of protein factors. Therefore, it could be concluded that the activation of adenylate cyclase in cultured OB astrocytes negatively regulates the expression of CNTF-mRNA by multiple mechanisms, namely, by the transcriptional inhibition of the CNTF gene and the destabilization of CNTF-mRNA.

The differential regulation of CNTF- and NGF-mRNAs by the activation of adenylate cyclase in vitro suggests either differential functions or differential modes of expression of these neurotrophic factors in vivo. The findings reported here should help to clarify the role of neurotrophic factors in the mechanism of the neurogenesis and regeneration of olfactory neurons.

Acknowledgments. This study was supported by a Grant-in-Aid for Scientific Research on Priority Areas (05261201) from the Ministry of Education, Science, and Culture, Japan.

References

1. Graziadei PPC, Monti-Graziadei AG (1992) Sensory reinnervation after partial removal of the olfactory bulb. J Comp Neurol 316:32–44
2. Guthrie KM, Gall CM (1991) Differential expression of mRNAs for the NGF family of neurotrophic factors in the adult rat central olfactory system. J Comp Neurol 313:95–102
3. Stöckli KA, Lillien LE, Näher-Noè M, Breitfeld G, Hughes RA, Raff MC, Thoenen H, Sendtner M (1991) Regional distribution, developmental changes, and cellular localization of CNTF-mRNA and protein in the rat brain. J Cell Biol 115:447–459

[1] Faculty of Pharmaceutical Sciences, Hokkaido University, N-12 W-6, Kita-ku, Sapporo, 060 Japan

Neural Responses of Olfactory Bulb and Pyriform Cortex to Odors of Essential Oils and Terpenes in Rats

Mari Kimoto, Yasuko Fukushima, and Taeko Yamada[1]

Key words. Leaf oil—Olfactory system

It has been suggested that the inhalation of vapors arising from forest trees has a physiological influence in various species of animals and in humans. To characterize the functional roles of leaf oil odors in the olfactory system, we studied the neuronal responses of the olfactory bulb (OB) and the pyriform cortex (PC) to four essential oil odorants and their components, five terpenes.

Male Wistar albino rats (weighing 230–560 g), anesthetized i.p. with Nembutal (60 mg/kg) were used. The odorants were the essential oils of *Abies sachalinensis Fr. Schm.*, *Chamaecyparis obtusa Endl.*, *Cinnamomum camphora Siebold*, and *Eucalyptus globulus*, and the terpenes, α-pinene, limonene, cineole, camphor, and bolnyl acetate. All odorants were diluted to 10^{-6}, 10^{-4}, 10^{-2}, and 1% with mineral oil. Four concentrations were used for the OB experiment, and two higher concentrations were used for the PC experiment. Thirty ml of the odorous vapor was presented for 5 s at 4-min interstimulus intervals. Extracellular recordings of single units in the OB (63 units) and PC(33) were made with tungsten microelectrodes (10 MΩ), which were inserted perpendicularly into the OB at a depth of 0.4–3.4 mm from the surface or into the cortex at a depth of 6–10 mm, 0 mm to the bregma, and 3.5–4.5 mm lateral to the midline. The responses were recorded 1 min before and after the stimulus onset. The relative magnitude (RM) of the response to odors was calculated, this being the ratio of the number of spikes obtained 0–30 and 31–60 s after stimulation to the number obtained 30 s before stimulation. The responses to each odorant in the OB and PC were classified as excitatory (RM > 1.2), inhibitory (RM < 0.8), or no response (0.8 ≤ RM ≤ 1.2). To determine the response profiles, the percentage of each response type was calculated in the recorded units. The largest and the smallest magni- tudes of response to each odorant for all units were plotted according to the depth of electrode insertion; in the OB, the depth from the dorsal surface, and in the PC, the depth from the ventral surface. A corre- lation matrix was constructed for the odorants, based on their RM responses in the PC.

In the OB, the responses of single units differed according to the odorants and their concentrations. Recorded units in the dorsal region (400–700 μm) predominantly responded to α-pinene and limonene, and in the ventral region (2000–3400 μm) to cineole and camphor. The response profiles of the essential oils and terpenes were similar. The percentage for no response to all odorants was higher than that for the other two categories (i.e., excitatory and inhibi- tory) and the excitatory responses to essential oils were higher than those to terpenes.

In the PC, the RM of the responses differed according to the odorants and their concentrations, but there were close correlations between *Abies sachalinensis Fr. Schm.* and bornyl acetate; *Cin- namomum camphora Siebold* and camphor; and *Eucalyptus globulus* and cineole. In the superficial layer (750 μm), *Abies sachalinensis Fr. Schm.* evoked an excitatory response in some units, while, in the deep layer (750–1050 μm) clear excitatory responses were evoked by *Cinnamomum camphora Siebold*, limonene, and camphor. The response profiles were divided into two groups; in one group, the percent excitatory response was higher than the percent inhibitory response and in the other group, the major category was no response.

These results show that: (1) essential oils and terpenes are adequate stimuli for the rat's olfactory system, (2) OB neurons can discriminate leaf oils and terpenes, (3) responses to various odor stimuli are more sensitive in the PC than in the OB, and (4) the PC can process odor information about the quality of essential oils and their components, terpenes.

Acknowledgments. This study was supported by Grants in aid-for Special Scientific Research (1990, 1991, 1992) on Agriculture, Foresty, and Fisheries of Japan.

[1] Physiological Laboratory, Japan Women's University, 2-8-1 Mejirodai, Bunkyo-ku, Tokyo, 112 Japan

Suppressive Interactions During Olfactory Bulb Circuit Response to Odor: Computer Simulation

MICHAEL MEREDITH[1]

Key words. Olfactory bulb—Computer simulation

The olfactory bulb response to electrical stimulation is fairly well understood and is determined largely by two laminae of lateral inhibitory interneurons, the periglomerular (PG) and granule (GR) cells. Responses to odor are more complex. Profound suppression, complex responses, and non-monotonic intensity response (I/R) functions occur (both with excitation at low intensity and suppression at high; E-S type; and the opposite S-E type [1]). Studies using 2-deoxyglucose (2DG) or *c-fos* mRNA mapping [2] suggest that odor input is spatially non-uniform, with isolated foci of activity in different areas for different odors. A computer model, based on the known anatomy and physiology of the bulb was, therefore, constructed to examine output responses to restricted input. It calculates interactions between all cells within a 30×30 matrix of glomerular units, re-iterating until excitation, inhibition, and output are in balance [3]. In the latest version, each glomerular unit has one mitral, one middle tufted, and one outer tufted cell with associated interneurons. Relay cell axon collaterals and PG inhibition of PG cells have been added. This more realistic circuit does not qualitatively alter responses compared to earlier models. A two-dimensional normal distribution of input, centered on the matrix, produces a similar output pattern at low intensity, but with an inhibitory surround. At higher intensity, the activated area expands as surrounding cells are recruited, but input to the highly sensitive center saturates. Convergent lateral inhibition from the periphery drives output at the center to zero. This *secondary* inhibition, not yet demonstrated experimentally, could contribute to the profound inhibition observed in some cells. Relay cells at fixed matrix positions will be excited at some intensities and inhibited at others. Some cells will show non-monotonic E-S type and others S-E type I/R functions. Synaptic connections between cell types are uniform throughout the network; differences in response arise from placement relative to the odor focus. Multiple foci (suggested by 2DG data) pro-duce intense *interstitial* inhibition between the foci, where cells are inhibited at all intensities ("S-type" response [4]). The model(s) are highly simplified, but synaptic connections do conform to available experimental data, and there are constraints even on synaptic weights. If set too low or high, the percentage of cells inhibited at high concentration does not match in vivo observations. In previous versions, no threshold was provided for interactions with interneurons, but threshold inclusion here made only minor changes to patterns of output (increased spontaneous level and decreased inhibition). Lateral inhibitory connections appear capable of producing the odor responses observed in vivo, but other mechanisms are possible. Tufted cells begin responding at lower intensities than mitral cells [5], and, if mutually inhibitory, these cells could show non-monotonic I/R functions, of opposite type. To explore possible *interlaminar* inhibition, the basic 30×30 matrix was subdivided into 5×5 subregions, each representing a glomerulus with 25 mitral and 50 tufted cells. Careful adjustment of synaptic weights in three populations of PG cells allowed outer tufted cells to be excited and mitral cells inhibited at low intensity, with opposite responses at higher intensity. Thus, both E-S and S-E types of non-monotonic I/R function resulted from interlaminar inhibition and appeared in cells in different laminae connected to the same glomerulus. However, the interlaminar inhibition model required more speculative assumptions and was much less robust than the lateral inhibition models. Small changes in parameters disrupt the former but not the latter. Thus, within the limitations of this type of model, it appears that the complex responses and non-monotonic I/R functions observed with odor stimulation are more likely the result of lateral inhibition than interlaminar inhibition.

Acknowledgments. Supported by NIH Grant DC-00906.

References

1. Meredith M (1986) Patterned response to odor in mammalian olfactory bulb: I Influence of intensity. J Neurophysiol 56:572–597

[1] Neuroscience Program, Florida State University, Tallahassee, FL 32306, USA

2. Guthrie K (1993) Odor-induced increases in *c-fos* mRNA expression reveal an anatomical "unit" for odor processing in olfactory bulb. Proc Natl Acad Sci USA 90:3329–3333
3. Meredith M (1992) Neural circuit computation: Complex patterns in the olfactory bulb. Brain Res Bull 29:111–117
4. Kauer JS, Shepherd GM (1977) Analysis of the onset phase of olfactory bulb unit response to odour pulses in the salamander. J Physiol (Lond) 272:495–516
5. Schneider SP, Scott JW (1983) Orthodromic response properties of rat olfactory bulb mitral and tufted cells correlate with their projection patterns, J Neurophysiol 50:358–378

Analysis of Dopamine and Its Effects in the Salamander Olfactory Bulb

K.A. Hamilton[1], M.R. Gurski[1], S.S. Foster[1], D.S. Knight[1], and G.A. Gerhardt[2]

Key words. Catecholamines—Olfaction

The rat olfactory bulb contains a large number of putative dopaminergic neurons. In response to olfactory receptor cell excitation, the neurons appear to modulate the synaptic input from the receptor cell axons to their postsynaptic targets in the glomeruli [1,2]. To identify the sites and mechanisms of dopamine action in the olfactory bulb of the tiger salamander, we have examined the distribution of catecholaminergic neurons, catecholamine tissue levels, and the effects of dopamine (DA) and a DA receptor antagonist on field potentials evoked by stimulating the olfactory nerve (ON) and lateral olfactory tract (OT).

Anatomical studies of the salamander olfactory bulb have shown that the caudal cell layers contain a large number of neurons that are immunoreactive for tyrosine hydroxylase, the rate-limiting enzyme in catecholamine synthesis [3]. Many of the neurons have dendrite-like processes that project to glomeruli. When levels of the DA precursor L-dihydroxyphenylalanine (L-DOPA) are experimentally enhanced, neurons with similar dendritic patterns can be identified in sections reacted with DA-antiserum. The neurons are not immunoreactive for L-DOPA or for enzymes that catalyze the synthesis of norepinephrine and epinephrine.

Measurements of catecholamine tissue levels by high pressure liquid chromatography have provided initial support for the hypothesis that the neurons are dopaminergic [4]. The salamander bulb appears to contain 2–3 times as much DA as norepinephrine. It also contains 5–20 times as much DA as epinephrine, L-DOPA, or 3,4-dihydroxyphenylacetic acid or homovanillic acid, which are DA metabolites.

Field potentials were recorded in the granule cell layer of isolated hemi-brain preparations that were superfused with normal medium, then with either DA or the dopamine receptor antagonist fluphenazine (FLU). Ascending, low concentrations (0.05–10 µM) of both DA and FLU depressed the amplitude of ON-evoked responses by as much as 67% relative to controls in normal medium, whereas the amplitude of OT-evoked responses either remained the same (FLU) or decreased by ≤23% (DA) [5]. Studies of preparations exposed to single DA or FLU concentrations showed that the effects depended on the concentration and exposure time. After 15- to 20-min superfusion with 5–50 µM DA, the ON- and OT-evoked responses decreased by ≤76% and ≤18%, respectively, and after 200 µM DA, the ON-and OT-evoked responses decreased by 83% and 38%, respectively. After 5–50 µM FLU, the ON- and OT-evoked responses decreased by ≤28% and ≤8%, respectively. After 100–300 µM FLU, however, large decreases (≤72%) were observed in both the ON- and OT-evoked responses. The effects of DA are similar to those previously observed in the isolated turtle olfactory bulb [6].

The stronger effects of 5–50 µM concentrations of DA and FLU on the ON- than OT-evoked responses in the salamander olfactory bulb suggest that both drugs target receptors in the glomerular layer which have a higher density or affinity for DA than receptors in the more caudal (deeper) layers. The results also suggest that dopamine binds to several types of receptors which have different distributions on the olfactory receptor cell axons and their postsynaptic targets. Studies with intracellular recording methods are planned to identify specific sites and mechanisms of DA action.

Acknowledgments. Supported by grants from the NIH, NSF, and Whitehall Foundation. Abstracts have appeared previously.

References

1. Nickell WT, Norman AB, Wyatt LM, Shipley MT (1991) Olfactory bulb DA receptors may be located on terminals of the olfactory nerve. Neuroreport 2:9–12
2. Sallaz M, Jourdan F (1992) Apomorphine disrupts odor-induced patterns of glomerular activation in the olfactory bulb. Neuroreport 3:833–836
3. Hamilton KA, Foster SS (1991) Immunocytochemical evidence for dopaminergic cells in salamander olfactory bulb. Soc Neurosci Abstr 17:637
4. Knight DS, Foster SS, Hamilton KA (1992) Analysis of catecholamine concentrations in the salamander olfactory bulb (abstract). Chem Senses 17:652
5. Hamilton KA, Gurski MR (1992) Effects of dopamine on responses to electrical stimulation in the salamander olfactory bulb: Initial results. Soc Neurosci Abstr 18:1199
6. Nowycky MC, Halász N, Shepherd GM (1983) Evoked potential analysis of dopaminergic mechanisms in the isolated turtle olfactory bulb. Neuroscience 8:717–722

[1] Louisiana State University Medical Center, Department of Cellular Biology and Anatomy, 1501 Kings Highway, Shreveport, LA 71130-3932, USA
[2] University of Colorado Health Science Center, Department of Psychiatry, Box C268-71, 4200 East 9th Avenue, Denver, CO 80262, USA

Modulation of Dendrodendritic Interactions in the Rat Olfactory Bulb by Locus Coeruleus Stimulation

Fumino Okutani, Hideto Kaba, Hideo Saito, and Katsuo Seto[1]

Key words. Locus coeruleus—Field potential

Observations that noradrenaline induces enduring enhancement of the efficacy of a second transmitter in central synapses are intriguing and have attracted considerable attention, as the process may serve as a neuronal substrate for learning and memory. This may be true for synapses in the olfactory bulb (OB), as the significance of noradrenergic centrifugal inputs to the OB in olfactory learning has been noted in a number of models [1]. Noradrenergic terminals in the OB originate in the locus coeruleus (LC) [2]. However, the role of LC noradrenergic neurons in OB function has not been fully clarified. We therefore examined the effect of stimulation of the LC on mitral-granule cell interactions in the rat OB.

Ovariectomized Wistar female rats ($n = 10$) were used for the experiments; the surgical procedure was carried out at least 4 weeks prior to the electrophysiological experiments. The rats, weighing 210–290 g, were anesthetized with urethane (0.8 g/kg, IP), placed in a stereotaxic instrument, and had holes drilled in the skull to allow access to the OB, lateral olfactory tract (LOT), and LC. A glass microelectrode filled with 0.5 M sodium acetate was lowered into the granule cell layer to record the potentials evoked by stimulation of the LOT. Coaxial bipolar stimulating electrodes were used; these were constructed from 25-gauge stainless steel tubing and stainless steel wire 0.2 mm thick. Constant current stimuli, consisting of 0.2-ms square pulses with an intensity of 1.7 mA or less, adjusted so as to elicit a near-maximal response, were delivered to the LOT. LOT stimulation produced a characteristically large, positive evoked field potential in the granule cell layer. Such a conditioning stimulus resulted in a prolonged inhibition of the same wave evoked by a second identical test stimulus. This paired-pulse inhibition is believed to be a measure of granule cell-mediated feedback inhibition onto mitral cells [3]. LC neurons were activated either by microinjecting glutamate (10 mM, 1.0 μl), or electrically (0.5 mA, 333 Hz, 50 pulses). Paired-pulse testing at an interpair interval of 20 ms was performed before and after glutamate or electrical activation of the LC. LC glutamate activation significantly decreased the amplitude of test responses without any obvious effect on the amplitude of conditioning responses. A significant change occurred 4 min after glutamate ejections. LC electrical activation, however, had no effect on the amplitude of either response.

These results suggest that the injections of glutamate in the vicinity of LC neurons strongly activate LC noradrenergic neurons, thereby potentiating granule cell-mediated feedback inhibition onto mitral cells in the OB. There is strong evidence that GABAergic granule cells act to control mitral/tufted cell excitability [3]. It is, therefore, hypothesized that the consequence of LC glutamate activation is a gating of incoming olfactory information at the OB level.

References

1. Gervais R, Holley A, Keverne B (1988) The importance of central noradrenergic influences on the olfactory bulb in the processing of learned olfactory cues. Chem Senses 13:3–12
2. Shipley MT, Halloran FJ, de la Torre J (1985) Surprisingly rich projection from locus coeruleus to the olfactory bulb in the rat. Brain Res 329:294–299
3. Shepherd GM (1972) Synaptic organization of the mammalian olfactory bulb. Physiol Rev 52:864–917

[1] Department of Physiology, Kochi Medical School, Nankoku, Kochi, 783 Japan

Determinants of Odor Mixture Perception in Squirrel Monkeys, *Saimiri sciureus*

Matthias Laska and Robyn Hudson[1]

Key words. Odor mixture perception—Squirrel monkeys

Primates are typically regarded as visual animals and the sense of olfaction has been traditionally considered of only minor importance in this order of mammals. However, it is becoming increasingly clear from studies of both humans and non-human primates that olfaction may, in fact, play a significant part in the regulation of a wide variety of primate behaviors. In addition to its more obvious role in food identification and selection, there is now evidence from a number of primate species for olfactory involvement in social behaviors as diverse as the establishment and maintenance of rank, territorial defense, recognition of the estrous state of female conspecifics, and recognition of species, group members, or even individuals. All these behaviors require that the animal be capable of discriminating fine differences between odor signals. It was therefore the aim of our study to investigate the ability of squirrel monkeys, a common New World primate species, to differentiate between odor mixtures similar in quality.

In a task designed to simulate olfactory-guided foraging, 1.5-ml Eppendorf flip-top reagent cups were attached to the arms of a climbing frame and equipped with absorbent paper strips impregnated with 10 µl of an odorant signalling either that they contained a peanut food reward (S+) or did not (S−). Eighteen such cups, 9 positive and 9 negative, were inserted in random order in holes along the horizontal bars of the climbing frame; the monkeys were trained to inspect the cups by sniffing closely and were allowed 1 min per trial to harvest as many of the baited cups as possible.

To test the monkeys' ability to discriminate subtle differences in the composition of two odors, an artificial 12-component odorant was presented as S+ together with various 3-, 6-, 9-, or 11-component submixtures systematically varied in composition, as S−.

A combination of factors was found to contribute to the animals' performance: (1) Discriminability generally decreased as the number of components in the submixture increased. (2) Submixtures did not contribute equally to mixture perception, and one component in particular disproportionately influenced stimulus discriminability. (3) Interactive effects between submixtures resulted in marked deviations from the general pattern of discriminability. (4) Changes in the relative concentration of submixtures could also influence discriminability. (5) Finally, individual differences in responsiveness to particular stimuli were apparent.

Human subjects tested on several of the same discrimination tasks performed very similarly to the monkeys, suggesting that the factors found to affect the performance of the squirrel monkeys may be characteristic of primates more generally. Our findings demonstrate that the interaction between components in odor mixtures can be complex, and that seemingly small changes in their composition may strongly affect the perception and, thus, the potential signal function of semiochemicals.

In future investigations of odor mixture perception in both humans and nonhuman primates, the method of systematically varying and combining submixtures may be particularly useful in defining the contribution of components to a signal.

[1] Institute of Medical Psychology, University of Munich Medical School, Goethestr. 31, 80336 Munich, Germany

Acknowledgments. Supported by the Deutsche Forschungsgemeinschaft (Hu 426/1).

Oxidative Energy Metabolism Supporting Evoked Potentials in Olfactory Cortex Slice

Toshiyuki Saito[1]

Key words. Evoked potentials—Oxidative energy metabolism

It is accepted that neuronal activity in the central nervous system (CNS) depends on oxidative phosphorylation. For the production of energy (adenosine triphosphate; ATP), glucose is utilized as the sole substrate which provides reducing equivalents in the mitochondrial respiratory chain.

Another cell type, the glial cell, surrounds neurons in the CNS. Glial activity such as K^+ uptake and degradation of amino acids also requires an energy supply.

In this study, the author investigated the metabolic activity required to support the evoked potential and participation of glial cells in a functionally intact brain preparation, olfactory cortex slices.

The olfactory cortex slices were prepared from the brains of Sprague-Dawley rats (200–300 g bw), and perfused with HEPES-buffered Ringer's solution at room temperature, as reported previously [1]. The oxidation-reduction (redox) state of the mitochondrial respiratory enzymes, cytochromes, was measured by reflectance spectrophotometry in the piriform cortex, in which the evoked potential was also recorded [1,2]. The oxygen tension in an oxygen uptake chamber was polarographically measured.

The evoked potential produced by afferent stimulation of the lateral olfactory tract (LOT) was diminished by DL-2-amino-4-phosphono-butyric acid (2-APB, 1 mM), an antagonist of the excitatory action of glutamic acid. The evoked potential also decreased in amplitude in slices exposed to 2-deoxy-D-glucose (2-DG; 10 mM) and aminooxyacetic acid (AOAA; 1 mM). In the presence of AOAA (1 mM), cytochromes were oxidized. On the addition of fluoroacetate (1 mM), an inhibitor of the glial Krebs cycle, the oxidation of cytochromes, especially that of aa_3, was marked. Oxygen consumption was decreased after treatment with each inhibitor. Morphological investigation by the immunoperoxidase technique revealed that glial fibrillary acidic protein (GFAP)-positive cells and fibers existed in the preparation.

These results indicated: (1) that evoked potentials and metabolic activity primarily depend on the supply of external glucose and oxygen, (2) that reducing equivalents in the mitochondrial respiratory chain are also derived from the metabolic pathway linked to the gamma-aminobutyric acid (GABA) shunt, and (3) that glial cells participate in oxygen-dependent metabolic activity in the olfactory cortex slice.

Acknowledgments. The author is grateful to Prof. Dr. T. Kanno for his thoughtful advice. Most of this work was done at the Department of Physiology, Faculty of Veterinary Medicine, Hokkaido University, Sapporo, Japan.

References

1. Saito T (1990) Glucose-supported oxidative energy metabolism and evoked potentials are sensitive to fluoroacetate, an inhibitor of glial tricarboxylic acid cycle in the olfactory cortex slice. Brain Res 535:205–213
2. Kobayashi S, Yoshimura M, Shibahara T, Nakase Y, Yaono S (1981) Scanning spectrophotometry for dynamic study of the organ oxidative metabolism. Biomed Res 2:390–397

[1] National Institute of Animal Industry, Laboratory of Bionics, 2 Ikenodai, Kukizaki-machi, Inashiki-gun, Ibaraki, 305 Japan

Localization of Nitric Oxide Synthase in Mouse Olfactory and Vomeronasal System: A Histochemical, Immunological, and In Situ Hybridization Study

J. Kishimoto[1], E.B. Keverne[3], J. Hardwick[4], and P.C. Emson[2]

Key words. Nitric oxide synthase—In situ hybridization

The recent discovery of the short-lived toxic gas, nitric oxide, as a novel intercellular messenger molecule in various tissues including the CNS [1], and the particularly high activity of nitric oxide synthase (NOS) in the main and accessory olfactory bulbs [2] prompted us to investigate the distribution of NOS in the mouse olfactory bulb and olfactory epithelia, including the vomeronasal organ. This was done using an anti-NOS antibody, NADPH diaphorase histochemistry [3], and in situ hybridization with NOS-specific antisense oligonucleotide probes [2]. A fusion protein expression system (pGEX vector system) [4] was utilized to obtain NOS antigen and, hence, anti-NOS polyclonal antibody. Interneurones containing NOS protein mRNA and exhibiting NADPH diaphorase activity were detected in the plexiform layer of the main olfactory bulb and the granule cell layer of the main and accessory olfactory bulbs. Periglomerular cells in the main olfactory bulb were also NOS-positive, with diaphorase activity and immunostaining for NOS. In contrast, no evidence for NOS expression was found either in the main olfactory epithelium or in the vomeronasal organ, despite the strong diaphorase staining of the surface of the main olfactory epithelium (presumably supporting cells). Polymerase chain reaction amplification experiments for the detection of NOS gene expression further indicated that NOS was expressed in the olfactory bulb, but not in either the main olfactory epithelium or in the vomeronasal organ. The use of an antibody raised against another enzyme, NADPH-P450 oxidoreductase, showed that this protein was strongly expressed in the olfactory epithelium. The activity of this enzyme may account for the diaphorase histochemical staining of the epithelia. The involvement of constitutive neuronal nitric oxide synthase in signalling in olfactory receptor neurons is therefore doubtful, although NOS is clearly expressed in neurons in both main and accessory olfactory bulbs.

[1] Biotechnology Group, Shiseido Basic Research Laboratories, 1050 Nippa-cho, Kohoku-ku, Yokohama, 223 Japan
[2] MRC Molecular Neuroscience Group, Department of Neurobiology, AFRC Babraham Institute, Babraham, Cambridge CB2 4AT, UK
[3] Sub-department of Animal Behaviour, University of Cambridge, High St., Madingley, Cambridge CB3 8AA, UK
[4] Department of Biochemistry, NEOUCOM, 4209, State Rt, 44, Rootstown, OH 44272, USA

References

1. Garthwaite J (1991) Glutamate, nitric oxide, and cell-cell signalling in the nervous system. Trends Neurosci 14:60–67
2. Bredt DS, Hwang PM, Glatt CE, Lowenstein C, Reed RR, Snyder SH (1991) Cloned and expressed nitric oxide synthase structurally resembles cytochrome P-450 reductase. Nature 351:714–718
3. Vincent SR, Kimura H (1992) Histochemical mapping of nitric oxide synthase in the rat brain. Neuroscience 46:615–624
4. Smith DB, Johnson KS (1988) Single-step purification of polypeptides expressed in *Escherichia coli* as fusions with glutathione S-transferase. Gene 67:31–40

Removal of the Vomeronasal Organ: Long-Term Studies of Male and Female Prairie Voles

John J. Lepri, Kimberly A. Veillette, and Kennedy S. Wekesa[1]

Key words. Vomeronasal organ—*Microtus*

Short-term experiments, usually ranging from 10 min to 3 days, have convincingly demonstrated that the surgical removal of the vomeronasal organ (VNX) in rodents leads to impairments in odor-guided social and reproductive behaviors [1]. For example, we found that VNX interferes with male-induced reproductive activation in nulliparous and primiparous female prairie voles, *Microtus ochrogaster* [2,3]. However, there is usually a small percentage, typically 20% or less, of VNX subjects which are *not* reproductively impaired by the surgery, suggesting that long-term tests of VNX animals could reveal whether or not they are merely slower to respond than are animals with intact vomeronasal organs. Accordingly, we conducted two experiments in which VNX male and female prairie voles were paired with intact mates for 8 weeks, thereby providing additional time for reproductive responses to be evoked. In the first experiment, we found that only 2 of 9 VNX males sired offspring during this interval, a considerable reduction in reproductive performance when compared to the breeding success observed in 9 of 12 sham-operated (SHAM) males. Notably, the body, testes, and seminal vesicles mass of the VNX and SHAM males were not significantly different at the end of the 8 weeks, suggesting that the reproductive impairment of the VNX males was the result of disrupted behavioral responses to females. In the second experiment, we found that only 1 of 12 VNX females produced pups during an 8-week interval, compared to 11 of 16 SHAM females. We also found that VNX females engaged in lower levels of perineal sniffing than did SHAM females (data not shown),

although there were no differences in a number of other behaviors, e.g., duration of naso-nasal contact. Previous studies to explore the mechanism of reproductive impairment in VNX females have focused on the ways that vomeronasal chemoreception influences the endocrine and neural control of reproductive behavior. To address this issue, we conducted a final study in which VNX and SHAM females received daily subcutaneous injections of 5 μg estradiol benzoate in 0.05 ml sunflower oil. Both groups readily developed sexual receptivity to males within 1 week: 4 of 5 SHAM and 5 of 6 VNX mated, as evidenced by the presence of sperm in vaginal smears. This result suggests that vomeronasal chemoreception activates the hypothalamic-pituitary-gonadal axis, resulting in the secretion of sufficient amounts of estrogen to promote the expression of sexual receptivity in female prairie voles; following VNX, there is apparently insufficient stimulation of this axis to generate the response. In summary, the importance of vomeronasal chemoreception for the chemosensory coordination of reproduction in prairie voles is more firmly established by the results of these long-term studies, and we have also implicated a potential mechanism related to the vomeronasal organ's role in the stimulation of hormone secretion in females.

References

1. Wysocki CJ, Lepri JJ (1991) Consequences of removing the vomeronasal organ. J Steroid Biochem Molec Biol 39:661–669
2. Lepri JJ, Wysocki CJ (1987) Removal of the vomeronasal organ disrupts the activation of reproduction in female voles. Physiol Behav 40:349–355
3. Wysocki CJ, Kruczek M, Wysocki LM, Lepri JJ (1991) Activation of reproduction in nulliparous and primiparous voles is blocked by vomeronasal organ removal. Biol Reprod 45:611–616

[1] Department of Biology, University of North Carolina at Greensboro, Greensboro, NC 27412, USA

Odor-Induced Puberty and Expression of *c-fos* in the Olfactory Bulb of Female Mice (*Mus musculus*)

RICHARD E. BROWN[1,2], HEATHER M. SCHELLINCK[1], CATHERINE SMYTH[2], and MICHAEL WILKINSON[2,3]

Key words. Puberty acceleration—*c-fos*—Mice

Juvenile female mice exposed to the odors of adult males show accelerated sexual maturation [1,2]. The pheromonal stimulus acts on the hypothalamic-pituitary-gonadal axis by modulating luteinizing hormone, prolactin, and estradiol secretion [3]. The accessory olfactory bulb (AOB) is important in initiating such primer effects [4–6] and the neuro-transmitters glutamate, gamma aminobutyric acid (GABA), and norepinephrine have been implicated in such pheromonal effects [7]. Precisely which neurons are involved in pheromone-induced sexual maturation, however, is unknown. We used fos-like immunoreactivity (FLI) in the olfactory bulb of juvenile female mice to detect cells activated by male odors and by peppermint odors [8].

Groups of juvenile female CD1 mice ($n = 3$ per group; age 28–30 days and weight, 17.5–18.5 g) were exposed to bedding from an adult male mouse, to fresh bedding sprayed with peppermint extract, or to their own bedding for 3 h. Following odor exposure, the mice were anesthetized with Somnotol (0.3 ml; i.p.), their uteri removed and weighed, and their brains removed and sectioned (40 μm coronal sections); immunohistochemistry was performed by standard procedure [9]. The FOS antibody-peroxidase complex was visualized with a diamin-obenzidine-nickel-glucose oxidase technique [10].

FLI was found throughout the granular and mitral cell layers of the AOB of juvenile female mice exposed to the bedding odors of adult males. There was also evidence of FLI in the granular cell layer of the main olfactory bulb immediately adjacent to the AOB in all of these animals. Extensive staining was found in the granular cell layer of the main olfactory bulb (MOB) of the mice exposed to peppermint odors, but little or no staining was found in the AOB of these mice. No mice exposed to their own bedding showed immunostaining for fos in the AOB, and virtually no fos staining was found in the granular cell layer of the MOB of these mice. The odors which activated fos in the AOB also increased uterine weights after 24–48 h of exposure.

The expression of fos in the accessory olfactory bulb was coincident with the exposure of juvenile female mice to the bedding odors of male adults, but not to peppermint odors. This apparent selective activation of *c-fos* by chemical odors and pheromones is consistent with the belief that different neural mechanisms are involved in the processing of socially relevant versus socially irrelevant odors [5,7]. These results indicate that the expression of *c-fos* can be used to detect cells in the olfactory bulb which are activated by specific olfactory stimuli; that *c-fos* expression in distinct cell groups within the olfactory bulb responds to specific odors (male urine vs peppermint), and that odor-activated neural pathways involved in the neuroendocrine changes underlying puberty acceleration may be traced via *c-fos* expression.

References

1. Vandenbergh JG (1983) Pheromonal regulation of puberty. In: Vandenbergh JG (ed) Pheromones and reproduction in mammals. Academic, New York, pp 95–112
2. Brown RE (1985) The rodents I: Effects of odours on reproductive physiology (primer effects). In: Brown RE, MacDonald DW (eds) Social odours in mammals. Clarendon, Oxford, pp 245–344
3. Bronson FH, Desjardins C (1974) Circulating concentrations of FSH, LH, estradiol, and progesterone associated with acute, male-induced puberty in female mice. Endocrinology 98:1101–1108
4. Keverne EB (1983) Pheromonal influences on the endocrine regulation of reproduction. Trends Neurosci 6:381–383
5. Wysocki CJ, Meredith M (1987) The vomeronasal system. In: Finger T, Silver W (eds) The neurobiology of taste and smell. Wiley, New York, pp 125–150
6. Li CS, Kaba H, Saito H, Seto K (1990) Neural mechanisms underlying the action of primer pheromones in mice. Neuroscience 36:773–778
7. Brennan P, Kaba H, Keverne EB (1990) Olfactory recognition: A simple memory system. Science 250:1223–1226
8. Schellinck HM, Smyth C, Brown RE, Wilkinson M (1993) Odor-induced sexual maturation and expression of *c-fos* in the olfactory system of juvenile female mice. Dev Brain Res 74:138–141
9. MacDonald MC, Robertson HA, Wilkinson M (1990) Expression of *c-fos* protein by N-methyl-D-aspartic acid in hypothalamus of immature female rats: Blockade by MK-801 or neonatal treatment with monosodium glutamate. Dev Brain Res 56:294–297
10. Shu S, Ju G, Fan L (1988) The glucose oxidase-DAB-nickel method in peroxidase histochemistry of the nervous system. Neurosci Lett 85:169–171

Departments of [1] Psychology, [2] Physiology and Biophysics, and [3] Obstetrics and Gynecology, Dalhousie University, Halifax, Nova Scotia, Canada, B3H 4J1

c-fos Expression in Vomeronasal Pathways During Mating Behavior in Male Hamsters

Gwen D. Fernandez and Michael Meredith[1]

Key words. Chemosensory—Accessory olfactory

Male hamsters are very dependent on chemosensory input for their mating behavior, especially input from the vomeronasal system. This accessory olfactory system projects to the accessory olfactory bulb (AOB), to the medial (Me) and posterior medial cortical nuclei (PMCN) of the amygdala, the bed nucleus of the stria terminalis (BNST), and the medial preoptic area (MPOA). These are relatively direct projections to central structures important in reproductive behavior, compared to the indirect access of the main olfactory system to these areas. The expression of immediate early gene, *c-fos*, was used to study the role of chemosensory input during mating behavior. Intact hamsters and those with vomeronasal organs removed (VNX) were stimulated with a receptive female, or female hamster vaginal fluid (HVF), or left unstimulated for 45 min. Their olfactory bulbs and brains were processed 45 min later for immunocytochemical detection of FOS, the protein product of *c-fos* and an indicator of neuronal activation. Since mated animals receive a variety of inputs, Fos activation could be attributed to chemosensory inputs, to other sensory inputs, to "mating-related" inputs, or to the motor and integrative aspects of copulatory performance. Contributions from these different sources were studied by examining *c-fos* expression patterns in animals with different sensory inputs and/or different behavior. In animals exposed to HVF, the stimulus is restricted to the vomeronasal and main olfactory systems. Activity related to mating performance and female cues other than HVF are eliminated here. Activation due to non-vomeronasal input was studied in VNX animals stimulated with a female or HVF. Most VNX animals do not mate, but do perform intense chemoinvestigatory behavior. Any FOS activation here would be non-vomeronasal. In this study, two VNX animals did mate and their *c-fos* expression pattern is attributed to mating-related activation and to non-VN inputs. Unstimulated animals placed in clean cages without a female or HVF had little activation and none attributable to mating-related stimuli. FOS activation was studied quantitatively in vomeronasal and main olfactory targets by counting FOS-positive nuclei in vomeronasal and main olfactory targets.

Central vomeronasal targets, AOB and the Me, appeared to be selectively activated in intact animals during mating behavior and HVF stimulation. VNX animals, both those that mated and those that did not mate, had no significant activity in the AOB but had a higher level of activity in the Me than unstimulated animals. VNX animals that did not mate had significantly lower activity in the Me than VNX animals that did mate, and these, in turn, were significantly less activated than intact stimulated animals. Unilateral VNX animals exposed to HVF also showed a decrease in Me activation on the VNX side compared to the intact side. In contrast, the main olfactory system did not appear to be selectively activated after mating or HVF stimulation. Both the stimulated and unstimulated animals had similar levels of activation in the main olfactory bulb and low activity in the posterior lateral cortical amygdala.

Mating animals, whether intact or the rare VNX-mating animal, showed intense activation in the MPOA and in the posterior medial BNST (pmBNST), both rostrally (pr) and caudally (pc). Non mating animals, i.e., animals stimulated with HVF, non-mating VNX animals, or unstimulated animals, showed little or no activation in the MPOA or mBNSTpr. Both mating and non-mating animals that were exposed to chemical stimuli and engaged in chemoinvestigatory behavior, whether mated or not, showed FOS activation in the mBNSTpc. Thus, it is possible that the activation in the MPOA and mBNSTpr is related to mating and the activation in the mBNSTpc is related to chemoinvestigatory behavior. This mBNSTpc activation appears to be non-vomeronasal and could include consequences of some motor/integrative aspect of chemoinvestigation. Mating animals, including VNX, displayed two characteristic clusters of activated nuclei in the posterior Me, possibly reflecting mechanosensory activation.

[1] Program in Neuroscience, Department of Biological Sciences, Florida State University, Tallahassee, FL 32306, USA

452

Social Environment Affects Synapses in the Accessory Olfactory Bulb of Adult Hamster: Quantitative Electron Microscopic Study

Masato Matsuoka[1,3], Yuji Mori[2], Kunio Hoshino[1], and Masumi Ichikawa[3]

Key words. Synaptic plasticity—Vomeronasal system

There is increasing evidence that the complexity of the environment to which animals are exposed has an effect on fine synaptic structure, and that the effect varies among different regions of the central nervous system [1,2]. The vomeronasal system, consisting of the vomeronasal organ, the accessory olfactory bulb (AOB), and the higher vomeronasal centers receiving efferents from the AOB, is now thought to play a critical role in the perception and processing of conspecific chemical signals (pheromones) in mammals [3]. We have reported that differential rearing affects the synapses in the granule cell layer of the AOB and the molecular layer of the medial amygdaloid nucleus in adult rats [4,5]. The AOB is the primary nucleus in the vomeronasal system. Since synapses in the glomeruli of the AOB are formed between axons of vomeronasal sensory cells and dendrites of mitral/tufted (M/T) cells [6], they are the first synaptic contact in the vomeronasal pathway. The present study was therefore carried out to examine whether or not different pheromonal environments could affect synaptic structure in the vomeronasal system.

At 30 days of age, six sets of male littermate hamsters (*Mesocricetus auratus*) were assigned to one of three experimental conditions: (1) isolated condition (IC), where a male animal was housed alone, (2) neighbor condition (NC), where one male was caged together with, but separated from, two females, by wire shields, and (3) social condition (SC), where two males and two females were placed together. The SC males were in unrestricted contact with the two females, and all SC females gave birth to pups. The NC males would perhaps be able to recognize females through visual, auditory, and olfactory senses, but could not come into contact with them. After 60 days of differential rearing, 17 (IC, 6; NC, 5; SC, 6) males were prepared for morphological examination. The area of the somata of the M/T cells in the AOB was measured. Fifty or more synapses in the glomeruli of the AOB in each animal were photographed with an electron microscope. The lengths of synaptic contact zones were measured on printed papers.

The AOB of the adult hamster consists of the vomeronasal nerve layer, glomerular layer, mitral/tufted cell layer, olfactory tract layer, and granule cell layer. The somal area of the M/T cells was $173 \pm 1.4\,\mu m^2$ (mean $\pm$ SEM, $n = 451$) in the IC group, $191 \pm 2.1\,\mu m^2$ ($n = 308$) in the NC group, and $185 \pm 1.7\,\mu m^2$ ($n = 525$) in the SC group. The somal area was larger ($P < 0.01$) in both the SC and NC groups than in the IC group, but there was no significant difference between the SC and NC groups. The lengths of the synaptic contact zones of synapses in glomeruli were $243 \pm 4.9\,nm$ ($n = 283$) in the NC group, $246 \pm 4.6\,nm$ ($n = 325$) in the IC group, and $287 \pm 6.2\,nm$ ($n = 325$) in the SC group. The length was significantly ($P < 0.01$) greater in the SC group than in the IC and NC groups. There was no significant difference between the IC and NC groups. These results demonstrate that exposure to different rearing conditions, in which the pheromonal environment can be substantially different, induces morphological changes in both synapses and neurons in the AOB of adult hamsters [7].

References

1. Calverley RKS, Jones DG (1990) Contributions of dendritic spines and perforated synapses to synaptic plasticity. Brain Res Rev 15:215–249
2. Greenough WT, Chang F-LF (1988) Plasticity of synapse structure and pattern in the cerebral cortex. In: Peters A, Jones EG (eds) Cerebral cortex, vol 7, development and maturation of cerebral cortex. Plenum, New York, pp 391–440
3. Wysocki CJ (1979) Neurobehavioral evidence for the involvement of the vomeronasal system in mammalian reproduction. Neurosci Biobehav Rev 3:301–341
4. Ichikawa M, Matsuoka M, Mori Y (1993) Effect of differential rearing on synapses and soma size in rat medial amygdaloid nucleus. Synapse 13:50–56

[1] Laboratory of Veterinary Reproduction, Tokyo University of Agriculture and Technology, 3-5-8 Saiwai-cho, Fuchu, Tokyo, 183 Japan
[2] Laboratory of Veterinary Ethology, The University of Tokyo, 1-1-1 Yayoi, Bunkyo-ku, Tokyo, 113 Japan
[3] Department of Anatomy and Embryology, Tokyo Metropolitan Institute for Neuroscience, 2-6, Musashidai, Fuchu, Tokyo, 183 Japan

5. Ichikawa M, Matsuoka M, Mori Y (1993) Effects of differential rearing on the structure of perforated synapses in the granule cell layer of the rat's accessory olfactory bulb. Neurosci Res 18:19–25
6. Barber PC, Raisman G (1974) An autoradiographic investigation of the projection of the vomeronasal organ to the accessory olfactory bulb in the mouse. Brain Res 81:21–30
7. Matsuoka M, Mori Y, Hoshino K, Ichikawa M (1993, in press) Social environment affects synaptic structure in the accessory olfactory bulb of adult hamster. Neurosci Res

The Action of Paraventricular Oxytocinergic Neurons on Mitral and Granule Cells of the Rat Olfactory Bulb

Guo-Zhong Yu and Hideto Kaba[1]

Key words. Oxytocin—Olfactory bulb

A significant release of oxytocin has been measured in the olfactory bulb of sheep during parturition [1]. The olfactory bulb has also been implicated in the control of maternal behavior in rodents and maternal recognition in sheep [2]. In order to clarify the role of oxytocin in the processing of olfactory information, we examined the effects of electrical stimulation of the hypothalamic paraventricular nucleus (PVN; 0.4–1.0 mA, 100 Hz, 5–30 pulses) on the spontaneous firing of mitral and granule cells in the main olfactory bulb in the rat.

Ovariectomized Wistar rats ($n = 66$) were used for the experiments. The animals were anesthetized with urethane (1.2 g/kg, IP), placed in a stereotaxic instrument, and had holes drilled in the skull to allow access to the olfactory bulb, lateral olfactory tract (LOT), and PVN and, in some cases, to the medial olfactory tract or medial forebrain bundle. Coaxial bipolar stimulating electrodes were guided into the LOT and the PVN. Single-unit extracellular recordings were made with glass micropipettes containing 0.5 M sodium acetate and Pontamine sky blue dye, in the mitral or granule cell layers. The criteria for antidromic invasion were constant latency, the ability to follow a stimulus pulse pair at greater than 100 Hz, and collision between a spontaneous action potential and a stimulus-evoked potential. Glass micropipettes and 30-gauge stainless steel tubing were used for the microinfusion of oxytocin or the selective oxytocin receptor antagonist, [d(CH$_2$)$_5$, Tyr(Me)^2, Orn^8]-vasotocin (MTOV), and the local anesthetic, lignocaine, respectively. Five-barrel micropipettes with tip diameters of 3–5 μm were used for the microiontophoretic application of drugs. The electrodes were glued to a single-barrel recording electrode so that the tip of the recording electrode protruded by 10–30 μm. One barrel was filled with 4 M NaCl for automatic current balancing to prevent tip polarization artifacts. The remaining barrels contained the following drug solutions: 0.5 mM oxytocin in 154 mM NaCl; 0.5 mM MTOV in 154 mM NaCl; and 154 mM NaCl. A retaining current of −5 nA was applied to drug barrels between periods of ejection.

Of 49 mitral cells tested, 32 (65.3%) showed inhibitory responses to PVN stimulation, with latencies ranging from 4 to 125 s. In contrast, 14 (66.7%) of 21 granule cells tested showed excitatory responses to PVN stimulation, with latencies ranging from 5 to 75 s. Both the inhibitory and excitatory responses were blocked by local infusions of MTOV (10 pmol) into the olfactory bulb. Infusions of 0.5% lignocaine into the medial olfactory tract and the medial forebrain bundle both failed to block mitral and granule cell responses to PVN stimulation. Mitral and granule cell responses to PVN stimulation were replicated by cerebroventricular infusions of oxytocin (100–400 ng, 1–2 μl). Mitral and granule cells were also tested for their sensitivity to the microiontophoretic application of oxytocin and MTOV. Oxytocin inhibited 17 out of 20 mitral cells, with a mean latency of 35 s, while it excited 8 out of 13 granule cells, with a mean latency of 32 s. Mitral and granule cell responses to oxytocin were both blocked by the oxytocin antagonist, MTOV, administered microiontophoretically.

Taken together with findings that specific binding sites for oxytocin were localized in the granule cell layer of the guinea pig olfactory bulb [3], the present results suggest that oxytocin released from nerve terminals originating in the PVN excites granule cells via the cerebrospinal fluid, thereby inhibiting mitral cells in the olfactory bulb.

References

1. Kendrick KM, Keverne EB, Chapman C, Baldwin BA (1988) Intracranial dialysis measurement of oxytocin, monoamine, and uric acid release from the olfactory bulb and substantia nigra of sheep during parturition, suckling, separation from lambs, and eating. Brain Res 439:1–10
2. Keverne EB (1988) Central mechanisms underlying the neural and neuroendocrine determinants of maternal behaviour. Psychoneuroendocrinology 13:127–141
3. Tribollet E, Barberis, Dubois-Dauphin M, Dreifuss JJ (1992) Localization and characterization of binding sites for vasopressin and oxytocin in the brain of the guinea pig. Brain Res 589:15–23

[1] Department of Physiology, Kochi Medical School, Nankoku, Kochi, 783 Japan

Effective Induction of Pregnancy Block by Electrical Stimulation of the Mouse Accessory Olfactory Bulb Coincident with Prolactin Surges

Hideto Kaba, Cheng-Shu Li, Hideo Saito, and Katsuo Seto[1]

Key words. Accessory olfactory bulb—Pregnancy block

Pheromones present in the urine of male mice have significant effects upon the reproductive physiology of female mice. Such effects include the acceleration of puberty, induction of estrus in grouped females, and the blocking of pregnancy in newly mated females. These effects are induced by activation of the accessory olfactory system, with receptors in the vomeronasal organ [1]. Our electrophysiological studies [2] in anesthetized female mice have shown that electrical stimulation of the accessory olfactory bulb (AOB) activates tuberoinfundibular dopaminergic (TIDA) arcuate neurons via the amygdala and, subsequently, the medial preoptic area. The present study examined the effectiveness and critical timing of AOB stimulation in inducing pregnancy block in newly mated female mice.

Adult male and virgin female mice of the Balb/c strain were singly housed with a reversed 12-h:12-h light cycle (lights on 21:00, lights off 09:00). The female mice had had bipolar stimulating electrodes permanently implanted stereotaxically into the bilateral AOBs, while they were anesthetized with Avertin. The electrodes were constructed from 10% iridium-platinum wires 125 μm thick, coated with Teflon except for the cut tip. The females were allowed to mate with a male, mating being confirmed by the presence of a vaginal plug at 21:00, which was designated day 0 of pregnancy. The females were then removed to a clean cage for the remainder of the experiment. Rectangular pulses of 0.2-ms duration and 0.2 mA were delivered at 0.33 Hz to the bilateral AOBs from an isolated stimulator at various times after mating, and for different durations. The stimulus frequency was the same as that previously used in in vivo electrophysiological studies [2]. To demonstrate pregnancy block, mice were killed by cervical dislocation 6 days after mating and the uterine horns were examined for implantation sites. The absence of implantation sites confirmed a positive block to pregnancy.

The olfactory block to pregnancy can be accounted for in terms of the pituitary hormone, prolactin, which is regulated by TIDA neurons [1]. Mating initiates the release of two daily surges of prolactin, of a nocturnal surge peaking about 1 h before lights on, and a diurnal surge peaking about 1 h before lights off. Therefore the timing of AOB stimulation to induce pregnancy block was investigated in relation to mating-induced prolactin surges. Stimulation for 4 h produced a significant level of pregnancy block, 70% ($n = 10$), providing that stimulation was for two 2-h periods (17:00–19:00 on day 1 and 08:00–10:00 on day 2) coincident with prolactin surges, compared to the non-stimulated group (20% pregnancy block, $n = 10$, $P < 0.05$). In contrast, stimulation for 4 h (01:00–03:00 and 14:00–16:00 on day 2) between prolactin surges, or two 1-h periods (17:00–18:00 on day 1 and 08:00–09:00 on day 2) coincident with prolactin surges was without effect. These results provide convincing evidence that prolactin is the hormone mainly responsible for pregnancy block. Our results, taken together with the results of previous studies [2], lead us to suggest that the following sequence of events occurs in newly mated female mice after exposure to the pheromones of a strange male. The chemosignal received via the vomeronasal organ is conveyed to the AOB mitral cells, the first relay neurons. The projections of the AOB mitral cells activate excitatory amino acid receptors within the amygdala, and from here the signal is channelled through the stria terminalis. The stria terminalis neurons release cholecystokinin (CCK), which acts on CCK-B receptors located in the medial preoptic area, thereby causing excitation of TIDA neurons as part of the final common pathway of the accessory olfactory system. The increased release of dopamine from the TIDA neurons into the hypophyseal portal circulation lowers plasma levels of prolactin, which, in the mouse, is luteotrophic. As a secondary consequence, progesterone secretion from the corpus luteum decreases and thereby prevents blastocyst implantation.

References

1. Keverne EB (1983) Pheromonal influences on the endocrine regulation of reproduction. Trends Neurosci 6:381–384
2. Kaba H, Li C-S, Keverne EB, Saito H, Seto K (1992) Physiology and pharmacology of the accessory olfactory system. In: Doty RL, Müller-Schwarze D (eds) Chemical signals in vertebrates, 6. Plenum, New York, pp 49–54

[1] Department of Physiology, Kochi Medical School, Nankoku, Kochi, 783 Japan

The Effect of Microinfusions of Drugs into the Accessory Olfactory Bulb on the Olfactory Block to Pregnancy

HIROYUKI NAKAZAWA[2], HIDETO KABA[1], TAKASI HIGUCHI[1], and SHINPEI INOUE[2]

Key words. Olfactory memory—Glycoprotein

On mating with a male, female mice form a long-term memory for the urinary pheromones of this male. This olfactory memory is of critical biological importance, since it prevents any subsequent exposure to the mate's pheromones from initiating the neuroendocrine mechanism that would terminate pregnancy. Pheromones from strange males, those for which no memory has been formed, activate the accessory olfactory system, thereby causing pregnancy block. It is now established that the formation of this memory is dependent on a neural mechanism within the accessory olfactory bulb (AOB) [1]. We have shown that the infusion of a protein synthesis inhibitor into the AOB blocks memory formation [2]. Our attention has been directed towards the synthesis of glycoproteins during memory formation, since glycoproteins are thought to be important in mediating changes in neuronal connectivity. Jork et al. [3] identified a specific inhibitor of fucoglycoprotein synthesis and a competitive inhibitor to galactose, 2-deoxygalactose, which prevented fucosylation in a galactose-fucose sequence. It has been shown that the central administration of 2-deoxygalactose produces amnesia for a number of learning tasks in chicks and rodents [4]. We therefore examined the significance of fucoglycoprotein synthesis in memory formation at the time of mating. Thirty-one female Balb/c mice had 25-gauge stainless steel cannulae permanently implanted stereotaxically into the AOBs while they were anesthetized with tribromoethanol. To investigate the local effect of the drug in the AOB on the olfactory block to pregnancy we infused 0.5-μl of the drug into females after they had mated with a male of the same strain, during the critical period of memory formation. The females were left with the stud males for 6 h following mating and were then removed to a clean cage. On the afternoon following mating, the stud male was reintroduced to the female's cage and left for 48 h. If pregnancy was blocked by the strange male but not by the familiar stud male, memory formation had clearly taken place. If pregnancy was blocked by the familiar male, then memory formation had not occurred. Bilateral infusions of 2-deoxygalactose (0.1 + 0.1 μmol/each bulb) into the AOB at 0 and 1.5 h, and 3 and 4.5 h after mating produced low levels of pregnancy block following re-exposure to the stud male, of 14% ($n = 7$) and 0% ($n = 5$), respectively. A higher dose (0.5 + 0.5 + 0.5 + 0.5 + 0.5 + 0.5 μmol) of the drug administered immediately following mating and at 1-h intervals still produced a low level of pregnancy block, 0% ($n = 8$), which was significantly lower than that produced by exposure to the strange male (55% pregnancy block, $n = 11$, $P < 0.05$). These results indicate that memory formation still occurred during 2-deoxygalactose administration; the results do not support the viewpoint that fucoglycoprotein synthesis is an essential step in the formation of olfactory memory.

References

1. Brennan P, Kaba H, Keverne EB (1990) Olfactory recognition: A simple memory system. Science 250: 1223–1226
2. Kaba H, Rosser A, Keverne EB (1989) Neural basis of olfactory memory in the context of pregnancy block. Neuroscience 32:657–662
3. Jork R, Grecksch G, Matthies H (1986) Impairment of glycoprotein fucosylation in rat hippocampus: Consequences on memory formation. In: Matthies HJ (ed) Learning and memory: Mechanisms of information storage in the nervous system. Pergamon Oxford, pp 223–228
4. Rose SPR (1991) How chicks make memories: Cellular cascade from c-fos to dendritic remodelling. Trends Neurosci 14:390–397

Departments of Physiology[1] and Neuropsychiatry[2], Kochi Medical School, Nankoku, Kochi, 783 Japan

Convergent Olfactory, Taste/Visceral and Other Sensory Inputs to the Monkey Orbital Cortex

S.T. Carmichael and J.L. Price[1]

Key words. Prefrontal cortex—Primary olfactory cortex

The olfactory pathway from the receptor neurons in the nasal cavity through the olfactory bulb to the primary olfactory cortex is relatively well known, both in rats [1] and in primates [2]. Similarly, the gustatory pathway from receptors on the tongue through the nucleus of the solitary tract and the parvicellular part of the ventroposteromedial thalamic nucleus (VPMpc) to the gustatory cortex has been well described [3]. On the other hand, these pathways do not provide for convergence of olfactory and gustatory sensory information, in spite of the common observation that the sensation of flavor is a combination of olfaction and taste. Furthermore, it is unclear how visceral afferent information, also relayed through the nucleus of the solitary tract, may interact with gustatory and olfactory sensory activity.

The most likely site of convergence of olfactory and gustatory activity is in the orbital cortex. Nine histological methods were used to define the architectonic subdivisions of this region. The caudal orbital surface is occupied by the agranular insular cortex, which can be divided into medial, intermediate, lateral, posteromedial, and posterolateral areas (Iam, Iai, Ial, Iapm and Iapl, respectively; Fig. 1). The primary gustatory cortex (G) lies just lateral to Ial and Iapl. Rostral to the agranular insula, area 13 can be divided into four areas, 13a, 13b, 13m, and 13l. Subdivisions of areas 10, 11, 12, and 14 occupy the more rostral, lateral, and medial parts of the orbital cortex.

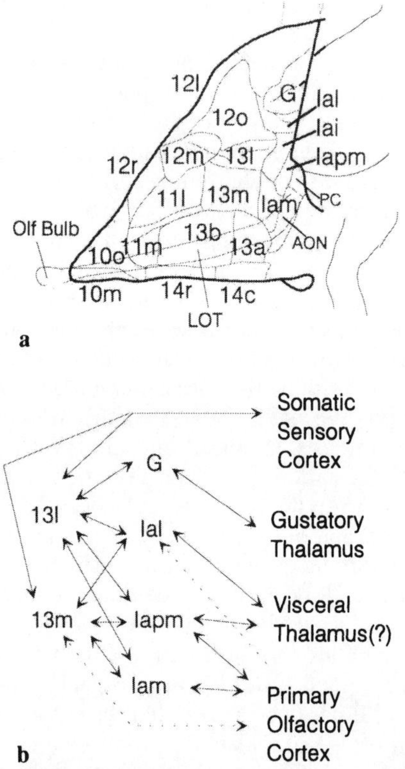

The connections of the primary olfactory cortex, VPMpc, and the orbital cortex in macaque monkeys were defined in experiments with injections of axonal tracers. Anterograde tracer injections into the primary olfactory cortex labeled substantial projections to areas Iam and Iapm (Fig. 1), and lighter projections to Ial, Iai, Iapl, 13m, 13a, 14c, and 25. Retrograde axonal tracer injections into the orbital cortex confirmed these projections, and demon-

Fig. 1. **a** An architectonic map of the orbital cortex in macaque monkeys (see text for abbreviations). **b** Summary of the major convergent connections from the olfactory, gustatory, visceral, and somatic sensory systems onto areas in the caudal and central orbital cortex

strated that they arise from several parts of the primary olfactory cortex, including the anterior olfactory nucleus (AON), piriform cortex (PC), and periamygdaloid cortex. Retrograde tracer experiments also confirmed projections from the dorsal part of the VPMpc to the gustatory cortex, and further demonstrated connections from the ventral part of the VPMpc to areas Ial and Iapm (Fig. 1). Other reports suggest that this ventral part of the VPMpc is a relay for visceral afferent activity [4–6].

Further axonal tracer experiments demonstrated that areas 13m and 13l receive convergent inputs from olfactory, visceral, and gustatory related areas. Area 13l receives input from all of these pathways, while 13m receives olfactory and visceral input.

[1] Department of Anatomy and Neurobiology, Washington University School of Medicine, Campus Box 8108, Washington University School of Medicine, 660 S. Euclid Avenue, St. Louis, MO 63110, USA

458

In addition, both areas are connected to ventral portions of somatic sensory areas 3b, 1–2, and SII that are related to the mouth and hand. Areas 12l and 12m also receive inputs from visual and somatic sensory cortical areas, respectively, and more rostral connections from these areas and from 13m and 13l converge on area 11l. Many of these areas are also connected with the amygdala and other limbic areas. This network appears to provide a mechanism for multisensory and hedonic appreciation of food, and possibly other affective stimuli.

References

1. Price JL (1987) The central olfactory and accessory olfactory systems. In: Finger TE, Silver WL (eds) Neurobiology of taste and smell. John Wiley and Sons, New York, pp 179–204
2. Price JL (1990) The olfactory system. In: Paxinos G (ed) The human nervous system. Academic, San Diego, pp 979–998
3. Norgren R (1984) Central mechanisms of taste. American Physiological Society, Washington DC
4. Blomquist AJ, Benjamin RM, Emmers R (1962) Thalamic localization of afferents from the tongue in squirrel monkey (*Saimiri sciureus*). J Comp Neurol 118:77–88
5. Rogers RC, Novin D, Butcher LL (1979) Electrophysiological and neuroanatomical studies of hepatic portal osmo- and sodium-receptive afferent projections within the brain. J Aut Nerv Syst 1:183–202
6. Beckstead RM, Morse JR, Norgren R (1980) The nucleus of the solitary tract in the monkey: Projections to the thalamus and brainstem. J Comp Neurol 190: 259–282

Emotional Behavior in Cats Induced Olfactorily by matatabi (*Actinidia polygama*)

KIYOAKI KATAHIRA[1]

Key words. Emotional behavior—Olfactory system

The odor of the plant known as matatabi in Japan (*Actinidia polygama*) attracts cats and causes a quick response, in which the animals reach an ecstatic state, sniffing, licking, salivating, rolling around, and rubbing the back against the floor. We designated these responding cats, R-cats. The signs of response seem to be similar to the normal sexual reflexes of cats. The specificity of responsiveness to the matatabi odorant varies widely. There is a minority of adult cats (we designated them, NR-cats) which exhibit little response. The biologically active substances, matatabi-lactone, actinidine, and beta-phenylethylalcohol, have been isolated from the plant [1].

In this study, we constructed a recording apparatus for matatabi response and we investigated the behavioral mechanism underlying the response by using the apparatus. We also investigated the effects of unilateral damage to the rhinencephalon upon matatabi response and sexual and emotional behavior in R- and NR-cats.

The matatabi response was clear in male and female adult R-cats and unclear in kittens. Kittens began responding from the age of about 8 months. The cats were presented with matatabi, and the specific movement of the R-cats was recorded with a static sensograph and an XY-recorder. The matatabi-induced movement ceased after olfactory deprivation.

In three NR- and four R-cats, we produced a unilateral lesion of the rhinencephalon, the amygdala, parts of the inferior temporal neocortex, the temporal pole, the hippocampus, and the pyriform cortex. NR-cats showed the matatabi response clearly 6–32 days after the operation, and responded to matatabi in all presentation trials over a 5-month period. The R-cats showed no change, their performance level being same as that shown preoperatively. When these animals received an aversive stimulus, i.e., electroshock (10–40 V, 50 Hz, AC) administered for 2.5 s through a grill floor during the matatabi response, they ceased responding to it. At higher voltages, the duration of matatabi response blocking in amygdalectomized NR-cats was longer than in control cats.

Fifteen R-cats of both sexes were divided into three groups: amygdalectomized, neocortical control, and unoperated controls. In the neocortical control group, lesions were produced unilaterally or bilaterally in the cortex of the suprasylvian gyrus. The weekly score for the appearance of sexual behavior, i.e., mounting or lordosis, as a ratio of the score on test trials, was referred to as sexual activity. Successive estrus in the amygdalectomized females was observed for a long period following the onset of estrus. In the males, sexual activity scores rose gradually from 2 to 11 weeks after the amygdalectomy. They also mounted other males, homosexually, as well as mounting females. In the control cats, sexual activity scores did not reach a high level.

The specificity of the matatabi response in NR-cats is, conceivably, not due to the smell itself, but to neural mechanisms. The above notion for the exclusion of false effects is supported by the findings of lack of response to matatabi in NR-cats preoperatively stage, and less effective response to matatabi in R-cats following unilateral amygdalectomy. The physiological finding [2] that substances isolated from matatabi influenced the electrical activity of the amygdala and hippocampus in cats also supports the above view.

The results of this study revealed that the mechanism responsible for the release of the matatabi response may be closely related to the central nervous and the olfactory system, regulation of emotional and sexual behaviors. Matatabi response in cats may be a good index for the study of sexual and emotional behavior.

References

1. Sakan T (1967) Matatabi (*Actinidia polygama Miq.*): The isolations and the structures of its biologically active components (in Japanese). Protein, Nucleic Acid and Enzyme 12:2–9
2. Yoshii N, Hano K, Suzuki Y (1963) Effect of certain substances isolated from 'Matatabi' on the EEG of cat. Folia Psychiatr Neurol jap 17:335–350

[1] Experimental Animal Laboratory, Fukushima Medical College, 1 Hikarigaoka, Fukushima, 960-12 Japan

9. Mechanisms of Learning and Memory Involving the Chemical Senses

Memory for Taste Aversion: Molecular and Cellular Mechanisms

STEVEN P.R. ROSE[1]

Key words. Memory—Taste aversion—Chick—Early genes—Glycoproteins—NMDA

Introduction

Young chicks peck spontaneously at small bright objects in their field of view. If the object is a colored bead dipped in a bitter-tasting liquid, the chick subsequently avoids pecking at even a dry bead of that color and shape, although its general pecking activity is unimpaired. It has thus learned to associate the sight of and pecking response to the bead with the aversive taste of the bitter liquid. This behavior forms the basis of the "one-trial passive avoidance" learning task introduced by Cherkin during the 1960s [1] and is an excellent model system in which to study the neural correlates of memory formation. Furthermore, it is not necessary to pair the beadpeck directly with the aversive taste; chicks peck spontaneously at a dry colored bead; and if some time later they are made sick by an intraperitoneal injection of lithium chloride, they subsequently avoid beads of similar color [2]. Retention for these tasks persists up to several days.

Learning to suppress pecking at the bitter bead initiates an intracellular cascade of cellular processes, which, beginning with pre and postsynaptic membrane transients and proceeding by way of genomic activation to the lasting structural modification of these membranes, occurs in identified regions of the chick forebrain. I believe that these synaptic modifications form in some way the neural representations of the aversive bead-pecking experience and encode the instructions for the changed behavior (avoid pecking a bead of these characteristics) that follows [3]. Here I review the key molecular steps identified as occurring in this cascade and discuss what these processes can reveal about the mechanisms and nature of memory storage in vertebrates.

Loci of Change

For the basic experimental design [3,4] day-old chicks are placed in small pens under controlled illumina-

tion and, after a period of equilibration, may be injected intracerebrally with appropriate precursors or potentially amnestic agents—a procedure made simple, without the need for anesthesia, by the chick's soft, unossified skull. The birds are then presented with a small chrome bead dipped either in water (W) or the bitter aversant methylanthranilate (M). At times ranging from 30 min to 24 h after training, chicks are tested and their brains removed for analysis. More than 80% of W birds peck on the test, and more than 75% of M birds avoid pecking. The percent avoidance among the M birds, by comparison with W birds, is taken as a measure of recall.

To localize the areas of the brain that might be involved in the response to pecking at the bitter bead, we gave chicks a 30-min pulse of $[^{14}C]$2-deoxyglucose just prior to or after training on the bead and compared autoradiograms of forebrains from M and W birds. Two regions in particular showed enhanced accumulation of radioactivity during the 30 min after training: the intermediate medial hyperstriatum ventrale (IMHV) and the lobus parolfactorius (LPO). Interestingly, and of considerable relevance to our subsequent studies, there was also evidence of lateralization, with the greatest changes being seen in the left hemisphere regions [5].

Synaptic Transients

Having identified IMHV and LPO as sites of enhanced neural activity during the minutes following training on the passive avoidance task, in subsequent experiments we followed biochemical, physiologic, and morphologic changes in these regions. The working hypothesis is that memory formation requires synaptic modulation, initially by way of changes in receptor activity and subsequently by lasting changes in synapse structure or number. The IMHV is a region of the chick brain rich in NMDA-glutamate receptors, and three lines of evidence point to these receptors being required for memory formation for the passive avoidance task. The first, most direct, is that within 30 min of training the chick there is a significant upregulation of NMDA receptors in the left IMHV [6] with a reciprocal change in accumulation of inositol phosphates via non-NMDA-activated second messenger pathways

[1] Brain and Behaviour Research Group, Open University, Milton Keynes MK7 6AA, UK

[7]. The third piece of evidence is that administration of the noncompetitive NMDA antagonist MK801 or 7 chlorokynurenine, an antagonist of the glycine binding site of the NMDA receptor, just before or just after training results in amnesia for the avoidance task; that is, chicks trained on the bitter bead subsequently peck at it rather than avoid it [8,9].

Upregulation of the NMDA receptor opens postsynaptic calcium channels; and if prisms of IMHV are cut 30 min after training the chicks and incubated in vitro, we can measure an increase in $^{45}Ca^{2+}$ uptake. This increase is probably mediated not through L-type but through conotoxin-sensitive calcium channels, as conotoxin (but not the L-channel blockers nifedipine or nimodipine) produces amnesia for the passive avoidance task if injected at the time of training and abolishes the increased Ca^{2+} flux [10].

Postsynaptic NMDA upregulation is assumed to result in some retrograde message being transmitted to the presynaptic side; and, as is well known, nitric oxide (NO) has become a powerful candidate for one such retrograde messenger. In accordance with this possibility we have shown that blocking NO production with nitroarginine prior to training results in amnesia for the avoidance, and that this amnesia can be alleviated by injecting excess arginine along with the nitroarginine [11]. It seems likely that the signal provided by the NO arriving at the presynaptic side also affects calcium flux at the presynaptic membrane because within the same time frame (30 min following training) there is a change in the phosphorylation state of an exclusively presynaptic, 52-kDa membrane protein immunologically identical to the phosphoprotein variously called B50, GAP 43, neuromodulin, or F1 [12]. The enzyme responsible for the phosphorylation of B50, protein kinase C (PKC), exists in a number of isoforms and is partially membrane-bound and partially cytosolic. One widely canvassed model for the regulation of phosphorylation of its membrane substrates is by way of translocation of the enzyme from cytosol to membrane [13]. We have used a specific antibody to the α/β (translocatable) forms of the enzyme to assay the enzyme in synaptic membranes and found a small but significant increase in the amount of membrane-bound PKC in the left IMHV 30 min after passive avoidance training [14]. Furthermore, if the phosphorylation step is essential for memory formation, it should follow that intracerebral injection of PKC inhibitors such as melittin or H7, which prevent the phosphorylation of B50, should result in amnesia for the passive avoidance. This situation indeed turns out to be the case; localized unilateral injection of melittin or H7 into left, but not right, IMHV just before or just after training results in amnesia in birds tested 6 to 24 h subsequently [14]. Thus within the forebrain the phosphorylation steps seem to be localized to the left IMHV. These processes are analogous to those that occur with long-term potentiation (LTP), where PKC inhibitors have been reported to affect the maintenance (though not the initiation) of the effect [15]. As with LTP, retention of weak memories (e.g., if the 100% methylanthranilate is replaced by a 10% solution of the aversant in alcohol) can be potentiated by the injection of activators of PKC such as phorbol esters around the time of training.

From Synapse to Nucleus

Conversion of such transient modifications of membrane properties into more lasting pre- or postsynaptic modulations of connectivity must depend on the synthesis of new membrane constituents; thus it has been shown in virtually every system in which it has been studied, including the chick [16,17], that the formation of long-term memory is prevented by inhibitors of protein synthesis. The molecular biological mechanisms involved in triggering the synthesis of such membrane proteins are assumed to involve the initial activation of members of the family of immediate early genes of which the protein oncogenes c-fos and c-jun are among the best known. Expression of c-fos and c-jun is believed to be initiated by signals emanating from the membrane, especially the opening of calcium channels and the activation of the PI cycle mediated by the phosphorylation steps described above [18]. Using Northern blotting and in situ hybridization, we have shown that 30 min after M-training c-fos and c-jun mRNAs are induced in IMHV and LPO [19]. This effect seems to be due to learning, not merely experience-related.

From Nucleus to Synapse

Whatever the intervening intracellular signals and genomic mechanisms, within an hour after training there is enhanced synthesis of a variety of proteins (notably tubulin [20,21]) intended for export from the cell body. Most of our attention has been directed toward the glycoproteins of the synaptic membrane because of the major role that several glycoprotein families (e.g., the N-CAMs and integrins) play as cell recognition and adhesion molecules in stabilizing intercellular connections. There is enhanced incorporation of radioactively labeled fucose into pre-and postsynaptic membrane glycoproteins for many hours following training, regulated by increased activity of the rate-limiting enzyme fucokinase [22,23]. Using double-labeling techniques, we have separated the glycoproteins on SDS gels and identified a number of fractions of interest. In particular, in IMHV and LPO a presynaptic component of molecular weight around 50 kDa and postsynaptic components of 100 to 120 kDa and 140 to 180 kDa seem particularly training-sensitive [24]—molecular weights that are interestingly close to those of the N-CAMs.

Training-related glycoprotein synthesis has been reported in several species and tasks; for instance, increased fucosylation has been found following brightness discrimination training in rats by Matthies' group in Magdeburg, who showed that the increased fucosylation occurred in two waves, separated by some 6 h: the first hippocampal, the second cortical in location [25]. We have become intrigued by the possible biochemical and behavioral significance of this "second wave."

As with the phosphorylation and protein synthesis steps of the cascade, we would expect to find that if the synthesis of glycoproteins was a necessary step in the formation of long-term memory inhibiting this synthesis should produce amnesia. 2-Deoxygalactose (2-DGal) is a competitive inhibitor to galactose which, once incorporated into the nascent glycoprotein chain, prevents terminal fucosylation. We have found that intracerebral administration of 2-DGal during a time window of up to an hour or so following training produces amnesia for passive avoidance in the chick [26]. We have gone further, however, and shown that there is a second time window, at some 5 to 8 h after training, at which injections of 2-DGal result in amnesia in chicks tested 24 h subsequently. Furthermore, injections of a monoclonal anti-NCAM, though without effect at the earlier time window, results in amnesia if injected 5.5 h after training [27].

The behavioral significance of these two waves of glycoprotein synthesis in the context of memory for taste becomes apparent from two further experiments. Methylanthranilate is a strong aversant. If it is replaced as the aversant by the less strong tasting quinine, chicks avoid the dry test bead for the first few hours after training, but memory subsequently declines and by 24 h is almost completely lost. Under these conditions, only the first wave of glycoprotein synthesis occurs [28]. A similar effect may be occurring in the conditioned taste aversion paradigm referred to above. Chicks pecking a dry green bead and made sick 30 min later by intraperitoneal injection of 1 M LiCl, avoid the bead on test 3 h after the LiCl injection, though they no longer avoid it if the presentation of the bead is delayed to 24 h. If, however, 2-DGal is injected intracerebrally at the time of the beadpeck (though not at the time of LiCl injection), the chicks are amnesic on test and peck the green bead. Thus even "weak" learning of an unpaired experience, such as pecking at a taste-neutral but conspicuous object, seems to require the first wave of glycoprotein synthesis [2].

Our explanation for these observations is that initial learning of the taste aversion is associated with a cascade of macromolecular processes occurring primarily in the left IMHV of the chick, including at least some synaptic restructuring that demands glycoprotein synthesis. However, if the memory trace is to be more permanently "stamped in," a more lasting synaptic reorganization is required—one that involves the mechanisms of pre-/postsynaptic recognition provided by the neural cell adhesion molecule and presumably other glycoproteins with similar functions, such as the integrins. A similar role has been postulated for the molluscan homolog of N-CAM ("ApCAM") by Kandel and his colleagues [29].

How might this intracellular biochemical cascade "translate" into altered pre- and postsynaptic morphology? Working on the hypothesis that changes in synaptic connectivity might be expressed by changes in the numbers or dimensions of axodendritic synapses, a series of studies by Stewart has quantitatively examined morphological parameters, at the light and electron microscope levels, in IMHV and LPO of chicks 24 h after training on the passive avoidance task [30]. At this time, there is a large (60%) increase in the density of dendritic spines on a class of large multipolar projection neurons of the left IMHV in the M chicks compared with W chicks and somewhat smaller increases in a similar class of cells in the LPO. This lateralized change is superimposed on a left–right asymmetry that already exists in control chicks and is accompanied by a significant increase in the spine head diameter [31].

Stereological analysis of the synapses of the IMHV and LPO has also shown training-related changes, the most striking of which include increases in synapse number of both left and right LPO and a 60% increase in the number of synaptic vesicles per synapse in the left IMHV and left LPO. Changes in synapse number in the IMHV can be detected as early as 1 h after training, but these changes are transient, presumably involving only the first wave of glycoprotein synthesis [32]. The changes in the LPO can be detected as early as 12 h after training. These changes are reminiscent of those found by Greenough and Bailey following exposure of rats to enriched versus impoverished environments [33] and Bailey and Kandel in the synapses involved in the gill withdrawal reflex in *Aplysia*; these authors have argued that changes in vesicle number are relatively transient, whereas those in synapse number are longer-lasting [34].

Physiologic Correlates

Among the structural changes that might be expected were memory for the training experience to be represented by an alteration in the physiologic connectivity of IMHV and LPO synapses. For instance, theoretic calculations show that spine synapses are more effective than shaft synapses in depolarizing the postsynaptic neuron; and the spine head diameter can limit the magnitude of the postsynaptic signal. The IMHV of the young chick shows interesting neurophysiologic plasticity, notably its capacity to express LTP-like phenomena in vitro [35]. In accord with the prediction that the structural modification of synapses induced during memory formation is associated with changes in electrical properties, extracellular recordings from the IMHV (and LPO) of anesthetized chicks during the hours after training with the bitter bead show dramatic time-dependent increases in the incidence of bouts of high-frequency neuronal firing–bursting activity [36].

It is of course essential to show that these substantial biochemical, morphologic, and physiologic changes are in some way directly related to memory formation and are not simply the sequelae of the combination of sensory, motor, and aversive experiences associated with pecking a bitter bead. To check for this possibility we made use of the fact that a brief subconvulsive transcranial electroshock given during the minutes after training on the passive avoidance task results in subsequent amnesia; birds peck on the test. If the electroshock is delayed until around 10 min after training, however, birds show recall on the test [37]. This phenomenon is presumably a consequence of the fact that the earliest phases of memory formation are dependent on transient ionic fluxes at the synapse, which the electroshock disrupts. In any event, this effect makes it possible for us to dissociate the sequelae of the experience of pecking the bead from those of memory for the avoidance by simply comparing our presumed biochemical, morphologic, or physiologic markers in birds that have all pecked the bead and been shocked but, because of the time of administering the shock, some showed recall whereas others did not. Using this paradigm we have shown that enhanced fucosylation, increased spine density, and neuronal bursting occur only in birds showing recall; the mere tasting of the bead is without effect on these markers [38].

Memory Storage: Localized or Distributed?

What do these data have to say about how memories for the taste aversion/passive avoidance response are stored in the brain? The biochemical and cellular cascade described implies, as do most cellular models of memory formation, that a linear sequence of processes in a pair of neurons or, more realistically, in a small ensemble of such neurons, results in lasting modification of synaptic connectivity within the ensemble, a modulation that is the brain representation of some association of events and experiences whose consequences are changed be-

haviors. Other results from our laboratory, including lesion studies (not presented in detail here), however, cast doubt on such a simplistic sequential model [39]. Memory traces are not, it appears, stably located within a single neuronal ensemble but are, at least during their early phases, dynamic and fluid, moving from site to site within the brain. Representations are multiple, and the concept of a fixed locus may be misleading. (See in this context the alternative, global "chaos" model of memory proposed by Skarda and Freeman [40].) During the processing and stabilizing of the memory trace for passive avoidance in the chick, there is sequential activation of the right IMHV and LPO. The second wave of glycoprotein synthesis, probably occurring in the LPO, is reponsible for producing the glycoproteins, including N-CAM, which stabilize the changed synaptic connectivities by creating new synapses or dendritic structures or altering the locations of preexisting ones. Understanding just how this stabilization is achieved now depends on interpreting the time dependencies, identifying the relevant glycoproteins and their cellular functions, and, perhaps more importantly, rethinking the overly simple associationist models of memory that have so far guided experimental approaches to memory mechanisms.

Acknowledgments. The experiments described here have been based on collaborative work by many members of the Brain and Behaviour Research Group over the past decade. I thank all who are cited and those others whose work forms part of the essential background to the results discussed here. Our research has been principally funded through the Open University, MRC, and SERC.

References

1. Cherkin A (1969) Kinetics of memory consolidation: role of amnesic treatment parameters. Proc Natl Acad Sci USA 63:1094–1100
2. Barber AJ, Gilbert DB, Rose SPR (1989) Glycoprotein synthesis is necessary for memory of sickness-induced learning in chicks. Eur J Neurosci 1:673–677
3. Rose SPR (1992) The making of memory. Bantam, Uxbridge
4. Lossner B, Rose SPR (1983) Passive avoidance training increases fucokinase activity in right forebrain tissue of day-old chicks. J Neurochem 41:1357–1363
5. Rose SPR, Csillag A (1985) Passive avoidance training results in lasting changes in deoxyglucose metabolism in left hemisphere regions of chick brain. Behav Neural Biol 44:315–324
6. Stewart MG, Bourne RC, Steele RJ (1992) Quantitative autoradiographic demonstration of changes in binding to NMDA-sensitive [^{3}H]glutamate and [^{3}H]MK801, but not [^{3}H]AMPA receptors in chick forebrain 30 min after passive avoidance training. Eur J Neurosci 4:936–943

7. Bullock S, Rose SPR, Pearce B, Potter J (1993) Training chicks on a passive avoidance task modulates glutamate stimulated inositol phosphate accumulation. Eur J Neurosci 5:43–48

8. Burchuladze R, Rose SPR (1992) Memory formation in day-old chicks requires NMDA but not non-NMDA glutamate receptors. Eur J Neurosci 4:533–538

9. Steele R, Stewart MG (1994, in press) 7-Chlorokynurenine, an inhibitor of the glycine binding site of NMDA receptors inhibits memory formation in day old chicks (*Gallus domesticus*). Behav Neural Biol

10. Clements M, Rose SPR (1994, in press)

11. Holscher C, Rose SPR (1992) An inhibitor of nitric oxide synthesis prevents memory formation in the chick. Neurosci Lett 145:165–167

12. Ali S, Bullock S, Rose SPR (1988) Phosphorylation of synaptic proteins in chick forebrain; changes with development and passive avoidance training. J Neurochem 50:1579–1587

13. Akers RF, Lovinger DM, Colley D, Linden D, Routtenberg A (1986) Translocation of protein kinase C activity after LTP may mediate hippocampal synaptic plasticity. Science 231:587–589

14. Burchuladze R, Potter J, Rose SPR (1990) Memory formation in the chick depends on membrane-bound protein kinase C. Brain Res 535:131–138

15. Reymann KC, Schulzeck K, Dase H, Matthies HJ (1988) Phorbol ester-induced hippocampal long term potentiation is counteracted by inhibitors of PKC. Exp Brain Res 71:227–230

16. Rosenzweig MR, Bennett EL, Martinez JL, et al. (1991) Stages in memory formation in the chick: findings and problems. In: Andrew RJ (ed) Neural and behavioural plasticity. Oxford University Press, New York, pp 394–418

17. Freeman F, Rose SPR (1993) Proceedings 10th BRA Meeting, St. Andrews

18. Chiarugi VP, Ruggiero M, Corradetti R (1989) Oncogenes, protein kinase C, neuronal differentiation and memory. Neurochem Int 14:1–9

19. Anokhin K, Mileusnic R, Shamakina I, Rose SPR (1991) Effects of early experience on c-fos gene expression in the chick forebrain. Brain Res 544:101–107

20. Mileusnic R, Rose SPR, Tillson PJ (1980) Passive avoidance learning results in changes in concentration of and incorporation into colchicine binding proteins in the chick forebrain. J Neurochem 34:1007–1015

21. Scholey AB, Bullock S, Rose SPR (1992) Passive avoidance learning in the young chick results in time- and locus-specific elevations of α-tubulin immunoreactivity. Neurochem Int 21:343–350

22. Sukumar R, Burgoyne RD, Rose SPR (1980) Increased incorporation of ³H-fucose into chick brain glycoproteins following training on a passive avoidance task. J Neurochem 34:1000–1007

23. Rose SPR (1989) Glycoprotein synthesis and post-synaptic remodelling in long-term memory. Neurochem Int 14:299–307

24. Bullock S, Zamani MR, Rose SPR (1992) Characterisation and regional localisation of pre- and post-synaptic glycoproteins of the chick forebrain showing changed fucose incorporation following passive avoidance training. J Neurochem 58:2145–2154

25. Pohle W, Rüthrich HL, Popov N, Mathies H (1979) Fucose incorporation into rat hippocampus structures after acquisition of a brightness discrimination. Acta Biol Med Germ 38:53–63

26. Rose SPR, Jork R (1987) Long-term memory formation in chick is blocked by 2-deoxygalactose, a fucose analogue. Behav Neural Biol 48:246–258

27. Scholey AB, Rose SPR, Zamani MR, Schachner M, Bock E (1993) A role for the neural cell adhesion molecule in a late consolidating phase of glycoprotein synthesis 6 hs following passive avoidance training of the young chick. Neurosci 55:499–509

28. Bourne RC, Davies DC, Stewart MG, Csillag A, Cooper M (1991) Cerebral glycoprotein synthesis and long-term memory formation in the chick (*Gallus domesticus*) following passive avoidance training depends on the nature of the aversive stimulus. Eur J Neurosci 3:243–248

29. Mayford M, Barzilai A, Keller F, Schacher S, Kandel ER (1992) Modulation of an NCAM-related cell adhesion molecule with long-term synaptic plasticity in *Aplysia*. Science 256:638–644

30. Stewart MG (1991) Changes in dendritic and synaptic structure in chick forebrain consequent on passive avoidance learning. In: Andrew RJ (ed) Neural and behavioural plasticity. Oxford University Press, New York, pp 305–328

31. Patel SN, Stewart MG (1988) Changes in the number and structure of dendritic spines, 25 h after passive avoidance training in the chick (*Gallus domesticus*). Brain Res 449:34–46

32. Doubell T, Stewart MG (1993) Short term changes in numerical density of synapses in the intermediate medial hyperstriatum ventrale of the chick following passive avoidance training. J Neurosci 13:2230–2236

33. Greenough WT, Bailey CH (1988) The anatomy of memory: convergence of results across a diversity of tests. Trends Neurosci 11:142–147

34. Bailey CH, Kandel ER (1993) Structural changes accompanying memory storage. Annu Rev Physiol 55:397–426

35. Bradley PM, Burns BD, Webb AC (1991) Potentiation of synaptic responses in slices from the chick forebrain. Proc R Soc Lond [Biol] 243:19–24

36. Mason RJ, Rose SPR (1987) Lasting changes in spontaneous multi-unit activity in the chick brain following passive avoidance training. Neuroscience 21:931–941

37. Benowitz L, Magnus JG (1973) Memory storage processes following one-trial aversive conditioning in the chick. Behav Biol 8:367–380

38. Patel SJ, Rose SPR, Stewart MG (1988) Training induced dendritic spine density changes are specifically related to memory formation processes in the chick, *Gallus domesticus*. Brain Res 463:168–173

39. Rose SPR (1991) Memory—the brain's Rosetta stone? Concepts Neurosci 2:43–64

40. Skarda CA, Freeman WJ (1990) How brains use chaos to make order. Concepts Neurosci 1:275–286

Neural Mechanisms of Conditioned Taste Aversion Revealed by Functional Ablation Procedures

Jan Bures[1]

Key words. Memory—Gustatory cortex—Amygdala—Parabrachial nucleus—Amnesia—Tetrodotoxin

Introduction

Conditioned taste aversion (CTA) is an acquired adaptive reaction protecting animals against repeated ingestion of flavors the intake of which has been followed by illness. The biological importance of CTA is reflected by the robustness of the phenomenon as well as by its ubiquitous presence in the animal kingdom. CTA was introduced into physiological psychology by Garcia [1], and studies have been reported in about 2000 articles [2] and scores of monographs [3] and symposial volumes [4,5]. The main attraction of CTA for neurobiologists is the unusual properties of the underlying learning process.

Multiple Stages of CTA Learning

The aim of CTA is the association between the taste of food (CS) and visceral consequences of its ingestion (US). Because US onset is necessarily delayed by the slow passage of the ingested food through the gastrointestinal tract, digestion, and resorption, the association is not directly formed between the CS and the US but between a CS trace and US. CTA learning proceeds, therefore, in three subsequent stages: (1) processing of gustatory signals and formation of the gustatory short-term memory (GSTM); (2) persistence of the GSTM trace; and (3) association of the GSTM with the visceral signals of poisoning and formation of the gustatory long-term memory (GLTM). The existence of these hypothetical stages cannot be demonstrated with the conventional lesion approach, which affects all stages of CTA acquisition and retrieval. A more adequate analysis must be based on functional ablation methods, that is, transient inactivation of a definite center or circuit for the duration of a particular stage of CTA learning, leaving the brain intact for retrieval testing.

Differential Sensitivity of CTA Stages to Disruption

The long evolutionary history of CTA learning suggests that both the GSTM and GLTM traces are formed in close proximity to or within the gustatory and visceral projections. The differential effect of a lesion on acquisition and retrieval mechanisms can only be demonstrated when it prevents acquisition of new CTAs but does not impair retrieval of old CTAs. The situation cannot be reversed: a lesion blocking CTA retrieval may leave the CTA acquisition intact, but this possibility cannot be ascertained in absence of retrieval. Again, this impasse can be solved by functional ablation procedures.

The first reports showing differential sensitivity of various CTA stages to disruption appeared in the early 1970s. Berger [6] demonstrated that rats anesthetized with pentobarbital shortly after saccharin drinking and poisoned with LiCl under deep anesthesia acquire strong CTA to saccharin. These findings, replicated in a number of laboratories with different general anesthetics and different CS and US [3], indicate that anesthesia does not disrupt the GSTM, does not impair processing of the poison-induced visceral signals, and prevents neither association of the GSTM with the visceral US nor consolidation of the GLTM.

On the other hand, formation of the GSTM was prevented when the taste stimulus was presented to anesthetized animals by i.p. administration of concentrated tastant, for example, 2% saccharin, 1% body weight [7]. Similar results were obtained with other systemically acting disruptive procedures such as hypothermia and anoxia and by various amnesia-eliciting interventions [3].

The foregoing experiments showed that the disruption-resistant stage of the CTA acquisition process does not require active support of arousal and motivational mechanisms. Because no systemic intervention is known that prevents the GSTM–poisoning association, it is possible that the neurotransmitter blockade achieved by systemic drug application was incomplete. This assumption has

[1] Institute of Physiology, Academy of Sciences, Prague, Czech Republic

467

stimulated attempts to find the critical regions by intracerebral application of drugs inducing a well-defined and localized blockade of a specific function.

Gustatory Cortex

Gustatory cortex (GC) lesions elicit CTA impairment [8], which cannot be ascribed, however, to interference with a particular phase of acquisition or retrieval. Early attempts to use cortical spreading depression to disrupt CTA learning showed that taste stimuli applied during bilateral functional decortication do not form the GSTM and are not associated with subsequent poisoning [9]. On the other hand, functional decortication elicited after administration of the gustatory CS and maintained during application of the US did not prevent CTA acquisition. These results indicated that the cerebral cortex is required for the initial processing of the taste signal and GSTM formation, but that it is not necessary for later stages of CTA learning. Cerebral cortex was also shown to contribute to CTA retrieval [10]: rats with strong CTA to saccharin did not display aversive oromotor reactions to infusion of the aversive fluid into the oral cavity through an indwelling cannula when their cortex was blocked by repeated waves of cortical spreading depression.

These results were recently confirmed in experiments with local inactivation of the gustatory cortex by microinjection of tetrodotoxin (TTX). Unlike the functionally decorticated rats, which do not drink spontaneously and have to be force-fed the CS flavor, rats with TTX blockade of the gustatory cortex have almost normal fluid intake. In spite of spontaneous consumption of the saccharin CS, no CTA was acquired by rats with predrinking TTX inactivation of GC. On the other hand, postdrinking GC blockade neither disrupted GSTM nor prevented its association with poisoning [11]. Because it is improbable that synaptic changes underlying memory trace formation can proceed in the TTX-inactivated neural tissue, GC is obviously the locus of neither the GSTM nor the GLTM.

Amygdala

Less clear is the role of amygdala in CTA acquisition and retrieval. Most lesion studies reported attenuation rather than complete disruption of CTA acquisition [12]. Amygdala lesions interfere with retrieval of previously learned CTA, but it is not clear whether the effect is caused by elimination of CTA engrams or to readout failure from interruption of corticoamygdaloid connections [13]. Results obtained with the TTX inactivation of amygdala show that amygdala, unlike GC, is not necessary for recording of the taste signal but that it significantly

contributes to the processing of the visceral CS and to its association with the GSTM [11]. Some amnesic effect can be elicited by posttrial TTX injection, but it disappears within 1.5 h after LiCl injection [14]. CTA retrieval was partly impaired by TTX blockade of amygdala, but the saccharin avoidance was still significant and comparable to that observed after GC blockade.

Parabrachial Nucleus

Another candidate for the locus of the CTA engram storage is the parabrachial nucleus (PBN), a mesencephalic relay center of the gustatory and visceral pathways [15]. PBN lesions were reported to disrupt CTA retrieval or prevent CTA acquisition [16–18]. Various explanations of CTA absence in PBN-lesioned rats have been discussed by Spector [18]:

a. Blockade of the conditioned aversive reaction (cessation of licking and attempts to remove the aversive fluid from the oral cavity)
b. Impairment of the gustatory signal processing
c. Impairment of the visceral signal processing
d. Failure of the integration of taste and visceral signals and of the formation of the corresponding CTA trace.

Some of the foregoing explanations can be directly rejected when using TTX blockade of PBN instead of irreversible damage. Ivanova and Bures [19,20] have shown that PBN blockade induced after saccharin drinking and overlapping with the duration of LiCl poisoning prevents formation of CTA, manifested by absence of CTA during preference testing performed 2 days later. Because the animals have fully recovered from the TTX block during the retention test, the absence of CTA could not be caused by impairment of aversive responding or by reduced taste sensitivity. The latter explanation cannot account for the CTA acquisition failure because the gustatory stimulus was applied to intact animals. PBN blockade could have prevented processing of the visceral signals of poisoning and thus disrupted the GSTM association with US, but CTA was also disrupted by TTX application performed 24 h after LiCl injection [20]. This leaves the memory explanation as the only alternative.

It seems that PBN or some adjacent centers are indispensable for the integration of the GSTM with the visceral US and for the subsequent storage of the CTA engram. The prolonged retrograde amnesia elicited by TTX inactivation of PBN suggests that the block has affected the storage site. The relative importance of PBN for CTA formation is supported by experiments showing that CTA was not prevented by lesions of the nucleus of the solitary tract and of the gustatory thalamus, but was fully suppressed by PBN lesions [18]. While thalamic lesions do not interrupt the gustatory input into the limbic system, CTA present in rats with NTS lesions must result from nongustatory detection of the CS. According to Yamamoto [21], lesions of the GC, amygdala, and gustatory thalamus cause partial impairment of CTA, which is completely blocked by PBN lesion or by combined lesions of GC and amygdala.

The pivotal role of PBN in CTA learning is supported by experiments [22] showing that postdrinking TTX blockade of PBN disrupts CTA elicited by various CS fluids (saccharin, NaCl, quinine) as well as by various US (LiCl, amphetamine, carbachol). Particularly important is the finding that PBN blockade disrupts both the LiCl-based and amphetamine-based CTA, which is or is not accompanied by orofacial rejection responses [23], respectively.

Interaction of PBN with GC and Amygdala

Both the reversible and irreversible PBN lesion studies point to this brain region as to the probable site of the CTA engram, but other experiments indicate that the convergence of the visceral and gustatory inputs in this region [24] is not a sufficient condition for CTA learning. The GC seems to be necessary during the initial presentation of the taste stimulus and amygdala during the action of the visceral stimulus. It is conceivable that during simultaneous activation of the PBN and cortical gustatory projections, corticofugal feedback may cause priming of the corresponding PBN input representing the GSTM. The primed PBN neurons are prepared for several hours to be associated with the visceral signals of poisoning, the effectiveness of which is perhaps also enhanced by centrifugal feedback from amygdala. The feasibility of such priming is supported by experiments showing that novocaine blockade of gustatory cortex modulates PBN responses to taste stimuli [25] and that taste-responsive PBN neurons are activated by electrical stimulation of the gustatory cortex [26].

Similarly, readout of the PBN trace cannot be mediated by direct gustatory input alone, but requires participation of GC, which contains the key required for correct interpretation of the new PBN code during retrieval.

Further insight into the mutual interaction of the GC, amygdala, and PBN during CTA acquisition and retrieval can be obtained in experiments with unilateral inactivation of these brain centers. The importance of ipsilateral connections between PBN and cerebral cortex was demonstrated in rats that drank saccharin during cortical spreading depression in the right hemisphere and received, 15 min later, TTX injection into the right or left PBN. Retention tested 2 days later with intact brain revealed CTA presence or absence in rats in which functional decortication and PBN inactivation were applied to

the same hemisphere or to opposite hemispheres [27]. This indicates that the corticoparabrachial interaction is mainly ipsilateral. Similar analysis of the amygdaloparabrachial and corticoamygdalar interaction during CTA acquisition suggests that it is also implemented by mainly ipsilateral connections.

Nature of the PBN Engram

It is conceivable that in a network the connectivity of which has been recently changed by learning, the TTX-induced impulse blockade interferes with the most recent modifications and that the persistence of such may require continuous input of mediators. Such plastic change was reported by Yamamoto [21] in c-*fos* immunoreactivity studies showing that although saccharin ingestion elicits activation of the central lateral subnucleus of PBN in normal rats, it is the ventral lateral subnucleus that is activated in CTA-trained rats. Switching of the saccharin projection can be caused by long-term synaptic potentiation or depression at the PBN level. Consonant with this assumption is the finding [28] of changing protein kinase C (PKC) concentration in the PBN of CTA-trained rats: the cytosolic PKC was increased and the membrane-bound PKC decreased in comparison with untrained rats. Attempts to disrupt the putative plastic change by PBN microinjection of amnesia-eliciting drugs failed: no attenuation of CTA was achieved by the cholinergic antagonist scopolamine, by the NMDA receptor blockers ketamine, APV, and MK 801, and by the protein synthesis inhibitor cycloheximide. This is not surprising if we take into account that these drugs can serve as US in CTA experiments at systemic dosages eliciting retrograde amnesia in passive or active avoidance paradigms [29–31]. They disrupt CTA acquisition, however, when administered before CS presentation [32].

Conclusion

CTA has been described as an example of a specialized adaptive form of learning [33] differing from the more conventional conditioned responses by the long CS–US delay and by the "preparedness" of the flavor–illness association. While these behavioral peculiarities were gradually absorbed by the current theory of associative learning [34], the evidence reviewed in this article indicates that the claim for CTA exceptionality is supported by neurobiological rather than by behavioral arguments. CTA acquisition in a sequence of stages differentially sensitive to disruption, the robustness of the GSTM–US association proceeding even in deep coma; the indispensability of forebrain for the formation and retrieval of the mesencephalic engram and the enigmatic nature of the US and of the GLTM are just the most striking features distinguishing CTA from conventional instances of learning. Much of the confusion in CTA research could be avoided by keeping in mind that a lesion or treatment preventing CTA acquisition may leave stages of the associative process unaffected. Only a detailed analysis of the mechanisms involved may accelerate the separation of trivialities from fundamental discoveries and warrant a more productive direction of the future CTA research.

Acknowledgments. This research was supported by the Czech Academy of Sciences grant 71140.

References

1. Garcia J, Kimeldorf DJ, Koeling RA (1955) A conditioned aversion towards saccharin resulting from exposure to gamma radiation. Science 122:157–159
2. Riley AL, Tuck DL (1985) Conditioned food aversions. A bibliography. Ann NY Acad Sci 443:381–437
3. Bures J, Buresova O, Krivanek J (1988) Brain and behavior. Paradigms for research in neural mechanisms. Wiley, Chichester
4. Barker LM, Best MR, Domjan M (eds) (1977) Learning mechanisms in food selection. Baylor University Press, Waco, TX
5. Braveman NS, Bronstein P (eds) (1985) Experimental assessment and clinical applications of conditioned food aversions. Ann NY Acad Sci 446:1–441
6. Berger BD (1970) Learning in anesthetized rat. Fed Proc 29:749
7. Bures J, Buresova O (1989) Conditioned taste aversion to injected flavor: differential effect of anesthesia on the formation of the gustatory trace and on its association with poisoning in rats. Neurosci Lett 98:305–309
8. Braun JJ (1990) Gustatory cortex: definition and function. In: Kolb B, Tees BC (eds) The cerebral cortex of the rat. MIT Press, Cambridge, MA, pp 407–430
9. Buresova O, Bures J (1973) Cortical and subcortical components of the conditioned saccharin aversion. Physiol Behav 11:435–439
10. Brozek G, Buresova O, Bures J (1974) Effects of bilateral cortical spreading depression on the hippocampal theta activity induced by oral infusion of aversive gustatory stimulus. Exp Neurol 42:661–668
11. Gallo M, Roldan G, Bures J (1993) Differential involvement of gustatory insular cortex and amydala in the acquisition and retrieval of conditioned taste aversion in rats. Behav Brain Res 52:91–97
12. Simbay LC, Boakes RA, Burton MJ (1986) Effects of basolateral amygdala lesions on taste aversions produced by lactose and lithium chloride in the rat. Behav Neurosci 100:455–456
13. Yamamoto T, Azuma S, Kawamura Y (1981) Significance of cortical-amygdalar-hypothalamic connections in retention of conditioned taste aversion in rats. Exp Neurol 74:758–768
14. Roldan G, Bures J (1991) Tetrodotoxin blockade of amygdala prevents acquisition of conditioned taste aversion (abstract). Regional Meeting of International Union of Physiological Sciences, Prague, p 39
15. Norgren R (1985) Taste and the autonomic nervous system. Chem Senses 10:143–161
16. DiLorenzo PM (1988) Long-delay learning in rats with parabrachial pontine lesions. Chem Senses 13:219–229
17. Flynn FW, Grill HJ, Schulkin J, Norgren R (1991) Central gustatory lesions: II. Effects on salt appetite, taste aversion learning and feeding behaviors. Behav Neurosci 105:944–955
18. Spector AS, Norgren R, Grill HJ (1992) Parabrachial gustatory lesions impair taste aversion learning in rats. Behav Neurosci 106:147–161
19. Ivanova SF, Bures J (1990) Acquisition of conditioned taste aversion in rats is prevented by tetrodotoxin blockade of a small midbrain region centered around the parabrachial nuclei. Physiol Behav 48:543–549
20. Ivanova SF, Bures J (1990) Conditioned taste aversion is disrupted by prolonged retrograde effects of intracerebral injection of tetrodotoxin in rats. Behav Neurosci 104:948–954
21. Yamamoto T (1993) Neural mechanisms of taste aversion learning. Neurosci Res 16:181–195
22. Bielavska E, Bures J (1992) Functional blockade of parabrachial nuclei disrupts conditioned taste aversion learning irrespective of the nature of the gustatory conditioned (CS) and visceral unconditioned stimulus (US). Physiol Res 41:P23
23. Parker LA (1988) Positively reinforcing drugs may produce a different kind of CTA than drugs which are not positively reinforcing. Learn Motiv 19:207–220
24. Hermann GE, Rogers RC (1985) Convergence of vagal and gustatory afferent input within the parabrachial nucleus of the rat. J Auton Nerv System 13:1–17
25. DiLorenzo PM (1990) Corticofugal influence on taste responses in the parabrachial pons of the rat. Brain Res 530:73–84
26. DiLorenzo PM, Monroe S (1992) Corticofugal input to taste-responsive units in the parabrachial pons. Brain Res Bull 29:925–930
27. Gallo M, Bures J (1991) Acquisition of conditioned taste aversion in rats is mediated by ipsilateral interaction of cortical and mesencephalic mechanisms. Neurosci Lett 133:187–190
28. Krivanek J, Novakova L (1993) Protein kinase C in parabrachial nucleus during conditioned taste aversion (abstract). 7th Meeting of Czech and Slovak Neurochemists, High Tatras, p 38
29. Bammer G (1982) Pharmacological investigations of neurotransmitter involvement in passive avoidance responding: a review and some new results. Neurosci Biobehav Rev 6:247–296
30. Jackson A, Sanger DJ (1989) Conditioned taste aversion induced by phencyclidine and other antagonists of NMDA. Neuropharmacology 28:459–464
31. Squire LR, Emanuel CA, Davis HP, Deutsch JA (1975) Inhibitors of cerebral protein synthesis: dissociation of aversive and amnesic effects. Behav Biol 14:335–341
32. Rosenblum K, Meiri N, Dudai Y (1993) Taste memory: the role of protein synthesis in gustatory cortex. Behav Neural Biol 59:49–56
33. Rozin P, Kalat JW (1971) Specific hungers and poison avoidance as adaptive specializations of learning. Psychol Rev 78:459–486
34. Rescorla RA (1988) Behavioral studies of Pavlovian conditioning. Annu Rev Neurosci 11:329–352

A Neural Model for Taste Aversion Learning

Takashi Yamamoto[1]

Key words. Conditioned taste aversion—C-fos—Saccharin—Learning—Immunohistochemistry—LiCl

Introduction

When ingestion of a food substance is followed by an internal malaise accompanying gastrointestinal disorders or nausea, a learned association between the taste of the ingested substance and the internal consequences is established. Thus animals remember the taste for a long time and thereafter reject foodstuffs with a similar taste. This phenomenon is called conditioned taste aversion (CTA).

CTAs are distinguished from classical conditioning in the following points [1,2]: (1) strong CTAs to novel taste stimuli can be acquired after only one pairing of conditioned stimulus (CS) and unconditioned stimulus (UCS); (2) successful CTAs can be obtained to the CS after delays of as long as several hours between exposure to the CS and delivery of the UCS; and (3) the association between the CS and the UCS can proceed even under conditions of deep anesthesia.

Although CTA is a fundamental feeding-related learned reaction with these unique characteristics, its brain mechanisms are still unresolved. The current study was performed to determine some basic neural substrates for CTAs, using an approach combining behavioral and immunohistochemical techniques.

Materials and Methods

Behavioral Experiment

Wistar male rats weighing 250–300 g were used. In the first week, the rats were deprived of water for 20 h and trained to drink water in a test box to which a single drinking tube was attached. On the conditioning day, the rats received an i.p. injection of 0.15 M LiCl (2% of body weight) soon after 20 min of exposure to 0.01 M sodium saccharin. The animals were then subjected to the same water deprivation,

and the amount of saccharin ingested in 20 min was recorded for the successive 4–5 days. The amount of fluid intake was assessed by measuring the weight difference of each bottle before and after the drinking period. Following the test session, each rat was given tap water for a period of 4 h and 40 min to avoid dehydration.

The rats were anesthetized by sodium pentobarbital (60 mg/kg) and received small amounts (0.2–0.5 µl) of 1% ibotenic acid in confined areas of the brain before or after the conditioning procedure. Lesion sites included both sides of the cortical gustatory area (CGA), thalamic taste area (parvicellular part of posteromedial ventral nucleus of the thalamus, VPMpc), parabrachial nucleus (PBN), hippocampus, amygdala, etc. Each rat had a 1-week recovery period in which water and food were available ad lib. When testing was completed, the rats were killed and perfused through the heart with physiological saline, followed by 10% formalin. Coronal sections of 50 µm were prepared, stained with cresyl violet, and examined microscopically for extent of the lesions.

Immunohistochemical Study

Wistar male rats weighing 150–200 g were used. They were deprived of water for 20 h and trained to drink water in a test box. On the experimental day, the animals were allowed to drink water or one of these palatable taste solutions: 0.5 M sucrose, 0.01 M sodium saccharin, or 0.2 M NaCl. Some animals were conditioned to avoid the saccharin solution with the same CTA procedure as described on the day before the day of the experiment. The aversive stimuli including the conditioned stimulus, 0.02 M HCl, and 0.001 M quinine hydrochloride were delivered to the mouth through an intraoral cannula.

After 1–2 h of presentation of one of the liquids, the rats were deeply anesthetized and perfused with 4% paraformaldehyde in 0.1 M phosphate-buffer following a perfusion with 0.02 M phosphate-buffered saline (PBS). The brain was removed, immersed in the same fixative for 4–24 h, then in ice-cold 30% sucrose: PBS overnight, and sectioned at a thickness of 50 µm on a freezing microtome. Sections were processed for detection of c-fos-like immunoreactivity by the peroxidase-antiperoxidase method [3].

[1] Department of Behavioral Physiology, Faculty of Human Sciences, Osaka University, 1-2 Yamadaoka, Suita, Osaka, 565 Japan

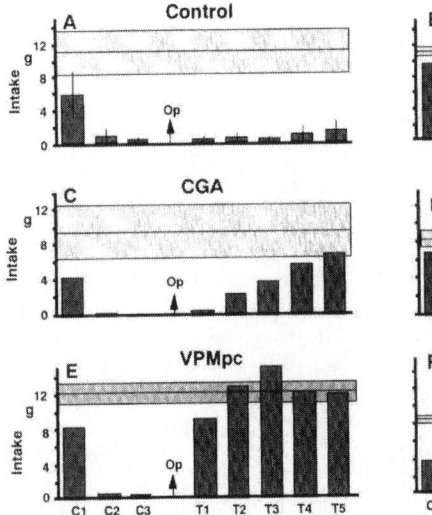
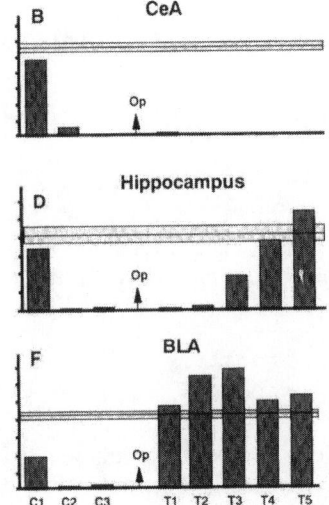

FIG. 1. Acquisition and retention of conditioned taste aversion (CTA) to saccharin in normal rats (**A**), and in rats with lesions of central amygdaloid nucleus (*CeA*) (**B**), cortical gustatory area (*CGA*) (**C**), hippocampus (**D**), parvicellular part of ventral posteromedial nucleus of thalamus (*VPMpc*) (**E**), and basolateral amygdaloid nucleus (*BLA*) (**F**). Each column in graph **A** shows the mean ± SD (*n* = 5) saccharin intake for 20 min on conditioning days (C1–C3) and 5 test days (T1–T5); other graphs are from single representative animals. Horizontal lines show mean ± SD of water intake during training session. Ibotenic acid lesions were made after acquisition of CTA (indicated by *Op*)

Results and Discussion

Behavioral Lesion Experiment

We have extensively surveyed the brain sites responsible for formation of the CTA by the lesion method. In the first behavioral experiment, ibotenic acid lesions were made before acquisition of the CTA. Rats with ibotenic acid (1%, 0.2–0.5 μl) lesions of the PBN (second-order taste relay nucleus) completely failed to establish the CTA. The second most effective site was the VPMpc (third-order taste relay nucleus), followed by the lateral amygdaloid nucleus (LA) and the basolateral amygdaloid nucleus (BLA). Lesions of the CGA and hippocampus induced moderate effects, but lesions of the other subnuclei of the amygdala, entorhinal cortex, bed nucleus of the stria terminalis, substantia innominata, lateral hypothalamic area, and ventromedial hypothalamus induced slight or no effects.

In the next behavioral experiment, ibotenic acid lesions were made after acquisition of the CTA. Figure 1 shows the mean volume of saccharin solution ingested for a 20-min administration period on the successive 3 conditioning days and the subsequent 5 test days in normal control and in brain-lesioned rats. On the first exposure conditioning day, the intake of saccharin in each rat was less than that of water, indicating a cautious intake of the novel taste stimulus (neophobia). As shown in Fig. 1A, the intake of saccharin in control rats was greatly suppressed by the three CTA acquisition procedures, and the aversion was well maintained during the five test sessions (20-min exposure per session; one session per day), although the volume of intake increased gradually from the first to the fifth day. No disruption of retention of the CTA was observed by lesions of the central amygdaloid nucleus (Fig. 1B). Lesions of the CGA or the hippocampus showed a gradual disruption of the retention (Fig. 1C,D), whereas lesions of the VPMpc or BLA showed a dramatic effect; that is, rats with these lesions ingested the saccharin solution as much as water from the first test session (Fig. 1E,F).

It is interesting that essentially the same results were obtained when the ibotenic acid lesions were made before or after CTA procedures. That is, the PBN, VPMpc, and BLA are essential for acquisition or retention of the taste aversion learning.

Immunohistochemical Experiment

To clarify the ascending routes and projection sites of gustatory and general visceral information concerning the CTA, we tried to localize c-*fos*-like protein in the brain as an anatomic marker for activated neurons. C-*fos* is a proto-oncogene that is expressed in neurons following voltage-gated calcium entry into the cell [4]. It is known that neuronal excitation induces immunoreactivity to c-*fos*-like protein in nuclei as a result of a rapid induction of c-*fos* [5,6]. Using this method, we showed that a functional segregation exists in the PBN. For example, intraperitoneal injection of

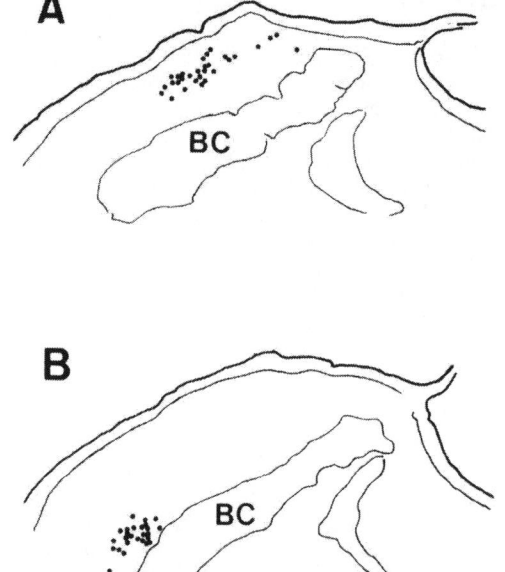

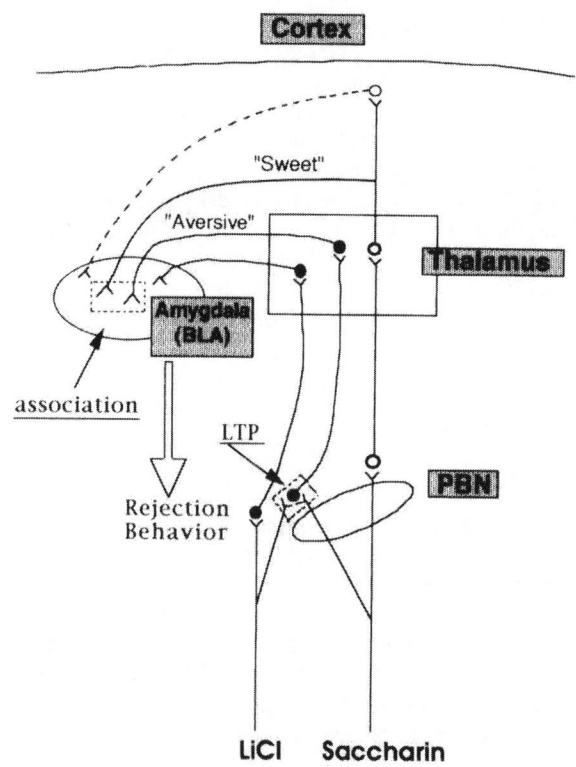

FIG. 2. Camera lucida drawings illustrate expression of c-*fos* neurons (*dots*) in dorsal lateral subnucleus (**A**) and external lateral subnucleus (**B**) of parabrachial nucleus after ingestion of 0.01 M sodium saccharin. Rats in both **A** and **B** received bilateral lesions in cortical gustatory area, thalamic taste area, and amygdala. **B** was obtained from a rat that had received acquisition procedure for conditioned taste aversion. *BC*, brachium conjunctivum

FIG. 3. Diagram of possible neural substrate for formation of conditioned taste aversion. *LTP*, long-term potentiation; *PBN*, parabrachial nucleus; *BLA*, basolateral amygdaloid nucleus

0.15 M LiCl (2% body weight) induced c-*fos*-like immunoreactive neurons (or c-*fos* neurons) dominantly in the external lateral subnucleus and moderately in the dorsal lateral subnucleus of the PBN [3]. The taste information of NaCl induces c-*fos* neurons exclusively in the central medial subnucleus of the PBN [7].

C-*fos* immunoreactivity studies [8–10] in normal rats showed that ingestion of saccharin and other palatable stimuli induced a remarkable activation of the dorsal lateral subnucleus of the PBN, while stimulation with quinine hydrochloride induced activation in the external lateral subnucleus and the lateral portion of the ventral lateral subnucleus of the PBN. On the other hand, rats with CTA to saccharin showed c-*fos* neurons predominantly in the external lateral subnucleus of the PBN after forced ingestion of saccharin. These results suggest that the recipient zone for saccharin taste shifts from the dorsal lateral to external lateral subnuclei after acquisition of CTA.

There is a possibility that activation of taste-hedonic neurons in the dorsal lateral and external lateral subnuclei of the PBN is caused by descending influences from the higher brain centers rather than ascending inputs. To elucidate this point, c-*fos* immunohistochemical study was performed in rats with combined large ibotenic acid lesions of the CGA,

VPMpc, and amygdala including LA and BLA. Blockade of these regions should shut out almost all the ascending and descending taste information from and to the PBN. As shown in Fig. 2A, when a lesioned rat drank the saccharin solution (10.5 ml/h), c-*fos* neurons appeared in the dorsal lateral subnucleus. Because rats with the combined lesions cannot acquire CTA [11], they drink the saccharin even after intake of this solution is paired with i.p. injection of LiCl. When such a rat ingested saccharin (9.0 ml/h), c-*fos* neurons were observed in the external lateral subnucleus of the PBN (Fig. 2B). It is of great interest that even though the rat drank saccharin solution, c-*fos* expression was observed in the region where a hedonically aversive feeling is concerned. These results suggest that hedonic expression of c-*fos* neurons in the PBN can be shown without involvement of the higher central nervous system. Moreover, since both categories of rats took the similar amount of saccharin, drinking behavior itself and its accompanying physiological functions have no essential roles in induction of c-*fos* expression in the PBN.

On the basis of these results showing that the PBN, VPMpc, and BLA are essential in both acquisition and retention of CTA, a possible neural substrate for CTA formation was diagrammed (Fig. 3). Neurons in the external lateral subnucleus of the PBN are concerned with aversive hedonics, and are assumed to receive collateral convergence from general visceral (LiCl) input and taste (saccharin) input. The existence of such convergent neurons was suggested by Hermann and Rogers [12]. Because the saccharin responses in these neurons are assumed to be enhanced by the visceral input in a long-term manner, the taste of saccharin is changed to be hedonically aversive, and this information is transmitted to the amygdala via the thalamus. Taste quality information of saccharin is also sent to the amygdala; thus, a learned association that "saccharin taste is aversive" may be established in the amygdala. According to this notion, therefore, LiCl information serves as an enhancer of saccharin responses in the external lateral subnucleus of the PBN. The importance of the lateral part of the PBN, which includes the external lateral subnucleus, in acquisition of the CTA has been demonstrated recently by Sakai et al. (chapter "Effects of lesions of the medial and lateral parabrachial nuclei on acquisition and retention of conditioned taste aversion", this volume) in behavioral lesion experiments.

Acknowledgments. A part of this work was supported by grants from the Japanese Ministry of Education (03304042, 04454465, 05267101), the Brain Science Foundation, and the Salt Science Research Foundation. I thank S. Azuma, T. Tanimizu, N. Ozaki, and N. Sakai for technical assistance in c-*fos* immunohistochemistry.

References

1. Bures J, Buresova O, Krivanek J (1988) Brain and behavior. Paradigms for research in neural mechanisms. Academia, Prague, pp 185–237
2. Bernstein IL (1991) Flavor aversion. In: Getchell TV, Doty RL, Bartoshuk LM, Snow JB (eds) Smell and taste in health and disease. Raven, New York, pp 417–428
3. Yamamoto T, Shimura T, Sako N, Azuma S, Bai W-Zh, Wakisaka S (1992) C-*fos* expression in the rat brain after intraperitoneal injection of lithium chloride. Neuroreport 3:1049–1052
4. Morgan JI, Curran T (1986) Role of ion flux in the control of c-*fos* expression. Nature (Lond) 322:552–555
5. Morgan JI, Cohen DR, Hempstead JL, Curran T (1987) Mapping patterns of c-*fos* expression in the central nervous system after seizure. Science 237:192–197
6. Mugnaini E, Berrebi AS, Morgan JI, Curran T (1989) Fos-like immunoreactivity induced by seizure in mice is specifically associated with euchromatin in neurons. Eur J Neurosci 1:46–52
7. Yamamoto T, Shimura T, Sako N, Sakai N, Tanimizu T, Wakisaka S (1993) C-*fos* expression in the parabrachial nucleus after ingestion of sodium chloride in the rat. Neuroreport 4:1223–1226
8. Yamamoto T (1992) Central mechanisms of food aversion learning. Neurosci Res [Suppl] 17:S32
9. Yamamoto T, Shimura T, Sako N, Tanimizu T (1992) Expression of C-fos-like protein in the parabrachial nucleus following taste stimulation in the rat. Neurosci Res [Suppl] 17:S265
10. Yamamoto T, Shimura T, Azuma S, Bai W-Zh, Mochizuki R, Wakisaka S (1992) Distribution of C-*fos*-like immunoreactivity in the rat brain following taste stimulation. Chem Senses 17:93
11. Yamamoto T (1993) Neural mechanisms of taste aversion learning. Neurosci Res 16:181–185
12. Hermann GE, Rogers RC (1985) Convergence of vagal and gustatory afferent input within the parabrachial nucleus of the rat. J Auton Nerv Syst 13:1–17

Nerve Growth Factor Accelerates Recovery of Conditioned Taste Aversion Learning by Insular Cortical Grafts

Federico Bermúdez-Rattoni and Martha L. Escobar[1]

Key words. Insular cortex—Conditioned taste aversion—Growth factors—Grafts—Fetal brain transplants—Learning—Memory

Introduction

In our laboratory we have used the conditioned taste aversion paradigm (CTA) to examine the effects of brain grafts on learning in cortically lesioned animals. In this paradigm, normal animals acquire aversion to taste when it is presented as the conditioned stimulus (CS) followed by gastric irritation as an unconditioned stimulus (US) [1]. The CTA has been demonstrated in many laboratories and with many different species of animals. The neuroanatomical pathways for CTA have been established with the use of anatomical, electrophysiological, and behavioral methods [1].

The nucleus solitarius (NTS, the first gustatory relay) receives visceral input as well as inputs from primary taste afferents from the entire tongue via nerves VII and IX and from the larynx and pharynx via nerve X. Neurons responding to both gustatory and visceral stimuli are found in the pontine taste area of the parabrachial complex (second gustatory relay). From the parabrachial nucleus, one major projection of fibers passes to ventral forebrain structures, including the amygdala, lateral hypothalamus, and the substantia innominata, and a second projection passes to the posterior ventromedial and ventromedial nuclei of the thalamus. The thalamic taste area projects to the insular cortex, a region 1 mm wide by 3 mm long located along the rhinal sulcus in the rat. The insular cortex (IC) (Krieg's areas 13 and 14) has been referred to as the gustatory or visceral cortex because it receives taste and visceral information from the thalamus. These anatomical connections suggest that the IC plays an integral role in the mediation of visceral reactions. Moreover, it has been postulated that the IC receives convergence of limbic input with primary sensory inputs that is not seen within any other sensory areas in the cortex [2]. The IC area has been involved in mediating the associations between taste and illness, but is not involved in the hedonic responses to taste. Thus, like normal rats, IC-lesioned animals prefer sucrose as well as low concentrations of sodium chloride over water and reject quinine and acid solutions. Also, taste responsiveness remained intact even in decerebrate or anesthetized rats [3,4].

Functional behavioral recovery from brain injury has recently been demonstrated using the fetal brain transplant technique in adult mammal brains. It has been established that the transplanted neurons differentiate and make connections with the host brain [5]. We have shown that the fetal brain transplants produced a significant recovery in the ability of IC-lesioned rats to acquire a CTA [6]. The possibility of spontaneous recovery was excluded, because the animals with IC lesions that did not receive transplants were unable to acquire the CTA at 8 weeks after the transplant, even with two acquisition series of trials [6]. The findings of another series of experiments recently carried out in our laboratory showed that the degree of functional recovery induced by fetal brain tissue grafts depends on the place from which graft tissue has been taken. Animals that received homotopic but not occipital cortical tissue recovered the CTA. Further, animals that received either tectal heterotopic tissue or no transplant showed no behavioral recovery. Results based on horseradish peroxidase (HRP) histochemistry revealed that cortical but not brain-stem grafts established connections with amygdala and with the ventromedial nucleus of the thalamus [7]. These results suggest that connectivity between an IC graft and the host tissue may play a role in graft-mediated behavioral recovery.

In another series of experiments, we established the recovery-time curve after fetal brain grafts. Thus, in one experiment rats with lesions of IC showing disrupted taste aversions received neocortical transplants and were retrained at 15, 30, 45, and 60 days after transplantation. The behavioral results showed almost complete functional recovery at 60 days, slight recovery at 45 and 30 days, and a poor recovery at 15 days post graft. HRP histochemistry

[1] Departamento de Neurociencias, Instituto de Fisiología Celular, Universidad Nacional Autónoma de México, 04510 Mexico City, Mexico

revealed the absence of HRP-labeled cells in the ventromedial nucleus or into the amygdala at 15 days. At 30, 45, and 60 days post graft the connections with the thalamus and with the amygdala were increasing and becoming almost as those seen in the controls. The behavioral recovery was correlated with increased acetylcholinesterase activity, which was detected histochemically, and with morphological maturation, revealed by Golgi staining [8]. In another experiment our biochemical analyses showed that the IC fetal grafts are able to release acetylcholine (ACh), gamma ammobutyric acid (GABA), and glutamate to K^+ depolarization. However, heterotopical grafts taken from the occipital cortex released GABA and glutamate but not ACh in response to K^+ depolarization at 60 days post graft [9,10].

The implication of these series of experiments is that, for the IC and CTA, functional recovery is related to the morphological maturation or cholinergic activity of the graft. Such findings suggest that if neurotrophic factors are involved they need to be associated with cortical homotopic grafts or with some time-dependent factor essential for producing functional recovery. In this regard, several experiments have shown that the nerve growth factor (NGF) produces a significant regeneration, regrowth, and penetration of cholinergic axons in the hippocampal formation [11,12]. It has been thus demonstrated that constant infusion of NGF in fimbria-fornix-lesioned animals with severe learning deficits produced functional and anatomical recovery [13]. Moreover, it has been demonstrated that aged rats with functional impairments that received constant intracerebral infusions of NGF improved drastically after several weeks of treatment [14]. We have evaluated the role of NGF in the recovery of the ability to acquire CTA induced by IC grafts at three periods after transplantation: 15, 30, and 60 days.

Materials and Methods

Animals

Male Wistar rats weighing about 200–225 g were assigned one per cage and were fed on standard laboratory chow. A 12-h light: dark cycle (lights on at 9:00 A.M.) was maintained throughout the experiment.

Lesions

Stereotaxic surgery was used to create bilateral electrolytic lesions in the IC. Animals were kept under deep pentobarbital anesthesia (50 mg/kg) during the procedure. A 0.5-mm monopolar stainless steel electrode coated with epoxy (except for the tip) was inserted using the following coordinates with respect to bregma: AP, +1.2; L, +5.5; and V, −5.5. The lesions were produced by passing a 2-mA anodal current through the electrode for 45 s.

IC Implants

Sixteen-day-old fetuses were removed from the abdominal cavities of pregnant rats under pentobarbital anesthesia and their brains were extracted. Tissue blocks of approximately 3 mm³ were dissected from the region corresponding to the temporoparietal area (above the rhinal sulcus) for IC implants. Using a 100-μl Hamilton syringe (1700 series, gastight), tissue was placed stereotaxically into the cavity produced by the lesion, using the same set of coordinates given above. Before the implants the tissues were embedded in a high-concentration solution (20 mg/ml) of NGF 7S (Sigma, St. Louis, MO) + DMEM (Dulbecco's modified Eagle's medium, GIBCO, Grand Island, NY)/0.25 BSA (bovine serum albumin) (Sigma) according to Otto et al. [15]. Following the implant, a cap of gelfoam embedded in the same solution was inserted into the cavity. Implant surgery started 10–11 days after the lesion procedure.

Conditioning

The CTA procedure used in this experiment has been described elsewhere [8,10]. In brief, animals were first deprived of water for 24 h and then given 10-min drinking sessions twice a day (9:00 A.M. and 4:00 P.M.) for 2 days. On the third day, was tap water replaced by a 0.1 M LiCl solution to induce taste aversion. There followed four drinking sessions with tap water before a retention test of the CTA using a 0.1 M NaCl solution, the latter being indistinguishable from LiCl by rats [16]. A second retention trial with NaCl was given after two further sessions with tap water. Fifteen days after the implant surgery the CTA procedure was repeated, this time with only one NaCl test. The analyses of all data were made with ANOVA, with post hoc analyses using the Sheffé test for comparisons among groups. P values are usually given only when significant to $P < 0.05$.

Procedure

Several groups of rats showing disrupted taste aversions because of IC electrolytic lesions were grafted as follows. One group received gelfoam + NGF ($n = 15$), the second group received fetal cortical graft + NGF ($n = 15$), the third group received gelfoam + vehicle ($n = 15$), and the fourth group received only fetal cortical graft ($n = 24$). Unoperated control animals were used as a control group (CON; $n = 18$). All the groups were sub-

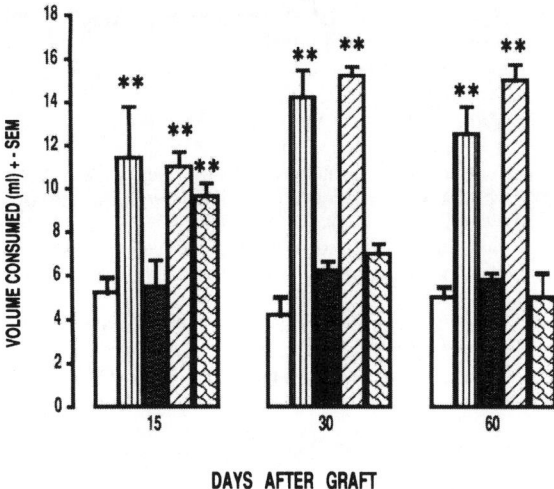

FIG. 1. Amount of NaCl consumed by control (*open bars*), gelfoam + NGF (*vertically hatched bars*), insular cortex (IC) graft + NGF (*shaded bars*), gelfoam + vehicle (*diagonally hatched bars*), and IC graft-alone (*shingled bars*) groups at 15, 30, and 60 days postgraft. **$P < 0.01$ (Sheffé test)

divided in three subgroups that were retrained for CTA at 15, 30, and 60 days, respectively, post graft.

Results

Figure 1 shows the volume of liquid consumed by each group in 15, 30, and 60 days post transplantation. The control groups showed strong taste aversions in all three postgraft times. The IC graft + NGF group showed a significant recovery of ability to acquire the taste aversions at the three postgraft times when compared with the control group. The IC graft groups showed a disrupted taste aversion at 15 days post graft ($P < 0.01$), and recovered the ability to acquire the taste aversion by days 30 and 60 post graft. The groups that received gelfoam + NGF or gelfoam + vehicle remained significantly impaired at the three postgraft times tested as compared with the control group ($P < 0.01$). The results indicate that the application of NGF with the cortical grafts accelerates the ability to acquire the CTA at 15 days post graft.

Discussion

The results presented here clearly shows that the application of NGF with insular cortical grafts accelerates recovery in the ability to acquire conditioned taste aversion learning. It is important to note that previous studies showed that in absence of NGF the insular cortical grafts started to induce the functional recovery at 30 days after implantation, so the

effects of NGF appear to be accelerating recovery. It is noteworthy that the tissue taken at various times after fetal brain transplants without NGF suplementation were at different stages of maturation. That is, at 15 days there was a little development of neurons and blood vessels in the implanted tissue. By 60 days a greater neuronal density had developed; with abundant vascularization, proliferation of glial cells was very apparent, and fibers increasingly crossed the border between the graft and the host tissue [8]. These results, together with those presented here suggested that the behavioral effects are related to the integration and maturity of the implanted tissue. In other studies, we tested the possibility that the combination of NGF with other brain tissue could produce functional recovery in cortical lesioned animals [17; Escobar et al., manuscript in preparation]. Thus, groups of IC-lesioned animals showing disrupted taste aversions received either insular or occipital cortical grafts + NGF or insular grafts + vehicle or remained lesioned as a control group. The results showed that the combination of NGF with insular, but not with occipital cortical grafts, produced recovery in the ability to acquire CTA at 15 days post graft.

The cholinergic neurotransmitter system has long been known to be involved in learning and memory processes [18]. In this regard, several studies have demonstrated that lesions affecting cholinergic innervation of the fimbria fornix produce severe impairments on different learning tasks [13]. Using this model, several authors have demonstrated that exogenous application of NGF can enhance and accelerate regrowth and penetration of cholinergic axons into a lesioned area [12,19], restore rhythmic theta activity in the denervated hippocampus [20], and stimulate the recovery of cholinergic functions [21]. As mentioned, biochemical analyses revealed that IC but not occipital fetal grafts released ACh in response to K^+ depolarization [10]. These results suggest that cholinergic transmission may play a role in graft mediated behavioral recovery.

In a later study we were able to demonstrate that choline acetyltransferase (ChAT) activity in the IC grafts with, but not without, NGF was similar to the normal IC activity of unoperated controls at 15 days post graft [17]. In addition, measurements of GAD activity in the same groups of animals remained similar. We concluded that the NGF associated with IC grafts induced recovery of learning abilities that correlate with the reestablishment of ChAT activity [17].

Acknowledgments. This research was supported by grants from DGAPA-UNAM IN204689, IN201893 and CONACyT 0178-N9107.

References

1. García J, Lasiter PS, Bermúdez-Rattoni F, Deems D (1985) General theory of aversion learning. Ann NY Acad Sci 443:8–20

2. Krushel LA, van der Kooy D (1988) Visceral cortex: integration of the mucosal senses with limbic information in the rat agranular insular cortex. J Comp Neurol 270:34–54

3. Bermúdez-Rattoni, F, Sánchez MA, Pérez J, Forthman D, García J (1988) Odor and taste aversions conditioned in anesthetized rats. Behav Neurosci 102:726–732

4. Kiefer SW (1985) Neural mediation of conditioned food aversion. Ann NY Acad Sci 443:100–109

5. Bjorklund A, Stenevi U (1981) Intracerebral neural implants: neural replacement and reconstruction of damaged circuitries. Brain Res 229:403–428

6. Bermúdez-Rattoni F, Fernández J, Sánchez M, Aguilar-Roblero R, Drucker-Colin R (1987) Fetal brain transplants induce recuperation of taste aversion learning. Brain Res 416:147–150

7. Escobar M, Fernández J, Guevara-Aguilar R, Bermúdez-Rattoni F (1989) Fetal brain grafts induce recovery of learning deficits and connectivity in rats with gustatory neocortex lesions. Brain Res 478:368–374

8. Fernández-Ruíz J, Escobar ML, Piña AL, Díaz-Cintra S, Cintra-McGlone L, Bermúdez-Rattoni F (1991) Time-dependent recovery of taste aversion learning by fetal brain transplants in gustatory neocortex-lesioned-rats. Behav Neural Biol 55:179–193

9. López-García JC, Bermúdez-Rattoni F, Tapia R (1990) Release of acetylcholine, gamma-aminobutyrate, dopamine, acid glutamate and activity of some related enzymes, in rat gustatory neocortex. Brain Res 523:100–104

10. López-García JC, Fernández-Ruíz J, Bermúdez-Rattoni F, Tapia R (1990) Correlation between acetylcholine release and recovery of conditioned taste aversion induced by fetal neocortex grafts. Brain Res 523:105–110

11. Gage FH (1990) NGF-dependent sprouting and regeneration in the hippocampus. Prog Brain Res 83:357–370

12. Hagg T, Vahlsing HL, Manthorpe M, Varon S (1990) Nerve growth factor infusion into denervated adult rat hippocampal formation promotes its cholinergic reinnervation J Neurosci 10:3087–3092

13. Varon S, Hagg T, Vahlsing L, Manthorpe M (1989) Nerve growth factor in vivo actions on cholinergic neurons in the adult rat CNS. In: Cañedo LE, Todd LE, Packer L, Jaz J (eds) Cell function and disease. Plenum, New York, pp 235–248

14. Gage FH, Bjorklund A (1986) Enhanced graft survival in the hippocampus following selective denervation. Neuroscience 17:89–98

15. Otto D, Frotscher M, Unsicker K (1989) Basic fibroblast growth factor and nerve growth factor administered in gelfoam rescue medial septal neurons after fimbria-fórnix transection. J Neurosci Res 22:83–91

16. Nachman M (1963) Learned aversion to the taste of lithium chloride and generalization to other salts. J Comp Physiol Psychol 56:343–349

17. Escobar ML, Jiménez N, López-García JC, Tapia R, Bermúdez-Rattoni F (1993) Nerve growth factor with insular cortical grafts induces recovery of learning and reestablishes graft choline acetyltransferase activity. J Neural Transplant & Plast 4:167–172

18. Decker SB, McGaugh JL (1991) The role of interactions between the cholinergic and other neuromodulatory systems in learning and memory. Synapse 7:151–168

19. Vahlsing HL, Hagg T, Spencer M, Connor JM, Manthorpe M, Varon S (1991) Dose-dependent responses to nerve growth factor by adult rat cholinergic medial septum and neostriatum neurons. Brain Res 552:320–322

20. Buzsaki G, Gage FH, Czopf J, Bjorklund A (1987) Restoration of rhythmic slow activity (theta) in the subcortically denervated hippocampus by fetal CNS transplants. Brain Res 400:334–347

21. Lapchak PA, Jenden DJ, Hefti F (1991) Compensatory elevation of acetylcholine synthesis in vivo by cholinergic neurons surviving partial lesions of the hippocampal pathway. Neuroscience 11:2821–2828

A Comparative Study of Neocortical Olfactory Neuron Responses in Very Young Pups and Adult Dogs

Norihiko Onoda and Kazuyuki Imamura[1]

Key words. Orbitofrontal cortex—Neocortical olfactory area—Odor response—Animal product odor—Dog

Introduction

Response selectivity of neocortical neurons to specific odors has been observed in adult mammals [1,2]. We wondered if neurons of this type were already present in newborn animals or if olfactory experience or learning is required. For technical reasons it is difficult to obtain data from naive animals, i.e., those without olfactory experience. In this study, neocortical neuron responses of very young pups to specific odor patterns were recorded and compared with results from adults.

Methods

Eleven adult dogs (4–7 kg) and ten very young dogs (0.8–3.0 kg) in the preclosure period of the skull, aged 3 months or less, were used. Surgical procedures have been described [1–3]. In brief, a tracheal cannula was inserted caudally for artificial ventilation and a postnasal cannula for artificial air intake was inserted into the rostral end of the trachea under sodium pentobarbital anesthesia (30 mg/kg, i.v.). The head was immobilized on a stereotaxic frame fixed to an overhead bar [4] and the dorsal surface of the frontal cortex was exposed. Fine teflon tubes were tightly inserted into the nasopalatine ducts on both sides. About 6 h after the initial application of sodium pentobarbital, the animal was paralyzed with bromine pancuronium (1 mg/kg/h, i.v.). Extracellular single unit responses of neocortical neurons were recorded by glass micropipettes filled with a dye of brilliant blue in 0.5 M sodium acetate. Air intake into the nasal cavity was induced by applying negative pressure pulses (300-msec duration, 1 Hz) and was monitored by a pressure transducer [4].

The animal's own fresh urine and feces, conspecific fresh urine and feces, dry food pellets for dogs (Oriental Kohbokogyo, Chibaken), and eight pure chemicals were used as odor stimuli. The pure chemicals were as follows: dl-camphor (CM), borneol (BL), cineole (CL), 1,2-dichloroethane (DE), methylcyclopentenolone (CLT), γ-undecalactone (UDL), isovaleric acid (VA), and isoamyl acetate (AA). The chemicals were diluted to a concentration of 10^{-2} in an odorless mineral oil to which a lack of response was verified. All stimulus substances were put into separate syringes. They were applied to the animal's nostrils through the teflon tubing for 3–5 s at about 1-min intervals. Each odor stimulation was repeated at least three times.

Results

We examined the neocortical response of the orbital gyrus to odors. In adult dogs, 24 units showed changes in their discharge rate during the application of at least one of the odors used in this experiment. These units are shown within dotted areas in Fig. 1. From histological checks, the unit location occupied the restricted region of the orbital gyrus and was mainly in the fifth layer.

Odor responses obtained from 24 neocortical neurons of adult dogs are summarized in Fig. 2. A large number of units (91.7%) showed a facilitatory response, 1 unit (4.2%) showed an inhibitory response, and another unit (4.2%) a mixed response. Although not all odor stimuli were tested on each unit, the number of units that responded to one odor was the largest (37.5%). From these results, it appears that in the adult dog orbitofrontal cortex there exist particular neurons, such as a urine neuron and a feces neuron, which respond exclusively to specific odors in a manner similar to neurons found in the rabbit neocortex [1,2]. Ten units (41.7%) responded to odors of pure chemicals. Further, 3 units (12.5%) showed a facilitatory response to only one pure chemical odor but no response to odors of animal products. In puppies, 24 units that responded to at least one of the odors examined were identified in the orbital gyrus. Of 24, 7 units (29.0%) responded to odors of pure chemicals whereas 21 units (87.5%) responded to those of animal products (Fig. 3).

Figure 4 shows a comparison between adult and very young dogs of the number of units that

[1] Department of Physiology, Kanazawa Medical University, Uchinada, Kahoku, Ishikawa, 920-02 Japan

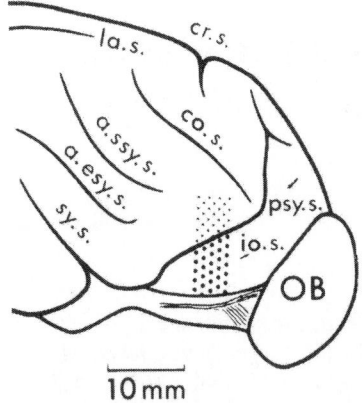

FIG. 1. Lateral view of dog hemisphere. Areas indicated by small and large *dots* show location of units that responded to odors. Small *dotted area* indicates dorsal bank of the orbital gyrus, which is hidden by the anterior compositus gyrus. *a. esy. s.*, anterior ectosylvian sulcus; *a. ssy. s.*, anterior suprasylvian sulcus; *co. s.*, coronal sulcus; *cr. s.*, cruciate sulcus; *io. s.*, intraorbital sulcus; *la. s.*, lateral sulcus; *OB*, olfactory bulb; *psy. s.*, presylvian sulcus; *sy. s.*, sylvian sulcus. (From [5], with permission)

responded to each odor. Of 21, 9 units in adults (42.9%) responded to the odor of strange urine whereas 1 of 13 (7.7%) in pups responded to the same odor.

Discussion

Selective neocortical responses to biologically significant odors were recorded in the middle part of the orbital gyrus of the dog. According to Kreiner's classification, unit location in this study belongs to ORB II a and b [6]. In the rabbit [1,2], 29% of neocortical neurons responded exclusively to odors of animal products and 31% of neurons responded to both odors of animal products and to pure odors, but no units showed any response to pure odors alone. In this study, however, 42% of units responded to both kinds of odors. Further, 3 units showed a facilitatory response to only one pure chemical odor.

Results obtained from the dog are similar to those from the rabbit. Because the number of units is

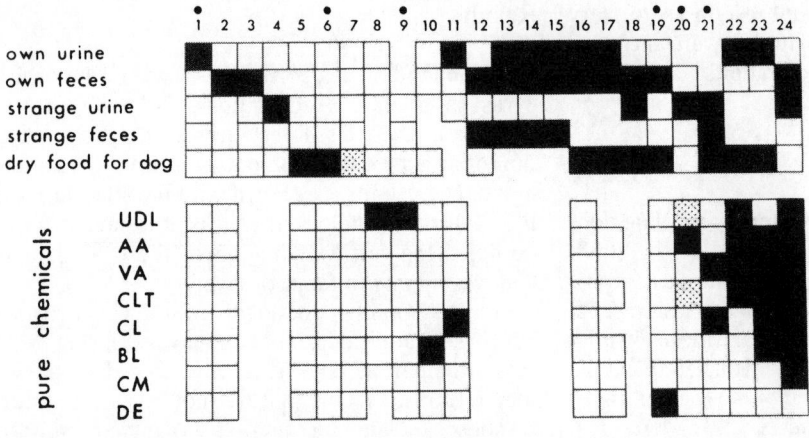

FIG. 2. Panel histogram from 24 units. Number at *top* indicates unit numbers arranged in order of number of odors to which it responded. *Dots* over unit numbers show that nasopalatine ducts on both sides were closed. *Solid squares*, facilitatory response; *open squares*, no response; *dotted squares*, inhibitory response; *CM*, *dl*-camphor; *BL*, borneol; *CL*, cineole; *DE*, 1,2-dichloroethane; *CLT*, methylcyclopentenolone; *UDL*, γ-undecalactone; *VA*, isovaleric acid; *AA*, isoamyl acetate

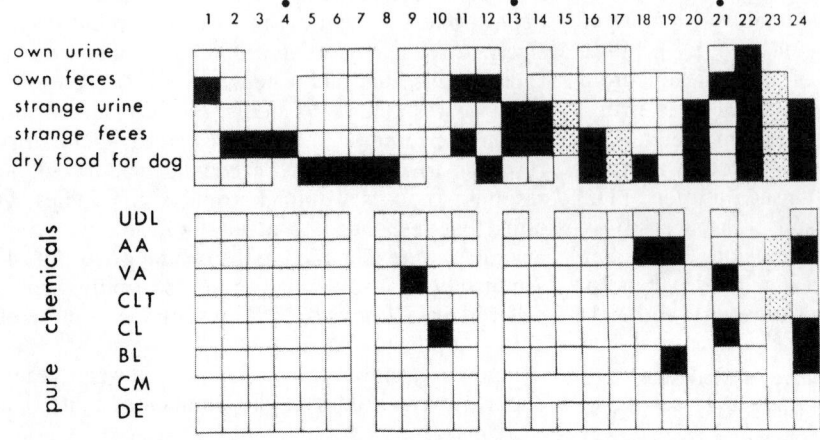

FIG. 3. Panel histogram of neocortical responses to odors obtained from 24 odor-sensitive neurons in very young pups. Number over histogram indicates unit number arranged in order of number of odors to which it responded. *Dots* over unit numbers show that nasopalatine ducts on both sides were closed. *Solid squares*, facilitatory response; *open squares*, no response; *dotted squares*, inhibitory response. (From [5], with permission)

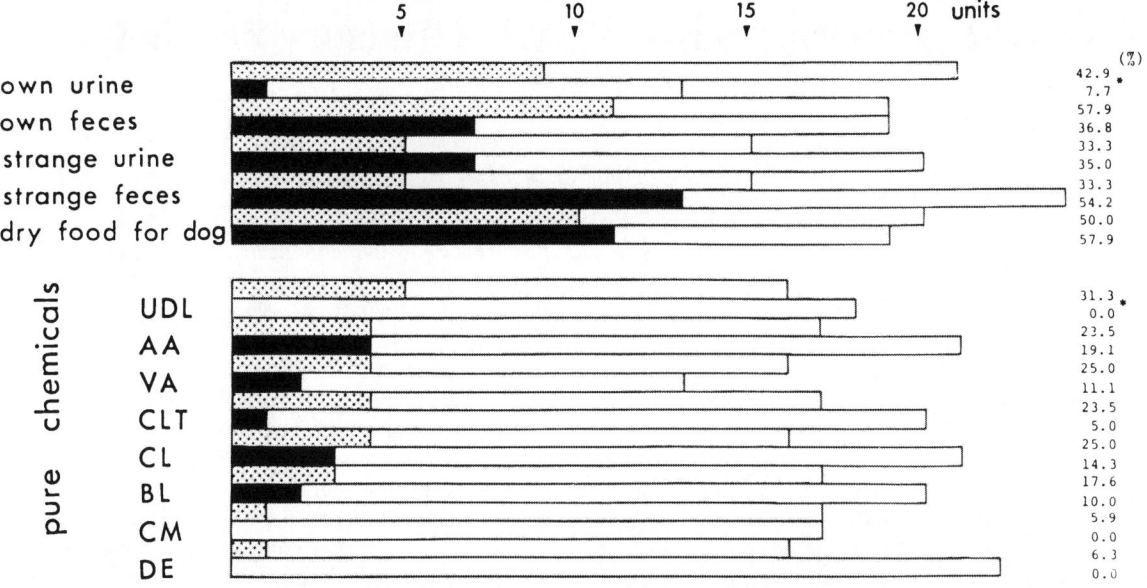

FIG. 4. Comparison of response characteristics of neocortical neurons between adult and very young dogs. *Open column*, number of units tested for each odor; *dotted column*, units in adult dogs responding to each odor; *filled column*, units in very young pups responding to each odor. Numbers in *right-hand column* indicate ratio (%) of number of units that responded to each odor to number of units tested for each odor

small, it is difficult to find a difference of neocortical responses between adult dogs and pups (see Fig. 4). It appears, however, that response selectivity to specific odors is present in newborn animals. It is suggested that the response selectivity to specific odors might be determined by inherent neuronal connections which might be modified by olfactory experience.

References

1. Onoda N, Iino M (1980) Selective responses to odors of animal products in the neocortex neurons of rabbits. Proc Jpn Acad 56:300–305
2. Onoda N, Imamura K, Obata E, Iino M (1984) Response selectivity of neocortical neurons to specific odors in the rabbit. J Neurophysiol 52:638–652
3. Onoda N, Ariki T, Imamura K, Iino M (1982) Neocortical response to odors of sex steroid hormones in the dog. Proc Jpn Acad. 58:222–225
4. Onoda N, Mori K (1980) Depth distribution of temporal firing patterns in olfactory bulb related to air-intake cycles. J Neurophysiol 44:29–39
5. Onoda N, Imamura K, Ariki T, Iino M (1981) Neocortical responses to odors in the dog. Proc Jpn Acad 57B:355–358
6. Kreiner J (1961) The myeloarchitectonics of the frontal cortex of the dog. J Comp Neurol 116:117–133

The Noradrenergic Basis of Early Olfactory Plasticity

Michael Leon[1]

Key words. Olfactory preferences—Development—Olfactory bulb—Early experience—Noradrenaline—Glomeruli

Introduction

A variety of stimuli can facilitate the acquisition of early olfactory preferences, including milk or sucrose ingestion, warmth, high humidity, tail pressure, and medial forebrain bundle stimulation [1–6]. In addition, a strong or prolonged odor exposure will induce such an olfactory preference in young rats [7]. The relatively nonspecific nature of the stimuli that can facilitate early olfactory preference acquisition suggests that there may be a common mechanism underlying these responses.

One possible common mechanism is the activation of the noradrenergic system in the olfactory bulb. A number of facts point toward this possibility. Olfactory bulb noradrenaline is involved in olfactory learning in adult rats, mice, and rabbits (see [8] for a review). There is a large noradrenergic projection to the olfactory bulb [9]. This pathway is present [10] and functional [11] during the first postnatal week. Moreover, tactile stimulation activates the noradrenergic neurons in the locus coeruleus in the first week of life [12], and this stimulation increases bulb norepinephrine during that time [13].

To test this hypothesis, β-noradrenergic antagonists were injected before odor preference training [14,15], and this procedure blocked the behavioral and neural changes characteristic of such learning. Specifically, the rat pups were not attracted to the trained odor, they did not evidence the enhanced uptake of the glucose analogue 2-deoxyglucose (2-DG), and they did not change the firing pattern of olfactory bulb mitral cells in response to the trained odor [14,15]. Placement of the antagonist directly into the bulb similarly blocked the neurobehavioral responses [16]. While activation of the β-noradrenergic receptors is required for the acquisition of the early olfactory preference, it is not required for the expression of either the behavioral preference or the enhanced 2-DG response [17]. Activation of the α-noradrenergic receptors is required for neither the development nor the expression of early olfactory preferences (Woo and Leon, unpublished observations).

If noradrenaline is the critical factor in the facilitation of the acquisition of early olfactory preferences, then it should be possible to substitute a noradrenergic agonist for the reinforcing stimulation and evoke a subsequent olfactory preference. Indeed, pairing the odor with a β-noradrenergic agonist in the absence of tactile stimulation produces the behavioral and neural changes characteristic of normally trained pups [14,15]. Amphetamine, a nonspecific adrenergic agonist, has similar properties when paired with an odor [18].

Both infant and adult rats [19] respond to noradrenaline with a "U"-shaped dose–response relationship. That is, both low and high levels of noradrenaline prevent the learned response. In the case of early olfactory learning, adding together reinforcing stimulation with additional noradrenaline blocks the learning. Specifically, tactile stimulation paired with exogenous noradrenaline and an odor induces neither a behavioral preference nor an enhanced neural response to the trained odor [14,15]. Similarly, combining reinforcing stimulation (tactile stimulation + high humidity [2] or pairing high levels of noradrenergic agonist with an odor prevents the development of the neurobehavioral response to the trained odor [14,15]. Pairing odor with combinations of high temperature, stroking, and noradrenergic agonists block early olfactory preference formation [18]. If the combination of odor + tactile stimulation + noradrenergic agonist blocks the neurobehavioral response by increasing noradrenergic stimulation in an additive manner, then reducing the amount of tactile and agonist should cause the noradrenergic levels to return to an optimal level for learning. Indeed, half the amount of tactile stimulation, combined with half the dose of the noradrenergic agonist, successfully induced the neural and behavioral responses when paired with an odor [14].

While the role of noradrenaline in the establishment of the neurobehavioral changes following early olfactory preference training seems clear, a critical question that remains is how the noradrenergic input affects glomerular layer responses when the noradrenergic input principally reaches subglomerular

---
[1] Department of Psychobiology, University of California, Irvine, CA 92717, USA

lamina in the bulb [9]. Beta receptors have been reported to be at low densities in the bulb [20], although there is a high density of alpha receptors in that structure [21]. β-Noradrenergic receptors drive the second messenger cAMP directly, while alpha receptors potentiate the beta-receptor effect [22,23]. The olfactory bulb has the highest level of alpha-receptor potentiation in the brain [24].

How can the beta antagonists and agonists be effective in the bulb if there are few beta receptors to mediate their action? One reason may be that the developing rat brain expresses β-noradrenergic receptors transiently, and the receptor localization studies have been done only with adult rats. Indeed, a transient expression of these β-noradrenergic receptors is seen in developing rat liver, with the adult liver subsequently developing a predominance of alpha receptors [25]. In fact, bulb α-adrenergic receptor density is low at birth with a gradual postnatal increase [21].

Another problem in understanding the action of β-noradrenergic mechanisms in the developing bulb is that early olfactory experience induces a variety of changes in the glomerular layer of the olfactory bulb [14,26–30], but few noradrenergic fibers reach the glomeruli [9]. To address this question, we determined the developmental pattern of receptor density and localization in the bulb.

β-Noradrenergic stimulation also affects mitral cell firing patterns, and such responses are altered by early olfactory learning [14,31]. Therefore, we wanted to determine whether noradrenergic control is exerted indirectly on the mitral cells via granule cells or directly on mitral cells themselves. To address this question, we localized the cells that express β-noradrenergic receptors in the olfactory bulb.

Materials and Methods

We used three methods to determine the developmental course of the β-noradrenergic receptors. First, we determined the density and binding affinity of β-noradrenergic receptors in olfactory bulb homogenates using the binding of the radiolabeled β-noradrenergic ligand (^{125}I)iodopindolol. The second method involved the use of autoradiography to localize the receptors in the bulb lamina during development. The third approach involved the use of in situ hybridization and Northern blots for β-noradrenergic mRNA using antisense riboprobes to localize the cells that express these receptors.

Results

Analyses of the homogenates revealed that a gradual increase in receptor density in the olfactory bulb from postnatal day 1 (PND 1) to PND 19 [32]. The binding affinity of the receptors remained relatively stable throughout this period. Receptor autoradiography initially showed diffuse labeling throughout the bulb for both the β$_1$- and β$_2$-noradrenergic subtypes, which were revealed by the use of the β$_1$ antagonist ICI 89,406 and the β$_2$ antagonist ICI 118,551 [32]. By PND 12, however, distinct regions of receptors can be seen in the bulb that correspond to the glomerular and granule cell layers, with some labeling in the external plexiform layer. Specific β$_2$ binding also is located in the internal plexiform layer. Discrete glomerular foci for both receptor types are seen by PND 12. These foci increase in number through PND 30 and correspond to the neuropil of individual glomeruli.

Northern blot analysis indicated that the mRNA for the β$_1$ noradrenergic receptor was present in bulb homogenates at a fairly constant level from PND 1 to PND 15 and increased somewhat by PND 19 [33]. In situ hybridization for the β$_1$-receptor mRNA indicated that these receptors were expressed by both juxtaglomerular cells and by cells within the mitral cell layer. Similar patterns of β$_1$ receptor expression were observed from PND 1 to PND 19 in the bulb. β$_2$-Receptor localization is currently being investigated.

Discussion

While the noradrenergic innervation to the glomerular layer of the olfactory bulb is minimal, there is a high density of β-noradrenergic receptors in specific glomerular foci that appear to be expressed by glomerular layer cells. The focal cell population in the glomerular layer increases in number in response to early olfactory learning [29]. Astrocytes are also present in the glomerular layer of the bulb [34], and these cells increase the size of their processes in response to early learning [28]. Astrocytes contain high levels of glycogen [35] and have noradrenergic receptors [36]. Glycogen levels are very high in the neonatal bulb (Coopersmith and Leon, manuscript in preparation) and olfactory bulb glycogen phosphorylase, which is the rate-limiting enzyme in the mobilization of glycogen into glucose is the highest in the brain [37]. Noradrenaline can mobilize glycogen from olfactory bulb, with beta-receptor activation being particularly effective in the first week of life and alpha-receptor activation subsequently most effective (Coopersmith and Leon, manuscript in preparation). It seems possible that the effect of noradrenaline on early olfactory plasticity may involve activation of this glucose store, either to provide a source of energy for the local neurons [38] or to provide energy for a modulating function of astrocytes in the glomerular layer [39].

The granule-layer β-noradrenergic receptor population appears to be expressed by mitral cells. Again, the activity of these cells is altered by early olfactory preference training [31]. The action of noradrenaline on early learning therefore may be mediated by beta receptors expressed by those cells that change in response to such early experiences.

Acknowledgments. This research has been supported in part by grants MH45353A from NIMH, N0001489-1960 from ONR, and HD24236 from NICHD.

References

1. Alberts JR, May B (1984) Nonnutritive, thermotactile induction of filial huddling in rat pups. Dev Psychobiol 17:161–181
2. Do JT, Sullivan RM, Leon M (1988) Behavioral and neural correlates of postnatal olfactory conditioning II: respiration during conditioning. Dev Psychobiol 21:591–600
3. Johanson IB, Teicher M (1980) Classical conditioning of an odor preference in 3-day-old rats. Behav Neural Biol 29:132–136
4. Kenny JT, Blass EM (1977) Suckling as incentive to instrumental learning in preweanling rats. Science 196:898–899
5. Sullivan RM, Brake SC, Hofer MA, Williams CL (1986) Huddling and independent feeding of neonatal rats can be facilitated by a conditioned change in behavioral state. Dev Psychobiol 21:625–635
6. Wilson DA, Sullivan RM (1991) Olfactory associative conditioning in infant rats with brain stimulation as reward: II. Norepinephrine mediates a specific component of the bulb response to reward. Behav Neurosci 105:843–849
7. Leon M, Galef BG, Behse J (1977) Establishment of pheromonal bonds and diet choice in young rats by odor pre-exposure. Physiol Behav 18:387–391
8. Gervais R, Holley A, Keverne B (1988) The importance of central noradrenergic influences on the olfactory bulb in the processing of learned olfactory cues. Chem Senses 13:3–12
9. Shipley MT, Halloran FJ, De la Torre J (1985) Surprisingly rich projection from locus coeruleus to the olfactory bulb in the rat. Brain Res 329:292–299
10. McLean JH, Shipley MT (1991) Postnatal development of the noradrenergic projection from the locus coeruleus to the olfactory bulb in the rat. J Comp Neurol 304:467–477
11. Wilson DA, Leon M (1988) Noradrenergic modulation of olfactory bulb excitability in the postnatal rat. Dev Brain Res 42:69–75
12. Nakamura S, Kimura F, Sakaguchi T (1987) Postnatal development of electrical activity in locus coeruleus. J Neurophysiol 58:510–524
13. Coopersmith R, Weihmuller F, Marshall J, Leon M (1990) Odor and tactile stimulation each increase olfactory catecholamine release. Soc Neurosci Abstr 16:102
14. Sullivan RM, McGaugh JL, Leon M (1991) Norepinephrine-induced plasticity and one-trial olfactory learning in neonatal rats. Dev Brain Res 60:219–228
15. Sullivan RM, Wilson DA, Leon M (1989) Norepinephrine and learning-induced plasticity in the infant rat olfactory system. J Neurosci 9:3998–4006
16. Sullivan RM, Zyzak DR, Skierkowski P, Wilson DA (1992) The role of olfactory bulb norepinephrine in early olfactory learning. Dev Brain Res 70:279–282
17. Sullivan RM, Wilson DA (1991) The role of norepinephrine in the expression of learned olfactory neurobehavioral responses in infant rats. Psychobiology 19:308–312
18. Pedersen PE, Williams CL, Blass EM (1982) Activation and odor conditioning of suckling behavior in 3-day-old albino rats. J Exp Psychol Anim Behav Processes 8:329–341
19. McGaugh JL (1983) Hormonal influence on memory. Ann Rev Psychol 34:297–323
20. Wanaka A, Kiyama H, Murakami T, Matsumoto M, Kamada T, Malbon CC, Tohyama M (1989) Immunocytochemical localization of β-adrenergic receptors in the rat brain. Brain Res 485:125–140
21. Battie CN (1986) Postnatal development of [^{3}H] yohimbine binding in the olfactory bulb of male and female rats. Dev Brain Res 30:129–132
22. Leblanc GG, Ciaranello RD (1984) α-Noradrenergic potentiation of neurotransmitter-stimulated cAMP production in rat striatal slices. Brain Res 293:57–65
23. Stone EA (1987) Central cyclic-AMP-linked noradrenergic receptors: new findings on properties as related to the actions of stress. Neurosci Biobehav Rev 11:391–398
24. Stone EA, Herrera AS (1986) α-Adrenergic modulation of cyclic AMP formation in rat CNS: highest level in olfactory bulb. Brain Res 384:401–403
25. McMillen MK, Schanberg S, Kuhn CM (1983) Ontogeny of rat hepatic adrenoreceptors. J Pharmacol Exp Ther 227:181–186
26. Coopersmith R, Leon M (1984) Enhanced neural response to familiar olfactory cues. Science 225:849–851
27. Coopersmith R, Weihmuller F, Kirstein CL, Marshall JF, Leon M (1991) Extracellular dopamine increases in the neonatal olfactory bulb during odor preference training. Brain Res 564:149–153
28. Matsutani S, Leon M (1993) Elaboration of glial cell processes in the rat olfactory bulb associated with early learning. Brain Res 613:317–320
29. Woo CC, Leon M (1991) Increase in a focal population of juxtaglomerular cells in the olfactory bulb associated with early learning. J Comp Neurol 305:49–56
30. Woo CC, Coopersmith R, Leon M (1987) Localized changes in olfactory bulb morphology associated with early olfactory learning. J Comp Neurol 263:113–125
31. Wilson DA, Leon M (1988) Spatial patterns of olfactory bulb single-unit responses to learned olfactory cues. J Neurophysiol 59:1770–1782
32. Woo CC, Leon M (1993) Distribution of beta adrenergic receptors in the olfactory bulb of young rats. Soc Neurosci Abstr
33. Ivans K, Neve RL, Ibrahim E, Leon M (1993) Localization of β-noradrenergic receptor RNA in the olfactory bulb of young rats. Soc Neurosci Abstr
34. Bailey MS, Shipley MT (1993) Astrocyte subtypes in the rat olfactory bulb: morphological heterogeneity and differential laminar distribution. J Comp Neurol 328:501–526
35. Ibrahim MZ (1975) Glycogen and its related enzymes of metabolism in the central nervous system. Adv Anat Embryol Cell Biol 52:1–85
36. Murphy S, Pearce B (1987) Functional receptors for neurotransmitters on astroglial cells. Neuroscience 22:381–394

37. Coopersmith R, Leon M (1987) Glycogen phosphorylase activity in the olfactory bulb of the young rat. J Comp Neurol 261:148–154
38. Tsacopoulos M, Evequoz-Mercier V, Perrottet P, Buchner E (1988) Honeybee retinal glia cells transform glucose and supply neurons with metabolic substrate. Proc Natl Acad Sci USA 85:872–873
39. Khayari A, Math F, Davrainville JL (1988) Evidence for an absence of K$^+$ spread in the glomerular layer of the rat olfactory bulb. Neurosci Lett 93:56–60

Odor Learning and Odor Memory in the Rat

Key words. Olfaction—Odor discrimination—Odor memory—Odor learning—Entorhinal cortex—Hippocampus

Introduction

For many years it was believed that olfaction played an important role in species-specific but not learned or cognitive behaviors. This view was based largely on the fact that the modality was phylogenetically old, that it appeared to project only to primitive (limbic) parts of the brain, and that olfactory bulbectomy produced profound deficits in reproductive and other social behaviors in macrosmatic species. However, beginning in 1965, the results of both anatomical and behavioral studies forced investigators to reevaluate this position. The initial report of olfactory cortical projections to the thalamus [1] was somewhat surprising because these projections extended to components of the mediodorsal and submedial nuclei, nuclei that projected to prefrontal cortex. Subsequent anatomical studies revealed more clearly the full range of projections from the olfactory bulb and olfactory cortex. Currently, it is clear that olfactory information projects to two well-defined areas of the forebrain that are known to be involved in learning and memory: the prefrontal cortex and the limbic system (amygdala, entorhinal cortex, and hippocampus). Interestingly, among sensory systems olfaction appears to have the most direct access to these forebrain areas.

Perhaps the first studies indicating that olfaction was important in learning were those showing that rats had a preference or ability to use olfactory cues in instrumental learning tasks despite the provision of other seemingly more salient discriminative stimuli [2,3]. Subsequent reports demonstrated that rats are not only adept in learning simple odor detection and discrimination tasks but could learn complex problems that required acquisition of cognitive strategies. The latter included acquisition of multiple problem and serial reversal learning sets, cross-modality transfer, and stimulus matching and nonmatching to sample tasks.

Olfactory Stimulus Control

Initial reports by Jennings and Keefer [4] and Langworthy and Jennings [5] indicated that rats tested in a maze or choice box rapidly learned odor discriminations and showed good interproblem transfer across a series of novel two-odor tasks. These outcomes were replicated and extended in a series of studies using operant conditioning and better control of stimulus variables. The results demonstrated that under appropriate training conditions rats could learn an initial odor discrimination task with few errors and were more attentive to odor cues than to visual or auditory cues [6]. When tested on sequential discrimination problems, rats showed excellent interproblem transfer and many were able to achieve essentially errorless learning. Thus, after sufficient training on odor reversal tasks or novel two-odor discriminations, most rats made no or only 1–2 errors in learning subsequent problems [7–9]. These results indicate that rats are able to acquire a "win-stay, lose-shift" response strategy similar to that obtained with primates tested on visual discrimination tasks. A particularly impressive demonstration of this learning ability is the rapid improvement in learning a series of eight-odor discrimination problems [10]. For rats trained on nine novel eight-odor sets, performance improved gradually and, despite the relatively large number of elements in each set, most rats made few errors in acquiring the last two or three discrimination tasks.

Other, equally important demonstrations of odor-based cognitive learning are those of cross-modality transfer [11] and matching to sample [12,13]. The exceptional performance of rats on such tasks provides an excellent opportunity for examining the neurobiological basis of olfactory learning and memory.

Long-Term Odor Memory

Although it is generally accepted that memories of past events can be evoked by odor stimuli (the "Proust effect"), there are few empirical studies of long-term odor memory in animals. Beauchamp and Wellington [14] used habituation of sniffing to demonstrate that guinea pigs could remember the odor of urine from a particular female for a period of at least 5 weeks. Slotnick and Pazos (unpublished study) found that, in rats trained on an operant task,

---

[1] Department of Psychology, The American University, Washington, DE 20016, USA

memory for an odor was intact after an 8-week rest period, and Slotnick and Risser [15] reported that rats trained on a eight-odor discrimination task (4 S+ and 4 S− odors) had virtually perfect memory of each stimulus element after a 2-week rest period.

In the multiple odor learning set study described [10], rats were retested with odors of set 3 and set 6 after having learned to discriminate among the initial nine different sets of eight odors. Despite the considerable potential for anterograde and retrograde interference effects in these tests, each of the 10 rats demonstrated good retention for the significance of odors used in these earlier problems.

To investigate the locus of the engram for such long-term odor memory, Slotnick and Risser [15] gave rats extensive training on an eight-odor discrimination task and tested for memory 2 weeks after transection of the posterior aspect of the lateral olfactory tract (LOT), destruction of the mediodorsal thalamic nucleus (MD), or combined lesions of the LOT and MD. Rats were trained using partial reinforcement and, to provide an unambiguous test of memory, tested under extinction. Rats with only MD lesions or with only transection of the LOT had virtually perfect retention. However, those with combined lesions performed at chance levels on the retention test. The results indicate that neither the olfactory thalamocortical nor the olfactory projections to the limbic system are essential for retrieval of well-established odor memories. The memory disturbance produced by combined interruption of both these pathways suggests that the olfactory neocortex (orbital frontal cortex) and posterior paleocortex are both involved (and may be equipotential) in the retrieval of odor memories.

Short-Term Odor Memory

A number of studies have demonstrated that rats have excellent short-term memory for the significance of odors they had sampled for only a few seconds. Thus, Staubli et al. [16] reported that rats trained in a maze on a series of novel two-odor tasks performed as well with a 3- to 10-min interval between trials as they did when trials were separated by 0–2 min. In both cases odor discriminations were acquired within 20 trials. This general outcome has been replicated several times under varying conditions in our own laboratory (Slotnick, Thanos, and Lovelace, unpublished study).

Perhaps the procedure most commonly used to assess short-term memory in animals is delayed matching- or nonmatching-to-sample. Although prior studies indicated that rats perform poorly on such tasks when visual stimuli were used [17,18], recent studies demonstrate that quite different results obtain when rats are trained with odors. Lu et al. [12] reported that rats readily learn an olfactory matching-to-sample task. Rats were reinforced for

responding to the second odor of a pair if and only if it was identical to the first odor presented. Rats performed equally well on two- and three-stimulus sets and, over a series of such problems, demonstrated acquisition of a matching-to-sample learning set. Performance was not disrupted by a 10-s delay between stimuli or by presentation of a strong irrelevant odor during this delay.

Longer delay functions were examined by Otto and Eichenbaum [13] using a continuous nonmatching-to-sample method. Rats were trained to respond in the presence of 1 odor if it was different from that presented on the last trial. When a set of 16 different odors was used rats achieved 80% correct responding after only three 100-trial training sessions and a terminal accuracy of 90%. Increasing the delay between stimuli produced a decrease in performance accuracy; however, accuracy was better than 75% with a 60-s delay.

Excellent memory for even longer delays after a brief odor sample was found by Lovelace and Slotnick (manuscript in preparation). Rats were first trained on a series of 16 novel two-odor discriminations using a 10-s intertrial interval (ITI) and then tested for acquisition of other novel odor problems using a 10-min and a 30-min ITI. Rats achieved errorless or near errorless performance by the end of the first 16 problems and were able to acquire the long ITI tasks with few or no errors. Thus, it appears that the rat can remember for at least 30 min whether an odor was associated with reward after a single brief (0.5–1 s) exposure to that odor.

The hippocampus has frequently been implicated in short-term memory and consolidation, and thus it is not surprising that most studies of olfactory short-term memory have concentrated on olfactory projections to the entorhinal cortex. Because the entorhinal cortex receives a direct input from the LOT and, in turn, provides a primary source of afferents to the hippocampus, lesions of the cortex would largely separate the hippocampus from the olfactory system. The results of three independent studies suggest that olfactory projections to the entorhinal cortex play a significant role in olfactory short-term memory. In an initial study, Staubli et al. [16] found that rats with entorhinal cortical lesions had no deficits in learning novel two-odor discrimination tasks when a short (0–2 min) intertrial interval was used but, unlike controls, performed at near chance levels when the intertrial delay was 3–10 min. Further, when these rats were trained on a new pair of odors and tested 1 h later on the reversal of this problem, control rats showed negative transfer but experimental rats rapidly acquired the reversal task. Staubli et al. [16] suggested that these results demonstrate that olfactory denervation of the hippocampus produces rapid forgetting of olfactory information.

These outcomes were largely replicated by Thanos, Wagner, and Slotnick (unpublished study).

Control rats and those with posterior transection of the LOT were trained in a maze on a series of two-odor discrimination problems using a 10-s ITI and then tested on a novel problem using a 10-min ITI. Experimental rats performed as well as controls on the 10-s ITI problems but made many more errors on the 10-min ITI task. However, with additional training, experimental rats were able to reach criterion levels of performance. The results of a reversal task were similar to that described in the Staubli et al. report [16]; memory for the initial learning task was demonstrated for controls but not for experimental rats. The posterior LOT transection used in this study served to deafferent both the amygdala and entorhinal cortex from olfactory input. However, on the basis of the Staubli et al. report and the fact that amygdala lesions have little or no effect on olfactory learning [8], we assumed that the deficits obtained were caused by olfactory denervation of the entorhinal cortex. It is less clear whether, as Staubli et al. suggest, the deficits represent rapid forgetting of olfactory information. This is because experimental rats in the Thanos et al. study (unpublished data) were able to learn an odor discrimination even when a 10-min ITI was used and because factors such as a lesion-induced disinhibition may have contributed to more rapid learning of a reversal task. It should also be noted that lesions used in these studies leave intact olfactory bulb projections to the dorsal hippocampal rudiment.

Perhaps clearer evidence for the role of the entorhinal cortex in short-term memory comes from a recent study by Otto and Eichenbaum [13]. Rats with orbitofrontal lesions, entorhinal lesions, or lesions of the fornix were tested on delayed-nonmatching-to-sample tasks. Rats with orbitofrontal cortex lesions (lesions that destroyed at least part of the frontal cortex receiving an input from the central or olfactory division of MD) performed poorly even under optimal training conditions (short delay, minimal interitem interference). Deficits in the group with entorhinal lesions emerged only with longer (30–60 s) delays between stimuli or a combination of long delays and an increase in opportunity for interitem interference. Interestingly, rats with lesions that destroyed the descending columns of the fornix performed as well as controls in all tasks. According to Otto and Eichenbaum, these results suggest that the orbital frontal cortex may play a role in the acquisition of rules that govern the olfactory nonmatching-to-sample task and that the entorhinal or parahippocampal cortex selectively impairs short-term recognition memory. Because transection of the fornix was without effect in this study, this implicates the parahippocampal cortex itself in mediating the recognition memory required when there is a significant delay between a sample and a comparison stimulus. This view is in accord with studies demonstrating that, in primates, hippocampal lesions which spare the parahippocampal cortex produce only modest deficits in delayed nonmatching-to-sample tasks while lesions which include that cortical area produce profound deficits [19].

Conclusions

In recent years, two sets of data have provided strong evidence that the olfactory system serves other than primitive or species-specific functions in the life of macrosmatic animals. The first is that, when trained with odors, the performance of rats rivals or is qualitatively similar to that of primates in a number of learning tasks including complex discriminations that require excellent memory and acquisition of abstract concepts. Second, the results of anatomical and electrophysiological studies demonstrate that the rat olfactory system has fairly direct projections to many areas of the forebrain which have been implicated in learning and memory in both rodent and primates. These data suggest but, despite the enthusiasm expressed in a number of recent papers (e.g., [13,20], do not yet demonstrate that rodent olfaction can provide a model system for computational neuroscience, studies of memory, and studies directed at better understanding the functions of such structures as the hippocampus and prefrontal cortex in neural plasticity. Certainly there are significant differences among species in anatomical and neurochemical organization among brain structures as well as in behavioral capacities and strategies. It remains unclear whether, for example, deficits in rat olfactory learning or memory stemming from hippocampal lesions will help us understand the impaired memory processes of primates with similar lesions. Before advocating olfaction as a model system we need to know more about olfactory memory itself and what role olfactory bulb targets play in the establishment, maintenance, and retrieval of olfactory memories. The few available studies reviewed here provide only a beginning in this quest. To determine whether the olfactory system provides an adequate model for a neurobiological analysis of cognitive function, the first step might be a concerted effort to better understand the sensory functions of olfactory bulbar projections and better characterize the deficits in learning produced by damage to their target structures. Finally, it must be noted that animal studies of olfaction pose unique problems in selection and control of the stimulus and in stimulus control of behavior. The methods used in prior studies varied widely in the choice of stimuli, stimulus concentration, adequacy of generating and controlling these stimuli, and in methods used to present stimuli to subjects. Future studies on the neural basis of olfactory learning should come closer to the standards generally considered acceptable in studies of other sensory systems.

References

1. Powell TPS, Cowan WM, Raisman G (1965) The central olfactory connexions. J Anat 99:791–793
2. Miller SD, Erickson RP (1966) The odor of taste solutions. Physiol Behav 1:145–146
3. Thorne BM, O'Brien AL (1971) The use of olfactory cues in solving a visual discrimination task. Behav Res Methods Instrum 3:240
4. Jennings JW, Keefer LH (1969) Olfactory learning set in two varieties of domestic rat. Psychol Rep 24:3–15
5. Langworthy RA, Jennings JW (1972) Oddball, abstract, olfactory learning in laboratory rats. Psychol Rec 22:487–490
6. Nigrosh B, Slotnick BM, Nevin J (1975) Reversal learning and olfactory stimulus control in rats. J Comp Physiol Psychol 80:285–294
7. Slotnick BM (1984) Olfactory stimulus control in the rat. Chem Senses 9:157–165
8. Slotnick BM (1990) Olfactory perception in animals. In: Stebbins W, Berkley M (eds) Comparative perception, vol 1: Basic mechanisms. Wiley, New York
9. Slotnick BM, Katz H (1974) Olfactory learning-set formation in rats. Science 185:796–798
10. Slotnick BM, Kufera A, Silberberg AM (1991) Olfactory learning and odor memory in the rat. Physiol Behav 50:555–561
11. Youngentob SL, Markert LM, Mozell MM, Hornung DE (1990) A method for establishing a five odorant identification confusion matrix task in rats. Physiol Behav 47:1053–1059
12. Lu X, Slotnick BM, Silberberg AM (1993) Odor matching and odor memory in the rat. Physiol Behav 53:795–804
13. Otto T, Eichenbaum H (1992) Complementary roles of the orbital prefrontal cortex and the perirhinal-entorhinal cortices in an odor-guided delayed-nonmatching-to-sample task. Behav Neurosci 102:762–775
14. Beauchamp GK, Wellington JL (1984) Habituation to individual odors occurs following brief, widely-spaced presentations. Physiol Behav 32:511–514
15. Slotnick BM, Risser J (1990) Odor memory and odor learning in rats with lesions of the lateral olfactory tract and mediodorsal thalamic nucleus. Brain Res 529:23–29
16. Staubli U, Ivy G, Lynch G (1984) Hippocampal denervation causes rapid forgetting of olfactory information in rats. Proc Natl Acad Sci USA 81:5885–5887
17. Pontecorvo MJ (1983) Effects of proactive interference on rats' continuous non-matching-to-sample performance. Anim Learn Behav 11:356–366
18. Rothblat LA, Hayes LL (1987) Short-term object recognition memory in the rat: non-matching with trial-unique junk stimuli. Behav Neurosci 101:587–590
19. Zola-Morgan S, Squire LR, Amaral DG, Suzuki WA (1989) Lesions of perirhinal and parahippocampal cortex that spare the amygdala and hippocampal formation produce severe memory impairment. J Neurosci 9:4355–4370
20. Davis JL, Eichenbaum H (eds) Olfaction. A model system for computational neuroscience. MIT Press, Cambridge

Olfactory Recognition Memory

ERIC B. KEVERNE[1], PETER A. BRENNAN[1], and KEITH M. KENDRICK[2]

Key words. Recognition memory—Bulbar synapses — Microdialysis — Norepinephrine — GABA —Glutamate

Mice have an olfactory (pheromone) recognition memory that is acquired with one-trial learning, contingent on norepinephrine release at mating [1]. This recognition is specific to the odor of the male that mates, and prevents pregnancy block on subsequent exposure to this male, or his pheromones. The pheromones from this male will, however, block the pregnancy of other females, and likewise his mated female can have her pregnancy blocked if exposed to the pheromones of an unfamiliar male [2,3]. Hence, by recognition of familiar male pheromones, pregnancy loss is normally avoided. Since familiar pheromones are likely to be around during the vulnerable period leading up to implantation, the biological relevance of this recognition is very clear.

It is now established that this olfactory recognition is formed during a critical period lasting a few hours immediately after mating and is dependent on noradrenergic innervation of the accessory olfactory bulb (AOB) [4]. However, removal of this noradrenergic innervation after memory formation is without effect, suggesting that noradrenergic dependent changes in synaptic connection strengths may occur among intrinsic accessory bulbar neurons at the time of mating. The question therefore arises as to whether the AOB itself or other projection sites of the system are required for the storage of the olfactory recognition memory to the mate's pheromones. To address this question we have temporarily disabled each relay site along the central neural projections of the vomeronasal system during the critical period immediately after mating, and examined whether recognition memory failed to occur [5]. It should be noted that any permanent lesion to these pathways would prevent effective recall of the recognition memory, because it would disrupt pathways to the neuroendocrine effector system. Infusion of the local anesthetic lignocaine into the AOB prevented recognition memory, and the male's pheromones blocked pregnancy in the female he had previously mated. When lignocaine was infused into the medial amygdala immediately after mating, there was no block to memory formation, and only strange males were able to disrupt the females' pregnancies. Since lignocaine infusions to the amygdala disabled transmission, but did not prevent memory formation, it is clear that the recognition memory must be occurring at some point in the pathway that is anatomically proximal to the amygdala. Taken together with the necessary control procedures which examine the nonspecific effects of infusions, these experiments point to the accessory olfactory bulb itself as an important site, not only for the recognition process, but also for the memory trace which prevents the familiar mate from blocking his own partner's pregnancy. Therefore, this structure became the focus for our attention in addressing plastic changes which enable the recognition memory to be formed.

Localized drug infusions into the accessory olfactory bulb have provided the basis for studying the transmitters and receptors of importance for this olfactory recognition. This model has the advantage that experiments are performed on freely behaving animals and, because of the established knowledge on the structure of the AOB, the effects of the drugs infused locally can be attributed to actions at specific synapses. Failure to form a memory gains an effect not revealed previously, and the test for "memory" formation involves an unambiguous physiological end-point, the animals are either pregnant or nonpregnant.

The synaptic circuitry of the AOB is comparatively simple and is very similar to that of the main bulb [6,7]. Mitral cells receive afferents from the vomeronasal nerve and project to the medial amygdala, forming the excitatory pathway to the hypothalamus for pheromonal signals received by the vomeronasal organ receptors. The mitral cells form reciprocal dendrodendritic synapses with granule cells, the main class of interneuron in the AOB. Granule cell synapses are depolarized by an excitatory amino acid input from the mitral cells, and in turn provide a feedback inhibition to the mitral cells via gamma-aminobutyric acid (GABA) release. This interaction between mitral and granule

[1] Sub-Department of Animal Behaviour, University of Cambridge, Madingley, Cambridge CB3 8AA, UK
[2] Department of Neurobiology, AFRC Babraham Institute, Babraham, Cambridge CB2 4AT, UK

cells, at the reciprocal synapse, regulates mitral cell activity in a negative-feedback manner. Since norepinephrine is thought to reduce the inhibition exerted by granule cells [8], then sustained excitation of mitral cells may be expected to occur at mating. This condition can be mimicked without mating by local infusions of the GABA-ergic antagonist, bicuculline, in the presence of male pheromones. Such a procedure results in the formation of connection strengths that are nonspecific to recognition memory, and no male pheromones are able to block pregnancy even though mating has not occurred [9]. Specificity is lacking because drug infusions spread throughout the AOB and cannot be directed at any subset of granule cells that distinguish each male's odorprint.

The converse is found and formation of a memory to the mate fails to occur when the excitatory arm of the dendrodendritic synapse is blocked with local infusions of a nonselective excitatory amino acid antagonist (D-glutamyl glycine) following mating [10]. However, recognition memory proceeds normally with infusions of specific N-methyl-D-aspartate (NMDA) receptor blockers D-2-amino-5-phosphonovalerate (APV or MK801) or a non-NMDA receptor blocker 6,7,-dinitroquinoxaline-2,3-dione (DNQX), but is prevented if equimolar infusions of each are given simultaneously. Hence, olfactory memory in the context of pregnancy block may occur by using either NMDA or voltage-gated Ca^{2+} channels, and both probably require blocking in order to prevent olfactory recognition memory from forming.

These studies implicate the granule to mitral cell dendrodendritic synapse in the formation of a recognition memory. Further support for such synaptic changes in olfactory recognition comes from other studies involving mother-infant recognition in sheep. Although the context is very different the mechanisms are remarkably similar. Both mate recognition in mice and offspring recognition in sheep depend upon vaginocervical stimulation (mating/ parturition) which promotes norepinephrine release, activating the critical period when olfactory recognition occurs. Olfactory recognition of lambs is accomplished in two phases starting with the completely naive, nonparturient ewe, which rejects all lambs. In the first few hours after parturition, all ewes accept any lamb, but after this period they become highly selective, recognizing and accepting only their own lamb. Ewes remain in the unselective phase if the olfactory bulbs are depleted of norepinephrine or if β-adrenergic receptors are blocked; they fail to make specific recognition, and therefore accept any lamb [11,12]. Following the first birth experience, the nonselective phase is much shorter and multiparous ewes rapidly develop selective recognition [13]. Clearly, olfactory changes established at the first birth are carried over into sub-

sequent births, and enable the specialist or selective recognition phase to be accomplished faster.

To further understand how the mitral cells increase their responsiveness to odors we used in vivo microdialysis to measure the effect of odors on the release of intrinsic GABA and glutamate as well as the release of centrifugal norepinephrine and acetylcholine in the olfactory bulb before and after olfactory recognition of lambs [14,15]. In this part of the study, 10 adult Clun Forest ewes were anesthetized and implanted bilaterally with 18-gauge stainless steel guide tubes aimed in the olfactory bulbs, which were placed by using X-ray guidance. The guide tubes were inserted through the nasal sinus and directed at right angles to the rostral pole of the bulb. In this way, the microdialysis probes (CAM-10, 4 mm membrane length, CMA/Microdialysis, Stockholm, Sweden) preferentially sampled the external plexiform layer and monitored the neurotransmitter interactions between granule and mitral cells. Neurotransmitter release in response to lamb odors was measured 4 to 24 h before parturition and again in the same ewes 24 h after parturition when a recognition memory had formed. Samples were collected at 5-min intervals before, during, and after a 10-min exposure to lamb odors. Neurotransmitter concentrations were measured by high-performance liquid chromatography. Behavioral tests were given to all the ewes 6 h after birth to confirm that they had all formed a selective recognition bond with their lambs.

Prior to birth, lamb odors produced no detectable change in the release of any bulbar transmitters. After parturition, when ewes had established a selective bond with their lamb, the odor of this lamb, but not that of an alien one, induced a significant increase in the release of both the excitatory amino acid glutamate and the inhibitory one, GABA, which are the major transmitters between mitral and granule cells at the dendrodendritic synapse. These changes in glutamate and GABA release only occurred during the first 5 min of exposure to the lamb odor, and the increase in GABA was significantly greater than that for glutamate. Basal release of GABA and glutamate was also significantly higher in the period after birth and, again, GABA release was significantly greater than that of glutamate. Because GABA-ergic granule cells are intrinsic bulbar neurons excited by mitral cells and provide feedback inhibition to the mitral cells by way of the reciprocal dendrodendritic synapses, then the proportionately higher release of GABA with respect to glutamate might be explained in terms of a changed efficacy of glutamate at these synapses in the period after birth.

This increased efficiency of glutamate in promoting GABA release was further examined by comparing the regression slopes for the two transmitters in the period before and after birth [14]. Glutamate and GABA release were correlated both before and after birth but the slopes for the two periods were significantly different. The displacement of the slope illustrated that, for a given level of glutamate recovered, more GABA was present in the sample. The overall increase in both glutamate and GABA release in response to lamb odors after birth is synonymous with more mitral cell activity, while the significant shift in the regression slope could only have resulted from the increased efficacy of glutamate in promoting GABA release. Such enhancement of neurotransmission at the mitral cell to granule cell dendrodendritic synapse after birth, and in response to lamb odors, would produce a change in the firing frequency of those neurons that are odor-coded for the lamb, and may therefore provide a bias in the network with respect to these odors.

These studies have gone some way to revealing the neurochemical changes which underlie olfactory recognition of lambs. Parturition induces massive increases in the release of norepinephrine (NE) and acetylcholine (ACh) in the olfactory bulb [14,16], a release which is enhanced by the peptides oxytocin and vasopression [17], also recovered in high concentrations from the olfactory bulb by microdialysis at parturition [17]. Although the release of transmitters from centrifugal projections contingent on stimulation of mitral cells by lamb odors is essential for recognition, there appears to be no direct interaction of ACh or NE on glutamate release [18]. Nor does the release of bulbar ACh and NE distinguish between familiar and strange lambs [14]. Hence, the main influence of ACh and NE appears to involve the periglomerular and granule cell interneurons. Once learning has occurred, there is a decrease in periglomerular dopamine release, and the resulting disinhibition at the glomerular level may well underlie the increased sensitivity of lamb odors in the early postpartum period. Selectivity in recognition is best explained by the changes which occur between mitral to granule cells and their interaction with centrifugal afferents. Changes in the granule cells make them more sensitive to glutamate, which would enhance feedback inhibition on the mitral cell. However, since there is also an enhancement of inhibiting influences on the granule cell via noradrenergic innervation and autoreceptor feedback inhibition, the overall outcome is a change in the oscillatory frequency of the mitral-granule cell unit [19], producing an increased frequency in mitral cell burst firing. This proposed mechanism is substantiated by the electrophysiological recordings following lamb recognition when more units respond to lamb odors, and some units differentiate between own and alien lambs with increased firing frequency [14].

In conclusion, these studies reveal the influence which mating and birth have on the processing of biological odors thereby facilitating their recognition on subsequent exposure. The capacity of the bulbar networks to undergo plastic changes bears a remarkable resemblance to the plasticity which occurs in the hippocampus as revealed by long term potentiation (LTP) [20]. Not only has evolution been conservative in its neural mechanisms for memory (the combination of noradrenergic, GABA-ergic, and excitatory amino acid transmitters frequently feature in mechanisms of memory [21]), but in the examples studied here these mechanisms occur at the first sensory relay. In the case of mate recognition in mice the olfactory memory acts upon a neuroendocrine readout. Since neuroendocrine neurons are frequency coded, any change in the burst firing of mitral cells would interfere with the efficacy of the olfactory signal producing a neuroendocrine response leading to pregnancy block. In this system the memory resides in the bulbar network. Recognition of lambs by the post-parturient ewe is more complex, although remarkably similar changes in the neural network of the olfactory bulb underlie this olfactory recognition. During the learning phase, both the periglomerular and granule cells come under strong inhibition. In response to lamb odors in the post-learning phase there is a relaxation of inhibition at the olfactory glomerulae and a tight coupling of feedback inhibition by the granule cell. The outcome is an increased sensitivity to lamb odors in general and a tightening of the granule-mitral cell oscillator in a subpopulation of mitral cells that respond to their own lamb. Hence, recognition involves a spatial (anatomical organization of mitral cells) as well as a temporal dimension (change in firing frequency of mitral cells) and in this system the memory depends on establishing a larger network with projection sites deep in the brain.

References

1. Keverne EB, de la Riva C (1982) Pheromones in mice: reciprocal interaction between the nose and the brain. Nature 296:148–150
2. Bruce HM (1960) A block to pregnancy in the mouse caused by proximity to strange males. J Reprod Fertil 1:96–103
3. Keverne EB (1983) Pheromonal influences on the endocrine regulation of reproduction. Trends Neurosci 6:381–384
4. Rosser AE, Keverne EB (1985) The importance of central noradrenergic neurones in the formation of an olfactory memory in the prevention of pregnancy block. Neuroscience 16:1141–1147
5. Kaba H, Rosser AE, Keverne EB (1989) Neural basis of olfactory memory in the context of pregnancy block. Neuroscience 32:657–662
6. Shepherd GM (1972) Synaptic organisation of the mammalian olfactory bulb. Physiol Rev 52:864–917
7. Mori K (1987) Membrane and synaptic properties of identified neurons in the olfactory bulb. Prog Neurobiol 29:275–320

8. Jahr CE, Nichol RA (1982) Noradrenergic modulation of dendrodendritic inhibition of the olfactory bulb. Nature 297:227–229

9. Brennan PA, Kaba H, Keverne EB (1990) Olfactory recognition: a simple memory system. Science 250: 1223–1226

10. Brennan PA, Keverne EB (1989) Impairment of olfactory memory by local infusions of non-selective amino acid receptor antagonists into the accessory olfactory bulb. Neuroscience 33:463–468

11. Pissonier D, Thiery JC, Fabre-Nys C, Poindron P, Keverne EB (1985) The importance of olfactory bulb noradrenalin for maternal recognition in sheep. Physiol Behav 35:361–364

12. Lévy F, Gervais R, Kindermann U, Orgeur P, Piketty V (1990) Importance of β-noradrenergic receptors in the olfactory bulb of sheep for recognition of lambs. Behav Neurosci 104:464–469

13. Poindron P, Le Neindre P (1980) Endocrine and sensory regulation of maternal behavior in the ewe. In: Rosenblatt JS, Hinde R, Beer C, Busnel MC (eds) Advances in the study of behavior, vol II. Academic, New York, pp 75–119

14. Kendrick KM, Lévy F, Keverne EB (1992) Changes in the sensory processing of olfactory signals induced by birth in sheep. Science 256:833–836

15. Keverne EB, Lévy F, Guevara-Guzman R, Kendrick KM (1993) Influence of birth and maternal experience on olfactory bulb neurotransmitter release. Neuroscience 56:557–565

16. Kendrick KM, Keverne EB, Chapman C, Baldwin BA (1988a) Microdialysis measurement of oxytocin, aspartate, gamma-aminobutyric acid and glutamate release from the olfactory bulb of the sheep during vaginocervical stimulation. Brain Res 411:171–174

17. Kendrick KM, Keverne EB (1992) Control of synthesis and release of oxytocin in the sheep brain. In: Pedersen CA, Caldwell JD, Jirikowski GF, Insel TR (eds). Oxytocin in maternal, sexual and social behaviors. New York Academy of Sciences. New York, pp 102–121

18. Lévy F, Guevara-Guzman R, Hinton MR, Kendrick KM, Keverne EB (1993) Effects of parturition and maternal experience on noradrenaline and acetylcholine release in the olfactory bulb of sheep. Behav Neurosci 107:662–668

19. Taylor JG, Keverne EB (1991) Accessory olfactory learning. Biol Cybernet 64:301–306

20. Morris RGM, Anderson F, Lynch GS, Baudry M (1986) Selective impairment of learning and blockade of long term potentiation by an N-methyl-D-aspartate antagonist. Nature 319:774–776

21. Artola A, Singer W (1987) Long term potentiation and NMDA receptors in rat visual cortex. Nature 330:649–652

22. Bliss TVP, Goddrad GV, Riives M (1983) Reduction of long term potentiation in the dentate gyrus of the rat following selective depletion of monoamines. J Physiol (Lond) 334:475–491

Modifiability of NaCl Taste Aversion Generalization Functions

Heather J. Duncan and David V. Smith[1]

Key words. Taste aversion—Salt

The conditioned aversion procedure has been used extensively to evaluate perceptual similarities among taste stimuli to experimental animals, allowing the categorization of chemical stimuli into groups with similar or dissimilar tastes. Using this technique to examine the effects of taste mixtures and salt modifiers on patterns of generalization to NaCl, water-deprived hamsters received pairings of a conditioned stimulus (CS) and LiCl-induced illness. In a generalization test, responses (mean licks per trial) were measured to the CS, two lower concentrations of the CS, and these three stimuli adulterated with another chemical, either of a different taste quality (e.g., citric acid), a similar taste quality (e.g., $NaNO_3$), or a taste modifier (L-arginine-L-aspartate; AA). After an aversion was conditioned to 0.1 M NaCl, hamsters generalized to 0.032 M and 0.01 M NaCl, yielding a concentration-response function reflecting stimulus intensity. $NaNO_3$ was chosen as an adulterant because both electrophysiological and behavioral data [1–3] from the hamster indicate that it is similar to NaCl. When added to a NaCl concentration series, animals responded to the increased Na^+ concentration; that is, the adulterated stimulus appeared to be more salty than NaCl alone. Monosodium glutamate (MSG) has also been shown to be similar to NaCl to the hamster [2] and our results, while less dramatic than for $NaNO_3$, support this conclusion. Stimuli of known perceptual *dissimilarity*, such as sucrose or citric acid, resulted in a decreased aversion when added to NaCl, which, at the least, indicates a perceptual dissimilarity to NaCl, and may represent a less salty perception. AA, about which effects in the hamster are not yet known, also *decreased* the NaCl aversion, suggesting that AA does not act as a saltiness enhancer for the hamster. Because some stimuli produce an increase in the aversion to NaCl, we may assume that these combined stimuli have increased "saltiness"; we suggest that a reduction in the conditioned aversion to NaCl with other adulterants may be indicative of decreased "saltiness" of these combined stimuli.

References

1. Smith DV, Travers JB, Van Buskirk RL (1979) Brainstem correlates of gustatory similarity in the hamster. Brain Res Bull 4:359–372
2. Yamamoto T, Matsuo R, Kiyomitsu Y, Kitamura R (1988) Taste effects of "umami" substances in hamsters as studied by electrophysiological and conditioned taste aversion techniques. Brain Res 451:147–162
3. Frank ME, Nowlis GH (1989) Learned aversions and taste qualities in hamster. Chem Senses 14:379–394

[1] Department of Otolaryngology, Head and Neck Surgery, 231 Bethesda Ave., ML 528, University of Cincinnati Medical Center, Cincinnati, OH 45267-0528, USA

Effects of Lesions of the Medial and Lateral Parabrachial Nuclei on Acquisition and Retention of Conditioned Taste Aversion

Nobuyuki Sakai, Tomoko Tanimizu, Noritaka Sako, Tsuyoshi Shimura, and Takashi Yamamoto[1]

Key words. Conditioned taste aversion—Parabrachial nucleus

When the consumption of novel flavored foods is followed by the administration of a toxin, the animals avoid ingesting this food on subsequent presentations. This alteration in behavior is referred to as a conditioned taste aversion (CTA). Understanding of the neural mechanisms subserving this taste aversion learning has been a goal of a variety of studies. It has been reported that confined lesions of the brain attenuate or abolish CTA learning in rats. The lesions included those in the area postrema, amygdala, gustatory cortex, ventral posteromedial nucleus of the thalamus, as well as chronic supracollicular decerebration.

Several investigators have reported that cytotoxic lesions of the whole parabrachial nucleus (PBN), which invade the taste and visceral recipient zones, disrupt acquisition of CTA in the rat [1] and that electrolytic lesions of the medial subnuclei of the PBN, which receive taste information, disrupt acquisition of CTA [2,3].

We aimed to examine whether or not the external lateral subnuclei of the parabrachial nucleus (PBel), where c-fos-like immunoreactive neurons were most densely observed after intraperitoneal injection of isotonic LiCl [4], were involved in the acquisition and retention of CTA.

In experiment 1, to establish CTA, 0.01 M Na-saccharin and i.p. injection of 0.15 M LiCl were used for the conditioned stimulus (CS) and unconditioned stimulus (UCS), respectively. Before conditioning, each rat received electrolytic lesions focusing on the parabrachial external lateral subnucleus (PBel). The lesioned rats were divided into three groups; (1) complete lesion group (CL); rats in this group received bilateral lesions of the PBel, and c-fos expression induced by i.p. injection of LiCl was not found, (2) partial lesion group (PL); lesions were unilateral or incomplete and partial c-fos expression

was found, and (3) a sham lesion group (SL); lesions were around the PBN and normal c-fos expression was found. The controls consisted of intact animals. The CL group did not show neophobia to saccharin and failed to establish CTA. Similar effects were shown in the PL group, to a lesser extent. The SL and control groups showed neophobia, and strong CTA was acquired. After this CTA test, a two-bottle preference test was performed for each rat, with taste stimuli consisting of 0.02 M quinine hydrochloride, 0.01 M HCl, 0.1 M NaCl, 20% Polycose, and 0.5 M sucrose solutions. It was found that PBel lesions had no significant effects on preference behavior.

In experiment 2, rats were conditioned to avoid saccharin first, then received the PBel lesions. The CL group showed a little faster extinction of CTA than the other groups. The two-bottle test showed no significant difference in taste preference among the four groups, except that control animals drank less sucrose than the CL group.

In experiment 3, rats were conditioned to avoid saccharin first, then received bilateral lesions of the medial PBN (PBm), which receives gustatory information. In contrast to the PBel-lesioned rats in experiment 2, the PBm-lesioned rats showed a complete extinction or no retention of CTA. The two-bottle preference test showed that PBm-lesioned rats consumed more QHCl and HCl solutions than the PBel-lesioned and control groups.

In experiment 4, the PBel-lesioned and PBm-lesioned rats were reconditioned to 0.2 M NaCl, and were examined for their ability to acquire CTA. Neither group acquired taste aversion learning.

The present study demonstrated that both the PBm (taste area for saccharin) and PBel (general visceral area) are crucial for the acquisition of CTA, and that the retention of CTA is severely impaired by PBm lesions, but only slightly so by PBel lesions.

References

[1] Department of Behavioral Physiology, Faculty of Human Sciences, Osaka University, 1-2 Yamadaoka, Suita, Osaka, 565 Japan

1. Yamamoto T, Fujimoto Y (1992) Brain mechanisms of taste aversion learning in the rat. Brain Res Bull 27:403–406

2. Spector AC, Norgren R, Grill HJ (1992) Parabrachial gustatory lesions impair taste aversion learning in rats. Behav Neurosci 106:147–161
3. Di Lorenzo PM (1988) Long-delay learning in rats with parabrachial pontine lesions. Chem Senses 13:219–229
4. Yamamoto T, Shimura T, Sako N, Azuma S, Bai W-Zh, Wakisaka S (1992) c-fos Expression in the rat brain after intraperitoneal injection of lithium chloride. Neuroreport 3:1049–1052

Parabrachial Lesions Disrupt Conditioned Taste and Odor Aversions, but not Learned Capsaicin Aversion or Flavor Preference

Tsuyoshi Shimura[1], Patricia S. Grigson[2], Steve Reilly[2], and Ralph Norgren[2]

Key words. Conditioned taste aversion—Parabrachial nucleus

Four experiments evaluated the process by which bilateral ibotenic acid lesions of the parabrachial nucleus (PBN) disrupt acquisition of a conditioned taste aversion.

Experiment 1

Conditioned taste aversion (CTA) and conditioned odor aversion (COA) learning was evaluated in ten rats that received bilateral ibotenic acid lesions ($0.2\,\mu l$; $20\,\mu g/\mu l$) of the PBN (PBNx) and eight unlesioned controls (SHAM). All rats received 15-min access to 0.3 M alanine (CTA) or to dH_2O + almond odor (COA). After a 15-min interval, they were injected IP with 0.3 M LiCl (1.33 ml/100 g bw). SHAM rats acquired the CTA following a single pairing, while PBNx rats failed to demonstrate any evidence for CTA learning. SHAM rats also acquired the COA, albeit more slowly than they acquired the CTA. PBNx rats appeared to acquire the COA, but subsequent experiments have confirmed that PBNx rats are, indeed, incapable of COA learning. Thus, an intact PBN is essential for both taste and odor aversion learning.

Experiment 2

We evaluated whether the failure to acquire a CTA in PBNx rats was due to a general inability to process gustatory cues. Six SHAM rats and five PBNx rats received alternating 24-h access periods to two flavors of Kool-Aid: one flavor presented in dH_2O (CS−), and the second flavor presented in 0.15% saccharin (CS+). Rats received 3 days' access to the CS− solution alternated with 2 days' access to the CS+ solution. Following a 24-h access period to plain dH_2O, a two-bottle preference test was conducted. During this test, rats had 24-h exposure to two bottles of water, one flavored with the CS− and the other with the CS+. The results showed that both SHAM and PBNx rats displayed evidence for gustatory processing by selectively increasing intake for the CS+ flavored dH_2O relative to the CS− flavored dH_2O.

Experiment 3

We examined whether the PBNx rats, which failed to predict the occurrence of illness using a gustatory cue, could predict the illness using a trigeminal cue. The same rats used in experiment 1 were given 15-min access to 0.01 mM capsaicin. After a 15-min interval, half of the rats were injected IP with either 0.15 M LiCl (1.33 ml/100 g bw) or with physiological saline. Both SHAM and PBNx rats, injected with LiCl, demonstrated a reliable aversion to capsaicin.

Experiment 4

Finally, we tested whether successful performance in the capsaicin aversion task was a consequence of recovery of function (i.e., capsaicin aversion training was conducted 5 months after the original CTA experiment). Thus, the same rats used in experiments 1 and 3 were simply trained on a new CTA, using 0.1 M NaCl as the conditioned stimulus. All rats were given 15-min access to 0.1 M NaCl and, after a 15-min interval, were injected IP with 0.15 M LiCl (1.33 ml/100 g bw). SHAM rats readily acquired the CTA to NaCl, while PBNx rats failed to show any evidence of learning a CTA over three trials.

Conclusions

(1) Cell bodies of the PBN, rather than fibers of passage, are critical for CTA and COA learning. (2) The failure to acquire a CTA in PBNx rats is not due to an inability to detect, process, or reject a gustatory cue. (3) The successful capsaicin aversion

[1] Department of Behavioral Physiology, Faculty of Human Sciences, Osaka University, 1-2 Yamadaoka, Suita, Osaka, 565 Japan
[2] Department of Behavioral Science, College of Medicine, Pennsylvania State University, Hershey, PA 17033, USA

was not a consequence of recovery of function and, therefore, must be taken as evidence that rats with lesions of the PBN can experience LiCl-induced illness and can predict the occurrence of illness if trigeminal, but not gustatory, cues are provided.

Thus, we conclude that rats with PBN lesions fail to acquire a CTA because they suffer from a deficit that renders them incapable of making a specific association between a gustatory conditioned stimulus and a LiCl unconditioned stimulus.

Olfactory Recognition of Five Species by Urine Odor in Tufted Capuchin (*Cebus apella*)

YOSHIKAZU UENO[1]

Key words. Primate—Species recognition

The tufted capuchin can distinguish the urine odors of conspecifics from those of other species. However, there is no knowledge that the tufted capuchin can distinguish among the urine odors of other species. The present study investigated how the tufted capuchin can discriminate and recognize the urine odors of five different species, on the basis of the paradigm of operant conditioning.

The subjects were two adult tufted capuchins, born and reared at the Primate Research Institute, Kyoto University. Therefore, they did not have any experience of direct social interaction with other species. These monkeys had previously been trained in odor-discrimination experiments. They were deprived of water for 4 h prior to the beginning of training, but could drink water without restraint after training. The odor stimuli were urine samples, collected from three species of New World monkeys (tufted capuchin, squirrel monkey, and cotton top tamarin) and two species of Old World monkeys (rhesus macaque and Japanese macaque). Ten stimulus sets of urine from two species were examined. The monkeys were trained on a successive odor-discrimination task with two odors in one session per day, which session consisted of 30 trials of each stimulus odor. One trial lasted 20 s, and the intertrial interval was 10 s. The response to a key was reinforced by 0.25 ml sweet water on the average of once in 20 s during one stimulus (S+), but not reinforced during another stimulus (S−). The number of responses to the key was counted on each trial, and the response latency to the first response in each trial was measured. The discrimination ratio and the mean response rate to the first reinforcement on both S+ and S− trials were calculated from the response number every session. The time for reinforcement on the S− trial was logically calculated. If there was no significant difference between the response latency on S+ and S− trials at the 15th session, the monkey was considered not to have acquired the discrimination and the training was stopped.

The monkeys could not discriminate between the urine from rhesus and Japanese macaque, but could discriminate the urine from other species. The mean of the response-rate ratio (RRR) for each stimulus set was calculated in order to compare the discrimination among stimulus sets. This score was highest for discrimination of conspecific odors, and decreased in the order of New World monkeys and Old World monkeys. Further, the similarity of urine odors calculated on the basis of the RRR was analyzed by multidimensional scaling (MDS) and cluster analysis. Two species of macaques belonged to the same cluster at the lowest part of the tree. They were also close to the center of the distribution on the two-dimensional MDS plot. The other species were placed on the periphery of the MDS plot. This suggests the following: (1) The tufted capuchin recognized distinctly the difference among New World monkeys, although they were phyletically closer to the tufted capuchin than to macaques. (2) On the other hand, for the tufted capuchins, the odors of macaques were not only similar to each other, but were ambiguous or neutral in comparison to the other species. As stated above, the subjects had never had social relations with other species. Therefore, the properties of olfactory recognition described here are due to difference of experience.

The distribution of the tufted capuchin overlaps that of other New World monkeys in the wild, but does not overlap that of Old World monkeys. An animal may, in general, tend to show high sensitivity or attention to conspecifics. This study demonstrates that tufted capuchins show high sensitivity not only to their conspecifics but also to other species with which they have a close ecological relationship. Consequently, it can be concluded that species recognition by odor in the tufted capuchin must reflect an innate difference in the relationship with other species, which arises from sympatry or allopatry.

[1] Center for Experimental Plants and Animals, Hokkaido University, N10 W8, Kita-ku, Sapporo, 060 Japan

Changes in Urinary Marking Behavior of Mice with Development

Key words. Urinary marking—Olfactory cue

Mice deposit urine marks on their home range, their marking patterns varying according to their sex, social rank, experience, and reproductive state [1]. In order to investigate when and how these differences appear in mouse development, we observed the marking behavior of five classes of mice classified according to sex and conditions of rearing after weaning. The five classes were isolated and sexually naive males, grouped and naive males, grouped and naive females, copulated males, and copulated females. The animals were observed every 3rd or 4th day in an experimental field ($30 \times 30\,\mathrm{cm}$), at various stages, from weaning (13 days of age) to maturation (71 days of age). At the center of the field, different urine marks of other mice had been deposited previously. The location of the mouse was recorded with a video camera, in regard to the deposited marks on substrate. The time spent in resting or investigating, time spent in locomotion, number of urine marks, and area of urine marks were measured in three parts of the experimental field; the central region with the previously deposited marks (the stimulus urine), the corner region, and the remainder, the peripheral region.

Generally, the amount of motion (total amount of active time) decreased slightly after weaning, but increased again at puberty, decreasing in the adults. The mice preferred staying in the corner of the field, but, corresponding to the general increase in motion, the time spent in the corner region was reduced during puberty. These differences were largest in the copulated males, whereas the isolated males did not reveal marked changes.

The total amount of deposited urine (marked area) increased gradually as the mice grew, but fell after maturation. The amount of urine deposited was inhibited by the stimulus of the previously deposited urine. Interestingly, the grouped females ceased to mark the field from an early stage, and the urine of sexually naive males suppressed marking by grouped naive males at puberty.

The number of deposited urinary marks did not change to a great extent with aging in females, but the size of each mark increased. In males, however, the number of marks increased with growth and the size decreased. Of particular note, copulated males deposited a number of urine spots during puberty; in early puberty, they responded to the urine stimuli of diestrous females, while in the later stage, they responded to these stimuli of estrus females.

Counter-marking (marking in the central region containing previously deposited stimulus urine) was marked at puberty, and strong responses to female urine were shown by both males and females. In the grouped naive females, however, these responses developed and diminished in the early stage.

The counter-marking rate (the area of the spot deposited in the center region as a proportion of the area deposited in all regions, divided by the time spent in the center region as a proportion of total observation time) for puberty stage mice increased in response to diestrous female urine stimuli. Grouped naive and copulated males had an interest in possible mates, females, while grouped naive and copulated females responded strongly to possible competitors. In isolated males, however, the responses to female urine were delayed.

Thus, urinary marking patterns peculiar to their sex were revealed in puberty in copulated males and females which seemed to be developing normally. Grouped or isolated mice that had received hormonal suppression during maturation or during the sexual cycle showed diminished transition or ceased marking. These results both confirmed that these mouse urinary marking patterns were controlled by sex hormones, and supported previous findings that injury of the vomeronasal system, which is responsible for the primer effect of pheromone, has a greater effect on urinary marking than injury of the main olfactory system [2].

References

1. Hurst JL (1990) Urine marking in a population of wild house mice *Mus domesticus* Rutty: I. Communication between males II. Communication between females, III. Communication between the sexes. Anim Behav 40:209–243
2. Hatanaka T (1992) Consequences of injuring the main or accessory olfactory system on counter urine marking behavior in mice (in Japanese). Proceedings of the 26th Japanese Symposium on taste and smell (3–4, Dec., 1992, Tokyo):257–260

[1] Department of Biology, Faculty of Education, Chiba University, 1-33 Yayoi-cho, Inage-ku, Chiba, 263 Japan

10. Nutrition and Brain Mechanisms of Food Recognition

Functional Correlation Between Feeding-Related Neurons and Chemical Senses

Y. Oomura[1,2], Z. Karadi[2], H. Nishino[2], S. Aou[2], and T.R. Scott[2]

Key words. Awake monkey—Lateral hypothalamic area—Glucose-sensitive neuron—Glucose-insensitive neuron—Gustatory stimulation—Olfactory stimulation

Introduction

Food and water intake are regulated by glucose-sensitive neurons (GSNs) in the lateral hypothalamus (LHA); the firing rates of GSNs decrease in response to glucose administered by electrophoresis [1], and to metabolites, hormones, cytokines, growth factors, and neurotransmitters [2]. About two-thirds of LHA cells are glucose-insensitive neurons (GISNs) and do not respond to these substances. The GISNs respond specifically to environmental visual and auditory signals in tasks to acquire food [3]. Thus endogenous chemical information is processed by GSNs, and external information is processed by GISNs to regulate feeding. Gustatory [4–6] and olfactory [7,8] projections to the LHA have been identified in primates, and both GSNs and GISNs respond to taste and olfaction. This report describes activity changes of single neurons in the LHA of alert rhesus monkeys during a conditioned bar-press feeding task and changes in these responses due to gustation, olfaction, and electrophoretic application of glucose, norepinephrine, and dopamine.

Methods

The results described here were obtained from 12 adult rhesus monkeys (*Macaca mulatta*) of both sexes. They were seated in primate chairs facing a panel 25 cm away. Details of preparation and testing are described elsewhere [6,7]. The monkeys were trained in a high fixed-ratio (FR) bar-press feeding task consisting of four phases: (1) a red cue light (CL) to signal the availability of food reward; (2) a 20- or 30-bar press (BP); (3) a short cue tone (CT)

[1] Institute of Bio-Active Science, Nippon Zoki Pharmaceutical Co., Ltd., Yashiro-cho, Hyogo, 673-14 Japan
[2] Department of Biological Control Systems, National Institute for Physiological Sciences, Okazaki, Aichi, 444 Japan

followed by food delivery; (4) pickup and ingestion of food (reward period, RW). Through a 2.5-mm-diameter hole in the skull a nine-barreled micropipette was inserted to the LHA. Electrophoretic chemical applications were made through a barreled glass micropipette, and single neuron activity was recorded through a 7 µm diameter carbon fiber in the central barrel. Each barrel was filled with one of the following solutions: 0.5 M monosodium L-glutamate, 0.5 M D-glucose, 0.5 M Na hydrochloride, 0.5 M dopamine hydrochloride, 0.25 M sotalol HCl, saturated phenoxybenzamine HCl, or saturated spiperon tartrate or 0.05 M naloxone HCl.

The five prototypical taste qualities—0.3–1.0 M glucose, 0.3–1.5 M NaCl, 3–30 mM HCl, 0.3–3.0 mM quinine hydrochloride, 1.8–9.0 mM monosodium-L-glutamate (MSG), 0.5–2.5 mM citric acid and 10%–50% orange juice were used as gustatory stimuli. Application of a tastant was always followed by one or two injections of deionized water. For odor stimulation, seven synthetic volatile odor substances prepared by Takasago Koryo Co. were used: 1% orange (ORA), 1% isoamyl acetate (AA), 0.01% skatole (SKAT), 1% borneol (BOR), 1% musk (MUSK), 1% 2,4,6-trimethylpyridine (TMP), and 1% trimethylamine (TMA). Each stimulus was followed by two injections of clean air from a separately prepared syringe.

Results

Activity Change During Feeding Task

Of 647 neurons tested with electrophoretically applied glucose, 181 (28%) were GSNs and 466 (72%) were GISNs. GSN activity was depressed during the BP and RW periods (Fig. 1,I,A). The BP response was blocked by electrophoretically applied sotalol, a β-adrenoceptor blocker [3], and the RW response was blocked by naloxone [1]. When food was salty, as shown in Fig. 2A, behavior did not change, although the response during the RW period disappeared; i.e., opioid receptors were not activated. In early trials the monkey was hungry and would eat even bitter (quinine-treated) bread [4]. The mechanism of the unresponsiveness in RW period is discussed later in the chapter. Responses of

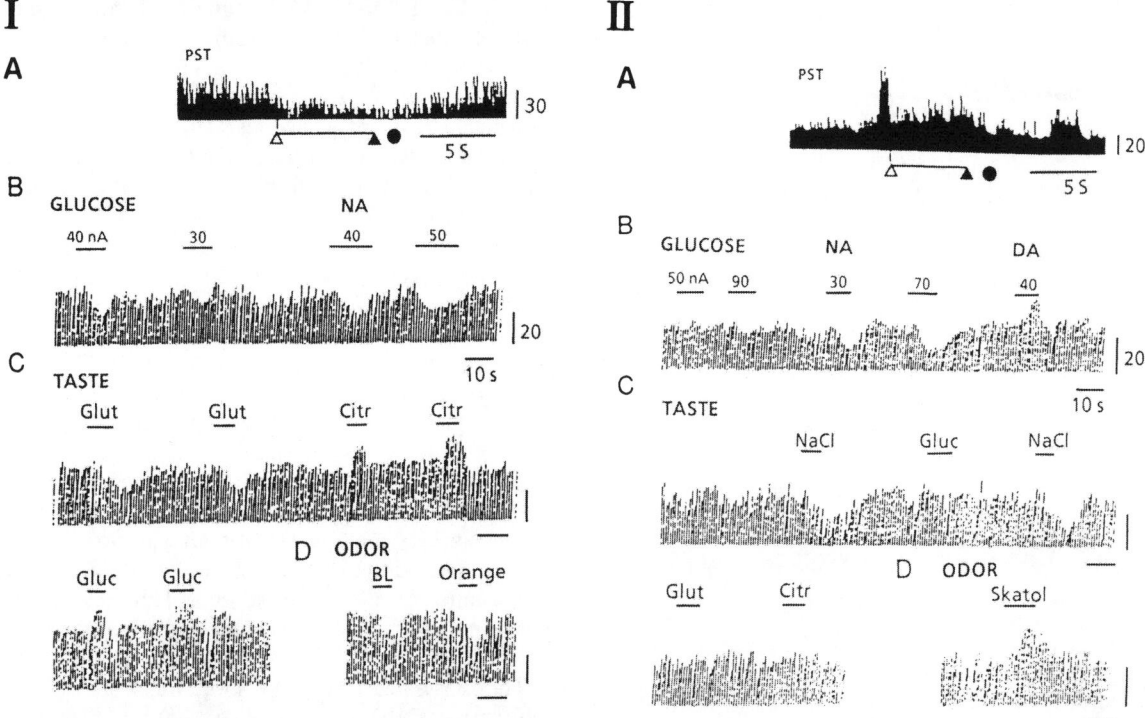

FIG. 1. Neuronal responses in feeding task (*A*), to electrophoretically applied glucose, NA and DA (*B*), and to taste (*C*) and odor (*D*) stimulations. **I** Glucose-sensitive neuron. *A*: Firing decreased during bar press (BP) and reward phases. Firing histogram (*PST*), impulses per 50 ms for 25.6 s, totaled five times, and lined up at first bar press. *Abscissa*, time (s); *ordinate*, impulses/s for *B–C*. *Horizontal bar* under PST, BP period. *Open triangle*, cue light on; *closed triangle*, cue tone. *B*: Electrophoretically applied glucose and NA inhibited firing dose-dependently. *C*: Glutamate (*Glut*) inhibited whereas citrate (*Citr*) and glucose (*Gluc*) excited firing. *D*: Bolneol (*BL*) and orange depressed firing. **II** Glucose-insensitive neuron. *A*: Firing increased at cue light and during bar press. *B*: Glucose had no effect; NA decreased and DA increased firing. *C*: NaCl decreased firing, but glucose, glutamate and citrate had no effect. *D*: Skatol increased firing [12]

GISNs were also closely correlated with feeding behavior but in functionally different ways. Responses were elicited by the CL, during the CL–BP interval, during the BP period, by the CT, or in any combinations of these times (Fig. 1,II,A). The CL responses and the response during the CL–BP interval could be blocked by spiperone, a dopamine 2 receptor blocker, and the CT response could be blocked by phenoxybenzamine, an α-adrenoceptor blocker [3].

Responses to Taste and Olfaction

Examples of taste and olfaction responses of GSNs and GISNs are shown in Fig. 1,I and II, respectively [6,7]. The gustatory responses are summarized in Fig. 3 (top). More GSNs (*n* = 33, 66%) than GISNs (*n* = 24, 39%; χ^2 test, $P < 0.01$) responded to at least one taste. Furthermore, among 50 GSNs, 24% of cells responded to one taste stimulation, and 42% of cells were significantly more broadly tuned to the basic gustatory stimuli; among 61 GISNs, these rates were 31% and 8%, respectively (χ^2 test, $P < 0.01$).

Thus GISNs were more taste discriminating than GSNs (χ^2 test, $P < 0.001$) (also see Fig. 1,I and II). The GSNs were slightly more responsive than the GISNs to all gustatory stimuli tested, except NaCl and orange juice. GISNs were significantly more inhibited than GSNs by these two tastants (χ^2 test, $P < 0.05$). Responses of the GSNs, however, were mainly facilitation. The one taste that was unique among those tested was quinine, which inhibited significantly more GSNs than GISNs (χ^2 test, $P < 0.05$) and excited almost all GISNs.

Figure 3 (bottom), summarizes the olfaction tests. The general pattern of olfaction responses was similar to the pattern of taste responses. Among 50 GSNs, 12% of cells responded to one odor stimulation and 76% were significantly more broadly tuned to odor stimuli; but among 56 GISNs, these rates were 29% and 23%, respectively (χ^2 test, $P < 0.01$). Thus GISNs were more discriminating in their responses (see also Fig. 1,I and II). ORA, AA, and SKAT odors tended to elicit responses more often in GSNs than GISNs (AA and SKAT: χ^2 test, $P < 0.05$). ORA and AA evoked excitation mainly, and

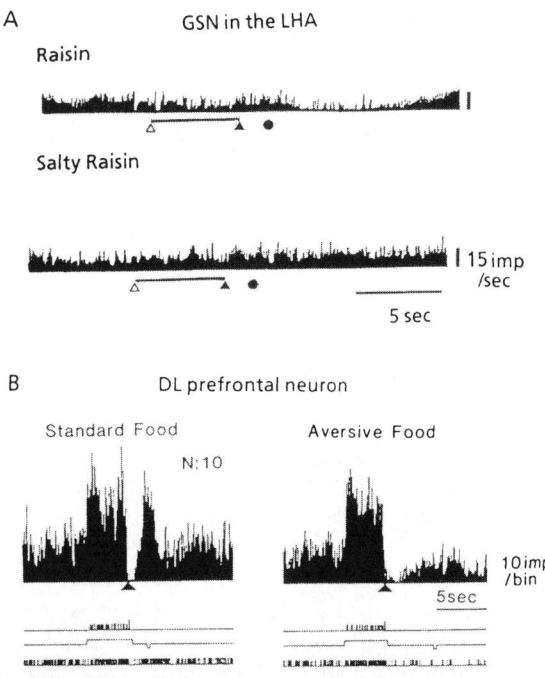

FIG. 2. Neuronal response to reward situation in bar press (BP) feeding task. **A** Firing rate of the same glucose-sensitive LHA neuron decreased during the reward (RW) phase after a normal raisin (*top*) and did not change after a salted raisin (aversive) was presented (*bottom*). **B** Firing rate of dorsolateral prefrontal cortex neuron increased during the BP period and the early phase of the normal food RW period (left) and did not during the RW period when food was aversive. *Middle line*, BP and cue tone (large); *lower line*, cue light on (upward deflection) and off, and food put into the mouth (short deflection down); *open triangle*, cue light on; *horizontal bar*, BP; *filled triangle*, cue tone; filled circle, food put into mouth [12,16]

SKAT inhibition, in GSNs and GISNs. GSNs were predominantly inhibited, whereas GISNs were mainly excited by MUSK, TMA, and TMP. BOR had special, intermediate attributes among the olfactory stimuli: the GSNs were also more responsive to BOR than the GISNs (χ^2 test, $P < 0.05$).

Discussion

More GSNs were sensitive and broadly tuned to tastants and odorants than GISNs, and the latter were more discriminating. Appetitive or palatable tastes (e.g., those of NaCl and ORA) as well as unpalatable, or aversive, tastes (e.g., quinine or TMA and TMP) evoked opposite responses in the two groups of cells. In contrast to the GSNs, the GISNs were excited by aversive tastes. Two sets of neurons displayed a similar dichotomy with respect to familiar and unfamiliar or unpleasant odorants [7]. Approximately 70% of GSNs responded to both gustatory and olfactory stimulation, in contrast to

only 21% of GISNs. Convergence of inputs from both chemical senses is reported in the nucleus tractus solitarii [9], the insula [10], orbitofrontal (OBF) cortex [11], and central nucleus of the amygdala [8,10] of macaques. The distinct chemosensitivity of GSNs and GISNs may be due to their different anatomic connections. In addition to the close behavioral and neurophysiologic relations between the LHA and dorsolateral prefrontal (DL) and OBF regions, results indicate strong connections between GISNs and the motor area and less strong connections between GISNs and the DL-OBF areas, as well as strong connections between GSNs and the DL-OBF areas (Fig. 4) [4,12]. Thus the prefrontal-OBF regions, which represent an advanced stage of chemosensory processing, project selectively to hypothalamic GSNs [4,7,12]. According to other data [8,13], most neurons in the caudolateral OBF area (the secondary gustatory relay, also directly interconnected with the LHA) are sharply and specifically tuned to taste stimuli. Thus integrated gustatory information to GSNs and discriminative chemical signals to GISNs may reach the LHA.

The GSNs were selectively suppressed by norepinephrine (NE) and dopamine (DA) [1,3]. For food acquisition, they were suppressed during BP and RW, and GISNs were excited by CL, BP, and CT. These results agree with previous reports that implicated DA circuits in sensorimotor integration, cue-motor coupling, and movement initiation, and associating NE systems with food acquisition [1,6,7], opiate-mediated reinforcement, and stimulus-reward coupling [1]. As shown in Fig. 4, GSNs and GISNs connect with other regions and systems through which these neurons receive information and then project signals to integrate and control feeding behavior. GSNs receive, in addition to gustatory and olfactory inputs, inputs from endogenous chemicals [2] and from the NE_β, opioid, and 5-hydroxytryptamine 1 (5-HT_1) systems [14]. GISNs receive similar gustatory and olfactory inputs plus stimulation through other modalities, as well as DA, NE_α and 5-HT_2 system inputs [14].

Cue-related responses are reward-coupled, which is one form of learning. CL-responding neurons responded only to the cue-related visual signal and not to light signals that were not associated with reward. Spiperon blocked CL-related responses, so the DA system and DA_2 receptors on GISNs could also be involved in the learning process. The same mechanism appears in GISNs not only in the LHA but also in the amygdala [15]. Light signals pass through the inferotemporal cortex and basolateral amygdala to GISNs in the central medial nucleus of the amygdala and to GISNs in the LHA. If a visual signal is coupled to a reward, it is recognized in the DL area [16]. This signal in turn activates the DA system through the ventral tegmental area or substantia nigra to open a pathway from the inferotem-

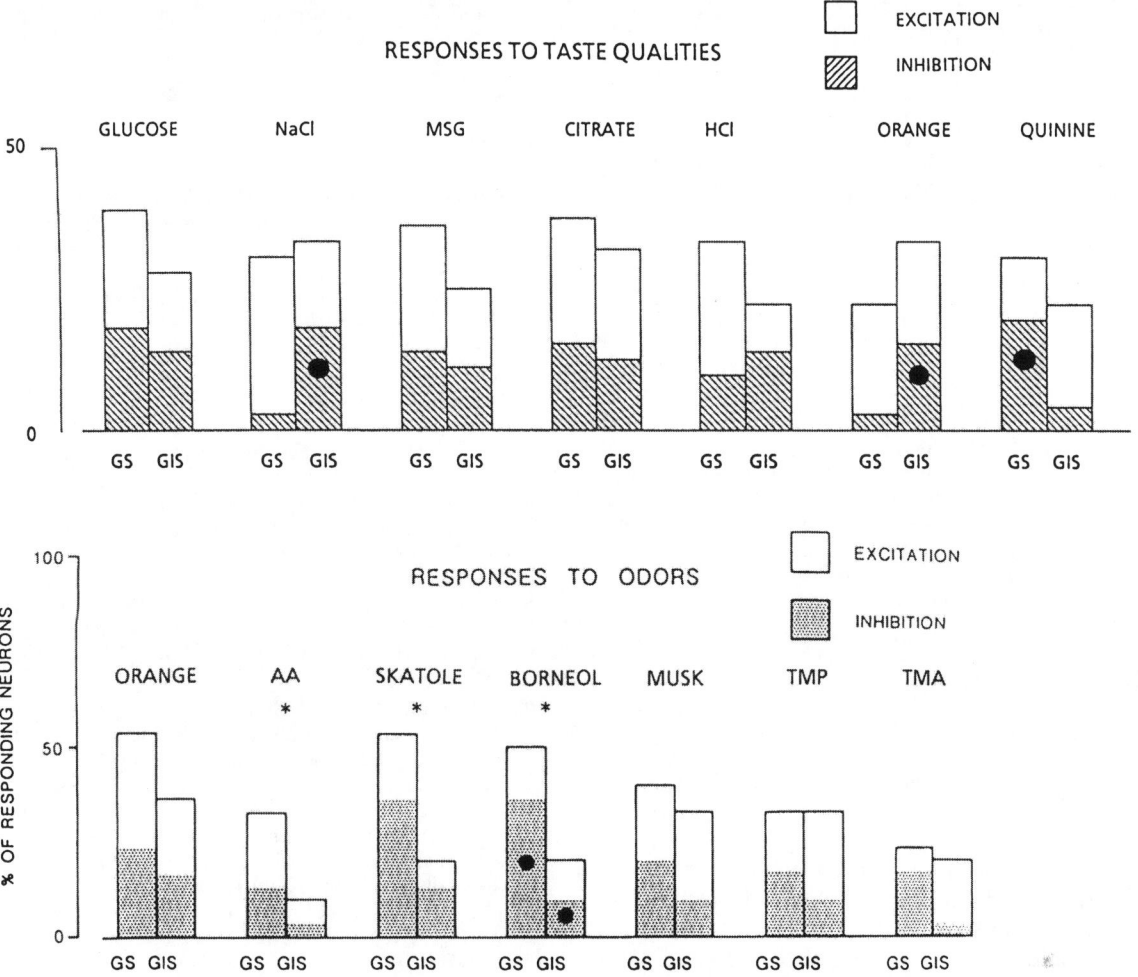

FIG. 3. Responses of LHA neurons to tastes (*top*) and odors (*bottom*). No significant differences in total responsiveness of glucose-sensitive neurons (*GS*) and glucose-insensitive neurons (*GIS*) to the various tastants. Inhibitory activity changed in response to NaCl and orange juice (*orange*) more frequently in GISNs. GS cells were mainly suppressed by quinine hydrochloride. Total responsiveness of GSNs was significantly higher than that of GIS cells to AA, skatole, and borneol odors. GSNs were more frequently inhibited by borneol than GISNs. There were opposite response characteristics of the two sets of neurons to MUSK, TMP, and TMA. *Filled circles*, $P < 0.05$, significantly different GSN and GISN response patterns; *$P < 0.05$, significantly different response to different odors [6,7,12]

poral cortex. The opening might be accomplished through presynaptic facilitation at the nerve terminals of the DA neurons that project to the GISNs.

The results in which the neuronal activity changed during the reward period can be explained in terms of functions in which the DL area recognizes ingestion of palatable food, ingestion of unpalatable food, or absence of ingestion (extinction). As shown in Fig. 2B, neurons in the DL area increase their activity when they recognize palatable food (a rewarding situation); but when the food is unpalatable or absent, they do not respond [16]. The response to palatable food activates the DA system, the NE system, and the opioid system. In the absence of prefrontal cortical responses (aversive food or no food), the RW period response, whether a decrease or increase in activity, is attenuated [4].

Our data support the existence of a complex gustatory and olfactory code in the monkey LHA, where chemical information is linked to both endogenous cues and exogenous perception and motivation. Gustatory and olfactory senses project to GSNs and GISNs, as shown in Fig. 4. GSNs integrate multimodal endogenous and exogenous chemosensory information. GISNs are driven by changes in the external environment and distinguish fewer but highly specific taste and odor cues in discriminative feeding regulation. A physiologically relevant, distinct organization of GSNs and GISNs is suggested in the LHA of the rhesus monkey for integrative processing of endogenous chemical signals and exogenous feeding-related cues.

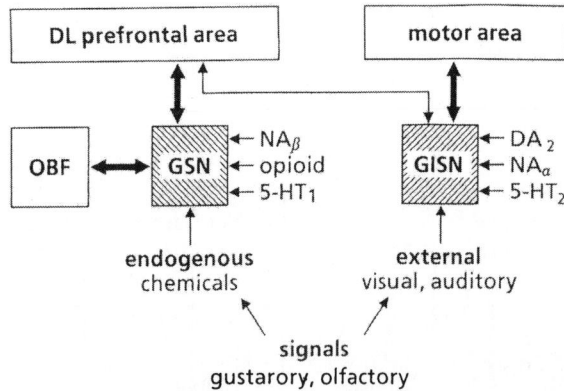

Fig. 4. Organization of glucose-sensitive neurons (*GSN*) and glucose-insensitive neurons (*GISN*), with emphasis on their functional separation in the feeding-related processing of endogenous and exogenous signals. Chemical senses appear to connect the internal and external milieus through these two sets of LHA cells. *Thick arrows* indicate, extensive mutual functional projections; *thin arrows* indicate, relatively limited connections [12]

Summary

Single neuron activity in the lateral hypothalamus (LHA) of awake monkeys was recorded during a high fixed-ratio, bar-press feeding task. Activity was recorded during electrophoretic application of various chemicals and gustatory and olfactory stimulations, and glucose-sensitive (GSN) and glucose-insensitive (GISN) neurons were compared. Of 647 neurons tested, the activity of GSNs (28%) was suppressed and that of GISNs (72%) did not change during electrophoretic glucose application. Because of the respective activation of β-adrenoceptors and opioid receptors, GSNs were significantly more often inhibited during bar pressing and food reward. GISNs responded more than GSNs to the food available cue light during the bar-press period and to the cue tone reward signal. Cue light and BP excitation were through activation of DA_2 receptors, and cue tone excitation was through α-adrenoceptors.

Of the GSNs, 66% responded to taste and 88% to odor; of the GISNs, 39% responded to taste and 52% to odor. Most GSNs responded to two or more taste and odor stimuli, and GISNs responded to only one. GSNs have dense mutual connections with the prefrontal area, and GISNs are connected with the motor area.

These two LHA neuron types have distinct functions in the acquisition of, and reward by, food and in processing chemical signals. GSNs monitor the internal milieu and use associative connections to integrate chemosensory information, including responses to taste and odor. GISNs respond to changes in the external environment and distinguish specific taste and odor. They receive gustatory and olfactory cues and may be important in discrimination during feeding control.

References

1. Oomura Y, Nishono H, Aou S, Lenard L (1986) Opiate mechanisms in reward-related neuronal responses during operant feeding behavior of the monkey. Brain Res 365:335–339
2. Oomura Y (1989) Endogenous chemical sense in control of feeding. In: Ottoson D, Perl ER, Schmidt RF, Shimazu H, Wills WD (eds) Progress in sensory physiology. Springer, Heidelberg, pp 171–191
3. Nishino H, Oomura Y, Aou S, Lenard L (1987) Catecholaminergic mechanisms of feeding related lateral hypothalamic activity in the monkey. Brain Res 405:56–67
4. Oomura Y (1987) Modulation of prefrontal and hypothalamic activity by chemical senses in the chronic monkey. In: Kawamura Y, Kare MR (eds) Umami: A basic taste. Dekker, New York, pp 481–509
5. Burton MJ, Rolls ET, Mora F (1976) Effects of hunger on the responses of neurons in the lateral hypothalamus to the sight and taste of food. Exp Neurol 51:668–677
6. Karadi Z, Oomura Y, Nishino H, et al. (1992) Responses of lateral hypothalamic glucose-sensitive and glucose-insensitive neurons to chemical stimuli in behaving rhesus monkey. J Neurophysiol 67:389–400
7. Karadi Z, Oomura Y, Nishino H, Aou S (1989) Olfactory coding in the monkey lateral hypothalamus: behavioral and neurochemical properties of odor-responding neurons. Physiol Behav 45:1249–1257
8. Takagi SF (1986) Studies on the olfactory nervous system of the old world monkey. Prog Neurobiol 27:195–250
9. Van Buskirk RL, Ericksom RP (1977) Odorant responses in taste neurons of the rat NTS. Brain Res 135:287–303
10. Mesulam MM, Mufson EJ (1982) Insula of the old world monkey. III. Efferent cortical output and moments on function. J Comp Neurol 212:38–52
11. Scott T-R, Giza BK (1987) The effects of physiological condition on taste in rats and primates. In: Kawamura Y, Kare MR (ed) Umami: A basic taste. Dekker, New York, pp 409–437
12. Oomura Y, Nishino H, Karádi Z, Aou S, Scott TR (1991) Taste and olfactory modulation of feeding related neurons in behaving monkey. Physiol Behav 49:769–779
13. Rolls ET, Sienkiewcz ZJ, Yaxley S (1989) Hunger modulates the responses to gustatory stimuli of single neurons in the caudolateral orbitofrontal cortex of the macaque monkey. Eur J Neurosci 1:53–60
14. Kai Y, Oomura Y, Shimizu N (1988) Responses of rat lateral hypothalamic neuron activity to dorsal raphe nuclei stimulation. J Neurophysiol 60:524–535
15. Jiang LH, Oomura Y (1988) The effects of catecholamine receptor antagonists on feeding related neuronal activity in the central amygdaloid nucleus of the monkey: a microiontophoretic study. J Neurophysiol 60:536–548
16. Inoue M, Oomura Y, Aou S, Nishino H, Sikdar SK (1985) Reward related activity in monkey dorsolateral prefrontal cortex during feeding behavior. Brain Res 326:307–312

Taste Responses During Ingestion

Ralph Norgren[1], Hisao Nishijo[2], and Kiyomi Nakamura[3]

Key words. Chronic recording—Neurophysiology—
Nucleus of the solitary tract—Parabrachial nucleus—
Quality coding—Taste

Introduction

The conventional view of central sensory nuclei is
epitomized in the term "relay," a system that faithfully reproduces an incoming signal and passes it
on to the next level with minimal distortion. This
concept is derived from the assumption that perception takes place in the neocortex and that the
task of the subcortical sensory nuclei is to transmit
the activity of the peripheral receptors to that sen-
sorium. For both vision and audition, this viewpoint
has been replaced by a more dynamic model in
which different parameters of the sensory message
are processed separately, in parallel. Although most
scientific attention remains focused on the neocortex,
the process of elaborating the sensory activity produced at the receptors is thought to take place all
along the neuraxis.

In the gustatory system the process of replacing
the old telephonic analogy with a newer one has
proceeded more slowly, perhaps because the data
from the chemical senses never really fit well into
what was essentially a spatiotemporal framework.
Although the code for gustatory quality remains at
issue, particularly within modalities, data from acute
studies indicated that taste neurons in the central
nervous system (CNS) differed only in minor ways
from their peripheral counterparts. This was particularly true for the first two central relays in the rodent
gustatory system, which are less suppressed by
general anesthesia than those in the forebrain [1].

Nevertheless, other evidence indicated that the
brainstem gustatory areas, the nucleus of the solitary
tract (NST) and the parabrachial nuclei (PBN), were
capable of supporting taste-guided behavior without
the benefit forebrain processing. Chronically decerebrate rats exhibit normal orofacial responses to
gustatory stimuli and even alter these responses as
a function of some need states [2]. In otherwise
normal rats, damage to the NST produced different
effects on taste-guided behavior than similar lesions
of the PBN. For example, bilateral lesions of the
NST reduce or eliminate ordinary taste preferences
and aversions but allow acquisition of a conditioned
taste aversion (CTA) and the expression of sodium
appetite [3]. Similar lesions in the PBN have only
modest effects on preference functions but routinely
block CTA and salt appetite [4]. This point suggests
that the brainstem gustatory nuclei may not only
support taste-guided behavior but also each have
functions that differ from one another.

To assess gustatory afferent activity in the brainstem of an awake, behaving animal, we exploited a
preparation perfected in the laboratory of Professor
Ono in Toyama. We have now tested several hundred
taste neurons in both the NST and PBN using a
variety of sapid stimuli. In the awake, behaving rat,
gustatory activity in the NST and PBN not only
differed from that in the anesthetized state but also
across the two levels. These data do not explain the
differing functions of the two areas, but they do
support the contention that, as in other sensory
systems, the central gustatory system transforms and
utilizes the afferent activity that passes through it on
the way to the neocortex.

Materials and Methods

These experiments use identical preparations and
essentially identical procedures. Only the recording
site and the sapid stimuli vary. The rats are ac-
customed to receiving most of their daily water
ration in a small, plastic chamber. Subsequently,
while anesthetized, an acrylic cap is attached to their
skull, several intraoral catheters are implanted, and
the intended recording site is identified electrophysiologically. The animals are then returned to the
water deprivation schedule and gradually adapted
to having their heads restrained by anchoring the
acrylic cap to the stereotaxic apparatus. When they
have acclimated to remaining in the apparatus for up
to 3 h, the recording experiment is begun. Typically,
each animal has two recording sessions per week for
up to 8 weeks. On the intervening days, the rats are
maintained on their deprivation schedule and, at
least part of the time, given their water ration in the

[1] Department of Behavioral Science, College of Medicine,
The Pennsylvania State University, Hershey, PA 17033,
USA
[2] Department of Physiology, Toyama Medical and Pharmaceutical University, Sugitani, Toyama, 930-01 Japan
[3] Department of Electronics and Informatics, Toyama Prefectural University, Kosugi-machi, Toyama, 939-03 Japan

recording apparatus to ensure their continued habituation to the situation.

During a recording session, a conventional hydraulic microdrive is used to position a glass-insulated tungsten microelectrode in either the NST or the PBN using coordinates established during the acute phase of the experiment. When a neuron is isolated in the vicinity of the taste area, it is screened for gustatory responsiveness, typically using a moderate concentration of NaCl or sucrose. If the cell responds to sapid stimuli and the amplitude and waveform remain stable, it is tested with a battery of chemicals according to a predetermined protocol. To date, with minor exceptions, each stimulus trial has consisted of a 50 µl infusion made over 1.0 s via an intraoral cannula, always preceded and followed by at least one similar infusion of distilled water. All fluids are at room temperature (24° ± 1°C). A sketch of the chronic recording setup has been published elsewhere [5].

Neural activity is archived on analog magnetic tape along with stimulus onset markers and voice commentary. Again with a few exceptions, the data consist of the number of action potentials that occur during the first 5.0 s after stimulus onset. We tabulate both the raw responses and the differential responses (stimulus minus water), but most of the subsequent analyses use the latter data. A significant response is defined arbitrarily as activity that differs from baseline by ±2.5 SD. For water responses, the baseline is a running average of spontaneous activity; for taste responses, it is a running average of water responses. For the experiments summarized here, all data analysis was done off-line using the taped records.

Two sets of virtually identical experiments have been carried out while recording from the nucleus of the solitary tract and the parabrachial nuclei. Initially in each area, taste neurons were tested with a concentration series of sucrose, NaCl, citric acid, and quinine HCl—chemicals chosen to serve as exemplars of the four standard gustatory modalities. Subsequently, another sample of cells from each area was tested with an extended array of chemicals, usually at only a single concentration. For comparison, this array contained a midrange concentration of the four standard stimuli plus 11 other chemicals (Table 1). Two other, smaller experiments also have been carried out, one dealing with gustatory responses to monosodium glutamate [6] and the other with intraoral infusions versus licking as a method of stimulus presentation [7]. The most detailed summary of the methods appears in the first publication in the series [8].

Results

Although the rats in these experiments are awake and behaving, their situation is hardly normal. One difference is that the stimuli are applied via intraoral catheters rather than through normal ingestion. In the first case, fluids are infused continuously; in the other, the animal itself applies them in a pulsatile fashion by licking. To assess these two methods of stimulation, we tested 39 PBN neurons with the same sapid chemicals applied by intraoral infusion and by licking [7]. On average, the route of stimulation had little effect on the magnitude of the neural responses. For individual neurons, the correlation between infusion and licking responses ranged from 0.68 to 0.9997, but the average for the sample was 0.95. This measure is based on response magnitude alone, but visual inspection of the neural data and the accompanying genioglossus electromyographic (EMG) activity revealed apparent correlation between the two for about 40% of the neurons

TABLE 1. Stimulus chemicals and their molar concentrations.

Stimuli	Standard (M)	Range ($\frac{1}{2}$ log steps)
Sucrose	0.3	0.01–1.00
Sodium chloride	0.1	0.01–1.00
Citric acid	0.01	0.0001–0.03
Quinine hydrochloride	0.0001	0.00001–0.003
Glycine	0.3	
Fructose	0.3	
Maltose	0.3	
Polycose	0.1	
Monosodium L-glutamate	0.1	
Sodium nitrate	0.1	
Potassium chloride	0.1	
Magnesium chloride	0.1	
Ammonium chloride	0.1	
Malic acid	0.01	
Hydrochloric acid	0.01	

The four standard stimuli are italicized.

tested. Although no attempt was made to quantify these relations, they do raise the possibility that, at least at the microstructural level, pulsatile stimulus application remains evident in the neural record three synapses away from the receptors. Nevertheless, because our data analysis so far is based on simple response magnitude over 5 s, the method of stimulus application is less important than it might be otherwise.

The largest sample of neurons from which we gathered data came from the nucleus of the solitary tract using the four standard stimuli [9]. Based on the best-stimulus classification, these cells were more or less equally divided among NaCl-best ($n = 25$), sucrose-best ($n = 42$), and citric acid-best ($n = 30$). Only 4 of the 101 taste neurons responded best to quinine. Within the categories, however, the specificity of the response profiles differed considerably. Acid-best neurons were the most specific (23 of 30 responded significantly only to citric acid), but they also had lower overall spontaneous rates and response magnitudes than either NaCl-best or sucrose-best cells. The NaCl-best cells, on the other hand, had the largest average response magnitude and the fewest specific neurons (3/25). In fact, using our standard concentrations, more than 70% of these cells responded nearly as well to sucrose as to NaCl. Sucrose-best cells were more numerous than any of the other categories. Their response profiles were more or less equally divided between sucrose-specific ($n = 18$) and those that also responded to NaCl ($n = 14$) or to citric acid ($n = 10$).

In fact, it is the degree of specificity that most distinguishes NST gustatory neurons in awake, behaving animals from their counterparts in anesthetized preparations. Using a common statistical criterion, 34% of the NST taste neurons in our sample responded to only one of the four standard stimuli, whereas in anesthetized rats the corresponding figure was only 5.5%. This difference was particularly noticeable for acid responses. In acute preparations, the same cell often responds well to acid and NaCl. In our sample, this combination was not common (9% of the NaCl- and acid-best cells). On other dimensions, the presence or absence of anesthesia appeared to have relatively little effect on NST gustatory cells. In awake animals, the spontaneous rate was almost twice that in the anesthetized preparation, but the variance was large in both situations. The relative frequency of best-stimulus categories was similar, as were the concentration response functions and the breadth of response measures.

In the parabrachial nuclei, the spectrum of responsiveness changes considerably with respect to NST taste cells in awake animals and to PBN cells in the anesthetized case [8]. The mean spontaneous rate of PBN taste neurons in our sample was at least twice that of acute preparations and three times that

of the chronic NST cells. Perhaps because of this higher spontaneous rate, only 17% of the PBN gustatory neurons also responded to water, but 40% did so in the NST. Within response categories, the biggest changes were in NaCl- and acid-best neurons. In the NST only 25% of the taste neurons were NaCl-best; in the PBN that figure rose to 71%. In addition, a much larger percentage of the pontine NaCl-best cells responded significantly only to salt (48% vs 12%). The prevalence of sodium sensitivity mimics that observed in acute preparations in both the medulla and the pons, but the response profiles differ. As mentioned above, in acute preparations Na^+ and acid sensitivity appear commonly in the same neurons. In behaving rats, pontine taste cells seldom respond to acid under any circumstances. Only 2 of 42 PBN neurons responded best to citric acid and only 4 of the NaCl-best cells (<10%) responded second best to acid.

When we replicated the NST and PBN experiments using an extended array of stimuli, the basic response patterns did not vary greatly, but some further complexity appeared ([10]; Nishijo and Norgren, manuscript in preparation). Most of the additional chemicals fit into classes represented by the four standard stimuli (i.e., sodium salts, non-sodium chloride salts, sugars, and acids). Three chemicals were picked for their unusual properties: glycine, which is sweet to humans, but not a sugar; Polycose, which is preferred by rats, but barely tastes sweet to humans; and monsodium glutamate (MSG), which is associated with a unique taste sensation (umami). With only four stimuli, multidimensional analyses of the neural data reveal little that is not evident from a simple best-stimulus categorization. The additional stimuli allowed for a check on these admittedly preconceived notions; and, in fact, the multidimensional analyses suggest that allowing only four perceptual categories may constrain the data more than is warranted.

Not surprisingly, in the NST sodium salts drove NaCl-best neurons robustly and nearly equally ($\bar{x} = 15.7$–20.8 spikes/s), whereas the nonsodium, chloride salts did not. Perhaps not so obviously, the sucrose-best cells responded as well to glycine and MSG as to sucrose ($\bar{x} = 13.7$–17.8), but only modestly to the other sugars ($\bar{x} = 8.4$ and 9.8, respectively). As previously, acid-best cells were numerous but relatively sluggish, particularly when HCl was the stimulus ($\bar{x} = 3.1$). When all these data (50 cells $\times$ 15 stimuli) were processed through a factor analysis, five factors accounted for almost 99% of the variance. Three of the factors appeared to correspond to standard taste qualities (sweet, salty, sour) because they included heavy loadings for stimuli that elicit those qualities in humans. The other two, however, did not fit obvious categories. One contained heavy loadings for the nonsodium, chloride salts; the other covered HCl, quinine, and Polycose. Only two stimuli, Polycose and MSG, were represented prominently in more than one factor.

In the PBN, the expanded array of stimulus chemicals revealed a response profile that was only hinted at in the NST. Based on the responses of NST taste cells, the nonsodium, chloride salts (KCl, $MgCl_2$, NH_4Cl) are, at best, moderate to poor sapid stimuli. In addition, the cells that were activated by these chemicals were more or less evenly distributed across best-stimulus classes. Nevertheless, several multidimensional analyses separated these three chemicals from the others in the array. In the pons, the same three stimuli became effective at driving about 50% of the NaCl-best taste cells. In fact, parabrachial NaCl-best neurons had two distinct response profiles. In one, Na^+ ions appeared to be the effective stimulus moiety because the neurons were driven equally well by equimolar NaCl, $NaNO_3$, and MSG ($\bar{x} > 60/s$). On average, these cells also responded to sucrose and glycine but at lower rates ($\bar{x} < 30/s$). In the other group of NaCl-best cells, Cl^- ions appeared to be the effective moiety because they were driven as well by the nonsodium, chloride salts as by NaCl, but not at all by either $NaNO_3$ or MSG. Neither quinine HCl or HCl itself were effective for this group, but the standard concentrations for these chemicals were much lower than for the salts.

Discussion

The awake, behaving rats used in these experiments provide a far more natural preparation than does an anesthetized animal, but the situation cannot be construed as ideal. Although familiar with the circumstances, the rats are away from their home cages, moderately water-deprived, and restrained. In addition, to increase the stability of the recording, we find it useful to infuse stimuli through intraoral cannulas rather than allow the rats to ingest fluids normally by licking. Based on the overall magnitude of the responses, these two methods of applying fluids appeared to be equivalent. A study of oral EMG responses supports this equivalence [11]. Intraoral infusions of sapid stimuli and spontaneous licking of the same fluids elicited virtually identical patterns of EMG activity from a variety of lingual and masticatory muscles.

Although this summary has concentrated on the differences between the data from awake and anesthetized preparations, at least in the medulla, many similarities exist as well. These similarities are all the more noteworthy because the preparations differ on a variety of dimensions in addition to the presence or absence of general anesthesia. In the chronic preparation, stimulus application results in a behavioral response, which then produces a barrage of intraoral somatosensory activity that is processed along with the gustatory neural message. The volume, distribution, and rate of stimulation vary considerably between the two situations. Our typical stimulus was 50 µl in 1.0 s, distributed throughout the oral cavity and swallowed. In acute preparations, the volumes are usually in excess of 1.0 ml applied for at least several s, but often to only one receptor field. The same is true for the amount of rinse used and the effects of saliva. Given all the differences in these parameters, often considered to be critical variables by gustatory researchers, the similarities between the responses of NST taste cells in the awake and anesthetized animals may be more remarkable than the differences.

Nevertheless, differences do exist. In the NST of awake animals, taste neurons respond more robustly, but to fewer stimuli than in the anesthetized condition, suggesting more active processing and perhaps better discrimination. In the pons, many more cells respond to NaCl, but much of this additional activity apparently results from neurons that extract information about Cl^- ions rather than Na^+ ions in the stimuli. In other words, the brainstem nuclei are not only relaying gustatory afferent activity to the forebrain but actively processing it as well.

Acknowledgments. The research reviewed here was performed at the College of Medicine of the Pennsylvania State University while Drs. Nishijo and Nakamura were on leave from their faculty positions in the Department of Physiology, Faculty of Medicine, Toyama Medical and Pharmaceutical University. We are grateful to Professor Taketoshi Ono for his support through all phases of this project. The research was supported by grants from the US NIH (DC 00240 and MH 43787), the Salt Science Research Foundation (no. 9351), and the Japanese Ministry of Education, Science, and Culture (04246105). Dr. Norgren is a recipient of an NIMH Research Scientist Award (MH 00653).

References

1. Travers J, Travers S, Norgren R (1987) Gustatory neural processing in the hindbrain. Annu Rev Neurosci 10:595–632
2. Norgren R, Grill H (1982) Brain stem control of ingestive behavior. In: Pfaff D (ed) Physiological mechanisms of motivation. Springer, New York, pp 99–131
3. Shimura T, Grigson PS, Norgren R (1992) Gustatory function after lesions of the rostral nucleus of the solitary tract in rats. Soc Neurosci Abstr 18:1040
4. Spector A, Grill H, Norgren R (1993) Concentration-dependent licking of sucrose and sodium chloride in rats with parabrachial nucleus lesions. Physiol Behav 53:277–283
5. Norgren R, Nishijo H, Travers S (1989) Taste responses from the entire gustatory apparatus. Ann NY Acad Sci 575:246–264

6. Nishijo H, Ono T, Norgren R (1991) Parabrachial gustatory neural responses to monosodium glutamate ingested by awake rats. Physiol Behav 49:965–971

7. Nishijo H, Norgren R (1991) Parabrachial gustatory neural activity during licking in rats. J Neurophysiol 66:974–985

8. Nishijo H, Norgren R (1990) Gustatory neural activity in the parabrachial nuclei of awake rats. J Neurophysiol 63:707–724

9. Nakamura K, Norgren R (1991) Gustatory responses of neurons in the nucleus of the solitary tract of behaving rats. J Neurophysiol 66:1232–1248

10. Nakamura K, Norgren R (1993) Taste responses of neurons in the nucleus of the solitary tract of awake rats: an extended stimulus array. J Neurophysiol 70:879–891

11. Travers JB, Norgren R (1986) An electromyographic analysis of the ingestion and rejection of sapid stimuli in the rat. Behav Neurosci 100:544–555

Role of Taste in Dietary Selection

Thomas R. Scott and Barbara K. Giza[1]

Key words. Taste—Feeding—Hunger—Rat—Diet—Electrophysiology

Introduction

The selection of an adequate diet places multiple demands on an omnivore. Diverse nutrients are called for to satisfy its complex biochemical requirements, and the mix among them may change with age, disease, or nutritional or reproductive status. Yet omnivores *do* select an appropriate diet over time, a capacity for which Curt Richter coined the term "body wisdom." The hypothesis we advance here is that body wisdom is achieved simply by maximizing hedonic reward—by consuming that which tastes best at the moment. The flexibility required to maintain the match between the needed and the desirable is provided by the taste system. Thus taste responses in the central nervous system (CNS) are influenced not only by the quality of the chemical in the mouth but also by the past experiences and momentary physiological needs of the animal. There are instances in which these factors have been shown to be responsible for modifying taste sensitivity such that chemicals that taste best are also nutritionally most beneficial for the animal at the moment.

Decisions about the acceptability of a potential food are based largely on the taste responses it evokes. That gustatory input has at least four consequences: (1) it activates *somatic reflexes* that control the immediate decision to swallow or reject; (2) if the former, it releases *autonomic reflexes* by which the impending digestive processes are facilitated; (3) it guides a *hedonic appraisal* of the stimulus, which, if positive, sustains feeding, extending a bite to a meal, a meal to a diet; and (4) it permits an *evaluation of quality and intensity*. Each of these processes may be mediated by a separate anatomic branch of the central gustatory complex, as described below.

The role of taste as a mediator between the external and internal chemical environments is reflected in its anatomy. In the hindbrain, the special visceral afferent fibers of taste terminate in close apposition to both somatosensory and visceral afferents, first in the nucleus tractus solitarius (NTS) and subsequently in the pontine parabrachial nuclei (PBN). Beyond the pons, taste axons bifurcate, sending one projection to the thalamocortical axis in conjunction with oral somesthesis and the other into ventral forebrain areas associated with autonomic processes.

Responses to Nutrients and Toxins

Somatic Reflexes

Clusters of neurons in the ventral region of gustatory NTS send their projections not rostrally to the PBN but ventrolaterally. These paths lead to salivatory and pharyngeal efferents in the reticular formation, to the hypoglossal nucleus, and toward the facial and ambiguus nuclei, through which acceptance-rejection reflexes may be orchestrated. These reflexes are fully integrated within the hindbrain. They are stereotypical to each of the basic taste qualities and are unaltered by the loss of tissue rostral to the midbrain. The orofacial reflexes are adaptive in dealing with the chemical in the mouth—swallowing if appetitive, rejecting if aversive—and in communicating a hedonic message to other members of the species. This hedonic monitor is innate and is neurally intact by the seventh gestational month in humans.

Autonomic Reflexes

A second projection from the ventral NTS runs caudally and ventrally, to ramify through viscerosensory NTS and the dorsal motor nucleus of the vagus (DMNX), through which it may influence the autonomic processes of digestion. There are several cephalic phase pancreatic and gastrointestinal reflexes associated with ingestion, the best understood being that which controls insulin release. Thousands of beta cells in the rat's islets of Langerhans are stimulated to release insulin by fewer than 200 fibers coursing through the hepatic and gastric branches of the vagus. These fibers originate in the rostromedial DMNX, which is overlaid by the gustatory NTS. Neurons in DMNX send apical dendrites into the NTS, effectively fusing the two structures and offering gustatory input direct influence over autonomic reflexes. Consequently, a sweet taste stimulus

[1] Department of Psychology, University of Delaware, Newark, DE 19716, USA

elicits insulin secretion, and this effect is blocked by vagotomy [1].

Hedonic Appraisal

From dorsal areas of the gustatory NTS, axons proceed rostrally to the PBN and then to thalamus and ventral forebrain. Parabrachial projections include the lateral hypothalamus, substantia innominata, central nucleus of the amygdala, and bed nucleus of the stria terminalis in their itinerary. Moreover, the connections are reciprocal, with forebrain areas returning centrifugal projections to brainstem taste nuclei. Therefore the rat's taste system communicates directly and reciprocally with structures associated with hedonics, feeding, and emotions.

It is not surprising, then, that taste input carries a hedonic component. The area most clearly involved in mediating gustatory-induced hedonics is the hypothalamus. Those tastes that arouse activity in lateral regions of the hypothalamus are appetitive; those that elicit activity in ventromedial hypothalamus are avoided. Moreover, areas of the lateral hypothalamus that sustain stimulus-bound feeding project to the PBN and NTS, evoking from taste-sensitive neurons responses similar to those they would give to taste stimuli [2]. Thus the hypothalamus has the means to impose a bias on hindbrain taste cells, presumably according to the animal's nutritional state about which it receives such detailed information.

The neurochemical basis for these processes is also being defined. Positive reinforcement is associated with increased dopamine (DA) release in the limbic system [3] and with reductions in acetylcholine (ACh) concentration in the nucleus accumbens [4] and serotonin (5-hydroxytryptamine; 5-HT) in the hypothalamus [5]. The release of DA, then, is associated with hedonically positive experiences, whereas the reverse is true for ACh and 5-HT. Saccharin placed on a rat's tongue evokes DA release in the nucleus accumbens, in accordance with its reinforcing value. If paired with nausea to create a conditioned taste aversion, however, saccharin, now aversive to the rat, causes a reduction in DA and, instead, releases ACh in accumbens and 5-HT in the lateral hypothalamus. Therefore the likely neurochemical basis for the reinforcing value of saccharin was lost. Conversely, when a rat is conditioned to prefer an originally neutral taste by pairing it with the intragastric infusion of nutrients, that acquired taste comes to elicit increased DA release in accumbens in accord with its newfound rewarding properties.

In sum, the anatomic, physiological, and neurochemical bases for an intimate interaction between gustation and both positive and negative reinforcement have been established.

Evaluation of Quality and Intensity

Humans make discriminations among the qualities and concentrations of gustatory stimuli independently of hedonic appreciation. Electrophysiological recordings in the primary taste cortex of the macaque suggest that quality and intensity information is represented here. Concentration-response functions for all basic stimuli except HCl are compatible with those reported in human psychophysical experiments. Taste quality, as indexed by the correlations among patterns of activity evoked by a range of tastants in the monkey cortex, matched well with the perceived similarity of the same stimuli, as reported in humans. Thus activity in primary taste cortex of the macaque appears to serve the gustatory discriminative capacity of the human observer.

Alterations in the Gustatory Code

The data cited above support the position that the gustatory system performs a differentiation of nutrients from toxins, and that the consequences of that process are manifested in reflexive behavior, hedonic evaluation, and cognitive appraisal. The capacity to perform this analysis is probably genetically endowed and so is consistent across individuals and over time. Although providing a broad, effective mechanism for ingesting nutrients and avoiding toxins, this innate process would not be sensitive to the idiosyncratic needs or allergies of the individual or to changes in those needs over time. If taste is to serve its host faithfully, its signals should be modifiable to reflect that variety of conditions. There is now a body of information showing that taste responses can indeed by influenced to accommodate the individual experiences and momentary needs of the animal.

Effect of Experience

An animal's experience has a profound and enduring effect on its reaction to taste stimuli. Experimental protocols have now been established for creating conditioned taste aversions and preferences by pairing gustatory experiences with nausea and nutritional repletion, respectively.

The effects on taste responsiveness of creating a conditioned preference have not yet been assessed, though neurophysiological studies are under way. The effects of developing a conditioned aversion, however, have been documented. Chang and Scott [6] recorded single unit gustatory activity from the NTS of naive rats and from those conditioned to avoid the taste of saccharin (the conditioned stimulus; CS). The response among saccharin-sensitive neurons to the CS in conditioned animals was 65% greater than in controls. Moreover, an analysis of the time course of responses in the two groups revealed that nearly the entire difference in discharge rate was attributable to a burst of activity that diverged from control levels at 600 ms and reached a peak at 900 ms following stimulus onset. Therefore the major consequence of the conditioning procedure was to increase responsiveness to the saccharin CS through a well-defined peak of activity.

What was the effect of this change in responsiveness on the quality code for saccharin? The activity profile evoked by saccharin in conditioned rats was more closely related to that elicited by quinine than is normally the case. The convergence of the neural signals for saccharin and quinine offers a neural basis for the similar behavioral reaction evoked by quinine and sweet chemicals to which an aversion has been conditioned. The altered gustatory code for the formerly sweet CS may also underlie the abolition of the cephalic insulin release observed after aversive conditioning, as well as the decline in dopamine release in the nucleus accumbens. Therefore the consequences of replacing a "sweet" with a "bitter" signal were manifested throughout the various components of the ingestive process. The remarkable plasticity of the gustatory afferent code at the level of the medulla was sufficient to explain the reflexive and hedonic consequences of a conditioned aversion.

Effect of Physiological Need

The momentary physiological status of an animal has a major impact on feeding behavior. Because physiological needs are complex and in constant flux, the hedonic value of the taste perceptions that guide feeding must be correspondingly labile.

Long-Term Needs

Many mammals express an innate preference for the taste of sodium, an adaptive response to diets that are salt-deficient. This preference is exaggerated under conditions of severe sodium deprivation and is thought to result from changes in the hedonic value of tasted salt: Concentrations of NaCl that had been rejected by sodium-replete rats are eagerly sought when these same animals are salt-deprived.

The total gustatory response to NaCl declined in salt-deprived rats. When this reduction was analyzed by taste neuron type, it was found that the responsiveness of salt-sensitive cells was profoundly suppressed, and that this effect was partially offset by a sharp increase in sensitivity to NaCl among sweet-sensitive neurons [7]. The net effect was to transfer the burden of coding sodium from salt- to sweet-sensitive cells. Accordingly, the activity profile evoked by NaCl shifted decisively toward those elicited by sugars. In conjunction with this shift, the aversive behavioral reactions normally associated with the taste of concentrated NaCl were replaced by acceptance reflexes, and dopamine was released in the nucleus accumbens.

Just as the formation of a conditioned taste aversion altered the behavioral, autonomic, and neurochemical consequences of a taste from positive to negative, so salt deprivation changed these same processes in the opposite direction in response to NaCl. In both the negative and positive cases, alterations in gustatory afferent activity in the hindbrain were sufficient to account for the attendant feeding behavior and the reward derived from it.

Transient Needs

Whereas the appreciation of salt levels occurs over a period of days, the availability of glucose is of almost hourly concern to a small, homeothermic mammal. There must be accommodation of the decision to swallow or reject and in the hedonic assessment that drives that decision, as the danger of malnutrition weighs more heavily against the risk of ingesting a toxin. Common experience reinforces the results of psychophysical studies: With deprivation foods become more palatable and with satiety less so. These effects may also be mediated by changes in gustatory activity.

When rats were made hyperglycemic by an intravenous glucose load, gustatory responsiveness in the NTS was reduced [8]. A behavioral study verified that the rats' perceived intensity of glucose solutions declined in parallel with this gustatory suppression [9]. One implication of this finding is that the reward associated with appetitive tastes might be lessened, making termination of the meal more likely. Related studies in macaques revealed taste-sensitive neurons in orbitofrontal cortex whose activity was related to the hedonic value of a tastant [10]. Thus, incentives for the initiation and maintenance of feeding may be modulated by momentary physiological needs and by their fulfillment.

Conclusions

The sense of taste is both exteroceptor and interoceptor. In the former capacity, taste signals the quality and intensity of chemicals through a spatiotemporal code devoted to the discrimination of nutrients from toxins. This analysis offers a first approximation of the appropriateness of consuming a potential food.

As interoceptor, taste is subject to a degree of plasticity that permits finer adjustments in food selection. The acceptability of a tastant must be verified by its physiological consequences. If a taste that arouses feeding reflexes and pleasure is followed by nausea, the gustatory signal is altered toward one of a toxin. With that gustatory modification, all subsequent processes are affected: Somatic reflexes reverse from acceptance to rejection, autonomic reflexes associated with digestion are suppressed, and the neurochemical substrates of reward are abolished. Conversely, an inherently unappealing taste that is paired with nutritional repletion gains hedonic appeal. Therefore the wisdom inherited with the species may be tailored to the biochemical needs of the individual.

Finally, within each individual, taste responses and their reflexive and hedonic sequelae are in flux to match momentary physiological needs. It follows that a diet selected purely according to the reinforcing value of its components is a varied one. The variability need not reside in hindbrain reflex arcs or in forebrain reward systems, but only in the form of the gustatory code that drives them. Thus the animal has one simple goal to fulfill: to maximize hedonism. What tastes best to it is what will best serve its physiological needs at the moment.

Acknowledgments. The authors were supported by research grants from the NIH (DK 30964) and the National Science Foundation (BNS 9001213).

References

1. Louis-Sylvestre J, Giachetti I, LeMagnen J (1983) Vagotomy abolishes the differential palatability of food. Appetite 4:295–299

2. Murzi E, Hernandez L, Baptista T (1986) Lateral hypothalamic sites eliciting eating affect medullary taste neurons in rats. Physiol Behav 36:829–834

3. Wise RA, Spindler J, DeWit H, Gerber GJ (1978) Neuroleptic-induced "anhedonia" in rats; pimozide blocks reward quality of food. Science 201:262–264

4. Rada P, Mark GP, Pothos E, Hoebel BG (1991) Systemic morphine simultaneously decreases extracellular acetylcholine and increases dopamine in the nucleus accumbens of freely-moving rats. Neuropharmacology 30:1133–1136

5. Stanley BG, Schwartz DH, Hernandez L, Liebowitz SF, Hoebel BG (1989) Patterns of extracellular 5-hydroxyindoleacetic acid (5-HIAA) in the paraventricular hypothalamus (PVN): relation to circadian rhythm and deprivation-induced eating behavior. Pharmacol Biochem Behav 33:257–260

6. Change F-CT, Scott TR (1984) Conditioned taste aversions modify neural responses in the rat nucleus tractus solitarius. J Neurosci 4:1850–1862

7. Jacobs KM, Mark GP, Scott TR (1988) Taste responses in the nucleus tractus solitarius of sodium-deprived rats. J Physiol (Lond) 406:393–410

8. Giza BK, Scott TR (1983) Blood glucose selectively affects taste-evoked activity in rat nucleus tractus solitarius. Physiol Behav 31:643–650

9. Giza BK, Scott TR (1987) Blood glucose level affects perceived sweetness intensity in rats. Physiol Behav 41:459–464

10. Rolls ET, Sienkiewicz ZJ, Yaxley S (1989) Hunger modulates the responses to gustatory stimuli of single neurons in the caudolateral orbitofrontal cortex of the macaque monkey. Eur J Neurosci 1:53–70

Changes of Taste Preference with Nutritional Disorder and Neural Plasticity in the Brain

Kunio Torii[1,2], Takuya Murata[2], Masato Mori[2], Kazumitsu Hanai[3], and Taketoshi Ono[4]

Key words. Preference—Nutritional disorder—Neural plasticity—Neural response—Neurotrophic factor—Lateral hypothalamic area

Introduction

The gustatory and anticipatory cephalic stimuli that are detected during a meal yield nutritional information and aid in the efficient digestion of food [1–3]. It is possible that animals, including humans, can detect the amount and quality of protein ingested during a meal and use this information, via cephalic relays, to initiate digestion. They must, of course, ultimately have a means of determining whether the level of dietary protein intake is sufficient for body needs. When the animal detects that protein intake is deficient or imbalanced, mechanisms of protein sparing as well as those associated with searching for the deficient materials can come into play.

Animals change their diet or liquid preference when some component of their diet is changed. For example when an L-lysine-deficient diet is offered, they preferentially ingest a previously aversive tasting solution of L-lysine (see below). Although there are reports of dependence of the content of amino acids in the brain on amino acid intake [4], few reports suggest neural control of amino acid intake [5,6]. Some studies suggest that this mechanism might include neurons of the lateral hypothalamic area (LHA) [6].

It is well known that the LHA is a feeding center that controls overt actions, such as eating and drinking [7,8]. It has been reported that when an animal is exposed to a diet deficient in the essential amino acids anorexia occurs, and the concentrations of essential amino acids in the brain and plasma decrease within a few hours [4]. These data suggest that the central nervous system, especially the LHA, may influence amino acid and protein metabolism

through control of amino acid ingestion. The present studies were undertaken to investigate the role of the LHA in the ingestion of amino acids.

This chapter discusses not only the biologic roles of learning preference but the neural plasticity of LHA neurons in awake rats to both central iontophoretic application of the essential amino acid L-lysine and its ingestion under a condition of L-lysine deficiency. Additionally, release and brain localization of possible neurotrophic factors assayed by behavioral changes of *Hydra japonica* are also discussed.

Induction of Learning Preference

Sprague-Dawley rats were exposed to prolonged essential L-amino acid deficiency. The diurnal patterns of L-amino acid concentration in plasma and brain were examined. In a study examining L-lysine deficiency, the diurnal patterns in plasma and brain L-lysine were comparable. In the L-lysine-deficient animals, plasma and brain L-lysine both dramatically declined subsequent to the onset of feeding (during the dark period), and both measures were elevated to near control levels by the end of the light period.

The daily patterns of other amino acids were essentially unchanged, regardless of the nutritional status of L-lysine. Initially, when the L-lysine-deficient diet was offered to rats, they ate it in a manner and amount similar to that of the control diet. Several hours after the onset of the dark period, however, they were reluctant to eat the L-lysine-deficient diet and began to ingest an L-lysine solution.

Two weeks later, animals displayed a preference for both L-lysine and monosodium L-glutamate (MSG) during the dark period and were growing normally. Rats ceased drinking the L-lysine solution immediately after the L-lysine-sufficient diet was offered, but preference for MSG was sustained.

In this case, the anorexic animal may seek out and develop a strong preference for the missing nutrient, e.g., L-lysine. Once normal growth and maintenance has been reestablished, preference for an umami taste substance is reestablished. This reestablishment of umami preference may be attributable to the reestablishment of normal levels of ammonia in

[1] Ajinomoto Co., Inc., Yokohama, 244 Japan
[2] Torii Nutrient-stasis Project, ERATO, Research Development Corporation of Japan, Technowave 100, 1-1-25 Shin-urashima-cho, Kanagawa-ku, Yokohama, 221 Japan
[3] Shiga Medical University, Shiga, 252 Japan
[4] Toyama Medical & Pharmaceutical University, Toyama, 930-01 Japan

plasma and, probably, brain [9]. It seems as though, with an adequate protein diet, the presence of MSG or preference for it results in lower plasma ammonia levels [9].

Neuronal Plasticity of Sensitivity to Nutrients

The single neuron activity of the LHA of awake rats, Wistar strain, was recorded using the paradigm previously reported [10].

Neuronal responses to cue tones were classified into two types: nondifferential and differential. All neurons that responded nondifferentially to every cue tone also responded nondifferentially to licking for any solution; and most nondifferential and differential neurons that responded to cue tones also responded to licking. These phenomena indicate that the neurons responding to cue tones were associated with licking, and those responding nondifferentially to cue tone kept the same characteristics in their responses to licking. The number of neurons that responded differentially to cue tones predicting MSG and arginine solution tended to be larger in control animals, and the number responsive to cue tone predicting L-lysine solution tended to be larger in L-lysine-deficient animals.

These data were consistent with the preference for amino acids by control and L-lysine-deficient animals in the behavioral study. It is suggested that these LHA neurons are related to integration of auditory stimuli and taste stimuli or reward [11,12]. In the present study any solution the rats obtained during licking behavior was also a reward after deprivation.

The LHA neurons generally responded nondifferentially during licking when animals were fed a diet containing a normal content of L-lysine However, when the diet was deficient in L-lysine, neurons responded differentially more to L-lysine than to other amino acids. Among the differential responses, MSG-specific neurons, which responded only to licking of MSG solution, were found only in controls, and L-lysine-specific neurons, which responded only to licking of L-lysine solution, were found only in animals with L-lysine deficiency, in 203 LHA neurons. These data suggest that the responsivity of LHA neurons may increase to promote ingestion of essential amino acids that are deficient in the diet.

Although it is difficult to determine whether MSG-specific neurons were the same as L-lysine-specific neurons, MSG-specific neurons appeared only in the control condition, and L-lysine-specific neurons appeared only with L-lysine deficiency. Such specific neurons that appeared in one condition only had similar characteristic relations; MSG was most preferred in controls, and L-lysine was most preferred during L-lysine deficiency in the behavioral study [4,13]. Moreover, both MSG- and L-lysine-

specific neurons were localized mainly in the dorsal and lateral LHA. These data strongly suggest that LHA neurons have plasticity to stimulation of a specific taste and integrate that taste with the physiologic need of that substance.

It is also plausible that these neurons might be related to a positive systemic reinforcement, such as "satisfaction," when rats ingest a solution that contains a substance necessary for the body (e.g., L-lysine during L-lysine deficiency), as some of these neurons responded during intracranial self-stimulation. We have found that the growth factors inhibin or activin, with or without the specific binding protein (follistatin), were released after intake of a control diet (L-lysine-sufficient) or a nonprotein diet, respectively [14]. These facts indicate that ingestion of L-lysine-deficient diet or nonprotein diet caused a change in serum (and possibly brain) levels of physiologic factors, including activin and inhibin. This release may elicit plasticity in the sensitivity of some LHA neurons to amino acids that could selectively drive ingestive behavior for particular amino acids (e.g., L-lysine) to maintain amino acid homeostasis.

Additionally, the immunohistochemical distribution of activin A in the brain was found in the nucleus tractus solitarius, the area postrema, and the arcuate nucleus and LHA. Increased oxygen consumption in the hypothalamus, LHA, and the arcuate nucleus was observed by a system of magnetic resonace spectrometry and imaging (MRSI) 30 min after L-lysine-deficient rats had been given intraperitoneal L-lysine. In addition to the fact that ingestion of an L-lysine-deficient diet or a nonprotein diet caused a change in serum levels of activin A as a possible neurotrophic factor, this release may elicit plasticity in the sensitivity of neurons to deficient amino acid in the nuclei that could selectively drive ingestive behavior for its particular amino acid (e.g., L-lysine) to maintain homeostasis. Our data strongly suggest that activin A is a possible candidate either to induce neuronal plasticity in the brain to allow prolonged adaptation to nutritional deficits or to elicit selective quantitative ingestion of a beneficial nutrient. It is possible that the ratio of inhibin and activin A could play an important role in notifying the brain as to whether the quantity of consumed food was nutritionally adequate.

Previous and present studies suggest that selective ingestion of a deficient amino acid depends on the blood level of the amino acid being lowered. It could trigger LHA neurons, directly or indirectly, and elevate the responsivity of LHA neurons to the deficient amino acid. The motor system would then carry out selective ingestion where possible. A positive systemic response would promote selective ingestion of the deficient amino acid—when the deficient substance is available. We also suggest that, based on data from studies previously published

[4,15,16], and from our iontophoretic experiments, the LHA is important in intrinsic regulation of amino acids (i.e., the monitoring of amino acids in the blood and cerebrospinal fluid). Our studies presently in progress indicate that LHA neurons respond more actively to iontophoretically applied amino acids during L-lysine deficiency, and some neurons respond to applied L-lysine during L-lysine deficiency. Animals fed the control diet do not display L-lysine-specific neurons. These findings suggest that the LHA might integrate intrinsic and extrinsic regulation of amino acids and undergo neuronal plasticity in terms of sensitivity to nutrients under malnutrition conditions.

References

1. Brand JG, Cagan RH, Naim M (1982) Chemical senses in the release of gastric and pancreatic secretions. Annu Rev Nutr 2:249–276
2. Naim M, Kare MR, Merritt AM (1978) Effects of oral stimulation on the cephalic phase of pancreatic exocrine secretion in dogs. Physiol Behav 20:563–570
3. Steffens AB (1976) Influence of the oral cavity on insulin release in the rat. Am J Physiol 230:1411–1415
4. Torii K, Mimura T, Yugari Y (1987) Biochemical mechanism of umami taste perception and effect of dietary protein on the taste preference for amino acids and sodium chloride in rats. In: Kawamura Y, Kare MR (eds) Umami: A basic taste. Dekker, New York, pp 513–563
5. Leung PMB, Rogers QR (1971) Importance of prepyriform cortex in food-intake response of rats to amino acids. Am J Physiol 221:929–935
6. Rogers QR, Leung PMB (1977) The control of food intake: when and how are amino acids involved? In: Kare MR, Maller O (eds) The Chemical sense and nutrition. Academic, Orlando, FL, pp 213–248
7. Anand BK, Brobeck JR (1951) Hypothalamic control of food intake in rats and cats. Yale J Biol Med 24:123–140
8. Ninomiya Y, Tanimukai T, Yoshida S, Funakoshi M (1991) Gustatory neural responses in preweanling mice. Physiol Behav 49:913–918
9. Mori M, Kawada T, Ono T, Torii K (1991) Taste preference and protein nutrition and L-amino acid homeostasis in male Sprague-Dawley rats. Physiol Behav 49:987–996
10. Tabuchi E, Ono T, Nishijo H, Torii K (1991) Amino acid and NaCl appetite, and LHA neuron responses of lysine-deficient rat. Physiol Behav 49:951–964
11. Nakamura K, Ono T (1986) Lateral hypothalamus neuron involvement in integration of natural and artificial rewards and cue signals. J Neurophysiol 55:163–181
12. Ono T, Nakamura K, Nishijo H, Fukuda M (1986) Hypothalamic neuron involvement in integration of reward, aversion, and cue signals. J Neurophysiol 56:63–79
13. Torii K, Mimura T, Yugari Y (1986) Effects of dietary protein on the taste preference for amino acids in rats. In: Kare MR, Brand JG (eds) Interaction of the Chemical Senses with Nutrition. Academic, Orlands, FL, pp 45–69
14. Torii K, Hanai K, Ono T, Wysocki CJ (1991) 21st Annual Meeting of Society for Neuroscience [abstract 535]
15. Panksepp J, Booth, DA (1971) Decreased feeding after injections of amino-acids into the hypothalamus. Nature 233:341–342
16. Wayner MJ, Ono T, De Young A, Barone FC (1975) Effects of essential amino acids on central neurons. Pharmacol Biochem Behav 3:85–90

Chemosensory Factors Affecting Acceptance of Food During Human Growth

Gary K. Beauchamp, Beverly J. Cowart, and Julie A. Mennella[1]

Key words. Infant—Taste—Smell—Flavor—Development—Experience

Introduction

The chemical senses—taste, smell, and chemical irritation (chemesthesis)—are critical to an organism's ability to select appropriate foods. Although all of these senses are involved in food recognition, their ontogeny and the relative roles they play in food acceptance during early human development vary. There has been little study of the development of chemesthetic sensitivity, so this topic is not considered further here except to state that the sense is important for future research.

In contrast to chemesthetic perception, there is now considerable research on the development of human taste and olfaction. Here, we describe some of our recent work in both of these areas. We first discuss development of human salt taste perception and preference and then describe studies on the transfer of flavor from foods consumed by nursing mothers into their milk, and how these flavors influence the suckling behavior of their infants.

Development of Salt Taste Perception and Preferences

Sweet taste perception and preference are well developed in the term [1] and even the preterm [2] infant; similarly, responsiveness to umami (glutamate) [3], sour acids [4,5], and some intensely bitter compounds [4] are evident in the human newborn. In contrast, newborns exhibit minimal responsiveness to salt; as outlined below, however, there is evidence for maturation of sensitivity during the first several months of life.

In the newborn, studies using measures of facial expression and of intake relative to water or weakly sweet diluents indicate an indifference to salt [6]. However, when sucking patterns have been evaluated, investigators report that salt elicits shorter bursts of sucking than water, suggesting that it is

relatively unpleasant [7]. No study demonstrates a preference or liking for salt in the newborn. In light of these ambiguous data and the evidence for a postnatal maturation of salt sensitivity in other species [review: 8], it seems likely that newborns are relatively insensitive to the saltiness of NaCl. At birth, Na-specific taste cells may be undeveloped, and the few responses to NaCl seen at this age may reflect its stimulation of nonspecific receptor channels (e.g., those also sensitive to KCl).

During the first few months of age, responsiveness to NaCl changes dramatically [9]. After 4 months of age, infants prefer salt solutions to water. It has been argued [10] that experience with salty tastes probably does not play a major role in the shift from apparent indifference to salt at birth to acceptance during later infancy; rather, this change in response may reflect postnatal maturation of central or peripheral mechanisms (or both) underlying salt taste perception, allowing expression of a largely unlearned preference for saltiness.

We [11] have replicated in newborns the finding that some measures of sucking are inhibited by salt (0.2 and 0.4M) relative to water. However, we found a substantial and significant change by 4 to 8 months of age: Infants tend to prefer or respond indifferently (depending on the measures evaluated) to 0.2M NaCl relative to water and to show significantly less rejection of 0.4M NaCl than do newborns (Fig. 1). In a second, longitudinal study in which responsiveness to salt added to familiar formula was assessed monthly in infants between 2 and 7 months of age, a marked developmental change was also observed (G.K. Beauchamp, B.J. Cowart, J.A. Mennella, and R.A. Marsh, unpublished data). On several sucking measures, the infants appear to be indifferent to salted (~0.15M) versus unsalted formula at 2 to 3 months of age. By 6 to 7 months, however, these same infants tended to reject the salted formula. Salting formula substantially suppresses its sweetness as judged by adult observers (unpublished observations). We propose that this suppression accounts for the older infants' rejection of the salted formula, whereas the younger infants did not perceive a suppression in sweetness owing to their inability to detect the salt.

Taken together, these data suggest that salt taste perception may be unique in developing postnatally

[1] Monell Chemical Senses Center, 3500 Market Street, Philadelphia, PA 19104, USA

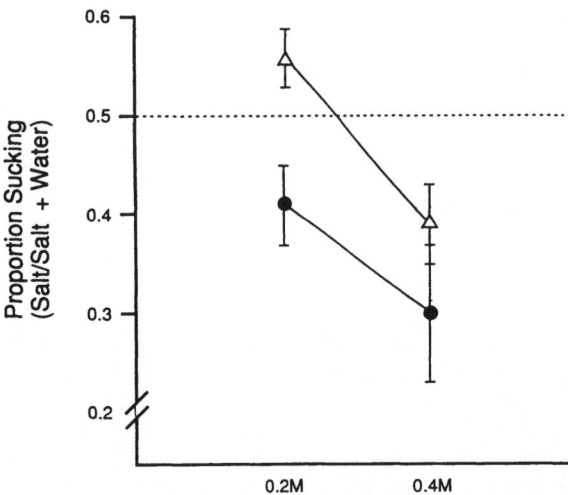

FIG. 1. Relative number of sucks (sucks in response to salt divided by total sucks) in newborn (*circles*) and 4- to 8-month-old infants (*triangles*) for two concentrations of salt. Each infant was tested only once using methods similar to those described by Maone et al. [2]. There was a significant ($P < 0.05$) effect of concentration (0.2 M vs 0.4 M) and age. (From [11], with permission)

in human infants. Interestingly, this postnatal maturation, likely due to addition of amiloride-sensitive channels in taste cells [8], parallels postnatal maturation of the kidney system, which also makes use of such channels to regulate sodium balance. It will be important for future studies to determine whether postnatal exposure to high or low salt levels during this developmentally labile period alters subsequent acceptance or preference for salt.

Transmission of Flavor (Odor) Cues from Mother to Infant

A number of early studies demonstrated that very young infants are responsive to volatile chemicals and can discriminate among them [12]. More recent studies, focusing on odors derived mainly from mothers or other individuals that may have biological relevance, have shown infants to be able to discriminate among individuals and to identify the odor of their mother [13].

Less work has been done with food odors. We have conducted a number of studies that indicate that flavors in foods nursing mothers consume are (1) transmitted through their milk and (2) detected by the infant, as shown by a change in the infant's response to the milk. In the first study [14], we showed that 1 to 3 hours after nursing mothers consume garlic, their milk has a distinctive garlic-like odor as determined by adult sensory panelists. Moreover, relative to a control day on which the

mother consumed placebo rather than garlic, infants nursed longer when the milk was garlic-flavored. This finding strongly suggests that the infants can detect the garlic flavor and that they find this flavor palatable.

In a second study [15], using the same methodology, we found that consumption of alcohol also flavors mothers' milk, but in this case intake was suppressed. The suppression probably was due either to the change in flavor of the milk (current studies are investigating preferences for alcohol-flavored milk compared with unflavored milk) or to an alcohol-induced suppression of the milk ejection reflex.

What role does experience play in altering the response of infants to milk-transmitted flavors? We found (J.A. Mennella and G.K. Beauchamp, unpublished data) that when mothers consumed substantial amounts of garlic either during the 3 days immediately before the test day or on the 3 days before this test period (followed by a 3-day washout before testing), the enhanced nursing following the mother's garlic consumption was eliminated. That is, prior exposure to garlic via mother's milk seemed to attenuate the infant's responses to that flavor relative to a control day when the milk was unflavored. A likely explanation for this observation is that the infants no longer found the garlic-flavored milk novel or arousing, and hence enhanced sucking was not induced. Further studies are needed to verify this hypothesis.

This research indicates that the chemosensory world of the breast-fed infant is rich, varied, and different from that of the bottle-fed infant. Bottle-fed infants, who experience a constant set of flavors from standard formulas, may be missing significant sensory experiences, which until recent times in human history were common to all infants.

Although much research is required to fully understand the effects of exposure to flavors in mother's milk on the infant's behavior, it is clear that human milk is not a food of invariant flavor. Rather, it provides the potential for a rich source of varying chemosensory experiences to the infant. Based on a variety of animal model studies [e.g., 16,17], the infant's prior exposure to flavors in mother's milk may actually increase the desirability of those flavors through familiarization. Unfamiliar foods, which are often not preferred by children, become preferred as a function of repeated presentation and increased familiarity [18]. Studies on other animals [19] also suggest that experience with a variety of flavors during breast-feeding, in contrast to the invariant flavor experience of formula feeding, predisposes the infants toward an increased willingness to accept novel flavors. Indeed, one study reported that breast-fed infants consumed more of a novel vegetable than did formula-fed infants [20].

Discussion

It is important to have a fuller understanding of the early development of the chemical senses and, in particular, better insight into the potential role of early exposure on long-term preference and aversion. Such information will be valuable when determining optimal diets for infants. Furthermore, this research may suggest strategies to assist in maintaining adequate but not excessive intake of nutrients such as salt, fat, protein, vitamins, and other minerals.

Acknowledgments. The research described here was supported in part by NIH grants RO1 DC-00882 and NRSA HD-07375 and a gift from the Mead Johnson Nutritional Group of Bristol-Myers Squibb.

References

1. Desor JA, Maller O, Turner RE (1977) Preference for sweet in humans: infants, children and adults. In: Weiffenbach JM (ed) Taste and development: The genesis of sweet preference. US Government Printing Office, Washington, DC
2. Maone TR, Mattes RD, Bernbaum JC, Beauchamp GK (1990) A new method for delivering a taste without fluids to preterm and term infants. Dev Psychobiol 23:179–191
3. Steiner JE (1987) What the neonate can tell us about umami. In: Kawamura Y, Kare MR (eds) Umami: A basic taste. Dekker, New York
4. Steiner JE (1977) Facial expressions of the neonate infant indicating the hedonics of food-related chemical stimuli. In: Weiffenbach JM (ed) Taste and development: The genesis of sweet preference. US Government Printing Office, Washington, DC
5. Desor JA, Maller O, Andrews K (1975) Ingestive responses of human newborns to salty, sour and bitter stimuli. J Comp Physiol Psychol 89:966–970
6. Rosenstein D, Oster H (1990) Differential facial responses to four basic tastes in newborns. Child Dev 59:1555–1568
7. Crook CK (1978) Taste perception in the newborn infant. Infant Behav Dev 1:52–69
8. Hill DL, Mistretta CM (1990) Developmental neurobiology of salt taste sensations. Trends Neurosci 13:188–195
9. Beauchamp GK, Cowart BJ, Moran M (1986) Developmental changes in salt acceptability in human infants. Dev Psychobiol 19:17–25
10. Cowart BJ, Beauchamp GK (1990) Early development of taste perception. In: McBride RL, MacFie HJH (eds) Psychological basis of sensory evaluation. Elsevier, London, pp 1–18
11. Beauchamp GK, Cowart BJ (1993) Development of salt taste responses in human newborns, infants and children. In: Pearl J (ed) Development, growth and senescence in the chemical senses. NIH Publication No. 93-3483. US Department of Health and Human Services, Public Health Service, National Institutes of Health, Bethesda, MD, pp 61–68
12. Beauchamp GK, Cowart BJ, Schmidt HJ (1991) Development of chemosensory sensitivity and preference. In: Getchell TV, Bartoshuk LM, Doty RL, Snow JB Jr (eds) Smell and taste in health and disease. Raven, New York, pp 405–416
13. Schaal B (1988) Olfaction in infants and children: Developmental and functional perspectives. Chem Senses 13:145–190
14. Mennella JA, Beauchamp GK (1991) Maternal diet alters the sensory qualities of human milk and the nursling's behavior. Pediatrics 88:737–744
15. Mennella JA, Beauchamp GK (1991) The transfer of alcohol to human milk: Effects on flavor and the infant's behavior. N Engl J Med 325:981–985
16. Galef BG, Henderson PW (1972) Mother's milk: A determinant of the feeding preferences of weaning rat pups. J Comp Physiol Psychol 78:213–219
17. Mainardi M, Poli M, Valesecchi P (1989) Ontogeny of dietary selection in weaning mice: Effects of early experience and mother's milk. Biol Behav 14:185–194
18. Birch LL (1979) Dimensions of preschool children's food preferences. J Nutr Educ 11:77–80
19. Capretta PJ, Petersik JT, Stewart DJ (1975) Acceptance of novel flavours is increased after early exposure of diverse tastes. Nature 254:689–691
20. Sullivan SA (1992) Infant experience and acceptance of solid foods. Ph.D. dissertation, University of Illinois at Urbana-Champaign

Sensory Preferences for Sweet High-Fat Foods: Evidence for Opiate Involvement

ADAM DREWNOWSKI[1]

Key words. Naloxone—Butorphanol—Obesity—Binge eating—Hedonic preference—Sugar—Fat

Introduction

Endogenous opiates influence the regulation of food intake in both humans and rats [1,2]. Food intake is generally reduced by opiate antagonists such as naloxone but increased by opiate agonists such as butorphanol [3–5]. These effects appear to be strongest for the best-tasting foods, notably those rich in sugar and fat. Whereas naloxone and naltrexone suppressed the intakes of sugar, fat, and a palatable cafeteria diet [4–6], morphine and butorphanol increased fat consumption in rats [7,8]. It has been proposed that the pleasure response to palatable foods is mediated by the central release of endogenous opiates [2]. Blockade of opiate receptors may therefore suppress taste preferences and so reduce overeating associated with exposure to palatable sweet or high-fat foods.

If opiate antagonists influence food intake throught their effects on food palatability, we should expect naloxone to have measurable effects on hedonic preferences as well as on food consumption. In previous studies, opiate antagonists were most effective in reducing intakes of sweet glucose, sucrose, or saccharin solutions by laboratory rats [1,2]. Conversely, morphine increased preferences for sweet taste and the intake of sweet solutions [1,2]. Central administration of mu and delta receptor agonists also increased intakes of saccharin solutions in rats [9]. In other studies, the consumption of chocolate and chocolate milk by rats led to the release of endogenous opiate peptides [10]. Sensory preferences for sweet and high-fat foods may therefore be under opiate control.

Clinical reports also indicate that sweet high-fat foods are the frequent target of food "cravings" [11,12]. Elevated preferences for chocolate, cookies, or ice cream have been reported by obese patients, bulimic women, or women diagnosed with the binge-eating syndrome. Although some investigators described preferences for chocolate as "carbohydrate cravings," the fact is that many sweet desserts largely consist of only two ingredients, sugar and fat. Preferences for such foods may also be influenced by endogenous opiates. In one clinical study, oral naltrexone reduced the acceptability of sweet glucose solutions and diminished the pleasantness of food odors [13].

However, another clinical study showed that the antagonist nalmefene selectively reduced the consumption of the most preferred foods during a lunch meal, including some that were neither sweet nor rich in fat [14]. It is therefore unclear whether opiate blockade reduces taste preferences for the best-liked foods, regardless of their nutrient composition, or only for those foods that are rich in fat, sugar, or both.

Given the adverse health consequences of high-fat diets, human preferences for high-fat foods have become a topic of particular interest [15–17]. Our study [18] was the first to examine the effects of opiate blockade on hedonic preferences for sugar–fat mixtures and the consumption of sweet and high-fat foods.

Materials and Methods

The subjects were 14 women binge-eaters, 18 to 40 years of age, and 12 nonbingeing controls. Of the 14 binge-eaters, 8 were obese. The subjects were given intravenous infusions of naloxone, butorphanol, and saline placebo using a double-blind within-subject design. Randomized testing sessions were spaced at least 2 days apart. The dose of naloxone used (6 mg bolus followed by 0.1 mg/kg per hour drip for 2.5 h) had been found in other studies to decrease the size of binge-eating episodes in female bulimics [19]. Butorphanol, a synthetic mixed agonist–antagonist, was administered as a 1 µg/kg bolus followed by an intravenous saline drip. This dose had been found to increase food intake [20].

Sensory stimuli were 20 sweetened dairy products of varying sugar and fat content, including whole milk (3.5% fat), Half and Half (10% fat), light cream (20% fat), and heavy cream (36% fat), each sweetened with 2%, 4%, 8%, 16%, and 32% sucrose (w/w) in a 4 × 5 factorial design [15,17]. The stimuli,

[1] Human Nutrition Program, School of Public Health, The University of Michigan, 1420 Washington Heights, Ann Arbor, MI 48109-2029, USA

TABLE 1. Foods available for consumption: a summary of food choices.

Food	Hedonic rating	Energy per portion (kcal)	Macronutrient content (kcal)		
			CHO	Protein	Fat
Category 1: low sugar/low fat					
Popcorn	7.19	101	79	12	9
Saltines	6.19	94	72	9	19
Breadsticks	5.69	99	79	12	9
Pretzels	6.65	109	84	7	8
Category 2: low sugar/high fat					
Fritos	5.69	98	41	4	55
Cream cheese + Saltines	4.50	101	20	8	72
Potato chips	6.57	101	38	4	58
Lorna Doones	6.16	96	58	4	38
Category 3: high sugar/low fat					
Jelly beans	4.84	100	100	0	0
Marshmallows	4.84	99	97	3	0
Jello	5.27	100	100	0	0
Jelly + Saltines	4.42	109	93	2	5
Category 4: high sugar/high fat					
Chocolate chip cookies	7.73	109	55	4	41
Snickers bar	6.73	105	48	10	43
M&M candies	7.77	100	54	5	41
Oreo cookies	6.61	96	58	4	38

From [18], with permission.

chilled to 5°C, were presented to subjects in plastic cups for sensory and hedonic evaluations. The subjects tasted each sample and rated its perceived sweetness, creaminess, fat content, and overall acceptability using 9-point category scales [15,17].

Sixteen snack foods, containing different amounts of sugar and fat, were divided into four categories: low sugar/low fat (category 1); low sugar/high fat (category 2); high sugar/low fat (category 3); and high sugar/high fat (category 4). Each subject was presented with eight foods, two from each category, during each experimental session. Nutrient composition and energy content per portion of the 16 foods are summarized in Table 1.

Results

Hedonic preference ratings for sugar–fat mixtures followed an inverted U shape, characterized by a sweetness optimum or "breakpoint" (Fig. 1). The sweetness breakpoint for binge-eaters (16% sucrose) was higher than for nonbingeing controls (8% sucrose), in accordance with previous results [15,17]. Naloxone reduced taste preference ratings relative to the saline condition equally for binge eaters and for nonbingeing controls. In contrast, butorphanol had no significant effects on taste preferences.

Naloxone reduced food consumption relative to the saline condition in binge-eating subjects, but not

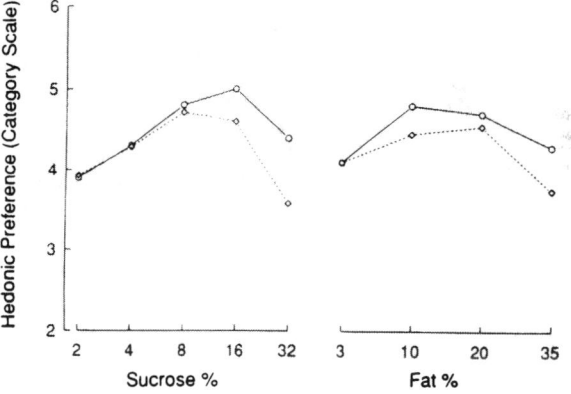

FIG. 1. Hedonic preference profiles of binge-eaters (*circles*; n = 14) and controls (*diamonds*; n = 12) at baseline as a function of stimulus sucrose and fat content. (From [18], with permission)

in nonbingeing controls (Table 2). The most pronounced effects of naloxone were observed for the sweet high-fat foods in category 4 [18]. Naloxone did not selectively decrease the intakes of the best-liked foods. Consumption of some of the most highly rated foods (e.g., popcorn) was actually increased by naloxone.

TABLE 2. Summary of food consumption patterns for binge-eaters and controls as a function of drug infusion.

Food	Butorphanol	Naloxone	Saline
Binge eaters			
Category 1: low sugar/low fat	128 (39)	87 (33)	95 (20)
Category 2: low sugar/high fat	157 (52)	144 (53)	226 (56)
Category 3: high sugar/low fat	70 (20)	65 (16)	112 (26)
Category 4: high sugar/high fat	161 (49)	121 (51)[a]	261 (69)
Total	518 (103)	417 (101)[a]	694 (121)
Normal controls			
Category 1: low sugar/low fat	87 (20)	107 (22)	105 (18)
Category 2: low sugar/high fat	193 (50)	140 (36)	195 (38)
Category 3: high sugar/low fat	72 (27)	49 (17)	55 (17)
Category 4: high sugar/high fat	143 (30)	102 (33)	143 (30)
Total	496 (83)	397 (76)	505 (66)

Data are expressed as mean (kcal/h) and SEM.
[a] $P < 0.05$; Dunnett's t-test.
From [18], with permission.

Discussion

Naloxone reduced taste preferences for sugar–fat mixtures and caused a decrease in the consumption of foods that were sweet, high in fat, or both. The reduction in taste preferences was observed in all subjects. However, the decrease in intake was significant only among female binge-eaters but not among nonbingeing controls. The strongest effects of naloxone were obtained for sweet high-fat foods such as M&M candies, Snickers bars, and Oreo and chocolate chip cookies (category 4). As noted previously, such foods are among the most common targets of food "cravings" and are often listed in reports of eating binges observed in obesity and bulimia nervosa [16,17]. Hunger and fullness ratings were not affected by naloxone infusions, arguing against the notion that opiate antagonists modulate hunger and satiety in normal eating.

Not all foods listed as highly preferred were affected by naloxone. The consumption of popcorn or breadsticks was actually increased. At the same time, naloxone effects were not macronutrient specific. In contrast to some animal studies [6], naloxone did not selectively reduce the intake of carbohydrate or fat. Instead, the present data suggest that some palatable foods are more sensitive to opiates than others. Although several past studies have shown opiate effects on the consumption of sandwiches and other lunch foods, our study [18] suggests that foods rich in fat, sugar, or both may be affected the most. The difference is most likely that of degree rather than specific food type. However, all foods in category 4 did contain chocolate. Chocolate cravings have been observed with obesity and eating disorders, seasonal affective disorder, as a function of the menstrual cycle, and during the course of antidepressant treatment [11,12,16,17]. The potential connection between opiate peptides and preferences for chocolate is the subject of continuing investigation.

The main therapeutic value of opiate antagonists may be in reducing binge-eating episodes and not in the long-term control of body weight. Obvious parallels have also been drawn between binge eating and drug addiction, as both behavioral syndromes involve intense cravings and loss of control. It may be that the same physiological mechanisms are involved in mediating food cravings and drug reward. Both clinical observations and anecdotal reports indicate that sweet cravings tend to be associated with opiate addiction [22], and opiate withdrawal is sometimes alleviated by sweets [22]. The present studies suggest that food cravings in obesity and eating disorders may be mediated by endogenous opiates and point to the key sensory role of dietary sugars and fat.

Acknowledgments. Supported by USPHS grants DA05471 and DK38073, and by PHS grant 5M01RR00042.

References

1. Olson GA, Olson RD, Kastin AJ (1992) Endogenous opiates: 1991. Peptides 13:1247–1287
2. Cooper SJ, Kirkham T (1990) Basic mechanisms of opioids' effects on eating and drinking. In: Reid LD (ed) Opioids, bulimia, and alcohol abuse and alcoholism. Springer, New York, pp 91–110
3. Gosnell BA, Levine AS, Morley JE (1985) The stimulation of food intake by selective agonists of mu, kappa and delta opioid receptors. Life Sci 38:1081–1088
4. Apfelbaum M, Mandenoff A (1981) Naltrexone suppresses hyperphagia induced in the rat by a highly palatable diet. Pharmacol Biochem Behav 15:89–91

5. Bertiere MC, Sy TM, Baigts F, Mandenoff A, Apfelbaum M (1984) Stress and sucrose hyperphagia: role of endogenous opiates. Pharmacol Biochem Behav 20:675–679

6. Marks-Kaufman R, Kanarek R (1981) Modifications of nutrient selection induced by naloxone in rats. Psychopharmacology (Berl) 74:321–324

7. Marks-Kaufman R, Kanarek RB (1980) Morphine selectively influences macronutrient intake in the rat. Pharmacol Biochem Behav 12:427–430

8. Levine AS, Morley JE (1983) Butorphanol tartrate induces feeding in rats. Life Sci 32:781–785

9. Gosnell BA, Majchrzak MJ (1989) Centrally administered opioid peptides stimulate saccharin intake in nondeprived rats. Pharmacol Biochem Behav 33:805–810

10. Dum J, Gramsch C, Herz A (1983) Activation of hypothalamic beta-endorphin pools by reward induced by highly palatable food. Pharmacol Biochem Behav 18:443–447

11. Tomelleri R, Grunewald KK (1987) Menstrual cycle and food cravings in young college women. J Am Diet Assoc 87:311–316

12. Paykel ES, Mueller PS, De la Vergne PM (1973) Amitriptyline, weight gain, and carbohydrate craving: a side effect. Br J Psychiatry 123:501–507

13. Fantino M, Hosotte J, Apfelbaum M (1986) An opioid antagonist naltrexone reduces preference for sucrose in humans. Am J Physiol 251:R91–R96

14. Yeomans MR, Wright P, Macleod HA, Critchley JAJH (1990) Effects of nalmefene on feeding in humans. Psychopharmacology (Berl) 100:426–432

15. Drewnowski A, Brunzell JD, Sande K, Iverius PH, Greenwood MRC (1985) Sweet tooth reconsidered: taste preferences in human obesity. Physiol Behav 30:629–633

16. Drewnowski A, Kurth CL, Holden-Wiltse J, Saari J (1992) Food preferences in human obesity: carbohydrates versus fats. Appetite 18:207–221

17. Drewnowski A (1990) Taste responsiveness in eating disorders. Ann NY Acad Sci 575:399–409

18. Drewnowski A, Krahn DD, Demitrack MA, Nairn K, Gosnell BA (1992) Taste responses and preferences for sweet high-fat foods: evidence for opioid involvement. Physiol Behav 51:371–379

19. Mitchell JE, Laine DE, Morley JE, Levine AS (1986) Naloxone but not CCK-8 may attenuate binge-eating behavior in patients with the bulimia syndrome. Biol Psychiatry 21:1399–1406

20. Morley JE, Parker S, Levine AS (1985) Effect of butorphanol tartrate on food and water consumption in humans. Am J Clin Nutr 42:1175–1178

21. Kaye WH, Pickar D, Naber D, Ebert MH (1984) Cerebrospinal fluid opioid activity in anorexia nervosa. Arch Gen Psychiatry 41:350–355

22. Morabia A, Fabre J, Chee E, et al. (1989) Diet and opiate addiction: a quantitative assessment of the diet of noninstitutionalized opiate addicts. Br J Addict 84:173–180

Taste Preference in Adrenalectomized Rats

Tsuyoshi Horio and Yojiro Kawamura[1]

Key words. Taste preference—MSG

Umami taste is widely considered to be one of the basic tastes [1,2], a representative umami substance being monosodium L-glutamate (MSG). In this experiment, using the two-bottle method, we evaluated the preference for umami taste in adrenalectomized rats, which prefer NaCl for survival [3].

A series of preference tests was performed in a free choice situation with two bottles. The subjects were 28 male Wistar albino rats. Sixteen rats were adrenalectomized (experimental group) and 12 were sham-operated (controls). The rats were allowed food ad libitum. A 5-day period was allowed for postoperative recovery before the experiment was begun; during this period all rats were given 0.15 M NaCl solution.

Na levels in the plasma of the adrenalectomized rats were almost the same value as in controls when sufficient NaCl was given. Na levels in urine after adrenalectomy were a little lower than those in controls; however, urine volume per day in the adrenalectomized animals was more than in the controls. Body weight after adrenalectomy increased a little less than in controls. Food intake did not differ between the adrenalectomized and control rats.

The drinking rates of 0.15 M MSG, 0.15 M NaCl, and 0.15 M NaNO₃ solutions, compared to those of distilled water, were significantly higher in adrenalectomized rats than in the controls. However, the volume of sodium salt solutions and water ingested did not differ, even in adrenalectomized rats, when the salts were dissolved in amiloride solution, which reduces the taste intensity of Na salts.

Both adrenalectomized and control rats drank more NaCl solution than NaNO₃ or MSG solution.

There were no differences between adrenalectomized and control animals in the volumes ingested of solutions of 0.5 M sucrose, 0.01 M tartaric acid, 0.0005 M quinine HCl and mixtures of 0.5 M sucrose + 0.05 M MSG, 0.01 M tartaric acid + 0.05 M MSG, 0.0005 M quinine HCl + 0.05 M MSG, using the one-bottle method.

These results indicate that adrenalectomized rats prefer MSG to water, to the some extent as they prefer other sodium salts.

References

1. Kawamura Y, Kare MR (eds) (1987) Umami: A basic taste. Dekker, New York
2. Kawamura Y, Kurihara K, Nicolaidis S, Oomura Y, Wayner MJ (eds) (1991) Umami: Proceedings of the second international symposium on umami. Physiol Behav 49:831–1028
3. Richter CP (1936) Increased salt appetite in adrenalectomized rats. Am J Physiol 115:155–161

[1] Faculty of Nutrition, Koshien University, 10-1 Momiji-gaoka, Takarazuka, Hyogo, 665 Japan

Appetite and Taste Preferences in Growing Rats in Various States of Protein Nutrition

Masato Mori, Richard L. Hawkins, Masashi Inoue, and Kunio Torii[1,2]

Key words. Taste preference—Protein nutrition

Cephalic gustatory stimuli received during a meal yield nutritional information and aid in the efficient digestion of food [1]. When animals detect a food with a bitter and/or sour taste, they exercise caution in ingesting something that might be toxic [2], whereas foods with a familiar or pleasant taste are swallowed without hesitation. In general, foods having a sweet, umami, or mildly salty taste are palatable [3].

This study focused on appetite for flavored food and on feeding behavior in growing male Sprague-Dawley rats in various states of protein nutrition. Three experiments were performed.

In experiment 1, rats (N = 6, in each group) were supplied with a 15% purified whole egg protein (PEP) diet and offered a choice of three sweet-tasting solutions, 4 mM sodium saccharin (saccharin), and 0.5 and 1.0 M glucose, as well as deionized distilled water as drinking water. In every experimental group, the ingestion of the sweet solutions (saccharin or glucose) exceeded that of water. A strong preference for both glucose and saccharin was observed immediately after food deprivation. Moreover, these preferences were sustained in rats ingesting glucose solution. However, the intake of the saccharin solution declined during the period of starvation. In fasted rats, endogenous protein degradation was suppressed by the ingestion of an amount of glucose that was sufficient to meet energy needs. This study clearly indicates that, at some point, animals are able to discriminate which of various sweet-tasting substance are an energy source and which are not.

In experiment 2, rats that received a 15% PEP diet were given access to both water and one of following sweet-tasting solutions: 1.0 M glucose, 0.5 M sucrose, 1.0 M fructose, 0.5 M maltose, 0.5 M lactose, or 4 mM saccharin as drinking water (N = 6, in each group). The total energy intake of each group which ingested sugars, except sucrose, was unchanged. Rats that ingested sucrose exceeded 115% of the total energy intake compared to the ingestion of saccharin as a control. These findings supported the concept that the degree of reduction in food intake paralleled the amount of sugar ingestion, except in the case of sucrose. Sucrose, which consists of a glucose and a fructose moiety, may not yield enough nutritional information as an energy source to affect the control of total energy intake in rats.

In experiment 3, rats (N = 6, occasionally N = 5, in each group) were supplied with a nonprotein or a 15% casein diet and water ad libitum for 3 weeks. Each of seven cages contained a diet without (control), or with a distinctive flavoring: 5.6% (w/w) monosodium L-glutamate (MSG; umami), 0.2% MSG + 0.5% guanosine 5'-monophosphate (GMP; umami), 6.3% citric acid (sour), 20% sucrose (sweet), 1.8% NaCl (salty), or 0.2% quinine hydrochloride (quinine; bitter). The growth of the rats and their consumption of the 15% casein diet with MSG, MSG + GMP, or sucrose was quite comparable to the growth and consumption in controls fed the diet without the taste materials. However, when rats were supplied with the diet containing NaCl, quinine, or citric acid, their food intake was reduced and their growth was retarded in this order of experimental groups. In the rats that received the non-protein diet, food consumption was significantly reduced and growth was significantly retarded in all groups, regardless of food flavoring. The diurnal pattern of feeding in rats fed a 15% casein or a nonprotein diet with flavoring was recorded using a strain gauge system. Nocturnal rhythm was sustained regardless of flavoring or protein restriction. When palatable taste materials, such as MSG, MSG + GMP, or sucrose, were added to the 15% casein diet, the meal size exceeded that when aversive compounds, i.e., quinine or citric acid were added. This phenomenon was not as marked when rats were fed a nonprotein diet with any flavoring.

The animals' primary concern was energy intake and their second concern was protein nutrition, regardless of flavoring.

[1] Torii Nutrient-stasis Project, ERATO, Research Development Corporation of Japan, Technowave 100, 1-1-25 Shin-urashima-cho, Kanagawa-ku, Yokohama, 221 Japan
[2] Ajinomoto Co., Inc, 214 Maeda-cho, Totsuka-ku, Yokohama, 244 Japan

References

1. Torii K, Mimura T, Yugari Y (1987) Biochemical mechanism of umami taste perception and effect of dietary protein on the taste preference for amino acids

and sodium chloride in rats. In: Kawamura Y, Kare MR (eds) Umami: A basic taste. Dekker, New York, pp 513–563

2. Beauchamp GK, Maller O (1977) The development of flavor preferences in humans. In: Kare MR, Maller O (eds) The chemical senses and nutrition. Academic, New York, pp 291–310

3. Rozin P, Kalat JW (1971) Specific hungers and poison avoidance as adaptive specializations of learning. Psychol Rev 78:459–485

Developmental Changes in Taste Response to and Preferential Ingestion of Various Sugars in Rats

Taeko Yamada, Mari Kimoto, and Motoko Nogami[1]

Key words. Sugar intake—Developing rat

In rats, the preference for various sugars of equivalent concentration differs, and the intake of sucrose is large. This study was performed to clarify the developmental changes in sugar intake preferences (experiment I) and to determine whether the difference in sugar intake was caused by taste (experiment II).

We examined Wistar albino rats in three developmental stages, preweaning (15- to 21-days-old), postweaning (22- to 28-days-old), and adult (10- to 14-weeks-old). The sugars used were sucruose, maltose, lactose, glucose, and fructose, dissolved in distilled water (DW).

In experiment I, we examined intake of increasing concentrations of saccharides and preferences shown. Rats of the three age periods were divided into five sugar groups, and were given a two-bottle 23-h preference test, the two bottles being filled with sugar solution and DW. The concentration of sugar was increased every 2nd day, from 0.025, to 0.05, 0.1, 0.4, and 0.8 M in the preweaning group, and every day, from 0.025, to 0.05, 0.1, 0.2, 0.4, 0.8, and 1.6 M in the postweaning and adult groups.

In the preweaning group, a preference of 70%–80% was shown for higher concentrations of four sugars, excluding lactose. There was little difference in the intake of these four sugars; the intake per body weight (bw) of these sugars was also similar. In the postweaning group, preferences for lower concentrations of the five sugars were greater than in the preweaning group. The preferences for 0.1–0.4 M sucrose, 0.05–0.8 M maltose, and 0.2–0.8 M glucose were 90%–100%. The largest intake, per 10 g bw was 4.8 ml of 0.2 M maltose and 5.4 ml of 0.4 M glucose, whereas for 0.05–0.4 M sucrose it was about 2.5 ml. In adults, preferences for sucrose at all concentrations were greater than in the pre- and postweaning groups. The intake per 10 g bw of 0.1 M sucrose was the largest (4.2 ml) of the three test groups, although the intake of maltose, glucose, and lactose decreased. No developmental change was found in intake per bw value for fructose in the three age groups.

In experiment II, we examined intake of sugars at concentrations for which nerve response magnitudes were equivalent. We were recorded the chorda tympani integrated responses to five sugars, at 0.05 to 1.0 M concentrations, in the pre- and postweaning and adult rats. The concentrations of sugars at which nerve response magnitudes were equivalent were obtained from the dose response curves. These concentrations were; suc-0.1, mal-0.45, lac-0.4, glu-0.4, fru-0.1 for preweaning; suc-0.1, mal-0.7, lac-0.8, glu-0.8, fru-0.4 for postweaning; suc-0.1, mal-0.4, lac-0.15, glu-0.45, and fru-0.3 M for adult. Six bottles, filled with five sugar solutions and DW, were presented to the rats and intake over 23 h was measured. The intake ratio of each sugar to total intake, was then calculated. For lactose, the intake ratio was greatest (9%–13%) in the preweaning group and it decreased according to age. For sucrose, the intake ratio was 19%–23% in the preweaning group, 7%–14% in the postweaning group, and 9%–24% in adults. For glucose, the intake ratio was greatest (42%) in animals that were 24 days old, and was 37% in 25-day-old animals. On the other days, the ratio was about 15%. For maltose, the intake ratio was 30%–40% in the preweaning group, 17%–40% in the postweaning group, and 50%–76% in adults. These values were higher than those for other sugars (except for glucose on days 24 and 25), throughout the three developmental stages. There was a reciprocity of intake ratio between maltose and glucose in the postweaning period and on the 2nd to 4th test days for adults; the 5th and 6th test days in adults, there was a reciprocity between maltose and sucrose.

These results indicate that: (1) In the preweaning group, the preference for sugars was obtained only at high concentrations and sugars were not discriminated, except for lactose. (2) In the postweaning group, there was selective consumption of sugars; intake volume varied in relation to the kind of sugar and the concentration. (3) the large intake of maltose in experiment II, appeared to be caused by a factor other than taste preference, and the large intake of sucrose compared to other sugars in adults appeared to be due to taste preference. (4) The reciprocity in intake volume of maltose and glucose, and maltose and sucrose indicates the presence of a mechanism that monitors the intake volume of sugar.

[1] Physiological Laboratory, Japan Women's University, 2-8-1 Mejiro-dai, Bunkyo-ku, Tokyo, 112 Japan

529

Gustatory Neural Input for L-Lysine in Mice

Hideaki Kajiura[1,2], Yuzo Ninomiya[1], Kazumichi Mochizuki[1], and Kunio Torii[3]

Key words. Gustatory nerves—L-lysine

Our previous study [1] demonstrated that the mouse glossopharyngeal nerve was highly responsive to umami and essential L-amino acids and quinine, compared with the chorda tympani (CT) nerve. In contrast, the CT nerve showed greater responses to NaCl and sugars. The differential responses of the CT and GL nerves suggest possible differences in physiological function between the two taste nerves. To study the physiological roles of these two taste nerves on food and fluid intake, we examined taste sensitivity to an amino acid, L-lysine (Lys). We then examined the effects of denervation of the GL and CT nerves on reduction of the aversive responses to Lys in Lys-deficient animals, to determine whether these responses would constitute an adjustment to the nutritional deficiency.

In the first series of experiments, to study taste sensitivity to Lys in mice, we examined behavioral responses, measured by using a conditioned taste aversion (CTA) paradigm and neural responses obtained from the GL and CT nerves. The experimental animals were male and female mice (C57BL/KsJ: body weight 20–35 g). In the behavioral experiments, the animals were divided into two groups: a GL-denervated and an intact group. The experimental schedule and procedures have been described in a previous study [2]. In the electrophysiological experiments, we recorded the integrated responses of CT and GL nerves to $0.1\,\mu M$–$0.1\,M$ Lys while the animals were anesthetized (pentobarbital 40–50 mg/kg). The experimental procedures have also been described previously [1].

When avoidance of 0.1 M Lys was conditioned, the animals showed a significantly lower number of licks for Lys at concentrations of $0.1\,\mu M$ or higher than for distilled water. This behavioral threshold for Lys increased to 1.0 mM when the bilateral GL nerves of animals were denervated. In correspondence with the behavioral thresholds, the threshold of neural responses to Lys was 3 log units lower in the GL (about $1.0\,\mu M$) than in the CT nerves (about 1.0 mM).

In the second series of experiments, we compared the effects of denervation of the GL or CT nerve on behavioral responses (aversive responses without CTA) to Lys in mice fed Lys-deficient food or a standard diet which contained Lys. Animals were divided into the following 6 groups: intact, CT-denervated, and GL-denervated; each consisted of two subgroups, one supplied with the standard diet containing Lys and the other with the Lys-deficient diet. Animals were given the experimental diet 5 days before the beginning of the training period. In this experiment, behavioral responses (number of licks) in one-bottle preference tests were measured, and avoidance to 0.1 M Lys was not conditioned. In the preference tests, test solutions were presented to animals for 10 s bottle by bottle. Other aspects of the experimental schedule and procedures were as described above.

The behavioral aversion threshold for Lys in intact control mice ($3.0\,\mu M$), obtained from the one-bottle preference test, was close to the neural threshold for Lys in the GL nerve. This threshold increased to a level ($300\,\mu M$) similar to the threshold of the CT responses to Lys when animals were fed the Lys-deficient diet. In the GL-denervated mice, the Lys deficiency did not change the aversion threshold for Lys. Both control and Lys-deficient groups had an aversion threshold ($300\,\mu M$) similar to the neural threshold of the CT nerve. The greatest increase in the aversion threshold for Lys produced by Lys-deficiency was observed in the CT-denervated mice (from $3.0\,\mu m$ to 10 mM). These results indicate that taste information from the GL nerve plays a more important role in changes in the behavioral response of mice to Lys caused by Lys deficiency than does taste information from the CT nerve. It is probable that the GL nerve conveys chemical information regarding nutrients that is important for maintaining the animal's homeostatic balance.

[1] Department of Oral Physiology, Asahi University School of Dentistry, 1851 Hozumi, Hozumi-cho, Motosu-gun, Gifu, 501-02 Japan
[2] Sensory Evaluation Group, R&D Division, Kirin Brewery Co., Ltd., 26-1 Jingumae 6-chome, Shibuya-ku, Tokyo, 150-11 Japan
[3] Ajinomoto Co., Ltd., 214 Maeda-cho, Totsuka-ku, Yokohama, 244 Japan

References

1. Ninomiya Y, Tanimukai T, Yoshida S, Funakoshi M (1991) Gustatory neural responses in preweaning mice. Physiol Behav 49:913–918
2. Ninomiya Y, Funakoshi M (1989) Behavioral discrimination between glutamate and the four basic taste substances in mice. Comp Biochem Physiol 92:365–370

Effects of Angiotensin II and ANP on Taste Nerve Activity

Tomio Shingai, Yoshihiro Takahashi, and Yozo Miyaoka[1]

Key words. Angiotensin II—Taste nerve

Angiotensin II (A II), a peptide hormone that plays an important role in the regulation of water and sodium balance, is involved in the mechanism of sodium retention and induces water drinking. Atrial natriuretic hormone (ANP), which is also a peptide, antagonizes these effects of A II, enhancing sodium excretion and reducing the drinking induced by A II. We have recently reported that afferent signals from the glossopharyngeal and superior laryngeal nerves also influence renal function [1,2]. It can be presumed that these peptides modulate afferent signals from the oral cavity and larynx. This study was designed to examine the effects of A II and ANP on afferent signals from taste nerves. For this purpose we used rabbits (2.0–2.5 kg) and rats (250–360 g) anesthetized with urethane (1.0 g/kg) or pentobarbital (40 mg/kg). The spontaneous activity of entire glossopharyngeal and chorda tympani nerves was measured before and after the intravenous injection or infusion of A II and ANP. The spontaneous activity of the glossopharyngeal nerve was increased by the injection of A II (2–4 µg/kg), but was decreased by ANP (4–5 µg/kg) injection. The spontaneous activity of the chorda tympani nerve was gradually enhanced by the continuous infusion of A II (0.1–0.2 µg/kg·min). The effects of A II and ANP on the response of the chorda tympani nerve to NaCl solutions were examined. The magnitude of the responses to these solutions was not affected by the continuous intravenous infusion of A II (0.2 µg/kg·min), or by adding A II to the stimulating solutions (1 µg/ml). These results suggest that A II and ANP may exert their effects through the central nervous system, presumably the circumventricular organs, and the autonomic nervous system. Changes in the spontaneous activity of taste nerves probably reflected the autonomic control of salivation and circulation in the tongue.

Acknowledgments. This study was partially supported by Grant No. 92050 from The Salt Science Research Foundation of Japan (T.S).

References

1. Shingai T, Miyaoka Y, Shimada K (1988) Diuresis mediated by the superior laryngeal nerve in rats. Physiol Behav 44:431–433
2. Shingai T, Miyaoka Y, Shimada K (1991) Afferent signals from sodium-responsive fibers in the glossopharyngeal nerve decrease urine flow and increase renal sympathetic nerve activity. In: Yoshikawa M, Uono M, Tanabe H, Ishikawa S (eds) New trends in autonomic nervous system research. Excerpta Medica, Amsterdam, p 502

---

[1] Niigata University School of Dentistry, 5274 Gakkochodori Niban-cho, Niigata, 951 Japan

Dietary NaCl Deficiency Affects Gustatory Neurons in the Nucleus of the Solitary Tract of Behaving Rats

KIYOMI NAKAMURA[1] and RALPH NORGREN[2]

Key words. Salt appetite—Nucleus of the solitary tract

Dietary sodium deprivation alters gustatory neural responses to sodium stimuli both on the periphery and in the central nervous system. This change is thought to influence the compensatory behavior that restores sodium balance. Prior investigations used anesthetized preparations, across subjects designs, and, in one instance, diets that differed in nutrient, as well as sodium, content [1]. In the present experiment, we used the same awake, behaving animals before and during dietary sodium deprivation, isolated single neurons from the nucleus of the solitary tract (NST), and tested their response to a battery of sapid stimuli.

Ten male rats were fitted with a cranioplastic cap and two intraoral cannulae and trained to receive their water while restrained in the recording apparatus. During the first recording session, they were maintained on a sodium-replete diet (sodium-deficient diet with 0.4% Na added) and 41 NST taste neurons were tested. Subsequently, the rats were switched to the sodium-deficient diet for a minimum of 10 days and then an additional 58 NST cells were tested. Finally, the rats were returned to a sodium-replete diet, and a third, small sample of data was collected ($n = 12$). The diet conditions were verified by analysis of urinary sodium concentration and by behavioral tests for ingestion of 0.51 M NaCl. The response criteria and the statistical analyses were the same as those described in our previous study [2].

During sodium deprivation, spontaneous activity increased to an average of 142%, but responses to water dropped to 72%. Under these conditions, taste responses to the four standard stimuli were reduced as follows. The mean response to NaCl decreased to 53% of its pre-deprivation level; that for sucrose dropped to 41%; citric acid to 68%, and quinine HCl to 84%. The other stimuli exhibited similar decreases—monosodium L-glutamate (MSG) dropped to 39%, glycine to 35%, and Polycose to 61%. Although the sample was small, these ratios appeared to revert toward previous values when sodium was added back to the diet.

Based on the best-response categories for the four standard stimuli, the response profiles of taste neurons were not changed by the dietary conditions. Within the NaCl-best category, however, the percentage of neurons that responded exclusively to NaCl increased from 27.2% (3/11) in the replete condition to 50% (10/20) in the Na-depleted state. Under both dietary conditions, multidimensional scaling of neurons produced a ring-like arrangement in a two-dimensional space [2]. In the replete condition, neurons in different best-stimulus categories did not intermingle in the space. In the deficient condition, however, sucrose-best neurons overlapped with NaCl-best cells on one side and with acid-best ones on the other.

Compared with a prior study in an anesthetized animals, the change in diet conditions in the present experiment failed to produce a shift in NaCl responsiveness from NaCl-best neurons to sucrose-best cells. In the Na-replete state, 61% of the activity elicited by NaCl occurred in Na-best cells; and 33% in sucrose-best neurons. In the depleted state, these figures were 60% and 26%, respectively. At higher concentrations, however, deprivation did alter the relative responsiveness of gustatory neurons to sucrose and NaCl. When the animals were sodium replete, in sucrose-best neurons, 1.0 M NaCl elicited only 60% as much activity as that produced by 0.3 M sucrose. When depleted, the response to strong salt was 101% that of sucrose. Similarly, for Na-best neurons, the response to 1.0 M sucrose was only 38% of that to the 0.1 M NaCl standard in the replete condition, but rose to 71% when the rats were sodium deprived.

Acknowledgments. Supported by The Salt Science Research Foundation 9351, PHS DC-00240, MH-43787, and MH-00653.

References

1. Jacobs KM, Mark GP, Scott TR (1988) Taste responses in the nucleus tractus solitarius of sodium-deprived rats. J Physiol (Lond) 406:393–410
2. Nakamura K, Norgren R (1991) Gustatory responses of neurons in the nucleus of the solitary tract of behaving rats. J Neurophysiol 66: 1232–1248

[1] Department of Electronics and Informatics, Toyama Prefectural University, 5180 Kurokawa, Kosugi-machi, Toyama, 939-03 Japan
[2] Department of Behavioral Science, College of Medicine, Pennsylvania State University, Hershey, PA 17033, USA

Taste Functions and Na$^+$-Appetite After Excitotoxic Lesions of the Parabrachial Nuclei in Rats

G. Scalera[1] and R. Norgren[2]

Key words. Parabrachial lesions—Taste functions

Bilateral lesions of the parabrachial nucleus (PBN) do not render rats ageusic, but they do alter responsiveness to sapid stimuli. The same lesions also eliminate the ability of rats to acquire a conditioned taste aversion (CTA) and prevent the expression of sodium appetite following acute sodium depletion. The present experiments extended these observations to rats with ibotenic acid lesions of the PBN by: (1) measuring their responsiveness to a concentration range of basic tastes, (2) assessing CTA when an aversive solution was used as a conditional stimulus (CS), and (3) testing sodium appetite following dietary sodium restriction. In anesthetized rats, ibotenic acid (0.2 µl; 20 µg/µl; PBNxlbo, $n = 5$) or the same amount of vehicle (PBS; pH, 7.40; VehicPBS, $n = 3$) was injected bilaterally into the electrophysiologically localized gustatory area of the PBN; three intact rats served as nonsurgical controls (NSC).

In a specially designed gustometer, water-deprived rats were trained for 3 days to lick a drinking spout to receive 10-s access to distilled water. During this training, as well as during the testing under deprived conditions, the rats had access to water for only 1 h immediately after gustometer tests. After water training, the rats were tested with water and one stimulus concentration series, presented randomly, for 30 min. The following solutions was tested for 1 day each under deprived conditions and another day under rehydrated conditions: sucrose, quinine, NaCl, citric acid, and L-alanine. If a rat failed to initiate at least one trial for each taste stimulus, its data were excluded. Unlike earlier experiments, as measured by this test, these PBNxlbo rats were rendered almost ageusic, although they still demonstrated concentration-dependent licking for citric acid.

Rats with lesions of the PBN fail to acquire a CTA. All previous tests of this effect have used normally preferred stimuli as the CS. Using a nonpreferred stimulus as a CS, however, actually facilitates acquiring a CTA. If the PBNx rats simply acquired a CTA more slowly than normal animals, then a mildly aversive CS might eliminate the effects of the lesions. The same rats used in the gustometer tests were given ad lib food and adapted to a deprivation schedule in which they received water for 15 min each morning and for 1 h each afternoon. The CS was 0.01 M malic acid (MA); the UCS, 0.15 M LiCl injected i.p. (1.33 ml/100 g body weight). On days 1, 4, 7, and 10, the rats were given MA to drink in the morning and, on the first 3 of these days, were immediately thereafter injected with LiCl. On the intervening days they were given water. For all groups, water intake during both access periods and body weights remained stable throughout the experimental period, but both measures were lower for the PBNxlbo rats than for the controls. After the first LiCl injection, both VehicPBS and the NSC rats essentially ceased ingesting MA, while the PBNxlbo group continued to ingest significant amounts of the same stimulus.

Rats with lesions of the PBN also fail to exhibit a sodium appetite when acutely depleted with furosemide. In order to assess the generality of this effect, the present group of PBNxlbo rats was placed on a sodium-deficient diet for 10 days and then given access to a strong NaCl solution. At the end of this period, urine Na$^+$ was less than 1.0 mmol/l. The baseline intake of water, 3% NaCl solution, and food before and after the Na$^+$-deficient diet did not differ between groups. On the test day, all rats had simultaneous access to 3% NaCl and water. Over the next 24 h, water intake was virtually identical for all three groups. During the first 15 min, the two control groups drank substantial amounts of NaCl, but the PBNxlbo rats, practically none. These differences persisted over the next 2 h, but then decreased somewhat over 24 h.

Acknowledgments. This study was supported by NIH grants DC00240, MH43787, and MH00653.

[1] Istituto di Fisiologia Umana, Via Campi, 287, 41100 Modena, Italy
[2] Department of Behavioral Science, College of Medicine, Pennsylvania State University, P.O. Box 850, Hershey, PA 17033, USA

Responses of Lateral Hypothalamic Neurons in Lysine-Deficient Rats

Takashi Kondoh[1], Tentcho Voynikov[1], Eiichi Tabuchi[1], Takashi Yokawa[1],
Taketoshi Ono[2], and Kunio Torii[1,3]

Key words. Lateral hypothalamus—Lysine deficiency

Appetite for food, and taste preferences for amino acids and NaCl in animals are affected by the state of protein nutrition [1–4]. Past experience is also thought to be important for food selection. Rats in a nutritionally well-balanced condition prefer monosodium L-glutamate (MSG) and L-arginine solutions to NaCl and other amino acids, this preference being directed toward L-lysine and NaCl when the animals are fed lysine-deficient diets. The central site at which preference is regulated, however, is unknown. One of the most important brain sites may be the lateral hypothalamic area (LHA). Some LHA neurons respond during feeding and to anticipatory cues after appropriate conditioning. The ingestion response of the monkey LHA is modulated both by palatability of food and satiation. Some rat LHA neurons respond selectively to the ingestion of lysine during lysine deficiency, but this does not occur in lysine-sufficient controls [4]. To evaluate the role of the LHA in the regulation of the rat preferences in different nutritive conditions, we recorded single neuron activity from the LHA during cue tone presentation and ingestion of various amino acids or NaCl, and compared the responses in controls with those in lysine-deficient rats that had learned to select amino acids and NaCl.

Thirty male Wistar strain rats, weighing about 100 g at the beginning of the experiments, were used. The animals were assigned to 6 groups, each of five animals. The five rats in each group were housed together and fed a control or a lysine-deficient diet with six kinds of solutions (0.2 M L-lysine HCl, 0.15 M MSG, 0.05 M L-arginine, 0.5 M glycine, 0.15 M saline, and distilled water) in a choice paradigm. We first confirmed that rats fed the control diet selectively ingested MSG and arginine, and that rats fed the lysine-deficient diet learned to selectively ingest lysine. When they grew to 200 g, the rats were held painlessly in a stereotaxic apparatus with a specially designed holder [5], and were trained to lick a spout to obtain various solutions that had been used in the earlier behavioral experiments. After the end of a 2-s cue tone, the spout was automatically extended close to the rat's mouth for 2 s. Each of the different solutions was signaled by its respective cue tone (1000–8750 Hz). In the recording sessions, rats were housed individually and fed the control or lysine-deficient diet previously received in their home cages. The activity of single LHA neurons was recorded through glass microelectrodes during cue tone and ingestion-related behavior. Response activity in the controls was compared with that elicited during lysine deficiency.

We analyzed 379 LHA neurons during conditioned cue tone stimuli and the subsequent ingestion of amino acids, saline, and water: 176 in the controls and 203 in lysine-deficient rats. Of the neurons tested in the controls, 44% (77/176) responded to the cue tones and 52% (92/176) responded during ingestion. During lysine deficiency, 46% (94/203) responded to the cue tones and 58% (117/203) responded during ingestion. Of the neurons tested, 9.7% (17/176) in the controls and 8.4% (17/203) during lysine deficiency differentiated cue tone or ingestion of taste solutions or both. During lysine deficiency, more LHA neurons responded to cue tone or ingestion or both for lysine or saline, and fewer neurons responded to the ingestion of arginine than in the controls. Typically, during lysine deficiency, neurons that responded to cue tone or the ingestion of lysine or MSG also responded to cue tone or the ingestion of saline; this relation was not evident in the controls. Neurons that differentiated cue tone or ingestion of the solutions, or both, were more numerous in the dorsal and lateral LHA both in controls and during lysine deficiency.

The results suggest that preference for at least some deficient amino acids and NaCl in states when they are deficient, and for MSG and arginine during periods of normal protein sufficiency, might be mediated through the LHA, and that learning and memory could affect these processes.

[1] Torii Nutrient-stasis Project, ERATO, Research Development Corporation of Japan, Technowave 100, 1-1-25 Shin-urashima-cho, Kanagawa-ku, Yokohama, 221 Japan
[2] Department of Physiology, Faculty of Medicine, Toyama Medical and Pharmaceutical University, 2630 Sugitani, Toyama, 930-01 Japan
[3] Ajinomoto Co., Inc., 214 Maeda-cho, Totsuka-ku, Yokohama, 244 Japan

References

1. Torii K, Mimura T, Yugari Y (1987) Biochemical mechanism of umami taste perception and effect of dietary protein on the taste preference for amino acids and sodium chloride in rats. In: Kawamura Y, Kare MR (eds) Umami: A basic taste. Dekker, New York, pp 513–563
2. Mori M, Kawada T, Torii K (1991) Appetite and taste preference in growing rats given various levels of protein nutrition. Brain Res Bull 27:417–422
3. Mori M, Kawada T, Ono T, Torii K (1991) Taste preference and protein nutrition and L-amino acid homeostasis in male Sprague-Dawley rats. Physiol Behav 49:987–995
4. Tabuchi E, Ono T, Nishijo H, Torii K (1991) Amino acid and NaCl appetite, and LHA neuron responses of lysine-deficient rat. Physiol Behav 49:951–964
5. Nakamura K, Ono T (1986) Lateral hypothalamus neuron involvement in integration of natural and artificial rewards and cue signals. J Neurophysiol 55: 163–181

Lateral Hypothalamic Neuron Response to Application of Amino Acids in Different Nutritive Conditions

E. Tabuchi[1], T. Kondoh[1], T. Voynikov[1], T. Yokawa[1], T. Ono[2], and K. Torii[1,3]

Key words. Amino acid—Lysine deficiency

Nutritive state can be indicated by behavior that reflects preferences for amino acids. Rats prefer monosodium L-glutamate (MSG) and arginine in a nutritively normal condition, but they prefer lysine and/or NaCl when in a lysine-deficient condition. We have found neurons in the lateral hypothalamic area (LHA) that responded specifically to ingestion of lysine solution in lysine-deficient rats, but not in lysine-sufficient controls. This may indicate that a given nutritive condition, such as deficiency of a certain amino acid, could enhance sensations, constituting a taste system to increase the drive to obtain the deficient substance. To examine the central function mediating preference-related behavior, we recorded single neuronal activity from the LHA during the iontophoretic application of amino acids in 17 male Wistar strain rats in the nutritively rich condition and the lysine deficient condition.

After recovery from an operation performed to permit painless restraint in a stereotaxic instrument, the rats were allowed to freely eat a lysine-deficient diet or a control (lysine-sufficient) diet, and to drink water ad lib. During a recording session, each rat was fixed in a stereotaxic apparatus by placing its head in a specially designed clamp; neuronal activity was recorded from the LHA during the iontophoretic application of various amino acids. The substances applied and their concentrations were: 0.05 M MSG, 0.05 M glycine, 0.05 M threonine, 0.05 M lysine HCl, 0.05 M arginine, and 0.15 M NaCl. These six solutions, and 4 M NaCl for current balance, were each placed in one barrel of a seven-barreled glass micropipette (22–80 MΩ), which was then attached to a glass recording microelectrode (1.4–3.5 MΩ) filled with 4 M NaCl. The tip of the recording electrode projected about 10 μm beyond the tip of the micropipette. DC current for iontophoretic application was in the range of 1–90 nA. Neuronal activity was recorded during the application of an amino acid and NaCl, taking strict care to avoid artifacts due to the iontophoretic application technique. If the distance between the tip of the recording electrode and that of the micropipette was more than 30 μm at the end of a recording session, all data obtained in that session were omitted, and data that indicated a current effect were omitted.

The activity of 250 LHA neurons (122 in controls, 128 in animals during lysine deficiency) was recorded during the application of amino acids. Of the neurons on which amino acids were applied, 106 (46 in controls, 60 during lysine deficiency) could be recorded during the ingestion of amino acid solutions. Of the 250 neurons, 163 (73 in controls, 90 during lysine deficiency) were analyzed and compared in the two nutritional conditions, since 87 neurons (49 in controls, 38 during lysine deficiency) were omitted according to our criteria for anomalous responses to iontophoresis. Of the 163 neurons analyzed, 112 (49 in controls, 63 during lysine deficiency) responded during the application of any amino acid, and 51 (24 in controls, 27 during lysine deficiency) did not respond. In controls, neurons responsive only to glutamate were often observed (19/49, 39%), and those responsive to any amino acid responded mostly to glutamate (37/49, 76%). Some neurons responded to glutamate plus threonine (7/49, 14%). The remaining neurons responded to fewer than three kinds of amino acids and did not display any particular pattern of responses. During lysine deficiency, neurons responsive only to glutamate were observed often (23/63, 36%), and those responsive to any amino acid also responded mostly to glutamate (47/63, 75%). However, some neurons responded specifically to lysine or lysine plus glutamate (7/63, 11%). Some neurons responded to glutamate plus threonine (7/63, 11%), and others responded to more than four kinds of amino acids (9/63, 14%).

During lysine deficiency, some LHA neurons responded to applied lysine, whereas, in a nutritively rich condition, most neurons did not respond to lysine. It is suggested that, in order to maintain homeostasis, an animal may have the potential to change its sensitivity to both external and internal stimuli in different nutritive conditions, to obtain nutrient substances that it may lack and which are necessary.

[1] Torii Nutrient-stasis Project, ERATO, Research Development Corporation of Japan, Technowave 100, 1-1-25 Shin-urashima-cho, Kanagawa-ku, Yokohama, 221 Japan
[2] Department of Physiology, Faculty of Medicine, Toyama Medical and Pharmaceutical University, 2630 Sugitani, Toyama, 930-01 Japan
[3] Ajinomoto Co., Inc., 214 Maeda-cho, Totsuka-ku, Yokohama, 244 Japan

Gustatory and Olfactory Responses
of Chemosensitive Pallidal Neurons

Zoltán Karádi, Béla Faludi, András Czurkó, Csaba Niedetzky, and László Lénárd[1]

Key words. Tasting and smelling—Glucose-monitoring system

The globus pallidus (GP), a key structure of the extrapyramidal motor system, is known to be intimately involved in the central regulation of body weight and food and fluid intake behaviors [1–3]. In addition to its roles in the motor control of eating and drinking [4,5], it is important in metabolic regulation [6,7], as well as in feeding-associated sensory-motor and perceptual processes [1,4,8]. These functions of the GP are, at least partially, mediated by catecholaminergic (CA) forebrain mechanisms, most importantly by the nigrostriatal and mesolimbic-mesocortical dopamine (DA) systems [1,4,9]. The close functional-morphological interrelations of the GP and the lateral hypothalamic area (LHA) [2,10] and the availability of information on related properties of neurons of the latter [11–13], indicate the necessity for investigating feeding-associated attributes of GP neurons, with particular regard to their sensitivity to glucose. Despite the abundance of related data in the literature, our knowledge is very poor concerning the chemical characteristics of GP cells. To date, information on the endogenous and exogenous chemosensitivity of pallidal neurons in the primate is very limited. The present studies, therefore, aimed to reveal the complex chemosensory properties of GP cells. To do so, the extracellular single neuron activity of the GP was recorded in anesthetized rats and anesthetized or alert rhesus monkeys, by means of carbon fiber multibarreled glass microelectrodes, during: (1) local, microelectrophoretic administration of chemicals (including glucose, catecholamines, acetylcholine, gamma aminobutyric acid [GABA], and N-methyl-D-aspartic acid [NMDA] enantiomers) and (2) gustatory and (3) olfactory stimulations. Seventeen male CFY rats (270–430 g; LATI, Gödöllö, Hungary) and 3 (1 male, 2 females) adult monkeys (*Macaca mulatta*) were used. Anesthesia, when employed, was introduced and maintained by either urethane (for rats; 20%, 1.5 ml/300 g, i.p.) or ketamine (for monkeys; 35 mg/kg; i.m.). The surgery, recordings,

electrophoretic applications, and data analysis were performed according to the standard methodology of our laboratory, details of which have been published elsewhere [11,14]. The technique of gustatory and olfactory stimulations in the rat (with appropriately lower volumes of tastants and odorants) was the same as that previously described for primates [11,15]. The activity of a total 196 neurons (101 and 95 cells in the rat and monkey, respectively) was recorded in these studies. Glucose was electroosmotically administered to 98 neurons in the rodent and primate GP. Firing rate changes, exclusively inhibitory ones, were observed in 12 of all cells tested (12%) and the proportion of these glucose-sensitive (GS) neurons did not vary significantly among species and treatments. The whole gustatory and olfactory stimulus array was applied to 95 pallidal neurons. Eight of all the GP cells examined (approximately 9%) responded, usually to two or more tastants, whereas the proportion of smell-responsive neurons was only 6% (6 cells). "Gustatory" and "olfactory" units were mainly found among the GS neurons of the GP: More than the half of all taste-responsive pallidal cells in the rat, and about the three-quarters of the "gustatory" GP neurons in the monkey, proved to be GS. (Discharge rate changes of a typical, taste- and smell-responsive GS cell in the ventral region of a monkey's GP are shown in Fig. 1.) Pallidal neurons also displayed distinct sensitivities when noradrenaline (NA), DA, acetylcholine, GABA, and NMDA enantiomers were applied microelectrophoretically. GS cells, with predominant responsiveness to tastes and smells, were more likely to change in activity (being mainly suppressed) by NA and DA applications than were the glucose-insensitive units. Our present results provide evidence for the existence of specific chemosensitive pallidal cells that may be integrant constituents of a hierarchically organized glucose-monitoring system along the central neuraxis [16,17]. We have also demonstrated that GP neurons, especially the GS ones, are intimately involved in central taste (and smell) information processing. The physiological importance of these cells in feeding control is substantiated by the direct interconnections of GP loci with GS units of the LHA in the rat [10], as well as by our most recent findings on alimentary task-related activity patterns of GP neurons in the monkey [18].

[1] Institute of Physiology, Pécs University School of Medicine, Pécs, Szigeti út 12, H-7643 Pécs, Hungary

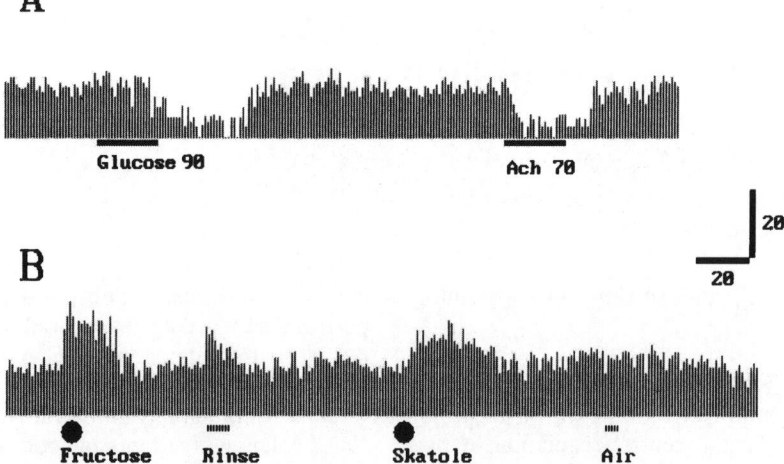

A

Glucose 90   Ach 70

B

Fructose   Rinse   Skatole   Air

FIG. 1**A,B.** Firing rate changes of a chemosensitive neuron in the ventral region of the globus pallidus of a rhesus monkey. **A** Activity decrease to glucose and to acetylcholine, both applied microelectrophoretically. **B** Facilitation of neuronal discharge rate in response to fructose taste and skatole odor. *Horizontal bars*, numbers, duration (s), and intensity (nA) of electrophoretic current ejections. *Solid circles*, Onset of gustatory and olfactory stimulation. Calibration, time in s, and firing rate in spikes/s, respectively

Acknowledgments. The present work was supported by the Ministry of People's Welfare of Hungary (ETT T-565/1990; L.L.), the Hungarian National Research Fund (OTKA/1404; L.L.), and the Hungarian Health Foundation (Z.K.).

References

1. Lénárd L, Karádi Z, Szabó I, Hahn Z (1982) Pallidal mechanisms in the organizations of feeding and sensorimotor integration. In: Lissák K (ed) Recent development of neurobiology in Hungary IX. Akadémiai Kiadó, Budapest, pp 79–113
2. Morgane PJ (1961) Alterations of feeding and drinking behavior of rats with lesions in globi pallidi. Am J Physiol 201:420–428
3. Sándor P, Hajnal A, Jandó G, Karádi Z, Lénárd L (1992) Microelectrophoretic application of kainic acid into the globus pallidus: Disturbances in feeding behavior. Brain Res Bull 28:751–756
4. Marshall JF, Richardson JS, Teitelbaum P (1974) Nigrostriatal bundle damage and the lateral hypothalamic syndrome. J Comp Physiol Psychol 87:375–395
5. Szabó I, Sarkisian J, Lénárd L, Németh L (1977) Pallidal stimulation in rats: Facilitation of stimulus bound chewing by food deprivation. Physiol Behav 18:361–368
6. Hahn Z, Karádi Z, Lénárd L (1988) Sex-dependent increase of blood glucose concentration after bilateral pallidal lesion in the rat. Acta Physiol Hung 72:99–101
7. Hahn Z, Lénárd L, Ruppert F (1979) The connection of plasma triiodothyronine levels with the sex-dependent body weight loss after bilateral pallidal lesion in rats. Acta Physiol Hung 53:17–21
8. Levine MS, Ferguson N, Kreinick CJ, Gustafson JW, Schwartzbaum JS (1971) Sensorimotor dysfunctions and aphagia and adipsia following pallidal lesions in rats. J Comp Physiol Psychol 77:282–293
9. Lénárd L (1977) Sex-dependent body weight loss after bilateral 6-hydroxydopamine injection into the globus pallidus. Brain Res 128:559–568
10. Oomura Y, Nakamura T, Manchanda SK (1975) Excitatory and inhibitory effects of globus pallidus and substantia nigra on the lateral hypothalamic activity in the rat. Pharmacol Biochem Behav 3(S1):23–36
11. Karádi Z, Oomura Y, Nishino H, Scott TR, Lénárd L, Aou S (1992b) Responses of lateral hypothalamic glucose-sensitive and glucose-insensitive neurons to chemical stimuli in behaving rhesus monkeys. J Neurophysiol 67:389–400
12. Oomura Y (1980) Input-output organization in the hypothalamus relating to food intake behavior. In: Morgane PJ, Panksepp J (eds) Handbook of the hypothalamus, vol 2. Marcel Dekker, New York, pp 557–620
13. Yamamoto T, Matsuo R, Kiyomitsu Y, Kitamura R (1989) Response properties of lateral hypothalamic neurons during ingestive behavior with special reference to licking various taste solutions. Brain Res 481:286–297
14. Czurkó A, Faludi B, Vida I, Niedetzky CS, Hajnal A, Karádi Z, Lénárd L (1991/1992) Effect of electrophoretically applied neurochemicals on activity of extrapyramidal and limbic neurons in the rat. Acta Biochim Biophys Hung 26:57–60
15. Karádi Z, Nishino H, Oomura Y, Scott TR, Aou S (1990) A method for gustatory stimulus delivery in awake rhesus monkeys. Brain Res Bull 25:665–668
16. Nakano Y, Oomura Y, Lénárd L, Nishino H, Aou S, Yamamoto T, Aoyagi K (1986) Feeding-related activity of glucose- and morphine-sensitive neurons in the monkey amygdala. Brain Res 399:167–172
17. Oomura Y, Yoshimatsu H (1984) Neural network of glucose monitoring system. J Autonom Nerv Syst 10:359–372
18. Karádi Z, Lénárd L, Faludi B, Czurkó A, Vida I, Sándor P, Niedetzky Cs (1993) Feeding-associated characteristics of globus pallidus neurons in the rhesus monkey. Neurobiology 1:170

Effects of IV Angiotensin II on Cortical Neurons During a Salt-Water Discrimination GO/NOGO-Task in Monkeys

M. Ohgushi, H. Ifuku, and H. Ogawa[1]

Key words. Angiotensin II—Task-related neurons

Intraventricular (i.v.) administration of angiotensin II (AII) elicits sodium appetite in animals and increases water intake [1]. In a previous experiment, we disclosed prolonged response latencies and a decreased fraction of correct responses after i.v. administration of AII during a NaCl-water discrimination GO/NOGO-task in monkeys [2]. These findings indicate that AII increased the threshold of monkeys to discriminate NaCl from water. Such changes in responses recovered within about 30 min after AII injection, while increased water intake was observed even 1 h after injection. To clarify the underlying neural mechanisms, we recorded the activities of task-related neurons in the frontal operculum (Fop) and orbitofrontal cortex (OFO) of monkeys during the task, and examined effects of i.v. injections of AII on these neurons.

Two Japanese monkeys (*Macaca fuscata*) were trained to perform a NaCl-water discrimination GO/NOGO-task. Various concentrations of 0.5 ml NaCl solution were provided as GO cues. When the monkey pressed the lever in response to NaCl he received 0.5 ml of 0.3 M sucrose as reward. Water was provided as a NOGO cue. When the monkey did not press the lever, he was rewarded. After the monkeys were found to have learned the task, a cannula for administration of AII to the lateral ventricle and an Evert's type cylinder to cover the Fop and OFO were aseptically implanted on the skull of the anesthetized animal (ketamine hydrochloride; 10 mg/kg, i.m.). Tungsten microelectrodes were inserted to record the activities of single neurons in these areas. When the neuron activities under study were stable for more than 10 min, 5 µg of AII was administrated through the cannula. Cues and reward were provided by a 16 bit microcomputer. The state of performance and neural activities were shown on-line on the display of the same microcomputer and recorded on mini-floppy disks every trial for later analysis. After the termination of the experiment, the recording sites were reconstructed histologically.

A total of 135 task-related neurons were recorded. In 22 neurons of these, the effects of i.v. injections of AII were examined. Based on the neural discharges in relation to cue presentation, neurons were classified into the following four types: (1) Cue-sensitive; these responded to both NaCl and water with the same magnitude and in a short latency to the cue onset. (2) Cue-responsive; these responded to both cues in a latency of longer than 200 ms. (3) Cue-differentiation; these responded differentially to two cues. And (4) cue-expectation; these showed the high spontaneous activity between trials, which was inhibited by cue presentation. Injection of AII altered firing rates in 12 neurons, changed the response pattern in 4, and shortened response latencies in 1 neuron. It also increased spontaneous activity in 7 neurons, including 2 cue-expectation neurons, and increased inhibition in 2 cue-expectation neurons. Two contradictory changes were seen in the response patterns. First, some cue-sensitive and cue-responding neurons were changed to the cue-differentiation type, indicating an increased discrimination of NaCl vs water. Second, the responses of cue-differentiation type neurons were decreased, indicating decreased discrimination. Such changes in the neural activity of task-related neurons lasted more than 30 min, except for 2 neurons which returned to a control level at 10 or 30 min after AII injections. Neural changes lasted longer than the behavioral effects of AII. Intraventricular administration of AII modulated task-related neurons so that NaCl discrimination was decreased or increased, depending on the neurons. It also increased expectation for cue presentation, probably corresponding to increase in errors in response to water (NOGO-task). AII is received at the SFO and at other periventricular organs [1] which, in turn, may modulate, directly and/or indirectly through the hypothalamus, the neural processing of gustatory information in the primary and secondary gustatory areas.

References

1. Simpson JB (1981) The circumventricular organs and the central action of angiotensin. Neuroendocrinology 32:248–256
2. Ohgushi M, Ifuku H, Ito S, Ogawa H (1991) Salt-water discrimination at increased sodium appetite condition in monkeys. Chem Senses 16:185

[1] Department of Physiology, Kumamoto University School of Medicine, Honjo 2-2-1, Kumamoto 860, Japan

11. Clinical Assessment of Taste

Influences of Aging on Taste Bud Distribution and Taste Perception in Humans and Rodents

INGLIS J. MILLER, JR.[1]

Key words. Aging taste buds—Human rodents variation

Many people experience subjective changes in their orosensory perception as they get older. Total loss of taste perception is rare [1–3], but both chemosensory [1] and electrogustometry thresholds [4] *seem* higher in older people than in younger ones. Figure 1, from Bartoshuk et al. [1], shows taste thresholds for NaCl and sucrose as a function of age. While thresholds for NaCl and sucrose appear to be *higher* for older people, their ability to detect salt and sugar seems to be preserved.

Electrogustometry thresholds reported by Tomita and others [4] are shown in Fig. 2. Like taste thresholds for chemicals, the average electrogustometry thresholds for older people seem to be elevated in comparison to younger ones. Notice the large variations in electrogustometry thresholds among older subjects.

In addition, some individuals experience exaggerated oral sensations, parageusias, and paresthesias, as they get older. Longevity among humans complicates the objective assessment of taste perception across an individual lifespan, but cross-sectional studies of taste thresholds among people of different ages are replete with individual variations [1–3]. Contemporary studies [5] show that taste perception in older people remains remarkably intact.

There is a relationship between the variations in anatomical distributions of taste receptor organs and in taste sensitivity. American university students [6] who had higher taste bud densities on the tongue tip also gave higher intensity ratings of taste stimuli than students with lower taste bud densities. Subjects from a clinical population [7] and dental students [8] also showed a relationship between relative taste sensitivity and the abundance of fungiform taste buds that they had. A summary of these observations is illustrated in Fig. 3. The young people with more fungiform papillae also had more fungiform taste pores on them. In addition, the students with more taste receptors gave higher taste intensity ratings,

and another group [7] had lower thresholds for propylthiouracil.

Studies of human cadaver tongues published in the 1930s from Japan [9,10] and the United States [11] reported that older individuals had fewer taste buds in vallate and foliate papillae than younger subjects. The observations by Arey et al. [11] and by Mochizuki [9,10] for vallate papillae are shown in Fig. 4. These graphs show that the number of taste buds per vallate papilla may reach its maximum in youth and diminish later in life. More recent studies of cadaver tongues [12,13] and excised papillae from living subjects [14] show a large variation in the prevalence of taste buds among subjects, regions of the tongue, and among individual papillae [14,15] from the same tongue. Differences in taste bud prevalence are as great among subjects of the same age group as they are among tongues from young adults, middle-aged adults, and older adults [15].

A persistent problem in the analysis of tissue specimens from aged human populations is to understand whether the anatomical conditions which prevail at the time of autopsy are due to conditions (a) in the more recent life of the subject, (b) from earlier events of the subject's life, or (c) from cumulative effects over a lifespan. The donors of anatomical specimens from the 1930s, for example, had lifespans which extended back to the middle of the 19th century. The oldest specimens from our own cadaver studies were born in the beginning of the 20th century. The point is that standards of nutrition, hygiene, diagnostics, and therapeutics have changed so much over the last 150 years that younger specimens differ in other ways from the older specimens than just in chronological age at the time of death. In an anatomical study [17] which compared tongues from young and old monkeys, healed lesions on the anterior part of the tongue influenced taste bud distributions in older animals, but it was not clear at what age the lesions occurred.

Rodents offer an alternative choice for the study of individuals across their entire lifespans. Since the lifespan of rats is 2–3 years, it is possible to follow them from birth until their natural deaths. It has been reported that small but significant differences are present in the numbers of taste buds that young adult and old [18,19] Fischer-344 rats have, and that the neural taste responses [20] between young and

[1] Department of Neurobiology and Anatomy, Bowman Gray School of Medicine of Wake Forest University, Winston-Salem, NC 27157-1010, USA

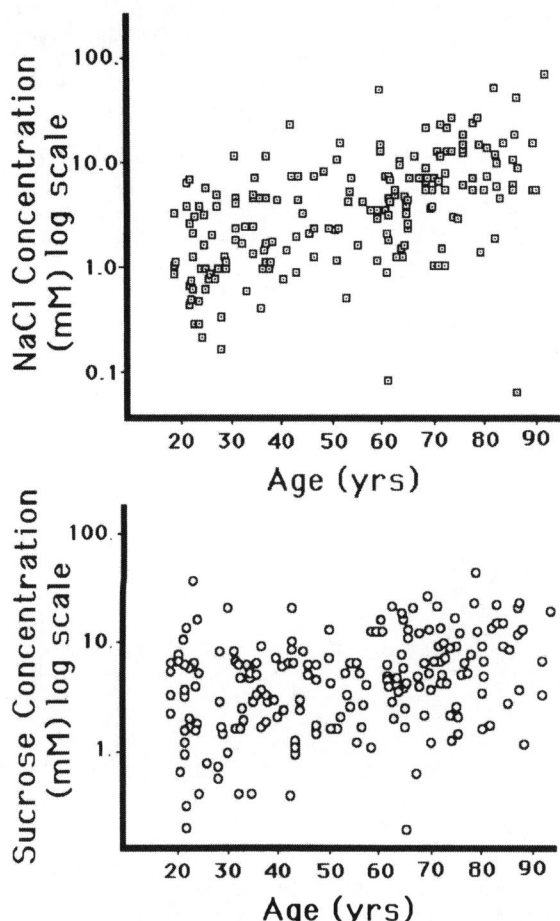

FIG. 1. Published taste thresholds for NaCl and sucrose as a function of age. (From Bartoshuk et al. [1] with permission)

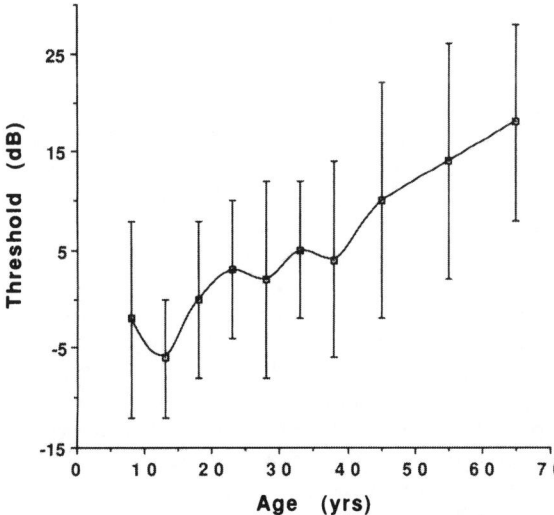

FIG. 2. Electrogustometry thresholds for human subjects. (From Tomita et al. [4] with permission)

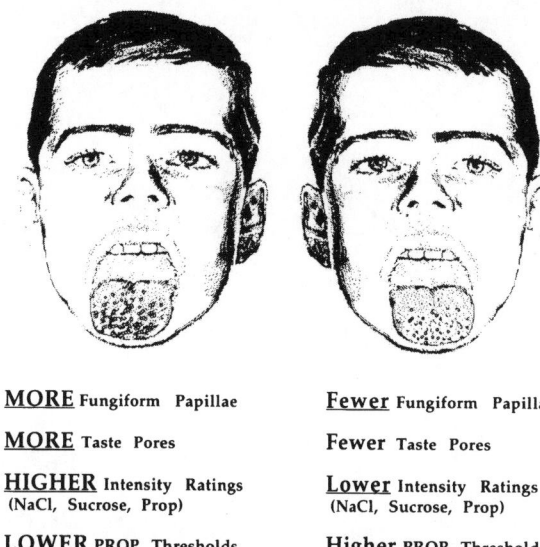

MORE Fungiform Papillae   <u>Fewer</u> Fungiform Papillae

MORE Taste Pores   Fewer Taste Pores

HIGHER Intensity Ratings (NaCl, Sucrose, Prop)   <u>Lower</u> Intensity Ratings (NaCl, Sucrose, Prop)

LOWER PROP Thresholds   <u>Higher</u> PROP Thresholds

FIG. 3. Diagram showing that young human subjects with more fungiform papillae and taste pores on their tongue tips also have lower taste thresholds and higher intensity ratings

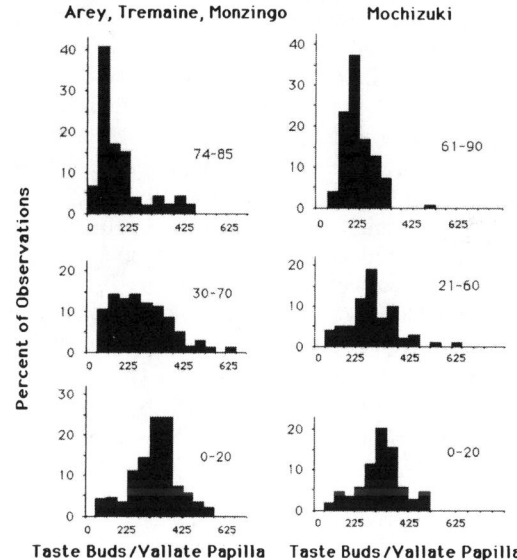

FIG. 4. Data from Arey et al. [11] (*left*) and from Mochizuki [9,10] (*right*) showing vallate taste bud numbers as a function of age

old animals also differ. Thus, like humans, rat taste responses seem to be different between youth and old age, but catastrophic taste loss seems not to be very prevalent.

It was hypothesized that rats, like humans, differ from one another in taste sensitivity, and that some differences in taste sensitivity among rats are related to the numbers of taste buds that individuals have. With J.C. Smith of Florida State University, we have participated in a collaborative project to study

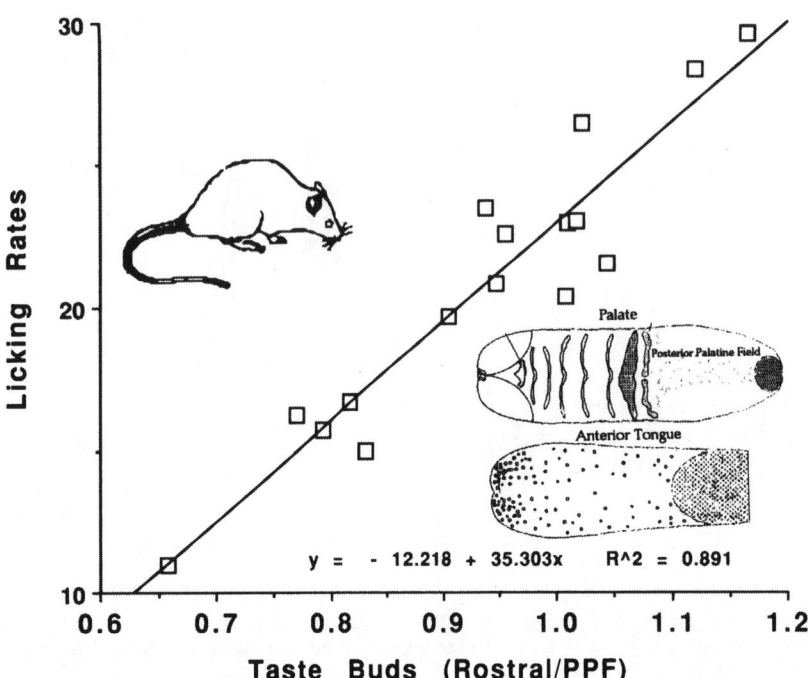

FIG. 5. The maximal licking rates of rats for sucrose are plotted for $n = 16$ aging rats as a function of a ratio of the numbers of taste buds that the rats have on their tongue tips and nasoincisor ducts, divided by the numbers of taste buds on the rats posterior palatine fields (*PPF*). (From [22], with permission)

the relationship between taste bud distribution and the ingestive preference for sucrose solutions in Fischer-344 rats. Smith studied the intake behaviors of rats from 2 months of age until their natural deaths at 24–36 months of age. He has [21] reported that the licking rate within drinking bouts was proportional to stimulus concentration with a slope which was similar to the concentration–response function for the integrated nerve response of the greater petrosal nerve in the rat.

Maximal licking rates for sucrose solutions by rats are related significantly to the numbers of taste buds on the rats' tongues and palates [21]. The licking rates per 6s are shown in Fig. 5 for each rat in relation to the ratio of the numbers of taste buds that each rat has in its rostral oral cavity and posterior palatine field. This relationship suggests that behavioral differences in the rats' ingestion of sucrose are related to taste bud distribution. As the animals age from 1 to 2 years old, their licking rates and consumption of sucrose increase [22]. There does not seem to be a dramatic decrease in the rats' sucrose intake prior to death. We conclude that taste perception in rats appears to be sustained until old age. Taste buds seem to be well preserved in old, healthy rats.

What useful information can be learned from rodents about aging in the taste system? One must postulate that the taste system serves both rodents and humans in a similar way across the lifespan. Taste receptor activity is necessary for oropharyngeal reflexes to reinforce ingestive behavior and to provide individuals with discriminative information about the content of certain sweet, salty, sour umami, bitter and, perhaps, other qualities. Two important features of discriminative taste behavior seem to be similar among humans and rodents: (1) both humans and rodents learn selective intake behaviors partly by taste perception; (2) genetics influences relative taste sensitivity among closely related individuals [24,25]. The impact of taste perception on ingestive behaviors is contingent on both inherent and environmental factors. It is much easier to control genetic and environmental influences in rodents than in humans. The available evidence suggests that the taste system remains intact and functional across most of the lifespans of both humans and rodents.

When episodes of incapacity, serious illness, or starvation compromise the health of older individuals, there can result a regressive phase of life with decreased ingestion, debilitation, and death. It appears that substantial loss of taste perception in both humans and rodents usually results from pathological conditions. Although the taste system regenerates continuously, regeneration is incomplete [26] and the capacity for regeneration seems to diminish over the lifespan. In a debilitated condition, altered taste perception may exacerbate anorexia, compromise nutrition, and contribute to terminal illness.

The following postulates, which are listed in Fig. 6, about aging, taste bud distribution, and taste perception are offered in summary. The first postulate is that the number and distribution of human taste buds are related to the person's taste sensitivity. While it has been assumed for a long time that variation in taste bud distribution is related to variations in taste perception, there are now substantive

1. **The number and distribution of human taste buds are related to the person's taste sensitivity.**

2. **Renewal and redundancy in the taste system preserves gustatory function in old age.**

3. **Unusual taste insensitivity in older individuals usually results from pathology or the side effects of medication.**

4. **Because of genetic diversity in the human population, quantitative measures of "NORMAL" taste perception yield a wide range of values.**

5. **Taste sensitivity and the prevalence of taste buds may be diminished in older individuals as a consequence of etiological factors which occurred much earlier in their lives.**

FIG. 6. Five postulates about aging, taste bud distribution, and taste perception

observations to support this postulate [6–8]. Important clinical observations have also supported the relationship between diminished taste perception and taste bud loss [4]. Contemporary findings [1–3] support the conclusion that taste perception is sustained in many older persons. Both renewal of taste buds [26] and redundancy in the taste system seem to contribute to the preservation of gustatory function. However, we need to study the etiology of taste dysfunction in those people in whom it occurs. Unusual taste insensitivity in older individuals usually results from pathology or the side effects of medication. Assessment of taste perception in older humans is complicated because of genetic diversity. Quantitative measures of "normal" taste perception yield a wide range of values. Since both taste sensitivity and the prevalence of taste buds may be diminished in older individuals as a consequence of etiological factors which occurred much earlier in their lives (trauma, surgery), it is necessary to devise new methods of differential diagnosis which discriminate between acute and chronic causes of dysgeusia and hypogeusia. Both controlled animal studies and new diagnostic imaging techniques may provide beneficial information.

Acknowledgments. This project was supported by NIH Grants DCD 00230 and AG 04932. David Sink and David Black provided important technical contributions. The author is indebted to Professors Hiroshi Tomita and Kenzo Kurihara, and to the Organizing Committee of ISOT-XI and JASTS XXVII, for support and kind assistance.

References

1. Bartoshuk LM, Rifkin B, Marks LE, Bars P (1986) Taste and aging. J Gerontol 41:51–57
2. Weiffenbach JM, Baum BJ, Burghauser R (1982) Taste thresholds: quality specific variation with human aging. J Gerontol 37:372–377
3. Cowart B (1989) Relationships between taste and smell across the adult life span. Ann NY Acad Sci 561:39–56
4. Tomita H, Ikeda M, Okuda Y (1986) Basis and practice of clinical taste examinations. Auris Nasus Larynx (Tokyo) [Suppl I]:S1–S15
5. Bartoshuk LM (1989) Taste: robust across the age span? Ann NY Acad Sci 561:65–75
6. Miller IJ Jr, Reedy FE Jr (1990b) Variations in human taste bud density and taste intensity perception. Physiol Behav 47:1213–1219
7. Reedy FE Jr, Bartoshuk LM, Miller IJ Jr, Duffy VB, Yanagisawa K (1993) Relationships among papillae, taste pores, and 6-n-propylthiouracil (PROP) suprathreshold taste sensitivity. Chem Senses 18:618
8. Englehardt R, Zuniga J, Davis S, Chen N, Phillips C, Schiffman S, Miller IJ Jr (1993) Taste performance of the anterior human tongue varies with fungiform taste bud density. Chem Senses 18:550
9. Mochizuki Y (1937) An observation of the numerical and topographical relation of taste buds to circumvallate papillae of Japanese. Okajimas Folia Anat Jpn 15:595–608
10. Mochizuki Y (1939) Studies of the papilla foliata of Japanese. Okajimas Folia Anat Jpn 18:355–369
11. Arey L, Tremaine M, Monzingo F (1935) The numerical and topographical relation of taste buds to human circumvallate papillae throughout the life span. Anat Rec 64:9–25
12. Arvidson K (1979) Location and number of taste buds in human fungiform papillae. J Dent Res 87:435–442
13. Miller IJ Jr (1986) Variation in human fungiform taste bud densities among regions and subjects. Anat Rec 216:474–482
14. Arvidson K, Friberg U (1980) Human taste: response and taste bud number in fungiform papillae. Science 209:807–808
15. Miller IJ Jr (1988) Human taste bud density across adult age groups. J Gerontol Biol Sci 43:B26–B30
16. Kullaa-Mikkonen A, Koponen A, Seilonen A (1987) Quantitative study of human fungiform papillae and taste buds: variation with aging and in different morphological forms of the tongue. Gerondontology 3:131–135
17. Bradley RM, Stedman HM, Mistretta CM (1985) Age does not affect numbers of taste buds and papillae in adult rhesus monkeys. Anat Rec 212:246–249
18. Mistretta CM, Baum BJ (1984) Quantitative study of taste buds in fungiform and circumvallate papillae of young and aged rats. J Anat 138:323–332
19. Mistretta CM, Oakley IA (1986) Quantitative anatomical study of taste buds in fungiform papillae of young and old Fischer rats. J Gerontol 41:315–318
20. McBride MR, Mistretta CM (1986) Taste responses from the chorda tympani nerve in young and old Fischer rats. J Gerontol 41:306–314
21. Smith J (1988) Behavioral measures of the taste of sucrose in the rat. In: Miller IJ Jr (ed) The Beidler Symposium on Taste and Smell. Book Service Associates, Winston-Salem, NC, pp 205–213
22. Miller IJ Jr, Smith JC (1991) Sucrose intake behavior is related to taste bud distribution in Fischer-344 rats. Chem Senses 16:558
23. Smith JC, Wilson LS (1989) Study of a lifetime of sucrose intake by the Fischer-344 rat. Ann NY Acad Sci 561:291–306
24. Miller IJ Jr (1991) Taste preference and taste bud prevalence among inbred mice. In: Wysocki C, Kare

M (eds) Chemical senses, vol 3, Genetics of perception
and communication. Dekker, New York, pp 263–266
25. Miller IJ Jr, Harder DB, Whitney G (1991) Taste bud
distribution and taste preference among mice. Chem
Senses 15:557

26. Oakley B (1985) Trophic competence in mammalian
gustation. In: Pfaff D (ed) Taste, olfaction, and the
central nervous system. A Festschrift honoring Carl
Pfaffman. The Rockefeller University Press, New
York, pp 92–103

Influence of Aging on Taste Threshold

S. Endo, M. Hotta, Y. Yamauchi, K. Ootsuka, and H. Tomita[1]

Key words. Taste threshold—Taste acuity—Whole mouth method—Electrogustometry—Scaling—Aging

Introduction

The effects of aging on visual and hearing acuity are well-established, and the effects on smell acuity are also well-known [1], but another chemical sense—taste—may have a different response to aging.

In a clinical setting, there have already been many reports investigating the effects of aging on taste acuity. It is generally agreed that the effect of aging on taste acuity is minimal. To confirm this, we investigated the changes in thresholds of taste with aging using three different methods:

1. The whole-mouth gustatory test
2. The electrogustometric test
3. Self-rating of physiologic function by multiple choice questionnaire

Method and Results

Whole Mouth Gustatory Test

We investigated taste thresholds with the whole-mouth method in over 500 normal nonsmoking subjects. The subjects consisted of students, volunteers visiting the health clinic or the senile communication center, and normal hospital staff. Before the taste examination, all subjects were examined for otolaryngologic disease and their medical history taken. Diabetics and patients who had undergone gastrectomy or had a history of central nervous system disease were excluded from this study. The taste solution was prepared as serial (1:1) dilutions. The stimulus was given in gradually ascending order and the threshold was measured by approaching the limit. More detailed information is presented in another chapter of this book [2].

The distribution of the detection thresholds of four tastants is shown in Fig. 1. The vertical axis is a logarithmic scale of concentration of the taste solu-

tion. For these four tastants, there was no remarkable increase in the threshold according to age. It is possible to apply a regression line on the threshold distribution, and this may indicate a slight increase in the threshold according to age, as we mentioned before [3]. Comparison of the threshold mean of each decade indicated some increase in the threshold with respect to age, but it was not significant because the higher limit of the young age group overlapped with the lower limit of the old age group; i.e., many young subjects had a higher detection threshold than the mean detection threshold of the elderly group.

The recognition thresholds, which were measured at the same time, were also stable with respect to aging.

The results of the whole-mouth gustatory test were as follows.

1. There were minimal increases in taste thresholds with aging.
2. The range of thresholds among individuals was wider than that of any variation according to age.

Electrogustometric Test

We studied the threshold of the electric taste in over 500 normal nonsmoking subjects. The test equipment was a commercially available electrogustometer (TR-06, Rion, Tokyo, Japan). This was modified to extend the dynamic range, from -20 to $34\,dB$ for normal sensitive subjects. Three pairs of locations were tested for thresholds. Gradually increasing levels of stimulation were given and so the threshold was measured by ascending toward the limit. More detailed information is presented in another chapter of this book [4].

The distribution of the threshold for each nerve area is shown in Fig. 2. As the side difference was not remarkable, only the right-side data are shown. Sex difference was not significant. There were many nonresponders to the strongest electric stimulus in the greater petrosal nerve area, so they were plotted at the level of 40 dB tentatively. There was a marked gradual increase in the threshold according to age in the chorda tympani and the greater petrosal nerve areas. In the glossopharyngeal nerve area the threshold was constant until age 60 and then there was a gradual rise according to age. The upper limit of the young age group was slightly higher than the lower limit of the old age group especially in the chorda

[1] Department of Otorhinolaryngology, Nihon University School of Medicine, 30-1 Ooyaguchi-kami-machi, Itabashi-ku, Tokyo, 173 Japan

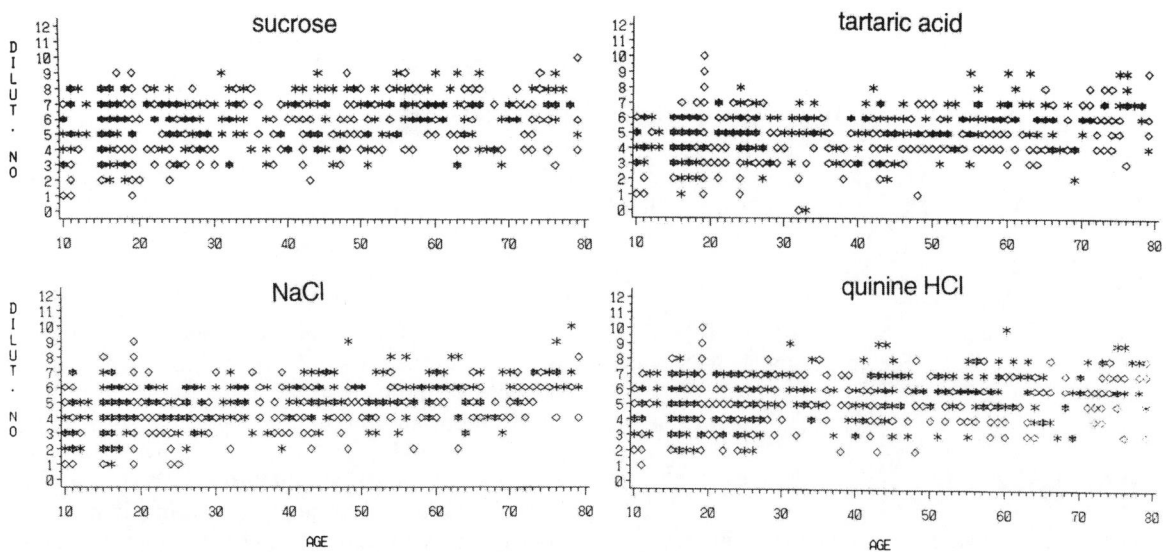

FIG. 1. Detection threshold using whole-mouth gustatory test. *Diamond*, female; *asterisk*, male; *Dilut. no*, serial 1:1 dilution number

RIGHT CHORDA TYMPANI NERVE AREA

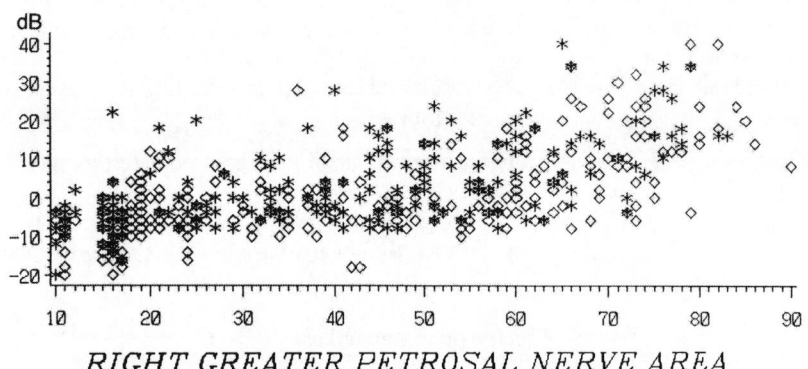

RIGHT GREATER PETROSAL NERVE AREA

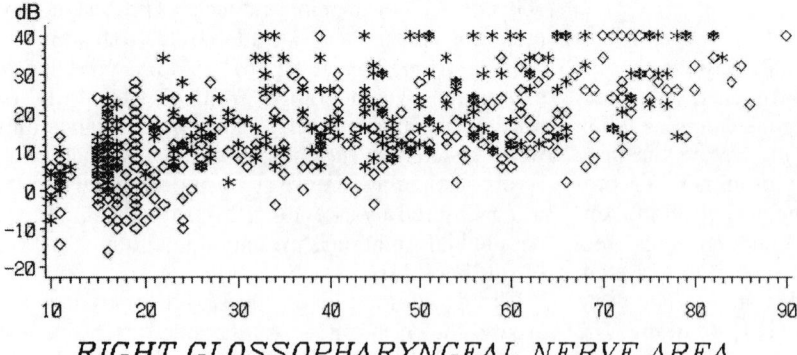

RIGHT GLOSSOPHARYNGEAL NERVE AREA

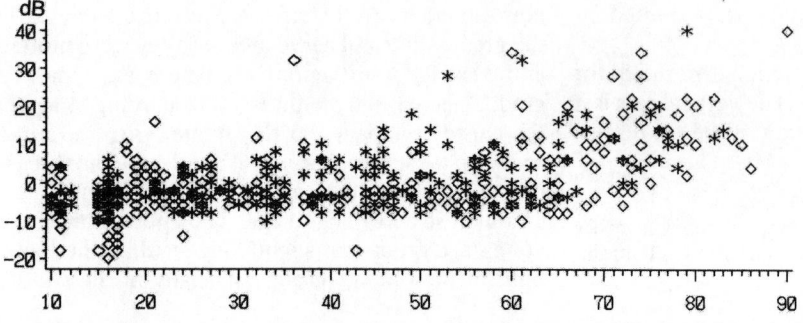

FIG. 2. Electrogustometric test. *Diamond*, female; *asterisk*, male

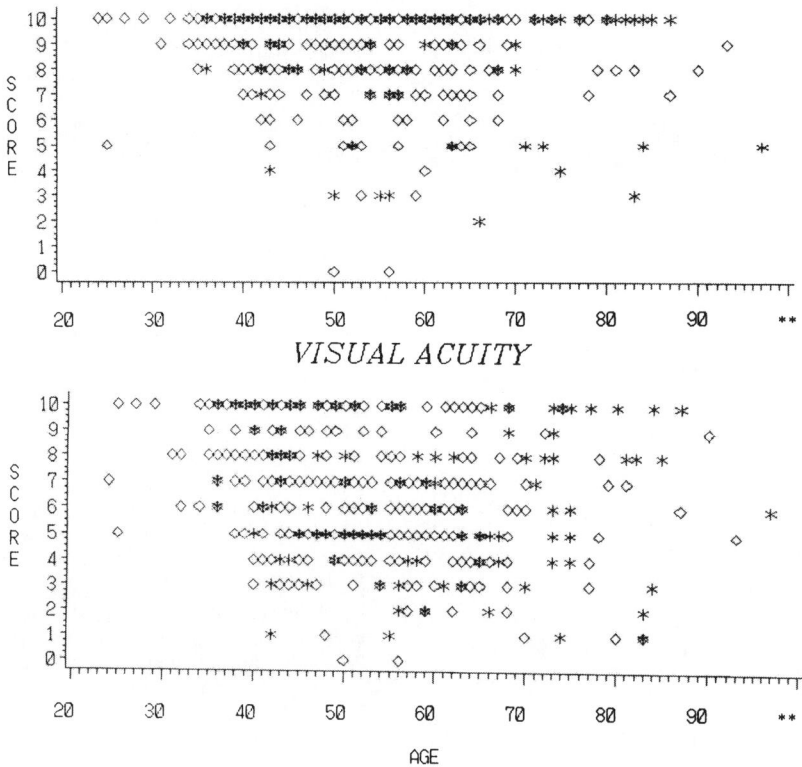

FIG. 3. Self-rating of physiologic function versus age. *Diamond*, female; *asterisk*, male

tympani nerve and the glossopharyngeal nerve areas. It was possible to apply a regression line to the distribution of the threshold as in our previous report [5]; however, in the glossopharyngeal nerve area it appeared to be inappropriate. The increase in the threshold according to age was significant. For a more appropriate statistical analysis, a more suitable model for the aging pattern of the threshold is now being researched.

The results of the electrogustometric test were as follows.

1. There were gradual increases in the thresholds in the chorda tympani nerve and the greater petrosal nerve areas.
2. The threshold was rather stable in the glosso-pharyngeal nerve area until age 60, but after that the threshold showed a considerable increase with age.

Self-Rating of Physiologic Function

We investigated the self-consciousness of the taste function by asking subjects for a self-rating on each physiologic function. The subjects were normal volunteers visiting the health clinic or the senile communication center. A multiple questionnaire test

was given in which they were asked to rate their physiological function by score from 0 to 10. A score of 10 means full sensation, as at the most sensitive younger age. A score of 0 means the complete loss of function.

The distributions of the scores from self-rating of visual and taste acuity are shown in Fig. 3. There was a wide variety of scores, with medians of 7 to 5 in the ratings of visual acuity. This indicated a perception of a decrease in visual acuity with aging. In contrast to visual acuity, most subjects rated their taste acuity as 10. This suggested stability of taste function according to age.

The result of the self-rating test was that the subjective assessment of taste acuity was robust to aging.

Discussion

There were already many reports investigating the taste threshold with the whole-mouth method or electrogustometry; these were reviewed by Murphy [6]. A number of reports favored an increase in the threshold of taste with respect to aging. However, with the whole-mouth method the reported increment was insufficient to discriminate completely the two groups, the young and the aged. In the Baltimore longitudinal study conducted by Weiffenbach and colleagues [7], a small decline with age in taste acuity was reported with NaCl and quinine HCl but not with sucrose or citric acid. Also, the reported increment was no more than the standard deviation. With the whole-mouth method, as mentioned before, any increment of the threshold is not statistically significant, so the thresholds measured using this method are said to be robust to aging.

Now it is thought that a more sophisticated method for obtaining the threshold of taste is the forced-choice protocol with a prestimulus water rinse. In our study, an ascending approach to the limit was used. According to Weiffenbach [8], our method was a measurement of a subjective threshold rather than an objective one. However, in the present study, determination of the range in normal subjects according to age was carried out using three methods, including a subjective method, and in a clinical setting subjective measurement is thought to be important.

Conclusions

1. In general, the taste threshold or acuity is robust to aging.
2. The local application of the threshold test showed a gradual increase in the taste threshold.

References

1. Doty RL (1989) Age-related alterations in taste and smell function. In: Goldstein JC, Kashima HK, Koopermann CF (eds) Geriatric otorhinolaryngology. American Academy of Otolaryngology—Head and Neck Surgery, Decker, Philadelphia, pp 97–104
2. Yamauchi Y, Endo S, Sakai F, et al. (1994) Influence of aging on taste threshold examined by whole mouth method
3. Yamauchi Y, Sakai F, Endo S, et al. (1991) Whole mouth gustation method of clinical taste examination. Proc Jpn Symp Taste Smell 25:141–144
4. Hotta M, Endo S, Tomita H (1994) Changes with age of the electrogustometric threshold
5. Hotta M, Endo S, Tomita H (1991) Changes with age of the electrogustometric threshold. Proc Jpn Symp Taste Smell 25:257–260
6. Murphy C (1987) Taste and smell in the elderly. In: Meiselman HL, Rivlin RS (eds) Clinical measurement of taste and smell. Macmillan, New York, pp 343–371
7. Weiffenbach JM, Baum BJ, Durghauser R (1982) Taste threshold: quality specific variation with human aging. J Gerontol 37:372–377
8. Weiffenbach JM (1989) Assessment of chemosensory functioning in aging: subjective and objective procedures. In: Murphy C, Cain WS, Hegsted DM (eds) Nutrition and the chemical senses in aging: recent advances and current research needs. Ann NY Acad Sci 561:56–64

Human Sensory Function for Taste Is Robust Across the Adult Life-Span

JAMES M. WEIFFENBACH[1]

Key words. Human—Taste intensity—Aging—Psychophysics—Taste loss—Taste complaints

The persistent belief that taste function declines with age encourages the study of the relation of human taste to aging. Differences in taste function between younger and older individuals are inherently as interesting as those between individuals of different sexes or who differ with respect to some other characteristic. Age differences are, however, of particular personal relevance because all of us are growing older. To specify age differences in sensory function for taste requires resolution of three issues. One is the distinction between age-related and non-age-related differences between older and younger people. Another is the distinction between differences in experience or performance that reflect sensory differences and those that do not. Finally, the tyranny of the idea that sensory declines with aging are universal and inevitable must be addressed.

Three views regarding the mechanisms of age-related change that can provide a framework for the study of taste and aging are already in place and may be readily appropriated from gerontology. First, is the common-sense view that decline in taste with age is universal. In this view, taste loss is a predictable feature of old age, a part of normal aging that will affect all of us. A second view similarly postulates that declines in function occur with advancing age but it differentiates between declines in function that are expected as a result of biological aging and other changes that can be attributed to an individual's experience of trauma or of disease and its treatment. In this view, the elderly may suffer declines not only due to having aged but also as a cumulative effect of what has happened to them during their lifetime. Because the elderly have lived for a longer period of time, they have had a longer period of vulnerability to illness and trauma. The elderly thus suffer declines as a joint function of their age and their past experience. A third and more recently developed view is radically different [1]. In this view, the focus shifts from the decline of function to the maintenance of function. Some individuals maintain function into advanced age. These individuals have aged successfully. Their performance is better than that of their age-mates and may be equivalent to, or better than, that of many younger individuals. Such individuals may have begun life with a fortunate genetic endowment, experienced minimal exposure to detrimental environmental insults, and been particularly healthy. Elderly subjects who demonstrate high levels of functioning may have experienced no loss of function with age, or may have experienced age-related declines from an even higher level of performance. Other older individuals who demonstrate less adequate performance may have experienced declines that reflect their vulnerability to either aging or life experience or both.

Differences in taste experience between groups composed of individuals of different ages are not necessarily due to aging. For example, Cohen and Gitman [2] found no complaints referable to taste among their younger individuals, whereas 33% of the individuals in an older group complained. However, the older group, as the authors point out, were residents of an "old folks' home." The differential frequency of complaints may be age-related but due to institutionalization rather than to aging. Institutions are likely to contain a higher proportion of individuals who are either functioning poorly or who are prone to complain than are found in the population at large. Since complaints about taste may arise as a result of a discrepancy between the memory of previous experiences with food compared to how one experiences food now, taste complaints may reflect memory changes—changes in the remembered joy of eating. Alternatively, the discrepancy that leads to a taste complaint may reflect changes in the present experience of foods. Such changes may arise as a consequence of dining in an institutional setting, or of realistic depression associated with current conditions or prospects. It is even possible that the institutional food is objectively less tasty than the home-cooked meals of decades past. Differences between groups composed of individuals of different ages and living in different circumstances may be due to influences that affect older people, or even that differentially affect older people, but are not a part of biological aging.

Older individuals, even from a generally healthy community-dwelling population, may demonstrate

[1] Clinical Investigations and Patient Care Branch, National Institute of Dental Research, National Institutes of Health, Bethesda, Maryland 20892, USA

an increased frequency of taste complaints. For example, participants in the National Institute on Aging's Baltimore Longitudinal Study of Aging (BLSA) demonstrated an age-associated rise in taste complaints [3]. The percentage of individuals reporting diminished taste rose from 5 to 15 to 20 for individuals of increasing age. A reanalysis of these data categorized individuals as "healthy" or of "questionable health status" [4]. Because all the individuals in the youngest group were healthy, this age group was excluded from the reanalysis. For persons of questionable health status, the percentage of middle-aged and old individuals with complaints of diminished taste (30 and 27) was much higher than that for the healthy group (11.6 and 12.5). Differences due to age were not significant but those for health status were. Aging may not be the only or even the most powerful influence on complaints. Health status and associated medications may have influenced taste complaints directly, or this effect may have been mediated by objectively measurable changes in the sensory function.

Although it is important to understand that age differences in taste experience may be related to influences that are only incidently related to age, it is equally important to determine if differences in the experience of taste actually reflect sensory differences. Much of what we know about age-related differences in taste experience comes from taste complaints or self-assessment of taste acuity. The validity of such subjective reports, however, depends on the assumption that individuals are willing and able to give an accurate account of their sensory experience. To determine if differences between the taste experience of different aged individuals reflect sensory differences requires objective measurement of taste functioning. This type of measurement is obtained when individuals are required to demonstrate their sensory competence by performing a task requiring information from taste stimuli [5].

A variety of different tasks are available for the assessment of different aspects of sensory competence. Threshold tasks address the question of how strong a taste stimulus must be to be detected or to be discriminated from a taste-absent stimulus. Threshold tasks employ weak, barely detectible stimuli and were the focus of other presentations at this symposium [see chapter "Influence of aging on taste threshold, in this volume]. The question addressed by direct scaling, "How intense is the taste?," is arguably more relevant to the everyday use of the sense of taste [5]. The direct-scaling task employs stimulus strengths covering the dynamic range of the sensory system and will be of concern here.

The direct-scaling task is well adapted for the study of taste intensity perception and has been used for that purpose since its development in the mid-1950s. For the direct scaling of taste intensity, subjects are presented with stimuli of different strengths in a random order. The subject judges the intensity of each presentation. Intensity judgments of stimuli of the same strength are averaged and the availability of sensory information to the subject is evaluated. As is the case for other sensory modalities, the relation of taste intensity judgments to tastant strength is commonly described mathematically by the slope of the function relating the logarithm of judgments to the logarithm of stimulus strength.

Studies from different laboratories that address the question of age-related variation in the perception of taste intensity will be reviewed. All the studies assess response to aqueous solutions representing four psychophysically defined taste qualities. The researchers, however, express different philosophies of design, draw their subjects differently and from different populations, and use different response measures and different analytic procedures. Different substances are used to represent bitter. Normalization or standardization is employed in some studies and not in others.

Bartoshuk et al. [6] compared an older group recruited from a home for the aged and a younger group recruited from the surrounding community. Each subject assigned numerical estimates to the intensity of stimuli at 10 levels of strength for each substance. At the same time, judgments of the intensity of low-frequency tones were also obtained. Judgments of tones were used to standardize the taste judgments. Comparison of the sensory functions generated by older and younger subjects indicated general concordance. Minor elevations at the lower concentrations differentiated the intensity functions of the older group from those of the younger group. The authors attributed this to a mild background taste elevating the judgments of the elderly.

Murphy and Gilmore [7] also compared an older with a younger group. All of the subjects reported excellent health and none had been hospitalized in the last year. Three elderly and six young subjects were eliminated from the study because they could not do the task. Taste intensity judgments were obtained for stimuli representing sweet, sour, salty, and bitter at three levels of stimulus strength. In addition, intensity judgments were obtained for mixtures of the substances. Each subject's taste judgments were standardized or normalized to their judgments of lifted weights. An analysis of variance (ANOVA) comparing normalized judgments for the unmixed qualities found no age differences for sweet and salty but significant age differences for the other two qualities. Comparisons for the mixed qualities found age-stability for all qualities except bitter.

Cowart [8] also obtained measures of intensity perception for all four qualities. The study sample included approximately equal numbers of individuals of each sex in each of three distinct age groups (19 to

35 years of age, 45 to 56 years of age, and 65 to 79 years of age). In this study, subjects made intensity judgments of stimuli at five concentration levels and communicated their judgments by extending the blade of a retractable tape measure. Intensity estimates were not normalized. Separate ANOVAs were carried out for each quality. The expected significant effects for stimulus concentration were observed for each quality. Critically, effects for age were not significant for any of the qualities. Two other measures were calculated for individual subjects. ANOVAs were carried out separately for each measure and each quality [9]. The first of these measures, slope, reflects the rise in logarithm of intensity relative to increases in the logarithm of stimulus concentration. For slopes there was no significant effect for age for any of the qualities. The other measure, the intraclass correlation coefficient (ICC), reflects the consistency of response from presentation to presentation. This measure showed no significant effect for age either.

Weiffenbach et al. [10] drew their study population from the National Institute on Aging's Baltimore Longitudinal Study of Aging. Participants are community-dwelling volunteers who return to the Gerontology Research Center on a regular schedule for 2 to 3 days of assessment. They are of middle to upper-middle socioeconomic class and appear to be generally healthy with good access to medical care. They are from 23 to 88 years of age and form four contiguous age groups (less than 40, 40 to 56, 56 to 70, and over 70) of approximately equal size. Stimuli representing each of four taste qualities were presented at seven levels of concentration. The findings of this study for group slopes and raw estimates of intensity are consistent with those already cited, in that average intensity estimates are robust across the adult age span. The slopes and ICCs calculated for individual subjects were submitted to ANOVA. A significant main effect indicated differences between qualities for the slope of the intensity function but, critically, no significant effect for age was observed. The measure of trial-to-trial repeatability, ICC, declined significantly with age for salty, sour, and bitter, but not for sweet. Analogous regression analysis for the ICC demonstrated a significant relation to age only for salty and bitter. Thus, the slope of the taste functions is age-stable while the consistency with which judgments are made differs with age for some but not all qualities.

The findings of age stability for taste intensity functions and of age-related decrements in the consistency of judgments of some qualities derive from cross-sectional studies in which different individuals represent each age. Recently, 42 of the 170 original participants in the BLSA study [10] were retested with the same procedures after an interval of 10 years. Thus, for the first time, it is possible to compare taste intensity judgments obtained from

the same individuals at different ages. Such longitudinal observations reflect on the life history of an individual's sensory function for taste across years. In longitudinal analysis, age changes are directly measured. They are not, as in cross-sectional analysis, an inference from differences between different aged groups. Longitudinal analysis of slopes demonstrates age stability that is consistent with that observed cross-sectionally. For each quality the proportion of individuals who generated higher slopes on retest did not differ from the proportion with lower slopes. The ANOVA for individual slopes demonstrated no significant main effect of age for any of the qualities. Thus, the slope of the intensity functions did not change significantly across the 10-year span. The quality-specific age differences observed for ICC cross-sectionally are likewise supported by the longitudinal findings. Comparisons of ICCs from the initial test to those obtained from the same individuals on subsequent retest demonstrate a significant decline for salt but not for the other qualities. Both the cross-sectional analysis from the original study and the longitudinal analysis of the subsample demonstrate the robustness of the rise in taste intensity with stimulus concentration and are consistent with a quality-specific decline in the trial-to-trial consistency of intensity judgments.

In summary, the first available longitudinal study of taste intensity leads to the same general conclusion derived from a diverse group of cross-sectional studies: taste is robust across the adult life-span.

References

1. Rowe JW, Kahn RL (1987) Human aging: usual and successful. Science 237:143–149
2. Cohen T, Gitman L (1959) Oral complaints and taste perception in the aged. J Gerontol 14:294–298
3. Baum BJ (1981) Characteristics of participants in the oral physiology component of the Baltimore longitudinal study of aging. Community Dent Oral Epidemiol 9:128–134
4. Weiffenbach JM (1984) Taste and smell perception in aging. Gerodontology 3(2):137–146
5. Weiffenbach JM (1989) Assessment of chemosensory function in aging: subjective and objective procedures. Ann NY Acad Sci 561:56–64
6. Bartoshuk LM, Rifkin B, Marks LE, Bars P (1986) Taste and aging. J Gerontol 41(1):51–57
7. Murphy C, Gilmore MM (1989) Quality-specific effects of aging on the human taste system. Perception Psychophys 45:121–128
8. Cowart BJ (1989) Relationships between taste and smell across the adult life span. Ann NY Acad Sci 561:39–55
9. Cowart BJ (1991) Derivation of an index of discrimination from magnitude estimation ratings. In: Bolanowski SJ (ed) Ratio scaling of psychological magnitude. Lawrence Erlbaum Associates, Hillside, New Jersey, pp 115–127
10. Weiffenbach JM, Cowart BJ, Baum BJ (1986) Taste intensity perception in aging. J Gerontol 41(4):460–468

Early Development of the Glossopharyngeal Nerve in Humans

SHUNTARO SHIGIHARA[1], HIROSHI TOMITA[1], and RAYMOND F. GASSER[2]

Key words. Glossopharyngeal nerve—Human embryo

Although many anatomical studies have been made on the development of the glossopharyngeal nerve alone or the tongue alone, little information is available on the dynamic relationship that exists between the two structures during their early gestational stages. We examined the relationship between the glossopharyngeal nerve and the tongue during the early embryonic period, using three-dimensional (3-D) computer reconstructions.

Materials and Methods

The human embryos had a crown-rump length of 2.4–21.5 mm and an estimated postovulation age of 24–52 days. 3-D reconstructions were made of the right glossopharyngeal nerve and the tongue region in six human embryos. The reconstructed images were made on a Silicon Graphic Iris computer, using the Skandha program that was written at the University of Washington, Department of Biological Structure, Seattle, Washington, for the Digital Anatomist project.

Results

At stage 13 (5.2 mm, 28 days), the primordium of the mucosa covering of the tongue can be identified as elevations in the floor of the oral cavity. Peripheral branches of the glossopharyngeal and facial nerves do not yet reach the tongue region. In stage 15 embryo (7.0 mm, 33 days), the lingual branch of the glossopharyngeal nerve grows medially at almost a right angle with its parent nerve. The chorda tympani nerve is the first recognizable branch of the facial nerve and courses rostrally and medially to the lateral swellings of the first arch. At stage 19 (18 mm, 47.5 days), the tongue primordium has enlarged and the peripheral lingual branches of each nerve have increased dramatically in number. In stage 21 embryo (21.5 mm, 52 days), some of these branches cross the midline to reach the mucosa of the other side where they communicate with nerve branches there. Lingual branches of the glossopharyngeal nerve are widely distributed to the posterior portion of the tongue mucosa. Chorda tympani branches are distributed mainly forward from the entry point of the nerve into the tongue. This point was considered as the junction (sulcus terminalis) between the lateral swellings of the tongue and the region medial to the third pharyngeal arch. Few chorda tympani branches were observed near the junction, whereas many glossopharyngeal nerve branches were present caudal to the junction and were found throughout from the midline to the lateralmost margin of the tongue. Primordia of taste buds were recognizable in the lingual mucosa at this stage, suggesting that peripheral growth of the nerve branches was nearing completion.

Conclusion

We have shown the manner in which the glossopharyngeal and facial nerves innervate the tongue in six stages, in serially sectioned human embryos, using 3-D computer reconstructions. Chorda tympani nerve branches are distributed mainly to the lingual mucosa anterior to the nerve's entry point, whereas glosso-pharyngeal branches form more rapidly and are more widely distributed to the mucosa posterior and lateral to the sulcus terminalis. The early, profuse distribution of glosso-pharyngeal branches to the taste bud region suggests that they probably innervate these taste buds by association.

Acknowledgments. Supported by the Louisiana Educational Quality Support Fund, No 90-93-RD-A-09.

[1] Nihon University, School of Medicine, Department of Otolaryngology, 30-1 Ohyaguchi-kamimach, Itabashi-ku Tokyo, 173 Japan
[2] Louisiana State University Medical Center, Department of Anatomy, New Orleans, LA, USA

Influence of Aging on Taste Threshold, Examined by Whole Mouth Method

Yuki Yamauchi, Sohei Endo, Fumitaka Sakai, and Hiroshi Tomita[1]

Key words. Whole mouth method—Aging

The whole mouth method is a procedure that is now extensively used for comprehensively evaluating the sensitivities of taste examinations. In this study, the authors evaluated the recognition thresholds of four basic taste qualities in normal subjects (male and female) to observe the changes caused by aging; sex differences were also investigated.

The subjects of the study were 225 males and 330 females ranging in age from 10–79 years. We divided these patients into the following groups: 10–14 years, 15–19, 20–29, and 70–79. Subjects who complained of taste disorder, who had full dentures, or who smoked were excluded from the study.

Sucrose, salt, tartaric acid, and quinine were used for identifying sweet, salty, sour, and bitter taste, respectively. A 13-rank concentration sequence was created by preparing a multiple dilution series for each taste quality. With a glass syringe, 1 ml of test solution was sprayed all over the mouth of a subject after they had abstained from eating for 1 h or more, thus allowing sufficient time before tasting. The subject then swallowed the solution and noted the taste sensed. The minimum threshold for sensing any kind of taste was defined as the detection threshold, and the minimum threshold for sensing the correct taste was defined as the recognition threshold. The subjects were instructed to rinse the mouth well with tap water between each of the tests, so that the taste perceived during the previous test would not affect the subsequent test.

Regression analysis was performed on a PC-SAS (version 6.04; NEC PC-9801, SAS Institute, Nihon University School of Medicine, Tokyo, Japan). We found that the recognition threshold rose slowly with age for each of the four taste qualities, but that the differences were not significant. There were also no differences according to sex. A slight variation in the median value was assumed to reflect fluctuations in individual data rather than to be attributable to the age factor.

Many studies have used the whole mouth method in attempts to elucidate the relationship between age and taste recognition; the results have been inconclusive. Our studies here demonstrated that the rise in the recognition threshold was not significantly affected by, aging.

Further, there were no sex differences in this threshold.

References

1. Endo S (1990) Taste localization in the whole mouth gustatory test (in Japanese). Nichidaiishi (J Nihon Univ Med Ass) 49(6):590–600
2. Oowada K (1972) The effect of age taste sensitivity (in Japanese). Nichieishi (Jpn J Hyg) 27:243–247
3. Glanville EV, Kaplan AR, Fischer R (1964) Age, sex, and taste sensitivity. J Geront 19:474–478

[1] Nihon University School of Medicine, Department of Otolaryngology, 30-1 Ooyaguchi-kami-machi, Itabashi-ku, Tokyo, 173 Japan

Changes with Age in the Electrogustometric Threshold

Mahoko Hotta, Sohei Endo, and Hiroshi Tomita[1]

Key words. Electrogustometric threshold—Aging

The clinical value of electrogustometry, which can measure taste threshold quantitatively for each innervation area, has been acknowledged for some time. Although on changes in normal values due to aging in the chorda tympani innervation area have been reported [1], to the best of our knowledge no report has been made thus far on the greater petrosal innervation area or the glossopharyngeal innervation area. We conducted electrogustometry in 516 normal men and women (all nonsmokers) in age groups ranging from teenagers to the late seventies to determine changes according to aging and sex.

The equipment used was an upgraded model of the Rion TR-06 type. Whereas the former model was capable of measuring a range from −6 dB to 34 dB, using 8 A as the 0 standard (0 dB), the upgraded model has the capacity to measure a range of −20 dB–34 dB. The stainless steel electrodes of the electrogustometric meter were applied to the points of the three innervation areas of the chorda tympani nerve, greater petrosal nerve, and the glossopharyngeal nerve. We first carried out stimulation with a current of 20 db and then had the subjects identify the taste qualities. We then used a circulated electric current, starting from −20 db and had the subjects push a button when they detected the same taste as before.

A PC-SAS (Version 6.04) was used for the actual calculations.

In the chorda tympani innervation area, the electrogustometric threshold values increased with age in both males and females. No significant difference was observed in the degree of elevation between males and females. In the greater petrosal innervation area and the glossopharyngeal innervation area also, the electrogustometric threshold values increased with age in both males and females.

We have already reported [1] that the electrogustometric threshold values of the chorda tympani area increased with age. The results of this study confirmed that finding. In the greater petrosal area, we found that the elevation of threshold values due to aging was much greater than that in the chorda tympani area. In contrast, the rise of threshold value in the glossopharyngeal area was smaller, suggesting that the electrogustometric threshold values in this area were better preserved against aging. We assume that the changes in threshold values due to aging are related to numbers of taste buds.

[1] Department of Otolaryngology, Nihon University School of Medicine, 30-1 Ooyaguchikamimachi Itabashi-ku, Tokyo, 173 Japan

References

1. (1971) The normal value of electrogustometry (in Japanese). J Otolaryngol Jpn 74:1580–1587

Taste Loss and Taste Phantoms: A Role of Inhibition in the Taste System

LINDA BARTOSHUK[1], JOHN KVETON[1], KEN YANAGISAWA[1], and FRANK CATALANOTTO[2]

Key words. Taste—Chorda tympani—Glossopharyngeal—Phantom—Dysgeusia—Inhibition

Introduction

Patients often fail to notice taste loss even when their taste systems are severely damaged. Pfaffmann [1,2] presented his own case of Ramsey-Hunt's syndrome (cranial nerve damage produced by the virus responsible for chicken pox) at a meeting of the Association for Chemoreception Sciences. The virus destroyed taste on the left side of his mouth yet he had no subjective awareness of loss in everyday taste experience.

Halpern and Nelson [3] proposed that the chorda tympani (a branch of VII) normally inhibits the glossopharyngeal (IX) nerve. They recorded from an area in the medulla receiving input from both VII and IX. Anestheia of VII increased neural responses when the area innervated by IX was stimulated. Halpern and Nelson proposed that the anesthesia abolished the normal inhibition of VII on IX, thereby enhancing responses from IX. Additional support for the existence of similar inhibition came from [4–7].

The term dysgeusia refers to a chronic taste that occurs in the absence of obvious stimulation [8]. The majority of clinical cases probably result from actual taste stimuli that are not apparent to the patient. For example, some chemotherapy patients report bitter tastes associated with drug administration [9]. These drugs may be tasted in saliva [10] or crevicular fluid [11] or directly from blood (venous taste phenomenon [12]. Postnasal drip, reflux, and bleeding into the mouth also produce dysgeusia. Furthermore, chronic tastes can originate from the nervous system, although this phenomenon is rarer. We suggest the term "taste phantom" for these cases. A prosthesis implanted in the middle ear stretched the chorda tympani nerve and produced bitter and metallic phantoms [13]. Bull [14] evaluated patients in whom the chorda tympani was sectioned or stretched during stapedectomy and noted, "In under half the patients with this symptom, the metallic sensation was on the side of the tongue on which the chorda tympani had been cut, but in the remaining cases the sensation was either felt all over the tongue or confined to the tip." Central nervous system (CNS) tumors can produce taste phantoms [15] and temporal lobe epilepsy occasionally produces taste auras preceding seizures [16]. Yamada and Tomita reported bitter taste phantoms after transtympanic anesthesia of the chorda tympani for the treatment of tinnitus [17].

This review presents studies we have done to delineate the interaction between the chorda tympani and glossopharyngeal nerves using anesthesia of the chorda tympani nerve and topical anesthesia of the mouth with both normal volunteers and patients with known lesions of the taste system.

Methods

Spatial Test

The stimuli ($1.0\,M$ NaCl, $1.0\,M$ sucrose, $0.032\,M$ citric acid, and $0.001\,M$ quinine hydrochloride, presented in that order) were applied with long-handled cotton-tipped applicators (i.e., Q-tips) to the front (fungiform papillae), foliate papillae, most anterior circumvallate papilla, and palate, with care taken not to cross the midline. Right and left sides were stimulated close in time to allow subjects to compare intensities. Subjects rinsed before each stimulus to prevent localization illusions [18]. The stimuli were rated with the Natick 9-point scale such that 0 = no taste, 1 = very weak, 5 = medium, 9 = very strong [19]. A particular stimulus was applied to all of the loci before the next stimulus was tested. After all loci were tested, subjects were asked to rinse, take a comfortable amount of solution in the mouth, swish it around, spit it out, swallow the remaining few drops, and then rate the highest intensity that occurred. This essentially stimulated all of the taste-sensitive areas in the mouth.

[1] Yale University School of Medicine, Department of Surgery, Section of Otolaryngology, PO Box 3333, New Haven, CT 06510, USA
[2] University of Medicine and Dentistry of New Jersey, New Jersey Dental School, Newark, NJ, USA

Topical Anesthesia Test

To evaluate the source of a phantom taste, patients were first asked to describe the quality of the sensation, localize it, if possible, and rate its intensity (on the 9-point scale). To anesthetize the mouth, patients held about 5 cc of 0.5% dyclonine and 0.5% diphenhydramine (in 0.9% saline) in the mouth for 60 s, spit it out, rested for 60 s, and then rinsed with water. For the duration of the anesthesia, patients were asked periodically to describe their experience providing quality, intensity, and localization information.

Injected Anesthesia

Anesthesia of the chorda tympani nerve was carried out much as it is in routine anesthesia of the ear canal. A 30-gauge needle was used to inject 0.2–0.4 cc of 1% xylocaine with 1:100 000 epinephrine into the posterior ear canal wall under direct visual guidance through a Zeiss operating microscope. Anesthesia was usually preceded by a slight sensation of paresthesia extending from the foliate papillae to the midline of the front of the tongue.

Lower concentrations of tastants were used as stimuli in the spatial tests with injected anesthesia to prevent ceiling effects. The order of the stimuli was varied and subjects used the method of magnitude estimation to rate perceived intensities (responses were normalized to the grand mean of all of the subject's responses before anesthesia).

Results and Discussion

Anesthesia of the Chorda Tympani via Lingual-Chorda Block

Our first study was intended to determine how much of the taste system had to be anesthetized before the loss was perceptible to the subject. Dental anesthesia (lingual block) offered a safe way to remove selected parts of the taste system temporarily. This anesthesia blocks both the trigeminal and chorda tympani nerves and so renders the affected side numb and insensitive to touch and thermal stimulation as well as taste. To our initial surprise, anesthesia on one side produced statistically significant increments in whole-mouth perceived intensity [20,21]. We ought not to have been surprised because a decade before, Halpern and Nelson had proposed that the chorda tympani normally inhibits the glossopharyngeal nerve [3].

Anesthesia of the Chorda Tympani Nerve via the Ear Canal

To determine whether the apparent release of inhibition seen in the human study came from IX, the next study stimulated individual loci [22,23] using both dental anesthesia and anesthesia of the chorda tympani via the ear canal. As expected, based on the Halpern-Nelson model, there was significant enhancement of taste on the circumvallate papillae (innervated by IX) when VII was anesthetized. Unexpectedly, the enhancement occurred for the side contralateral to the anesthetic. Since taste appears to project ipsilaterally [24], a contralateral effect must occur in the central nervous system. We concluded, as did Halpern and Nelson, that VII must normally inhibit IX in some fashion. Anatomical work [25,26] suggests that the inhibition may be produced via descending pathways from the cortex to the lower taste structures. The enhancement was greatest for quinine, slight for sucrose and citric acid, and failed to occur for NaCl. In fact, the intensity with NaCl was slightly decreased on the ipsilateral circumvallate papilla.

A second experiment by Yanagisawa et al. [27] included anesthesia of both chorda tympani nerves via the ear canals. After the first injection, as in the previous experiment, the quinine intensity was enhanced on the contralateral circumvallate papilla and that of NaCl was decreased on the ipsilateral circumvallate papilla. The slight enhancement of sucrose and citric acid intensities seen in the first experiment did not occur. After anesthesia of the second side, enhancement at the circumvallate papilla was equal on both sides and was the same magnitude as the enhancement with anesthesia of one side only.

Release-of-Inhibition Taste Phantoms

In Yanagisawa's experiment [27], a subject noticed a spontaneous salty taste on the side contralateral to the anesthesia. Subsequently, we asked subjects to report any similar occurrences. Eight out of seventeen subjects reported phantoms that were contralateral to the anesthesia and of the same quality in repeat sessions. The quality of the phantoms varied across subjects (salty, bitter, sour, metallic, and sweet, in order of frequency). Application of a topical anesthetic to the phantom abolished it. Since anesthesia abolishes spontaneous activity [4], it might be the source of the phantoms. We suggest calling these phantoms "release-of-inhibition phantoms," to distinguish them from phantoms produced by direct stimulation of a nerve (i.e., nerve-stimulation phantoms).

Bull [14] described phantoms resulting from damage to the chorda tympani that were not clearly localized to the side of the damage. These may have included release-of-inhibition phantoms. Similarly, the bitter phantoms reported by the subjects receiving transtympanic anesthesia of the chorda tympani may also have been release-of-inhibition phantoms [17].

Patients Undergoing Surgery for Acoustic Neuromas

Acoustic neuromas (tumors associated with VIII) or their surgical removal can damage VII. The acoustic neuroma patients who experience unilateral loss of the chorda tympani branch of VII can be compared to the subjects who had anesthesia of the chorda tympani. A series of seven acoustic neuroma patients in whom VII was damaged were given the spatial taste test. Patients were included in this series only if their tumors were small or if testing with cold at the foliate papillae showed no asymmetry, since the IXth nerve is close enough to be damaged by the tumor or the surgery in some patients. The acoustic neuroma patients [28] showed taste asymmetry at IX: responses from the contralateral circumvallate papilla were greater than those from the ipsilateral papilla. We interpret the larger responses on contralateral IX as release of inhibition from the effects of the tumor/surgery on VII.

Patients with Taste Phantoms

Damage to IX Via a Tonsillectomy

RY (female) developed a bitter phantom after a tonsillectomy that was localized to the back of her mouth on both sides. Spatial testing showed essentially no function at the circumvallate papillae and abnormal function at the foliate papillae (most stimuli were called bitter). The topical anesthesia test increased the intensity of her phantom.

We suggest that the phantom arose from bilateral damage to IX during surgery [29]. Topical anesthesia of the mouth of RY effectively blocked her chorda tympani nerves and released inhibition on IX just as the anesthesia of the chorda tympani did in the experiments discussed previously.

Damage to All Taste Branches of VII

Childhood surgery damaged the branches of VII innervating EO's palate. EO (female) underwent right and left mastoidectomies that damaged her chorda tympani nerves. After her first surgery, she experienced a salty phantom roughly localized to the area innervated by the damaged chorda tympani. The topical anesthesia test increased the intensity of her phantom [23]. This observation was replicated six times throughout her surgeries with the same results.

We suggest that her phantom was produced because the damage to VII stimulated it in some way. The topical anesthesia of IX released inhibition on this nerve-stimulation phantom from VII. Thus, this case provides evidence that the glossopharyngeal nerve normally inhibits neural responses from the chorda tympani nerve just as the anesthesia experiments suggest that the chorda tympani nerve normally inhibits neural responses from the glossopharyngeal nerve.

Damage to VII

JR (male) experienced a salty taste phantom after surgery to repair a torn tympanic membrane. The phantom was localized to the same side as the damaged nerve, but was surprisingly far back near the foliate papillae. We attempted to anesthetize his chorda tympani via the ear canal central to the presumed site of damage in the middle ear. The anesthesia was partially successful and reduced the intensity of the phantom; however, to the patient's surprise, a new phantom appeared on the side contralateral to the damaged nerve near the base of the tongue. The new phantom was salty as well and was more intense than the original phantom. During the next several minutes, both phantoms diminished somewhat in intensity. About 20 min after the injection, we anesthetized JR's mouth topically. The new phantom was immediately abolished and the original phantom became much more intense. As the anesthesia wore off, the original phantom returned to the same intensity that it had at the beginning of the session.

We suggest that the original phantom was a nerve-stimulation phantom produced by damage to the chorda tympani. The anesthesia injected into the ear canal may have partially anesthetized the chorda tympani central to the damage. The anesthesia also produced a release-of-inhibition phantom like those produced in Yanagisawa's [27] experiment. In our previous studies, topical anesthesia in the mouth enhanced nerve-stimulation phantoms (through release of inhibition) and abolished release-of-inhibition phantoms (perhaps by abolishing spontaneous activity). Thus, we expected the topical anesthesia in the mouth to abolish JR's new phantom and intensify his original phantom, just as it did.

Conclusions

Inhibition

Studies of anesthesia of VII suggest that it inhibits IX at some locus in the central nervous system. This conclusion is supported by patient RY whose bitter nerve-stimulation phantom arose from IX. Anesthesia of VII released inhibition of IX thereby intensifying her phantom.

The data from patient EO suggest that IX inhibits VII as well. EO's salty nerve-stimulation phantom arose from VII. Anesthesia of IX released inhibition of VII thereby intensifying her phantom.

Phantoms

We suggest that there are two types of taste phantoms that originate from the nervous system: nerve-stimulation phantoms and release-of-inhibition phantoms. These phantoms can be distinguished by their localization and their response to topical anesthesia of the mouth. Nerve-stimulation phantoms tend to localize to the area innervated by the damaged nerve and are intensified by anesthesia of the mouth. Release-of-inhibition phantoms tend to localize to a site different from (and possibly contralateral to) the site of damage and are abolished by anesthesia of the mouth.

Clinical Implications

If anesthesia of the mouth intensifies a taste, then the taste clearly does not originate via normal stimulation of taste receptors in the mouth. Either venous taste or a neural origin should be suspected. If anesthesia of the mouth abolishes a taste, then the taste may reflect the actual presence of a stimulus (e.g., something in saliva) or it may be a release-of-inhibition phantom. To test for the presence of an actual stimulus, attempt to rinse it away. If the taste disappears, even for a few seconds, an actual stimulus for the taste should be sought. To test for a release-of-inhibition phantom, test spatially to see if there is a damaged area that might be responsible for the release-of-inhibition.

Acknowledgments. This work was supported by NIH grant DC-00283.

References

1. Pfaffmann C, Bartoshuk LM (1989) Psychophysical mapping of a human case of left unilateral ageusia. Chem Senses 14:738
2. Pfaffmann C, Bartoshuk LM (1990) Taste loss due to herpes zoster oticus: an update after 19 months. Chem Senses 15:657–658
3. Halpern BP, Nelson (1965) Bulbar gustatory responses to anterior and posterior tongue stimulation in the rat. Am J Physiol 209:105–110
4. Norgren R, Pfaffmann C (1975) The pontine taste area in the rat. Brain Res 91:99–117
5. Ninomiya Y, Funakoshi M (1982) Responsiveness of dog thalamic neurons to taste stimulation of various tongue regions. Physiol Behav 29:741–745
6. Ogawa H, Hayama T (1984) Receptive fields of solitario-parabrachial relay neurons responsive to natural stimulation of the oral cavity in rats. Exp Brain Res 54:359–366
7. Sweazey RD, Smith D (1987) Convergence onto hamster medullary taste neurons. Brain Res 408:173–184
8. Snow JB, Doty RL, Bartoshuk LM, Getchell T (1991) Categorization of chemosensory disorders. In: Getchell T, Doty RL, Bartoshuk LM, Snow JB (eds) Smell and taste in health and disease. Raven, New York, pp 445–447
9. Fetting JH, Wilcox PM, Sheidler VR, Enterline JP, Donehower RC, Grochow LB (1985) Tastes associated with parenteral chemotherapy for breast cancer. Cancer Treat Rep 69:1249–1251
10. Stephen KW, McCrossan J, Mackenzie D, Macfarlane CB, Speirs CF (1980) Factors determining the passage of drugs from blood into saliva. Br J Clin Pharmacol 9:51–55
11. Alfano M (1974) The origin of gingival fluid. J Theor Biol 47:127–136
12. Bradley RM (1973) Electrophysiological investigations of intravascular taste using perfused rat tongue. Am J Physiol 224:300–304
13. Jones RO, Fry TL (1984) A new complication of prosthetic ossicular reconstruction. Arch Otolaryngol 110:757–758
14. Bull TR (1965) Taste and the chorda tympani. J Laryngol Otol 79:479–483
15. El-Deiry A, McCabe BF (1990) Temporal lobe tumor manifested by localized dysgeusia. Ann Otol Rhinol Laryngol 99:586–587
16. Hausser-Hauw C, Bancaud J (1987) Gustatory hallucinations in epileptic seizures. Brain 110:339–359
17. Yamada Y, Tomita H (1989) Influences on taste in the area of chorda tympani nerve after transtympanic injection of local anesthetic (4% lidocaine). Auris Nasus Larynx 16:S41–S46
18. Todrank J, Bartoshuk LM (1991) A taste illusion: Taste sensation localized by touch. Physiol Behav 50:1027–1031
19. Kamen JM, Pilgrim FJ, Gutman NJ, Kroll BJ (1961) Interactions of suprathreshold taste stimuli. J Exp Psychol 62:348–356
20. Östrum KM, Catalanotto FA, Gent JF, Bartoshuk LM (1985) Effects of oral sensory field loss of taste scaling ability. Chem Senses 10:459
21. Catalanotto FA, Bartoshuk LM, Östrum KM, Gent JF, Fast K (1993) Effects of anesthesia of the facial nerve on taste. Chem Senses (in press)
22. Lehman C (1991) The effect of anesthesia of the chorda tympani nerve on taste perception in humans. PhD thesis, Yale University
23. Bartoshuk LM, Kveton J, Lehman C (1991) Peripheral source of taste phantom (i.e., dysgeusia) demonstrated by topical anesthesia. Chem Senses 16:499–500
24. Norgren R (1990) Gustatory system. In: Paxinos G (ed) The human nervous system. Academic, New York, pp 845–861
25. London JA, Donta TS (1991) Cortical projections to the rostral pole of the hamster NTS exhibit bilateral differences in strength and area of origin. Chem Senses 16:514–515
26. DiLorenzo P (1990) Corticofugal influence on taste responses in the parabrachial pons of the rat. Brain Res 530:73–84
27. Yanagisawa K, Bartoshuk LM, Karrer TA, Kveton JF, Catalanotto FA, Lehman CD, Weiffenbach JM (1992) Anesthesia of the chorda tympani nerve: Insights into a source of dysgeusia. Chem Senses 17:724
28. Kveton J, Bartoshuk LM (1993) The effect of unilateral chorda tympani damage on taste. Laryngoscope (in press)
29. Rieder C (1981) Eine seltene Komplikation: Geschmacksstörung nach Tonsillektomie. Laryng-Rhinol 60:342

Drug-Induced Taste Disorders

MIGUEL CIGES, FRANCISCO FERNANDEZ-CERVILA, JULIO ORTEGA, and JOSÉ MANUEL RUIZ[1]

Key words. Ageusia—Dysgeusia—Phantogeusia—Taste iatrogenia—Taste biochemistry—Thiols

Introduction

Drug-induced taste disorders have been recognized only recently and have added a new dimension to the study of this much-neglected sense by demonstrating the complexity of its biochemical regulation. Although a great many papers dealing with this subject have been published, the true number of both the disorders and the responsible drugs is probably higher than reported. It is due not only to the fact that taste disorders may go unnoticed by the patient, especially as a complete recovery is often made, but also that the prescribing physician does not systematically test for taste alterations.

Iatrogenic Taste Disorders

There are many drugs known to modify the sense of taste; they have been classified into pharmacologic groups by Uziel and Smajda [1], who studied 16 examples taken from a wide pharmacologic spectrum.

The most frequent iatrogenic taste disorder is hypogeusia or ageusia, often only partial at the onset but with a rapid evolution resulting in total ageusia, although permanent ageusia occurs only after prolonged treatment or high dosage. Parageusia or dysgeusia sometimes arises, and is either metallic or, less frequently, bitter in nature.

The metallic taste has been described as phantogeusia by Rollin [2], and this special type of dysgeusia may also occur with bitter and salty sensations. Such qualitative disorders are sometimes associated with true hypo- or ageusia.

The perception of sweetness seems to be affected first, and frequently it is the most extensively reduced in comparison to other disorders. In our experience, the alteration in sweetness is followed by anomalies in bitterness perception; furthermore, when the causative drug is discontinued, more time is needed for the sense of sweetness to return completely to normal.

Smell is never affected, but patients sometimes complain not only of tastelessness of food but also of alteration in flavors. Tactile, thermic, and pain sensations in the tongue and the rest of the mouth are normal.

Before a taste disorder can be properly considered as iatrogenic, it is necessary to eliminate other causes, such as the disease itself for which the drug is being administered or other pathophysiological phenomena different from the intrinsic taste mechanism. Smell disorders must be excluded, especially in patients complaining of poor detection of flavors. Saliva disorders must also be carefully considered, for although in iatrogenic dysgeusias the patient frequently complains of a dry mouth we have not found any such association with hypo- or ageusias. As tastants are dissolved in the saliva and it is not known if quantitative or qualitative alterations in saliva production affect taste, we consider that dysgeusia must be excluded if true hyposialia exists, as it may be due to the alteration in saliva production rather than a disorder of the intrinsic taste mechanism. A new area of taste pathophysiology would be necessary to include taste disorders of salivary origin.

It is impossible to pinpoint the actual drug dosage that alters taste acuity, as it varies not only from one patient to another but from one drug to another. The most accurate criterion for classifying a taste disorder as iatrogenic is the relation between the drug and the disorder and the recovery made on discontinuation of the drug.

Drug-induced taste disorders affect not only the tongue but also the palate and pharynx. A functional examination must be carried out by both chemical and electrical methods, although the former is more important. The number of drugs capable of altering taste is enormous; however, an especially interesting chemical group exists—that of drugs containing the thiol groups, with R-SH in their molecule. This group encompasses antirheumatic drugs such as penicillamine, mucolitics such as acetyl homocysteine, and antihypertensive drugs such as captopril. Although the origin and pharmacologic use of the drugs in this group is diverse, taste disorders have been associated with all of them. Furthermore, these disorders are always hypo- or ageusia.

Taste disorders have been most frequently described with the administration of D-penicillamine; alterations in taste are associated with as many as

[1] Department of Otorhinolaryngology, University Hospital, University of Granada School of Medicine, Avenida Doctor Oloriz, 18012 Granada, Spain

30% of cases. We have found the second most frequent drugs associated with taste disorders to be acetylhomocysteine, which is used as a mucolitic in mild respiratory infections, and captopril, a hypotensive agent. When these drugs are administered for more than 20 days, the incidence of resulting taste disorders is 5%.

Drugs not containing thiols are less often associated with a taste disorder; and when these disorders do occur they are more likely to be dysgeusia, frequently related to hyposialia, than hypo- or ageusia. Indeed, we have detected hypo- or ageusia only in isolated cases of administration of phenylbutazone or carbamacepine and not at all with antineoplastic drugs. However, taste disorders have been described associated with treatment with clofibrate, lincomycin, 5-mercapyridoxal, phenindione, oxyphedrine hydrochloride, acetylsulfosalicylic, griseofulvin, and tranquilizers, among others [3].

Pathophysiology of Drug-Induced Taste Disorders

As mentioned, there are a large number of drugs with varying chemical composition that can cause taste disorders, making classification difficult. As we have seen, however, drugs containing thiols form a specific group that are particularly toxic for taste, similar to the way aminoglucosides affect hearing or other drugs affect smell. Apart from this group, there are other drugs causing alterations in the perception of taste that act by a different mechanism.

Henkin et al. [4] proposed that drugs containing thiols (R-SH) affect the preneural phase. They elaborated an interesting molecular theory for the regulation of taste acuity by studying, clinically and experimentally, the action of D-penacillamine and its relation with transitional metals. They had noticed that serum ceruloplasmin and copper concentration fell to below normal levels in some patients treated with D-Penicillamine and that it was associated with a significant decrease in taste acuity. This finding suggests that copper is involved in the regulation of taste acuity, as diminished serum copper levels are related to the onset of hypogleusia. This hypothesis has been confirmed experimentally. Furthermore, when copper was added to the diet of ageusic patients, taste acuity returned to normal within a week. Other transitional metals, such as zinc and nickel, were also tested, with positive results.

It may be concluded that taste acuity is regulated by both thiols and metals—thiols decreasing and metals increasing it. When a drug rich in thiols is administered, the balance is altered, resulting in ageusia. Although a chelation mechanism of the R-SH on the metals by the recognized crab-claw action may be considered, it has been demonstrated that certain drugs rich in SH groups, such as 5-

mercaptopyridoxal, produce ageusia without altering cupremia.

We studied copper concentrations in various organs and fluids of ageusic guinea pigs treated with D-Penicillamine and acetylhomocysteine [5]. Organs and fluids were chosen for either their direct involvement in the process or for a control. Copper concentration was analyzed in the tongue mucosa (in the zone rich in taste papillae), saliva, submandibular gland serum, liver, muscle, and heart. For both drugs, a significant decrease in copper concentration was found only in the liver. In the tongue and serum, the values were practically unchanged; a slightly higher level was found with acetylhomocysteine and a virtually normal level with D-penicillamine. However, it has clearly been shown that D-penicillamine reduces serum copper concentration. Our apparently paradoxical findings resulted from the fact that the guinea pigs always died before complete ageusia was achieved. We therefore investigated only the simple hypogeusic level and presume that if we had been able to increase the dose of D-penicillamine we would have found a reduction in serum copper levels. With acetylhomocysteine complete ageusia was reached without hypocupremia; it also occurs with 5-mercaptopyridoxal.

Copper metabolism is carried out mainly by the liver, and its homeostasis is maintained primarily and almost exclusively by bilary excretion. Both of the drugs we studied reduced copper concentration in the liver, and D-penicillamine is known to cause hypocupremia; however, neither drug altered the concentration of copper in the tongue membrane, mucosa, or saliva. If a direct relation did exist between taste acuity and copper, a decrease in copper concentration would be expected in the tongue or saliva. We believe, rather, that the thiols have a direct action on the preneural phase of taste, and that metals simply act as an antagonist without any direct effect on the taste process. Although there appears to be an evident interaction between metals and thiols, the active mechanism is difficult to understand because in diseases such as Wilson's disease, where serum copper levels are high, hypergeusia has never been reported.

That thiol drugs act on the preneural phase is demonstrated by the fact that the process is reversible and the taste buds are apparently normal. If the lingual surface is completely stained with osmium tetroxide, the papillae of ageusic animals stain just as deeply as those of normal ones, suggesting that the neural part of the bud is preserved. According to Henkin, thiols could act on the *gatekeeper proteins* of the taste pore, reducing its diameter. The action of the transitional metals such as copper or zinc would not be an unfolding of these proteins but rather act directly on the thiols. When we studied the pore and pit region of the taste buds [6] we observed that their permeability depends to a large

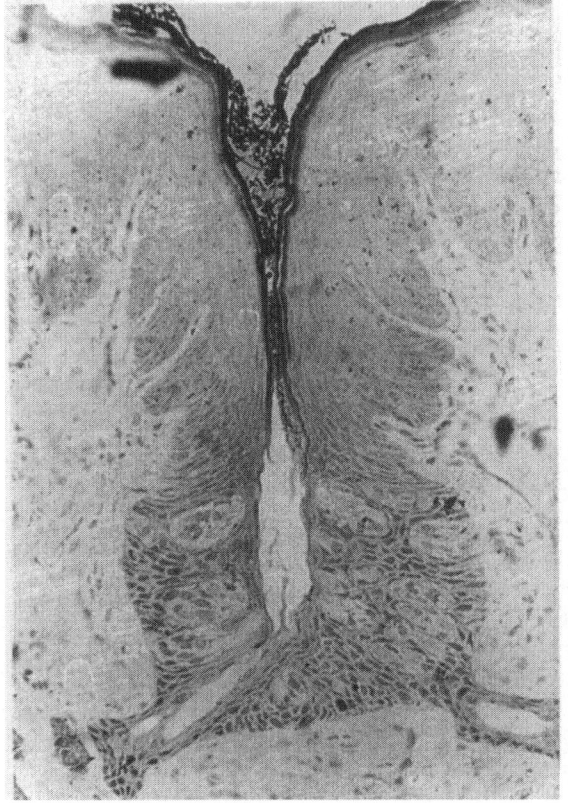

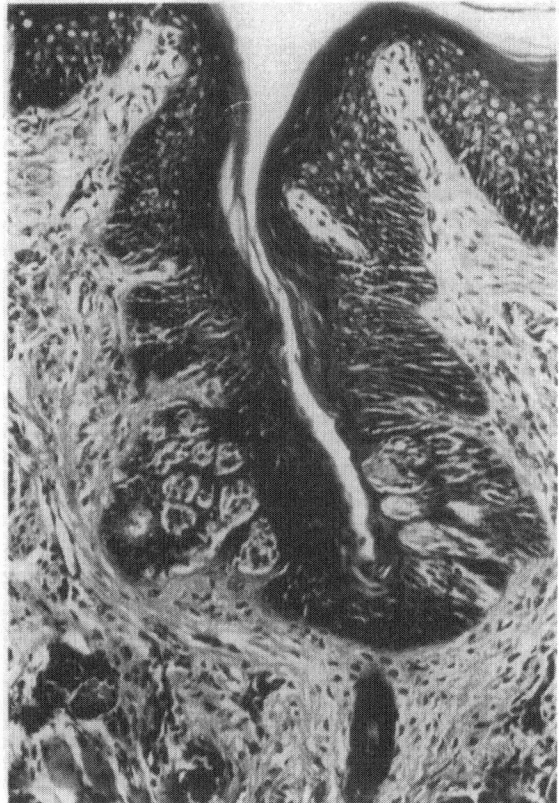

FIG. 1. Plug of amorphous and granular substance is seen in the entrance of the papillary groove. Taste buds are apparently normal. Semithin section, toluidine blue. ×40

FIG. 3. Normal taste papilla for comparison with the previous figures. H&E. ×40

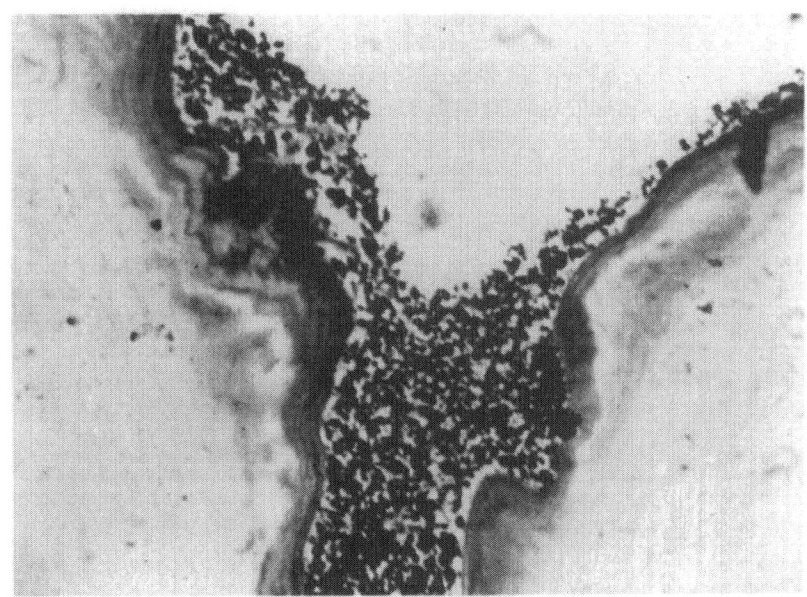

FIG. 2. Amorphous and granular substance of the plug in another papilla. Semithin section, toluidine blue. ×100

extent on the desquamation of their peripheral cells. Pores from which the peripheral cells have been desquamated are wider, allowing the amorphous substance found in the pit region to flow freely out of the pore. Thus the pore is both wider and emptier, facilitating contact between the cilia of the sensory cells and tastants.

Thiol-containing drugs may act by affecting the regulation of pore size by altering its desquamation process. Henkin et al. [4] found many more closed taste pores in D-penicillamine-treated rats than in controls. Moreover, we have observed [5] a plug of granular and amorphous substances at the entrance to the papillary groove in animals ageusic due to acetylhomocysteine. (Compare Figs. 1 and 2 with the normal papillae in Fig. 3.) The taste buds themselves were apparently normal, but the plug blocked the entrance of the tastant into the groove, preventing its contact with the bud. As the plug is always found at the entrance of the groove and the bottom region where the buds lie is normal, it is probable that the blocking substances originates from outside. To date, however, our results are not representative enough to draw any final conclusions.

A general hypothesis for pathophysiology of drugs other than the thiol group is almost impossible to offer as their chemical structure and pharmacologic properties vary greatly, although many do contain sulfur in a chemical form different from that of R-SH. We consider that they act on the neural phase, deranging the taste cell or its afferent nerves by neurotropic action. Finally, we believe that many of these drugs, particularly those producing dysgeusia, primarily affect salivary flow, and the taste bud is only indirectly altered. Lingual homeostasis is basically maintained by the saliva, and therefore any alteration, either quantitative or qualitative, would affect the biology of the taste buds themselves as well as the conditions under which the tastant makes contact with the receptors.

References

1. Uziel A, Smadja JG (1985) Exploration fonctionnelle et trobles du goût. In: Encyclopédie médico-chirurgicale. Paris, p 10
2. Rollin H (1978) Drug-related gustatory disorders. Ann Otol 87:37–42
3. Henkin RL (1971) Griseofulvin and dysgeusia: implications? Ann Intern Med 74:795–796
4. Henkin RI, Graziadei PPG, Bradley DF (1969) The molecular basis of taste and its disorders. Ann Intern Med 71:791–821
5. Ciges M, Morales J (1980) Experimental ageusia. Acta Otolaryngol (Stock) 89:240–248
6. Ciges M, Gonzalez M, Ceballos A (1978) Desquamation on taste buds. Acta Otolaryngol (Stock) 85:290–295

Clinical Assessment of Taste Disorders and Evaluation of Zinc Therapy

M. Ikeda and H. Tomita[1]

Key words. Electrogustatory test—Filter-paper disk test—Taste disorder—Zinc—Therapy of taste disorder

Introduction

The cause of taste disorders varies among patients. Hypogeusia may be due to damage of various parts of the gustatory tract. In practice, we try to clarify the cause and severity of the hypogeusia in each patient by conducting clinical examinations and then selecting an appropriate therapeutic regimen.

In this chapter we discuss taste disorders mainly from a clinical viewpoint. The first section concerns the clinical examinations we conduct on patients with hypogeusia. Second, clinical analysis concerning the causes and the incidence of hypogeusia is briefly explained. The third section relates to therapy and the clinical results achieved in our clinic.

Clinical Examinations

Clinically, it is important to identify the damaged region of the gustatory tract and the severity of the damage. Based on this understanding, we determine or suspect the cause of hypogeusia, select the therapy, and evaluate the efficacy of the treatment. To evaluate taste function, we perform three tests: the electrogustatory test, the filter-paper disk test, and most recently the whole-mouth test.

Electrogustatory Test

It is well known that a metallic taste is identified when a positive direct-current (DC) electrode touches the tongue. This taste is thought to be identified through the electrolysis of saliva or direct stimulation of the gustatory nerve endings by electricity.

The positive stimulating electrode of the gustometer (TR-06, Rion, Tokyo, Japan) we use is stainless steel with a 5 mm diameter. The strength of the stimuli is expressed in decibels because the magnitude of the sensation of the electrical taste correlates well with the strength of the electrical current on a logarithmic scale. There are 21 levels of stimuli in the range -6 dB ($4\,\mu A$) to 34 dB ($400\,\mu A$). The threshold for the chorda tympani in young healthy individuals is set as 0 dB (Table 1). The normal threshold for the glossopharyngeal nerve area is 4 dB, and that of the greater petrossal nerve is 8 dB. If there is a 6 dB or larger difference between the left and right sides, it is regarded that the threshold is significantly increased.

The electrogustatory test is especially useful for distinguishing the presence of damage to the gustatory nerve because of its fine scaling.

Filter-Paper Disk Test

The filter-paper disk (FPD) test uses taste solutions and examines the function of each taste nerve. The test kit is named Taste-Disc (Sanwa Kagaku, Nagoya, Japan), and it is commercially available. The kit consists of four solutions, with each solution having five concentrations (Table 2). The median for healthy persons is solution 2, and the upper limit of the normal range is solution 3.

If the patient can taste solution 1, he or she can be regarded as having hypergeusia, though it does not necessarily reflect an abnormal condition. Tasting solutions 2 and 3 indicate normal function. Tasting solution 4 indicates mild hypogeusia, and tasting solution 5 suggests moderate hypogeusia; those who cannot identify the taste even in solution 5 have severe hypogeusia. Almost 60% of patients show mild disturbance; 35% have moderate hypogeusia and 5% severe hypogeusia.

Whole-Mouth Taste Test

We have begun to use the whole-mouth test. It consists of a series of tastes in solutions of 13 concentrations. Solution 0 is distilled water and solution 6 is considered the median value for normal subjects. The test is performed by spreading 1 ml of the test solution around the mouth using a syringe.

Results of the Three Tests

Changes of the thresholds of the three examinations were compared in 50 patients whose hypogeusia was

[1] Department of Otolaryngology, Nihon University School of Medicine, 30-1 Oyaguchi, Itabashi-ku, Tokyo, 173 Japan

TABLE 1. Stimulus level of decibel units in electrogustatory test (electrogustometer TR-06).

dB	μA
−6	4
−4	5
−2	6.4
0	8
2	10
4	13
6	16
8	20
10	25
12	32
14	44
16	50
18	64
20	80
22	100
24	136
26	170
28	210
30	260
32	320
34	400

TABLE 2. Concentrations of test solutions for FPD test.

Solution	Concentration (%)				
	1	2	3	4	5
Sucrose	0.3	2.5	10	20	80
NaCl	0.3	1.25	5	10	20
Tartaric acid	0.02	0.2	2	4	8
Quinine	0.001	0.02	0.1	0.5	4

improved subjectively after treatment. About 86% of the patients showed improvement according to the whole-mouth test, 90% by the FPD test, and 64% by the electrogustatory test. Hence it was considered appropriate to use the whole-mouth and FPD tests to observe the course of the disease.

Clinical Analysis of Patients with Taste Disorders

An examination of the distribution of age and sex of 1498 patients who visited our clinic from 1981 to 1987 revealed that elderly patients (50–69 years old) markedly predominated, and women (925 cases) outnumbered men (573 cases). The incidence of taste disorders increases remarkably in those over 50 years of age.

Table 3 summarizes the causes of hypogeusia and the incidence of patient visits to our clinic. We have concluded that drugs, zinc deficiency, and idiopathic causes are important etiologies of hypogeusia.

TABLE 3. Classification and incidence of taste disorders.

Disorder	Incidence (%)
Primary taste disorders	
Congenital	0
Peripheral nerve	3.0
Central nervous system	1.9
Simultaneous disorder of taste and smell	5.1
Senile	1.9
Idiopathic	14.7
Secondary taste disorders	
Oral diseases	8.1
Internal diseases	17.7
Zinc deficiency	22.9
Drug-induced	24.4
Head trauma	1.4
Flu	1.4
Irradiation	0.4
Flavor disorder	6.6
Psychogenic	9.5

Some cases were classified into more than two categories.

The number of patients with hypogeusia increases markedly with age. Therefore it is necessary to investigate whether any geriatric changes relate to the taste disorders of the elderly. We studied 257 hypogeusic elderly patients (>70 years old) and compared them with 406 younger patients (<49 years old) with hypogeusia. According to our study, the incidence of idiopathic hypogeusia was significantly lower in elderly patients. It was only 6.6%, whereas it was 18.0% in the patients below age 40. On the other hand, in elderly patients hypogeusia related to drug intake (33.9%) or to internal (21.8%) or oral cavity (11.7%) problems was, significantly higher than in the younger age group. We thus believe that the number of patients who could be diagnosed as having "geriatric hypogeusia" is small.

Animal Experiments on Zinc-Related Taste Disorders

Zinc is one of the essential trace elements and involves major metabolic actions through utilization of approximately 200 or more enzymes. It is known that zinc deficiency induces taste dysfunction in addition to growth suppression, disorders of the skin and its appendages, and anorexia [1].

In an animal model in our study, hypogeusia appeared in 30% of young rats and 70% of elderly rats [2]. Those rats showed such conditions as disappearance of dense substances in the taste pore region, a marked decrease in the dark granules of taste cells, and the appearance of intracellular microvacuolation and torn microvilli of taste cells [3]. In addition, when compared with normal rats, zinc-deficient rats showed flattening and a decreased number of gustatory papillae on the soft palate [4].

We believe these findings indicate the importance of zinc to the function and morphology of the taste organ.

Treatment of Hypogeusia

As the first choice of therapy for hypogeusia, we prescribe zinc preparations, not only for zinc-deficient cases but also for idiopathic cases. Zinc sulfate ($ZnSO_4 \cdot 7H_2O$) 100 mg is administered three times a day during meals. By administering it with meals, any adverse effects on the digestive system can be effectively prevented.

Clinical efficacy of zinc therapy is high, especially in the categories of zinc-deficiency hypogeusia (73.7%) and idiopathic hypogeusia (75.8%). Overall efficacy was 66.7%. Because those with idiopathic hypogeusia were improved by zinc preparations, we presume that subclinical zinc deficiency is responsible for the appearance of hypogeusia in these patients.

The efficacy of zinc therapy depends on the time of presentation of the patient. Therapy is more effective in patients who present early in the course of the hypogeusia. Among our patients seen within 1 month of its onset, 79.6% responded well to zinc administration.

Zinc therapy also shows good efficacy in elderly patients. We studied 106 hypogeusic elderly patients (>70 years old) and compared them with 117 younger hypogeusic patients (30–49 years old). When zinc administration was started at an early stage of the disease, the efficacy was high in both groups. Among patients who were seen within 6 months of the onset, the efficacy rate in elderly patients was 78.6% and in younger patients 72.3%. Zinc administration is continued for as long as symptoms remain, even when the monitored serum zinc level is sufficiently high, because the serum zinc level does not accurately reflect the tissue zinc level.

Double-Blind Study of Zinc Gluconate

In a double-blind study of zinc gluconate [5] the subjects were divided into two groups and were given zinc gluconate or placebo for up to 4 months. Improvement of the taste examination and subjective symptoms were then analyzed in 65 of the subjects.

The overall efficacy in the group who underwent zinc therapy was 77.1% and that in the placebo group 60.0%. We could detect no significant difference between the two groups. However, when we considered only patients with zinc-deficiency and idiopathic hypogeusia, the efficacy of zinc gluconate treatment was 82% and that in the placebo group 54%. Hence zinc gluconate was significantly superior to placebo. Moreover, alleviation of the hypogeusia was compared between these two groups, and again there was significant therapeutic superiority of zinc gluconate.

We believe that zinc administration should be regarded an effective therapeutic regimen for hypogeusia, especially in patients with a low serum zinc level and in those with idiopathic hypogeusia.

Conclusion

Causes of hypogeusia vary among patients. Therefore when treating hypogeusia it is important to understand the cause and severity of the disease in order to, first, select the most appropriate therapy for the patient and, second, evaluate the efficacy of that therapy.

Several clinical studies have reported that deficiency of zinc may induce hypogeusia in humans; and in several animal experiments an important relation has been demonstrated between zinc and taste. Therefore zinc may be important in the treatment of hypogeusia. The efficacy of zinc administration in the treatment of hypogeusia was first reported almost 30 years ago. Our clinical results support the efficacy of zinc administration for taste disorders. We intend to continue our clinical and experimental studies to clarify the relation of zinc and taste disorders.

References

1. Cunnane SC (1988) Zinc: Clinical and biochemical significance. CRC Press, Boca Raton
2. Hasegawa H, Tomita H (1987) Assessment of taste disorders in rats by simultaneous study of the two-bottle preference test and abnormal ingestive behavior. Auris Nasus Larynx 13[Suppl I]:33–41
3. Kobayashi T, Tomita H (1986) Electron microscopic observation of vallate taste buds of zinc-deficient rats with taste disturbance. Auris Nasus Larynx 13[Suppl I]:25–31
4. Naganuma M, Ikeda M, Tomita H (1988) Changes in soft palate taste buds of rats due to aging and zinc deficiency. Auris Nasus Larynx 15:117–127
5. Yoshida S, Endo S, Tomita H (1991) A double-blind study of the therapeutic efficacy of zinc gluconate on taste disorder. Auris Nasus Larynx 18:153–161

Concepts of Therapy in Taste and Smell Dysfunction: Repair of Sensory Receptor Function as Primary Treatment

ROBERT I. HENKIN[1]

Key words. Taste—Smell—Taste disease treatment—Olfactory disease treatment—Receptor pathology—Receptor disease treatment

Introduction

Most physicians consider dysfunctions of taste and smell minor clinical problems, primarily aesthetic in nature. Physicians consider treatment practical if correctable by surgical means or if adrenal cortical steroids are effective. If these approaches are unsuccessful, further treatment is usually considered impossible. This approach is incorrect, as neither surgery nor steroids can correct abnormalities of taste and smell function that are primarily biochemical in nature. The purpose of this chapter is to describe effective treatment for pathologies of taste and smell based on biochemical principles that underlie these sensory systems.

Physiology of Taste and Smell Function

The sensory systems for taste and smell may be considered to be composed of three parts (Fig. 1): (1) receptor; (2) sensory nerve; and (3) central nervous system (CNS). Each system part is regulated by biochemical activation and inhibition through short and long feedback loops. The initial functional event involves stimulus binding to receptor. At the receptor a complex series of events occur from which generator and action potentials are derived. The action potential is then transmitted along the sensory nerve to the CNS for integration. The focus of this discussion is on receptor events of taste/smell function, their pathology, and their treatment.

Molecular Biology of Receptor Function

The receptor, the initial component of taste/smell perception, is similar to the seven-loop, four-

transmembrane chemical moeity previously described for rhodopsin in the visual system (Fig. 2) [1]. Taste receptors reside on cilia of taste bud type III cells, which comprise only 5% of bud cells [2]. These cells have no blood vessels, no lymphatics, and few if any mitotic figures [2], yet they turn over rapidly [3]. Basal cells in the bud have the potential to differentiate into type III cells under appropriate conditions. Smell receptors reside on cilia of mitral cells [4], which also comprise a small percentage (~5%) of cells of olfactory epithelium. These cells also have no blood vessels, no lymphatics, and few if any mitotic figures [4] yet turn over rapidly [5]. Basal cells in the epithelium can differentiate rapidly into mitral cells under appropriate conditions. Primary nutrition of both taste and smell receptors depends on proteins and growth factors that are secreted into saliva and nasal mucus from supporting cells [4,6,7].

Following stimulus-receptor binding, a series of receptor-coupled events occur. For the "on" mode (Fig. 3), receptor binding institutes a conformational change at the receptor that initiates G-protein activity, resulting in the generator potential with subsequent effector activation through either adenylyl cyclase [8,9] or phospholipase C with subsequent cyclic 3,5-adenosine monophosphate (cAMP) and inositol-1,4,5-triphosphate (IP$_3$) activity, opening of Na and Ca channels, and generation of an action potential transmitted along sense specific cranial nerves with subsequent CNS integration. For some tastes, receptor-coupled events may be directly mediated by ion channel activity, bypassing G-protein and effector activation. System failure results primarily in loss of sensory acuity.

Termination of receptor binding, with cessation of sensory signal or the "off" mode response (Fig. 4), is critical for normal sensory function so a given stimulus action is terminated or the binding "erased" and a new stimulus appreciated. Sensory cessation can occur in several ways. Because normal stimulus-receptor binding is weak (~Kd 10^{-3} [10]) a novel stimulus can readily replace a previously bound stimulus based on its hierarchical Kd or another stimulus can bind to another receptor class, initiating a new set of receptor-coupled events. Decreased receptor number or receptor injury commonly initiates tighter receptor binding resulting in sensory

[1] The Taste and Smell Clinic, Center for Molecular Nutrition and Sensory Disorders, 5125 MacArthur Boulevard, Washington, DC 20016, USA

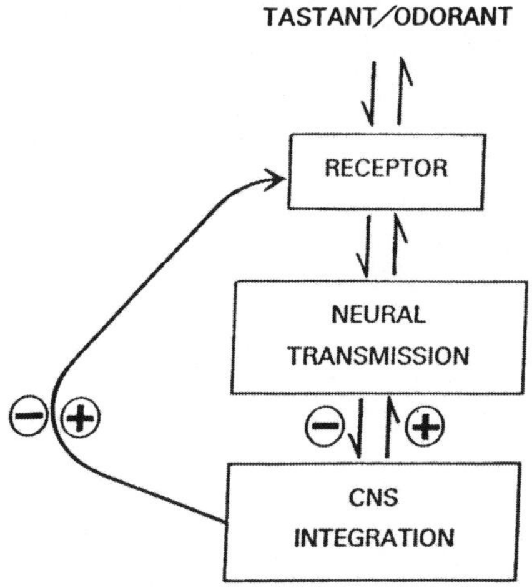

TASTANT/ODORANT

FIG. 1. Components of taste/smell systems. Short and long feedback loops control activity at several levels of biochemical and neurophysiological activity

Pathology of Taste and Smell Function

Pathology results in two major functional abnormalities of these systems: loss of acuity [14,15] and distortion of function (Table 1) [4,16,17]. These abnormalities may occur alone or simultaneously, or one may precede or follow the other. In patients with taste and smell loss, most (98%) exhibit hypogeusia, hyposmia, or both; most of those with hypogeusia (90%) exhibit type II hypogeusia. In patients with hyposmia, most (about two-thirds) exhibit type II hyposmia with half as many (about one-third) exhibiting type I hyposmia. Fewer than 5% of patients with hyposmia or hypogeusia exhibit either type III hyposmia or hypogeusia, or anosmia or ageusia [4].

About 50% of patients exhibit distortions of taste or smell usually weeks to months after acuity loss [4,16]. Patients commonly exhibit both aliageusia [16] and aliosmia [4], rarely aliageusia or aliosmia alone. Torquegeusia [4,17] and torquosmia [4] are the most common types of these distortions, with about half as many patents exhibiting cacogeusia or cacosmia. Mixed types are not uncommon, but parageusia and parosmia are rare.

Patients with aliosmia or aliageusia may also exhibit phantogeusia and phantosmia. It is more common for patients to exhibit either phantogeusia or

perseveration [4]. An "off" mode of receptor events also normally occurs with activation of several receptor kinases [11], Gi proteins [11,12], P450 proteins [13], or other proteins and chemical moieties that assist in termination of receptor sensory signals (Fig. 4). System failure results primarily in distortion of sensory function.

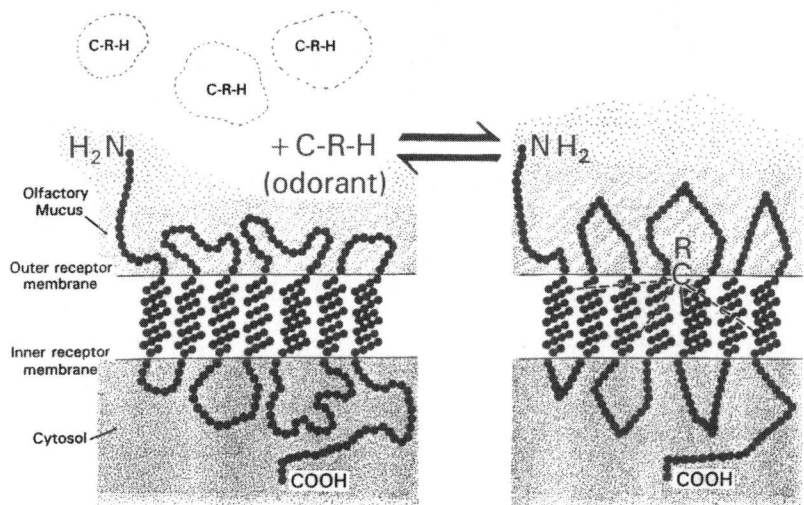

FIG. 2. Model of taste/smell receptor based on the model of the rhodopsin receptor [1]. This is a two-dimensional representation of a three-dimensional model consisting of a seven-transmembrane α helical structure composed of a single polypeptide chain with about 350 amino acids (similar to other G-protein receptors that may contain as many as 450 amino acids). The polypeptide chain transverses the cellular lipid bilayer seven times, the seven-transmembrane helices being connected by loops, with the two termini, the NH_2 and the carboxy termini, on opposite sides of the membrane. In a manner similar to a light photon, a tastant/odorant molecule, C-RH, probably binds to a specific tastant/odorant receptor molecule on the transmembrane helices, i.e., to a molecule similar to retinal, such that it is covalently linked to several amino acids on several of the predicted transmembrane helices (for convenience shown here for only one helix). Then, hydrogen bonding occurs between this (these) active site(s) and other portions of the receptor molecule, represented in the figure for convenience, by three of the other transmembrane helices (similar to the binding of inhibitors to the active site on aspartyl proteinases, e.g., pepsin or renin). Through this binding a conformational change in the receptor occurs through which this stimulus "information" initiates the next series of steps in the sensory process involving the receptor-coupled systems

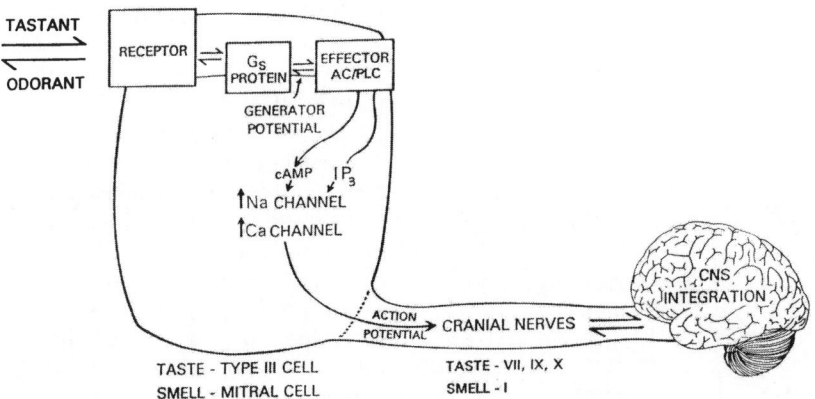

FIG. 3. Model of the receptor-coupled system in the "on" mode related to perception of taste/smell acuity. After tastant/odorant receptor binding (Fig. 1) a G-protein (Gs, Gq) and the effector system (*AC*, adenyly cyclase; *PLC*, phospholipase C) are activated, which activate cAMP or IP_3. The latter activate Na and Ca channel activity, result-ing in initiation of the action potential, which is transmitted along sense-appropriate nerves to the CNS for sense-appropriate integration. For taste, direct ion channel activity may mediate some responses, bypassing the G-protein and effector activation

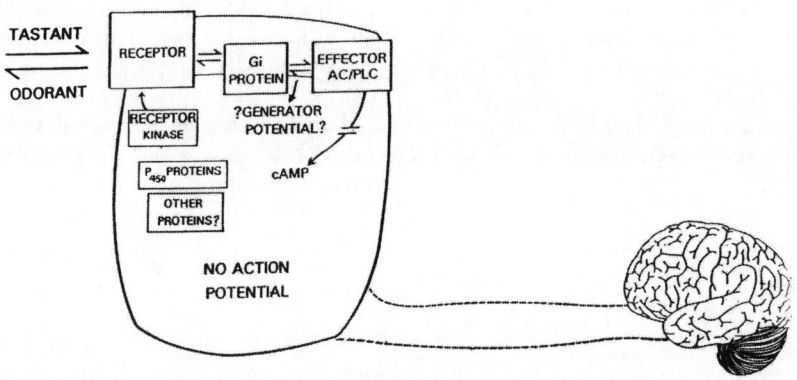

FIG. 4. Model of receptor-coupled system in the "off" mode related to taste/smell perseveration (i.e., abnormal persistence of taste/smell sensations) or initiation of taste/smell distortion. Following stimulus binding to a receptor, an inhibitory G-protein (*Gi*) is activated or receptor kinases are activated to inhibit the signal emanating from receptor. Other proteins, including P450 proteins, may also be activated; they inhibit receptor activation, which terminates receptor activity

TABLE 1. Classification of loss and distortion of taste and smell function in relation to anatomy, physiology, and pathology of taste and smell.

Dysfunction	Taste	Smell	Definition
Loss	Ageusia	Anosmia	Inability to detect or recognize any stimuli
	Hypogeusia	Hyposmia	Decreased ability to detect or recognize any stimuli
	Type I	Type I	Absent stimulus recognition
	Type II	Type II	Decreased ability to detect or recognize stimuli
	Type III	Type III	Decreased ability to judge stimulus intensity
Distortion	Aliageusia	Aliosmia	From environmental stimuli
	Cacogeusia	Cacosmia	Rotten, decayed, fecal
	Torquegeusia	Torquosmia	Chemical, metallic, bitter, burned
	Mixed	Mixed	Both caco- and torque
	Parageusia	Parosmia	Skewed but normal (onions, like roses)
	Phantogeusia	Phantosmia	From endogenous stimuli
	Cacogeusia	Cacosmia	Rotten, decayed, fecal
	Torquegeusia	Torquosmia	Chemical, metallic, bitter, burned
	Mixed	Mixed	Both caco- and torque-

See Henkin [4] for details.

phantosmia than to see both symptoms together. Torquegeusic and torquosmic types of phantoms are more than twice as common as cacogeusic or cacosmic types, which are about as common as mixed types [4].

Location of Pathology

CNS

Injury to the CNS accounts for less than 5% of taste and smell pathology related mainly to cerebral vascular accidents, brain tumors, demyelinating, metabolic and nutritional disorders, drug effects [18], genetic diseases, infections, Parkinson's disease, and dementias of several types including Alzheimer's disease.

Sensory Nerves

Disorders of neural transmission also contribute less than 5% of taste and smell pathology, reflected mainly in infectious diseases, demyelinating, metabolic, and nutritional disorders, drug effects, and surgical nerve injury or interruption.

Receptor Pathology

Receptor dysfunction comprises more than 90% of taste and smell dysfunction. Direct receptor pathology contributes about half the pathology; receptor destruction produces a type I defect; injury or impaired stimulus binding produces a type II defect and normal functioning; and inadequate receptor number produces a type III defect (Table 1). Inhibition of receptor function may follow infectious processes, head injury, drug effects, allergen inhibition, congenital receptor absence or genetic mutation, vitamin deficiencies (A, B_{12}), heavy metal (Cd, Pb, Hg) toxicity, or vapor-phase toxicant exposure (e.g., formaldehyde or halogenated hydrocarbons). Molecular mechanisms underlying these events may involve alteration of primary, secondary, or tertiary protein structure with subsequent receptor down regulation, N-terminal cleavage, decreased receptor sensitivity, inhibition of receptor turnover, irreversible occupation of the N-terminal receptor site, receptor mutations, N-terminal inhibition, receptor/G-protein decoupling, G-protein/effector decoupling, or the presence of peptides that inhibit receptor activity or inhibit receptor activation.

Pathology of receptor-coupled events in the "on" mode relates primarily to loss of sensory acuity and involves inhibition of G-protein activity or synthesis or effector-mediated impairment. G-protein impairment has been associated with hormonal abnormalities (e.g., type Ia pseudohypoparathyroidism [19,20] and Turner's syndrome [21,22]); decreased gustin synthesis [23]; binding or storage; trace metal

(Zn, Cu) deficiency [24–27]; or viral inactivation of the system [28]. Pathology of effector inhibitory events involve hormonal abnormalities including abnormalities of carbohydrate-active steroids (Addison's disease, Cushing syndrome [29]); parathyroid hormone (pseudohypoparathyrodism, type 1a [19,20,29]); thyroid hormone (hypothyroidism [30], use of antithyroid drugs [30] or radioactive iodine [31]); gonadal steroids (Turner syndrome, hypogonadotrophic hypogonadism [29]); or drug effects (inhibition of adenylyl cyclase, cAMP, phospholipase C, P450 antagonists, or calmodulin deficiency [32]). Pathology of ion channel activity may also affect taste function directly.

Pathology of receptor-coupled events in the "off" mode relates primarily to taste and smell distortions and involves both direct receptor pathology and abnormalities of Gi proteins, inhibition of receptor kinase activity, cAMP inhibition, P450 protein inhibition, or inhibition of other proteins and peptides involved in the termination of receptor-mediated events.

Treatment of Taste and Smell Dysfunction

CNS

Treatment includes excision of offending tumors, use of antibiotics, or drug cessation.

Sensory Nerves

Treatment includes antibiotics and hormones to restore functional neural integrity. In patients with trace metal deficiency, replacement usually restores normal function. Surgical repair of severed nerves by graft or suture has also been of benefit.

Nasal Lesions

Treatment of local nasal pathology by nasal polypectomy or nasal turbinectomy can restructure airway system integrity. However, correction of these local lesions are usually associated with only transient correction of smell function, and a more fundamental, commonly receptor pathology must be kept in mind.

Receptors

Treatment of receptor pathology per se involves inhibition of infectious agents, hormone, trace metal or vitamin replacement, or heavy metal chelation with EDTA, D-penicillamine [25], or triene.

Treatment of abnormalities of the "on" mode of the receptor-coupled system involves stimulation of G (Gq, Gs)-protein synthesis or activity accomplished by either zinc administration in patients in whom gustin synthesis is inhibited by decreased availability of zinc cofactor [23,33] or by gustin replacement when this protein becomes available as an approved pharmaceutical [34]. Effector system stimulation can be corrected by increasing the endogenous cAMP concentration through administration of theophylline (which inhibits phosphodiesterase metabolism of cAMP [4,35]) or with F, Al or Mg, which directly stimulate cAMP synthesis.

Treatment of receptor-coupled events in the "off" mode involves inhibition of receptor overactivity [36,37] through use of dopaminergic agonists (e.g., haloperidol, thioridizine), calcium channel blockers (e.g., pimozide, nimodipine), GABAergic agonists (e.g., thioridizine, valproic acid, bacolfen), dopamine receptor blockers (e.g., tetrabenazine), benzodiazepenes (e.g., diazepam, klonopin), tricyclics (e.g., desipramine), serotonin-uptake inhibitors (e.g., fluoxitine), or α-adrenergic receptor agonists (e.g., clonidine). Other treatment modalities include stimulation of Gi protein activity, stimulation of receptor kinase activity, P450 stimulation, and stimulation of other inhibitory peptides.

Discussing the pathology and subsequent treatment of the most common cause of taste and smell dysfunction, postinfluenza hyposmia and hypogeusia (PIHH) [4,28], may be useful. Pathology of this syndrome may be divided into three phases: (1) edema of nasal mucous membranes with hyposmia related to obstruction of vapor passage to olfactory receptors in the epithelium; (2) direct viral destruction of sensory receptors; and (3) secondary viral effects inducing autoimmune changes; these autoimmune effffects may be secondary to antigen-antibody reactions at the receptor or to immune system-mediated receptor impairment as occurs in multiple sclerosis models (Theiler's disease, experimental allergic encephalomyelitis). If receptor destruction is complete, a type I defect occurs; if only injury occurs, a type II defect results. Recovery may be spontaneous, although the longer the loss the less likely is recovery.

Treatment of PIHH involves correction of abnormalities associated with each disease phase: (1) Usually self-limited, therapeutic diminution of mucous membrane edema (antihistamines or local nasal steroid sprays). (2) Control of acute viral infection. Treatment is not usually clinically effective (α-interferon, systemic acyclovir, other antiviral agents). (3) Direct receptor stimulation or indirect activation of the receptor-coupled system or effector stimulation. This treatment is usually effective.

Treatment of "on" receptor-coupled events increases sensory acuity. Zinc can induce receptor activity through direct T-cell stimulation [4], as a cofactor in enhancing Gs/Gq protein synthesis [4,33], or through gustin modulation of phosphodiesterase activity [38]. Theophylline modulates immune system function and can indirectly increase cAMP con-

centration via inhibition of phosphodiesterase [4]; F, Al, and Mg can stimulate cAMP directly. Systemic carbohydrate-active steroids can inhibit autoimmune responses, induce transformation of sensory basal cells (thereby promoting receptor synthesis and activity, and inhibit cellular apoptosis.

Treatment of "off" receptor-coupled events inhibits dysgeusia and dysosmia. Drugs (dopaminergic agonists, calcium channel blockers, GABAergic agonists) can inhibit several components of receptor overactivity, as can activation of Gi proteins or other inhibitory peptides.

Conclusions

Most pathology of taste and smell dysfunction involves taste/smell receptors and affects both processes of dysfunction, causing loss of acuity and sensory distortion. Receptor pathology can be defined by hypogeusia or hyposmia (types I–III) or dysgeusia or dysosmia (alia-, phanto-, caco-, or torquegeusia, or mixed variety). Receptor treatment usually involves systemic medical therapy, which corrects receptor malfunction either at the receptor or in the receptor-coupled system in the "on" or "off" mode. Drugs or chemical moieties that activate G-proteins or effector systems involving enhanced activity of adenylyl cyclase, cAMP, phospholipase C, IP_3, or ion channels are useful for restoring taste/smell acuity. Drugs that inhibit receptor overactivity or enhance Gi protein activity are effective in inhibiting dysgeusia and dysosmia.

This systematic approach defines medical therapy that corrects induced biochemical abnormalities of chemosensation.

References

1. Finley JB (1990) The structure of G-protein linked receptors. Biochem Soc Symp 56:1
2. Henkin RI (1976) Taste in man. In: Harrison D, Hinchliffe R (eds) Scientific foundation of otolaryngology. Heinemann, London, pp 468–483
3. Beidler LM, Smallman RL (1965) Renewal of cells within taste buds. J Cell Biol 27:263
4. Henkin RI (1993) Evaluation and treatment of human olfactory dysfunction. In: English G (ed) Otolaryngology, vol 2. Harper & Row, New York, pp 1–93
5. Moulton DG (1975) Cell renewal in the olfactory epithelium of the mouse. In: Denton D, Koghland JP (eds) Olfaction and taste, vol 5. Academic Press, Orlando, FL, pp 111–114
6. Henkin RI (1978) Saliva, zinc and taste: interrelationships of gustin, nerve growth factor, saliva and zinc. In: Hambidge KM, Nichols BL (eds) Zinc and copper in clinical medicine. Spectrum, Jamaica, NY, pp 35–48
7. Henkin RI (1984) Zinc in taste function: a critical review. Biol Trace Elem Res 6:263
8. Reed RR, Bakalyar HA, Dhallan RJ, Reed RR (1991) GTP-binding proteins, adenylyl cyclase and ion channels in olfactory transduction. Biophys J 59:391a

9. Dhallan RS, Yam KW (1990) Primary structure and functional expression of a cyclic nucleotide-activated channel from olfactory neurons. Nature 347:184

10. Lum CKL, Henkin RI (1976) Sugar binding to purified fractions from bovine taste buds and epithelial tissue: relationships to bioactivity. Biochim Biophys Acta 421:380

11. Hausdorf WP, Caron MG, Lefkowitz RL (1990) Turning off the signal: desensitization of β-adrenergic receptor function. FASEB J 4:2881

12. Dawson TM, Arriza IL, Jaworsky DE, et al. (1993) β-Adrenergic receptor kinase-2 and β-arrestin-2 as mediators of odorant-induced desensitization. Science 259:825

13. Lazard D, Zupke K, Poria Y, Lancet D (1991) Odorant signal termination by olfactory UDP glucuronosyl transferase. Nature 349:790

14. Henkin RI (1967) The definition of primary and accessory areas of olfaction as the basis for a classification of decreased olfactory acuity. In: Hayashi T (ed) Olfaction and taste, vol 2. Pergamon, London, pp 235–252

15. Hoye RC, Ketcham AS, Henkin RI (1970) Hyposmia after paranasal sinus exenteration and laryngectomy. Am J Surg 120:485

16. Henkin RI (1992) Phantogeusia. In: Taylor RB (ed) Difficult diagnosis, vol 2. Saunders, Philadelphia, pp 348–356

17. Henkin RI, Frazier, KB (1989) Torquegeusia: a new distinct and common form of dysgeusia. FASEB J 3:A338

18. Henkin RI (1986) Drug effects on taste and smell. In: Pradhan SN, Maikel RP, Dutta SN (eds) Pharmacology in Medicine: Principles and Practice. SP Press, Bethesda, pp 748–753

19. Henkin RI (1968) Impairment of olfaction and of the tastes of sour and bitter in pseudohypoparathyroidism. J Clin Endocrinol Metab 28:624

20. Farfel Z, Bourne HR (1982) Pseudohypoparathyroidism: mutation affecting adenylyl cyclase. Miner Electrolyte Metab 8:227

21. Henkin RI (1967) Abnormalities of taste and olfaction in patients with chromatin negative gonadal dysgeusia. J Clin Endocrinol Metab 27:1436

22. Zisman E, Henkin RI, Ross GT, Bartter RC (1970) A biochemical similarity between chromatin negative gonadal dysgenesis and pseudohypoparathyroidism. Acta Endocrinol Panam 1:39

23. Henkin RI, Lippoldt RE, Bilstad J, Edehoch H (1975) A zinc containing protein isolated from human parotid saliva. Proc Acad Sci USA 72:488

24. Henkin RI, Patten BM, Re P, Bronzert D (1975) A syndrome of acute zinc loss. Arch Neurol 32:745

25. Henkin RI, Keiser HR, Jaffe IA, Sternlieb I, Scheinberg H (1967) Decreased taste sensitivity after D-penicillamine reversed by copper administration. Lancet 2:1268

26. Tomita H (1990) Zinc in taste and smell disorders. In: Trace elements in clinical medicine. Springer, Berlin Heidelberg New York, Tokyo, pp 15–37

27. Brosvic GM, Slotnick BM, Henkin RI (1992) Decreased gustatory sensitivity in zinc deprived rats. Physiol Behav 52:527

28. Henkin RI, Larson AL, Powell RD (1975) Hypogeusia, dysgeusia, hyposmia, and dysosmia following an influenze-like infection. Ann Otol Rhinol Laryngol 84:672

29. Henkin RI (1975) The role of adrenal corticosteroids in sensory processes. In: Blasckho A, Sayers G, Smith AD (eds) Handbook of physiology, vol VI: Endocrinology. American Physiological Society, Washington, DC, pp 209–230

30. McConnell RJ, Menendez CW, Smith FR, Henkin RI (1975) Defects in taste and smell in patients with hypothyroidism. Am J Med 59:354

31. Varma VM, Grollman EF, Dai WL, Henkin RI (1991) Taste loss after I131 treatment of thyroid cancer. Clin Res 40:417A

32. Law JS, Henkin RI (1986) Low parotid saliva calmodulin in patients with taste and smell dysfunction. Biochem Med Metab Biol 36:118

33. Shatzman AR, Henkin RI (1981) Gustin concentration changes relative to salivary zinc and taste in human. Proc Natl Acad Sci USA 78:3867

34. Coon HG, Henkin RI (1992) Continuous culture of human parotid gland cells secrete the major salivary proteins: amylase, gustin and lumicarmine. Clin Res 40:261A

35. Henkin RI, Aamodt RL, Babcock AK (1981) Treatment of abnormal chemoreception in human taste and smell. In: Morris DM (ed) Perception of behavioral chemicals. Elsevier, Amsterdam, pp 229–253

36. Henkin RI (1990) Hyperosmia and depression following exposure to toxic vapors. JAMA 264:2803

37. Henkin RI (1991) Salty and bitter taste. JAMA 256:2253

38. Law JS, Nelson N, Watanabe K, Henkin RI (1987) Human salivary gustin is a potent activator of calmodulin-dependent brain phosphodiesterase. Proc Natl Acad Sci USA 84:1674

Is the Whole Mouth Method an Appropriate Index of Taste Disorders?

Yuki Yamauchi, Eikichi Tokunaga, Sohei Endo, and Hiroshi Tomita[1]

Key words. Whole mouth method—Therapeutic index

Electrogustometry, the filter-paper disc test, and the whole mouth method are the major test methods we employ for determining the effects of treatment at our clinic for taste disorders. In this study, we examined whether the whole mouth method was as an appropriate index for assessing these therapeutic effects.

Seventy of the patients with taste disorders who visited our clinic between March 1991 and August 1993 were the subjects of this study. Subjectively, 50 improved and 20 did not. As the basic taste qualities examined (sweet, salty, sour, bitter), we used sucrose, salt, tartaric acid, and quinine, respectively.

One ml of each basic taste substance, diluted into 13 ranked concentrations, was sprayed into the patient's mouth. We began with the lowest concentration, increasing concentrations when the patient recognized the correct taste quality. After giving the patient sufficient time to taste it, we let them swallow the liquid and taste the quality he or she had sensed. We made sure that the patients rinsed their mouth well with tap water to eliminate any lingering taste before tasting a different taste.

We examined the change in recognition thresholds by the whole mouth method before and after treatment. According to this method, thresholds were significantly improved ($P < 0.01$) in all taste qualities in the subjectively improved group, while improvements were nonsignificant in the subjectively-non-improved group. Therefore, we suggest that the whole mouth method can be an appropriate index to examine the effects of treatment for taste disorders. To compare improvement as determined by the three test methods, we used the following criteria for judging improvement for each taste:

1. Whole mouth method: those patients who demonstrated improvement by one rank or more in two or more taste qualities out or four. According to these criteria 86% of patients were improved.

2. Filter-paper disc test: those patients who demonstrated improvement by one rank or more, in two or more taste qualities out of four, or three or more locations in the right and left areas innervated by the chorda tympani and glossopharyngeal nerves (90% improvement).

3. Electrogustometry: those patients who demonstrated improvement by 6 dB or more in two or more locations in the right and left areas innervated by the chorda tympani and glossopharyngeal nerves (64% improvement).

Patients who showed a subjective improvement in symptoms invariably demonstrated improvement in at least one of the test methods. None of the patients showed improvement in electrogustometry alone. Electrogustometry was least correlated with the subjective improvements; this is probably because the mechanism of taste differs between electrogustometry and the other two taste methods. The values first examined for the whole mouth method were often within the normal range. It would be appropriate to judge the effects of treatment in terms of improvements in values for the concentration series. When the whole mouth method is used, even patients suffering from severe taste disorder will not have measurable results on the filter-paper disc test; this will seldom be the case when electrogustometry is used, however. Consequently, the whole mouth method enables differentiation of the degree of serious symptoms that are otherwise difficult to measure. We conclude that the results of the whole mouth method can serve as an appropriate index for the follow up of the effects of treatment for taste disorders.

References

1. Takumida M (1983) Correlation of gustatory tests (in Japanese). Jibirinshou [Practica Otologia (Kyoto)] 76(11):2905–2910
2. Ikeda M (1992) Study of a simplified method of taste assessment and evaluation of zinc therapy (in Japanese). Koukuuintouka (Stomato-pharyngology) 4(2):51–57

[1] Department of Otolaryngology, Nihon University School of Medicine, 30-1 Ooyaguchi-kamimachi, Itabashi-ku, Tokyo, 173 Japan

Secretory Function of the Salivary Glands in Patients with Taste Disorders: Correlation Between Zinc Deficiency and Secretory Function of Salivary Glands

Masami Tanaka and Hiroshi Tomita[1]

Key words. Secretory function of the salivary glands—Zinc deficiency

Introduction

Patients with xerostomia and taste disorders are frequently observed clinically. We inferred that zinc deficiency was one of the causes of xerostomia, and carried out studies to elucidate the correlations between zinc deficiency and salivary secretory function.

Subjects and Methods

Studies were carried out on 93 patients (27 males and 66 females; age range 39–83 years, mean, 64 years) who were suffering from hypogeusia and/or xerostomia. The following tests were conducted:

1. Dynamic salivary scintigraphy, using ^{99m}Tc-pertechnetate. The patients first received an intravenous injection of 3 mCi of 99 mTc-pertechnetate. Fifteen min later, they were given 1 g of Cinal (vitamin C) orally; this was given to stimulate salivary secretion. We used the maximum secretory ratio as an index for salivary secretory function.
2. Taste examination. We carried out electrogustometry and a filter-paper disc test.
3. Hemo-biochemical examination. The patients were classified into three groups, each group consisting of a male and female sub-group: a normal zinc group; serum zinc and copper levels were 70–110 μg/dl and 80–130 μg/dl, respectively, zinc deficiency group; serum zinc level was below 70 μg/dl, and group suspected of zinc deficiency; serum copper level was above 130 μg/dl.

Results

Correlations Between Serious Taste Disorders and Salivary Secretory Function

No clear correlation was observed between serious taste disorders and salivary secretory function.

Correlation Between Salivary Secretory Function and Zinc Deficiency

In the female zinc deficiency group (including the group suspected of zinc deficiency), salivary secretory function declined compared with the normal zinc group. Comparison between the parotid gland (PTG) and the submandibular gland (SMG) revealed a significant decline ($P < 0.05$) for the SMG. In the male group, no significant difference was observed in salivary secretory function between the zinc deficiency and the normal zinc groups.

Efficacy of Zinc Treatment for Xerostomia

Zinc sulfate was prescribed for 38 of 63 patients with xerostomia, in 22 of whom (57.9%) the condition disappeared or significantly improved.

Case Report

A typical case in which xerostomia improved significantly through zinc treatment was that of a 70-year-old woman who had xerostomia and hypogeusia. Her serum Zn was 65 μg/dl and serum Cu, 110 μg/dl. She received 300 mg/day of zinc sulfate, taken orally, for 6 months. The dynamic salivary sintigram showed a remarkable improvement in the function of each salivary gland as a result of the treatment.

Discussion

It has been reported that, where zinc decreased the secretory function of the salivary gland, alkaline phosphatase (AP) activity and carbonic anhydrase (CA) activity in the SMG of zinc-deficient rats were reduced to 26% and 44% of values in their respective control groups [1,2]. Both AP and CA are metalloenzymes of zinc [3] and are chiefly localized in the myoepithlial cells (MEC) and the cells of the duct system. Zinc deficiency hinders the growth of these cells and causes functional impairment.

[1] Department of Otolaryngology, Nihon University School of Medicine, 30-1 Oyaguchi, Itabashi-ku, Tokyo, 173 Japan

Conclusion

A decline in the secretory function of the salivary gland was observed in patients with zinc deficiency, particularly in the submandibular gland of female patients. We concluded that xerostomia can be a symptom of zinc deficiency.

References

1. Chaudry IM, Meyer J (1979) Response of submandibular gland of the rat to nutritional zinc deficiency. J Nutr 109:316–320
2. Gandor DW, Fanslow DJ, Meyer J (1981) Effects of zinc deficiency on developmental changes in alkaline phosphatase and carbonic anhydrase activities in the submandibular gland of the rat. Arch Oral Biol 28: 609–614
3. Cunnane SC (1988) Enzyme. In: Zinc: Clinical and biochemical significance. CRC Press, Florida, pp 609–614

Clinical Experience of Zinc Picolinate in the Treatment of Patients with Taste Disorders

YOICHIRO YAMADA, KOICHI ISHIYAMA, KEN-ICHI WATANABE, SOHEI ENDO, and HIROSHI TOMITA[1]

Key words. Zinc picolinate—Clinical experience

The clinical efficacy of zinc picolinate in the treatment of taste disorders was assessed in an open trial.

Materials

We obtained serum from a total of 34 patients with taste disorders. The sera were classified according to the etiology of the condition: zinc-deficient, serum zinc level $<70\,\mu g/dl$, not having any other cause, $n = 11$; idiopathic, serum zinc level $\geq 70\,\mu g/dl$, not having any other overt cause, $n = 3$; drug-induced, $n = 13$; and others, $n = 7$.

Methods

Ten to 20 mg of Zn was given t.i.d. over a 4-month period. The threshold of recognition of the four basic taste qualities was determined by the filter paper disc (FPD) method at the test point of the bilateral chorda tympani and glossopharyngeal nerve fields. The average score (AS) and that at the first examination (AS 1) and at the end of the trial (AS 2) were measured. We defined the effectiveness of zinc picolinate as shown below (DA indicates duration of administration).

AS 2 − AS 1 < 1 and AS 2 < 4 or AS 2 − AS 1 ≥ 1, effective

AS 2 − AS 1 < 1 and AS 2 ≥ 4 (DA < 90 days), dropout

AS 2 − AS 1 < 1 and AS 2 ≥ 4 (DA ≥ 90 days), ineffective

$$AS = \frac{\text{All the sums of each score of the four basic taste qualities examined on the above two nerve fields bilaterally}}{16}$$

The serum zinc levels at the first examination (Zn 1) and at the end of the trial (Zn 2) were also measured and evaluated.

Wilcoxon non-parametric statistical analysis with a PC SAS Ver.6.04 was used.

Results

Zinc picolinate was effective in 15/27 subjects in this study (55.6%). In patients in whom this was effective, and the zinc picolinate was given for 90 days or more, the change of serum zinc level (Zn 2 > Zn 1) was 69.2% (9/13), while in those who received the drug for less than 90 days, the change of serum zinc level (Zn 2 > Zn 1) was 50.0% (3/6). Side effects (abdominal pain, diarrhea, and loss of appetite) were observed in 3/27 (11.1%) of patients.

Conclusion

We considered that zine picolinate would be more effective than zinc sulfate due to its better absorbability. However, this could also be one reason for the side effects being worse than with zinc sulfate.

[1] Nihon University School of Medicine, Department of Otolaryngology, 30-1 Kami-cho Ooya-guchi Itagbashi-ku, Tokyo, 173 Japan

Double-Blind Study of the Therapeutic Efficacy of Zinc Picolinate in Taste Disorder: Preliminary Report

Fumitaka Sakai, Shinya Yoshida, Sohei Endo, and Hiroshi Tomita[1]

Key words. Zinc picolinate—Double-blind study

In an earlier study, we investigated the therapeutic efficacy of orally given zinc gluconate in taste disorders in a double-blind study [1]. No significant difference was detected between zinc gluconate and placebo. However, a significant superiority of zinc gluconate to placebo was observed in patients with idiopathic and zinc-deficient taste disorders. Here, we studied the therapeutic efficacy of zinc in patients with taste disorders limited to idiopathic and zinc-deficient groups. The test drug was zinc picolinate. We used this drug since Barrie et al. [2] reported that picolinic acid promoted the absorption of zinc in humans, and that it was superior to gluconate. The efficacy was assessed in a double-blind study; the degree of improvement in taste examination, and subjective symptoms, were evaluated.

Materials and Methods

The subjects were 50 patients with taste disorders who were diagnosed clinically as having idiopathic or zinc-deficient taste disorders. They had visited the ENT clinic of the Itabashi Hospital of Nihon University School of Medicine between October 1991 and March 1993. They were randomly assigned to either of two 25-patient groups, one receiving zinc picolinate and one placebo, and were given capsules for up to 3 months. The test drug capsule contained 12 mg of zinc, and lactose, as an additive; the placebo contained lactose alone. Both preparations were given t.i.d. Of the 50 patients enrolled in the study, 44 (24 on zinc picolinate and 20 on placebo) were eligible for analysis. The mean age of these patients was 53.0 year (53.5 for the zinc picolinate group and 52.6 for the placebo group), and the male-to-female ratio was 1:1.9. Improvements in taste examination and subjective symptoms were analyzed. The whole

mouth method, filter-paper disk (FPD) method, and electrogustometry (EGM) were conducted for the taste examination.

Results

Subjective symptoms were rated in five grades of severity, from normal taste sense of 5 to ageusia, 1. The degree of subjective recovery was defined as the difference between the initial and final subjective symptoms elicited by questionnaires. The grade of improvement in the taste examination was defined as the difference between the initial and final value in each gustatory test. Our findings were:

1. There was a significant difference between the zinc picolinate and placebo groups in the degree of total subjective recovery ($P < 0.05$, Wilcoxon test).
2. There was no significant difference between the two groups in the grade of improvement in four basic taste qualities (sweet, salty, sour, bitter) in the whole mouth method (Wilcoxon test).
3. There was no significant difference between the zinc picolinate and placebo groups in the grade of improvement of the four taste qualities in the FPD method and EGM at the chorda tympani and glossopharyngeal nerve area (Wilcoxon test).

Conclusion

Thus far, the patients who have received zinc picolinate have shown no significant improvement. However, we will continue the study.

References

1. Yoshida S, Endo S, Tomita H (1990) A double-blind study of the therapeutic efficacy of zinc gluconate on taste disorder. Auris Nasus Larynx 18:153–161
2. Barrie SA, Wright JV, Pizzorno JE, Kutter E, Barron PC (1987) Comparative absorption of zinc picolinate, zinc citrate, and zinc gluconate in humans. Agents Actions 21:223–228

[1] Department of Otolaryngology, Nihon University School of Medicine, 30-1 Oyaguchi-kamicho, Itabashi-ku, Tokyo, 173 Japan

Xanthurenic Acid and Zinc

Yukio Shibata, Takao Ohta, Masahiro Nakatsuka, and Yazo Kotake[1]

Key words. Triptophan—Vitamin B_6—Xanthurenic acid—Zinc—Taurine—Methionine

Prof. H. Tomita of the Department of Otorhinolaryngology, Nihon University, has been carrying out experiments to determine the relationship between Zn^{2+} and taste abnormality in diabetes mellitus. Normal taste capacity was recovered by the administration of $ZnSO_4$ to these patients. K. Okamoto of Kinki University [1] discovered oxine (8-hydroxy quinoline) diabetes; its mechanism might be due to chelation with Zn^{2+}.

Furthermore, Kotake, Inada, and Matsumura [2] have performed experiments to determine the relationship between xanthurenic acid (4,8-dihydroxyquinoline 2-carboxylic acid; XA) and diabetes. H. Okamoto of Tohoku University [3] has since investigated the inhibition of proinsulin synthesis by quinoline compounds, using isolated pancreas cells.

Against the background of these experiments, we have carried out the following investigations.

Change of Blood Sugar in Diabetic Rats

Blood sugar was increased by the injection of quinoline compounds only and not by isoquinoline compound. The $-OH$ group in the 8-position of the quinoline ring was active, but the $-OCH_3$ group was not. The $-COOH$ group in the 2-position of the quinoline ring was active, but the $-COOC_2H_5$ group was not. The N atom in the quinoline ring was also active, but N-oxide was not.

These results suggest that the quinoline compounds, being able to chelate with Zn^{2+}, were important for the change of blood sugar.

Kynureninase Activity

Xanthurenic acid is metabolized from 3-OH-kynurenine, so it depends on liver kynureninase (LKA) activity. LKA decreased in vitamin B_6-deficient rats and also in streptozotocin diabetic rats. The importance of the $-SH$ group on the LKA

apoenzyme was reported by Takeuchi and Izuta [4,5], and the active center of the kynureninase apoenzyme was almost completely clarified (Asp-Phe-Ala-Cys-Trp-Cys-Ser-Tyr-Lys-Tyr).

XA Formation Decreased by Taurine Administration

Taurine was administered to rats at a concentration of 4.6% per 150 g of standard diet. With excess methionine (Met) administration, much XA excretion from tryptophan (Trp) was observed, but with excess taurine administration, XA excretion was not increased.

Chelation of XA and Zn^{2+}

Many experiments indicate that XA might be chelated with Zn^{2+} (Kotake [6], Murakami [7], Ogasawara [8], Shiraishi [9], and von Butenant [10]). Thus, we concluded that the administration of Zn^{2+} and taurine was important for the recovery of normal taste in diabetes mellitus patients.

References

1. Okamoto K, Kadota I (1949) Acta scholae med Univ Kioto 27:43
2. Kotake Y, Inada T, Matsumura Y (1954) J Biochem 42:355
3. Okamoto H (1975) Acta Vitamin Enzymol 29:227–231
4. Takeuchi F, Izuta S, Tsubouchi R, Shibata Y (1991) J Nutri 1366–1373
5. Takeuchi F, Tsubouchi R, Nakatsuka M, Shibata Y (1992) Advances in Tryptophan Research 1992 (Proceedings Seventh Meeting of the International Study Group for Tryptophan Research), pp 435–438
6. Kotake Y, Sato Y, Shibata Y (1990) Proceedings of 24th Japanese Symposium Taste and Smell. Tsu, Mie, Japan, p 33
7. Murakami E (1975) Acta Vitamin Enzymol 29:240–242
8. Ogasawara M, Hagino Y, Kotake Y (1962) J Biochem 52:162–166
9. Shiraishi S, Ohta T, Fuse H, Yamada S (1988) Trace Nutrients Research, no. 4, pp 127–132
10. Butenandt A (1955) Lecture in Wakayama Medical College

[1] Department of Biochemistry, Aichi Medical University, Ooaz Yazako, Nagakute-cho, Aichi-gun, Aichi, 480-11 Japan

Clinical Aspects of Phantogeusia

A. Ikui, M. Ikeda, I. Tahara, I. Ito, S. Endo, and H. Tomita[1]

Key words. Phantogeusia—Spontaneous dysgeusia

The most common complaint of patient with taste disturbance is usually hypogeusia or ageusia. However, in some cases, patients complain of an abnormal taste sensation which occurs spontaneously, for example, a bitter or astringent taste. The authors named these symptoms phantogeusia and spontaneous dysgeusia, or nonstimulated (no oral stimulus required to induce the sensation) dysgeusia, and clinically analyzed 132 patients with these symptoms. The patients were selected from 1000 patients who visited the ENT Clinic of Nihon University Hospital between 1976 and 1983.

The most common taste complained of by the patients was bitterness ($n = 53$; 40%) and the second most common was saltiness ($n = 31$; 23%). Some patients complained of an astringent taste ($n = 17$; 13%), of acidity ($n = 15$; 11%), or of sweetness ($n = 10$; 8%).

Two-thirds of the patients who complained of bitterness were female; however, there were no sex differences in the patients who complained of the other tastes. More than half of the patients who complained of bitterness were in their fifties or sixties. None of the patients who complained of sweetness were older than their sixties.

The most common etiology of these symptoms seemed to be drug-related ($n = 38$; 29%), and insufficiency of zinc was considered to be the second most common cause of this abnormality ($n = 28$; 21%). General internal diseases or local conditions were other causes of these symptoms.

The taste function was examined by the filter paper disc method. Bitterness was examined with quinine hydrochloride. There were many patients with normal response in both the areas innervated by the chorda tympani nerve and those innervated by the glossopharyngeal nerve. Saltiness was examined with NaCl. Many patients showed no response in the areas innervated by the chorda tympani and the glossopharyngeal nerves. Acidity was examined with tartaric acid. Many patients showed a good response to areas innervated by the chorda tympani and the glossopharyngeal nerves. Sweetness was examined with sucrose; there were patients with a poor response in the areas innervated by the chorda tympani and glossopharyngeal nerves.

Ninety-one patients with taste disturbance were selected from among 500 of the original 1000 patients who had visited the ENT Clinic of Nihon University Hospital between 1981 and 1983. Blood urea nitrogen (BUN), Na, K, Cl, Ca, P, and Mg were examined in the saliva of those patients and in normal controls. The amount of magnesium in saliva in the patients differed significantly from that in the controls; however, no difference was observed for the other components. Twenty of the 91 patients 22% had reduced serum zinc levels.

The patients who showed a normal response to the taste function test were not treated. One-third of the patients ($n = 32$; 35%) were treated with zinc sulfate and of these, 22 (69%) had a good prognosis (no symptoms of phantogeusia or spontaneous dysgeusia after treatment).

[1] Department of Otolaryngology, Nihon University School of Medicine, 30-1 Ooyaguchi, Itabashi-ku, Tokyo, 173 Japan

Effect of Muscle Relaxation on Chronic Pain Patients Who Have Taste Dysfunction

Iwao Saito[1], Satoshi Okuse[2], Nobuyoshi Yashiro[2], Yoshio Inamori[3], and Yasuko Saito[4]

Key words. Taste dysfunction—Muscle relaxation

Introduction

Taste dysfunction is not rare in chronic pain patients. In the treatment of chronic pain, we employ multidisciplinary interventions such as pharmacological treatment, relaxation techniques such as autogenic training and biofeedback training (BFT), and soft massage for home practice. We noted that complex symptoms of the eyes, ears, tongue and nose improved on treatment for chronic pain. It seemed that relaxation training on tongue-related muscles improved the capacity to taste sweetness. This study was performed to examine whether muscle relaxation had any subjective or physiological effects on taste dysfunction.

Patients

The subjects were 15 patients (5 male and 10 female; aged between 23 and 71 years, mean 42 years) who suffered from chronic muscle contraction headache.

Methods of Muscle Relaxation

Thirteen patients practiced soft massage on tongue-related muscles and 2 patients practiced electromyogram biofeedback training (EMG-BFT).

Soft Massage

Patients were encouraged to massage softly or stroke on the depth of the jaw where the tongue muscle group was located. They had learned muscle relaxation. Using the tips or the backs of their fingers, the patients lightly stroked the surface of the target muscle, stroking slowly to just move the skin for 3 min.

EMG-Biofeedback Training

For the EMG-biofeedback training, we used apparatus P-303 (Cyborg, Illinois, U.S.A.) and a computerized UT-201 (Unique Medical, Tokyo, Japan). The EMG level was monitored from the skin surface under which the glossal muscles were located.

Evaluation of Sweetness

Patients were asked to evaluate the intensity of sweetness on a scale of 0 to 100 when a therapist touched their tongue with a candy. Each touch lasted 3 s. The examination was done at four locations: the tip, right and left sides of the middle, and the center of the tongue. The procedure was: (1) evaluation of the intensity of sweetness, (2) muscle relaxation (soft massage or EMG-BFT), and (3) reevaluation of the intensity of sweetness. The results was examined with a paired t-test.

Results

The mean sweetness values for 13 patients are shown in Table 1. Two patients showed better sensation of sweetness in the tip and the left side of the tongue with EMG-BFT. Five of the total 15 patients showed improvements in taste and improved anorexic complaints as assessed clinically after the hospital and home training.

Discussion and Summary

In painful or tight muscles, particularly in myodolor, the intramuscular circulation becomes sluggish, and plasma lactic acid, bradykinin, and prostaglandin E_2 are increased due to anaerobic metabolism; noxious effects are thus engendered. Muscle relaxation training, however, helps to improve intramuscular circulation, to expel pain-inducing substances [1,2], and to ameliorate emotional state [5]. In a previous study, we found that soft massage of the face en-

[1] Muroran Institute of Technology, 27-1 Mizumoto-cho, Muroran, Hokkaido, 050 Japan
[2] Sapporo Meiwa Hospital, 1-10 Tsukisamu-nishi, Toyohira-ku, Sapporo, 062 Japan
[3] Hyogo Medical College, 1-1 Mukogawa-cho, Nishinomiya, Hyogo, 663 Japan
[4] Hokkaido University School of Medicine, W7 N15 Kitaku, Sapporo, 060 Japan

582   I. Saito et al.

TABLE 1. Mean sweetness values.

	Tip	L-middle	R-middle	Center
Before soft massage	34.7 ± 7.1	34.2 ± 7.3	29.9 ± 6.8	19.0 ± 6.1
After soft massage	50.3 ± 7.1	39.3 ± 7.1	40.0 ± 7.6	23.4 ± 6.4
	(P < 0.01)	NS	(P < 0.05)	NS

hanced positive subjective feelings and decreased
EMG values in the shoulder [3,4]. Although the
sensation of sweet taste has been discussed mainly in
relation to the central nervous system, this simple
intervention suggests that evaluation of sweetness
also seems to be influenced by peripheral conditions
such as excessive muscle tension.

References

1. Saito I, Takaoka K (1993) Facial massage for stress
   control. In: Institute of Shiseido Beauty Sciences (ed)
   The psychology of cosmetic behavior. Fragrance
   Journal (Tokyo), pp 365–389
2. Saito I (1990) A soft massage for chronic pain and
   discomfort. Workshop presented at 21st annual meeting
   of Association for Applied Psychophysiology and
   Biofeedback (AAPB): AAPB tape collection, No.
   90-8, Washington, DC
3. Mott A, Mills L, Pepper E (1992) Muscle tension,
   massage, and EMG: A comparison of pre- and post-
   intervention measures. Biofeedback Self Regul 17:333
4. Saito I (1990) Does EMG-biofeedback training have
   some cardiovascular effect in the treatment of chronic
   pain patients? In: AAPB (ed) Proceedings of 21st annual meeting of AAPB
5. Tassinary LG, Ccioppo JT, Geen TR (1989) A psychometric study of surface electrode placements for
   facial electromyographic recording. Psychophysiology
   26:1–16

Reinnervation After Severance of the Chorda Tympani During Middle Ear Surgery

Takehisa Saito, Yasuhiro Manabe, Takechiyo Yamada, and Hitoshi Saito[1]

Key words. Chorda tympani—Reinnervation

The chorda tympani, which crosses the tympanic cavity of the middle ear, must be severed in some cases of middle ear surgery. However, it is not clear whether the chorda tympani becomes reinnervated after being severed. We followed 22 patients, whose recovery courses were evaluated by electrogustometry (EGM), for several years after surgery. The chorda tympanis of these patients were cut during middle ear surgery, and the separated nerves were not approximated to each other. Conditions necessary for reinnervation were investigated.

EGM was performed preoperatively, and then postoperatively at intervals for 2–5 years. At the final EGM examination the findings of the fungiform papillae of the tongue and ear drum were observed under an operating microscope. The patients were classified into three groups according to course of recovery of electrogustation. The thresholds of the first group ($n = 7$) recovered within 2 years; those in the second group ($n = 6$) did not recover within 2 years, but recovered thereafter, while those of the third group ($n = 9$) showed no recovery. Finally, seven patients showed complete recovery, six showed incomplete recovery, and nine showed responses beyond the scale. On the other hand, after surgery, 35% of the patients complained of subjective gustatory disorders. However, none of the patients complained of any disorders during the long follow-up period. The fungiform papillae of the tongue were observed in order to indirectly assess whether the chorda tympani had been reinnervated. In the complete recovery group, evaluated by EGM, two patients showed complete regeneration of the fungiform papillae, three showed incomplete regeneration, and two had white papillae. In the incomplete recovery group, five patients showed incomplete

regeneration and one had white papillae. In the nine patients with responses beyond the scale, only one patients showed incomplete regeneration and the others had either white papillae or no fungiform papillae. Thus, there was a correlation between the recovery demonstrated by EGM and the regeneration of fungiform papillae. In the incomplete recovery group, as determined by EGM, there were two patients whose reinnervated chorda tympani were detected inside the ear drum during the second surgery. The periods between their first and second surgeries were 7 years and 1 year and 2 months. After both second surgeries, the thresholds of the EGM were elevated and responses were beyond the scale. However, the responses gradually recovered and reached the preoperative thresholds 2 and 4 months later, respectively. These findings indicate that the nerve had the potential to be reinnervated in some of the patients whose chorda tympani was cut.

As for the conditions necessary for reinnervation of the chorda tympani, two factors are considered likely. One is the regrowth of connective tissue such as granulation or scar around the tympanic isthmus where the chorda tympani passes. This also leads to retraction of the attic. The other is a thickening of the reconstracted ear drum. We suspect that these conditions play important roles as contact guidance which leads to reinnervation of the nerve.

We also investigated the correlation between the findings of the ear drum and the electrogustometrical results. Thickening of the ear drum was observed in all patients. On the other hand, postoperative attic retraction due to stenosis of the tympanic isthmus was also observed in all groups classified by EGM. Therefore, these two factors are not sufficient for promoting the regeneration of the nerve. That is to say, the direction of the remaining part of the nerve after being severed and the distance between its central and peripheral ends are important factors for reinnervation. If the cut ends are approximated to each other during surgery, the probability of reinnervation might increase.

[1] Department of Otolaryngology, Fukui Medical School, Matsuoka-cho, Yoshida, Fukui, 910-11 Japan

12. Clinical Assessment of Olfaction

Analysis of Neurogenesis and Aging of Olfactory Epithelium in the Rat

YASUYUKI KIMURA[1]

Key words. Immunohistochemistry—N-CAM—MAP2—MAP5—Aging—Neurogenesis

Introduction

Among nerve cells of vertebrates, the olfactory elements are uncommon in their capacity to turn over through life. However, little is known about the initial stage of neurogenesis, maturation, and aging of olfactory cells. It has become possible to trace the neurogenesis by staining using several antibodies. The purpose of this study is to investigate the olfactory epithelium of embryonic, neonatal, adult, and aged rats systematically using immunohistochemical methods.

Materials and Methods

Animals

Wistar rats were used. The ages of the animals were embryonic day 19, postnatal day 3, postnatal 12 weeks as adult rats, and postnatal 2 years as aged rats.

Methods

Under pentobarbitone anesthesia, the animals were perfused through the left ventricle with Zamboni's solution. The mucosa of the nasal septum was resected and then cryoprotected in 25% sucrose solution. The specimen was rapidly frozen in Tissue-Tek O.C.T. (Lab-Tek division, Miles Lab., Elkhart, IN, USA) with liquid nitrogen and cut in 10 μm thick axial sections on a cryostat. Anti-N-CAM (neural cell adhesion molecule, ×400, Chemicon, Temecula, CA, USA), anti-MAP5 (microtubule-associated protein 5, ×400, Chemicon) and anti-MAP2 (×400, BioMakor, Rehovot, Israel) antibodies were used as the primary antibody for 24 h at room temperature. Next, the specimens were incubated for 1 h with the HRP conjugated secondary antibodies (rat HRP-Ig

for detecting N-CAM, mouse HRP-Ig for MAP5 and MAP2). DAB was used as chromogen.

N-CAM

Neural cell adhesion molecules (N-CAMs), which are classified as cell surface molecules, have been implicated in cell interactions that underlie formation of the nervous system. N-CAM plays a role in fasciculation of neurites and is important in the directed growth of fascicles toward nerve growth factor [1,2].

MAP

Microtubules, the most abundant cytoskeletal elements found in neuronal processes, are composed of tubulin as well as a variety of microtubule-associated proteins (MAPs). MAPs are related to polymerization of tubulin to microtubules. MAP5 is important in modulating microtubule function during the formation of neuronal processes, and MAP2 is important in stabilizing microtubules [3,4].

Results

Adult Rat

In the adult rat, N-CAM immunopositive staining was found exclusively in the basal layer of the epithelium and axons in the lamina propria. The formations of the fascicle were clearly seen (Fig. 1). MAP5 immunopositive staining was also found in the basal layer of the epithelium, and the middle layer was weakly stained. The axons in the lamina propria were moderately stained (Fig. 2). In contrast, MAP2 immunopositive staining was observed in the middle layer of the epithelium and was not found in the basal layer, supporting cell layer, or axons (Fig. 3).

Embryonic Rat

N-CAM and MAP2 were not detectable in the olfactory mucosa of the embryonic rat. However, olfactory axons that arise from the olfactory bulb, proceed through the lamina cribrosa, and separate in the

[1] Department of Otolaryngology, School of Medicine, Kanazawa University, Takara-machi 13-1, Kanazawa, Japan 920

FIG. 1. Tissue section of olfactory mucosa from adult rat, stained with anti-N-CAM antibody. Staining was seen exclusively in the basal layer of the epithelium and axons in the lamina propria. ×200

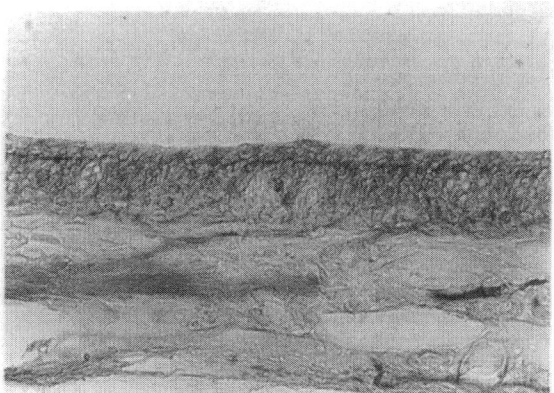

FIG. 4. Section from neonatal rat, stained with anti-MAP5 antibody. Staining was seen in the axons and several layers closer to the surface of the epithelium. ×200

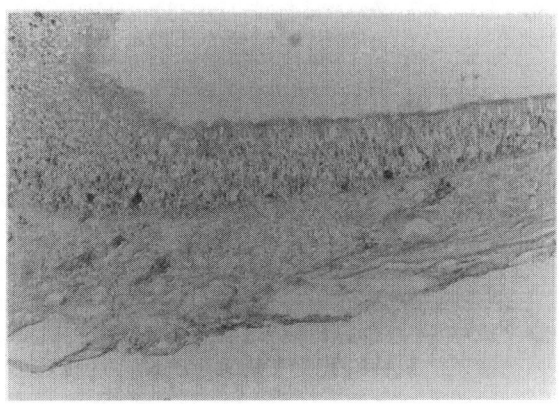

FIG. 2. Section from adult rat, stained with anti-MAP5 antibody. Staining was seen in the basal layer of the epithelium and in the lamina propria. The middle layer of the epithelium was also weakly stained. ×200

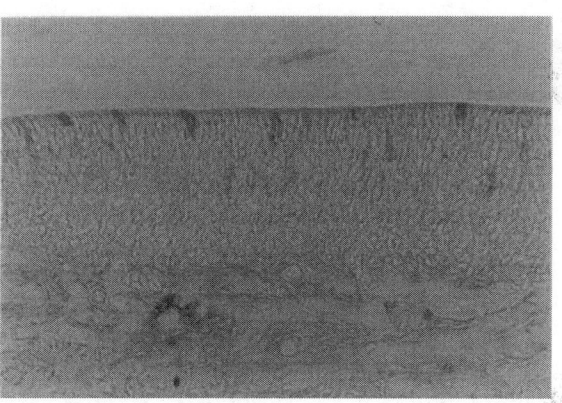

FIG. 5. Section from neonatal rat, stained with anti-MAP2 antibody. Because of its flask-like shape, these cells might be the so-called microvillar cells. ×400

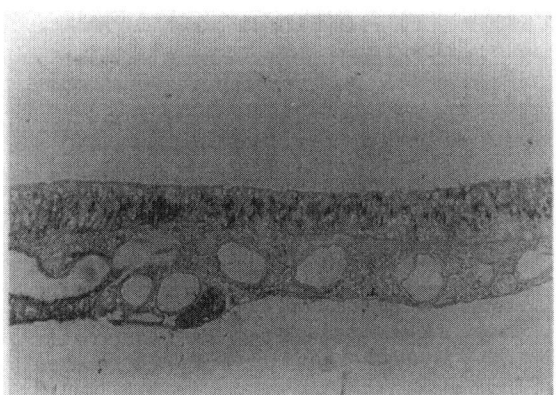

FIG. 3. Section from adult rat, stained with anti-MAP2 antibody. Staining was observed in the middle layer of the epithelium but not in the basal or supporting cell layer. ×200

lamina propria were labeled by the antibody to MAP5. In the epithelium, the staining was located in several layers closer to the free surface.

Neonatal Rat

In the postnatal day 3 rat, N-CAM-immunopositive staining was observed in the lamina propria but not in the epithelium. In the case of MAP5, positive staining was located in the olfactory cell layer of the epithelium, occupying a position closer to the epithelial surface. The axons in and under the epithelium were also stained (Fig. 4). MAP2 was detectable only near the free surface of the epithelium, and its axons extended downward. Because of its flask-like shape, these cells might be the so-called microvillar cells (Fig. 5).

TABLE 1. Staining grades for each antibody.

Stain	Grade in various tissues			
	Epithelium			
	Basal	Middle	Superficial	Axon
N-CAM				
E19	–	–	–	–
P3	–	–	–	±
Adult	++	–	–	++
Aged	+	–	–	+
MAP5				
E19	–	–	++	++
P3	–	–	++	++
Adult	++	+	–	+
Aged	–	–	–	+
MAP2				
E19	–	–	–	–
P3	–	–	+	–
Adult	–	++	–	–
Aged	–	++	–	–

++, strong; +, moderate; ±, weak; –, negative.

Aged Rat

Weak immunoreactivity to anti-N-CAM was found in the basal layer of the epithelium and in axons in the lamina propria of the aged rat. The axons immunopositive to anti-N-CAM were found in the epithelium in places. This finding was not observed in embryonic, neonatal, or adult rats. MAP5 immunopositive staining was observed in the axons under the epithelium. In contrast, no components in the epithelium reacted to MAP5. Immunoreactivity to anti-MAP2 was almost the same as in the adult rat. Positive stainings were found in the middle layer of the epithelium.

Discussion

In the case of MAP5, the staining in the epithelium was different between adult and neonatal rats. In the adult rats, staining was found mainly in the basal layer. In the neonatal and embryonic rats, positive staining was observed in the layers closer to the surface. In neonatal rats, the cells that differentiated during the embryonic period emigrate toward the free surface. The grade of maturation of MAP5-positive cells is thought to be low. I consider that the cells in the basal and middle layers were less de-

veloped than the MAP5-positive cells, although the morphologic structure appeared almost fully developed, as the functioning of the neuroepithelium was immature.

Table 1 shows the grades for positive staining. The antibodies I used are related to neurogenesis. However, there is a subtle difference among them in terms of the times of appearance and their role. It is said that MAP5 appears earlier than MAP2 and appears during the initial phase when a neuron grows its neuronal processes [5]. In the adult rats, the cells in the basal layer were positive to anti-MAP5 and N-CAM but did not react to anti-MAP2 antibody. Therefore the cells in the basal layer were thought to be immature. I conclude that it is possible to observe the process of differentiation and aging using the method of functional morphology staining with these monoclonal antibodies.

The immunoreactivity to anti-MAP2 in the aged rats was similar to that in the adult rats. Thus I speculate that aged rats may have maintained their olfactory acuity. However, axons in the epithelium were found in places that indicated an age-related change. Furthermore, in the epithelium, the immunoreactivity to anti-N-CAM was weak, and no reactivity to anti-MAP5 was found. Therefore the olfactory epithelium of the aged rats may have maintained its function, but the supply of new neurons was not as plentiful as in the younger rats.

Acknowledgments. This reserch was supported by the fund for medical treatment of the elderly, School of Medicine, Kanazawa University, 1992.

References

1. Ederman GM, Chuong CM (1982) Embryonic to adult conversion of neural cell adhesion molecules in normal and staggerer mice. Proc Natl Acad Sci USA 79: 7036–7040
2. Chung WW, Lagenaur CF, Yan Y, Lund JS (1991) Developmental expression of neural cell adhesion molecules in the mouse neocortex and olfactory bulb. J Comp Neurol 314:290–305
3. Matus A (1988) Microtubule-associated proteins: their potential role in determining neuronal morphology. Annu Rev Neurosci 11:29–44
4. Tucker RP, Binder LI, Matus AI (1988) Neuronal microtubule-associated proteins in the embryonic avian spinal cord. J Comp Neurol 271:44–55
5. Riederer B, Cohen R, Matus A (1986) MAP5: a novel brain microtubule-associated protein under strong developmental regulation. J Neurocytol 15:763–775

Pathology of Olfactory Mucosa in Patients with Age-Related Olfactory Disturbance and Dementia

Masuo Yamagishi[1], Yoichi Ishizuka[1], and Kohji Seki[2]

Key words. Alzheimer's disease—Age-related olfactory disturbance—Olfactory mucosal biopsy—Immunohistochemistry—Tau protein—Amyloid-β protein

Introduction

Two types of sensorineural olfactory loss have been recognized in elderly patients. One is an olfactory disturbance that occurs in normal elderly persons, with olfactory loss the only major complaint [1]. The other type of olfactory loss is combined with a neurodegenerative disease such as Alzheimer's disease (AD) or Parkinson's disease [2,3]. The mechanisms underlying these disorders have not yet been elucidated, and the pathologic changes in the olfactory pathway remain to be clarified.

In this study, we have examined the pathologic changes in the peripheral olfactory mucosa in patients with these disorders using the olfactory mucosal biopsy. The possibility of making a definitive diagnosis of AD by olfactory mucosal biopsy was also explored.

Materials and Methods

Olfactory mucosa was obtained from six patients who had been clinically diagnosed as having AD and three patients with olfactory disturbance due to aging. One of the AD patients was diagnosed as having familial AD. In each case, two samples were obtained from the tegmen of the olfactory cleft without anesthesia using the special biopsy instruments we routinely use in our smell clinic. Consent was obtained from the AD patients' families because all six were severely demented. Control specimens were obtained from three age-matched patients who had olfactory disturbance due to chronic sinusitis

and upper respiratory viral infection, or depression without dementia.

The routine streptavidin-biotin peroxidase method was used for immunohistochemistry. Specimens for amyloid-β protein-immunostaining were pretreated with formic acid for 5 min to enhance immunoreactivity. Rabbit anti-polyclonal tau protein (p-Tau, 1:1500) antiserum, mouse anti-monoclonal tau protein (m-Tau, 1:1000) antibody, rabbit anti-polyclonal ubiquitin (p-Ubq, 1:150) antiserum, and mouse anti-monoclonal amyloid-β protein (m-ABP, 1:100) antibody were used. p-Tau and p-Ubq were used for markers of neurofibrillary tangles, and m-Tau was used that of normal tau protein localization. M-ABP was used to identify senile plaque. P-tau antiserum was a gift from Dr. Ihara (Brain Institute, Tokyo University), m-Tau antiserum was obtained from Sigma Immunochemicals (St. Louis, USA), and p-Ubq and m-ABP antisera were obtained from Dakopatts (Glostrup, Denmark).

Results

Alzheimer's Disease

p-Tau Localization

Three kinds of p-Tau-immunoreactive patterns were observed in the olfactory epithelium of clinically typical AD cases. In five AD cases, p-Tau immunoreactivity was seen in the dendrite and perikarya of the olfactory receptor cells (Fig. 1A). In one case, immunoreactivity was restricted to the dendrite. In addition to these changes, in two cases an extracellular mass was immunoreactive in the epithelium (Fig. 1B). In the lamina propria, which was contained in four cases, olfactory nerve bundles also displayed immunoreactivity for p-Tau in most cases.

m-Tau Localization

m-Tau immunoreactivity was observed only in the dendrite of the olfactory receptor cells in the epithelium. In the lamina propria, m-Tau accumulations were also observed in the olfactory nerve bundles in three cases. In a specimen from the familial AD case, an extracellular mass displayed immunoreactivity for m-Tau.

[1] Department of Otolaryngology, Mizonoguchi Hospital, Teikyo University School of Medicine, Mizonoguchi 74, Takatu-ku, Kawasaki, Kanagawa, 213 Japan
[2] Clinic of Neurologic Disease, Mishima Hospital, Mishima-machi, Mishima-gun, Niigata, 940-23 Japan

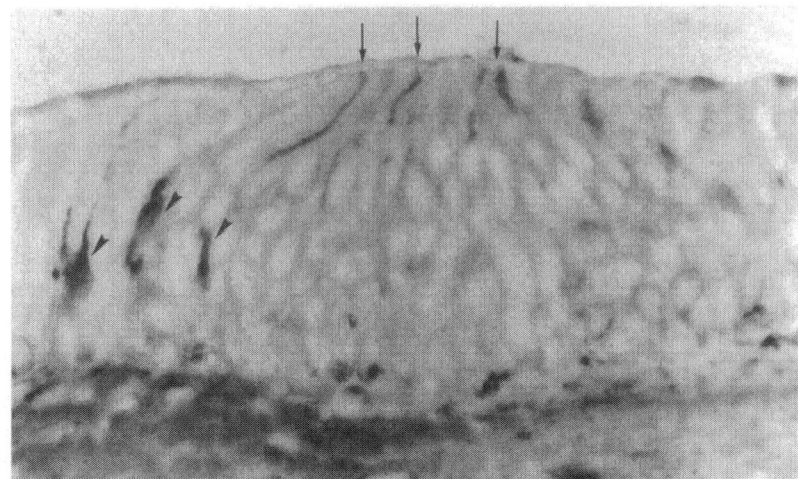

FIG. 1. AD olfactory epithelium stained with p-Tau antiserum. **A** p-Tau immunoreactivity is seen in the dendrite (*arrows*) and perikarya (*arrowheads*) of the olfactory receptor cells. ×400. **B** Abnormal extracellular mass (*arrow*) in the epithelium reacts to p-Tau antiserum. ×200

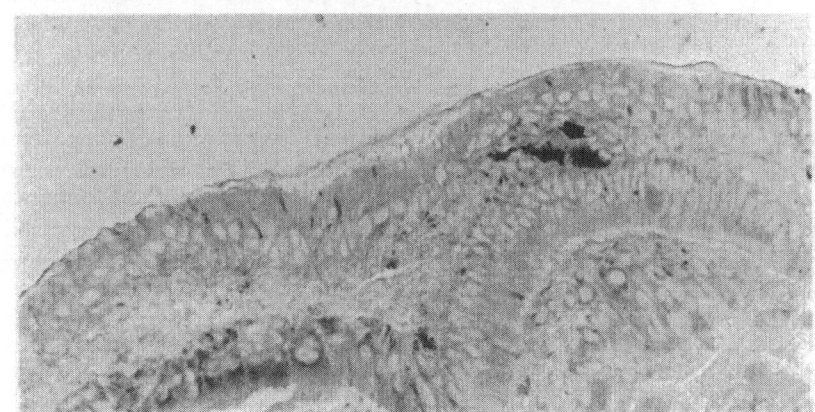

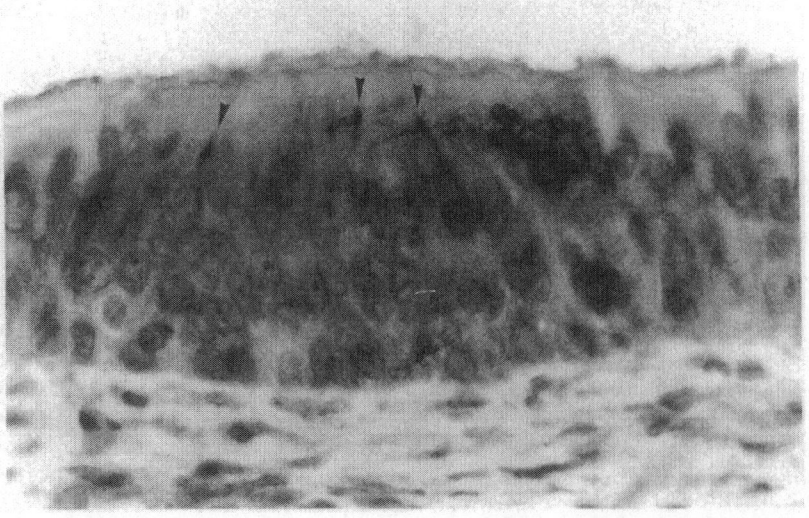

FIG. 2. p-Ubq immunoreactivity in the olfactory mucosa. Diffuse cytoplasmic immunoreaction appears in the olfactory receptor cells, especially strongly in the dendrites (*arrowheads*). ×400

p-Ubq Localization

Diffuse p-Ubq immunoreactivity was found throughout the olfactory receptor cells weakly, with strong localization in the dendrite (Fig. 2).

m-ABP Localization

The extracellular mass, which was observed in the familiar AD case, reacted weakly to anti-ABP antiserum. No immunoreactivity could be seen in other specimens.

TABLE 1. Immunoreactivity of anti-Tau and amyloid β-protein antibodies.

No	Age/Sex	Clinical diagnosis	p-Tau				m-Tau				m-Ubq				m-ABP			
			DR	PK	ON	EM	DR	PK	ON	EM	DR	PK	ON	EM	DR	PK	ON	EM
1	85/M	AD	+	−	/*	+	+	−	/	−	±	±	/	−	−	−	/	−
2[+]	58/F	AD	+	+	/	+	−	−	/	+	+	−	/	−	−	−	/	±
3	75/F	AD	±	+	−	/**	−	−	−	/	−	−	−	/	−	−	−	/
4	87/F	AD	+	+	+	/	+	−	+	/	−	−	−	/	−	−	−	/
5	63/F	AD	+	+	+	/	−	−	+	/	+	−	−	/	−	−	−	/
6	73/F	AD	+	+	±	/	−	−	+	/	+	±	−	/	−	−	−	/
7	58/F	Anosmia (Aging)	Squamous metaplasia															
8	80/M	Anosmia (Aging)	−	−	+	/	−	−	−	/	−	−	−	/	−	−	−	/
9	58/M	Anosmia (Aging)	+	+	+	/	−	+	+	/	−	−	−	/	−	−	−	/
10	70/M	Hyposmia (Sinusitis)	±	−	−	/	−	−	−	/	−	−	−	/	−	−	−	/
11	72/M	Hyposmia (Viral)	−	−	−	/	−	−	−	/	−	−	−	/	−	−	−	/
12	56/M	Dysosmia (Depression)	−	−	+	/	−	−	−	/	−	−	−	/	−	−	−	/

DR, dendrite; PK, perikarya; ON, olfactory nerve; EM, extracellular mass.
[+] Familiar Alzheimer's disease; * lamina propria was not contained in the specimen; ** extracellular mass was not found in the specimen.

Olfactory Disturbance Due to Aging

Of the patients with olfactory disturbance due to aging, one was found to have squamous metaplasia of the neuroepithelium. In two other cases the basic nuclear arrangement was retained, but the area of the olfactory epithelium was reduced and nuclei of the receptor cells had decreased. P-Tau immunoreactivity was found in the dendrite and perikarya in only one case and in the olfactory nerve bundles in two cases.

Controls

In age-matched control patients without dementia, p-Tau immunoreactivity was found in the dendrite and in the olfactory nerve in each case. No p-Ubq or m-ABP immunoreactivity was found in the olfactory mucosa.

Detailed immunostaining data are shown in Table 1.

Discussion

As noted by Waldton [2], it is well known that the sense of smell is compromised in AD patients. This loss is present early in the disease process and is considered to be caused by disruption of the central olfactory pathway due to the neuritic plaques and neurofibrillary tangles that are sometimes found in disproportionate numbers in the olfactory bulb and olfactory cortex [4]. In 1989 a group of US neuro-

logists, Talamo et al., reported that dystrophic olfactory neurites, pathologic changes characteristic of AD, were present in the olfactory mucosa of autopsy cases [5]. In the study of Talamo et al., the olfactory receptor cells were decreased in some areas and showed degeneration, and the phosphorylated neurofilament protein heavy subunit and Alz50 were identified by immunohistochemical staining. On the other hand, Trojanowski et al. reported that dystrophic neurites, which were immunostained by neurofilament protein, MAP5, synaptophysin, and peripherin, appeared in the olfactory epithelium not only in AD but also in Parkinson's disease and other neurodegenerative disorders, and even in normal elderly controls. They concluded that detection of dystrophic olfactory neurites with the antibodies described in their study was unlikely to be useful for discriminating AD from other neurodegenerative diseases [6].

In our examination, p-Tau immunoreactivity of the dendrite in the olfactory receptor cells and olfactory nerve bundles was observed not only in AD cases but also in controls. These findings were not characteristic in the AD patients. Accumulation of abnormal tau protein in the dendrite and olfactory nerve bundles may occur in elderly persons, as can normal tau protein. On the other hand, immunoreactivity of perikarya in the receptor cells was observed in most of the biopsied mucosal specimens from AD patients but in only one of the anosmia cases due to aging; control cases did not display such immunoreactivity. This unusual neurite mass is assumed to be analogous to the neurofibrillary tangles in the AD brain, all of which stain for p-Tau; and it is characteristic in AD olfactory mucosa. Furthermore p-Ubq immunoreactivity was seen in only AD cases. Ubiquitin has been shown immunohistochemically to coexist with abnormal tau protein in the neurofibrillary tangles. This fact proves the presence of abnormal tau protein indirectly. In certain cases of AD, an extracellular mass was observed; and it is suspected to be senile plaque, though it did not react to p-ABP antibody. From these results, it became clear that pathologic changes appeared in the peripheral olfactory mucosa in AD and they played some part in the olfactory disturbance in AD.

Next, the idea is conceived that olfactory mucosal biopsy may be useful for making a definitive diagnosis of AD. AD is a central nervous system (CNS) disorder characterized by the presence of neuritic plaques and neurofibrillary tangles in susceptible areas of the brain [7]. To diagnose this disease definitively, it is necessary to examine the patient's brain tissue histopathologically. Brain biopsy is impossible during life and is performed only at autopsy. However, if it is possible to obtain specimens containing neuronal tissue, and if these specimens show degenerative findings closely related to intracranial pathologic changes, it will facilitate the definitive diagnosis of AD. Our examination indicates that it may be possible to diagnose AD definitively using the immunoreactivity of p-Tau and p-Ubq antisera. Although the olfactory mucosa in patients with other forms of dementia that accompany neurodegenerative diseases has not yet been examined, the olfactory mucosal biopsy, which can be easily and safely performed by otolaryngologists, is useful for evaluating pathologic changes in the brain and facilitating a definitive diagnosis of AD.

In contrast to AD olfactory mucosa, neurofibrillary tangle-like dystrophic neurites were found in only one specimen from patients with olfactory disturbance due to aging. In these patients, only decreased olfactory receptor cells and squamous metaplasia were found. With this type of olfactory disturbance, it is suspected that impaired neurogenesis or misdirection of the differentiation of olfactory receptor cells takes place, although in animals renewal of receptor cells occurs quickly and continuously.

References

1. Schiffman S (1977) Food recognition by the elderly. J Gerontol 32:586–592
2. Waldton S (1975) Clinical observations of impaired cranial nerve function in senile dementia. Acta Psychiatry Scand 50:539–547
3. Anisari KA, Johnson A (1975) Olfactory function in patients with Parkinson's disease. J Chronic Dis 28:493–497
4. Esiri M, Wilcock GK (1984) The olfactory bulbs in Alzheimer's disease. J Neurol Neurosurg Psychiatry 47:56–60
5. Talamo BR, Rudel RA, Kosik KS, et al. (1989) Pathological changes in olfactory neurons in patients with Alzheimer's disease. Nature 337:736–739
6. Trojanowski JQ, Newman PD, Hill WD, Lee VMY (1991) Human olfactory epithelium in normal aging, Alzheimer's disease, and other neurodegenerative disorders. J Comp Neurol 310:365–376
7. Alzheimer A (1911) Über eigenartige Krankheitsfälle des späteren Alters. Z Ges Neurol Psychiatry 4:356–385

Rise and Fall of Olfactory Competence Through Life

WILLIAM S. CAIN and JOSEPH C. STEVENS[1]

Key words. Olfaction—Threshold—Odor identification—Aging—Children—Psychophysics

Introduction

Human olfactory functioning declines during adulthood, a matter that is reflected in virtually all measures, including absolute sensitivity, quality discrimination, perceived intensity, and so on. The decline apparently occurs during the course of normal aging. The focus on changes that occur during adulthood, however, has left any changes that occur before then poorly explored. The functioning of the child theoretically has as much relevance to understanding chemosensory aging as does the functioning of the middle-aged or elderly subject. This investigation extended the range of measured functioning to children as well as to adults from young to old.

Of the tasks used to assess olfactory functioning in the laboratory or clinic, some, especially that of detection, are simple in that they require relatively limited cognitive and verbal skill from the subject. Others, such as odor identification, require experience with the olfactory world [1,2]. For exploration of olfactory functioning throughout the life span, it seemed potentially enlightening to explore such complementary tasks. Because odor identification makes cognitive as well as sensory demands, a comparable task of cognitive skill, but with visual material, offers a suitable backdrop for performance. For the present investigation, we assessed odor detection, odor identification, and picture identification with a well established test (Boston Naming Test) [3].

Experiment 1: Odor Identification, Threshold, and the Boston Naming Test

Method

Subjects

A group of 125 persons participated: 25 children (8–14 years of age), 25 young adults (18–28 years), 25 young middle-aged adults (33–47 years), 25 older middle-aged subjects (48–63 years), and 25 senior citizens (65–90 years).

Procedure

The subjects participated in three tasks. The first was the Boston Naming Test (BNT) of picture identification, which comprised 60 drawings of items, some that occur in everyday life with high frequency (e.g., a tree) and others that occur with low frequency (e.g., an abacus).

The second task involved measurement of the detection of 1-butanol vapor via a two-alternative forced-choice procedure used in clinical testing [2]. A test series comprised 12 successive threefold dilutions from a stock of 4% (v/v) in deionized water. Testing began with dilution step 12 versus a water blank. The subjects squeezed the bottles of a pair successively under the nostrils and chose the one with the stronger odor. If correct in the choice, subjects underwent the same step until they made five consecutive correct choices, defined then as the threshold point. If incorrect in the choice, subjects received the next higher concentration.

The third task entailed identification of 20 common odorants [baby powder, banana, bubble gum, butterscotch, coconut, chocolate, coffee, crayons, dirt (soil), disinfectant (Lysol), grape jelly, honey, onion, orange, peanut butter, perfume, potato chips, rubber, soap (Ivory), and Vicks] chosen largely to be easy for children to identify. The experimenter presented the items in irregular order. When subjects guessed wrong on an item, they had the opportunity to choose a name from a prompt card that contained four possibilities.

Results

Sensitivity

Threshold, depicted in Fig. 1, varied with age ($F[4,120] = 45.1$, $P < 0.0001$) by a span of 3.2 threefold dilution steps (i.e., 30-fold). Children performed essentially as well as young adults and better than the older middle-aged and elderly adults by Newman-Keuls tests ($P < 0.05$ here and for subsequent post hoc comparisons). From young adulthood onward, performance declined progressively.

[1] John B. Pierce Laboratory, 290 Congress Avenue, New Haven, CT 06519, USA

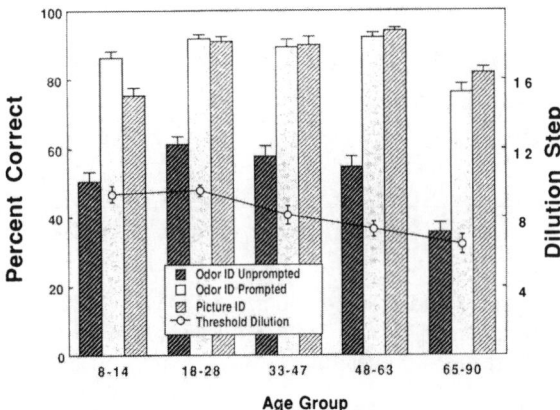

FIG. 1. Performance (± SE) in threshold detection expressed as a dilution step. Bars show performance in free odor identification (unprompted), prompted odor identification, and free picture identification in the Boston Naming Test

Odor Identification

For unprompted odor identification, performance first improved and then declined over ages (F[4,120] = 13.2, P < 0.0001). The elderly differed from all other groups, and the children differed from the young adults. With prompting, all five age groups improved. Although the differences among groups shrank with prompting, total performance still varied with age (F[4,120] = 14.2, P < 0.0001), with the elderly falling below all other groups and the children falling below the older middle-aged group.

Aside from differences in net performance, the various groups showed reasonable similarity in how well they identified the items. The groups displayed an average intercorrelation (r) of 0.82 in unprompted odor identification across items (range 0.73–0.90, all P < 0.001 in all cases). The generally good agreement among the groups implied the existence of a general factor of performance.

Picture Identification

Performance in picture identification exceeded that in corresponding odor identification (87% vs 52%) but varied with age similarly (F[4,120] = 18.4, P <

0.0001). Young adults and middle-aged adults could name more than 90% of the pictures, whereas children and the elderly could identify only 75% and 82%, respectively. Performance of the children fell below all groups except the elderly, and the performance of the elderly fell below all groups except the children.

Interestingly, the performance of 87% on the BNT equaled that of prompted performance on odor identification. A comparison accordingly yielded no main effect of task (F[1,120] = 0.67, n.s.). Not surprisingly, it yielded a effect of age (F[4,120] = 25.7, P < 0.0001), but more importantly it yielded an interaction of age by task (F[4,120] = 6.0, P < 0.0002). Analysis of simple effects isolated the interaction to the performance of the children (F[1,120] = 19.1, P < 0.001) and the elderly (F[1,120] = 5.3, P < 0.025) across the two identification tasks. The children performed relatively better at the odors and the elderly at picture identification. This outcome suggests that a discriminative deficit in part controlled the olfactory performance of the elderly.

Correlation Among Performance Indices and Demographic Variables

With individual subjects rather than groups as the object of analysis, age of *adults* proved the most robust correlate of performance on all tasks (Table 1). Age correlated significantly with both odor and picture identification and with odor threshold. Among the adults, threshold correlated modestly with odor identification.

In a multiple regression of performance in odor identification against the variables of age, education, and performance on the BNT for the adults, age and performance on the BNT entered as significant variables (F ≥ 3.5), accounting for 31% of the variance (multiple R = 0.56), whereas education did not (partial correlation 0.07).

Experiment 2: Threshold of Children and Young Adults Reexamined

The finding that children had about the same absolute sensitivity as young adults merited further scrutiny

TABLE 1. Intercorrelations between age, threshold, and performance at odor and picture identification.

Parameter	Age	Odor ID (unprompted)	Odor ID (prompted)	Picture ID	Threshold
Age	—	−0.53[a]	−0.51[a]	−0.40[a]	−0.43[a]
Odor ID (unprompted)		—	−0.55[a]	0.36[a]	0.31[a]
Odor ID (prompted)			—	0.33[a]	0.14
Picture ID				—	0.12

[a] P < 0.01.

since: (1) Strauss [4] had reported better sensitivity in children than in young adults (to be sure, though, she used a questionable blast-injection mode of stimulus delivery, did not control for response criterion, and could thereby have confounded criterion with sensitivity); (2) about one-third of our children and young adults yielded thresholds within one step from the highest available dilution (lowest concentration), which may have truncated and thereby obscured differences in the distributions; and (3) our procedure, derived principally for use in the clinic with its priority on speed, penalized errors and lapses of attention heavily (possibly more commonly in children) and could therefore have given them a falsely higher average threshold. Accordingly, we chose to measure thresholds of young adults and children with a wider range of concentrations and with a procedure that would penalize errors less severely and allow a continuous estimate of threshold throughout testing.

Method

Subjects

Twenty-five children (8–14 years of age) and 26 young adults (18–28 years) participated in two threshold tests.

Procedure

As in experiment 1, the task involved detection of 1-butanol vapor via a two-alternative forced-choice procedure. The test series in this instance comprised 15 successive dilutions (in contrast to only 12 in experiment 1) from the stock of 4% butanol.

For the measurement of a threshold, a succession of 30 forced-choice trials was given according to a protocol, called the "step method," devised by Simpson [5] to give an efficient estimation of threshold (<40 trials). The step method is one of a class of "adaptive" measures that rely on a subject's ongoing performance to decide which stimulus levels to present on the next trial. Many such methods generate an up–down staircase-like pattern over trials from which to calculate a threshold. Unlike the regular staircase procedures, the present one can call for a change of several stimulus steps on a subsequent trial; such large changes are frequent at the beginning of the staircase. With more and more information available as trials accumulate, the change called for tends to narrow, and the same step may be called for on several subsequent trials.

Results

Children and young adults yielded approximately the same thresholds at the end of the 30 trials: dilution steps 8.04 ± 1.65 (SD) and 8.46 = 1.46

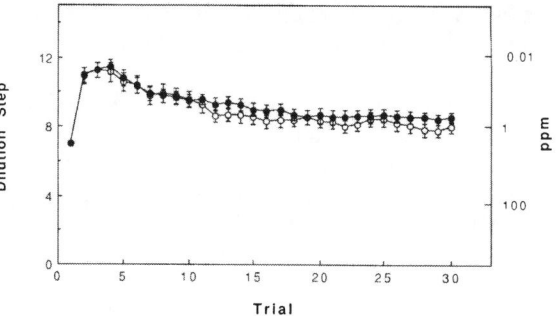

Fig. 2. Average dilution step (± SE) estimated as threshold by the algorithm of the step method for the 30 trials of the test (average of two sessions). *Open circles*, children; *closed circles*, young adults

(SD), respectively (F[1,47] = 0.90, n.s., for age; F[1,47] = 0.63, n.s. for sex). A trial-by-trial view revealed similar performance by the two groups throughout (Fig. 2). Most subjects in both groups detected the starting level of step 7, which led to probe stimuli at high dilutions. As subjects then began to miss at the higher dilution steps, their average performance came back to levels close to their final averages.

Discussion

Both experiments implied that youngsters halfway through childhood have about the same absolute sensitivity as young adults. Presumably, the process of olfactory decline seen in various studies of adults begins only during adulthood. From previous results, it was expected that children might evince poorer odor identification than adults. Although in principle such performance could have arisen from poorer absolute sensitivity in children, it clearly did not. Any disadvantage for children seems to have risen from lack of odor-specific knowledge.

During the life span, odor identification seems to face three major limitations: the need to associate odors with their names; the need to retrieve names from existing associations; and the need to discriminate among stimuli. Early in life, children presumably learn to associate names with smells, one substance at a time. Some odors appear more frequently than others and therefore offer more opportunities for learning. Just as children add more words per month to their vocabulary than do adults, they probably learn more odors per month. In addition to simple frequency of occurrence, some odors (e.g., chocolate) have special salience for children, who may overlearn them, much as they overlearn the names of common objects in the visual world. It is somewhat surprising, though, how little overlearning occurs. Although most young adults, our best group, could identify such prevalent odors as bubble gum, onion, orange, and soap, one in five subjects failed on each of these odors until helped by prompting.

The gain in odor identification that occurred with prompting emphasizes how far odors are from being overlearned but at the same time suggests the presence of a storehouse of information that people fail to retrieve on command. On the surface, it may appear that the difference between unprompted performance (recall memory) and prompted performance (recognition memory) measures only retrieval. The prompted task also eases discriminative burden. The four choices (e.g., coffee, baby powder, lemon, and cloves for the item coffee) essentially define the degree of precision necessary to identify an item correctly. Use of distractors with odors similar to that of the correct choice undoubtedly depresses performance and enlarges the differences between the performance of the elderly and that of younger groups. Prompted performance should nevertheless reflect discriminative differences better than does unprompted performance. Hence the poorer performance of the elderly at prompted identification presumably reflects principally the deterioration of discrimination.

Whereas the BNT seems to measure facility at word retrieval in adult subjects, it appears to measure principally word knowledge or vocabulary in children, in whom performance increases progressively into adolescence. The outcome for children on the BNT and on unprompted odor identification presumably have the same source. Previous studies of the BNT have indicated that elderly subjects rate below nonelderly adults by about 10%. The present investigation found the same, an outcome that implies some contribution of word-retrieval to unprompted identification in our elderly. Irrespective of any cognitive handicap, the elderly do possess a sensory handicap in their elevated olfactory threshold. Although the elderly have had the bene-fits, such as they are, of many decades of odor learning, such learning ultimately fails to overcome deterioration caused by the sensory apparatus.

Although middle-aged adults also yielded elevated thresholds, the data on prompted odor identification failed to permit the conclusion that the middle-aged have discriminative losses that would impair functioning. Nevertheless, during early middle age the accumulation of knowledge and the sensory capacity to use that knowledge may intersect to produce optimum performance. When Eskenazi et al. [6] fitted a quadratic function to composite data on olfactory performance, they found it to reach a maximum at age 34 [see also 1,2]. When the present data were fitted, the function reached a maximum at age 33.5 years.

Acknowledgments. Supported by grants AG 04287 and DC 00284 from the US National Institutes of Health.

References

1. Doty RL, Shaman P, Appelbaum SL, et al. (1984) Smell identification ability: changes with age. Science 226:1441–1443
2. Cain WS, Gent JF, Goodspeed RB, Leonard G (1988) Evaluation of olfactory dysfunction in the Connecticut Chemosensory Clinical Research Center. Laryngoscope 98:83–88
3. Kaplan EF, Goodglass H, Weintraub S (1983) The Boston Naming Test (2nd edn). Lea and Febiger, Philadelphia
4. Strauss EL (1970) A study of olfactory acuity. Ann Otol Rhinol Laryngol 79:95–104
5. Simpson WA (1989) The step method: a new adaptive psychophysical procedure. Percept Psychophys 45:572–576
6. Eskenazi B, Cain WS, Friend F (1986) Exploration of olfactory aptitude. Bull Psychonom Soc 24:203–206

Olfactory Dysfunction in the Elderly and in Alzheimer's Disease

Richard L. Doty[1]

Key words. Aging—Olfaction—Alzheimer's disease —Dementia—Smell—Psychophysics

Introduction

The sense of smell, along with its sister sense of taste, serves to monitor the intake of environmental chemicals and nutrients into the body and determines, to a large degree, the flavor and palatability of foods and beverages. Importantly, this sense warns of spoiled foods, leaking natural gas, polluted air, and smoke. As indicated in this chapter, decreased ability to smell is common in older age and can be an early sign of a number of age-related neurodegenerative disorders, including Alzheimer's disease [1].

Changes in the ability to smell during the later years of life are not inconsequential. Thus a disproportionate number of elderly die from accidental gas poisoning [2], and many elderly report that their food is tasteless [3]. The latter problem, which can lead to decreased motivation to eat, explains at least some cases of age-related nutritional deficiencies. As shown in clinical studies [4], decreased "taste" perception during eating largely reflects decreased retronasal stimulation of the olfactory receptors [5]. Whole-mouth sweet, sour, bitter, salty, and (possibly) umami perception is more resilient to major age-related changes, although marked age-related changes in localized taste perception are now clearly documented (Matsuda and Doty, manuscript in preparation). In addition, distortions in taste perception, including persistent sour, bitter, and salty sensations, are common during old age and are debilitating and difficult to manage [4,6].

In this chapter I discuss the changes in structure and function of the olfactory system that occur in many elderly persons and in individuals with Alzheimer's disease. The reader is referred elsewhere for exhaustive treatises on this topic [1,7–10].

Age-Related Changes in Odor Perception

A wide variety of olfactory tests reveal age-related deficits, including tests of odor identification, detection, discrimination, adaptation, and suprathreshold intensity perception [7]. In addition, age-related declines have been noted in the ability to detect airborne irritants that stimulate free nerve endings of the trigeminal nerve (CN V) within the nasal mucosa [11]. Both linear and quadratic components are present in the functions that describe the average age-related declines in odor detection and identification [12], although large individual differences are present. Olfactory dysfunction is most noticable after age 60 and men, on average, evidence earlier age-related declines in odor perception than do women. These phenomena are well illustrated by scores on the University of Pennsylvania Smell Identification Test (UPSIT), a widely used clinical olfactory test [13] (Fig. 1).

As is apparent from the data in Fig. 1, age-related deficits in olfactory function are profound. Indeed, using the criterion of an UPSIT score of ≤19 as indicative of major impairment (10 is chance performance), more than 50% of individuals between the ages of 65 and 80 years evidence such impairment, whereas more than 75% of those over the age of 80 years do so [3]. Indeed, age is a much more important factor than either gender or smoking for the ability to smell [3,13]. The age-related decline and gender difference noted above appear to be universal, as they occur in individuals representing a wide variety of cultures [14].

Although the decline in olfactory function may occur more for some odorants than for others (depending on such factors as the odorant's threshold and the form of the function relating odorant concentration to perceived intensity), such decline is present for a wide spectrum of odorants and reflects a generalized phenomenon. It is now well documented, for example, that, among both young and elderly individuals, persons who evidence comparatively low sensitivity to one odorant typically evidence low sensitivity to others, whereas those who evidence comparatively high sensitivity to an odorant typically evidence high sensitivity to others. Therefore, with the possible exception of so-called specific anosmias (i.e., decreased ability to detect

[1] Smell and Taste Center, Hospital of the University of Pennsylvania, and Department of Otorhinolaryngology—Head and Neck Surgery, School of Medicine, 3400 Spruce Street, Philadelphia, PA 19104, USA

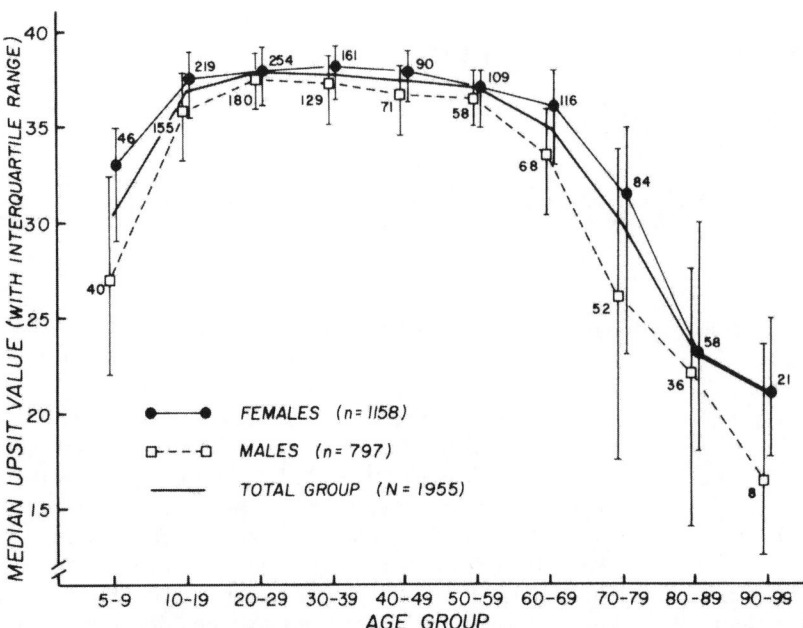

Fig. 1. Scores on the University of Pennsylvania Smell Identification Test as a function of age in a large heterogeneous group of subjects. Numbers by data points indicate sample sizes. (From [3] with permission, copyright American Association for the Advancement of Science 1984)

one or a few compounds in the presence of normal sensitivity to most odorants), a "general olfactory acuity" factor appears to exist, analogous to the general intelligence factor derived from items of intelligence tests [15,45].

Underlying Causes of Age-Related Changes in Olfactory Function

The olfactory receptor neurons are more or less directly exposed to the outside environment, making them susceptible to insult from bacteria, viruses, toxins, and other nosogenic agents. For this reason, it is not surprising that environmentally induced damage to the receptor epithelium appears to be a primary cause of age-related alterations in the ability to smell. Cumulative destruction of the olfactory neuroepithelium has been described over the entire life span, with metaplasia from respiratory epithelium occurring as islands within the membrane [16]. Age-related physiologic or structural changes may also directly damage the olfactory epithelium or predispose it to damage from environmental insults. Included in such changes are reduced protein synthesis or metabolic insufficiency (as with hypothyroidism) [17], loss of neurotrophic factors [18], occlusion of cribriform plate foramina [19], decreased intramucosal blood flow [20], changes in the vascular elasticity of the epithelium [21], altered airway patency [22,23], increased nasal mucus viscosity [24], atrophy of secretory glands and lymphatics [24], and, potentially, decreases in enzyme systems that deactivate xenobiotic materials within the olfactory mucosa [25].

Despite the fact that the olfactory receptor cells have the capacity to reconstitute themselves at periodic intervals [26], this plasticity is likely altered by age-related processes, as demonstrated by studies of rodent and amphibian olfactory epithelia [27]. For example, in older animals the ratio of dead or dying cells to the number of receptor cells increases, suggesting that such receptor cells have less mitotic activity than receptor cells from younger animals. Repair of the olfactory epithelium following its destruction by zinc sulfate or methylformininomethylester is slower or nonexistent in older than in younger animals, showing that the neurogenic process is altered with age [28].

In general, degeneration of glomeruli of the olfactory bulb is observed after destruction of the receptor elements of the olfactory neuroepithelium. This observation was used by Smith [29] to estimate age-related losses of human olfactory receptors. In this early study, Smith assessed the number and form of glomeruli present in 205 olfactory bulbs of 121 individuals at autopsy and concluded that loss of olfactory nerves begins soon after birth and continues to occur throughout life at approximately 1% per year. However, reevaluation of Smith's data (Fig. 1 in [29]), using medians rather than means, indicates that, on average, major loss does not occur until the fifth decade of life. No sex differences were apparent in this pioneering study, although considerable variability in the destruction of the glomeruli was seen at all ages examined.

Widespread age-related alterations in bulbar structures have been clearly demonstrated in rats. For example, Hinds and McNelly [30] measured, in Sprague-Dawley rats, the volume of main olfactory bulb components (including the glomerular, external plexiform, internal granular, and olfactory nerve layers) at 3, 12, 24, 27, and 30 months of age. The size and number of mitral cells were determined in both the main and accessory bulbs. Even though developmental increases in the volume of the examined bulb and brain layers occurred during the first 24 months of age, decreases occurred after that time. Importantly, a marked decline in the number of mitral cells was found, accompanied by an increase in the size of the mitral cell bodies and nuclei, and in the volume of individual mitral cell dendritic trees.

In a subsequent study, Hinds and McNelly [31] examined age-related changes in both bulbar and neuroepithelial tissue. A decline in the number of epithelial receptor cells appeared to have occurred several months before the bulbar alterations, implying that the latter were secondary to destruction of the epithelial elements. In the oldest group examined, the number of synapses per receptor cell was increased, conceivably reflecting a compensatory increase in the relative numbers of synapses per cell in the surviving receptor cell population.

Unlike the olfactory neuroepithelium and bulb, the piriform cortex of the rat evidences little age-related change. Curcio et al., for example, carefully examined the cells and synapses of the piriform cortices of rats ranging in age from 3 to 33 months [32]. No meaningful changes were observed in the volumes of cortical laminae Ia and Ib or in the numerical and surface densities of the synaptic apposition zones in layer Ia (which are formed mainly by mitral cell axons). Although no age-related changes in nuclear volume, soma volume, or numerical density of layer II neurons were found, a modest (18%) decline in the proportion of layer Ia occupied by dendrites and spines was observed, accompanied by an increase in the proportion of glial processes, but not by an alteration in the proportion of axons and terminals.

Even though the aforementioned studies suggest that peripheral factors seem to be paramount in olfactory system pathology, alterations in more central pathways may also occur. For example, the olfactory bulbs of more than 40% of *nondemented* persons 50 years of age and older exhibit neurofibrillary tangles (NFTs) [33]; and, as indicated below, considerable destruction of central pathways have been described in Alzheimer's disease.

Olfactory System Function and Morphology in Alzheimer's Disease

Relative to age-matched controls, decreased olfactory function is present in individuals with Alzheimer's disease (AD) (see [34] for review). In general, the magnitude of the dysfunction is weakly related (Pearson rs are typically <0.40) to the severity of the dementia [35–39]. Although investigators from one laboratory [38] argued that this association is present only for tests of odor identification (and therefore, by questionable inference, central rather than peripheral olfactory structures), other laboratories have demonstrated such an association for other measures (including olfactory thresholds) [e.g., 35,37], and the basis for such an inference cannot, at the present time, be empirically justified. Importantly, the possible influence of dementia on subtle aspects of nonolfactory components of a number of olfactory tests makes such findings somewhat enigmatic.

Although AD-related olfactory dysfunction may reflect, in part, the physiologic changes associated with normal aging, additional factors presumably also come into play. Thus even early-stage AD patients with mild dementia score much more poorly on olfactory tests than do age-matched controls [40], and AD-related pathophysiologic changes occur rather specifically in olfactory-related brain structures. For example, in AD patients high numbers of NFTs and neuritic plaques (NPs) are found in the anterior olfactory nucleus, olfactory bulb, hippocampal formation, periamygdaloid nucleus, prepiriform cortex, entorhinal cortex, prefrontal cortex, and dorsomedial thalamic nucleus (for review, see [8]), and damage to the olfactory neuroepithelium has been reported [41]. Whether such damage is related to the transit of environmental agents, such as viruses or toxins, from the nasal cavity into the central nervous system via the olfactory fila is unknown; however, this hypothesis is currently receiving widespread attention [1,8,34,42]. Another hypothesis, put forward by Kurtz et al. [43], is that degeneration of the olfactory system per se is the basis of the cognitive decline in AD, as rats that had been anosmic for some time had considerable difficulty learning an active avoidance learning task, unlike those whose anosmia was of recent origin. The involvement of olfactory structures in AD was summarized by Pearson et al. [44] in 1985 as follows:

The invariable finding of severe and even maximal involvement of the olfactory regions in Alzheimer's disease is in striking contrast to the minimal pathology in the visual and sensorimotor areas of the neocortex and cannot be without significance. In the olfactory system, the sites that are affected—the anterior olfactory nucleus, the uncus, and the medial group of amygdaloid nuclei—all receive fibers directly from the olfactory bulb. These observations at least raise the possibility that the olfactory pathway is the site of initial involvement of the disease [p. 4534].

Summary

Studies have demonstrated that the ability to smell is markedly decreased in a number of older persons.

Such diminution is observed on a variety of olfactory tests and is exaggerated in persons with Alzheimer's disease. Both peripheral and central nervous system olfactory structures may be involved, although additional research is needed to establish the relative roles of a variety of olfactory structures in producing the dysfunction.

Whatever the physiologic basis for the decreased olfactory ability, such alterations are often profound and place many elderly persons at considerable risk for food and gas poisoning. Hopefully, future research will lead to a better understanding of age-related olfactory disorders and will provide gerontologists with a means of reversing such changes and ensuring enhanced quality of life for their patients.

Acknowledgments. Supported by grant PO1 DC 00161 from the National Institute on Deafness and Other Communication Disorders and by grant RO1 AG08148 from the National Institute on Aging.

References

1. Doty RL (1991) Psychophysical measurement of human olfactory perception. In: Laing DG, Doty RL, Breipohl W (eds) The human sense of smell. Springer, Berlin, pp 95–134
2. Chalke HD, Dewhurst JR, Ward CW (1958) Loss of sense of smell in old people. Public Health (Lond) 72:223–230
3. Doty RL, Shaman P, Applebaum SL, et al. (1984) Smell identification ability: changes with age. Science 226:1441–1443
4. Deems DA, Doty RL, Settle RG, et al. (1991) Smell and taste disorders: a study of 750 patients from the University of Pennsylvania Smell and Taste Center. Arch Otolaryngol Head Neck Surg 117:519–528
5. Burdach K, Doty RL (1987) Retronasal flavor perception: influences of mouth movements, swallowing and spitting. Physiol Behav 41:353–356
6. Cohen T, Gitman L (1959) Oral complaints and taste perception in the elderly. J Gerontol 14:294–298
7. Doty RL (1990) Aging and age-related neurological disease: olfaction. In: Goller F, Grafman J (eds) Handbook of neuropsychology. Elsevier, Amsterdam, pp 211–226
8. Ferreyra-Moyano H, Barragan E (1989) The olfactory system and Alzheimer's disease. Int J Neurosci 49:157–197
9. Murphy C (1986) Taste and smell in the elderly. In: Meiselman HS, Rivlin RS (eds) Clinical measurement of taste and smell. Macmillan, New York, pp 343–371
10. Weiffenbach JM (1984) Taste and smell perception in aging. Gerodontology 3:137–146
11. Stevens JC, Cain WS (1986) Aging and the perception of nasal irritation. Physiol Behav 37:323–328
12. Deems DA, Doty RL (1987) The nature of age-related olfactory detection threshold changes in man. Trans Penn Acad Ophthalmol Otolaryngol 39:646–650
13. Doty RL, Shaman P, Dann M (1984) Development of the University of Pennsylvania Smell Identification Test: a standardized microencapsulated test of olfactory function. Physiol Behav 32:489–502
14. Gilbert AN, Wysocki CJ (1987) The smell survey results. Natl Geogr Mag 172:514–525
15. Yoshida M (1984) Correlation analysis of detection threshold data for "standard test" odors. Bull Fac Sci Eng Chuo Univ 27:343–353
16. Nakashima T, Kimmelman CP, Snow JB Jr (1984) Structure of human fetal and adult olfactory neuroepithelium. Arch Otolaryngol 110:641–646
17. Mackay-Sim A, Beard MD (1984) Hypothyroidism disrupts neural development in the olfactory epithelium of adult mice. Dev Brain Res 36:190–198
18. Appel SH (1981) A unifying hypothesis for the cause of amyotrophic lateral sclerosis, parkinsonism, and Alzheimer's disease. Ann Neurol 10:499–505
19. Krmpotic-Nemanic J (1969) Presbycusis, presbystasis, and presbyosmia as consequences of the analogous biological process. Acta Otolaryngol 67:217–223
20. Hasegawa M, Kern EB (1977) The human nasal cycle. Mayo Clin Proc 52:28–34
21. Somlyo AP, Somlyo AV (1968) Vascular smooth muscle: I. Normal structure, pathology, biochemistry, and biophysics. Pharmacol Rev 20:197–272
22. Doty RL, Frye R (1989) Nasal obstruction and chemosensation. Otolaryngol Clin North Am 22:381–384
23. Nishihata S (1984) Aging effect in nasal resistance. Nippon Jibiinkoka Gakkai Kaiho 87:1654–1671
24. Koopmann CF Jr (1989) Effects of aging on nasal structure and function. Am J Rhinol 3:59–62
25. Dahl AR (1988) The effect of cytochrome P-450-dependent metabolism and other enzyme activities on olfaction. In: Margolis RL, Getchell TV (eds) Molecular neurobiology of the olfactory system. Plenum, New York, pp 51–70
26. Graziadei GA Monti, Graziadei PPC (1979) Studies on neuronal plasticity and regeneration in the olfactory system: morphologic and functional characteristics of the olfactory sensory neuron. In: Meisami E, Brazier MAB (eds) Neural growth and differentiation. Raven, New York, pp 373–396
27. Breipohl W, Mackay-Sim A, Grandt D, Rehn B, Darrelmann C (1986) Neurogenesis in the vertebrate main olfactory epithelium. In: Breipohl W (ed) Ontogeny of olfaction. Springer, Berlin, pp 21–33
28. Matulionis DH (1982) Effects of the aging process on olfactory neuron plasticity. In: Breipohl W (ed) Olfaction and endocrine regulation. IRL Press, London, pp 299–308
29. Smith CG (1942) Age incidence of atrophy of olfactory nerves in man. J Comp Neurol 77:589–595
30. Hinds JW, McNelly NA (1977) Aging of the rat olfactory bulb: growth and atrophy of constituent layers and changes in size and number of mitral cells. J Comp Neurol 171:345–368
31. Hinds JW, McNelly NA (1981) Aging in the rat olfactory system: correlation of changes in the olfactory epithelium and olfactory bulb. J Comp Neurol 203:441–454
32. Curcio CA, McNelly NA, Hinds JW (1985) Aging in the rat olfactory system: relative stability of piriform cortex contrasts with changes in olfactory bulb and olfactory epithelium. J Comp Neurol 235:519–528
33. Kishikawa M, Iseki M, Nishimura M, Sekine I, Fujii H (1990) A histopathological study on senile changes in the human olfactory bulb. Acta Pathol Jpn 40:255–260
34. Doty RL (1991) Olfactory dysfunction in neurodegenerative disorders. In: Getchell TV, Doty RL,

Bartoshuk LM, Snow JB Jr (eds) Smell and taste in health and disease. Raven, New York, pp 735–751

35. Knupfer L, Spiegel R (1986) Differences in olfactory test performance between normal, aged, Alzheimer-, and vascular-type dementia individuals. Int J Geriatr Psychiatry 1:3–14

36. Moberg PJ, Pearlson GD, Speedie LJ, et al. (1987) Olfactory recognition: differential impairments in early and late Huntington's and Alzheimer's diseases. J Clin Exp Neuropsychol 9:650–664

37. Murphy C, Gilmore MM, Seery CS, Salmon DP, Lasker BR (1990) Olfactory thresholds are associated with degree of dementia in Alzheimer's disease. Neurobiol Aging 11:465–469

38. Serby M, Larson P, Kalkstein D (1991) The nature and course of olfactory deficits in Alzheimer's disease. Am J Psychiatry 148:357–360

39. Waldton S (1974) Clinical observations of impaired cranial nerve function in senile dementia. Acta Psychiatr Scand 50:539–547

40. Doty RL, Reyes P, Gregor T (1987) Presence of both odor identification and detection deficits in Alzheimer's disease. Brain Res Bull 18:597–600

41. Jafek BW, Filley CM, Eller PM, et al. (1991) Abnormal olfactory epithelium in biopsies of Alzheimer's patients. Presented at the Thirteenth Annual Meeting of the Association for Chemoreception Sciences, Sarasota, Florida

42. Roberts E (1986) Alzheimer's disease may begin in the nose and may be caused by aluminosilicates. Neurobiol Aging 7:561–567

43. Kurtz P, Schuurman T, Prinz H (1989) Loss of smell leads to dementia in mice: is Alzheimer's disease a degenerative disorder of the olfactory system? J Protein Chem 8:448–451

44. Pearson RCA, Esiri MM, Hiorns RW, Wilcock GK, Powell TPS (1985) Anatomical correlates of the distribution of the pathological changes in the neocortex in Alzheimer's disease. Proc Natl Acad Sci USA 82:4531–4534

45. Doty RL, Smith R, McKeown D, Raj J (1994, in press) Tests of human olfactory function: Principal components analysis suggests that most measure a common source of variance. Percept Psychophysics

Differential Expression of Cell-Specific Molecules in Olfactory Receptor Neurons of Humans of Different Ages

Marilyn L. Getchell[1,2] and Thomas V. Getchell[1,2,3]

Key words. Olfactory marker protein—Apoptosis

Human olfactory receptor neurons (ORNs) express the neuron-specific molecules neuron-specific enolase (NSE; [1]) and protein gene product (PGP) 9.5 [2], as well as the olfactory-specific molecule olfactory marker protein (OMP [3–5]). To determine the relative frequency of expression of these markers throughout the human life span, to investigate the relationship between the expression of these markers and ongoing cell death, and to initiate studies on immune involvement in ORN death, double-staining immunofluorescence techniques with antibodies to NSE or PGP 9.5 and OMP were applied to tissue obtained at autopsy from 16 subjects (10 males, 6 females). Subjects included a 16-week-old fetus, a 24-week-old fetus, a 10-week-old infant, 4 young and middle-aged subjects (24–59 years old; mean, 45 years), 5 old subjects (63–90 years old; mean, 75 years old), and 4 subjects (52–85 years old; mean, 70 years old) in whom the diagnosis of Alzheimer's disease (AD) was confirmed by brain histopathology. Post-mortem intervals (range, 2–22 h; median, 8 h) did not differ significantly. The numbers of OMP-immunoreactive and NSE- or PGP 9.5-immunoreactive ORNs in each microscopic field that contained at least one OMP-immunoreactive ORN were counted; the densities of neurons expressing each marker and the ratios of OMP:NSE or PGP 9.5-immunoreactive neurons were calculated. Statistical analyses were performed with Kruskal-Wallis and Dunn's multiple comparison tests.

Mean densities of OMP-immunoreactive neurons increased from 3/100-μm length of olfactory epithelium (OE) in the 16-week-old fetus ($n = 9$ fields) to 8–9/100 μm in the 24-week-old fetus ($n = 8$), the 10-week-old infant ($n = 7$), and the young to middle-aged subjects ($n = 24$). In old subjects ($n = 26$) and those with AD ($n = 35$), mean densities declined to 5–6/100 μm. In contrast, mean densities of NSE- or PGP 9.5-immunoreactive neurons decreased from 15–20/100 μm in the 16- and 24-week-old fetuses and 10-week-old infant to 8–10/100 μm in young to middle-aged and old subjects and those with AD. Neuronal densities between fields in each subject varied considerably. The corresponding ratios of OMP:NSE or 9.5-immunoreactive neurons were more consistent. Mean ratios increased through mid-life, from 0.19 ± 0.07 in the 16-week-old fetus to 0.42 ± 0.08 in the 24-week-old fetus to 0.60 ± 0.08 in the 10-week-old infant, to a high of 0.83 ± 0.10 in the young and middle-aged subjects. In old subjects and those with AD, the mean ratios declined to 0.58 ± 0.16 and 0.57 ± 0.16, respectively, which differed significantly ($P < 0.001$) from that in young and middle-aged subjects.

Because OMP and NSE have comparable cellular half-lives (10 vs 9.3 days [6]), this decline was not attributable to differences in turnover rate. Hypothesizing that the decline represented cessation of OMP expression before that of neuronal markers in dying ORNs, we investigated the mode of ORN death. DNA staining with bis-benzamide demonstrated numerous clusters of apoptotic bodies in the OE of the two fetuses and the infant, as well as in old subjects and those with AD, indicating that ORNs undergo apoptosis [7], which occurred more frequently during development (fetal and neonatal ages) and aging. Staining with an antibody to a human macrophage marker using ABC-peroxidase techniques identified macrophages in the OE of all subjects, with the greatest numbers in old subjects and those with AD, suggesting that some characteristic(s) of the surfaces of dying ORNs change so that they become targets of phagocytosis by macrophages.

These studies demonstrate that the number of mature olfactory receptor neurons relative to the total number of neurons changes with age, and suggest a higher rate of cell death with the involvement of cells of the immune system in the developing and aged human olfactory epithelium.

Acknowledgments. This research was supported by NIH grants DC-01715 (MLG) and DC-00159 (TVG). We thank F.L. Margolis, Ph.D., Roche Institute of Molecular Biology, Nutley, NJ, USA, for supplying antiserum to OMP; and W.R. Markesbery, M.D., and D.R. Wekstein, Ph.D., for pro-

---

[1] Division of Otolaryngology—Head Neck Surgery, Department of Surgery, [2] Sanders-Brown Center on Aging, and [3] Department of Physiology and Biophysics, University of Kentucky College of Medicine, Lexington, KY 40536, USA

viding access to human tissue through the Alzheimer's Disease Research Center of the Sanders-Brown Center on Aging, University of Kentucky, which is supported by NIH grant P50-AG05144.

References

1. Takahashi S, Iwanaga T, Takahashi Y, Nakano Y, Fujita T (1984) Neuron-specific enolase, neurofilament protein, and S-100 protein in the olfactory mucosa of human fetuses. An immunohistochemical study. Cell Tissue Res 238:231–234
2. Takami S, Getchell ML, Chen Y, Monti-Bloch L, Berliner DL, Stensaas LJ, Getchell TV (1993) Vom-eronasal epithelial cells of the adult human express neuron-specific molecules. NeuroReport 4:375–378
3. Nakashima T, Kimmelman CP, Snow JB Jr (1985) Olfactory marker protein in the human olfactory path-way. Arch Otolaryngol 111:294–297
4. Getchell ML, Mellert TK (1991) Olfactory mucus secretion. In: Getchell TV, Doty RL, Bartoshuk LM, Snow JB Jr (eds) Smell and taste in health and disease. Raven, New York, pp 83–95
5. Getchell ML, Chen Y, Ding X, Sparks DL, Getchell TV (1993) Immunohistochemical localization of cyto-chrome P-450 isozyme in human nasal mucosa: Age-related trends. Ann Otol Rhinol Laryngol 102:368–374
6. Kream RM, Margolis FL (1984) Olfactory marker pro-tein: Turnover and transport in normal and regenerat-ing neurons. J Neurosci 4:868–879
7. Wyllie AH, Kerr JFR, Currie AR (1980) Cell death: The significance of apoptosis. Int Rev Cytol 68:251–306

Olfactory Acuity of American and Japanese Subjects Tested with a T and T Olfactometer

C. Maetani[1], I. Takemoto[1], K. Shimada[1], I. Koizuka[1], T. Matsunaga[1], E.B. Kern[2], and T.V. McCaffrey[2]

Key words. Sense of smell—Clinical examination

No standard method has yet been established to evaluate the sense of smell in patients with olfactory disorders. In Japan, we have been using the T and T olfactometer (Takasago, Tokyo, Japan) as a standard for this purpose since 1975. There have been very few studies of differences in olfactory acuity between American and Japanese subjects using the same olfactory test. In this study, we compared the olfactory acuity of American and Japanese subjects, using the T and T olfactometer.

Our subjects were 154 healthy adults aged between 17 and 68 years; they had no abnormality in the nasal cavity as confirmed examination performed by one of the authors. Seventy-seven subjects were American, and 77, Japanese. The details for the American subjects were: 37 males, mean age 38.5 ± 11.4 years, and 40 females, mean age 40.7 ± 13.8 years. For the Japanese, there were 37 males, mean age 38.6 ± 11.6 years, and 40 females, mean age 40.7 ± 13.0 years. Seventy-three (95%) of the American subjects were Caucasian, the other 4 were Hispanic.

We used five odorants in the T and T olfactometer. These were: A, β-phenyl ethyl alcohol (the odor of roses); B, methyl cyclopentenolone (the odor of caramel); C, iso-valeric acid (the odor of dirty socks); D, γ-undecalactone (the odor of canned peaches); and E, skatole (the odor of excrement). Various concentrations of odorants were used, the highest concentration being denoted 5; there were eight concentration levels 5, 4, 3, 2, 1, 0, −1, and −2 in descending order. The subjects smelled the odorant liquid sequentially, from the lowest concentration to the highest. The number "0" represents the threshold concentration for olfactory sensitivity in a normal young adult. The detection and recognition thresholds were measured by the same procedure at both the Mayo Clinic and Osaka University.

The detection thresholds (mean $\pm$ SE) of odors A, B, C, D, E, respectively, were: American males, 0.41 ± 1.50, 0.59 ± 1.00, 0.32 ± 1.40, 0.65 ± 1.53, and 0.22 ± 1.49; Japanese males, 0.89 ± 1.37, 0.41 ± 0.75, 0.19 ± 0.77, 0.43 ± 0.75, and 0.46 ± 0.86; American females, 0.18 ± 1.16, 0.53 ± 1.24, 0.10 ± 1.14, 0.68 ± 1.08, and 0.05 ± 1.12; Japanese females, 0.50 ± 0.87, 0.35 ± 0.61, 0.08 ± 0.61, 0.25 ± 0.62, and 0.18 ± 0.67.

The recognition thresholds (mean $\pm$ SE) of odors A, B, C, D, E, respectively, were: American males, 1.35 ± 1.85, 1.03 ± 1.20, 1.03 ± 1.44, 2.11 ± 2.01, and 1.03 ± 1.70; Japanese males, 2.68 ± 1.88, 1.97 ± 1.37, 0.70 ± 0.77, 1.97 ± 2.01, and 1.54 ± 1.39; American females, 0.83 ± 1.36, 1.28 ± 1.10, 0.63 ± 1.41, 1.50 ± 1.36, and 0.90 ± 1.45; Japanese females, 2.40 ± 1.55, 1.45 ± 1.26, 0.70 ± 1.05, 1.95 ± 1.76, and 1.83 ± 1.82.

There was no significant difference in the odor detection threshold between American and Japanese subjects. There was, however, a significant difference in the recognition threshold for odor A ($P < 0.01$ in males; $P < 0.005$ in females), and the recognition threshold for odor B was higher in Japanese males than in American males ($P < 0.01$). Since there was no significant difference in detection threshold between American and Japanese subjects, we concluded that there was no essential difference in the detection of odors between American and Japanese subjects. However, the recognition threshold of odors A and B was significantly different in Japanese and American subjects.

This difference may be due to differences in life style; it is well known that the recognition of tastes and odors may vary depending on the culture and on other circumstances. Thus, the recognition threshold is probably related to the subjects' environment, both physical and cultural.

References

1. Takagi SF (1989) Standardized olfactometries in Japan —a review over ten years. Chem Senses 14:25–46
2. Takagi SF (1989) Olfactory Tests. In: Takagi SF (ed) Human olfaction. University of Tokyo Press, Tokyo, pp 35–69

[1] Department of Otorhinolaryngology, Osaka University Medical School, 2-2 Yamadaoka, Suita 565, Osaka, Japan
[2] Department of Otorhinolaryngology, Mayo Clinic, 200 First Street, Rochester, MN 55905, USA

Clinical Evaluation of Disorders of the Sense of Smell

James B. Snow Jr.[1]

Key words. Chemosensory—Olfaction—Evaluation—Psychophysical—Differential diagnosis—Sensorineural

Introduction

Each year, more than 200000 persons in the United States consult physicians for smell or taste problems. Many more smell and taste disturbances go unreported. How these disorders are evaluated has a direct impact on how they are managed and treated.

The clinical evaluation of the patient with a chemosensory disorder should consist of four basic components: history, physical examination, psychophysical assessment, and medical imaging procedures. Each of the components provides valuable information on an individual basis, but it is only when they are viewed collectively that accurate anatomic and etiologic diagnoses can be made and proper management and treatment can be provided. The focus of this chapter is the evaluation of olfactory disorders.

History

Obtaining a pertinent patient history is the essential first step for successful evaluation of an olfactory disorder. The history provides valuable insight into the disorder and often sets the tone for the remainder of the evaluation. The process of obtaining a history should begin by allowing the patient to describe his or her complaints without interruption. This technique provides useful information about the patient's perception of his or her olfactory disorder and also identifies the aspects of the olfactory problem that are most troubling for the patient. Following the patient's description, the examiner should ask questions that reveal additional relevant information.

It is necessary to know if there was illness or injury that was temporally related to the onset of the symptoms. Related viral infections and head injury, for example, are of great importance. It is necessary to inquire if the sense of smell is completely lost

or only diminished, and whether the loss is for all odorants or only a few. The effects of strong trigeminal stimulants as well as relatively pure olfactory stimulants also must be known. Was the onset sudden or gradual? Were there qualitative as well as quantitative changes in the symptoms as the disorder evolved?

It is also necessary to know if the loss is continuous or intermittent because some patients report temporary recovery of the sense of smell under certain circumstances, such as with exercise, changes in the relative humidity and temperature, and treatment with corticosteroids. The ability of the disorder to change under these circumstances suggests interference with the transport of the odorant to the olfactory neuroepithelium, as in allergic, infectious or noninfectious rhinitis, rather than a sensorineural deficit. Additional history regarding symptoms of nasal stuffiness or airway obstruction, clear or purulent rhinorrhea, sneezing, nasal dryness, epistaxis, crusting, pain or itching, or headache is also helpful to determine if allergies might be a cause.

The examiner must inquire about the perversion of the sense of smell. In certain states of degeneration or recovery of olfactory function, there may be the perception of a foul odor in the presence of a normally pleasant odorant or with environmental changes in temperature and relative humidity. Furthermore, it is useful to know if there is the perception of a foul odor in the absence of an odorant, as this problem may occur in psychotic states and some forms of epilepsy.

The history should attempt to determine if a loss of taste accompanies the olfactory disorder, if there has been occupational exposure to threatening substances, if substance abuse including alcohol and tobacco is involved, or if there is significant family history for chemosensory disorders. Pertinent medical history (e.g., dietary history, medications or drugs the patient is taking or has taken, radiation therapy to the head and neck, neurosurgical or otolaryngologic operations, renal or hepatic disease, diabetes and thyroid, parathyroid or adrenal diseases) also contributes to determining the cause of the olfactory disturbance.

Physical Examination

Once a comprehensive history is completed, it is necessary for a physician to conduct a physical

[1] National Institute on Deafness and Other Communication Disorders, Building 31, Room 3C-02, National Institutes of Health, Bethesda, MD 20892, USA

examination. For most patients information derived from the history coupled with results from the physical examination provides the basis for the anatomic and etiologic diagnoses.

The physical examination often provides key etiologic information and may focus attention on a local factor in the nose to explain the development of the chemosensory disorder. Lack of evidence for nasal pathology, however, does not exclude the possibility of a process in the past that could have resulted in the olfactory disorder.

The physical examination must begin with close inspection of the ears, upper respiratory tract, and head and neck in general. It includes a neurologic evaluation that emphasizes the cranial nerves. A small fiberoptic endoscope should be used to directly visualize the olfactory neuroepithelium and to detect any structural abnormalities that would result in a nasal obstruction, such as deviation of the nasal septum, polyps, and masses. Although this process is difficult even with the smallest endoscopes, it is useful for detecting any abnormality that would interfere with the transport of odorants to the olfactory neuroepithelium or structural anomalies that could be contributing to the olfactory disorder. Following this initial inspection, a vasoconstrictor should be applied to enhance visualization.

The nasal mucous membrane should be evaluated for color, surface texture, swelling, inflammation, atrophy, epithelial metaplasia, exudate, erosion, and ulceration. Paleness of the mucous membrane is usually due to edema in the lamina propria and is suggestive of allergy. Epithelial metaplasia suggests the inhalation of environmental or industrial pollutants, as do swelling, inflammation, exudate, erosion, and ulceration. Unusual spaciousness, dryness, and crusting suggest atrophy of the lamina propria, such as occurs with atrophic rhinitis. Evidence of trauma should be sought, and nasopharyngeal masses should be detected.

Any purulent rhinorrhea should be noted, as should its site of origin. If the rhinorrhea occurs generally throughout the nasal cavity, a rhinitis is suggested; however, a discharge that emanates from the middle meatus suggests maxillary, anterior ethmoidal, or frontal sinusitis. Discharge from the superior meatus suggests posterior ethmoid or sphenoid sinusitis.

Psychophysical Evaluation

The patient's olfactory complaints are corroborated by the psychophysical evaluation. Because a disorder of one chemosense can affect the function of the other (frequently without the patient's awareness), the psychophysical examination must include evaluations of both the olfactory and the gustatory systems. Serious errors in diagnosis occur if only one of the two senses is evaluated psychophysically.

A number of procedures are available for assessing the patient's ability to smell. The most common clinical tests examine the ability to detect and identify odors. Both unilateral and bilateral olfactory testing should be performed, even though patients rarely notice or seek help for unilateral olfactory deficits and clinically significant unilateral deficits are rarely observed. Unilateral testing may identify a rare tumor or lesion involving only one olfactory bulb, tract, or associated projection region.

Odor identification tests that have found wide clinical usage require the patient to select, from a set of alternatives, a term that best describes his or her odor experience. Such tests are highly reliable and generally correlate well with the patient's olfactory complaint. Forced-choice formats are used to minimize confounding of detection and discrimination threshold measures with response biases. Forced-choice formats range from simple yes–no tests, such as "Does this odor smell like a lemon?" to multiple-choice tests in which a number of alternatives are presented. The most popular clinical smell identification test provides four response alternatives for each of 40 odorants located on microencapsulated strips. A patient's score on this test is compared to that of normal subjects of equivalent age and gender using standardized norms that allow determination of a percentile value.

A popular odor detection threshold test is often referred to as the "method-of-limits" procedure. During this assessment technique the concentration of an olfactory stimulus is increased or decreased incrementally until the patient is just barely able to perceive it. Most commonly, log-step or half-log-step concentrations of an odorant are presented to a patient using "sniff" or squeeze bottles. For example, on a given trial a patient may be asked to report which of two successively presented stimuli smells stronger. One is a given concentration of an odorant dissolved in a liquid medium such as light mineral oil and the other the undiluted liquid medium. The presentation order of the two stimuli is randomized from trial to trial, and successively higher stimulus concentrations are presented until a reliable detection is made. Staircase procedures, in which odorant concentrations are increased or decreased as a function of the correctness of the subject's responses, have gained in popularity and provide a reliable measure of threshold with a minimum number of trials. Because threshold tests are time-consuming, usually a patient's sensitivity to only one or two odorants is measured.

The olfactory system is capable of detecting and discriminating among thousands of odorants, and a threshold test can sample only a small portion of the potentially detectable odorants. Therefore the question arises as to whether the patient's complaint reflects insensitivity to a stimulus domain other than the one being sampled. Fortunately, it appears that individuals who are insensitive to one compound usually evidence insensitivity to others. This common insensitivity has been inferred through clinical studies that have found relatively high correlations among threshold values obtained from different compounds within the same individuals. Nevertheless, instances of "smell blindness" occur in which only a small number of stimuli are involved. For this and other reasons, most clinics use odor identification tests in addition to threshold tests for assessing olfactory ability.

Other, less popular olfactory tests have also been applied in the clinic. These tests include tests of: (1) differential threshold, where the concentration of an odorant is increased until it is just noticeably different from another above-threshold odorant; (2) suprathreshold buildup in odor perception using either odor intensity rating scales or the techniques of magnitude estimation; (3) adaptation; (4) the ability to make fine distinctions among odors of varying quality, as measured by techniques such as multidimensional scaling; and (5) odor memory, which requires the subject to smell a target odor and then select, after an interval, the same odor from a set of others.

Psychophysiologic measures that reflect autonomic nervous system activity, resulting from stimulation of nonolfactory afferents such as the trigeminal nerve, should be avoided, as should tests where the odorant is directly absorbed into the bloodstream. Despite this variety of tests, most current psychophysiologic tests are subjective. Considerable progress, however, is being made in both the development and understanding of some objective measures of olfaction, such as odor-evoked potentials, which may soon be available for clinical use. Such tests may help determine whether the problem is traceable to the central nervous system or the peripheral nervous system, the disorder is based on the olfactory nerve or trigeminal nerve, or it is attributable to a sensorineural defect or airflow occlusion.

Medical Imaging Procedures

The final component of the comprehensive evaluation for olfactory disorders is the use of medical imaging procedures. Until recently the ability of a physician to image the olfactory pathways was limited. Technologic advances now allow physicians to examine the integrity of the olfactory pathways and permit identification of many treatable causes of olfactory dysfunction. Most importantly, the use of medical imaging procedures allows the physician to detect serious or life-threatening disorders that may cause olfactory symptoms.

Computed tomography provides detailed structural information about the nasal cavities, particularly in the superior portions of the nasal cavities, the cribriform plates, and the anterior cranial fossa. It provides definitive information on the presence or absence of sinusitis. It is a valuable adjunct for ruling in or out neoplasms of the nose, paranasal sinuses, and cranial cavity. Particularly with the use of enhancement with intravenous iodinated compounds, computed tomography is an effective means of excluding intracranial neoplasms (e.g., meningiomas of the anterior cranial fossa) that may affect the olfactory system. Unsuspected trauma to the cribriform plate and the base of the skull may be found in this manner. Magnetic resonance imaging is particularly helpful when evaluating the contents of the cranial cavity. It can differentiate neoplasms from mucoceles exceedingly well and allows diagnosis of extramucosal fungal disease. For the initial evaluation, however, computed tomography takes precedence because of the bony detail it provides.

Differential Diagnosis of Transport and Sensorineural Chemosensory Disorders

Once the clinical evaluation of the chemosensory disorder is completed, the examiner can proceed with the differential diagnosis, which occurs on three levels. The first differentiation is whether one or both chemosensory systems are involved. This determination rests almost entirely with the psychophysical evaluation. The second step is to determine the site of the lesion that causes the olfactory complaint. This diagnosis is essentially an anatomic one. It is important to know whether the anatomic diagnosis of the disorder is a transport loss, a sensory loss, or a neural loss. Transport losses result from lesions of the nasal cavities; sensory losses result from lesions of the olfactory neuroepithelium; and neural losses may be either peripheral or central, resulting from lesions of the olfactory nerves or the central olfactory pathways. The third differentiation is to determine the cause of the anatomic abnormality.

The history can help determine whether the loss is a transport loss or a sensorineural loss. It is best evidenced by knowledge of events temporally related to the onset of the loss or distortion of the sense of smell. Bacterial infections of the nose and paranasal sinuses, specific and nonspecific rhinitis, and allergic rhinitis, for example, suggest a transport olfactory loss. Viral infections suggest a sensorineural olfactory loss, although they may, of course, produce a transport olfactory loss during the acute phase. Trauma to the maxillofacial area or head suggests a sensorineural olfactory loss resulting from transection of the olfactory fila. Nasal trauma, particularly that extending to the frontoethmoid complex, may produce a transport olfactory loss. Drugs affecting cell turnover are more likely to produce a sensory lesion than a neural lesion. However, neurotoxic agents may produce either a sensory or a neural lesion.

Positive findings during the physical examination in general suggest a transport olfactory loss. Positive findings in the medical imaging of the nose and paranasal sinuses usually suggest a transport olfactory loss, although malignant neoplastic processes originating in the nasal cavities and paranasal sinuses may destroy the olfactory neuroepithelium and result in a sensory or neural lesion. Positive findings inside the cranial cavity on medical imaging, of course, suggest a neural olfactory loss.

The anatomic differentiation reduces the number of etiologic possibilities for the olfactory disorder because only a limited number of causes can produce a lesion at a given anatomic site. Knowledge of the known possible causes of olfactory disorders coupled with the psychophysical evaluation, the anatomic possibilities, and positive findings from the history, physical examination, imaging, and perhaps biopsy data makes it possible to deduce a credible etiologic diagnosis that can be used as the basis for rational treatment.

Conclusion

The knowledge and methods now available to conduct a comprehensive clinical evaluation of chemosensory disorders were made possible by the hard work of physicians, psychophysical clinicians, and scientists devoted to research in the chemical senses. Although international interest in the chemical senses has increased scientific opportunities and advancements, there is an ever constant need for more quality research. With international commitment to the study of chemosensory disorders, accelerated progress is not only possible but an exciting and challenging probability. The future promises to provide unparalleled benefits from the discoveries that result from our worldwide quest to improve the quality of life for all with chemosensory disorders.

Psychophysical Assessment of Chemosensory Disorders in Clinical Populations

CLAIRE MURPHY[1,2], JUDITH A. ANDERSON[2], and STACY MARKISON[1]

Key words. Children—Aging—Dementia—Nasal disease—Odor threshold—Odor identification

Introduction

This chapter focuses on two groups in whom clinical problems of smell function occur: the very young and the old. The intent here is to examine several of these problems and their psychophysical manifestations and then to examine what this work reveals about normal function and psychophysical assessment.

After encountering mediocre scores when testing children's odor identification ability with tests designed for assessing adults, our laboratory has been developing a Children's Test of odor identification, focusing on two of the problems children experience in this task: unfamiliarity with many of the odors easily identified by adults and, particularly in the youngest children, inability to read and thereby perform in the response mode used for adult testing. A report describing these experiments is currently in preparation.

If the test is to be useful in the clinic, normal children need to perform well on the test. Beginning with a large group of odorants and testing their identifiability in several groups of children ranging from 3 to 8 years of age, a battery of eight items on which the average 5-year-old, prereading, preschooler can manage a 90% correct score on identification was identified. To minimize problems of the prereader and early reader, a picture response board was refined to a cue sheet of line drawings to which the child could point. Practice with identifying the pictures was incorporated into the protocol.

The battery is currently being tested on children with diseases in which olfactory impairment is suspected. It is a working component of the battery of testing used at the Nasal Dysfunction Clinic at the University of California, San Diego (UCSD) Medical Center and is administered to every child under 12 who presents with nasal disease, allergic rhinitis, or smell loss.

Method

Stimuli

For threshold assessment *N*-butyl alcohol was prepared in a series of tertiary dilutions beginning with 4% v/v in deionized water.

Odor identification testing employed three odor identification batteries. The Children's Test utilized baby powder, bubble gum, cinnamon, chocolate, coffee, mustard, peanut butter, and Play-Doh. The Adult Test utilized ammonia, baby powder, chocolate, cinnamon, coffee, Ivory soap, mothballs, peanut butter, Vicks, and wintergreen. A subset of subjects was given the items on the University of Pennsylvania Smell Identification Test (UPSIT) [1], described below.

Procedure

All threshold assessments, regardless of the age of the subject, employed a two-alternative, forced-choice, ascending procedure [2]. Subjects reported which of two samples smelled stronger. Incorrect responses led to more concentrated solutions. Five successive correct responses determined threshold.

For the Children's Test, identification testing was introduced as a smell game. Children were first presented with the picture cue sheet and prompted to identify all 20 line drawings. If the child was unsure of a picture, the tester provided the correct label. The tester ensured that the child was able to inhale properly when instructed. The child then held a blindfold over the eyes. The tester held an odorant under the nostrils for approximately 5 s and instructed the child to sniff. The child then removed the blindfold and pointed to a picture on the cue sheet to identify the odorant. For older children, a verbal response was accepted. Odorants were presented in random order.

For the Adult Test, the blindfolded subject smelled the contents of an opaque jar and then attempted identification from a cue sheet containing odor names and distractors.

[1] Department of Psychology, San Diego State University, San Diego, CA 92182-0551, USA
[2] Division of Head and Neck Surgery, University of California Medical Center San Diego, San Diego, CA, USA

The UPSIT consists of 40 microencapsulated odors [1]. Subjects scratch a microencapsulated odor and then attempt to identify it from a list of four odor names.

Subjects

A total of 92 children, including normal children, allergic children, cystic fibrosis patients, and adults with Down's syndrome were compared. In the first three groups the youngest children were 5 years old and the mean ages were 7, 10, and 10 years, respectively. The Down's syndrome subjects were 21 to 46 years (mean 31 years).

Results

Children

Several experiments were conducted to examine the reliability, validity, and appropriateness of the Children's Test. The percent correct identifications for each subject in each patient group are shown in Fig. 1. Groups differed in performance. The mean performance score was significantly poorer in the cystic fibrosis (CF) and Down's syndrome subjects than in the controls and allergy patients.

Subjects were tested twice, with a mean delay of 5.4 days. Test-retest reliability was good throughout the range of function represented by the different groups, $r = 0.86$.

A subset of 23 normal children with a mean age of 6.5 years were tested on both the Children's Test and the adult odor identification test. Children could identify a significantly greater percentage of items on the Children's Test (mean = 76) compared with the Adult Test (mean = 39).

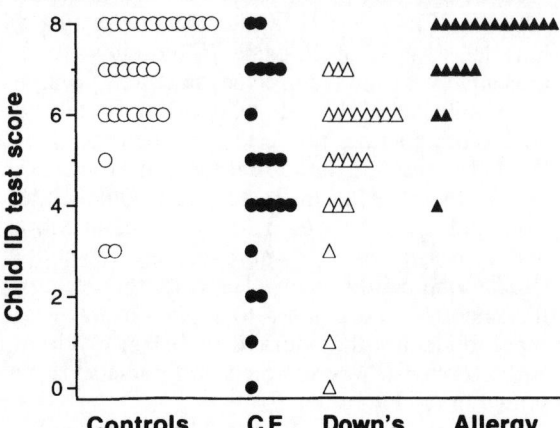

FIG. 1. Percent correct odor identification on the Children's Test for each child in each of the groups

Cystic Fibrosis Patients

Further examination of the data from the CF patients provides information concerning the utility of the Children's Test. Patients with CF have been tested as they present to the Cystic Fibrosis/Pulmonary Clinic, UCSD, for their annual evaluations. These patients exhibit varying levels of CF disease severity as defined by Schwachman/Kulczycki clinical disease scores. Adults underwent the adult version of the odor identification test and children the children's version. Scores were computed as percent correct of the total in the respective test. The mean percent correct for the CF patients was significantly lower (mean, 66) than for the controls (mean, 88). Odor identification was particularly impaired in patients in the adult group.

Performance of the normal control children on the Children's Test was comparable to the performance of normal adults on the Adult Test. Differences were not significantly different on statistically testing. There was a trend toward better performance for children on the Children's Test than for teens on the Adult Test.

Threshold scores confirmed the olfactory impairment in CF patients relative to controls. Older adults with CF had poorer thresholds than controls and than young adults, teens, or children with CF (Fig. 2, top panel). These older adults with CF whose Schwachman/Kulczylcski scores indicated more severe clinical disease also had more severe odor identification impairment (Fig. 2, bottom panel). These data suggest that olfactory function and disease status of CF patients may be more closely linked than previously thought.

One of the aims of these studies was to determine whether odor identification testing could substitute for the threshold because maintaining attention of young children in a threshold task is difficult. Lapses in attention result in thresholds that suggest children are less sensitive than they are. As shown in Fig. 2, the data from the CF children do suggest that olfactory impairment measured with the Children's Odor Identification Test mirrors olfactory impairment measured by painstaking threshold assessment.

The data from all the patient groups underscore the need for an odor identification test specifically developed for children. Children with normal olfactory function tested with both the Adult and Children's Identification Tests identified a greater percentage of items on the Children's Test.

The psychometric properties of the test appear to be sound. The test shows good test-retest reliability, differentiates between groups with varying levels of olfactory sensitivity, and is related to threshold scores. It may be more useful than threshold tests for children, as they show greater variability than adults in olfactory threshold scores. Difficulty maintaining attention on the lengthy threshold task prob-

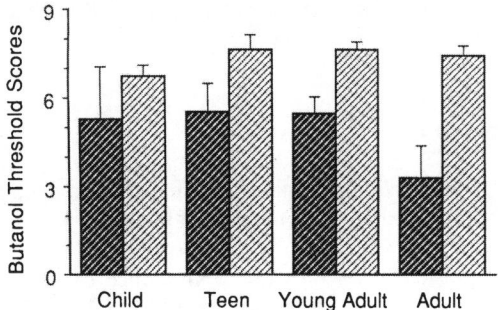

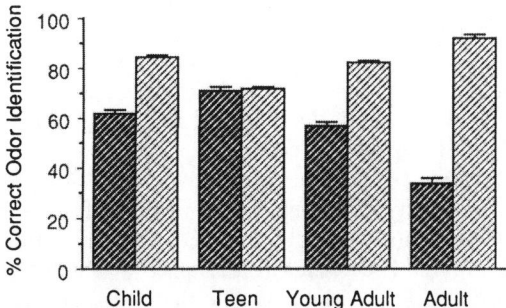

FIG. 2. Butanol threshold (*top*) and odor identification (*bottom*) scores for normal controls and cystic fibrosis (CF) patients of different ages (and with increasing duration of the disease). *Dark hatched bars*, CF; *light hatched bars*, controls

ably underlies some of this variability. Further research is needed on this point. The shorter and less tedious odor identification test is appealing. Its brevity renders it useful in situations where threshold testing is impractical.

Demented Patients

Alzheimer's disease (AD) is a debilitating, progressive brain disorder characterized by dementia; it is definitively diagnosed on autopsy or biopsy. Patients with AD exhibit cell loss and neurofibrillary tangles and neuritic plaques, most concentrated in association areas, particularly the hippocampus. Increased plaques and tangles, cell loss, and granulovacuolar degeneration occur in entorhinal cortex, prepiriform cortex, and the anterior olfactory nucleus. These areas mediate olfactory information [3–6], suggesting olfactory impairment in AD. Characterizing the extent of olfactory dysfunction in AD patients has presented a psychophysical challenge. The lessons learned from studying these patients with psychophysical methods facilitate the characterization of olfactory function in other patients in whom cognitive abilities, memory, and attention are concerns.

Early attempts to describe olfactory function in AD patients focused largely on odor identification ability [7–10]. A typical odor identification task involves sampling the odors presented, reading the multiple choices of possible odor names (labels), and then choosing the label. Performance requires the sensory ability to detect an odor and the short-term memory and attention to recognize an odor and select the closest identification from the multiple choices offered. The latter ability implies enough facility at odor memory to recall what those multiple choices smell like. In most studies of odor identification in AD and in most taste and smell clinics where olfactory assessment takes place, no attempt is made to discern the role cognitive function or dementia plays in the patients' ability to perform an odor identification task. Without this information, it is risky to attribute performance on such tests to olfactory function. Patients unable to name odors perform poorly regardless of smell function.

Olfactory threshold testing might appear to present less difficulty, as patients need not name the odor; however, threshold tasks are not without memory demands. To perform in a two-alternative odor threshold procedure, for example, the patient must hold in memory the perception of one sample while sniffing the second and then choose which is stronger. Whether the olfactory system or the memory demands of the task dictate performance is an important question in the demented, as in children. The approach we have taken with many of these groups has been to study simultaneously the threshold for another sensory system not implicated in the disease and to make the subject's task as similar as possible: to choose from each pair presented the stronger of the two stimuli.

In the study described briefly below, olfactory and taste thresholds were assessed in AD patients and normal elderly controls. Taste stimuli were a series of concentrations of sucrose in deionized water. Complete details regarding this study can be found in Murphy et al. [11]. The AD patients had significantly higher olfactory thresholds than controls, but taste thresholds did not differ. Thus patients were able to perform a two-alternative, forced-choice taste threshold task. It was concluded that the deficits seen on olfactory threshold testing in AD patients do truly indicate olfactory impairment in the disease.

Correlations used to examine the relation between olfactory threshold and dementia for the AD patients indicated that patients who make more errors on the dementia scales also show decreased sensitivity to odor (Fig. 3). Indeed, the yearly decrement in dementia scores correlates equally well with the yearly decrement in butanol threshold scores.

There are instances where criterion- or response-free methods of assessing olfactory function that are noninvasive would be useful. The olfactory evoked potential techniques may ultimately prove useful for the assessment of AD patients, as the ability to respond with a psychophysical judgment is not re-

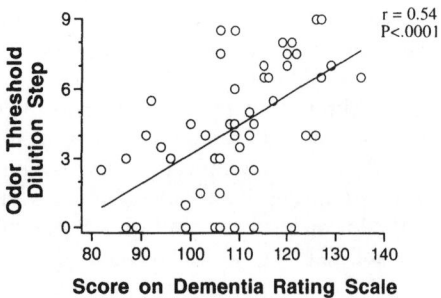

FIG. 3. Butanol thresholds for Alzheimer's disease patients as a function of dementia scores

quired. We have used this technique with normal elderly subjects and young controls and demonstrated age-associated differences in amplitude that are correlated with the olfactory threshold [12; see also Kobal and Hummel, this volume].

Normal Aging

Olfactory loss in the elderly is reflected in threshold, intensity, perceptual quality and pleasantness of odors, odor identification, and odor memory [13–18]. In each case, the elderly show greater variability than young adults. How does the older person who maintains normal function differ from most older persons who show an age-related decline? It seems natural to include otorhinolaryngologic examinations in studies of chemosensory aging, but this has rarely been done. The degree to which nasal disease affects olfactory function, both acutely and cumulatively over the life span, seems to be an important potential influence on the loss observed in the elderly. Alzheimer's disease and other dementias in which olfactory impairment has been demonstrated affect older people. Thus investigating cognitive function is important in order to differentiate between olfactory impairment due to aging from other etiologies.

An experiment from our laboratory drew a random sample of elderly persons who were of high socioeconomic status and prescreened them with a thorough rhinologic examination for nasal-sinus disease, allergic rhinitis, head trauma, and upper respiratory disease; they were also prescreened for dementia [19]. Those with problems in these areas did not meet the criteria for participation and were excluded from the study. Although odor thresholds did show significant impairment relative to the young, the elderly in that study showed better olfactory function than was previously considered possible in this age group [20]. Thus information derived from testing clinical patients with nasal disease and those with dementia has facilitated identification of several factors that play a role in the variability seen

in olfactory function in older subjects that derives from something other than normal aging.

Summary

Work with clinical populations suggests the need for screening subjects in normal populations for nasal disease and dementia. It is important to tailor psychophysical testing to the competence of the subjects, and the interpretation of the data should carefully incorporate a clear understanding of exactly which functions are necessary and sufficient for performing a given psychophysical task. What has been learned from testing clinical populations has shed light on chemosensory disorders in those populations and, at the same time, served to sharpen psychophysical assessment techniques.

Acknowledgments. Supported by NIH grants AG04085 and AG08203 from the National Institute on Aging. The technical assistance of Jill Razani, Steven Nordin, Maritess Mauricio, Ruel Canada, Carlo Quiñonez, Lisa Maxwell, and Michael Madowitz is gratefully acknowledged.

References

1. Doty RL, Shaman P, Dann M (1984) Development of the University of Pennsylvania Smell Identification Test: a standardized microencapsulated test of olfactory function. Physiol Behav 32:489–502
2. Cain WS, Gent JS, Catalanotto FA, Goodspeed RB (1983) Clinical evaluation of olfaction. Am J Otolaryngol 4:252–256
3. Averback P (1983) Two new lesions in Alzheimer's disease. Lancet 19:1203
4. Esiri MM, Wilcock GK (1984) The olfactory bulbs in Alzheimer's disease. J Neurol Neurosurg Psychiatry 47:56–60
5. Ohm TG, Braask H (1987) Olfactory bulb changes in Alzheimer's disease. Acta Neuropathol (Berl) 73:365–396
6. Reyes PF, Golden GT, Fagel PL, et al. (1987) The prepiriform cortex in dementia of the Alzheimer type. Arch Neurol 44:644–645
7. Waldton S (1974) Clinical observations of impaired cranial nerve function in senile dementia. Acta Psychiatr Scand 50:539–547
8. Serby M, Corwin J, Novatt A, Conrad P, Rotrosen J (1985) Olfaction in dementia. J Neurol Neurosurg Psychiatry 48:848–849
9. Warner MD, Peabody CA, Flattery JJ, Tinklenberg JR (1986) Olfactory deficits and Alzheimer's disease. Biol Psychiatry 21:116–118
10. Koss E (1986) Olfactory dysfunction in Alzheimer's disease. Dev Neuropsychol 2:89–99
11. Murphy C, Gilmore MM, Seery CS, Salmon DP, Lasker BR (1990) Olfactory thresholds are associated with degree of dementia in Alzheimer's disease. Neurobiol Aging 11:465–469
12. Murphy C, Nordin S, de Wijk R, Cain WS, Polich J (1994) Olfactory evoked potentials: assessment of

young and elderly and comparison to psychophysical threshold. Chem Senses 19:47–56

13. Murphy C (1983) Age-related effects on the threshold, psychophysical function, and pleasantness of menthol. J Gerontol 38:217–222

14. Doty RL, Shaman P, Applebaum SL, et al. (1984) Smell identification ability: changes with age. Science 226:1441–1443

15. Murphy C, Cain WS, Gilmore MM, Skinner RB (1991) Sensory and semantic factors in recognition memory for odors and graphic stimuli: elderly versus young persons. Am J Psychol 104:161–192

16. Schiffman SS (1986) Age related changes in taste and smell and their possible causes. In: Meiselman HL, Rivlin RS (eds) Clinical measurements of taste and smell. Macmillan, New York, pp 326–342

17. Murphy C (1986) Taste and smell in the elderly. In: Meiselman HL, Rivlin RS (eds) Clinical measurements of taste and smell. Macmillan, New York, pp 343–371

18. Murphy C (1989) Aging in chemosensory perception of and preference for nutritionally significant stimuli. In: Murphy C, Cain WS, Hegsted DM (eds) Nutrition and the chemical senses in aging: recent advances and current research needs. New York Academy of Sciences, New York, pp 251–266

19. Murphy C, Davidson TM (1992) Geriatric issues: special considerations. J Head Trauma Rehabil 7:76–82

20. Markison S, Murphy C (1993, submitted) Olfactory threshold comparison in young and elderly persons prescreened for nasal disease and dementia

Clinical Evaluation of Olfactory Dysfunction at Kagoshima University

SHIGERU FURUTA[1], RIKAKO HIROTA[1], MASAHIKO EGAWA[1], MASARU OHYAMA[1], and ETSURO OBATA[2]

Key words. Smell—Olfactory function test—Sinusitis—Common cold—Head trauma

Introduction

The sense of smell is of considerable importance to humans. Individuals lacking smell ability are subject to the consequences of being unable to detect escaping gas, dangerous fumes, and fires in the home, automobile, and workplace; and they frequently complain of loss of enjoyment from eating and drinking. It is clear that a large number of individuals experience olfactory dysfunctions following accidents, disease states, medical interventions, aging, and exposure to a number of environmental pollutants. Although the degree of overall impairment from olfactory loss is less than that produced by major losses in the other senses, smell disturbances are of considerable significance to persons experiencing them. According to the previous clinical evaluation in patients with the olfactory disturbance, we noted that there was a distinct difference among the results of the etiologic classes of olfactory disturbance in Japan and in the other countries [1,2]. In this chapter, we describe the clinical evaluation and the results from patients with olfactory disturbances in our clinic.

Materials and Methods

Subjects

The study group was composed of 132 male and 154 female patients with olfactory disturbance who visited the Smell and Taste Clinic at the Hospital of Kagoshima University over a 28-month period (mean ± SD age 49.52 ± 17.35 years; age range 8–87 years).

[1] Department of Otolaryngology, Faculty of Medicine, Kagoshima University, 8-35-1 Sakuragaoka, Kagoshima, 890 Japan
[2] Nagahama Hospital, 904-1, Shiromoto, Oonejime-cho, Kimotsuki-gun, Kagoshima, 893-23 Japan

Olfactory Function Test

The subjects underwent an olfactory function test in an air-conditioned room. Four primary psychophysical tests were used to assess olfactory function: (1) the smell identification test (SIT) [3,4], a standard 12-stimulus microencapsulated "scratch and sniff" odor identification test; (2) a forced-choice single staircase odor detection threshold test using phenylethyl alcohol (PEA) [5], a rose-like odor compound with comparatively little intranasal trigeminal stimulative properties; (3) a standard olfactory acuity test using the T&T olfactometer [6], including five odorants; and (4) a intravenous olfactory test using fursultiamine [7], a garlic-like odor compound.

Statistical Analysis

We used ANOVA for the analysis of the relations between olfactory function and the clinical data of each patient.

Results

Based on data from the medical histories and examinations, each patient was classified into one of primary probable categories. Because of the problems of concurrent disorders, medication usage, and incomplete data, the reliability of such assignments could not always be exact. The most common olfactory-related complaint was that of reduced smell (hyposmia). Twenty-five percent of patients reported total loss of smell function (anosmia). Eleven patients complained of unilateral olfactory disturbance. As can be seen in Table 1, a variety of primary probable etiologies was associated with olfactory disorders. Sinusitis (44.4%), the common cold (20.2%), and nasal allergy (10.4%) accounted for about 75% of cases. About 60 of the persons reporting smell loss believed that they had at least partial taste loss. Sinusitis, common cold, nasal allergy, and head trauma served as the focus of analyses to be described herein. Three-fourths of patients with head trauma reported total loss of smell as their chief complain. However, patients with upper respiratory infections (URIs) complained of

TABLE 1. Selected patient information for 281 patients evaluated at Kagoshima University Hospital.

Probable eitology	Frequency	Male/female	Taste loss
Sinusitis	127 (44.4%)	77/50	16 (12.5%)
Common cold	58 (20.2%)	15/43	21 (36.2%)
Nasal allergy	30 (10.4%)	12/18	5 (6.6%)
Head trauma	8 (2.8%)	4/4	1 (12.5%)
Psychiatric disorder	8 (2.8%)	0/8	2 (25.0%)
Nasal disease	5 (1.7%)	2/3	0
CNS disease	5 (1.7%)	2/3	1 (20.0%)
Nasal tumor	5 (1.4%)	3/2	1 (20.0%)
Other sinus disease	3 (1.0%)	1/2	0
Operation-related	3 (1.0%)	2/1	1 (33.3%)
Age-related	2 (0.7%)	2/0	1 (50.0%)
Toxic chemicals	2 (0.7%)	0/2	0
Congenital	1 (0.3%)	1/0	1 (100%)
Irradiation	1 (0.3%)	1/0	1 (100%)
Other	28 (9.7%)	10/18	9 (32.1%)
Total	286	132/154	60 (20.9%)

reduced smell more than anosmia. Unilateral olfactory dysfunction was found in a few cases with URIs.

The SIT scores in patients with head trauma were lower than those in patients with URIs. Their olfactory function was severe by our classification [4]. Patients with nasal allergy had the highest SIT scores, 6.28 ± 0.63, among the four etiologic groups (Fig. 1). The olfactory functions in patients with URIs were moderate. The results of detection thresholds test were similar to those of the identification test (Fig. 2). Both detection and recognition thresholds of T&T olfactometry were significantly elevated in those with head trauma, followed by those with sinusitis, the common cold, and nasal allergy, in that order. The difference between the thresholds of T&T olfactometry was by only one step of the concentration of each odorant, and dissociation between the two thresholds was not observed (Fig.

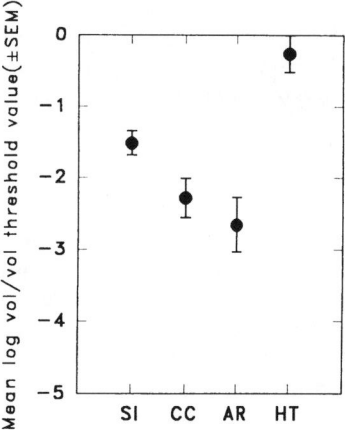

FIG. 2. Average values of the detection threshold (± SEM) by the phenylethyl alcohol detection test in four groups

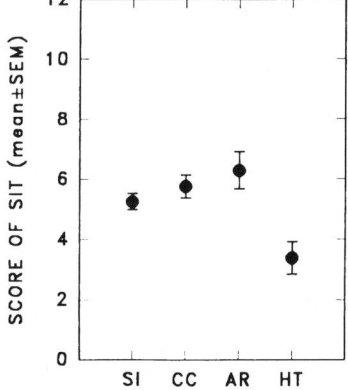

FIG. 1. Average composite score (mean ± SEM) of the smell identification test (*SIT*) in four groups. *SI*, chronic sinusitis; *CC*, common cold; *AR*, nasal allergy; *HT*, head trauma

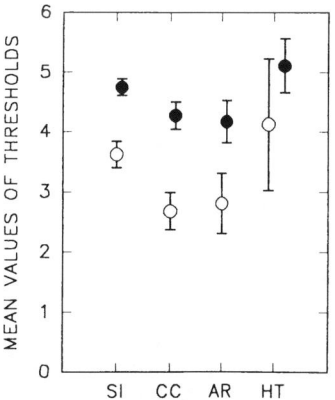

FIG. 3. Average composite values (mean ± SEM) of the detection threshold (*open circles*) and recognition threshold (*filled circles*) of T&T olfactometry in four groups

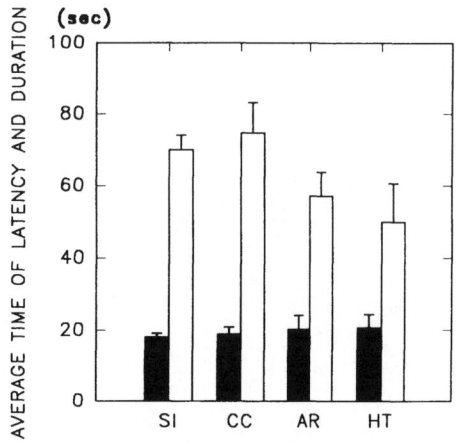

FIG. 4. Average composite time (mean ± SEM) of the latency (*filled bars*) and duration (*open bars*) by the detection of a garlic-like odor after intravenous injection of fursultiamine in four groups

3). The latencies of the intravenous olfactory test extend in all groups. Its short duration was found in patients with nasal allergy and head trauma (Fig. 4).

Discussion

The most common cause of olfactory disturbance is nasosinus disease blocking airflow to the olfactory mucosa. Severe bilateral airway obstruction, such as chronic sinusitis, nasal polyposis, allergic rhinitis, vasomotor rhinitis, malignant disease of the paranasal sinuses with extension into the nasal cavities, and hyperplasia of turbinates can influence smell function by restricting airway patency.

The anosmia or hyposmia due to polyps, allergies, and sinusitis has been cured by using prednisolone [8] as well as endonasal surgery [9], including polypectomy, endonasal ethmoidectomy, and septoplasty. In the case of chronic sinusitis, the surface of the olfactory region was apparently replaced by respiratory epithelium with ciliated columnar and goblet cells and extensive tracts of dense collagenous tissue in the lamina propria indicative of scar tissue [10,11]. The authors have reported that severe olfactory dysfunction in patients with sinusitis may be due to damaged olfactory epithelium rather than obstruction of the olfactory cleft. This conclusion was based on radiologic findings concerned with the relation of the olfactory cleft and olfactory functions [12].

Airway blockage resulting from viral infections such as the common cold can distort olfactory function. However, with the common cold a period of either hyperosmia or hyposmia can ensue, even after airway patency has returned to normal. In such cases it is not clear whether the state of the mucous membrane is at fault or the receptors and associated cilia are incapacitated or damaged in some fashion. Although in most instances the sense of smell returns to normal after several weeks of such symptoms, it is not always the case. The onset of symptoms seems to be related to the age or associated disease of the patients in these cases [13].

The sole biopsy study reported in the literature of the olfactory mucosa of a patient exhibiting long-term anosmia after a bout of influenza evidenced severe loss of neuronal elements in the subepithelial tissue and scar tissue in adjacent regions. Yamagishi et al. reported that morphologic and immunohistochemical examinations on the olfactory mucosa of eight patients with olfactory disturbances caused by the common cold revealed three patterns of degeneration: (1) an olfactory epithelium that was of adequate thickness with a basic arrangement of supporting cells, olfactory receptor cells, and basal cells but slightly fewer than normal receptor cells; (2) olfactory cells were decreased; and (3) the olfactory mucosa was thin and atrophic with no receptor cells or olfactory vesicles, and only the supporting and basal cells remaining [14].

The olfactory disturbances caused by head trauma have increased [15]. The physical trauma is produced by frontal fractures of the skull or by blows to the occipital region. The olfactory filaments passing through the cribriform plate are fragile and can be damaged or torn following shearing actions produced by blows and whip-like forces, which tear the bulbs away from the cribriform plate [16].

It appears that a conservative estimate of the incidence of anosmia is associated with the severity of the trauma. Judging from the few available accounts, the prognosis is poor, as it is with severe cases caused by the common cold. We should develop a new treatment for severe cases with a poor prognosis.

References

1. Doty RL (1979) A review of olfactory dysfunctions in man. Am J Otolaryngol 1:57–79
2. Asaga H (1990) Diagnosis of olfactory disorders. Otolaryngol Head Neck Surg (Tokyo) 62:727–730
3. Doty RL, Shaman P, Dann MS (1984) Development of the University of Pennsylvania Smell Identification Test. Physiol Behav 32:489–502
4. Furuta S, Egawa M, Ozaki M, Ohyama M (1992) Clinical application of the smell identification test. J Otolaryngol Jpn 95:1339–1344
5. Doty RL, Gregor TP, Settle RG (1986) Influence of intertrial and sniff-bottle volume on phenyl ethyl alcohol odor detection thresholds. Chem Senses 11:259–264
6. Shiokawa H (1975) The clinical studies on the olfactory test using standard odorous substances. J Otolaryngol Jpn 78:1258–1270
7. Furukawa M, Kamide M, Miwa T, Umeda R (1988)

Significance of intravenous olfaction test using thiamine propyldisulfide (alinamin) in olfactometry. Auris Nasus Larynx (Tokyo) 15:25–31

8. Goodspeed RB, Gent JF, Catalanotto FA, Cain WS, Zagraniski RT (1986) Corticosteroids in olfactory dysfunction. In: Meiselman HL, Rivlin RS (eds) Clinical Measurement of Taste and Smell. Macmillan, New York, pp 514–518

9. Hosemann W, Goertzen W, Wohlleben R, Wolf S, Wigand ME (1993) Olfaction after endoscopic endonasal ethmoidectomy. Am J Rhinol 7:11–15

10. Egawa M, Matsune J, Hanamure Y, Furuta S, Ohyama M (1992) Relationships between olfactory and respiratory epithelium in experimentally induced sinusitis of rabbit. JASTS 26:373–377

11. Ohyama M, Chung SK, Hanamure Y, Egawa M, Tsurumaru H (1991) Infuence of experimental sinusitis on olfactory mucosal morphology in the rabbit. J Clin Electron Microsc 24:793–794

12. Furuta S, Moriyama I, Hirase H, Ishikawa T, Nishimoto K (1992) Relationship between radiological finding and olfactory function in patients with chronic sinusitis. JASTS 26:377–380

13. Kimura Y, Sakumoto M, Yamamoto T, Yachi N, Furukawa M (1991) Clinical observation on olfactory disturbance following influenza-like infection. Otol Rhinol Laryngol (Tokyo) 34:647–652

14. Yamagishi M, Nakamura H, Suzuki S, Hasegawa S, Nakano Y (1990) Relationships among olfactory epithelium appearance, olfactory test results and prognosis. Pract Otol (Kyoto) 83:383–390

15. Zusho H (1982) Posttraumatic anosmia. Arch Otolaryngol 108:90–92

16. Costanzo RM, Becker DP (1986) Smell and taste disorders in head injury and neurosurgery patients. In: Meiselman HL, Rivlin RS (eds), Clinical measurement of taste and smell. Macmillan, New York, pp 565–578

Diagnosis and Treatment
of Distorted Olfactory Perception

Donald Leopold[1] and Gary Meyerrose[2]

Key words. Phantosmia—PET scan—Olfactory hallucination—Hyposmia—Troposmia—Odorant

Introduction

There are three types of human olfactory dysfunction: (1) decreased ability to perceive odorants (hyposmia, anosmia); (2) distorted quality of perceived odorants (troposmia, derived from *trop*, "to turn or react," and *osmia*); and (3) perceived odor when no odorant stimulus is present (phantosmia, hallucination). Patients with only hyposmia or anosmia generally complain of decreased ability to smell odorants and appreciate foods, but the quality of the perceived odors is "normal" for the patient. Although patients with troposmia, phantosmia, and hallucinations may have decreased ability to perceive odorants, their main concern is that perceived odors are distorted or different from what they remember.

Diagnosis

The clinical task when evaluating patients with complaints of olfactory distortion is to determine the etiology of the problem and then recommend treatment (if possible). The methods to accomplish this goal include the clinical history, physical examination, imaging studies, and testing.

Clinical History

While obtaining the clinical history, it is important to be sympathetic, as many of these patients are anxious about this symptom, which is difficult to diagnose and sometimes associated with psychiatric disease. Initially, three general questions must be answered: (1) Does the patient have a decreased sense of smell, a distorted one, or both? (2) Is the complaint a perceived taste or a perceived smell? (3) Is the perceived smell a distortion of an existing odorant (troposmia), or can it be present without a chemosensory stimulus (phantosmia or hallucination)? If the patient is complaining of a distorted smell, a description of that smell should be elicited, such as "burned rubber" or "rotted meat." It should also be determined if the perceived smell masks other environmental odorants and if it is described as a "body odor" or something other than a body odor.

Further details about the onset and character of the distorted smell aid in the diagnosis. A real smell due to a metabolic problem (e.g., trimethylaminuria) is present throughout the patient's life but is initially noticed by the patient during the teenage years. It is also noticed more during exhalation. In contrast, most other complaints of olfactory distortion begin at least a decade later, during the young adult years, and are noticed during inhalation. Smells that are perceived in only one nostril are usually classified as troposmia or phantosmia, whereas most hallucinations are bilateral or cannot be localized. Similarly, events that cause loss of olfactory ability, such as an upper respiratory infection or head trauma, are sometimes associated with the onset of phantosmia or troposmia, whereas the onset of hallucinations is usually associated with a psychiatric or neurologic disease.

Physical Examination

The physical examination includes a thorough head and neck evaluation. Healthy tympanic membranes should have no abnormalities near the chorda tympani nerve, and the normal tongue should have a plentiful display of taste papillae on each side. Nasal endoscopy should reveal the nasal cavity to be lined by pink, moist mucosa with adequate airflow pathways for respiration and olfaction. Trigeminal nerve function must be intact and symmetric. Finally, the patient's general health must be assessed, along with the general psychiatric status.

Imaging Studies

Computed tomography (CT) scanning is the mainstay of imaging for olfactory dysfunctions because it provides the best images of the nasal and sinus cavities, the olfactory cleft, and the bony skull base.

[1] Department of Otolaryngology—Head & Neck Surgery and [2] Department of Radiology, Division of Nuclear Medicine, Johns Hopkins Medical Institutions, Baltimore, MD 21205, USA

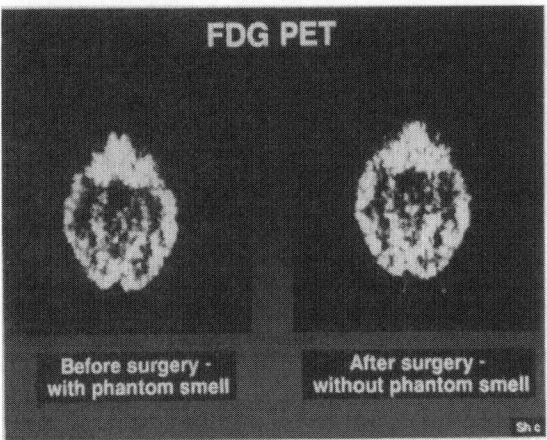

FIG. 1. PET scan images of the brain of a patient who had phantosmia in the right nostril. Note the increased activity (brightness) in the left frontal, insular, and temporal regions, which decreased after excision of the olfactory epithelium from the right nasal cavity. Top is anterior, and the *right* side of each image represents the patient's left side

Many nasal cavities do not allow gentle endoscopic examination of the olfactory cleft, and CT scanning can ensure their patency. Magnetic resonance imaging (MRI) is also a useful tool, as it can give more detailed information on the soft tissues of the brain and the status of the cranial ventricles. Positron emission tomography (PET) scanning using fluorodeoxyglucose (FDG) is a research imaging technique that has revealed asymmetries in temporal lobe function (Fig. 1) [8]. Further studies are needed to define its role in the workup of patients with distortions of olfactory ability.

Clinical Testing

Chemosensory testing is useful for isolating the site of dysfunction. Smell and taste tests of threshold or identification ability often locate a deficit, especially when uninasal testing is performed. These areas of deficit are usually the site of the problem (e.g., phantosmia isolated to the left nares). When the location of the problem is still not clear, regional local anesthesia of the tongue, chorda tympani, and nasal cavity with topical and injected medications can be even more helpful. Patients who have suspected seizure activity should have appropriate electroencephalography (EEG) performed. Similarly, the diagnosis of those with suspected psychiatric disease may benefit from psychometric or personality testing.

Specific Diagnostic Categories

In the sections that follow, the diagnostic categories for olfactory distortion are presented, with attention ·

to how the diagnosis can be made and what treatment may be available.

Internally Generated Odorants

Rare individuals have conditions whereby they actually generate an odor that they smell, and others can also smell it if in close proximity. Trimethylaminuria is an example of this type of disorder that can masquerade as a smell distortion [1]. Because they are concerned that others might also be able to smell this unpleasant odor, the social lives of these individuals are usually bleak. Other historical clues to this condition are that these patients have great difficulty in deciding whether their problem is a taste or a smell dysfunction, and they notice it more during exhalation. Special metabolic testing is available to confirm the diagnosis.

Troposmia

Troposmia is a condition whereby the patient has distorted olfactory perception of existing odorants. This distortion can be of two types. One is a blanket contamination of all odorants by the same smell, which may range from a slight discoloration to almost complete replacement. With the second type of troposmia, there is a different distortion for each odorant, and all odorants have distinct smells. Patients with this second type often learn a new smell "vocabulary."

In the literature, the word "parosmia" seems to be used to describe both troposmia and phantosmia, resulting in some confusions [2,3]. Troposmias have been reported in patients who have had olfactory losses after head trauma, after an upper respiratory infection, and after exposure to inhaled toxins. They have rarely been noted in individuals who are on certain antibiotic or cardiovascular drugs, and in those who have cancers of the upper aerodigestive tract [2–5]. The pathophysiology of troposmia is unknown, although some defect in coding due to decreased neural input is possible, as most individuals with this problem have decreased olfactory ability in at least one nostril.

Treatment for this problem is unclear. In many patients the problem resolves with time, sometimes paralleling an improving or declining olfactory ability. Zilstorff [3] has suggested using topical cocaine hydrochloride drops on the olfactory mucosa. Unfortunately, no other reports of this therapy have appeared in the literature. We have identified a patient with a troposmia that was temporarily eliminated with uninasal cocainization. Although partial excision of the olfactory epithelium on this side of the nose has been partially successful in improving this patient's troposmia, there is still some distortion present.

Phantosmia

Phantosmia is the intermittent or continuous perception of an odor when no odorant stimulus is present. It is a rare condition and is found mostly in women. The typical history begins with a woman in the second or third decade of life who experiences an unpleasant odor for about 20 min with no apparent inciting event. She then experiences the same smell 2 to 3 weeks later, again for no apparent reason. Over the next several months the odor is experienced more and more frequently, until it appears almost every day and may last most of the day. Sudden changes in nasal airflow, such as with sniffing or snorting, are reported to trigger the problem in most patients, and they learn to avoid this practice. For the first few months or years, patients sometimes abort the odor experience by performing a modified Valsalva maneuver. Most patients report that the bad odor is in only one nostril and can be blocked by occluding that nostril. Sometimes the same odor or another develops in the opposite nostril after several months or years.

During the evaluation of patients with suspected phantosmia, it is useful to rule out seizure activity with an EEG study. MRI or CT scans of the brain is important to eliminate a major brain abnormality.

During the clinical evaluation, each nostril should be blocked independently to note the effect on the phantom smell. Then each olfactory area should be anesthetized. If the phantom smell can be eliminated along with the olfactory ability, the diagnosis can be made.

The pathophysiology of this disease is unclear [9]. Because it often occurs in only one nostril and can be triggered by nasal airflow, the neural epithelium in the upper nasal cavity may be the site of a neuron or group of neurons that generate an abnormal signal that the brain interprets as a smell.

I have been able to reduce or eliminate the phantom smell in five patients over the past 4 years by excising some or all of the olfactory epithelium on the affected side of the nose. Others have reported successful treatment for phantosmia with a craniotomy to excise the olfactory bulb [7]. There is other information to suggest that the problem is in the brain, and that the primary olfactory neurons merely trigger this process. As previously noted, EEG abnormalities have been noted in one-third of the patients reported by Potoliccio et al. [6], although our patients have not shown these changes. Leopold and associates [8] have noted an asymmetry in temporal lobe activity on PET scans of a patient with a unilateral phantosmia that resolved after successful surgery (Fig. 1). Clearly further studies are needed before a definite etiology can be determined.

The treatment of phantosmia is still evolving. It has been suggested that antiepileptic drugs such as valproic acid or antipsychotic medications such as the butyrophenones are effective [6]. I have found that patients on these treatments either do not experience relief or they are bothered by the side effects. The surgical treatment involves potentially dangerous operations, but the patients seem to have a more permanent result and a more pleasant lifestyle afterward.

Olfactory Hallucinations

Hallucinations, as defined here, refer to olfactory sensations that are not associated with any odorant stimulation and are not dependent on nasal airflow in one or both nostrils. They tend to be associated with either psychiatric or neurologic disease.

Schizophrenia

Since at least 1934 it has been recognized that patients with thought disorders and incongruity of affect can also have olfactory hallucinations [10]. These odors are described as coming from an external source (extrinsic), such as "poisonous vapors," "smells of holiness," or the "odor of interplanetary visitors" [11]. The duration of the odor sensation is limited to minutes, and it does not bother the patient much. Approximately half of these patients have had psychiatric care, and many cannot be employed because of their schizophrenia.

Olfactory function has been noted to be diminished in schizophrenia; and this assessment, along with other tests, is used to classify patients into either an orbitofrontal or a dorsolateral profile of central nervous system involvement [12–14]. After appropriate evaluations and diagnosis, prescribed drug therapy can be useful for treating the symptoms of schizophrenia, including the hallucinations.

Alcoholic Psychosis

Patients with chronic alcoholism have symptoms of olfactory sensation similar to those seen with schizophrenia. These patients are different, however, in that alcoholics are generally older, more often male, have a history of more psychiatric care, and have more frequent delusions. Chronic alcoholics without hallucinations have been shown to have deficits in odorant matching and identification skills, related to reduced brain size [15,16]. Treatment for these patients therefore is directed toward both the alcoholism and the psychosis.

Depression

Although depressive episodes may accompany many other psychiatric conditions, patients who have only depression can have olfactory hallucinations associated with it. These hallucinations generally last minutes and greatly bother the patient; rarely are there other sense deceptions [17]. The odor is characteristically described as emanating from the patient's own body, and the patient is often deeply ashamed, embarrassed, and self-abasing. These patients tend to wash excessively, change their clothes often, and annoy people around them with their unusual behavior. Because nondepressed patients with an actual internally generated odorant from a metabolic problem can have similar symptoms, care must be taken to distinguish the two.

Patients with this type of olfactory hallucination must be diagnosed by their obvious depressive symptoms, as imaging and EEG studies are generally normal. Understanding the pathophysiology and clinical status of this process is hampered by contradictory reports of olfactory ability [13,18].

Olfactory Reference Syndrome

A group of clinically distinct patients have olfactory hallucinations similar to patients with depression [17]. They think an odor is emanating from them, and they are concerned how others perceive the odor. They are different, however, in that they do not have symptoms of depression and are generally younger. Fortunately, these patients, said to have olfactory reference syndrome, respond to antidepression medication.

Epileptic Seizures

From clinical experience with brain tumors and electrical brain stimulation, it has become clear that abnormalities in the inferior and medial regions of the temporal lobes can result in the perception of smells [19]. Temporal lobe seizures are characterized by an initial aura, which in one series was olfactory in 8% of the patients [20]. The smells lasted only seconds and were described as coming from outside the person's body, such as "bad," "rotten," "sickening," or like "burning food." Most patients then developed some type of tonic motor activity, usually involving head or mouth movements.

In patients suspected of having seizures, the workup includes brain imaging and EEG. Sometimes the EEG must be performed utilizing specialized techniques to elicit the abnormality. Treatment is usually successful using antiseizure medications. For those few patients who are unable to be controlled with medication, surgical techniques have been devised to excise the abnormal area(s). In general, part of the temporal lobe can be excised with few sequelae.

Conclusion

In general, patients who experience distortions of olfactory ability are concerned and unhappy and have difficulty adjusting to the distortion. With careful history taking, examination, and testing, a precise description of the patient's symptoms and diagnosis of the problem can be established. Only after it is completed can a plan for treatment be initiated. The important factor when evaluating these patients is to consider each one individually and to listen carefully to their concerns, no matter how ridiculous they sound.

References

1. Leopold DA, Preti G, Mozell MM, Youngentob SL, Wright HN (1990) Fish-odor syndrome presenting as dysosmia. Arch Otolaryngol Head Neck Surg 116: 354–355
2. Scott AE (1989) Clinical characteristics of taste and smell disorders. Ear Nose Throat J 68:297–315
3. Zilstoff K (1966) Parosmia. J Laryngol 80:1102–1104
4. Henkin RI, Larson AL, Powell RD (1975) Hypogeusia, dysgeusia, hyposmia, and dysosmia following influenza-like infection. Ann Otol 84:672–682
5. Markley EK, Mattes-Kulig DA, Henkin RI (1983) A classification for dysgeusia. Perspect Pract 83:578–580
6. Potolicchio SJ, Lossing JH, O'Doherty DS, Henkin RI (1986) Partial seizures with simple psychosensory symptomotology (cyclic phantosmia). Clin Res 34: 635A
7. Kaufman MD, Lassiter KRL, Shenoy BV (1988) Paroxysmal unilateral dysosmia: a cured patient. Ann Neurol 24:450–451
8. Leopold DA, Meyerrose GE, Szabo Z, Sostre S (1993) PET scan representation of central olfactory processing in phantosmia. Chem Senses 18:587
9. Leopold DA, Schwob JE, Youngentob SL, et al. (1991) Successful treatment of phantosmia with preservation of olfaction. Arch Otolaryngol Head Neck Surg 117:1402–1406
10. Bromberg W, Schilder P (1934) Olfactory imagination and olfactory hallucinations. Arch Neurol Psychiatry 32:467–492
11. Pryse-Phillips W (1975) Disturbance in the sense of smell in psychiatric patients. Proc R Soc Med 68: 26–28
12. Hurwitz T, Kopala L, Clark C, Jones B (1988) Olfactory deficits in schizophrenia. Biol Psychiatry 23: 123–128
13. Serby M, Larson P, Kalkstein D (1990) Olfactory sense in psychoses. Biol Psychiatry 28:829–830
14. Seidman LJ, Talbot NL, Kalinowski AG, et al. (1992) Neuropsychological probes of fronto-limbic system dysfunction in schizophrenia: olfactory identification and Wisconsin Card Sorting Performance. Schizophr Res 5:55–65
15. Kesslak JP, Profitt BF, Criswell P (1991) Olfactory function in chronic alcoholics. Percept Motor Skills 73:551–554
16. Shear PK, Butters N, Jernigan TL, et al. (1992) Olfactory loss in alcoholics: correlations with cortical and subcortical MRI indices. Alcohol 9:247–255
17. Pryse-Phillips W (1971) An olfactory reference syndrome. Acta Psychiatr Scand 47:484–509
18. Amsterdam JD, Settle RG, Doty RL, Abelman E, Winokur A (1987) Taste and smell perception in depression. Biol Psychiatry 22:1477–1491

19. Penfield W, Jasper H (1954) Epilepsy and the functional anatomy of the human brain. Little, Brown, Boston

20. King DW, Marsan CA (1977) Clinical features and ictal patterns in epileptic patients with EEG temporal lobe foci. Ann Neurol 2:138–147

Endoscopic Observations for the Diagnosis of Olfactory Disturbance

M. Shirakura[1] and H. Asaka[2]

Key words. Olfactory disturbance—Diagnosis—Endoscope—Selfox lens

Since 1968 we have observed the olfactory mucous membrane in about 8000 cases using a narrow, rigid endoscope with a diameter of 1.7 mm. It is known that observation of the olfactory mucosa is not possible using anterior rhinoscopy. In this chapter we report findings concerned with olfactory disturbance, mainly derived from our olfactory mucosal observations.

Observation Method of Olfactory Mucous Membrane

Figure 1 shows the endoscope we use to examine the olfactory mucous membrane. Images are transmitted through a special lens called the selfox lens, the diameter of which is 1.0 mm and the length 200 mm. Unlike the fiberscope, this instrument allows clear images even with a diameter of 1.0 mm.

Figure 2 shows the positions of the examiner and patient during an olfactory mucosal examination.

FIG. 1. Endoscope used to examine the olfactory mucous membrane

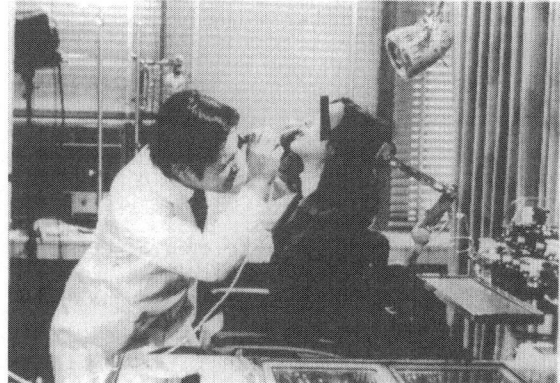

FIG. 2. Examination of the olfactory mucosa

The patient remains in a sitting position without local anesthesia (there is no pain). The method of inserting the instrument is as follows: The tip of the endoscope is held softly with the left hand, fixed at the anterior nares, and the scope itself is held by the right hand. Once the tip of the endoscope enters the nasal vestibule, the endoscope is gradually inserted while the examiner look into it, being careful not to touch the nasal septum and concha. The inside of the nasal cavity is of considerable width, so it should be easier to insert the endoscope into the nasal cavity without touching the nasal septum and mucosa. If the endoscope is inserted in this manner, observation of the olfactory mucosa is possible unless a large nasal polyp or a severely curved septum is present.

Figure 3 shows the olfactory mucosa of a person with normal olfaction. Anatomically, the connective tissue from the mucous epithelium that coats the bone of the olfactory mucosa is thin. For this reason

[1] Department of Otorhinolaryngology, Yamanashi Medical College, Shimogato 1110, Tamaho-cho, Nakakoma-gun, Yamanashi, Japan
[2] Department of Otorhinolaryngology, Takatsu-Chuou Hospital, Mizonokuchi 211, Takatsu-ku, Kawasaki, Kanagawa, Japan

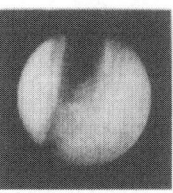

FIG. 3. Olfactory mucosa of a person with normal olfaction

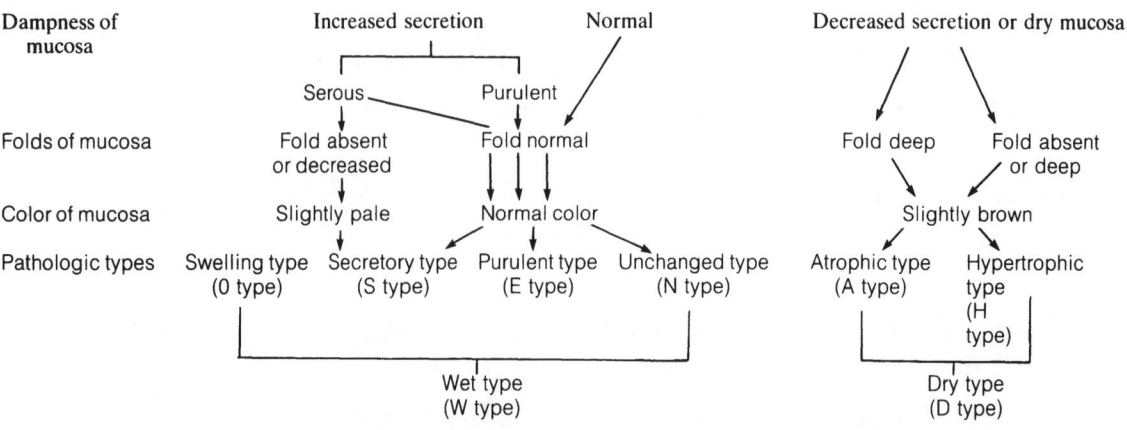

FIG. 4. Classification of pathology of the olfactory mucosa

the folds of the bones at the area of the lamina cribrosa are in a plate-like pattern on the surface of the olfactory mucosa. It is not clear in the picture, but the surface of the mucosa is richly damp.

Pathologic Types of Olfactory Mucosa

The pathologic types of olfactory mucosa are classified in reference to three conditions: dampness, folds, and color tone (Fig. 4). The dampness of the olfactory mucosa is roughly classified into four groups, the folds into three types, and the color tone into three types. There is a relation among the three conditions of dampness, folds, and color tone, and criteria have been established for classifying the pathologic types.

There are two types of mixed case: a swelling plus secretory type, and swelling plus purulent type. We categorize the atrophic and hypertrophic types together and generally call it a "dry type." The swelling, secretory, and purulent types are categorized together as well and generally are called the "wet type."

The five typical types (swelling, secretory, swelling + purulent, atrophic, and hypertrophic) are shown in Fig. 5a–e.

Among 1596 olfactory disorder patients seen at our clinic over a 3-year period, the causes for the disorder were as follows: chronic sinusitis, 626 cases; nasal allergy, 193 cases; and other causes, 777 cases.

Therapeutic Methods and Results of Treatment

Basic treatment for the patient in whom an olfactory mucosa lesion is found is local application of a steroid hormone agent with the patient in the head-down position (2 or 3 drops of 0.1% beta-methasone

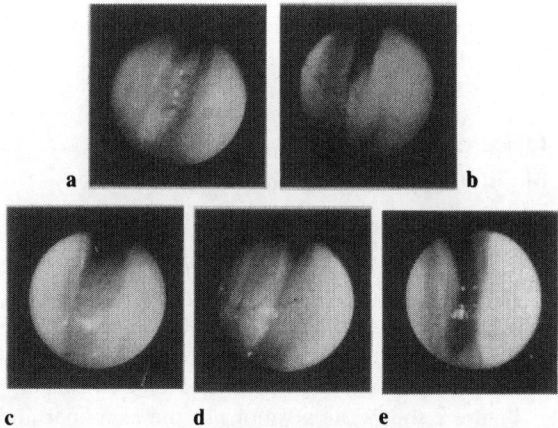

FIG. 5. Typical appearances of pathologic olfactory mucosa. **a** Swelling. **b** Secretory disturbance. **c** Swelling plus purulence. **d** Atrophy. **e** Hypertrophic mucosa

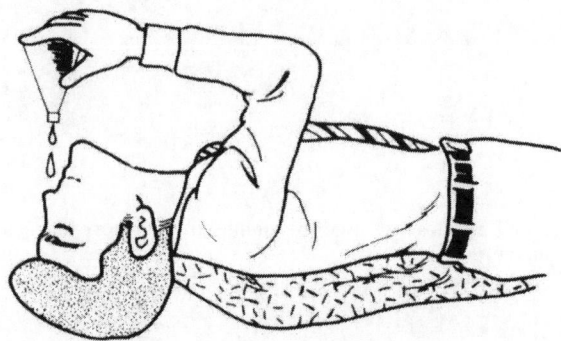

FIG. 6. Application of a steroid hormone to a patient in whom a nasal mucosal lesion has been found

or 0.1% dexamethasone liquid for topical use) (Fig. 6). Concurrent treatment for chronic sinusitis is given, according to its severity. With the treatment just described, we achieved the following results: cured 37.4%; alleviated 35.8%; unchanged 26.8%.

Conclusions

Observation of olfactory mucosa is important for the diagnosis and treatment of olfactory disturbances. The endoscopic findings in the olfactory mucosa are especially important for evaluating the results of treatment. We have obtained clear images using a narrow, rigid endoscope.

Suggested Reading

Asaka H (1981) Disease type on olfactory mucous membrane. Adv Neurol Sci 25:272–278

Asaka H, Fukushima Y, Fujii S (1986) Actual conditions of patients with dysosmia. Proceedings of the Japanese Symposium on Taste and Smell 20:61–64

Takagi SF (1989) Human olfaction. University of Tokyo Press, Tokyo, pp 71–100

Olfactory Event-Related Potentials in the Diagnosis of Smell Disorders

GERD KOBAL[1]

Key words. Olfactory event-related potentials—Chemosensory event-related potentials—Chemoreception—Pain—Olfaction—Anosmia

Introduction

In 1978 Kobal and Plattig [1] introduced a technique that allowed chemosensory stimulation of the nasal cavity without simultaneously altering mechanical or thermal conditions at the mucosa. Thus after the previous endeavors of Finkenzeller [2] and Allison and Goff [3], it finally became possible to record olfactory event-related potentials (OERPs) in humans [4].

By choosing the appropriate stimulant it is possible to selectively activate either the trigeminal or the olfactory system, both of which contribute to the sensation induced by most odorants. Substances such as vanillin or hydrogen sulfide stimulate the olfactory system specifically [5,6], whereas others, such as carbon dioxide, stimulate the trigeminal system predominantly if not solely [7,8]. Investigations of the OERPs' topographic distribution revealed that stimulation of the trigeminal nerve with carbon dioxide produces a pattern that is distinctively different from the pattern elicited by odorous stimulation with substances such as vanillin [9]. Huttunen et al. [10] established, by means of magnetoencephalographic recordings, that late near-field event-related potentials to intranasal chemical stimulation with carbon dioxide represent neocortical activity in the secondary somatosensory area (S II), which is closely connected to the stimulation of nocisensors and the sensation of pain [11]. New recordings with a novel 37-channel magnetoencephalographic device (Krenikon Siemens, Germany) revealed that the olfactory generators of the OERPs are located in the temporal lobe (Kobal et al., unpublished data). In addition, relations of the chemosensory event-related potentials) (CSERPs) to flow rate, stimulus concentration, and ratings of intensity have been thoroughly investigated (for review see [12]).

Thus in addition to the recording of electroolfactograms [13], OERPs provide another electrophysiologic correlate of odorous sensations in humans. Because electroolfactograms are difficult to obtain, OERPs appear to be the method of choice for general electrophysiologic investigations of patients with olfactory disorders. In these clinical situations, reliable parameters of patients' ability to smell are needed, that is, where it is preferable not to depend on subjective responses alone, particularly in malingering patients. Therefore the aim of this study was to test the reliability of OERP recordings in anosmic patients.

Material and Methods

Patients

A group of 45 anosmic patients (mean age 44 years, range 17–69 years) were investigated in our laboratory. Table 1 gives a summary of the most probable causes of the anosmias diagnosed in the Ear-Nose-Throat Clinic and the Department of Neurology of the University of Erlangen-Nürnberg. Diagnosis was primarily based on the patients' complaints and their history. Within this context it is important to state that none of the patients received any financial advantage when he or she was diagnosed as anosmic. The study was conducted in accordance to the Helsinki/Hong Kong declaration.

Stimulation and Testing Procedure

For chemosensory stimulation, an apparatus (Olfactometer, OM4, Burghart, Germany) was employed that delivered the chemical stimulants without altering the mechanical or thermal conditions at the mucosa [4]. This monomodal chemical stimulation was achieved by mixing pulses of the stimulants in a constantly flowing air stream with controlled temperature and humidity (36.5°C; 80% relative humidity). The air stream was led into both nasal cavities by way of teflon tubing (6 cm length, 4 mm outer diameter). The total flow rate was 140 ml/s. Stimulus duration was 200 ms with a rise time below 20 ms.

Stimuli were applied nonsynchronously to breathing. To avoid respiratory flow of air in the nose, the

[1] Department of Experimental and Clinical Pharmacology and Toxicology, University of Erlangen-Nürnberg, Krankenhausstr. 9, 91054 Erlangen, Germany

TABLE 1. Most probable causes of anosmia in investigated patients.

Cause	No. of patients		
	Male	Female	Total
Trauma	18	8	26
Postviral reaction	3	4	7
Kallmann syndrome	3	0	3
Postoperative (rhinoseptal surgery) aftermath	1	1	2
Intoxication (solvent)	1	0	1
Idiopathic	3	3	6
Total	29	16	45

patients were requested to breathe through the mouth. They were comfortably seated in an acoustically and electrically shielded, air-conditioned chamber. White noise, of approximately 50 dB SPL, was used to mask switching clicks of the stimulator.

The three substances—carbon dioxide (52% v/v), hydrogen sulfide (0.78 ppm), and vanillin (2.06 ppm)—were tested during each session. To our knowledge, carbon dioxide is a stimulant that largely, if not exclusively, excites intranasal chemoreceptors of the trigeminal nerve [7,8]. Vanillin and hydrogen sulfide are both considered to be primarily olfactory stimulants [5,6].

At each session, all three stimulants were applied 15 times to the left and the right nostrils. The interstimulus interval was approximately 40 s. The sessions, including preparation, lasted 80 min.

Olfactory and Chemosensory Event-Related Potentials

The electroencephalogram (EEG) (filtering 0.2–70.0 Hz, electrode impedance 2–4 kΩ) was recorded from six positions of the international 10/20 system (Fz, Cz, Pz, C3, C4, and Fp2, all referenced to A1+A2). EEG records of 2048 ms duration were digitized (sampling frequency 250 Hz) and averaged in six groups according to the three stimulants and the binasal stimulation. Data were evaluated by OFFLAB programs [14]. EEG records contaminated by eye-blinks (Fp2/A1+A2) or motor artifacts were discarded from the average [4].

Psychophysical Testing of Olfactory and Trigeminal Function

In addition to the recording of chemosensory event-related potentials, the sense of smell was tested by

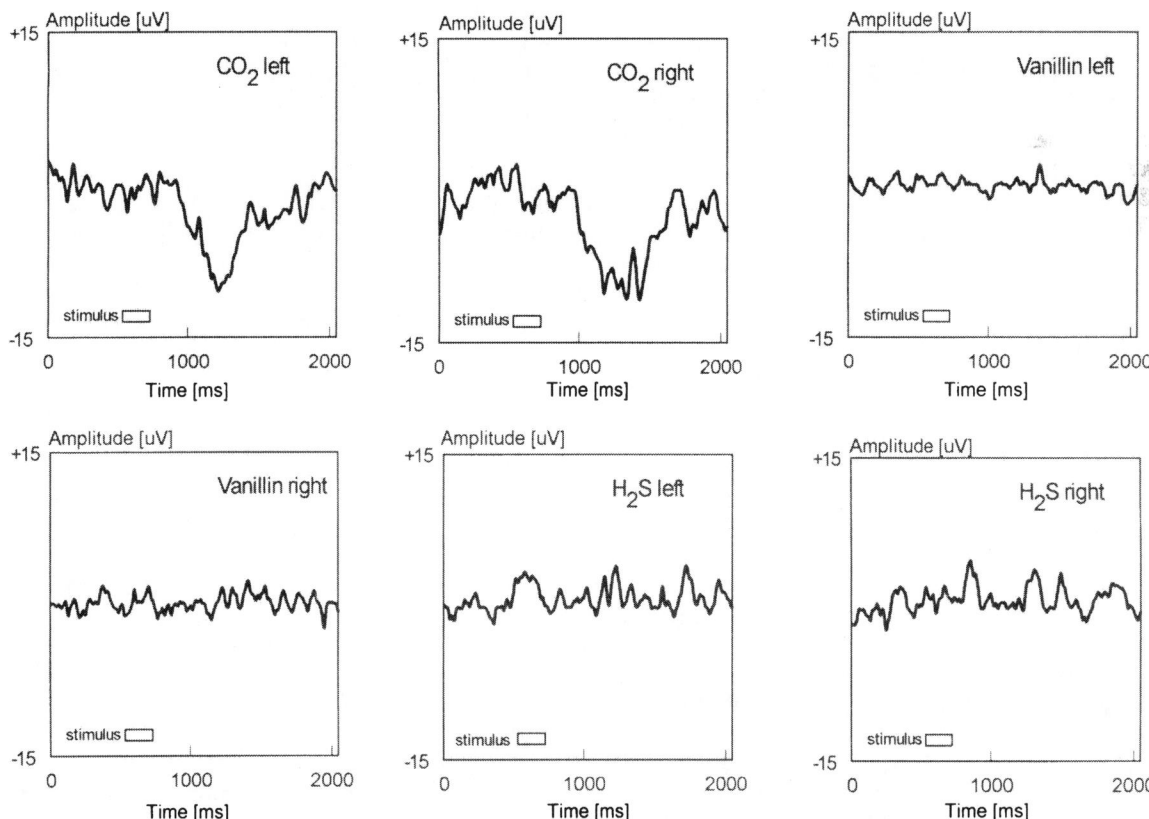

FIG. 1. Chemosensory event-related potentials of a patient suffering from Klinefelter's syndrome. There are chemosensory event-related potentials to the trigeminal stimulant carbon dioxide, but no olfactory event-related potentials to the odorants vanillin and hydrogen sulfide (anosmia)

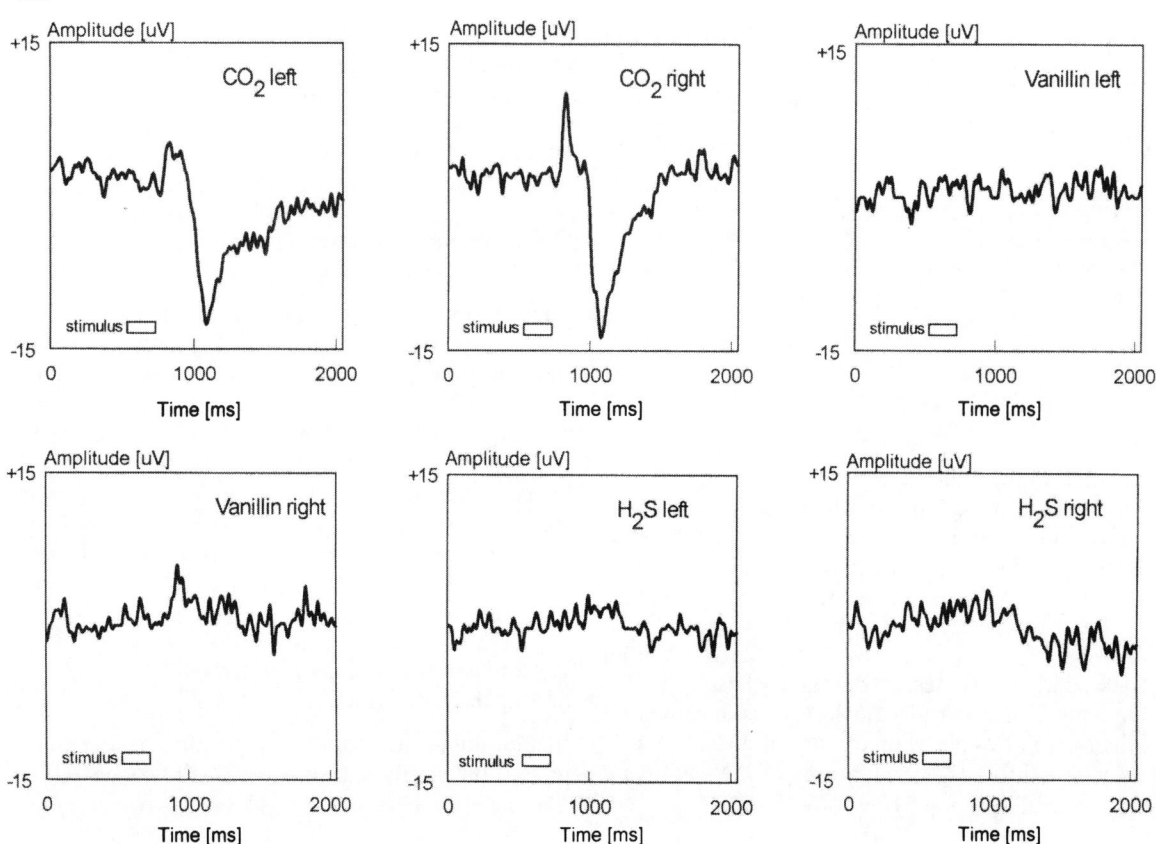

FIG. 2. Chemosensory event-related potentials of a patient suffering from olfactorius meningioma (*left side*). There are chemosensory event-related potentials to the trigeminal stimulant carbon dioxide, but no olfactory event-related potentials to the odorant vanillin and only a very small response to hydrogen sulfide on the affected left side (hyposmia)

Results and Discussion

In all investigated anosmic patients intranasal CSERPs could be obtained after stimulation with carbon dioxide, a substance that produces a stinging, painful sensation; i.e., it excites fibers of the trigeminal nerve [7,8]. In contrast, no OERPs could be detected after stimulation with the odorants hydrogen sulfide and vanillin (Figs. 1 and 3). In all probability, those two odorants exclusively excite receptors of the olfactory nerve [5,6].

Developments have opened new fields for electrophysiologic investigation of human olfaction. Using an "odd-ball paradigm," Durand-Lagarde and Kobal [15] succeeded in demonstrating that a late positive component can be observed in response to olfactory stimuli. This late positivity, often labeled "P_{300}" or "P3," is recorded in human event-related potentials during a discrimination task where subjects are requested to discriminate low probability stimuli from

means of an odor identification and discrimination test. Thresholds were also measured. Results of the psychophysical findings will be published elsewhere.

high probability stimuli [16]; and it appears only after application of the low probability stimulus. Considering the large body of literature in P_{300} research (for review see Renault et al. [17]) OERPs appear to provide a promising approach with which to gain new information about cognitive events in olfactory disorders.

In conclusion, it is possible to independently investigate the two sensory systems primarily responsible for the perception of odorants by means of event-related potentials. This technique may be of advantage in many clinical situations, such as the monitoring of olfactory disorders (Fig. 2). Specifically, the testing of malingering patients complaining of anosmia no longer needs to be the difficult task with which clinicians are presently confronted.

Summary

The aim of this study was to demonstrate the usefulness of olfactory event-related potentials (OERPs) and chemosensory event-related potentials (CSERPs) in the diagnosis of anosmia. Forty-five patients participated in the study. CSERPs were

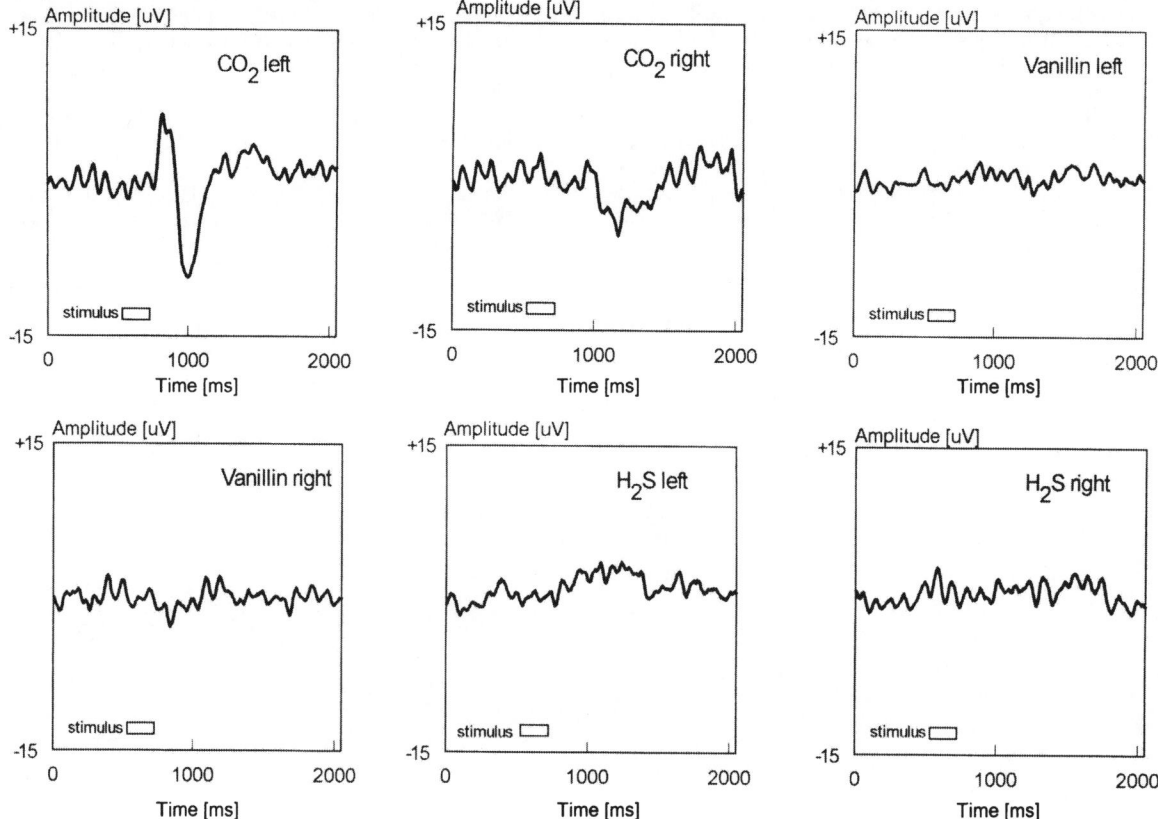

Fig. 3. Chemosensory event-related potentials of a patient suffering from chronic exposure to solvents. There are no olfactory event-related potentials to the odorants vanillin and hydrogen sulfide (anosmia)

recorded after stimulation of the trigeminal nerve with carbon dioxide. For the recording of OERPs, vanillin and hydrogen sulfide were used as stimulants. In addition, patients performed a simple odor identification test.

After olfactory stimulation, OERPs could not be obtained in any of the anosmic patients, indicating the complete loss of the sense of smell. However, all patients responded to stimulation of the trigeminal nerve with carbon dioxide. Thus it could be demonstrated that OERPs provide important information for the assessment of olfactory dysfunction.

Acknowledgments. This work was supported by grant Ko812/5-1 of the Deutsche Forschungsgesellschaft.

References

1. Kobal G, Plattig KH (1978) Methodische Anmerkungen zur Gewinnung olfaktorischer EEG-Antworten des wachen Menschen (objektive Olfaktometrie). Z Electroencephalogr Electromyogr 9:135–145
2. Finkenzeller P (1966) Gemittelte EEG-Potentiale bei olfactorischer Reizung. Pfluegers Arch 292:76–80
3. Allison T, Goff WR (1967) Human cerebral evoked responses to odorous stimuli. Electroencephalogr Clin Neurophysiol 23:558–560
4. Kobal G, Hummel C (1988) Cerebral chemosensory evoked potentials elicited by chemical stimulation of the human olfactory and respiratory nasal mucosa. Electroencephalogr Clin Neurophysiol 71:241–250
5. Doty RL, Brugger WPE, Jurs PC, et al. (1988) Intranasal trigeminal stimulation from odorous volatiles: psychometric responses from anosmic and normal humans. Physiol Behav 20:175–185
6. Kobal G, Van Toller S, Hummel T (1989) Is there directional smelling. Experientia 45:130–132
7. Cain WS, Murphy C (1980) Interaction between chemoreceptive modalities of odour and irritation. Nature 284:255–257
8. Thürauf N, Friedel I, Hummel C, Kobal G (1991) The mucosal potentials elicited by noxious chemical stimuli: is it a peripheral nociceptive event? Neurosci Lett 128:297–300
9. Kobal G, Hummel T, Van Toller S (1992) Differences in human chemosensory evoked potentials to olfactory and somatosensory chemical stimuli presented to left and right nostrils. Chem Senses 17:233–244
10. Huttunen J, Kobal G, Kaukoronta E, Hari R (1986) Cortical responses to painful CO_2-stimulation of nasal mucosa; a magnetencephalographic study in man. Electroencephalogr Clin Neurophysiol 64:347–349
11. Chudler EH, Dong WK, Kawakami Y (1985) Tooth pulp evoked potentials in the monkey: cortical surface and intracortical distribution. Pain 22:221–223

12. Kobal G, Hummel T (1991) Olfactory evoked potentials in humans. In: Getchell TV, Doty RL, Bartoshuk LM, Snow JB Jr (eds) Smell and taste in health and disease. Raven, New York, pp 255–275

13. Kobal G, Hummel T (1992) Human electro-olfactograms and brain responses to olfactory stimulation. In: Laing DG, Doty RL, Breipohl W (eds) The human sense of smell. Springer, Berlin, pp 135–151

14. Kobal G (1981) Elektrophysiologische Untersuchungen des menschlichen Geruchssinnes. Thieme, Stuttgart, p 172

15. Durand-Lagarde M, Kobal G (1992) P300: a new technique of recording a cognitive component in the olfactory evoked potentials. Chem Senses 16: 379

16. Sutton S, Braren M, Zubin J (1965) Evoked-potential correlates of stimulus uncertainty. Science 150:1187–1188

17. Renault B, Kutas M, Coles MGH, Gaillard AWK (1989) Event-related potentials investigations of cognition. North-Holland, Amsterdam, p 362

Ultrastructural Analysis of Olfactory Mucosa in Patients with Congenital Anosmia

Seog-In Paik[1], Soon-Il Park[1], Allen M. Seiden[2], and David V. Smith[2]

Key words. Congenital anosmia—Ultrastructure

Congenital anosmia is a rare condition which can be diagnosed by taking a careful personal history of lifelong anosmia. Although mucosal biopsies of the neuroepithelium have been obtained by several investigators, it is not clear whether the olfactory mucosa is totally absent in cases of congenital anosmia. Only two of these previous reports have demonstrated the presence of olfactory epithelium, which contained immature olfactory receptor cells [1,2]. Other investigators, however, have shown only respiratory epithelium [3,4].

In the present study, in order to investigate the ultrastructural characteristics of the mucosa, we obtained mucosal specimens from the olfactory area in four patients with congenital anosmia, using the biopsy instrument developed by Lovell et al. [5]. Since 1991, two Korean patients, one with Kallmann's syndrome and one isolated type, have been diagnosed at the Smell and Taste Clinic of Wonju College of Medicine. Two Caucasian women, one familial type and one isolated type, were diagnosed at the University of Cincinnati Taste and Smell Center between 1989 and 1991.

In each of the four, olfactory epithelium was demonstrated on light microscopic examination of semithin sections and on ultrastructural examination with transmission electron microscopy. Morphologic changes included decreased numbers of olfactory receptor cells with few or no cilia and disintegration of cellular organelles in most olfactory receptor cells and microvillar cells. The basal bodies were usually located in the distal part of dendrites and the olfactory vesicles rarely protruded above the surface, which seems to be a finding of immature olfactory receptor cells. The fila olfactoria in the lamina propria was decreased in number and showed axonal degeneration. Intraepithelial axonal collections were observed in the patient with Kallmann's syndrome and in the isolated type Caucasian woman patient. These findings imply that there might be different morphologic manifestations in congenital anosmia, depending upon its possible heterogeneous pathophysiology.

References

1. Douek E, Bannister LH, Dodson HC (1975) Olfaction and its disorders. Proc R Soc Med 68:467–470
2. Leopold DA, Hornung DE, Schwob JE (1992) Congenital lack of olfactory ability. Ann Otol Rhinol Laryngol 101:229–236
3. Igarashi Y, Oka Y, Ishizuka Y (1986) Electron microscopic observations of the nasal olfactory area in Kallmann's syndrome. Prac Otol (Kyoto) 79:2041–2048
4. Jafek BW, Gordon ASD, Moran DT, Eller PM (1990) Congenital anosmia. Ear Nose Throat J 69:331–337
5. Lovell MA, Jafek BW, Moran DT, Rowley JC III (1982) Biopsy of human olfactory mucosa. Arch Otolaryngol Head Neck Surg 108:247–249

[1] Department of Otolaryngology, Yonsei University Wonju College of Medicine, 162 Ilsan-Dong, Wonju 220-701, Korea
[2] Department of Otolaryngology and Maxillofacial Surgery, University of Cincinnati Medical Center, 231 Bethesda Ave., Cincinnati, OH 45267-0528, USA

Morphological Study of Olfactory Dysfunction in Patients with Sinusitis

SHOJI MATSUNE, MASAHIKO EGAWA, SHIGERU FURUTA, and MASARU OHYAMA[1]

Key words. Olfactory dysfunction—Sinusitis

Chronic sinusitis is the most common cause of olfactory dysfunction seen at the ENT outpatient clinic. In our department, from January 1991 to April 1994, the most common etiologic class of olfactory disturbance was chronic sinusitis; this accounted for 128 of the 286 patients (44.4%) seen during that period. The important causative factors in these patients were considered to be the inflammatory damage to the olfactory epithelium and the obstruction of the olfactory clefts by the swollen nasal mucosa. That the operation to relieve the patency of the nasal cavity does not always improve olfactory ability would seem to explain the etiological importance of epithelial damage. Also, even after the surgical removal of the inflammatory foci of sinusitis and/or after conservative chemotherapy, epithelial damage in the olfactory portion may tend to persist. In order to observe the morphological recovery of the olfactory epithelium in sinusitis, we created an experimental sinusitis model.

Experimental sinusitis was prepared in eight rabbits by injecting bacteria (*Staphylococcus aureus* 209P strain) into the paranasal sinuses after sensitization with egg albumin. A purulent nasal discharge was produced by the infection, after which an antibiotic (ofloxacin) was injected into the paranasal sinuses once a day for 3 days to four rabbits (group 1) and for 7 days to the remaining four rabbits (group 2). After the antibiotic administration was completed, all the rabbits were sacrificed under anesthesia and specimens were harvested from both respiratory and olfactory areas for light and scanning electron microscopic (SEM) studies.

[1] Department of Otolaryngology, Kagoshima University School of Medicine, 8-35-1 Sakuragaoka, Kagoshima, 890 Japan

In control rabbits, the mean number of olfactory vesicles was 11.0 ± 4.7 at a magnification of $\times 5000$ under SEM observation. Accordingly, we assumed that the normal range of the number of olfactory vesicles was more than 6 per field at the same magnification. Further, the border between the respiratory and the olfactory areas was quite clear as a normal pattern under SEM observation. Even in group 2 animals, however, the number of olfactory vesicles was significantly decreased. The olfactory area seemed to be invaded by respiratory components such as ciliated cells. In the respiratory area, on the other hand, although the ciliated epithelium was still damaged in group 1, the normal structure was completely restored in group 2. On the basis of these findings, we suggest that the damage to the olfactory epithelium in patients with sinusitis persists in the period during which conservative therapy is administered for the sinusitis.

We hypothesized that differences in the cell-proliferation cycle between the respiratory and olfactory epithelium were an important causative factor in this persistence. Accordingly, to analyze cell-proliferation capacity in the inflammatory respiratory and olfactory areas, we investigated the uptake of bromodeoxyuridine (BrdU) to cell nuclei histochemically after the intravenous injection of BrdU. The number of BrdU-positive cells at the inflammatory stage was increased in comparison with that at the non-inflammatory stage in the respiratory area, but not in the olfactory area. The cell turnover of respiratory and olfactory epithelium is generally thought occur within periods of about 2 and 4 weeks, respectively. This cell kinetic study with BrdU indicated that inflammation seemed to induce more activated cell kinetics and faster epithelial turnover in the respiratory than in the olfactory epithelium. This is likely to be another reason that olfactory epithelial damage often persists even during effective conservative therapy for sinusitis.

Clinical Features of Olfactory Disorder Following Common Cold

Tsunemasa Aiba, Yoshiaki Nakai, Midori Sugiura, and Jyunko Mori[1]

Key words. Olfactory disorder—Virus

There is an olfactory disorder syndrome that occurs after the common cold is characterized by a relatively severe olfactory disorder that is not associated with rhino-sinusitis. This syndrome is now recognized as a disorder of the olfactory epithelium caused by viral infection, but its detailed pathogenesis still remains obscure. The purpose of this study was to define the clinical features of this syndrome and thereby to clarify its true pathology.

The subjects were outpatients seen at the Olfaction Clinic of the Osaka City University Medical School Hospital during the past 11 years (1982–1992). Of the 796 patients involved in the study, 162 had a syndrome that met the following diagnostic criteria:

1. A history of olfactory disorder occurring after symptoms of common cold or upper respiratory tract inflammation
2. Patency of the olfactory cleft noted on rhinoscopy, endoscopy, and X-ray examination
3. No evidence of rhinitis or sinusitis
4. Exclusion of any other causes of olfactory disorder

We investigated the distribution of the patients according to sex and age, the results of olfactory function tests, and the accompanying abnormal smell sensation, and we investigated the month of onset of the common cold symptoms, as the trigger of this syndrome; this was done to identify the causative virus of this syndrome epidemiologically.

The male to female ratio of patients with this syndrome in this series was 26 : 136; 137 (85%) of the patients were in the forties to sixties. It was thus obvious that the syndrome occurred most frequently in middle-aged women, suggesting a relationship between olfaction and climacteric or sex hormone disorders.

Olfactory function tests yielded the following results: On the standard olfaction test using the T & T olfactometer (Daiichi Yakuhin Sangyo Co.), the mean detection threshold for patients with this syndrome was 2.87 and the mean identification threshold was 4.20, indicating that the decrease of olfactory acuity in patients with this syndrome was relatively mild compared with that arising from other causes. On the intravenous prosulthiamine injection test, however, 55 patients (34.8%) showed no response. For responders, the mean latency time was almost normal, being 9.41 s, while the mean duration was relatively short, at 30.03 s (normal: >45 s). These results can be interpreted as indicating that the lesion responsible for this syndrome is on the olfactory epithelium and that symptoms are caused by olfactory epithelial dysfunction and easy fatiguability.

This syndrome is often accompanied by abnormal smell perception in comparison with olfactory disorders due to other causes. Twenty-nine patients (17.9%) complained of abnormal smell perception at the onset of the syndrome, while 20 (12.3%) complained of it after the initiation of treatment. In many patients, the parosmia was characterized by the invariable penception of a certain, usually bad, smell, regardless of the odor stimulus. The mechanism responsible for this phenomenon is not clear. Probably, it occurs with a disorder in any part of the odor recognition system. Thus, it may be caused by the abnormal function of an olfactory epithelium damaged by viral infection.

An analysis of the monthly incidence of the common cold as a trigger of this syndrome revealed that it occurred mostly in spring and summer. Therefore, on comparison with the monthly incidence of viral diseases or the frequencies with which viruses were detected it follows that influenza virus and Respiratory Syncytial virus are extremely unlikely to be the causative agent because they are prevalent in winter, while para-influenza virus, the enterovirus group, adenoviruses, and other viruses are rather more likely to be the causative agents of the disorder.

---

[1] Department of Otolaryngology, Osaka City University Medical School, 1-5-7 Asahimachi, Abeno-ku, Osaka, 545 Japan

Monorhinal Smell Identification Ability Before and After Temporal Lobectomy

Sumio Ishiai[1], Morihiro Sugishita[1], Hiroyuki Shimizu[2], Ichiro Suzuki[2], and Buichi Ishijima[2]

Key words. Smell identification—Temporal lobectomy

Patients who have had a temporal lobectomy have been reported to have normal olfactory detection thresholds, but to show an impairment in olfactory identification [1] and a deficit in smell discrimination in the nostril ipsilateral to the excision [2]. These impairments were not severe, and were rather mild. The patients were not tested before the operation. In this study, we examined the monorhinal smell identification ability of patients before and after temporal lobectomy.

The subjects were 18 patients with intractable temporal lobe epilepsy: 11 patients underwent right lobectomy and 7 underwent left lobectomy. Seventeen of the patients were right-handed. The remaining 1 was left-handed, but an intracarotid sodium Amytal test showed left hemisphere dominance for speech. In the right lobectomy, the anterior half of the hippocampus, almost all the amygdala, and the anterior 4 cm of the middle and inferior temporal, and the fusiform and parahippocampal gyri were excised. In the left lobectomy, the extent of excision was the same as in the right lobectomy, except that the middle temporal gyrus was preserved. Two patients, one with right excision and the other with left excision, underwent amygdalohippocampectomy, which left the neocortical gyri. All patients gave their informed consent for testing.

For the smell identification test, we used ten microencapsulated fragrances that represent the smell of mint and those of nine fruits familiar to Japanese. These fragrances were coated onto labels, as in the University of Pennsylvania smell identification test [3] and, in testing, the examiner scratched the label with sandpaper. We asked the subjects to sniff each of the smells monorhinally and to identify it by choosing among four alternatives. The subjects held the other nostril closed with a finger. The order of presentation of smells was counterbalanced across nostrils.

Fourteen patients did not show a significant difference between the numbers of correctly identified smells before and after temporal lobectomy, in either nostril. One patient who had had a right lobectomy and three who had had a left lobectomy showed a significant decrease in the number of correctly identified smells in the nostril ipsilateral to the excision. In three of these four patients, the extent of temporal lobectomy was comparable to those of the other patients with intact smell identification ability. In the remaining patient, who had had a left lobectomy, the lesion was small, as he had undergone amygdalohippocampectomy.

On magnetic resonance imaging (MRI) examination, the brains of the patients with intact smell identification ability were normal, except for the excised temporal lobe. The four patients with smell identification impairment in the nostril ipsilateral to the excision also had pathological changes on the contralateral side: atrophy of the temporal lobe or the cerebral hemisphere, a cystic lesion in the posterior temporal white matter, and an abnormal intensity lesion in the temporal stem.

It has been shown that the olfactory tract projects ipsilaterally from the olfactory bulb to the prepiriform cortex, the uncus, amygdala, and the entorhinal area [4]. However, our results suggest that monorhinal smell identification may be functionally processed not only in the ipsilateral temporal lobe but also in the contralateral temporal lobe. Identification of smells may be impaired in the nostril ipsilateral to the side of temporal lobectomy only when the function of the contralateral temporal lobe is also impaired.

References

1. Jones-Gotman M, Zatorre RJ (1988) Olfactory identification deficits in patients with focal cerebral excision. Neuropsychologia 26:387–400
2. Zatorre RJ, Jones-Gotman M (1991) Human olfactory discrimination after unilateral frontal or temporal lobectomy. Brain 114:71–84
3. Doty RL, Shaman P, Dann M (1984) Development of the University of Pennsylvania Smell Identification Test: A standardized microencapsulated test of olfactory function. Physiol Behav 32:489–502
4. Eslinger PJ, Damasio AR, Van Hoesen GW (1982) Olfactory dysfunction in man: Anatomical and behavioral aspects. Brain Cogn 1:259–285

[1] Department of Rehabilitation, Tokyo Metropolitan Institute for Neuroscience, [2] Department of Neurosurgery, Tokyo Metropolitan Neurological Hospital, 2-6 Musashidai, Fuchu, Tokyo, 183 Japan

Anosmic Patients Can Separate Trigeminal and Nontrigeminal Stimulants

David E. Hornung, Daniel Kurtz, and Steven L. Youngentob[1]

Key words. Anosmia—Trigeminal

Anosmia is the inability of the first cranial nerve pathway to detect airborne odorant molecules. In humans, however, information about airborne odorant molecules comes from a variety of sensory systems in addition to the first cranial nerve. As a first step in determining the degree of involvement of other sensory systems in the smell deficits of anosmic patients, a retrospective study was undertaken to determine if these patients could separate trigeminal and nontrigeminal odorants.

The data for this study come from 143 anosmic patients seen in The Smell and Taste Disorders Clinic at the Health Science Center in Syracuse, N.Y. The olfactory ability of these patients was measured using the Odorant Confusion Matrix (OCM) [1], a ten-odorant forced-choice olfactory test in which patients are presented with a bottle containing one of the ten test odorants and ask to pick from a list of the ten odorants the name which best describes what they smell. Each odorant is presented once in a block of ten, and ten blocks (100 trials) are presented in a session. The ten odorants in the OCM are: ammonia, cinnamon, licorice, mint, mothballs, orange, rose, rubbing alcohol, vanilla, and vinegar. Ten correct responses out of 100 represents the chance level of performance, whereas 80 or above indicates that the patient is normosmic. With a score of 20 or below the patient is considered to be anosmic.

For the anosmic patients, the percent correct for each odorant was not statistically different from that for any other odorant or from the chance level of performance.

Because the odorants are presented within a confusion matrix, it is possible to analyze patterns of response in addition to the overall percent correct score. The patterns of responses were thus analyzed with a 2×2 chi square design to determine if the patients used odorant names associated with trigeminal stimulants when presented with ammonia and vinegar. Since ammonia and vinegar are the only two odorants in the OCM with strong trigeminal components, the use of names associated with trigeminal stimulants in response to these odorants suggests an ability to detect trigeminal stimulants in general. The results of this analysis demonstrated ($P < 0.00005$) that anosmic patients could separate ammonia and vinegar from the other eight odorants of the OCM, even though these odorants are usually not correctly identified. Likewise, these patients were able to separate the other eight OCM odorants from ammonia and vinegar ($P < 0.0005$) even though correct identification of specific odorants was different from chance.

We next placed the anosmic patients in one of eight diagnostic groups by evaluating the records from The Smell and Taste Disorders Clinic. The groups reflected what was the most likely cause of the anosmia. The eight groups were: mucosal swelling, post upper respiratory infection, congenital, head trauma, idiopathic, polyposis, exposure to hazardous substances, and aging. The diagnostic groups did not differ from each other in the overall percent correct. The results of an analysis of variance suggest that the ability to separate trigeminal from nontrigeminal odorants was not different for the different diagnostic groups.

The conclusions of this study are:

1. Even though anosmic patients cannot often correctly name the odorants in the OCM, these patients are still quite facile at separating the sting from the two stronger trigeminal odorants from the sensation of the other eight odorants.
2. The ability to make these trigeminal discriminations is not affected by the etiology of the olfactory loss.

Reference

1. Wright HN (1987) Characterization of olfactory dysfunction. Arch Otol 113:163–168

---

[1] Department of Physiology, SUNY Health Science Center, Syracuse, NY 13210, USA

Olfactory Function in Patients with Hypogonadotropic Hypogonadism: An All-or-One Phenomenon?

Matthias Laska[1], Thomas Berger[1], Robyn Hudson[1], Babett Heye[2], Jochen Schopohl[3], Joachim Stemmler[4], Thomas Vogl[4], Adrian Danek[5], and Thomas Meitinger[2]

Key words. Clinical testing—Hypogonadotropic hypogonadism

Hypogonadotropic hypogonadism (HH) refers to a heterogeneous group of disorders of the hypothalamic-pituitary-gonadal axis typically associated with deficient release of gonadotropin-releasing hormone (GnRH) from the hypothalamus, consequent failure of the pituitary to secrete gonadotropins sufficient for full gonadal development, and subsequent failure of the patient to undergo puberty and achieve reproductive capacity. One of the best known forms of HH is Kallmann's syndrome (KS), which is characterized by aplasia or hypoplasia of the olfactory bulbs and consequent anosmia or severely impaired olfactory function. It apparently results from a disruption in the migration of neurons from the olfactory placode to the bulb and hypothalamus early in development, and so provides a unique opportunity to investigate olfactory function in human subjects with congenitally incomplete peripheral systems. KS and idiopathic hypogonadotropic hypogonadism without anosmia (IHH) are both highly variable in phenotypic expression and may be associated with a variety of neurological, urinary, skeletal, and other anomalies.

As part of a clinical study, 37 patients with HH were tested on a modified version of the Munich Olfaction Test (MOT) and compared to 37 age-matched controls. The MOT is based on the sniff-bottle method and includes tests of: (a) odor quality discrimination, (b) intensity discrimination, (c) detection thresholds, and (d) tests of recognition, hedonic evaluation, and identification ability for odors. Furthermore, all patients and part of the control group underwent magnetic resonance imaging (MRI) in the coronal and sagittal planes in order to assess the possible relationship between olfactory dysfunction and pathology of structures.

It was the aim of this study to evaluate:

1. Whether the considerable variation in the phenotypic expression of HH corresponds to a continuum of olfactory capacity or whether the groups of KS and IHH can be clearly separated
2. The suitability of the MOT for differential diagnosis, in this case of KS and IHH
3. To what degree anosmic subjects can rely on the common chemical sense and to estimate the contribution of the trigeminal nerve to olfaction in both impaired and unimpaired subjects

We found that the HH patients could be divided into two distinct groups differing significantly on all four subtests of the MOT and showing no overlap in performance: 20 anosmics, conforming to KS, and 17 apparently normosmics (IHH patients) whose performance, though sometimes slightly poorer, was not significantly different from that of the controls. Thus, there was no correlation between endocrinological impairment and olfactory dysfunction. MRI findings reflected the sharp distinction between patient groups and the similarity between IHH patients and controls. The MOT proved to be a useful tool for assigning patients with HH to one of the two groups, especially as patients with congenital anosmia are usually unaware of olfactory impairment. The considerable variation in the expression of HH did not correspond to a continuum of olfactory capacity. These findings are consistent with a neurodevelopmental defect rather than with a mutant GnRH gene being responsible for the occurrence of anosmia in KS.

Acknowledgments. Supported by the Deutsche Forschungsgemeinschaft (Hu 426/1).

[1] Institut für Medizinische Psychologie, [2] Abt. Pädiatr. Genetik der Kinderpoliklinik, [3] Medizinische Klinik Innenstadt, [4] Radiologische Klinik, [5] Neurologische Klinik, University of Munich Medical School, Goethestr. 31, 80336 Munich, Germany

13. Noninvasive Measurements of Human Chemosensory Response

Gustatory-Evoked Potentials in Man: A Long History and Final Results

KARL-HEINZ PLATTIG[1], PRODROMOS BEKIAROGLOU[1,2], and SUSANNE SCHELL[1]

Key words. Taste stimulation—Gustatory-evoked potentials—Electrophysiological methods—EEG—Tastants in life span—Minidrop

Introduction

For eliciting gustatory-evoked potentials (GEPs) from electroencephalograms (EEGs), the same evaluation methods are used as in other sensory modalities [compare with visual-evoked potentials, or the objective audiometric methods of evoked response audiometry (ERA) and brainstem-evoked response audiometry (BERA)]. The main principle is the "averaging" of many stimulus-linked EEG responses to produce "event-related potentials" or correlation of EEG sweeps, leading to auto- or cross-correlograms.

The main difficulties in studying the chemical senses of smell and taste are with stimulation. To elicit GEPs, we have applied tastants to the tongue in an open jet and via a closed chamber system. A third, completely new method of gustatory stimulation is our new "minidrop method."

Material and Methods

Both the open jet and closed chamber stimulation make use of the principle that small doses of sapid tastants must replace a steady stream of neutral water on its way to the human tongue [1]. This replacement has to be performed without any turbulence or oscillations, because these would add a tactile stimulation to the gustatory one. For minimization of these tactile artifacts, we developed several magnetic valves usable both for the open jet application (open system) and for the chamber application (closed system). For the latter, several chambers also had to be constructed [2].

For application of tastants to the rear parts of the tongue, our "method of one standard minidrop" was developed, an open system using a syringe com- monly used for injections of insulin or tuberculin. It is equipped with a shortened butterfly infusion set tubing (25 gauge) of which only 2–3 cm remain. With this fairly stiff and stable hose with an anterior opening usually of 0.3 mm precisely, tiny doses of about 10 µl can be delivered; these doses correspond to 5–15 mg in weight, as evaluated in 50 weight trials for each of the four taste qualities and for distilled water (used for control as "fifth quality").

It is to be kept in mind that these "minidrops" are not really drops, defined as depending on surface tension, but "doses," which can be delivered by slightly wiping the outlet tip of the syringe on the tongue (or any other surface). This wiping or touching certainly causes an additional somatosensory response that must be subtracted from the gustatory one. To keep the somatosensory influence minimal, it is important that the mucosa is touched not by the outlet of the syringe but only by the fluid itself.

Equipping an insulin syringe with a micrometer drive developed by the mechanical department of our institute according to the ideas of our student Thomas Weber makes application even more precise in the doses: turning the screw at a certain angle will advance the piston a small, very constant distance, and thereby a tiny amount of fluid is released to the tongue.

To trigger an averaging computer, a thin platinum wire is embedded in the outlet of the syringe or the initial tubing, respectively. The triggering device consists of an RC circuit in a box near the subject's hands, connected to the subject's head at F3 of the international 10/20 system of EEG electrodes and to the platinum electrode in the syringe as mentioned before. The condenser can be charged by the subject by pressing a pushbutton attached to a 9 V-battery, both located at or in the box. If the tongue is touched by the minidrop, the electric current from the RC circuit will mark the time of stimulation precisely by producing a "spike" in F3 with a sharp onset and a flatter exponential decrease. The onset flank can nicely trigger the computer. The experimenter must be grounded to the eight-channel EEg recorder (Mingograph, manufactured by SIEMENS-ELEMA, Solna, Sweden) to keep the EEG as free of noise artifacts as possible.

GEP computing was performed by the Erlangen program of topographical EEG evaluation, Brain-

[1] Institute for Physiology and Biocybernetics, University of Erlangen-Neurnberg, Universitätsstrasse 17, 91054 Erlangen, Germany
[2] On leave from Aristoteles University, Thessaloniki, Greece

Star, supplied by U. Brandl, on an AT-IBM-compatible 386 or 486 computer system.

In the main psychophysical experiments, there were 4 female subjects aged 14, 26, 49, and 77 years, while for "psychophysical tongue mapping" 10 females and 10 males (aged 20–30 years) were used. EEG and GEP experiments were performed on 8 females and 6 males (aged 20–30 years).

Results

Subjective taste thresholds produced by minidrops in four female subjects, aged 14, 26, 49, and 77 years, on the anterior and posterior parts of the tongue are shown in Figs. 1 (sodium chloride), 2 (tartaric acid), 3 (quinine-HCl), and 4 (sucrose). In the ordinates the millimolar concentrations are given in logarithmic divisions; in the abscissae are the number of hits of eight repeated presentations of the same concentration to the same place on the tongue.

Figure 5 represents a GEP of a 22-year-old woman elicited by 300 mmol NaCl applied as a minidrop to her right-anterior tongue portion (after zero voltage correction and slight smoothing), and Fig. 6 shows the response of a man also aged 22 years to plain water applied as a minidrop to his right anterior tongue portion. Figure 7 shows the GEP of a 22-year-old woman elicited by 5 mmol tartaric acid

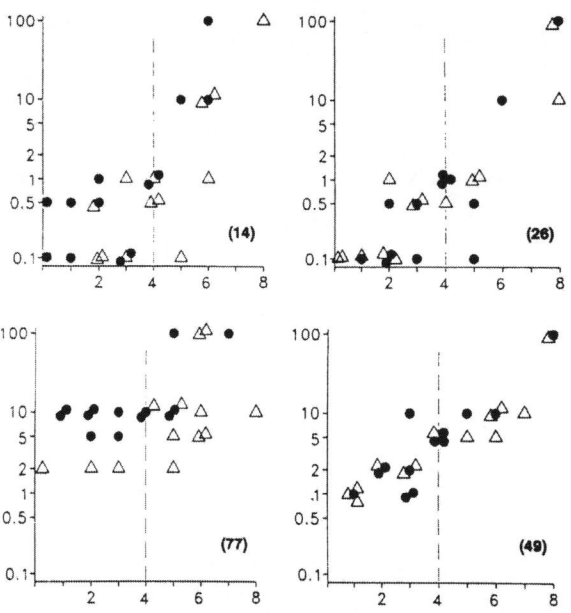

FIG. 2. Same experimental conditions as Fig. 1 using tartaric acid

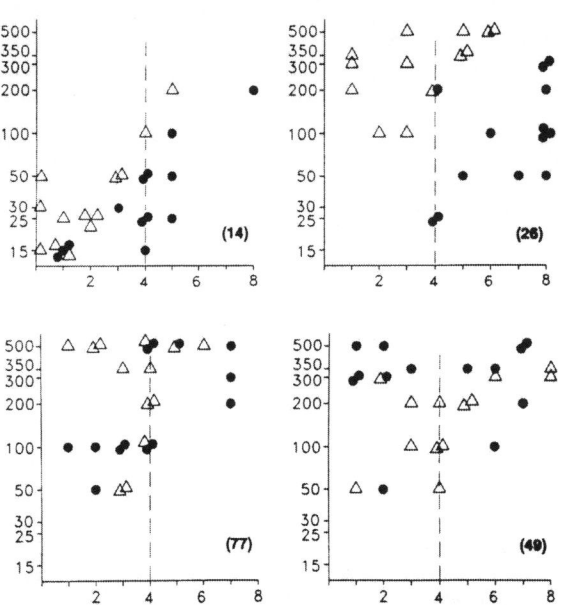

FIG. 1. Subjective taste thresholds gained by minidrops of sodium chloride applied to anterior (*closed circles*) and posterior (*open triangles*) parts of the tongue of four female subjects aged 14, 26, 49, and 77 years. In the ordinates, millimolar concentrations are given in logarithmic divisions; in the abscissae, the number of hits of eight repeated presentations of same concentration to same place on the tongue

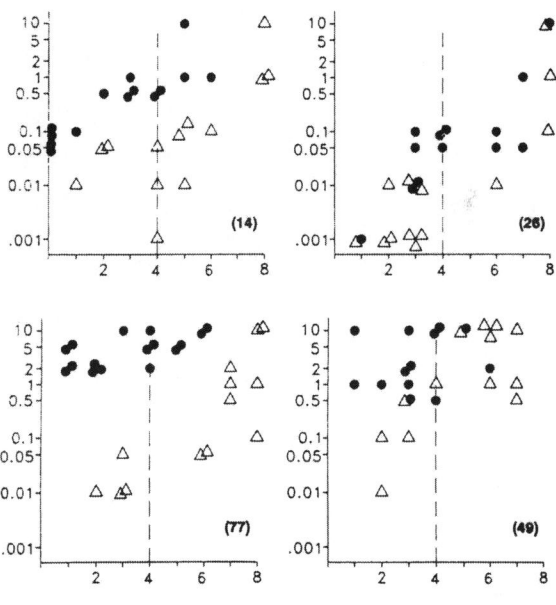

FIG. 3. Same experimental conditions as Fig. 1 using quinine-HCl

applied as a minidrop to her right anterior tongue portion. Responses to quinine-HCl and sucrose are similarly small as to tartaric acid.

All GEP pictures compare the five 10/20 leads, Cz, C3, C4, P3, and P4 versus A1+A2. The time scale is 4 s; in the ordinate, 2 μV indicates the y-scale (negativity up), and the number of averaging steps used is marked at the bottom of each figure.

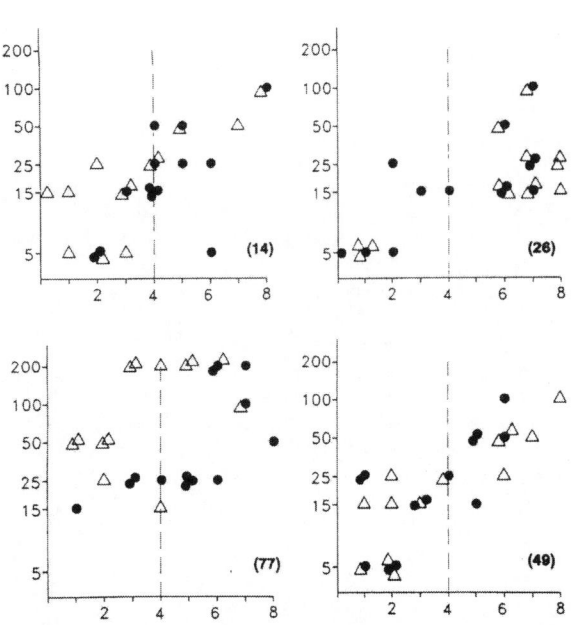

FIG. 4. Same experimental conditions as Fig. 1 using sucrose

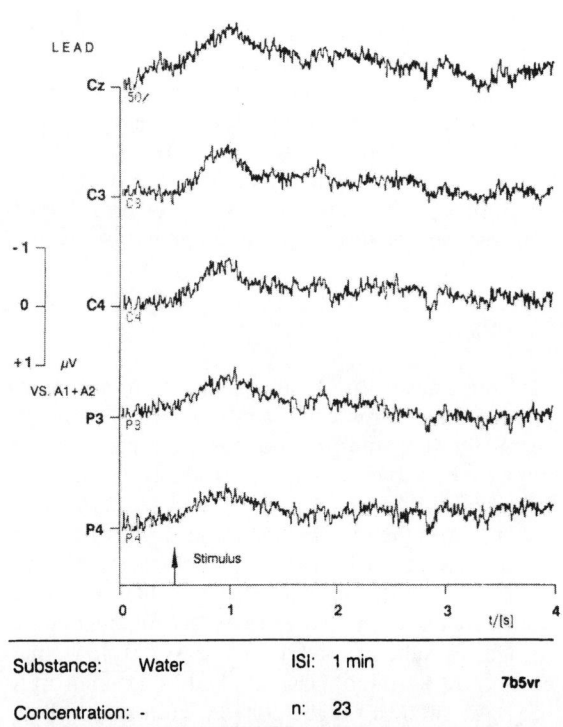

| Substance: | Water | ISI: | 1 min | |
| Concentration: | - | n: | 23 | 7b5vr |

FIG. 6. Response of 22-year-old male to plain water, also applied as minidrop to the right-anterior tongue portion (zero voltage correction, no smoothing); 23 averaging steps

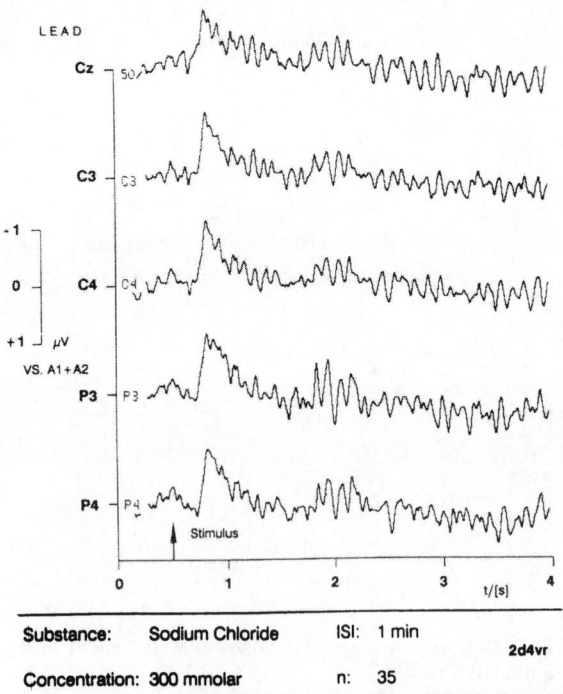

| Substance: | Sodium Chloride | ISI: | 1 min | |
| Concentration: | 300 mmolar | n: | 35 | 2d4vr |

FIG. 5. GEP of a 22-year-old female elicited by 300 mmol NaCl applied as minidrop to the right-anterior tongue portion (after zero voltage correction and slight smoothing); 35 averaging steps

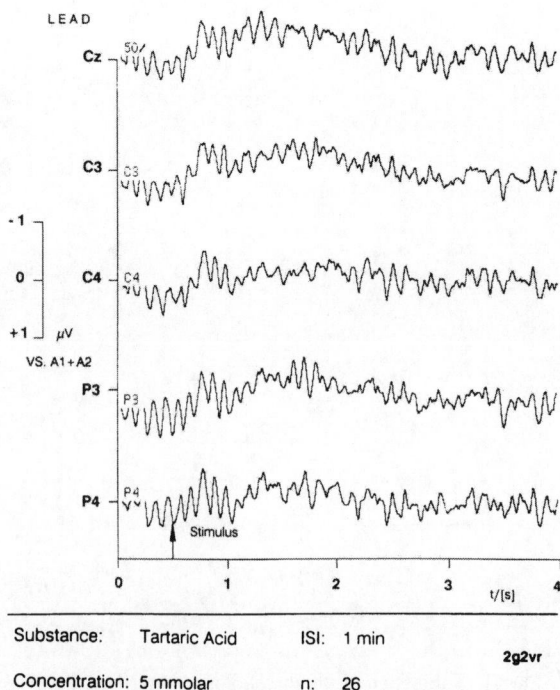

| Substance: | Tartaric Acid | ISI: | 1 min | |
| Concentration: | 5 mmolar | n: | 26 | 2g2vr |

FIG. 7. GEP of a 22-year-old female elicited by 5 mmol tartaric acid applied as minidrop to the right-anterior tongue portion (zero voltage correction, no smoothing); 26 averaging steps

Discussion

In the four females aged 14, 26, 49, and 77 years, thresholds were determined for the right- and left-anterior and rear portions of their tongues. The results (Figs. 1–4) show the well-known broad variance. Eight presentations of each tested concentration of the four substances (quinine-HCl, sucrose, tartaric acid, and sodium chloride, and also distilled water) were performed in randomized order, but only the hits for the four substances were registered.

For bitter (see Fig. 3), as has long been known, sensitivity is high in the rear (hits between 0 and 8 on the abscissa, marked by full circles for anterior and by open triangles for posterior tongue). A certain value for a threshold by calculated regression curves cannot be given, but looking at the scattered hits we can find some orientation. There will be no good regression curves before much more data are available, but if these curves were elaborated, they could likely represent oblique lines of different steepness. The concentration where they cross the vertical line at four hits (representing the 50% score) could be taken as a measure for the "actual threshold value." Our results also suggest that between the 14th and 77th years of age the bitter thresholds rise markedly at the tip, but less at the rear, of the tongue.

For sweet (see Fig. 4) there are about equal anterior–posterior tongue thresholds in the youth and a distinct elevation in the rear of the tongue at age 77, but in midlife the conditions appear similar to those in the youth.

For sour (see Fig. 2) there might again be about equal anterior–posterior thresholds in the youth, both slightly rising during midlife, but in age there seems to be a marked rise of the anterior threshold. The same may hold for salty (see Fig. 1), except that the rise in greater age is more pronounced and occurs in both anterior and posterior positions; however, in both qualities variances are also higher than in bitter and sweet, and the subjects are much more inclined to mix up salty and sour.

Stimulation of the anterior tongue by the minidrop system is very similar to that performed by our chambers with the exception that it is now again an open system. Application of standard minidrops to the rear of the tongue is still not easy: to recognize the targeted foliate and fungiform papillae a fiberglass illuminator must be used and to avoid the retching and vomiting reflexes of the subject, his tongue must be touched gently.

Disruptive somatosensory influences of the application of the minidrops were checked by application of plain water (see Fig. 6). Water causes a slow negative wave of 0.5–1 µV total amplitude with short latency of 100–200 ms. It is terminated about 1–1.5 s after stimulation.

In contrast, 300 mmol sodium chloride elicits a sharp rise 250–400 ms after stimulation (see Fig. 5).

The place of stimulation is encoded in these pictures in the lower right: the first and second numbers and figures bear infermation about the subjects, the third position gives the quality (1: sweet; 2: sour; 3: bitter; 4: salty; 5: tasteless water), and v, h, l, and r are the position of stimulation (v/h is anterior/posterior and l/r is left/right). In this case of right-side anterior stimulation the contralateral P3 amplitude appears slightly, but not significantly, higher than the ipsilateral P4.

Results for sweet, sour, and bitter (see Fig. 7 for sour) so far are confusing: only for tartaric acid is a distinct short-latency onset present, although still quite disturbed by noise and motion artifacts, while for sweet and bitter the longer latency of onset that we had demonstrated by chamber stimulation [2] cannot be seen. In this regard we have to consider the very small amounts of tasting substances in our minidrops: 10 µl contains 3.10^{-6} or 2.10^{-6} mole of sodium chloride or sucrose if a 300- or 200 mmol solution is used (0.2 mg NaCl for 300 mmol, 0.7 mg sucrose for 200 mmol), while the respective values for 5 mmol tartaric acid are 5.10^{-8} mol or 7 µg per minidrop, and even for a 0.5 mmol solution of quinine-HCl, only 5.10^{-9} mol or 2 µg per minidrop. Also, the termination of stimulation is not notable when using the minidrops.

Conclusions

So far this is the only method known to us for stimulating human foliate and vallate papillae to record electrophysiological (EEG) responses. This new method of minidrops might become a valuable tool that should be developed further for stimulating the rear of the tongue and certainly also other areas in the mouth. Stimulation via the chamber is apparently now already a valuable tool, but is still being improved so as to analyze how men and women are delighted by the aroma, taste, and smell of fine things.

Acknowledgments. We are grateful to our medical student Thomas Weber for planning an improved dosage device for the minidrops and to Gerald Kramp, head of the mechanical workshop in our institute, for realizing it.

References

1. Plattig KH, Dazert S, Maeyama T (1988) A new gustometer for computer evaluation of taste responses in men and animals. Acta Otolaryngol (Stockh) Suppl 458:123–128
2. Plattig KH, Eisentraut U, Eichner R (1990) Human gustatory evoked EEG responses—methodological improvements. Biomed Res (India) 1:55–66

Brain Event-Related Potentials to Primary Tastes

JOHN PRESCOTT[1]

Key words. Event-related Potentials—Taste—Topography—Intensity

Introduction

For nearly three decades, the recording of human brain event-related potentials (ERPs) to visual, auditory, and somatosensory stimuli has provided an effective tool for investigating sensory and cognitive processing, and also has been an increasingly useful clinical tool (see, for example, [1]). While there is now a rapidly increasing research literature on olfactory ERPs [2], there has been very little research on gustatory ERPs. This stems at least partly from the difficulties in accurately producing discrete stimuli. As a result, there is no consensus on the nature of the ERP components associated with taste stimuli, their topographical distribution, or their responsiveness to physical or psychological manipulations of stimulus or response parameters.

Plattig [3], using sophisticated gustatory stimulation that delivered taste solutions to the tongue surface, described a negative (410 ms)–positive (1200 ms) complex produed by solutions of NaCl. Effects of stimulus intensity and duration on potential amplitude and latency were demonstrated, as was a dependence on the tastant used. Responses to air saturated with acetic acid sprayed onto the tongue were demonstrated by Kobal [4], who described a sequence of four major components: a positivity (300 ms), a negativity (410 ms), a positivity (660 ms), and a negativity (860 ms). Maetani et al. [5] reported a triphasic waveform: N1 (200 ms)–P1 (350 ms)–N2 (1000 ms), using NaCl as the stimulus. They noted that NaCl-free stimuli also produced N1 and P1, suggesting that only N2 was specifically related to taste.

The research to date is thus equivocal regarding the nature of gustatory ERPs, although it does suggest that the method of stimulation may be an important determinant of the responses that are seen. The aim of the research presented here was to describe the potentials and their scalp distribution that occur following gustatory stimulation and to relate peak amplitudes and latencies to ratings of tastant intensity.

[1] Sensory Research Centre, CSIRO Division of Food Science and Technology, Sydney, NSW 2113, Australia

Methods

The taste delivery system (Fig. 1) used in these experiments was based on that described by Kelling and Halpern [6]. It consists of a closed delivery system in which the subject's anterior tongue seals an elliptical opening (8 mm × 5 mm) in the wall of a tube through which the tastant flows. A pair of electrodes situated within the tube opposite the opening lead to a circuit that detects changes in electrical conductivity, relative to a rinse of deionized water, which are produced by the arrival of the tastant at the tongue. The tastants and rinse are supplied from pressurized containers via Teflon tubes to solenoid valves that, under computer control, allow either one of the tastants or the rinse to flow at about 6 ml/s to the subject.

The stimuli were solutions of NaCl plus a control solution of ordinary water. Between stimuli, subjects received a continuous rinse of deionized water. Forty stimuli of each concentration of NaCl and water were presented randomly in blocks of 5 with a random interstimulus interval (ISI) of 20–28 s, a nominal duration of either 300 or 500 ms, and a rise time of about 80 ms.

One group of subjects rated solutions of 0.2, 0.4, and 0.8 M NaCl and water; for another group, 0.4, 0.8, and 1.6 M NaCl and water were passively received in one session and rated in a second session. Subjects were asked to keep their eyes open and fixate on a cross on a computer screen 1 m in front of them at eye level. In conditions in which subjects rated taste intensity, a graphic rating scale with endpoint descriptors of *not at all salty* and *extremely salty* appeared on the computer screen 5 s after the stimulus. Subjects used a computer mouse to place a mark at any point on this scale.

EEGs were recorded from electrodes at Fz, F3, F4, Cz, C3, C4, T3, T4, Pz, P3, P4, T5, T6, O1, and O2 (10–20 system), referred to the right earlobe. An electrode on the forehead was used as a ground, while another above the right eye was used to monitor eye movement. The EEG was sampled at 125 Hz and amplified using a Grass Model 12 Neurodata System with a gain of 50 000 and a bandpass of 0.1–30 Hz. The conductivity changes were used as event markers to form 4000-ms epochs following stimulus onset. Averages for each taste concentration were obtained after rejecting trials that were severely affected by artifacts.

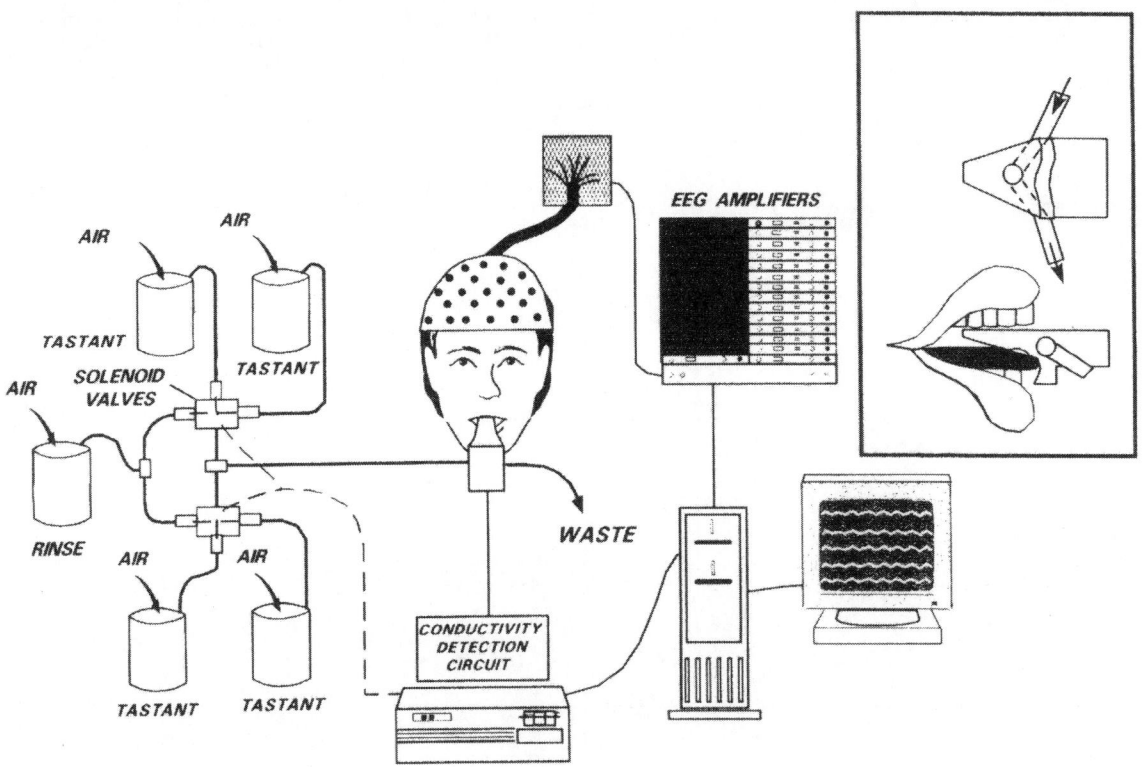

FIG. 1. Schematic depiction of gustatory stimulus delivery and EEG recording system. Inset: detailed views of mouthpiece; *top*, (from underside of mouthpiece) shows aperture, which is sealed by subject's tongue, and *bottom*, positioning of mouthpiece within subject's mouth

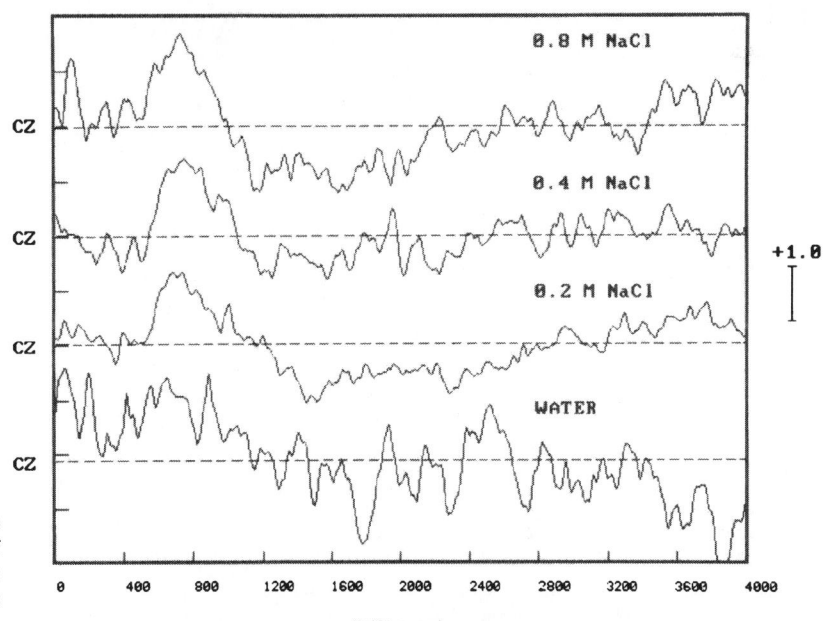

FIG. 2. Group average ERPs to three different concentrations of NaCl and water in a condition in which subjects rated stimulus intensity

Results

Figure 2 shows ERPs to three different concentrations of NaCl plus water in a condition in which subjects rated the stimulus intensity. The most pro-minent component is a positive peak occurring at approximately 700 ms. In individual averages, this potential frequently had a double peak, which was masked in the group averages. The potential had maximum amplitude over the parietotemporal areas,

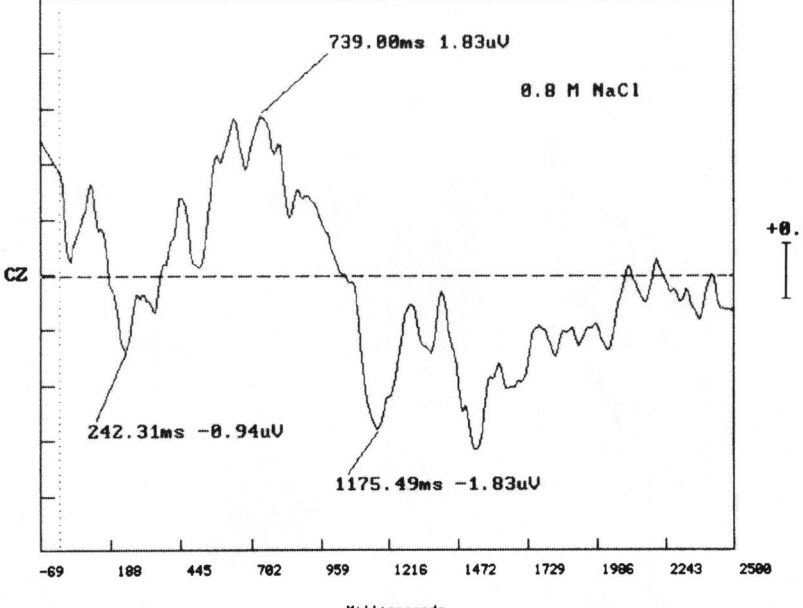

FIG. 3. Detailed view of one subject's ERPs to solution of 0.8 M NaCl shows the three components most commonly seen with their latencies and amplitudes

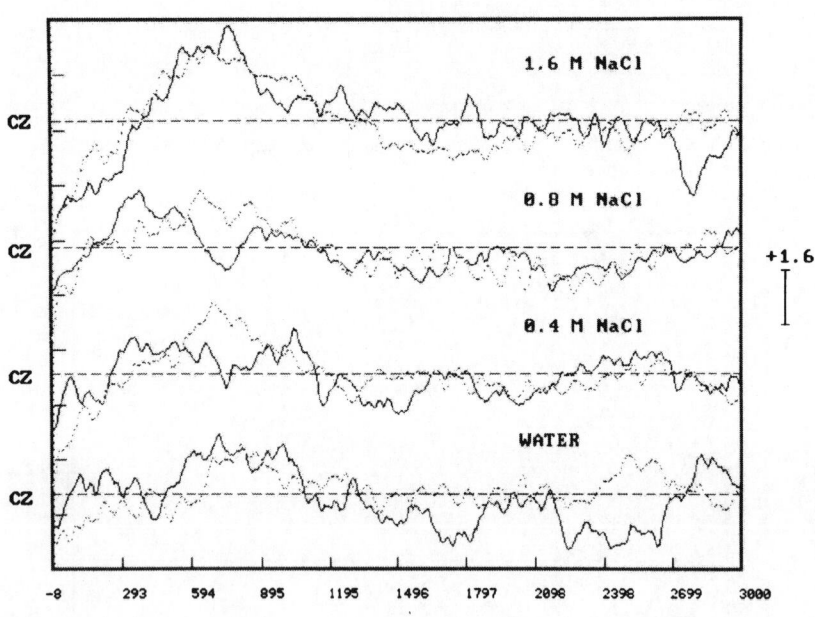

FIG. 4. Group average ERPs to three different concentrations of NaCl and water in conditions in which same subjects rated stimulus intensity (*dotted traces*) or produced no response (*solid traces*)

with a slight right-hemisphere bias, and also fronto-centrally. This was followed by a broad negativity, whose onset and maximum amplitude (~1200 ms) seem to be intensity dependent. An early negativity may also occur at 200–300 ms, although this is not always discernible. These potentials, shown in detail in Fig. 3, have very small amplitudes, particularly when compared to those described in previous research on gustatory ERPs. In contrast to the NaCl solutions, the response to water (which had a slight taste relative to the rinse) shows less distinct components.

Figure 4 shows data recorded in conditions in which subjects rated the intensity of stimuli (dotted traces) or passively received stimuli (solid traces). In these traces, the amplitude of the positive component at about 700 ms is largest with the highest intensity stimulus (1.6 M NaCl), although there is little distinction between 0.8 M and 0.4 M NaCl solutions. There is some indication that the peak is delayed relative to the passive condition for 0.8 M and 0.4 M NaCl solutions. The temporoparietal and frontocentral topography of the component does not appear to be significantly altered by the inclusion of a rating task.

The amplitude of the positive component and intensity ratings showed only a modest association ($R^2 = 0.36$). In fact, the intensity ratings showed both high variability and poor discrimination between stimuli. Correlations between intensity ratings and latencies were effectively zero.

Discussion

Compared to the research of Plattig [3], who used very similar stimuli (although of greater duration) delivered as a liquid, the present results show little agreement on component polarity or latency. Where Plattig [3] showed a prominent negative peak at approximately 400 ms, the most prominent component in the present data was a positivity at about 700 ms.

The most obvious distinction between the two sets of results is the differing methods of stimulation, because not only do the modes of delivery differ but also the extent of stimulation on the tongue. The present data were generated by stimulation of a discrete area of the anterior tongue, while Plattig's method of stimulation suggests a more extensive and less well-defined stimulus area. The restricted stimulus area used in the present research may also account for the small amplitudes of the ERP components. Such inconsistencies are perhaps to be expected at this stage in the study of gustatory ERPs. However, they do emphasize the need to reach agreement regarding the optimal stimulus parameters and methods.

Acknowledgments. I would like to thank Andrew Eddy, who provided computer programming and technical assistance.

References

1. Regan D (1989) Human brain electrophysiology. Elsevier, New York
2. Kobal G, Hummel T (1991) Olfactory evoked potentials in humans. In: Getchell TV, Bartoshuk LM, Doty RL, Snow JB (eds) Smell and taste in health and disease. Raven, New York, pp 255–275
3. Plattig K-H (1991) Gustatory evoked brain potentials in humans. In: Getchell TV, Bartoshuk LM, Doty RL, Snow JB (eds) Smell and taste in health and disease. Raven, New York, pp 277–286
4. Kobal G (1985) Gustatory evoked potentials in man. Electroencephaologr Clin Neurophysiol 62:449–454
5. Maetani C, Notake N, Takemoto I, Koizuka I, Ogino H, Matsunaga T, Yoshimura S, Tonoike M (1989) Gustatory evoked potentials by taste solution. In: Proceedings of JASTS XXII, Fukuoka, Japan, November 8–9, 1988. Chem Senses 14(2):311
6. Kelling ST, Halpern BP (1986) The physical characteristics of open flow and closed flow taste delivery apparatus. Chem Senses 11(1):89–104

EEG Topography of Affective Response to Odors

Shizuo Torii[1], Yoshiro Okazaki[2], and Jeffrey D. Winchester[3]

Introduction

Researchers in the past few decades have successfully employed electroencephalograms (EEGs) in their investigations of cerebral lateralization of mental functions [1–3]. For example, when the brain waves of a subject engaged in a spatial-oriented task (e.g., manipulating blocks to reconstruct a pattern from memory) are monitored, a decrease in alpha waves in the right hemisphere is seen, indicating activation of the right side of the brain during that task. Similarly, the left hemisphere usually shows decreased alpha-wave activity during a task that is more verbally focused. Also, the left and right cerebral cortex contribute differentially to the regulation of emotion. For example, greater left-frontal activation has been associated with reports of more intense positive affect in response to positive film clips, whereas greater right-frontal activation has been associated with more intense reports of negative affect in response to negative film clips [4].

EEG topographic observation has also helped reveal more discretely the localization of some mental functions [5–8]. Because alpha waves are presumed to reflect an "idling" state of cortical tissue, decreased alpha waves in a particular area in the cortex can be taken to indicate activation of that area. Using topographic mapping of alpha waves (7.6–9.4 Hz), we undertook to identify a typical alpha-wave reduction patterns during sniffing odors, and in doing so found a novel topographic pattern presumed to reflect affective arousal.

Subjects and Methods

The subjects were 10 volunteers, 4 men and 6 women between the ages of 24 and 43. All were right-handed; the four male subjects were professional perfume sniffers. The female volunteers all expressed an interest in scents. We found in a pilot study that the subjects' highly developed sensitivity to odors aids in obtaining clear objective and subjective data.

EEGs were recorded using 16 electrodes with a linked ear reference, employing the international 10–20 system of electrode placement. EEG activity was analyzed in real time and displayed on a signal processor 7T18 (Nihondenki-Sanei, Tokyo, Japan) by obtaining 10-s epochs of EEG stored and transformed through a fast Fourier transformation in numerical data representing rhythm power. Alpha-wave power was displayed; black designated areas of higher amplitude and white designated areas of lower amplitude.

The subjects sat in a comfortable highbacked chair in a well-ventilated room adjacent to the room containing the EEG amplifiers. They were instructed to keep their eyes closed throughout the experiment. Subjects rested for 5 min and EEGs were recorded for the last 1 min of this resting period. Next, EEGs were recorded while the subject sniffed perfume for 1 min. Following the completion of EEG recording, topographic color maps were obtained for the power at alpha frequency (7.6–9.8 Hz) from each 5-s period of recording during sniffing odors.

Each subject sniffed his or her favorite perfume. The reason for using the subjects's favorite perfume instead of individual essential oils is because favorite perfumes were found in a pilot study to induce much more affective arousal.

Results and Discussion

During the rest period, the topographic pattern of alpha waves for each 5-s period of recording varied a great deal within, as well as between, individuals. However, while sniffing perfume, the topographic patterns seen were rather consistent. Therefore, representative findings obtained in one subject at 5-s intervals are described here in detail.

Before undertaking this experiment, however, the difference in topographic patterns between hedonic positive perfume and hedonic neutral perfume was examined. Figure 1 shows that hedonic-positive perfume (right column) produced much more complicated patterns than did hedonic-neutral perfume (left column). Consequently, the subjects were asked to sniff hedonic-positive perfumes.

One of the male subjects, I.N., provided objective alpha-wave data and collaborative subjective reports

[1] Professor emeritus, Toho University, 2-6-20 Shin-Ishikawa, Midori-ku, Yokohama, Japan
[2] Takasago Central Research Laboratory, 1-4-11 Nishi-Yahata, Hiratuka, Kanagawa, Japan
[3] School of Medicine, Toho University, 5-21-16 Omori-nishi, Ota-ku, Tokyo, Japan

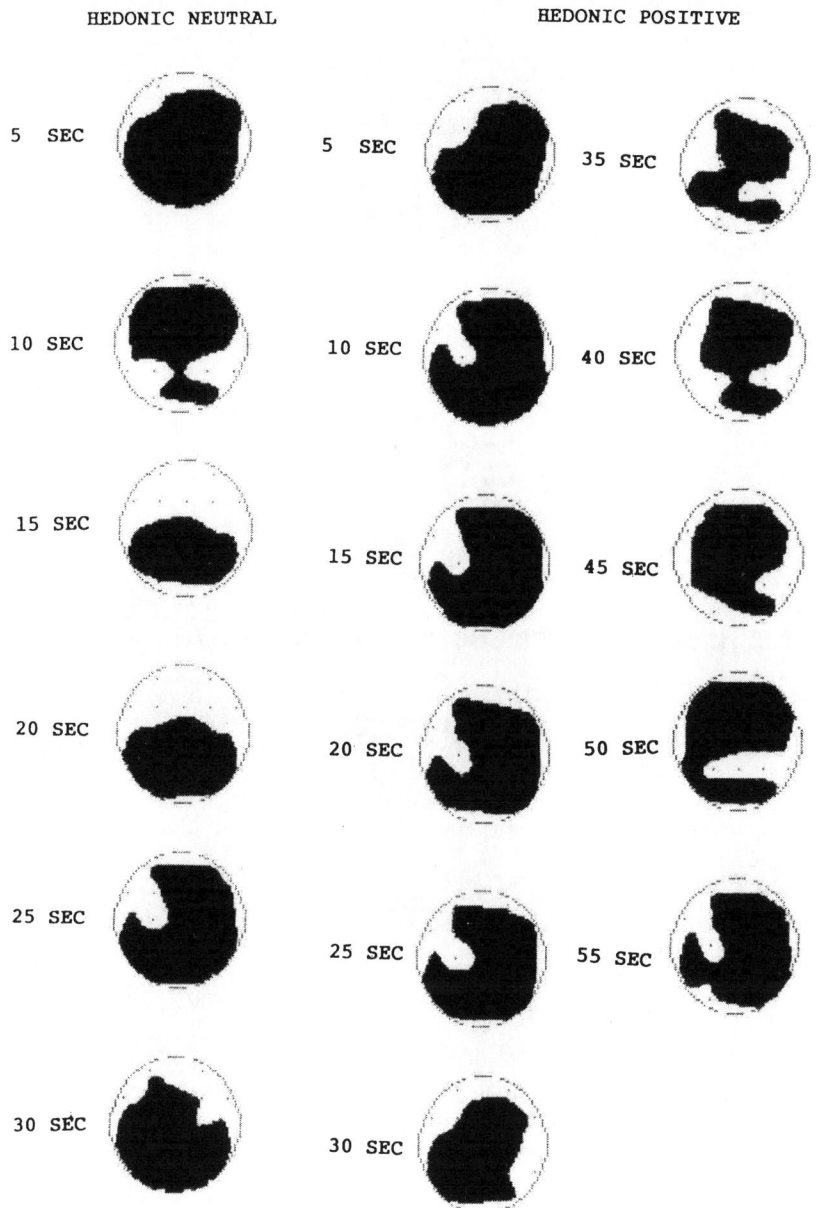

FIG. 1. Topographic maps of the cerebral cortex of subject I.N. while sniffing his favorite perfume. Each map represents one 5-s interval. *Left column*, hedonic-neutral pefume; *right column*, hedonic-positive perfume. Localized *white areas* indicate activation of that area. *Black dots* indicate positions of the 16 electrodes (according to the international 10–20 system)

that were typical of all 10 subjects. For the first 25 s after starting to sniff the perfume, a clear lack of alpha waves (white area) was seen at the left frontal and central area. The subject reported that he was analyzing the odors contained in the perfume. At 30 s, an alpha-wave dearth was seen at the right-parietal area, but at 35 s, it switched over to the left-central area. The subject reported having difficulty getting a clear mental image. At 40 s, alpha waves were gone from both sides of the parietal area. Although the stimuli that may possibly evoke this

pattern are as of yet unknown, it might be possible to associate this alpha-wave pattern with the subject's report, as 45 s, which follows.

At 45 s, the alpha-wave deficit was seen in the right-parietal area, which is known to be involved with mental visual images. The subject reported seeing the image of a female figure in his mind. At 50 s, the alpha waves were gone from the midline as well as from both sides of the parietal area. The subject reported feeling sexual arousal. This was a very unusual pattern; a search of the literature failed

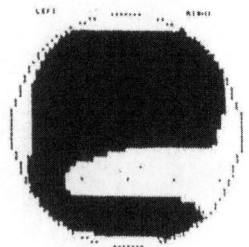

I.N.(male) SMELLING HIS FAVORITE PERFUME AND FEELING
SEXUAL AROUSAL

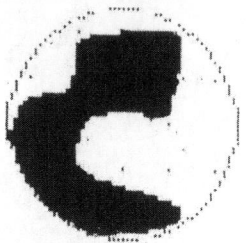

S.A.(female) LISTENING TO EXCITING MUSIC

H.H.(female)   HAVING FUN SOLVING MENTAL ARIHMETIC
PROBLEMS

FIG. 2. Topographical maps of cerebral cortex of three subjects during three activities causing affective arousal. An almost identical pattern of alpha-wave activity reduc- tion is seen in midline and parietal areas on both sides in all three subjects

to uncover other reports of this pattern. At 55 s, the lack of alpha waves had resumed in the left side of the frontal and central areas. The subject reported feeling satisfied with the experience and clear-minded. Thus, by using EEG topographic maps, we were able to clearly visualize and record areas of the brain involved in olfactory information processing.

It is interesting to note that the topographic pattern this subject exhibited during the time he reported sexual arousal (at 50 s) was the same as that seen other subjects who reported being "excited" while listening to their favorite music and "having fun" doing mental arithmetic problems (Fig. 2). To the best of our knowledge, this pattern has not been previously identified. It may be possible that this pattern of reduced alpha-wave activity in the midline and the parietal areas of both sides of the brain is an indicator of affective or emotional arousal. All the other subjects exhibited a similar pattern when sniffing their favorite perfumes.

There is supportive evidence for this assumption. We know from clinical EEG studies that the electrical activity of the medial part of the cerebral cortex, known to be involved in affection, may be recorded in the midline of the cerebral cortex. Of course, such a small sample is insufficient for drawing any solid conclusions. Moreover, obtaining the subjects' introspective reports after the actual recordings makes it difficult to claim clear correlation between the alpha-wave activity patterns identified and the subjects' thoughts and feelings. As similar experiments are repeated and replicated, we hope to build a database that will allow us to confirm these correlations.

Summary

Changes occurring in EEGs while sniffing perfumes were studied in a group of 10 healthy subjects. Alpha-wave mappings yielded the following results:

1. Hedonic-positive perfume produced more complicated topographic alpha wave patterns than did hedonic-neutral perfume.
2. Analyzing perfume odors was accompanied by suppression of alpha-wave activity in the left-frontal area.
3. Mental visual imagery produced a decrease of alpha-wave activity in the right-parietal area.
4. Affective arousal produced decreases of alpha-wave activity in the midline and both sides of the parietal areas.

References

1. Robbins KI, McAdam DW (1974) Interhemispheric alpha asymmetry and imagery modes. Brain Lang 1: 189–193
2. Furst CJ (1976) EEG alpha asymmetry and visuospatial performance. Nature (Lond) 260:254–255
3. Gevins AS, Zeitlin GM, Doyle JC, Schaffer RE, Callaway E (1979) EEG patterns during cognitive tasks. II. Analysis of controlled tasks. Electroencephalogr Clin Neurophysiol 47:704–710
4. Wheeler RE, Davidson RJ, Tomarken AJ (1993) Frontal brain asymmetry and emotional reactivity: a biological substrate of affective style. Psychophysiology 30:82–89
5. Van Toller S, Behan J, Howells P, Kendal-Reed M, Richardson A (1993) An analysis of spontaneous human cortical EEG activity to odours. Chem Senses 18:1–16
6. Kano K, Nakamura M, Matsuoka T, Iida H, Nakajima T (1992) The topographical features of EEGs in patients with affective disorders. Electroencephalogr Clin Neurophysiol 83:124–129
7. Grillon C, Buchsbaum MS (1987) EEG topography of response to visual stimuli in generalized anxiety disorder. Electroencephalogr Clin Neurophysiol 66: 337–348
8. Pfurtscheller G, Berghold A (1989) Patterns of cortical activation during voluntary movement. Electroencephalogr Clin Neurophysiol 72:250–258

Olfaction and the EEG:
Old Paradigms and New Approaches

TYLER S. LORIG[1]

Key words. Olfaction—EEG—ERP—Odor labeling

Introduction

Electrophysiology has been applied to the study of many human senses with outstanding success. Vision, audition, the somatosensory system, and even the vestibular system have been studied extensively and fruitfully using the electroencephalogram (EEG) and event-related potentials (ERPs). The application of these techniques to taste and smell has similar promise. Unfortunately, chemosensory stimuli can be difficult to control and, because precise stimulus control is essential for ERP studies, data of this type have been difficult to obtain in many clinical settings. Alternative electrophysiological techniques are available that offer flexibility and good correlations with traditional psychophysical measures.

Review

Figure 1 illustrates several electrophysiological techniques that have been applied to the study of olfactory processing (a somewhat similar list may be constructed for gustatory stimuli). The most simple application is to record EEG during the presentation of odors (olfactory EEG). The first such recordings were demonstrated by Moncrieff [1]. More recently, van Toller [2], Lorig and Schwartz [3], Klemm et al. [4], and van Toller et al. [5] have applied sophisticated quantitative analysis to the EEG of subjects exposed to different odors. The results of these analyses have indicated consistent differences between groups for some odorants, but the differences are not sufficiently large to provide a sensitive clinical tool at this time.

Olfactory ERPs (see Fig. 1) are obtained by repeatedly administering a brief olfactory stimulus and recording the resulting changes in EEGs. These EEG segments are averaged to create an evoked potential (ERP). While a number of investigators have used this approach, Kobal and Hummel [6,7] have been the most active in promoting the technique. The advantage of this approach is that the entire cortical response to the stimulus may be captured in the time domain, thus revealing the transduction and sequential neural processing of the olfactory stimulus. Additionally, ERPs have very good signal-to-noise ratios because they represent the average of many stimuli. The difficulty in this approach stems from problems in controlling the precise onset of the olfactory stimulus. Variability in stimulus delivery leads to "smearing" of the resulting waveform and degradation of the amplitude of the response. While it is an exceptionally promising technique, few labs have constructed or purchased olfactometers capabel of providing the stimulus precision necessary for the collection of these data.

There are other ways, however, to use ERP techniques to study olfaction. Torri and colleagues [8] described the use of olfactory stimuli to modulate more traditional electrophysiological recordings. They presented odors while subjects were engaged in a contingent negative variation (CNV) task. The CNV paradigm requires a subject to respond to a stimulus (S2) after a warning stimulus has been provided (S1). Torri et al. found that some odors enhanced the negativity that arises in the S1–S2 interval, a finding later replicated by Lorig and Roberts [9]. The ability of odors to modulate the CNV and other types of ERPs can be a powerful tool for understanding olfactory processing even though it cannot establish the series of neural events associated with odor reception.

Our laboratory has built upon the work of Torri and associates by using odors to modulate ERPs to auditory and visual stimuli. Lorig et al. [10] found that increasing concentrations of galaxolide affected the amplitude of auditory P300 differentially for rare versus frequent tones. Amplitude of the P300 was greatest during the rare tones presented during the administration of the highest concentrations of galaxolide. The P200 component also changed similarly. We have found odor-modulated ERP effects in both visual and auditory "oddball" paradigms and for several different odors [11].

[1] Department of Psychology, Washington and Lee University, Lexington, VA 24450, USA

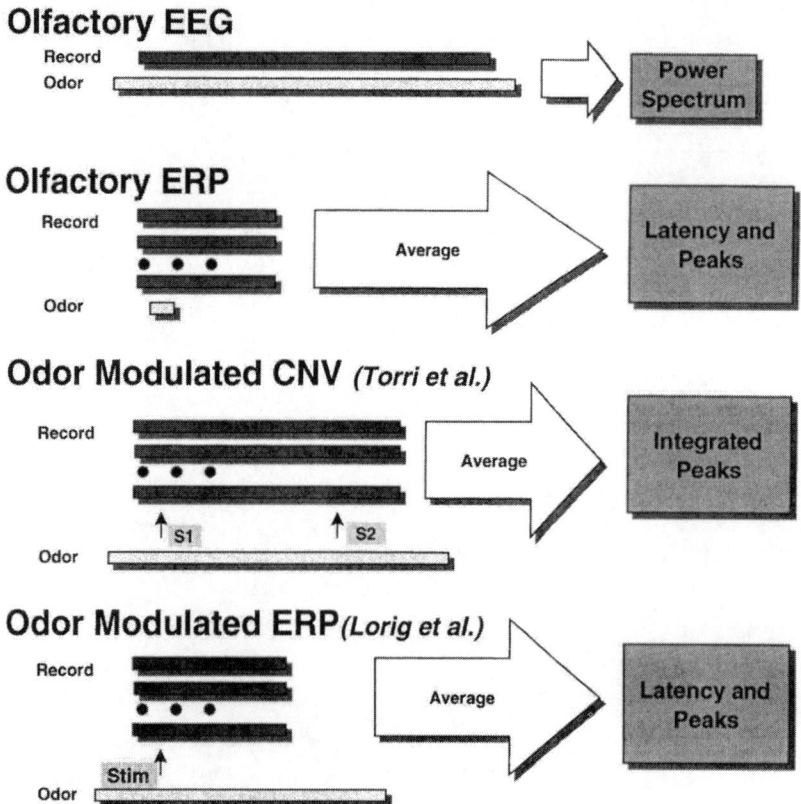

FIG. 1. Diagram of various paradigms using EEG and ERP techniques to study olfaction

Most recently, our work has centered on ERPs produced as subjects attempt to label odors. In our first experiment [12], an odor was presented and followed by a visual label that correctly identified the odor on 75% of the trials. ERPs were collected to the visual label. An analogous visual condition was used as a control condition. Thus, subjects smelled or saw a stimulus such as SOAP (depending on the condition), which was then followed by a visual stimulus such as "SOAP." We hypothesized that subjects with the best olfactory ability would be most certain of incongruent labels and would have the largest P300. While the areas of greatest P300 amplitude were the most positively correlated with olfactory ability, a large negativity over the frontal areas (especially left-frontal) overlapped the P300 (Fig. 2). This negativity was found to be significantly correlated with subject olfactory ability ($r = -0.74$ for the left-frontal electrode, $P < 0.01$).

While providing very interesting information, this study was not designed to assess the left-frontal negativity related to odor labeling. A second study was completed to more thoroughly evaluate this phenomenon. In this experiment, complex odors were presented and followed by arbitrary labels. Data were collected during the entire S1–S2 interval (4 s) from a 32-electrode montage. Additionally, the

visual labeling control condition was changed so that subjects were required to label more complex visual images (polygons). As in the previous experiment, the S2 was always a visual label ("A", "B", or "C"). Results of the experiment replicated the previous study. A left-frontal negativity arose during the S1–S2 interval that was highly correlated with olfactory ability ($r = -0.76$; $P < 0.01$). The left-frontal negativity was atypical of a traditional CNV because it was highly asymmetrical and tended to diminish as S2 approached (Fig. 3).

Conclusions

The purpose of this paper has been to acquaint investigators with a variety of techniques for the noninvasive study of olfactory processing. Old techniques common to the clinic and laboratory may be used to investigate odor processing and may be useful for clinical diagnosis because these objective electrophysiological measurements can correlate highly with more subjective psychophysical testing.

Acknowledgments. The author wishes to acknowledge the support of International Flavors and Fragrances and NIH grant DC01323-02.

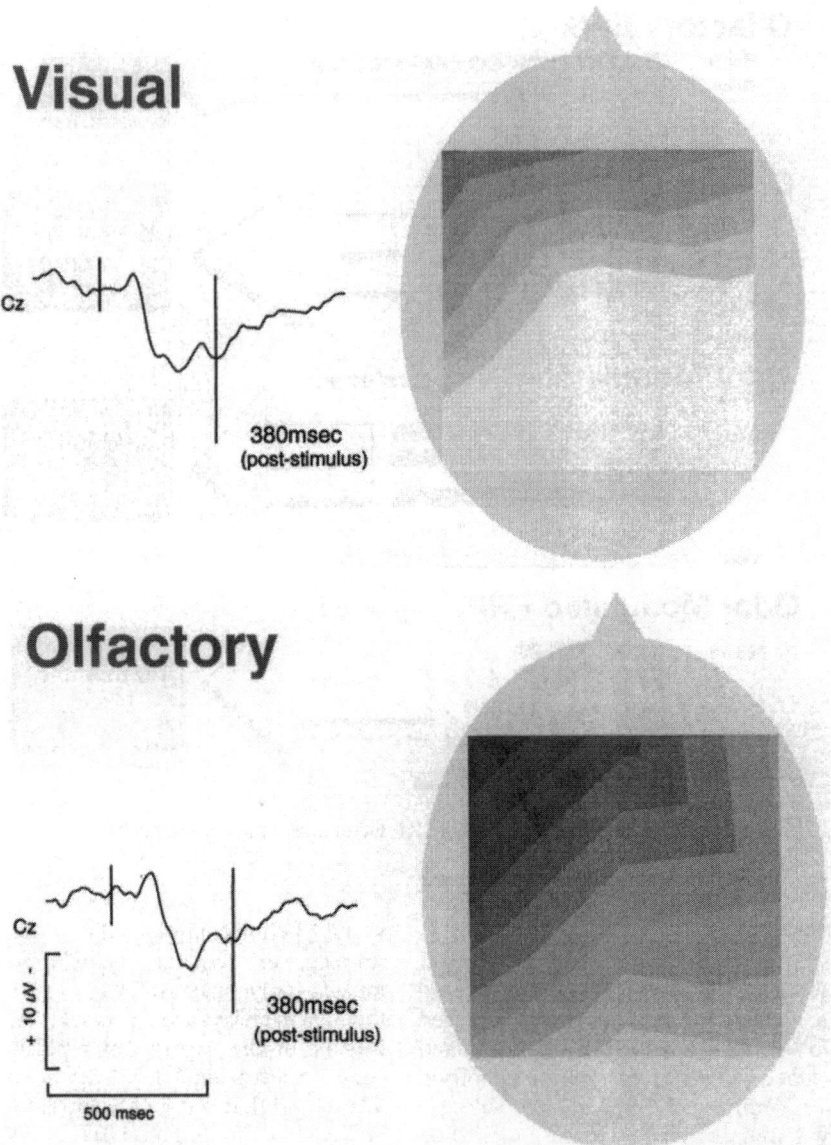

FIG. 2. Grand mean waveforms ($n = 16$) for visual and olfactory phases of odor labeling experiment. Scalp topography maps (nine channels) for 380-ms time slice. Negativity is given in *darker shades*

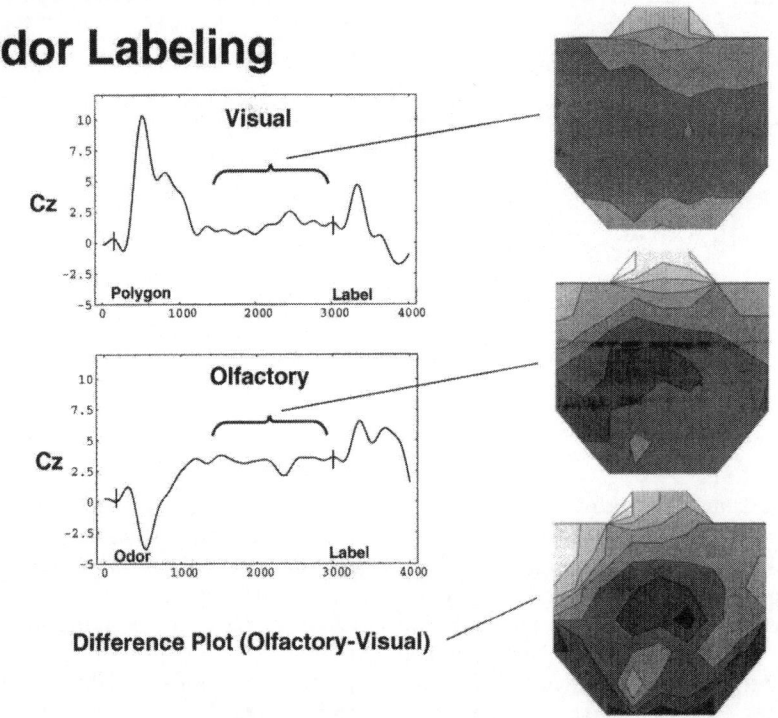

FIG. 3. Grand mean waveforms ($n = 15$) for visual and olfactory phases of second odor labeling experiment. Waveforms represent 4000 ms of data; odors were presented asynchronously with breathing. Scalp topography maps (32 channels) of integrated voltage from 1000 to 3000 ms. Note: Negativity is given in lighter shades for these data

References

1. Moncrieff RW (1962) The effect of odours on EEG records. Perfum Essen Oil Rec 53:757–760
2. Van Toller S (1988) Emotion and the brain. In: Van Toller S, Dodd G (eds) Perfumery: The psychology and biology of fragrance. Chapman Hall, London
3. Lorig TS, Schwartz GE (1988) Brain and odor I: alteration of human EEG by odor administration. Psychobiology 16:281–284
4. Klemm WR, Lutes SD, Hendrix DV, Warrenburg S (1992) Topographical EEG maps of human responses to odors. Chem Senses 17:347–361
5. Van Toller S, Behan J, Howells P, Kendal-Reed M, Richardson A (1993) An analysis of human cortical EEG activity to odors. Chem Senses 18:1–16
6. Kobal G, Hummel C (1988) Cerebral chemosensory evoked potentials elicited by chemical stimulation of the human olfactory and respiratory nasal mucosa. Electroencephalogr Clin Neurophysiol 71:117–119
7. Kobal G, Hummel T (1991) Olfactory evoked potentials in humans. In: Getchell T, Doty R, Bartoshuk L, Snow J (eds) Smell and taste in health and disease. Raven, New York, pp 387–398
8. Torri S, Fukuda H, Kanemoto H, Miyanchi R, Hamauzu Y, Kawasaki M (1988) Contingent negative variation and the psychological effect of odor. In: Van Toller S, Dodd G (eds) Perfumery: The psychology and biology of fragrance. Chapman Hall, London
9. Lorig TS, Roberts M (1990) Odor and cognitive alteration of the contingent negative variation. Chem Senses 15:537–545
10. Lorig TS, Huffman E, DeMartino A, DeMarco J (1991) The effects of low concentration odors on EEG and behavior. J Psychophysiol 5:69–77
11. Lorig TS (1991) Chemosensory modulation of visual and auditory event-related potentials. Proc IEEE Eng Med Biol 13:548–549
12. Lorig TS, Mayer TS, Moore FH, Warrenburg S (1993) Visual event-related potentials during odor labeling. Chem Senses 18

Do the Brain Hemispheres React Differently to Odors? A Study Using Pleasant and Unpleasant Odors

Takuji Yamamoto[1]

Key words. Odors—Feeling—Emotion—CNV—Hedonics—Hemispheric reactivity

Introduction

Stimuli of any modalities may provoke various kinds of feelings, emotions, and moods. However, the use of terms such as feelings, emotions, and moods, is rather confusing in daily life and also in the literature. It is also true that each of these states switches over to the other and that each psychological domain characterized by these terms is difficult to differentiate from another. To avoid these confusing situations, it may be reasonable to define the terms temporarily as follows: feeling is conceived as a subjective response to a variety of conditions aroused by both internal and external stimuli, in which mental images are also included, and emotion is defined as a psychic activity with salient behavior expressions resulting from significance evaluation of the feeling. A mood is an affective state that lasts longer and can be shared with others.

Each sensory modality may have its own characteristics for these psychic activities. The idea that olfactory modality and feeling are closely linked has become commonplace in both popular and scientific discussions of the sense of smell. However, the feelings or emotions aroused by olfactory modality have less dimensionality compared to the feelings or emotions aroused by other modalities. This is because the affective meanings of sensory stimuli of the other modalities involve more cognitive processes. Through clinical studies of brain-damaged patients or studies of normal subjects in which the subjects respond to emotionally evocative stimuli, there is no doubt that the two brain hemispheres differ with regard to their role in feelings and emotions. Most of these studies, however, have been done in the domain of a modality other than olfactory, although in the other modalities also the stimuli are believed to be hemispherically distinct in processing their meanings.

Ehrlichman [1] reported that unpleasant odors to the right nostril were rated as more unpleasant than the same odors in the left nostril. He also reported that, in contrast, no such difference were found for pleasant odors. In his study, however, the task was only to rate the odors on a pleasantness–unpleasantness scale. It is claimed that the influence of olfactory stimuli on feelings and emotions takes two forms: nonspecific and specific. The nonspecific claim is based on hedonics, and seems to be the minimal demands of experiencing an odor. The specific concept is that certain odorants have particular effects on certain aspects of feelings or emotions.

As the olfactory modality seems to demand the minimal cognitive processes of experiencing the stimuli as pleasant or unpleasant, a nonspecific form of olfactory influence would seem to provide odors well suited for addressing questions of different involvement in feeling between the two hemispheres. Two experiments are described here, and nonspecific aspects of the sense of smell are discussed with special reference to the function of the right and left brain hemispheres.

Methods

Two groups of young healthy females participated in experiments as paid volunteers. They ranged from 20 to 24 years of age (mean age, 21.8 years) and all were right-handed. No subject showed any signs of neurological and psychiatric involvements. Written informed consents were obtained for the unpleasant odor experiment.

Experiment 1 used lemon oil as an example of pleasant odorants and isovaleric acid as an example of unpleasant ones; 12 subjects participated. In experiment 2, two groups of odorants were prepared: favored (lemon oil, grapefruit, orange), and ill–favored, or less enjoyable (cis-jasmon, eugenol, styrallyle); 8 subjects participated. In both experiments, each of these odorants was placed in the small receptacle in front of the subject's nose throughout the odorant session during which ERPs were recorded. After each experimental session, six categories of psychological states were rated for feeling profile on a three-point differential scale. These categories were "elated," "depressed," "favored," "ill-favored," "relaxed," and "irritated."

[1] Department of Psychophysiology, Tokyo Instutute of Psychiatry, Setagaya-ku, Tokyo, 156 Japan

EEG were recorded from 19 Ag/AgCl electrodes placed on the scalp according to the 10/20 electrode system and referenced to linked ear lobes with a time constant of 5.0 s. ERPs were evoked by S1–S2 and S1–S2 + R paradigms in which S1 (a tone pip) followed S2 (white noise) by 2.5 s. In the latter paradigm, subjects were asked to press the button in their right hand as quickly as possible when S2 was presented. Both ERPs were analyzed by multivariate analysis of single-trial ERPs. Methods of EEG data processing have been described elsewhere [2]. The area of the largest ERP basis wave in each paradigm was measured in $\mu V \cdot ms$, and the increase ratio of this basis wave was obtained at each electrode site. Then, mean hemispheric increase ratios and the Rt/Lt (right to left) ratio of these hemispheric means were obtained.

Results

Experiment 1

Results are summarized in Figs. 1 and 2. Figure 1 depicts ERP basis waves recorded by S1–S2 and S1–S2 + R under control and two odorant conditions. Figure 2 shows the feeling profile, increase ratio, and Rt/Lt ratio for the lemon oil and isovaleric acid presentations.

Lemon oil induced a feeling profile that has rather relaxed and favored characteristics, while isovaleric acid produced an irritated and ill-favored one. Numbers on the bar graphs (Fig. 2) in the middle column indicate only mean hemispheric increase ratios of each condition. The mean of the left-hemispheric increase ratio increased in the lemon oil condition compared to the control condition, but the ratio did not increase remarkably at the right hemisphere. This was just the reverse of the isovaleric acid condition, in which the hemispheric increase ratio increased more in the right hemisphere than in the left hemisphere.

In the far right column (see Fig. 2), Rt/Lt ratios are illustrated. The Rt/Lt ratio was smaller in the lemon oil condition than in the control condition, while the ratio increased in the isovaleric acid condition compared to that of the control condition.

These results implied two possibilities:

1. Lemon oil, which in this case means pleasant odor, increases the reactivity of the left hemisphere but not of the right hemisphere, resulting in a lower Rt/Lt ratio compared to the control condition.
2. Isovaleric acid, which here means unpleasant odors, increases the reactivity of the right hemisphere, but not of the left hemisphere, resulting in a higher Rt/Lt ratio.

Experiment 2

The second experiment was designed to test the possibilities just listed. One group of odorants consisted of lemon oil, grapefruit, and orange, and the other of cis-jasmon, eugenol, and styrallyle. From these two groups, the subject was asked to choose one each as her favored and ill-favored. Five of eight subjects picked lemon oil as their favored, and two chose grapefruit and one orange. For their ill-favored, four chose cis-jasmon, three eugenol, and one styrallyle.

Feeling profile, increase ratio, and Rt/Lt ratio are shown in Fig. 3. In the lemon oil group, the feeling profile was quite similar to that of the first group, but the increase ratio did not duplicate the results of experiment 1. The Rt/Lt ratio remained the same throughout the control and experiment conditions.

In the cis-jasmon group, the feeling profile revealed a similar profile to that of the isovaleric one, but again the results of the mean hemispheric increase ratio were not as expected. In this case, the increase ratio decreased on both hemispheres, although the Rt/Lt ratio increased compared to that of the control condition. The eugenol group was between these two groups for profile, increase ratio, and Rt/Lt ratio.

From these two experiments, it may be safely concluded, as far as the Rt/Lt ratio is concerned, that when the feeling induced by odor tended to be unpleasant or ill-favored, the Rt/Lt ratio increased, showing more reactivity on the right hemisphere than on the left hemisphere.

Discussion

There can be little doubt that the experience of odors linked to hedonic tone and that odors directly influence feeling, emotion, and mood. From a number of studies on the neural substrates for emotions, there is substantial evidence that the two cerebral hemispheres in humans differ in their emotional propensities. It has been also claimed that there might be functional asymmetry for experiencing odors.

There have been several reports supporting these claims. Lorig and Schwartz [3] reported, on the basis of data obtained at F7, F8, T5, and T6 electrodes, that there was a significant difference in the amount of EEG theta activity produced by smelling the different odorants compared with smelling the no-fragrance control. They also noted that odor conditions were found to produce differences in the amount of alpha activity between right and left hemispheres. For the theta activity, different odors produced differing patterns of hemispheric activity during the time EEG data were collected. Their additional finding was that even odors of high per-

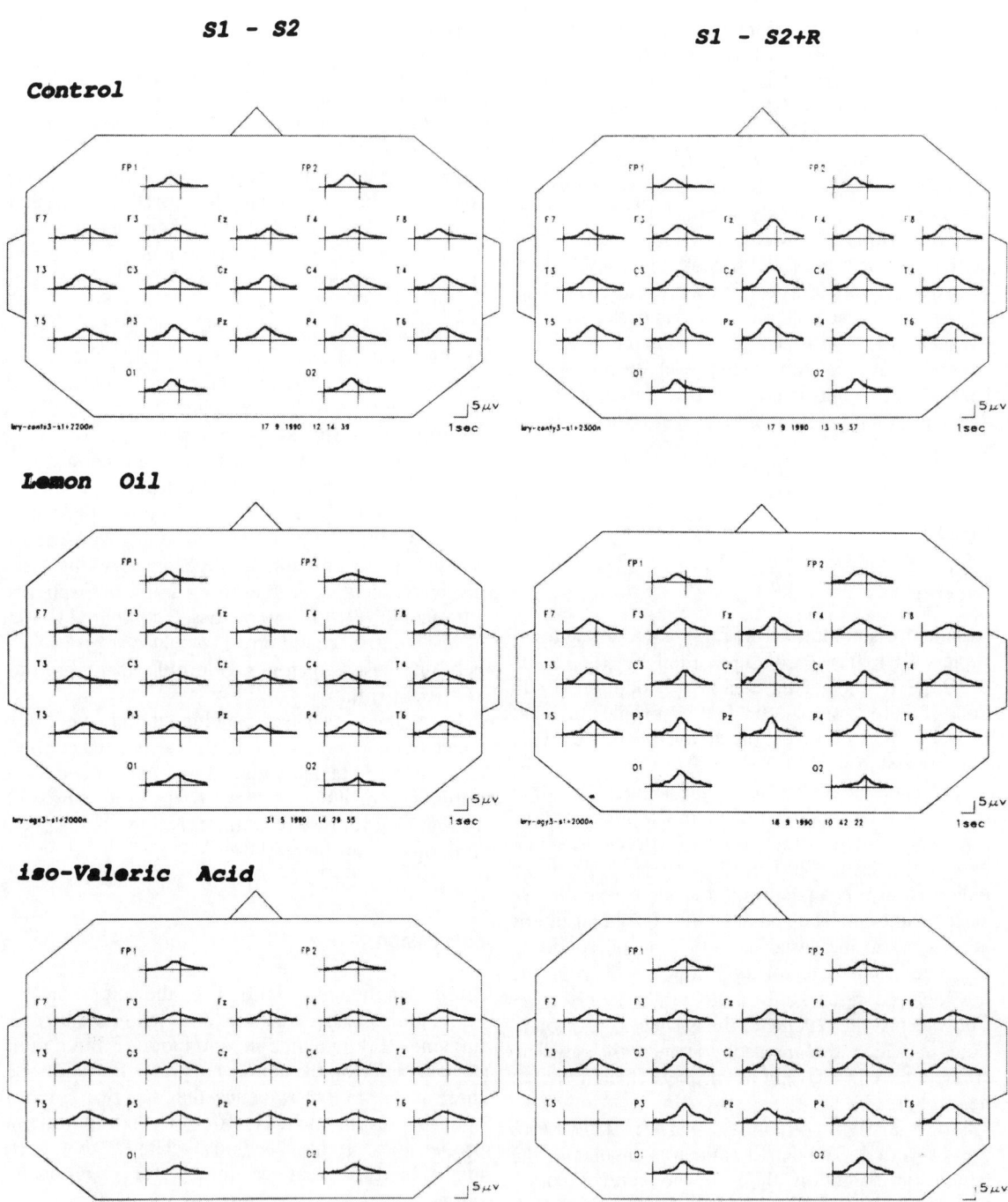

FIG. 1. ERP basis waves extracted from the ERPs recorded in control and in lemon oil and isovaleric acid conditions of experiment 1

ceptual similarity can produce very different patterns of neurophysiological activity. By using 19 electrodes, Klemm and Warrenburg [4] found specific and reproducible EEG changes across subjects in certain odors, particularly in the theta band in the left temporal region. According to their report, birch tar, jasmine, lavender, and lemon odors caused a statistically significant increase in the theta activity,

whereas this decreased during no-odor control trials.

It is clear from these studies that odor stimuli modify EEGs asymmetrically and that the EEG changes seem to correlate well to functional asymmetry of experiencing feeling or emotions induced by olfactory stimuli. In the studies mentioned, the term "emotion" as commonly used seems to imply two conditions: a way of acting and a way of feeling.

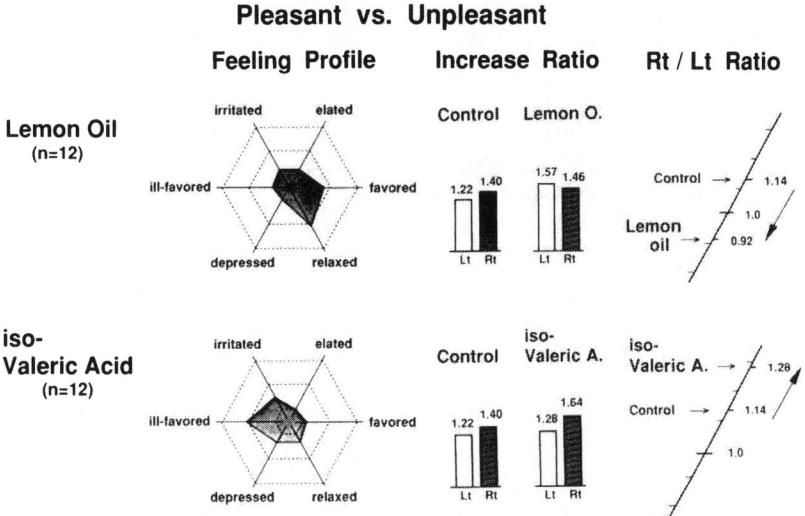

FIG. 2. Feelings profile, increase ratio, and Rt to Lt ratio of lemon oil and isovaleric acid in experiment 1

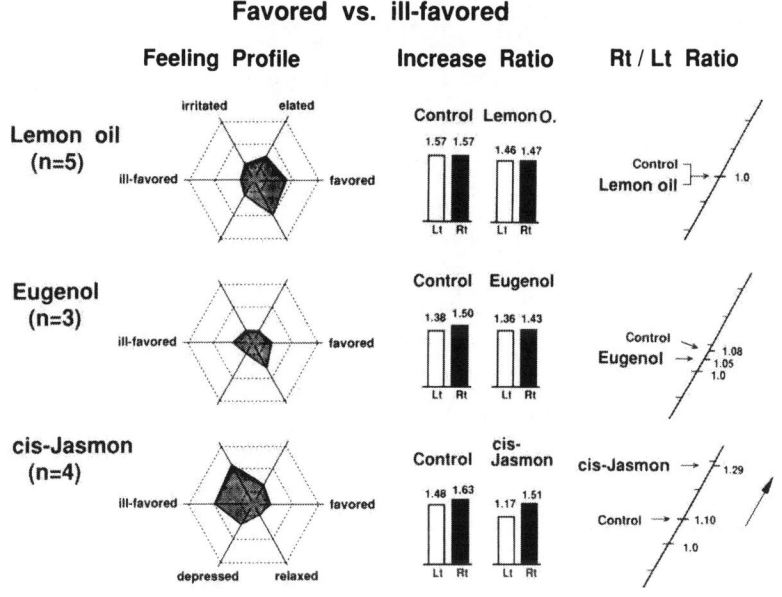

FIG. 3. Feelings profile, increase ratio, and Rt to Lt ratio of lemon oil, eugenol, and *cis*-jasmon in experiment 2

Torii et al. [5] measured CNV amplitude while subjects were presented with various essential oils, and their results supported claims for the stimulating effects of jasmine and the sedative effect of lavender. They measured CNV amplitude of the early CNV component on the midline frontal region and associated their results with a rather specific aspect of olfactory sensation.

In this study, the nonspecific aspect was treated as an attribute of experiencing odors and the responses of the right and left hemispheres were studied by effects on the largest CNV basis wave, which is distributed widely over the scalp. When the Rt/Lt

ratio or the mean of increase ratio are taken into consideration as an index of brain reactivity, a pleasant type of feeling profile is related to lower reactivity of the right hemisphere and an unpleasant type of feeling profile to higher reactivity of the right hemisphere. These relationships seem to depend on the feeling profile, but are not dependent on odorant substance and subjects.

It is assumed from these results that the brain hemispheres react differently according to the state of one's feelings or emotions provoked by olfactory stimuli. The results obtained here seem to be compatible with the results obtained in the study of clinical patients with cerebral lesions and in experimental, psychopharmacological, and psychological studies. Differential involvement of the cerebral hemispheres to olfactory stimuli has been in many occasions interpreted by predominant ipsilateral projection of the olfactory nerves and by a model in which the right hemisphere is specialized for mediating negative affect. However, as it is known that the right and left olfactory nerves reciprocally communicate through the anterior commisurre, a differing state of hemispheric reactivity might underlie the different feeling profiles.

Acknowledgments. The author thanks Drs. S. Ayabe-Kanamura (Tukuba University), S. Ogata (Tokyo Institute of Psychiatry), Y. Saito (Asikaga Institute of Technology), and Y. Takashima (Takasago Inteinational Corp.) for their support.

References

1. Ehrlichman H (1986) Hemispheric asymmetry and positive-negative affect. In: Ottoson D (ed) Duality and unity of the brain: unified functioning and specialization of the hemispheres. Macmillan, London, pp 194–206
2. Saito Y, Yamamoto T (1990) Multivariate analysis of CNV-paradigm-evoked ERP and topographic brain mapping of ERP basis waves. In: Brunia CHM, Gaillard AKK, Kok A (eds) Psychophysiological brain research. Tilburg University Press, Tilburg, The Netherlands, pp 87–93
3. Lorig TS, Schwartz GE (1988) Brain and odor: I. Alteration of human EEG by odor administration. Psychophysiology 16:218–284
4. Klemm WR, Warrenburg S (1992) Topographical EEG maps of human responses to odors. Psychophysiology 29:s11 (SPR abstracts)
5. Torii S, Fukuda H, Kanemoto H, Miyauchi R, Hamauzu Y, Kawasaki M (1988) Contingent negative variation (CNV) and the psychological effects of odor. In: Van Toller S, Dodd GH (eds) Perfumery: the psychology and biology of fragrance. Chapman and Hall, New York, pp 107–120

Chemosensory Event-Related Potentials: Effects of Dichotomous Stimulation with Eugenol and Dipyridyl

Thomas Hummel and Gerd Kobal[1]

Key words. Olfactory nerve—Trigeminal nerve —Event-related potential—Emotion—Smell— Laterality

Introduction

It is often suggested that there is a close link between olfaction and emotion. Thus, it is surprising to note that there are only a few studies on olfactory-induced emotions in humans. It is possible that this discrepancy lies in the difficulty associated with the handling of chemical stimulants (e.g., precise control of delivery of odorous stimuli [1]).

Kobal and coworkers [2] published an evoked potential study that suggested that the hedonic quality of an odorous stimulus may be assessed by examining the differences between chemosensory event-related potentials (CSERP) after stimulation of the left or right nostrils. In that study they found that the pleasant smell of vanillin produced significantly longer CSERP latencies when applied to the left nostril compared to right-sided stimulation. In contrast, this pattern was reversed when subjects were stimulated with the unpleasant odor of rotten eggs (hydrogen sulfide). To verify these results, the follow-up study [3] used acetaldehyde in two concentrations; the lower concentration produced pleasant sensations whereas acetaldehyde at the higher concentration was rated as unpleasant. In addition, the pleasant smell of phenyl ethylalcohol and, again, the unpleasant smell of hydrogen sulfide were investigated. The results of this study confirmed the earlier findings in that the pleasant stimuli [acetaldehyde (low concentration), phenyl ethylalcohol] produced longer latencies when applied to the left nostril compared to right-sided stimulation; the inverse pattern was found for the two unpleasant stimuli.

These results could not be explained by differences in subjectively perceived intensities; that is, none of the stimulants elicited significantly different estimates for stimulation of the left or right side. It was therefore hypothesized that the observed

differences were related to the processing of emotional information within the left or right hemispheres. In line with this, previous studies revealed that positive emotions are predominantly processed by the left hemisphere whereas negative emotions are processed to a larger extent by the right [4,5]. It is thus conceivable that the emotional quality of an olfactory stimulus may be evaluated differently depending on the hemisphere in which it is initially (or predominantly) processed.

The aim of this study was to investigate the processing of dichotomously presented stimuli that not only produce olfactory sensations but also excite fibers of the trigeminal nerve [6,7]. Further, a stimulant should be tested that produced hedonically neutral sensations. Thus, eugenol was chosen as a mixed trigeminal–olfactory stimulus and dipyridyl was used as a hedonically indifferent stimulus.

Materials and Methods

Thirty-nine healthy volunteers, 19 female and 20 male (21–34 years old; mean 26 years) participated in the study. Right-handedness was ascertained by means of a German translation of the Edinburgh Inventory [8]. Subjects provided written informed consent. The study was performed in accordance with the Declaration of Helsinki/Hong Kong.

The concentrations of the two stimulants (eugenol, 43 ng/ml; dipyridyl, 0.5 ng/ml) were adjusted before testing so as to be of approximately equivalent intensity. Subjects were familiar with a breathing technique (velopharyngeal closure [9]) whereby respiratory flow inside the nasal cavity is avoided. A specially devised apparatus ensured that stimulus presentation did not activate mechano- or thermoreceptors in the nasal mucosa [1]. Subjects were seated in an electrically shielded, ventilated chamber. White noise of approximately 50 dB SPL prevented them from hearing the switching process.

All subjects underwent a training session during the course of which they were acquainted with the experimental procedure. Subsequently, each subject participated in one experiment composed of two identically designed blocks separated by a 5-min recovery period. During that period, the outlets of the stimulators were switched to balance possible

[1] Department of Pharmacology and Toxicology, University of Erlangen-Nürnberg, Universitätsstr. 22, 91054 Erlangen, Germany

differences between the two stimulators used. The two stimulants were randomly delivered 16 times to the left or right nostril, resulting in a total of 64 stimuli (duration 200 ms; interval ≈60 s).

Tracking Performance

The testing chamber was equipped with a computer screen and a joystick. During the session, two squares of different sizes appeared on the screen. Subjects were instructed to keep the smaller square (controlled by the joystick) inside the larger one (moving unpredictably). Performance was checked by counting how often, and by measuring how long, subjects lost track of the independently moving square (0%–100% success). Data from each experiment were averaged separately for the two stimulants according to stimulation of the left or right nostril.

Ratings

Following each presentation, subjects were required to indicate both the overall stimulus intensity and the side of stimulation. The stimulus intensity was compared to the intensity of the first stimulus presented, eugenol, which served as a standard (100 estimation units = 100 EU). Ratings were made by moving the joystick to adjust the length of a bar on a visual analogue scale (VAS) displayed on the screen. As with the tracking performance, intensity estimates were averaged according to the four groups of stimuli. Subsequently, on another screen, the subjects indicated the stimulated nostril by moving a pointer to either the left or right.

After each session subjects evaluated the hedonic quality of the stimulants in two different ways. First, subjects estimated stimulus quality by adjusting a pointer superimposed on a VAS. The left-hand end of the scale indicated a most unpleasant sensation (0 EU), and the right a most pleasant sensation (100 EU). In addition, subjects analyzed the hedonic quality by means of a questionnaire (translated from Dravnieks et al. [10]). They were at liberty to choose from a total of 140 verbal descriptors, each of which corresponded to a numerical value ranging between −4 (most unpleasant) and +4 (most pleasant).

CSERP

The EEG (bandpass, 0.2–70 Hz) was recorded from seven positions of the 10/20 system referenced to linked earlobes (A1+A2; Mingograf, Siemens, Erlangen, Germany). Possible blink artifacts were registered from an additional site (Fp2/A1+A2). Analogue to digital conversion of the stimulus-linked EEG segments of 2048 ms started 540 ms before stimulus onset. Single responses contaminated by eyeblinks were discarded from the average, which yielded late nearfield ERPs. The amplitudes, A-N1

and A-P2, and the latencies, T-N1 and T-P2, were measured [9].

Statistical Analyses

Data were analyzed by means of the SPSS/PC+ program package. Results of both tracking performance and intensity estimates were submitted to analysis of variance [ANOVA; "stimulant" (df, 38/1) and "side" (df, 38/1) as within subject factors; interaction "stimulant" by "side" (df 38/1)]. CSERP were investigated separately for the two stimulants [ANOVA; factors "recording position" (df, 228/6) and "side" (df, 38/1), interaction "recording position" by "side" (df, 228/6)]. In addition, a second analysis was performed separately for the seven recording positions [ANOVA; factors "stimulant" (df, 38/1) and "side" (df, 38/1), interaction "stimulant" by "side" (df, 38/1)].

Results

For both eugenol and dipyridyl the side of stimulation could be easily localized (eugenol: 77% correct, 11% incorrect, 12% no decision; dipyridyl: 74% correct, 12% incorrect, 14% no decision).

Tracking Performance

The tracking performance did not differ for the two stimulants neither for left- nor right-sided stimulation (Table 1).

Ratings

There was no significant difference for intensity estimates of both eugenol and dipyridyl after stimulation of the left or right nostril (Table 1). Despite balancing stimulus intensities before the experiment, eugenol produced more intense sensations compared to stimulation with dipyridyl (F = 50.3; P < 0.001).

Estimates of the stimulants' hedonic quality was performed in two different ways at the end of each experiment. Data assessed by the VAS indicated that the majority of subjects perceived eugenol as pleasant (M = 61.7; SD = 20.5) whereas dipyridyl was estimated to be slightly unpleasant (M = 43.0; SD = 15.2). When analyzing data of the verbal item scale, again eugenol was rated as pleasant (M = 1.1; SD = 0.8). In contrast to the VAS ratings, on the verbal item scale, dipyridyl was described as neither pleasant nor unpleasant (M = 0.0; SD = 1.3).

CSERP Amplitude A-N1

After stimulation of the left and right nostrils with dipyridyl, amplitude A-N1 was found to be largest at position Cz (Table 1). Similarly, eugenol produced

TABLE 1. Means (M) and standard deviations (SD) of investigated parameters ($n = 39$).

Stimulus	Tracking performance, %		Intensity estimates (estimation units)	
	M	SD	M	SD
Eugenol (left)	87.6	5.6	96.6	27.6
Eugenol (right)	87.2	6.7	90.6	27.3
Dipyridyl (left)	87.6	5.6	84.9	33.5
Dipyridyl (right)	87.7	5.5	77.3	29.6

ERPs (stimulus)	A-N1 (µv)		A-P2 (µv)		T-N1 (msec)		T-P2 (msec)	
	M	SD	M	SD	M	SD	M	SD
Eugenol (left)	−5.4	3.7	12.0	4.7	327	48	592	107
Eugenol (right)	−4.7	4.1	11.3	5.0	326	68	578	137
Dipyridyl (left)	−4.3	3.2	11.6	4.6	352	66	625	114
Dipyridyl (right)	−3.9	3.5	11.1	5.6	343	68	582	117

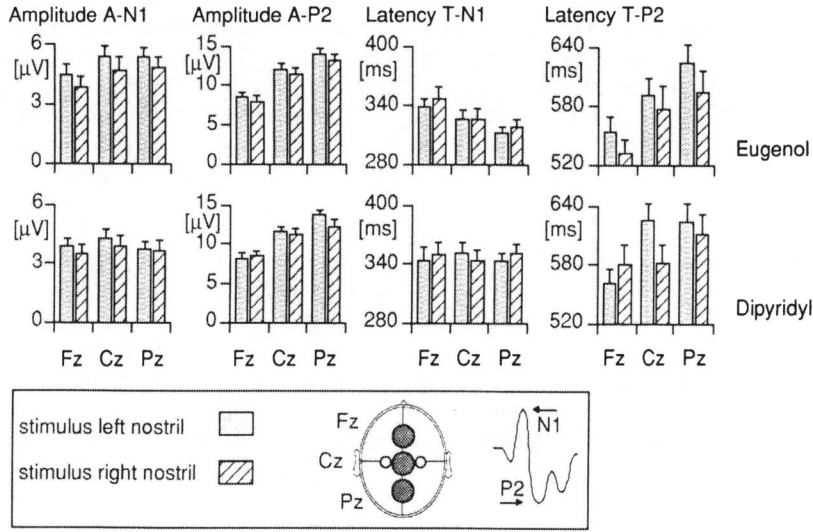

FIG. 1. Means and SE of CSERP amplitudes and latencies ($n = 39$) at midline recording positions (Fz, Cz, Pz) after stimulation of left (*dotted bars*) and right (*shaded bars*) nostrils. Responses to eugenol were larger compared to CSERP amplitudes after stimulation with dipyridyl. In general, amplitudes A-N1 were largest at Cz. In contrast, amplitude A-P2 had its maximum at Pz. Latencies T-P2 were prolonged after stimulation of left nostril compared to right-sided stimulation. This effect was more pronounced for CSERP in response to eugenol. Note different scaling of Y axes. Inset: Schematic drawing of location of recording positions and CSERP peaks N1 and P2

amplitudes A-N1 peaking at Cz/Pz (factor "recording position": eugenol, F = 5.6, $P < 0.001$; dipyridyl, F = 2.36, $P < 0.05$). In general, amplitudes A-N1 in response to eugenol were larger than responses elicited by dipyridyl (factor "stimulant": C3, F = 9.39, $P < 0.01$; C4, F = 14.52, $P < 0.01$; Pz, F = 10.02, $P < 0.01$). A significant interaction between the factors "stimulant" and "side" was observed at position F3 (F = 5.74, $P < 0.05$). That is, eugenol produced larger amplitudes after stimulation of the right nostril compared to stimulation of the left

side while dipyridyl elicited an inverse pattern of responses.

CSERP Amplitude A-P2

Both stimulants produced largest amplitudes A-P2 at Pz (Fig. 1) after stimulation of either the left or right nostril (factor "recording position": eugenol, F = 47.43, $P < 0.001$; dipyridyl, F = 50.54, $P < 0.001$). Additionally, for dipyridyl an interaction emerged between factors "side" and "recording position" (F = 4.04, P < 0.01). That is, while stimulation of the right nostril produced larger amplitudes at the left-sided recording positions (F3, C3) compared to stimulation of the left nostril, this pattern was reversed at the right-sided recording positions (F4, C4).

CSERP Latency T-N1

In general, latencies T-N1 were longest at frontal recording positions and shortest at Pz. This became significant for the CSERP to eugenol (factor "recording position": F = 4.67, P < 0.001). In addition, an interaction between the factors "side" and "recording position" emerged for both stimulants (eugenol, F = 2.62, P < 0.05; dipyridyl, F = 3.42, P < 0.01), that is, when applied to the left nostril at F3 and C3 the two stimulants produced longer latencies compared to stimulation of the right nostril, while at positions F4 and C4 latencies were prolonged after right-sided stimulation. The comparison between responses to the two stimulants revealed that eugenol elicited shorter latencies T-N1 at centroparietal recording sites (factor "stimulant": C3, F = 7.55, P < 0.01; Cz, F = 9.71, P < 0.01; C4, F = 14.93, P < 0.001; Pz, F = 31.85, P < 0.001). In addition, a significant effect of the factor "side" was found at F4 (F = 10.83, P < 0.01); that is, CSERP had longer latencies after stimulation of the right nostril compared to stimulation of the left nostril.

CSERP Latency T-P2

For both stimulants, longest latencies T-P2 were found at Pz, and they were shortest at the frontal sites (eugenol, F = 9.01, P < 0.001; dipyridyl, F = 3.20, P < 0.01). Responses to dipyridyl generally had longer latencies compared to those elicited by eugenol (factor "stimulant": F3, F = 6.94, P < 0.05; Fz, F = 11.99, P < 0.01; C3, F = 4.76, P < 0.05; C4, F = 5.22, P < 0.05). Moreover, statistical analyses revealed differences between CSERP recorded after left- or right-sided stimulation at positions Cz and C4 (Cz, F = 4.77, P < 0.05; C4, F = 4.13, P < 0.05). That is, stimulation of the left nostril with either one of the two stimulants produced significantly longer latencies compared to stimulation of the right side.

Discussion

The selection of stimulants met experimental requirements. That is, results of the localization task suggested that both eugenol and dipyridyl were exciting fibers of the trigeminal nerve, that is, the stimulated nostril could be easily identified [7]. In addition, dipyridyl was rated on the verbal item scale to be neither pleasant nor unpleasant, and its ratings on the VAS also were close to the point of indifference. Because the objective of the study was to investigate CSERP after dichotomous stimulation, it is likewise important to note that (1) there was no significant difference for intensity estimates after stimulation of the left or right nostrils [2,3], and (2) the tracking performance did not differ for the four groups of stimuli, suggesting that the subject's attention was stable.

Activation of the trigeminal system by both stimulants was reflected in the topographical distribution of CSERP amplitude A-N1 confirming earlier findings [2,11], that is, A-N1 was found to be largest at the vertex while amplitude A-P2 peaked at Pz. Using magnetoencephalographic techniques [12], research is currently underway to examine the different cortical sources being involved in the generation of CSERPs [13].

The current study also revealed differences in CSERP latencies after left- or right-sided stimulation. After right-sided stimulation, T-N1 was significantly prolonged at F4 for both stimulants. Because this pattern was inversed, in terms of mean latencies, at left-sided recording positions for both stimulants, this result has to be regarded as an isolated statistical finding. In contrast, for latency T-P2, the data were more consistent with significantly longer latencies T-P2 at Cz and C4 after left-sided stimulation for both stimulants. Regarding mean latencies after stimulation with eugenol, this effect was observed at all recording positions. However, when dipyridyl was used as the stimulant, the pattern was reversed at frontal recording sites.

The current data only partly confirmed the hypothesis that pleasant stimuli such as eugenol would produce longer latencies after left-sided stimulation compared to stimulation of the right nostril [2,3]. For both stimulants this might result from trigeminal excitation, which is processed contralaterally to the stimulation site as opposed to the ipsilateral processing of olfactory information. Thus, it appears as if the complex effects of mixed trigeminal/olfactory stimulants within this model are more difficult to interpret compared to stimuli that predominantly excite the olfactory system. The other important finding was that dipyridyl did not elicit major changes in the CSERPs relative to the dichotomous stimulation. It may be hypothesized that stimulants estimated as hedonically neutral produce a "balanced" pattern of response latencies, that is, emotionally indifferent stimulants applied to the left or right nostrils will not produce uniformly different response latencies in relationship to the stimulated side as it is observed for distinctively pleasant or unpleasant odorants.

Acknowledgments. This research was supported by DFG grant Ko812/2-2. We thank Dr. Elisabeth Pauli, Dept. of Neurology, University of Erlangen-Nürnberg, for statistical analysis of the data, and Dr.

Steve Meller, Dept. of Pharmacology, University of Iowa, USA, for helpful suggestions during preparation of the manuscript.

References

1. Kobal G (1985) Pain-related electrical potentials of the human nasal mucosa elicited by chemical stimulation. Pain 22:151–163
2. Kobal G, Hummel T, Van Toller S (1992) Differences in chemosensory evoked potentials to olfactory and somatosensory chemical stimuli presented to left and right nostrils. Chem Senses 17:233–244
3. Kobal G, Hummel T, Pauli E (1989) Correlates of hedonic estimates in the olfactory evoked potential. Chem Senses 14:718
4. Davidson JD (1984) Affect, cognition, and hemispheric specialization. In: Izard CE, Kagan J, Zajonc RB (eds) Emotions, cognition, and behavior. Cambridge University Press, Cambridge, pp 320–365
5. Dimond SJ, Farrington L, Johnson P (1976) Differing emotional response from right and left hemispheres. Nature (Lond) 261:690–692
6. Doty RL, Brugger WPE, Jurs PC, Orndorff MA, Snyder PJ, Lowry LD (1978) Intranasal trigeminal stimulation from odorous volatiles: psychometric responses from anosmics and normal humans. Physiol Behav 20:175–185
7. Kobal G, Van Toller S, Hummel T (1989) Is there directional smelling? Experientia (Basel) 45:130–132
8. Oldfield RC (1971) The assessment and analysis of handedness: the Edinburgh inventory. Neuropsychologia 9:97–113
9. Kobal G, Hummel T (1991) Olfactory evoked potentials in humans. In: Getchell TV, Doty RL, Bartoshuk LM, Snow JB Jr (eds) Smell and taste in health and disease. Raven, New York, pp 255–275
10. Dravnieks A, Masurat T, Lamm RA (1984) Hedonics of odors and odor descriptors. J Air Pollut Control Assoc 34:752–755
11. Hummel T, Kobal G (1992) Differences in human evoked potentials related to olfactory or trigeminal chemosensory activation. Electroencephalogr Clin Neurophysiol 84:84–89
12. Huttunen J, Kobal G, Kaukoronta E, Hari R (1986) Cortical responses to painful CO_2-stimulation of nasal mucosa; a magnetoencephalographic study in man. Electroencephalogr Clin Neurophysiol 64:347–349
13. Kobal G, Hummel T, Kettenmann B, Pauli E, Schüler P, Stefan H (1994, in press) Chemosensory event-related potentials in temporal lobe epilepsy and first recordings of olfactory event-related magnetic fields. Chem Senses

Olfactory Event-Related Potentials and Olfactory Neuromagnetic Fields in Humans

Mitsuo Tonoike[1]

Key words: Olfactory event-related potentials—Topographical mapping system—Chemosensory event-related potentials (CHERPs)—Olfactory neuromagnetic fields—DC-SQUID gradiometer—Trigeminal nervous system

Introduction

It is very important to measure human olfactory responses noninvasively to determine their origins and deficiencies accurately. Olfactory evoked potentials have been investigated since 1966–1976 by Finkenzeller [1] and Allison and Goff [2]. Plattig [3], Kobal [4], and Hummel [5] have reported recording chemosensory evoked potentials. We [6,7] have also measured olfactory event-related potentials using an odorant pulse synchronized with subject respiration. However, these studies have provoked controversial discussion.

In this experiment, the most important problem is whether the observed response is induced by the stimulation of the olfactory or the trigeminal nervous terminal. The aim of this study is to investigate noninvasively the origin of chemosensory event-related potentials (CHERPs) by olfactory event-related potentials and olfactory neuromagnetic fields in humans.

Materials and Methods

Olfactory Event-related Potentials

In the first experiment we used a pulsed odorant generator in which three-way solenoid valves were controlled under synchoronization with subject respiration. The odorant gas was administered directly into the olfactory cleft by inserting a thin Teflon tube into the nasal cavity (we call this is the "blast method"). We measured the olfactory event-related potentials recorded with the topographical mapping system (NEC-San-Ei EEG system, Tokyo, Japan) using odorant pulse stimuli synchronized with subject respiration. Amyl acetate, skatole, and other odorants were administered through a thin Teflon tube (3.0 mm outer diameter) inserted into the right or left nostril. Flowrate was controlled at 3 l/min, stimulus duration was 100 ms, the high-cut off filter was 120 Hz, and the time constant was 0.3 ms. In this experiment the responses were averaged over 20 stimuli. The subjects sat with their eyes opened in a dimly light, isolated, shielded room. White noise of approximately 60 dB SPL, delivered through headphones, prevented subjects from hearing the switching sound and other noises.

Olfactory Neuromagnetic Fields

In addition to the evoked potentials, we measured the responses of olfactory neuromagnetic fields at the frontal region using a 3-channel DC-SQUID (superconducting quantum interference device) gradiometer with the first-order pickup coil. The first experiment was carried out in the magnetically shielded room in ETL-Tsukuba. A special type of olfactory stimulator was made of nonmagnetized materials. A respiration mask with an optical fiber sensor detected subject repiration state, and the odorant pulse was given to the subject's nose under synchronization with his or her respiration using an air pressure valve. In this experiment, neuromagnetic fields were averaged over a total of times with a 500-ms odorant pulse. Time response of the airvalve was calibrated by an S_nO_2 gas sensor using ethanol gas.

In the second experiment, our DC-SQUID sensor was designed to the most suitable dimension (pick-up coil, 30 mm Φ; baseline length, 60 mm) to catch the olfactory magnetic responses. Neuromagnetic fields were measured in a new magnetically shielded room with 5-cm-thick aluminium panels constructed in ETL-Osaka. Odorant pulses were controlled more precisely and quickly by water pressure valves of nonmagnetic materials. We could eliminate the magnetic and auditory noises of these water pressure valves. Olfactory neuromagnetic responses were measured with olfactory evoked potentials on the vertex C_z at the same time.

[1] Life Electronics Research Center, Electrotechnical Laboratory, Amagasaki, Hyogo, 661 Japan

Results

Results of Olfactory Event-Related Potentials

Olfactory event-related potentials were clearly obtained by the blast method using 100-ms odorant pulse stimuli synchronized with subject respiration. Some negative and positive peaks (N1, P1, N2, P2 and P3) were observed (Fig. 1). Latencies of each peak were about 156, 187, 255, 351, and 441 ms, respectively. Deodorized fresh air was used as a control, and we obtained N1, P1, and P2 peaks except for N2 and P3 at an air stimulation. This is also shown in Fig. 1. In our experiments, olfactory event-related potentials were able to obtain topographical mapping from 16 points in the 10–20 system. N2 and P3 peaks in the waves were observed at the frontal and the vertex region, respectively (T-test analysis; significance, $p < 0.05$). From the results of contralateral dominance of N1, P1, and P2 peaks for air stimulation, these responses were estimated to be trigeminal responses. On the other hand, N2 and P3 peaks in the olfactory event-related potentials suggest that the origin of these responses is the olfactory nervous terminal. Recent investigations by our research group [7] for the P3 component suggested this peak may be a P300 response (with experiments using an oddball paradigm).

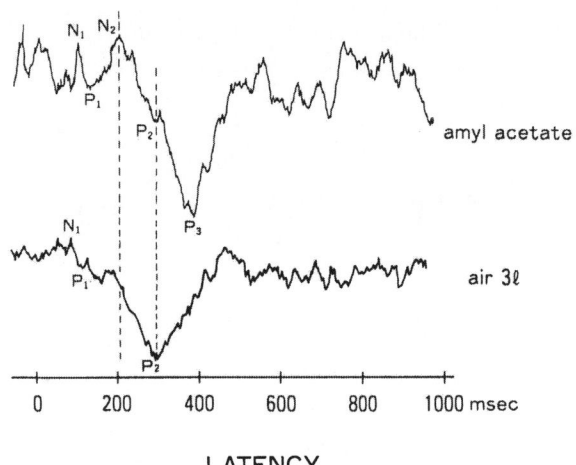

LATENCY

Fig. 1. Olfactory event-related potentials evoked by odorant and air stimulation. Flowrate, 3 l/min; odorant, amyl acetate. Each peak of evoked potentials was named as N1, P1, N2, P2, P3, respectively. N2 and P3 peaks were not observed in air stimulation

Results of Olfactory Neuromagnetic Fields

In the first DC-SQUID experiments, the odorant pulse was given to the subjects using air pressure valves. These magnetic fields showed a large but

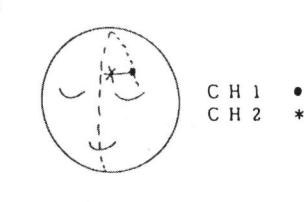

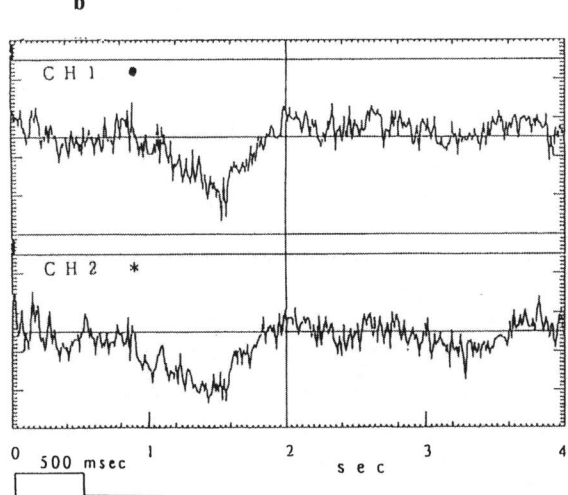

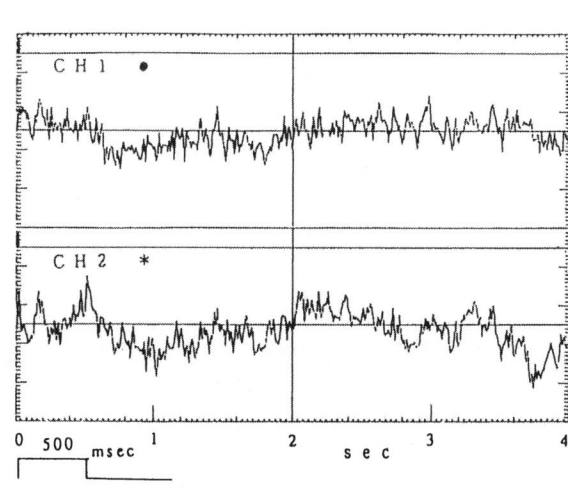

Fig. 2. Neuromagnetic responses evoked by odorant and air stimulation. **a** Measuring regions of frontal area on head surface. **b** Response wave of neuromagnetic field for odorant stimulation by amyl acetate (500-ms duration). **c** Response wave of neuromagnetic field for odorless air stimulation

Magnetically Shielded Room

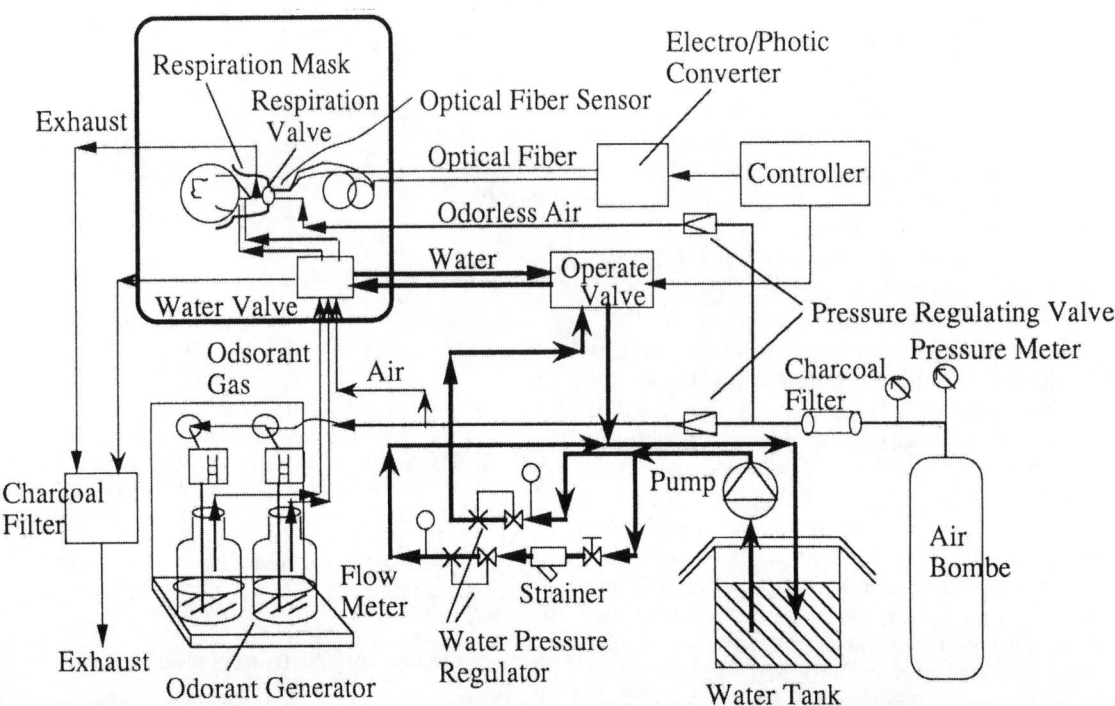

FIG. 3. Olfactory stimulator controlled by water pressure valve. Water pressure valve system constructed with non-magetized materials. Respiration mask can detect subject's respiration by optical fiber sensor. Odorant pulse was quickly and sharply given to subject nose in magnetically shielded room

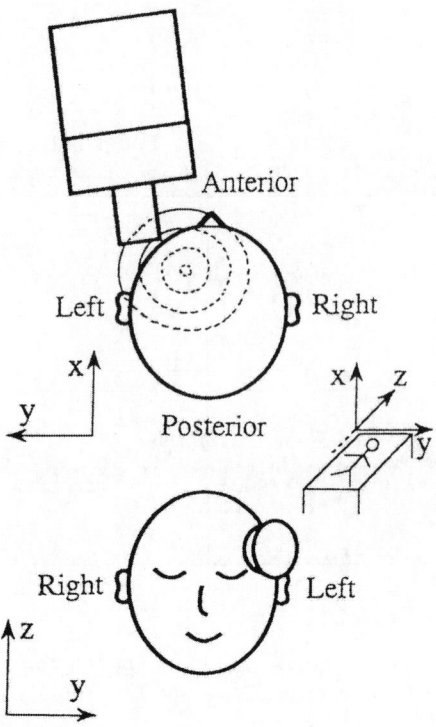

slow response (Fig. 2) having about 1.0- to 1.8-s latencies that were included with the rising time of air-pressure valves. However, olfactory event-related potentials and olfactory neuromagnetic fields were also discriminated from the response obtained by using the odorless air stimulation.

In the second DC-SQUID experiments, water pressure valves were used as shown in Fig. 3. We could obtain very quick responses and present odorant stimuli more sharply using water pressure valves.

Location of SQUID dewar and measuring regions on the head were determined at the frontal area (Fig. 4) from previous physiological information. We found that a technique of noise elimination from the valve system was most important for the measurements of olfactory neuromagnetic fields.

FIG. 4. Location of DC-superconducting quantum interference device (SQUID) gradiometer in measurement of olfactory evoked fields (OEFs). Location of SQUID dewar and measuring regions on head were determined at frontal area in which magnetic flux was estimated from olfactory nervous center in brain

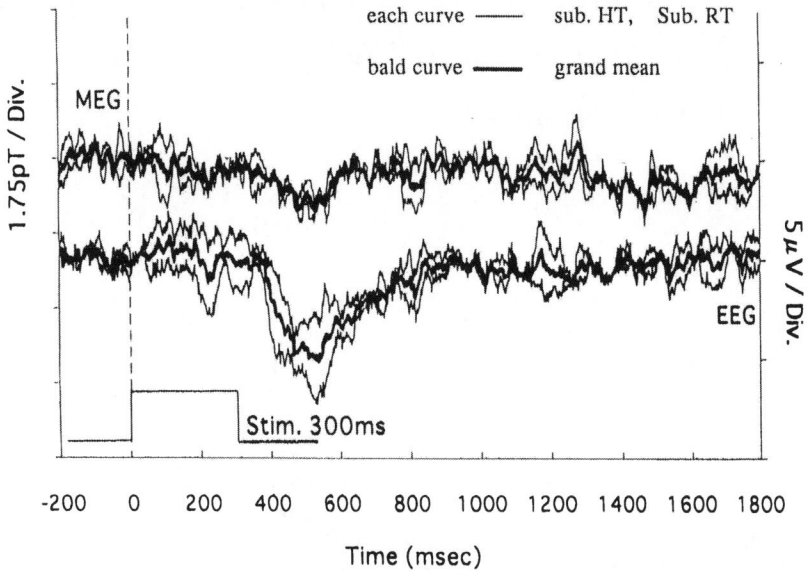

FIG. 5. Olfactory evoked fields and olfactory evoked potentials. *Upper* curves, olfactory evoked fields (MEG); *lower* curves, olfactory evoked potentials (EEG). Each curve was sub HT and sub RT, respectively; bald curve was grand mean (amyl acetate, 300 ms). Both OEFs and OEPs were measured at same latency time

Figure 5 shows neuromagnetic responses obtained with the responses of olfactory event-related potentials on the vertex C_z at the same latency time.

Discussion

In the current study, we were able to obtain a more definite olfactory event-related potential by the blast method and also olfactory neuromagnetic fields during the same experiments using a three-channel DC-SQUID gradiometer. From the results of the odorant and odorless air stimulation, early components of N1, P1 and P2 peaks were considered to be trigeminal responses and later components of N2 and P3 peaks to be olfactory nervous responses.

In this study, olfactory neuromagetic measurements were tried first to record the origin of chemosensory event-related potentials (CHERPs) using the DC-SQUID gradiometer noninvasively. The results suggest that the neuromagnetic measurements are effective and very important in analyzing the location of the source for the olfactory nervous center.

References

1. Finkenzeller P (1966) Gemittelte EEG-potentiale bei olfactorischer reizung. Pflügers Arch 292:76–80
2. Allison T, Goff WR (1967) Human cerebral evoked responses to odorous stimuli. Electroencephalogr Clin Neurophysiol 23:558–560
3. Plattig KH, Kobal G (1979) Spatial and temporal distribution of olfactory evoked potentials and techniques involved in their measurement. In: Lehmann D, Callaway E (eds) Human evoked potentials. Plenum, New York, pp 285–301
4. Kobal G (1991) Olfactory evoked potentials in human. In: Getchell TV, Doty RL, Bartoshuk LM, Snow JB (eds) Smell and taste in health and disease. Raven, New York, pp 255–275
5. Hummel T, Kobal G (1992) Differences in human evoked potentials related to olfactory or trigeminal chemosensory activation. Electroencephalogr Clin Neurophysiol 84:84–89
6. Tonoike M (1987) Response characteristics of olfactory evoked potentials using time-varying filtering. In: Roper SD, Atema J (eds) Olfaction and taste IX. Ann NY Acad Sci 510:658–661
7. Tonoike M, Seta N, Maetani T, Koizuka I, Takebayashi M (1990) Measurements of olfactory evoked potentials and event related potentials using odorant stimuli. In: Pedersen PC, Onaral B (eds) Proceedings of 12th Annual International Conference of IEEE Engineering in Medicine and Biology, vol 12, pp 912–913

Electroencephalography and Its Fractal Analysis During Olfactory Stimuli

KIMIKO KAWANO[1]

Key words. EEG—Fractal dimensional map—Correlation dimension—Graph dimension—Olfactory stimulation

Introduction

Electroencephalograph (EEG) band power analysis is an approximate linear analysis that uses the fast Fourier transform (FFT) method. However, physical functions are essentially nonlinear, so a fractal analysis of EEGs [1] as a nonlinear oscillation has been studied. Fractal dimensions were calculated using two methods, one having a correlation dimension and the other a graph dimension. Topographical, two-dimensional maps were then compiled based on the values from 12 channels on the scalp.

It is difficult to analyze the olfactory response by ordinary EEG. Some reports state that the α-wave, especially α_2, appears on the frontal region during olfactory stimulation [2]. However, according to the author's observation, this frontal α-wave is not very clear on general subjects.

The application of the fractal dimensional maps to olfactory experiments revealed clear differences in their patterns from those in the resting state, especially around the frontal-midline (Fz) region of the graph dimensional map. Among the several odor stimuli, cacao or chocolate had a clearer effect than the fragrant oils such as lemon or jasmine generally used. In this report, the results obtained from the application of these fractal dimensional maps to olfactory experiments are discussed.

Methods and Analysis

EEG was measured with 12 channels on the scalp according to the international 10–20 system (Fig. 1). For the power analysis, EEG signals were relayed to the signal processor (7T-18, NEC-Sanei, Tokyo Japan) at a sampling time of 5 ms with 1024 points per unit epoch, and analyzed for each epoch using FFT from 0.2 to 80 Hz with a 0.2-Hz step. The average of six epochs was then available for the data

[1] Information Processing Center for Medical Sciences, Nippon Medical School, Bunkyo-ku, Tokyo, 113 Japan

summary. The frequency bands used for the analysis in this report are 8.0–12.8 Hz for α-waves and 13.0–29.8 Hz for β-waves.

For the dimensional analysis, EEGs were fed into a minicomputer (Facom S-3300, Tokyo, Japan) through a general purpose interface bus (GPIB) at a sampling clock of 1.28 ms with 32 767 points. The fractal dimensions were then calculated using the two methods.

Graph Dimension

A graph dimension corresponds to the dimensional analysis on the fractal nature of a coastline reported by Mandelbrot [3]. It was calculated using the cumulative wavelength [4], which is shown by the following equation:

$$L_{\Delta t} = \Sigma \, | X_{i+1} - X_i |$$

The $L_{\Delta t}$ value can be easily calculated from the data for a time interval Δt. With the same data set for Δt, we can then calculate $L_{2\Delta t}$, $L_{3\Delta t}$, etc. If $L_{n\Delta t} \propto (n\Delta t)^{-\beta}$ and β can be appropriately determined, it then indicates the fractal nature of EEG and the power β can be regarded as the fractal dimension of the EEG graph.

Correlation Dimension

A correlation dimension corresponds to the dimension of the attractor. It was estimated by a correlation integral [5]. For a time series of data $(x_0, x_1, \ldots, x_M)$, where x_i $(i = 0, 1, \ldots, M)$ is the ith data of the EEG amplitude and M is the total number of data points obtained with the time interval Δt (i.e., 32 767 and 1.28 ms in this report, respectively), we can calculate the following correlation integral:

$$C(r) = (1/N^2) \, \Sigma\{\Sigma H(r - \, \| \, X_i - X_j \, \| \,) - 1\}$$

where the $H(z)$ function is the step function, which is 1 for a nonnegative z and 0 for a negative z. $\| \cdot \|$ is a proper norm for the d-dimensional space. N is the total number of d-dimensional vectors X_i $(i = 1, 2, \ldots, M - d + 1)$, which are constructed from $(x_0, x_1, \ldots, x_M)$ as $X_i = (x_i, x_{i+1}, x_{i+2}, \ldots, x_{i+d-1})$ $(i = 1, 2, \ldots, M - d + 1)$ [6]. If the correlation integral $C(r)$ satisfies $C(r) \sim r^D$, then the power D can be regarded as a fractal dimension, that is, "correlation

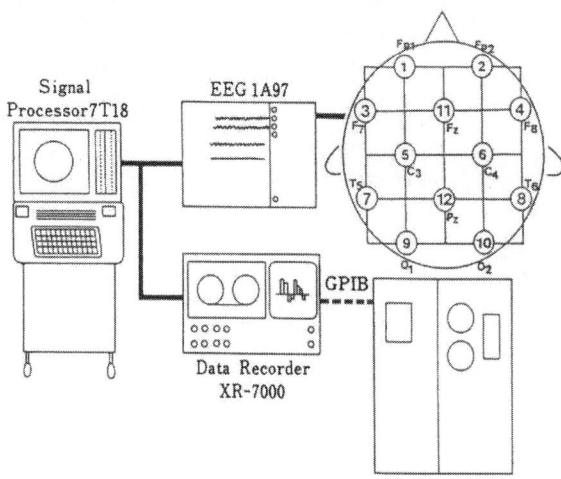

FIG. 1. EEG recording system with placement of electrodes

dimension." A more detailed explanation is given in our previous report [7].

Fractal Dimensional Maps

After the dimension calculations of the 12 crosspoints presented in Fig. 1, additional values on some lattice points were calculated, averaging the given values of the neighboring points. These values were then transferred to the graphic software DeltaGraph, and its three-dimensional surface chart produced the resulting map.

Results

Figure 2a shows the fractal dimensional maps of a student in the resting state with eyes closed. The left map is that of the correlation dimension, and the right is that of the graph dimension. The upper end of each map denotes the frontal region. The values of the dimension are shown on the right scale. In the resting condition, both maps showed almost symmetrical patterns. The correlation dimension was rather higher for the anterotemporal regions and lower for the occipital region. The graph dimension was higher for the occipital and lower for the frontal region.

As a reference, the conventional power maps of the α-wave (left) and the β-wave (right) are shown in Figure 2b. The numerals under each map indicate the voltage value of one step in 11 shades of gray. These maps also showed the symmetric, normal pattern.

Figure 3a shows maps obtained from the same student while smelling cacao. The graph dimensional map obviously showed a different pattern for the frontal-midline (Fz) region than that in the resting state. The correlation dimensional map showed a lateral difference. However, in the conventional power maps, only a slight change was observed from the central- to the frontal-midline of the α-band topography (Fig. 3b).

For comparison, the maps of a professional perfumer are shown in Fig. 4. Figure 4a presents the dimensional maps, and Figure 4b is the α and β power maps in the resting state. The α-waves of the odor specialist spread toward the frontal region, and

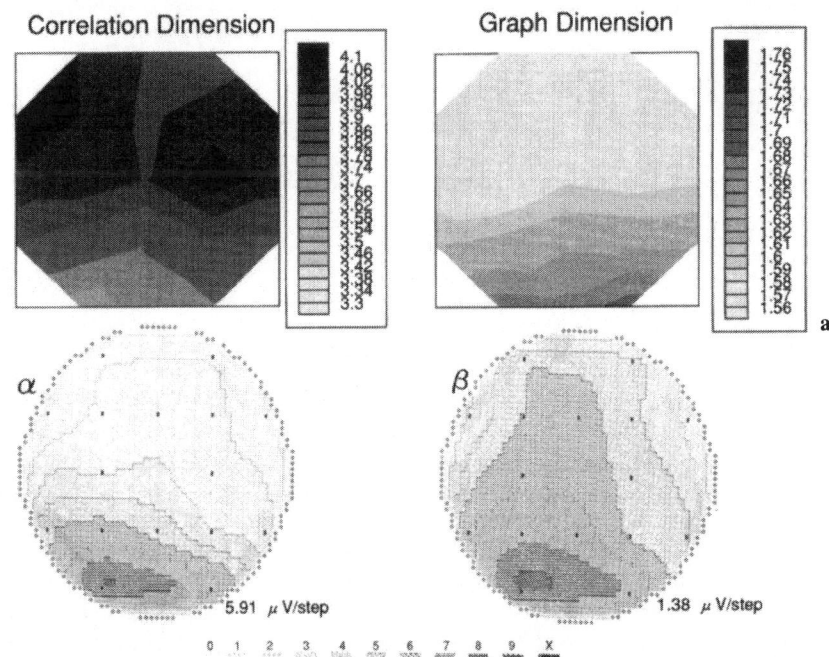

FIG. 2a,b. Maps of student subject in resting state with eyes closed. a *Left*, correlation dimensional map; *right*, graph dimensional map. b *Left*, α power map; *right*, β power map. All maps show almost symmetric patterns

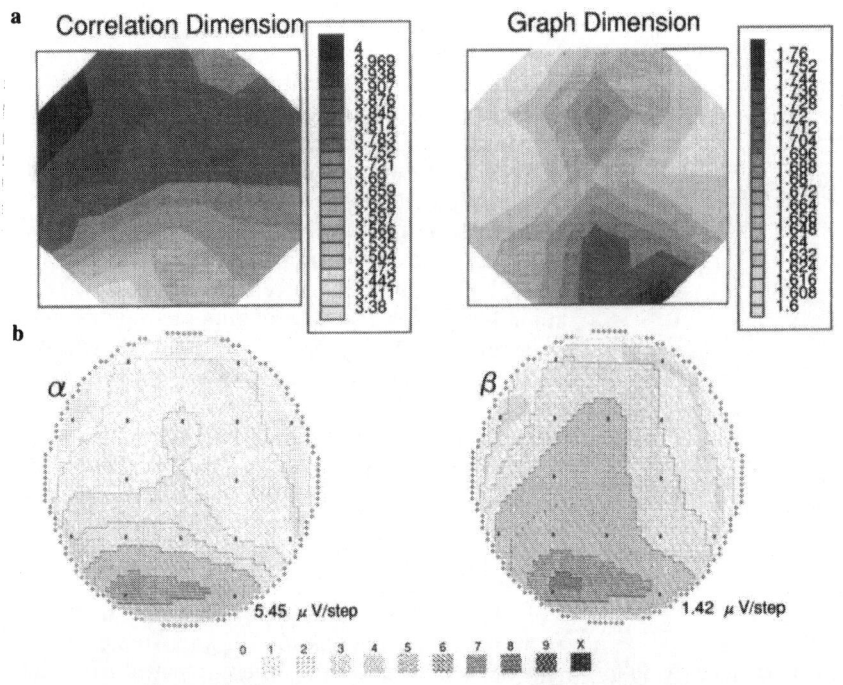

FIG. 3. Maps while smelling cacao (obtained from same student as in Fig. 2). While subject is smelling, correlation dimensional map becomes slightly asymmetric and graph dimensional map shows higher Fz pattern

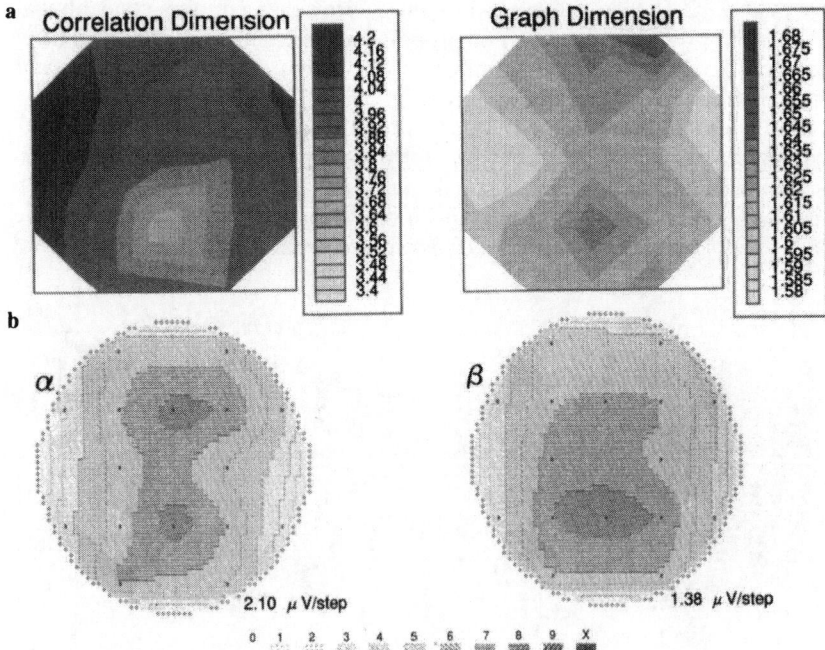

FIG. 4. Maps of professional perfumer in resting state with eyes closed. α-Waves of this perfumer spread toward frontal region, and graph dimensional map shows higher Fz pattern even in resting state

the region for his graph dimensional map showed a higher frontal pattern even in the resting state.

In Fig. 5, maps of the same perfumer, Fig. 5a shows those while smelling lemon and Figure 5b while smelling chocolate. The patterns of Fig. 5a showed a similar change with those of the student while smelling cacao (Fig. 3a). While smelling chocolate, the graph dimensional values were larger

and the correlation dimensional map showed a more asymmetrical pattern.

Figure 6 shows the fractal maps of another student in the resting state (Fig. 6a), while smelling two types of chocolate flavor, C1 and C2 (Figs. 6b and 6c, respectively). Instead of a very small difference in the flavors, the graph dimensional map showed a tendency for a higher Fz pattern with the C1 flavor

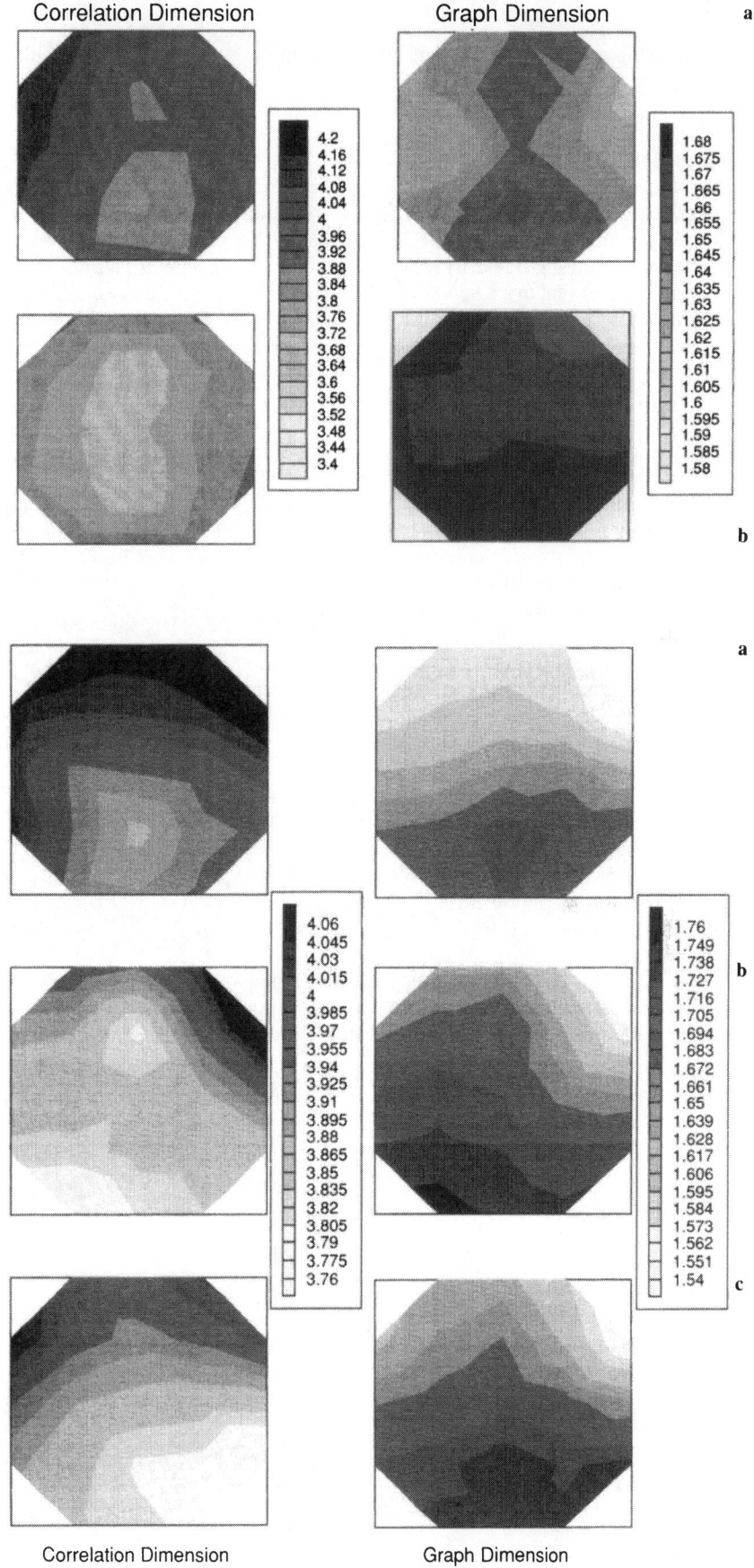

FIG. 5. Fractal dimensional maps of perfumer while smelling lemon (**a**) and chocolate (**b**). Chocolate has a stronger effect than other fragrant oils

Correlation Dimension

Graph Dimension

FIG. 6. Fractal dimensional maps of another student subject when **a** in resting state; **b** and **c** smelling two types of chocolates, C1 and C2, respectively. Flavor C1 shows a slight tendency for the Fz pattern

Correlation Dimension

Graph Dimension

embedded 8

rather than with C2 throughout the experiments using several students.

Discussion

These fractal dimensional maps provide information completely different from that provided by the power maps. During olfactory stimulation, the correlation dimensional maps sometimes became asymmetric. The correlation dimension may reflect the movement of the mind in its asymmetric pattern.

During olfactory stimulation, especially by good smells, the graph dimensional maps showed a specific pattern around Fz. However, as shown in Fig. 4, the map of the perfumer presented a higher Fz pattern even in the resting state. This Fz pattern from the resting condition was observed not only for perfumers but also for many kinds of specialists, such as Japanese abacus specialists, professional Japanese chess players, Chinese Qigong masters, and others who must have a very high concentration ability.

The conventional power maps describe the condition of the neural activities of the brain at the time of measurement, while the fractal dimensional maps may additionally show some inherent characteristics of the subject. On nonspecialist subjects, this Fz pattern was not obtained during the resting state, but during olfactory stimulation or, often, while concentrating on some task such as mental calculations. The degree of concentration was measured by another indicator that has been previously reported [8], and a relationship to this Fz pattern was observed. This pattern may be related to a person's concentration.

The Fz pattern observed during olfactory stimulation could be considered as a result of concentration on the odor. However, it might indicate that odors could directly stimulate a deep part of the brain. As an odor stimulant, cacao or chocolate had a clearer effect compared to the usual fragrant oils. Further only the small differences in chocolate flavors resulted in differences in the patterns of the fractal maps, as shown in Fig. 6.

It has been reported that the chaotic activity of the EEG is very important to the rapid judgment or recognition of sensory information [9]. More precise study of this analysis reveals the effects on neural activity caused by small changes in olfactory stimuli.

References

1. Babloyantz A, Salazar JM, Nicolis C (1985) Evidence of chaotic dynamics of brain activity during the sleep cycle. Phys Lett 111A:152–156
2. Harada H, Shiraishi K, Kato T, Soda T (1988) Observation of the topographic EEG during olfactory stimulation. In: Proceedings of the 7th Japanese Conference of Topographic Electroencephalography, pp 153–158, Saitama
3. Mandelbrot BB (1982) The fractal geometry of nature. Freeman, San Francisco
4. Higuchi T (1988) Approach to an irregular time series of the basis of the fractal theory. Physica D 31:277–283
5. Takens F (1980) Detecting strange attractors in turbulence. In: Rand DA, Young LS (eds) Dynamical systems and turbulence. Springer, Berlin Heidelberg New York
6. Grassberger P, Procaccia I (1983) Estimation of Kolmogorov entropy from a chaotic signal. Phys Rev 28A:2591–2593
7. Shinagawa Y, Kawano K, Matsuda H, Seno H, Koito H (1991) Fractal dimensionality of brain wave. Forma 6:205–214
8. Shinagawa Y, Kawano K (1992) Concentration and EEG. Clin Electroencephalogr 34:168–173 (in Japanese)
9. Freeman WJ (1991) The physiology of perception. Sci Am 34:34–41

Evoked Potentials Produced by Electric Stimulation on the Tongue

I. Takemoto[1], I. Koizuka[2], C. Maetani[2], K. Shimada[2], K. Doi,[2] T. Matsunaga[2], and M. Tonoike[3]

Key words. SEP—Taste

Introduction

In a previous study [1], we reported that two kinds of responses were elicited by stimulation with a salt solution on the tongue, one being evoked by taste stimulation and the other resulting from the mechanical stimulus of the application of the solution. When a taste solution is used for stimulation, a lag time between the trigger and the stimulation may be unavoidable. Electrical stimulation, on the other hand, even with an artifact on evoked responses, demonstrates measurable and constant latency, and has a certain advantage as it can be readily applied with most signal processors that are available clinically. When electric stimulation is applied to the tongue, a metallic taste is induced together with the tactile sensation. Therefore, it may be assumed that not only the trigeminal nerve but also the chorda tympani is stimulated with electric stimulation applied to the anterior two-thirds of the tongue. In this study, we attempted to identify the component via the trigeminal nerve and via the chorda tympani.

Methods and Materials

Informed consent was obtained from all subjects. In the first protocol, 17 healthy volunteers, aged from 23 to 46 years old (mean age, 30.6 years), were examined. Evoked potentials were recorded by silver plated electrodes on the scalp (C3, Cz, C4, according to the 10–20 system), as well as by silver ball electrodes on the posterior part of the annulus tympani. Reference electrodes on both auricles were short-circuited and the general ground was placed on the glabella. Using a 5-mm diameter electrode, we applied electric stimulation for 1 ms to the tongue and the upper gum, at a frequency of 0.9 Hz and an intensity 0.1 mA greater than a pre-obtained threshold. One hundred responses were averaged in each trial, using a Synax 1000 signal processor (NEC-Sanei Co. Ltd., Tokyo, Japan). Sampling time was 100 ms, high-cut filter was 1.5 Khz and low-cut filter was 0.5 Hz.

In the second protocol, we investigated five patients with otosclerosis, aged from 21 to 48 years. In these patients, the evoked potentials were recorded directly from the chorda tympani by needle electrode, as well as being recorded on C3, Cz, and C4 during stapedotomy under general anesthesia. The method used was the same as that in the first protocol. However, electric stimulation was applied only to the tongue.

Results

In the first series, the positive components recorded in C3 and C4, as well as Cz, at 43.8 ms by stimulation in the upper gum and 44.6 ms in the tongue were almost identical and well reproduced. These responses were obtained from a wide range of the scalp. Of the five components (N10, N24, P44, N48, and P75) obtained in Cz by stimulation on the tongue, P44 with a large amplitude was noted in all subjects. N10 was also well reproduced, but with a rather small amplitude. Potentials recorded in the external auditory canal consisted of N38 and P46. Responses in Cz elicited by gingival stimulation were made by N12, N24, P43, N50, P65, and N85. P43 with a large amplitude was recorded in all subjects, as was P44 on tongue stimulation. N24, N50, and N85 showed poor reproducibility. P65 with a large amplitude was characteristic of stimulation to the gingiva and was observed in 70% of the subjects.

In the second protocol, polyspikes were apparent at around 6 ms after the electrical stimulation. In the patients under general anesthesia, the responses in C3, C4, and Cz observed in the awake subjects were no longer identified.

Discussion

Both P44 evoked by tongue stimulation and P43 evoked by stimulation of the gingiva demonstrated

[1] ENT Department, Izumisano Municipal Hospital, 3-2-1 Ichiba-nishi, Izumisano, Osaka, 598 Japan
[2] ENT Department, Osaka University, 2-2 Yamadaoka, Suita, Osaka, 565 Japan
[3] Osaka-LERC Electrotechnical Laboratory, 3-11-46 Jyakuoji, Amagasaki, Hyogo, 661 Japan

almost the same latency and topographical allocation. Thus, it is possible that they are induced via the trigeminal nerve, which distributes over these areas.

Reference

1. Takemoto, et al. (1988) Gustatory evoked potential evoked by salt solution. J Otolaryngol Jpn 91:1637

Topographic Analysis of Hemispheric Differences in Chemosensory Event-Related Potentials

MITSUO TONOIKE[1], NORIKO ITO[2], MARI NAKAMURA[1], CHIKAHIDE MAETANI[2], IZUMI KOIZUKA[2], TORU MATSUNAGA[2], and MASAMINE TAKEBAYASHI[1]

Key words. Chemosensory event-related potentials (CHERPs)—Topographic analysis

The aim of this study was to analyze chemosensory event-related potentials (CHERPs), to show the origin of their peaks, and also to examine hemispheric differences in responses to odorants and differences in subjects by using topographic mapping. CHERPs were measured by our blast method, in which an odorant gas is administered directly into the nasal cavity by inserting a thin Teflon tube in synchronization with the subject's respiration [1,2].

With this blast method we were able to obtain more obvious waveforms of N1, P1, N2, P2, and P3 peaks for our six normal subjects than we obtained with an open method [3] in our previous study. In the present study, we obtained the CHERP responses by odorant pulse stimulation with amyl acetate, skatole, beta-phenylethyl alcohol, and other odorants, administered into the right or left nasal cavity by the one-side stimulation method. However, N2 and P3 peaks were not obtained on stimulation with odorless air. Two patients whose olfactory epithelium had been removed showed the same response patterns as those of the odorless air stimulation in the six normal subjects. These results suggested that the N1, P1, and P2 peaks in the CHERP wave were trigeminal responses; these were clearly discriminated from the N2 and P3 peaks, which we regarded as olfactory nervous responses.

In this experiment, we analyzed topographical patterns from the mapping data of 16 points on the scalp by the international 10–20 recording method. N1, P1, and P2 peaks in the odorless air stimulation clearly suggested the dominance of the contralateral hemisphere for all subjects except one, who had right dominance. On the other hand, later components of N2 and P3 were clearly obtained; they were considered as CHERPs from the changing of the N2-P3 peak for dose rate response. The P3 component showed dominance around the vertex area.

The N2 component showed some unique patterns for each odorant and each subject. For example, dominance of the ipsilateral hemisphere was shown in response to amyl acetate (pleasant odor) in two of the six subjects, while dominance of the contralateral hemisphere was shown in response to skatole (unpleasant odor) in an other two of the six subjects. However, one other subject showed a reversal of the above-mentioned hemispheric dominance in response to amyl acetate and skatole, and other subjects showed right and left dominance in response to some characteristics. Only one subject always showed right dominance for all odorants. These results suggest that hemispheric dominance in response to odorant stimuli is an individual characteristic.

References

1. Tonoike M (1987) Response characteristics of olfactory evoked potentials using time-varying filtering. In: Roper SD, Atema J (eds) Olfaction and taste IX. Annals of The New York Academy of Sciences, vol 510, New York, pp 658–661
2. Tonoike M, Seta N, Maetani T, Koizuka I, Takebayashi M (1990) Measurements of olfactory evoked potentials and event related potentials using odorant stimuli. In: Pedersen PC, Onaral B (eds) Proceedings of the 12th annual international conference of the IEEE engineering in medicine and biology society, New York, NY, USA, vol 12, no 2, pp 912–913
3. Tonoike M, Kurioka Y (1971) Precise measurements of human olfactory evoked potentials for odorant stimuli synchronized with respirations (in Japanese). Jpn J Electroencephalogr Electromyogr 9(3):214–223

[1] Life Electronics Research Center, Electrotechnical Laboratory, 3-chome, 11-46 Nakoji, Amagasaki, Hyogo, 661 Japan
[2] Department of Otorhinolaryngology, Osaka University Medical School, 2-2 Yamadaoka, Suita, Osaka, 565 Japan

Effects of Odor Pleasantness on Facial EMG

MAYA KATO[1] and AKIHIRO YAGI[2]

Key words. Facial EMG—Pleasantness of odor

This study employed facial electromyography (EMG) as an index of subtle changes of facial expression due to changes in feelings assumed to be caused by odors. Facial EMG has been confirmed to discriminate positive feelings from negative ones, even when facial expressions are too faint to be observed. Most previous facial EMG studies [1,2] have used imagery strategy to create an emotional context. However, in this study, we measured EMG using olfactory stimuli.

Thirteen subjects (6 male and 7 female students) perceived three different odors (pleasant, neutral, and unpleasant odor conditions): The pleasant odor was that of 3.33% lemon oil solution (500 µl), the neutral odor was that of distilled water (1 ml), and the unpleasant odor was that of 2.44% butyric acid solution (250 µl). The solvent was propylene glycol. Each odor solution was sealed in a 300-ml flask with two glass tubes on the top. The subjects applied the odors themselves by pushing a silicon bulb to release air from the flask, almost 20 ml being released at a time. Each odor appeared three times, and the order was pseudo-randomized not to be followed by the same odor.

EMGs were recorded in three muscle regions: (1) in the corrugator supercilii region (BROW) for its sensitivity to depressive feelings; (2) in the zygomaticus major region (CHEEK) for its sensitivity to positive feelings; and (3) in the levator labii superioris alaeque nasi region (NOSE) for its sensitivity to disgusting or unpleasant stimuli.

The subjects closed their eyes and relaxed while an EMG baseline was recorded in each trial. After hearing a signal, they released the odor, by pushing the pump six times, and inhaled twice. After each trial, the subjects were asked to rate the pleasantness of the odor. The pre-odor EMG recording period (baseline) and the post-odor EMG recording period were both 11 s. The inter-trial interval varied, but in each case was at least 90 s. When the EMG activity returned to the baseline level, the EMG baseline for the following trial was recorded.

[1] Research Institute of Human Engineering for Quality Life, 1-2-5 Dojima, Kita-ku, Osaka, 530 Japan
[2] Kwansei Gakuin University, Department of Psychology, 1-155 Uegahata Ichiban-cho, Nishinomiya, Hyogo, 662 Japan

EMG change was calculated as follows: the integration of the rectified EMG during the pre-odor period (base line level) was subtracted from the integration of the rectified EMG during the post-odor period, and this was divided by the base line level to minimize individual differences in baseline EMG activity. Data with obvious noise were discarded. For all but two trials, subjects rated the odors in the predicted manner. Data from these two trials were also not used.

As predicted, the different conditions had a significant effect on EMG changes in BROW and NOSE ($P < 0.01$), but there was no significant effect in CHEEK. While the unpleasant condition elicited greater EMG change than neutral and pleasant conditions in these two regions, there was no significant difference between the EMG response to the neutral and pleasant conditions in any region. Data in the unpleasant condition were divided into two groups based on the pleasantness ratings (high unpleasant and low unpleasant groups). It seemed that EMG change was greater for the high unpleasant group in NOSE and CHEEK than for the low group. However, the difference was not significant in any region, probably due to the limited amount of data due to the elimination of noisy data.

In conclusion, EMG changes in BROW and NOSE regions reflect subjective feelings of unpleasantness of odor. Although the changes were not significant, it is believed that facial EMG can reflect the subjective degree of unpleasantness of odor. Pleasantness of odor might not evoke EMG change easily. Finally, although it is possible that the response was not unpleasantness-specific, but stimulus-specific, the following finding rules out this possibility: When a subject felt a lemon odor to be pleasant, the EMG did not respond, but when the subject felt the same odor to be unpleasant, there was a dramatic EMG response. This confirms that the results are reliable in reflecting the subjective appraisal of odor.

Acknowledgments. This research was commissioned by the agency of Industrial Science and Technology under the Ministry of International Trade and Industry through New Energy and Industrial Technology Development Organization as a part of an Industrial Science and Technology Frontier Program entitled "Human Sensory Measurement Application Technology."

References

1. Schwartz GE, Brown SL, Ahern GL (1980) Facial muscle patterning and subjective experience during affective imagery. Psychophysiology 17:75–82

2. Vrana SR (1993) The psychophysiology of disgust: Differentiating negative emotional contexts with facial EMG. Psychophysiology 30:279–286

Different Feeling Profiles in Response to Odor: Relationship to Hemispheric Reactivity

Y. SAITO[1], T. YAMAMOTO[2], S. OGATA[2], Y. TAKASHIMA[3], and S. AYABE-KANAMURA[4]

Key words. Odors—Contingent negative variations (CNV)—Multivariate analysis

It is known empirically that scented environments induced by favored or unfavorable odor stimuli affect the emotional aspects of human behavior. These effects seem to be reflected in such psychological factors as mood and feeling. Several studies have addressed the effects of odors on human psychological function by using event-related potentials (ERPs) [1,2]. In our previous studies, also [3,4], we found that both favored and unfavorable odor stimuli affected the basis waves of ERPs extracted by multivariate analysis. In this study, right/left (rt/lt) ratios measured from basis waves of ERPs obtained from 20 subjects were examined in different "feeling" profiles induced by the same odor stimulus.

Lemon oil, which was evaluated as favored by most subjects, was used to produce a scented environment. The odor stimulus was presented continuously during the ERP recording session. A mood check test was used to check the subjective feelings in the control conditions and during odor presentation. This test consists of six categories: favored, unfavorable, relaxed, irritated, elated, and depressed. For visual investigation, feeling profiles were made in accordance with these six categories.

ERPs were recorded between the Ag/AgCl electrode attached on the scalp (10/20 electrode system) and a reference connecting both ear lobes. ERPs were evoked by the S1–S2 and S1–S2 + R paradigms (S1–S2 interval was 2.5 s). S1 was a pip tone of 1.5 kHz and S2 was white noise. Response (R) was pressing the button in the right hand. There were 36 trials of the S1–S2 paradigm and 36 trials of S1–S2 + R in an experimental session. Two ERP recording sessions were done before odor presentation (control condition) and another two were done during odor presentation.

In order to estimate the basis waves, a principal component analysis (PCA)—varimax rotation—multiple regression analysis (MRA) strategy was applied to single trial ERPs at each respective electrode site. Hemispheric mean areas of the largest basis waves (Ns1 + 2000 – 2200) were obtained in the S1–S2 and S1–S2 + R conditions and an increase ratio was obtained due to the paradigm shift. The increase ratios for the right and left hemisphere were then compared, resulting in the rt/lt ratio.

The 20 subjects were divided into three groups ($n = 4$, $n = 8$, and $n = 8$) according to the results of the mood check test. The four subjects in group 1 scored high points in the categories of unfavorable and irritated. In this group, the rt/lt ratios were 0.92 in the control condition and 1.02 in the odor presentation condition. While subjects in groups 2 and 3 showed similar feeling profile patterns, group 3 scored higher than group 2 in the categories of relaxed, favored, and elated. The rt/lt ratios of group 2 were 1.00 in the control condition and 0.84 during odor presentation, while the rt/lt ratios of group 3 were 1.00 in the control condition and 0.91 during odor presentation.

In this study, although the same odor stimulus was used, there were correspondences between the feeling profiles and the increase ratio of electrical activity in the right and left hemispheres. When feeling profiles showed higher scores for the categories of unfavorable and irritated, the increase ratio was greater in the right hemisphere than in the left one. On the other hand, the higher scores for the categories relaxed, favored, and elated corresponded to a greater increase ratio in the left hemisphere than in the right one.

These results were compatible with the results previously obtained in a pleasant-unpleasant odor experiment [3] and a favored-unfavorable experiment [4].

[1] Department of Systems and Engineering, Ashikage Institute of Technology, 268 Ohmaecho, Ashikaga, Tochigi, 326 Japan
[2] Department of Psychophysiology, Tokyo Institute of Psychiatry, 2-1-8 Kamikitazawa, Setagaya-ku, Tokyo, 156 Japan
[3] Takasago International Corporation, 1-4-11 Nishihachiman, Hiratsuka, Kanagawa, 254 Japan
[4] Department of Psychology, University of Tsukuba, 1-1-1 Tennodai, Tsukuba, Ibaraki, 305 Japan

References

1. Torii S, Fukuda H, Kanemoto H, Miyauchi R, Hamauzu Y, Kawasaki M (1988) Contingent negative variation (CNV) and the psychological effects of odor. In: Van

Toller S, Dodd GH (eds) Perfumery: The psychology and biology of fragrance. Chapman and Hall, New York, pp 107–120

2. Koga Y, Takeuchi H (1989) The effect of odors on brain functions; mental and physiological discussions. Fragrance J 9:20–27

3. Saito Y, Yamamoto T, Kanamura S (1990) Scented environment and ERP basis waves (2) (in Japanese).

Proceedings of the 24th Japanese symposium on taste and smell, Japanese Association for the Study of Taste and Smell, Tsukuba, pp 51–54

4. Saito Y, Yamamoto T, Kanamura S (1991) Scented environment and ERP basis waves (3) (in Japanese). Proceedings of the 25th Japanese symposium on taste and smell, Japanese Association for the Study of Taste and Smell, Tsukuba, pp 117–120

Influences of Blink Response and Eye Movement on Olfactory Evoked Potentials

Shigeru Furuta, Hirofumi Nishizono, Rikako Hirota, and Masaru Ohyama[1]

Key words. Olfactory evoked potentials—Blink response

Human cerebral evoked responses to odorous stimuli have been reported [1]. Blinks or eye movements were excluded from the analysis of olfactory evoked potentials (OEP) as an artifact [2,3]. However, when the OEP were not used, the phenomenon of the blink reflex and eye movement associated the stimuli of odor were useful as an objective olfactory function test [4]. Here, we discuss the influence of blink responses or eye movements on OEP potentials.

Forty normosmic subjects participated in this study. The odorants, phenyl ethyl alcohol, isoamyl acetate (IAA), isovalic acid, methyl cyclopentenelone (MCP), were presented with a new apparatus developed by our department. An electronically controlled valve was activated by means of a respiration monitor attached to the subject so that delivery of odor could be synchronized to occur only during inspiration. Electrical evoked potentials were measured, simultaneously with the surface recordings, from the midline electrodes Fz and Cz, referred to the right earlobe. Vertical eye movements and blinks were monitored with electrodes above and below the left eye referred to electrodes below the right eye.

An evoked potential of Fz or Cz for the odorant presented a positive peak in a mean time of about 250 ms after the odorant stimulus and recovered gradually. The eye movement was observed from about 100 ms after the stimulus, and the blink response was observed earlier than the eye movement. There was a blink response for each odorant in 25%–35% of the subjects, this being more than that for air, which response occurred in 18% of the subjects. The results sugested that the afferent source of the blink response for an odorant might be not only the olfactory nerve but also the trigeminal nerve. Although the incidence of eye movements for each odorant was less than that of the blink response,

there were significant differences between the incidence of eye movements for each odorant and for air. The blink responses for IAA and MCP decreased with an increasing number of stimuli, indicating the phenomenon of odorant fatigue in olfactory receptors. No significant difference was found in the latency of the blink responses and the eye movement for each odorant. Five subjects who had a high sensitivity for olfactory stimuli served in a further study, in which measurements were made before and after anesthesia of nasal mucosa was produced by 4% lidocaine. The incidence of the blink response and the eye movement decreased significantly ($P < 0.001$) after local anesthesia, while the latency did not change for either measurement.

Attempts to quantify human responses to odorous stimuli have included measuring the influence of odorants on pupillary dilation, respiration rate, blood flow in the extremities, and electrical conductivity of skin, and measuring odorant-induced changes in electroencephalograms, as well as psychophysical measures of various kinds. In this study, the blink response for olfactory stimulation occurred in 30% of the subjects. This is in agreement with the findings of Ichihara et al. [4]. Although responses greater than 40 μV in the eye channel have been excluded from the analysis of olfactory evoked potentials as an artifact [2], we suggest that these responses may be useful as an objective olfactory function test.

References

1. Allison T, Goff WR (1967) Human cerebral evoked responses to odorous stimuli. Electroenceph Clin Neurophysiol 23:558–560
2. Hummel T. Kobal G (1992) Diffrences in human evoked potentials related to olfactory or trigeminal chemosensory activation. Electroenceph Clin Neurophysiol 84:84–89
3. Tonoike M, Kurioka Y (1981) Precise measurements of human olfactory evoked potentials for odorant stimuli synchronized with respirations. Electroencephalograph Myograph 9:214–223
4. Ichinara M, Komatsu A, Ichihara F, Asaga H, Taira K, Kubota T (1967) Osmatic test with optic-winking as an objective response (in Japanese). Otolaryngol (Tokyo) 39:947–953

[1] Department of Otolaryngology, Faculty of Medicine, Kagoshima University, 8-35-1 Sakuragaoka, Kagoshima, 890 Japan

Olfactory Evoked Potentials Using Blast Method for Odor Stimulation

K. Shimada[1], N. Seta[2], I. Takemoto[2], I. Koizuka[2], M. Tonoike[3], S. Yoshimura[4], and T. Matsunaga[2]

Key words. OEP—P_{300}

Since Finkenzeller [1] (in 1966) and Allison and Goff [2] (in 1967) reported olfactory evoked potentials (OEPs) for the first time, this phenomenon has been the subject of much research. Although the precise origin of OEPs is still unknown, their study is one of the most useful methods for objectively evaluating odor sensation in humans.

We have examined OEPs, using a blast method for administering odorant gases [3]. We separated the OEPs into two components and concluded that the early component was the chemo-somatosensory evoked potential (CSSEP), evoked by trigeminal nerve stimulation, and the late component was the chemo-sensory evoked potential (CSEP), which originated from the olfactory system.

In this study, we investigated the significance of the late positive component (P_3) in our OEP, the latency of which was approximately 300 ms, using an oddball paradigm. Here, we discuss whether the P_3 component in our OEP is equivalent to P_{300}.

For stimulation, we used a pulsed stimulator developed by one of the authors [4]. The odorant gas was applied once per two breaths directly into the nasal cavity via double-barreled silicon tubes 2 mm in diameter. For each stimulant gas these were inserted approximately 1 cm into the nasal cavity. The application of gas was synchronized with the subjects' respiration; duration of stimulation was 200 ms.

We used an NEC-San-Ei multipurpose electro-encephalograph (EEG) type 1A97 for EEG recording and a 7T18 signal processor (NEC, Tokyo, Japan) for accumulation and EEG analysis.

The high-cut filter was 120 Hz and the time constant was 0.3 s. The responses to 20 rare stimuli and 20 of 80 frequent stimuli were averaged. Silver-silver chloride electrodes were placed on the scalp at three points (Fz, Cz, and Pz) according to the international 10–20 electrode system.

We used a pair of stimulants, amyl acetate and skatole, as odorant gases. Our subjects were one healthy female and one healthy male, who both showed normal olfactory recognition and thresholds tested by an objective olfaction test. The subjects sat with their eyes open in a dimly lit sound proof room with no visual cues. White noise of approximately 60 dB SPL, delivered through headphones, prevented the subjects from hearing the switching process.

When the odorant gas was administered into the nasal cavity by the blast method, both the olfactory and the trigeminal nerve systems were stimulated simultaneously. In a previous study [3], we found that the early components (N_1,P_1) were representative of changes in potential due to trigeminal nerve stimulation, while the middle components (N_2,P_2) were representative of changes due to olfactory nerve stimulation. Since the latency of the late component, P_3, was long, we assumed that the P_3 could be an event-related potential.

In this study, we compared the P_3 components of each OEP obtained by the pair of odorants, amyl-acetate and skatole, using an oddball paradigm. When the odorants were applied as a rare stimulus, the P_3 components had a much higher amplitude than those evoked when the odorants were administered as a frequent stimulus. No obvious difference due to differences of the odorant gases was observed in this study. Therefore, we concluded that the P_3 component in our OEP may be equivalent to P_{300}.

References

1. Finkenzeller P (1966) Gemittelte EEG-Potentiale bei Olfactorischer Reizung. Pfluegers Arch 292:76–80
2. Allison T, Golf WR (1967) Human cerebral evoked response to odorous stimuli. Electroenceph Clin Neurophysiol 23:558–560
3. Notake N, Maetani T, Koizuka I, Ogino H, Matsunaga T, Yoshimura S, Sakurai K, Tonoike M (1989) Olfactory evoked potential by blast method and significance of its component. Proceedings of the 23rd Japanese symposium of taste and smell 23:43–46
4. Tonoike M, Kurioka Y (1982) Precise measurements of human olfactory evoked potentials for odorant stimuli synchronized with respiration. Bul Electrotech Lab 46,(11):622–633

[1] Department of Otolaryngology, Otemae Hospital, 1-5-34 Otemae, Chuo-ku, Osaka, 540 Japan
[2] Department of Otolaryngology, Osaka University Medical School, 2-2 Yamadaoka, Suita, Osaka, 565 Japan
[3] L.E.R.C., 3-11-46 Wakaoji, Amagasaki, Hyogo, 661 Japan
[4] Asuka Electronic Co. Ltd., 8-8-11 Tenjinbashi, Kita-ku, Osaka, 531 Japan

Wavelet Analysis of Event-Related Potentials (ERPs) Under Ambient Odors

Yoshihiko Koga[1], Kazuhiko Yanai[2], and Yoshikazu Shutara[1]

Key words. Wavelet analysis—Event-related potentials (ERPs)

Introduction

Event-related potentials (ERPs) have recently been shown to be a useful tool for investigating the processing of sensory input in the human brain. Of the ERP components, P300 is considered to reflect the stage of stimulus evaluation. The peak latency of the P300 component is thought to provide an index of stimulus evaluation time, while the amplitude of the P300 is thought to indicate the amount of information processing resources allocated to conduct stimulus evaluation. In order to study the influence of odors on the function of information processing in the human brain from a physiological aspect, we recorded ERPs, under conditions of ambient odors, using an oddball paradigm. We then applied wavelet analysis to further investigate the ERP waveforms. Wavelet analysis is a method for the time-frequency analysis of time series data. Different from Fourier analysis, wavelet analysis is able to demonstrate changes of the frequency of ERP components within a time-frequency window.

Method

The subjects were ten right-handed normosmic adults (four male) with a mean age of 23.0 years. For each subject, ERP was elicited by using an oddball paradigm in six experimental blocks while different odors, i.e., distilled water, ethanol, immature whiskey, immature whiskey without alcohol, mature whiskey, and mature whiskey without alcohol were presented respectively.

The stimuli used for the oddball task consisted of two rectangular-shaped figures of different colors (red and blue). The red one was the target, with a probability of 0.2. The stimuli were projected in random order on a cathode ray tube (CRT) screen. The glass containing the odorous sample was set below the subject's nose. The subject was instructed to count the target stimuli while breathing as usual to feel the ambient odor. EEG was recorded from 16 scalp sites referred to the linked ears. Twenty trials without artifact were averaged. The ERP waveforms of the six experimental blocks were calculated for each subject. The ERPs for the most favorite odor, the least favorite odor, and distilled water were collected for all the subjects, and grandmean waveforms of these three conditions were computed. The grandmean ERP waveforms were subjected to wavelet analysis by using the Gabor function as a wavelet base.

Results

The P300 latency was significantly shorter for the most favorite odor condition than for the least favorite. However, no significant differences were found among the P300 amplitudes of the three conditions. The results of the wavelet analysis showed that in every condition the large component, which corresponded to the P300 of the original wave, appeared clearly during the period between about 150–400 ms. In conditions of the most favorite odor and distilled water, the frequency band of this component was 2–8 Hz, while it was 2–6 Hz in the least favorite odor condition. The shape of this component in the most favorite odor condition was sharper to some degree than that in the other two conditions. Furthermore, the components of 10 Hz rhythmically appeared throughout the analysis period in the least favorite odor condition. In the distilled water condition, these components also appeared until 300 ms. They were not observed in the most favorite odor condition.

Discussion

The findings for the P300 latency suggest that the rate of visual information processing is increased when the ambient odor is favorable to the subject. With respect to the results of wavelet analysis, the lack of an element with a relatively high frequency

[1] Department of Neuropsychiatry, School of Medicine, Kyorin University, 6-20-2 Shinkawa, Mitaka, Tokyo, 181 Japan
[2] Department of Human Sciences, Kao Institute for Knowledge and Intelligence Sciences, Kao Corporation, 1-14-10 Nihonbashi-Kayaba, Chuo-ku, Tokyo, 103 Japan

band during the period of the P300 in the least favorite odor condition indicates that the mode of allocation of information processing resources may be different from that in the other two conditions. The rhythmic 10 Hz components observed by wavelet analysis in the least favorite odor condition may be due to superimposed alpha waves which cannot be distinguished by visual inspection of the original ERP waveform. The appearance of these 10 Hz waves could be caused by the subjects' decreased vigilance level in that condition.

In the light of these results, we consider that wavelet analysis is useful for finding the dominant frequency band of the ERP component throughout the period during which it appears. Moreover, this analysis can elucidate the structure of the ERP waveform, which consists of several subcomponents.

14. Artificial Sensing Devices

The Taste Sensor

Key words. Multichannel taste sensor—Lipid membrane—Taste discrimination—Quantified taste—Taste map

Introduction

Taste is classified into five kinds of basic qualities, elicited by many chemical substances that affect each other and change the taste strength. Sourness is produced by the hydrogen ions of hydrochloric acid, acetic acid, etc. Bitterness is elicited by many alkaloids as quinine, caffeine, etc. Electrolytes such as NaCl and KCl produce saltiness, while many nonelectrolytes, such as sucrose and glucose, show sweetness. Umami is characteristic of monosodium glutamate (MSG) [1]. The intensity of saltiness is weakened by coexistent bitter substances and strengthened by sour substances. In the same way, umami substances enhance each other. These are called suppression and synergistic effects, respectively.

Conventional chemical sensors cannot reproduce the human taste sense, because only the quantities of contained taste substances are analyzed. It may be difficult to correlate the output signals with the taste sense. The physical value measured for example, by a pH meter to measure sourness or a refractometer to measure sweetness is also affected by extra substances (e.g., starch) that do not elicit taste sensations in humans. In addition, the value cannot express the interactions between taste substances.

Lipid membranes have been investigated in detail as a candidate for transducers of the taste sensor [2–4]. The lipid membranes were found to discriminate between basic taste qualities and to detect interactions between taste substances, such as a synergistic effect.

Based on these facts, a multichannel taste sensor has been developed that utilizes lipid membranes [5,6]. This sensor has greater sensitivity and reliability than those of humans.

[1] Department of Electronics, Faculty of Engineering, Kyushu University 36, Fukuoka, 812 Japan

Material and Methods

Multichannel Electrode

Eight kinds of lipid analogues were used for membranes [5,6]. The lipid membrane, which was cast with polyvinyl chloride and dioctyl phenylphosphonate, was a transparent, colorless, soft film with about 150 μm thick.

The electric potential across the membrane was detected by a silver wire in the multichannel electrode and a reference electrode. The selected electric signal from the sensor was converted to a digital code and routed to a computer.

Measurement of Beer

The taste solutions studied were various kinds of beer. The initial precondition was made by preserving the electrode in a certain standard beer (named K1).

A taste map is proposed for expressing the taste quality of beer [7]. This map is based on sensory tests made by humans, and is composed of the abscissa axis expressing "rich taste" or "soft taste" and the ordinate axis expressing "sharp touch" or "smooth touch." The rich or soft taste may be mainly related to the concentration of wheat, whereas the sharp or smooth touch may arise from the concentrations of alcohol, hops, and so on. We tried to express these taste qualities quantitatively by transforming the output pattern of the taste sensor by means of principal component analysis.

Results

Figure 1, which indicates the sensor responses to taste substances [5], shows taht taste was measured in a fashion similar to the human gustatory sensation because the sensor showed a similar pattern in the same group of tastes and quite different patterns for different taste groups. In fact, similar outputs were obtained for NaCl, KCl, and KBr as eliciting saltiness, for example, and their patterns also differed clearly from those for other basic taste qualities. Further, this result suggests a possibility of molecular recognition; careful selection of membranes will enable us to detect minute differences between molecules and thus classify them.

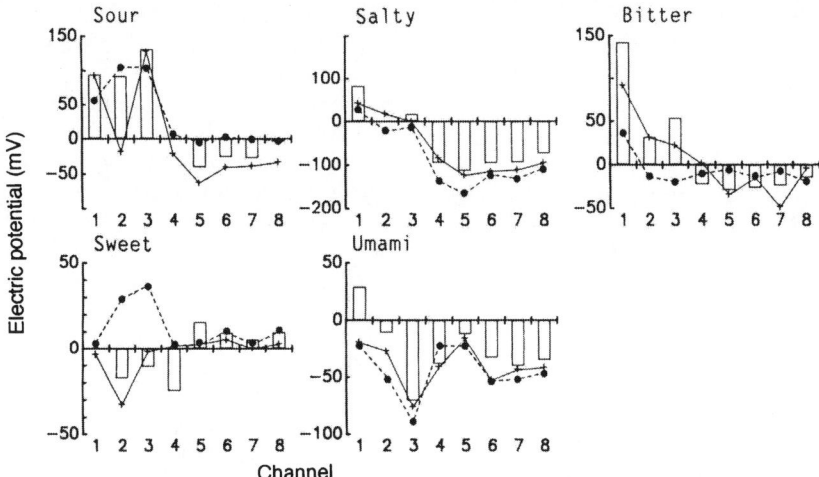

FIG. 1. Response patterns for taste substances belonging to five kinds of taste qualities. *Bar graph*, *plus sign*, and *solid circles* denote, respectively, 3 mM HCl, 30 mM citric acid, and 30 mM acetic acid, for sourness; 300 mM each NaCl, KCl, and KBr, for saltiness; 3 mM quinine, 30 mM MgSO$_4$, and 3 mM phenyltiourea, for bitterness; 1 M each sucrose, glucose, and fructose for sweetness; and 100 mM MSG, 1 mM monosodium inosinate, and 100 mM guanylate for umami

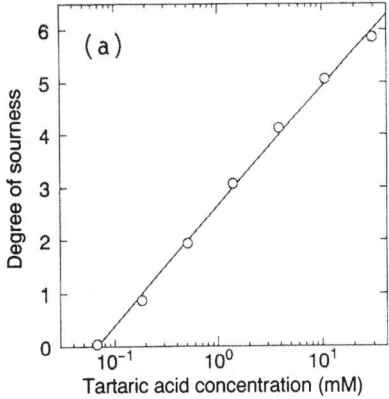
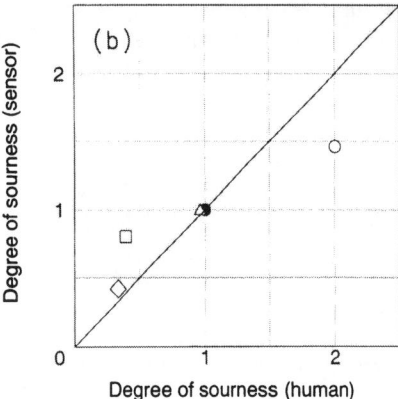

FIG. 2. Degree of sourness defined by sensor outputs (**a**) and correspondence between intensities obtained from the sensor and human sense (**b**). *Solid circle*, tartaric acid; *open circles*, hydrochloric acid; *diamond*, acetic acid; *triangle*, citric acid; *square*, lactic acid; all of same concentration

Intensities of sourness and saltiness were quantified using the taste sensor; in Fig. 2A, "degree of sourness" is introduced as a function of tartaric acid concentration [8]. Once the *degree* of sourness is defined by means of the response electric potential of the sensor, it can be applied to other sour substances. Figure 2B shows the correspondence between this sensor and the sense of human taste. The sensor and human sense matched, indicating that this sensor reproduces the human taste sensation. When tartaric acid and NaCl were used together, their tastes enhanced each other [9]. This phenomenon was reproduced using the sensor-defined degree of sourness [8]; thus, the taste sensor expresses the taste interactions experienced by humans.

Certain kinds of commercial beverages, such as coffee, beer, and aqueous ionic drinks, were easily

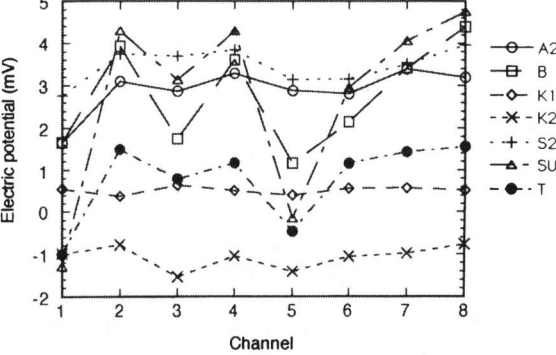

FIG. 3. Output electric potential patterns for seven brands of beer

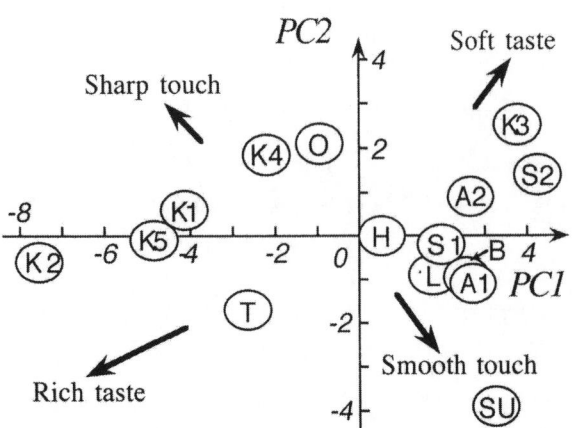

FIG. 4. Taste map obtained from output electric potential patterns

discriminated using the taste sensor [10]. Figure 3 shows the response patterns to 7 brands of beer among the 15 brands listed in Fig. 4 [11]. Although the difference of electric potential between different brands was within a few millivolts, each beer was easily distinguishable because the SD were about 0.1. The results in Fig. 4 agree with our taste "feelings" on beer as reported [7,12], if each axis is interpreted as named in Figure 4. The electric potential pattern of beer was drastically affected by temperature, implying that the taste as usually experienced is considerably altered by temperature.

Discussion

The taste sensor has higher sensitivity, reproducibility, and reliability than those of humans. The taste quality was quantified, and the taste interactions were reproduced.

The foregoing method quantified taste by the sensor to extract the characteristics of output patterns using such analyses as principal component analysis. Another method is to compare output patterns between test solution and the mixed solutions by performing many measurements of various mixed solutions. We tried to reproduce the taste of a commercial aqueous beverage by blending four basic taste substances (HCl, NaCl, quinine, sucrose) so that the response pattern would most resemble that of the aqueous drink. With this attempt, the best combination of the concentrations of basic taste substances was obtained using 2 mM HCl, 50 mM NaCl, 0.2 mM quinine, and 100 mM sucrose. This mixed solution produced almost the same taste as the aqueous drink.

The concentration of taste substances can be transformed to the taste strength [13]. The mixed solution just mentioned is composed of 4.04 sourness, 2.03 saltiness, 5.01 bitterness, and 2.24 sweet-

ness in terms of the τ scale. Therefore, the taste of any drink can be quantified using the multichannel taste sensor. Quantitative measurements of taste by the sensor are more accurate than those determined by human sense. It is very important to note that this method automatically incorporates the interactions between taste substances.

Tomatoes were also studied [14]; different brands of tomatoes were easily distinguished using the taste sensor. The taste was expressed quantitatively in terms of basic taste qualities. Taste of amino acids was also studied using the taste sensor [15].

The taste sensor can be used for quality control in the food industry and will help automation of the production. The sense of taste is imprecise, and taste sensations largely depend on subjective factors of human feelings. If we compare the standard index measured by means of the taste sensor with the sensory evaluation, we will be able to assess taste objectively. Moreover, the mechanism of information processing of taste in the brain as well as the reception at taste cells [16] will also be clarified by developing a taste sensor with output similar to that of the biological gustatory system.

References

1. Kawamura Y, Kare MR (eds) (1987) Umami: a basic taste. Dekker, New York
2. Hayashi K, Matsuki Y, Toko K, Murata T, Yamafuji Ke, Yamafuji Ka (1989) Sensing of "umami" taste and synergistic effects with synthetic lipid membranes. Sens Mater 1:321–334
3. Iiyama S, Toko K, Yamafuji K (1987) Taste reception in a synthetic lipid membrane. Membrane (Maku) 12:231–237
4. Iiyama S, Toko K, Hayashi K, Yamafuji K (1989) Effect of sweet substances on electric characteristics of a dioleyl phosphate membrane. Agric Biol Chem 53:675–681
5. Hayashi K, Yamanaka M, Toko K, Yamafuji K (1990) Multichannel taste sensor using lipid membranes. Sens Actuators B2:205–213
6. Ikezaki H, Hayashi K, Yamanaka M, Tatsukawa R, Toko K, Yamafuji K (1991) Multichannel taste sensor with artificial lipid membranes. Trans IEICE Jpn J74-C-II:434–442
7. Kirin Brewery Co Ltd (1992) Biru no umasa wo saguru [Study of deliciousness of beer]. Shokabo, Tokyo, Chap. 7
8. Murata T, Hayashi K, Toko K, Ikezaki H, Sato K, Toukubo R, Yamafuji K (1992) Quantification of sourness and saltiness using a multichannel sensor with lipid membranes. Sens Mater 4:81–88
9. Bartoshuk LM (1975) Taste mixture: is mixture suppression related to compression? Physiol Behav 14:643–649
10. Toko K, Hayashi K, Yamanaka M, Yamafuji K (1990) Multichannel taste sensor with lipid membranes. Tech Dig 9th Sens Symp 9:193–196
11. Toko K, Murata T, Matsuno T, Kikkawa Y, Yamafuji K (1992) Taste map of beer by a multichannel taste sensor. Sens Mater 4:145–151

12. Tanaka F (1991) Axis of taste, vol 10, no 7. Quark, Kodansha, Tokyo, pp 54–59
13. Indow T (1969) An application of the τ scale of taste: interaction among the four qualities of taste. Percept Psychophys 5:347–351
14. Kikkawa Y, Toko K, Yamafuji K (1993) Taste sensing of tomato with a multichannel taste sensor. Sens Mater 5:83–90
15. Kikkawa Y, Toko K, Matsuno T, Yamatuji K (1993) Discrimination of taste of amino acids with a multichannel taste sensor. Jpn J Appl Phys 32:5731–5736
16. Kurihara K, Yoshii K, Kashiwayanagi M (1986) Transduction mechanisms in chemoreception. Comp Biochem Physiol 85A:1–22

Intelligent ChemSADs for Artificial Odor-Sensing of Coffees and Lager Beers

Julian William Gardner[1] and Philip Nigel Bartlett[2]

Key words. Electronic nose—Odor sensors—Coffee—Beer

Introduction

The ability to analyze aromas for identification, product blending, and process control is an essential requirement of the food and drinks industry. Conventional analytical methods of aroma analysis, such as gas chromatography (GC) and GC/mass spectrometry, are expensive and rather slow. In addition, not only is the sensitivity often too low to detect the presence of some key compounds but it can also be difficult to relate the chemical composition to its flavour impact. Moreover, the training and implementation of a sensory or organoleptic panel is an even more expensive and slower process and the results are only qualitative in nature. The principal difficulty arises from the complex nature of odor-sensing. For example, there are about 670 flavor constituents of coffee [1] with the largest classes being the furans, pyrazines, pyrroles, and ketones. Similarly, studies on beer flavor have shown that there are more than 100 separately identifiable flavor compounds of which only 15 or so of the 39 key ones can be explained with confidence. It is therefore evident that the development of a low-cost, fast, electronic instrument to monitor aromas and flavors is highly desirable.

Considerable effort has been directed in the past 10 years at Warwick University into the general development of chemical sensoric array devices (ChemSADs) for both artificial odor sensing and gas mixture analysis. A ChemSAD consists of the combination of a chemical sensor array and a pattern recognition technique (e.g., discriminant function analysis or an artificial neural network); when applied specifically to odor sensing it may be referred to as an "electronic nose." A comprehensive review of artificial odor sensors and electronic noses has been published [2].

[1] Department of Engineering, University of Warwick, Coventry, CV4 7AL, UK
[2] Department of Chemistry, University of Southampton, Southampton, SO9 5NH, UK

Materials and Methods

Figure 1 shows a general arrangement of the Warwick–Southampton artificial odor-sensing system. Two systems have been constructed. The first consists of an array of 12 metal oxide gas sensors (Figaro Engineering, Osaka, Japan) in which thick sintered tin oxide layers form the active material. The second is an array of 12 polymeric chemoresistors (in-house) using microelectronic technology and electrochemistry (Fig. 2). The electrode gap, across which the black electroactive polymers are deposited, is $10 \mu m$ wide and $1 mm$ long. Table 1 gives details of the sensors employed in each of the two systems. Each array of 12 sensors was inserted into a 2-l glass vessel and surrounded by a heated water bath into which the measurand was introduced. The measurand was either $20 cm^3$ of headspace for the metal oxide sensor or $100 cm^3$ of liquid for the polymer array. The conductance $G_{ij}(t)$ of all the odor sensors was transduced into a dc voltage $V_{ij}(t)$ via basic analogue op-amp circuitry (Fig. 1). The voltage generated by each sensor was then fed into an a/d converter within a microcomputer system. The dynamic response of each sensor i to odor j was then processed to define a time-independent response parameter x_{ij}. The choice of sensor-processing parameter has been shown to affect the performance of the pattern recognition engine [3,4]. Using the fractional change in conductance x_{ij} helps reduce the temperature dependence of the sensor and, in the case of metal oxide only, linearize the concentration dependence,

$$x_{ij} = \frac{G_{ij}^{odor} - G_{ij}^{air}}{G_{ij}^{air}} \qquad (1)$$

where G_{ij}^{odor} is the steady-state conductance in odor j and G_{ij}^{air} is the steady state conductance in air of sensor i. The array response is thus a 12-dimensional vector X_j. In dilute odor mixtures, the performance of the pattern recognition (PARC) engine may be further enhanced by normalizing the response vector of the sensor array as they conform to a principle of linear superposition:

$$X_j' = \frac{X_j}{\| X_j \|} \equiv \frac{k_{ij}}{\sqrt{\Sigma_i k_{ij}^2}}$$

where $x_{ij} \approx k_{ij} C_j$

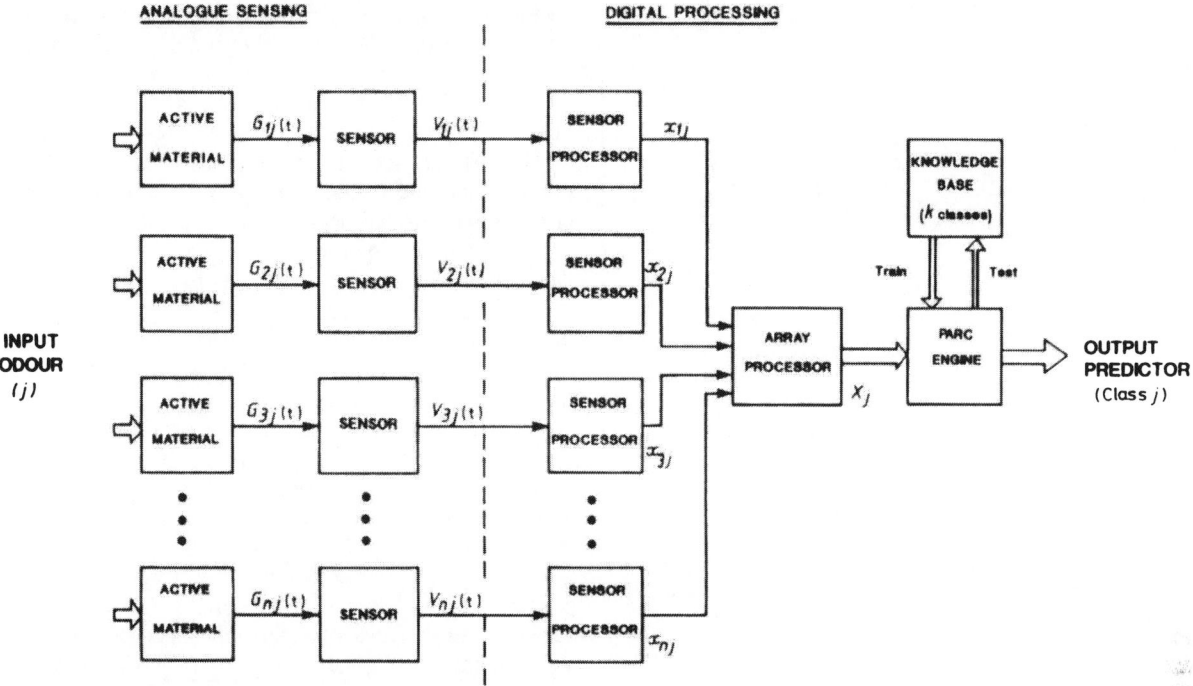

FIG. 1. Architecture of artificial odor-sensing system

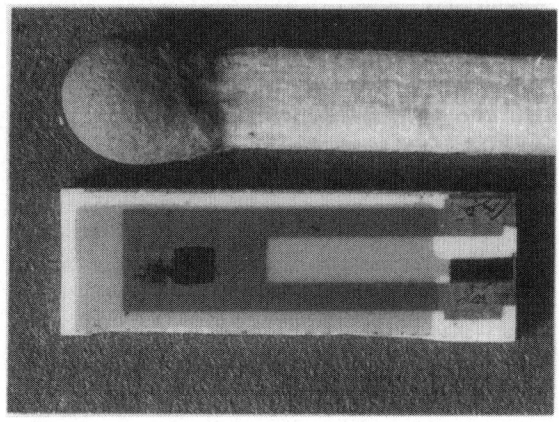

FIG. 2. Photograph of odor-sensitive polymer chemoresistor

TABLE 1. Details of odor sensors used in electronic noses.

Sensor	Oxide	Polymer
1	TGS 803	PPy-BSA
2	TGS 831	PPy-PSA
3	TGS 815	PPy-HxSA
4	TGS 816	PPy-HpSA
5	TGS 823	PPy-OSA
6	TGS 800	PPy-DSA
7	TGS 842	PPy-TSA(Na)
8	TGS 882	PPy-TSA(m)^a
9	TGS 881	PPy-TEATS^a
10	TGS 883	PPy-TEATS
11	TGS 825	PAN-NaHSO$_4$
12	TGS 821	P3MT-TEATFB^a

^{a}Nonaqueous solutions.

Finally, the response vector is fed into the PARC engine. Under a supervised training scheme (e.g., principal components analysis or discriminant function analysis), the multivariate input vector is initially mapped onto a knowledge base containing crisp data (e.g., brand of coffee) or fuzzy data (e.g., organoleptic). Then unknown input vectors can be fed into the unsupervised (predictive) part of the PARC engine and classified according to established rules. Various PARC techniques have been applied to electronic nose data, such as discriminant function analysis and artificial neural paradigms. The success of each technique depends on not only the nature of the sensor mechanism but also the sampling method and odor complexity. Classification space of odors can be not only non linear but also unstable, exemplified by ground coffee, which oxidizes in air, and beer, which stales in air.

Results

An investigation was carried out to analyze the response of an array of 12 commercial sensors to the headspace of coffees (two different blends and two different roast levels) [5]. Figure 3 shows the

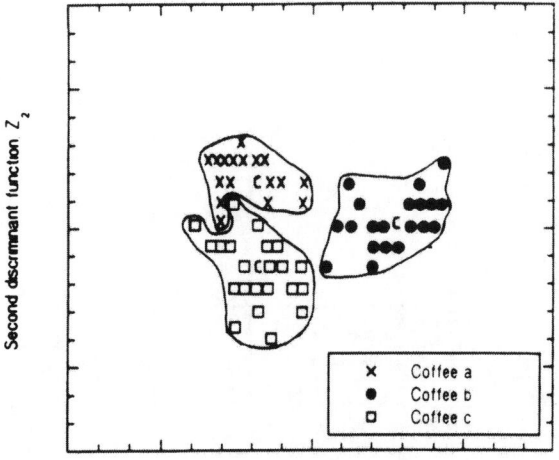

FIG. 3. Discrimination of different coffee blends and roasts by metal oxide electronic nose. Cluster centroids are marked by letter 'c'

results of discrimination function analysis on 30 samples of each coffee type. Type (a) is a medium roast of blend 1, (b) a dark roast of blend 1 and (c) a dark roast of blend 2. The discrimination of the roasting level appears to be slightly better than that of the blend using this procedure. A classification success rate of 89.9% was found for the standard response vectors X_j, which increased to 95.5% after array normalization. More recently, the response of both electronic noses to Brazilian and Colombian

coffee beans has been made using principal components analysis. Although the metal oxide sensor array was only able to discriminate the samples at 90% level, the polymer array gave a clear distinction (100%) (Fig. 4).

Analysis of beer headspaces showed that the metal oxide electronic nose was capable of discriminating between beer, lagers, and spirits using a back-propagation neural networking technique [6]. A more comprehensive study has been carried out on the application of a polymer electronic nose to lager beer aromas. (See [7] for full details.) As Fig. 5 shows, a Euclidean cluster analysis is able to discriminate between a control lager (cluster A) and a artificially staled (diacetyl) lager (cluster B).

Summarizing, the application of two types of electronic noses (metal oxide and polymer) to coffee and beer aromas has been discussed. The response time, sensitivity, and stability of the metal oxide nose is generally better but the discriminating power of the polymer nose to coffees and lagers is superior. Both types of electronic nose are under commercial development. Figure 6 shows a six-element metal oxide electronic nose manufactured by Alpha MOS (Toulouse, France) that is currently available, and a 12-element polymer electronic nose (Neotronics, Bishop's Stortford, UK), which is under trial in industry.

Acknowledgments. We wish to thank Campden Food and Drinks Research Association, Neotronics Ltd., Bass Plc, and the Department of Trade and Industry for their financial and technical support.

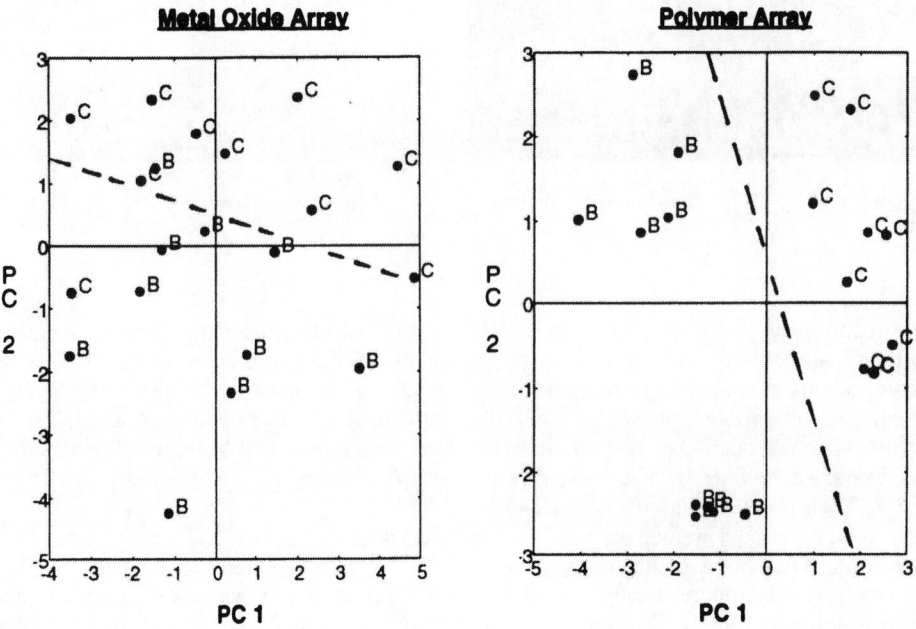

FIG. 4. Discrimination of Brazilian (B) and Colombian (C) coffee beans by metal oxide and polymer electronic noses

Fig. 5. Discrimination of control (*aC*) and tainted (*aT*) lager beers by polymer electronic nose. Two clusters, *A* and *B*, are apparent from data

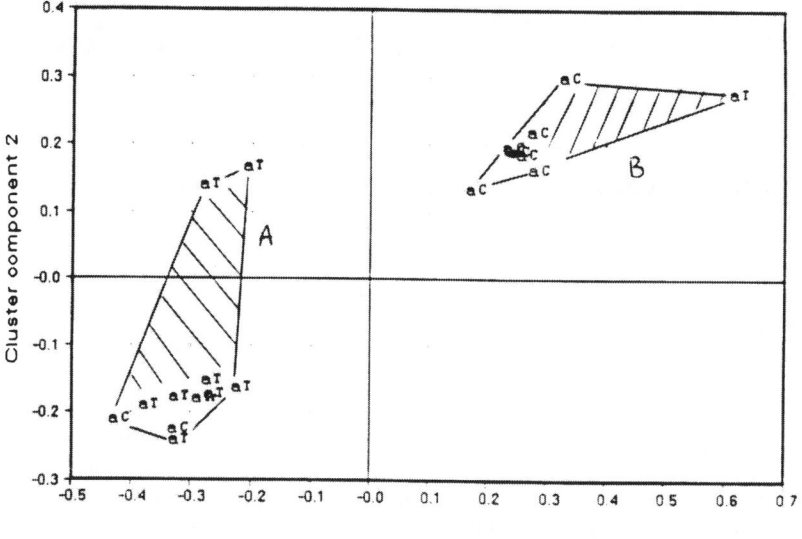

Fig. 6. Photographs of Fox 2000 Intelligent Electronic Nose (Alpha MOS) and Polymer Beer Monitor (Neotronics)

References

1. Dodd GH, Bartlett PN, Gardner JW (1992) Odours—the stimulus for an electronic nose. In: Gardner JW, Bartlett PN (eds) Sensors and sensory systems for an electronic nose. Kluwer Academic Press, Dordrecht, pp 1–13
2. Gardner JW, Bartlett PN (1992) Sensors and sensory systems for an electronic nose. Kluwer Academic Press, Dordrecht, pp 1–327
3. Gardner JW (1991) Detection of vapours and odours from a multisensor array using pattern recognition. I: Principal component and cluster analysis. Sensors Actuators B4:109–116
4. Gardner JW, Hines EL, Tang HC (1992) Detection of vapours and odours from a multisensor array using pattern recognition. II: Artificial neural networks. Sensor Actuators B4:133–142
5. Gardner JW, Shurmer HV, Tan TT (1992) Application of an electronic nose to the discrimination of coffees. Sensors Actuators B6:71–75
6. Gardner JW, Bartlett PN (1992) Pattern recognition in odour sensing. In: Gardner JW, Bartlett PN (eds) Sensors and sensory systems for an electronic nose. Kluwer Academic Press, Dordrecht, pp 161–179
7. Pearce TC, Gardner JW, Friel S, Bartlett PN, Blair N (1993) Electronic nose for monitoring the flavour of beers. Analyst 118:371–377

Odor-Sensing System Using QCM Gas Sensors and an Artificial Neural Network

Toyosaka Moriizumi[1], Takamichi Nakamoto[1], and Yuichi Sakuraba[1]

Key words. Odor sensor—QCM—Artificial neural network

Introduction

In various fields, odor sensors or odor-sensing systems are widely in demand. Gas chromatography is an established way of analyzing odors and can reveal their detailed molecular components. Coffee flavor, for example, was found to be composed of more than 400 kinds of molecules. However, a gas chromatographic measurement cannot identify an odor, at least not directly. Moreover, it has the disadvantage of long measurement time (usually a few hours). In contrast, a human can identify an odor sort (coffee flavor, for example) as soon as it is smelled. Biomimicking would be one of the best ways to realize odor sensors. Odorant molecules are adsorbed at a bilayer lipid membrane in an olfactory receptor, which does not show strong specificity. Output patterns from many receptors with slightly different characteristics are processed by an olfactory neuron network for odor identification.

If we assume that odors have a "molecular spectrum," as shown in Fig. 1b, the sensing and recogmition mechanism in olfaction and the principle of biomimetic odor sensors can be understood schematically. In Fig. 1, the abscissa is the axis for discriminating molecules, for example, molecular weight, retention time in the case of gas chromatography, etc., while the ordinate is that of molecular concentration. S_1, S_2, and S_3 are the response spectra of receptor cells or artificial odor sensors. If those sensing curves cover the entire spectrum range of the odor to be detected, the overall spectrum can be caught by the group of sensors S_1, S_2, and S_3, and can be reproduced from their output signals. Figure 1a shows a gas sensor. Let us assume gas M_1 is an objective gas to be detected. However, as the selectivity of a gas sensor is generally low, another kind of gas molecule, such as M_2, is also detected by S_1. To subtract the signal M_2 from S_1, another sensor, such as S_2, is necessary, but results in signals

M_3 and M_4. Consequently, a third sensor, S_3, is required. In conclusion, recognition of the output pattern from an array of gas sensors with partially overlapped specificity is effective for both gas and odor identification.

The first study following this scheme was reported with semiconductor gas sensors [1], and other research was conducted to identify gas sorts using quartz resonators (in other words, QCM, quartz crystal microbalance) [2], SAW sensors [3], and electrochemical cells [4]. The recognition algorithm in those studies was that of multivariate analysis. We developed a system for odor identification using a QCM array and artificial neural network (ANN) pattern recognition, and showed that odors of alcoholic beverages [5–7], fruit flavors, and fragrances can be discriminated with high recognition probabilities [8]. Moreover, we developed a new ANN algorithm named fuzzy learning vector quantization (FLVQ) [9]. The application of FLVQ to an odor recognition system made it possible to achieve a high recognition rate and distinguish an unlearned category from learned ones [7]. We are now studying the expression of odor sensory quantity by FLVQ. In this report, our achievements in both odor-sensing systems and the FLVQ algorithm to express sensory quantity are reviewed.

Odor-Sensing System

System Fundamentals

Our system, shown schematically in Fig. 2, consists of a standard air supplier, odor sampling part, temperature stabilizer, QCM sensors, electronic circuits, and computers. Odorous vapor in the headspace of a sample bottle is carried by the clean air flow into the sensor cell, where eight QCM sensors are installed. Two kinds of automatic systems have been developed for odor sampling. In the first version, a flow-pipe network with solenoid valves was set over several sample bottles, and clean and odorous air flows were obtained alternately by switching the valves under computer control [10]. In the next version, a robot arm with two needle tubes is used to sample odorants automatically by inserting the needles into sample vial bottles [11]. The alternating

[1] Department of Electrical and Electronic Engineering, Faculty of Engineering, Tokyo Institute of Technology, Meguro-ku, Tokyo, 152 Japan

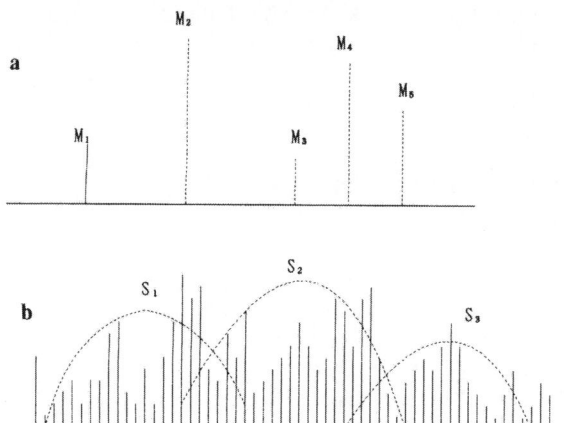

FIG. 1. Schematic diagram of "molecular spectrum" for gas (**a**) or odor (**b**)

flow changes of clean air and odorous vapor are also performed to refresh the sensors by odorant desorption.

The quartz resonators used here are AT cut with fundamental resonant frequencies of about 10 MHz and with sensing films coated over the electrode surfaces. During the sensing process, the films adsorb odorant molecules, giving rise to decreases in resonant frequency. The films are deposited by a casting method, and the thickness is also checked by the resonance frequency decrease. The values of resonator Q were kept above 5400 in this study.

The film materials are those of gas chromatographic stationary phase, cellulose, and lipid. Okahata et al. reported that a lipid film was effective in detecting odorant materials [12]. We have also found lipid films promising but not universally suited for odor discrimination. The materials were selected so that a high recognition probability could be obtained when a group of samples is to be discriminated. The selection utilized the statistical

index, partial F of Wilks' lambda, to evaluate pattern separation [13], or a hierarchical clustering method.

The QCM sensors were connected to oscillation circuits and their frequencies were measured using frequency counters. A personal computer read out the frequency data through an I/O interface. The ANN was realized using software, and its program was installed into the workstation. The network (back propagation, BP [14]) consisted of eight input units, seven hidden ones, and five or six output units. First, every sample was measured ten times, and the data were transferred from the personal computer to the workstation, where the network was initially trained 5000–20000 times using those data. During the identification measurements, the adaptive training (500 times) was performed regularly to compensate the data drift [15].

After initial studies [5], various kinds of measurement system modifications were performed [6,16]: air-flow and temperature (22°C) stabilizations, cleaning of the flow network, and raising the reproducibility of sample vapor supply. After those modifications, the coefficient of variation of the measurement data, which is defined as the ratio of the standard deviation to the average, was reduced from 3% to 0.5%.

Experiments

Identification experiments were performed on the following groups of samples.

Whisky aroma identification. The five Japanese whiskies that were tested have aromas too similar for most people to distinguish easily [10]. Liquors are mainly composed of ethanol and water, and the rest of the components (<1% by weight) contributes to the identification. Therefore, one sample signal was used as a reference signal and the data were input to the neural network after subtracting the

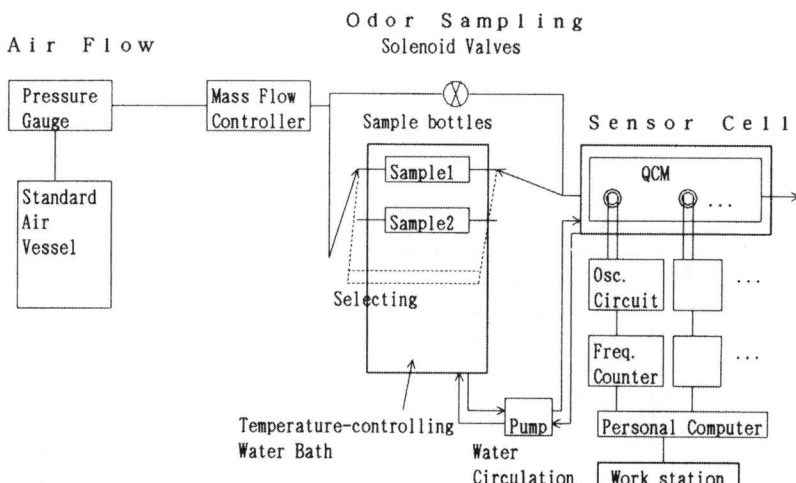

FIG. 2. Schematic diagram of odor-sensing system. *QCM*, quartz crystal microbalance

reference signal from the original data so that the subtle differences between the samples could be enhanced. When ten measurements were performed for every sample, the average recognition probability, as high as 94%, was obtained. The details have been described elsewhere [6,7].

Fragrance and Flavor Identification. Typical perfumes such as citrus cologne, floral bouquet, chypre, modern bouquet, and oriental, and flavor samples such as orange, strawberry, apple, grape, and peach were also examined by this system. The results of principal component analysis showed that the differences between the output patterns from the sensor array were large for both perfume and flavor samples, and the coefficient of variation of the data was as small as 0.34%. In agreement with the results, the recognition probability by the neural network was 100% [8].

Experiments for discriminating subtle differences between flavors were performed using orange flavors with small amounts of a foreign substance (2-butylidene cyclohexanone, 2-BC) as samples. The detection limit of 2-BC concentration was found to be 0.5 vol% from both ANN and principal component analyses. The sensor with an ethylcellulose film showed the highest contribution to the discrimination, and its output is interestingly proportional to 2-BC concentration (about $-20\,\mathrm{Hz/vol\%}$) [10]. This result revealed that our system can also be used for an odor quantitative measurement (something like an odor scale).

Essential Oil. The study of various essential oils by this system has been performed and published [11].

Discussion

The improvements of the recognition probability have so far been carried out by both enlargement of the distance between data categories and suppression of data scattering. They can be achieved by film selection and the various system modifications, respectively. The details have been discussed in [7].

It should be pointed out that the present system needs appropriate reference samples to discriminate samples with subtle odor differences, and the measurement is not absolute but relative. This situation is analogous to human olfaction. However, there is a clear difference between human and artificial "noses." In human olfaction, odors are recognized and discriminated according to our memory if the odor difference is large. If it is small, however, we cannot discriminate plural samples just by sniffing; we sniff and compare them several times. Subtle odor differences cannot be recognized using our long-term memory, but must be discriminated by means of sensory memory, which is kept for only a few tenths of a second. In contrast, the signals from the sensor array in the present system can be

kept without any deterioration almost permanently and processed precisely for recognition in a computer. This is the absolute advantage of an artificial odor-sensing system, which should be emphasized.

Expression of Odor Sensory Quantity by FLVQ

Learning vector quantization (LVQ) is known as a network to extract data topology by supervised learning using an Euclidean distance [17]. The model of FLVQ is based on a neuron in the same way as that of LVQ. However, a FLVQ neuron deals with a fuzzy quantity using a membership function, and has a reference vector to output the fuzzy similarity between the reference vector and an input vector. If there are several categories in the input data space, each category consists of several neurons. The category output is the maximum value among neuron outputs in the same category, and refers to the similarity between the input and category data. When all the category outputs are zero, the input vector is considered to belong to an unlearned category. This is the way to distiguish an unlearned category from learned ones [9].

In addition to its categorization capability, FLVQ has been found suitable for relating the sensor output with human olfactory quantity [18]. The sensory test was performed to investigate human sensory quantity. Although there are several methods of sensory tests, we chose the semantic differential method, and enumerated the relationship between samples. Ten females in their teens and twenties were tested using 25 kinds of natural essential oils as the odor samples. Subjects smelled each flavor and graded it on a five-point scale in terms of 23 characteristic adjectives. After the sensory test, the similarities between the flavor samples were estimated from the scores to the adjectives using a conventional similarity equation.

FLVQ had 25 categories, each of which consisted of 25 neurons (in total, 625 neurons). The network was trained 200 times to input the sensor signals and use the flavor similarities for the desired outputs. For comparison with FLVQ, a BP network was also used. The BP network was composed of 8 neurons for the input layer, 10 neurons for the hidden layer, and 25 neurons for the output layers; the number of learning processes was 5000. We determined the number of neurons for the hidden layer experimentally to optimize the network performance. The root mean square error of FLVQ was 0.147 per input data whereas that of BP was 0.317. For example, the result on spearmint after the learning process is shown in Fig. 3; the height of each bar corresponds to the similarity of another flavor to spearmint. Figure 3 confirms that the estimation accuracy of FLVQ was better than that of BP, and

FIG. 3. Similarity of various odors to spearmint obtained from fuzzy learning vector quantization (*FLVQ*) after supervision by human sensory tests. *Solid bars*, desired output; *open bars*, FLVQ; *hatched bars*, backpropagation (*BP*)

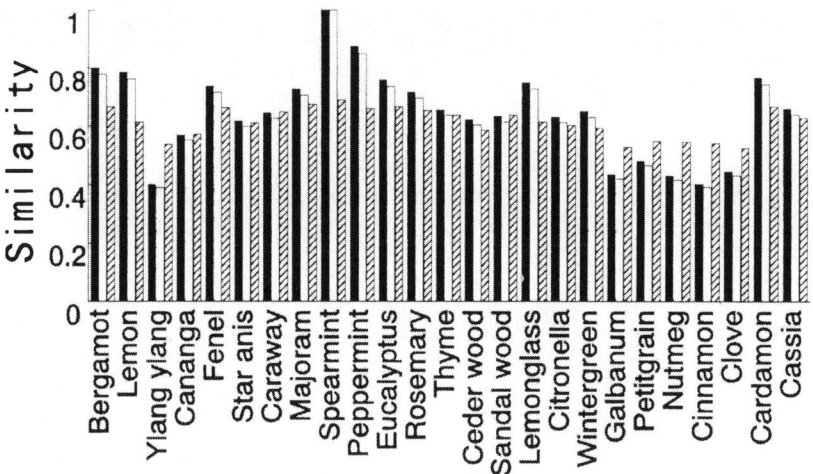

odor similarity mapping or construction of odor space might be possible following this approach. In other words, FLVQ can realize mapping from the sensor data space onto odor sensory space.

Conclusion

Several simple odor-sensing systems have already been commercialized by several companies. The authors believe that the present system is more sophisticated in its measurement reliability and intelligence because of the ANN algorithm. Further progress in the development of these systems will require additional studies on sensing materials and devices, flow system designing, and algorithms.

Acknowledgments. The authors express many thanks to J. Ide of T. Hasegawa Co., Ltd., for offering us perfume and flavor samples and giving us helpful advice. We are also grateful to the Japanese Ministry of Science and Education for Grant-in-Aid 03302036.

References

1. Persaud KC, Dodd GH (1982) Analysis of discrimination mechanisms in the mammalian olfactory system using a model nose. Nature (Lond) 299:352–354
2. Carey WP, Beebe KR, Kowalski BR (1986) Selection of adsorbates for chemical sensor arrays by pattern recognition. Anal Chem 58:149
3. Ballatine DS, Rose SL, Grante JW, Woltjen H (1986) Correlation of surface acoustic wave device coating responses with solubility properties and chemical structure using pattern recognition. Anal Chem 58:3058
4. Stetter JR, Jurs PC, Rose SL (1986) Detection of hazardous gases and vapors: pattern recognition of data from an electrochemical sensor array. Anal Chem 58:860
5. Ema K, Yokoyama M, Nakamoto T, Moriizumi T (1989) Odour-sensing system using quartz-resonator sensor array and neutral-network pattern recognition. Sens Actuators 18:291–296
6. Nakamoto T, Fukuda A, Moriizumi T, Asakura Y (1991) Improvement of identification capability in odor sensing system. Sens Actuators B3:221–226
7. Moriizumi T, Nakamoto T, Sakuraba Y (1992) Pattern recognition in electronic noses by artificial neural network models. In: Sensors and Sensory Systems for an Electronic Nose. Kluwer Academic Publishers, Netherlands, pp 217–236
8. Nakamoto T, Fukuda A, Sasaki S, Moriizumi T (1993) Perfume and flavour identification by odour-sensing system using quartz-resonator sensor array and neural-network pattern recognition. Sens Actuators B10:85–90
9. Sakuraba Y, Nakamoto T, Moriizumi T (1991) New method of learing vector quantization using fuzzy theory. Syst Comp Jpn 22(13):93–103 (translated from Japanese)
10. Nakamoto T, Fukunishi K, Moriizumi T (1990) Identification capability of odor sensor using quartz-resonator array and neural-network pattern recognition. Sens Actuators B1:473–476
11. Ide J, Nakamoto T, Moriizumi T (1993) Development of odor sensing system using an auto-sampling stage. In: Sens Actuators B13–14:351–354
12. Okahata Y, Shimizu O, Ebata H (1990) Detection of odorous substances by using a lipid-coated quartz-crystal microbalance in the gas phase. Bull Chem Soc Jpn 63(11):3082–3088
13. Nakamoto T, Sasaki S, Fukuda A, Moriizumi T (1992) Selection method of sensing membranes in odor sensing system. Sens Mater 4(2):111–119
14. Rumelhart DE, Hinton GE, Williams RJ (1986) Learning representations by back-propagating errors. Nature (Lond) 323:533–536
15. Nakamoto T, Moriizumi T (1988) Odor sensor using quartz-resonator array and neural-network pattern recognition. Proc IEEE Ultrason Symp 1988:613–616
16. Nakamoto T, Fukuda A, Moriizumi T (1993) Perfume and flavor identification by odor sensing system using

quartz-resonator sensor array and neural-network pattern recognition. Sens Actuators B10:85–90

17. Kohonen T (1987) Self-organization and associative memory, 2nd edn. Springer, Berlin Heidelberg New York

18. Sakuraba Y, Nakamoto T, Moriizumi T (1993) Neural network to express odor sensory quantity. Technical Digest of World Congress on Neural Network '93, INNS Press, pp 70–73

Chemical Vapor Sensor Using a Lipid-Coated SAW Resonator Oscillator

Eiichi Tamiya[1] and Isao Karube[2]

Key words. Biosensor—SAW device—Odorant-sensor—Natural lipids—LB membrane—Neural-network pattern recognition

Introduction

The sensitivity of piezoelectric detectors and surface acoustic wave (SAW) devices is directly proportional to the square of the resonant frequency, and is inversely proportional to the surface area. Therefore, a SAW device is considered to be an excellent transducer for a chemical vapor sensor because the oscillation frequency of a SAW device can be two or three orders of magnitude higher than that of an AT-cut quartz crystal resonator. Moreover, the surface area of a SAW resonator is very small, and thus an increase in sensitivity can be expected for SAW devices. As well as being sensitive, SAW devices can be miniaturized with precise and reproducible characteristics using photolithographic techniques. In addition, the lithographic fabrication capability easily permits a complex circuit to be present on the same crystal surface.

Most SAW devices used as physical, chemical, or biological sensors are based on delay line oscillators. However, a resonator oscillator offers advantages over the SAW delay line, particularly at higher oscillator frequencies [1,2]. In view of these advantages, we applied SAW resonator oscillators to detect odorants from coated lipids on the surface of devices.

In this study the LB technique was applied to produce uniform layers sufficiently thin to ensure that the insertion loss from the mass loading remained low. Phospholipids and a fatty acid were used for the thin coating of SAW devices. The loading was controlled by the frequency shift and the insertion loss. We investigated the properties of a SAW resonator as an odorant sensor [3–5].

[1] Japan Advanced Institute of Science and Technology (JAIST), 15 Asahidai, Tatsunokuchi-cho, Nomi-gun, Ishikawa, 923-12 Japan
[2] RCAST, The University of Tokyo, 3-8-1 Komaba, Meguro-ku, Tokyo, 153 Japan

Materials and Methods

The SAW oscillators fabricated on Y-cut (cut angle, 36°), X-propagation quartz using standard photo lithographic techniques were obtained from Toshiba (Tokyo, Japan). The device consisted of a two-port resonator. The electrodes were metallized using aluminium. The interdigital transducer (IDT) consisted of 55 pairs with a half-wavelength finger. The space between the IDTs was 65 µm because the insertion loss is least at this distance. The resonant frequency of the SAW device, 310 MHz, was determined using a Network/Spectrum analyzer.

The lipids were first dissolved in chloroform (2 mM). The solution was then spread onto a clean water surface. After evaporation of the chloroform, the molecules of the film were compressed by means of a movable barrier until they formed a close-packed structure. A SAW device was then passed through the air–water interface, and a monolayer of lipid was deposited onto the surface of the SAW device. The deposition of monolayers was performed at a surface pressure of 15 mN m^{-1} by the horizontal lifting method. The quality of phospholipids was checked by measuring the isotherms using an LB trough. The speed of the film left was kept at 6 mm min^{-1}.

In this study, a three-layer neural network was used to recognize the various odorants automatically with the back-propagation algorithm. A sigmoid is used as an activation function in the back-propagation algorithm because the sigmoid function compresses the sum of input signals so that the output lies between zero and one. Coefficients for the back-propagation algorithm can be chosen, but here the values as in the NEC Neuro-07 software (NEC, Tokyo, Japan) were adopted [6]. The structure of the neural network is shown in Fig. 1. There were six input layer units, which number corresponds to the number of sensors, and seven output layer units, which corresponds to the number of odorants measured. The teaching signals were composed of character symbols and seven-order data, which correspond to the number of odorants measured. The number of intermediate-layer (called hidden-layer) units is decided by a trial-and-error method to accelerate and improve the convergence of the learning process, but seven units were used.

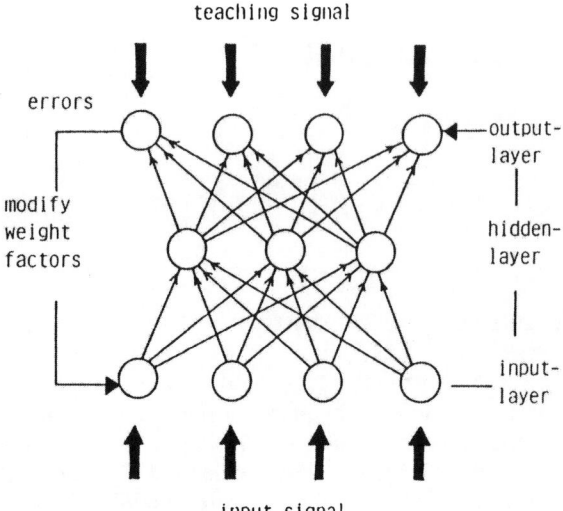

FIG. 1. Hierarchical model of neuron network for back-propagation algorithm

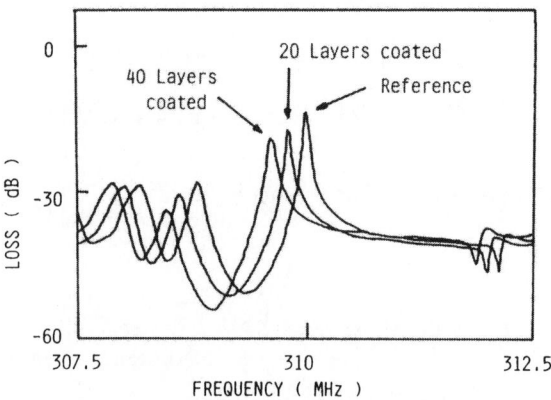

FIG. 2. Typical insertion loss versus frequency for two-port SAW resonator

The initial values of the connecting weight factors in the neural network were random numbers between −1 and 1.

Results and Discussion

The deposition of monolayers was performed by the horizontal lifting method. With this method, it was possible to cover the SAW device and to make it reproducible for as many as 40 monolayers of the X-type structure. The mass loading could be determined with high precision by the frequency change of the SAW device. Figure 2 shows that the typical frequency characteristics are dependent on the number of monolayers deposited. The frequency decreases continuously, while the insertion loss from the dissipation of energy increases with the number of monolayers deposited. If the number of monolayers reached a critical value (between 20 and 30, depending on the surface pressure and molecular weight of the material deposited), then the loss became so high that the frequency characteristics changed markedly. Therefore, by applying the LB technique it was possible to deposit the optimal number of layers, which was found to be between 10 and 20 when the frequency shift was about 200 kHz.

Figure 3 shows a typical response profile obtained from two consecutive on–off exposures to 148 ppm (by volume) of acetoin. From this result, it was shown that the response is quite typical and could repeatedly be duplicated.

Figure 4 shows the correlation between the resonant frequency shift and the odorant concentration of a phosphatidylcholine-coated SAW resonator. The results demonstrate that the lowest

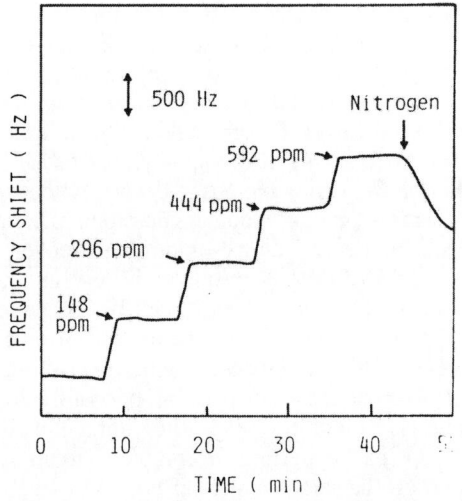

FIG. 3. Typical response of phosphatidylcholine-coated SAW resonator exposed to step increase of 148 ppm of acetoin

concentration required to give a measurable frequency change differs for the various odorants. The sensitivity represents the slope obtained from a linear least-squares fit of replicate data set of responses. The values are about 3 ppm and 66 Hz ppm^{-1} for menthone, 24 ppm and 6.3 Hz ppm^{-1} for amyl acetate, and 35 ppm and 4.0 Hz ppm^{-1} for acetoin, respectively.

The frequency shift is about 20 times that obtained by 9 MHz AT-cut crystal in our laboratory, but the minimum concentration is almost the same. These results were thought to be caused by the high frequency and the small active surface area. However, there are good correlations between these results and the olfactory threshold values in the biological cell [7–9]. The surface pressure of lipid monolayers or the membrane potential of liposome changes with the addition of various odorants in a liquid phase. The structure of odorants is extremely

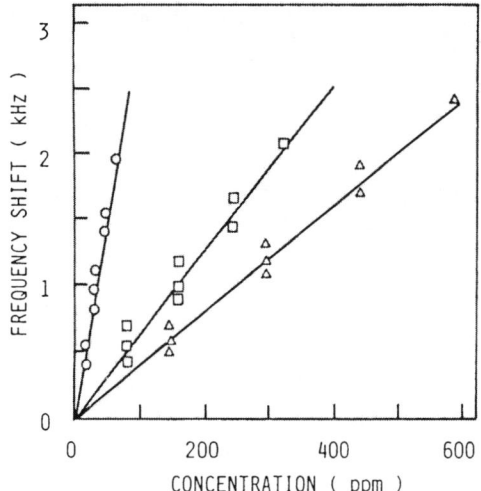

FIG. 4. Correlation between odorant concentration and resonant frequency shift for phosphati dylcholine-coated SAW resonator. *Circles*, menthone; *squares*, amylacetate; *triangles*, acetoin

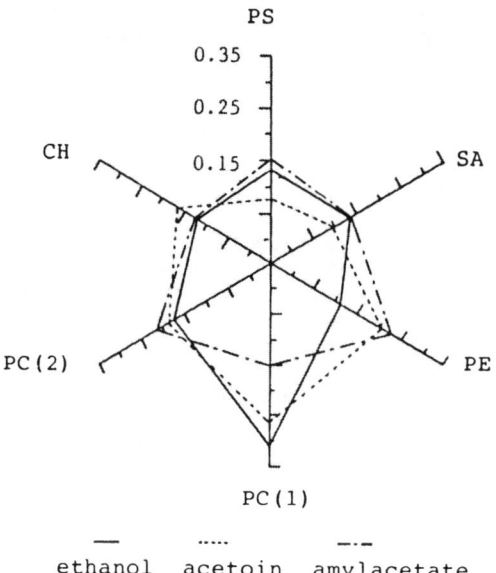

FIG. 5. Normalized patterns of resonant frequency shifts to various odorants. *SA*, stearicacid; *PE*, phosphatidylethanolamine; *PS*, phosphatidylserine; *CH*, cholesterol; *PC*(1), phosphatidylcholine (dicaproyl); *PC*(2), phosphatidylcholine (dipalmitoyl)

diverse, and it is difficult to find the molecular recognition mechanism by lipid layers in terms of chemical structure. However, it was suggested in these reports [7–9] that the changes were induced by odorant adsorption on the hydrophobic region of the membrane, the resultant conformation change causing variations in the surface charge environment. It is considered that these results reflect the equilibrium between the lipid membrane and the gas

or liquid phase. From these results, it follows that there is a good similarity between the mechanism of lipid monolayer surface pressure change, liposome membrane potential change, and the resonant frequency change of the SAW resonator.

Other lipids were also coated on the SAW device and used as odorant sensors. The responses were different for each lipid; therefore, the frequency shift of the different lipids for each odorant was represented by a pattern. The patterns cannot be compared directly with each other because the vapor concentrations differ. Normalization is necessary for comparison. The results after the normalization procedure are shown in Fig. 5. The pattern itself is specific and distinct for each odorant, so the normalized pattern can be used to identify odorants. On the basis of the monolayer properties of phospholipids and fatty acids, an explanation can be given for the differences in odorant affinity. However, the structure of the odorants is extremely diverse and it is difficult to establish the mechanism of molecular recognition by the lipid layers on the chemical structure level.

From the foregoing, it follows that a lipid-coated SAW resonator can monitor different odorants. Using a number of different lipids for coating the surfaces of SAW resonators, odorants can be identified by a computerized pattern recognition algorithm.

On the basis of these normalized patterns, the patterns were processed using a three-layer neural network. The procedures and results of the learning process are shown in Fig. 6. The normalized patterns are transferred to the input layer as input signals, and the teaching signals are composed of character symbols and seven-order data, which are "on" (1) at the position of the odorant concerned, and "off" (0) at other positions. The position of methanol is "on" and other positions are "off" in Fig. 6. The shaded areas indicated by the squares for the units are proportional to the activity levels, which are accumulated by the conversion function of the last layer. The input signals and teaching signals are processed and connected into the neural network repeatedly, and the neural network is ordered so that the output layer indicates the accurate character symbol as a result of training. As shown in Fig. 7, the output layer at first indicates a random symbol, but eventually indicates the accurate symbol according to the learning processes.

When the input signals of the seven kinds of odorants using responses of seven pure compounds were placed in the network, the output layer showed the activity levels and were indicated by the character symbols. The fact that the relatively large squares are aligned on the diagonal and that the relatively accurate symbols are indicated means that this system can be used for the pattern recognition of odorants.

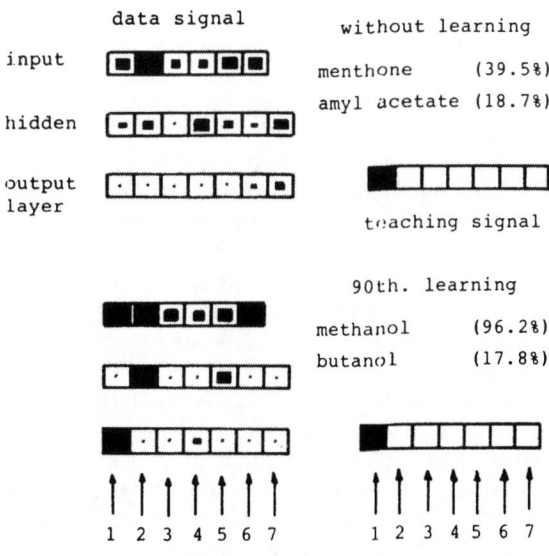

FIG. 6. Schematic diagram of learning process: 1, methanol; 2, ethanol; 3, *n*-propanol; 4, *n*-butanol; 5, acetoin; 6, amylacetate; 7, menthone

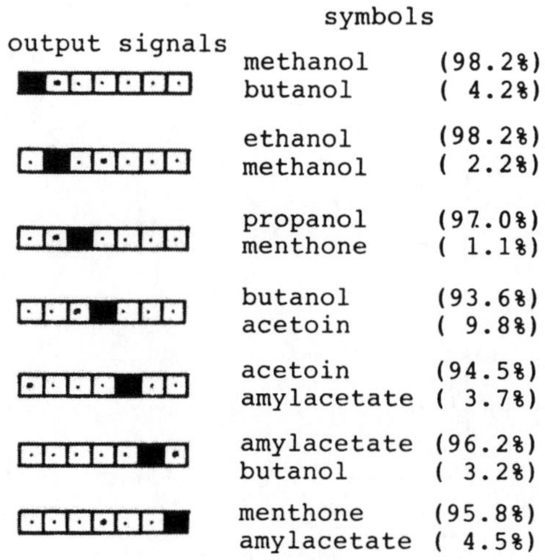

output signals	symbols	
▪░·░·░░	methanol	(98.2%)
	butanol	(4.2%)
░▪░·░·░	ethanol	(98.2%)
	methanol	(2.2%)
░░▪░·░·░	propanol	(97.0%)
	menthone	(1.1%)
░·░▪·░·░	butanol	(93.6%)
	acetoin	(9.8%)
░·░·▪░·░	acetoin	(94.5%)
	amylacetate	(3.7%)
░·░·░▪░	amylacetate	(96.2%)
	butanol	(3.2%)
░·░·░·▪	menthone	(95.8%)
	amylacetate	(4.5%)

FIG. 7. Results of output layer units after 250 learning processes

Using this algorithm, odorants could be recognized at greater than 90% probability, which was primarily caused by the deviation of the sensor responses even if pure compounds were used. From these results, it follows that a multichannel lipid-coated SAW devices can monitor different odorants. Using a number of different lipids for the coating of surfaces of SAW, odorants can be identified by a neural-network pattern recognition algorithm, where transformation of a nonrecognizable signal to a recognizable symbol is called pattern recognition.

References

1. Ash EA (1978) Fundamentals of signal processing devices. In: Oliner AA (ed) Acoustic surface waves. Springer, Berlin Heidelberg New York, pp 115–123
2. Morgan DP (1985) Surface-wave devices for signal processing, Elsevier, Amsterdam
3. Chang SM, Tamiya E, Karube I, Sato M, Masuda Y (1991) Odorant sensor using lipid-coated SAW resonator oscillator. Sensors Actuators B5:53–58
4. Change SM, Tamiya E, Karube I (1991) Chemical vapour sensor using a SAW resonator. Biosens & Bioelectronics 6:9–14
5. Change SM, Ebert B, Tamiya E, Karube I (1991) Development of chemical vapour sensor using SAW resonator oscillator incorporating odorant receptive LB films. Biosens & Bioelectronics 6:293–298
6. Chang SM, Iwasaki Y, Suzuki M, Tamiya E, Karube I (1991) Detection of odorants using an array of piezoelectric crystals and neural-network pattern recognition. Anal Chim Acta 249:323–329
7. Nomura T, Kurihara K (1987) Liposomes as a model for olfactory cells: Changes in membrane potential in response to various odorants. Biochemistry 26:6135–6140
8. Nomura T, Kurihara K (1987) Effects of changed lipid composition on responses to liposomes to various odorants: possible mechanism of odor discrimination. Biochemistry 26:6141–6145
9. Muramatsu H, Tamiya E, Karube I (1989) Detection of odorants using lipid-coated piezoelectric crystal resonators. Anal Chim Acta 225:399–408.

A Lipid-Coated Quartz Crystal Microbalance as an Olfaction Sensor

Y. Okahata[1]

Key words. Olfaction sensor—Lipid-cast films—Quartz crystal microbalances—Frequency changes—Adsorption behavior—Olfaction threshold

Introduction

The olfactory reception of various odorants and perfumes is thought to occur when molecules are received at the olfactory cell membrane, thus producing electric signals in our bodies that travel to the brain. [1,2]. However, the molecular mechanism of the reception of chemical substances and the translation process in olfactory cells are not well understood at present. The chemical structures of odorants are extremely diverse, and it is difficult to find any one chemical structure that is common to those substances.

Odorant molecules are relatively hydrophobic, and the threshold concentration of general odorants is roughly determined by the partition coefficients between water and oil [2]. Kurihara and coworkers reported that the surface pressure of the lipid monolayer at the air–water interface from bovine olfactory epithelium [3] from synthetic phospholipids is selectively changed by the addition of various odorants in the aqueous phase. Their response magnitude strongly correlates with olfactory reception in humans [3–5]. These results suggest that the first step of olfactory reception takes place on the adsorption of odorous substances at a lipid bilayer matrix without any specific receptor proteins in biological cells [1].

These biological results prompted us to study the partition process of various odorants and perfumes in a lipid matrix by using a quartz crystal microbalance (QCM) coated with a synthetic multibilayer-immobilized film in the gas phase. QCMs are known to provide very sensitive mass-measuring devices at nanogram levels because the resonance frequency changes sharply on deposition of a given mass on the electrode of the quartz plate. The experimental setup is shown in Fig. 1.

[1] Department of Biomolecular Engineering, Tokyo Institute of Technology, Nagatsuda, Midori-ku, Yokohama, 227 Japan

Materials and Methods

Preparation of a polyion complex-type synthetic bilayer-forming amphiphile, dimethyldioctadecylammonium poly(styrene-4-sulfonate) $(2C_{18}N^+2C_1/PSS^-)$, has been reported [6,7].

Lipid-Coated QCM

A quartz crystal microbalance (8 mm diameter, AT cut, 9 MHz) was connected to a homemade oscillator designed to drive quartz at its resonance frequency and was driven at 5 V dc [6–10]. The frequency of the vibrating quartz was followed by a frequency counter (model SC7201, Iwatsu, Tokyo, Japan) attached to a microcomputer system (model PC9801, NEC, Tokyo, Japan). Calibration of the QCM used in our experiments obeyed a Sauerbrey equation [13] that a frequency change of 1 Hz corresponds to a mass increase of 1.05 ± 0.01 ng on the electrode of QCM [6–10].

$$\Delta m = -(1.05 \pm 0.01) \times 10^{-9} \, \Delta F \qquad (1)$$

A chloroform solution of synthetic $2C_{18}N^+2C_1/PSS^-$ or naturally occurring dipalmitoylphosphatidylcholine (DPPC) bilayer-forming amphiphiles was cast on electrodes on both sides of QCM, dried in air, and aged in a hot-water bath at 60°C for 1 h to prepare well-oriented multilamellar structures in the film. Biological olfactory cells were cast from an aqueous dispersion of human olfactory epithelium. The membrane-coated QCM was set in a 60-ml closed vessel that had been saturated with the vapor of odorous substances by the injection of 2 μl of odorants; the frequency change of the QCM caused by the adsorption of odorants in a lipid matrix was followed with time in the gas phase under stirring.

Results and Discussion

Adsorption Behaviors of Odorants

Figure 2 shows the frequency changes of a $2C_{18}N^+2C_1/PSS^-$ multibilayer film-coated QCM (20 μg, 1.0 μm thick) when the QCM was set in saturated vapor of β-ionone in a 60-ml closed vessel. The concentration of β-ionone was calculated to be

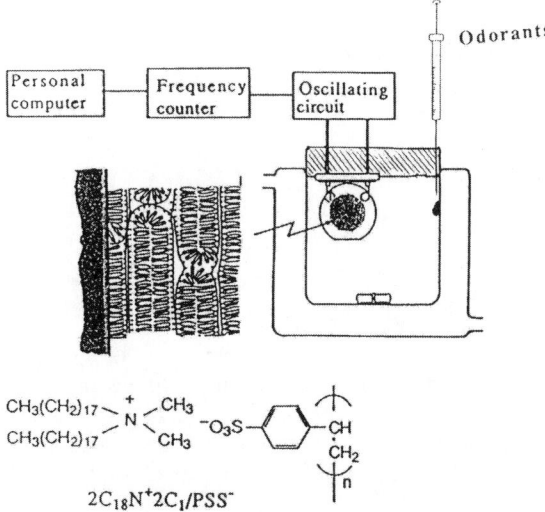

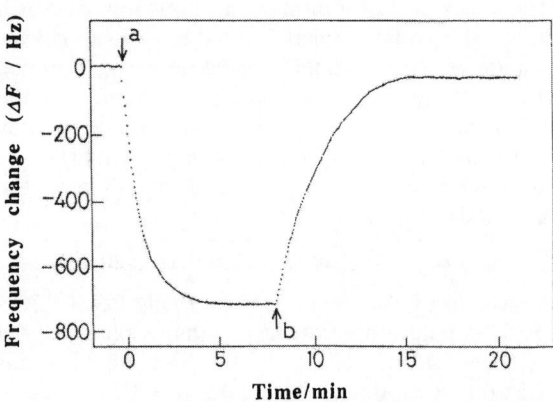

FIG. 1. Experimental setup for frequency measurements of $2C_{18}N^+2C_1/PSS^-$ multibilayer film-coated quartz crystal microbalance (QCM) in gas phase. DPPC-cast film or olfactory epithelium was also employed as membrane on the QCM

TABLE 1. Adsorption amounts (Δm) and partition coefficients (P) of β-ionone in membrane-coated quartz crystal microbalances (QCMs) in gas phase at 25°C[a].

Membranes on QCM	$\Delta m/ng$	$P^b/10^3$
Uncoated	8 ± 2	4.1
$2C_{18}N^+2C_1/PSS^-$ cast film	760 ± 10	390
DPPC-cast film[c]	540 ± 10	280
Olfactory cell membrane	640 ± 10	330
Polystyrene	74 ± 4	38
Poly(vinyl alcohol)	26 ± 3	13
Poly(methyl L-glutamate)	42 ± 5	22
Bovine plasma albumine	32 ± 5	16
Kelatine	28 ± 4	14

[a] Obtained in the saturated vapor of β-ionone ($6.12 \mu g$) in 60 ml of air. The cast amount of membranes was $20 \pm 2 \mu g$ on the QCM.
[b] Contains $\pm 5\%$ of experimental errors.
[c] Dipalmitoylphosphatidylcholine.

FIG. 2. Typical frequency changes of $2C_{18}N^+2C_1/PSS^-$ film-coated QCM (at $20 \mu g$) responding to setting in saturated vapor of β-ionone ($6.12 \mu g/60$ ml of air) at *arrow a* and brought in atmosphere at *arrow b*

$6.12 \mu g/60$ ml of air from the saturated vapor pressure (9.9×10^{-3} mm Hg at 25°C). The frequency of the QCM decreased immediately and reached to the equilibrium ($\Delta F = 720 \pm 5$ Hz) within 5 min, which corresponded to an adsorption of 760 ± 5 ng in a lipid matrix on the QCM from equation 1.

The frequency reverted to the original value after the QCM was removed from the vessel to the atmosphere (*arrow* b in Fig. 2), indicating desorption of β-ionone from the lipid matrix. These reversible adsorption and desorption phenomena could be repeated many times without damaging the membrane and were observed for other odorants and perfumes. The adsorbed amount in the $2C_{18}N^+2C_1/PSS^-$ film on the QCM increased linearly with increasing concentration ($1.0-30 \mu g/60$ ml in air) of odorants. Partition coefficients (P) of the odorants in the lipid matrix were obtained when the concentration (g/l) of the adsorbed substance in the lipid matrix was divided by the concentration (g/l) of substances in the vessel calculated from the saturated vapor pressure.

The adsorption amounts and partition coefficients of β-ionone to various membrane-coated QCMs are summarized in Table 1. β-Ionone adsorbed only sparingly onto the uncoated QCM and hydrophobic [polystyrene and poly(methyl L-glutamate)] or hydrophilic [poly(vinyl alcohol)] polymer-coated QCMs. The partition coefficients in proteins such as albumin and keratin were also very small. In contrast, β-ionone adsorbed specifically onto the lipid bilayer matrix of synthetic $2C_{18}N^+2C_1/PSS^-$ and naturally occurring DPPC-cast films. A similar large adsorption was observed on the olfactory cell membranes in humans on the QCM. These results indicate that odorants tend to adsorb into the lipid bilayer matrices, but not onto proteins, and that the partition behaviors hardly depend on the detailed structure of the hydrophilic head groups; further, the well-packed dialkyl bilayer structure is important for adsorbing odorous substances. Although it is difficult to compare the adsorption behaviors onto simple proteins such as albumin and keratin with that onto membrane proteins in a cell membrane, it seems that the adsorption amount of odorants onto the proteins in the cell membrane is very small compared with that of the lipid matrices; the odor intensity is mainly determined by the adsorption behavior to the lipid bilayers.

Adsorption Behaviors of Perfumes

Adsorption experiments were carried out on six typical perfumes (1-undecanol, p-anisaldehyde, anethol, citral, phenethyl acetate, and benzyl acetate) by using the synthetic $2C_{18}N^+2C_1/PSS^-$ film-coated QCM and the human olfactory epithelium-coated QCM in the gas phase at 25°C. Partition coefficients (P) were calculated by dividing the adsorption amount in the lipid matrix by the concentration of odorants in the vessel. Figure 3 shows a correlation between partition coefficients obtained by QCM methods and the perfume intensity proposed originally by Appell. Appell determined the perfume intensity by smelling vapors from solutions of standard concentration and normalized by setting the intensity of citral to unity [11,12]. Because Appell's intensities were not considered to be the true concentration of perfume in air, we corrected Appell's values by dividing them by the partial vapor pressure of each perfume. We found a good correlation between the log P values of perfumes in the lipid matrix, or the log P values in the olfactory epithelium, and the modified perfume intensity. Perfume with stronger intensities showed higher partitions to the lipid matrix or the olfactory cell membrane on the QCM.

These results agree with the proposal [1,3–5] that the first step of olfactory reception takes place on the adsorption of odorous substances at a lipid bilayer matrix, without any specific receptor proteins in the biological cells. The intensity of olfactory reception is mainly determined by the partition amounts in the lipid matrix of biological olfactory cells. The partition coefficients in the olfactory epithelium are smaller than those in the synthetic

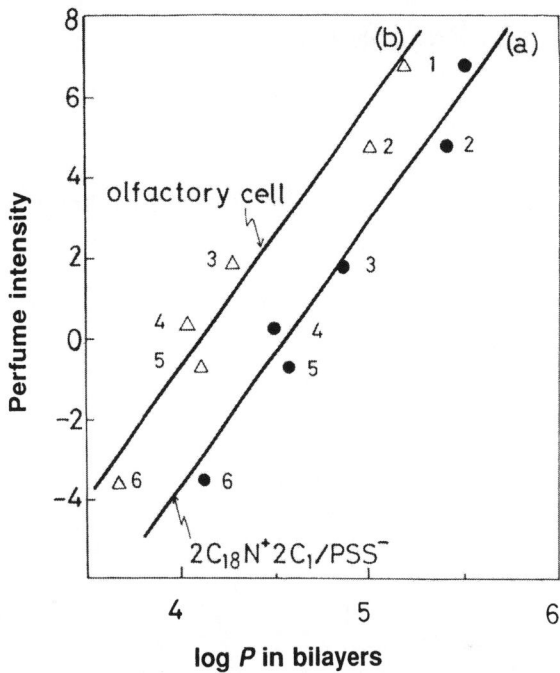

FIG. 3. Relationships between partition coefficients of perfumes [in (a) synthetic $2C_{18}N^+2C_1/PSS^-$ film or (b) olfactory epithelium on QCM] and their perfume intensities in humans. Numbers: 1, 1-undecanol; 2, p-anisaldehyde; 3, anethol; 4, citral; 5, phenethyl acetate; 6, benzyl acetate

$2C_{18}N^+2C_1/PSS^-$ membranes, probably because of the smaller amount of lipid matrix in the cell membrane as compared with a synthetic multibilayer film. These substances adsorbed very little onto the protein matrix on the QCM (see Table 1).

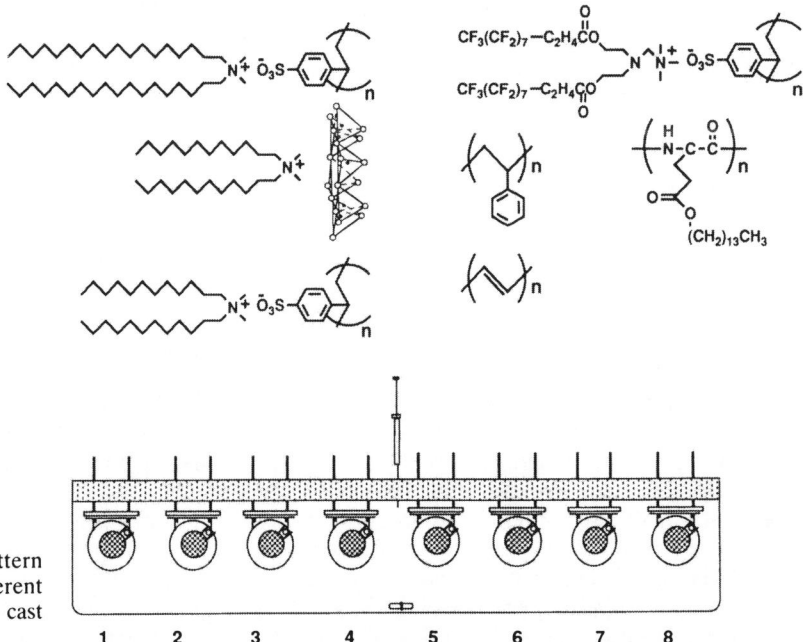

FIG. 4. Eight-QCM system for pattern recognition of odorants. A different lipid or polymer membrane was cast on each QCM

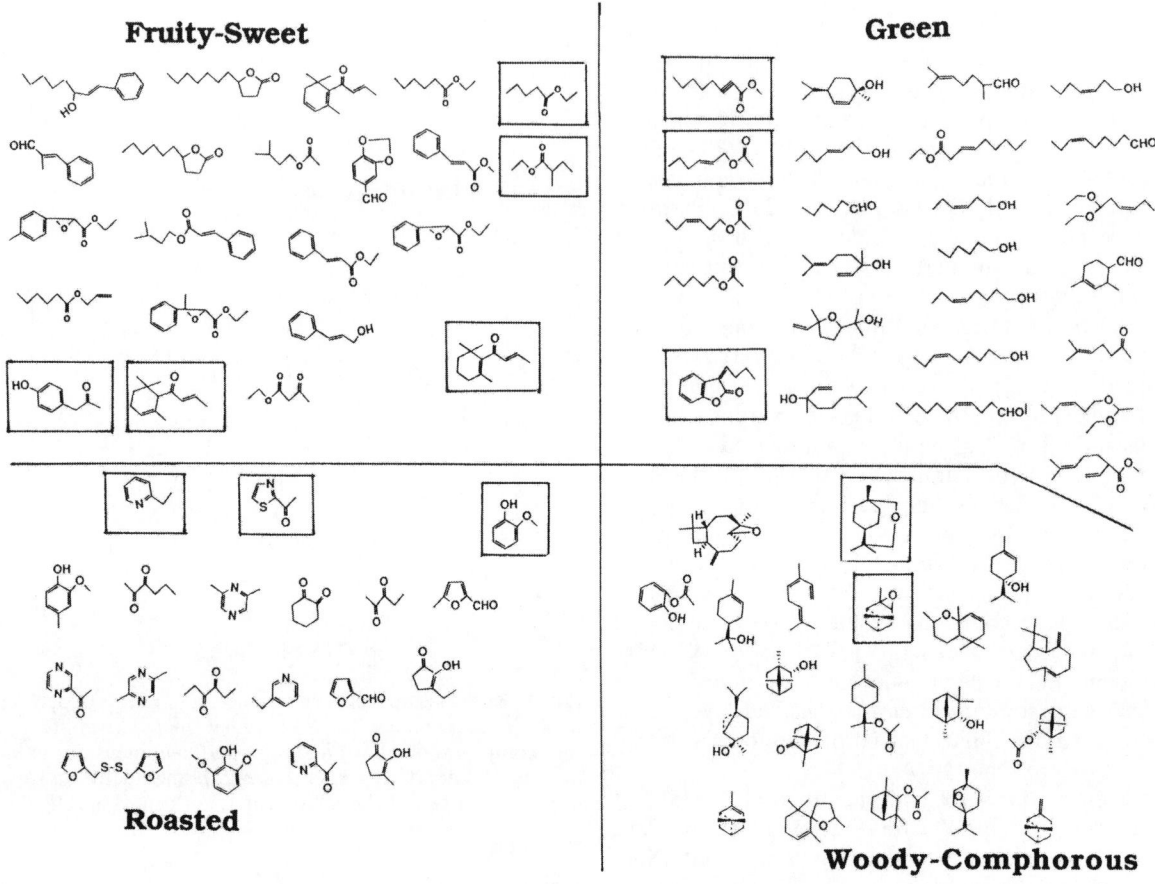

FIG. 5. Chemical structures of 87 perfumes divided into four categories by professional perfumers

TABLE 2. Discriminate analyses for four categories of 87 perfumes by using eight QCMs, each with different membrane[a].

Category	Number of perfumes	Number correct	Percent correct (%)	Number of perfumes classified			
				Fruity-sweet	Green	Roasted	Woody-camphorous
Fruity-sweet	23	16	70	16	2	4	1
Green	25	16	64	5	16	1	3
Roasted	19	16	84	2	1	16	0
Woody-camphorous	20	16	80	2	2	0	16

[a] Chemical structures of membranes and perfumes are shown in Figs. 4 and 5, respectively.

Pattern Recognition of Four Kinds of Odorants

We could estimate the olfaction intensity with partition coefficients to a lipid matrix by using a QCM system. In the second step, we moved to pattern recognition of various odorants depending on chemical structures. We prepared an eight-QCM system with a different lipid membrane cast on each (Fig. 4). We employed 87 kinds of various odorants, which were divided into four kinds of odors such as fruity-sweet (23 kinds), green (25 kinds), roasted (19 kinds), and woody-camphorous (20 kinds) by pro-

fessional perfumers. The chemical structures of odorants are shown in Fig. 5. We obtained partition coefficients of 87 perfumes to each of eight different lipid membranes in the same manner.

The data were analyzed by discriminate analysis using a personal computer (Table 2). Our eight-QCM system can discriminate 87 odorants to four categories in the range of 64%–80% correct. Chemical structures that were incorrectly discriminated by our eight-QCM system are shown in squares in Fig. 5. Green odorants seemed to be unrecognized relatively often in QCM systems because their chemical structures are very similar. When students in our laboratory tried to discriminate these 87 odorants to four categories, their correct percentages ranged from 20% to 75%, and the correct percentages for green odorants were also relatively low compared with other categorites.

Conclusion

The partition coefficients of various odorants and perfumes in the lipid matrix can be easily and quantitatively obtained by using the synthetic multibilayer film-coated QCM in the gas phase. The intensity of various odorants and perfumes can be explained by the adsorption amount to a lipid matrix. The lipid-coated QCM is physically stable and reusable, and provides a new sensor system to determine the intensity and discriminate odorants and perfumes in the gas phase.

References

1. Kurihara K, Yoshii K, Kashiwayanagi M (1986) Transduction mechanisms in chemoreception. Comp Biochem Physiol 85A:1
2. Beildler LM (1971) Handbook of sensary physiology, volumes IV-1. Springer, Berlin, p 322
3. Koyama N, Kurihara K (1972) Effect of odorants on lipid monolayers from bovine olfactory epithelium. Nature (Lond) 236:402
4. Nomura T, Kurihara K (1987) Effects of changed lipid composition on responses of liposomes to various odorants: possible mechanism of odor discrimination. Biochemistry 26:6135, 6141
5. Nomura T, Kurihara T (1987) Liposomes as a model for olfactory cells: changes in membrane potential in response to various odorants. Biochemistry 26:6135
6. Okahata Y, En-na G, Ebato H (1990) Synthetic chemoreceptive membranes: sensing bitter or odorous substances on a synthetic lipid multibilayer film by using quartz-crystal microbalances and electric responses. Anal Chem 62:1431
7. Okahata Y, Ebato H (1991) Adsorption behaviors of local anesthetics into synthetic lipd membranes coated on a quartz-crystal microbalance and relations with their anesthesia. J Chem Soc Perkin Trans 2:457
8. Okahata Y, Shimizu O, Ebato H (1990) Detection of odorous substances by using a lipid-coated quartz-crystal microbalance in the gas phase. Bull Chem Soc Jpn 63:3082
9. Okahata Y, Ebato H (1991) Adsorption of surfactant molecules on a lipid-coated quartz-crystal microbalance: an alternative to eye-irritant tests. Anal Chem 63:203
10. Okahata Y, Ebato H (1992) Detection of bioactive compounds by using a lipid-coated quartz-crystal microbalance. Trends Anal Chem 11:344
11. Appell L (1964) Physical foundations in perfumery. Am Perfum Cosm 79:29
12. Appell L (1969) Physical foundations in perfumery: part VII. Odor tonality 84:45

Odor Evaluation of Foods Using Conducting Polymer Arrays and Neural Net Pattern Recognition

Krishna C. Persaud, Ahmad A. Qutob, Paul Travers, Anna Maria Pisanelli, and Stefan Szyszko[1]

Key words. Conducting polymers—Odor sensing—Food quality—Neural networks

Introduction

An increasing need exists for chemical sensing systems that mimic biological olfaction. Examples of potential applications of such systems are in quality control of foods and beverages and environmental monitoring.

Organic, electrically conducting polymers derived from aromatic or heteroaromatic compounds have been used as chemical sensors [1-3]. It has been observed that many conducting polymers show reversible changes in conductance when chemical molecules adsorb and desorb from the surface. The current understanding of the mechanisms by which odorants or other polar chemicals modulate changes in conductivity of electroactive polymers is still poor. It is believed that these may function as reversible dopants which may modulate structural conformation or charge transfer in the polymers. An array of such materials modified to have differing chemical adsorption characteristics may be applied to generate a pattern that can be used to "fingerprint" different odorants.

A conducting polymer sensor array, consisting of 20 conducting polymer sensing elements was coupled with neural network-based pattern recognition and used to assess the odor quality of raw food materials such as corn and green coffee with respect to human perception.

Methods

The design of the odor-sensing system has been previously reported [3]. A small pump pulls clean air or the odor sample across the sensor array. For the odor headspace of raw food materials, 10–50 g was placed in 250-ml Duran bottles that were left to equilibrate at constant temperature. The odor head-space was then sampled via a two-way valve switching between clean air and the odor sample. The changes in DC resistance of each sensor element in the array are measured at 50-ms intervals via a microprocessor-based data acquisition system. After some preprocessing for signal-to-noise-level optimization, data are transmitted via a serial RS232 link to an external IBM-compatible personal computer (PC) for further processing and display at 1-s intervals. A normalized pattern of responses of each sensor element relative to the entire array was used as the input to a neural network for real-time odor recognition. A strip chart display was used for odor intensity.

Neural Network Pattern Recognition

After evaluating a number of neural network architectures, two-layer neural networks were found to be a good compromise between speed of training and odor discrimination capability. All units in a network are divided into two groups of input units Q_j, which receive the input patterns directly, and output units O_i, which have associated teaching or target inputs. In our application these teaching inputs specify the desired odor classes. Inputs are connected to the outputs via a matrix of weights W_{ij}; this matrix represents the strength of interconnections in a network. The outputs are biased and a vector of biases B_i must be added. Training of a network relies on an appropriate adjustment of the matrix of weights W_{ij} and the vector of biases B_i. The delta rule was applied for calculating the necessary weights at each training epoch.

Results and Discussion

Figure 1a shows the averaged response profile of the entire sensor array when odor from a green coffee was sampled at a rate of $150\,ml\,min^{-1}$. The y-axis measures the averaged percentage change of resistance of the sensor elements. A stable signal was achieved after about 15 s and remained constant as long as the odor stimulus was applied. When clean air was again switched into the system, the

[1] Department of Instrumentation and Analytical Science, UMIST, Manchester M60 1QD, England

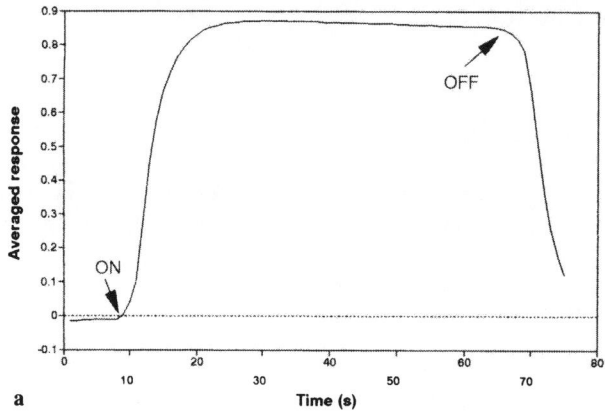

FIG. 1. **a** Kinetic response profile of conducting sensor array to odor of green (unroasted) coffee beans. Plot shows averaged percentage change in resistance observed.

b Relative response pattern obtained for green coffee sample that is used for training neural network to discriminate between different coffee types

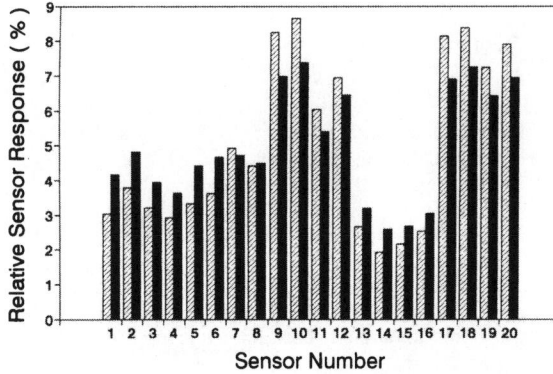

FIG. 2. Comparison of patterns obtained from corn samples evaluated as good (*solid bars*) and bad (*hatched bars*) by odor panel

return to baseline occurred in about 20 s. Relative response patterns were generated at 5-s intervals over the stable part of the odor response profile, and Fig. 1b shows the average pattern obtained for this particular sample. All sensors respond to some extent, but it is clear that some groups of sensors have higher affinities for some odor components in the green coffee sample.

Other types of raw food materials have also been measured, including wheat, corn, and soya. Attempts have been made to correlate the sensor array response with human evaluation of the odor quality. We have observed that differences in the odor patterns exist that can indeed be used effectively. Figure 2 shows the different patterns obtained from two samples of corn, classified respectively as

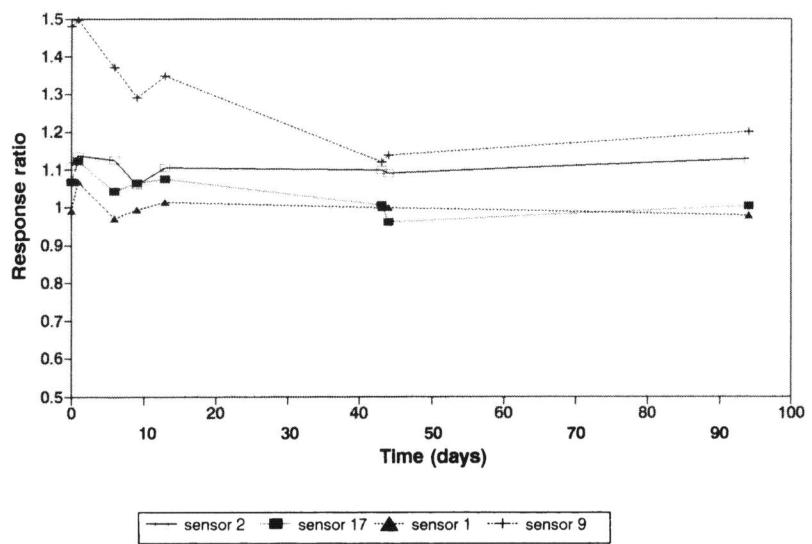

FIG. 3. Corn pattern stability during 3-month period relative to methanol response (standard). For clarity, only normalized responses of 4 sensors of 20-sensor array are shown

good or bad by an odor panel. While the example shown is clear-cut, in general we need to distinguish subtle nuances between samples and to correlate these to human odor perception. To achieve this, we found that it was necessary to collect data from a variety of food material samples over long periods of time and under varied sampling conditions in which the parameters of temperature and humidity were changed. For each odor class to be discriminated, about 150 patterns were collected, and these were used for training a two-layer neural network as described. Generally, each odor descriptor obtained from a human panel was used as an output node of the neural network.

Because it was essential that the trained system be capable of being used over long periods of time, we investigated means of maintaining pattern stability even if sensor drift or changes in temperature and humidity were occurring. Many raw food materials change in odor characteristics with time, and it was necessary to distinguish these changes from any sensor-related changes. At each odor measurement session, the pattern obtained from methanol vapor was recorded and used as a standard. Figure 3 illustrates the record of the response of four sensors in the array to corn odor relative to the methanol response during a 3-month period, the corn samples being stored at 4°C between measurements. Larger changes are observed in the responses of some sensor elements than others, but the patterns generated after normalization in this way are generally stable and any changes in the sample itself are easily followed.

Conclusion

Measurements of the odor headspace of raw food materials or final products are extremely difficult because a large number of chemical components are present. Subtle changes in the relative concentrations of such components may cause large changes in the human perception of odor quality. The system that we have constructed now allows a rapid objective assessment of odor quality to be achieved using instrumental methods. Robust, real-time odor classification can be obtained over long periods. This system will be of use in preliminary screening of raw materials for the food industry.

Acknowledgments. This work was supported in part by the Science and Engineering Research Council, UK.

References

1. Pelosi P, Persaud KC (1988) Gas sensors: towards an artificial nose. In: Dario P (ed) Sensors and sensory systems for advanced robots, NATO ASI Series vol F43. Springer, Berlin, pp 49–70
2. Persaud KC (1992) Electronic gas and odour detectors that mimic chemoreception in animals. Trends Anal Chem 11:61–67
3. Persaud KC, Travers P (1991) Multielement arrays for sensing volatile chemicals. In: Intelligent Instruments and Computers. New York, Elsevier, pp 147–154

Discrimination of Food Aromas by Applying Chemometric Pattern Recognition to Gas Sensor Array

Tetsuo Aishima[1]

Key words. Gas sensor array—Aroma—Food—Chemometrics—Pattern recognition—Artificial neural network

Introduction

Although pattern recognition applied to gas chromatography (GC) data can objectively discriminate aromas, as is widely used in food analysis, the mammalian olfactory system performs a similar function without separating individual components. All sensors developed so far are nonselective in nature; however, if different gas sensors are integrated to make a sensor array, a function similar to the mammalian olfactory system can be expected because of the intrinsic nonselectivity in our odor receptors. Since the first successful report on discriminating odor compounds using a multisensor system by Persaud and Dodd [1], some attempts to discriminate aromas using gas sensor arrays and subsequent pattern recognition have been reported, using quartz resonator arrays (Ema et al. [2] and Nakamoto et al. [3]) or semiconductor gas sensor arrays (Aishima [4,5] and Gardner [6]).

In this study, an array composed of six TGS gas sensors was installed in a glass flask to construct an "artificial nose." Purge-and-trap concentration of aroma components using porous polymers such as Tenax TA or Porapack Q has been widely used in aroma analysis. By using such polymers, most aroma components in the headspace can be adsorbed but water, ethanol, and methanol vapors pass through them. The adsorbed volatiles can be reversibly desorbed by heating the polymers. Thus, we can concentrate food aromas without using any solvents. To prevent the effects of overwhelming ethanol content on sensing and to standardize the analytical conditions, aroma components adsorbed on Tenax TA were introduced into the sensing flask. Statistical and artificial neural network (ANN) pattern recogni-tion techniques were applied to the resulting sensor responses.

Materials and Methods

Gas Sensing System

The scheme of the gas sensing system is shown in Fig. 1. The six TGS semiconductor gas sensors, TGS813, TGS812, TGS711, TGS800, TGS815D, and TGS712D, were supplied by Figaro Sensors, Inc. (Minoo, Osaka, Japan). One sensor array was installed in the three-necked 5-l sample flask and another array was used as reference sensors. The circuit voltage for all sensors was held at approximately 2V; the heater temperature was about 350°C. The resistance decreased when reducing gas contacted the surface of the gas sensors. Amplified differences between sample sensors and reference sensors were recorded. After each measurement was finished, the flask was ventilated with purified air.

Materials

Essential oils of 14 fruits

Coffee: ground coffee beans (*Coffea arabica* and *C. robusta*), and spray-dried and freeze-dried instant coffees

Liquors: shochu (Japanese spirits), German white wine, French red wine, Japanese lager beer, sake (rice wine), gin, bourbon whiskey, Canadian whiskey, Scotch whiskey, Japanese whiskey, and cognac

Sample Pretreatment for Sensing

A Tekmar LSC 2000 headspace concentrator was used for the pretreatment of all sample aromas. Sample aroma was purged with N_2 gas for 15 min at 40 ml/min at room temperature. The volatiles trapped on a mixture (~150 mg) of Tenax TA and silica gel were dry purged with N_2 gas for 35 min at 40 ml/min. Aroma desorption from the trap was performed at 180°C for 4 min. The desorbed aroma was introduced into the flask with N_2 gas at 40 ml/min

[1] Research and Development Division, Kikkoman Corporation, 399 Noda, Noda, Chiba, 278 Japan

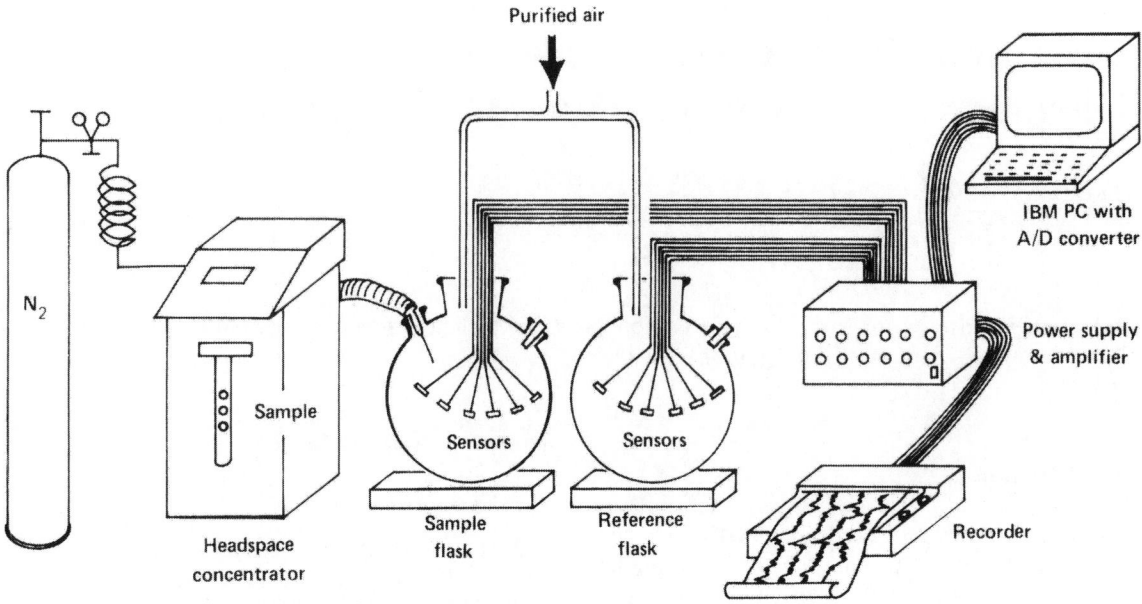

FIG. 1. Aroma sensing system using gas sensor arrays. (From [4], with permission)

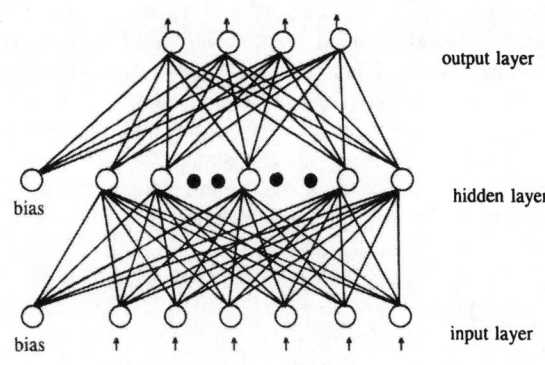

FIG. 2. Three-layer network applied for pattern recognition of gas sensor responses

through a fused silica capillary column heated at 100°C.

Pattern Recognition

The mean heights of sensor responses for 30 min were calculated for each of the six sensors. Statistical and ANN pattern recognition analyses were applied for the resulting six-dimensional data matrix using SPSS PC+ Ver 4.0 and NeuralWorks Professional I software, respectively, on an IBM PS/55 5551T computer with a math coprocessor 80387. For the ANN pattern recognition, the back-propagation algorithm was utilized on three-layer networks (Fig. 2).

Results

Sensor Responses

The six sensors responded immediately when desorbed volatiles were injected into the sample flask. The responses from the six sensors were similar to each other. No apparent differences were observed in their pattern comparison among liquors or coffee samples. However, large differences are not a necessary condition for discriminating patterns so long as their repeatability is within the reliable range.

Coffee

Cluster analysis was applied to the normalized data matrix of sensor responses. The resulting dendrogram suggested clear separation of four groups, that is, *Coffea arabica*, *C. robusta*, and freeze-dried and spray-dried instant coffees. Each of six cultivars in *C. arabica* and two cultivars in *C. robusta* also formed individual clusters (Fig. 3).

TGS812 was selected as the most effective sensor to discriminate coffee aromas when stepwise linear discriminant analysis (LDA) was applied for ground coffees and instant coffees. By use of TGS812 alone, 78% correct sample classification was attained. When TGS813 was added at step 2, the ratio of correct responses increased to 96.7%. Perfectly correct classification was attained by adding TGS800 and TGS815D at steps 3 and 4, respectively. Thus a combination of only three or four sensors can discriminate two ground coffees and two instant coffees.

After 6000 iterations with 10 units in the hidden layer in the training process, a ratio of 96.7% correct

FIG. 3. Clustering of samples in eight coffee cultivars. *Coffea arabica*: *M*, *T*, *B*, *S*, *A*, and *G*; *C. robusta*: *P* and *W*

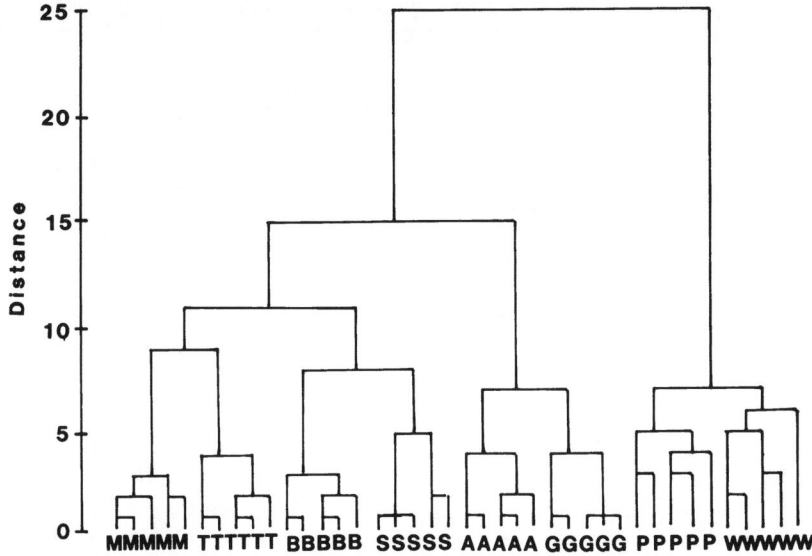

classifications was attained. By increasing the number of iterations to 21 000, all samples were correctly assigned.

Liquors

Cluster analysis was applied to eight different liquors, each composed of three replicate response patterns (Fig. 4). First, three sensor responses of each liquor made their own clusters, then they merged to make three large clusters. The four liquors in the first cluster, cognac, shochu, and red and white wines, contain a considerable amount of terpenes. Three liquors in the second cluster, beer, sake, and whiskey, are made from grains. Considering its peculiar aroma, the uniqueness of gin may be understandable.

Cluster analysis and LDA was applied to the responses for bourbon, Scotch, Canadian, and Japanese whiskeys, and cognac. Clusters corresponding to each of five spirits were observed. TGS815D was selected at the first step in LDA followed by TGS813, TGS712D, TGS800, and TGS812. Although aroma characters were not similar, one Japanese whiskey was falsely assigned to Canadian whiskey.

After 10 000 iterations in the training process, the correct classification ratio was 100% (50:50) for the training set and the ratio in the test set was 92% (23:25) (Table 1).

Essential Oils

Cluster analysis applied for sensor responses for the 14 different essential oils suggested the existence of two distinct groups (Fig. 5). One cluster is composed of only citrus essential oils; another contains apple, peach, melon, banana, and strawberry essential oils.

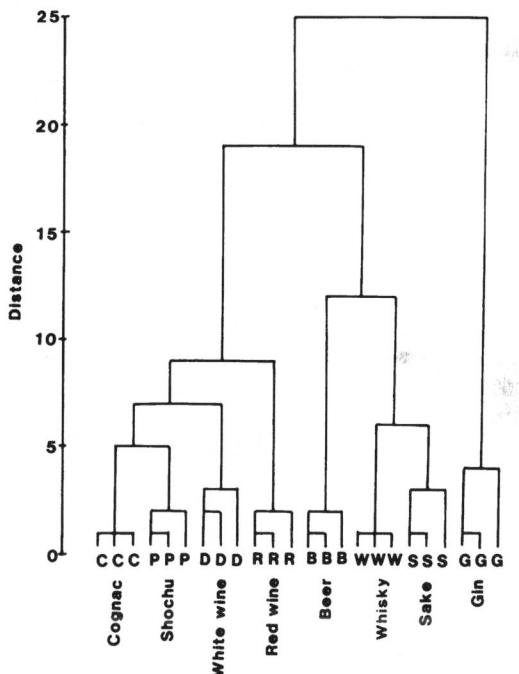

FIG. 4. Clustering of eight liquors based on responses from six sensors. (From [4], with permission)

We cannot statistically assess the appropriateness of the clustering results but this classification seems acceptable based on aroma characteristics of essential oils.

Discussion

Ethanol itself does not have a smell detectable by the human nose, but all gas sensors sensitively

TABLE 1. ANN pattern recognition results of five spirits.

	Classification ratios				
	Cognac	Canadian	Scotch	Bourbon	Japanese
Cognac	10 (5)[a]	0 (0)	0 (0)	0 (0)	0 (0)
Canadian	0 (0)	10 (4)	0 (1)	0 (0)	0 (0)
Scotch	0 (0)	0 (0)	10 (5)	0 (0)	0 (0)
Bourbon	0 (1)	0 (0)	0 (0)	10 (4)	0 (0)
Japanese	0 (0)	0 (0)	0 (0)	0 (0)	10 (5)

[a] Test samples. Numbers represent correct classification ratios for training set; test set ratios are in parentheses.

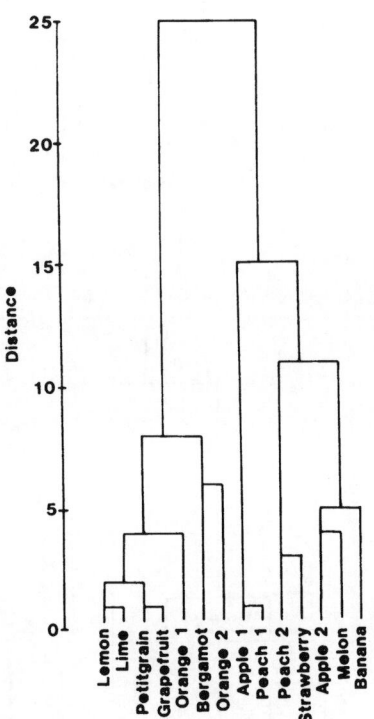

FIG. 5. Clustering of essential oils based on responses from six sensors. (From [5], with permission)

respond to it with their sensing mechanisms. Therefore, the overwhelming content of ethanol in most foods, especially fermented ones, disrupts the ability of an "artificial nose" to sense the genuine aroma components. The use of porous polymers may be one solution to avoid ethanol effects on gas sensing, but continuous aroma monitoring cannot be performed if such a pretreatment step must be incorporated. Thus, no instrument that can sense food aromas has previously been developed. All instruments applied to aroma analysis so far were merely measuring ionized compounds, electric impedance,

or weight changes caused by aroma compounds. The subsequent data analysis made these instrumental outputs correlate with human response. Thus, only the human nose can detect "aromas," but so far, in spite of many efforts, the detecting mechanism in the olfaction has not been known and a device mimicking the olfaction mechanism in the true sense has not been developed. If the recently discovered set of proteins is the true odorant receptors [7], then a biomimic olfaction device may soon appear.

Acknowledgments. I thank Figaro Sensors, Inc. for their kind supply of gas sensors. The courtesy of samples supplied by Soda Aromatic Co., Ltd. and Tokyo Allied Coffee Roasting Co., Ltd. is gratefully acknowledged.

References

1. Persaud K, Dodd G (1982) Analysis of discriminant mechanisms in the mammalian olfactory system using a model nose. Nature (Lond) 299:352–255
2. Ema K, Yokoyama M, Nakamoto T, Moriizumi T (1989) Odour-sensing system using a quartz-resonator sensor array and neural-network pattern recognition. Sens Actuators 18:291–296
3. Nakamoto, T, Fukunishi K, Moriizumi T (1990) Identification capability of odor sensor using quartz-resonator array and neural-network pattern recognition. Sens Actuators B1:473–476
4. Aishima T (1991) Discrimination of liquor aromas by pattern recognition analysis of responses from a gas sensor array. Anal Chim Acta 243:293–300
5. Aishima T (1991) Aroma discrimination by pattern recognition analysis of responses from semiconductor gas sensor array. J Agric Food Chem 39:752–756
6. Gardner JW (1991) Detection of vapours and odours from a multisensor array using pattern recognition. Part 1: Principal component and cluster analysis. Sens Actuators B4:109–116
7. Buck L, Axel R (1991) A novel multigene family may encode odorant receptors: a molecular basis for odor recognition. Cell 65:175–187

Fish Freshness Detection by Semiconductor Gas Sensors

Makoto Egashira, Yasuhiro Shimizu, and Yuji Takao[1]

Key words. Semiconductor gas sensor—Trimethylamine—Dimethylamine—Ammonia—Fish smell—Freshness

Introduction

Objective measurement of the freshness of fish is important from the viewpoint of food sanitation. The most widely used method of testing for freshness is the K value method, and the K value is defined as the percentage of inosine and hypoxanthine among six kinds of adenosine triphosphate-(ATP)- related compounds. The method is based on chemical analysis of breakdown products of ATP in the fish muscle, employing complicated pretreatment procedures and expensive instruments. Thus, the method is destructive analysis and requires much effort and time. The amounts of volatile bases such as trimethylamine (TMA), dimethylamine (DMA), and ammonia in the fish muscle are also often used as the freshness index, but this method again is destructive and complicated.

On the other hand, it is well known that fish gives off a peculiar smell during deterioration after death. The smell contains TMA, DMA, ammonia, and so on [1,2]. If these components are monitored by special gas sensors, the freshness can be easily determined by measuring the resistance changes of the sensors. In this work, fundamental experiments have been carried out to develop sensors that detect the important odor components with high sensitivity and selectivity. Some results of preliminary tests for freshness are also described.

Material and Methods

It is well known that surface oxygen plays the most important role in the gas detection by semiconductor metal oxides, which is now widely utilized for the detection of urban gas and LP gas leakage. Oxygen can adsorb in the forms of O_2^-, O^-, O^{2-}, etc. Formation of these negatively charged oxygen adsorbates should be accompanied by electron transfer from the oxide to the adsorbates. As a result, electric conductivity of the oxide decreases in the case of n-type oxides. If a part of these oxygen adsorbates is consumed by inflammable molecules such as hydrogen and methane, the trapped electrons are returned to the oxide, resulting in high conductivity. The smell from fish should be detected in a similar manner because TMA, DMA, and NH_3 are inflammable.

Various n- and p-type oxides were examined as the sensing material. Powder of a sensor material mixed with a small amount of water was applied on the surface of an alumina tube 3.5 mm × 1.5 mm⌀ on which a pair of gold electrodes had been printed. The sensing layers, partially sintered at 800°C, were porous enough for gaseous molecules to permeate into the inner region where the electrodes were located. A small nichrome heater was insterted inside the alumina tube to preheat the sensor as well as to control the operating temperature at 300°–650°C. The concentration of testing gases was mostly 300 ppm in air. The gas sensitivity, k, was defined as the ratio of electric resistance in air R_a to that in a sample gas R_g for n-type oxides, and as the reverse ratio for p types.

Results and Discussion

Trimethylamine Sensor

The key technology of the present method for freshness detection is to develop highly sensitive and selective sensors to each odor component in fish smells. First, changes in electrical resistance of various n- and p-type oxides were measured when they were exposed to gaseous TMA molecules [3,4]. All the oxides examined showed substantial TMA sensitivity. The most sensitive was TiO_2, the electrical resistance of which decreased by a factor of 59 on exposure to 300 ppm TMA in air.

Another interesting feature was that In_2O_3 exhibited relatively high sensitivity despite its very low resistance in air. If the bulk conductivity of a sensor is thus high, the gas sensitivity becomes small in general, because any change in conductivity at the surface region does not cause any large variation in total conductivity of surface plus bulk. In other words, reducing the bulk conductivity with valency control may cause increased gas sensitivity. This consideration prompted us to examine the effect of

[1] Department of Materials Science and Engineering, Faculty of Engineering, Nagasaki University, 1-14 Bunkyomachi, Nagasaki, 852 Japan

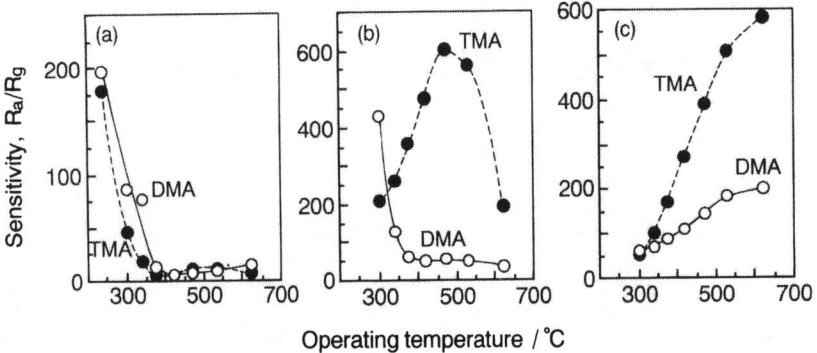

FIG. 1. Sensing characteristics of In₂O₃–MgO (5 mol %) specimens loaded with (a) Pd, (b) Pt, and (c) Au at 3.0 wt % to 300 ppm TMA and DMA in air

substitutional doping of divalent magnesium and zinc ions into the trivalent In_2O_3 lattice [4]. With doping of MgO up to about 5 mol%, the resistance of the In_2O_3 speciments in air markedly increased, and correspondingly the TMA sensitivity markedly increased by a factor of 7–8. Doping of ZnO also resulted in enhancement of the sensitivity, although the solubility limit appeared smaller than that of MgO. Thus, it was suggested that it is of great importance to decrease the concentration of donor electrons in the bulk to increase the sensitivity.

In the field of semiconductor gas sensors, noble metal sensitizers are often utilized to enhance sensitivity [5]. Using this technique, the TiO_2 specimen was loaded with a small amount of various noble metals. The TMA sensitivity was markedly increased by loading In, Ru, and Au, while addition of Pt, Pd, and Rh resulted in a considerable decrease in sensitivity [6]. Such variations in sensitivity were found to be intimately related with the catalytic activity of the sensor materials for the reaction between TMA molecules and oxygen adsorbates [7]. If the sensing material was catalytically very active, as in the cases of Pd-, Pt-, an Rh-doped TiO_2, the sensitivity became very small. This is probably because TMA molecules to be detected were almost completely consumed at the outer region of the sensor and did not permeate into the inner region where the electrodes were located. It is natural that the materials having small catalytic activity exhibit small sensitivity, because oxygen adsorbates are not consumed as much. Thus, high TMA sensitivity can be realized with the sensing materials having mild or appropriate catalytic activity, because TMA molecules can migrate into the inner region of the sensor, where the electrodes for resistance measurement are located, to be oxidized there.

DMA Sensor

The TMA sensors developed based on In_2O_3 and TiO_2 exhibited relatively high sensitivity to DMA as well, although the values were smaller than those to TMA. Sensors selective not only to TMA but to DMA are necessary for precise determination of the freshness of fish. The enhancement in the DMA selectivity has therefore been explored by doping the In_2O_3-based sensor with a small amount of noble metal sensitizers. Some typical results are shown in Fig. 1. The sensor loaded with 3 wt % Pd exhibited almost comparable sensitivity to TMA and DMA, the sensitivity values being high at low temperatures, while the Au-loaded sensor was more sensitive to TMA than to DMA, especially at high temperatures. In contrast, the Pt-loaded sensor was highly sensitive and fairly selective to TMA at about 500°C, but at 300°C it was more sensitive to DMA than to TMA.

Enhancement in the DMA selectivity of the Pt-loaded sensor was further examined by changing the sensor structure; namely, the sensing layer was covered with another porous layer of TiO_2. As a result, the sensor became preferentially sensitive to DMA, the value being extremely large (k > 9500) (Fig. 2). The reason for this variation in the sensing property is not clear at present. Permeation of TMA and oxygen molecules into the inner sensing layer may have been suppressed by the filtration or catalytic effect of the porous TiO_2 layer. However, this modified sensor is very attractive as a selective DMA sensor.

NH₃ Sensor

Figure 3 shows the transient response to 300 ppm NH_3 of an In_2O_3–MgO sensor with electrodes at the innermost region [8]. Only a very slight decrease in resistance was observed on exposure to ammonia at 300°C. At 530°C, however, a very complicated response was observed; an abrupt decrease in resistance followed by a gradual increase when exposed to NH_3, and an abrupt increase followed by a gradual decrease after removal of NH_3. If negatively charged oxygen adsorbates are consumed by NH_3 oxidation, the sensor resistance decreases. On the other hand,

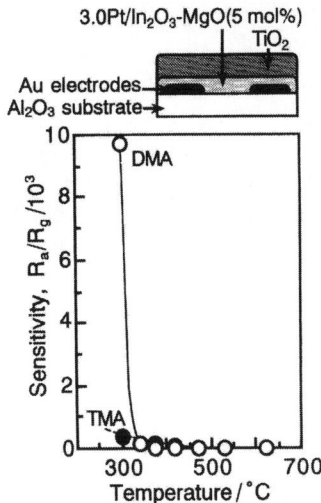

FIG. 2. Effect of coating with porous TiO₂ layer on TMA and DMA sensitivity of In₂O₃–MgO (5 mol %) sensor loaded with Pt (3.0 wt %)

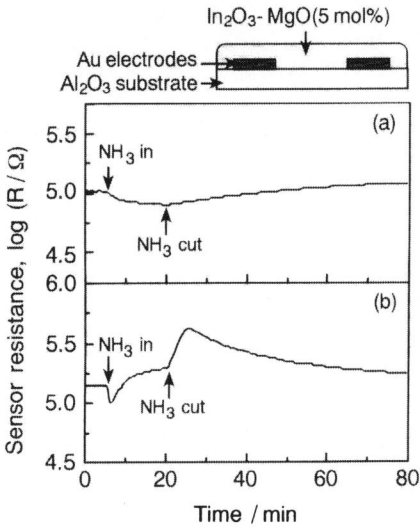

FIG. 3. Transient response of In₂O₃–MgO (5 mol %) sensor to 300 ppm NH₃ at (a) 300°C and (b) 530°C. Schematic drawing shows sensor structure

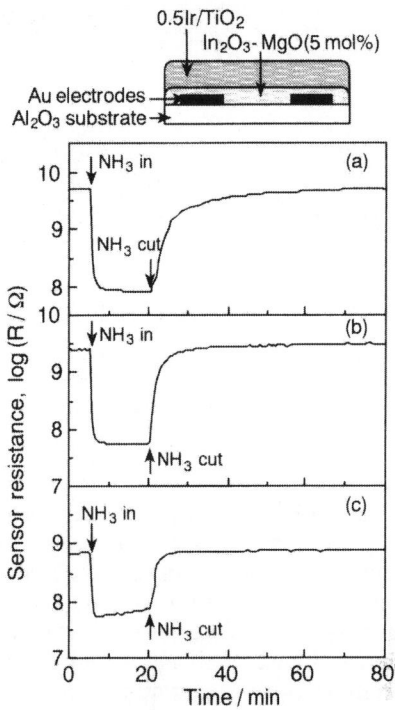

FIG. 4. Transient response of In₂O₃–MgO (5 mol %) sensor coated with catalyst layer Ir(0.5 wt %)/TiO₂ to 300 ppm NH₃ at (a) 470°C, (b) 530°C, and (c) 620°C. Schematic drawing shows sensor structure

formation of negatively charged adsorbates such as NO^-, NO_2^-, and NO_3^- leads to an increase in resistance. Competitive occurrence of those reactions is probably the main cause for the very complicated response observed.

The interference from NO_x was more pronouncedly observed when the sensing layer was covered with a Pt catalyst layer [8]. At 300°C, a relatively large normal response or a decrease in resistance is observed. At 370°C, however, the resistance remained almost constant on exposure to NH_3, whereas it increased when introduction of NH_3 was cut off. More dramatically, the resistance

abruptly increased when exposed to NH_3 at 620°C. There is no doubt that these enhanced interferences from NO_x species arose from the high catalytic activity of Pt for NH_3 oxidation.

When other nobel metals were employed as the catalyst layer, the interference from NO_x was considerably smaller, especially in the case of iridium catalyst; that is, the response to NH_3 appeared principally as a decrease in resistance at every temperature. However, it seemed that the interference from NO_x species still existed at higher temperatures, because the resistance after removal of NH_3 was larger than the initial value. To eliminate the interference, the position of electrodes was changed from the innermost region of the sensing layer to the interface between the sensing and catalyst layers [8] so that the sensor could restore its initial resistance after removal of NH_3 (Fig. 4). This one-directional response characteristic is essential for a sensor. The sensitivity value was greater than 60 at 470°C.

Preliminary Tests for Fish Freshness

Table 1 summarizes promising sensors that have been developed. Note that high sensitivity (400–500) was achieved to 300 ppm TMA with TiO₂ sensors loaded with Ru and In. Sensitivity of 100–500 was also realized with In₂O₃–MgO sensors, depending

TABLE 1. Prospective sensors for TMA, DMA, and NH$_3$.

Target	Sensor materials[a]	Sensitivity, R_a/R_g [b]	Operating temperature (°C)
TMA	0.5Ru/TiO$_2$	400	560
	2.0In/TiO$_2$	480	380
	0.5Au/TiO$_2$	170	470
	In$_2$O$_3$–MgO	100–500	420
DMA	3.0Pt/In$_2$O$_3$–MgO coated with TiO$_2$	9700	300
NH$_3$	In$_2$O$_3$–MgO coated with 0.5Ir/TiO$_2$	60–100	470

[a] The loading level of metal sensitizers is given in wt%; In$_2$O$_3$ was doped with 5 mol% MgO.
[b] To 300-ppm level in air.

on the preparation conditions. A sensor with extraordinarily high sensitivity and selectivity to DMA was developed. As for ammonia, the sensitivity of the developed sensor was relatively small, compared with the TMA and DMA sensors, but it is still attractive. It is expected that a multisensor array utilizing these promising sensors can discriminate the freshness of any kind of fish. However, such experiments have not been extensively made yet. Only preliminary tests were carried out in which resistance changes of some single sensors were monitored when exposed to actual odors from some fish samples.

In an experiment, variations in resistance of a Ru:TiO$_2$ sensor were continuously monitored when it was exposed to the smell from a fish (Japanese saurel) stored in a closed box at room temperature [9]. The variations were well correlated with those in the biochemical freshness index, the K value, which was determined on the basis of the chemical analysis of ATP-related compounds at given intervals. In another experiment, the response of an In$_2$O$_3$–MgO sensor was measured to the smell from cuttlefish stored at 8°C. It is well known that in the case of cuttlefish very large K values are often observed even when it is very fresh. Thus, the K value is not believed to be useful as a freshness indicator, and the amount of TMA–nitrogen in the muscle is the more useful index [10]. Variations in the sensor resistance corresponded closely to the freshness deterioration based on the latter index. Figure 5 is another example indicating the potential advantages of the present sensor [11]. Note that there is a good linear correlation in a log–log plot between the sensor response and the concentration of TMA in various kinds of fish muscles stored under different conditions.

Thus, it is obvious that the semiconductor gas sensor method is quite useful for freshness detection even when a single sensor was employed, compared with the biochemical K value or volatile amine method. Very simple, very cheap, nondestructive,

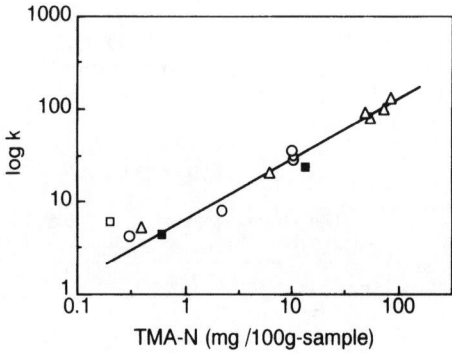

FIG. 5. Sensitivity to fish smells of In$_2$O$_3$–MgO (5 mol %) sensor at 420°C and concentration of TMA-nitrogen atoms in fish muscles stored under different conditions. *Circles*, sardine; *triangles*, cod; *open square*, yellowtail (ordinary muscle); *closed squares*, yellowtail (bloody muscle)

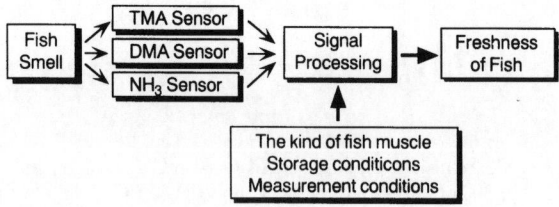

FIG. 6. A multisensor system for discriminating freshness of fish

and continuous measurement is possible. For more precise determination of freshness, however, we have to set up a multisensor system combined with sophisticated signal processing, as illustrated in Fig. 6. The system of course must be applicable to any kind of fish muscle, irrespective of the storage and measurement conditions. We hope the sensors developed in our work will find practical application in such an artificial olfactory system.

Acknowledgments. This work was partly supported by grants-in-aid for scientific research from the

Japanese Ministry of Education, Science, and Culture.

References

1. Miwa K, Tokunaga T, Iida H (1976) Studies on protecting methods of occurrence of bad odors and their removing methods in fisheries processing factories: II. Cooking odor and drying odor of fish. Bull Tokai Reg Fish Res Lab 86:7–27
2. Katayama T, Sugimoto K (1966) Chemical studies on aromatic and odoriferous components of fisheries products: I. Odoriferous components of fish accompanying lowering of freshness in Japanees *Saurel trachurus*. Mem Fac Fish Kagoshima Univ 15:19–26
3. Egashira M, Shimizu Y, Takao Y (1990) Trimethylamine sensor based on semiconductive metal oxides for detection of fish freshness. Sens Actuators B1:108–112
4. Takao Y, Miya Y, Tachiyama Y, Shimizu Y, Egashira M (1990) Improvement in trimethylamine sensitivity of In_2O_3 and Cr_2O_3 sensors by valency control. Denki Kagaku 58:1162–1168
5. Yamazoe N, Kurokawa Y, Seiyama T (1983) Effects of additives on semiconductor gas sensor. Sens Actuators 4:283–289
6. Takao Y, Iwanaga Y, Shimizu Y, Egashira M (1993) Trimethylamine-sensing mechanism of TiO_2-based sensors: 1. Effects of metal additives on trimethylamine-sensing properties of TiO_2 sensors. Sens Actuators B10:229–234
7. Takao Y, Fukuda K, Shimizu Y, Egashira M (1993) Trimethylamine-sensing mechanism of TiO_2-based sensors: 2. Effects of catalytic activity of TiO_2-based specimens on their trimethylamine-sensing properties. Sens Actuators B10:235–239
8. Takao Y, Miyazaki K, Shimizu Y, Egashira M (1993) Semiconductor ammonia gas sensor with double-layered structure. In: Digest of Technical Papers, 7th International Conference on Solid-State Sensors and Actuators, Yokohama, pp 360–363
9. Shimizu Y, Takao Y, Egashira M (1988) Detection of freshness of fish by a semiconductive $Ru:TiO_2$ sensor. J Electrochem Soc 135:2539–2540
10. Takao Y, Ohashi E, Shimizu Y, Egashira M (1992) Detection of cuttlefish freshness by semiconductor trimethylamine gas sensors. Sens Mater 3:249–259
11. Ohashi E, Takao Y, Fujita T, Shimizu Y, Egashira M (1991) Semiconductive trimethylamine gas sensor for detecting fish freshness. J Food Sci 56:1275–1278, 1286

Taste Recognition Using Multichannel Fiber-Optic Sensors with Potential-Sensitive Dye Coatings

SHINZO YAMAKAWA[1]

Key words. Potential-sensitive dye coating—Fiber-optic taste sensor

To our knowledge, there is, as yet, no published work on taste recognition using optical response patterns from multichannel fiber-optic sensors, although electrode taste sensors with artificial lipid membranes have already been developed [1]. Here, we propose a new taste recognition system using optical response patterns from a fiber-optic sensor array that has potential-sensitive dye/polymer coatings.

The dye/polymer coatings of the fiber-optic sensors were formed by dip-coating the fiber end faces in ethanol solutions of dye and polymer. The sensor structure was a transmitted light type. The optical fiber was plastic, and the coating polymer was silicone and poly(methyl methacrylates).

The potential-sensitive dye coatings give large changes in dye absorption spectra when they are immersed in various taste solutions. Also, it has been found that different dye coating show different responses to the same taste solution, while the same dye coating shows different responses to different taste solutions. Accordingly, a sensor array consisting of these dye coatings can be used for taste pattern recognition. Six different dyes, which gave large changes in dye absorption, were selected and used for a six-channel fiber-optic taste sensor array. The absorption spectra change data were processed by multiple discriminant analysis and neural networks, using a back-propagation algorithm. The optical taste sensor system has been found to discriminate salty (NaCl), bitter (quinidine-H_2SO_4), sweet (sucrose), sour (HCl), and *umami* (monosodium glutamate; MSG) substances.

To compare the fiber-optic taste sensors with electrode taste sensors, coated wire Ag electrodes with the same dye coatings were used. The dye-coated electrodes also showed large changes in membrane potential when they were immersed in the five basic taste solutions. A four-channel electrode sensor array consisting of four different dye coatings was also used for taste pattern recognition for comparison with the fiber-optic sensors. The taste concentration dependence of dye absorbance at a particular wavelength was compared with that of the membrane potential. The results indicate that the fiber-optic sensors are more sensitive to taste solutions than the electrode sensors.

[1] Department of Computer Engineering, Toyama National College of Maritime Technology, Shin-minato, Toyama, 933-02 Japan

References

1. Hayashi K, Yamanaka M, Toko K, Yamafuji K (1990) Multichannel taste sensor using lipid membranes. Sens Actuators B2:205–213

Halide Anion Sensor

Izumi Kubo[1], Atsushi Seki[1], Hiroyuki Sasabe[2], and Hiroaki Tomioka[2]

Key words. Halorhodopsin—Ion-sensitive field effect transistor (ISFET)

Introduction

We studied biosensors that consisted of an electrochemical device and functional biomolecules. We recognized that membrane proteins, which play a role in energy conversion, environmental sensing, and information transmission, could be candidates for bio-recognition elements in the biosensor. In this study, halorhodopsin (hR), a light-driven chloride anion pump in the cytoplasmic membrane of *Halobacterium halobium* [1], was chosen as a halide anion recognition element, and it was coupled with an ion-sensitive field effect transistor (ISFET) so as to construct an anion-sensing system. In this system, hR-containing vesicles were immobilized on the surface of the ISFET. Under yellow light illumination, this hR-ISFET sensor showed a halide anion-dependent response.

Materials and Methods

Preparation of Vesicle Containing hR

Vesicles containing hR were prepared from a *Halobacterium halobium*, strain OD_2S, a bacteriorhodopsin-deficient mutant. The cells, harvested by centrifugation, were disrupted by a freeze-thaw method. The membrane fractions were collected and dialysed against 1 mM piperazine-N,N'-bis(2-ethanesulfonic acid) (PIPES) buffer containing 1 M Na_2SO_4 to remove chloride anion.

Flash-Induced Absorbance Change of hR

The photochemical activity of hR was measured by flash-photolysis [2]. Potassium halide was added to the membrane suspension and the absorbance change at 580 nm induced by red-flash ($\lambda > 600$ nm) was measured.

Fabrication of hR-Immobilized ISFET

An ISFET was covered with polyvinylbutyral (PVB) resin according to a method reported previously [3]. The PVB-ISFET was immersed in the membrane suspension to attach the hR-containing vesicles in the PVB matrix. The hR-immobilized ISFET (hR-ISFET) was immersed in the 1 M Na_2SO_4-1 mM PIPES buffer. While the hR-ISFET was illuminated with a halogen lamp passed through a yellow filter ($\lambda > 500$ nm), potassium halide was added to the buffer solution. The output voltage was measured with a source-follower circuit [4], which is a voltmeter for the gate output voltage of ISFET. An Ag/AgCl electrode was used as a reference electrode, with a salt bridge of potassium nitrate to prevent contamination by choride anion.

Results and Discussion

Halide Anion Dependency of Photochemical Activity of hR

The photochemical activity of hR was measured as flash-induced absorbance change. This absorbance change is tightly coupled with anion pumping [5]. We monitored the absorbance change at 580 nm (ΔA_{580}), ΔA_{580} being dependent on the halide anion. We used three kinds of halide anions, Br^-, Cl^-, and I^-. The amplitude of ΔA_{580} increased according to increases in halide anion concentration. The sequence of effectiveness of halide anions on ΔA_{580} was $Br^- > Cl^- > I^-$. This sequence in sulfate buffer agreed well with the sequence in phosphite buffer [5].

Response of hR-ISFET

hR-ISFET responded to the addition of halide anion under stationary illumination with yellow light; however, the ISFET itself did not respond to the halide addition. The amplitude of the hR-ISFET response was dependent on the concentration of halide anion, the sequence of the amplitude of the response to halide anions being $Br^- > Cl^- > I^-$. This sequence is correlated with that of the

[1] Department of Bioengineering, Faculty of Engineering, Soka University, 1-236 Tangi, Hachioji, Tokyo, 192 Japan
[2] Laboratory for Nano-Photonics Materials, Frontier Research Program, The Institute of Physical and Chemical Research (RIKEN), 2-1 Hirosawa, Wako, Saitama, 351-01 Japan

photochemical activity of hR. These results indicate that hR immobilized in the resin layer on the surface of the ISFET retained its pumping activity and that the ISFET detected halide anion pumping by hR.

References

1. Schobert B, Lanyi JK (1982) Halorhodopsin is a light-driven chloride pump. J Biol Chem 257:10306

2. Otomo J, Tomioka H, Sasabe H (1992) Bacterial rhodopsin of newly isolated halobacteria. J Gen Microbiol 138:1027

3. Gotoh M, Tamiya E, Karube I, Kagawa Y (1986) A microsensor for adenosine-5′-triphosphate pH-sensitive field effect transistors. Anal Chim Acta 187:287

4. Kitagawa Y, Tamiya E, Karube I (1987) Microbial-FET alcohol sensor. Anal Lett 20:81

5. Hazemoto N, Kamo N, Kobatake Y, Tsuda M, Terayama Y (1984) Effect of salt on photocycle and ion-pumping of halorhodopsin and third rhodopsinlike pigment of *Halobacterium halobium*. Biophys J 45:1073

Pattern Analysis of Odors by Multiple Metal Oxide Semiconductors: Odor Analyzer with Human Sense of Smell

KATSUO EHARA[1] and RENZO HATTORI[2]

Key words. Odor sensor—Anthropomorphic sensing

To carry out a pattern analysis of odors, we used an odor analyzer with a human sense of smell; this analyzer indicates not only the odor quality, i.e., pleasantness and unpleasantness in the pattern, but also the odor intensity. Conventional systems indicate only the intensity level of an odor, irrespective of its quality. This newly developed system can discriminate offensive from inoffensive odors by the shapes of graphic patterns. From a hexagonal radar pattern made by six unique semiconductor sensors, odor quality and intensity are obtained, the sharpness of radar pattern indicating the odor quality, while the area indicates odor intensity. An odor map can then be made from the odor quality and intensity obtained. Thus, various odors can be visually identified, and this procedure, rather than the human sense of smell, can be applied to classify various types of odors.

The odor-sensing unit of the analyzer is a sensor assembled with a unique metallic oxide semiconductor; it has a platinum coil heater made by the vacuum evaporation method. The six selected metallic oxide sensors have characteristic properties for determining nitrogen, sulfur, hydrocarbons, oxygen, and alcoholic and aromatic compounds. These are the principle components of natural odorous materials. The test sample is placed in an odor box with the sensor for 10 min for vaporization.

When the sensor coil is heated to about 300°C, the electrical resistance of the sensor circuit drops due to the reductive reaction of the attachment of odorant molecules. Variations of the resistance value are measured and processed with a computer. Hexagonal radar charts are formed, their characteristics depending on the quality and intensity of the odors. The measurement of various types of odors shows that odors pleasant to humans have a hexagonal radar chart resembling a spearhead, while unpleasant odors tend to have an indeterminate form. By identifying pleasant odors in the radar pattern, a value for psychological pleasantness can be determined. An odor map created from the odor quality of psychological pleasantness and intensity is very convenient for the estimation of an agreeable and tasty zone of volatile materials.

To visualize various odor qualities by making good use of semiconductor sensors without GC (gas chromatograph) and MS (mas spectrometer) or other apparatus and to apply this odor sensor to anthropomorphic sensing fields, four kinds of SnO_2 and two kinds of ZnO sensors were assembled.

By using the combination of these six metallic oxide sensors, we found interesting results in that: (1) Enhancement and cancellation effects could be determined for the quality of a mixed odor, (2) there was a dependency of odor quality on humidity changes, (3) the maturity of alcoholic beverages could be estimated, (4) the freshness of fish could be estimated, (5) the quality of the aroma of roasted coffee beans could be assessed, and (6) pleasant and unpleasant odors could be evaluated.

Acknowledgments. We thank Mrs. H. Tanahashi (Japan Women's University) for her work in this study.

[1] Tokyo Institute of Technology, 12-1 Ookayama 2, Meguro-ku, Tokyo, 152 Japan
[2] Takasago International Corporation, 19-22 Takanawa 3-chome, Minato-ku, Tokyo, 108 Japan

Novel Intelligent Gas Sensor
Based on Informational Source of Nonlinear Response

You Kato and Kenichi Yoshikawa[1]

Key words. Olfactory model—Intelligent gas sensor

We are currently studying dynamic model systems of taste and olfactory sensation [1–3]. It is well known that receptor cells in taste and olfaction show rather broad spectra to chemical substances. However, living organisms can actually distinguish thousands of chemical species in the environment. We have been stressing the importance of nonlinear dynamic response on the very initial stage of perception. We have shown that, based on the information of nonlinear response, one can distinguish and quantitate taste compounds with high reliability [1,2]. As an extension of these studies, we report our study on the dynamic response of a gas sensor. It will be shown that a conventional ceramic sensor, with a broad chemical spectrum, affords us enough information to distinguish gas molecules when one analyses its dynamic response [3].

We measured the dynamic response of an SnO_2 sensor accompanied by the repetitive change of the temperature in the sensor body. The logarithm of the time-dependent output resistance was Fourier-transformed into complex frequency. From the theoretical point of view, this procedure corresponds to the following expansion of the resistance, $R(t)$:

$$R(t) = R_0 \exp[I_1 \sin(\omega t + \Phi_1) + I_2 \cos(2\omega t + \Phi_2) + \ldots] \quad (1)$$

On the other hand, the temperature dependence of the resistance is given with the usual Arrhenius relationship.

$$R(T) = R_0 \exp(eV_s/kT) \quad (2)$$

Where V_s is the surface barrier potential. When the temperature is time-dependent, $V_s(T)$ may be expanded as follows:

$$V_s(T) = a_1 T + b_1 \left(\frac{dT}{dt}\right) + a_2 T^2 + b_2 T \left(\frac{dT}{dt}\right)$$
$$+ c_2 \left(\frac{dT}{dt}\right)^2 + \text{higher order} \quad (3)$$

From the relationship given by the combination of Eqs. 2 and 3, it is apparent that the complex harmonics in Eq. 1 imply the information on the dynamic interaction between sensor and gas molecules. The parameters a_1 and a_2 are related to the thermo-equilibrium state and the parameters b_1, b_2, and c_2 are concerned with the dynamic of adsorption, desorption, and chemical reaction. In the actual experiments, we have made it clear that evaluation of the concentration of individual gases is quite possible for a mixed gas sample, even with a single detector.

Recently, there have been many studies of models of taste and olfactory sensation, utilizing informational processing with artificial neural networks [4,5]. Most of these studies are based on the fundamental idea of using the information of the linear responses with multi-detectors. However, this framework breaks down in general, because: (1) sensors exhibit rather large nonlinearity, such as in the saturation effect, (2) the additive rule for different chemical species does not hold, and (3) sensors show marked time-dependent characteristics, such as aging or hysterisis effects. In contrast, the nonlinear dynamic response gives us a rich variety of information concerning the chemical environment.

We believe that the inclusion of the nonlinear response is essentially important in understanding the mechanism of recognition in living organisms.

Acknowledgments. The authors thank Dr. S. Nakata, and Mrs. Y. Kaneda, S. Akakabe (Nara University of Education) for their experimental support and useful discussions.

References

1. Yoshikawa K, Matsubara Y (1984) Chemoreception by an excitable liquid membrane: Characteristic effects of alcohols on the frequency of electrical oscillation. J Am Chem Soc 106:4423–4427
2. Yoshikawa K, Yoshinaga T, Kawakami H, Nakata S (1992) New strategy of informational processing: Utilization of nonlinear dynamics in chemical sensing. Physica 188A:243–250
3. Nakata S, Nakamura H, Yoshikawa K (1992) New strategy for the development of a gas sensor based on dynamic characteristics: Principle and preliminary experiment. Sensors and Actuators 8B:187–189
4. Erdi P, Gröbler T, Barna G, Kaski K (1993) Dynamics of the olfactory bulb: bifurcations, learning, and memory. Biol Cybern 69:57–66
5. Freeman WJ (1987) Simulation of chaotic EEG patterns with a dynamic model of the olfactory system. Biol Cybern 56:139–150

[1] Graduate School of Human Informatics, Nagoya University, Furo-cho, Chikusa-ku, Nagoya, 464-01 Japan

Neural Network Processing of Responses to Odorants by a Biological Nose and a Sensor Array

Graham A. Bell[1], Donald Barnett[1], Fan Ng[1], Junni Zhan[2], and David C. Levy[3]

Key words. Neural networks—Olfaction

A neural network is a processing device, either an algorithm or actual hardware, whose design was motivated by biological neural functions. It can be trained to operate as a classifier. The application of neural networks to the study of olfactory processing in vivo and to identifying and classifying complex chemical mixtures from the outputs of chemical sensor arrays is the subject of this study.

A previous study [1], using the radioactive 2-deoxyglucose (2-DG) method of metabolic monitoring, demonstrated mixture suppression in odor-specific patterns of metabolic activity in the glomerular layer of the rat main olfactory bulb following stimulation of the olfactory system with controlled dilutions of pure chemical odorants. The question then arises whether similar metabolic effects can be observed in the olfactory mucosa of a mammal. An attempt to reconstruct autoradiographs from rats' noses in three dimensions proved logistically impracticable [2]. The septal wall of the nose of mouse has now been studied by flat-mounting the mucosa from the septum, after injection of the animals with [3H] 2-DG, followed by stimulation with diluted vapor of propionic acid, or of ethylacetoacetate, or by clean air.

Data from 2-DG autoradiographic images derived from mouse noses after stimulation by odorants, and data from voltage outputs from an array of Figaro chemical sensors were used as inputs to a self-organizing neural network which was trained using a Kohonen algorithm.

Results show that neural networks can be used to demonstrate that the mammalian nasal epithelium is responding in a complex way, which, nevertheless, supports a classifier of chemical inputs.

Complex odor mixtures generated from the head spaces of containers each holding a sample of one of four kinds of Australian cheese were delivered to a sensor array. The array consisted of eight Figaro sensors housed in a stainless steel chamber. After a period of stabilization, the outputs from the sensor array were digitized and analyzed, using a three-layer neural net and a back-propagation algorithm. The net learned to correctly classify the four types of cheese, to a 100% criterion, thereby demonstrating that identification of very complex odor mixtures can be achieved with minimal prior knowledge of the response capabilities of each sensor in an array.

A five-layer net designed to incorporate connection patterns of the olfactory bulb also learned the classification. A high degree of specialization in the proportional output strengths developed in certain neurons, particularly in the third hidden layer. After five of the low-output neurons were pruned, the net again learned the task to 100% criterion.

Neural network processing assisted in demonstrating that both the animal nose and the sensor array produce information upon which a classification of the odors can be based. In the case of the biological nose, single pure odorants were identified from a pre-processed reduction of spatial information generated by odor stimulation. This confirmed the impression of the image data by human observers that the odorants produced odor-specific activity patterns across the nasal septa of the mice. Classification by neural nets of biological image data obtained from a mammalian nose, and of data from sensor array outputs, assisted in the confirmation of the functions of both systems. Neural network processing can assist in the interpretation of complex patterns of neural activity such as occur in the olfactory system, and should expedite the development of sensor arrays for identification of complex chemical mixtures of interest to a number of industries.

[1] CSIRO, Sensory Research Centre, Division of Food Science and Technology, P.O. Box 52, North Ryde, Australia 2113
[2] Department of Anatomy, Monash University, Clayton, Australia, 3168
[3] Department of Electronic Engineering, University of Natal, Durban, Republic of South Africa

References

1. Bell GA, Laing DG, Panhuber H (1987) Odour mixture suppression: Evidence for a peripheral mechanism in human and rat. Brain Res 426:8–18
2. Bell GA, Zhan J (1991) Metabolic activity in the rat and mouse olfactory epithelium and olfactory bulb following odour stimulation. Chem Senses 16(5):500–501

Measurement of Halitosis
with a Zinc Oxide Thin Film Semiconductor Sensor

Masayo Shimura, Yoko Yasuno, Masaki Iwakura, Seizaburo Sakamoto, Yoshinori Shimada, Sai Sakai, and Kengo Suzuki[1]

Key words. Halitosis—Sensor

Halitosis, defined as unpleasant oral odor, is a health concern among the general public. Halitosis has been conventionally diagnosed by organoleptic examination and by gas chromatographic analysis of the main source of halitosis, i.e., volatile sulfur compounds, such as H_2S and CH_3SH. Since gas chromatography requires a large-scale system and a long running time, we investigated the use of a zinc oxide thin film semiconductor sensor for measuring trace volatile sulfur compounds in mouth air. Mouth air samples collected in Teflon bags from 7 patients and 22 volunteers were analyzed by the three methods. The readings of the sensor were correlated with the values of the volatiles measured by gas chromatography ($r = 0.712$, $P < 0.01$), and also with the organoleptic scores given by three judges ($r = 0.714$, $P < 0.01$). The organoleptic scores were correlated with the gas chromatographic values ($r = 0.703$, $P < 0.01$).

These results suggest that this zinc oxide thin film semiconductor sensor may be used for the diagnosis of halitosis; its small size and simplicity of handling may enable us to use it for routine chair-side study and field surveys of halitosis.

[1] Department of Preventive Dentistry, Tohoku University School of Dentistry, 4-1 Seiryo, Aoba-ku, Sendai, 980 Japan

Development of Odor-Sensing System Using an Auto-Sampling Stage and Identification of Natural Essential Oils

Junichi Ide[1], Takamichi Nakamoto[2], and Toyosaka Moriizumi[2]

Key words. Quartz-resonator—Pattern recognition—Sensing system—Cyclodextrins—Optical isomer

Introduction

An odor-sensing system is desirable for use in a broad variety of fields, from the food, drink, and cosmetics industries, to use in environmental and other protection fields. The human sensory tests now utilized to discriminate odors in those fields are inevitably affected by the inspector's state of health and mood. It is therefore essential that an objective evaluation method be developed for practical use.

The present authors developed an odor sensing system using a quartz-resonator sensor array and neural-network pattern recognition, by which the odor of alcoholic beverages could be identified [1]. Since then, it was found that selection of the sensing films by multivariate analysis, and the use of a measurement system modification for reducing data variation were effective in heightening the identification capability [2,3]. However, the number of samples was limited (up to six) due to the increase in the number of solenoid valves, since two valves per sample were used for controlling sample vapor supply.

A new system for measuring many samples with an automatic sampling stage has been developed to overcome this problem, and this has been applied to flavor discrimination. The results of discrimination of many kinds of essential oils by principal component analysis are reported here. Modified cyclodextrins were tested as sensing films. Since the cyclodextrin molecule has an internal cavity, it is thought to be effective for magnifying differences of aroma based upon molecular size and stereochemical structure. The experimental results showed that cyclodextrins were effective in detecting odorant materials with benzene rings and in discriminating optical isomers. The details are described in the latter half of this article.

Principle of System

A quartz-resonator sensor, consisting of a resonator coated with a sensing film, was used here. The resonance frequency decreases when odorant molecules are adsorbed onto the film, and the frequency recovers after desorption. This phenomenon is called the mass-loading effect [4–6] and the shift in frequency is proportional to the total mass of adsorbed odorant molecules. As the responses to odors of each sensor coated with a different film vary slightly, the output pattern from the sensor array can be used to identify odors.

Measurement System

A schematic diagram of the authors' system is shown in Fig. 1. A vapor-flow system is used, and the sensor output can be repeatedly measured within a short time. A sample vapor is injected into a sensor cell where eight sensors are installed, and then dry air is supplied to refresh the sensors. After recovery of the sensor responses, the next sample vapor is injected.

In the present system, liquid odor samples are poured into vials that are then sealed with rubber stoppers. A mechanical stage (Sankyo SKIRAM SR5383, purchased from Sankyo Seiki MFG. Co., Ltd., Tokyo, Japan) is used; this moves along XYZ axes, and two syringe needles are placed on the vials, and inserted into them through the rubber stoppers. Three solenoid valves also move together with the syringe needles After penetration of the needle, dry air, with a flow rate of 120 ml/min, controlled by a mass flow controller, is supplied through one needle for 30s, and the sample vapor flows out through the other needle into the sensor array. The needles are then moved to an empty vial, and dry air flows in the same way to refresh the sensors. These operations are repeatedly controlled

[1] Kawasaki Research Center, T. Hasegawa Co., Ltd., 335-Kariyado, Nakahara-ku, Kawasaki, Kanagawa, 211 Japan
[2] Faculty of Engineering, Tokyo Institute of Technology, 2-12-1 Ookayama, Meguro-ku, Tokyo, 152 Japan

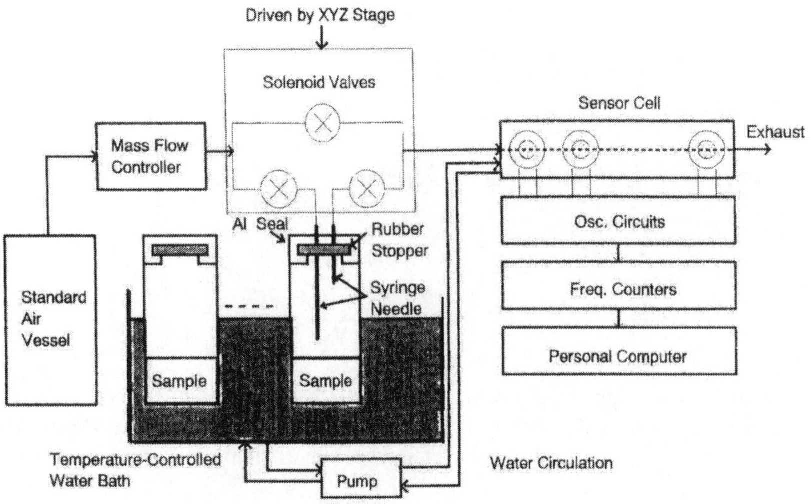

Fig. 1. Odor-sensing system using an auto-sampling stage

by a personal computer so that many samples can be measured automatically.

In the same way as in the previous system [3], the quartz resonators used here are AT-cut with fundamental resonance frequencies of about 10 MHz. The eight resonators with different film materials are connected to Complementary Metal Oxide Semiconductor (CMOS) oscillation circuits and their frequencies are measured in parallel, using frequency counters. A personal computer (NEC 9801 DX2; NEC, Tokyo, Japan) reads out the frequency data through an Input/Output (I/O) interface.

Experimental

The measurement data are eight-dimensional, since eight sensors are used. Thus, principal component analysis (PCA), a dimensional reduction technique, is used for studying the pattern separation among samples. The data are analyzed after the normalization of $X_i/\Sigma_{j=1}^{8} X_j$, where X_i is the signal of the i-sensor.

Essential Oil-Aroma Separation

The sensing films used here are lipid, mixed-lipid, and cellulose, which have been selected for discriminating citrus essential oils (Table 1) [7]. Okahata et al. [8] reported that a lipid film was effective in detecting odorant materials. The authors [9] have previously reported experiments on typical flavors and on citrus flavors with common substantial aroma components. Following that study, 20 typical essential oils were measured, these oils being of various notes, such as citrus, flower, mint, spicy, and so on, as listed in Table 2. The scattering diagram by PCA is shown in Fig. 2. The pattern separation among these samples was improved in comparison with that in the previous study, since the films were modified, using those shown in Table 1, and the data

TABLE 1. Films used in the sensor array.

No.	Sensing film		Classification
1	Lecithin	67%	Lipid
	Cholesterol	33%	Sterol
2	Lecithin		Lipid
3	Ethyl cellulose		Cellulose
4	Cholesterol		Sterol
5	Phosphatidyl choline	63%	Lipid
	Sphingomyelin 37%		Lipid
6	Dioleyl phosphatidyl serine		Lipid
7	Sphingomyelin (egg)		Lipid
8	Dioleyl phosphatidyl choline		Lipid

TABLE 2. Twenty typical essential oils.

Note	Oil
Citrus	Bergamot
Flower	Ylang ylang
	Cananga
Anise	Fennel
	Star anise
	Caraway
Mint	Marjoram
	Spearmint
	Peppermint
Resin	Galbanum
Rustic	Eucalyptus
	Rosemary
	Thyme
Woody	Cedar wood
	Sandalwood
Citronella	Lemongrass
Medical	Wintergreen
Spicy	Nutmeg
	Cardamon
	Cassia

normalization described above was performed. The coefficient of variation of the data was reduced to 1.6% after normalization, whereas it was around 3% without normalization. It is essential to suppress

1st axis (67%)

2nd axis (29%)

Coefficient
of variation (1.6%)

F-value (119)

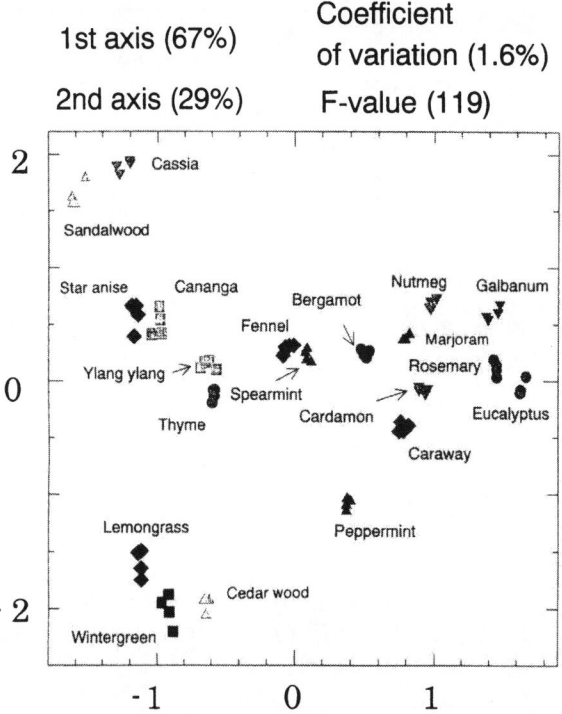

FIG. 2. Scattering diagram of 20 typical essential oils by principal component analysis (*PCA*)

TABLE 3. Chemicals and their vapor pressure.

Samples	Vapor Pressure (18°C)
(D)-Rose oxide	0.7 mmHg
(L)-Rose oxide	0.7 mmHg
Hexane	110.5 mmHg
Heptane	31.9 mmHg
Benzene	67.9 mmHg
Toluene	19.4 mmHg
Methoxybenzene	2.4 mmHg
Cyclohexane	70.4 mmHg
Benzaldehyde	0.5 mmHg
Hexanal	35.3 mmHg
Heptanal	1.6 mmHg

data variation and to select suitable sensing films for the identification of these types of flavors.

Sensor Responses Using Cyclodextrins

Separation of Nine Chemicals

Three types of modified cyclodextrins were used for sensing films:

1. Modified α-cyclodextrin:(Hexakis-2,3,6-tri-O-methyl)-α-cyclodextrin
2. Modified β-cyclodextrin:(Heptakis-2,3,6-tri-O-methyl)-β-cyclodextrin and
3. Modified γ-cyclodextrin:(Octakis-2,3,6-tri-O-methyl)-γ-cyclodextrin

After the modified cyclodextrin (designated CD hereafter) was dissolved in an organic solvent such as acetone and chloroform, the solution was painted on the electrode of a quart resonator, in a manner similar to that reported in our previous study [3]. The sensing films No. 1, 2, and 3 (listed in Table 1) were changed into CD films, and nine of the chemicals listed in Table 3 were measured. The vapor pressure of the chemicals (at 18°C) is also shown in Table 3. Figure 3 shows the outputs of the sensors with four kinds of films that showed larger responses than the others. A lipid mixture of phosphatidylcholine 63% and sphingomyelin 37% gave rise to large responses to all the chemicals, in agreement with our previous work [10]. Of three chemicals six

carbon atoms listed in Table 3, hexane had the highest vapor pressure, and cyclohexane and benzene had comparable vapor pressures. In spite of having the same carbon number and in spite of the vapor pressure difference, only benzene gave the β- and γ-CD sensors very large frequency shifts. Other chemicals with benzene rings and aldehyde functional groups gave larger responses to the sensor with β- and γ-CD, whereas normal chain chemicals, such as hexane and heptane, showed lower outputs. It seems evident from these results that there is some relation of host-guest chemistry between the CD cavity and the molecular structure of adsorbed odorants.

Although benzaldehyde has the lowest vapor pressure, it gave a higher frequency shift than heptanal, which is an aliphatic compound with the same carbon number as benzaldehyde. It is suggested that the interaction (which might be due to electron affinity) between cyclodextrin and the benzene ring influenced the frequency shift.

Optical Isomer Separation

The experiments were performed on the optical isomers, (D)-rose oxide and (L)-rose oxide, 2-(2-methyl-1-propenyl)-4-methyltetrahydropyran, shown in Fig. 4. The isomers are abbreviated as (D) and (L) hereafter. In the first experiment, the samples were measured at the same temperature (18°C) as in the previous section. Since the sensor responses were small, the sample temperature was raised to 24°C so that the vapor concentration could be increased. The experiment was performed five times.

The average sensor outputs and their standard deviations were: i.e., the frequency shifts of (L) were 110 ± 1, 554 ± 14, and 127 ± 7 for sensors with films of α-, β-, and γ-CD, respectively. On the other hand, the frequency shifts of (D) were 261 ± 23, 1045 ± 24, and 223 ± 8 for the sensors of α-, β-, and γ-CD, respectively. It was found from these experiments that the (D) sample output about twice as large a response as the (L) sample from these CD

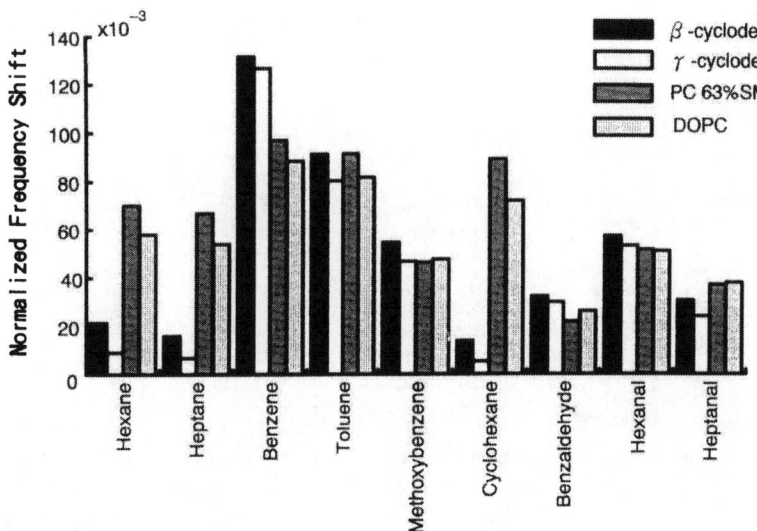

Fig. 3. Normalized frequency shift to four kinds of film. *PC*, phosphatidylcholine; *SM*, sphingomylelin; *DOPC*, dioleylphosphatidylcholine

Rose oxide

$C_{10}H_{18}O = 154.25$

Fig.. 4. Structure of rose oxide

sensors, while the other sensors showed smaller response differences (about 1.2 times). Further, the responses of the β-CD sensor were significantly larger than those of the other two CD sensors. The coefficient of variation was estimated to be about 2%, except for the response of (D) to α-CD. On gas chromatography, using the modified cyclodextrin as the stationary phase material, (D) was eluted more slowly than (L), suggesting that (D) has a larger partition coefficient than (L). This finding agrees with the present result.

Conclusions

The authors developed an odor sensing system using an automatic sampling stage, and found that this system showed considerable efficiency in measuring multiple samples. Twenty samples of essential oils were successfully separated, and the coefficient of variation of the data was reduced to 1.6%.

Modified cyclodextrins were tested as sensing films; the results suggested that these molecules were effective in detecting odorant materials with a benzene ring. It was also found that optical isomers

were separated with the modified cyclodextrin. The responses of the β-CD sensor were significantly larger than those of the other two CD sensors.

Acknowledgment. The authors wish to thank Mr. M. Hayashi of Meidensha Corp. for providing the quartz resonators.

References

1. Nakamoto T, Moriizumi T (1988) Odor sensor using quartz-resonator array and neural-network pattern recognition. Proc IEEE Ultrasonics Symp, Chicago, pp 613
2. Nakamoto T, Fukunishi K, Moriizumi T (1990) Identification capability of odour sensor using quartz-resonator array and neural-network pattern recognition. Sensors and Actuators B1:473–476
3. Nakamoto T, Fukuda A, Moriizumi T, Asakura Y (1991) Improvement of identification capability in an odour-sensing system. Sensors and Actuators B3:221
4. Sauerbrey G (1959) Verwendung von Schwingquartzen zur Wägung dünner Schichten und zur Mikrowägung. Z Phys 155:206–222
5. King WH (1964) Piezoelectric sorption detector. Anal Chem 36:1735
6. Nakamoto T, Moriizumi T (1990) A theory of a quartz crystal microbalance based upon a Mason equivalent circuit. Jpn J Appl Phys 29:963
7. Ide J, Nakamoto T, Moriizumi T (1993) (in Japanese) Nippon Nogeikagaku Kaishi
8. Okahata Y, Ebato H, Taguchi K (1987) Specific adsorption of bitter substances on lipid bilayer-coated piezoelectric crystals. J Chem Soc Chem Commun 1363
9. Ide J, Nakamoto T, Moriizumi T (1993) Development of odour-sensing system using an auto-sampling stage. Sensors and Actuators B13-14:351–354
10. Nakamoto T, Sasaki S, Fukuda A, Moriizumi T (1992) Selection method of sensing membranes in odor-sensing system. Sensors and Materials. 42:111–119

A Sensor for the Detection of p-Cresol in Air

D. Barnett[1], S. Skopec[1], C.A. Clarke[1], G.G. Wallace[2], and D.G. Laing[3]

Key words. Sensor—p-Cresol

The odor of piggeries is unpleasantly distinctive and derives from the volatile components of pig feces, principally p-cresol, skatole, and hydrogen sulfide. With the increasingly urban location of piggeries, nuisance odor is a cause of concern for neighbors and operators alike. Odor control by piggery staff is possible, but because of tolerance, a more objective indicator is needed.

This article describes a sensor for airborne p-cresol based upon the use of antibody-coated piezoelectric crystals [1]. The sensor system is capable of indicating when problem levels of odor are present and it could be used to generate a warning signal or for feedback control of washing.

The antibody was developed in rabbits, to a chemically synthesized analogue of p-cresol. For small entities (haptens), which constitute most odoriferous substances, it is necessary to couple the hapten to an immuno-reactive protein such as bovine serum albumin (BSA). The p-cresol analogue (carboxylic acid) was activated by esterification with N-hydroxy succinimide. The activated ester coupled spontaneously with BSA and the conjugate was injected into rabbits in Freund's adjuvant. Activity of the antibodies was confirmed by immunoprecipitation and their specificity was examined by competitive enzyme-linked immunosorbent assay (ELISA). Piezoelectric crystals with gold electrodes (9 MHz, AT cut) were coated with antibody and dried. The resultant frequency shifts upon exposure to p-cresol of the specifically-coated crystals were detected as the output from an oscillator/mixer [2,3].

In this study, the feasibility of using an antibody to a fecal odor component as a sensor receptor element is indicated. It would appear that the interaction of antibody and antigen in air is reversible and specific. The sensor for p-cresol has a rapid and linear response over the range 5–300 ppm. One earlier report of a similar sensor device detecting the insecticide parathion is the only other example of the specific immuno-detection of a hapten in the vapor phase and the reversibility of antigen/antibody binding in such a situation [1]. Use of a dual crystal detector enables cancellation of temperature and humidity effects [2]. Further study is required to examine in detail the interferences, long-term stability, and reproducibility of the sensor.

Acknowledgments. This work was supported in part by the Pig Research and Development Corporation of Australia. Thanks is also due to Peter Dencher for invaluable support services.

[1] CSIRO, Sensory Research Centre, North Ryde, Australia
[2] Department of Chemistry, University of Wollongong, Northfields Avenue, Wollongong, NSW, 2522, Australia
[3] Faculty of Food Science and Technology, University of Western Sydney, Bourke Street, Richmond, NSW, 2753, Australia

References

1. Ngeh-Ngwainbi J, Foley J, Kuan PH, Guilbault GG (1986) Parathion antibodies on piezoelectric crystals. J Am Chem Soc 108:5444–5447
2. Karmarkar KH, Guilbault GG (1974) A new design and coatings for piezoelectric crystals in measurement of trace amounts of sulphur dioxide. Anal Chim Acta 71:419–424
3. Guilbault GG, Hock B, Schmid R (1992) A piezoelectric immunobiosensor for atrazine in drinking water. Biosensors Bioelectron 7:411–419

Revolutionary Device to Identify Smell —The Development of an Olfactory Receptor-Coated Biosensor

Tzong-Zeng Wu and Hsi-Hua Wang[1]

Key words. Electronic nose—Biosensor

A model system of an artificial nose is proposed in this study. Here we describe the approach toward the development of an electronic nose, by mimicking the sensory pathway of the human olfactory recognition mechanism. Molecular recognition components—olfactory receptor proteins (ORPs), isolated from bullfrogs (*Rana* sp.), were coated onto the surface of a piezoelectric (PZ) electrode. The PZ crystal served as a signal transducer. The ORPs and PZ were combined to form a multi-array PZ-olfactory biosensor. When the odorants bind with the receptor proteins, which are immobilized on the surface of the PZ crystal, a response of decreasing frequency follows. The sensor head is composed of multi-array receptor-coated PZ crystals, so that each one will generate an individual response signal. These signals are then processed by an appropriate mathematical method and are expressed in a discrimination pattern in a hexagonal diagram. The shape of the pattern is regarded as the particular "fingerprint" of the particular odorant under detection. This particular odor pattern can also be further processed by numerical cluster analysis to form a simplified datum (also called coding) and can then be stored in the library of the computer memory. For known odorants, we call the above measurement path the "learning process." Thus, for a particular unknown odorant, the computer will compare the resultant coding with the library database. If a match is found, the nature of the unknown odorant will be identified. If no match is found, the computer will point out the most similar attributions of an odorant by using the method of numerical adjacency. Therefore, this model system clearly suggests how an "intelligent" artificial nose could be built, and, also, both the qualification and quantification of odorants can be easily implemented. Preliminary results indicated a reversible and long-term (more than 5 months) stable response to the volatiles of alcoholic beverages such as n-caproic acid, isoamyl acetate, n-decyl alcohol, β-ionone, and n-octyl alcohol. The sensitivity of receptor protein-coated crystals ranged from 10^{-7} to 10^{-8} g, equivalent to the olfactory threshold value of humans.

[1] National Taiwan University, Department of Agricultural Chemistry, Laboratory of Applied Microbiology, Taipei, Taiwan 10704, ROC

References

1. Fesenko EE, Novoselvo VI, Bystrova MF (1988) Properties of odour-binding glycoproteins from rat olfactory epithelium. Biochim Biophys Acta 937:369–378
2. Ngeh-Ngwainbi J, Suleiman AA, Guilbault GG (1990) Piezoelectric crystal biosensors. Biosensors Bioelectron 5:13–26

15. Chemoreception in Aquatic Organisms

Enhancing Effects of Betaine on the Taste Receptor Responses to Amino Acids in the Puffer *Fugu pardalis*

SADAO KIYOHARA[1], HIROO YONEZAWA[2], and IWAO HIDAKA[3]

Key words. Taste—Enhancing effect—Betaine—Amino acids—Synergism—Fish

Introduction

It is well known that there exists a synergism between monosodium glutamate and nucleotides in human taste sensation [1,2]. A similar synergism is known for these compounds in the peripheral taste nerve responses in rats [3–5]; the neural response is synergistically enhanced by applying them together.

In the taste receptors of fish, synergistic effects have been reported for combinations of acids and salts, pairs of amino acids, and amino acids and other compounds [6–8]. We have shown in the puffer that taste responses to alanine or glycine are greatly enhanced when each amino acid is applied to the receptors after a prior application of betaine [9]. Two types of independent taste units for the amino acid and betaine are found in this species: amino acid-sensitive and betaine-sensitive units. Such an enhancement between alanine and betaine is observed only in the amino-acid-sensitive units, suggesting that betaine behaves as an enhancer for the amino acid receptor responses [9].

To analyze this enhancement of amino acid responses of the puffer by betaine, the effects of betaine on the dose-response curves for amino acids and the structure-activity relations between the amino acids and betaine in the synergism were studied by recording the neural responses from the facial nerve innervating the anterior palate.

Material and Methods

A total of 45 puffers (*Fugu pardalis*) weighing 60 to 270 g were used in the experiments. The fish was lightly anesthetized with MS222 and then im-

mobilized with gallamine triethiodide. After removing the eye, the ramus palatinus facialis was isolated. The electrical activity of the nerve was recorded as the summated response using an electronic integrator.

For successive application of two chemical solutions, a stimulant delivery system with two solenoid valves was used [9]. One solution was applied to the palate for 4 s, and then the other solution was applied for the same period or longer. The palate was irrigated with artificial sea water (SW) throughout the experiment. Amino acids, betaine, and their various derivatives, dissolved in SW, were used as taste stimuli. The pH of the solutions was adjusted from 7.6 to 8.0 using hydrochloric acid or hydroxide.

Results

Figure 1 shows a typical example of the summated whole nerve responses of the palatine nerve to 5×10^{-4} M alanine (Ala) in SW and in 5×10^{-4} M betaine (Bet) after an application of 5×10^{-4} M Bet. The response to Ala at 5×10^{-4} M was small but becomes much greater when Ala is applied after the application of 5×10^{-4} M Bet. There was no significant difference in response magnitude (phasic component) between the Ala single solution applied after Bet, the mixture of Ala and Bet applied after Bet, and that applied without pretreatment, though the tonic component was more distinct in the latter two than in the former. The effects of the application time of Bet on the response to Ala were also studied at 0.5, 1, 2, 4 and 8 s. In every case, the response to Ala was enhanced with no appreciable differences in response magnitude among the five application periods of Bet. In the following experiments, solutions of single amino acids and other chemicals were used to analyze the effect of the prior application of Bet on the responses to them.

Figure 2 shows effects of various concentrations for Bet on the responses to 5×10^{-5}, 5×10^{-4}, and 5×10^{-3} M Ala. In this figure the relative magnitude of the response to Ala is plotted as a function of logarithmic concentration of Bet taking the magnitude of the response to 10^{-2} M Ala without pretreatment with Bet as unity. For each of the three concentrations of Ala, the increase of the response

[1] Department of Biology, College of Liberal Arts and Sciences, Kagoshima University, Kagoshima, 890 Japan
[2] Department of Chemistry, Faculty of Science, Kagoshima University, Kagoshima, 890 Japan
[3] Faculty of Bioresources, Mie University, Tsu, 514 Japan

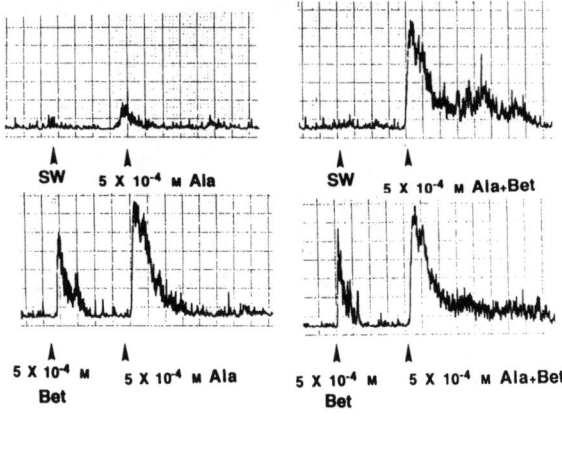

FIG. 1. Typical examples of the summated whole nerve responses to 5×10^{-4} M alanine in sea water and in 5×10^{-4} M betaine after application of 5×10^{-4} M betaine. *Arrowheads* under each record show the onset of stimuli

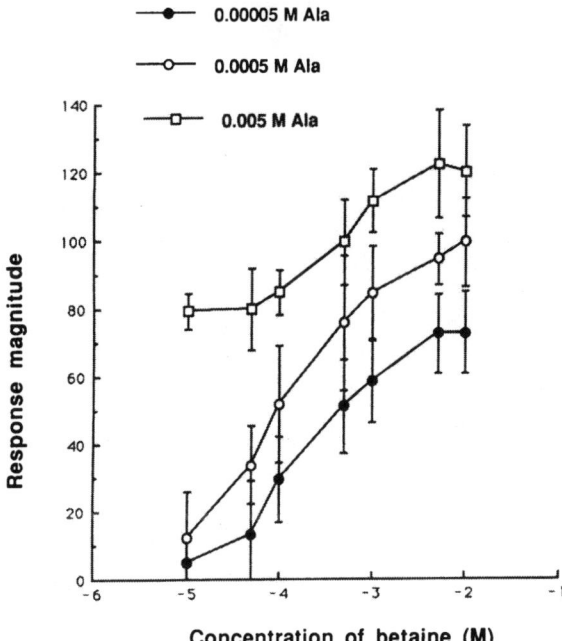

FIG. 2. Effects of betaine on the responses to alanine. Summated responses to three concentrations of alanine are recorded after application of various concentrations of betaine, and their magnitudes are plotted as a function of betaine concentration. The magnitude of response to 10^{-2} M alanine by itself is taken as unity. Each point represents the mean of five preparations. *Vertical bars* represent the standard deviation

was significant at 10^{-4} M of Bet and became larger with increasing concentrations of Bet, attaining a saturation level at around 10^{-2} M.

Dose-response relations for glycine (Gly), serine (Ser), and proline (Pro) after application of 10^{-2} M Bet were also examined, and the results are shown in Fig. 3. A marked enhancing effect of Bet was found for Ser: For example, 10^{-3} M Ser did not elicit any detectable response by itself but yielded a noticeable response when applied after Bet. The dose-response curve for Ser shifted to the left by about 2 log-units, as seen in Fig. 3C: A similar shift of the response curve by Bet was also yielded for Ala (Fig. 3A), and Bet enhanced the responses to Gly over a wide concentration range (Fig. 3B) but did not alter its threshold significantly. In contrast with these three amino acids, the response to Pro was not affected by Bet. Bet also had an enhancing effect on the responses to cysteine, histidine, methionine, monosodium asparatate, and phenylalanine (Table 1).

To understand the mechanism of the synergism between Bet and Ala, the structure-activity relations for both substances were studied using a variety of their derivatives. First, the effect of 5×10^{-4} M Bet on the responses to various derivatives of Ala, Gly, and Ser were examined at 5×10^{-4} M and the results are summarized in Table 1. All the chemicals in Table 1 except Ala, Gly, Pro, and N-methylalanine were ineffective at 5×10^{-4} M by themselves. The results in Table 1 show that replacement of the amino or carboxyl group by $-OH$ or $-H$ (lactic acid, acetic acid, propionic acid, methylamine, ethanolamine) always yields no responses, indicating that both the amino and carboxyl groups are essential for the enhancement by Bet. Removal of a charge from the amino group by formylation or acylation (formylalanine, acetylalanine) abolishes the responses. Introduction of one methyl group (N-methylalanine) exerts no influence on the enhancing effect of Bet, whereas that of two or three (N-di- or N-trimethylalanine) abolishes it almost completely. Shifting the position of amino group from α to β position (β-alanine) also abolished the responses. When the carboxyl group is esterified (alanine methyl or ethyl ester), a marked response is yielded when the ester is applied after Bet, though by itself it is not effective. Amidation of the carboxyl group almost eliminates the enhancement (alanine amide). D-Ala is less effective than L-Ala in the enhancement. When the hydrogen at the α-C atom was replaced by a methyl group (α-aminoisobutylic acid), no detectable response was observed. When the side chain was changed, enhancements were observed for some amino acids as mentioned above.

To reveal which parts of the Bet molecule participate in the enhancement of Ala response, the effects of a variety of derivatives of Bet on the response to Ala was examined (Table 2). The

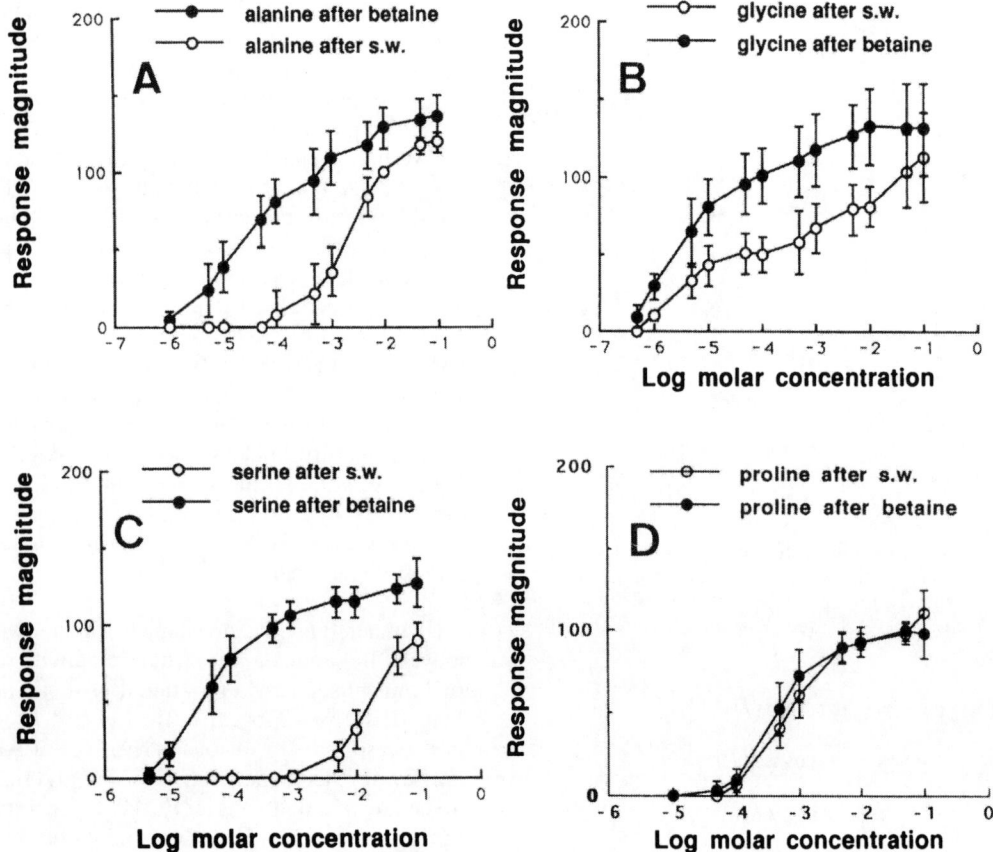

FIG. 3. Relative magnitude of response to various concentrations of alanine (**A**), glycine (**B**), serine (**C**), and proline (**D**) after application of sea water (*s.w.*) or 10^{-2} M betaine. The magnitude of response to 10^{-2} M alanine by itself is taken as unity. Each point represents the mean of five preparations. *Vertical bars* represent the standard deviation

number of methyl groups at N is important: with three, markedly effective; with two, moderately effective; and with one, ineffective. Amidation and esterification of the carboxyl group weakened the enhancing effect to a certain extent but not completely. Substitution of the methyl group at N by the ethyl group (triethylglycine) abolished the enhancing effect. The hydrogen in the side chain is essential. Replacement of the H by CH_3 or CH_2OH (trimethylalanine, trimethylserine) caused complete reduction of the enhancement. Neither acetylcholine nor N-trimethylamine was effective. Replacement of the carboxyl group by CH_2OH (choline) also yielded almost no enhancement.

Discussion

The results described above show that Bet has an enhancing effect on the responses of puffer taste receptor responses to several amino acids. Our previous single fiber analyses and cross-adaptation experiments using Ala, Gly, Pro, sarcosine, dimethylg-lycine (DMG) and Bet suggested the presence of three groups of receptor sites for these chemicals: (1) Ala sites for Ala, Gly, and sarcosine; (2) Pro sites for Pro and DMG; and (3) Bet sites for Bet and DMG. It was also found in the previous study that marked enhancement of the response to Ala or Gly resulted from prior application of Bet to the receptors. This synergism between Ala or Gly and Bet was observed with the amino-acid-sensitive fibers but not with Bet-sensitive ones, showing that the synergism is related to the Ala sites, and Bet enhances the Ala receptors. The profound cross-adaptation of the neural response between Ala and Gly led us to the conclusion that both amino acids might stimulate the same receptor sites. The present finding that the responses to the two amino acids were similarly enhanced by Bet is in accord with the previous conclusion. The enhancing effect of Bet on the responses to 5×10^{-5}, 5×10^{-4}, and 5×10^{-3} M Ala is a function of Bet concentration and becomes saturated at around 5×10^{-3} M, giving different maximum responses depending on the concentration of Ala.

TABLE 1. Effectiveness of amino acids and derivatives at a concentration of 5×10^{-4} M after application of 5×10^{-4} M betaine.

Compound	Relative response magnitude (mean ± SD)	No. of fish
Lactic acid	—	5
Acetic acid	—	5
Propionic acid	—	5
Methylamine	—	5
Ethanolamine	—	5
L-Formylalanine	—	5
L-Acetylalanine	—	9
L-N-Methylalanine	98.9 ± 12.8	5
	(29.8 ± 3.8)	5
L-N-Dimethylalanine	—	5
L-N-Trimethylalanine	—	5
L-Alanine amide	2.1 ± 4.0	9
L-Alanine methylester	76.6 ± 10.4	5
L-Alanine ethylester	73.4 ± 12.4	5
β-Alanine	—	6
D-Alanine	45.6 ± 7.2	7
α-Aminoisobutylic acid	—	5
L-Alanine	92.2 ± 16.1	15
	(17.6 ± 6.8)	15
L-Arginine	—	5
L-Aspartic acid	30.2 ± 9.7	6
L-Cysteine	34.4 ± 12.3	5
L-Glutamine	9.3 ± 9.2	6
L-Glutamic acid	2.5 ± 3.9	10
Glycine	110.4 ± 14.0	5
	(65.4 ± 13.7)	5
L-Histidine	43.4 ± 15.9	5
L-Leucine	0.8 ± 1.6	5
L-Lysine	—	5
L-Methionine	44.4 ± 14.1	5
L-Phenylalanine	48.2 ± 7.7	5
L-Proline	50.2 ± 10.2	5
	(43.4 ± 15.6)	5
L-Serine	66.0 ± 7.5	5
Taurine	—	
L-Threonine	0.8 ± 1.6	5
L-Tryptophan	9.0 ± 10.8	5
L-Varine	2.6 ± 5.2	5

Blanks in this column show no responses to the chemical at 5×10^{-4} M after the application of 5×10^{-4} M betaine. Numbers in parentheses indicate the magnitude of response to the chemical by itself.

TABLE 2. Enhancing effect of betaine and its derivatives at 5×10^{-4} M on the response to 5×10^{-4} M alanine.

Compound	Relative response magnitude (mean ± SD)	No. of fish
Betaine	92.2 ± 16.1	15
Triethylglycine[a]	26.8 ± 14.7	5
N-Dimethylglycine	59.6 ± 9.4	5
N-Methylglycine	8.4 ± 7.4	5
Betaine amide[a]	81.8 ± 22.7	5
Betaine methylester[a]	68.2 ± 10.5	5
Betaine ethylester[a]	62.0 ± 10.3	5
L-N-Trimethylalanine[a]	24.2 ± 12.9	5
D-N-Trimethylalanine[a]	19.8 ± 10.2	5
L-N-Trimethylserine[a]	17.4 ± 2.7	5
Trimethylamine	16.6 ± 2.4	5
Choline chloride	25.0 ± 13.6	5
Acetylcholine	21.2 ± 8.0	5
Control	17.8 ± 6.8	15

Each number indicates the magnitude of response to alanine after the application of the compound except for the control, which represents the magnitude of response to alanine by itself.

[a] Synthesized by H.Y.

the affinity of Ala or Ser for Ala sites so that both amino acids become as effective as Gly at the lower concentration range. On the other hand, the response curves for the three amino acids become saturated at around 10^{-2} M and attain similar magnitudes, which seems to indicate that the total number of the Ala sites does not vary by the application of Bet. A similar mechanism has been proposed for the synergism between sodium glutamate and nucleotides in the taste receptors of rats [5].

Proline stimulates the same fibers or taste units that are stimulated by both Ala and Gly. However, the previous cross-adaptation experiments suggested that the receptor sites for Pro are different from those for the latter two. The present result—that no appreciable enhancement of the response to Pro was observed over a wide concentration range of Pro—is also in accord with the previous conclusion.

The present study of the structure-activity relationship in the synergism between Bet and Ala shows that a molecule must have both the amino group and the carboxyl group to exert the enhancing effect of Bet. Formylation or acylation of the amino group results in no response. It can be explained by either removal of positive charges from the amino group or the steric hindrance due to the modification of the amino group. On the other hand, introduction of one methyl group to the amino group maintains the synergistic effect, suggesting that there is some allowance in the space of the receptor site for the amino group. A negative charge at the carboxyl group is not necessary, as the response to alanine methyl or ethyl ester is enhanced considerably by Bet. Some amino acids were effective at 5×10^{-4} M

In the present study, it was found that Ser and several other amino acids are also involved in the synergism with Bet (Table 1). This finding suggests the possibility that these amino acids also stimulate the Ala sites. In this connection, comparison of the effects of Bet on the dose-response curves for Ala, Gly, and Ser in Fig. 3 shows noticeable tendencies among the three amino acids: First, the stimulatory effectiveness of the three amino acids with decreasing concentrations comes closer to each other by pretreatment of the receptors with Bet, though their original thresholds are more separated from each other. This finding seems to indicate that Bet alters

after the application of 5×10^{-4} M Bet, though they were not effective by themselves even at two orders or more higher concentrations. This finding suggests that Bet binds the taste receptor membrane that is associated with the Ala sites, leading to a change in the membrane conformation, which may in turn remove the steric hindrance to the side chain of the amino acid.

The results obtained in the synergistic effects of Bet derivatives also show that the structural requirement for Bet is rather strict: Only modifications of the carboxyl group, such as esterification, amidation, or reduction of the methyl group at N from three to two, are possible to maintain the enhancing effect at a fairly high levels.

References

1. Kuninaka A (1960) Studies on taste of ribonucleotide derivatives. J Agric Chem Soc Jpn 34:487–492
2. Kuninaka A (1966) Recent studies of 5'-nucleotides as new flavor enhancers. In: Gould RF (ed) Chemistry. American Chemical Society, Washington, DC, pp 261–274
3. Adachi A, Okamoto J, Hamada T, Kawamura Y (1967) Taste effectiveness of mixtures of sodium 5'-inosinate and various amino acids. J Physiol Soc Jpn 29:65–71
4. Hiji Y, Sato M (1967) Synergism between 5'-GMP and amino acids on taste receptors of rats. J Physiol Soc Jpn 29:274–275
5. Yoshii K, Yokouchi C, Kurihara K (1986) Synergistic effects of 5'-nucleotides on rat taste responses to various amino acids. Brain Res 367:45–51
6. Hidaka I (1970) Effects of salts and pH on fish chemoreceptor response. Nature 228:1102–1103
7. Yoshii K, Kamo N, Kurihara K, Kobatake Y (1979) Gustatory responses of eel palatine receptors to amino acids and carboxylic acids. J Gen Physiol 74:301–317
8. Marui T, Harada S, Kasahara Y (1986) Multiplicity of taste receptor mechanisms for amino acids in the carp, Cyprinus carpio L. In: Kawamura Y, Kare MR (eds) Umami: A basic taste. Dekker, New York, pp 185–199
9. Kiyohara S, Hidaka I (1991) Receptor sites for alanine, proline, and betaine in the palatal taste system for the puffer, Fugu pardalis. J Comp Physiol [A] 169:523–530

Sorting Food from Mud: Vagal Gustatory System of Goldfish (*Carassius auratus*)

THOMAS E. FINGER[1]

Key words. CGRP—Glutamate—Swallowing—Nucleus of the solitary tract—Nucleus ambiguus—Vagal lobe

Introduction

Adult goldfish (*Carassius auratus*) and many related carps feed by taking in mouthfuls of substrate and, utilizing specialized oropharyngeal structures, manipulate the substrate to retain food particles while rejecting the inedible material [1]. In these fishes, the posterior part of the oropharynx includes a large, muscular palatal organ and a pharyngeal chewing organ. The palatal organ is studded densely with taste buds and plays a crucial role in sorting food from substrate particles.

This entire oropharyngeal apparatus is innervated by branches of the vagus nerve that provide both sensory and motor innervation to these organs. Previous work [2] has shown that the oropharynx is represented in a mapped (orotopic) fashion within the sensory and motor nuclei of the brainstem of goldfish. The pharyngeal chewing organ is represented within the lateral general visceral nucleus, and the palatal organ is represented in a dorsal evagination of the caudal medulla called the vagal lobe. The vagal lobe consists of a superficial sensory layer homologous to the gustatory part of the nucleus of the solitary tract and a deeper, motor layer homologous to part of the nucleus ambiguus of other vertebrates. Both the sensory and motor layers are organized into mapped representations of oral space. These sensory and motor maps are in register so a given point in the oral cavity is represented in a column within the sensory layer and receives motor innervation from the radially subjacent motor neurons (Fig. 1) [2,3]. A radially inward-directed reflex fiber system connects the sensory to the motor layers in a point-to-point fashion [3,4].

Primary gustatory sensory fibers terminate in three of the ten layers of the sensory zone [2]. The clearly laminated nature of the vagal lobe makes it an ideal system in which to study the neurophysiology and neurotransmitter systems involved in transmitting and processing gustatory inputs to the medulla. The studies described herein are representative of the multiple approaches we are utilizing to study gustatory functions in the goldfish. In vitro recording and histological experiments were used to test whether glutamate plays a role as a neurotransmitter in the lobe; immunohistochemical studies were used to examine neuromodulatory inputs.

Materials and Methods

The studies described below were carried out on common goldfish. *Carassius auratus*, ranging in size from 5 to 15 cm (standard length). For all invasive procedures, the fish were anesthetized with a 1:10000 solution of tricaine methanesulfonate. For immunohistochemical studies, anesthetized animals were perfused transcardially with teleost Ringer's solution followed by a fixative solution of 4% paraformaldehyde in phosphate buffer (0.1 M, pH 7.2) with or without the addition of 0.05% to 0.30% glutaraldehyde. Sections were cut either on a vibratome or, after cryoprotection in 20% buffered sucrose, on a cryostat.

Primary antibodies employed were directed against calcitonin gene-related peptide (CGRP) (C. Sternini, UCLA), cholecystokinin (CCK) (Cambridge Research Biochemicals, Wilmington, DE), and glutamate (J. Madl, CSU, Ft. Collins, CO). Biotinylated secondary antibodies were obtained from Jackson Laboratories (West Grove, PA); standard avidin-biotin-complex peroxidase procedures were used for ultimate localization of immunoreactivity.

In vitro recording was made from transverse slices (approximately 500 μm thick) of the vagal lobe obtained from anesthetized fish. The slices were allowed to recover for at least 1 h in artificial fish CSF placed in a slice recording chamber [5] prior to recording. A stimulating bipolar electrode was placed on the incoming vagal root fibers; a glass recording electrode was situated in an area of maximum evoked activity in the sensory layers. Drugs were applied into the recording chamber through a perfusion system that constantly supplied bathing medium to the chamber at a flow rate of 2 ml/min.

[1] Department of Cellular & Structural Biology, University of Colorado School of Medicine, 4200 E. Ninth Avenue, Denver, CO 80262, USA

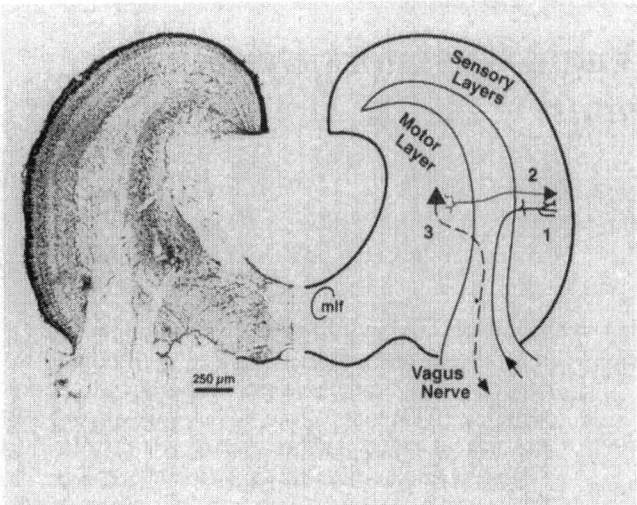

FIG. 1. Transverse section through the vagal lobe of a goldfish. (*Left*) Photomicrograph shows the pattern of lamination as seen with a Nissl stain. (*Right*) Line drawing shows the sensory and motor layers separated by the central fiber layer, which is continuous with the vagal nerve root. The essential reflex circuitry is shown: (*1*) primary gustatory afferent fiber; (*2*) reflex interneurons connecting to the motor layers; and (*3*) motoneurons that innervate the palatal musculature. *mlf*, medial longitudinal fasciculus

For these studies on glutamate, the antagonist kynurenic acid was added to the bathing medium at final concentrations of 50 or 100 μM.

Results

Glutamate in the Primary Gustatory Center

Immunocytochemistry

Glutamate-like immunoreactivity was present in neurons, fibers, and puncta within the vagal lobe. Studies by other investigators have shown that with the fixation and reaction conditions employed, punctate immunoreactivity is indicative of glutamatergic terminals; axonal and somatic reactivity is a less reliable indicator of glutamatergic systems. Punctate immunoreactivity is present within most of the sensory layers except layers iv and vi (Fig. 2B). These layers are those that receive heavy input from the primary gustatory afferent fibers (Fig. 2A). This inverse correlation indicates that primary gustatory afferent fibers do *not* utilize glutamate as a neurotransmitter.

The motor layer of the vagal lobe also is replete with punctate immunoreactivity surrounding the motoneurons (Fig. 2C). In addition, numerous radially oriented fascicles of immunoreactive fibers course between the sensory and motor layers of the lobe (Fig. 2B). Similarly, immunoreactive fibers enter the ascending secondary gustatory tract and terminate in the superior secondary gustatory nucleus (pontine taste area). These findings are consonant with the interpretation that glutamate is utilized as a neurotransmitter by second- and higher-order neurons of the vagal lobe, including those neurons projecting to brainstem reflex systems and to the pontine gustatory nuclei.

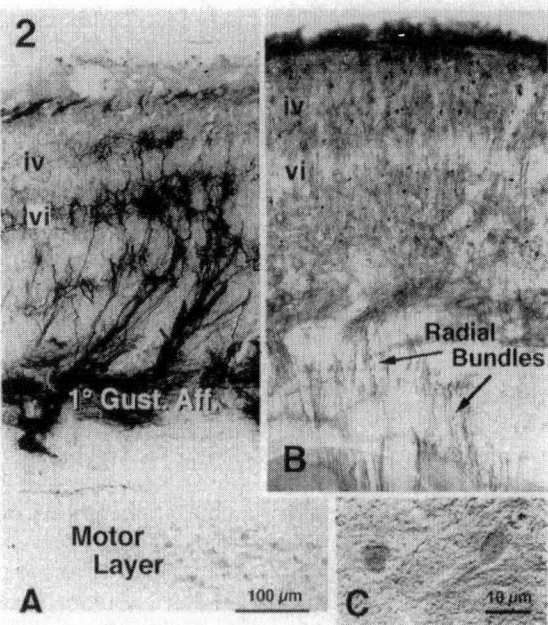

FIG. 2. **A** Photomicrograph of a section through the vagal lobe in which the primary afferent fibers had been filled with HRP (*black*). These fibers terminate mostly in layer vi with lesser inputs to layers iv and ix. **B** Glutamate-like immunoreactivity in the sensory layers of the vagal lobe. Note the absence of punctate immunoreactivity (*clear area*) corresponding to layers vi and iv. Also, radial bundles of immunoreactive fibers cross the central fiber layer to enter the motor layer. Magnification as for **A**. **C** Nomarski optics view of punctate immunoreactivity around the motoneurons of the motor layer

Electrophysiology

The evoked population response following electrical stimulation of the vagal nerve root is a complex waveform containing three negative-going spikes

[5]. The first negative potential is attributed to the incoming primary afferent fiber volley; the second spike is a synaptic response due to activation of the primary afferent fibers; and the third spike is a higher-order synaptic response presumably due to activation of second-order neurons. With application of 50 or 100 µM kynurenic acid, the second and third negative waves are abolished; the first wave is unaffected. These findings indicate that the first-order synapse does utilize an excitatory amino acid or other kynurenate-sensitive neurotransmitter.

In summary, primary gustatory afferent terminals do utilize a kynurenate-sensitive neurotransmitter despite the absence of glutamate immunoreactivity. Other neurons within the vagal taste system, including those that provide reflex connection to the brainstem motor nuclei and those conveying gustatory information to pontine centers, do appear to utilize glutamate.

Neuromodulators in the Vagal Lobe

Both CGRP and CCK have been implicated in the central regulation of feeding behavior. Immunocytochemical studies were carried out to determine whether neural systems utilizing either of these neuromodulators provide input to the vagal lobe.

CGRP

Immunoreactivity to CGRP antisera was present within the superficial fiber plexus in the rostral half of the vagal lobe (Fig. 3A). Vagal rhizotomy shows that these superficial immunoreactive fibers are primary afferents innervating branchial structures.

CCK

The CCK-immunoreactive fibers are present within both the sensory and motor zones of the vagal lobe. Those in the sensory zone lie in deep layers as well as in the superficial fiber layer. CCK-immunoreactive fibers in the motor zone lie mostly superficial to the motoneuron somata and are dorsally continuous with the immunoreactive fibers of the deep sensory layers (Fig. 3B). The CCK-immunoreactive fibers appear to arise from a longitudinal fiber system that courses along the ventrolateral margin of the brainstem tegmentum. The origin of this system is unknown, although numerous CCK-immunoreactive cell bodies of the pontine raphe system appear to be continuous with the ventrolateral fiber bundle.

Discussion

The regular, laminated structure of the vagal lobe of goldfish is conducive to anatomic and physiologic

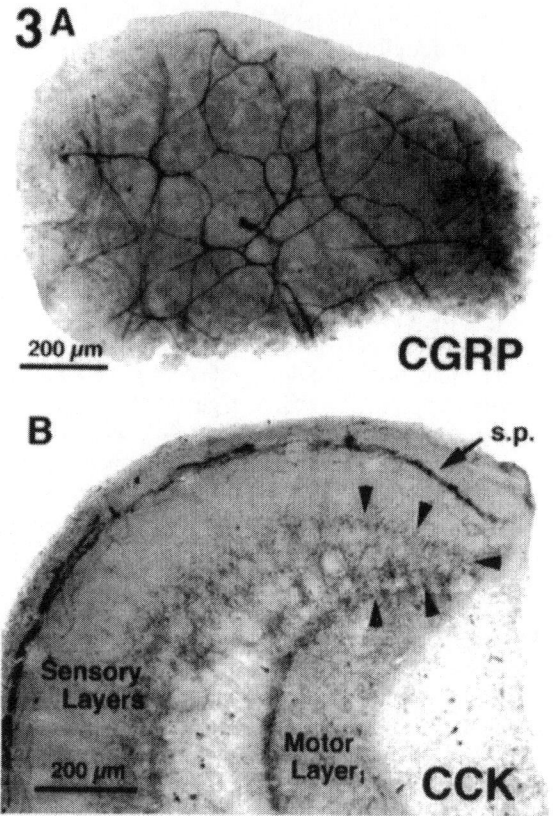

FIG. 3. Immunocytochemistry of the vagal lobe. **A** CGRP-like immunoreactivity in a tangential section skimming through the superficial plexus of fibers showing the fine net-like appearance of this layer. **B** CCK-like immunoreactivity in a transverse section shows reactivity in the superficial plexus (*s.p.*) as well as a band of punctate labeling (*arrowheads*) reaching from the deep sensory layers to the motor layer

experiments on the role of different neurotransmitters and neuromodulators in processing gustatory information. The experiments described above indicate that: (1) glutamate is used as a neurotransmitter by interneurons and projection neurons of the vagal lobe but not by the primary gustatory afferents; (2) capsular root fibers of the vagus nerve contain CGRP, but they are not primary gustatory afferent fibers; and (3) CCK-containing fibers originating from elsewhere in the central nervous system (CNS) provide substantial input to the vagal lobe.

Glutamate

Glutamate is an excitatory neurotransmitter at many synapses within the CNSs of all vertebrates. In particular, glutamate is utilized by several primary afferent systems, including those of the somatosensory system [6]. some evidence also exists that glutamate may serve as a neurotransmitter in at least some primary afferent fibers of the vagus nerve [7,8]. To test whether glutamate serves as a neurotransmitter for primary vagal gustatory fibers, both immunohistochemical and pharmacologic approaches were utilized. The anatomic experiments showed extensive glutamate-like immunoreactivity within the vagal lobe but not in the principal layers of termination of the primary gustatory fibers (layers iv and vi). Conversely, in in vitro recording experiments, kynurenic acid, a broad-spectrum glutamate antagonist, does block the synaptic potential attributed to activation of the primary gustatory fibers. Thus, while glutamate does not seem to be a neurotransmitter for the primary gustatory nerve terminals another kynurenate-sensitive transmitter is. These results are consonant with biochemical results showing that glutamate levels in the vagal lobe are not changed following vagal rhizotomy [9].

CGRP

The neuropeptide CGRP is present in many fine-caliber somatosensory fibers [10] as well as being distributed in many visceral sensory nuclei within the CNS [11]. Immunohistochemical results show that although CGRP is present in primary sensory fibers entering the superficial root of the vagal lobe, this peptide is absent from the primary gustatory fibers. The type of peripheral receptor innervated by these CGRP-immunoreactive afferents is unknown but will be the object of further study.

CCK

The neuroactive peptide CCK has been implicated in central control of feeding and satiety [12]. Accordingly, we examined the vagal lobe to determine if any CCK-containing central inputs reached the primary gustatory centers. A prominent CCK-immunoreactive fiber system extends from the ventrolateral brainstem into both the sensory and motor layers of the vagal lobe. The location of the cells of origin and the possible role of this system in modulating vagal lobe neuronal activity remain to be determined.

Conclusion

The studies described demonstrate the role of glutamate in the vagal lobe and establish the presence of two potential neuromodulatory systems. Future experiments will attempt to determine the neurotransmitter of primary gustatory afferent terminals and the physiologic role of CCK and CGRP in gustatory processing.

Acknowledgments. The immunocytochemical studies were carried out in collaboration with K. Cullen-Dockstader, B. Böttger, and C. Adams; in vitro physiologic studies were done with T.V. Dunwiddie and C.F. Lamb. This work supported by NIH grant DC 00147 to T.E.F.

References

1. Sibbing FA (1982) Pharyngeal mastication and food transport in the carp (Cyprinus carpio L.): a cineradiographic and electromyographic study. J Morphol 172: 223–258
2. Morita Y, Finger TE (1985b) Topographic representation and laminar organization of the vagal gustatory system in the goldfish, Carassius auratus. J Comp Neurol 238:187–201
3. Finger TE (1988) Sensorimotor mapping and oropharyngeal reflexes in goldfish, Carassius auratus. Brain Behav Evol 31:17–24
4. Goehler L, Finger TE (1992) Functional organization of vagal reflex systems in the brainstem of the goldfish, Carassius auratus. J Comp Neurol 319:463–478
5. Finger TE, Dunwiddie TV (1992) Evoked responses from an in vitro slice preparation of a primary gustatory nucleus: the vagal lobe of goldfish. Brain Res 580: 27–34
6. DeBiasi S, Rustioni A (1988) Glutamate and substance P coexist in primary afferent terminals in superficial laminae of spinal cord. Proc Natl Acad Sci USA 85:7820–7824
7. Dietrich WD, Lowry OH, Loewy AD (1982) The distribution of glutamate, GABA and aspartate in the nucleus solitarius of the cat. Brain Res 237:254–258
8. Schaffer N, Pio J, Jean A (1990) Selective retrograde labeling of primary vagal afferent cell bodies after injection of [^{3}H]D-aspartate into the rat nucleus tractus solitarii. Neurosci Lett 114:253–258
9. Contestabile A, Villani L, Bissoli R, Poli A, Migani P (1986) Cholinergic, GABAergic and excitatory amino acidic neurotransmission in the goldfish vagal lobe. Exp Brain Res 63:301–309
10. Kruger L, Mantyh PW, Sternini C, Brecha NC, Mantyh PW (1988) Distribution of calcitonin-gene-related peptide immunoreactivity in relation to the rat central somatosensory projection. J Comp Neurol 273:182–213
11. Yasui Y, Saper CB, Cechetto DF (1989) Calcitonin gene-related peptide immunoreactivity in the visceral sensory cortex, thalamus and related pathways in the rat. J Comp Neurol 290:487–501
12. Antin J, Gibbs J, Holt J, Young RC, Smith GP (1975) Cholecystokinin elicits the complete behavioral satiety sequence in rats. J Comp Physiol Psychol 89:784–790

Amino Acids and Other Stimuli for Catfish Feeding, Escape, and Aversion Behaviors

Tine Valentinčič[1]

Key words. Taste—Olfaction—Feeding—Reflexes —Behavior patterns—Conditioning

Introduction

Catfish detect amino acids by taste and olfactory systems [1–3]. Olfactory and taste information are fed into different areas of the brain, suggesting that in the control of behavior the roles of olfaction and taste are different. It was indicated that the taste system directly releases and guides the feeding behavior of fish [4–6], whereas olfactory mechanisms control feeding arousal and excitatory states and the formation of the search image [7]. In catfish, gustatory information is transmitted to the enlarged facial and vagal lobes of the medulla [8–10]. This structural separation of sensory inputs and of the segmentally organized facial and vagal lobes [11,12] is expected to result in a differential sensory control of motor patterns of feeding behavior during the orienting response and the appetitive and consummatory phases of feeding behavior.

Escape behavior has a high survival value, and the tendency to escape is always present in organisms. In addition, the escape behavior has priority over feeding and other behaviors. Even under carefully selected experimental conditions, stimuli that release escape behavior cannot be totally avoided, and the escape behavior frequently occurs during experiments. Therefore knowledge of the stimuli that release escape behavior is required to successfully conduct experiments in feeding behavior. Escape behavior in channel catfish is released by visual stimuli, such as objects shown from above [13], whereas the escape behavior of bullhead catfish is released by low-frequency vibrations (Valentinčič and Pečenko, unpublished data).

Materials and Methods

Dilutions and distributions of stimulus eddies were established by injecting dopamine and measuring its concentrations with carbon fiber redox electrodes

positioned at various locations in the test aquaria [14]. Super-VHS recordings were analyzed to reveal how chemical, visual, and low frequency vibrational stimuli influenced the behavior of catfish. Escape responses of channel catfish (*Ictalurus punctatus*) to visual stimuli, such as views of darkly colored substrate and large objects, were tested experimentally. To determine the properties of low frequency vibration that release escape behavior and cardiac arrests, electrodes were chronically implanted into the pericardium of a bullhead catfish (*Ictalurus nebulosus*), which was restrained in a Plexiglas chamber in an 80-liter aquarium positioned near the concrete block used for seismometry by the Ljubljana Geological Survey station. Cardiac arrests and ensuing bradycardias [15–17] were monitored parallel to the seismologic recordings. Orienting, appetitive, and consummatory behavior patterns of the of both bullhead and channel catfish feeding behavior were recorded with a super VHS camera after stimulation with amino acids and derivatives. Discrete behavioral events (behavior patterns) were counted either directly or from video recordings and their occurrences compared under different experimental conditions using nonparametric procedures (Wilcoxon rank and Fischer's exact tests).

In experiments investigating the responsiveness to taste stimuli, either single or both olfactory organs together with the entire nasal epithelium were removed under MS-222 (3-aminobenzoic acid ethyl ester in concentration 1:8000) anesthesia. Anosmic catfish were tested several months after the surgery when the surgical wounds were regrown with connective tissue and skin. Both animal species showed normal behavioral responsiveness to taste stimulation from 1 month to several years after the surgery. Olfactory responsiveness was tested in intact catfish, in unilaterally osmic and in anosmic catfish. Bullhead and channel catfish were conditioned during 40 to 50 trials to single amino acids with a food reward given either 90 s (channel catfish) or 120 s (bullhead catfish) after the amino acid stimulus injection. One milliliter of a 10 mM amino acid solution was injected into a test aquarium yielding an expected contact concentration of 0.001 to 0.010 mM. After 50 conditioning trials, a short discrimination training (5–10 trials) was applied successively using the conditioned and a noncondi-

[1] Department of Biology, University of Ljubljana, Aškerčeva 12, 61000 Ljubljana, Slovenia

tioned amino acid. Later the conditioned amino acid and several other amino acids and analogs were tested in the same fish.

Results

The escape and feeding behavior conflict arises when stimuli that release the two opposing behaviors occur simultaneously. The central excitatory escape state was triggered in channel catfish by white substrate, such as white gravel, which inhibits feeding behavior. Channel catfish did not feed in room light, whereas the same animals exhibited normal feeding behavior less than 60 s after the room lights were switched off. Escape behavior by channel catfish was also triggered by visual stimuli, such as large objects presented from above. Dummy bird models and paper strips were equally effective in releasing escape behavior. In bullhead catfish, the major key stimuli for escape behavior were low frequency vibrations of 0.1 to 10.0 Hz. Such vibrations originated from nearby events, such as the opening and closing of doors, city traffic, industrial noise, and low frequency vibrations originating in earthquakes of magnitudes greater than 2.2 Richter scale with epicenter distances of up to several hundred kilometers from the Ljubljana Geological Survey Station.

In experiments investigating the responsiveness to chemical stimuli, key stimuli that release escape behavior by either of the two catfish species were carefully avoided. The following behavior patterns of feeding behavior occurred after stimulation with feeding stimuli: barbel twitching, orienting response, searching swim, turning to food, biting-snapping, opercular movements, mastication, and swallowing. The most effective stimuli that elicited snapping behavior in anosmic channel catfish were L-alanine and L-arginine at low concentrations (10^{-6} to 10^{-4} M) and L-alanine and L-proline at concentrations above 10^{-3} M. The number of bite-snaps, which varied from 1 to 12 per single stimulation, was dependent mainly on the number of eddies containing the amino acid stimulus at concentrations greater than the behavioral threshold for single amino acids and not on the total concentration of the amino acid within a single eddy. The most effective stimuli (L-alanine, L-arginine, L-proline) for releasing biting-snapping behavior in intact animals were previously determined to be the most potent stimuli tested electrophysiologically [1,2]. The number of bite-snaps was influenced by the swimming activity that brought the catfish into and out of the areas containing the stimulus eddies.

The conditioning paradigm resulted in increased swimming activity in intact, semiosmic, and to a lesser extent anosmic catfish. The swimming activity would in nature increase the chances of the animal to encounter the source of stimulation. After 50 conditioning trials and a short discrimination training (5–15 trials) comprising alternating conditioning trials and presentations of the nonconditioned stimulus presented without a food reward, the channel catfish started to swim more to the conditioned amino acid than to the nonconditioned ones.

In channel catfish, the only compound that released the swimming response as effectively as the conditioned stimulus, L-proline, was L-pipecolic acid. All other compounds cross-tested with L-arginine, L-proline, L-alanine, and L-lysine conditioning stimuli were shown to be significantly less effective than the conditioned stimuli. In nearly all cases the amount of swimming activity after stimulation with the conditioned amino acids was approximately twice that of the swimming activity after stimulation with other compounds.

Bullhead catfish that resided in shelters (plastic tubes) did not leave the shelters in cross-tests when the applied amino acid was essentially different from the conditioning amino acid. Amino acids with properties similar to the conditioned amino acid, such as short-chain neutral amino acids, which are similar to L-alanine, released considerable swimming activity. This swimming activity was in most cases significantly less than the swimming activity released by the conditioned stimulus. The following conditioning stimuli were tested: L-alanine, L-cysteine, L-arginine, and L-norleucine. In bullheads, D-alanine was equally stimulatory in releasing swimming behavior as the conditioned stimulus, L-alanine. Glycine also provoked searching activity. Importantly, it was shown that in most cases even substances similar to the conditioned stimulus. L-norleucine, such as L-leucine, L-isoleucine, L-valine, and L-norvaline, were discriminated from L-norleucine.

Unilaterally, osmic channel catfish conditioned to L-alanine showed less swimming activity than intact channel catfish. However, their swimming activity to stimulation with a conditioned amino acid was significantly greater than that after stimulation with other amino acids or their analogs. Anosmic channel catfish also increased their general swimming activity after they were repeatedly stimulated with amino acids and given a food reward at the end of the test. These anosmic catfish did not swim more to the amino acid presented repeatedly with a food reward than to a novel amino acid stimulus. Thus at least one intact olfactory organ was necessary for discriminating conditioned from nonconditioned stimuli.

Aversion to previously harmful food protects animals from consuming the same toxic food repeatedly. LiCl-soaked food induced a short-term aversion in bullhead catfish that did not last beyond the duration of the malaise behavior. After only a few unpleasant experiences with an undigestible material that increased its volume in the stomach before it was regurgitated, the aversion lasted for several weeks. In initial tests, all inexperienced catfish chewed and ingested the plastic material; however, within 24 to 72 h after ingestion the fish regurgitated and rejected this material. A strong aversion to the plastic material developed during the first six repeated presentations. Later, the aversive response consisted of either rejection of the plastic object when it touched the lips or barbels of the catfish or the expectoration of it after the material was taken into the mouth. When the plastic material soaked in 1 M L-alanine solution yielding an expected concentration of less than 1 mM upon contact with the bullhead catfish, it provoked turning toward the object and biting-snapping activity even during tests when the plastic material was rejected following tactile contact.

Discussion

The taste system directly controls the feeding behavior of catfish. Different parts of the medulla are involved to a different extent in driving the sequential behavior patterns of the complex feeding behavior. Thus barbel movements, orienting posture, search swimming activity, and turning to food, which are released by either chemical or mechanical stimuli, are largely controlled by the facial lobes. The later consummatory phases of the feeding behavior are controlled either by both facial and vagal lobes (biting-snapping activity) or by the vagal lobe exclusively, such as opercular movements, mastication, and swallowing.

In contrast, the olfactory system does not contribute to the direct release of feeding behavior in catfish. Olfaction serves an arousal function and enables the animals to learn to discriminate chemical stimuli, such as amino acids and their mixtures. The increased food-searching activity by catfish in part results from decreased inhibition from the escape behavior and its central excitatory escape state (fear); however, the specific increase in food-searching activity observed after stimulation with a conditioned amino acid is a result of learning, which increases the expectation of a food reward. It is important to point out that both channel and bullhead catfish discriminated the conditioned stimulus from most of the chemical stimuli cross-tested in our experiments. Such a result was expected based on the findings of multiple olfactory receptor sites in catfish and other species that were demonstrated electrophysiologically [18–21] and with molecular cloning [22–24] and behavioral techniques [11,25,26].

Conclusions

Catfish detect amino acids with facial and vagal taste and olfactory systems. The facial taste system is a skin taste system that collects information from barbels and other taste buds, whereas the vagal taste system is the intraoral system involved in control of initial food processing within the mouth and swallowing. The two taste informations are fed into enlarged areas of medulla: the large facial (typical for catfish) and vagal lobes, respectively. This separation of sensory and central structures enables separate control of behavior patterns during the consummatory phase of feeding. In channel catfish (*Ictalurus punctatus*) L-alanine, L-proline, L-arginine at low concentrations (0.001–1.000 mM), and some other short-chain neutral amino acids release reflexes and behavior patterns (turning, snapping/biting, and opercular movement behaviors), whereas L-arginine (>10 mM) at high concentrations releases rhythmic movements of the hyoid probably identical to mastication. Any of these behavior patterns can be released independently or as a part of the sequence of the feeding behavior patterns. In addition, intermittent taste stimulation derived from the patchy distribution of stimuli in eddies maintains the central excitation for feeding. In contrast, the olfactory system is not involved in direct release of single behavior patterns. Its main function is maintenance of feeding arousal; and became of its plasticity, it enables the conditioning to environmental odors including single amino acids and complex olfactory stimuli. The searching image of a single amino acid, such as L-alanine, L-arginine, L-isoleucine, or L-cysteine, enables the catfish to remain excited after stimulation with the conditioned amino acids significantly more than after any other amino acid stimulation.

References

1. Caprio J (1978) Olfaction and taste in the channel catfish: an electrophysiological study of the responses to amino acids and derivatives. J Comp Physiol 123: 357–371
2. Caprio J (1988) Peripheral filters and chemoreceptor cells in fishes. In: Atema J, Fay RR, Popper AN, Tavolga WN (eds) Peripheral filters in chemoreceptor cells in fishes. Springer, New York, pp 313–338
3. Marui T, Caprio J (1992) Teleost gustation. In: Hara TJ (ed) Fish chemoreception. Chapmar & Hall, London, pp 171–198
4. Sato M (1937) Further studies on the barbels of a Japanese goatfish, Upenoides bensasi (Temminc & Schlegel). Sci Rep Tohoku Imp Univ 11:297–302
5. Bardach JE, Todd JH, Crickmer RK (1967) Orientation by taste in fish of genus Ictalurus. Science 155: 1276–1278
6. Atema J (1971) Structures and functions of the sense of taste in the catfish (Ictalurus natalis). Brain Behav Evol 4:273–294

7. Atema J, Holland K, Ikehara W (1980) Olfactory responses of yellowfin tuna (Thunnus albacares) to prey odors: chemical search image. J Chem Ecol 6: 457–465

8. Finger TE (1987) Gustatory pathways in the bullhead catfish. II. Facial lobe connections. J Comp Neurol 180:691–706

9. Kanwal JS, Caprio J (1987) Central projections of the glossopharyngeal and vagal nerves in the channel catfish, Ictalurus punctatus: clues to differential processing of visceral inputs. J Comp Neurol 264:216–230

10. Hayama T, Caprio J (1989) Lobule structure and somatotopic organization of the medullary facial lobe in the channel catfish Ictalurus punctatus. J Comp Neurol 285:9–17

11. Goehler LE, Finger TE (1992) Functional organization of vagal reflex systems in the brain of the goldfish, Carassius auratus. J Comp Neurol 319:463–478

12. Finger TE (1993, in press) Sorting food from mud: the vagal gustatory system of goldfish. In: 11th International Symposium on Olfaction and Taste

13. Valentinčič T, Caprio J (1994) Chemical and visual control of feeding and escape behaviors in channel catfish, Ictalurus punctatus. Physiol Behav 55:845–855

14. Moore PA, Gerhardt GA, Atema J (1989) High resolution spatio-temporal analysis of aquatic chemical signals using microelectrochemical electrodes. Chem Senses 14:829–840

15. Holland KN, Teeter JH (1981) Behavioral and cardiac reflex assays of the chemosensory acuity of channel catfish to amino acids. Physiol Behav 27:699–707

16. Little EE (1981) Conditioned cardiac response to the olfactory stimuli of amino acids in the channel catfish, Ictalurus punctatus. Physiol Behav 27:691–697

17. Morin P-P, Dodson JJ, Dore FY (1987) Heart rate conditioning to L-cysteine and other chemical stimuli in young Atlantic salmon (Salmo salar). Can J Zool 65:2710–2714

18. Caprio J, Byrd RP (1984) Electrophysiological evidence for acidic, basic, and neutral amino acid olfactory receptor sites in the catfish. J Gen Physiol 84: 403–422

19. Ohno T, Yoshii K, Kurihara K (1984) Multiple receptor types for amino acids in the carp olfactory cells revealed by quantitative cross-adaptation method. Brain Res 310:13–21

20. Sveinsson T, Hara TJ (1990) Multiple olfactory receptors for amino acids in arctic char (Salvelinus alpinus) evidenced by cross-adaptation experiments. Comp Biochem Physiol [A] 97:289–293

21. Kang J, Caprio J (1991) Electro-olfactogram and multiunit olfactory receptor responses to complex mixtures of amino acids in the channel catfish, Ictalurus punctatus. J Gen Physiol 98:699–721

22. Buck L, Axel R (1991) A novel multigene family may encode odorant receptors: a molecular basis for odor recognition. Cell 65:175–187

23. Ngai J, Chess A, Dowling MM, et al. (1993) Coding of olfactory information: topography of odorant receptor expression in the catfish olfactory epithelium. Cell 72:667–680

24. Ngai J, Dowling MM, Buck L, Axel R, Chess A (1993) The family of genes encoding odorant receptors in the channel catfish. Cell 72:657–666

25. Zippel HP, Voight R, Knaust M, Luan Y (1993) Spontaneous behavior, training and discrimination training in goldfish using chemosensory stimuli. J Comp Physiol 172:81–90

26. Valentinčič T, Wegert S, Caprio J (1994) Learned olfactory discrimination versus innate taste responses to amino acids in channel catfish, Ictalurus punctatus. Physiol Behav 55:865–873

Specific Amino Acid Binding to Membrane Fractions of Carp Taste Epithelium

Yuji Amagai and Takayuki Marui[1]

Key words. Fish taste receptor—Amino acid binding

The gustatory system of the carp, *Cyprinus carpio*, is highly sensitive to amino acid stimuli. The amino acid receptor mechanisms have been characterized in previous electrophysiological studies to have thresholds in the nM range. Self- and cross-adaptation experiments have indicated independent taste receptors for the L-isomers of alanine, glutamate, aspartate, proline, and cysteine, and for glycine betaine [1]. Receptor adaptation with each of these amino acids did not affect taste responses to the other amino acids. This study of Marui et al. [1] suggested the presence of six different types of high affinity amino acid receptors in the taste cell membranes of the carp.

To elucidate the putative stimulus-receptor interactions, we investigated the binding of [³H]-labeled amino acids to a membrane fraction prepared from the carp taste epithelium, using the methods of Krueger and Cagan [2] and Kalinoski et al. [3]. Briefly, barbels and lips containing taste buds were dissected from the carp; they were then washed, minced, and homogenized in 1 mM $NaHCO_3$ and 1 mM $CaCl_2$. The homogenate was centrifuged at $2000 \times g$, and the resulting supernatant was centrifuged at $7300 \times g$. The precipitate was washed three times with rinse buffer (10 mM Tris-HCl, 1 mM $CaCl_2$; pH 7.8) and used as the membrane fraction. Binding reactions were carried out by incubating the membrane fraction (25–50 μg) in a rinse buffer with the [³H]-labeled amino acid (L-alanine, L-glutamate, L-aspartate and L-arginine, respectively) at 0°C for an appropriate time. The incubation was stopped by filtering the reaction mixture rapidly through a nitrocellulose membrane filter (pore size; 0.22 μm). The radioactivity that remained on the membrane was then measured. Non-specific binding was obtained in the presence of an excess of the non-labeled amino acid. The amount of [³H] L-alanine binding was proportional to the amount of protein added, and optimal binding occurred at pH 7.5. The time course of the binding was rapid; half-maximal binding was achieved within 5 min of incubation and a plateau was reached at 30 min. Specific binding was augmented depending on the ligand concentration. Scatchard analysis indicated the presence of both high- and low-affinity binding sites. Ligand binding was specific for taste organ membrane, indicated by our findings that binding was negligible in membrane fractions prepared from head skin, where there are no taste buds. In the presence of a self-adapting ligand at 300 μM, the binding of L-alanine, L-glutamate, and L-arginine decreased to 25%, 37%, and 17% of control, respectively; for L-aspartate binding, the addition of 1 mM of non-labeled ligand decreased the binding to 33% of control. In contrast, the binding of each of the amino acids was not affected by the presence of any of the other amino acids at concentrations of up to 300 μM. These results are indicative of the binding of amino acids to membrane fractions derived from taste receptor membranes. The binding of each amino acid was not competitive with the other amino acids used and was consistent with previous electrophysiological results that indicated the presence of specific taste receptor binding sites for these amino acids in the facial taste system of the carp.

References

1. Marui T, Harada S, Kasahara Y (1987) Multiplicity of taste receptor mechanisms for amino acids in the carp, *Cyprinus carpio* L. In: Kawamura Y, Kare MR (eds) Umami: A basic taste. Dekker, New York, pp 185–199
2. Krueger JM, Cagan RH (1976) Biochemical studies of taste sensation VI. Binding of L-[³H]alanine to a sedimentable fraction from catfish barbel epithelium. J Biol Chem 259:88–97
3. Kalinoski DL, Bryant BP, Shaulsky G, Brand JG, Harpaz S (1989) Specific L-arginine taste receptor sites in the catfish, *Ictalurus punctatus*: Biochemical and neurophysiological characterization. Brain Res 488: 163–173

[1] Department of Oral Physiology, Ohu University School of Dentistry, 31-1 Aza Misumido, Tomita-machi, Koriyama, Fukushima, 963 Japan

L-Lactate Sensitivity of Fish Taste Receptors

Jun Kohbara, Kazuhiro Oohara, and Iwao Hidaka[1]

Key words. Taste—Lactic acid

Organic acids are abundant in the tissue extracts of fish and other aquatic organisms. However, our knowledge of their stimulatory effect on feeding and their stimulatory effect on the taste receptors of fish is still fragmentary. In a previous study [1], the palatal taste receptors of the yellowtail *Seriola quinqueradiata* were found to be sensitive to some organic acids, especially to L-lactic acid. This suggests the biological importance of these acids associated with the gustatory sense of fish, since the tissues of potential food organisms contain these acids at varying concentrations as cell metabolites. Moreover, it seems interesting to determine whether lactic acid stimulates the receptor sites for amino acids, since its chemical structure has some resemblance to alanine.

In this study, we examined, electrophysiologically, lactic acid sensitivity and specificity in the facial taste system of two marine teleosts. The stimulatory effects of 24 organic acids and related chemicals on the taste receptors of the yellowtail were examined by recording the electrical responses from the palatine branch of the facial nerve. The yellowtail was found to be highly sensitive to L-lactic acid at neutral or weakly alkaline pH. Some other organic acids, i.e., DL-glyceric, pyruvic, glycolic, D-2-phosphoglyceric, oxalacetic, and propionic acids were also stimulatory at neutral or weakly alkaline pH. The threshold for L-lactic acid, the most potent stimulus tested, was around 10^{-6} M, as low as that of L-proline, the most potent amino acid for the yellowtail. However, the magnitude of response to L-lactic acid was less than half of that to L-proline at 10^{-2} M. D-lactic acid was not effective at 10^{-2} M. Similar to the yellowtail, the scorpionfish *Sebastiscus marmoratus* showed a high sensitivity to L-lactic acid. Pyruvic and propionic acids were also stimulatory at 10^{-2} M. The threshold for L-lactic acid, located at around 10^{-6} M, was as low as that of L-proline, the most potent amino acid for the scorpionfish. The magnitude of response to L-lactic acid was about 1.7 times higher than that of L-proline at 10^{-2} M. D-lactic acid was not effective for this teleost either.

Single taste fiber analysis was performed on the yellowtail. Most of the L-lactate-sensitive fibers were also responsive to L-alanine. However, the relative responses to the two chemicals differed in magnitude from fiber to fiber, suggesting independent receptor sites for the two compounds in taste units. Cross-adaptation experiments were also conducted for a combination of L-lactic acid and L-alanine. In both teleosts, a reciprocal but incomplete cross-adaptation between the two chemicals was observed in the summated responses. This suggests that some common receptor mechanisms are shared for L-lactic acid and L-alanine.

Reference

1. Zeng C, Hidaka I (1989) Gustatory responses to organic acids in the yellowtail *Seriola quinqueradiata*. Proc 23rd Jpn Symp Taste and Smell 23:221–224

[1] Faculty of Bioresources, Mie University, 1515 Kamihama, Tsu, 514 Japan

NH$_3$ Blocks Gustatory Responses of Frogs and Eels

CHIKAKO YOTSUI and KIYONORI YOSHII[1]

Key words. pH$_{in}$—Electrophysiology

We investigated the pH dependence of taste responses in the frog, *Xenopus laevis*, and in the eel, *Anguilla japonica*, to amino acids, CaCl$_2$, and caffeine, by recording integrated neural responses. The taste substances were dissolved in Tris or 2-(N-Morpholino) ethanesulfonic acid (MES) solutions adjusted to pH 7.0–9.0. The receptors were adapted to the respective buffer solutions before stimulation to eliminate responses to these solutions. The pK$_a$ of the amino acids used were out of the pH range of the buffer solutions. The following results were obtained: (a) The responses of frogs to 0.1 mM Pro, Trp, Tyr, 1 mM Val, and 10 mM caffeine, potent stimuli of frog taste receptors, were practically independent of the pH of the stimulating solutions at pH 7.0–9.0. (b) The addition of 10 mM NH$_4$Cl to the stimulating solutions led to remarkable and reversible pH dependence of response. In the presence of NH$_4$Cl, the responses to Pro were completely blocked at pH 9.0. The blocking effect decreased with decreasing pH and was unclear at pH 7.0. (c) The blocking effect decreased with decreasing NH$_4$Cl concentration at pH 9.0. (d) Both the pH- and concentration-dependent blocking effects were expressed by a single curve as a function of NH$_3$ concentration in the stimulating solution, calculated from the Henderson-Hasselbalch equation:

$$[NH_3] = \frac{Ka}{Ka + [H^+]}$$

where Ka is a dissociation constant of NH$_4^+$. (e) NH$_4$Cl, at pH 9.0, blocked the responses to Trp, Tyr, and Val, but did not block these responses at pH 7.0, indicating that NH$_3$ blocks these responses. (f) Similarly, the responses to caffeine and CaCl$_2$ were also blocked in the presence of NH$_3$. (g)

Trimethylamine also blocked the responses to Pro and caffeine. (h) The addition of NaCl, tetramethylammonium chloride, or choline chloride to the stimulating solutions at pH 9.0 did not block the responses. These results indicate that the blocking effect was due neither to an increase in ionic strength nor to nonspecific binding of ions to the receptor membranes. These results, rather, suggest that NH$_3$ and trimethylamine, the nonionized forms, blocked the responses. (i) NH$_3$ shifted the concentration-response relation for Pro to higher concentrations. (j) The taste responses of eels to Arg, as well as those of the frogs, were blocked as a function of NH$_3$ concentration.

There are two possible interpretations of the present results. The first is that the receptor sites for the taste substances examined are similar. Namely, binding sites for different taste substances in the frog and eel preserve subsites to which NH$_3$ and trimethylamine bind. This hypothesis, however, does not agree with the conventional understanding that these taste substances stimulate different receptors.

NH$_3$ and trimethylamine rapidly diffuse into cells where most of them combine with protons, resulting in an increase in intracellular pH [1]. Changes in intracellular pH modify enzyme activity and the gating of gap junctions or the surface potential of plasma membranes [2] among other effects. In taste cells, these factors are thought to play a role in the taste transduction mechanism. Thus, we considered another possibility, that both NH$_3$ and trimethylamine increase the intracellular pH of the taste cells to block the taste responses. This action would explain the present results of a nonselective blocking effect on taste responses.

References

1. Roos A, Boron WF (1981) Intracellular pH. Physiol Rev 61:296–434
2. Hille B (1968) Charges and potentials at the nerve surface: Divalent ions and pH. J Gen Physiol 51:221–236

[1] Department of Biochemical Engineering and Science, Kyushu Institute of Technology, 680-4 Kawazu, Iizuka, Fukuoka, 820 Japan

Responses to Binary Mixtures of Amino Acids in the Facial Taste System of the Channel Catfish

K. Ogawa and J. Caprio[1]

Key words. Mixtures—Amino acids

Prior studies indicated that olfactory receptor responses to mixtures of amino acids in the channel catfish, *Ictalurus punctatus*, were enhanced only if the components bound to independent receptor sites [1,2]. Here we report the responses of 32 multiunit preparations and 55 single facial taste fibers to binary mixtures of amino acids. Tested were (a) L-alanine and L-arginine; L-alanine and L-proline; L-arginine and L-proline, and L-arginine and L-glutamate (group I), amino acids indicated to bind to independent taste receptor sites, and (b) L-alanine and glycine; L-alanine and L-methionine; L-arginine and L-alpha-amino beta guanidino proprionic acid, and L-alanine and L-glutamate, amino acids indicated to bind to the same or highly cross-reactive taste receptor sites (group II). All component stimuli were adjusted in concentration to provide approximately equal response magnitude- height of integrated multiunit activity (IMA) or action potentials generated/3 s of response time/single taste fiber (STF). The mixture discrimination index (MDI), defined as the response to the mixture divided by the average of the responses to the component stimuli, was calculated for each binary mixture. All group I binary mixtures resulted in enhanced taste activity [MDI: (IMA) = 1.16 ± 0.12 (SD), $n = 44$; (STF) = 1.17 ± 0.20 (SD); $n = 116$], whereas all group II binary mixtures showed no significant enhancement [MDI: (IMA) = 1.06 ± 0.15 (SD), $n = 41$; (STF) = 1.03 ± 0.14 (SD); $n = 34$]. There were no significant differences between the IMA and STF MDIs for group 1 and group 2 mixtures, respectively, nor were there differences between the MDIs of the two major fiber types, the alanine and arginine fibers [3].

Acknowledgments. This study was supported by NSF BNS-8819772.

References

1. Caprio J, Dudek J, Robinson JJ II (1989) Electro-olfactogram and multiunit olfactory receptor responses to binary and trinary mixtures of amino acids in the channel catfish, *Ictaluruspunctatus*. J Gen Physiol 93: 245–262
2. Kang J, Caprio J (1991) Electro-olfactogram and multiunit olfactory receptor responses to complex mixtures of amino acids in the channel catfish, *Ictalurus puctatus*. J Gen Physiol 98:699–721
3. Kohbara J, Michel W, Caprio J (1992) Responses of single facial taste fibers in the channel catfish, *Ictalurus punctatus*, to amino acids. J Neurophysiol 68:1012–1026

[1] Louisiana State University, Department of Zoology and Physiology, Baton Rouge, LA 70803-1725, USA

Gustatory Responses to Amino Acids in Salmonids: Phylogenetic Considerations

Toshiaki J. Hara[1], Yasuyuki Kitada[2], and Robert E. Evans[1]

Key words. Fish—Gustatory receptor

In a recent study [1], we showed that three char species (*Salvelinus alpinus*, *S. fontinalis*, and *S. namaycush*), each with a distinct pattern of taste bud distribution, responded specifically to L-proline (L-Pro), hydroxy-L-proline (L-Hpr), and L-alanine (L-Ala). Cross-adaptation experiments further demonstrated that the three char species possessed only one main palatal gustatory amino acid receptor, that for L-Pro. Recognizing the considerable phylogenetic implications, we emphasized the need for further investigation in other fish species so that a definitive statement could be made about phylogeny of receptor distribution. In order to enhance the basis of functional evidence for phylogenetic relationships of the gustatory system of salmonids, we examined the distribution pattern of palatal taste buds and their characteristic electrical responses to amino acids in five salmonid species: Arctic grayling (*Thymallus arcticus*), lake whitefish (*Coregonus clupeaformis*), Atlantic salmon (*Salmo salar*), brown trout (*Salmo trutta*), and kokanee (*Oncorhynchus nerka*).

In these fish, the densest taste bud populations were found on ridges in the area around the palatine teeth. With the exception of lake whitefish, the ridges adjacent to the teeth on the head and shaft of the prevomer were also dense in taste bud content, which taste buds may be largely responsible for amino acid detection. All species, except for lake whitefish, responded well to L-Pro and L-α-amino-β-guanidinopropionic acid (L-AGPA), with thresholds at 10^{-8}–10^{-7} and 10^{-5}–10^{-4} M, respectively. These species also responded to varying degrees to L-Hpr and L-Ala, believed to be agonists for L-Pro receptors. In addition, kokanee detected L-phenylalanine, L-leucine, and betaine. Of the six lake whitefish examined, only one specimen responded to L-Pro, L-AGPA, and L-arginine. In a separate experiment, nine more lake whitefish investigated for gustatory responses to amino acids also failed to register any responses from the palatine nerve. The general low sensitivity to some and total lack of sensitivity to other amino acids in lake whitefish may be due, in part, to the sparse distribution or lack of taste buds on the ridges adjacent to the teeth on the head and the shaft of the prevomer. Of most interest is the finding that response spectra for amino acids are narrow and may exhibit a species or group specificity. The only amino acids commonly detected by all five species are L-Pro and its agonists, L-Hpr and L-Ala. The specificity of the Pro receptor varies from species to species, depending upon the degree of its interaction with L-Hpr and L-Ala. Specificity indices (1-[L-Hpr or L-Ala response]/L-Pro response) for both L-Hpr and L-Ala were highest in the Arctic grayling and lake whitefish, and lowest in rainbow trout [2].

We conclude that, in these five salmonid species, L-Pro and L-AGPA receptors play a major role in amino acid gustation, and we further propose the hypothesis that salmonids may have initially evolved these two gustatory receptors for amino acids. With phylogenetic advancement, they gained greater response capabilities by (i) acquiring new receptor types, and (ii) by losing the specificity of existing receptors.

Acknowledgments. This study was supported, in part, by the JSTF fund (061-92-047).

References

1. Hara TJ, Sveinsson T, Evans RE, Klaprat DA (1993) Morphological and functional characteristics of the olfactory and gustatory organs of three *Salvelinus* species. Can J Zool 71:414–423
2. Marui T, Evans RE, Zielinski B, Hara TJ (1983) Gustatory responses of the rainbow trout (*Salmo gairdneri*) palate to amino acids and derivatives. J Comp Physiol A 153:423–433

[1] Canada Department of Fisheries and Oceans, Freshwater Institute, 501 University Crescent, Winnipeg, Manitoba, R3T 2N6, Canada
[2] Department of Physiology, Okayama University School of Dentistry, 2-5-1 Shikata-cho, Okayama, 700 Japan

Development of the Facial Lobe in the Sea Catfish, *Plotosus lineatus*

A. Shito[1], S. Kiyohara[1], and J. Kitoh[2]

Key words. Taste—Facial lobe

There is a distinct topographical arrangement in the facial lobe (FL) of *Plotosus*. In the anterior two-thirds of the FL, five distinct lobules are seen, and these constitute five longitudinal columns. The first four lobules, from medial to lateral, receive projections from the medial mandibular, lateral mandibular, maxillary, and nasal barbels, respectively. The fifth lobule receives projections from the body. Electrophysiological study has shown that the tip to base axis of each barbel is represented in the rostro-caudal axis of each barbel lobule, while the antero-posterior axis of the face and flank is represented in the dorsal lobule in the posteroanterior axis [1]. The aim of this study was to examine the development of the barbel lobules, with special attention being paid to the proliferative zones and migratory pathway of newly formed cells in the developing lobules.

Larvae of every post hatched-day up to 25 days (P25d) were fixed, and serial sections were made and stained. A three-dimensional image analyzer (Cosmozone 2SA, Nikon, Tokyo, Japan) linked to a computer was used for the three-dimensional reconstruction of the barbel lobules in each larval stage. In order to detect DNA synthesis in the FL, we kept larvae from various hatched days in sea water containing BrdU (2 mg/ml) for 1 or 12 h and then fixed the specimens. Some of the larvae were allowed to survive for varying numbers of days before fixation. The larvae were embedded in paraffin and serially sectioned in the coronal plane at 5 or 10 µm. They were processed by BrdU immunohistochemical procedures.

At P5d, the FL grew dorsad and mediad because of an increase in the number of cells and fibers and the expansion of neuropile areas. Lobulation was first observed laterally, forming the nasal barbel lobule at P6d. At P12d, the four lobules were located in the anterior one-third of the FL, and the fifth lobule appeared to be located laterodorsal to the other four lobules. At P26d, the four barbel lobules occupied the anterior half of the FL, indicating that the relative growth of the lobules was greater than that in the other part of the FL.

In the first experiments with BrdU, it was applied for 1 h to larvae at various post hatched days, and they were immediately fixed to explore the location of the proliferative zone. Cells labeled with BrdU were clearly observed in the medial or dorsal margin of the FL; they were greatest in number at P8d and were observed up to P25d. After this time, few labeled cells were found in the facial lobe, indicating that neurons and glial cells in the FL were formed within this period. At P4-6d, labeled cells were observed only in the mediodorsal margin of the FL anteroposteriorly. When the lobules became distinct, labeled cells were distributed not only in the medial margin but also along the entire border of the anterior portion of the FL. Labeled cells increased in number anteriorly. This suggests that cell division was active in the anterior direction. The density of the labeled cells was highest in the anterior tip region of the FL among various regions of the FL.

In the second experiment, BrdU was applied for 1 h to the larvae and they were allowed to survive for various periods. Sections of these specimens showed that labeled cells were scattered in the FL, indicating the migration of cells that were formed in the proliferative zone of the FL.

[1] Department of Biology, College of Liberal Arts and Sciences, Kagoshima University, Kohrimoto 1-21-30, Kagoshima, 890 Japan
[2] Institute for Laboratory Animal Research, Nagoya University, School of Medicine, Syouwa-ku, Nagoya, 466 Japan

Reference

1. Marui T, Caprio J, Kiyohara S, Kasahara Y (1988) Topographical organization of taste and tactile neurons in the facial lobe of the sea catfish, *Plotosus lineatus*, Brain Research 446:178–182

Determination of Palatability in Goldfish

Charles F. Lamb and Thomas E. Finger[1]

Key words. Feeding behavior—Gustatory system

Sensory information conveyed along the glossopharyngeal and vagal nerves is used to determine the palatability of items within the oral cavity. The glossopharyngeal/vagal system is especially well developed in particulate-feeding cyprinid fishes (e.g., carp and goldfish), which manifest complex sorting behaviors to separate food from non-food items in the oral cavity [1]. These fishes are characterized by specializations of this system peripherally, in the oropharyngeal cavity, and centrally, in the prominant vagal lobe of the medulla. The dorsal wall of the pharynx contains a muscular palatal organ, a structure used to trap food particles against the floor of the cavity during feeding [1]. The palatal organ is lined with an epithelial layer containing high densities of taste buds. Sensory fibers from the palatal organ terminate in the vagal lobe, in a topographic organization reflecting the surface of the palatal organ [2]. The topographic organization of sensory information within the vagal lobe, and a specific reflex circuitry between the sensory and motor layers of the lobe, provide a neuroanatomical substrate for the feeding patterns displayed by cyprinid fishes.

We devised a behavioral paradigm to test whether particular feeding behaviors could be driven by gustatory stimulation of pharyngeal taste buds in goldfish. Taste stimuli included quinine and amino acids, because of their effectiveness in numerous fish species (for review, see [3]).

Goldfish were trained to feed on gelatin pellets ($4 \times 4 \times 5$ mm) and then tested with pellets containing different concentrations of quinine (10^{-6}–10^{-3} M), or quinine plus L-alanine (10^{-2} M), L-proline (10^{-2} M), or ground food particles (1%, 5%, or 10%, by weight). Behavioral patterns were characterized and rates of ingestion were recorded for each combination of stimulants (e.g., 10^{-5} M quinine alone, 10^{-3} M quinine plus 10^{-2} M L-alanine, 10^{-6} M quinine plus 5% food, etc.).

Quinine produced a dose-dependent rejection (spitting out) of pellets taken into the oral cavity, with behavioral threshold at approximately 10^{-5} M

and complete rejection at 10^{-3} M. Pellets with low concentrations of quinine (none, 10^{-6}, or 10^{-5} M) were ingested within 5–10 s after being sucked into the oral cavity, while pellets with high concentrations (10^{-3} M) were typically rejected immediately after the pellet reached the palatal organ. The addition of either L-alanine or L-proline to the quinine-flavored pellets markedly increased ingestion rates at quinine concentrations of 10^{-4} M ($2\times$) and 10^{-3} M ($10\times$). Both L-alanine and L-proline stimulated food-sorting behaviors when mixed with quinine. With 10^{-4} M quinine and either of the amino acids, pellet manipulation time increased to 20–30 s; the behavior was characterized by repeated open protrusion of the mouth. When the quinine concentration was 10^{-3} M, the addition of amino acids produced both open and closed protrusion of the mouth, resulting in a robust pumping of water through the oral cavity for 45–60 s.

Quinine is an aversive feeding stimulant for goldfish. The combination of quinine and either L-alanine or L-proline resulted in an increase in the ingestion rate over pellets with quinine alone, providing evidence that both of these amino acids are appetitive tastants. The mixture of aversive and appetitive stimulants also increased the time spent sorting the pellets within the oropharynx (associated with an increase in both open- and closed-mouth protrusion). Goldfish use these two protrusion behaviors, and the resulting orobranchial irrigation, to sample the gustatory qualities of food items in the oral cavity and to selectively hold those that are deemed palatable. The response to the combination of both appetitive and aversive stimulants illustrates that this sorting process is important for the determination of palatability.

References

1. Sibbing FA, Osse JWM, Terlouw A (1986) Food handling in the carp (*Cyprinus carpio L.*), its movement patterns, mechanisms, and limitations. J Zool (London) 210(A):161–203
2. Morita Y, Finger TE (1985) Topographic and laminar organization of the vagal gustatory system in the goldfish, *Carassius auratus*. J Comp Neurol 238:187–201
3. Caprio J (1988) Peripheral filters and chemoreceptor cells in fishes. In: Atema JA, Fay RR, Popper AN, Tavolga WN (eds) Sensory biology of aquatic animals. Springer, New York, pp 313–338

[1] Department of Cellular and Structural Biology, University of Colorado Health Science Center, B-111, 4200 E. Ninth Ave., Denver, CO 80262, USA

Ultrastructure of Taste Buds in the Spotted Dogfish *Scyliorhinus caniculus* (Selachii)

KLAUS REUTTER[1]

Key words. Taste buds—Selachians—Ultrastructure

The taste buds (TBs) of animals of different systematic groups differ considerably in structure [1], but our knowledge of these sensory organs across the vertebrate lineage is currently too limited to derive a TB phylogenetic tree. Since only a single report exists on gustatory receptors of selachians [2], the present study of the structure of TBs of the spotted dogfish, *Scyliorhinus caniculus*, was initiated to learn more of the interrelationships among TBs of vertebrates. Spotted dogfish were immersion-fixed in a glutaraldehyde-paraformaldehyde-picric acid solution. Pieces of skin removed from different parts of the head and oral cavity were processed conventionally for transmission electron microscopy. In the spotted dogfish, TBs were rare and occurred exclusively at the tip of epidermal papillae within the oral mucosa. In longitudinal section, the TBs were pear-shaped, were approximately 50 μm high and 30 μm wide, were located in the upper third of the squamous non-keratinized epithelium, and were surrounded by marginal cells, the normal epidermal cells. The base of each TB rested on the marginal cells and not on the underlying dermal (corium) papilla, which projected into the basal portion of the epithelium. Each TB was innervated by a nerve that ran through the corium papilla, penetrated the basal lamina of the epithelium, and ascended between the epithelial cells to the TB. Blood capillaries were present in the dermal papilla, but did not reach the TB. The sensory epithelium within the TB consisted of a total of about 25 slender light and dark cells bearing microvilli and situated parallel to the TB's longitudinal axis. *Light cells*, which stained less electron dense than dark cells, had a cytoplasm rich in mitochondria, smooth endoplasmic reticulum (sER), and microtubules, and their apical ends terminated in three to five large and divided microvilli. Intermediate (tono)-filaments were rarely observed. *Dark cells* occurred in two forms. In form one, the

cell ended apically in long, slender, and divided microvilli. This dark cell was rich in mitochondria, sER, and dark-staining secretory vesicles of about 200 nm diameter, and contained numerous intermedate filaments. Form two of the dark cell showed long, thin, and undivided microvilli. These form two dark cells contained numerous mitochondria, numerous profiles of the sER, and solitary intermediate filaments. Also, small filament bundles occurred in the basal portions of the cell. Some of the basal marginal cells, which were also rich in organelles, appeared to be basal stem cells, and were different from those of teleosts [1,3,4]. In the spotted dogfish, basal cells comparable to the ones in teleosts did not form the base of the TB, but lay directly on top of the corium papilla. The TB was innervated by unmyelinated nerve fibers, and presynaptic 130-nm dense core and 40-nm clear vesicles occurred mainly at the bases of both the light and dark (form 1) cells. Efferent synapses were not observed. Thus, the TBs of the spotted dogfish differed in the following ways from those of other vertebrates: (1) the TBs were not located directly on top of the corium papillae; (2) the TBs contained at least three types of elongated cells which reached the epithelial surface; and (3) the TBs did not contain basal cells analogous to those of teleosts and amphibians [1,3,4].

References

1. Reutter K, Witt M (1993) Morphology of vertebrate taste organs and their nerve supply. In: Simon SA, Roper SD (eds) Mechanisms of taste transduction. CRC, Boca Raton, pp 29–82
2. Pevzner RA (1976) Electron microscope study of the taste buds of Elasmobranchs, *Trigon pastinaca* and *Raja clavata* (in Russian). Tsitologiia 18:560–566
3. Reutter K (1971) Die Geschmacksknospen des Zwergwelses *Amiurus nebulosus* (LeSueur). Morphologische und histochemische Untersuchungen. Z Zellforsch 120: 280–308
4. Reutter K (1986) Chemoreceptors. In: Breiter-Hahn J, Matoltsy AG, Richards KS (eds) Biology of the integument, vol 2, vertebrates. Springer, Berlin Heidelberg New York Tokyo, pp 586–604

[1] Institute of Anatomy, University of Tübingen, Österbergstrasse 3, 72074 Tübingen 1, Germany

Cell Proliferation and Differentiation in the Olfactory Organ of the Embryonic and Larval Zebrafish, *Brachydanio rerio*

Anne Hansen and Eckart Zeiske[1]

Key words. Olfactory organ—Zebrafish

The olfactory organ of the zebrafish, *Brachydanio rerio*, a standard model in fish embryology, has been studied by light and electron microscopy [1], but little is known about the proliferation of placodal cells during the ontogeny of this organ. The aim of the present study was to show the proliferation sites in relation to the developmental stages of the olfactory organ of zebrafish. Eggs and larvae were obtained from our breeding colony, where adult fish are kept at a 14-h light/10-h dark cycle at a temperature of 26.5°C. The age of the animals studied ranged from 24 h after fertilization (AF) to hatching (3–4 days AF). We used standard methods of light and scanning electron microscopy to visualize the development of the olfactory organ, and we applied immunocytochemical methods to show bromodeoxyuridine (BrdU) incorporation and proliferating cell nuclear antigen (PCNA) immunoreactivity of proliferating cells. In addition to paraffin embedding, some animals were embedded in Quetol 651, an epoxy resin of low viscosity, since only thin sections give a precise idea of the placode in very young embryos. Scanning electron microscopic pictures of the head of the embryonic zebrafish at 24 h AF show that the epidermis covering the olfactory placode is still completely closed. Within the placode, the cells in the rostromedial region differentiate first. About 30–32 h AF, the epidermal cells separate to form the olfactory pit. Pools of round, undifferentiated cells in the rostral and caudal ends of the placode remain covered by the epidermis, even after hatching. Twenty-four h AF, tests with antibodies against BrdU or PCNA show stained nuclei all over the placode. There is no distinct proliferation center. Thirty-four to 36 h AF, the distribution of immunoreactive nuclei is different for the antibodies against PCNA and BrdU. BrdU-ir is seen in the basal region and occasionally in the upper part of the placode. In Quetol sections, however, PCNA-ir is seen in almost all nuclei, ranging from weak to strong staining. After hatching (3–4 days AF), the labeled nuclei show the same distribution for both antibodies. In the rostral and caudal end of the placode, nearly all nuclei are labeled by both antibodies in all stages examined. These findings confirm that, apart from the basal cells, there are regions of cells that retain their capacity to proliferate during the embryonic and larval development of the olfactory organ. The difference in the staining pattern of BrdU and PCNA may be caused by the long half-life of PCNA, so that PCNA could be immunologically detectable in cells that have already left the cell cycle. Moreover, Quetol sections are much clearer than paraffin sections, thus making small amounts of PCNA visible, which amounts escape detection in paraffin sections.

Acknowledgments. This study was supported by DFG (Deutsche Forschungsgemeinschaft) (Ze 141/8-2).

Reference

1. Hansen A, Zeiske E (1993) Development of the olfactory organ in the zebrafish, *Brachydanio rerio*. J Comp Neurol 333:289–300

[1] Zoological Institute and Zoological Museum, University of Hamburg, Martin-Luther-King-Platz 3, 20146 Hamburg, Germany

Specific Regeneration of Peripheral and Central Olfactory Pathways in Goldfish

CHRISTIAN VON REKOWSKI and HANS PETER ZIPPEL[1]

Key words. Regeneration—Olfactory system

In a first series, ten groups of two goldfish were trained to discriminate Arg 10^{-6} M (rewarded stimulus) from Gln 10^{-6} M (not rewarded). After successful discrimination training, significant positive responses were recorded when the rewarded stimulus was applied in 80-fold lower concentration (Arg 5×10^{-8} M vs Gln 4×10^{-6} M) and during contamination of the rewarded stimulus up to 50% with the unrewarded stimulus (50% Arg 10^{-6} M + 50% Gln 10^{-6} M vs Gln 10^{-6} M). Thereafter, in five groups of fish, the olfactory nerves and in five groups, the lateral olfactory tracts were intracranially and bilaterally dissected. Immediately after this operation, both collectives were unable to discriminate concentration differences and contaminated stimuli. Two weeks after receptor axotomy and lateral tractotomy, following functional regeneration, both collectives again were able to discriminate concentration differences and contaminations, as they did before operation. The regeneration of the olfactory nerves (peripheral regeneration) and the lateral olfactory subtracts (central regeneration) therefore is highly specific, even if the discriminative task is at the threshold of the discriminative ability in intact goldfish [1]. Following functional regeneration, intrabulbar and telencephalic information processing results in a complete return of the preoperative discriminative ability.

In a second series, olfactory bulbs were crossed (i.e., the olfactory nerves were intracranially dissected and the left olfactory bulb was located in the former position of the right bulb, and vice versa) after discrimination training (Ala vs Arg, 10^{-6} M) in six groups of two goldfish. In this case, a functional regeneration (positive responses during application of Tubifex food odor) was apparent 10 days postoperatively. A specific regeneration, however, was recorded only during application of Ala vs Arg in similar concentrations (10^{-6} M). Difficulties in discriminating concentration differences and contaminations of stimuli were evident; these theoretically demonstrate functional differences of the lateral and the medial olfactory bulb.

Acknowledgments. Supported by DFG Zi 112/3-2.

Reference

1. Rekowski C, Zippel HP (1993) In goldfish the qualitative discriminative ability for odors rapidly returns after bilateral nerve axotomy and lateral olfactory tract transection. Brain Research 618:338–340

[1] Physiologisches Institut der Universität Göttingen, Humboldtallee 23, 37073 Göttingen, Germany

Information Processing in the Olfactory Bulb of Goldfish: Interaction of Mitral and Granular Cells

Hans Peter Zippel, Jörg Rabba, and Pinar Aksari[1]

Key words. Olfactory bulb—Single cells

The responses of goldfish olfactory bulb relay and granular neurons were analyzed by extracellular recordings. Four amino acids (Arg, Lys, Ala, and Gly; 10^{-6} M) were applied to the olfactory mucosa for 30 s repetitively, and recordings from bulbar mitral cells were analyzed. In contrast to electro-olfactogram (EOG) recordings in which hierarchical dose-response EOG amplitudes were regularly apparent (Arg > Lys > Ala > Gly; [1]), the discriminative ability of bulbar relay neurons was less pronounced. Sixty percent of 47 uniformly responding bulbar neurons remained indifferent during amino acid stimulation, and in 2% excitatory, in 8% inhibitory, and in 30% (14 cells), discriminative responses were recorded. In 22% of neurons (16 cells), the activity either abruptly changed during repetitive runs or slowly changed during repetitive stimulus applications. In goldfish it is possible to simultaneously record relay and granular cell activities because in lower vertebrates these different cell types are less strictly layered. It is evident that change in relay neuron activity is directly correlated to granular cell responses: the relay neuron activity decreases when the granular cell activity increases during olfactory mucosal stimulation, and vice versa. During repetitive odor stimulation, decreasing excitatory or inhibitory effects recorded from relay neurons correlate with opposite responses of granular cells and result in more or less indifferent odorant discriminative ability of relay neurons with respect to the preceding water phase. In 65 neurons, stimulus-free intervals, after a continous laminary tap water current (1 ml/min) resulted in abrupt or slow changes of relay neurons' "resting" activity. In 25% of neurons (i.e., 16 of 65) a long-term excitation was recorded during the stimulus-free interval, while in 46%, long-term inhibition was recorded. In 29%, the activity remained similar, or became similar within 5 min, to the activity recorded during the preceding tap water flow. In contrast to amino acid stimulation, application of permanent tap water stimuli resulted in long-term changes of the bulbar "spontaneous" activity. It therefore is evident that the olfactory receptor neurons are multimodal, and chemo- and mechano-sensitive.

Acknowledgments. Supported by DFG Zi 112/3-2.

Reference

1. Zippel HP, Lago-Schaaf T, Caprio J (1993) Ciliated olfactory receptor neurons in goldfish (*Carassius auratus*) partially survive nerve axotomy, rapidly regenerate and respond to amino acids. J Comp Physiol A 173:537–547

[1] Physiologisches Institut der Universität Göttingen, Humboldtallee 23, 37073 Göttingen, Germany

Olfactory Imprinting and Homing Mechanisms in Salmonids

Hiroshi Ueda[1], Munetaka Shimizu[1,2], Hideaki Kudo[1,3], Akihiko Hara[4], Kenzo Kurihara[5], and Kohei Yamauchi[3]

Key words. Salmonid olfactory system-specific protein—Migratory behaviors

It has been confirmed in many behavioral and electrophysiological studies [1,2] that maternal stream odorants are imprinted in the olfactory system (olfactory epithelium, olfactory nerve, and olfactory bulb) of juvenile salmon during downstream migration, and that the olfactory discrimination of the maternal stream odorants causes adult salmon to home during upstream spawning migration. However, few attempts have been made to investigate the biochemical aspects of the olfactory system in any salmonid species in terms of detecting olfactory system-specific molecular markers for studying olfactory imprinting and homing mechanisms.

Using sodium dodecyl sulfate-polyacrylamide gel electrophoresis (SDS-PAGE), we have identified an olfactory system-specific 24 kDa protein (N24) in kokanee salmon (*Oncorhynchus nerka*; land-locked sockeye salmon) by the electrophoretic comparison of proteins restricted to the olfactory system with those found in other parts of the brain. N24 appeared as a single acidic protein on analysis of two-dimensional PAGE (2D-PAGE). Application of an ultracentrifugation technique showed that N24 was present mainly in the cytosol fraction.

A specific polyclonal antiserum to N24 was generated in a rabbit [3]. The antiserum recognized only the 24 kDa protein in the olfactory system, this protein being absent from other parts of the brain, not only in kokanee but also in sockeye (*O. nerka*), masu (*O. masou*), and chum salmon (*O. keta*), as determined by Western blotting analysis. In various species of teleosts, N24 immunoreactivity was found in the olfactory systems of species migrating between sea and river, such as smelt (*Spirinchus lanceolatus*) and Japanese eel (*Anguilla japonica*), but not in carp (*Cyprinus carpio*), rosyface dace (*Leuciscus ezoe*), or tilapia (*Oreochoromis nilotics*), which exhibit no migratory behavior.

Western blotting analysis showed that, both at the time of imprinting of the maternal stream odorants in masu salmon, and at the time of homing to the maternal stream in chum salmon, the immunoreactivity of N24 in fish in the maternal stream was stronger than that in fish in seawater. Immunocytochemical and immunoelectron microscopic observations revealed that N24 positive immunoreactivity occurred exclusively on 50-nm neurosecretory vesicles in ciliated and microvillus neuroepithelial cells, and in the olfactory nerve and the olfactory bulb where the olfactory nerve had penetrated, indicating that N24 has a possible role in neurotransduction from neuroepithelial cells to the olfactory bulb.

These results suggest that N24 may be important in imprinting and homing mechanisms in salmonids, and in migratory behavior in teleosts. Moreover, interesting questions arise as to whether N24 might be present in the olfactory systems of other migratory animals such as sea turtles and migratory birds.

[1] Toya Lake Station for Environmental Biology, Faculty of Fisheries, Hokkaido University, 122 Tsukiura, Abuta, Hokkaido, 049-57 Japan
[2] Research Institute of North Pacific Fisheries, [3] Department of Biology, Faculty of Fisheries, Hokkaido University, 3-1-1 Minato-cho, Hakodate, Hokkaido, 041 Japan
[4] Nanae Fish Culture Experimental Station, Faculty of Fisheries, Hokkaido University, 498 Sakura-cho, Nanae, Hokkaido, 041-11 Japan
[5] Department of Pharmaceutical Sciences, Faculty of Pharmaceutical Sciences, Hokkaido University, N12 W6, Sapporo, Hokkaido, 060 Japan

References

1. Hasler AD, Scholz AT (1983) Olfactory imprinting and homing in salmon. Springer, Berlin, pp 1–134
2. Stabell OB (1992) Olfactory control of homing behaviour in salmonids. In: Hara TJ (ed) Fish chemoreception. Chapman and Hall, Hants, pp 249–270
3. Shimizu M, Kudo H, Ueda H, Hara A, Shimazaki K, Yamauchi K (1993) Identification and immunological properties of an olfactory system-specific protein in kokanee salmon (*Onocrhynchus nerka*). Zool Sci 10:287–294

Olfactory Responses of Rainbow Trout Were Unchanged When Fish Was Adapted from Fresh Water to Seawater

Takayuki Shoji, Ken-ichi Fujita, and Kenzo Kurihara[1]

Key words. Olfaction—Seawater adaptation

Various species of fish migrate between river and sea. Salinity on the olfactory receptor membranes varies greatly in such migration. In order to explore olfactory transduction mechanisms in fish, it is interesting to know how olfactory responses are affected by such large changes in environmental salinity.

In this study, we used rainbow trout, *Oncorhynchus mykiss* adapted to pond water (freshwater rainbow trout) and seawater (seawater rainbow trout). Rainbow trout is a species of euryhaline fish that can easily adapt to sea water. We used rainbow trout (400–500 g) that had been reared in sea water for more than 4 weeks; they had completely acquired osmoregulatory ability in sea water. For comparison, we used the marine fish, jacopever, *Sebastes schlegeli* (90–120 g), which cannot adapt to fresh water. The olfactory responses of rainbow trout and jacopever were recorded form the olfactory nerve, according to the method of Sveinsson and Hara [1].

Seawater rainbow trout responded to various amino acids dissolved in pond water in the same manner as freshwater rainbow trout. The minimum concentrations (thresholds) to induce responses in seawater rainbow trout agreed with those in freshwater rainbow trout reported by Hara [2]. In both seawater and freshater rainbow trout, the magnitudes of responses to amino acids in artificial sea water (ASW) were practically equal to those in artificial pond water (APW). Unlike rainbow trout, the olfactory responses of jacopever to amino acids in APW were greatly diminished. This seems to be because jacopever cannot acquire osmoregulatory

ability in fresh water. The dependence of responses to 0.1 mM L-alanine on the concentration of NaCl and $CaCl_2$ was also examined with freshwater and seawater rainbow trout. The responses were diminished in the absence of the salts, and were restored by the addition of 0.1 mM NaCl or 0.01 mM $CaCl_2$ to the stimulating solution. There was no difference in ion dependence between seawater and freshwater rainbow trout. That is, the responses of rainbow trout to amino acids had a high tolerance for large changes in ionic environment, while the responses of the marine fish, jacopever, were affected by ionic changes.

In seawater, the olfactory epithelium of fish is exposed to salt water containing about 500 mM NaCl and 10 mM $CaCl_2$. Application of 500 mM NaCl to the epithelium induced large responses in both seawater and freshwater rainbow trout, which did not adapt to a spontaneous level. On the other hand, application of seawater elicited only a small response in both seawater and freshwater rainbow trout, both of which were easily adapted. This was due to suppression of the response to NaCl by Ca ions in seawater. In fact, the response to 500 mM NaCl was completely suppressed in the presence of 10 mM $CaCl_2$. Mg ions showed no suppressive effect. Thus, Ca ions may play a role in reducing the fatigue of olfactory receptor cells in seawater.

References

1. Sveinsson T, Hara TJ (1990) Analysis of olfactory responses to amino acids in arctic char (*Salvelinus alpinus*) using a linear multiple-receptor model. Comp Biochem Physiol 97A: 279–287
2. Hara TJ (1982) Structure-activity relationships of amino acids as olfactory stimuli. In: Hara TJ (ed) Chemoreception in fish. Elsevier, New York, pp 135–157

[1] Faculty of Pharmaceutical Sciences, Hokkaido University, N-12 W-6 Kita-ku, Sapporo, 060 Japan

Spatial Coding of Odor Information in Salmonids

Toshiaki J. Hara[1] and Chunbo Zhang[2]

Key words. Fish—Olfaction

In teleost fishes, the olfactory epithelium is raised from the floor of the naris into a series of folds or lamellae to form a rosette. The arrangement, shape, and degree of development of the lamellae vary considerably among species, and the sensory epithelium shows various distribution patterns [1]. Two types of morphologically and ontogenetically distinct receptor cells (ciliated and microvillar) in the olfactory epithelium detect and encode chemical signals; however, their functional differentiation is not clear [2]. The olfactory nerve, unmyelinated axons of the receptor neurons, courses to the olfactory bulb where the axons make a synaptic contact with the second-order bulbar neurons in the form of glomeruli.

An electrophysiological study of salmonids suggests significant functional segregation of axons in the olfactory bulb, where the lateral and medial regions respond differentially to amino acids and fish odors [3]. However, recent anterograde (horseradish peroxidase; HRP) and retrograde (fluorescent latex beads) studies reveal that the primary olfactory neurons in rainbow trout (*Oncorhynchus mykiss*) lack point-to-point or regionally topographic organization and that the entire extent of the olfactory epithelium contributes axons to each region of the glomerular layer [4].

We investigated the spatial coding of olfactory information by recording the electroencephalogram (EEG) responses to amino acids and bile salts from six different positions on the surface of the olfactory bulb, while simultaneously monitoring electroolfactogram (EOG) responses from the olfactory epithelium in Atlantic salmon (*Salmo salar*), rainbow trout, and Arctic char (*Salvelinus alpinus*). These two groups of olfactory stimuli are believed to be detected through separate receptor mechanisms in the salmonid olfactory system. The responses were further examined in fish that had had their olfactory lamellae partially removed (partial lamellectomy). We hypothesized that if olfactory receptors for amino acids and bile salts showed distinct localizations over the olfactory epithelium, the EEG responses thus recorded should be affected differentially as a result of partial lamellectomy.

EEG responses to L-arginine, L-glutamate and L-serine, representing, respectively, basic, acidic, and neutral amino acids, were centered on the latero-posterior portion of the bulb, while responses to a bile salt, taurocholate (TCA) centered on the mid-portion, especially medial and lateral. Results were qualitatively similar in the three species. When the olfactory rosette was partially removed, EOG responses to amino acids, bile salt, and their mixtures were considerably reduced, regardless of the portion of the rosette that had been removed (medial, lateral, anterior, or posterior half). Following partial lamellectomy, EEG responses to L-serine, recorded at the lateroposterior portion, and responses to TCA at the mid-lateral portion, were reduced by 25%–40% and 30%–40%, respectively. No significant differences were found in the degree of reduction due to partial lamellectomies.

We conclude that at least two types of olfactory neurons, one specific for amino acids and one for bile salts, are distributed over the entire extent of the olfactory epithelium and project topographically to their respective coding centers in the bulb.

References

1. Zeiske E, Theisen B, Breucker H (1992) Structures, development, and evolutionary aspects of the peripheral olfactory system. In: Hara TJ (ed) Fish chemoreception. Chapman and Hall, London, pp 13–39
2. Zielinski B, Hara TJ (1988) Morphological and physiological development of olfactory receptor cells in rainbow trout (*Salmo gairdneri*) embryos. J Comp Neurol 271:300–311
3. Thommesen G (1978) The spatial distribution of odour induced potentials in the olfactory bulb of char and trout (Salmonidae). Acta Physiol Scand 102:205–217
4. Riddle DR, Oakley B (1991) Evaluation of projection patterns in the primary olfactory system of rainbow trout. J Neurosci 11:3752–3762

[1] Canada Department of Fisheries and Oceans, Freshwater Institute, 501 University Crescent, Winnipeg, Manitoba, R3T 2N6, Canada
[2] Department of Zoology, University of Manitoba, Manitoba, Winnipeg, R3T 2N2, Canada

Involvement of Gonadotropin-Releasing Hormone in Olfactory Function in Salmonids

Hideaki Kudo[1,2], Susumu Hyodo[3], Hiroshi Ueda[1], Katsumi Aida[4], Akihisa Urano[5], and Kohei Yamauchi[2]

Key words. Salmon-type gonadotropin-releasing hormone (sGnRH)—Homing migration

Gonadotropin-releasing hormone (GnRH) is a decapeptide neuroendocrine hormone that is considered to play important roles in regulating teleost reproduction, mainly by stimulating the release of gonadotropin from the pituitary gland. At least two GnRH molecules, salmon-type (sGnRH) and chicken II-type (cGnRH-II), have been identified in salmonid brains. Several immunocytochemical studies have reported that sGnRH-immunoreactive neurons are distributed in the preoptic area, the telencephalon, and the olfactory bulb, while cGnRH-II-immunoreactive neurons are found mainly in the mesencephalon. However, little is known about their functional differences, especially in the olfactory system.

Since the olfactory imprinting and homing hypothesis was introduced to explain homing migration in salmonids, many behavioral and electrophysiological experiments [1,2] have suggested the importance of olfactory function during spawning migration (homing migration). To date, however, few studies have examined the physiological changes of any particular molecule during such migration. In the present study, we examined cytophysiological changes of sGnRH-producing neurons in chum salmon (Oncorhynchus keta) during homing migration, both by means of immunocytochemical techniques with specific antiserum to sGnRH, and by in situ hybridization techniques with mRNA encoding the sGnRH precursor.

Female 4-year-old adult chum salmon were caught at the coastal sea stage and at the maternal river stage in Ishikari Bay off Atsuta, and from the stock pond of Chitose Branch, Hokkaido Salmon Hatchery, respectively, in the middle of October, 1992. The olfactory bulb, including the olfactory nerve bundle and the telencephalon, were fixed with 4% paraformaldehyde and embedded in either OCT compound or Histosec. For immunocytochemistry, frozen sagittal sections were immunoreactivated with anti-sGnRH serum (1:7000) followed by the strept-avidin-biotin complex method. For in situ hybridization, a sGnRH precursor mRNA 44 mer oligonucleotide probe was labeled at the 3′ tail with [^{35}S] dATP. The details of the hybridization techniques and radioautography have been previously described [3].

sGnRH immunoreactive neurons and neurons showing signals for sGnRH precursor mRNA were observed in the following six areas; the dorsal portion of the olfactory nerve bundle (ONB), the transitional area between the olfactory nerve and the olfactory bulb (ON-OB), the ventral olfactory bulb (OB), the area between the olfactory bulb and the telencephalon (OB-T), the ventral telencephalon (VT), and the preoptic area (POA). sGnRH neurons in the ONB were observed at the coastal sea stage, but not at the maternal river stage. Although sGnRH neurons in the ON-OB, OB, OB-T, VT, and POA were always observed during homing migration, the number of sGnRH neurons in the ON-OB was more abundant at the coastal sea stage than at the maternal river stage, while the signals of sGnRH neurons in the VT and POA were stronger at the maternal river stage than at the coastal sea stage.

The coastal sea stage is an important period for the discrimination of maternal river odorants, and the abundance of sGnRH-producing neurons in the ONB and, probably, in the ON-OB at this crucial stage suggests that they might be involved in the olfactory recognition of the maternal river.

[1] Toya Lake Station for Environmental Biology, Faculty of Fisheries, Hokkaido University, 122 Tsukiura, Abuta, Hokkaido, 049-57 Japan
[2] Department of Biology, Faculty of Fisheries, Hokkaido University, 3-1-1 Minato-cho, Hakodate, Hokkaido, 041 Japan
[3] College of Arts and Sciences, The University of Tokyo, 3-8-1 Komaba, Meguro-ku, Tokyo, 153 Japan
[4] Department of Fisheries, Faculty of Agriculture, The University of Tokyo, 1-1-1 Yayoi, Bunkyo-ku, Tokyo, 113 Japan
[5] Division of Biological Sciences, Graduate School of Science, Hokkaido University, N10 S8, Sapporo, 060 Japan

References

1. Hasler AD, Scholz AT (1983) Olfactory imprinting and homing in salmon. Springer, Berlin, pp 1–134
2. Stabell OB (1992) Olfactory control of homing behaviour in salmonids. In: Hara TJ (ed) Fish chemoreception. Chapman and Hall, Hants, pp 249–270
3. Urano A, Hyodo S (1990) In situ hybridization techniques in the study of endocrine secretions. In: Epple A, Scanes CG, Stetson MH (eds) Progress in comparative endocrinology, vol 342. Wiley-Liss, New York, pp 309–314

Chemical Alarm Signals in Adult Lake Whitefish

Gunnar Bertmar[1]

Key words. Alarm signals—Chemosensory—Orientation—Salmonids—Schooling—Schreckstoff—Whitefish

Introduction

Alarm substances and fright reactions have been described in many enjured fish species, but as yet, have not been reported in salmonid fishes [1]. However, as part of an investigation on chemosensory attraction to conspecifics by adult lake whitefish *Coregonus clupeaformis* (Mitchill), it was recently found that this salmonid species seems to have reactions to home water containing chemical cues from a slightly injured fish of the same population [2]. This is unique behavior, for a salmonid species, and its biological significance is reported here.

Material and Methods

The 50 mature 4-year-old fish used for the present experiments were from an Ontario wild stock and were reared at the Biology Department, York University, Toronto, Canada. These adult fish, ranging from 26 to 35 cm in total length, had been kept for a long time in a 600-l fiberglass home tank with running dechlorinated tap water. This home water was used in the experiments. The fish had the diurnal rhythm and schooling behavior typical for this salmonid species. The study was made in May–June 1990. Blank tests were first made on the orientation behavior to tap water only, and then the fish were given a choice between tap water and home water. Such water was pumped into a $120 \times 32 \times 22$ cm test trough. Test water was introduced in one end and tap water in the other end of the trough, and in the next session, the position of inflow was reversed. When more than half the fish passed the middle of the trough, it was noted as one passage. The fish in the trough could not see the fish in the home tank or aquarium. Statistical tests of the chi squared type ($P \leq 0.05$) were made. The rest of the behavioral bioassay and testing procedure has been described earlier, and a significant attraction to home water

was found as compared to tap water, as well as water from adult rainbow trout, a potential predator [2].

Altogether, 8 fish were taken at random from this population of 50 fish and tested as to chemical clues from a slightly injured fish. Two fish were injured by scraping an area of about $1 \, cm^2$ in the epidermis between the dorsal fin and the lateral line. After these fish were tested, both were cut about 2 mm into the dermis in the same scratch. This operation now caused a slight bleeding. The behavioral effect of one injured fish was tested at a time. The fish was kept either together with the others in the tank, or together with four or six other fish of the same population in a 200-l home aquarium.

Results

The first test fish (No. 6) was tested on water containing chemical cues from a scraped fish that was kept together with four other fish in the home aquarium. There was, at first, an insignificant orientation towards this home water, but after reversal of inflow of the trough, there was instead a significant preference for tap water, so that when the results were put together there was an insignificant preference for tap water (Fig. 1, No. 6). A scraped and cut fish was then kept together with six fish in the home aquarium, and fish No. 10 was tested. When summarized, these results showed that there was an insignificant preference for tap water, both before and after reversal of the trough inflow (Fig. 1, No. 10). A scraped fish was then kept together with the other fish of the same population in the tank, and the orientation of fish No. 11 was studied. This time the results showed an insignificant preference for home water, both before and after reversal of inflow (Fig. 1, No. 11).

Finally, a scraped and cut fish was kept alone in the aquarium and a group of five fish was tested. Also this time, a summary of the observations showed that there was an insignificant preference for home water (Fig. 1, Nos. 18–22).

In total, the eight test fish oriented 457 times towards the inlet of home water compared to 421 times towards that of tap water. This difference, of only about 4%, was also very insignificant.

[1] Department of Animal Ecology, Umeå University, S-901 87 Umeå, Sweden

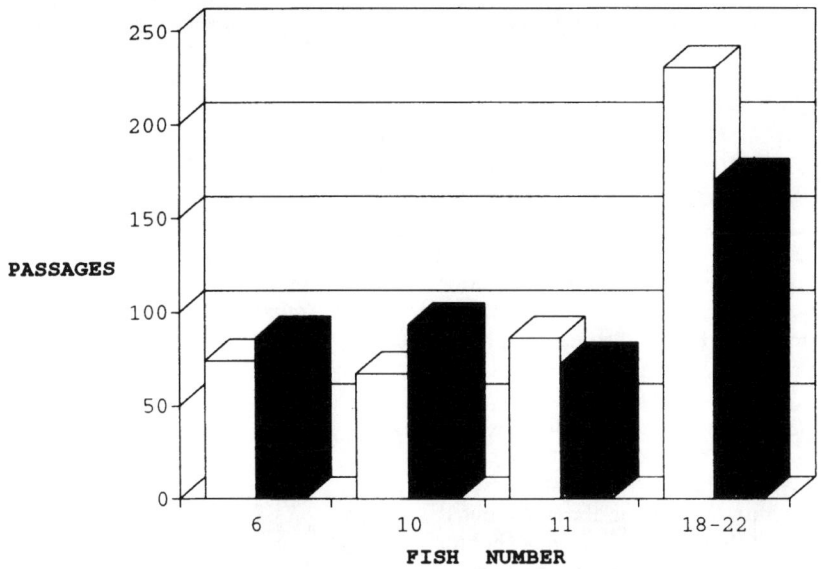

FIG. 1. Chemosensory orientation of adult lake whitefish to tap water (*solid columms*) and home water (*outlined columns*), the latter containing chemical cues from injured fish of the same population

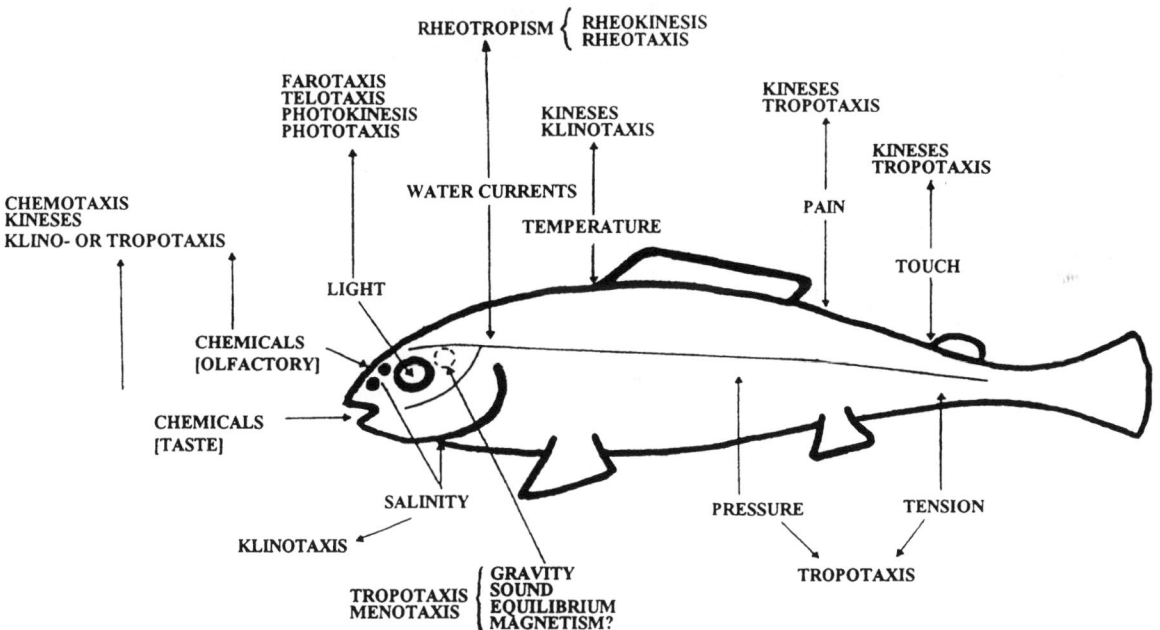

FIG. 2. Sensory ecology gives a sensory perspective of the entire organism in relation to its environment. Different external stimuli are the basis for various direct (*taxes*), indirect (*kineses*) or other types of orientation movements, all of which are structured according to the principle of Integrated Hierarchial Orientation [3]

Discussion

These four experiments showed that the test fish were neither attracted nor repelled by the home water. These results differ markedly from the earlier results of the chemosensory orientation to conspecifics of the same population of this species [2]. The 27 fish used in those experiments oriented 1396 times towards the inlet of home water but only 670 times towards that of tap water. However, both biotic and abiotic factors were the same in the two kinds of experiments: the same fish and temperature, the same test apparatus and technique were used, and the experiments were made during the same season, etc. It is therefore reasonable to suppose that the only difference was that the home water of the present experiments contained other chemical cues than those in the first type of experiments [2].

In the Schreckstoff system, and in other alarm pheromone systems in fishes as well, the active chemical(s) are released by mechanical damage to the fish, and the receiving fish responds with behavioral changes, often a fright reaction, that may reduce the receiver's vulnerability to predation [1]. The sending of alarm substances occurs only after damaging contact with the predator, which places these alarm pheromones with the distress calls given by birds and mammals when they are actually being attacked.

The present experiments on lake whitefish showed that the normal strong preference responses to home water [2] disappeared when the test fish were exposed to home water containing alarm chemical cues (probably a kind of Schreckstoff) from a slightly injured fish of the same population, irrespective of whether the injured fish was only scraped or slightly bleeding, or whether it was kept in an aquarium with 4 or 6 other fish of the same population, or in the home tank with 50 other fish and a bigger water volume. As the behavioral effect disappeared within 24 h, and as there were no avoidance reactions either, it seems that *the biological significance of this*

kind of alarm substance is to dissolve the fish school when attacked; the individuals may then hide.

The studies on lake whitefish also showed that, although they orientated to visual and temperature cues as well, their orientation drive towards chemical cues was stronger [2]. Orientation to these different cues in lake whitefish is therefore another example of the principle of Integrated Hierarchial Orientation that was suggested to guide, for example, homing salmonid fishes such as the Baltic salmon *Salmo salar* and sea trout *Salmo trutta* [3] (Fig. 2).

References

1. Smith RJF (1992) Alarm signals in fishes. Rev Fish Biol Fisheries 2:33–63
2. Bertmar G (1992) Chemosensory orientation to conspecifics and rainbow trout in adult lake whitefish *Coregonus clupeaformis* (Mitchill). In: Doty RL, Müller-Schwarze D (eds) Chemical signals in vertebrates VI. Plenum, New York
3. Bertmar G (1982) Structure and function of the olfactory mucosa of migrating Baltic trout under environmental stresses, with special reference to water pollution. In: Hara TJ (ed) Chemoreception in fishes, vol 8. Elsevier, Amsterdam

Avoidance of Pesticides by Olfaction in Carp

Y. Ishida[1], H. Yoshikawa[2], and H. Kobayashi[1]

Key words. Olfaction—Pesticide

Widespread use of pesticides in agricultural areas has caused aquatic pollution. Lethal concentrations of chemical pollutants are not usually present in nature. However, downpour or long-term rainfall effuses toxic concentrations of pesticides from farmland into fish habitats. If fish can detect these materials in water, they may be able to avoid toxic concentrations, and escape to safer areas. Avoidance behavior is the first step in the defense against toxic effects, and is a useful defense strategy. The ability of fish to avoid aquatic pollutants has been documented in many behavior tests [1,2]. The avoidance response is induced by chemoreception in fish. Olfaction and taste in fish have highly sensitive receptors. The importance of olfaction in the avoidance response was demonstrated by using anosmic fish [3]. We examined the role of olfaction in the avoidance behavior of carp, using three kinds of pesticide emulsion.

Electrophysiological study of the olfactory bulbar responses to pesticides indicated high sensitivities to isoprothiolane, at a threshold of $6.8 \times 10^{-3}\,\mu g/l$, and to benthiocarb, at $1.5 \times 10^{-2}\,\mu g/l$, but low sensitivity to fenitrothion, at $50\,\mu g/l$. Gustatory nerve responses were insensible to all tested pesticides.

Behavioral tests, using a Y-maze, showed an avoidance response to isoprothiolane and benthiocarb at low concentrations, $6.8\,\mu g/l$ and $1.5\,\mu g/l$, respectively. However, carp were unable to avoid fenitrothion at a low concentration, $5 \times 10^2\,\mu g/l$. Since the threshold of olfactory bulbar response to sodium lauryl sulfate (SLS) surfactant occurred at a remarkably low concentration, $10^{-3}\,\mu g/l$, we added SLS (1%) to fenitrothion. The modified fenitrothion caused a decrease in the avoidance threshold, to $5 \times 10^{-1}\,\mu g/l$.

The threshold concentration of isoprothiolane in avoidance behavior was higher than that in the olfactory bulbar response. To explain the reason for the difference in the two threshold concentrations, we examined the avoidance reaction of carp conditioned by electric shock. The threshold concentration of the reaction was $6.8 \times 10^{-2}\,\mu g/l$ to isoprothiolane.

The role of taste in the avoidance behavior of carp was examined by investigating the conditioned reaction to stimulation of the gustatory receptor only. The result showed that taste stimulation was independent of avoidance reaction. This result coincided with the lack of gustatory nerve response in carp.

[1] Kinki University, Faculty of Agriculture, 3327-204 Naka-machi, Nara, 631 Japan
[2] Interdisciplinary Research Institute of Environmental Sciences, 540 Higashi-yanagi-cho, Nishi-iru, Hichihon-matsu, Itsutsuji-dori, Kamigyo-ku, Kyoto, 602 Japan

References

1. Giattina JD, Garton RR (1983) A review of the preference avoidance responses of fishes to aquatic contaminants. Residue Rev 87:44–90
2. Beitinger TL, Freeman L (1983) Behavioral avoidance and selection responses of fishes to chemicals. Residue Rev 90:35–55
3. Hidaka H (1989) Avoidance by olfaction in a fish, Medaka (*Oryzias latipes*), to aquatic contaminants. Environment Pollution 56:299–309

Chemoreception in *Paramecium*

Judith L. Van Houten[1]

Key words. Chemoreception—Ciliate

Introduction

Aquatic organisms have the capability of detecting and responding to chemical stimuli. The stimulus can signify the presence of food, mates, predators, or a place to settle and continue development. Unicellular aquatic organisms are not exceptions and as an example, I discuss chemoresponses of *Paramecium tetraurelia*, a ciliated eukaryotic microorganism.

Paramecia swim through pond or sea water by beating their cilia, which they also use to sweep bacteria into a gullet where a food vacuole is forming. Paramecia have complementary mating types but appear to follow soluble chemical cues to locate food and not to communicate about mating or to avoid predators. One of *Paramecium*'s predators, *Didinium*, takes advantage of *Paramecium*'s attraction to food and follows the chemical cues from bacteria to locate the paramecia, which are feeding on the bacteria at the end of the stimulus trail [1].

Chemosensory signal transduction pathways of paramecia begin at the cell surface membrane and ultimately converge on the ciliary apparatus to produce a response, a change in swimming behavior (see Van Houten [2,3] for reviews). There are at least three chemosensory signal transduction pathways in *P. tetraurelia*, known in varying amounts of detail. Each pathway is discussed in turn, but first a short description of *Paramecium* physiology is in order.

Paramecia swim by beating their cilia toward the posterior in metachronal waves. The cells can modulate their swimming speed and can make abrupt turns by transiently reversing the power stroke of the ciliary beat and moving backward for a body length or so before righting the beat and moving off in a new direction (see Machemer [4] for review). From some elegant studies of *Paramecium* physiology, we know that the resting membrane potential correlates with the frequency of ciliary beating, and that the abrupt turn is due to a calcium action potential that transiently increases Ca^{2+} in the ciliary compartment [4–6]. Even more is now known about many conductances of the *Paramecium* membrane and there are Mendelian mutants available with defects in individual or multiple conductances [7–10]. These defects become apparent in the altered swimming behavior of the cell.

Paramecia are neuron-like excitable cells, and their membrane potential can be manipulated by changing extracellular cations. Chemical stimuli also ultimately affect membrane potential but indirectly through signal transduction pathways. Some chemical stimuli cause hyperpolarizations that in turn cause the cell to beat cilia faster and, because there are fewer spontaneous action potentials, to turn less frequently. Generally repellent stimuli cause the cell to beat cilia more slowly and turn frequently owing to frequent action potential episodes. These changes translate into rather fast and smooth swimming in attractants and slightly slower movement and frequent turning in repellents [2,3,11]. Although this discussion is an oversimplification of the membrane potential changes that different classes of stimuli can cause, it serves as a starting point for understanding the three signal transduction pathways. The changes in movement of the cell do not cause a "chemotaxis," that is, an oriented movement directly up or down a gradient of stimulus. Instead, the cells accumulate or disperse by a biased random walk, a chemokinesis, that also requires adaptation for most efficient results [2].

Pathway One

The chemical stimuli folate, acetate, and cyclic AMP (cAMP) fall into one class of stimuli that are attractants and hyperpolarize the cell, probably by activation of a plasma membrane calcium pump. The receptors for these stimuli have been characterized to varying degrees by binding and mutant studies (see Van Houten [2,12] for reviews). In particular, the cAMP receptor has been isolated and studied for its biochemical properties. There are two closely related proteins (48 kDa) that elute together from a cAMP affinity column and are inseparable by the many high-pressure liquid chromatography (HPLC) methods that we have tried. Both proteins are glycosylated [13], and cyanogen bromide cleavage products are identical except for one pair of bands ([12]; B. Cote, personal communication).

[1] Department of Zoology, University of Vermont, Burlington, VT 05405, USA

Despite the similarities, the N-termini are different (B. Cote, personal communication). Polyclonal antibodies produced against the proteins render whole cells unable to respond to cAMP in T-maze tests of chemoresponse, although responses to all other attractants tested remain intact [13].

The stimuli of this first signal transduction pathway initiate a hyperpolarization [14,15], but the characteristics are not consistent with channel activity [14]. Additional studies of the effects of LiCl on the chemoresponse and of the double mutant K-shyA/B in attractant stimuli gave us insight into this puzzle. LiCl profoundly affects the chemoresponse to only some stimuli (which included acetate, folate, and cAMP) and reduces calcium efflux from normal resting cells by about 50% [16]. The K-shy mutant has a yet undetermined alterations in calcium homeostasis [17], and is not attracted to the same class of stimuli that are affected by LiCl [15]. These observations led us to search for a plasma membrane calcium pump. Consequently, we described a Ca^{2+} APTase activity of the pellicle, a preparation of surface membranes that include the plasma membrane, cytoskeleton, and underlying calcium-sequestering organelles, alveolar sacs [18]. We have shown that this Ca^{2+} ATPase activity is characteristic of a plasma membrane pump and is distinct from intracellular organelle calcium pump activities [19,20]. There is a protein from the same preparton that likewise has the characteristics of the plasma membrane pump. Specifically, it forms a phosphoenzyme intermediate under the expected conditions, and it binds calmodulin, as is expected for the plasma membrane pump [19,20]. An increase in pump activity of less than 1% could account for the small hyperpolarizing current associated with chemical attractants, such as cAMP, acetate, and folate. Such a small difference in activity is within the variability of our assay, making a direct test of the pump in the signal transduction pathway impossible. We are therefore taking a more indirect route and are characterizing the pump and correlating its characteristics with chemoresponse [20]. Additionally, the gene for the plasma membrane pump of *P. tetraurelia* is almost completely cloned, with 3.1 kb of approximately 3.6 kb sequenced (Elwess, personal communication).

Pathway Two

Glutamate and inosine monophosphate (IMP) are stimuli that appear to utilize yet another signal transduction pathway. Cells swim rapidly and smoothly in glutamate, characteristic of a hyperpolarizing stimulus, and swim more slowly with frequent turns in IMP, characteristic of a depolarizing stimulus. Previously Preston and Usherwood [21] described an attractant response to nanomolar concentrations

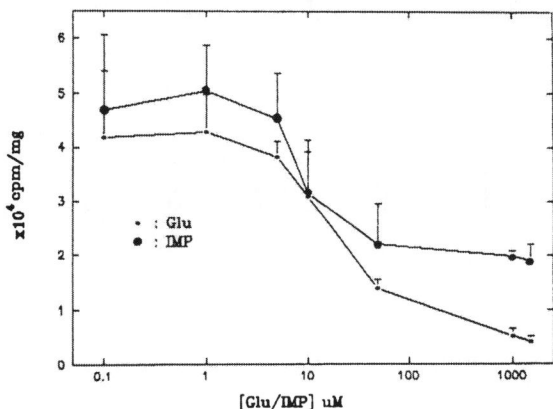

FIG. 1. Displacement of ^{3}H-glutamate binding to whole cells. Counts are normalized between experiments for specific activity. Unlabeled glutamate or unlabeled IMP is added in increasing amounts to displace binding. Approximately 50% glutamate binding in displaced by IMP. IMP > GMP > AMP ≫ CMP or cAMP. Approximately 50% glutamate-specific binding is not displaced except by quisqualate (not shown). (See [24] for details of methods)

of glutamate, but the response described here, as other chemoresponses of *P. tetraurelia*, requires much higher amounts. Studies of glutamate binding to whole cells has revealed two classes of glutamate binding sites: those that can be displaced only by glutamate or quisqualic acid and those that can be displaced by IMP, or to a lesser extent by GMP or AMP [22,23]. Figure 1 shows a displacement curve with glutamate or IMP as the unlabeled displacing ligand. These results parallel the behavioral studies, which show that glutamate can interfere with IMP response, but not vice versa, and that AMP or GMP but not CMP or cAMP can interfere with the response to IMP [25]. We infer that the exclusively glutamate binding sites are associated with attraction to glutamate and that the sites affected by IMP mediate repulsion from IMP.

Glutamate and IMP rapidly affect intracellular cAMP levels. Glutamate increases intracullar cAMP about threefold, roughly equivalent to the increase in cAMP caused by transferring cells from 10 mM KCl to 1 mM KCl buffer (Fig. 2) [26–28]. IMP, on the other hand, causes a decrease in cAMP levels by about 50%, roughly equivalent to that caused by a depolarization from transferring cells from 1 mM KCl to 10 mM KCl (Fig. 2).

We do not yet know the cause-and-effect relations of this pathway. It is possible that the hyperpolarization caused by the stimuli induces the cAMP change, which in turn affects ciliary beating. Bonini and Nelson [27] and Nakaoka and Ooi [29] have shown that cyclic nucleotides affect the ciliary beating of permeabilized cells, whose membrane potential can play no part in ciliary beat control. However, the results of Hennessey et al. [30] imply that

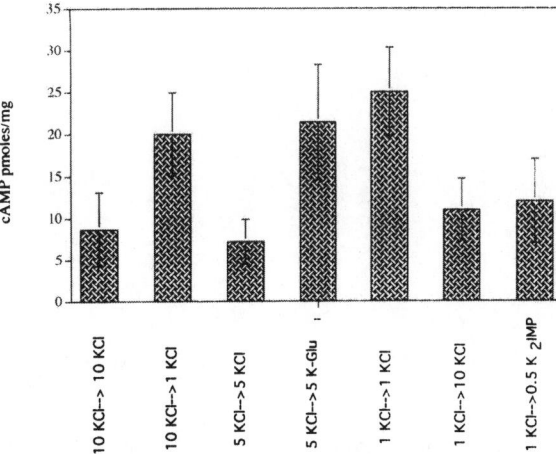

FIG. 2. Measurements of cAMP by radioimmunoassay. Cells are incubated in salt buffers and transferred to a new buffer from which cells are sampled in duplicate. The cells are mixed with EDTA, homogenized, boiled, and centrifuged to remove debris. The supernatant is then assayed by RIA [34] (Amersham, Arlington Heights, IL, USA) for cyclic AMP. Samples are also obtained for protein analysis. Cells are incubated in 10 mM KCl and transferred to 10 mM KCl and 1 mM KCl for control and hyperpolarizing conditions, respectively. Cells are then transferred from 1 mM KCl to 1 mM KCl or 10 mM KCl for control and depolarizing conditions, respectively. For chemical stimuli, cells are incubated in 5 mM KCl and transferred to 5 mM KCl or 5 mM potassium glutamate or incubated in 0.5 mM KCl and transferred to 0.5 mM KCl or 0.5 mM K-IMP. Note that the transfer to potassium glutamate increases cyclic AMP about as much as a hyperpolarization transfer to low KCl and that transfer to K-IMP decreases cyclic AMP about as much as a depolarization in high KCl. Glutamate stimulates an approximately three-fold increase in cAMP by 30 s. IMP reduces cAMP by approximately 50% by 30 s. Control cells hyperpolarized by changing K_o increase their cAMP by about threefold. Control cells depolarized by changing K_o decrease their cAMP by about 50%

Pathway Three

There is yet another class of stimuli that appear to utilize yet a third chemosensory signal transduction pathway. NH_4Cl is the best example of a stimulus for this pathway. The mutant K-shyA/B is attracted to NH_4Cl, and LiCl has little effect on the chemo-response to NH_4Cl [16]. Measurements of cAMP show no change of the NH_4Cl in cells relative to that in cells in control salt buffer (W.Q. Yang, personal communication). These results distinguish this stimulus from those of the other two pathways.

After a long search, we found no related compound that competes or interferes with the chemo-response to NH_4Cl, implying that there is either extreme receptor specificity or that NH_4Cl did not utilize a cell surface receptor to initiate its response (M. Gagnon and J. Van Houten, personal communication). It is possible that NH_4Cl crosses the membrane as NH_3 and there sets up an equilibrium with NH_4Cl, which alkalinizes the cell [32] and perhaps results in altered channel activity.

To study the changes in intracellular pH, we utilize an ion-sensitive fluorescent dye, BCECF [32,33]. This dye partitions into the cells as a membrane permeant acetoxymethyl ester; once inside the cell, it is cleaved to its trapped pH-sensitive membrane-impermeant form. Preliminary studies with BCECF have shown that the cells in 5 mM NaCl have an intraclleular pH of 6.80 ± 0.05 and that cells in 5 mM NH_4Cl have a lower intracellular pH of 6.36 ± 0.05. This acidification follows an initial alkalin-ization (D. Davis and J. Van Houten, data not shown).

We are at the beginning of these studies of intracellular pH. Because NH_4Cl hyperpolarizes the cells relative to NaCl, it will be interesting to determine the physiologic mechanism by which intracellular pH affects membrane potential [35].

Summary

There appear to be at least three signal transduction pathways in *P. tetraurelia* (Fig. 3), all of which converge on the membrane potential of the cell. The membrane potential, in turn, affects ciliary beating and, with adaptation, ultimately the accumulation or dispersal by a chemokinetic mechanism results. Second messengers (Ca^{2+}, cAMP, pH_i) are different for each pathway. A great deal of experimental detail has necessarily been omitted to avoid obscuring the picture of multiple signal transduction pathways. There is much left to accomplish in protein biochemistry, gene cloning, and electrophysiology to characterize these three pathways. It is also important to keep in mind that the division of signal transduction into separate pathways is most likely an oversimplification, and minimally there will be cross talk between the pathways.

the cylic nucleotide increases precede the hyper-polarization, which then directly affects ciliary beat. Additionally, Schultz et al. showed that the adenylyl cyclase that would produce the cAMP associated with hyperpolarization has intrinsic potassium conductance properties of a channel [31]. We know that for some attractants, such as acetate, intracellular cAMP levels do not seem to play a part in the hyperpolarization-based ciliary beat change [2]. Thus it is not clear how the chemosensory hyper-polarizations, only some of which seem to induce change in cAMP, relate to this potassium channel and other hyperpolarizing mechanisms in the cell.

Fig. 3. Three signal transduction pathways (I, II, III) in *Paramecium tetraurelia*

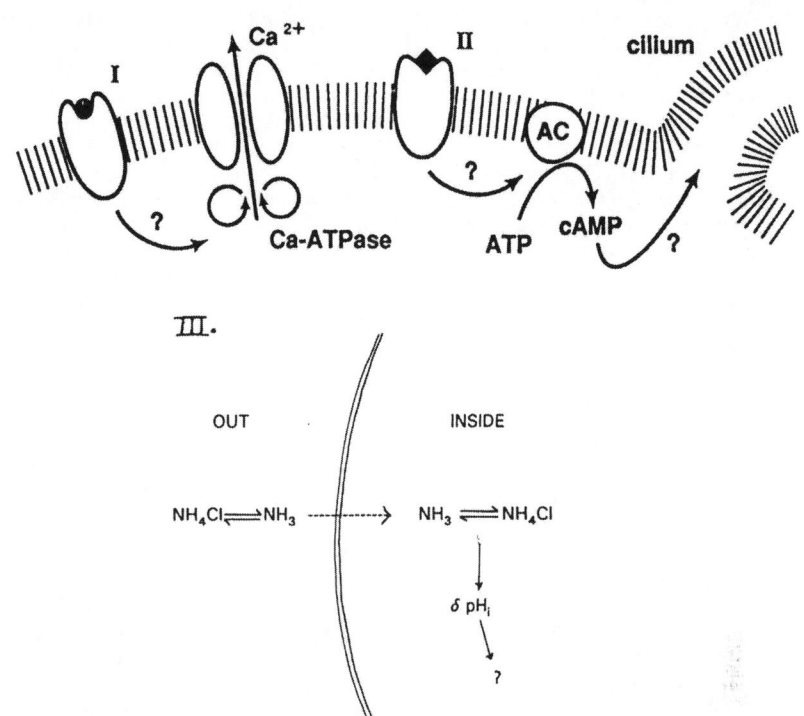

Acknowledgments. The work presented here is currently supported by NIH and the VCCC. The citations include a great deal of work done by past laboratory members, and I have attempted to acknowledge our current members for their valuable contributions. Thanks to T. Evans, D. Nelson, and C. Kung for mutants.

References

1. Antipa G, Martin K, Rintz M (1983) A note on the possible ecological significance of chemotaxis in certain ciliates protozoa. J Protozool 30:55–58

2. Van Houten J (1990) Chemosensory transduction in *Paramecium*. In: Armitage JP, Lackie JL (eds) Biology of the Chemotatic Response. University of Cambridge Press, Cambridge, pp 297–321

3. Van Houten J (1992) Chemosensory transduction in eukaryotic microorganisms. Annu Rev Physiol 54:639–663

4. Machemer H (1989) Cellular behavior modulated by ions: electrophysiological implications. J Protozool 36:463–487

5. Naitoh Y, Eckert R (1968) Electrical properties of *Paramecium caudatum*: all-or-none electrogenesis. Z Vergl Physiol 8:453–472

6. Eckert R (1972) Bioelectric control of ciliary activity. Science 176:473–481

7. Kung C, Preston RR, Maley ME, et al. (1992) In vivo *Paramecium* mutants show that calmodulin orchestrates membrane responses to stimuli. Cell Calcium 13:413–425

8. Preston RR (1990) Genetic dissection of Ca^{2+}-dependent ion channel function in *Paramecium*. Bioessays 12:273–281

9. Preston RR, Kink JA, Hinrichsen RD, Saimi Y, Kung C (1991) Calmodulin mutants and Ca^{2+}-dependent channels in *Paramecium*. Annu Rev Physiol 53:309–319

10. Saimi Y, Kung C (1987) Behavioral genetics of *Paramecium*. Annu Rev Genet 21:47–65

11. Van Houten J (1978 Two mechanisms of chemotaxis in *Paramecium*. J Comp Physiol [A] 127:167–174

12. Van Houten J (1993, in press) In: Kurjan J, Taylor B. (eds) Paramecium chemosensory transduction. Academic Orlaudo, FL, pp 309–327

13. Van Houten JL, Cote BL, Zhang J, Baez J, Gagnon ML (1991) Studies of the cyclic adenosine monophosphate chemoreceptor of *Paramecium*. J Membr Biol 119:15–24

14. Preston RR, Van Houten J (1987) Chemoreception in *Paramecium*: acetate- and folate-induced membrane hyperpolarization. J Comp Physiol 160:525–536

15. Van Houten J (1979) Membrane potential changes during chemokinesis in *Paramecium*. Science 204:1100–1103

16. Wright MV, Frantz M, Van Houten J (1991) Lithium fluxes in *Paramecium* and their relationship to chemoresponse. Biochim Biophys Acta 1107:223–230

17. Evans TC, Hennessey T, Nelson DL (1987) Electrophysiological evidence suggests a defective Ca^{2+} control mechanism in a new *Paramecium* mutant. J Membr Biol 98:275–283

18. Stelly N, Mauger J-P, Claret M, Adoutte A (1991) Cortical alveoli of *Paramecium*, a vast submembranous calcium storage compartment. J Cell Biol 113:103–112

19. Wright MV, Van Houten J (1990) Characterization of a putative Ca^{2+}-transporting Ca^{2+}-ATPase in the pellicles of *Paramecium tetraurelia*. Biochim Biophys Acta 1029:241–251

20. Wright MV, Elwess N, Van Houten JL (1993) Ca^{2+} transport and chemoreception in *Paramecium*. J Coump Physiol [B] 163:288–296

21. Preston RR, Usherwood PNR (1988) L-Glutamate-induced membrane hyperpolarization and behavioral responses in *Paramecium tetraurelia*. J Comp Physiol [A] 164:1–8

22. Yang WQ, Van Houten JL (1993) *Paramecium* chemoreception of IMP and glutamate. Soc Neurosci Abstr 19:1429

23. Yang WQ, Van Houten JL (1993) IMP and glutamate as stimuli in *Paramecium* chemoreception. Chem Senses 18:653

24. Smith R, Preston RR, Schulz S, Van Houten J (1987) Correlation of cyclic adenosine monophosphate binding and chemoresponse in *Paramecium*. Biochim Biophys Acta 928:171–178

25. Van Houten J, Yang WQ (1992) IMP as stimulus in *Paramecium* chemoreception. Soc Neurosci Abstr 19:845

26. Bonini NM, Gustin MC, Nelson DL (1986) Regulation of ciliary motility by membrane potential in *Paramecium*: a role for cyclic AMP. Cell Motil Cytoskel 6:256–272

27. Bonini NM, Nelson DL (1988) Differential regulation of *Paramecium* ciliary motility by cAMP and cGMP. J Cell Biol 106:1615–1623

28. Schultz JE, Grunemund R, von Hirschhausen R, Schonefeld U (1984) Ionic regulation of cyclic AMP levels in *Paramecium tetraurelia* in vivo. FEBS Lett 167:113–116

29. Nakoaka Y, Ooi H (1985) Regulation of ciliary reversal in triton-extracted *Paramecium* by calcium and cyclic adenosine monophosphate. J Cell Sci 77: 195

30. Hennessey T, Machemer H, Nelson DL (1985) Injected cyclic AMP increases ciliary beat frequency in conjunction with membrane hyperpolarization. Eur J Cell Biol 36:153–156

31. Schultz JE, Klumpp S, Benz R, Schurhoff-Goeters WJC, Schmid A (1992) Regulation of adenylyl cyclase from *Paramecium* by an intrinsic potassium conductance. Science 255:600–603

32. Anwer MS, Engelking LR (1993) Intracellular pH regulation. In: Tavoloni N, Berk PD (eds) Hepatic transport and bile secretion: Physiology and pathophysiology. Raven, New York, pp. 351–361

33. Welti R, Mullikin LJ, Helmkamp GM, Yoshimura T (1984) Partition of amphiphilic molecules into phospholipid vesicles and human erythrocyte ghosts: measurements by ultraviolet difference spectroscopy. Biochemistry 23:6086–6091

34. Schultz JE, Klumpp S (1993) Cyclic nucleotides and calcium signaling in *Paramecium*. In: Shenolikar S, Nairn AC (eds) Advances in Second Messenger and Phosphoprotein Research (vol 27). Raven, New York, pp 25–45

35. Van Houten J, Preston RR (1988) Chemokinesis. In: Gortz H-D (ed) *Paramecium*. Springer, Berlin Heidelberg, pp 282–300

Response to Reduced Glutathione of *Hydra* Cultured in Different Places: Involvement of Protease Activities in Modulation of the Behavioral Response

KAZUMITSU HANAI and YASUJI MATSUOKA[1]

Key words. Feeding behavior—Glutathione chemo-reception—Glutathione-binding protein—Modulation by protease—*Hydra*

Introduction

We have reported elsewhere the tentacle ball formation in *Hydra japonica* as its feeding response to reduced glutathione [1]. Glutathione-induced response also has been characterized by other behaviors by other workers [2,3]. The tentacle ball formation has interesting features, easy quantitative determination, and sensitive modification by many biologically active substances, including biologically active peptides, platelet proteins [4], and growth factors from mammals [5]. All these studies were done in a laboratory in Fukuoka. The same animal, however, showed a weak response when it was cultured in a different place, Otsu, located about 600 km away from Fukuoka despite the same culture method with the same prey. We became curious and so have studied why *Hydra* cultured in different places responds differently to glutathione. We found that short-term treatment of live animals at low concentration with chymotrypsin was enough to induce strong tentacle ball formation in response to reduced glutathione.

Materials and Methods

Hydra japonica was used in all experiments. The animals were cultured in a medium containing $NaHCO_3$ 0.1 g/L, KCl 0.015 g/L, $CaCl_2$ 0.05 g/L, $MgCl_2$ 0.015 g/L, and EDTA·4Na 0.05 g/L (basal medium).

Behavioral Response

Ten animals were transferred into 2 ml of 1 mM piperazine-*N*,*N'*-bis(2-ethanesulfonic acid)/1 mM $CaCl_2$, pH 6.2 (PIPES buffer) from the stock cul-

ture. Concentrated *S*-methyl-glutathione (GSM), 3–10 μl, solution was applied in final concentrations of 0.1, 0.3, 3.0, 10.0, and 50.0 μM 5 min after a rinse with PIPES buffer. *Hydra* responded with different susceptibilities to biologically active substances at these different concentrations of GSM [4,6]. The features of tentacle ball formation have been described elsewhere [1].

The animals showing a response were counted from 6 to 10 min after the start of the stimulation. The fraction of animals that showed a response were summated during this observation period in order to determine the degree of the response. When the animals responded completely during this observation period, the response score was 5.

Protease Treatment of Live Animals

Ten animals were incubated with chymotrypsin 10 ng/ml in 2 ml of 1 mM $NaHCO_3$/1 mM $CaCl_2$. After 5 min of incubation, the buffer was changed to PIPES buffer, and the response was examined.

Protease Activities of *Hydra* Homogenate

A few hundred *Hydra* were homogenized in a polytron homogenizer in 2 ml of 20 mM Tris 150 mM NaCl, pH 7.4 (TBS) and then centrifuged at $10^4 g$ for 10 min. The supernatant was used as an enzyme source. The amount of protein was determined by a dye-binding assay [7] (BioRad, Richmond, CA). Protease activities were examined with substrates carbobenzoxy-Phe-Arg-4-methyl-coumaryl-7-amide (ZFR-MCA, Peptide Institute, Osaka, Japan) [8], succinyl-Leu-Leu-Val-Tyr-4-methyl-coumaryl-7-amide (SucLLVY-MCA, Peptide Institute) [9,10], and Arg-4-methylcoumaryl-7-amide (R-MCA, Peptide Institute) [11]. These substrates are used for trypsin-type protease, chymotrypsin-type proteases, and cathepsin H-type protease, respectively. The substrate (1 μM) was incubated with the homogenated protein (3 μg) in 1 ml TBS at 30°C. After 10 min of incubation, 1 ml stopping solution (0.1 M CH_3COOH/0.1 M $ClCH_2COOH$, pH 4.4) was added to the mixture. The released 7-amino-4-methyl-coumarin was then determined by its strong fluorescence at 440 nm with excitation at 340 nm (RF1500

[1] Division of Neurochemistry, Institute of Molecular Neurobiology, Shiga University for Medical Science, Otsu, 520-21 Japan

Fluorophotometer, Shimadzu, Kyoto, Japan) [11]. The blank reaction was determined by adding the stopping solution before incubation and subtracted from the test reaction.

Protein Blotting Analysis of Glutathione-Binding Proteins

The membrane fraction was precipitated by centrifugation at $10^4 g$ for 20 minutes from *Hydra* homogenate prepared from about 10^6 animals with use of a teflon homogenizer in the presence of protease inhibitors [12]. The membrane fraction was solubilized by 1% octylglucoside in TBS. The solubilized membrane fraction was passed through an affinity column, which was prepared from epoxy-activated Sepharose (Pharmacia, Uppsala, Sweden) coupled with reduced glutathione. After extensive washing with TBS containing 0.1% Nonidet P-40, the bound protein was eluted with 50 mM $K_2CO_3/8$ M urea. The eluted fraction was analyzed by protein blotting analysis as described elsewhere [12].

Results

First we evaluated chemicals that might be contaminants in a normal laboratory environment. We examined the response of the animal after 4 days of culture in a medium to which trace amounts of various chemicals (1 ppm) were added. When a toxic effect was observed at this concentration, the concentration was reduced. The animals were transferred to this medium after feeding and were then fed once, on the second day after transfer. Chemicals examined were amino acids, essential metals, vitamins, and metabolites. We found that the animals cultured in the medium supplemented with ammonium acetate or pyroglutamic acid showed a stronger response more frequently (Fig. 1). Such a response was observed for only a few days after transfer to the supplemented medium. After this short period, only a poor response was resumed even after renewal of the supplemented medium (unpublished observation).

We examined the protease activity of the *Hydra* homogenate and the response of *Hydra* cultured in different media to see if there was a relation between these two. When the protease activities of the homogenate were high, we more frequently observed a strong response (Fig. 2). This tendency was observed for all three substrates and for all stimulatory conditions. When the response data were divided into two groups according to the protease activities at an appropriate value, the response observed in the animals with high protease activities was significantly higher than that with low protease activities by the paired Student's *t*-test (Table 1).

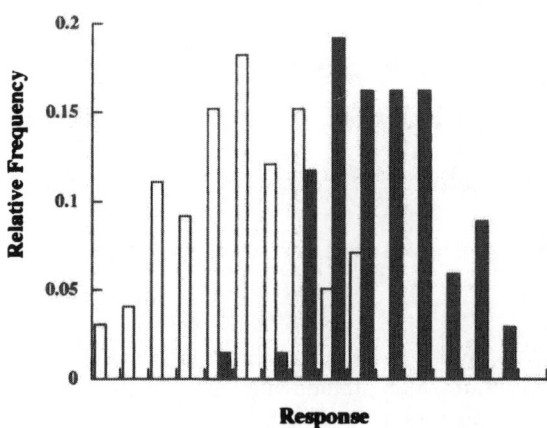

FIG. 1. Histogram of the response of *Hydra* cultured in various media. One group (*open bars*) was cultured in the basal medium and another group (*closed bars*) in the medium supplemented with ammonium acetate 1 ppm 4 days before behavioral examination. The response of these animals to 10 μM GSM was examined. We evaluated 100 tests for those cultured in the basal medium and 68 tests for those in the supplemented medium

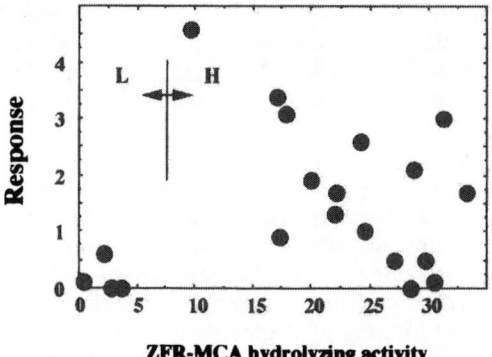

FIG. 2. Relation between the response of the animal and the protease activity (ZFR-MCA hydrolyzing activity). The animals were cultured 4 days in four groups: (1) in basal medium; (2) in medium supplemented with ammonium nitrate 1 ppm; (3) in medium with pyroglutamic acid 1 ppm; (4) in medium with ammonium nitrate 1 ppm and pyroglutamic acid 1 ppm. ZFR-MCA hydrolyzing activity (pmol/μg protein/min) and the response to 3 μM GSM was examined for each group of the animals

We examined the glutathione-binding protein, which binds to the glutathione receptor [13], by the protein blotting analysis with our monoclonal antibodies (mAbs) J245 and J5 [6,12]. A receptor candidate was characterized by a common antigen for these two mAbs and binds glutathione. We have reported elsewhere that 180- and 220-kDa proteins are the receptor proteins in the animal cultured in Fukuoka [12]. The proteins with these features in *Hydra* cultured in Otsu were 240 and 250 kDa, a little larger than those in Fukuoka (Fig. 3). This difference in the size of the glutathione-binding pro-

TABLE 1. Statistical analyses of the relation between the response and protease activities in the homogenate of *Hydra* cultured in different media.

| GSM (μM) | High protease activity group | | Low protease activity group | | |
	Mean ± SEM	No.	Mean ± SEM	No.	P
Substrate Z-FR-MCA					
0.1	1.713 ± 0.24	16	0 ± 0	4	<0.05
3.0	1.775 ± 0.326	16	0.175 ± 0.144	4	<0.05
10.0	1.831 ± 0.324	16	0.25 ± 0.132	4	<0.05
Substrate Suc-LLVY-MCA					
0.1	1.933 ± 0.252	9	0.909 ± 0.35	11	<0.05
3.0	2.189 ± 0.477	9	0.855 ± 0.281	11	<0.05
10.0	2.211 ± 0.401	9	0.945 ± 0.355	11	<0.05
Substrate R-MCA					
0.1	1.875 ± 0.276	12	0.613 ± 0.314	8	<0.05
3.0	2.025 ± 0.394	12	0.6 ± 0.258	8	<0.05
10.0	2.1 ± 0.355	12	0.637 ± 0.34	8	<0.05

The response data were divided into two groups according to the homogenate protease activities of the animals in different media. One such example is shown in Fig. 2. Statistical significance was examined between these two groups for each response by the paired Student's *t*-test.

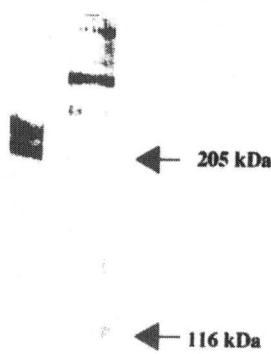

A   B

FIG. 3. Western blotting analysis of the glutathione-binding proteins of *Hydra* cultured in various places. The glutathione-binding protein was extracted from the animals cultured in Fukuoka (**A**) and Otsu (**B**) by a glutathione affinity column. The protein was electrophoresed in 5% sodium dodecylsulfate (SDS) polyacrylamide gel and transferred to polyvinylidenefluoride membrane. The protein was visualized by J5 antibody. *Arrows* show marker positions

tein may be responsible for the different behavioral responses to glutathione.

Because all the results suggested close relations between the response and protease activities, we

TABLE 2. Activation of the tentacle ball formation in *Hydra* treated with chymotrypsin.

Treatment	Response[a]		SEM
Control (*m* = 5)	1.46	±	0.14
Chymotrypsin (*m* = 9)	3.79	±	0.121[c]
Chymotrypsin + STI[b] (*m* = 6)	1.65	±	0.126

[a] After treatment of the live animals with chymotrypsin 1 ng/ml for 5 min, the response to 10 μM *S*-methylglutathione was examined.
[b] STI is soybean trypsin inhibitor, which was included in the reaction mixture at 50 ng/ml.
[c] $P < 0.01$ by one-way ANOVA.

examined the effect of protease treatment of live animals on the behavioral response. *Hydra* showed strong tentacle ball formation in response to GSM at all concentrations examined immediately after treatment with chymotrypsin for 5 min (Table 2). This activation by chymotrypsin was eliminated by inclusion of soybean trypsin inhibitor.

Discussion

We examined the effect of the trace impurities of more than 50 chemicals contained in the culture medium on the behavioral response as a possible source of different behavior of the animals cultured in different places. Among these chemicals, ammonium salts and pyroglutamic acid most frequently induced tentacle ball formation in the animals stimulated by GSM. The effect was complex, however.

Sometimes these chemicals decreased the response, especially when the animals in the basal medium showed a fair response. The response of the animals in the medium supplemented with these chemicals was not always significantly stronger than that of the animal in the basal medium. We observed a significant difference only when we compared the response of the animal groups divided on the basis of their protease activities. It is interesting to note that these chemicals were closely involved in nitrogen metabolism in the metabolic pathways of the biologic systems.

We observed a close relation between all the protease activities and the response to all stimulatory conditions examined. This binding suggests that a broad range of proteases activate the response because the examined substrates are susceptible to different proteases. The glutathione-binding proteins from the animals cultured in different places were of different molecular weights. Different proteolytic processing was thus suggested to yield different behavioral response in the animals cultured in different places.

Direct chymotrypsin treatment of live animals produced a strong response to GSM. Activation of the response was rapid, requiring only mild treatment, and was eliminated by soybean trypsin inhibitor, which is a potent inhibitor of chymotrypsin. Taking all observations into account, partial proteolytic digestion of the glutathione-binding protein appears to induce the response in the animal stimulated with GSM.

Acknowledgments. This study was supported by a Grant-in-Aid from the Ministry of Education, Science, and Culture of Japan, and by the Mishima-Kaiun Foundation.

References

1. Hanai K (1981) A new quantitative analysis of the feeding response in Hydra japonica: stimulatory effects of amino acids in addition to reduced glutathione. J Comp Physiol 144:503–508
2. Lenhoff H (1961) Activation of the feeding reflex in Hydra littoralis. I. Role played by reduced glutathione and quantitative assay of the feeding reflex. J Gen Physiol 45:331–344
3. Rushforth N, Hofman F (1972) Behavioral and electrophysiological studies of Hydra. III. Components of feeding behavior. Biol Bull 142:110–131
4. Hanai K, Kato H, Matsuhashi S, et al. (1987) Platelet proteins, including platelet-derived growth factor, specifically depress a subset of the multiple components of the response elicited by glutathione in Hydra. J Cell Biol 104:1675–1681
5. Hanai K, Oomura Y, Kai Y, et al. (1989) Central action of acidic fibroblast growth factor in feeding regulation. Am J Physiol 256:R217–R223
6. Sakaguchi M, Hanai K, Ohta K, et al. (1991) Monoclonal antibodies that depress a specific subset of multiple components of the glutathione-induced response of Hydra. J Comp Physiol 168:409–416
7. Bradford M (1976) A rapid and sensitive method for the quantitation of microgram quantities of protein utilizing the principle of protein-dye binding. Anal Biochem 72:248–254
8. Morita T, Kato H, Iwanaga S, et al. (1977) New fluorogenic substrates for alpha-thrombin, factor Xa, kallikrein, and urokinase. J Biochem 82:1495–1498
9. Sawada H, Yokosawa H, Hoshi M, Ishii S (1983) Ascidian sperm chymotrypsin-like enzyme; participation in fertilization. Experientia 39:377–378
10. Tsukahara T, Ishiura S, Sugita H (1988) An ATP-dependent protease and ingensin, the multicatalytic protease, in K562 cells. Eur J Biochem 177:261–266
11. Barrett A, Kieschke H (1981) Cathepsin B, cathepsin H, cathepsin L. Methods Enzymol 80:535–561
12. Ohta K, Hanai K, Morita H (1992) Glutathione-binding proteins identified by monoclonal antibodies which depress the behavioral response of Hydra. Biochim Biophys Acta 1117:136–142
13. Bellis S, Grosvenor W, Kass-Simon G, Rhoads D (1991) Chemoreception in Hydra vulgaris (attenuata): initial characterization of two distinct binding sites for L-glutamic acid. Biochim Biophys Acta 1061:89–94

Mechanisms of Detection and Discrimination of Mixtures in the Olfactory System of Spiny Lobsters

Charles Derby, Michele Burgess, Kirby Olson, Ted Simon, and Andrew Livermore[1]

Key words. Olfaction—Mixture—Transduction—Coding—Discrimination—Lobster

Introduction

A general problem facing animals is to extract from the many chemicals in their environment information about those that are behaviorally important. Important information includes the types of chemicals present. One question that my research group has been studying is how peripheral olfactory systems are organized to enable detection and discrimination of mixtures. We have focused on mixtures for two reasons. One is that most of the chemical signals that are behaviorally relevant to animals are mixtures. The second is that the response of an animal or neuron to a mixture is often unpredictable, even when the responses to the mixture's components are known [1]. These nonlinearities in response to mixtures are called mixture interactions, and they include mixture suppression (where the response to the mixture is less than expected) and mixture enhancement (where the response to the mixture is greater than expected). Here I discuss our work on the Caribbean spiny lobster *Panulirus argus* on two topics related to these issues: How do mixture interactions in olfactory receptor neurons affect neural coding and behavioral discrimination of mixtures? What cellular mechanisms may explain mixture interactions?

Materials and Methods

In studies of the Caribbean spiny lobster *Panulirus argus*, we have used the same set of chemical stimuli that are known to be behaviorally and physiologically active on *P. argus*. They include adenosine-5'-monophosphate (5'AMP), ammonium chloride, betaine, L-cysteine, L-glutamate, DL-succinate, taurine, and binary mixtures of these compounds. We think that using this set of chemicals is an important feature of our studies, as it enables us to

follow processing of the same olfactory information from receptor binding to behavior.

Our various approaches to the research questions require the use of different techniques. We have studied the ability of spiny lobsters to discriminate behaviorally among mixtures and their components by using an aversive conditioning paradigm [2,3]. Previous studies have suggested that across-neuron patterns (ANPs) could be used by the peripheral olfactory system of spiny lobsters to code the quality of odorant compounds [2,4,5]. Thus to evaluate the ability of the antennular olfactory pathway of *P. argus* to discriminate among odorant compounds and their binary mixtures, we have recorded single-unit extracellular responses of olfactory receptor neurons (ORNs) to mixtures and their components and then analyzed and compared ANPs of populations of these ORNs [6,7]. To examine odor-activated conductances in ORNs, we have been using whole-cell patch clamp recordings of ORNs in an in situ preparation, as developed by Ache and colleagues [8]. Binding of odorant molecules to their receptor sites has been studied with a radiochemical assay using a tissue fraction enriched in dendritic membrane from olfactory (aesthetasc) sensilla [9].

Results and Discussion

Mixture Interactions Resulting in Contrast Enhancement of Odorant Quality

Mixture interactions are common in the peripheral olfactory system of *P. argus*. For any given pair of compounds, there exists a diversity of types and degrees of mixture interactions across the individual members of a population of ORNs [6,10]. These observations suggested to us that such mixture interactions could create an ANP for a binary mixture that is different from the ANPs for the components of that mixture and different from the ANP predicted for that mixture from the responses to the components. This suggestion turns out to be the case for some mixtures [7]. It suggests that mixture interactions can improve the contrast between the neural responses to behaviorally important compounds and their mixtures. Given that it is so, then it should follow that the contrast enhancement in

[1] Department of Biology, Georgia State University, PO Box 4010, Atlanta, GA 30302-4010, USA

ANPs due to mixtures interactions should be expressed behaviorally. This statement is the case, at least for the three binary mixtures tested so far [3]. For example, the mixture interactions between 5'AMP and glutamate create an ANP for this binary mixture that does not resemble the observed ANP for either component or the predicted ANP for this binary mixture [6,7]. Behavioral expression of this neural mixture interaction for 5'AMP + glutamate is evident, as the perceived quality of this binary mixture is different from that of the individual components or the predicted quality of the mixture [3]. 5'AMP and taurine, which show no neural mixture interactions, also show no mixture interactions at the behavioral level [3,6,7]. Thus mixture interactions that affect the initial coding of the quality of a mixture can influence the subsequent perception and behavioral discrimination of that mixture such that there is enhancement of the contrast between that mixture and its components as well as other chemostimuli. We are presently extending this analysis to include additional binary mixtures.

Mechanisms of Mixture Suppression

Mixture suppression in the olfactory pathway of *P. argus* has been shown to be generated in the peripheral and the central nervous systems [10]. The physiologic basis for one type of peripherally generated mixture suppression has been demonstrated by Michel et al. [8] using whole-cell patch clamping of ORNs of *P. argus*. They showed that the depolarization evoked by a fish food (Tetramarin, Tetra Werke, Melle, Germany), which activated an inward current, can be inhibited by adding a single compound (L-proline, L-cysteine, or L-arginine) that activates an opposing outward (K$^+$) conductance. As a result, the depolarization produced by the mixture of Tetramarin plus the single compound is less than that produced by Tetramarin alone [8].

We have been examining mechanisms of mixture interactions in the peripheral system (i.e., in ORNs) of *P. argus* using this whole-cell patch clamp techniques [8] with the same set of compounds as used in our earlier behavioral and electrophysiologic studies: 5'AMP, NH$_4^+$, betaine, cysteine, glutamate, succinate, or taurine. To date, we have not seen mixture suppression with binary mixtures of these compounds as a result of the two components activating opposing ionic conductances. However, from both extracellular [6] and whole-cell patch clamp studies (T. Simon, unpublished data), we have observed mixture suppression in two other types of ORN: (1) where one compound of a binary mixture is excitatory and the other is neither excitatory nor inhibitory; and (2) where the two components of a binary mixture are both excitatory.

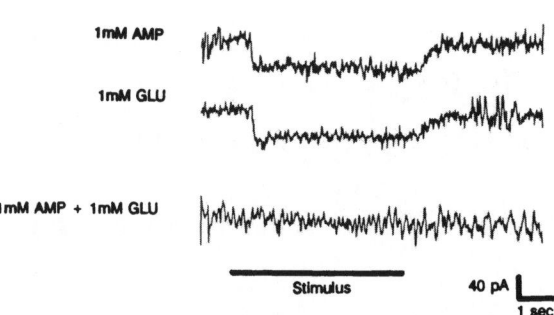

FIG. 1. Mixture suppression between two excitatory compounds for an olfactory receptor neuron excited by 5'AMP and L-glutamate. Current traces show inward currents in response to 1 mM 5'AMP, 1 mM L-glutamate, and 1 mM 5'AMP + 1 mM glutamate. This neuron was clamped to −40 mV. (Courtesy of T. Simon, unpublished data)

Examples of the latter case include mixtures of taurine + glutamate and mixtures of 5'AMP + glutamate (Fig. 1). One possible mechanism for such mixture suppression in ORNs is that some compounds in a mixture inhibit binding of other compounds to their receptor sites. For example, the mixture suppression in Fig. 1 could be due to 5'AMP and glutamate reciprocally inhibiting binding to their respective receptor sites. We have been examining such a mechanism by characterizing inhibition of binding at three olfactory receptor sites of *P. argus*: 5'AMP receptor sites, taurine receptor sites, and glutamate receptor sites (K. Olson and M. Burgess, unpublished data). Most odorant compounds tested tend to inhibit binding of 5'AMP, glutamate, and taurine to their respective receptor sites in a concentration-dependent fashion, often with 20% to 50% inhibition at higher concentrations. For example, 100 µM 5'AMP inhibits binding of 1 µM glutamate to its receptor sites by approximately 40%, and 100 µM glutamate inhibits binding of 1 µM 5'AMP to its receptor sites by about 20% (Fig. 2).

In all cases where we have identified that one odorant compound suppressed the electrophysiologic responses of a population of ORNs to taurine or 5'AMP [6], that odorant also inhibited binding of taurine or 5'AMP to its receptor sites (K. Olson, unpublished data). However, the ability of a compound to inhibit binding of taurine or 5'AMP is poorly correlated with the magnitude of mixture suppression as seen in electrophysiologic studies of ORNs. Thus in some but by no means all cases, mixture suppression of the responses of ORNs may be due to the fact that suppressing compounds inhibit binding of the excitant to its receptor sites. These binding events, along with intracellular events [8], may interact to generate mixture suppression in ORNs.

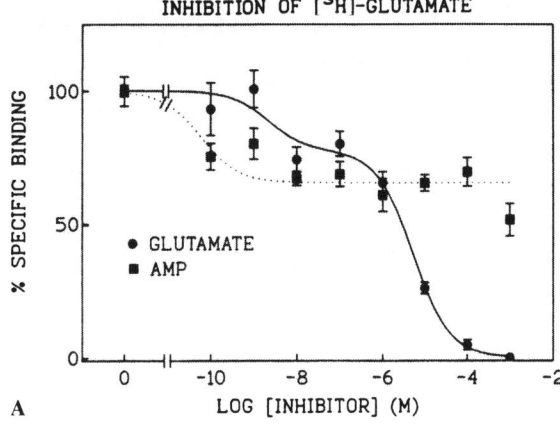

A

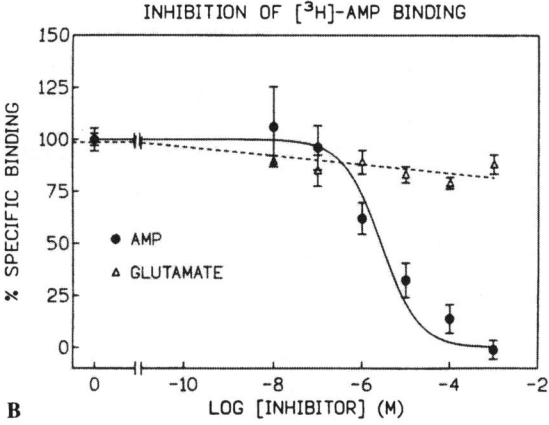

B

FIG. 2. Inhibition of binding of odorants to their receptor sites. **A** Inhibition of [³H]L-glutamate binding by 5'AMP. Olfactory tissue was incubated with 1 μM [³H]L-glutamate alone or with 1 μM [³H]L-glutamate plus varying concentrations of an inhibitor—either unlabeled L-glutamate (i.e., self-inhibition) or 5'AMP. The glutamate inhibition data were best fit by a two-site model, and the 5'AMP inhibition data were best fit by a one-site model (M. Burgess, unpublished data). **B** Inhibition of [³H]5'AMP binding by glutamate. Olfactory tissue was incubated with 1 μM [³H]5'AMP alone or with 1 μM [³H]5'AMP plus varying concentrations of an inhibitor—either unlabeled 5'AMP (i.e., self-inhibition) or L-glutamate. The 5'AMP inhibition data were best fit by a one-site model, and the glutamate inhibition data were best fit by linear regression (K. Olson, unpublished data). Values in both **A** and **B** are means ± SEM for three experiments run in triplicate

Conclusions

Mixture interactions in olfactory receptor neurons of spiny lobsters (Fig. 1) may result from various mechanisms, including mixture components either inhibiting the binding of excitants to their receptor sites (Fig. 2) or activating opposing ionic conductances [8]. Mixture interactions in olfactory receptor neurons can affect neural coding of mixture quality by the peripheral olfactory system by improving the contrast in the across-neuron patterns for mixtures and their components. Such neural mixture interactions appear to influence chemosensory perception and behavior, as similar contrast enhancement in olfactory discrimination is expressed at the behavioral level.

Acknowledgments. We thank the present and past members of our research group who have contributed to this and related work: David Blaustein, Peter Daniel, Jackie Fine, Nadia Girardot, Sandy Levett, Tripp Lynn, Liz Meyer, Michiko Nishikawa, Robert Simmons, Dae-Yong Sung, Marc Weissburg, and Debbie Wood. This research has been supported by NSF Grant IBN 9109783 and NIH Grants R01 DC00312 and K04 DC00002.

References

1. Laing DG, Cain WS, McBride RL, Ache BW (1989) Perception of Complex Smells and Tastes. Academic Press, Sydney
2. Fine-Levy JB, Daniel PC, Girardot MN, Derby CD (1989) Behavioral resolution of quality of odorant mixtures by spiny lobsters: differential aversive conditioning of olfactory responses. Chem Senses 14:503–524
3. Fine-Levy JB, Derby CD (1992) Behavioral discrimination of binary mixtures and their components: effects of mixture interactions on coding of stimulus intensity and quality. Chem Senses 17:307–323
4. Derby CD, Ache BW (1984) Quality coding of a complex odorant in an invertebrate. J Neurophysiol 51:906–924
5. Girardot MN, Derby CD (1990) Independent components of the neural population response for discrimination of quality and intensity of chemical stimuli. Brain Behav Evol 35:129–145
6. Derby CD, Girardot MN, Daniel PC (1991) Responses of olfactory receptor cells of spiny lobsters to binary mixtures. I. Intensity mixture interactions. J Neurophysiol 66:112–130
7. Derby CD, Girardot MN, Daniel PC (1991) Responses of olfactory receptor cells of spiny lobsters to binary mixtures. II. Pattern mixture interactions. J Neurophysiol 66:131–139
8. Michel WC, McClintock TS, Ache BW (1991) Inhibition of lobster olfactory receptor cells by an odor-activated potassium conductance. J Neurophysiol 65:446–453
9. Olson KS, Trapido-Rosenthal HG, Derby CD (1992) Biochemical characterization of independent olfactory receptor sites for 5'-AMP and taurine in the spiny lobster. Brain Res 53:262–270
10. Ache BW (1989) Central and peripheral bases for mixture suppression in olfaction: a crustacean model. In: Laing DG, Cain WS, McBride RL, Ache BW (eds) Perception of Complex Smells and Tastes. Academic Press, Sydney, pp 101–114

Turbulent Odor Dispersal, Receptor Cell Filters, and Chemotactic Behavior

Jelle Atema[1]

Key words. Odor dispersal—Chemotaxis—Temporal filter—Flicker fusion—Adaptation—Lobster

Introduction

Odor dispersal in the environment is driven by advection (air or water flow) and molecular diffusion. Diffusion is the only process that allows odor to penetrate the boundary layers that surround chemoreceptor organs and to reach the receptor cell surface. Understanding diffusion is therefore a critical component of understanding receptor cell function. For animals and olfactory organs larger than a few millimeters, however, diffusion distances become so large that they result in diffusion times exceeding several seconds, too slow to make the quick behavioral decisions necessary to compete successfully for food. This makes it imperative to understand advective flow as the basis of odor dispersal [1–4].

Advective flow is almost always turbulent under natural conditions. Over the course of 500 million years of evolution, animals and their receptors have faced turbulent conditions of odor dispersal. We assume that these natural stimulus conditions have placed constraints on receptor cell function, receptor organ morphology, and chemotactic behavior. It is likely that animals, particularly those that depend on chemical information for their survival and reproductive success, have found ways in which to extract information from odor plumes. This information includes, of course, recognition of the identity of the odor source, such as food odor mixture or pheromone. We propose that it also includes recognition of the spatiotemporal dynamics of odor plume dispersal [3,4].

To investigate the spatiotemporal dynamics of natural odor dispersal that are important physiologically and behaviorally, large aquatic arthropods offer significant advantages. Because of the greater density of water, the relevant scales of turbulence are about an order of magnitude smaller than in air.

This facilitates behavioral experiments and odor dispersal measurements. Also, crustacean chemoreceptor organs are cuticular pegs that form distinct boundaries with the surrounding fluid, uncomplicated by mucus transitions. These boundaries can be mimicked at the proper size scale with electrochemical microelectrodes [5]. In addition, electrochemical detection of oxidizable molecules in aquatic solution is a process of molecular counting that is fast enough to match and even exceed chemoreceptive integration. Finally, we chose the lobster *Homarus americanus* for its size and dependence on chemical information for survival.

This chapter reviews four experimental approaches that together argue in favor of a temporal analysis function of lobster olfaction. The experiments include high-resolution measurements of turbulent odor dispersal and lobster sampling behavior, electrophysiologic recording of in situ single cell responses to controlled and chaotic stimuli, and behavioral analysis of orientation and localization of odor sources.

Spatial Gradients in Odor Plumes

Because turbulent odor plumes simultaneously contain a broad suite of eddy sizes, the first problem is to decide which range of frequencies to measure and at which spatiotemporal scale to resolve the signal–noise complex. For spatial resolution, we scaled our odor tracer sensor (carbon-filled glass microelectrode) to the physical dimension of the lobster's 30 μm diameter olfactory receptor sensillum. For temporal resolution, we chose 5 ms (IVEC, Medical Systems, Gerhardt custom package), exceeding both the lobster's olfactory flicker-fusion frequency of 4 Hz (see below) and laboratory plume frequencies, which have little energy above 40 Hz. We studied odor dispersal patterns resulting from biologically scaled, constantly emitting jet sources in slow background flow. From high-resolution plume measurements using dopamine as a tracer, we described the spatial distribution of encounter probabilities of eddy features such as peak concentration, concentration gradients at their leading edge, and intermittency [6]. Some of these stimulus features showed spatial gradients that could be used to track and

[1] Boston University Marine Program, Marine Biological Laboratory, Woods Hole, MA 02543, USA

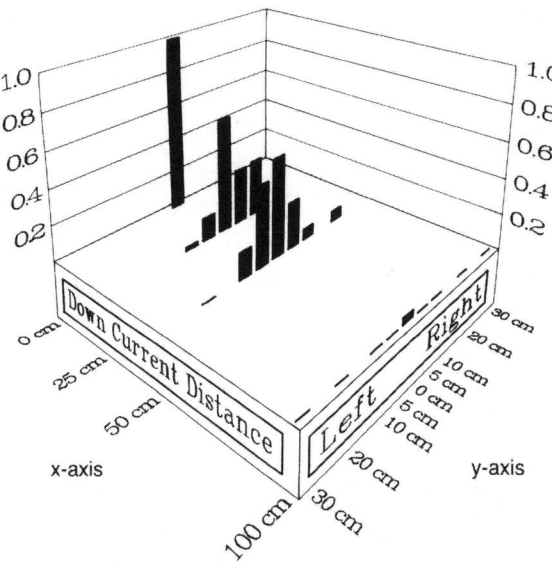

FIG. 1. Spatial gradient of odor patches in a plume. A constantly emitting point source is injecting a turbulent plume of food odor into a slow seawater carrier flow in a flume tank. *Bar height* shows probability of encountering patches with steep ($>10\,\mu M/s$) concentration slopes in different parts of the plume. At source location ($x = 0$, $y = 0$) $p = 1.0$ for reference. (After [6], with permission)

locate an odor source (Fig. 1). It demonstrates that, at the measuring scale of animal sensors, purely chemoreceptive information can be extracted from which the direction of a distant odor source can be estimated.

Orientation and Navigation in Odor Plumes

Subsequent behavioral experiments in the same laboratory flume [7] showed that lobsters indeed located the source of food odor plumes, whereas prior lesion experiments had shown that distance orientation is guided by olfaction [8] and the final approach of about 30 cm by taste [9]. Smell and taste are defined functionally [10]. The olfactory orientation was composed of an initial scanning phase (characterized by low walking speed and large heading angles) followed by an increasingly fast and accurate approach (Fig. 2). Both speed and accuracy reached a maximum that was maintained until walking legs (taste) took over the final search. Curiously, when not engaged in chemotactic search, lobsters can walk twice as fast. Unlike some insects [11,12] lobsters do not appear to use innate motor programs, such as counter turning. Instead, the individuality of their tracks (not shown here) and their relatively slow top speed during chemotactic approach suggest that they monitor and follow the variable spatial gradients characteristic of turbulent plumes. If walking animals such as lobsters use odor patch information for plume navigation, free-swimming animals without ground reference are even more likely to extract directional cues from turbulent odor dispersal patterns.

Odor Sampling Behavior

Feature extraction is the primary function of sensory systems. To this end, sense organs almost always use

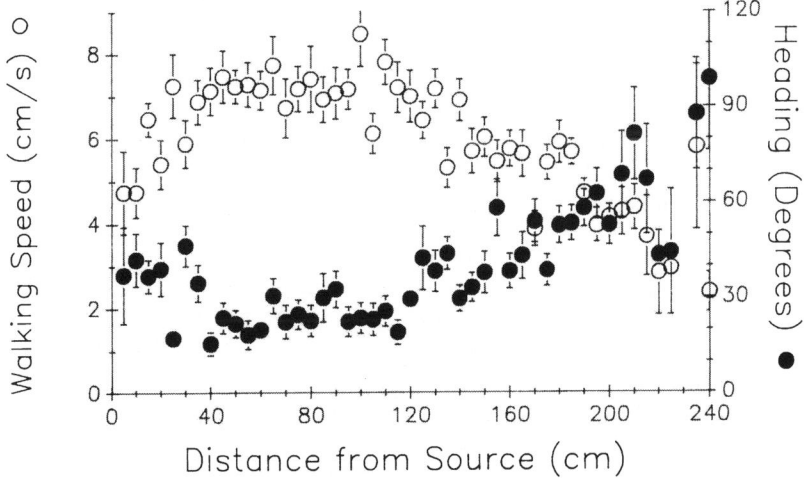

FIG. 2. Analysis of chemotactic search paths. Search path from right to left, source at $x = 0$. Initially, at $>200\,cm$ away from the odor source (food odor, flow conditions as in Fig. 1), lobsters walk slowly (3 cm/s) in many directions ("headings" of 90°) and with great individual variability (large standard error of the mean; N = 26 trials). Their performance in all respects improves gradually: Halfway toward the source walking speed is 7 cm/s, headings are 25°, and variability is relatively low. Close to the source (within 40 cm) search behavior deteriorates. It reflects a new behavior based on local "random" search with the walking legs. (From [7], with permission)

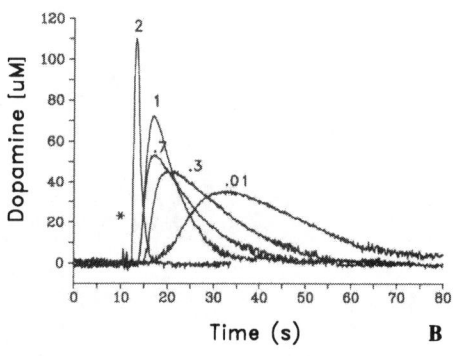

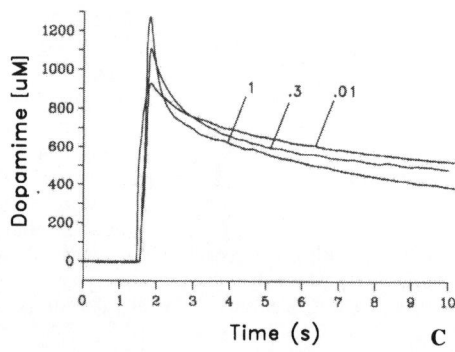

FIG. 3. Odor sampling behavior of lobster antennules: effect of "flicking." **A** Many aesthetasc (olfactory) sensilla and a few tall guard hairs on the lateral antennule of the lobster form dense clusters into which odor penetrates efficiently only through fast flow such as created by flicking, a rapid (>1 cm/100 ms) downstroke of the antennule. **B** Without flicking, a carrier flow brings an odor pulse quickly over the tips of the guard hairs (2 mm). The same pulse measured at the tips of the aesthetasc sensilla (1 mm; probe position in A) is delayed and reduced. Deeper into the hair cluster stimulus delay and reduction grow gradually and reach their maximum near the base of the sensilla (0.01 mm) where the odor peak arrives 30 s later at 30% of the original peak concentration. (From Moore et al. [5], with permission.) **C** During a simulated flick, odor has quick access to all levels of the sensilla (three levels shown) but remains trapped among the cluster until a new flick (not shown) washes it away as fast as it entered. Odor is measured as dopamine concentration with IVEC 30 μm carbon electrodes at 200 Hz sampling rate. (From Moore et al. [5], with permission)

three levels of signal filtering: a physical filter based on receptor organ morphology and associated behavior, a receptor cell filter based on biophysical and biochemical properties of transduction and adaptation, and a neural filter based on network connectivity in the central nervous system (CNS). We investigated the first two filters for lobster olfaction. The olfactory sensilla ("hairs") known to be critical for efficient orientation behavior [8] form a dense "toothbrush" at the distal half of the first antenna (Fig. 3). Video analysis and high-resolution electrochemical measurements showed that under low flow conditions (<5 cm/s) this brush forms a dense boundary layer that traps existing odor and shields the receptors from rapid odor access, thus making smelling virtually impossible [5]. Lobsters must flick their antennules (i.e., sniff) to smell. Flicking behavior drives water at high velocity (>12 cm/s) through the brush and causes the hairs to tremble in their sockets, which allows rapid odor exchange around the entire 1 mm long shafts of all the hairs (Fig. 3). Flick rates of up to 4 Hz occur in excited lobsters [13].

Temporal Resolution of Olfactory Receptor Cells

Lobster chemoreceptor cells show a great diversity of filter properties. Electrophysiological measurements show that each receptor cell is tuned not only

to one or a few preferred compounds [14, chapter "Comparison of tuning properties of five chemoreceptor organs of the American lobster: "Special filters", this volume] but also to a preferred frequency [15, chapter "Tuning properties of chemoreceptor cells of the American lobster: Temporal filters", this volume]. Temporal resolution in chemoreception comprises two components: rate of stimulus concentration increase (slope) of a single odor pulse and pulse repetition rate. Pulse slope corresponds to the arrival of an odor-flavored eddy; repetition rate represents arrival of different eddies. These dynamic properties are determined by two somewhat independent cellular processes: adaptation and disadaptation (or recovery). A cell's adaptation rate determines its preferred stimulus slope, and its recovery rate determines its flicker-fusion frequency. Flicker-fusion occurs at 4 Hz for the fastest receptor cells [chapter "Tuning properties of chemoreceptor cells of the American lobster: Temporal filters", this volume]. Some insect receptor cells can follow 4- to 10-Hz pulse rates [16–18]. Unfortunately, the non-linear dynamics of adaptation and disadaptation processes preclude a simple solution for determining meaningful transfer functions. However, results from experiments in which odor and single cell responses were measured simultaneously and with high spatio-temporal resolution, indicate that a steep pulse slope and large interpulse intervals are important excitatory stimulus features (Fig. 4).

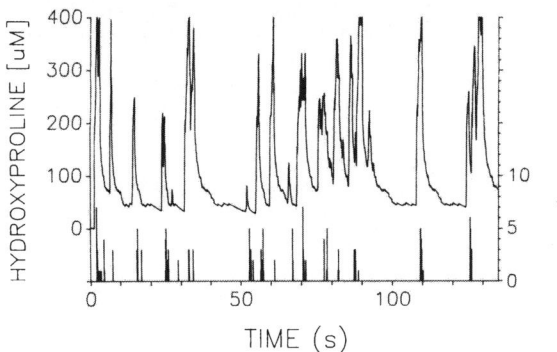

FIG. 4. Temporal receptor cell filter. Odor concentration patterns during 2 min of a simulated turbulent plume (such as in Fig. 3A) and response of a single aesthetase receptor cell. The stimulus pattern cannot be reconstructed from a single cell response pattern. Certain stimulus features, such as onset slope and time between odor pulses, are amplified preferentially, whereas pulses of long duration are deemphasized in the response

Discussion

The results demonstrate that (1) in turbulent odor plumes purely chemical spatial gradients can be calculated when measuring with sensors scaled to lobster olfactory organs; (2) rapid odor access to the lobster's olfactory organs is accomplished by flicking; (3) the maximum observed flick rate of 4 Hz corresponds to the neurophysiologically determined flicker-fusion frequency of olfactory receptor cells; (4) receptor cells are tuned not only to specific compounds but also to different temporal features of odor; and (5) lobsters locate odor sources with search paths, walking speed, and turning behavior suggestive of chemotaxis based on patchy odor distributions.

These results lead to the hypothesis that turbulently dispersed odor patches from a single source contain spatial gradients of signal features and that these features are recognized by olfactory temporal filters providing information for chemotactic navigation in odor plumes. Thus far we do not know the exact nature of these signal features or the sampling regimens and signal processing required to lead to efficient navigation. We are approaching these questions with behavioral, physiological, computational, and robotics methods.

Finally, lobsters are excellent models to learn about the natural world of odor dynamics, but they are probably not unique: They are opening our minds to possibilities not yet explored with other animals. The odor environment is richer and more complex than we know today, and lobsters are showing us a thus far unexplored dimension: navigation based on the characteristics of rapid odor signals.

Acknowledgments. I thank my superb and enthusiastic collaborators for their extensive contributions to this research, prominent among them Greg Gerhardt, George Gomez, Paul Moore, Nat Scholz, and Rainer Voigt. The long-term financial support of the National Science Foundation (grant BNS-8812952) is gratefully acknowledged.

References

1. Murlis J, Jones CD (1981) Fine-scale structure of odour plumes in relation to insect orientation to distant pheromone and other attractant sources. Physiol Entomol 6:71–86
2. Murlis J, Elkinton JS, Carde RT (1992) Odor plumes and how animals use them. Annu Rev Entomol 37:505–532
3. Atema J (1985) Chemoreception in the sea: adaptations of chemoreceptors and behavior to aquatic stimulus conditions. Soc Exp Biol Symp 39:387–423
4. Atema J (1988) Distribution of chemical stimuli. In: Atema J, Popper AN, Fay RR, Tavolga WN (eds) Sensory biology of aquatic animals. Springer, New York, pp 29–56
5. Moore PA, Atema J, Gerhardt GA (1992) Fluid dynamics and microscale odor movement in the chemosensory appendages of the lobster, *Homarus americanus*. Chem Senses 16:663–674
6. Moore PA, Atema J (1991) Spatial information contained in three-dimensional fine structure of an aquatic odor plume. Biol Bull 181:408–418
7. Moore PA, Scholz N, Atema J (1991) Chemical orientation of lobsters, *Homarus americanus*, in turbulent odor plumes. J Chem Ecol 17:1293–1307
8. Devine DV, Atema J (1982) Function of chemoreceptor organs in spatial orientation of the lobster, *Homarus americanus*: differences and overlap. Biol Bull 163:144–153
9. Derby CD, Atema J (1982) Function of chemo- and mechanoreceptors in lobster (*Homarus americanus*) feeding behaviour. J Exp Biol 98:317–327
10. Atema J (1977) Functional separation of smell and taste in fish and crustacea. In: Le Magnen J, MacLeod P (eds) Olfaction and taste (vol VI). Information Retrieval Ltd., London, pp 165–174
11. Akers RP (1989) Counterturns initiated by decrease in rate of increase of concentration: possible mechanism of chemotaxis by walking female *Ips paraconfusus* bark beetle. J Chem Ecol 15:183–208
12. David CT, Kennedy JS (1987) The steering of zigzagging flight by male gypsy moths. Naturwissenschaften 74:194–196
13. Berg K, Voigt R, Atema J (1992) Flicking in the lobster *Homarus americanus*: recordings from electrodes implanted in antennular segments. Biol Bull 183:377–378
14. Voigt R, Atema J (1992) Tuning of chemoreceptor cells of the second antenna of the American lobster (*Homarus americanus*) with a comparison of four of its other chemoreceptor organs. J Comp Physiol [A] 171:673–683
15. Gomez G, Voigt R, Atema J (1992) High resolution measurement and control of chemical stimuli in the lateral antennule of the lobster *Homarus americanus*. Biol Bull 183:353–354
16. Marion-Poll F, Tobin TR (1992) Temporal coding of pheromone pulses and trains in *Manduca sexta*. J Comp Physiol [A] 171:505–512

17. Rumbo E, Kaissling K-E (1989) Temporal resolution
    of odour pulses by three types of pheromone receptor
    cells in *Antheraea polyphemus*. J Comp Physiol [A]
    165:281–291

18. Christensen T, Hildebrand J (1988) Frequency coding
    by central olfactory neurons in the sphinx moth
    Manduca sexta. Chem Senses 13:123–130

Taste Discrimination in Salamanders

Takatoshi Nagai[1] and Hiro-aki Takeuchi[2]

Key words. Taste—Salamander—Aquatic animals —Behavior—Afferent nerve—Taste discrimination

Introduction

Salamanders are excellent experimental models for studying cellular mechanisms of taste reception because their taste receptor cells are large and form large taste buds of simple structure [1–3]. Intracellular recordings were made from taste receptor cells of the mudpuppy *Necturus maculosus* [4–7] and the tiger salamander *Ambystoma tigrinum* [8,9]. The chemosensitivity of the glossopharyngeal nerve (nIX) innervating a large area of the salamander lingual epithelium was studied in the mudpuppy [10,11] and the Mexican salamander *Ambystoma mexicanum* [12]. The ultrastructure of taste cells and associated synapses was studied in the mudpuppy [2,13,14] and the Mexican salamander [1,3]. Although there has been a wealth of physiological and anatomical studies, behavioral studies of the salamander's ability to discriminate taste are few [15,16]. Therefore we designed behavioral experiments to examine if salamanders can discriminate the chemicals that are commonly used in physiological experiments as taste stimuli.

Adequate stimulation of chemosensory receptor cells releases feeding behavior of animals. The receptor cells, such as those in the barbels of fish or in crustacean sensilla, are directly exposed to an aquatic environment. Thus relatively ready access to stimulants is achieved by simply placing them in an aquarium [17]. However, in aquatic salamanders the initial phase of feeding behavior (i.e., opening of the mouth) must be initiated by other sensory stimuli. The Mexican salamander, axolotl, snaps at a prey with highly successful feeding performances [18] when the prey induces water vibration and activates the lateral line nerve [19,20]. We utilized the stereotyped feeding behavior of axolotls to let them ingest the prey containing known concentrations of stimuli and then measured their ability to discriminate taste quantitatively.

Materials and Methods

Animals

Juvenile axolotls, *Ambystoma mexicanum*, were obtained from the Indiana University Axolotl Colony and raised in our laboratory until approximately 1 year of age (body length 13–18 cm; body weight 17–56 g). Animals were fed tubifex worms and dog food pellets.

Behavioral Experiments

Taste stimuli were presented to the animals as pellets (ca. $3 \times 3 \times 4$ mm) gelled by 3% agar, 1% locust bean gum, and 1% xanthan gum. Gel pellets contained either distilled water (DW), NaCl (Na: 0.1, 0.5, 1.0, 2.0 M), KCl (K: 0.1, 0.5, 1.0, 2.0 M), quinine hydrochloride (Q: 0.001, 0.01, 0.1, 0.3 M), citric acid (H: 0.001, 0.01, 0.1, 0.15 M) or sucrose (S: 0.1, 0.5, 1.0, 1.5 M). Binary mixtures of Q and Na (Q-Na) and Q and S (Q-S) were also tested to study possible interactions of two taste qualities.

Each animals was placed in a lucite chamber (size $6 \times 23 \times 8$ cm, water depth 5 cm) more than 1 hour before application of taste stimuli. The gel pellet was dropped into the water surface anterodorsal to the mouth of animals to elicit feeding behavior, i.e., snapping [20]. Snapping at the gel pellet that occurred highly successfully in each animal was succeeded by either a swallowing or rejecting response. The percentage of animals that showed the rejecting response was calculated as a measure of taste discrimination by the animal. However, satiety of the animal may also have induced rejection of the pellets. To check such a possibility, a dog food pellet that the animal had been usually fed was presented to the animal after the gel pellet was rejected. If it swallowed the dog food pellet, satiety in the animal was unlikely. Therefore the rejection that had occurred previously was included in the statistics. If it was rejected, the rejection of the gel pellet might have been induced by satiety and thus was not included.

Care was given to maintain an appetite of the animal throughout experiments: The behavioral test was performed on every third day; and after the test, the animal was fed only a small amount of tubifex worms. Daily test sessions included an ascending concentration series of taste stimuli (e.g., 0 M >

[1] Department of Physiology, Teikyo University School of Medicine, Kaga 2-11-1, Itabashi-ku, Tokyo, 173 Japan
[2] Department of Biology, Faculty of Science, Shizuoka University, Shizuoka, 422 Japan

0.1 M > 0.5 M > 1.0 M > 2.0 M NaCl); a single animal was exposed sequentially to the ascending concentrations unless it showed satiety.

The differences among the percentages of animals that showed rejection of each taste stimuli were analyzed by the chi-square test or Fisher's exact probability test (two-tailed).

Electrophysiological Experiments

Animals were anesthetized with 0.2% tricaine methanesulfonate (MS222; Sankyo, Tokyo, Japan), immobilized with 0.1 mg pancuronium bromide (Myoblock; Sankyo), laid in a recording chamber, and immersed in water except for the head region. The lingual branch of the ramus posttrematicus of the glossopharyngeal (nIX) nerve was exposed just anterior to the first gill by the same surgery as used for the mudpuppy [10,11]. The nerve was cut centrally, hooked onto bipolar platinum wire electrodes, and covered in liquid paraffin.

The overall activity of the nerve was differentially amplified and fed into a magnetic tape recorder (TEAC RD-101T) together with the output from the integrator (time constant 0.3 s) and a signal for stimulation time. Action potentials from the nerve and integrated nerve responses were recorded on a thermal array recorder (Nihon Kohden, Tokyo, Japan). All responses were quantified by measuring the area under the output from the integrator (10 s after the onset of stimulation) with a computer-assisted digitizer.

Taste stimuli (KCl, NaCl, quinine hydrochloride, citric acid, sucrose) and distilled water rinse were alternatively presented to the tongue through a peristalic pump with a flow rate of 10 to 12 ml/min. The stimulation paradigm consisted of applying the 0.5 M KCl standard followed by four to seven test solutions, each lasting 20 s and interspersed by 100-s water rinses; it was terminated by 0.5 M KCl. Distilled water was applied in the same manner as the taste stimuli to eliminate the tactile response from the integrated response. However, care was given to minimize tactile responses as much as possible.

In the mudpuppy the magnitude of the responses is known to be variable over time in a single animal and from animal to animal [10,11]. The gustatory response in the axolotl was also variable. When the response to the second 0.5 M KCl application was reduced by more than 30%, data were discarded for subsequent quantitative analysis. All responses were normalized by comparing each to the response to 0.5 M KCl recorded prior to the test solutions.

Results

Behavioral Response to Chemicals

After snapping at the gel pellet, either swallowing or rejection was exhibited as a consummatory feeding behavior. Rejection generally increased with an increase in the concentration of the stimulant, except sucrose (Fig. 1). It increased sigmoidally with citric acid and KCl, with a steeper slope for the latter. With quinine it reached more than 60% even at a threshold concentration for afferent nerve response (1 mM) and reached a plateau at higher concentrations. NaCl induced much lower rejections throughout the concentrations, with a slower increase. Sucrose did not induce any obvious rejection of the gel pellet. Rejection occurred to some extent when the pellet containing no sapid solution (plain pellet) was presented to axolotls (Fig. 1) at 0 M concentration: Q 4.0%; H 9.7%; K 6.6%; N 10%; S 7.5%.

Dose Response of the IX Nerve Response

To study the correlation between feeding behavior and neurophysiological data, the dose-response curve for the stimulants was recorded from the IX nerve of axolotls (Fig. 2). Sucrose was not an effective stimulus for any of the concentrations tested, but with the other four stimuli the neural response progressively increased with concentration.

Behavioral and Neural Responses to Mixtures

NaCl and sucrose induced few rejections (Fig. 1); in other words, these stimuli may have had potential to induce swallowing, which is the behavior opposite to rejection. However, such potential was not clearly shown in the paradigm to study the dose-rejection curve shown in Fig. 1 because rejections of plain pellets were also low and thus statistically not different from those of NaCl and sucrose. Therefore we mixed low-rejection-inducing stimuli and the stimulus that induced a high ratio of rejection in the hope of revealing their potential for swallowing.

When 100 mM NaCl was mixed with the pellet containing 1 mM quinine, rejection was significantly reduced, showing that 100 mM NaCl had potential for changing feeding behavior from rejection to swallowing. Sucrose, another stimulus inducing low rejection, did not have such potential. On the other hand, the modulation of opposite direction (i.e., from swallowing to rejection) was induced when 10 mM quinine was mixed with the pellets containing NaCl.

Discussion

The present study clearly showed that the axolotl had the ability to disciminate bitter taste from salty

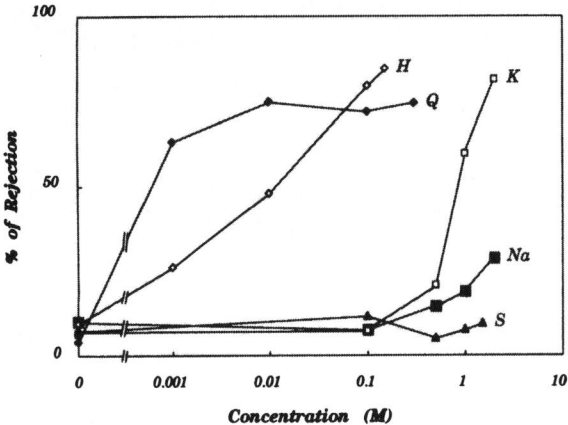

FIG. 1. Gel pellet rejection ratio as a function of the concentration of taste stimuli: KCl (K); NaCl (Na); sucrose (S); citric acid (H); quinine HCl (Q). Taste stimuli, except sucrose, induced higher rejections with an increase in concentration

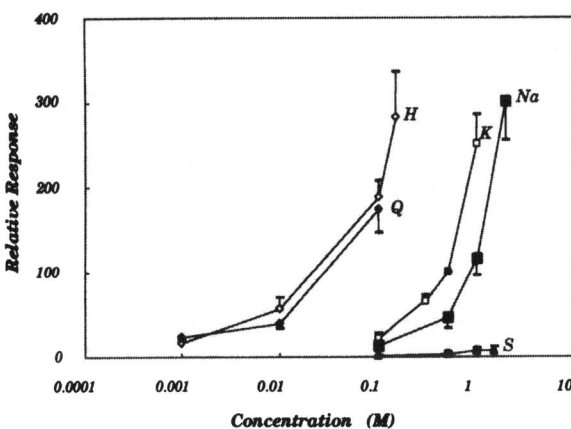

FIG. 2. Dose response curves for KCl (K), NaCl (Na), sucrose (S), citric acid (H) and quinine HCl (Q). Data points are means $\pm$ SEM of the responses normalized to the 0.5 M KCl response during each stimulus run

taste. Moreover, the axolotl feeding behaviors released by two types of salt proved to be different: KCl induced rejection, whereas NaCl at low concentrations induced swallowing. Because both KCl and NaCl are excitatory to the same extent in the overall afferent activities (Fig. 1), the difference in the behaviors may have been caused by depolarization in a different group of taste receptor cells [4] that subsequently excited a different set of peripheral and central neurons. KCl and NaCl must be perceived by the axolotl as different taste qualities.

Using the same paradigm as in our study, Bowerman and Kinnamon [15] also found that NaCl at a concentration of 1 M induced positive feeding behavior (i.e., swallowing) in the mudpuppy. However, in the axolotl, feeding behavior changed into higher rejection with an increase in concentration of NaCl

(Fig. 1). The different dose dependencies in two species of salamanders is interesting in light of possibly different mechanisms of NaCl taste transduction. In the tiger salamander (*Ambystoma tigrinum*), a species close to the axolotl, the amiloride-sensitive Na^+ channel plays a major role in depolarization in response to NaCl [9]. On the other hand, this channel is not involved in taste transduction in the mudpuppy [7,11].

Acknowledgments. This work was supported by the Sasakawa Scientific Research Grant from the Japan Science Society to H.T. We thank The Indiana University Axolotl Colony for the continuous supply of axolotls.

References

1. Fährmann W (1967) Licht- und elektronenmikroskopische Untersuchungen an der Geschmacksknospe des neotenen Axolotls (*Siredon mexicanum* Shaw). Z Mikrosk Anat Forsch 77:117–152
2. Farbman AI, Yonkers JD (1971) Fine structure of the taste bud in the mud puppy, *Necturus maculosus*. Am J Anat 131:353–370
3. Toyoshima K, Miyamoto K, Shimamura A (1987) Fine structure of taste buds in the tongue, palatal mucosa and gill arch of the axolotl, *Ambystoma mexicanum*. Okajimas Folia Anat Jpn 64:99–110
4. West CHK, Bernard RA (1978) Intracellular characteristics and responses of taste bud and lingual cells of the mudpuppy. J Gen Physiol 72:305–326
5. Roper S (1983) Regenerative impulses in taste cells. Science 220:1311–1312
6. Kinnamon SC, Roper SD (1987) Passive and active membrane properties of mudpuppy taste receptor cells. J Physiol (Lond) 383:601–614
7. Kinnamon SC, Roper SD (1988) Membrane properties of isolated mudpuppy taste cells. J Gen Physiol 91:351–371
8. Sugimoto K, Teeter JH (1990) Voltage-dependent ionic currents in taste receptor cells of the larval tiger salamander. J Gen Physiol 96:809–834
9. Sugimoto K, Teeter JH (1991) Stimulus-induced currents in isolated taste receptor cells of the larval tiger salamander. Chem Senses 16:109–122
10. Samanen DW, Bernard RA (1981) Response properties of the glossopharyngeal taste system of the mud puppy (*Necturus maculosus*). I. General organization and whole nerve responses. J Comp Physiol 143:143–150
11. McPheeters M, Roper SD (1985) Amiloride does not block taste transduction in the mudpuppy, *Necturus maculosus*. Chem Senses 10:341–352
12. Nagai T (1989) Chemoresponses of the glossopharyngeal nerve of the axolotl (*Ambystoma mexicanum*) [abstract]. Chem Senses 14:315
13. Cummings TA, Delay RJ, Roper SD (1987) Ultrastructure of apical specializations of taste cells in the mudpuppy, *Necturus maculosus*. J Comp Neurol 261:604–615
14. Delay RJ, Roper SD (1988) Ultrastructure of taste cells and synapses in the mudpuppy *Necturus maculosus*. J Comp Neurol 277:268–280

15. Bowerman AG, Kinnamon SC (1992) The effect of K$^+$ channel blockers on mudpuppy feeding behavior [abstract]. Proc Assoc Chemorecept Sci 14:44

16. Bowerman AG, Kinnamon SC (1993) The effect of minnow extract components on mudpuppy feeding behavior [abstract]. Proc Assoc Chemorecept Sci 15: 164

17. Atema J, Fay RR, Popper AN, Tavolga WN (eds) (1988) Sensory Biology of Aquatic Animals. Springer, Berlin Heidelberg New York

18. Lauder GV, Shaffer HB (1985) Functional morphology of the feeding mechanism in aquatic ambystomatid salamanders. J Morphol 185:297–326

19. Coombs S, Görner P, Münz H (eds) (1989) The Mechanosensory Lateral Line. Springer, Berlin Heidelberg New York

20. Takeuchi H, Nakamura S, Nagai T (1991) Participation of visual and lateral line systems in the feeding behavior of the axolotl [abstract]. Proc Int Congr Comp Physiol Biochem 3:101

Comparison of Tuning Properties of Five Chemoreceptor Organs of the American Lobster: Spectral Filters

Rainer Voigt and Jelle Atema[1]

Key words. Lobster—Chemoreception

The chemoreceptor organs of the American lobster, *Homarus americanus*, have been defined behaviorally as olfactory or gustatory organs. The lateral flagella of the first antenna (lateral antennules) act as olfactory organs and play a major role in long-distance orientation, while walking legs and third maxillipeds are gustatory organs important for feeding behavior. The behavioral function of medial antennules and second antenna as chemoreceptor organs remains unclear. We determined the spectral tuning properties of chemoreceptor cells of these five organs with amino acids and other compounds common in lobster prey. The distribution of differently tuned cells in the five organs determines the tuning of each organ.

Standard extracellular recording techniques were used to record from single receptor cells, which were identified with a mixture containing: alanine, arginine, aspartic acid, glutamine, glutamic acid, glycine, hydroxyproline, leucine, lysine, proline, serine, taurine, NH_4Cl, and betaine. Cells were tested with single compounds between 10^{-4} M and 10^{-7} M.

At lower concentrations, the chemoreceptor cells of all five organs were relatively narrowly tuned. The antenna was the most narrowly tuned organ [1]. It was dominated by a cell population which responded best to hydroxyproline (85% of all cells). In comparison, the lateral antennule contained a prominent Hyp-best cell population accounting for 46% of all cells, followed by Tau (13%) and Glu (10%) cell populations [2,3]. The medial antennule revealed Hyp (26%), Arg (21%), and Tau (24%) as prevalent best cell populations [4]. In the chelated (i.e., first two pairs of) walking legs Glu (38%), Hyp (16%), NH_4 (15%), and Bet (11%) formed the most numerous best cell populations [5]. Maxillipeds were similar to legs, with Glu (28%), Bet (15%), Hyp (13%), and Tau (13%) as major best cell populations [6]. These comparative results show similarities between the organ-level tuning of (cephalic) antenna and the lateral and medial antennules in contrast with the (thoracic) legs and maxillipeds.

Our cell tuning studies argue for independent receptors for all amino acids tested. We conclude that diversity of receptor cell tuning is created by cell-specific blends of receptors, just as differences in organ tuning result from different blends of receptor cells. Each organ represents a differently tuned spectral filter which allows the animal to extract different information from its chemical environment. Resemblance in spectral tuning could indicate functional overlap of chemoreceptor organs, and the tuning differences between cephalic and thoracic organs corresponds to the behavioral separation of olfaction and taste in the lobster. Their tuning may reflect behavioral function: Some compounds are most likely preferred long (olfactory)-distance signals (because of their longevity in seawater and their signal to background ratio), while other compounds could allow a refined evaluation of mixture quality near the source of release and/or upon contact.

Acknowledgments. This study was supported by NSF grant (BNS-8812952) to JA.

References

1. Voigt R, Atema J (1992) Tuning of chemoreceptor cells of the second antenna of the American lobster (*Homarus americanus*) with a comparison of four of its other chemoreceptor organs. J Comp Physiol A 171: 673–683
2. Johnson BR, Atema J (1983) Narrow-spectrum chemoreceptor cells in the antennules of the American lobster, *Homarus americanus*. Neurosci Lett 41:145–150
3. Weinstein A, Voigt R, Atema J (1990) Spectral tuning of lobster olfactory cells and their response to defined mixtures and natural food extracts. Chem Senses 15: 651–652
4. Tierney AJ, Voigt R, Atema J (1988) Response properties of chemoreceptors from the medial antennule of the lobster *Homarus americanus*. Biol Bull 174:355–363
5. Johnson BR, Voigt R, Borroni PF, Atema J (1984) Response properties of lobster chemoreceptors: Tuning of primary taste neurons in walking legs. J Comp Physiol 155:593–604
6. Corotto F, Voigt R, Atema J (1992) Spectral tuning of chemoreceptor cells on the third maxilliped of the lobster, *Homarus americanus*. Biol Bull 183:456–462

[1] Boston University Marine Program, Marine Biological Laboratory, Woods Hole, MA 02543, USA

Tuning Properties of Chemoreceptor Cells of the American Lobster: Temporal Filters

George Gomez, Rainer Voigt, and Jelle Atema[1]

Key words. Lobster—Chemoreceptor

Chemical stimuli in the environment are distributed in a turbulent fashion that results in a chaotic series of patches of varying concentrations. The distribution of these patches changes with increasing distance from the odor source [1]. A patch of odor moving over a stationary point is measured as a pulse of odor occurring in time [2,3]. Animals such as lobsters that orient to distant odor sources extract spatial information from the temporal pulse patterns. We hypothesize that their chemoreceptors should be designed to encode pulsatile information as it occurs at varying stimulus concentrations and frequencies in an odor plume. This study investigates the temporal filter properties of chemoreceptor cells of the lateral antennule (which mediates distance orientation) of the lobster *Homarus americanus*.

To accurately study temporal properties of chemoreceptors in situ, it is important to quantify the dynamics of the stimulus in the intact chemoreceptor organ, preferrably with millisecond and micrometer resolution. Aesthetasc sensilla containing hydroxyproline-sensitive receptors were localized using a concentric pipette, focal stimulation system. We measured the exact time course of each stimulus pulse on-line with a microelectrode and electrochemical detection of a tracer (dopamine) with a spatial and temporal resolution of $30\,\mu M$ and $5\,ms$, respectively (for details, see [4]). Responses were recorded extracellularly representing the sum of all perireceptor (e.g., diffusion) and receptor (e.g., transduction, spike generation) events. Receptor cells were given ten 100-ms pulses of 10^{-3}, 10^{-4}, and $10^{-5}\,M$ hydroxyproline (Hyp) at 0.5, 1, 2, 3, and $4\,Hz$.

Cells responded to single "square pulses" of Hyp with a phasic spike burst and rapidly adapted to the stimulus pulse. At a lower stimulus concentration ($10^{-5}\,M$), mean responses of 16 cells followed (i.e., gave discrete increases in spike activity to) individual pulses at 3 Hz but their responses fused at 4 Hz. At $10^{-4}\,M$ stimulus concentration, cells resolved pulse frequencies at 2 Hz. At $10^{-3}\,M$, cells only resolved

pulses at 1 Hz and their responses fused at 2 Hz. All stimulus pulses remained clearly separated at all frequencies and concentrations (determined with on-line stimulus measurement). Thus frequency resolution (as expressed in flicker fusion frequencies) decreased with increasing stimulus concentration.

The cumulative effect of adaptation was seen in the decline in the number of spikes given to succeeding pulses in a series. With increasing concentration, cells responded to the first pulse in each series with an increased number of spikes. However, succeeding pulses in a series resulted in a decline in the number of spikes given by the cells. At low (0.5 Hz) stimulation frequencies, responses to the second to the tenth pulse in a series were typically 50% in magnitude compared to the first pulse. At 2 Hz, the number of spikes given in response to succeeding (i.e., 3rd to 10th) pulses in a series was the same across all concentrations tested. Thus higher stimulation frequencies resulted in a loss of intensity discrimination by the receptor cells.

On-line stimulus measurement allowed us to closely investigate temporal filter properties of individual cells and attribute observed variability to cell properties, not to stimulus variability. We used the Synchronization Coefficient [5] to quantify capabilities of cells to encode pulse frequencies. Synchronization values of individual cells were spread over a continuum rather than in discrete classes, indicating an inherent diversity in frequency filter capabilities of individual cells. This effect was especially pronounced at 2 Hz stimulation. We hypothesize that dynamic response diversity is a feature of lobster olfaction that allows the animal to extract specific features present in an odor plume by "tuning" to different frequencies. This is reminiscent of auditory and mechanoreceptor systems that perform frequency analyses of natural stimuli.

Acknowledgments. This study was supported by NSF grant (BNS-882952) to JA.

References

1. Moore PA, Atema J (1991) Spatial information in a three-dimensional fine structure of an aquatic odor plume. Biol Bull 181:408–418

[1] Boston University Marine Program, Marine Biological Laboratory, Woods Hole, MA 02543, USA

2. Atema J (1985) Chemoreception in the sea: Adaptation of chemoreceptors and behavior to aquatic stimulus conditions. Soc Exp Biol Symp 39:387–423
3. Atema J (1987) Chemoreceptor adaptation: A patch in space is a pulse in time. Chem Senses 12(1):189–190
4. Gomez G, Voigt R, Atema J (1992) High resolution measurement and control of chemical stimuli in the lateral antennule of the lobster *Homarus americanus*. Biol Bull 183(2):353–354
5. Goldberg JM, Brown PB (1969) Response of binaural neurons of dog superior olivary complex to dichotic tonal stimuli: Some physiological mechanisms of sound localization. J Neurophysiol 32:613–636

Integration of Olfactory Input by Lobster Olfactory Receptor Neurons

WILLIAM C. MICHEL[1] and BARRY W. ACHE[2]

Key words. Olfaction—Inhibition

Approximately 50% of lobster olfactory receptor neurons (ORNs) support both odor-evoked excitatory and inhibitory conductances. In the present study we show that inhibition rapidly and reversibly diminishes both the intensity and the time course of the output of lobster ORNs within the natural sampling interval. We conclude that the dynamic interplay of opposing, odor-evoked inputs must be considered as a potentially important feature of coding in these neurons.

The olfactory organ of the Caribbean spiny lobster, *Panulirus argus*, is a dense tuft of olfactory sensilla (aesthetascs) located on the lateral filament of the antennule. Methods for preparation of tissue and recording have been considered in detail elsewhere [1,2]. The olfactory competence of an ORN was assessed by challenging the cell with a complex odor mixture, TET (an aqueous extract of a commercially available fish food), and recording a response concomitant with the time course of stimulation. Cells expressing an inhibitory conductance were identified by challenging them with 1mM proline and recording either a reduction in spontaneous discharge or an outward current in response to the odor.

In cells supporting an inhibitory conductance, 1 mM proline (added to TET) significantly delayed the time to and decreased the maximum instantaneous frequency of the response to TET tested alone. The average time to reach the maximum frequency of discharge increased from 371 ± 55 ms to 432 ± 58 ms ($n = 10$). The average maximum frequency decreased from 44 ± 5.1 Hz to 32.2 ± 4.2 Hz ($n = 10$). One mM proline did not significantly change either of these parameters in three cells that failed to express an inhibitory conductance.

The effect of proline supplementation was concentration-dependent. Increasing the added concentration of proline from 10^{-5} to 10^{-3} M decreased the peak instantaneous frequency of the response from 69 to 29 Hz. The nature of the decrement in the magnitude of the response elicited by increasing the amount of inhibitory input differed from the nature of the decrement in the magnitude of the response obtained by stimulating with a lower concentration of TET.

As little as 1 μM proline was sufficient to attenuate the magnitude of the intracellular excitatory response to TET. One mM proline significantly reduced the average magnitude of the inward current evoked by TET from −20.3 ± 2.4 pA to −12.6 ± 2.4 pA ($n = 12$). One mM proline significantly reduced the average rate of rise of the depolarizing current evoked by TET from −129 ± 26 pA/s to −76 ± 24 pA/s ($n = 12$). Correspondingly, 1 mM proline significantly decreased the rate of depolarization evoked by TET from 74.2 ± 18.9 mV/s to 22.6 ± 6.6 mV/s ($n = 6$). As little as 1 μM proline decreased the rate of rise of the receptor potential without any obvious decrease in the peak magnitude of the response, while 1 mM proline decreased both the rate and the magnitude of depolarization.

The average latencies of the outward currents (460 ± 60 ms, $n = 24$) evoked by proline were longer than the corresponding values for the inward currents (370 ± 50 ms, $n = 24$) evoked by TET. The slower onset of the inhibitory responses, however, may reflect their smaller magnitude, which would charge the cell membrane less rapidly than would the larger, excitatory responses. Overall, the average magnitude of the intracellular excitatory responses evoked by TET, (−20.6 ± 3.3 pA, $n = 24$) were larger than the average magnitude of the intracellular inhibitory responses evoked by proline (5.5 ± 0.6 pA, $n = 24$).

These results allow us to ascribe functional significance to having more than one transduction pathway in olfaction. Odor discrimination is generally thought to be based on neural pattern discrimination. Given that natural odors are complex mixtures with a high probability of co-activating inhibitory and excitatory inputs, inhibition would be realized through its effect on the (net) excitatory response. By including inhibition, the range for coding information across the population of receptor cells would be increased relative to that of a purely excitatory system. The result of having opposing transduction pathways, therefore, would be to en-

[1] University of Utah School of Medicine, Department of Physiology, 410 Chipeta Way, Salt Lake City, UT 84108, USA
[2] University of Florida Whitney Lab, 9505 Ocean Shore Blvd., St. Augustine, FL 32086, USA

hance the resolution of the across fiber pattern which is thought to be the basis of odor discrimination. This function could be expected to be fundamental to how olfactory information is acquired and processed.

References

1. Michel WC, McClintock TS, Ache BW (1991) Inhibition of lobster olfactory receptor cells by an odor-activated potassium conductance. J Neurophysiol 65:446–453
2. Michel WC, Ache BW (1992) Cyclic nucleotides mediate an odor-evoked potassium conductance in lobster olfactory receptor cells. J Neurosci 12:3979–3984

Study of Feeding Sensitivity in the Prawn, *Penaeus chinensis* (O'sbeck)

CHEN NANSHENG and SUN HAIBAO[1]

Key words. *Penaeus chinensis*—Feeding

Penaeus chinensis (O'sbeck) is widely cultivated in the northern coastal provinces of China. Currently, there are increasing problems of disease and malnutrition arising with the expansion of the cultivation area. It has been demonstrated that the nutritional status is responsible for most of those problems. As nutrition is obtained through feeding, and the optimal foraging theory (OFT) has been well received, we carried out a series of experiments on the feeding strategies and feeding mechanisms in the prawn during its developmental stages, so as to determine the mechanism underlying the selection of appropriate diets.

Chemoreception plays an important role in the feeding behavior of aquatic animals [1–4]. Here, we deal with chemosensitivity in larval, postlarval, and adult prawns.

Larvae and postlarvae were bred in the laboratory, and adult prawns were bought from the Huangdao Shrimp Farming Plant (about 10 km from Qingdao). Four experiments were carried out to elucidate the feeding sensitivity of larvae (Stages Zoea$_{1-3}$: Z_{1-3}; Mysis$_{1-3}$: M_{1-3}) and early postlarvae (Stage PL_{1-10}): (a) chemotaxis; (b) antennule flicking and body turning rates; (c) sand particle ingestion; (d) live and dead rotifer and artemia ingestion rates.

Results from the first two experiments showed that (a) Z_{1-3}, M_{1-3}, and PL_{1-10} did not have chemosensitivity to such stimuli as extracts of clam worm and artemia exudate, and their antennule flicking and body turning rates did not seem to be affected by these stimuli; (b) Z_{1-3}, M_{1-3}, and PL_{1-10} all ingested sand particles from water with a sand suspension; (c) the live rotifer ingestion rate was always greater than that for the dead ones, while the dead artemia ingestion rate of the Z_{1-3}, and M_{1-3} stages was greater than that for live ones, and the dead artemia ingestion rate of PL_{1-10} was less than that for live ones.

Two experiments were carried out to test the chemosensitivity of adult prawns: (a) searching behavior and (b) ingestion behavior. The first experiment was carried out in large tanks (about 500 l). The number of bites on a stimulus delivery pipe was used to evaluate the relative attractiveness of each stimulus to the prawn. The results showed that the extracts of clam worm and small fish had strong attractiveness, the extracts of mussel had relatively weak attractiveness, and the single amino acids had almost no attractiveness.

The second experiment was conducted in small tanks that held one prawn. The responses were divided into four categories: (1) no response; (2) walking, showing searching behavior; (3) walking, showing strong searching and grasping behavior; and (4) walking, showing strong grasping behavior and moving to the stimulus delivery pipe. About 20 amino acids at five different concentrations were used. A feeding sensitivity spectrum was obtained at the concentration of 10^{-4} M/l. Asp, Glu, Arg, Lys, Ser, were shown to have rather strong attractiveness.

From the above results, we can conclude that Z_{1-3}, M_{1-3}, and PL_{1-10} stage prawns do not utilize chemosensitivity, but rather, mechanosensitivity, to detect and ingest food; while adult prawns posesses chemosensitivity, which is utilized to detect and ingest food.

References

1. Carr WES (1988) The molecular nature of chemical stimuli in the aquatic environment. In: Atema J, Fay RR, Popper AN, Tavolga WN (eds) Sensory biology of aquatic animals. Springer, New York Berlin Heidelberg London Paris Tokyo, pp 3–27
2. Nakamura K (1987) Chemoreceptive property in feeding of the prawn (*Penaeus japonicus*). Mem Fac Fish Kagoshima Univ 36:201–205
3. Carr WES, Derby CD (1986) Chemically stimulated feeding behavior in marine animals. J Chem Ecol 12:989–1011
4. Hindley J (1975) The detection, location and recognition of food by juvenile banana prawn *Penaeus merguiensis de Man*. Mar Behav Physiol 3:193–210

[1] Department of Experimental Zoology, Institute of Oceanology, Academia Sinica, 7 Nanhai Road, Qingdao, 266071 People's Republic of China

Growth Enhancement in a Freshwater Prawn Through the Use of a Chemoattractant

Sheenan Harpaz[1]

Keywords. Chemoreception—Growth enhancement

Chemoreception is an important mechanism governing the process of food searching in many aquatic animals. Numerous investigations have demonstrated that chemostimulants, infused into the controlled environment of several decapod crustacean species, trigger almost identical feeding behaviors to those which are displayed by these organisms when a food item is presented [1,2]. Crustaceans are attracted to artificial food pellets by following an attractant "plume" originating from the food source [1]. Behavioral analyses of feeding responses were carried out on *Macrobrachium rosenbergii* prawns, showing that food searching, as well as substrate probing, were greatly enhanced in the presence of aqueous solutions of identified chemoattractants [3]. In a laboratory experiment using *M. rosenbergii* prawns, it was found that after 2 h, virtually no attractants for this prawn could be detected when aqueous extracts of the pellets were tested in a serial manner. In the present study, the effect of chemoattractant introduction, 2 h after food administration, on the growth of juvenile *M. rosenbergii* freshwater prawns was tested under laboratory conditions.

A total of 500 juvenile *M. rosenbergii* prawns, weighing around 0.1 g each, were individually weighed and placed in ten 60-l aquaria. The water was well aerated and kept at $28° \pm 2°C$. At the end of the 6-week growth trial, each of the experimental prawns was individually weighed. Prawns were fed, 6 days a week, with an artificial pelleted feed containing 25% protein. The feeding rate was set at 10% of the average body weight per day and the feed was administered twice a day. In the experimental groups receiving attractant augmentation, the treatment was as follows: 2 h after the feed was given, a pipette containing 6 ml of betaine-HCl solution, at a concentration of 10^{-3} M, was emptied by moving it to and fro above the aquarium water in order to achieve (as much as possible) even distribution of the attractant in the water volume. The average growth attained in the aquaria which had betaine solution augmentation was 17% better than that attained by the control group in the aquaria to which no attractant was added. It is interesting to note that in the aquaria receiving betaine augmentation, although the bulk of the population exhibited better growth rate, a certain proportion lagged behind. It has been shown that an aqueous solution of betaine infused into the water causes *M. rosenbergii* prawns to vigorously search for food [2,3]. The addition of an attractant to the water, at the stage at which the artificial pellet is no longer stimulating the prawns to search for food, leads to an additional burst of food-searching activity. The results clearly show that this enhances growth. The use of one attractant over long periods is, however, not recommended, since it might lead to habituation.

References

1. Atema J (1988) Distribution of chemical stimuli. In: Atema J, Fay RR, Popper AN, Tavolga WN (eds) Sensory biology of aquatic animals. Springer, Berlin, Heidelberg, pp 29–56
2. Harpaz S, Steiner JE (1990) Analysis of betaine-induced feeding behavior in the prawn: *Macrobrachium rosenbergii* (de Man, 1879) (Decapoda, Caridea). Crustaceana 58:175–185
3. Harpaz S, Kahan D, Galun R, Moore I (1987) Responses of the freshwater prawn *Macrobrachium rosenbergii* to chemical attractants. J Chem Ecol 13: 1957–1966

[1] Agricultural Research Organization, Fish and Aquaculture Research Unit, P.O. Box 6, Bet-Dagan 50250, Israel

Sensitivity of *Xenopus* Chemoreceptors to Astringent Compounds

Satoru Yamashita[1], Yoshiaki Moriyama[1], Masanobu Ohno[2], and Yukihiko Hara[2]

Key words. Astringency—Taste

Since Kawamura et al. [1] described the electrophysiological responses in a pioneering study of rat chorda tympani and glossopharyngeal nerves to tannic acid, an astringent, the neural mechanism of astringent taste has not been sufficiently studied [2]. In the present study, we examined the sensitivity of taste receptors to astringent compounds in the aquatic toad, *Xenopus laevis*, by recording electrical responses from the glossopharyngeal nerve; here we present the results and discuss the peripheral neural mechanisms of astringent taste.

Xenopus weighing 30–50 g were anesthetized with intraperitoneal injections of urethane, at doses of 0.25 g/100 g body weight. The neural activity detected by Ag-AgCl electrodes was integrated. Test solutions, all prepared with reagent grade chemicals and deionized water, were delivered to the oral cavity for 5 or 10 s, at a flow rate of 0.3 ml/s. The response to tannic acid was produced at approximately 10^{-5} M and increased with increasing concentration, up to 10^{-2} M, at which no saturation of the response was detected. To eliminate the contribution of the pH component to the response, we adjusted the pH of the tannic acid solution to 5.0–5.5, i.e., almost equal to the pH (5.6) of the rinsing solution, and compared the responses with those to non-pH-adjusted tannic acid solutions. The responses to pH-adjusted tannic acid solutions were rather small at higher concentrations. We also examined the dose-response function of gallic acid. The threshold concentration was similar to that for tannic acid, but, at higher concentrations, the magnitude of the response was about half that of the response to tannic acid. In contrast to the response to tannic acid, the responses to pH-modified gallic acid (pH 5.5) were larger than those to the pH-

unmodified gallic acid at higher concentrations. The enhancement of response at higher concentrations appeared to be due to the increasing ionic dissociation of gallic acid (pK_a, 3.0). Of five catechin compounds tested, epicatechin gallate and epigallocatechin gallate were highly stimulatory. The threshold lay between 10^{-6} and 10^{-5} M, and the magnitude of response reached 50% of the standard response (response to 10^{-3} M L-proline) at 10^{-2} M.

Two polyvalent cationic salts known to be astringent, $AlNH_4(SO_4)_2$ and $AlK(SO_4)_2$, were also used as test stimuli. Both had similar thresholds, but $AlNH_4(SO_4)_2$ produced responses of a greater magnitude at higher concentrations. The characteristic properties of the responses to the astringent compounds tested were: (1) the response consisted of the phasic component alone, (2) the response did not appear to be saturated at 10^{-2} M, at which concentration an abrupt decrease in response is often detected with repetitive application, and (3) after the application of the astringent compounds, responses to all other taste stimuli tested were also extremely depressed. Depression of the response to L-proline caused by stimulation with tannic acid was recovered by treating the oral epithelium containing the taste buds with HCl, L-methionine, and acetone. These substances, however, were not effective in recovering the response to tannic acid. The decrease in response to tannic acid caused by repetitive stimulation was, however, recovered by treatment with dimethyl sulfoxide (DMSO). DMSO may effectively remove molecules of astringent compounds that probably bind to the membrane-bound proteins associated with taste transduction.

References

1. Kawamura Y, Funakoshi M, Kasahara Y, Yamamoto T (1969) A neurophysiological study of astringent taste. Jpn J Physiol 19:851–865
2. Schiffman SS, Suggs MS, Sostman AL, Simon SA (1991) Chorda tympani and lingual nerve responses to astringent compounds in rodents. Physiol Behav 51:55–63

[1] Department of Biology, College of Liberal Arts and Sciences, Kagoshima University, Kagoshima, 890 Japan
[2] Food Research Laboratories, Mitsui Norin Co. Ltd., Fujieda, Shizuoka, 426 Japan

16. Chemoreception in Insects

Functional Organization of Insect Taste Hairs

KAI HANSEN[1]

Key words. Insects—Contact chemoreception—Taste receptors—Functional model—Signal transduction

Introduction

The research on insect contact chemoreception has many facets. It is based on a wealth of details known from ecoethologic observations regarding the biological significance and the diversity of the perceived stimuli. It has been possible, using electrophysiological methods, to characterize the stimulus specificities of single receptor cells and to identify binding sites of these cells [1–3]. Another area of interest is the functional organization of taste hairs. Our present knowledge here is based mainly on the work of Morita and coworkers [3,4]. They succeeded in measuring dose-response curves and developed kinetic interpretations. General functional aspects of insect sensilla, regarding especially their epithelial character and the importance of the transepithelial voltage, have been provided by Thurm and Küppers [5].

In this chapter we have tried to obtain further insights into the functional organization of taste hairs using a modeling technique and the available fine structural and electrophysiologic data, as reviewed in the next section. Special emphasis is placed on the signal transfer along the thin outer dendrites of the taste hair in flies and the influence of the transepithelial potential on the voltage distribution along the dendrite.

Structural Prerequisites and Localization of Functional Processes

All of the following transmission electron microscopy (TEM) [6] and electrophysiologic [7] data concern the "largest" taste hairs, with a length of 350 to 440 μm on the labellum of the fly *Protophormia terraenovae* Rob.-Desv [1]. These hairs contain four axially oriented chemoreceptor cells, one mechanoreceptor cell, and four sheath cells (Fig. 1A). The chemosensory outer dendrites are of ciliary origin and contain microtubules as the only cytoplasmic component. The dendrites extend through the canal (CI) of the chitinous hairshaft up to its terminal pore, through which stimulating molecules reach the membranes of the dendritic tips from outside; it is here that the transduction process takes place. The dendrites are conical in shape and differ in their dimensions. Thin dendrites have an apical diameter of 0.13 μm and a basal diameter of 0.6 μm; the corresponding values for thick dendrites are 0.22 μm and 1.0 μm, respectively. The extracellular space around the dendrites is rather small (Fig. 1, A2). In addition to canal I, the hairshaft bears canal II, which ends distally below the tip; in this region the inner hair wall is permeable (Fig. 1, A1, pC) [4]. Below the cuticle the outer dendrite is connected via a ciliary neck to the inner dendrite (1 μm diameter, 20 μm long). The latter leads to the ellipsoidal receptor cell body (10 μm diameter, 12 μm in length), from which the axon rises with a diameter of 0.3–0.7 μm. Only the membranes of the receptor cell parts lying proximal to the ciliary neck are capable for spike generation in adult flies. The sheath cells envelop the receptor cells concentrically in a staggered manner. The two outermost ones (the accessory, not shown in Fig. 1A, and the tormogen cell) are distally tightly attached to the cuticle. The innermost thecogen cell surrounds the dendritic inner segments and receptor cell bodies. The "apical receptor lymph space" is subdivided by the dendritic sheath into (1) a large outer part (oLS) confined by the apical membranes of the tormogen and trichogen cells and in open connection with canal II; and (2) an inner part (iLS) surrounding the dendrites. The dendritic sheath is assumed to be permeable for small ions. The lateral cell membranes of all sheath and receptor cells are connected by septate junctions, which constitute the diffusion barrier that separates the apical lymph space from the hemolymph space, surrounding the outside of the thecogen cell. Thus the sensillum represents a "miniature epithelium" [5], consisting altogether of nine cells. The apical membranes of the sheath cells exhibit a high potential of about 100 mV built up by an electrogenic proton pump (Fig. 1, A1, BSHa) coupled with an K^+/H^+ antiport [9,10]. Together with the basal sheath cell potential (BSHp), a "transepithelial potential" of about 45 mV results. Its primary function is enhancement of the receptor current [5].

[1] Zoological Institute, University of Regensburg, 93040 Regensburg, Germany

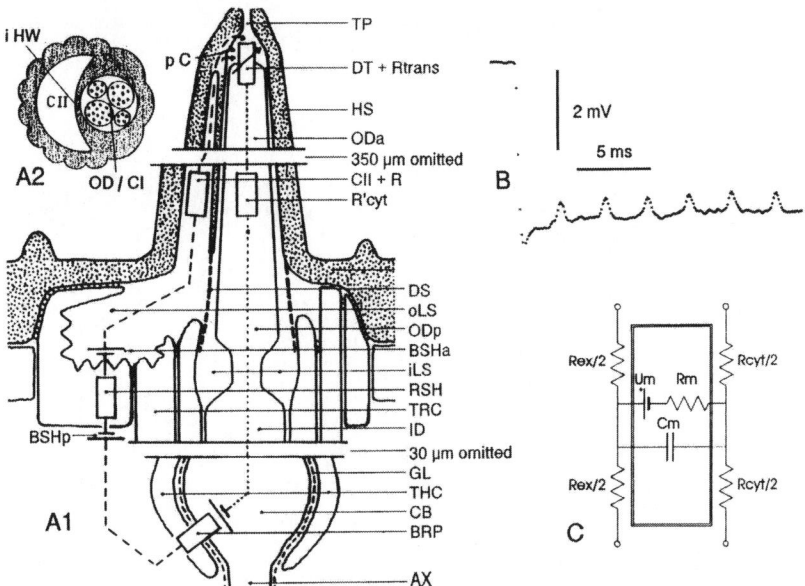

FIG. 1. **A1** Organization of a taste hair in a longitudinal section and the superimposed electrical equivalent circuit. The intracellular current pass is marked by a *dotted line* and the extracellular one by a *hatched line*. Only one receptor cell is shown; the cuticle is stippled. **A2** Cross section through the hair shaft showing the dendrite-free canal II and the four dendrites within canal I. **B** Electrophysiologic DC recording with 1 M potassium nitrate as stimulus showing the short latency (between onset of stimulation and the first spike) of 3.6 ms and a high subsequent spike frequency; the distance between two measuring points corresponds to 0.1 ms. **C** Electrical components of a single segment used for modeling; note that those components belonging to the membrane are framed. *AX*, axon; *BSHa BSHp*, batteries of the *apical*/proximal sheath cell membranes, here located in the tormogen cell; *CB*, receptor cell body; *CII + R*, canal II of the hairshaft and its resistance; *DS*, dendritic sheath; *DT + Rtrans*, distal tip of the dendrite as the membrane area of transduction and its stimulus-controlled resistance; *GL*, glia; *HS*, hairshaft; *ID*, inner dendrite; *iHW*, inner hair wall insulating CI from CII; *iLS*, apical inner receptor lymph space (basal labyrinth of the THC is omitted); *oLS*, outer lymph space; *ODa, ODp*, apical/proximal outer dendrite; *pC*, permeable cuticular region of the iHW; *BRP*, resting potential battery of CB and its resistance; *R'cyt*, total longitudinal resistance of the OD; *RSH*, transepithelial resistance of the sheath cells; *THC, TRC*, thecogen, trichogen cells; *TP*, porus at the hair tip

Materials and Methods

The TEM database is derived mainly from eight series of serial sections through hair tips of chemically fixed material [6]. Approximately circular diameters have been derived from the circumferences of the irregularly oval cross sections.

The transduction process is represented by the variable resistor (Rtrans), which reflects the occupancy of the receptor site population. The outer dendrite was subdivided into 30 segments of increasing diameter and length corresponding to the TEM dimensions, with the aim of approximating isopotentiality between adjacent segments. The electrical constituents of a segment are shown in Fig. 1C.

1. *Intracellular cytoplasmic resistance* (Rcyt) was calculated with a specific cytoplasmic resistivity of 100 Ohm∗cm [11].

2. *Extracellular resistance* (Rex) was set to the same value as that of Rcyt of the same segment; this assumption corresponds to (1) the finding that both canals within the hairshaft, except at the tip, are electrically insulated from each other [4]; and (2) the narrow extracellular spaces within canal I, as obtained from the TEM micrographs. Rcyt and Rex were subdivided for symmetric coupling of the segments [12].

3. *Membrane resistance* (RM) of the outer dendrite was calculated for a set of different specific membrane resistivities (specific RM) between $10\,k\Omega cm^2$ and $1\,M\Omega cm^2$. The membrane capacitance and the nernstian voltage across the dendritic outer membrane (Um) were not included in the modeling; Um was set to zero, taking into account the elevated potassium concentration in the receptor lymph [8].

The inner dendrite and the cell body were represented by five segments only, as they are electrically compact. The axon was modeled by 20 segments (length 60 µm long, diameter 0.5 µm). For all these segments a resting potential of −65 mV and a specific membrane resistance of $15\,k\Omega cm^2$ was assumed. The total input resistance of the inner dendrite, receptor cell body, and axon amounts to 1.6 GΩ. This figure

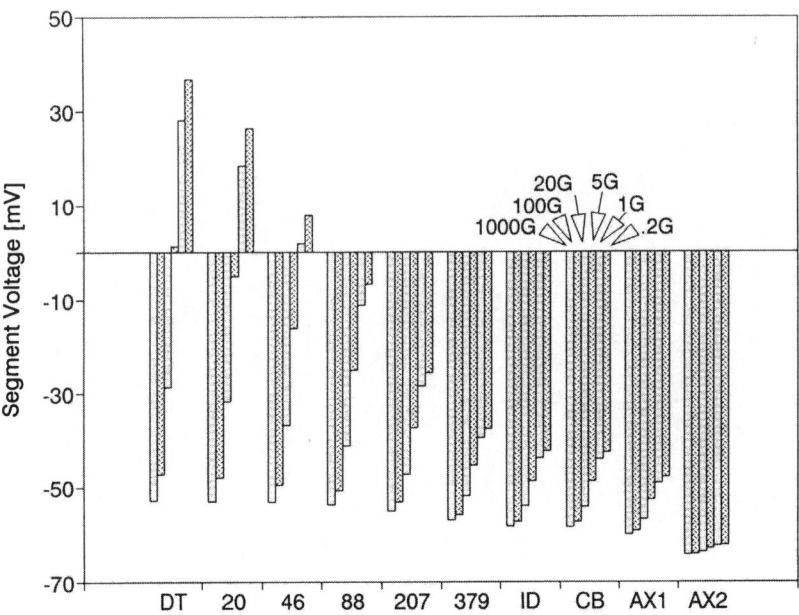

FIG. 2. Distribution of the voltage (y-axis) along the outer dendrite (small type) and the other parts of a receptor cell, assumed to be measured by an imaginary intracellular electrode with reference to the hemolymph space. Each group of bars represents six different resistances of the transduction membrane (*Rtrans*), as indicated for the bars belonging to the cell body. The depolarization is maximal at the dendritic tip (*DT*); it decreases toward the inner dendrite (*ID*) and cell body (*CB*) and disappears along the axon. *AX1*, axonal segment adjoining the cell body; *AX2*, approximately 1200 μm away from the cell body. The numbers on the x-axis represent the cumulative length of the outer dendrite in micrometers. Only 5 of 30 segments of the outer dendrite used for the calculation are displayed

is in fair agreement with the resistance obtained from dissociated taste receptor cells in the whole-cell recording mode (Kijima, personal communication). The network was computed with PSPICE 5.2 (Microsim Corp./Irvin, CA, Thomatronix/Rosenheim, Germany).

Results

Basis of the Electrical Equivalent Circuit of the Taste Hair

The cellular elements of the sensillum can be reduced to an equivalent circuit consisting of resistances and batteries, as shown superimposed in Fig. 1A. Because we have knowledge of the structural dimensions, the values of the electrical components can be determined. The receptor current is controlled mainly by the stimulus-dependent increase of the conductance (or the decrease of the resistor Rtrans, respectively) and tunes the spike generator located at the inner dendrite or at the cell body. Differing from other sensory neurons, the receptor current here is not caused simply by the dendritic resting potential but by a combination of the resting potential, restricted to the inner dendrite and soma, and the transepithelial voltage generated by the sheath cells. The receptor current path outside the receptor cell is also more complex than usual. It passes (1) the sheath cells and (2) two extracellular spaces: the hemolymph space and that of the apical receptor lymph.

Longitudinal Resistance of the Outer Dendrite

Calculation of the longitudinal cytoplasmic resistance (Fig. 1, A1, R'cyt) of the outer dendrite by summing the resistances Rcyt of the 30 segments results in values of 2 GΩ (dendrites of large size) to 6 GΩ (small size). Because of the conical shape of the outer dendrite, half of these resistances belong to the most distal 20% of the total dendritic length. This resistance is much higher than the corresponding one of the inner dendrite and is about two orders of magnitude larger than that of the extracellular path. In the following section all data refer to the "small dendrite."

Voltage Distribution Along the Receptor Cell

Figure 2 shows the intracellular voltages of selected segments of the outer dendrite and of other parts of the receptor cell. Because we chose hemolymph as the reference, the voltages across the membrane segments of the outer dendrite are higher by roughly the transepithelial voltage. The voltages are calculated in the steady state for six values of the transduction resistance (Rtrans; for a specific RM of $100 \, k\Omega cm^2$, see below). The first ($1000 \, G\Omega$) and last ($0.2 \, G\Omega$) values represent the resting ($1 \, pS$) and maximal ($5 \, nS$) transduction conductances, the values in between reflect different stimulus concentrations as well as numbers of simultaneously open channels: $100 \, G\Omega$ or $10 \, pS$ correspond to one open channel of $10 \, pS$ (an often observed value), $20 \, G\Omega$ ($50 \, pS$) to five channels, $5 \, G\Omega$ ($200 \, pS$) to 20 channels, and $1 \, G\Omega$ ($1 \, \mu S$) to 100 open channels.

If Rtrans is high, the potential at the tip is determined mainly by the resting potential of the cell body ($-65 \, mV$); if Rtrans is low, the potential of the apical lymph space ($+45 \, mV$) surrounding the dendrite determines the dendritic potential. Thus at maximal stimulation depolarization (as a positively moving voltage) of $96 \, mV$ absolutely occurs, starting from $-53 \, mV$ and going up to $+43 \, mV$. Along the dendrite the depolarization decreases to $17 \, mV$ at the inner dendrite and cell body and represents the voltage that controls the spike generator. Finally, the depolarization is minimal at the axon $1200 \, \mu m$ away from the cell body. Note that the sensitivity in terms of voltage change per resistance change is higher in the range between $100 \, G\Omega$ and $1 \, G\Omega$ than for the first and last resistance values ($1000 \, G\Omega$, $0.2 \, G\Omega$); this middle range represents the dynamic range of the "dose-response curve."

Permutation of the Specific Membrane Resistance

For another set of simulations, the specific RM was varied from $10 \, k\Omega cm^2$ to $1 \, M\Omega cm^2$ in order to study the influence on the transfer behavior.

1. In the resting state (Rtrans = $1000 \, G\Omega$) the voltage at the tip is similar to that of the soma only for high specific RM values (for $\geq 100 \, k\Omega cm^2$ U $> -53 \, mV$), whereas at $20 \, k\Omega cm^2$ it is reduced to $-25 \, mV$; thus the driving voltage for the receptor current in the latter case is drastically reduced. Moreover, at this low specific RM value the resting potential at the cell body is already depolarized ($-41 \, mV$) under the influence of the transepithelial voltage. Owing to the same effect at high specific RM values, the stimulus-dependent currents and depolarizations are higher than at low values: At $100 \, k\Omega cm^2$, for example, eight open channels lead to a receptor current through the membranes of the inner dendrite and the soma of $1.8 \, pA$ (without the part flowing into the axon), whereas at $20 \, k\Omega cm^2$ a current of only $0.9 \, pA$ is obtained.

2. Leakage currents (Ileak) increase with decreasing specific RM values. As additive components, they influence low receptor currents more markedly than higher ones and sum up at the proximal parts of the outer dendrite: At $100 \, k\Omega cm^2$ Ileak

is $0.1 \, pA$ at the dendritic tip and $1.8 \, pA$ at the soma or equal to a receptor current obtained by the action of eight open channels. At $20 \, k\Omega cm^2$ Ileak is four times higher, whereas the receptor current is reduced by a factor of 2.

For the modeling, a specific RM of $100 \, k\Omega cm^2$ (as used in Fig. 2) is the lowest value at which a concentration-dependent operation of the receptor cell seems plausible.

Discussion

The main problem here is how the transduction signal can be transferred along the outer dendrite over a distance of $400 \, \mu m$ to the spike-generating region at the inner dendrite or cell body. From the short latencies of 3 to $4 \, ms$ (Fig. 1C) following stimulation with high salt concentrations and the diffusion behavior of the salts, it can be determined that only the extreme tip of the dendrites with a membrane area of a few square micrometers is involved in the transduction process. Moreover, the signal transfer must occur electrotonically because active conduction along the outer dendrite must be excluded in adult flies owing to the polarity of the recorded action potentials. To estimate the loss of signal amplitude one must consider two factors: (1) The large cytoplasmic resistance of the thin outer dendrite, which lies in series with the one of transduction, reduces the receptor current; and (2) the resistance of the whole nontransducing membrane, being 100 times larger than the transduction area, lies in parallel with the transduction resistance and should increase leakage currents considerably.

The model presented here provides three main conclusions to this problem.

1. As a consequence of the high cytoplasmic resistance, only low currents can flow through the dendrite with a tolerable voltage drop. At $6 \, G\Omega$ resistance, a current of $20 \, pA$ results in a voltage drop of $120 \, mV$, which is in the same range as the available driving voltage (Fig. 2, $88 \, mV$). Such a current is obtained if, for example, 24 transduction channels of an assumed single channel conductance of $10 \, pS$ are open simultaneously. Consequently, the dynamic range of the signal transfer seems to be restricted to currents through a small number of channels.

2. As shown, a driving voltage of $100 \, mV$ exists at rest across the whole membrane area of the outer dendrite. Remarkable leakage currents are expected unless the membrane does not exhibit an extremely high specific resistivity. A value of $100 \, k\Omega cm^2$ results as a lower limit with respect to an acceptable ratio of transduction to leakage current.

3. The leakage currents behave in a manner similar to that of the true receptor currents and exert

a depolarizing influence on the membrane of the cell body at rest. This effect is highly undesirable because at resting voltages of more than $-50\,mV$ the spike generator reduces its activity because of the increasing inactivation of the sodium channels. This is an additional reason for postulating high specific membrane resistivity. It is not possible to compensate this effect by lowering the specific resistivity of the cell body membrane, as less effective coupling of the outer dendrite as sensor would result and hence reduce the sensitivity.

In two respects the model presented here is better adapted to taste receptors, than Thurms and Küpper's well-established model of the receptor circuit [5]. Although their model was developed for mechanoreceptors, it is often interpreted to be valid for other sensilla of insects. In the model presented here, the high longitudinal resistance of the outer dendrite and the resistance of the dendritic membrane are included. Although they are not involved in transduction, they are important in bridging the distance from the hair tip to the hair base. These parameters may play only a minor role in mechanoreceptors with short dendrites. Furthermore, the model shows that the influence of the transepithelial voltage is important and should be emphasized more strongly than was done in the model reported by Morita and Shiraishi [3].

Acknowledgments. The author thanks Drs. E. Hansen-Delkeskamp and M. Schnuch for critical comments on the manuscript.

References

1. Dethier VG (1976) The hungry fly. Harvard University Press, Cambridge, MA
2. Hansen K (1978) Insect chemoreception. In: Hazelbauer GL (ed) Taxis and behavior. Series B, vol 5. Receptors and recognition. Chapman & Hall, London, pp 232–292
3. Morita H, Shiraishi A (1985) Chemoreception physiology. In: Kerkut GA, Gilbert LI (eds) Comprehensive Insect Physiology, Biochemistry and Pharmacology (vol 6). Pergamon, Oxford, pp 133–170
4. Morita H (1969) Electrical signs of taste receptor activity. In: Pfaffmann C (ed) Proceedings, III International Symposium on Olfaction and Taste. Rockefeller University Press, New York, pp 370–381
5. Thurm U, Küppers J (1980) Epithelial physiology of insect sensilla. In: Locke M, Smith D (eds) Insect Biology in the Future. Academic Press, New York, pp 735–763
6. Waldhorst G (1988) The fine structure of gustatory sensilla in flies [in German]. Thesis, University of Regensburg
7. Schnuch M, Hansen K (1990) Sugar sensitivity of a labellar salt receptor of the blowfly *Protophormia terraenovae*. J Insect Physiol 38:671–680
8. Gödde J, Krefting ER (1989) Ions in the receptor lymph of the labellar taste hairs of *Protophormia terraenovae*. J Insect Physiol 35:107–111
9. Wieczorek H (1982) A biochemical approach to the electrogenic potassium pump of insect sensilla: potassium sensitive ATPases in the labellum of the fly. J Comp Physiol 148:303–311
10. Wieczorek H, Putzenlechner M, Zeiske W, Klein U (1991) A vacuolar-type proton pump energizes K^+/H^+ antiport in an animal plasma membrane. J Biol Chem 266:15340–15347
11. Rall W, Burke RE, Holmes WR, et al. (1992) Matching dendritic neuron models to experimental data. Physiol Rev 72:159–186
12. Segev I, Fleshman JW, Miller JP, Bunow B (1985) Modeling the electrical behavior of anatomically complex neurons using a network analysis program: passive membrane. Biol Cybern 53:27–40

Multiple Receptor Sites of Insect Taste Cells

ICHIRO SHIMADA[1]

Key words. Fleshfly—Taste—Chemoreception—
Multiple sites—Nucleotide site

Introduction

Multiple receptor sites for sweet taste have recently
been elucidated in various animals. Some of the
earliest direct evidence has been given by a pharmacological method with the sugar receptor of the
fleshfly, based on electrophysiological recordings of
single taste cell responses [1]. Four receptor sites
were identified by treatment of the sugar receptor
with some specific blocking reagents. They are
the pyranose, furanose, aryl, and alkyl sites. Stereo-
specificities of the four receptor sites have successfully been clarified by examining the structure—
activity relationships for stimulants. They are quite
rigid and different from each other [2].

Reactions at different receptor sites in a sugar
receptor cell, however, evoke the activity in the
same taste cell and send the same information of
taste quality to the central nervous system (CNS) of
the fly. What, then, is the biological meaning of the
presence of multiple receptor sites? The following
two functions can be predicted: (1) responsiveness
to various stimulants, and (2) rigorous discrimination. With only one type of receptor site, it would
be rather difficult for the two functions to coexist.
The more stimulants to which the putative receptor
is responsive, the less distinguishable the stimulants
are likely to be. With the four specific receptor sites,
the fly can respond to various stimulants in nature
and, at the same time, exclude deleterious substances by virtue of their rigid stereospecificity. Such
characteristics of the receptor sites may be advan-
tageous to the fly for its adaptation and evolution.

Recently, some sugar binding sites have been
found in the water and salt receptors of the fly
[3–6]. We give here a short review of such sugar
receptor sites in the other taste cells and compare
recent results suggesting the presence of some nucleotide binding sites in the sugar and salt receptors
of the fly [7,8]. The biological meaning of the
presence of such new multiple receptor sites is
discussed.

---

[1] Biological Institute, Faculty of Science, Tohoku University, Kawauchi, Aoba-ku, Sendai, 980 Japan

Materials and Methods

The fleshflies, *Boettcherisca peregrina*, 4–5 days old
and raised in our laboratory, were kept at room
temperature and given a 3% sucrose solution just
before the experiment. Recordings were made from
the chemosensory setae corresponding to the largest
hair (no. 10) on the left side of the labellum of
Phormia regina Meigen [9]. Impulses were recorded
from the side wall of the seta [10]. The methods
have been described in detail elsewhere [11]. Most
chemicals were dissolved in double-distilled water,
but 5'-AMP was dissolved in phosphate buffer (pH
5–7). The ambient temperature was $22° \pm 1°C$ and
the relative humidity was kept at 62%–80%.

Results

Anhydro Sugar Binding Site of the Salt Receptor

Most sugars possess reducing properties due to their
anomeric centers and are generally unstable for the
study of structure–activity relationships in taste.
Eight 1,6-anhydro-β-D-hexopyranoses in Fig. 1,
which have stable conformations without free
anomeric centers, were examined for their effectiveness in stimulating the labellar taste receptors.
1,6-Anhydro-β-D-idopyranose (*ido*) with three successive equatorial hydroxyl groups was found to be
the most effective on the sugar receptor, which
supports a model proposed for the pyranose site [1].
1,6-Anhydro-β-D-galactose (*gal*), 1,6-anhydro-β-D-
altrose (*altro*), 1,6-anhydro-β-D-talose (*talo*), and
1,6-anhydro-β-D-gulose (*gulo*) were found to stimulate the salt receptor though they are completely
uncharged. This was supported by comparing the
shape of evoked impulses and analyzing the effects
of mixed stimuli with NaCl and other anhydro sugars
on the salt and sugar receptors. These findings suggest the presence of a new specific anhydro sugar
receptor site on the salt receptor [5].

Nucleotide Binding Site of the Salt Receptor

In the course of the study of signal transduction of
the taste cell, various nucleotides have been found
to stimulate the salt receptor of the fleshfly. Most
nucleotides examined are more stimulatory than
NaCl at the corresponding concentrations. Figure

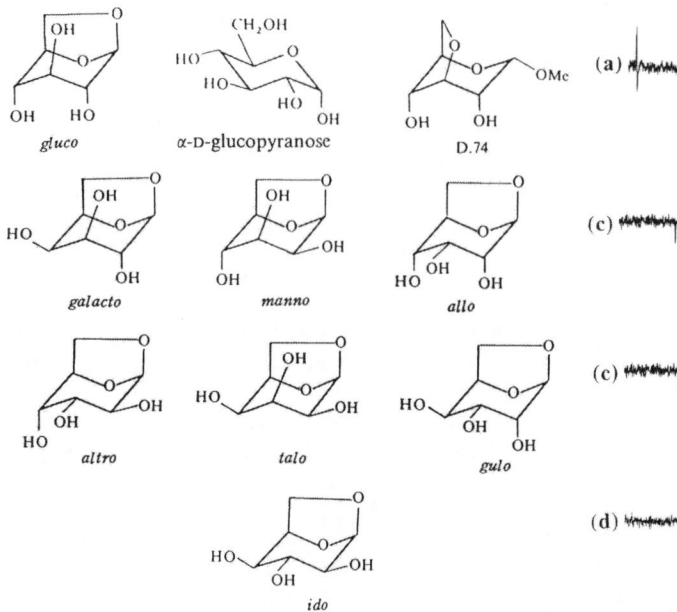

FIG. 1. Structures of eight 1,6-anhydro-β-D-hexopyranoses

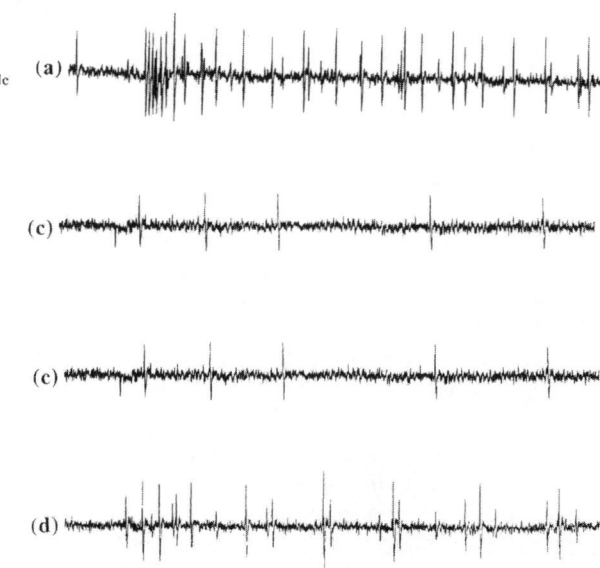

FIG. 3. Records of responses of taste receptor cells of the fleshfly. **a** Response to 100 mM sucrose; **b** water; **c** 500 mM NaCl; **d** 100 mM 5'-AMP (dissolved in water); **e** 300 mM 5'-AMP (in phosphate buffer)

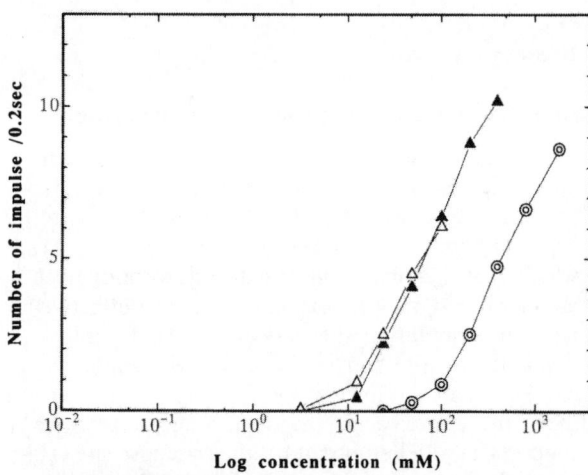

FIG. 2. Concentration–response curves for cyclic nucleotides and NaCl. The ordinate is the mean value of the responses of the salt receptor. ▲ *Closed triangles*, cAMP; △ *open triangles*, cGMP; ◎ *double circles*, NaCl

could also stimulate the salt receptor. These results suggest the presence of a nucleotide binding site on the salt receptor.

Stimulatory Effect of 5'-AMP on the Sugar Receptor

Among nucleotides, 5'-AMP is a rather peculiar stimulant. It can stimulate not only the salt receptor but also the sugar receptor.

In phosphate buffer ($\frac{1}{15}M$, pH 7.2), the excitability of both the salt receptor and the water receptor was markedly depressed while the sugar receptor was hardly affected. Figure 3 shows the records of responses to 5'-AMP in phosphate buffer and in distilled water. 5'-AMP in phosphate buffer primarily stimulated the sugar receptor. Under conditions of depressed salt response, a response–concentration relationship for 5'-AMP was obtained. Over the whole range of concentrations examined, the relative response to 5'-AMP was much smaller than to sucrose, as shown in Fig. 4, though the magnitude of the maximum response to 5'-AMP was almost the same as that to stimulatory amino acids [12,13]. Furthermore, concentration dependence for the response to 5'-AMP was clearly observed. On the other hand, the stimulatory effect on the salt receptor was not as clear as with other stimulatory nucleo-

2 shows the results of a comparative study on response–concentration relationships with cyclic nucleotides (cGMP and cAMP) and NaCl. Both nucleotides are disodium salts, but they were in fact stimulatory themselves on the salt receptor since considerable effect still remained after subtracting the magnitude of the response due to sodium ion. The remaining response increased as the concentration increased up to 200 mM. Owing to such concentration dependency, the cyclic nucleotides may be considered adequate stimuli to the salt receptor. 5'-GMP as well as dibutyryl cyclic GMP (dbcGMP)

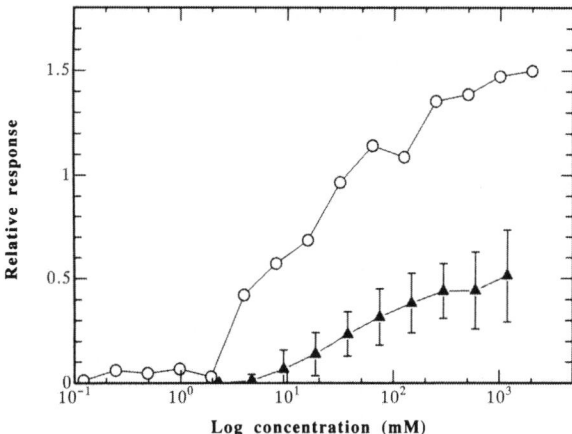

FIG. 4. Concentration–response curve for 5'-AMP and sucrose. The ordinate represents the ratio of the magnitude of response to each chemical, to the response to 100 m*M* sucrose. ○ *Circles*, sucrose; ▲ *triangles*, 5'-AMP

substances, mixed stimuli, but scarcely encounter the pure substances usually examined in the laboratory. Multiple receptor sites may present molecular bases for discrimination of such complex substances through various interactions and adaptations among them. They may even create new taste information different from that of pure substances owing to such interactions as inhibition, activation, cooperativity, adaptation, and so on under conditions of mixed stimuli. More studies, however, are necessary to prove such functions for multiple receptor sites.

Acknowledgments. I thank my students for their help with the experiments.

tides. These results suggest that there is also a nucleotide binding site on the sugar receptor.

Discussion

Our results with anhydro sugars can be compared with the results of Schnuch and Hansen [6] who reported the stimulatory effect on the salt receptor of some sugars that are supposed to bind to the pyranose site of the sugar receptor. Both results suggest the presence of specific receptor sites for uncharged sugars on the salt receptor and raise a question about the accepted idea of a classic "salt" receptor specific for alkali ions.

The stimulatory effect of certain nucleotides reported here indicates the presence of a further specific binding site on the salt receptor. On the other hand, 5'-AMP alone can stimulate the sugar receptor. These results with nucleotides contrast with those for the blowfly [7,8]. Most nucleotides examined in the latter studies stimulated the sugar receptor cell of the blowfly. Such a difference may be due to a species difference. Nucleotide receptor sites, however, may function in the feeding of the fly since carrion which serves as food for the fly contains a lot of nucleotides.

Various receptor sites recently discovered further support the concept of two functions indicated by multiple sites on the sugar receptor, mentioned in the introduction, and extend this concept to other receptors. In nature, flies often encounter complex

References

1. Shimada I, Shiraishi A, Kijima H, Morita H (1974) Separation of two receptor sites in a single labellar sugar receptor of the fleshfly by treatment with *p*-chloromercuribenzoate. J Insect Physiol 20:605–621
2. Shimada I (1987) Stereospecificity of the multiple receptor sites in the sugar taste receptor cell of the fleshfly. Chemical Senses 12:235–244
3. Wieczorek H, Koppl (1978) Effects of sugars on the labellar water receptor of the fly. J Comp Physiol 126:131–136
4. Wieczorek H, Shimada I, Hopperdietzel C (1988) Treatment with pronase uncouples water and sugar reception in the labellar water receptor of the blowfly. J Comp Physiol A 163:413–419
5. Shimada I, Ohrui H, Meguro H (1990) The stimulating effect of 1,6-anhydro-hexopyranoses on the salt receptor of the fleshfly. In: Døving KB (ed) ISOT X, p 359
6. Schnuch M, Hansen K (1990) Sugar sensitivity of a labellar salt receptor of the blowfly *Phormia terraenovae*. J Insect Physiol 36:409–417
7. Amakawa T, Ozaki M, Kawata K (1990) Effects of cyclic GMP on the sugar taste receptor cell of the fly *Phormia regina*. J Insect Physiol 36:281–286
8. Amakawa T, Kawata K, Ozaki M (1992) Nucleotide receptor site on the labellar sugar receptor cell of the blowfly *Phormia regina*. J Insect Physiol 38:365–371
9. Wilczek M (1967) The distribution and neuroanatomy of the labellar sense organs of the blowfly *Phormia regina* Meigen. J Morphol 122:175–201
10. Morita H, Yamashita S (1959) Generator potential of insect chemoreceptor. Science 130:922
11. Shimada I, Tanimura T (1981) Stereospecificity of multiple receptor sites in a labellar sugar receptor for amino acid and small peptides. J Gen Physiol 77:23–39
12. Shiraishi A, Kuwabara M (1970) The effects of amino acids on the labellar hair chemosensory cells of the fly. J Gen Physiol 56:768–782
13. Shimada I (1978) The stimulating effect of fatty acids and amino acid derivatives on the labellar sugar receptor of the fleshfly. J Gen Physiol 71:19–36

Immunocytochemistry of Odorant-Binding Proteins

R.A. STEINBRECHT[1], M. LAUE[1], S.-G. ZHANG[2], and G. ZIEGELBERGER[1]

Key words. Pheromone-binding protein—General Odorant-binding protein—Immunogold labeling—Pheromone-sensitive sensillum trichodeum—*Sensillum basiconicum*—*Antheraea polyphemus*—*Autographa gamma*—*Bombyx mori*—*Heliothis armigera*—*Spodoptera littoralis* (Insecta)

Introduction

Odorant-binding proteins are small water-soluble proteins that have been detected in the perireceptor compartment of olfactory receptor cells of vertebrates and insects [1,2]. Although their definite physiological role in olfaction is still unclear, similar functions have been proposed in stimulus transport or inactivation [1–6].

The odorant-binding proteins of insects are divided into two groups: the *pheromone-binding proteins* (PBPs [7]), which are predominant in the antennae of male moths, and the so-called *general odorant-binding proteins* (GOBPs [8,9]) which are antenna-specific but not sex-specific. Nine PBPs and nine GOBPs have been identified in seven moth species (for an overview, see Table 1 in ref. [10]). The PBPs, and even more so the GOBPs, are highly conserved between different species, but amino acid sequence homology between PBPs and GOBPs is only about 30% [8,9]. Moreover, odorant-binding experiments have proved extremely difficult in insects, and binding data are available only for the PBP of *Antheraea polyphemus* [7,11], the PBP of *Bombyx mori* [10], and the PBPs of *Lymantria dispar* [12]. In all other cases this function was inferred on the basis of sequence homology. Except for *A. polyphemus*, in which PBP was identified in drops of sensillum lymph collected from the pheromone-sensitive sensilla trichodea [13], nothing was previously known about the localization of odorant-binding proteins in insect antennae.

Immunocytochemistry using antibodies against odorant-binding proteins offers the possibility of localizing these molecules in sections of moth antennae with high selectivity, sensitivity, and spatial resolution. Thus, questions about the biosynthesis of odorant-binding proteins can be answered as well as questions concerning the correlation between the presence or absence of a certain binding protein and the specific olfactory function of a given sensillum. The fact that our antibodies against PBP and GOBP of *Antheraea polyphemus* cross-react with odorant-binding proteins of related species opens a further fruitful dimension to this kind of work.

Using a polyclonal antiserum against PBP of *Antheraea polyphemus*, Steinbrecht et al. [14,15] observed exclusive labeling of the pheromone-sensitive sensilla trichodea of male *Antheraea polyphemus* and *Bombyx mori*, while in *Autographa gamma* only a subpopulation of morphologically similar sensilla trichodea was labeled. It could be shown that biosynthesis of PBP is going on in the tormogen and trichogen cells of the sensilla trichodea. Moreover, there are indications of a degradative pathway as well. In this paper we report on an extension of this study to other species of moths and, in addition, on first immunolabeling data using a new polyclonal antiserum against GOBP purified from antennae of female *Antheraea polyphemus*.

Methods

The purification of PBP of *Antheraea polyphemus* and the production of the antiserum [anti-PBP(Apo)] as well as our immunolabeling protocol is described in detail in Steinbrecht et al. [15]. In brief, rabbits were injected three times with purified PBP plus Freund's adjuvant. Blood serum was obtained 13 and 17 days after the last injection, stored at −25°C, and used in dilutions of 1:3000 or more without further treatment. Immunolabeling was performed in a two-step postembedding protocol on sections of freeze-substituted antennae using antirabbit IgG coupled to 10-nm colloidal gold as a second antibody with subsequent silver intensification.

For the antiserum against the GOBP of *A. polyphemus* [anti-GOBP(Apo)], the same strategy was followed except that the GOBP purified from female *A. polyphemus* was used to immunize the rabbits.

[1] MPI für Verhaltensphysiologie, 82319 Seewiesen, Germany
[2] Institute of Zoology, Academia Sinica, Beijing, China

Results

Labeling with Anti-PBP(Apo)

General Observations

In labeled sensilla, very strong labeling was always observed in the sensillum lymph, in the hair lumen as well as in the sensillum-lymph cavity below the hair base. Intracellular label was mainly observed in the trichogen and tormogen cells of labeled sensilla, in the rough endoplasmatic reticulum, Golgi apparatus, and in particular in dense granules that may represent secretory granules. In the thecogen and in the receptor cells, organelles of protein biosynthesis were not labeled. However, the cell border between the receptor cell somata and the thecogen cell was weakly labeled. Moreover, labeled granules and coated pits and vesicles were observed in these cells. These observations were made mainly in male *Antheraea polyphemus* and *Bombyx mori*, but appear to be valid for the other species studied as well.

Cross-Reactions with PBP of Other Species

In male *Antheraea polyphemus* and in male *B. mori*, all long sensilla trichodea were labeled (Fig. 1a). In males of the other species studied, only a fraction of the long sensilla trichodea was labeled. A fairly large proportion of long sensilla trichodea was labeled in male *Heliothis armigera* and *Spodoptera littoralis*. However, in male *Autographa gamma*, the percentage of labeled sensilla trichodea was hardly more than 20%. Labeled and nonlabeled sensilla so far could not be distinguished by their morphology. Nevertheless, labeling was either very strong or absent in the sensillum lymph of these sensilla; intermediate reactions so far have not been observed.

In female *B. mori*, which in contrast to *Antheraea polyphemus* also have long sensilla trichodea on their antennae, PBP labeling was not observed initially [15] but an extensive search revealed that very few sensilla (<10%) that belong to the groups of medium-sized sensilla trichodea and of sensilla basiconica are labeled. Long sensilla trichodea were not labeled. In female *Spodoptera littoralis* and *Heliothis armigera*, however, the percentage of labelled sensilla trichodea was fairly large.

Labeling with Anti-GOBP(Apo)

So far, immunocytochemistry with anti-GOBP(Apo) has been performed on antennal sections of male *Antheraea polyphemus* and male and female *B. mori*. In males of both species the sensilla basiconica were labeled (Fig. 1b). The long sensilla trichodea were not labeled in male *A. polyphemus* and *B. mori*. In female *B. mori*, a large percentage of the sensilla basiconica and the medium-sized sensilla

trichodea were labeled and, in addition, all long sensilla trichodea (Fig. 1c).

The distribution of label in labeled sensilla basiconica resembled the picture obtained with anti-PBP (Apo) in the pheromone-sensitive sensilla trichodea.

Discussion

Although immunocytochemistry of odorant-binding proteins is only in its infancy, several intriguing issues are already apparent. First and foremost, there is now direct proof that PBP is localized in pheromone-sensitive sensilla while GOBP is localized in other olfactory sensilla that are known to respond to a large variety of general odours, as shown for *Antheraea* spp. in the classical electrophysiological paper of Schneider et al. [16]. Thus, Vogt et al. [9] were correct when they coined the term general odorant-binding protein and predicted the GOBPs to be associated with different classes of olfactory receptor neurons as compared to PBP.

The observation of PBP-labeled sensilla in the females of *B. mori* came as a surprise, because so far there are no electrophysiological data of pheromone receptor cells in female silk moths. The number of labeled sensilla, however, is so small that they could have well been overlooked, in particular because they appear to belong to another morphological subtype, the medium-sized sensillum trichodeum. Likewise, it is not surprising that so few sensilla do not create a measurable electroantennogram. The presence of PBP alone of course cannot prove pheromone sensitivity.

Nevertheless, it is now quite possible that the notion has to be abandoned that female silkmoths cannot smell their own sex attractant pheromone. This old dogma was recently challenged in another moth species, *Spodoptera littoralis*, by Ljungberg et al. [17] who could demonstrate by single-cell recording that pheromone-sensitive receptor cells are not at all rare in these females. They list also a few other cases of female responses to their pheromones shown by either behavior reactions, electroantennograms, or single-cell recordings; for example, females of *Heliothis armigera* are repelled by conspecific females, maybe to ensure spacing of egg-laying [18], while females of *Trichoplusia ni* are attracted by their main pheromone component [19]. It also appears likely in those moth species in which the females had been believed insensitive to their own pheromone that a few pheromone-sensitive sensilla persist which would suffice to monitor the emission of the pheromone gland.

In any event, it is now necessary to resume single-cell recording from female *B. mori*, in particular from the thus-far neglected medium-sized sensilla trichodea to prove that presence of PBP in the sensilla is coincident with pheromone sensitivity of

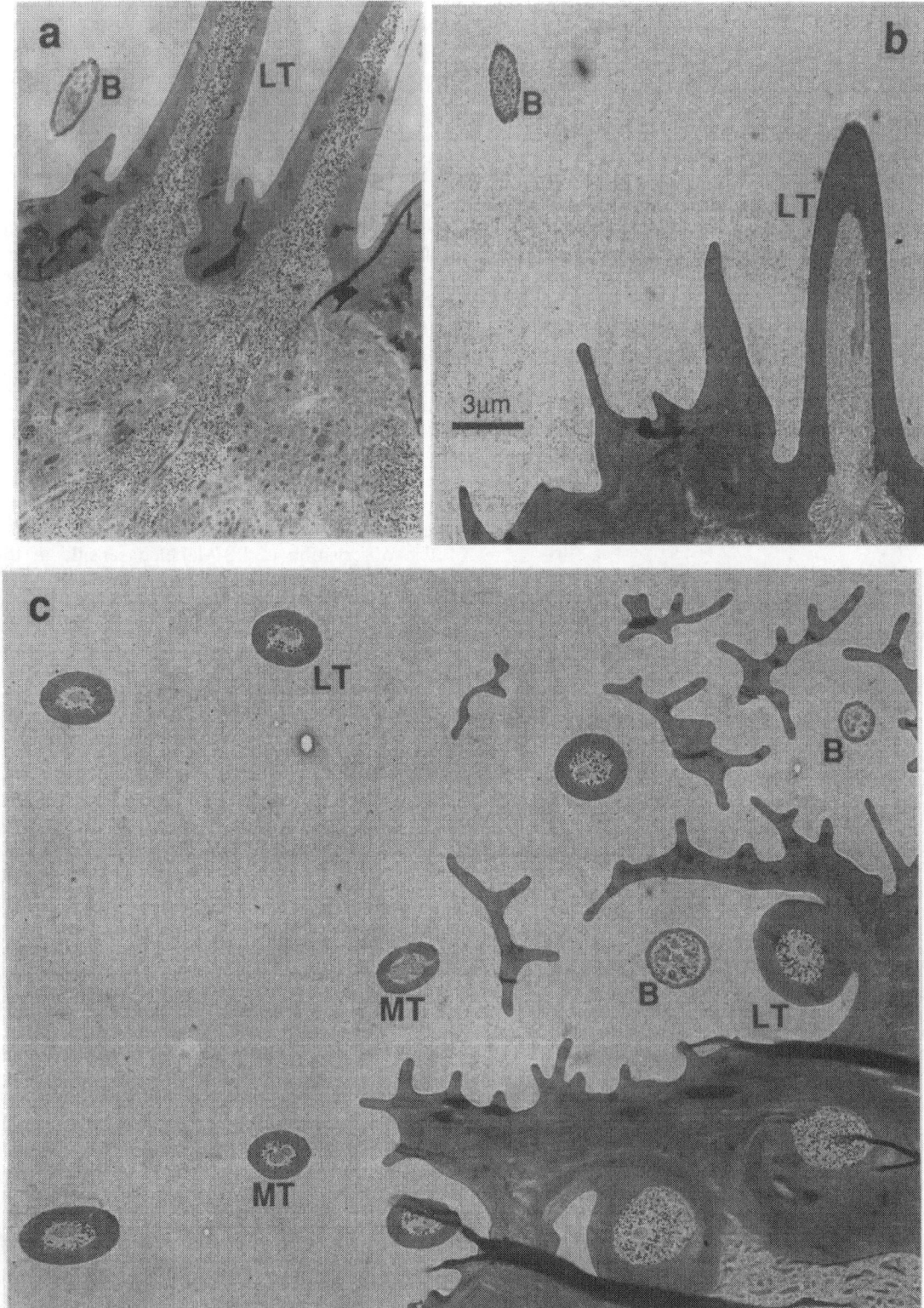

Fig. 1. Immunolabeling with antennal sections of male (**a,b**) and female (**c**) silkmoths, *Bombyx mori*. **a** Labeling with anti-PBP(Apo) demonstrates pheromone-binding proteins (PBP) in the sensillum lymph of long sensilla trichodea (*LT*); sensilla basiconica (*B*) are not labeled. **b** Labeling with anti-GOPB(Apo) shows the opposite labeling pattern: now sensilla basiconica (*B*) are labeled but not the long sensilla trichodea (*LT*). **c** In female *B. mori*, however, GOBP label is also present in all long sensilla trichodea (*LT*). Note that some medium-sized sensilla trichodea (*MT*) and sensilla basiconica (*B*) are not labeled. Scale bar: 3 μm.

the receptor cells. The best hint that presence or absence of PBP is highly correlated with pheromone sensitivity of the sensillum comes from the long sensilla trichodea in female *B. mori*. Unlike those moth species with a strong sexual dimorphism of the antennae (e.g., *Antheraea*, in which long sensilla trichodea are absent in females), in *B. mori* the long sensilla trichodea are present in both sexes but with changed specificity. In the male, they contain the pheromone receptor cells; in the female, they respond to linalool and benzoic acid [20,21]. Our immunolabeling experiments show that concomitant with the change in specificity from pheromone perception to general olfaction, these sensilla changed the odorant-binding protein from PBP to GOBP.

The intracellular distribution of label supports the notion that both odorant-binding proteins are synthesized in the tormogen and trichogen cell along typical pathways of protein biosynthesis. For PBP, we observed signs of a degradation pathway as well [15]. Whether the same is true for GOBP needs further investigation.

Acknowledgments. We wish to thank Anka Günzel and Barbara Müller for expert assistance and Leslie D. Williams for correcting the English style. This work was partially supported by a grant from Deutsche Forschungsgemeinschaft (DFG, Ste501/3–). One of us, Shan-gan Zhang, gratefully acknowledges support by K.C. Wong Education Foundation, Hong Kong. Dr. Bill Hansson, Lund, kindly provided pupae of *Spodoptera littoralis*.

References

1. Pelosi P, Maida R (1990) Odorant binding proteins in vertebrates and insects: similarities and possible common function. Chem Senses 15:205–215
2. Vogt RG, Rybczynski R, Lerner MR (1990) The biochemistry of odorant reception and transduction. In: Schild D (ed) Chemosensory information processing. NATO ASI Ser H, vol 39. Springer, Berlin, pp 33–76
3. Pevsner J, Snyder SH (1990) Odorant-binding protein: odorant transport function in the vertebrate nasal epithelium. Chem Senses 15:217–222
4. Kaissling K-E (1986) Chemo-electrical transduction in insect olfactory receptors. Annu Rev Neurosci 9:121–145
5. Vogt RG (1987) The molecular basis of pheromone reception: its influence on behavior. In: Prestwich GD, Blomquist GJ (eds) Pheromone biochemistry. Academic, Orlando, pp 385–431
6. Van den Berg MJ, Ziegelberger G (1991) On the function of the pheromone binding protein in the

7. Vogt RG, Riddiford LM (1981) Pheromone binding and inactivation by moth antennae. Nature (Lond) 293:161–163
8. Breer H, Krieger J, Raming K (1990) A novel class of binding proteins in the antennae of the silk moth *Antheraea pernyi*. Insect Biochem 20:735–740
9. Vogt RG, Prestwich GD, Lerner MR (1991) Odorant-binding-protein subfamilies associate with distinct classes of olfactory receptor neurons in insects. J Neurobiol 22:74–84
10. Maida R, Steinbrecht RA, Ziegelberger G, Pelosi P (1993) The pheromone binding protein of *Bombyx mori*: purification, characterization and immunocytochemical localization. Insect Biochem Mol Biol 23:243–253
11. Kaissling KE, Klein U, De Kramer JJ, Keil TA, Kanaujia S, Hemberger J (1985) Insect olfactory cells: electrophysiological and biochemical studies. In: Changeux JP, Hucho F, Maelicke A, Neuman E (eds) Molecular basis of nerve activity. De Gruyter, Berlin, pp 173–183
12. Vogt RG, Köhne AC, Dubnau JT, Prestwich GD (1989) Expression of pheromone binding proteins during antennal development in the gypsy moth *Lymantria dispar*. J Neurosci 9:3332–3346
13. Klein U (1987) Sensillum-lymph proteins from antennal olfactory hairs of the moth *Antheraea polyphemus* (Saturniidae). Insect Biochem 17:1193–1204
14. Steinbrecht RA, Keil TA, Ozaki M, Maida R, Ziegelberger G (1991) Immunocytochemistry of pheromone binding protein. In: Elsner N, Penzlin H (eds) Synapse—Transmission modulation, Proceedings of the 19th Göttingen Neurobiology Conference. Thieme, Stuttgart, p 172
15. Steinbrecht RA, Ozaki M, Ziegelberger G (1992) Immunocytochemical localization of pheromone-binding protein in moth antennae. Cell Tissue Res 270:287–302
16. Schneider D, Lacher V, Kaissling K-E (1964) Die Reaktionsweise und das Reaktionsspektrum von Riechzellen bei *Antheraea pernyi* (Lepidoptera, Saturniidae). Z Vergl Physiol 48:632–662
17. Ljungberg H, Anderson P, Hansson BS (1993) Physiology and morphology of pheromone-specific sensilla on the antennae of male and female *Spodoptera littoralis* (Lepidoptera: Noctuidae). J Insect Physiol 39:253–260
18. Saad AD, Scott DR (1981) Repellency of pheromones released by females of *Heliothis armigera* and *Heliothis zea* to females of both species. Entomol Exp Appl 30:123–127
19. Mitchell ER, Webb JR, Hines RW (1972) Capture of male and female cabbage loopers in field traps baited with synthetic sex pheromone. Environ Entomol 1:525–526
20. Priesner E (1979) Progress in the analysis of pheromone receptor systems. Ann Zool Ecol Anim 11:533–546
21. Heinbockel T, Kaissling K-E (1990) Sensitivity and inhibition of antennal benzoic-acid receptor cells of female silkmoth *Bombyx mori* L. Verh Dtsch Zool Ges 83:411

olfactory hairs of *Antheraea polyphemus*. J Insect Physiol 37:79–85

Transduction Ion Channels in Insect Taste Cells

HIROMASA KIJIMA, SATOSHI YAMANE, TOMOKI KAZAWA, HIROO NAKANO, SHINOBU GOSHIMA, and SHOKO KIJIMA[1]

Key words. Taste—Ion channel—Sugar receptor—Insect—Transduction—Chemoreception

Introduction

We have been studying the transduction mechanisms of insect taste cells, using mainly the sugar receptor neuron in the largest labellar chemosensory hair (the taste hair). This taste hair consists of two lumina: the inner and outer lumen. The inner lumen contains four sensory dendrites of taste neurons of primary sensory cell types: sugar receptor, water receptor, and two kinds of salt receptors. The outer lumen has a much larger cross section and is filled only with the receptor lymph secreted from the folded membrane of the supporting cells [1]. When the taste hair tip is stimulated by stimulant solutions, the receptor current flows within the outer lumen from the base toward the tip and passes the ion-permeable cuticle layer at the tip; it probably flows into the sensory dendrite of the stimulated neuron to cause the receptor potential and then evoke action potentials [2–4].

We report here three aspects of studies on the sugar-activated ion channels responsible for the transduction of taste sensation by the sugar receptor neuron. One is the very short latency (about 1 ms) of receptor current flow after stimulation of the taste hair tip; the second is the receptor current fluctuation analysis; and the third is the patch clamp study both on the isolated taste neuron and on the sensory dendrites grown out from the cut end of the taste hair. These results suggest that there are several kinds of transduction ion channels of a receptor–channel complex type on the receptor membrane of the sugar receptor neuron.

Materials and Methods

Measurement of the Receptor Current and Its Fluctuation

The adult of the fleshfly *Boettcherisca peregrina*, 2–8 days after emergence, was used. The head was separated from the body, mounted on the top of a glass capillary filled with fly Ringer solution, containing 112 mM NaCl, 5.5 mM KCl, 1.2 mM NaHCO$_3$, 0.08 mM NaH2PO$_4$, 1.8 mM CaCl$_2$, 0.8 mM MgCl$_2$, and 10 mM HEPES-Na, pH 7.1. Two methods of sidewall recording were employed [5]. The latency of the DC receptor current was measured with the two-sidewall method (Fig. 1b). Receptor current fluctuation was measured with the tip-sidewall method (Fig. 1a) during the time when transepithelial voltage was maintained to the level above +30 mV (Kijima et al., manuscript in preparation). Impulse generation was suppressed by applying 0.1 mM TTX to the taste hair tip for 10 min. The fluctuation was analyzed by computing its auto-correlation fluctuation (ACF) and power spectrum as described previously [5]. Sugar was dissolved in 0.1 M Na citrate solution (pH 6.9, [Na$^+$] = 0.28 M), and this solution was used for stimulation.

The Patch Clamp Experiment

The labellar taste neurons were isolated from the labellum of pupa about 2 days before emergence. The proboscis was isolated from the head and the labellum was cut as thin as possible leaving all the taste hairs on the labellum. The labellum was incubated with a mixture of 2 mg/ml chitinase (Sigma), 1 mg/ml cholagenase (Sigma, Type IA), and 2 mg/ml dispase (Boehringer, Mannheim, Germany) for 40 min at 23°C. Then, the bundle of sensory nerves with taste neurons contained in the thecogen cells at the end of nerves were taken out from the labellum with tweezers and extended on the surface of polylysine-coated glass bottom of a chamber filled with a 7:3 mixture of Schneider's culture medium (Gibco, Grand Island, NY, USA) for *Drosophila* and fly Ringer solution. The isolated taste neuron was incubated for about 2 h before the experiment.

The labellum isolated from the pupa 0.5–1 day before emergence was used for patch clamping on the sensory process extruded out from the cut end of the taste hair. The taste hair was cut in the middle by scissors and put on the bottom of the chamber containing either fly Ringer or Schneider's medium. The sensory processes had grown out within about 1 h, and some of their tip ends had spherical shapes with a diameter of 2–10 μm. The spherical ends

---

[1] Department of Physics, Faculty of Science, Nagoya University, Nagoya, 464-01 Japan

FIG. 1. Schemes of two recording methods of receptor current and its fluctuation that flows through outer lumen when labellar taste neuron is stimulated. **a** Tip-sidewall method adopted for fluctuation measurement. **b** Tip-sidewall method adopted for measurement of latency of DC receptor current.

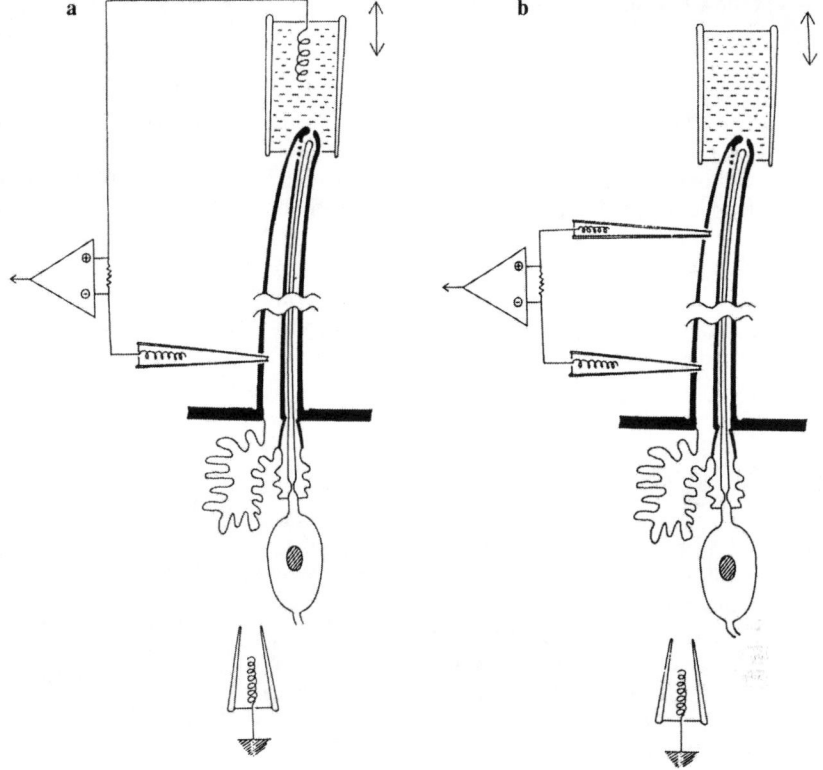

could be patch clamped by glass pipettes with a resistance of about 15 MΩ filled with fly Ringer containing 0.1 M sucrose plus 50 mM fructose.

Results

Very Short Latency of Receptor Current Flow After Stimulation at Tip of Taste Hair

The receptor current that flowed through the outer lumen of the largest type of labellar taste hair was recorded with the two-sidewall method (Fig. 1b). As shown in Fig. 2a, the receptor current was observed as the potential drop of the distal microelectrode with reference to the proximal one [2,4]. When the taste hair tip was stimulated by 1 M sucrose in 0.25 M citrate-Na (pH 6.9), the receptor current developed rapidly. When the time scale was expanded (Fig. 2b), it was clearly observed that the receptor current began to flow about 1.2 ms after the touch of sucrose solution to the taste hair tip (observed as the artifact of electrostatic discharge from the stimulation capillary; thick arrow) and the current reached the steady-state value within 8 ms. The very short response latency, about 1 ms, shows that transduction mechanism via the intracellular second messenger system is very unlikely.

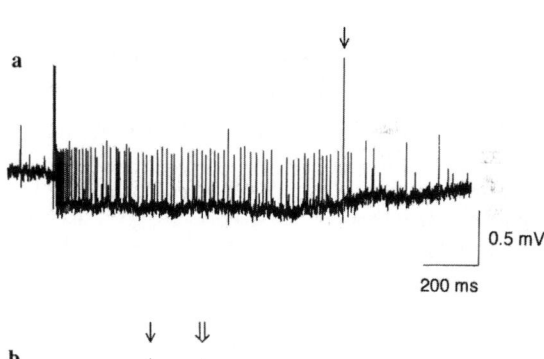

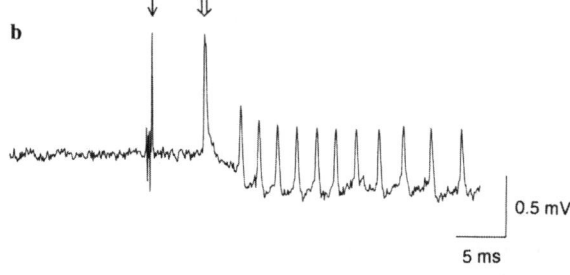

FIG. 2. Receptor current induced by stimulation with 1 M sucrose in 0.25 M citrate-Na solution, recorded by two-sidewall method. Downward voltage change shows current flow in outer lumen from base toward tip. **a** Whole feature of receptor current and sugar receptor impulses. Small spikes were from water receptor. *Arrow* shows artifact from turnoff of electromagnetic switch at end of stimulation. **b** Magnification of time scale at the start of stimulation to show latency. *Thick arrow* shows the artifact of electrostatic discharge from stimulation capillary when it is touched to taste hair tip. *Thin arrow* is artifact of electromagnetic switch when it was turned on

Receptor Current Fluctuation Analysis

The following results were obtained in the former study [5]:

a. The fluctuation developed in parallel with the development of the receptor current. It was observed only when the tetrodotoxin (TTX) treated taste hair was stimulated by effective stimulants for the sugar receptor.

b. The ACFs of the net fluctuation increase were well approximated by a single exponential term, whcih means that the ion channels obey approximately a simple open–shut transition scheme I:

$$ C \underset{\alpha'}{\overset{\beta'}{\rightleftharpoons}} O \qquad \text{scheme I} $$

where α' and β' are apparent rate constants.

c. The variance of fluctuation attained a maximum at a certain sugar concentration, in contrast with monotonic increase of DC receptor current with sugar concentration. The time constant, which shows the inverse of the sum of rate constants of open-and-close transition $(\alpha' + \beta')$, of fructose (5–10 ms) was larger than those of sucrose and maltose (2–4 ms). They all have a tendency to decreases with the increase of sugar concentration, which is consistent with the state transition scheme I of ligand-gated ion channels, in which β' increases with the ligand concentration.

d. The time constants of ACF differed with various sugar species and their concentrations.

These results as a whole indicate that the receptor current fluctuation reflects open–shut dynamics of the sugar-activated transduction ion channels. It is known that there are at least four receptor sites (receptor molecules) with definite specificities on the sugar receptor membrane [6]. We examined the transduction ion channels associated with these four receptor sites (they may form receptor–channel complexes) to determine whether they differed from each other, comparing their single channel currents (i) obtained by the relationship

$$ i = \sigma_I^2 / \{I(1 - p)\} \qquad (1) $$

where σ_I^2 is the variance of the receptor current fluctuation, I is the mean receptor current, and p is the channel open probability. σ_I^2 has the maximum value when p is 0.5. Thus, σ_I^2/I at the stimulant concentration to give the maximum value of σ_I^2 gives the relative magnitude of single channel current i. Instead of I, we measured sugar impulse frequency, which is proportional to I [2]. The following shows the relative values of a single-channel current obtained by using five kinds of stimulants; sucrose (P), maltose (P), fructose (F), L-valine (R), and L-phenylalanine (AR), where the letter in each pair of parentheses

is the receptor site each stimulant activates: sucrose, 1 (standard of relative value); maltose, 1.39 ± 0.24; fructose, 0.33 ± 0.05; L-valine, 0.64 ± 0.09; and L-phenylalanine, 0.59 ± 0.09 $(n = 7)$ (Kazawa et al., manuscript in preparation).

These results suggest that there are at least three different ion channels, which is consistent with the previous result [7] showing that both sucrose and maltose stimulate the pyranose site (P). That the time constants of sucrose and maltose differed from one another, when compared at the sugar concentrations at which the same frequencies of sugar impulses were evoked, suggests that the pyranose site forms the receptor–channel complex. The very short latency after stimulation strongly supports this view. The ion channel associated with the F site had the smallest single-channel current. The other two sites had ion channels with intermediate single-channel currents.

We are now investigating the ion specificities of the transduction ion channels. When addition of an ion into the stimulant sugar solution increases the sugar impulse frequency, there can be two causes: one is that the ion permeates through the ion channel and increases the single-channel current i, and the other possibility is that the ion increases the channel-open probability p. When the stimulant concentration is such as to give the maximum variance of the receptor current fluctuation σ_I^2, addition of ion increases σ_I^2 in the former case while it decreases or causes little change in the latter. Preliminary results showed that the P-site channel can permeate Na^+ and $choline^+$ but not the $Tris^+$ ion.

Patch Clamp Studies

To study the sugar-activated transduction ion channels at the single channel level, the patch clamp study was tried on the two kinds of preparations. One was single taste cell isolated from the labellum of pupa about 2 days before eclosion. The isolated taste cell was bipolar and spindle shaped with a truncated sensory process and an axon. Those cells belonging to the taste hairs were larger (about 8 × 15 μm) than those belonging to the pseudotracheal papillae (about 6 × 10 μm). About one-third of both types of isolated taste cells examined by the cell-attached patch pipette were found to evoke spontaneous action potentials as shown in Fig. 3a. The resting potential was estimated to be from −35 to −50 mV as judged by the zero current voltage at the instant of establishment of the whole-cell clamp. Under the whole-cell current clamp condition, the current injection-produced impulse trains. In two cells, 50 mM sucrose perfusion caused depolarization and impulse generation (Fig. 3b). Under whole-cell voltage clamp condition at −40 mV, three cells

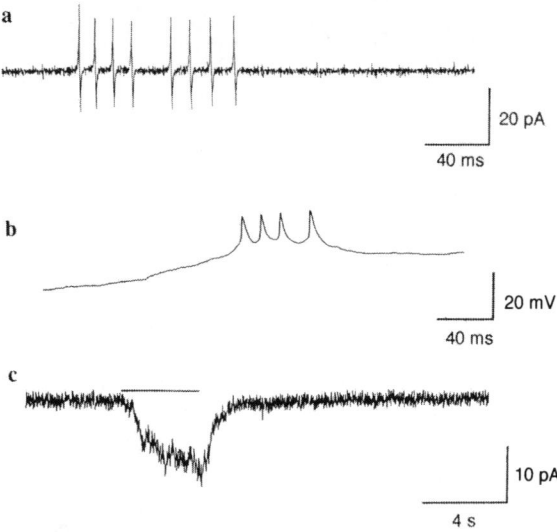

FIG. 3. Patch clamp records on isolated taste neurons obtained from labellum of pupa about 2 days before eclosion. **a** Spontaneous spike potentials of taste neuron recorded by cell-attached patch clamp pipette (voltage in pipette, 0 mV). **b** Depolarization and impulse initiation in whole-cell current clamped taste neuron induced by application of 50 mM sucrose. Perfusion solution was changed from fly Ringer containing 50 mM α-methyl mannoside (nonstimulant) to that containing 50 mM sucrose about 5 s before this record. Pipette potential at start of record was −40 mV at clamp current of −36 pA. **c** Inward current induced by puff application of 50 mM sucrose from pipette about 100 μm downstream of perfusion flow from isolated taste cell; whole cell clamped at −40 mV. *Bar* shows puff application of sucrose

caused inward current of 10–20 pA, responding to puff-applied 50 mM sucrose (Fig. 3c). The reversal potential of these sucrose-induced currents was about +15 mV, suggesting that permeability of Na$^+$ may be larger than that of the K$^+$ ion.

Another preparation for a patch clamp study is the sensory process of taste cells grown out from the cut end of the taste hair on the labellum isolated from the pupa about 12–24 h before eclosion, prepared in a way similar to Zufall and Hatt [8]. We succeeded in cell-attached patch clamp on about 20 sensory processes using the patch pipette filled with fly Ringer solution containing 0.1 M sucrose and 50 mM fructose, and could record the current of ion channels. However, we have not yet found sugar-activated single channels.

Discussion and Conclusion

The results presented here suggest the existence of a few types of sugar- or amino acid-activated

transduction ion channels of a receptor–channel complex type on the sensory process membrane of the labellar sugar receptor neuron. However, many problems remain to be resolved. First, the ionic mechanism and other properties of the transduction ion channels must be elucidated at the single-channel level. Second, the cellular modification mechanisms to these channels such as adaptation must be clarified. Ozaki and Amakawa [9] have reported that Ca^{2+} and IP$_3$ introduced into the taste cell from the detergent-treated end of the sensory process increased the rate of adaptation. Inflow of Ca^{2+} ion may be common in the adaptation process of various types of other sensory cells, such as the photoreceptor cell, the hair cell, and the olfactory cell. Amakawa et al. [10] also found that di-butylyl cGMP stimulated the sugar receptor cell. They suggested that there may be a transduction pathway in which cGMP is included as the second messenger, in parallel with the pathway by the receptor–ion channel complex. The third requirement is to clarify whether the transduction pathways mediated by the second messengers exist in parallel with the pathway of the receptor–ion channel complex.

References

1. Hansen K (1994) Functional organization of insect taste hairs (this volume)
2. Morita H, Shiraishi A (1985) Chemoreception physiology In: Kerkurt GA, Gilbert LI (eds) Insect physiology, biochemistry and pharmacology, vol 6. Pergamon, Oxford, pp 133–170
3. Dethier VG (1990) Chemosensory physiology in an age of transition. Annu Rev Neurosci 13:1–13
4. Morita H (1992) Transduction process and impulse initiation in insect contact chemoreceptor. Zool Sci (Tokyo) 9:1–16
5. Kijima H, Nagata K, Nishiyama A, Morita H (1988) Receptor current fluctuation analysis in the labellar sugar receptor of the fleshfly. J Gen Physiol 91:29–47
6. Shimada I (1994) Multiple receptor sites of insect taste cells (this volume)
7. Shimada I, Shiraishi A, Kijima H, Morita H (1974) Separation of two receptor sites in a single labellar sugar receptor of the fleshfly by treatment with *p*-chloromercuribenzoate. J Insect Physiol 20:605–621
8. Zufall F, Hatt H (1991) Dual activation of a sex pheromone-dependent ion channel from insect olfactory dendrites by protein kinase C activators and cyclic GMP. Proc Natl Acad Sci USA 88:8520–8524
9. Ozaki M, Amakawa T (1992) Adaptation-promoting effect of IP$_3$, Ca^{2+}, and phorbol ester on the sugar taste receptor cell of the blowfly, *Phormia regina*. J Gen Physiol 100:867–879
10. Amakawa T, Ozaki M, Kawata K (1990) Effects of cyclic GMP on the sugar receptor cell of the fly *Phormia regina*. J Insect Physiol 35:233–237

Elementary Receptor Potentials of Insect Olfactory Cells

This report summarizes studies of elementary receptor potentials (ERPs), which are considered as responses of receptor cells to single pheromone molecules except when they occur spontaneously. Possible mechanisms of ERP generation are discussed in view of biochemical and electrophysiological investigations of intracellular messengers in the moth *Antheraea polyphemus*.

Elementary receptor potentials (ERPs) are the first electrical responses to single pheromone molecules that can be recorded extracellularly from olfactory hairs of moth antennae [1,2]. ERPs are transient potential changes reaching a few tenths of a millivolt in transepithelial recordings and lasting several tens of milliseconds. Often they are followed by single action potentials or, sometimes, by groups of two or three action potentials. It is commonly assumed that receptor potentials reflect current flow that elicits action potentials in the inner dendrite and soma region [3]. According to an equivalent circuit diagram of the sensillum, based on morphological [4,5] and electrophysiological studies [2,6], the depolarization of the cell soma membrane during an ERP is expected to be in the range of 1 mV [7], which is much less than needed for axonal generation of action potentials. Most spontaneous action potentials are accompanied by ERPs (unpublished observation), which suggests that receptor molecules can be activated without pheromone stimuli.

ERPs can be observed best with pheromone stimuli eliciting up to a few action potentials per second (Fig. 1) [7,8]. ERPs can appear as discrete "bumps" without, or with, one or two action potentials (Fig. 1a,b). Often two or more bumps occur in groups (Fig. 1c), and sometimes bursts are formed in which single bumps can hardly be distinguished (Fig. 1d,e). Finally, there are bursts in which bumps are superimposed on each other (Fig. 1e).

Clearly these events occur in a very irregular manner; their distribution in time varies from stimulus to stimulus as one would expect for stimuli consisting of a small number of pheromone molecules per sensillum. Summation of many such responses reveals a fluctuating signal that corresponds to the overall receptor potential as obtained with a single stimulus of stronger intensity. This receptor potential can be as large as 30 mV, for example, in transepithelial recordings from sensilla trichodea of the moth *Antheraea polyphemus* [7,9]. Notably, certain pheromone derivatives produce overall receptor potentials that fluctuate much less than those produced by the pheromone [1]. This suggests that the underlying ERPs must be smaller and/or shorter than those produced by the pheromone. Curiously, the time course of the overall receptor potentials induced by these pheromone derivatives shows quicker transients at the beginning and end of stimulation; this means that the latencies of the ERPs caused by the derivatives are shorter than those for the pheromone [1,7].

The aim of this paper is to discuss the locus and mechanism of generation of ERPs, their distribution in time, and their function. The following questions are discussed here:

1. (a) Are ERPs generated by receptor cells or by auxiliary cells of a sensillum? (b) Where is the exact location of the generator of the ERPs?
2. Are ERPs produced by openings of single or several ion channels?
3. Are ERPs the result of direct coupling between pheromone receptor and ion channel or are intracellular messengers involved?
4. Are the latencies of ERPs mainly caused by the transport of the pheromone molecules from the hair surface toward the receptor cell or are they mainly caused by processes following the stimulus arrival at the receptor cell?

Although most of these questions cannot be answered, the current status of the investigations is summarized here.

Question 1a. Transepithelial recordings from single sensilla reflect properties of the receptor cells, mostly two or three per sensillum, but also of auxiliary cells. For instance, the transepithelial potential (TEP) largely depends on the activity of the electrogenic potassium pump located in the apical membranes of auxiliary cells [10]. An argument in favor of the receptor cells as generators of the ERPs is the fact that the time course of the ERP shows characteristic differences for each receptor cell of a sensillum [1,2]. In *Bombyx mori*, the action potentials fired by the bombykal receptor cell are accompanied by much shorter ERPs than the action potentials of the bombykol receptor cell. Similarly, the ERPs of the

[1] Max-Planck-Institut für Verhaltensphysiologie, 82319 Seewiesen/Starnberg, Germany

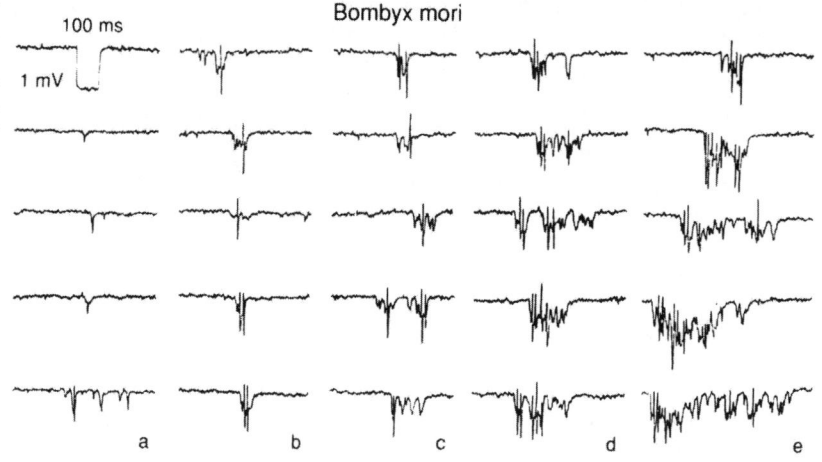

Elementary receptor potentials in response to Bombykal.

two major receptor cell types of *Antheraea polyphemus* differ in duration (unpublished observation).

Question 1b. Most likely the ERPs are generated in the outer dendritic segment of the receptor cell rather than in the inner segment or the soma region. This is expected from their negative polarity; when the membrane conductance of the outer dendrite is increased, the equivalent circuit of the sensillum produces negative potential changes (with the recording electrode in contact with the apical sensillum lymph space) [7]. The receptor cell dendrite is sensitive to pheromone stimuli along its entire length and can be locally adapted by pheromone stimuli [11]. Furthermore, local application of narcotic gases to restricted section of the hair sensillum can suppress a response to pheromone applied at the same locus (G. Stange and K.-E. Kaissling, unpublished data). It remains unclear whether adaptation and suppression act on the receptor site or affect the ion channels responsible for stimulus-induced conductance changes.

Question 2. The equivalent circuit model of the sensillum is compatible with an increase of dendritic membrane conductance in the range of 30 pS for a single ERP [2,7]. This means that ERPs could be generated by opening of single ion channels. It is, however, not excluded that the ERP is produced by simultaneous openings of several ion channels with smaller conductance per channel. This could, for instance, be performed by an intracellular signal cascade. The idea of several but smaller ion channel openings generating the pheromone-induced ERP is compatible with the aforementioned inference that pheromone derivatives cause smaller ERPs.

Interestingly, second messengers were found to open ion channels of 56 pS in inside-out patches excised from dendrites of pheromone receptor cells in the moth *Antheraea polyphemus* [12]. These channels were opened after stimulation with micro-

molar concentrations of cGMP or dioctanoyl glycerol (DOG), an analog of diacyl glycerol (DAG), and in the presence of MgATP. However, the characteristics of these channels measured in cell-attached patches after pheromone stimulation were different from the characteristics of the ERPs. The latter potentials have average durations of 10 ms or more whereas the open times of the ion channels in patch damp studies had two much shorter time constants (0.14 $\pm$ 0.02 ms and 1.48 $\pm$ 0.21 ms, respectively) and a mean burst length of 5.15 $\pm$ 1.42 ms [12]. Therefore, it seems to early to conclude that single channel openings of this type underly the ERP. It would also be difficult to understand if an intracellular cascade serves to open a single ion channel only. The possibility exists that the 56 pS channels have much smaller conductances in situ, for instance, because of partial blocking by Ca^{2+} [19] and that several such channels cooperate for inducing an ERP.

Question 3. The patch clamp experiments mentioned previously, as well as recent biochemical studies, suggest an important role of intracellular signal cascades for transduction in pheromone receptor cells. Two second messengers were formed on pheromone stimulation in antennal preparations of male moths of *A. polyphemus*.

First, an increase of cGMP concentration (but not of cAMP) was observed in antennae of living *A. polyphemus* after pheromone application for 1 to 60 s [13]. Because no cGMP increase was found in hairs isolated after pheromone stimulation, very likely cGMP is formed in the receptor cell soma only. Assuming that the total increase of cGMP (by 34%) occurred only in the pheromone receptor cells (A cells), the average increase was calculated to be about 3.6 μM (or 5×10^5 cGMP molecules per cell) [13]. This increment of cGMP corresponds to the concentration eliciting a just detectable response (opening of the 56 pS channels) in the patch clamp experiments [12]. Interestingly, cGMP (10 µM) effectively suppressed opening of Ca^{2+}-activated nonspecific cation (CAN) channels (48 ± 1.5 pS) found near and at the cell soma region [14].

Second, antennal homogenates of male *A. polyphemus* mixed with the sex pheromone component E,Z-6,11-hexadecadienyl acetate (1 pM) in stopped flow experiments produced a rapid and transient IP_3 response [15]. This pheromone component is perceived by one of the two or three receptor cells in each sensillum trichodeum (A cell) [16]. The kinetics of the IP_3 signal were very similar to those observed in cockroach and vertebrate preparations [17]; IP_3 rose from the basal level (80 pmol/mg protein) to a concentration of 400 pmol/mg protein within 25 msec [15]. After 50 ms, the IP_3 signal had already decayed to a lower level. No significant changes in the IP_3 concentrations were detected when living antennae were exposed to continuous pheromone stimuli for 10 s, although the receptor potential remained elevated during such stimuli [18]. This observation confirms the transient nature of the IP_3 response.

Based on these data it can be calculated how many IP_3 molecules were formed per hit of pheromone molecule within 25 ms using the following formula [15]:

$$IP_3 \text{ molecules/pheromone hit} \times 25 \text{ ms} = \frac{R_{IP_3} \times P_{ant}}{c_{ph} \times D_{ph} \times 4\pi a_v \times N_v}$$

Where R_{IP_3} (= 320 pmol IP_3 per mg protein and per 25 ms) is the measured maximum rate of IP_3 formation [15], P_{ant} (= 50 µg protein) is the amount of protein per antenna [13], c_{ph} (= 1 pM) is the initial pheromone concentration after mixing with the antennal homogenate [15], D_{ph} (= 10^{-5} cm²/s) is the assumed diffusion coefficient of the pheromone in the homogenate, a_v (= 0.5 mm) is the radius of dendritic vesicles if one assumes that the mechanical treatment during homogenization causes the dendrite of the receptor cell to disintegrate into 100 pieces, each forming a membrane vesicle [15], and N_v (= 6×10^6) is the assumed number of such vesicles formed per antenna (for 6×10^4 cells sensitive to the pheromone E,Z-6,11-hexadecadienyl acetate [16]).

The formula represents the situation in which the pheromone molecules in the homogenate need a finite time to diffuse to the sensitive dendritic membrane, assuming they are reflected by any targets. According to this calculation, 1.7×10^7 IP_3 molecules are formed per pheromone molecule within 25 msec [15]. This corresponds to a rate of 6.8×10^8 IP_3 molecules per pheromone molecule and per second. For comparison, one light quantum leads to hydrolysis of several tens of thousands of cGMP molecules in visual cells at a maximum rate of 10^6 s (before onset of adaptation processes) [19].

In an alternative situation, every pheromone molecule is assumed to be adsorbed at any membrane in the homogenate already during the mixing phase. The number of IP_3 molecules formed per each pheromone molecule adsorbed by a piece of pheromone-sensitive membrane of the receptor cells would then be 1×10^7 per 25 milliseconds (or 4×10^8 per second).

An amplification more than a hundredfold higher in olfactory cells than calculated for the transduction cascade in visual cells seems difficult to interpret. For the unlikely case that all pheromone molecules in the homogenate hit pheromone-sensitive membrane the amplification would be 100 fold smaller and still be greater than that for visual transduction. It seems unlikely that the enormous IP_3 production per pheromone molecule is used for generating a single ERP. IP_3 would reach 100 µM concentrations if 1.7×10^7 IP_3 molecules are formed per pheromone molecule in one receptor cell (volume, 240 µm³) [4,5], but micromolar concentrations of IP_3 were ineffective in patch clamp recordings [12]. An alternative function of IP_3 could be transient opening of the aforementioned CAN channels via Ca^{2+} release in the cell soma region [14,15].

In addition to cGMP and IP_3, DAG might play a role, as it is formed by phospholipase C at the same rate as IP_3. Less than micromolar concentrations of the analogue DOG caused 56 pS channel openings in excised inside-out patches. This concentration is astonishingly low if the in situ rate of DAG formation after pheromone stimuli corresponds to that found for IP_3 in the antennal homogenate. With 100 µM concentrations of DAG, one would expect substantial depolarization of the receptor cell unless the channel conductance in situ is much smaller than 56 pS.

Question 4. ERPs occur after pheromone stimulation with average latencies of several 100 ms [7,8]; their distribution, however, varies over a wide range. Latencies can be less than 10 ms as implied by the latency of receptor potentials at extremely strong stimulus intensity [15]; they also can reach several seconds. If the latencies mainly represent the time the pheromone molecules need for the transport from the hair surface to the receptor cell membrane, their diffusion coefficient would be in the range of 5×10^{-9} cm²/s [2,7]. This implies that they would migrate about as slowly as molecules in cell membranes. On the other hand, the latency could mainly represent delays caused by a second messenger cascade. So far, there is no direct experimental evidence to answer this question.

It is evident that our understanding of the ERPs and the role of intracellular messengers in *Antheraea* pheromone receptors is less than satisfactory. In addition to a signal cascade, direct coupling of pheromone receptors to ion channels seems possible. This could involve a special type of G-protein, as discussed for various transmitter receptors and their action on K^+ and Ca^{2+} channels (see ref. [20]). In conclusion, it seems clear that several mechanisms in each receptor cell are involved in olfactory transduction in moths. It should be noted that even related species show differences with respect to second messenger action [21].

References

1. Kaissling KE (1977) Structures of odour molecules and multiple activities of receptor cells. In: LeMagnen J, MacLeod P (eds) Int Symp Olfaction and Taste, information retrieval, London, pp 9–16
2. Kaissling KE, Thorson J (1980) Insect olfactory sensilla: structural, chemical and electrical aspects of the functional organization. In: Sattelle DB, Hall LM, Hildebrand JG (eds) Receptors for transmitters, hormones and pheromones in Insects. Elsevier, Amsterdam, pp 261–282
3. Morita H (1972) Primary processes of insect chemoreception. Adv Biophys 3:161–198
4. Keil TA (1984) Reconstruction and morphometry of silkmoth olfactory hairs: a comparative study of sensilla trichodea on the antennae of male *Antheraea polyphemus* and *Antheraea pernyi* (Insecta, Lepidoptera). Zoomorphology (Berl) 104:147–156
5. Gnatzy W, Mohren W, Steinbrecht RA (1984) Pheromone receptors of *Bombyx mori* and *Antheraea pernyi*. II. Morphometric data. Cell Tissue Res 235:35–42
6. DeKramer JJ (1985) The electrical circuitry of an olfactory sensillum in *Antheraea polyphemus*. J Neurosci 5:2484–2493
7. Kaissling KE (1987) R.H. Wright lectures on insect olfaction, Colbow K (ed) Simon Fraser University, Burnaby, B.C., Canada
8. Kaissling KE (1986) Chemo-electrical transduction in insect olfactory receptors. Annu Rev Neurosci 9: 121–145
9. Zack C (1979) Sensory adaptation in the sex pheromone receptor cells of saturniid moths. Dissertation Fak Biol Ludwig-Maximilians-Universität, Munich
10. Thurm U, Kueppers J (1980) Epithelial physiology of insect sensilla. In: Locke M, Smith DS (eds) Insect

biology in the future. Academic, New York, pp 735–763
11. Zack Strausfeld C, Kaissling KE (1986) Localized adaptation processes in olfactory sensilla of saturniid moths. Chem Senses 11:499–512
12. Zufall F, Hatt H (1991) Dual activation of sex pheromone-dependent ion channel from insect olfactory dendrites by protein kinase C activators and cyclic GMP. Proc Natl Acad Sci USA 88:8520–8524
13. Ziegelberger G, Van den Berg MJ, Kaissling KE, Klumpp S, Schulz JE (1990) Cyclic GMP levels and guanylate cyclase activity in pheromone sensitive antennae of the silk moth *Antheraea polyphemus* and *Bombyx mori*. J Neurosci 10:1217–1225
14. Zufall F, Hatt H, Keil TA (1991) A calcium-activated nonspecific cation channel from olfactory receptor neurons of the silkmoth *Antheraea polyphemus*. J Exp Biol 161:455–468
15. Kaissling KE, Boekhoff I (1993) Transduction and intracellular messengers in pheromone receptor cells of the moth *Antheraea polyphemus*. In: Wiese K, Gribakin FG, Popov AV, Renninger G (eds) Sensory systems of arthropods. Birkhäuser, Basel, pp 489–502
16. Meng LZ, Wu CH, Wicklein M, Kaissling KE, Bestmann HJ (1989) Number and sensitivity of three types of pheromone receptor cells in *Antheraea pernyi* and *Antheraea polyphemus*. J Comp Physiol 165: 139–146
17. Breer H, Boekhoff I, Tareilus E (1990) Rapid kinetics of second messenger formation in olfactory transduction. Nature (Lond) 345:65–68
18. Ziegelberger G (1990) Cyclic nucleotides in the olfactory system of the moths *Antheraea polyphemus* and *Bombyx mori*. In: Doving K (ed) International Symposium on Olfaction and Taste X, GCS A/S, Oslo, pp 85–91
19. Kaupp UB, Koch KW (1992) Role of cGMP and Ca^{2+} in vertebrate photoreceptor excitation and adaptation. Annu Rev Physiol 54:153–175
20. Brown AM, Birnbaumer L (1990) Ionic channels and their regulation by G protein subunits. Annu Rev Physiol 52:197–213
21. Stengl M (1993) Intracellular-messenger-mediated cation channels in cultured olfactory receptor neurons. J Exp Biol 178:125–147

Two Types of Sugar-Binding Protein in the Labellum of the Fly: Putative Taste Receptor Molecules for Sweetness

MAMIKO OZAKI, TAISAKU AMAKAWA, KOICHI OZAKI, and FUMIO TOKUNAGA[1]

Key words. Sweetness—Taste receptor protein

Flies have taste receptor cells that are particularly sensitive to sweetness. It has been suggested that these cells possess two types of receptor sites, a pyranose (P site) and a furanose site (F site), covering the receptive field of sweetness [1]. The pyranose and furanose sites were competitively inhibited by a starch, a polysaccharide consisting of glucopynanose and by levan, and a polysaccharide consisting of fructofuranose, respectively [2]. By affinity electrophoresis with these site-specific inhibitory polysaccharides, two types of sugar binding protein corresponding to the sugar receptor sites were isolated from a labellar extract of the fly [3].

Since polysaccharides are huge molecules with no electric charges, they themselves do not electrophoretically migrate in polyacrylamide gel. However, protein interactive with starch (P protein) or levan (F protein) was retarded in migration speed in polyacrylamide gel uniformly containing the polysaccharide and was easily separated from other proteins. The dissociation constant from starch or levan for either protein was identical to the inhibition constant of starch or levan for the response of the sugar receptor cell, respectively.

Sugars have affinities with these proteins in the following order; sucrose>maltosel>L-fucose>D-glucose≫D-fructose, and D-fucose for the P protein, and D-fucose>D-fructose≫sucrose, L-fucose, and D-glucose for the F protein. The corresponding receptor sites showed exactly the same order of electrophysiological sensitivity to these sugars.

The dissociation constant of the protein-sugar complex varied between 100 and 400 mM. According to the calculation based on the dissociation constant, the free energy change for the protein-sugar complex formation was estimated to be -1.3 to -0.5 kcal. This energy change corresponds to the partial affinity force due to one or two hydrogen bonds. Such a weak receptor-stimulant interaction occurs so that the animal can readily recognize the newly coming stimulant by washing the receptor membrane with fluid or sensillar lymph. The molecular weight of the P protein is 31 000 or 32 000, while that of the F protein is 27 000. Both proteins were water-insoluble and they were also detected in the isolated chemosensillum, which exclusively contains the sensory processes, i.e., the receptive region of the taste cells, as the cellular components. Thus, these sugar-binding proteins are likely to act as the sugar taste receptor molecules in the fly.

Acknowledgments. This work was supported by a grant from Nissan Science Fund to M. Ozaki.

References

1. Shimada I, Shiraishi A, Kijima H, Morita H (1974) Separation of two receptor sites in a single labellar sugar receptor of the fleshfly. J Insect Physiol 20: 605–621
2. Hara M (maiden name of Ozaki M) (1983) Competition of polysaccharides with sugar for the pyranose and the furanose sites in the labellar sugar receptor cell of the blowfly, *Phormia regina*. J Insect Physiol 29:113–118
3. Ozaki M (1988) A possible sugar receptor protein found in the labellum receptor of the blowfly, *Phormia regina*. Zool Sci 5:281–290

[1] Department of Biology, Faculty of Science, Osaka University, 1-1 Machikaneyama-cho, Toyonaka, Osaka, 560 Japan

Main Soluble Protein in the Taste Organ of the Fly: Localization and cDNA Sequence

Kazuyo Morisaki, Mamiko Ozaki, Koichi Ozaki, and Fumio Tokunaga[1]

Key words. Taste—Fly

The isolation and characterization of odorant binding proteins (OBPs) in vertebrates and pheromone binding proteins (PBPs) in insects led to the hypothesis that these proteins function as necessary cofactors in olfactory transduction by delivering lipophilic stimulants to the receptors. The expression of soluble proteins similar to OBPs (VEG proteins) has been found in the von Ebner's glands of vertebrates; their role in taste transduction was estimated by homology to OBPs [1].

We isolated an acidic soluble protein from the taste sensillum of the fly and determined the primary structure from the cDNA. It showed some homology to the insect PBPs. The taste organ, as well as the olfactory organ, of insects is a type of sensillum. The sensillum is filled with the receptor lymph and contains the sensory processes of the taste cells. Taste sensillar protein (TSP) was abundant in the isolated taste sensillum; its distribution was in proportion to the distribution of the taste sensillum. Thus, the TSP seemed to be a major component of the receptor lymph perfusing the receptor membrane surface, acting as a cofactor in insect taste transduction. The fly might use a pheromone mediated by the TSP and detected by the taste receptor, or the fly taste cell may respond to some fatty acids, which may be carried by the TSP to the receptor.

The molecular weight of the TSP, calculated from the amino acid composition, was 13 600; that determined by mass spectrometry was 13 900. This difference was due to phosphorylation. By SDS-PAGE with β-melcaptoethanol, however, the apparent molecular weight was 25 000, while it was 10 000 without β-melcaptoethanol. This suggested that the S—S bonds are indispensable in forming the functional structure of TSP. Actually, the TSP has six cysteine residues whose positions are conserved in the PBPs. The isoelectric point of TSP was estimated to be 4.5, from its amino acid composition, 25% of which was acidic residues.

Comparing the primary structure of insect PBPs and vertebrate OBPs reveals no significant similarities [2]. It is presumed that insect and vertebrate odorant binding proteins are, rather, functional analogues, and represent an evolutionary convergence. The TSP has some considerable homology to PBPs, but not to the VEG proteins. Thus, this study confirmed the evolutionary convergence between insects and vertebrates in regard to taste soluble proteins.

Acknowledgments. We would like to thank Professor Takekiyo Matsuo for his help with the mass spectrometry and Mr. Wataru Idei for his help with SDS-PAGE.

This work was supported by a grant from the Nissan Science Fund (to Mamiko Ozaki).

References

1. Shumale H, Holtgreve-Grez H, Christensen H (1990) Possible role for salivary gland protein in taste reception indicated homology to lipophilic-ligand carrier proteins. Nature 343:366–369
2. Krieger J, Raming K, Breer H (1991) Cloning of genomic and complementary DNA encoding insect pheromone binding proteins: Evidence for microdiversity. Biochim Biophys Acta 1008:277–284

---

[1] Department of Biology, Faculty of Science, Osaka University, 1-1 Machikaneyama-cho, Toyonaka, Osaka, 560 Japan

Receptor Current Fluctuation Analysis on the Taste Receptor of the Fleshfly

Tomoki Kazawa and Hiromasa Kijima[1]

Key words. Taste—Insect

The receptor current of labellar taste cells in the fleshfly, *Boettcherisca peregrina*, was recorded as a voltage drop between the taste hair tip and an electrode inserted into the outer lumen of the largest-type taste hairs. The fluctuations of this current reflected the open-shut transitions of the transduction ion channels at the tip of the sensory process of the sugar receptor, since the fluctuation increased markedly only when effective stimulants were applied to the hair. The fluctuations, recorded alternately coupled (AC) after the taste hair tip was treated with tetrodotoxin (TTX) to suppress impulse generation, were analyzed by computing autocorrelation functions. Net fluctuations in the transduction ion channel were obtained as differences from the autocorrelation functions of the control fluctuations when the solutions without stimulants were applied.

It is known that there are four or more different receptor sites on the labellar sugar receptor membrane: a P(pyranose) site, F(furanose) site, R(alkyl) site, and an Ar(aryl) site [1]. Five stimulants were tested for the four receptor sites: maltose (reacts with P site), sucrose (P), fructose (F), L-valine (R), and L-phenylalanine (Ar).

Stimulants were dissolved in 0.1 M Na citrate (Na^+ 0.28 M, pH 6.9) to suppress the response for water. Autocorrelation functions were approximately fitted with a single exponential term. This means that the channels obey approximately a simple two-state transiton scheme between an open and a shut state. Then the autocorrelation function $C_I(t)$ is expressed as

$$C_I(t) = \sigma^2_I exp(-t/\tau), \qquad \sigma^2_I = i \langle I \rangle (1 - p),$$

where σ^2_I is the variance of fluctuation, τ the time constant, i the single channel current, p the open probability, and $\langle I \rangle = iNp$ the mean receptor current where N is a number of channels.

In this scheme, σ^2_I changes little with a small change of p where it has the maximum value. There-fore, recording maximum fluctuations is easy and useful for obtaining information about transduction ion channels.

The maximum fluctuations induced by stimulants which activated each of the four sweet receptor sites were then recorded, respectively. Impulse frequency, which was proportional to the amplitude of receptor currents, was recorded before TTX treatment. Relative values of single ion currents were calculated from σ^2_I and relative values of $\langle I \rangle$. The results were: sucrose 1 (standard of relative value); maltose 1.39 ± 0.24 (n = 6); fructose 0.33 ± 0.05 (n = 7); L-valine 0.64 ± 0.99 (n = 6); and L-phenylalanine 0.59 ± 0.09 (n = 13), suggesting that there are at least three different ion channels on the sugar receptor membrane. If each of these ion channels is a ligand-gated ion channel, there being some evidence for this (H. Kijima et al., Transduction Ion Channels in Insect Taste Cells, this volume), it is natural to assume that these four receptor sites are different ligand-gated channel molecules.

To investigate this idea, mixtures of four stimulants, except for maltose were tested. If every site is an independent receptor-channel complex molecule, the variance for the mixture should be the sum of variance for each stimulant. The values of $\sigma^2_I(\times 10^{-4} mV^2)$ were: for sucrose 25 mM[S25], 19.6 ± 3.1 (n = 7); for fructose 100 mM[F100], 3.7 ± 0.5 (n = 6); for L-phenylalanine 2.5 mM[P2.5], 5.3 ± 0.4; for L-valine 5 mM[V5], 4.4 ± 0.9 (n = 7); for S25 and P2.5, 15.1 ± 2.5 (n = 7); for S25 and F100, 9.1 ± 1.1; for S25 and V5 (n = 7), 14.4 ± 3.0; for P2.5 and V5, 4.46 ± 1.9; for P2.5 and F100, 3.4 ± 1.0; for F100 and V5, 4.4 ± 1.2; for P2.5 and F100 and V5, 3.1 ± 1.1 (n = 7). In contrast to our prediction, variances for mixtures were reduced, indicating that each site is not a completely independent receptor-channel complex molecule. Details of these interactions will be studied.

References

1. Shimada I (1987) Stereospecificity of the multiple receptor sites in the sugar receptor cell of the fleshfly. Chemical Senses 12:235–244

[1] Department of Physics, Faculty of Science, Nagoya University, Furo-cho, Chikusa-ku, Nagoya, 464-01 Japan

The Effects of Cyclic Nucleotides on the Labellar Salt Receptor and Its Adaptation in the Fleshfly

MASAYUKI KOGANEZAWA and ICHIRO SHIMADA[1]

Key words. Insect taste—Second messenger

It has recently been suggested that cyclic nucleotides play an important role in taste cell transduction in vertebrates. Amakawa et al. [1] found that dibutyryl cyclic GMP (dbcGMP) had a stimulatory effect on the labellar sugar receptor cell of the blowfly; they suggested that cyclic GMP (cGMP) played an important role as a second messenger in sugar taste cell transduction. Why was the effect, then, restricted to the sugar receptor? Kijma et al. [2] suggested the presence of sugar-gated ion channels, based on receptor current fluctuation analysis. These findings have raised the question: Which type of receptor molecule is really necessary for taste cell transduction?

We tested the effects of several nucleotides on the taste receptor cell of the fleshfly. Most nucleotides we examined had stimulatory effects on the salt receptor cell. DbcGMP was more effective than cGMP, the magnitude of the response to DbcGMP being nearly equal to the sum of cGMP and 8-bromo cyclic GMP (8bcGMP) at 100 mM. The response to

dbcGMP was, then, interpreted to have two components, that is, membrane impermeable cGMP type, and permeable 8bcGMP type. 8bcGMP elicited responses at very low concentration, around $50\,\mu M$, which may be compatible with its permeability and possible function as a second messenger. Further, the slow adaptation curve for dbcGMP can also be regarded as consisting of two components, adaptive cGMP and slow adaptive 8bcGMP. The remarkable difference in responses to 8bcGMP and 8bcAMP suggests that cGMP, and not cAMP, plays a role as a second messenger in the salt receptor cell of the fly. Our results with the salt receptor, as a whole, resemble those for the sugar receptor of the blowfly reported by Amakawa et al. [1]. These findings suggest that cGMP may play a common role as a second messenger in taste cell transduction in the fly.

References

1. Amakawa T, Ozaki M, Kawata K (1990) Effects of cyclic GMP on the sugar taste receptor cell of the fly *Phormia regina*. J Insect Physiol 36:281–286
2. Kijima H, Nagata K, Nishiyama A, Morita H (1988) Receptor current fluctuation analysis in the labellar sugar receptor of the fleshfly. J Gen Physiol 91:29–47

[1] Biological Institute, Faculty of Science, Tohoku University, Kawauchi, Aoba-ku, Sendai, 980 Japan

Taste Cells Isolated from Adult Labella of the Blowfly

Taisaku Amakawa, Kaori Tanaka, and Kyoko Hirota[1]

Key words. Taste cell—Fly

Yamane and Kijima first succeeded in isolating taste cells from the pupal labella of the fleshfly and found ion-channels regulated by sugar [1]. The results of our previous tip-recording experiments showed that most of the taste cells of the pupal labella barely adapted to a sugar or salt stimulus and their responses were unstable [2]. The taste cells of the adult fly seemed more suitable for the study of the adaptation mechanism, although as taste cells are covered with thecogen cells which are tough and not easily ruptured, few cells were isolated. We did, however, obtain taste cells from the labella of the adult fly.

The labellar lobes of an adult blowfly (*Phormia regina*), about 1–8-days-old, were dissected and incubated in protease/chitinase solution for 45 min at 30°C, according to the method of Yamane and Kijima [1] with some modifications. A mixture of collagenase, dispase (1 mg/ml; Boehringer, Mannheim, Germany) and chitinase (2 mg/ml; Sigma, St. Louis, MO, USA) was used. Brief sonication treatment (45 kHz 100 w; 1–2 s) was carried out before the incubation. After this enzyme treatment, the samples were soaked in Ca^{2+}-free Waterhouse's solution for 10 min. Each lobe was then slipped onto a collagen-coated petri dish (60-mm diameter; Koken, Tokyo, Japan) so as to disperse the taste cells on the collagen membrane. The cells isolated were cultured in Schneider's solution for more than 2 h. Patch-clamp experiments were carried out at the taste cell body [3]. In some experiments, isolated cells were preincubated with saline solution for 5 min, then stained with nitroblue tetrazolium chloride (NBT) to visualize the metabolically active cells [4].

By our simple method described above, we obtained intact taste cells. The dimensions of the large cells were $8 \times 5\,\mu m$. Each cell had a large nucleus. Without sonication, the yield of isolated cells was reduced to one-third to one-half of that for the isolation method including sonication treatment, and those isolated had a rough surface at which gigaseal formation was usually difficult.

Several types of ion channels were found by cell-attached or inside-out patch-clamp. There were possible anomalous K-channels, voltage-dependent cation channels (inward- or outward-going current), and maxi K-channels (Ca^{2+}-dependent K-channels) which last was the most commonly seen (210 pS calculated from the I–V curve).

Not all the cells were stained with NBT when preincubated with 0.5 M NaCl. This NBT method seems to be suitable for identifying receptor cells of different qualities, as has been determined for photoreceptor cells [4].

Thus, the isolated taste cells showed physiological response such as ion-channel gating and metabolic activity detected by the NBT method. This preparation would be useful for further studies of insect taste reception.

References

1. Yamane S, Kijima H (1989) Sugar response of the labellar taste cell isolated from the fleshfly pupa. JASTS 23:95–98
2. Miki T, Amakawa T (1992) Responsiveness of the blowfly taste organ before and after eclosion (in Japanese). Seibutubuturi 30:[Suppl] S105
3. Hamill OP, Marty A, Nehr E, Sakmann B, Sigworth FJ (1981) Improved patch-clamp techniques for high resolution current recording from cells and cell-free membrane patches. Pflugers Arch 391:85–100
4. Marc RE, Sperling HG (1976) Color receptor identities of goldfish cones. Science 191:487–489

[1] Division of Sciences for Natural Environment, Department of Human Environment, Faculty of Human Development, Kobe University, 3-11 Tsurukabuto, Nada-ku, Kobe, 657 Japan

Oviposition Behavior of Insects Influenced by Chemoreceptors

ERICH STÄDLER[1]

Key words. Oviposition—Behavior—Insects—Chemoreception—Contact-chemoreception—Olfaction

Ecological Relevance of Oviposition

The choice of the ovipositing site by the female is very important for the survival of a species because most insect larvae have a very limited capability to locate alternative host plants. Several investigations have shown that the progeny develops best on host plants chosen by their mother insect (reviewed by [1]). Visual and mechanical properties can influence the oviposition behavior, but chemicals of the environment seem in almost all insects to be of paramount importance. Since nutritional and allelochemical compounds determine successful development, it is not surprising that chemoreceptors play a crucial role during oviposition.

Our knowledge about chemoreceptors of ovipositing females has long been very limited, as pointed out by earlier reviewers [2,3]. This may seem unimportant, because studies of other receptors involved in the perception of sexual pheromones, feeding stimulants, or deterrents are more numerous. General physiological principles are probably similar for all chemoreceptors, but the perception of host plants or oviposition sites is clearly more complex than, for example, that of sexual pheromones. Characteristic ecological features that most likely have an influence on oviposition behavior and chemoreception are: (a) interaction between different species and ecological levels (exception: marking pheromones); (b) perception of undamaged, dry leaf surfaces in most species; and (c) specialization of the majority of herbivore insects for specific plants or families.

Hypotheses About Chemoreception

Several partially mutually exclusive hypotheses about the nature of stimuli and receptors have been put forward in the past to explain the influence of chemical compounds on oviposition in general and host plant selection in particular. The behavioral aspects have been reviewed recently by [1] and the related chemoreception by [4]. The antithetical views can be summarized as follows:

Hypothesis 1. (a) Secondary plant compounds, which can be attractants, repellents, stimulants, or deterrents, are crucial in host-plant selection, and labelled-line coding is the "adapted" chemoreception mechanism; (b) Qualitatively and quantitatively, nutrients play only a minor role.

Hypothesis 2. (a) Host-plant recognition even in oligophagous (specialized) insects does *not* depend on the presence of secondary chemicals characteristic of the host taxa; (b) primary metabolites are part of complex mixtures and across-fiber pattern coding is the most likely sensory mechanism.

Chemoreceptors Involved in Oviposition

The data listed in Tables 1 and 2 show that in the last ten years significant progress has been made in the identification of compounds of plant and insect origin which influence oviposition behavior and the corresponding receptors. The results can be summarized as follows:

1. High specificity (labelled-line coding) has been found in some of the receptors for marking pheromones (Fig. 1). This is analogous to the sexual pheromones and underlines the importance of intraspecific communication.

2. Secondary plant compounds are important for oviposition and host-plant selection both as attractants/stimulants and repellents/deterrents. Depending on the species the compounds may act as volatiles and/or as contact-stimulants. Remarkable receptor specificity for some compounds has been reported.

3. Primary plant metabolites have now also been proven to play a role in the oviposition of some species. Receptors for these compounds were identified in polyphagous insects but not (yet?) in more specialized species.

4. The sensitivity of some contact-chemoreceptors was, contrary to the prevailing views, shown to be very high with thresholds estimated to be as low as $\sim 10^{-11} M$ (Fig. 1). This is many orders of magnitude

[1] Eidgenössische Forschungsanstalt, Schloss 334, CH 8820 Wädenswil, Switzerland

TABLE 1. Review of compounds of host environment influencing oviposition behavior and receptors identified.

Source	Compounds	Behavioral reaction[1]	Species and organs	Sensitivity threshold[2]	Authors
Cruciferae	Allylisothiocyanates	Attraction, partially stimulation	*Delia radicum* (Dipt., Anthomyiidae) antennae	EAG: 0.2 µg	[5]
	Glucobrassicin, glucobrassicanapin . . .	Stimulation	*Delia floralis* (Dipt., Anthomyiidae) tarsi: 1 cell, proboscis: 1 cell	$10^{-9} M$	[6]
	Glucobrassicin, glucobrassicanapin . . .	Stimulation	*Delia radicum* tarsi: 1-cell	$10^{-9} M$	[7]
	Glucotropaeolin		*Pieris brassicae* (Lep., Pieridae) tarsi: 1 cell	$10^{-5} M$	[8]
	Glucobrassicin, gluconasturtiin . . .		*P. rapae* (Lep., Pieridae) tarsi: 2 cells	$<10^{-7} M$	[9]
	"CIF" (tentative name until structure confirmed by a synthesis)	Stimulation	*Delia radicum* (Dipt., Anthomyiidae) tarsi: 1 cell	$\sim 10^{-11} M$	[10]
Cruciferae (nonhosts)	Cardenolide glycosides: Erychroside	Deterrency	*Pieris brassicae* (Lep., Pieridae) tarsi: 1 cell	$\ll 7 \times 10^{-7} M$	[11,12]
	Cardenolide glycosides: Erychroside, . . .	Deterrency	*P. rapae* (Lep., Pieridae) tarsi: 1 cell	$\sim 10^{-7} M$	[9]
Gramineae	Terpenoids: phenyl-acetaldehyde . . .	Attraction, stimulation	*Ostrinia nubilalis* (Lep., Pyralidae) antennae	EAG: $\ll 200$ µg	[13,14]
	Fructose, malic acid	Stimulation	*Ostrinia nubilalis* (Lep., Pyralidae) ovipositor: 1 cell?	$10^{-4} M$	[15] Baur et al., unpublished
	Sucrose	Stimulation	*Maruca testulalis* (Lep., Pyralidae) ovipositor	$10^{-3} M$	[16]
Liliaceae	Dipropyl-disulfide	Attraction, stimulation	*Delia antiqua* (Dipt., Anthomyiidae) antennae	EAG: 0.1 µg	[17]
	Propyl propane, thiosulfinate . . .	Attraction, stimulation	*Acrolepiopsis assectella* (Lep., Yponomeutidae) antennae	EAG: $4 \times 10^{-8} M$ (solution)	[18]
Pinaceae	Linalool, myrcene	Attraction, stimulation	*Ips pini* (Col., Scolytidae)	source: 0.5 µg	[19]
	α-pinene, β-pinene, limonene, myrcene	Attraction, stimulation	*Hylobius abietis* (Col., Curculionidae), antenna: several relatively specific cells	GC-single sensilla: $\ll 5$ ng	[20]
Pinaceae + Dendroctonus pheromone	α-pinene, ipsenol	Attraction? stimulation	*Coeloides pissodis*, (Hym., Braconidae)	EAG: 0.5 µg	[21]
Rosaceae	Propyl hexanoate . . .	Attraction, stimulation	*Rhagoletis pomonella* (Dipt., Tephritidae) antenna	EAG: 0.1 µg	[22]
	Guaiacol, methylbenzoate, 1-phenyl-1,2-propanedione	Attraction, stimulation	*Synanthedon pictipes* (Lep., Sesiidae)	EAG: 0.1 µg	[23]
	Glucose, malic acid	Stimulation	*Rhagoletis pomonella* (Dipt., Tephritidae) ovipositor	? $10^{-2} M$	[24]
Rutaceae	Naringin	Stimulation	*Papilio protenor* (Lep., Papilionidae) tarsi	?	[25]

TABLE 1. *Continued*

Source	Compounds	Behavioral reaction[1]	Species and organs	Sensitivity threshold[2]	Authors
Umbelliferae	Propenylbenzenes, furanocoumarins, polyacetylenes	Attraction, stimulation	*Psila rosae* (Dipt., Psilidae) antennae	EAG: ~1 µg	[26]
	4-terpineol, bornyl acetate, (*E,Z*)-sabinene hydrate . . .	Attraction, stimulation	*Papilio polyxenes* (Lep., Papilionidae) antennae	GC-EAG: ≪5 ng	[27]
	7-*O*-(6″-*O*-malonyl)-β-D-glucopyranoside, . . .	Stimulation	*Papilio polyxenes* (Lep., Papilionidae) tarsus: 1 cell, others for other compounds	~10^{-6} M	[28] Städler and Schöni, unpublished

[1] Behavioral reaction: as defined by [34].
[2] Sensitivity threshold: lowest dose at which significant responses were estimated or measured.
. . . , additional compounds were tested and found to be active; EAG, electroantennogram, source on filter paper or mixed in paraffin oil; GC, capillary gas-liquid-chromatograph; GC-EAG, EAG coupled with a GC in which compounds are injected, vaporized, separated (purified), and split GC-effluent used as stimulus source.

TABLE 2. Review of pheromones influencing oviposition behavior and receptors identified.

Pheromone source	Compounds	Behavioral reaction[1]	Species and organs	Sensitivity threshold[2]	Authors
Egg rafts on polluted water surface	(-)-(5*R*,6*S*)-6-acetoxy-5-hexadecanolide	Attraction, stimulation	*Culex quinquefasciatus* (Dipt., Culicidae) antennae	EAG: 0.1 µg	[29]
Female feces on fruit surface	*N*[15*R*(β-gluco-pyranosyl)-oxy-8*RS*-hydroxypalmitoyl]-taurine	Deterrency	*Rhagoletis cerasi* (Dipt., Tephritidae) tarsi, one receptor cell	2 × 10^{-10} M	[30,31]
Larval feces	*n*-nonanal, nerolidol, . . .	Repellency	*Spodoptera littoralis* (Lep., Noctuidae) antennae	EAG: 1 µg	[32]
Eggs	*trans*-2-[3-(3,4,5-trihydroxy-phenylpropenoyl)-amino]-3,5-dihydroxybenzoic acid	Deterrency	*Pieris brassicae* (Lep., Pieridae) tarsi, two receptor cells?	? ~10^{-7} M	[33,11, Van Loon et al., unpublished]

[1] Behavioral reaction: as defined by [34].
[2] Sensitivity threshold: lowest dose at which significant responses were estimated or measured.
. . . , additional compounds were tested and found to be active; EAG, electroantennogram, source on filter paper or mixed in paraffin oil; GC, capillary gas-liquid-chromatograph; GC-EAG, EAG coupled with a GC in which compounds are injected, vaporized, separated (purified), and split GC-effluent used as stimulus source.

more sensitive than the classical taste receptors (sugars, salt) known so far.

5. The sensitivity of olfactory chemoreceptors is certainly high too, but can be less well documented mainly because the stimulus quantification was made via the odor source only and no single sensilla recordings (only electroantennography) were used.

6. Conclusions about sensory coding remain tentative because the number and type of compounds tested is still limited. However, a high specificity (labelled-line coding) as well as elements of across-fiber coding have recently been found (e.g., *Pieris rapae*). Thus, high sensitivity does not exclude across-fiber coding.

7. Some insects ovipositing on the same host plants rely on the same (secondary) plant cues and have similar receptors. This is true for herbivores of Cruciferae, Liliaceae, and Pinaceae, but it is not true for the two known herbivores living on Umbelliferae which have receptors for differing compounds.

8. Parasitoids (third trophic level) seem to have receptors for similar plant compounds (first trophic level) as their hosts (second trophic level).

Conclusions

The impressive progress in the isolation of oviposition stimulants and their receptors has only been possible through multidisciplinary research involving chemistry, behavioral studies, and sensory physiology. The original hypotheses about host-plant selection and chemoreception have to be modified in order to take account of ovipositing females which mostly contact only the leaf surface and not the wet interior.

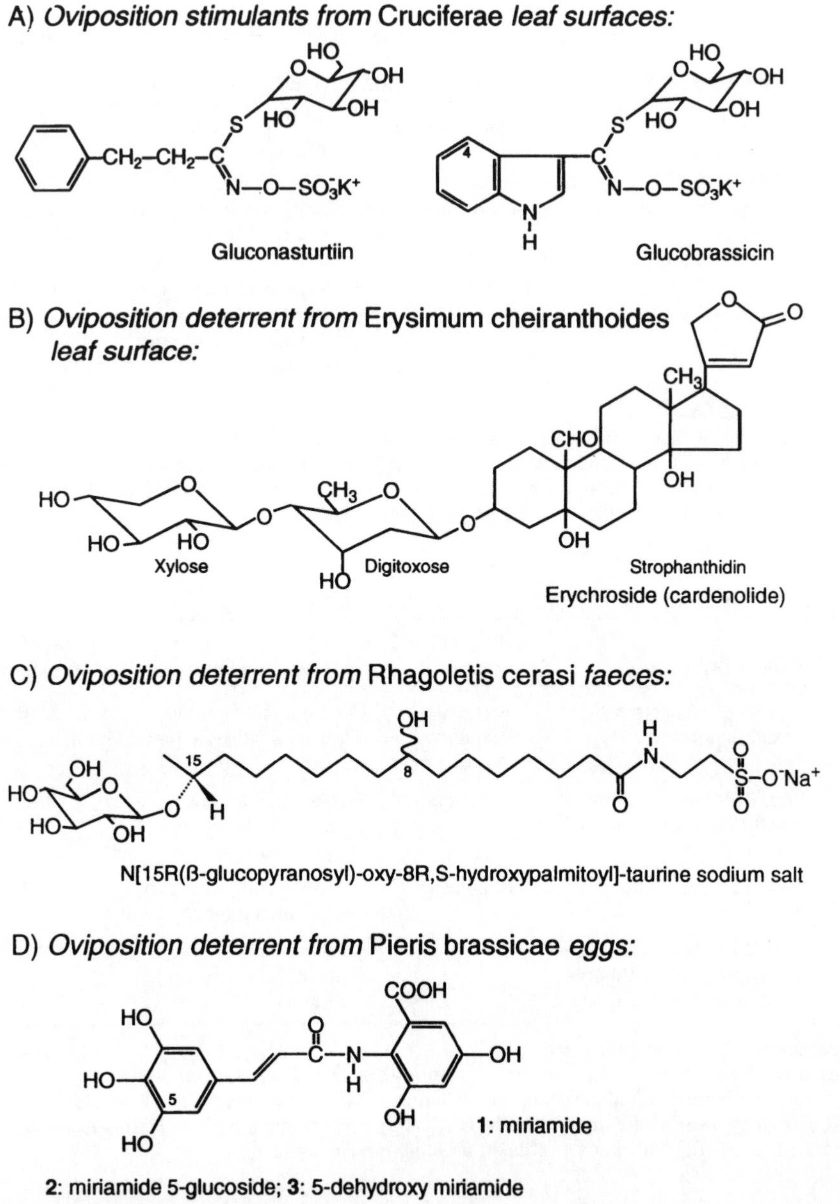

A) *Oviposition stimulants from* Cruciferae *leaf surfaces:*

Gluconasturtiin   Glucobrassicin

B) *Oviposition deterrent from* Erysimum cheiranthoides *leaf surface:*

Xylose   Digitoxose   Strophanthidin

Erychroside (cardenolide)

C) *Oviposition deterrent from* Rhagoletis cerasi *faeces:*

N[15R(ß-glucopyranosyl)-oxy-8R,S-hydroxypalmitoyl]-taurine sodium salt

D) *Oviposition deterrent from* Pieris brassicae *eggs:*

1: miriamide

2: miriamide 5-glucoside; 3: 5-dehydroxy miriamide

FIG. 1. Examples of compounds and sensitive contact-chemoreceptors influencing oviposition behavior. [References in Table 1]. **A** Two glucosinolates isolated from the host-plant leaf surface perceived by receptor cells in the sensilla of the butterfly *Pieris rapae* and the two vegetable flies *Delia radicum* and *D. floralis*. **B** Cardenolide isolated from a nonhost leaf surface perceived by a separate receptor cell in the tarsal sensilla of the butterflies *Pieris rapae* and *P. brassicae*. **C** Host marking pheromone deposited with feces by the cherry fruit fly on the fruit surface as a marking trail after oviposition inhibiting additional ovipositions into the same fruit, perceived by one type of tarsal receptor cell. **D** Host marking pheromone components of the butterfly *Pieris brassicae*, isolated from eggs, probably perceived by two tarsal receptor cells

Correlations between receptor activity and oviposition behavior were found to be significant and allowed initial conclusions about: (a) the relative importance of different receptors and compounds for the behavior, (b) differences between receptor cell types, and (c) assumed and possible interactions of different compounds on the same receptor cell and/or between cells in the same sensillum. The available results support the idea that a continuum between the two extreme types of coding, "labelled lines" and "across-fiber patterns," does exist in receptors involved in oviposition. The outlook for further investigations seems promising, since in a few insects the behavioral reactions and the receptors for the complex "real environment" have been successfully elucidated.

Acknowledgments. I thank Dr. R. Baur for discussions and improvements of the manuscript, Dr. J.J.A. Van Loon for the permission to use unpublished results, and the Schweizerisches Nationalfonds (grant # 31-30059.90 to E.S.) for financial support.

References

1. Städler E (1992) Behavioral responses of insects to plant secondary compounds. In: Berenbaum MR, Rosenthal GA (eds) Herbivores: their interaction with secondary plant metabolites, 2nd edn, vol II, Academic, New York, pp 45–88
2. Schoonhoven LM (1968) Chemosensory bases of host plant selection. Annu Rev Entomol 13:115–136
3. Städler E (1984) Contact chemoreception. In: Bell WJ, Cardé RT (eds) Chemical ecology of insects, Chapman and Hall, London, pp 3–35
4. Frazier JL (1992) How animals perceive secondary plant compounds. In: Berenbaum MR, Rosenthal GA (eds) Herbivores: their interaction with secondary plant metabolites, 2nd edn, vol II, Academic, New York, pp 89–134
5. Wallbank BE, Wheatley GA (1979) Some responses of cabbage root fly (*Delia brassicae*) to allylisothiocyanate and other volatile constituents of Crucifers. Ann Appl Biol 91:1–12
6. Simonds MSJ, Blaney WM, Mithen R, Birch ANE, Lewis JA (1994) Behavioural and chemosensory responses of the turnip root fly (*Delia floralis*) to gucosinolates. Entomol Exp Appl 71:41–57
7. Roessingh P, Städler E, Fenwick GR, Lewis JA, Nielsen JK, Hurter J, Ramp T (1992) Oviposition and tarsal chemoreceptors of the cabbage root fly are stimulated by glucosinolates and host plant extracts. Entomol Exp Appl 65:267–282
8. Ma W-C, Schoonhoven LM (1973) Tarsal contact chemosensory hairs of the large white butterfly, *Pieris brassicae* and their possible rôle in oviposition behaviour. Entomol Exp Appl 16:343–357
9. Städler E, Renwick JAA, Radke CD, Sachdev-Gupta K (1994) Ovipositional and sensory responses of tarsal sensilla of *Pieris rapae* (Lep., Pieridae) to stimulating glucosinolates and deterring cardenolides. Entomol Exp Appl, submitted
10. Roessingh P, Städler E, Hurter J, Ramp T (1992) Oviposition stimulant for the cabbage root fly: Important new cabbage leaf surface compound and specific tarsal receptors. In: Menken SBJ, Visser JH, Harrewijn P (eds) Proc 8th Int Symp Insect–Plant Relationships, Kluwer, Dordrecht, pp 141–142
11. Rothschild M, Alborn H, Stenhagen G, Schoonhoven LM (1988) A strophanthidine glycoside in the Siberian Wallflower: A contact deterrent for the large white butterfly. Phytochem 27:101–108
12. Sachdev-Gupta K, Renwich JAA, Radke CD (1990) Isolation and identification of oviposition deterrents to the cabbage butterfly, *Pieris rapae*, from *Erysimum cheiranthoides*. J Chem Ecol 16:1059–1067
13. Lupoli R, Marion-Poll F, Pham-Delègue MH, Masson C (1990) Effet d'émissions volatiles de feuilles de maïs sur les préférences de ponte chez *Ostrinia nubilalis* (Lepidoptera: Pyralidae). C R Acad Sci 311:225–230
14. Marion-Poll F (1986) La Chimioréception chez la pyrale du maïs (*Ostrinia nubilalis* Hbn.): approche anatomique et électroantennographique. Thèse INRA, Paris-Grignon, pp 1–158
15. Derridj S, Fiala V, Barry P, Boutin JP, Robert P, Poessingh P, Städler E (1992) Role of nutrients found in the phylloplane, in the insect host plant selection for oviposition. In: Menken SBJ, Visser JH, Harrewijn P (eds) Proc 8th Int Symp Insect–Plant Relationships, Kluwer, Dordercht, pp 139–140
16. Waladde SM, Ochieng SA (1991) Tarsi and ovipositor gustatory sensilla of *Maruca testulalis*: coding properties and behavioural responses. In: Szentesi Á, Jermy T (eds) Proc 7th Int Symp Insect–Plant Relationships, Symposia Biologica Hungarica 39:551–552
17. Ishikawa Y (1988) Electroantennogram responses of the onion fly, *Hylemya antiqua* Meigen (Diptera: Anthomyiidae) to oviposition stimulants. Appl Ent Zool 23:388–395
18. Lecomte C, Pouzat J (1986) Electroantennogram study of olfactory stimuli of plant origin in *Acrolepiopsis assectella*. Entomol Exp Appl 40:13–24
19. Mustaparta H, Angst ME, Lanier GN (1979) Specialization of olfactory cells to insect- and host-produced volatiles in the bark bettle *Ips pini* (SAY). J Chem Ecol 5:109–123
20. Wibe A, Mustaparta H (1992) Specialization of receptor neurons to host odours in the pine weevil, *Hylobius abietis*. In: Menken SBJ, Visser JH, Harrewijn P (eds) Proc 8th Int Symp Insect–Plant Relationships, Kluwer, Dordrecht, pp 127–128
21. Salom SM, Ascoli-Christensen A, Birgersson G, Payne TL, Berisford CW (1992) Electroantennogram responses of the southern pine beetle parasitoid *Coeloides pissodis* (Ashmead) (Hym, Braconidae) to potential semiochemicals. J Appl Entomol 114:472–479
22. Averill AL, Reissig WH, Roelofs WL (1988) Specificity of olfactory responses in the tephritid fruit fly, *Rhagoletis pomonella*. Entomol Exp Appl 47:211–222
23. Andersen JF, Mikolajczak KL, Reed DK (1987) Analysis of peach bark volatiles and their electroantennogram activity with lesser peachtree borer, *Synanthedon pictipes* (Grote and Robinson). J Chem Ecol 13:2103–2114
24. Crnjar R, Angioy AM, Pietra P, Stoffolano JG Jr, Liscia A, Tomassini Barbarossa I (1989) Electrophysiological studies of gustatory and olfactory responses of the sensilla on the ovipositor of the apple maggot fly, *Rhagoletis pomonella* Walsh. Boll Zool 56:41–46
25. Kusumi K, Shibuya T (1989) Response to tarsal contact chemoreceptor of the citrus feeding swallowtail burrerfly releasing the oviposition behavior. Zool Sci 6:1076
26. Städler E, Roessingh P (1991) Perception of surface chemicals by feeding and ovipositing insects. In: Szentesi Á, Jermy T (eds) Proc 7th Int Symp Insect–Plant Relationships, Symposia Biologica Hungarica 39:71–86
27. Baur R, Feeny P, Städler E (1993) Oviposition stimulants for the black swallowtail butterfly: identification of electrophysiologically active compounds in carrot volatiles. J Chem Ecol 19:919–937
28. Roessingh P, Städler E, Schöni R, Feeny P (1991) Tarsal contact chemoreceptors of the black swallowtail butterfly, *Papilio polyxenes*: responses to phytochemicals from host- and non-host plants. Physiol Entomol 16:485–495
29. Mordue AJ, Blackwell A, Hansson BS, Wadhams LJ, Pickett JA (1992) Behavioural and electrophysiological evaluation of oviposition attractants for *Culex quin-*

quefasciatus Say (Diptera: Culicidae). Experientia 48: 1109–1111

30. Städler E, Ernst B, Hurter J, Boller E, Kozlowski M (1992) Tarsal contact chemoreceptors of the cherry fruit fly, *Rhagoletis cerasi* (Dipt., Tephritidae): specificity, correlation who oviposition behavior, and response to the synthetic pheromone. In: Menken SBJ, Visser JH, Harrewijn P (eds) Proc 8th Int Symp Insect–Plant Relationships, Kluwer, Dordrecht, pp 143–145

31. Anderson P, Hilker M, Hansson BS, Bombosch S, Klein B, Schildknecht H (1993) Oviposition deterring components in larval frass of *Spodoptera littoralis* (Boisd.) (Lepidoptera: Noctuidae): a behavioural and electrophysiological evaluation. J Insect Physiol 39: 129–137

32. Städler E, Ernst B, Hurter J, Boller E (1994) Tarsal contact chemoreceptors for the host marking pheromone of the cherry fruit fly, *Rhagoletic cerasi*: responses to natural and synthetic compounds. Physiol Entomol, 19:in press

33. Blaakmeer A, van Beek TA, de Groot Æ, van Loon JJA, Schoonhoven LM (1994) Butterflies perceiving the presence of competitor's eggs: communication through novel chemicals. J Nat Prod, in press

34. Dethier VG, Barton Browne L, Smith CN (1960) The designation of chemicals in terms of the responses they elicit from insects. J Econ Entomol 53:134–136

Olfactory Mechanisms Underlying Sex-Pheromonal Information Processing in Moths

JOHN G. HILDEBRAND and THOMAS A. CHRISTENSEN[1]

Key words. Brain—Glomerulus—Insect—Inter-neuron—Olfaction—Pheromone

Introduction

Although they arise from lines of evolution that diverged hundreds of millions of years ago, insects and vertebrates possess differentiated olfactory systems that exhibit similarities far more striking than their differences [1]. In both groups of animals, olfactory receptor cells (ORCs), which arise and reside in an epithelium and have receptor dendrites exposed to the environment, transduce qualitative, quantitative, and temporal information about chemical stimuli into patterns of action potentials across the population of ORCs. The information encoded in those impulses is transmitted along the axons of the ORCs to the first synaptic way-station of the olfactory system in the brain, the vertebrate olfactory bulb (OB) or the insect antennal lobe (AL).

In both vertebrates and invertebrates, these primary olfactory centers are characterized by their remarkable arrays of glomeruli, which are spheroidal, condensed neuropil structures to which the ORC axons project and within which first-order synaptic processing of olfactory information takes place. Each olfactory glomerulus also receives neurites of intrinsic neurons [e.g., periglomerular and granule cells in the vertebrate OB, local interneurons (LNs) in the insect AL] and principal neurons [mitral and tufted cells in the vertebrate OB, projection neurons (PNs) in the insect AL]. In vertebrate and insect alike, each ORC axon projects to a single glomerulus, and large numbers of ORC axons converge on relatively few glomeruli, such that the primary-afferent "input" to each glomerular principal neuron reflects a convergence ratio on the order of 10^3. Even at the level of synaptic interactions among the various types of neurons associated with the olfactory glomeruli, extraordinary similarities between vertebrate and insect olfactory systems have been increasingly appreciated.

The outputs of the glomeruli are carried by the axons of their principal neurons, which project to higher-order olfactory centers in the brain for further synaptic processing and integration with other modalities. Among those higher centers are structures characterized by massively parallel fiber arrays (the pyriform cortex of the vertebrate, the mushroom body of the insect brain). On the basis of such similarities of neural organization, many investigators have studied the insect olfactory system as a valuable model to reveal mechanisms of olfaction in general.

Moth Olfactory System

We and our coworkers study the olfactory system of the experimentally favorable giant sphinx moth *Manduca sexta* with the aid of anatomical, neurophysiological, biochemical, cytochemical, developmental, and cell-culture methods. This brief review focuses on some of our findings and speculations about certain types of neurons and their interconnections in the AL of *M. sexta*; more comprehensive reviews of our work have been presented elsewhere [2–4]. A principal goal of our research is to understand the neurobiological mechanisms through which information about a specific olfactory signal, the female's multicomponent sex pheromone, is encoded, processed, and integrated with inputs of other modalities in the male moth's brain and how the message ultimately initiates and controls the male's characteristic behavioral responses. In pursuing that goal, we expect to learn much about how the brain processes chemosensory information and uses it to shape behavior.

The axons of antennal ORCs project into the ipsilateral AL in the deutocerebrum of the brain. The AL of *M. sexta* comprises a distinctive neuropil with a characteristic array of 64 ± 1 spheroidal, anatomically identifiable glomeruli bordered by lateral, medial, and anterior groups of neuronal somata [4,5]. These glomeruli are condensed neuropil structure $50-100\,\mu m$ in diameter, containing terminals of primary-afferent axons, arborizations of AL neurons, and synapses between sensory axons and AL neurons and among AL neurons [6,7]. Each ORC axon projects to a single glomerulus and pro-

[1] ARL Division of Neurobiology, University of Arizona, Tucson, AZ 85721, USA

vides excitatory, apparently cholinergic synaptic input to neurites of AL neurons innervating that glomerulus [8]. In particular, axons of male-specific receptor cells tuned to the two key components (the C_{16} aldehydes E10,Z12–16:AL and E10,E12,Z14–16:AL) of the female's sex pheromone project to a prominent male-specific neuropil structure in each AL, the macroglomerular complex (MGC), which comprises two "glomeruli"—one doughnut-shaped (the "toroid") and the other multilobate and globular (the "cumulus") [6,9]. Cells responding to E10,Z12–16:AL innervate the toroid, while those responding to E10,E12,Z14–16:AL innervate the cumulus [9]. Thus, the male's olfactory system consists of two parallel subsystems: a sexually dimorphic "labeled-line" subsystem specialized to detect and process information about the sex pheromone, and a more complex, sexually isomorphic subsystem that processes information about plant (and perhaps other environmental) odors encoded in "across-fiber" patterns of physiological activity. In the AL, the functions of both subsystems depend on neural circuitry involving the two main classes of AL neurons, LNs and PNs [4,6,10–12].

AL Projection Neurons

Uniglomerular PNs typically have their cell bodies in the medial and anterior AL cell groups, while multiglomerular PNs have somata in a particular cluster in the lateral group [4,11]. The axons of different types of PNs project in characteristic patterns into the protocerebrum via particular AL output tracts [4,11]. On the basis of their responses to pheromonal stimulation of the ipsilateral antenna, uniglomerular MGC PNs have been classified in two broad categories: pheromone generalists and pheromone specialists [10,13,14]. Pheromone generalists are neurons that respond similarly to stimulation of either the E10,Z12–16:AL input channel or the E10,E12,Z14–16:AL input channel and do not respond differently when the complete, natural blend is presented to the antenna. In contrast, we refer to neurons that can discriminate antennal stimulation with E10,Z12–16:AL from stimulation with E10,E12,Z14–16:AL as pheromone specialists. We have recognized several types of pheromone specialists, some of which receive input exclusively from either the dienal channel or the trienal channel. These cells therefore preserve information about individual components of the species-specific blend. These observations are consistent with behavioral studies and suggest that information about specific components of the blend, and not only the complete blend, is relevant to chemical communication in these animals.

An important subset of pheromone-specialist PNs found in male *M. sexta* receives input from both E10,Z12–16:AL and E10,E12,Z14–16:AL channels, but the physiological effects of the two inputs are opposite [10,13]. That is, if antennal stimulation with the dienal excites the PN, then stimulation with the trienal inhibits it, and vice versa. Simultaneous stimulation of the antenna with both E10,Z12–16:AL and E10,E12,Z14–16:AL elicits a mixed inhibitory and excitatory ($-/+/-$) response in these PNs. Thus, these neurons can discriminate between the two inputs based upon how each affects the spiking activity, and these cells also respond uniquely to the female's pheromone blend. Only this subset of pheromone-specialist neurons can follow intermittent pheromonal stimuli occurring at natural frequencies of up to 10 Hz [15] (see the following).

While most of the olfactory interneurons encountered in male *M. sexta* ALs respond preferentially to E10,Z12–16:AL and E10,E12,Z14–16:AL, some neurons also respond to other C_{16} components in the female's pheromone blend, such as the isomeric trienal E10,E12,E14–16:AL [13,14].

AL Local Interneurons

Primary-afferent synapses in the AL glomeruli usually involve LNs [4,6,16]. Thus, LNs are typically interposed between primary-afferent inputs to and PN outputs from the glomeruli. Some of the physiological properties and functions of these LNs have been revealed through the use of intracellular recording and staining, immunocytochemistry, pharmacology, and other methods [6,12,17]. All of the LNs studied to date have been spiking interneurons that respond to olfactory stimulation of the ipsilateral antenna with monosynaptic excitation and/or polysynaptic excitation or inhibition.

Insights about the roles of LNs in the AL have come from studies of the pheromone-specialist "$-/+/-$" MGC PNs that can discriminate between the two key components of the female's sex-pheromone blend [10,13,14]. The inhibitory input responsible for the early inhibitory postsynaptic potential (IPSP) in these neurons appears to be due to chemical-synaptic transmission mediated by γ-aminobutyric acid (GABA), one of the most prominent neurotransmitters in the ALs [18]. The IPSP in these PNs is easily reversed by injection of hyperpolarizing current, and ion-substitution experiments indicate that the IPSP is mediated by an increased Cl^- conductance [18]. This IPSP can be inhibited reversibly by picrotoxin, which blocks GABA-receptor-gated Cl^- channels, and by bicuculline, a blocker of vertebrate $GABA_A$ receptors. Furthermore, applied GABA also hyperpolarizes the postsynaptic neuron, and this response can be blocked reversibly by bicuculline, supporting the idea that bicuculline directly blocks GABA receptors. Blockade with bicuculline has demonstrated that the GABAergic synaptic transmission underlying the early IPSP is essential to the critical ability of the specialized MGC PNs to follow intermittent pheromonal stimuli, as occur in natural odor plumes [15].

To reveal which AL neurons contain GABA and might be responsible for inhibitory synaptic input to PNs, we used immunocytochemical methods employing GABA-specific antisera [17]. All of the GABA-immunoreactive neurons in the AL have somata in the large lateral cell group of the AL. There are approximately 350 GABA-immunoreactive LNs and 110 GABA-positive PNs (i.e., about 30% of the neurons in the lateral cell group appear to be GABAergic) [4,17]. Most of the LNs are GABA-immunoreactive, and these cells presumably are inhibitory interneurons. Further immunocytochemical studies have shown, moreover, that many of the LNs contain one or more putative neuropeptides colocalized with GABA [4,19]. The chemical identities and physiological functions of neuropeptides in the AL largely remain to be explored.

Synaptic Interactions Between AL Neurons

To test the idea that inhibition of PNs is mediated through LNs, we have begun to examine synaptic interactions between pairs of AL neurons [16]. In such experiments to date, using intracellular techniques, we passed current into one neuron while monitoring postsynaptic activity in the other. None of the PN–PN pairs examined to date showed any current-induced interactions, but a significant proportion of the LN–PN pairs exhibited such interactions, all of which were unidirectional. That is, LN activity could influence PN activity, but not vice versa. Depolarizing current injected into an LN, causing it to produce spikes, was associated with a cessation of firing in the PN. Spike-triggered averaging, based on spontaneous LN spikes, revealed a weak, prolonged IPSP in the PN. This result indicates that even background activity in LNs is sufficient to hyperpolarize PNs, and this action could help to improve signal contrast in the system. When the simultaneous responses of the two neurons to olfactory stimulation of the ipsilateral antenna with the sex-pheromone component E10,Z12–16:AL were recorded, a brief period of inhibition was observed in the LN, and this was followed shortly thereafter by a transient increase in the firing frequency in the PN. This suggests that the excitation in the PN is mediated through disinhibition [16]. That is, the LN that synaptically inhibits the PN may itself be inhibited by olfactory inputs, probably through another interneuron.

New evidence indicates, however, that not all LNs in the AL are inhibitory. These studies make use of simultaneous intracellular recordings from two AL neurons, as well as a new method for differentially labeling the two impaled neurons for subsequent examination in the laser-scanning confocal microscope and the transmission electron microscope [20; X.J. Sun, L.P. Tolbert, and J.G. Hildebrand, manuscript in preparation]. In one experiment involving simultaneous intracellular recordings from two LNs, one LN strongly excited the other, with a very short and constant synaptic latency suggesting a monosynaptic connection [X.J. Sun, unpublished]. Although that experiment was carried out in the AL of a female moth, the results support the hypothesis that there are excitatory LNs in the male AL as well.

Higher-Order Processing of Pheromonal Information in the CNS

Information about sex pheromone and other odors is relayed, by way of patterns of action potentials shaped by synaptic processing in the glomeruli, by the axons of AL PNs to higher centers in the protocerebrum. Toward the goal of understanding how pheromoal information controls the behavior of male moths, physiological and morphological studies have been launched to characterize neurons in the protocerebrum that respond to stimulation of the antennae with sex pheromone or its components [21,22]. That pilot work has revealed that many pheromone-responsive protocerebral neurons have arborizations in the lateral accessory lobes (LALs) of the protocerebrum [21]. Each LAL is linked, by neurons that innervate it, to the ipsilateral superior protocerebrum as well as the lateral protocerebrum, where axons of AL PNs terminate [10–12]. The LALs are also linked to each other by bilateral neurons with arborizations in each LAL. Neuropil adjacent to the LAL contains branches of many neurons that descend in the ventral nerve cord. Local neurons link the LAL to this adjacent neuropil. Some descending neurons also have arborizations in the LAL. Thus, the LAL appears to be interposed in the pathway of olfactory information flow from the AL through the lateral protocerebrum to descending neurons.

All protocerebral neurons to date that responded to antennal stimulation with pheromone were excited. Although brief IPSPs were sometimes elicited in mixed inhibitory/excitatory responses, sustained inhibition was not observed. Certain protocerebral neurons show "long-lasting excitation" (LLE) that sometimes outlasts the olfactory stimuli by up to 30 s. In some other protocerebral neurons, pheromonal stimuli elicit brief excitations that recover to background firing rates <1 s after stimulation. LLE is more frequently elicited by the sex-pheromone blend than by E10,Z12–16:AL or E10, E12,Z14–16:AL. LLE responses to pheromonal stimuli were observed in >50% of the bilateral protocerebral neurons sampled that had arborizations in the LALs. Fewer than 10% of the protocerebral local neurons examined exhibited LLE in response to similar stimuli. AL PNs responding to pheromone components do not show LLE [10,12]. Thus, in *M. sexta* males, LLE appears not to be produced at early stages of olfactory processing in the AL, but to occur first at the level of the protocerebrum.

These findings suggest that the LAL is an important region of convergence of olfactory neurons from other regions of the protocerebrum. Synaptic interactions in the LAL may mediate integration of both ipsilateral and bilateral olfactory information prior to its transmission to the bilateral pool of descending neurons. LLE appears to be one important kind of physiological response that is transmitted to thoracic motor centers. How this LLE might participate in the generation of the male moth's characteristic behavioral response is a subject for future research.

Acknowledgments. We thank our many present and former coworkers and collaborators who contributed to the work reviewed in this chapter. Our research in this area is currently supported by NIH grants AI-23253, DC-00348, and NS-28495.

References

1. Boeckh J, Distler P, Ernst K-D, Hösl M, Malun D (1990) Olfactory bulb and antennal lobe. In: Schild D (ed) Chemosensory information processing. NATO ASI Series, vol 39, Springer, Berlin Heidelberg, pp 201–227

2. Christensen TA, Hildebrand JG (1987) Functions, organization, and physiology of the olfactory pathways in the lepidopteran brain. In: Gupta AP (ed) Arthropod brain: its evolution, development, structure and functions. Wiley, New York, pp 457–484

3. Hildebrand JG (1985) Metamorphosis of the insect nervous system: Influences of the periphery on the postembryonic development of the antennal sensory pathway in the brain of *Manduca sexta*. In: Selverston AI (ed) Model neural networks and behavior. Plenum, New York, pp 129–148

4. Homberg U, Christensen TA, Hildebrand JG (1989) Structure and function of the deutocerebrum in insects. Annu Rev Entomol 34:477–501

5. Rospars JP, Hildebrand JG (1992) Anatomical identification of glomeruli in the antennal lobes of the male sphinx moth *Manduca sexta*. Cell Tiss Res 270:205–227

6. Matsumoto SG, Hildebrand JG (1981) Olfactory mechanisms in the moth *Manduca sexta*: Response characteristics and morphology of central neurons in the antennal lobes. Proc Royal Soc Lond B 213:249–277

7. Tolbert LP, Hildebrand JG (1981) Organization and synaptic ultrastructure of glomeruli in the antennal lobes of the moth *Manduca sexta*: A study using thin sections and freeze-fracture. Proc Royal Soc Lond B 213:279–301

8. Waldrop B, Hildebrand JG (1989) Physiology and pharmacology of acetylcholinergic responses of interneurons in the antennal lobe of the moth *Manduca sexta*. J Comp Physiol A 164:433–441

9. Hansson B, Christensen TA, Hildebrand JG (1991) Functionally distinct subdivisions of the macroglomerular complex in the antennal lobes of the sphinx moth *Manduca sexta*. J Comp Neurol 312:264–278

10. Christensen TA, Hildebrand JG (1987) Male-specific, sex pheromone-selective projection neurons in the antennal lobes of the moth *Manduca sexta*. J Comp Physiol A 160:553–569

11. Homberg U, Montague RA, Hildebrand JG (1988) Anatomy of antenno-cerebral pathways in the brain of the sphinx moth *Manduca sexta*. Cell Tiss Res 254:255–281

12. Kanzaki R, Arbas EA, Strausfeld NJ, Hildebrand JG (1989) Physiology and morphology of projection neurons in the antennal lobe of the male moth *Manduca sexta*. J Comp Physiol A 165:427–453

13. Christensen TA, Hildebrand JG (1990) Representation of sex-pheromonal information in the insect brain. In: Døving KB (ed) ISOT X. Proceedings of Tenth International Symposium on Olfaction and Taste. University of Oslo, Norway, pp 142–150

14. Christensen TA, Hildebrand JG, Tumlinson JH, Doolittle RE (1989) The sex-pheromone blend of *Manduca sexta*: Responses of central olfactory interneurons to antennal stimulation in male moths. Arch Insect Biochem Physiol 10:281–291

15. Christensen TA, Hildebrand JG (1988) Frequency coding by central olfactory neurons in the sphinx moth *Manduca sexta*. Chemical Senses 13:123–130

16. Christensen TA, Waldrop BR, Harrow ID, Hildebrand JG (1993) Local interneurons and information processing in the olfactory glomeruli of the olfactory glomeruli of the moth *Manduca sexta*. J Comp Physiol A 173:385–399

17. Hoskins SG, Homberg U, Kingan TG, Christensen TA, Hildebrand JG (1986) Immunocytochemistry of GABA in the antennal lobes of the sphinx moth *Manduca sexta*. Cell Tiss Res 244:243–252

18. Waldrop B, Christensen TA, Hildebrand JG (1987) GABA-mediated synaptic inhibition of projection neurons in the antennal lobes of the sphinx moth *Manduca sexta*. J Comp Physiol A 161:23–32

19. Homberg U, Kingan TG, Hildebrand JG (1990) Distribution of FMRFamide-like immunoreactivity in the brain and suboesophageal ganglion of the sphinx moth *Manduca sexta* and colocalization with SCP_B-, BPP-, and GABA-like immunoreactivity. Cell Tiss Res 259:401–419

20. Sun XJ, Tolbert LP, Hildebrand JG (1993) Synaptic organization of projection neurons of the antennal lobe of *Manduca sexta*: A confocal and electron-microscopic study. Soc Neurosci Abstr 19:127

21. Kanzaki R, Arbas EA, Hildebrand JG (1991) Physiology and morphology of protocerebral olfactory neurons in the male moth *Manduca sexta*. J Comp Physiol A 168: 281–298

22. Kanzaki R, Arbas EA, Hildebrand JG (1991) Physiology and morphology of descending neurons in pheromone-processing olfactory pathways in the male moth *Manduca sexta*. J Comp Physiol A 169:1–14

Olfactory Processing Pathways of the Insect Brain

Lateral Accessory Lobe System in the Protocerebrum Produces Olfactory Flip-Flopping Signals in *Bombyx mori*

RYOHEI KANZAKI[1]

Key words. Insect—Pheromone—Flip-flop—Descending interneuron—Brain—Olfactory pathways

Introduction

Male moths, *Bombyx mori*, display oriented zigzag walking with fluttering in response to the sex-attractant pheromone blend released by conspecific females. Even though they vibrate their wings, their body is too heavy to fly. The zigzag walking can even be initiated by a synthetic primary pheromone component (*E*10,*Z*12–16:OH, bombykol).

As proposed in many other flying moth species [1], including *Manduca sexta* [2], we have suggested that the *Bombyx* walking toward the pheromone source is controlled by a self-generated zigzag turning program which is triggered by intermittent pheromonal stimulation applied to the antennae [3].

Olfactory receptor neurons in trichodeal hairs on the antennae transmit information about pheromones via encoded firing patterns to the brain. In the brain of moths, olfactory information passes through several stages of synaptic processing involving the first-order olfactory center, the antennal lobes (ALs), and higher centers in the protocerebrum (PC). After local synaptic processing in the ALs and PC, the integrated information is relayed by PC neurons which have an axon descending (DNs) in the ventral nerve cord (VNC) to thoracic motor circuitry to generate and control the pheromone-modulated zigzag behavior.

We have found that in *Bombyx* many olfactory PC neurons, some restricted to the PC, and DNs, innervate the lateral accessory lobes (LALs) [4,5]. The LALs exist on both sides of the PC and are lateral to the central body. Each LAL is also linked with each other by protocerebral bilateral neurons (PBNs) with arborizations in each LAL [5]. The LALs appear to be important for processing olfactory information in the higher olfactory centers in the PC. The activity of these PC neurons, especially

of DNs, is of particular interest because the information they carry represents the integrated output of the brain circuits that may act on thoracic motor circuitry to affect olfactory modulated behavior.

Olberg [6] has found that some DNs of *Bombyx* males show characteristic state-dependent activities, like an electronic "flip-flop", which have two distinct firing frequencies, high and low. Switching back and forth between the two states occurs upon sequential pheromonal stimulation. It is suggested that the flip-flop signal may be important in controlling the pheromone-modulated zigzag walking of *Bombyx* males [6]. Neither the morphology of such flip-flop interneurons, nor the neural system required to produce their activity has been characterized.

In the present paper, we will show that in *Bombyx*, the LALs may be interposed in the pathway of olfactory information flow from the first-order olfactory center, the AL, through the lateral PC to the DNs. The findings presented in this report provide a substrate for understanding the neural system required to produce flip-flop activities, which may be an important signal to control zigzag walking of *Bombyx*.

Materials and Methods

Insect

Bombyx mori (Lepidoptera: Bombycidae) were purchased as pupae from Katakura Industries, Saitama, Japan. The pupae were kept in an incubator under a 12-h light and 12-h dark photo period regimen at 26°C. Adult males were used within 2 to 4 days after eclosion, which showed zigzag walking with fluttering in response to pheromonal stimulation applied to the antennae.

Physiology and Anatomy

Methods of intracellular recording, dye injection, and morphological reconstruction were similar to those described in Kanzaki and Shibuya [4]. Methods of suction recording, Co^{2+}-staining, and morphology of DNs in the VNC were similar to those described

---

[1] Institute of Biological Sciences, University of Tsukuba, Tsukuba, Ibaraki, 305 Japan

in Kanzaki et al. [7]. The electrode was filled with saline solution [in mM: 140 NaCl, 5 KCl, 7 CaCl$_2$, 1 MgCl$_2$, 4 NaHCO$_3$, 5 Trehalose, and 5 N-*tris* [hydroxymethyl]methyl-2-aminoethanesulfonic acid (TES), pH 6.8].

Olfactory Stimulation

An air-puff system similar to those described by Kanzaki et al. [3] was used to deliver pheromones to the antennae. Olfactory stimulation was applied separately to either antenna.

Two olfactory stimuli were used in the experiments: (1) "Female extract" was prepared by washing the tip of a virgin-female moth's abdomen with *n*-hexane, and (2) synthetic $E10,Z12$–16:OH (or bombykol), the principal pheromone component of *Bombyx mori*.

Two stimulant cartridges (1 ml plastic syringe barrels) were positioned about 1 cm from each antenna. Each stimulation from the cartridge (I.D. 5 mm) could be aimed selectively at one antenna or the other. From the other end of the cartridge an air puff (100 ms) was introduced into one of the stimulant cartridges by switching the continuous air stream (1 l/m or approximately 1 m/s) with a solenoid-driven valve to deliver the odorant to the antenna. After passing over the preparation, odorants were removed by gentle suction into an exhaust tube positioned nearby.

Constant Light Stimulation

Constant white light was presented to frontal regions of both compound eyes via light guide (intensity: approximately 3,500–7,000 lx; visual angle: approximately 45°), in what was defined as "light-on" conditions. "Light-off" conditions were less than 1 lx.

Results

Lateral Accessory Lobes of *Bombyx*

The spheroidal neuropile structure of the lateral accessory lobe (LAL, 150 × 100 μm) exists on both sides of the protocerebrum (PC) and lateral to the central body and posterior to the antennal lobe. Each LAL is linked by adjacent neuropiles and is also linked to each other via the commissure of the LALs [5].

Protocerebral Bilateral Neurons Innervating the LALs

Intracellular recording and staining of PC neurons revealed that some neurons innervated both LALs with heavily branched arborizations linked by a single major neurite which crossed the midline of the brain through the LAL commissure (protocerebral bilateral neurons or PBNs). These PBNs typically showed a characteristic, long-lasting excitation (LLE) which continued beyond the stimulation period. In case of a stimulation with 500 ng bombykol, firing levels >70% of the peak frequency lasted >20 s [5].

Descending Interneurons Innervating LALs

Cobalt staining of either the left or right connective, or a small bundle split off of one of the connectives, revealed DN morphology. The DNs which have a projection in the LAL fell into two classes according to the position of their cell bodies in the brain: (1) DNs which have a cell body in the cluster ventral to the medial cell cluster of the AL neurons (group I). Two to three cell bodies usually made up the cluster. They had widely spread arborizations throughout the LAL ipsilateral to the cell body and the neuropile ventral to the LAL. The major neurite projected to the contralateral LAL through the commissure of the LAL and descended with the contralateral connective of the VNC. (2) Those which have a cell body adjacent to the α-lobe of the mushroom body (group II). Ten to fifteen cell bodies usually made up the cluster. The major neurite projected to the ipsilateral LAL where it had sparse arborizations and descended with the ipsilateral connective of the VNC. Cell bodies of each group existed in bilateral clusters in the brain.

Antiphasic Flip-Flop Activities

Multiple suction recordings were made simultaneously from both connectives of the VNC. The flip-flop responses have been defined by Olberg's criteria, that is, a state transition is at least a doubling or a halving of the prestimulus firing frequency which persists for at least 5 s after the end of the stimulus [6]. Typical flip-flop activities were recorded from the dorsal part of the VNC (Fig. 1A,B). Even when a whole left or a whole right connective was picked up in the suction electrode, some of the largest spikes showed typical flip-flop activities in response to brief sequential pheromonal stimulation (100 ms, Fig. 1C,D). When two electrodes were applied to each of a whole connective, the largest spikes consistently showed antiphasic flip-flop responses. We classify the flip-flop activity pattern shown by some of the largest spikes into "FF" type. When "FF" activities were simultaneously recorded from the left and the right connectives, each of the "FF" activities had an antiphase relationship. The antiphasic flip-flop responses were also recorded within an individual connective. As shown in Fig. 1A,B, when two electrodes were applied to two different areas of the dorsal part of the left connective, pheromonal stimulation elicited antiphasic flip-flop

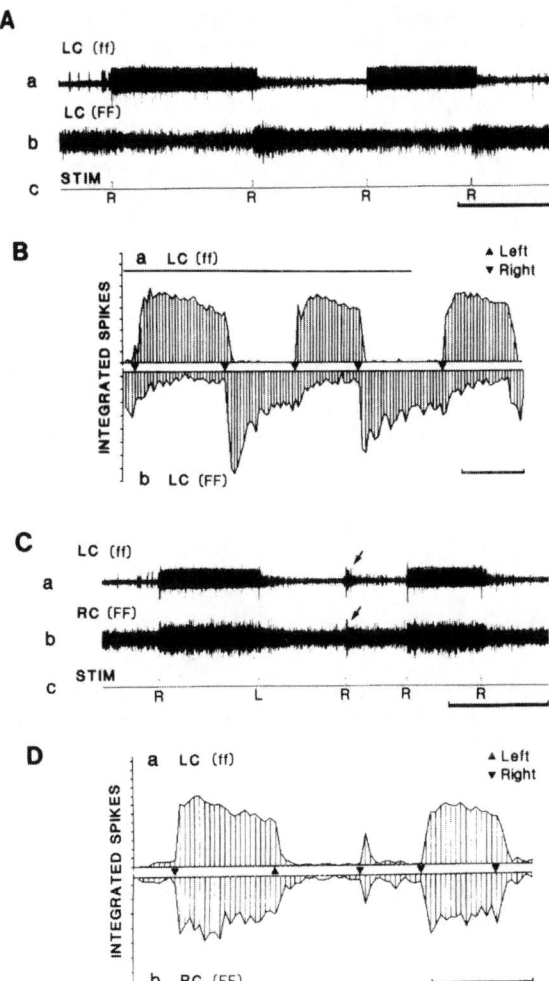

FIG. 1. Antiphasic flip-flop activities recorded from the connective of the ventral nerve cord (VNC). **A** Electrical activities were recorded simultaneously from two small bundles split off the dorsal part of the left connective (*LC*) in response to antennal stimulation (*STIM*) with *E*10, *Z*12–16:OH (1 ng). **B** Integrated spike histograms (0.5 s bin) of **A** are shown. The *bar* above the *histograms* indicates the period shown in traces in **A**. **C** Another electrode was applied to the whole right connective (*RC*) in the same preparation. The synchronized flip-flop activity pattern was observed between LC (same recording as **Aa**) and RC. Even when flip-flopping was not elicited in response to the stimulation, synchronized phasic firing was recorded (*arrows* in **a, b**). **D** Integrated spike histograms (0.5 s bin) of **C**. *L*, left; *R*, right; *scale bars*, 10 s

responses similar to those shown between the left and right connectives. We classified the antiphase flip-flop activities of the "FF" recorded within an individual connective into "ff" type. In order to clarify the flip-flop response patterns, i.e., "FF" or "ff", one suction electrode was applied to a whole left or right connective in most preparations. In some examples, "FF" and "ff" responses were simultaneously recorded by a single electrode sucking a small bundle split off a connective.

State transition of flip-flop activities was usually elicited in response to bombykol less than 10 ng. Sequential stimulation of a high concentration, e.g., 100 ng, hardly caused the state to drift. Once the state drifted from the low state to the high state after the high concentration stimulation, such a concentration elicited only a brief inhibition (approximately 500 ms) but did not alter the state much.

All of the flip-flop activities recorded were influenced by constant light stimulation. Although FF state transitions were typically observed under constant light-on conditions, peak firing rates were lower when the light was turned off. Extra spikes were elicited during the constant light stimulation. In some other preparations, flip-flopping interneurons showed only phasic activities in response to sequential olfactory stimulation in the dark.

After recording of FF activity, the DNs were stained by passing a current through a suction electrode containing 250 m*M* cobalt-lysine. In all cases (*n* = 5), group II DNs were stained. In 4 out of these 5 examples, a few other DNs which did not innervate the LALs were also stained. In one case, only group II DNs were stained in the brain. It is possible that some FF interneurons are contained in group II DNs.

Discussion

Hypothesis of the Functional Links Required for Flip-Flop Activities in *Bombyx*

Figure 2 schematically illustrates the functional links between LALs by DNs and PBNs of the male *Bombyx*. Morphologically similar DNs and PBNs, except for group-II-like DNs, have been characterized in *Manduca* [8,9]. Although group-II-like DNs have not yet been morphologically characterized in *Manduca*, similar functional links in the LAL system may exist in *Bombyx* and *Manduca*.

Physiological and morphological studies on DNs of *Bombyx* lead us to the following hypothesis on the functional links required for flip-flop activities. Through one LAL, group II DNs may conduct a "FF" signal to the ipsilateral connective of the VNC, and group I DNs conduct a "ff" signal to the contralateral one. Moreover, mirror symmetrical signal transmission would occur through the other LAL. When the "FF" type and "ff" type flip-flop activities were simultaneously recorded from left and right connectives, the activity patterns were synchronized as shown in Fig. 1C,D. Thus, "FF" in the left connective and "ff" in the right connective, or vice versa, transmitted a synchronized signal pattern. These results suggest that both "FF" and "ff" DNs running in each connective may receive a similar activity pattern (e.g., flip-flop activities) in the same neuropile, maybe in the LAL. Since it is strongly

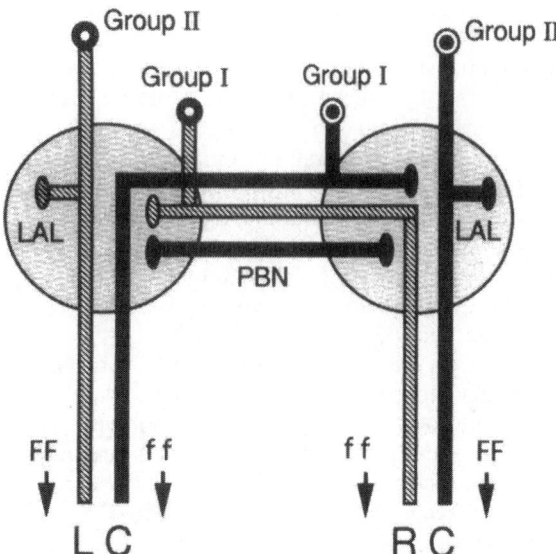

FIG. 2. Schematic representation of lateral accessory lobe (LAL) olfactory pathways in the brain of male moths, *Bombyx mori*. The two *lightly shaded circles* depict the left and right lateral accessory lobes (LALs) in the protocerebrum of the brain. *Lines with oval terminals* show group I and group II descending neurons (DNs) and protocerebral bilateral neurons (*PBNs*). The *ovals* indicate the arborizations of these neurons. Although group I DNs also had arborizations in the ventral protocerebrum (VP, adjacent to the LAL), the VP is not shown here. *LC*, left connective of the VNC; *RC*, right connective of the VNC. *FF, ff,* flip-flop response patterns

directed turns in the upwind approach of the male *Bombyx* to the female. We know that a male whose LAL commissure, linking each LAL, has been microsurgically dissected between the left and right LALs, shows only straight-line walking (Kanzaki et al., unpublished observation). The flip-flopping activity pattern was also consistent with the activity pattern of the neck motor neurons. These results suggest that the flip-flop signal may be important in controlling the pheromone-modulated zigzag walking of *Bombyx* [7].

Acknowledgments. This research was supported in part by a Grant-in-Aid for General Scientific Research from the Ministry of Education, Science and Culture of Japan (No. 05640761) and by a grant from Chemical Materials Research and Development Foundation to RK.

suggested that group II DNs show "FF" activities and their arborizations were restricted to the LAL, group II DNs may receive the "FF" signal in the LAL. On the other hand, when "FF" activities were simultaneously recorded from the left and right connectives, each of the "FF" activities had an antiphase relationship (Fig. 1). Therefore, flip-flop activities which had an antiphasic relationship were transmitted to each "FF" DN through the left and right LAL. These "FF" activities may be transmitted to group II DNs, and "ff" activities may be transmitted to group I DNs (Fig. 2).

Each LAL is linked with each other by protocerebral bilateral neurons (PBNs), some of which showed a long-lasting excitation [5]. In these functional links between LALs, PBNs or other bilateral PC neurons which link the two LALs, might have reciprocal inhibitory connections to produce antiphasic flip-flop activities.

Olberg [6] suggested that flip-flopping DNs carry left and right turning instructions for pheromone-

References

1. Baker TC (1990) Upwind flight and casting flight: Complementary phasic and tonic systems used for location of sex pheromone sources by male moth. In: Døving KB (ed) Proc 10th Int Symp Olfaction and Taste, Oslo, pp 18–25
2. Willis MA, Arbas EA (1991) Odor-modulated upwind flight of the sphinx moth *Manduca sexta*. J Comp Physiol A 169:427–440
3. Kanzaki R, Sugi N, Shibuya T (1992) Self-generated zigzag turning of *Bombyx mori* males during pheromone-mediated upwind walking. Zool Sci 9(3): 515–527
4. Kanzaki R, Shibuya T (1986) Descending protocerebral neurons related to the mating dance of the male silkworm moth. Brain Res 377:378–382
5. Kanzaki R, Shibuya T (1992) Long-lasting excitation of protocerebral bilateral neurons in the pheromone-processing pathways of the male moth *Bombyx mori*. Brain Res 587:211–215
6. Olberg RM (1983) Pheromone-triggered flip-flopping interneurons in the ventral nerve cord of the silkworm moth, *Bombyx mori*. J Comp Physiol 152:297–307
7. Kanzaki R, Ikeda A, Shibuya T (1992) Flip-flop descending interneurons controlling the olfactory triggered instinct behavior of insect (in Japanese with English summary). In: Arai S (ed) Proc 26th Jpn Symp Olfaction and Taste, Tokyo, pp 249–252
8. Kanzaki R, Arbas EA, Hildebrand JG (1991) Physiology and morphology of protocerebral olfactory neurons in the male moth *Manduca sexta*. J Comp Physiol A 168:281–298
9. Kanzaki R, Arbas EA, Hildebrand JG (1991) Physiology and morphology of descending neurons in pheromone-processing olfactory pathways in the male moth *Manduca sexta*. J Comp Physiol A 169:1–14

Pheromone-Modulated Flight Behavior of the Sphinx Moth, *Manduca sexta*

EDMUND A. ARBAS and MARK A. WILLIS[1]

Key words. Pheromone—Moth—Olfaction—Flight —Orientation—Behavior

Introduction

A complex pattern of zigzagging flight coupled with optomotor anemotaxis brings male moths up a plume of sexual pheromones to locate a female for mating. This behavior has attracted a great deal of interest because of its importance in the life cycle of many species of Lepidoptera, some of them of considerable economic importance as agricultural pests. The upwind zigzagging flight also provides an experimentally accessible model for the study of multimodal sensory integration and control of locomotion.

Previous studies of orientation to pheromones by flying insects have established a tradition of reducing filmed or videotaped flight tracks to a succession of movement vectors using the triangle of velocities method. The simple geometric equations of the triangle of velocities, together with the known wind speed and direction, the ground speed, and track angles measured from the flight tracks, can be used to calculate approximations of the animal's steering and air speeds. These track and flight parameter values are then typically averaged across part or all of the performance of an individual, and then lumped with matching values from many individuals in a particular experimental group. Values obtained in this way have fueled controversies about underlying mechanisms of flight control. While this approach enables an understanding of the average performance of a population of animals under particular conditions, it does not permit interpretations about neural mechanisms governing the behavior of individual insects, moment-to-moment, as they fly up the pheromone plume. To understand the latter, we have characterized the full range of individual performances that make up the population means, on a time scale that permits resolution of flight performance wingbeat-by-wingbeat. The results indicate considerable flexibility in the pheromone-modulated flight behavior by which male moths find females for mating, and several ways in which actual behavior

does not meet the assumptions of models based on average behavior.

Materials and Methods

Experimental procedures for eliciting, recording, and analyzing pheromone-modulated flight of the male *Manduca sexta* were very similar to those employed in our previous studies [1,2]. These are summarized briefly here.

Insects

Moths, *Manduca sexta*, were reared on artificial diet (modified from [3]) from eggs generated in our laboratory colony, or obtained from the U.S. Department of Agriculture (Beltsville, MD). Larvae were maintained on a 17:7 light:dark (LD) cycle at 25°C and 50%–60% relative humidity (RH). Pupae were separated by sex and adults allowed to emerge. Adult males were maintained on a 16:8 LD cycle at 28°C and 80% RH, and provided with a 5% solution of sucrose in water. All males flown in these experiments were 2–5 days old, unmated, and flown to pheromone between 2–6h into the scotophase.

Odor Sources

Extracts of the female pheromone gland (referred to as "pheromones") were made by excising the tip of the female abdomen and dipping the exposed pheromone gland in ca. 100 µl of hexane for 30 s. Several female glands were extracted in each 100 µl aliquot, and the final extract solution consisted of 10 female glands in 200 µl of hexane, thus yielding a 1 female equivalent/20 µl solution. Odor sources were made by applying 20 µl (1 female equivalent) of gland extract to a 0.7-cm diameter filter paper (Whatman No. 1, Whatman International Ltd., Maidstone, England) disk, and allowing the source to aerate for one hour in a fume hood. The extract was stored at −10°C when not in use.

Wind Tunnel and Data Recording

M. sexta males were flown in a Plexiglas flight tunnel with the dimensions 119 cm × 64 cm × 64 cm. The flight tunnel was designed to be fitted into a labora-

[1] ARL Division of Neurobiology, 611 Gould-Simpson Bldg., University of Arizona, Tucson, AZ 85721, USA

tory fume hood with sufficient draw through a set of nylon fabric baffles to produce a nearly laminar air current of approximately 100 cm/s. Odors are evacuated from the tunnel by the hood.

A floor pattern consisting of randomly arranged 10-cm diameter red dots on a white background [4] was positioned ca. 5 cm below the clear Plexiglas floor of the wind tunnel. Since red light bulbs were used as background illumination, these dots appeared nearly white in black-and-white video recordings, thus facilitating later tracing of flight tracks [4]. The room light intensity during experiments was ca. 2 lux and provided by one 40-watt incandescent white light diffused by reflection from a matte-white ceiling, and two 25-watt red incandescent bulbs.

Tracks of moths flying upwind in pheromone were recorded in plan view from above, using a Panasonic WV-1550 low-light video camera (Panasonic, Yokohama, Japan) attached via a Panasonic WJ-810 Time/Date generator, to a Sony SIV-373UC video cassette recorder (Sony, Tokyo, Japan). The camera was positioned 107 cm above the tunnel ceiling, pointed down toward the floor. This positioning resulted in a field of view of 80 × 60 cm at plume altitude with the upwind end of the field of view 10 cm downwind from the odor source.

Data Analysis

Flight tracks of individual moths were viewed frame by frame (30 frames/s) on a Panasonic WV 5410 black and white monitor and the location of each moth was marked every 0.03 s on transparent acetate sheets placed on the screen of the monitor. Flight tracks were analyzed as needed by methods described previously [1]. Course angles, airspeeds, and drift angles were calculated for each 0.03-s vector of the flight path using ground speed and track angle measured from the acetate tracing and the known wind speed and direction. The orientations of the body axes of flying moths were also measured directly from the same flight tracks used for this analysis.

For some of the analysis, we divided zigzagging flight tracks into two components, straight legs and turns. We defined straight legs as the middle 50% of the track leg between the apices of two turns. Turns were defined as the middle 50% of a track between the midpoints of two straight legs.

Results and Discussion

Analysis of the ground speeds across individual flight tracks indicates that, although *average* ground speeds were maintained near values identified in previous population studies, ground speeds of individual moths were modulated substantially along their upwind flight track. Moths typically achieved their fastest ground speeds along the straight parts of their tracks (between turns), and reached their slowest ground speeds at the apices of the turns.

Rapid changes in track angles marked the turns during flight, as expected. During some straight legs, track angles were maintained at relatively constant values with only small fluctuations superimposed. Small fluctuations atop a relatively constant distribution of track angles suggest that corrective maneuvers, perhaps mediated by optomotor feedback, may be used to stabilize the track leg. This suggests that steering may be governed by a specific set point during certain individual straight legs. Such stabilization around a set value of track angle may enable the moth to gauge the wind speed and direction, and adjust its own locomotor performance appropriately [5–7]. In other straight legs, the track angles varied over a wide range, suggesting that such straight legs may not be governed by a single set point for orientation.

Our data set includes an analysis of 69 inter-turn straight legs from 11 individual moths. More than two-thirds of these exhibited coefficients of variation ≤20% for track angles, indicating control of track angles to very close tolerances. When moths control the flight tracks, do they produce a "preferred track angle" near 60° as suggested by the population studies? They do not. The magnitude of the specific track angle maintained, varies from straight leg to straight leg across a flight performance, and from individual to individual.

If moths truly do monitor environmental parameters only during straight legs in which track angles are tightly regulated, then our own data suggests that the moth may make do with only periodic sampling of such information.

How would such information be gathered by the insect? One mechanism would be by breaking down the amount of upwind progress and drift that they undergo into transverse and longitudinal vectors of image slip over the compound eye [5,8–10]. Many studies estimate the values of transverse and longitudinal image slip from calculations of course angle using the triangle of velocities (see refs. in [2]). A major assumption of this approach is that the course angle, i.e., the angle of thrust during flight, corresponds to the longitudinal body axis. We calculated the course angles of male moths flying in pheromone plumes using this method, but then also measured the body axis directly from the video recordings for the same flight tracks and compared the two values. Deviation of the longitudinal body axis from the calculated course angle represents the angle of side slip experienced by the moth. Our data show that moths tolerate significant amounts of slip across their flight tracks. Indeed, only rarely does the body axis align with the axis of thrust. Some of the slip angle revealed by measurements of the body axis may be corrected by the moths' turning their heads in a compensatory response. We find that the limit of head turning in tethered animals is about 18°. Moths in our data sample frequently exhibited slip angles that exceeded this limit to compensation by head turning. To what extent moths actually do turn their heads to limit possible visual slip during free behavior remains to be determined.

These observations also indicate that estimates of longitudinal and transverse slip obtained by calculating the triangle of velocities may produce misleading values. Calculated values must be adjusted by the appropriate corrections for slip, and for head rotation [5].

Acknowledgments. Supported by NIH grants DC00348, NS07309 and NSF grant IBN 9216532.

References

1. Willis MA, Arbas EA (1991) Odor-modulated upwind flight of the sphinx moth, *Manduca sexta* L. J Comp Physiol (A) 169:427–440
2. Arbas EA, Willis MA, Kanzaki R (1993) Organization of goal-oriented locomotion: pheromone-modulated flight behavior of moths. In: Beer R, Ritzmann R, McKenna T (eds) Biological neural networks in invertebrate neuroethology and robotics. Academic, Orlando, pp 159–198
3. Bell RA, Joachim FA (1976) Techniques for rearing laboratory colonies of tobacco hornworms and pink bollworms. Ann Entomol Soc Am 69:365–373
4. David CT (1982) Competition between fixed and moving stripes in the control of orientation by flying *Drosophila*. Physiol Entomol 7:151–156
5. Marsh D, Kennedy JS, Ludlow AR (1978) An analysis of anemotactic zigzagging flight in male moths stimulated by pheromone. Physiol Entomol 3:221–240
6. Kennedy JS (1983) Zigzagging and casting as a programmed response to wind-borne odour: a review. Physiol Entomol 8:109–120
7. Cardé RT (1984) Chemo-orientation in flying insects. In: Bell WJ, Cardé RT (eds) Chemical ecology of insects. Sinauer Associates, pp 111–124
8. Ludlow AR (1984) Application of computer modelling to behavioral coordination. PhD thesis, University of London
9. David CT (1986) Mechanisms of directional flight in wind. In: Payne TL, Birch MC, Kennedy CEJ (eds) Mechanisms in insect olfaction. Clarendon, Oxford, pp 49–57
10. Kaissling KE, Kramer E (1990) Sensory basis of pheromone-mediated orientation in moths. Verh Dtsch Zool Ges 83:109–131

Behavioral Reaction Times of Male Moths to Pheromone Filaments and Visual Stimuli: Determinants of Flight Track Shape and Direction

Thomas C. Baker and Neil J. Vickers[1]

Key words. Sex pheromone—Moths—Plume—structure—Anemotaxis—Orientation—Behavioral latency

Introduction

Over the years, there have been several explanations for the existence of zigzagging upwind flight in male moths responding to sex pheromone [e.g., 1–7]. Far from being a frivolous argument about the superficial shape of the flight tracks, the root of the discussion involves the very mechanisms that the moths use to maneuver and reach the source of pheromone in wind. The behavioral mechanisms need to be precisely understood if we are to make sense of the underlying neuronal responses at the sensory, central nervous system (CNS), and motor levels and create a robust neuroethological knowledge about this powerful and agriculturally important biological process called attraction.

At the last meeting of the International Symposium on Olfaction and Taste (ISOT), Baker [7] proposed, using the results of previous studies performed predominantly with *Grapholita molesta* and *Heliothis virescens*, that each contact with a filament of pheromone would produce an upwind surge and high frequency of counterturning, and each pocket of clean air, if long enough, would produce a decrease in counterturning frequency and a subsiding of the surge. Sustained upwind flight would occur under conditions favoring the stringing together, or reiterations of, the upwind surging response to appropriately frequent contacts with filaments. The upwind surge was viewed as involving predominantly the anemotactic response system and was hypothesized to be tied to underlying blend-enhanced phasic neuronal pathways demonstrated in other species [8–10]. Casting flight was viewed as involving predominantly the counterturning program known to be switched on by pheromone and independent of the anemotactic system [11]. Counterturning was viewed as being driven by underlying, blend-dependent tonically firing neuronal elements dem-onstrated in other species [12–14]. It was envisioned that an increased frequency of exposure to filaments relative to the reaction time to the loss of pheromone would result in more straight-upwind flight with little zigzagging, because more frequent upwind surges would occur with less time for the counterturning program to subside in frequency and allow wide casting flight to be expressed fully [7]. Conversely, less frequent exposure to filaments would result in a greater degree of zigzagging in tracks due to the more visible expression of casting during lower-frequency counterturning occurring in the longer periods of clean air [7].

Recently, Willis and Arbas [15] found that their *Manduca sexta* males did not increase their rate of counterturning as the windspeed increased. They reasoned that the higher airspeeds generated by the moths in order to maintain constant groundspeed in elevated windspeeds should have produced more frequent contacts with filaments and resulted in a higher frequency of counterturning (and a narrower upwind path). Since their males did not exhibit increased counterturning, they argued that their results did not support the Baker model [7] and offered an alternative model in which an overall average level of stimulation by filaments, not reactions to individual filaments or the frequency thereof, would determine both the rate at which counterturning is performed and the intensity with which the anemotactic program is expressed.

Windspeed and Filament Frequency

Further examination of the supposition [15] that the moth's frequency of contact with filaments will increase with the moth's airspeed as windspeed increases reveals that it is likely to be flawed. Electroantennographic (EAG) measurements of the pheromone plume for *G. molesta* [16] showed that when windspeed was increased from 30 cm/s to 100 cm/s, the frequency of filaments contacting a stationary antenna 3 m from a point source increased only from 2.0 filaments/s to 2.6 filaments/s, a significant increase but far from the tripling of frequency one might assume would occur with this increase in airspeed over the antenna. Interestingly, when the

[1] Department of Entomology, 411 Science II Building, Iowa State University, Ames, IA 50011, USA

EAG preparation was pushed up the tunnel in order to increase the antenna's airspeed from 30 to 50 to 80 cm/s (the windspeed held constant at 30 cm/s), a more proportionate, expected increase in filament frequency occurred with readings of 1.6, 2.2, and 3.8 filaments/s at those three airspeeds, respectively [16].

These results suggest that if an increase in windspeed does not result in a significant increase in the frequency with which filaments are *generated*, then *in order to increase the frequency of filament contact when wind increases, a flying moth must increase the proportion of its groundspeed contributing to the airspeed*, as in the way that pushing the EAG up the wind tunnel did increase the frequency of filament contacts [16]. However, in no case have moths been shown to do this as a response to increased wind velocity. Rather, they maintain a constant groundspeed, and this is the now classical anemotactic response shown to be used by moths when they are exposed to pheromone [1,15,17–22].

To better visualize the situation, imagine a puffer device (e.g., as in [23]) generating filaments at 3/s, and the windspeed is doubled, then doubled again. Despite the quadrupling of windspeed, a stationary antenna downwind of the source will still only record 3 filaments per second, because this is the rate at which they were generated. A snapshot of the filaments will show them to be spaced more widely apart as the wind moves faster (Fig. 1), but nevertheless, increasing the stationary moth's "airspeed" by increasing the windspeed alone does absolutely nothing to increase the frequency of filaments arriving on the antenna. Next, assuming a known rate of filament generation from a typical pheromone point source, again, say 3/s [16], one can

see that the calculated contact rate with filaments by a moth flying at 40 cm/s upwind at each windspeed would not increase, and will actually decrease substantially, under conditions in which its airspeed increase is due entirely to an increase in windspeed. In fact, it would take ca. a 50% increase in the filament generation rate with each doubling of the windspeed just to keep the rate of contact from diminishing (Fig. 1).

The data of Willis and Arbas [15] show that although their *M. sexta* males tripled their airspeeds with a quadrupling of the windspeed, they did this by maintaining groundspeeds of ca. 40 cm/s upwind (50 cm/s overall) and thus, as in every other moth species studied thus far, the increased airspeed was due entirely to the increased windspeed (as in Fig. 1) [17–20]. Given this information, the data of Willis and Arbas [15] actually support the Baker model [7]; the observed lack of increase in counterturning frequency is expected by the model, given the probability that filament contact frequency may not have increased significantly, let alone proportionally, with windspeed. The rate with which filaments contacted an antenna in the plume was never measured by the authors. Thus, it should now be clear that the filament generation rate, not the speed with which the filaments move through the air, is the key variable that must increase proportionately with windspeed if a moth flying upwind is to contact filaments more frequently.

Olfactory Reaction Times

Other recent results from orientation studies are supportive of the model [7] and not of other models

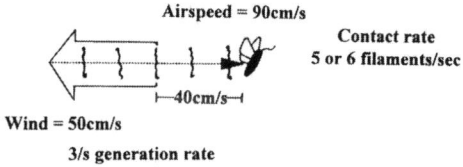

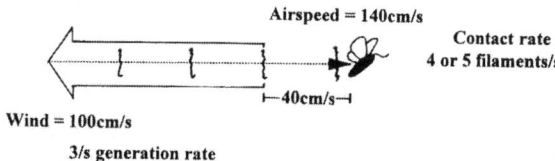

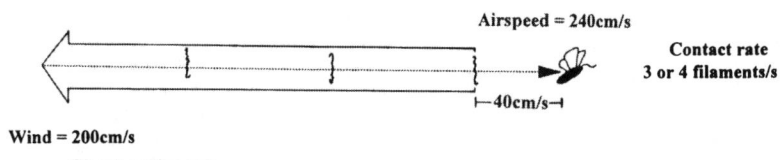

FIG. 1. Relationship between a moth's airspeed, groundspeed, and the frequency with which the moth contacts pheromone filaments as windspeed increases if the filament generation rate remains at 3/s and the moth maintains a preferred groundspeed of 40 cm/s

that have proposed that a counterturning program is not used by moths during pheromone-mediated flight [cf. 3,4]. Vickers [24] showed that male *H. virescens* do in fact respond to single filaments of pheromone by surging upwind. The reaction times of the males to both pheromone onset and offset are 0.23 s (−0.3 range) and 0.30 s (−0.4 range), respectively. This would translate into a frequency of contact of about 4 filaments per second if sustained upwind flight were to occur reliably, and in fact, frequencies of filaments puffed from the puffing device needed to exceed 4/s in order to evoke significant upwind flight and source contact [23]. The fact that casting across the windline occurred [24], always after flight into clean air, is evidence that a counterturning program is present and expressed in this species after emergence into clean air; any change in the direction of track legs to more crosswind must thus be viewed as the beginning of casting, and not as an error in the anemotactic system [3].

In other studies, the first-ever EAG's performed on flying male moths have demonstrated that encounters with filaments of pheromone from a standard rubber septum point source need to exceed 4 per second in order to sustain upwind flight [25]. Reaction times measured from EAG's of flying *H. virescens* males were consistent with those from single filament studies, with contact evoking an upwind surge 0.23 s (±0.11 SD) later in males that had been casting. Loss of pheromone caused crosswind flight 0.30 (±0.17 SD) seconds later, as measured by in-flight EAG's. Increased groundspeed up the windline produced, as predicted, an increase in the rate with which filaments contacted the antennae of the flying males (see foregoing discussion).

Visual Reaction Time

In a shifting wind-field, as under natural conditions in the field, the shape of a male's flight track will be determined by the latency of the male's reaction to the change in wind-direction plus the concomitant loss of contact with pheromone filaments [cf. 2,6,26]. The optomotor anemotactic system is necessarily a visual one [17,21,22], and thus the latency of the male's visually-mediated reaction to a change in wind direction also must play a part in shaping the track. For instance, when the wind direction shifted in the field, *H. virescens* males seemed to favor the correct direction, i.e., toward the newly displaced plume along the new windline, in which to make their first long cast [24]. Likewise, the published flight tracks of male gypsy moths during shifting wind direction also reveal a tendency for the males to make their first long cast in the correct direction toward the shifted plume [6,26].

In order to measure the visual reaction latency and try to explain this tendency, we created a nearly instantaneous "shift" in wind direction with which we challenged male *H. virescens* flying upwind in a plume of their pheromone. A rubber septum loaded with the six-component blend at 100 µg was placed on a metal platform at the upwind end of the wind tunnel in a windspeed of 50 cm/s. As the male reached the field of view of a video camera positioned looking down on the male's flight path, a dotted pattern located on the floor was abruptly moved sideways, 90° across the windline from the moth's left to its right at 50 cm/s (Fig. 2). The combination of 50 cm/s wind from the fan at the front of the tunnel plus the visual equivalent of wind from the dots moving 90° sideways at 50 cm/s below the moth instantly created a new windline coming at 45° from the right—a wind-shift that the moth, now flying to the left of the windline, must respond to by turning to the right, using this new visual feedback. The reaction time until the moth made a rightward response revealed the latency of the visual reaction to the wind-shift. These data show that males took, on average, 0.41 s (±0.17 SD; n = 14 males) to react to the visual stimulation by exhibiting a significant rightward change in the track direction (Fig. 2).

One key way a moth loses contact with pheromone is to fly into a large pocket of clean air caused by a

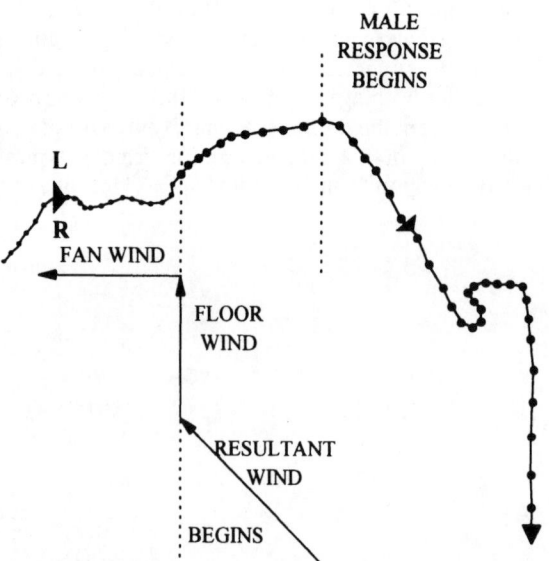

FIG. 2. The flight track of a male *H. virescens* that was flying upwind toward the pheromone source when the ground pattern was moved to create a new resultant wind direction at 45° from the moth's right while the plume remained unchanged in the center of the tunnel. *Larger dots* (every 1/30 s) on the flight track indicate the time period during which the ground pattern was moving and hence the period when the moth should have been responding to the new windline by steering to its right. The beginning of the rightward turn is indicated by male response begins

wind-shift [2,6,26]. Imagine a moth that in the field, is flying straight up the windline and encounters clean air at the shift-point of a plume that is being rapidly displaced to the left. During the 0.30 s that it takes the moth to begin responding to the loss of pheromone and begin casting [24], the inability of the moth to respond visually to the new windline for 0.41 s will allow it to be pushed off the windline *to the left, toward where the plume has gone*, and will facilitate recovery of contact with plume filaments. Thus, the similarities between olfactory [24] and visual reaction times, and indeed, the very existence of such time-lags, may explain why the flight tracks of moths seem to be biased toward the displaced plume in shifted wind-fields.

References

1. Kennedy JS (1983) Zigzagging and casting as a programmed response to wind-borne odour: a review. Physiol Entomol 8:109–120
2. Baker TC, Haynes KF (1987) Manoeuvres used by flying male oriental fruit moths to relocate a sex pheromone plume in an experimentally shifted wind-field. Physiol Entomol 12:263–279
3. Preiss R, Kramer E (1986) Mechanism of pheromone orientation in flying moths. Naturwissenschaften 73:555–557
4. Witzgall P, Arn H (1990) Direct measurement of the flight behavior of male moths to calling females and synthetic sex pheromones. Z Naturforsch 45c:1067–1069
5. David CT, Kennedy JS (1987) The steering of zigzagging flight by male gypsy moths. Naturwissenschaften 74:194–196
6. David CT, Birch MC (1989) Pheromones and insect behaviour. In: Jutsum AR, Gordon RFS (eds) Insect pheromones in plant protection. Wiley, New York, pp 17–35
7. Baker TC (1990) Upwind flight and casting flight: Complementary phasic and tonic systems used for location of sex pheromone sources by male moths. In: Døving KB (ed) Proc 10th Int Symp Olfaction and Taste, GCS A/S, Oslo, pp 18–25
8. Christensen TA, Hildebrand JG (1988) Frequency coding by central olfactory neurons in the sphinx moth *Manduca sexta*. Chem Senses 13:123–130
9. Christensen TA, Hildebrand JG, Tumlinson JH, Doolittle RE (1989) Sex pheromone blend of *Manduca sexta*: responses of central olfactory interneurons to antennal stimulation in male moths. Arch Insect Biochem Physiol 10:281–291
10. Christensen TA, Mustaparta H, Hildebrand JG (1989) Discrimination of sex pheromone blends in the olfactory system of the moth. Chem Senses 14:463–477
11. Kuenen LPS, Baker TC (1983) A non-anemotactic mechanism used in pheromone source location by flying moths. Physiol Entomol 8:277–289
12. Christensen TA, Mustaparta H, Hildebrand JG (1991) Chemical communication in heliothine moths. II. Central processing of intra- and interspecific olfactory messages in the male corn earworm moth *Helicoverpa zea*. J Comp Physiol A 169:259–274
13. Kanzaki R, Arbas EA, Hildebrand JG (1991a) Physiology and morphology of protocerebral olfactory neurons in the male moth *Manduca sexta*. J Comp Physiol A 168:281–298
14. Kanzaki R, Arbas EA, Hildebrand JG (1991b) Physiology and morphology of descending neurons in pheromone-processing olfactory pathways in the male moth *Manduca sexta*. J Comp Physiol A 169:1–14
15. Willis MA, Arbas EA (1991) Odor-modulated upwind flight of the sphinx moth, *Manduca sexta* L. J Comp Physiol A 169:427–440
16. Baker TC, Haynes KF (1989) Field and laboratory electroantennographic measurements of pheromone plume structure correlated with oriental fruit moth behaviour. Physiol Entomol 14:1–12
17. Marsh D, Kennedy JS, Ludlow AR (1978) An analysis of anemotactic zigzagging flight in male moths stimulated by pheromone. Physiol Entomol 3:221–240
18. Murlis J, Bettany BW, Kelley J, Martin L (1982) The analysis of flight paths of the male Egyptian cotton leafworm moths, *Spodoptera littoralis*, to a sex pheromone source in the field. Physiol Entomol 7:435–441
19. Kuenen LPS, Baker TC (1982) Optomotor regulation of ground velocity in moths during flight to sex pheromone at different heights. Physiol Entomol 7:193–202
20. Willis MA, Cardé RT (1990) Pheromone-modulated optomotor response in male gypsy moths, *Lymantria dispar* L: Upwind flight in a pheromone plume in different wind velocities. J Comp Physiol A 167:699–706
21. Kennedy JS (1940) The visual responses of flying mosquitoes. Proc Zool Soc Lond 109:221–242
22. Kennedy JS, Marsh D (1974) Pheromone-regulated anemotaxis in flying moths. Science 184:999–1001
23. Vickers NJ, Baker TC (1992) Male *Heliothis virescens* maintain upwind flight in response to experimentally pulsed filaments of their sex pheromone (Lepidoptera: Noctuidae). J Insect Behav 5:669–687
24. Vickers NJ (1992) Pheromone-mediated olfactory and behavioral mechanisms in the tobacco budworm, *Heliothis virescens* (F.). PhD dissertation, University of California, Riverside
25. Vickers NJ, Baker TC, Willis MA (1992) Correlation between electroantennogram activity and pheromone-mediated flight in the tobacco budworm. Annual Meeting of the Entomological Society of America, Baltimore, MD, USA
26. David CT, Kennedy JS, Ludlow AR (1983) Finding of a sex pheromone source by gypsy moths released in the field. Nature 303:804–806

Correlation of Chemoreception with Choice of Oviposition Site in Blowflies, *Phormia regina*

Azusa Nakagawa, Akifumi Iwama, and Atsuo Mizukami[1]

Key words. Blowfly—Oviposition

We investigated the correlation of chemoreception with the choice of oviposition site in blowflies, *Phormia regina*. Blowflies had to choose an oviposition site by odor or taste. As the odor source, we used decaying meat, and as the taste stimulant, 0.1 M sucrose solution. We made three combinations of odor and taste, that is, odorous or not, odorous or tasty and tasty or not. Flies laid eggs on the odorous site or the sucrose site. Besides, they also laid eggs between the meshes of the top cover of an oviposition cage. For the choice between odorous or not, 67.3% of eggs were found on the odorous site, 0.3% on the non-odorous site, and 32.4% between the meshes of the cage. Between odorous and tasty, 67.3% of the eggs were on the odorous site, 4.8% on the sucrose site, and 27.9% between meshes. Without the odor, flies laid 71.5% of eggs between the meshes and 16.4% on the sucrose site. These findings suggest that olfaction plays the principal role in the choice of oviposition site. Taste for sucrose may play a small role. We observed that flies explored the meshes, inserting their ovipositors, indicating that ovipositor mechano-reception may be related with the choice of a site.

Ablation of antennae reduced the proportion of eggs laid on the odorous site and increased the proportion on the other sites. For the choice between odorous or not, 26.3% of eggs were found on the odorous site. Between odorous and tasty, 20.6% were on the odorous site and 15.2% on the sucrose site. The proportion of eggs between the meshes was about 65% with any combination of odor and taste, being much larger than the proportion of eggs at other sites. These results suggest that the three modalities interact.

With both ablated antennae and maxillary palps, which have olfactory organs in many insects, the proportion of eggs on the odorous site decreased still further, while the proportions at other sites increased. The proportion at the sucrose site was larger than at the odorous site. However, these changes were not significant. Thus, we assumed that the maxillary palps function as an olfactory system, but are much less important than antennae. For the choice between odorous or not, eggs were still laid on the odorous site, suggesting that blowflies have an other olfactory system apart from the antennae and maxillary palps.

There may be several stages in the choice of oviposition site. Olfaction is probably involved in the first and following stages. In the final stage, ovipositor mechano-reception may be essential to induce egglaying. There may be a certain priority in which information from each modality is processed.

The above results indicate that antennae play an important role in the choice of oviposition site. Further, ablation of the antennae influenced the number of eggs laid. In the presence of the odor, flies with ablated antennae laid far fewer eggs than intact flies ($P < 0.05$ in the Mann-Whitney U-test), indicating that olfactory stimuli are likely to be important in determining the egg numbers. In the absence of the odor, on the other hand, the ablation of antennae had little effect on egg numbers; thus, the decrease that occurred with the odor may have been due not only to olfactory pathways from the antennae but also to some other olfactory pathway(s), which may function in regulating egg numbers. We do not completely understand how olfactory stimuli influence egg numbers. However, it is possible that olfactory input affects the endogenous physiological state in flies. Whether exogenous sensory inputs influence these states is a matter of some interest.

[1] Tsukuba Research Center, SANYO Electric Co., Ltd., 2-1 Koyadai, Tsukuba, Ibaraki, 305 Japan

Food Distribution Dependence in Fractal Feeding Behavior of *Drosophila*

Ichiro Shimada[1], Yoshinori Hayakawa[2], and Hiroaki Hara[3]

Key words. Fractal behavior—Feeding behavior of *Drosophila*

With the aid of our new system for the automatic analysis of the feeding behavior of *Drosophila*, we obtained a great deal of data for the locomotive activity and dwelling time on food in paired dense (60 wells) and sparse (8 wells) food distribution experiments.

A temporal fractal was clearly shown in a self-similar pattern of locomotive velocity in the feeding behavior of *Drosophila*. During the 1 h immediately after the fly was placed onto a new feeding place (micro test plate), the dwelling time on food showed a transition from a fractal, that is, an inverse power law distribution, to a nonfractal exponential [1]. In a choice situation between 100 mM sucrose (100 S) and 10 mM (10 S) in the dense food distribution, the dwelling time on 10 S showed an inverse power distribution. The dwelling time on 100 S, on the other hand, had a distribution close to exponential. In the sparse food distribution, a more striking difference between 100 S and 10 S was observed than with the dense food distribution, and the dwelling time on 10 S again showed an inverse power distribution. Under these choice situations, flies fed exclusively on 100 S but scarcely ingested 10 S, despite considerable stimulation of the peripheral sugar receptors in the legs. Such fractal behaviors may reflect situations in which the flies were exerting all their energy for searching, sampling, escaping, making decisions, etc., to adapt themselves to the new and complex surroundings.

Fractal behaviors are adaptive and intrinsic and reflect complex and high levels of information processing in the CNS of *Drosophila*. On the other hand, most behaviors that are expressed by exponential or Gaussian distribution may indicate rather simple information processing for reflex behavior or information processing that occurs after adaptation is attained.

Acknowledgments. We thank our students for their help with the experiments.

[1] Biological Institute, Faculty of Science, [2] Research Institute of Electrical Communication, [3] Graduate School Information Science, Tohoku University, Aoba-ku, Kawauchi, Sendai, 980 Japan

References

1. Shimada I, Kawazoe Y, Hara H (1993) A temporal model of animal behavior based on a fractality in the feeding of *Drosophila melanogaster*. Biol Cybern 68: 477–481

Elevation of Locomotor Activity After Sucrose Intake in the Blowfly, *Phormia regina* M

AKIO SHIRAISHI[1] and TOMOKO YAMAMOTO[2]

Key words. Feeding behavior—Dopamine

Feeding behavior in the blowfly was subdivided into component responses and the duration of these was measured. The relationship between the duration of the responses and the content of dopamine in the sub- and supraesophageal ganglion was investigated.

Two- or 3-day-old female blowflies, *Phormia regina* M, were used throughout the experiments; the flies did not ingest any food or tap water.

A transparent acryl cylinder, 1.5-mm-thick, 45-mm-external diameter, and 100-mm-high, was prepared. The top of the cylinder was covered with nylon net. A circular acryl plate 1.0-mm-thick and 45-mm-diameter was prepared. A hollow, 0.5-mm-deep and 5.0-mm-diameter, was made at the center of the plate. One volume of sucrose solution, $2\,\mu l$–$30\,\mu l$, was dropped into the hollow with a micropipette. One fly was placed in the cylindrical vessel. After 1 h on more, the vessel was put on the plate, and the duration of the fly's feeding responses was measured. Ambient temperature was $21°C \pm 1°C$ and relative humidity was more than 80%. While the fly was feeding in the cylindrical observation vessel air cooled with liquid nitrogen was passed through the vessel. The sub- and supraesophageal ganglia were then removed from the anesthetized fly and were homogenized with formic acid buffer. The homogenate was centrifuged at 15 000 rpm for 30 min and the supernatant was injected into a high-performance liquid chromatography electrochemical detector (HPLC-ECD) to measure the content of dopamine and 5-hydroxytryptamine. Phosphate buffer, pH adjusted to 3.5, was the mobile phase, the rate of flowing being 1 m/min. The voltage applied to the ECD was 750 mV.

Feeding behavior in the fly was divided into three component responses, volume intake; lapping the hollow with the labella, opening and shutting the labellar lobes; and moving around the hollow, extending and contracting the proboscis repeatly, respectively. These three component responses occurred sequentially. Volume intake was increased with increased sucrose concentration. The duration of the responses of lapping the hollow and opening and shutting the labellar lobes was constant, irrespective of the concentration of sucrose. The behavior of moving around the hollow has been called the "Fly Dance" [1]. The duration of this behavior increased when a high concentration and small volume of sucrose solution was given, but was decreased with large volumes of sucrose. The dopamine content in the sub- and supra-esophageal ganglion increased markedly with beginning of sucrose intake, this increase being maintained during sucrose intake. When the fly lapped the hollow with the labella, the dopamine content in the ganglion decreased considerably. During the occurrence of the Fly Dance, dopamine content in the ganglion again increased.

The increased locomotion after the intake of a high concentration and small volume of sucrose, a central excitatory state or Fly Dance also occurred after the intake of a large volume of sucrose, indicating that this is a kind of stereotyped behavior in the feeding of the fly. Dopamine content in the sub- and supraesophageal ganglion was shown to be related to both the duration of sucrose intake and the increased locomotor activity after feeding.

[1] Department of Biology, Faculty of Science, Kyushu University, 6-10-1 Hakozaki, Higashi-ku, FuKuoka, 812 Japan
[2] Chikushi Jogakuen Junior College, Ishizaka Dazaifu, 818-01 Japan

References

1. Dethier VG (1976) The hungry fly. Harvard University Press, Cambridge, pp 4–33

Central Projection of the Antennal Cold Receptor Axons of the Cockroach

Michiko Nishikawa, Fumio Yokohari, and Takaaki Ishibashi[1]

Key words. Antennal cold receptor—Cockroach

The antennae of the cockroach *Periplaneta americana* bear two types of cold receptive sensillia [1]. One type (cold/hygro sensillum) has smooth surface without pores; it includes four receptor cells: one cold, one dry, one moist cells and one whose modality is unknown. The other type of sensillum (cold/olfactory sensillum) is grooved, double-walled, and has pored surface. It includes one cold and three olfactory receptor cells. The cold receptors in both types of sensilla have been examined electrophysiologically, and differences in their static activities were revealed [2]. Do the afferents from these cold receptors project into the same target? We examined the central projections of the receptor axons in each type of cold receptive sensillum.

After electrophysiological identification of the sensillum, the receptor axons in a single sensillum were extracellularly stained by the iontophoretical injection of cobalt lysine. The pathways of the receptor axons in the brain were traced in whole mounted specimens. Four axons were usually observed in a specimen stained from a single cold/hygro sensillum. One axon branched in the ipsilateral dorsal lobe of the deutocerebrum. One branch passed through the tritocerebrum and terminated in the suboesophageal ganglion, and the other branches terminated in the ipsilateral dorsal lobe. Other axons terminated in the ipsilateral antennal lobe. Three or four axons were observed in a specimen stained from a single cold/olfactory sensillum. All axons terminated in the ipsilateral antennal lobe.

To reveal the terminal details of each axon, specimens embedded in Araldite were serially sectioned at 10–30 μm after being observed in whole mounted preparations. The terminal profiles of the four axons from the cold/hygro sensillum were as follows: One axon ran along three glomeruli lined up in the ventro-posterior and marginal region of the antennal lobe, and made a long brush-like cluster of arborizations with varicose terminals in each glomerulus (CH-1). The second and third ones also had terminal profiles similar to (CH-1). The second, however, ran along three glomeruli lined up in the ventro-posterior and lateral region of the antennal lobe (CH-2) and the third innervated two glomeruli located in the ventro-posterior region and the ventro-central region of the antennal lobe, respectively (CH-3). These eight glomeruli were located close to each other. The last one had several branches in the dorsal lobe, one of which terminated in the suboesophageal ganglion with varicose terminals, but which had no remarkable arborizations, while others ended in the lobe without terminal arborizations (CH-4).

The terminals of the four axons from a cold/olfactory sensillum were as follows: The profile of one axon (CO-1) appeared identical to that of (CH-1). Each of the other axons had short brush-like arborizations with varicose terminals in a glomerulus different from one another in the ventro-anterior and medial region of the antennal lobe (CO-2, 3, 4).

As the cold receptors are located in the antenna [2] and the cold-sensitive interneurons are located in the antennal lobe [3], it is feasible that the cold receptor axons may project to the antennal lobe. In addition, (CH-1) and (CO-1) had very similar profiles of axonal terminals, and innervated the same glomeruli. Therefore (CH-1) and (CO-1) must be the cold receptor axons. In the same context, (CH-2) and (CH-3) are putative hygroreceptor axons, and (CO-2), (CO-3), and (CO-4) are putative olfactory axons.

References

1. Altner H, Loftus R (1985) Ultrastructure and function of insect thermo- and hygroreceptors. Ann Rev Entmol 30:273–295
2. Nishikawa M, Yokohari F, Ishibashi T (1992) Response characteristics of two types of cold receptors on the antennae of the cockroach, *Periplaneta americana* L. J Comp Physiol A 171:299–307
3. Nishikawa M, Yokohari F, Ishibashi T (1991) Deutocerebral interneurons responding to thermal stimulation on the antennae of the cockroach, *Periplaneta americana* L. Naturwissenschaften 78:563–565

[1]Department of Biology, Faculty of Science, Fukuoka University, 8-19-1 Nanakuma, Jonan-ku, Fukuoka, 814-80 Japan

Portable Devices for Single Sensillum (SSR) and Electroantennogram (EAG) Recording from Insect Antennae in the Field

Jan N.C. van der Pers[1]

Key words. Sensor—Antennal receptors

In order to obtain more fundamental knowledge about the dispersion of behavior-modifying airborne chemical signals under natural conditions, it is necessary to measure these signals directly. These measurements require the use of the olfactory receptors of the target insect itself, because conventional physico-chemical detectors do not offer the required sensitivity and selectivity. Therefore, portable devices for single sensillium recording (SSR) and electroantennogram (EAG) recording have been developed, in which the insects' antenna is used as a sensor element.

The portable SSR device enables recording from individual antennal olfactory receptors by means of glass micropipettes or tungsten electrodes. The momentary air velocity over the antenna is recorded simultaneously. The signals and spoken comments are stored on a portable audio tape recorder and subsequently analyzed in the laboratory. Application of the instrument in orchards treated with pheromone dispensers for mating disruption of tortricid moths [1] showed that the instantaneous firing frequency of the receptor cells was strongly modulated by the air velocity. Due to the dense vegetation and the high release rate of the dispensers, the pheromone flux at the antenna is mainly determined by the air velocity. On the other hand, in open terrain where the antenna was exposed to artificial pheromone sources, there was little or no relation of the receptor response to the air velocity [2]. The receptors produce short periods of firing activity at irregular intervals. This response pattern of shorter and longer bursts was characteristic for the recordings made in the open field.

The responses obtained in the field can be quantified by comparison with responses of similar antennal preparations during exposure to well defined amounts of synthetic pheromone components. This was done by coupled gas chromatography-single sensillum recording (GC/SSR). Average pheromone concentration in the pheromone-treated area and the pheromone flux of pheromone plumes in open terrain were in the order of 50–150 pg/l [2].

The portable device for EAG recording is suitable for insect species in which the antennal receptors are not accessible for SSR, or in cases where SSR is too complicated to be utilized. Because an EAG signal represents the combined response of various types of antennal receptors, among which are mechano- and temperature receptors, EAG recording requires the antenna to be flushed with a conditioned filtered air current. For this reason, the portable EAG recording device is supplied with miniature air pumps and an activated charcoal filter, as well as cartridges containing standard stimuli for calibration. EAG responses are elicited by switching the antennal flow from filtered air to samples of unfiltered ambient air. Reference stimuli are presented at regular intervals. Samples can be taken on command of the operator or automatically, by means of a built-in timer circuit. The signals are stored on a portable audio cassette tape recorder, also useful for recording spoken comments. The signals are replayed in the laboratory and analyzed using specially developed software tools.

The advantage of the portable EAG recording apparatus is the extremely easy procedure for preparing the antennal sensor element. The sensitivity is comparable to that obtained with the portable SSR device. The SSR signal, however, has the same high time resolution and selectivity as the olfactory receptors themselves.

Both the portable EAG and the SSR recording devices are also suitable for applications in the laboratory (e.g., coupled GC/EAD, coupled GC/SSR), for measurements in wind tunnels, and for calibration purposes.

References

1. Van der Pers JNC, Minks AK (1993) Pheromone monitoring in the field using single sensillum recording. Entomol Exp Appl 68:237–245
2. Van der Pers JNC, Minks AK (1993) Field and laboratory single cell recordings of pheromone receptors of *Adoxophyes orana* and *Pandemis heparana*. In: Sommeijer MJ, van der Blom J (eds) Proc Exper Appl Entomol 4

[1] VDP Laboratories, P.O. Box 1547, NL-1200 BM Hilversum, The Netherlands

Sensitivity and Responsiveness of Neurons Associated with Male-Specific Trichoid Sensilla in the Beet Armyworm, *Spodoptera exigua*, to Volatile Emissions of Sympatric *Spodoptera* spp. and Plant Odors

Joseph C. Dickens[1]

Key words. Olfaction—Receptor neuron

Chemical signals emitted by insects (e.g., pheromones) may aggregate both sexes for feeding and mating as in Coleoptera [1,2], or may be gender-specific mating messages as in moths [3], and may be enhanced by plant odors [1,4–6]. Detection of olfactory chemical signals by insects involves primary sensory neurons which are housed within modified cuticular structures, called sensilla, on the insect's antenna [3]. These highly specialized receptor neurons in moths are thought to possess only one kind of membrane receptor molecule. However, selected plant volatiles stimulate pheromone receptor neurons in small ermine moths [7], the turnip moth [8], and the beet armyworm, *Spodoptera exigua* [9]. In my current study, the activity of volatile emissions of not only *S. exigua* females [10] but also those of two closely related sympatric species, *S. frugiperda* [11] and *S. eridania* [12], were tested. As in previous studies [9,13], only Z9–14:OH and Z9,E12–14:AC elicited appreciable activity from neurons associated with the type I sensilla on the proximal one-half of male antennae. In nearly all instances, the neuron giving the small amplitude spike was activated by Z9–14:OH, while the neuron with the large amplitude spike was selectively activated by Z9,E12–14:AC. Increases in action potential frequency from receptor neurons associated with male-specific type I sensilla were also elicited by 100-µg stimulus loads of several green leaf volatiles and benzaldehyde among 11 plant odors tested. Mean responses of the neuron with the small spike amplitude were significantly greater than responses of the neuron with the large amplitude spike. Dose-response curves revealed that the large amplitude spike responds only to Z9,E12–14:AC, while the small amplitude spike is responsive to Z9–14:OH, and the plant volatiles, E2–6:OH and 5:OH. I conclude that: (1) *Male-specific trichoid sensilla* house two neurons: one with a large amplitude spike, which responds to Z9,E12–14:AC, and one with a small amplitude spike, which responds to Z9–14:OH, both essential components (major and minor, respectively) of the conspecific female volatile emission. Increases in spike frequency are not elicited from either neuron by other components of volatile emissions of two sympatric *Spodoptera* spp. (2) *Plant odors* selectively activate a small amplitude spike. Since neurons characterized by either the small amplitude spike or the large spike are housed in the same sensillum and bathed in the same receptor lymph, it seems likely that either substantial differences exist in membrane *receptor mechanisms* for the neurons, or the receptor neuron with the small spike has more than one type of receptor site. (3) Since pheromone-sensitive sensilla probably evolved from sensilla primitively sensitive to plant volatiles [14], the observed responses of the pheromone receptor neurons to plant volatiles may be due to receptor sites for them remaining on these neurons [8,9]. Selective responses to plant odors by the neuron giving the small spike may indicate that detection of the minor component of the pheromone came after specialization of the neuron giving the large spike for the major pheromone component. (4) A *functional role* for the differential sensitivity and responsiveness of the pheromone-sensitive neurons to plant volatiles may be unlikely, due to the relatively high stimulus loads needed to elicit spikes from the small neuron. However, since plant volatiles differentially and selectively stimulate pheromone receptor neurons, thus altering the quality of the perceived pheromonal blend, and green leaf volatiles enhance behavioral responses to sex attractants in *S. exigua* and other Lepidoptera [5,6], plant odors could be involved in the cessation of pheromone-mediated flight, as has been previously shown for individual pheromone components [15].

[1] U.S. Department of Agriculture, Agricultural Research Service, Boll Weevil Research Unit, Mississippi State, MS 39762, USA

References

1. Byers JA (1989) Chemical ecology of bark beetles. Experientia 45:271–283

2. Tumlinson JH, Hardee DD, Gueldner RC, Thompson AC, Hedin PA, Minyard JP (1969) Sex pheromones produced by male boll weevils: Isolation, identification, and synthesis. Science 166:1010–1012

3. Kaißling KE (1987) Insect olfaction. In: Colbow K (ed) R.H. Wright lectures on insect olfaction. Biering, Munich, p 75

4. Dickens JC (1989). Green leaf volatiles enhance aggregation pheromone of boll weevil, *Anthonomus grandis*. Entomol Exp Appl 52:191–203

5. Dickens JC, Smith JW, Light DM (1993) Green leaf volatiles enhance sex attractant pheromone of the tobacco budworm, *Heliothis virescens* (Fabricius) (Lepidoptera: Noctuidae). Chemoecol 4

6. Light DM, Flath RA, Buttery RG, Zalom FG, Rice RE, Dickens JC, Jang EB (1993) Host plant green leaf volatiles synergize the synthetic sex pheromones of corn earworm and codling moth. Chemoecol 4

7. Van Der Pers JNC, Den Otter CJ (1980) Interactions between plant odors and pheromone reception in small ermine moths (Lepidoptera: Yponomeutidae). Chem Senses 5:367–371

8. Hansson BS, Van Der Pers JNC, Löfqvist J (1989) Comparison of male and female olfactory cell response to pheromone compounds and plant volatiles in the turnip moth. Physiol Entomol 14:147–155

9. Dickens JC, Visser JH, Van Der Pers JNC (1993) Detection and deactivation of pheromone and plant odor components by the beet armyworm, *Spodoptera exigua* (Hübner). J Insect Physiol 39:503–516

10. Tumlinson JH, Mitchell ER, Sonnet PE (1990) Analysis and field evaluation of volatile blend emitted by calling virgin females of the beet armyworm moth, *Spodoptera exigua* (Hübner). J Chem Ecol 16:3411–3423

11. Tumlinson JH, Mitchell ER, Teal PEA, Heath RR, Mengelkoch LJ (1986) Sex pheromone of fall armyworm, *Spodoptera frugiperda* (J.E. Smith). Identification of components critical to attraction in the field. J Chem Ecol 12:1909–1026

12. Teal PEA, Mitchell ER, Tumlinson JH, Heath RR, Sugie H (1985) Identification of volatile components released by the southern armyworm, *Spodoptera eridania* (Cramer). J Chem Ecol 11:717–728

13. Mochizuki F, Shibuya T (1991) Antennal single sensillum responses to sex pheromone in male beet armyworm, *Spodoptera exigua*, Hübner (Lepidoptera: Noctuidae). Appl Entomol Zool 26:409–411

14. Lanne BS, Schlyter F, Byers JA, Löfqvist J, Leufven A, Bergström G, Van Der Pers JNC, Unelius R, Baeckström P, Norin T (1987) Differences in attraction to semiochemicals present in sympatric pine shoot beetles. J Chem Ecol 13:1045–1067

15. Baker TC, Hansson BS, Löfstedt C, Löfqvist J (1988) Adaptation of antennal neurons in moths is associated with cessation of pheromone-mediated upwind flight. Proc Natl Acad Sci USA 85:9826–9830

Sex Pheromone Perception by Males
of *Periplaneta* and *Blatta* Species

Shozo Takahashi[1], Keisuke Watanabe[2], Shigeru Saito[2], and Yoshiharu Nomura[3]

Key words. *Periplaneta* sex pheromone— Periplanone-D

The six cockroach species of the genera *Periplaneta* and *Blatta* are distributed worldwide, from tropical to temperate zones. Since the discovery of the sex pheromonal activity of germacrene-D to male *P. americana*, we have been studying the relationship between chemical structure and sex pheromone activity toward males in the six species, by evaluating behavioral and electroantennogram (EAG) responses. Periplanone-A, -B (PB), and -J elicited behavioral responses not only to males of the conspecifics, but also to other species of the genera, except for *P. fuliginosa*, whose sex pheromone has not been investigated. We identified the *P. fuliginosa* pheromone, as well as estimating its sex pheromonal activity.

Volatiles from *P. fuliginosa* females were absorbed on Tenax TA (ENKA Research Institute, Paris) and purified by high-performance liquid chromatography (HPLC) to yield the isolated sex pheromone. Its mass spectrum (MS) was identical with that of periplanone-D (PD), a minor component of *P. americana* pheromones, obtained by Biendl et al. [1] and by us (work in progress). The behavioral responses of the male *P. fuliginosa* to PD ($+++$ at 10^{-8} g dose) were around 100 times higher than those of male *P. americana* or *P. japonica* to PD. On the other hand, EAG responses from the antennae of male *P. fuliginosa* to PD were much higher than those from the antennae to PB.

[1] Pesticide Research Institute, Faculty of Agriculture, Kyoto University, Kitashirakawa, Sakyo-ku, Kyoto, 606 Japan
[2] Takarazuka Research Center, Sumitomo Chemical Co., Takatsukasa, Takarazuka, Hyogo, 665 Japan
[3] Earth Chemical Co., Sakoshi, Akoh, Hyogo, 678-01 Japan

References

1. Biendl M, Hauptmann H, Sass H (1989) Periplanon D1 und periplanon D2—Zwei neue biogisch aktive Germacranoide Sesquiterpene aus *Periplaneta americana*. Tetrahedron Letters 30:2367–2368

Motor Patterns Underlying Pheromone-Modulated Flight in Male Moths, *Manduca sexta*

Mark A. Willis and Edmund A. Arbas[1]

Key words. Pheromone—Flight

Male moths typically find mates by flying a characteristic zigzag path up a plume of sex-pheromones released by a conspecific female. This upwind path results from a combination of self-steered maneuvers and continuous reactions to visual, olfactory, and mechanosensory inputs during flight. To characterize motor patterns underlying these flight maneuvers, we have simultaneously video-recorded the orientation tracks, and recorded electromyograms (EMGs) from flight muscles of male *Manduca sexta* flying freely in a plume of female pheromones. EMG activity was synchronized with video recordings of zigzag flight tracks in a laboratory wind tunnel. The activity of up to five muscles was recorded: (1) the left dorsal longitudinal muscle (DLM), the principal wing-depressor; (2 and 3) the left and right third axillary muscles (AxM), thought to be "steering" muscles; and (4 and 5) the left and right first basalar muscles (BaM), which generate lift and thrust by altering the angle of attack of the forewing. Parameters measured from the EMGs were: burst period (ms), and burst duration (ms) for all five muscles; and the time of onset of AxM and BaM activity relative to DLM burst onset (defined as the phase of onset).

The variability of motor patterns underlying even the most similar appearing maneuvers made it necessary for us to examine the flight tracks maneuver-by-maneuver. During pheromone-modulated upwind flight, the ground speed of the moths was modulated continuously, with the slowest speeds typically being reached at or near the apices of the turns. During some turns where continuous upwind progress was made throughout the turn, the DLM burst frequency decreased, with the lowest DLM burst frequency being correlated with the lowest ground speed. Alternatively, during some turns where continuous upwind progress was not maintained and some downwind movement occurred (i.e., maneuvers that may have incorported a brief period of hovering), the DLM frequency increased. During these maneuvers, the highest DLM burst frequency occurred during the period of lowest ground speed. In most moths, the relationship between DLM burst frequency and ground speed was variable, changing on a turn-by-turn basis. A few moths showed no association between DLM burst frequency and ground speed.

Coincident with the modulation of ground speed and DLM activity, the phase of AxM and BaM activity shifted symmetrically to characteristic relationships with respect to the DLM. Symmetrical advances in the phase of burst onset in bilateral AxMs may be strongly correlated with decreases in DLM burst frequency and the correlated changes in ground speed during some flight maneuvers. Alternatively, symmetrical phase delays in the activity of the AxMs and BaMs can be associated with increased DLM burst frequency, and with increases in ground speed. During other maneuvers, symmetrical phase changes of AxM activity can become dissociated from modulation of DLM burst frequency, but retain association with ground speed modulation, or can even vary on a maneuver-by-maneuver basis.

Characteristic asymmetries in the activity of bilateral AxMs were identified during zigzag turns in free flight. The most typical was the appearance of an additional spike at a later phase of the wing beat cycle in the AxM on the inside of the turn. We have also observed this pattern in the BaMs. These may occur during only one, or a few wing beat cycles in a turn. Asymmetric changes in AxM and BaM firing have been associated with changes in the body axis of the moth during large and small adjustments of the course. Thus, these bilateral asymmetries in the firing of AxMs and BaMs may contribute both to the minor adjustments in steering observed, and to the control of the larger temporally regular turns that make up pheromone-modulated flight.

We conclude from these data that *M. sexta* employ a rich complexity of motor strategies during the execution of their complex pheromone-guided flight.

Acknowledgments. This study was supported by NIH grants: DC00348, NS07309, and NSF grant IBN 9216532.

[1] ARL Division of Neurobiology, 611 Gould-Simpson Bldg., University of Arizona, Tucson, AZ 85721, USA

Morphology and Physiology of Pheromone-Triggered Flip-Flopping Descending Interneurons of the Male Silkworm Moth, *Bombyx mori*

Ryohei Kanzaki and Akira Ikeda[1]

Key words. Insect—Pheromone—Flip-flopping descending interneuron

Male silkworm moths, *Bombyx mori*, display oriented zigzag walking in response to the sex-attractant pheromones of females (bombykol). We have suggested that the walking toward the pheromone source is controlled by a self-generated zigzag turning program which is triggered by intermittent pheromonal stimulation applied to the antennae [1]. In the brain of the moths, olfactory information passes through several stages of synaptic processing involving the first-order olfactory center, the antennal lobes (ALs), and higher centers in the protocerebrum (PC). We have found that, in *Bombyx*, many olfactory PC neurons, some restricted to the PC, and others which have an axon descending (DNs) in the ventral nerve cord (VNC), innervate the lateral accessory lobes (LALs) [2, 3]. The LALs are also linked with each other [3].

The DNs that have a projection in the LAL fall into two classes according to the position of their cell bodies in the brain: (1) DNs which have a cell body in the cluster ventral to the medial cell cluster of the AL neurons (group I) and (2) those which have a cell body adjacent to the α-lobe of the mushroom body (group II).

Olberg [4] reported that, in response to sequential pheromonal stimulation, some DNs show state-dependent activities, like an electronic 'flip-flop (FF)', which have two distinct firing frequencies, high and low. In the present study, olfactory responses were recorded by glass suction-electrodes positioned against the descending, severed end of the small bundles or a whole connective of a VNC. We recorded two different types of FF activities simultaneously from each of the connectives, which had an antiphase relationship relative to each other. Typical FF activities were usually recorded from small bundles of the dorsal part of the connective. The FF activity pattern was also consistent with the activity pattern of the neck motor neurons, suggest-

ing that the signal may be important in controlling the pheromone-modulated zigzag walking of *Bombyx* [4]. State transition of FF activities was usually elicited in response to less than 10 ng bombykol. Sequential stimulation of a high concentration, e.g., 100 ng, hardly caused the state to drift. Once the state drifted from low to high when stimulated by the high concentration, such a concentration elicited only a brief inhibition, but did not alter the state much. All of the FF activities recorded in this study were influenced by constant light stimulation. Although FF state transitions were typically observed under constant light-on conditions, peak firing rates were lower when the light was turned off.

After the recording of FF activity, the DNs were stained by passing a current through a suction electrode containing 250 mM cobalt-lysine. In all cases ($n = 5$) group II DNs were stained. In four of these five examples, a few other DNs that did not innervate the LALs were also stained. In one case, only group II DNs in the brain were stained. It is possible that some FF interneurons are contained in group II DNs.

Acknowledgments. This research was supported, in part, by a Grant-in-Aid for General Scientific Research from the Ministry of Education, Science and Culture of Japan (No. 05640761), and a grant from the Chemical Materials Research and Development Foundation.

References

1. Kanzaki R, Sugi N, Shibuya T (1992) Self-generated zigzag turning of *Bombyx mori* males during pheromone-mediated upwind walking. Zool Sci 9(3): 515–527
2. Kanzaki R, Shibuya T (1986) Descending protocerebral neurons related to the mating dance of the male silkworm moth. Brain Res 377:378–382
3. Kanzaki R, Shibuya T (1992) Long-lasting excitation of protocerebral bilateral neurons in the pheromone-processing pathways of the male moth *Bombyx mori*. Brain Res 587:211–215
4. Olberg RM (1983) Pheromone-triggered flip-flopping interneurons in the ventral nerve cord of the silkworm moth, *Bombyx mori*. J Comp Physiol 152:297–307

[1] Institute of Biological Sciences, University of Tsukuba, 1-1-1 Tennodai, Tsukuba, Ibaraki, 305 Japan

List of Contributors

Abe, K. 65
Ache, B.W. 178, 196, 790
Adachi, M. 418
Adusumalli, M.D. 36
Aiba, T. 633
Aida, K. 761
Aihara, K. 418
Aishima, T. 711
Akaishi, T. 424
Aksari, P. 757
Amagai, Y. 747
Amakawa, T. 816, 820
Amasaka, J. 307
Anderson, J.A. 609
Andreini, I. 73
Aoki, J. 68
Aou, S. 502
Arai, S. 65
Arbas, E.A. 835, 850
Asaka, H. 340, 623
Asakoshi, R. 329
Asakoshi, T. 327, 331
Asanuma, N. 24
Atema, J. 778, 787, 788
Ayabe-Kanamura, S. 334, 345
Azen, E.A. 231

Baker, T.C. 838
Bakin, R.E. 139
Bard, J. 233
Barlett, P.F. 51
Barnett, D. 725, 731
Barry, P.H. 168
Bartlett, P.N. 690
Bartolomei, J.C. 425
Bartoshuk, L. 557
Beauchamp, G.K. 233, 239, 519
Becker, D.C. 396
Bekiaroglou, P. 638
Bell, G.A. 314, 725
Ben-Arie, N. 122
Berger, T. 636
Bertmar, G. 762
Bezrukov, S.M. 104

Blackshaw, S. 77
Boyse, E.A. 233
Brand, J.G. 73, 82, 93, 104, 135
Breer, H. 132
Breipohl, W. 27
Brennan, P.A. 490
Brown, R.E. 451
Bruch, R.C. 45
Bryant, B.P. 92, 104, 117
Buchholz, J.A. 45
Buck, L.B. 127
Bures, J. 467
Burgess, M. 775
Bushell, G.R. 51
Buttery, R.G. 271
Böttger, B. 322

Cain, W.S. 593
Calof, A.L. 36
Caprio, J. 750
Carlson, R.J. 222
Carmichael, S.T. 458
Catalanotto, F. 557
Caulliez, R. 376
Chaudhari, N. 382
Cho, H.H. 139
Christensen, T.A. 827
Ciges, M. 561
Cinelli, A.R. 433
Clarke, C.A. 731
Corrigan, C.J. 288
Cowart, B.J. 519
Curran, M. 233
Czurkó, A. 537

Danek, A. 636
Danho, W. 141
Dawson, T.M. 139
DeHamer, M.K. 36
DellaCorte, C. 135
DeMyer, S. 82
Derby, C. 775
Dermietzel, R. 32

DeSimone, J.A. 100
Di Lorenzo, P.M. 402
Dickens, J.C. 847
DiNardo, L.A. 422
Doi, K. 673
Doty, R.L. 597
Drewnowski, A. 522
Duncan, H.J. 494
Døving, K.B. 188

Egashira, M. 715
Egawa, M. 614, 632
Ehara, K. 723
Emson, P.C. 449
Endo, S. 547, 555, 556, 574,
       577, 578, 580
Enomoto, Shoji 90
Enomoto, Shuichi 152
Escobar, M.L. 475
Etoh, M. 113
Evans, R.E. 751

Fadool, D.A. 178, 196
Faludi, B. 537
Farbman, A.I. 45, 141
Faurion, A. 88, 301
Feigin, A.M. 104
Fernandez, G.D. 452
Fernandez-Cervila, F. 561
Finger, T.E. 322, 739, 753
FitzGerald, L. 135
Foster, S.S. 445
Froloff, N. 88
Fujikura, M. 341
Fujisawa, K. 385
Fujita, H. 323
Fujita, K. 323, 759
Fujita, T. 2, 5, 25
Fujiwara, M. 327, 329, 331
Fujiyama, R. 108, 109
Fukatsu, S. 282
Fukazawa, O. 340
Fuke, S. 357

Key Word Index